AF777411

Climate Change Impacts and Adaptation: Interdisciplinary Perspectives—Volume I

Climate Change Impacts and Adaptation: Interdisciplinary Perspectives—Volume I

Topic Editors

Cheng Li

Fei Zhang

Mou Leong Tan

Kwok Pan Chun

MDPI

Basel • Beijing • Wuhan • Barcelona • Belgrade • Novi Sad • Cluj • Manchester

Topic Editors
Cheng Li
College of Plant Protection
Yangzhou University
Yangzhou
China

Fei Zhang
College of Geography and
Environmental Sciences
Zhejiang Normal University
Jinhua
China

Mou Leong Tan
School of Humanities
Universiti Sains Malaysia
Pulau Pinang
Malaysia

Kwok Pan Chun
CATE School of Architecture
and Environment
University of the West of
England
Bristol
UK

Editorial Office
MDPI AG
Grosspeteranlage 5
4052 Basel, Switzerland

This is a reprint of the Topic, published open access by the journals *Agronomy* (ISSN 2073-4395), *Applied Sciences* (ISSN 2076-3417), *Climate* (ISSN 2225-1154), *Forests* (ISSN 1999-4907), *Remote Sensing* (ISSN 2072-4292), *Sustainability* (ISSN 2071-1050) and *ISPRS International Journal of Geo-Information* (ISSN 2220-9964), freely accessible at: https://www.mdpi.com/topics/2578M074VX.

For citation purposes, cite each article independently as indicated on the article page online and as indicated below:

Lastname, A.A.; Lastname, B.B. Article Title. *Journal Name* **Year**, *Volume Number*, Page Range.

Volume I
ISBN 978-3-7258-5819-4 (Hbk)
ISBN 978-3-7258-5820-0 (PDF)
https://doi.org/10.3390/books978-3-7258-5820-0

Set
ISBN 978-3-7258-5817-0 (Hbk)
ISBN 978-3-7258-5818-7 (PDF)

© 2025 by the authors. Articles in this book are Open Access and distributed under the Creative Commons Attribution (CC BY) license. The book as a whole is distributed by MDPI under the terms and conditions of the Creative Commons Attribution-NonCommercial-NoDerivs (CC BY-NC-ND) license (https://creativecommons.org/licenses/by-nc-nd/4.0/).

Contents

About the Editors

Cheng Li

Cheng Li is an Associate Professor at the Department of Ecology, School of Plant Protection, Yangzhou University, China. He earned his Ph.D. in Applied Meteorology from the Nanjing University of Information Science and Technology. With a strong academic background, Dr. Li has authored approximately 60 peer-reviewed articles in high-impact journals and has presented his research at numerous prestigious conferences. He serves as a Topic Editor, Guest Editor, and Topical Advisory Panel Member for several international journals. Dr. Li has been honored with the Young Elite Scientists Sponsorship Program of Jiangsu Province and the National Scholarship for Doctoral Students in China.

Fei Zhang

Fei Zhang is a professor at the College of Geography and Environmental Sciences, Zhejiang Normal University. He is mainly engaged in remote sensing research on eco-environments in arid zones, with a focus on remote sensing research on the water environment, remote sensing image classification, urbanization dynamic detection, and image processing. To date, he has published more than 100 journal papers as the first author or corresponding author. He is a Guest Editor of journals such as *Water* and *Remote Sensing*, and he is also an Editorial Board Member of journals such as the *Journal of Lake Science, Spectroscopy and Spectral Analysis* and *the International Journal of Geomatics*.

Mou Leong Tan

Mou Leong Tan is an Associate Professor at the Geography Section, School of Humanities, Universiti Sains Malaysia. His research focuses on hydrology, climatology, remote sensing, and geographic information systems (GISs). Dr. Tan currently serves as the President of the Malaysian Geographical Association (GeogMalaysia), Vice President of Water Watch Penang (WWP), and is a professional member of the Institution of Geospatial and Remote Sensing Malaysia (IGRSM). He has led more than ten national and international research projects and has published over 150 research papers, with an H-index of 38 and over 5,000 citations. He has received several prestigious awards, including the "Malaysia Water Award for Young Water Leadership", the "Top Research Scientist Malaysia" award, and the "Malaysia Research Star Award". From 2021 to 2024, he was listed among the top 2% of the world's most highly cited scientists under environmental engineering by Stanford University.

Kwok Pan Chun

Kwok Pan Chun is a Senior Lecturer in Environmental Management and Geography at the School of Architecture and Environment, University of the West of England. His core research interests include nonstationary spatiotemporal processes, numerical modelling for hydroclimate, and land use management. Dr. Chun is currently leading a UK Research Council-funded project titled Land Use Change Investigation and Regional Climate (LIRIC), collaborating with researchers in South America and Asia to advance Sustainable Development Goal 13: Climate Action. He also leads an International Association of Hydrological Sciences (IAHS) working group that challenges existing power dynamics and colonial legacies, aligning with UN frameworks and contributing to the broader goals of IAHS cross-cutting initiatives.

Preface

The idea for *Climate Change Impacts and Adaptation: Interdisciplinary Perspectives* took shape in 2022, when record-breaking heatwaves, floods and droughts were simultaneously affecting every continent. As the young and responsible editors we had spent the preceding decade documenting regional climate signals and their cascading impacts, yet we were struck by how rarely disciplinary insights were assembled into a coherent, scale-relevant body of evidence that decision-makers could act upon. Against this backdrop, we issued an open call to the global research community in 2023, inviting submissions of original work that crossed traditional boundaries between climatology, ecosystem science, geospatial analysis and social practice. Ultimately, we presents a curated selection of 69 research articles and reviews that collectively address the pressing challenges posed by climate change across a variety of sectors and regions.

We organised the contributions into four overlapping streams. Part I includes observed regional climate changes, demonstrating how advanced detection techniques can now resolve sub-national signals. Part II quantifies biophysical and socio-economic impacts of climate change on water, food, health and infrastructure systems, emphasising compound and cascading risks. Part III evaluates adaptation measures ranging from drought-tolerant cultivars and urban green infrastructure to community-based early-warning systems, offering empirical evidence on costs, co-benefits and limits to implementation. Part IV explores the application of new technologies and methodologies, including remote sensing, machine learning, statistical modeling, and participatory approaches.

Our purpose is twofold: to provide researchers with a consolidated reference of cutting-edge, interdisciplinary methods that reduce uncertainty in impact assessment; and to offer practitioners a portfolio of rigorously tested adaptation options that can be transferred across similar socio-ecological contexts. We address the volume to graduate students beginning their careers at the climate–society interface, to scientists engaged in regional down-scaling or risk-modelling exercises, and to planners tasked with translating global targets into local investments. By integrating remote-sensing products, citizen-science data, machine-learning algorithms and participatory narratives, we hope that the findings and methodologies presented here exemplify the epistemic diversity that effective climate response now demands.

We thank the reviewers for their meticulous critiques, and the editorial office for sustaining an efficient publication process. Finally, we dedicate this volume to the communities whose lived experience of climate extremes motivates the search for robust, just and timely adaptation solutions.

Cheng Li, Fei Zhang, Mou Leong Tan, and Kwok Pan Chun
Topic Editors

sustainability

Article

Analyzing Rainfall Trends Using Statistical Methods across Vaippar Basin, Tamil Nadu, India: A Comprehensive Study

Manikandan Muthiah [1,*], Saravanan Sivarajan [2], Nagarajan Madasamy [3], Anandaraj Natarajan [4] and Raviraj Ayyavoo [5]

1 Agricultural Research Station, Tamil Nadu Agricultural University, Kovilpatti 628501, Tamil Nadu, India
2 VIT School of Agricultural Innovations and Advanced Learning, Vellore Institute of Technology, Vellore 632014, Tamil Nadu, India; saravanan.siva@vit.ac.in
3 Agricultural Engineering College and Research Institute, Tamil Nadu Agricultural University, Kumulur, Trichy 621712, Tamil Nadu, India; nagarajan.m@tnau.ac.in
4 Dr. M. S. Swaminathan Agricultural College and Research Institute, Tamil Nadu Agricultural University, Eachangottai, Thanjavur 614902, Tamil Nadu, India; anandswc@yahoo.co.in
5 Agricultural Engineering College and Research Institute, Tamil Nadu Agricultural University, Coimbatore 641003, Tamil Nadu, India; rraj@tnau.ac.in
* Correspondence: manikandan.m@tnau.ac.in

Abstract: The Vaippar basin in southern India is economically important for rainfed and irrigated agriculture, mainly depending on the northeast monsoon (NEM) during October–December, and any changes in rainfall patterns directly affect crop ecosystems. This study aimed to analyze spatio-temporal rainfall changes using the monthly data from 13 scattered rain gauge stations in the Vaippar basin, India. They were converted into gridded rainfall data by creating 26 equally spaced grids with a spacing of 0.125° × 0.125° for the period between 1971 and 2019 through interpolation technique. Three methods, namely Simple Linear Regression (SLR), Mann–Kendell/modified Mann–Kendell (MK/MMK), and Sen's Innovation trend analysis (ITA), were employed to detect trends and magnitudes for annual and seasonal gridded rainfall series. The results showed significant trends at 2.3%, 7.7%, and 44.6% of grid points using SLR, MK/MMK, and ITA methods, respectively. Notably, ITA analysis revealed significant trends in annual and NEM rainfall at 57.69% and 76.92% of the grid points, respectively, at a 5% significance level. The southwestern and central parts of the basin exhibited a higher number of significant upward trends in annual rainfall. Similarly for the NEM season, the south-eastern, central, and extreme southern parts experienced significant upward trend. The western part of the basin exhibited significantly upward trend with a slope value of 2.03 mm/year, while the central part showed non-significant downward trend with a slope value of −1.89 mm/year for the NEM series. This study used the advantage of ITA method, allowing for exploration of monotonic/non-monotonic trends, as well as subtrends of low, medium, and high rainfall segments within the series. The key findings of this study serve as a scientific report from a policy perspective, aiding in the preparation and management of extreme climate effects on land and water resources in the Vaipaar basin.

Keywords: gridded rainfall; Mann–Kendell; innovative trend analysis; regression; magnitude; spatial interpolation; subtrend

1. Introduction

Whether it is rainfed agriculture or irrigated agriculture, rainfall remains the primary source of water for crop production. In countries like India, which heavily rely on agriculture, achieving grain self-sufficiency has been a significant accomplishment. However, the production is resource intensive, focused mainly on cereals, and biased towards specific regions, all while facing increasing stress on water resources [1]. The state Tamil Nadu, one of the major contributors to food production in India, includes 17 major river basins, with approximately 2.4 million hectares irrigated by surface water through major, medium, and minor schemes [2]. According to a recent study, the rainfed cropland area in Tamil Nadu was estimated to be 2.57 million hectares, accounting for 19.66% of the total geographical area of the state. The Virudhunagar (0.14 million hectares) and Thoothukudi district (0.17 million hectares), located within the Vaippar basin, contributed to higher rainfed cropland areas [3], and have been considered in this research work.

In the context of rainfall variability, uncertain distribution of rainfall poses a serious obstacle to agriculture [4]. The availability of rainwater for both rainfed and irrigated agriculture is becoming scarce due to uncertain rainfall patterns. Moreover, extreme climate-induced hazards such as droughts and floods are becoming more common due to hydro-climatological variability [5]. The trends of these events have been linked to changes in rainfall patterns [6]. Therefore, understanding the historical pattern of rainfall variation and referring to scientific reports on rainfall trends are crucial for planning water conservation efforts and formulating mitigation measures for extreme climate events [7]. Additionally, these details are essential for minimizing underestimation or overestimation of design parameters for water infrastructure [8].

In the recent years, trends in hydro-meteorological variables have gained considerable attention in different parts of the world [4]. Different statistical methods are presently available for detecting trends of hydro-meteorological variables and used by many researchers. Each method has its own merits and demerits. Generally, the trend detection methods are divided into parametric and non-parametric methods. Many scientists use non-parametric methods for trend analysis because these methods are not sensitive to outliers [9]. They can be particularly useful for analyzing non-normally distributed series with missing values [10], and they do not rely on any assumptions about the nature of the data [11]. The non-parametric methods such as the Mann–Kendall (MK) test, modified MK test, Spearman rank order correlation (SRC), Kendall rank correlation (KRC), and innovative trend analysis (ITA) are commonly followed by many researchers for trend detection at different time scales and significance levels [12–15]. Equally, parametric methods such as simple linear regression (SLR) method are extensively used for detecting the monotonic long-term trend in the rainfall time series [16,17]. The main advantage of SLR method is that it measures the statistical significance for testing hypothesis on the estimated slope and also provides the magnitudes of the parameters considered for analysis [18]. Most of the researchers used non-parametric tests for trend assessment [19] and some of the studies explored the comparison of non-parametric and parametric tests for trends [20]. The magnitude of rainfall trend (mm/year) can be determined using Sen's Slope estimator (SSE) and Simple Linear Regression (SLR) tests [15].

Numerous studies have been conducted on spatio-temporal trends of hydro-meteorological variables, such as stream flow [21,22]; rainfall [23]; temperature and potential evapotranspiration [10,24]; reference evapotranspiration [25] and other climate variables. Gridded satellite data and gauged climate data have been used for trend analysis [26,27]. Trend detection studies are available in regional and basin scale in India such as Kerala [28]; Ganga-Brahmputra-Meghna river basins of India [29]; Parambikulam Aliyar sub basin in Tamil Nadu [30]; Betwa Basin in Central India [31]; Godavari River basin in Southern Peninsular India [32,33]; Sindh river basin in India [34]; upper Cauvery Basin [35]; Lower Bhavani basin in Tamil Nadu [36]; Thamirabharani River Basin in Tamil Nadu [27]; Indian river basins [37]; Indravati river basin [26]; states such as Jharkhand [38,39]; Chhattisgarh State [40,41]; Maharashtra and Karnataka [42]; Gujarat [13,43] Central India—Madhya Pradesh and Chhattisgarh [44]; Parts of Rajasthan [12,45–47]; parts of Odisha [48]; Uttarakhand [15]; parts of Andhra Pradesh [49]; Kashmir Valley [50].

Some global studies on trend analysis of hydro-metrological variables include countries such as Ethiopia [51]; Turkey [52]; Ghana [53]; Netherlands [54]; Iran [55,56]; Mediterranean regions [57,58]; Tanzania coast [59]; China [60–64]; Egypt [65]; United States [66]. The spatial variation of trends could be identified under ArcGIS environment through the inverse distance weighting (IDW) [34,67,68] and Kriging method [40]. Some studies used

Thiessen polygon method to identify the area of influence of point rain gauges [15,31]. A standard methodology in a GIS environment using different trend methods was developed for analyzing the spatial pattern [13].

Recently, the innovative trend analysis (ITA) method developed by Sen (2012) [8] has gained more attention around the world and it was test verified by many researchers [14,17,44,47,52,63]. The main advantage of ITA method is analyzing trends by providing graphical forms of presentation and without any limitations, such as non-normality, serial correlation, and size of data in the time series. In addition, ITA method provides a robust and powerful result with minimum error. It is also possible to detect the monotonic and non-monotonic trends in a way that time series are divided into different subcategories of the time series such as high, medium, and low zones [69,70].

This study aims to explore the rainfall variations, spatial patterns of trends, and their magnitudes of annual and seasonal rainfall series of the Vaippar basin at micro-level using classical rainfall statistical methods. The uniqueness in this research work is the use of spatially interpolated gauge rainfall data applied at micro-level for trend detection and identification of subtrends within the rainfall series over space and time.

2. Materials and Methods

2.1. Study Area

The Vaippar River Basin, located in the southern part of Tamil Nadu, India is a significant river basin in the region. Situated between latitudes 8°97′ N and 9°78′ N and longitudes 77°24′ E and 78°37′ E, it encompasses a total catchment area of 5320 km^2. The basin is bounded on the north by the Vaigai and Gundar basins, on the south by the Tamaraparanibasin, on the west by the Western Ghats, and on the east by the Gulf of Mannar (Bay of Bengal). The basin area spans across four districts, with Virudhunagar accounting for 68%, Thoothukudi for 20%, Madurai for 7%, and Tirunelveli for 5% of the total area. The Vaippar River originates from the Echamalai mottai, Neduntheri mottai, and Kiladiparai hill ranges of the Western Ghats, situated near Sivagiri in Tirunelveli district. It begins at an elevation of 165 m above mean sea level and flows predominantly in an easterly and southeasterly direction for a distance of 146 km before joining the Gulf of Mannar. The catchment area of the Vaippar basin encompasses hilly regions such as Kodaliparai mottai, Vasudevanallur reserve forest, Periyasudangi malai, and others. These mountain ranges lie in the rain shadow regions of the Western Ghats, resulting in relatively low rainfall. The entire catchment area of the Vaippar basin is within the boundaries of Tamil Nadu state. The basin has been further divided into 13 sub-basins, namely: (1) Nichabanadhi, (2) Kalingalar, (3) Deviar, (4) Nagariyar, (5) Sevalperiyar, (6) Kayalkudiar, (7) VallampattiOdai/Uppodai, (8) SindapalliUppodai, (9) Arjunanadhi, (10) Kousiganadhi, (11) Uppathurar, (12) Senkottaiyar, and (13) Vaippar. The location of the Vaippar basin is depicted in Figure 1.

The basin experienced frequent drought with a range from four to eight years per drought [71]. Agricultural land covers approximately 74% of the total geographical area, while forested areas account for 10% of the area. Wasteland occupies 8% of the total geographical area, while settlements and water bodies together cover less than 8% of the basin's total geographical area. Out of the total agriculture area in the basin, cultivable land represents 43%. This cultivable land is primarily utilized for cultivating water-intensive crops such as paddy, sugar cane, and banana [72]. Additionally, cotton, non-paddy, and dry crops are also grown in the basin. The basin's total irrigated area accounts for 24% of the cultivable land. The remaining 76% of the cultivable land relies mainly on rainwater for irrigation [71].

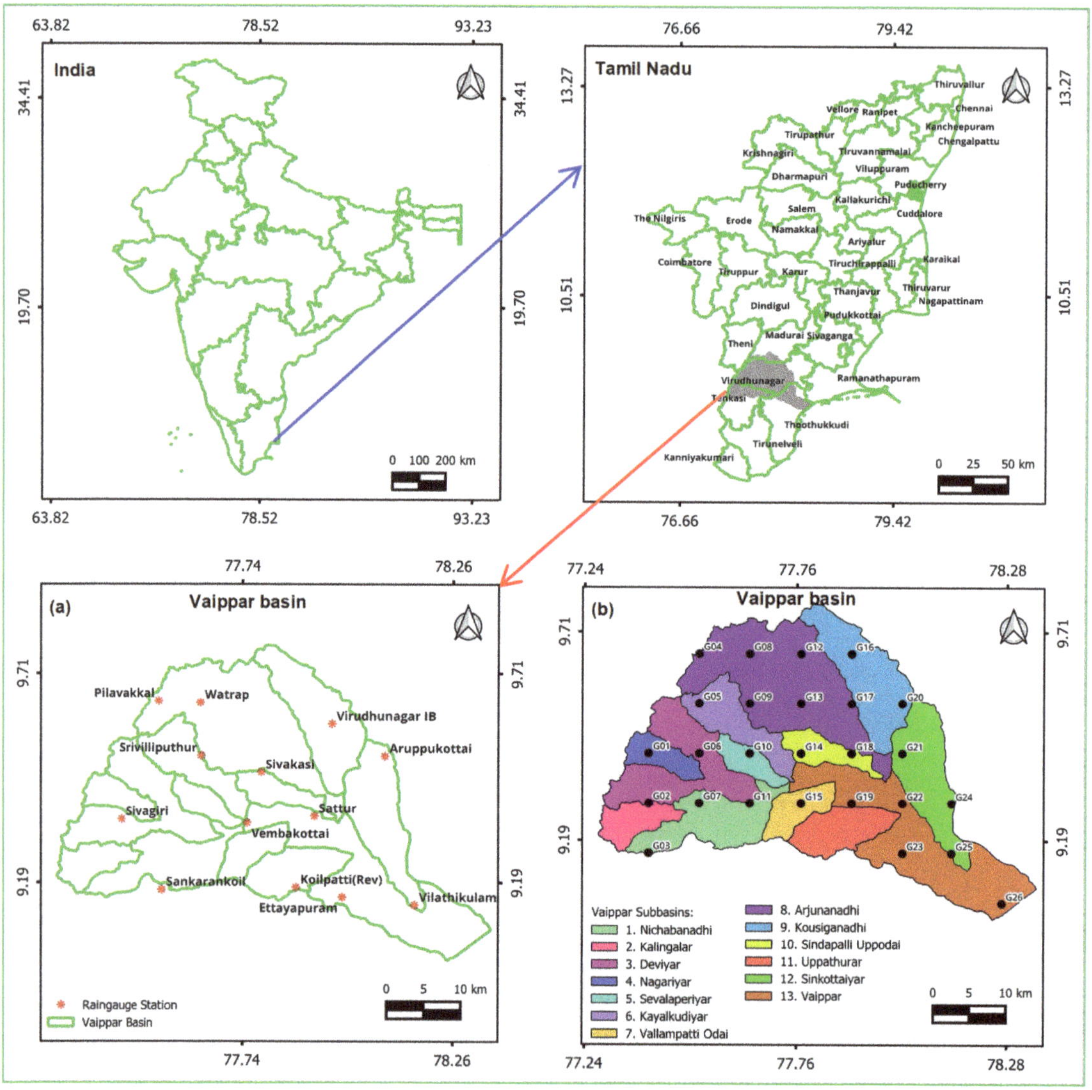

Figure 1. Geographical location of study area along with (**a**) rain gauge stations and (**b**) grid points in subbasin.

2.2. Data Used

Daily rainfall data from 13 rain gauge stations distributed across the Vaippar basin were collected for this study. The data were obtained from the State Ground and Surface Water Resources Data Center, Public Works Department, and Water Resources Organization in Chennai. The location details of each rain gauge station and the period of data utilized in this study are presented in Table 1. Standard quality control measures were undertaken to identify outliers and errors in rainfall data for all 13 rain gauges. Any potential outliers, including missing data and instrument errors, were checked and corrected. To facilitate the analysis, the daily rainfall data for each station was processed to derive monthly data. Subsequently, the monthly rainfall series was further organized into seasonal rainfall series. The seasons considered for this analysis were as follows: Southwest Monsoon (SWM) from

June to September, Northeast Monsoon (NEM) from October to December, winter season in January and February, summer season from March to May, and an annual rainfall series. The space–time correlation analysis of the mean monthly rainfall data with the rain gauge stations was conducted and is presented in Table A1. A strong correlation exists within the mean monthly rainfall series among the rain gauge stations.

Table 1. Details of rain gauge station geographical location and data used.

Sl. No	Rain Gauge Station	Latitude	Longitude	Elevation, m	Data Availability
1	Aruppukottai	09°30′09″	78°05′44″	123	1971–2019
2	Virudhunagar IB	09°35′00″	77°57′50″	120	1971–2019
3	Sattur	09°21′08″	77°55′13″	91	1971–2019
4	Sivakasi	09°27′44″	77°47′14″	127	1973–2019
5	Srivilliputhur	09°30′14″	77°38′08″	146	1971–2019
6	Watrap	09°38′11″	77°38′01″	73	1974–2019
7	Pilavakkal	09°38′25″	77°31′45″	141	1982–2019
8	Koilpatti (Rev)	09°10′27″	77°52′28″	130	1971–2019
9	Vilathikulam	09°07′48″	78°10′05″	47	1971–2019
10	Sankarankoil	09°10′04″	77°32′12″	143	1971–2019
11	Sivagiri	09°20′47″	77°26′01″	165	1971–2019
12	Vembakottai	09°20′06″	77°45′04″	111	2000–2019
13	Ettayapuram	09°08′53″	77°59′26″	60	1999–2019

2.3. Methodology

In this study, the monthly rainfall data collected from the 13 rain gauge stations were spatially interpolated to generate gridded rainfall data. A grid with a spacing of 0.125° in both latitude and longitude was employed to create 26 individual grids that spanned the entire basin area. The rainfall data available for each location in each year during the period specified in Table 1 underwent spatial interpolation. Using spatially interpolated monthly rainfall data, gridded rainfall data was prepared. Using the gridded rainfall data, trends in the monthly, annual, and seasonal rainfall series were estimated. Three different methods were employed for trend analysis: Mann–Kendall (MK) test, simple linear regression (SLR), and Sen's innovative trend analysis (ITA) method. The trends were evaluated by estimating the slopes using Sen's slope, SLR, and ITA methods to quantify the magnitude and direction of the trends in the rainfall data, and the percentage change of magnitude of trends from mean rainfall. The overall methodology used in this study is depicted as a flowchart in Figure 2.

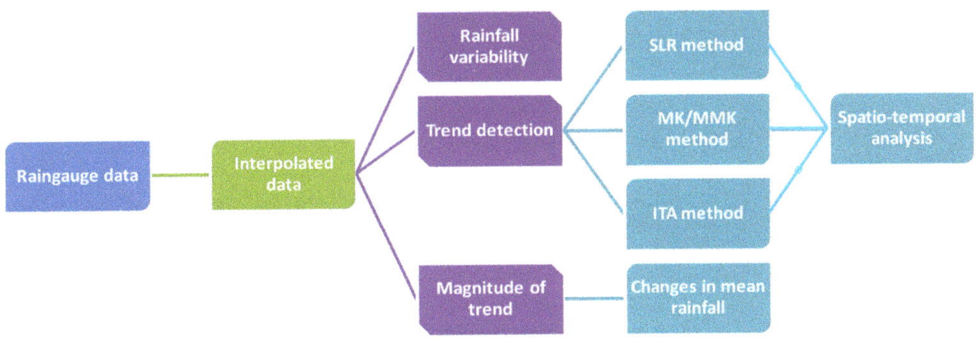

Figure 2. The overall process involved for rainfall characterization.

2.4. Spatial Interpolation of Rainfall

Spatial interpolation techniques play a crucial role in estimating values at locations where no observed data is available, using known data values. Several methods, such as inverse distance weighting (IDW), splines, and kriging, are commonly employed for spatial analysis of various variables [73–75]. In the present study, the IDW approach was specifically utilized for spatially interpolating rainfall data across the Vaippar basin. This method considers the proximity of known data points to the location of interest and assigns weights accordingly, resulting in an interpolated surface that provides valuable insights into the spatial distribution of rainfall within the basin.

The IDW interpolation technique assigns weights to each control point based on the inverse of their distances from the interpolated point. This means that closer control points have a greater influence on the interpolated value compared to those farther away. The technique assumes that each control point has a local influence that diminishes as distance increases. In IDW interpolation, the output value at a given location is determined by a specified number of nearest points or all points within a specified radius. The power parameter of the IDW method controls the significance of the surrounding points in determining the interpolated value. A higher power value reduces the influence of distant points, resulting in a stronger emphasis on the values of nearby points in the interpolation process [76].

The general form of IDW approach [77,78] is given in Equation (1):

$$zn_i = \frac{\sum_{j=1}^{m} z_j \times d_j^{-p}}{\sum_{j=1}^{m} d_j^{-p}} \tag{1}$$

where zn_i is the new value for grid i; z_j is the value of m nearest neighbours; d_j is the distance to m nearest neighbours; p is the exponent of distance. For this study, the exponent of distance was taken as two for spatial interpolation of the rainfall.

This study employed the Inverse Distance Weighting (IDW) technique instead of the Thiessen polygon method to estimate the gridded rainfall across the basin. The Thiessen polygon method often results in a crude approximation of rainfall spatial variation due to attribute variations associated with each rain gauge station, ranging from 14 to 30% [79,80]. To overcome this limitation, the Vaippar basin was divided into smaller square grids. Given the constraints of rainfall data availability and the uneven distribution of rain gauge stations within the basin, spatial interpolation of data at smaller grids was necessary to address these challenges and obtain a more accurate representation of rainfall patterns.

The total area of the Vaippar basin was divided into 26 grids, each with dimensions of 0.125° (205 km^2). These grids accounted for approximately 3.85% of the total area, equivalent to 5320 km^2. Monthly rainfall data recorded at 13 stations were spatially interpolated using the IDW method in QGIS 3.30.2 resulting in gridded rainfall data for the period from 1971 to 2019. The gridded rainfall data was then utilized to analyze the spatial patterns of significant trends in rainfall over the study period.

2.5. Rainfall Variability

In this study, the variability of rainfall has been assessed using the coefficient of variation (CV). The CV is a statistical measure that represents the ratio of the standard deviation to the mean of a rainfall data series. It is often used to quantify the relative variability or dispersion of rainfall. A higher CV indicates greater variability, while a lower CV suggests more stability or consistency in the data [27,81]. The CV for the rainfall events is expressed as a percentage (%). As mentioned by Hare (2003) [82], the CV can be used to classify the degree of variability of rainfall events into three categories: (i) when the CV is less than 20, rainfall events are considered to have less variability or consistent pattern of rainfall; (ii) when the CV is between 20 and 30, rainfall events are classified as having moderate variability or fluctuation; (iii) when the CV exceeds 30, rainfall events are categorized as having high variability indicating a less stable or more unpredictable rainfall.

2.6. Simple Linear Regression

Simple linear regression (SLR) analysis is a widely used parametric model for identifying monotonic long-term trends in monthly, seasonal, and annual rainfall time series [83]. SLR analysis establishes a relationship between two variables, often employed to determine the slope of hydro-meteorological variables over time. A positive slope indicates an upward trend, while a negative slope suggests a downward trend. The primary advantage of SLR analysis is its ability to provide a measure of significance through hypothesis testing on the slope, along with quantifying the rate of change. To obtain the percentage change during the specified period, the slope values are multiplied by the duration of the study period in years [16].

In this study, the simple linear regression models for the annual and seasonal time series were developed between the gridded rainfall data and time for trend detection [40,84,85]. The test statistic "t", which follows a Student's t-distribution with $(n - 2)$ degrees of freedom (where "n" represents the length of the data points in the time series), was calculated to examine the significance of trend. The null hypothesis (Ho) of a zero slope is rejected when the calculated test statistic "t" value surpasses the critical value "$t_{\alpha/2}$" at a given significance level.

2.7. Mann–Kendall (MK) Trend Analysis

The non-parametric Mann–Kendall (MK) test is a commonly used method for exploring trends in hydro-meteorological data. The trends identified by this test in the time series are monotonic, meaning they can be increasing or decreasing without assuming linearity. The test was originally proposed by Mann (1945) [86] and the test-statistic distribution was subsequently derived by Kendall (1975) [87]. The null hypothesis for this test assumes that the data is independent and randomly ordered. One advantage of this test is that it does not require any assumption of normality. However, it only indicates the direction of significant trends, without providing information about their magnitude [86,87]. The MK test statistic S indicates the number of positive differences minus the number of negative differences for all the considered differences. Using S value, its mean, and variance, the standardized test statistic Z was calculated for annul and seasonal rainfall series. A positive or negative Z value indicates an upward/downward trend [88]. The Z statistic follows a normal distribution. To test for statistical significance, compare the calculated Z value with critical values obtained from the standard normal distribution tables at two significance levels (α = 5% and α = 10%). The null hypothesis (Ho) of no trend is rejected if the calculated value of Z is greater than the table value of $Z_{1-\alpha/2}$ [13,30,89].

2.8. Modified Mann–Kendall Test

In the Modified Mann–Kendall (MMK) test proposed by Hamed and Rao (1998) [90], the autocorrelation coefficient (ACC) of the rainfall series is computed and then subjected to testing at various levels of significance. If the ACC value is determined to be significant, the MMK test is subsequently [13,23,89].

Initially, the modified variance was computed by integrating the Auto-Covariance Correction (ACC) at lag-i. Subsequently, this modified variance was utilized in the original Mann–Kendall (MK) test, leading to the calculation of the modified Mann–Kendall Z_c statistic. To assess the significance of the trend, the modified Mann–Kendall Z_c statistic was subjected to testing against threshold levels. For instance, significance levels of 10% and 5% were represented by threshold values of 1.645 and 1.96, respectively.

2.9. Autocorrelation Analysis of Time Series

The autocorrelation (serial correlation coefficient, r) analysis is a useful tool for evaluating the presence of randomness and periodicity within a time series at different lag periods [91]. To calculate autocorrelation coefficients (ACC), the normalized anomaly of the rainfall series, obtained from gridded rainfall data, along with the long-term average and standard deviation of annual and seasonal rainfall, can be utilized. Positive or negative

ACC values are generally influenced by the trends identified in the time series [90,92,93]. If the ACC is close to zero for different time lags, it suggests that the data points in the time series are randomly distributed, indicating no dependence. On the other hand, the presence of serial correlation, as examined through autocorrelation analysis, is an important preliminary test before performing the Mann–Kendall (MK) trend analysis. This is because the MK analysis requires the input data to be serially independent. Otherwise, the presence of positive serial correlation in the data can lead to an overestimation of trend significance. The lag-1 ACC specifically detects the serial correlation in the data series. The statistical significance of serial correlation can be tested by employing a normally distributed statistic at a significance level of "α", taking into account the lag period and the length of the series [94].

The ACC (r_k) of the rainfall time series at lag-k can be computed using the rainfall data of the time series and the total length of the time series, where "k" represents the time lag. The value of r_k ranges between +1 and -1. A value of zero for r_k signifies that the series is random for all lag-k values [15]. To evaluate the statistical significance of the ACC in the time series data, testing should be conducted at both upper and lower confidence limits. If the ACC turns out to be statistically insignificant, the Mann–Kendall (MK) test can then be applied to the original time series [95,96]. The null hypothesis (Ho) suggests the absence of serial correlation within the time series, while the alternative hypothesis (Ha) suggests the presence of some serial correlation rather than pure randomness. In this study, the hypothesis was examined using the lag-1 autocorrelation coefficient. The upper and lower confidence limits can be calculated [45,97] at a significance level denoted as α (e.g., for α = 10%, Z = ±1.645, and for α = 5%, Z = ±1.96) [45]. When the r_k value falls within the range of the critical values {(r_k) upper < r_k < (r_k) lower}, the null hypothesis Ho: $r_k = 0$ is rejected, while Ha: $r_k \neq 0$ is accepted. This indicates that the time series is not random and exhibits some level of persistence or serial correlation [11].

2.10. Sen's Slope

The magnitude of trends in monthly, seasonal, and annual time series was determined using Sen's slope, a non-parametric method introduced by Sen (1968) [98]. Sen's slope requires equally spaced data in a time series [11]. One of the main advantages of Sen's slope over simple linear regression (SLR) is its resilience to data errors, outliers, and extreme observations [15]. This method is particularly useful when assuming a linear trend line.

The slope estimate Q for each pair of data values is calculated using $Q_i = median\frac{x_j - x_k}{j-k}$ where x_j and x_k are the data values at time j and k, respectively ($j > k$). If there are n data values x_j in the time series, there will be a total of $N = n(n-1)/2$ slope estimates Q_i. The N values of Q_i are then ranked from the smallest to the largest. The Sen's slope is determined based on whether N is odd or even: if N is odd, the Sen's slope is $= Q_{[\frac{(N+1)}{2}]}$; if N is even, the Sen's slope is $Q = \frac{1}{2}\left(Q_{[\frac{N}{2}]} + Q_{[\frac{(N+2)}{2}]}\right)$. A positive value of Q indicates an upward trend, while a negative value represents a downward trend. A Q value of zero indicates no trend in the time series [15,30].

2.11. Innovative Trend Analysis

The Innovative Trend Analysis (ITA) method has been successfully utilized for trend detection in hydro-meteorological variables [8,52]. This method is simple, allowing for easy identification and visualization of trends in high, medium, and low data set on the trend line [70]. Unlike non-parametric trend identification tests, the ITA method does not require restrictive assumptions such as data series independence, normality, or data length [8,47,99]. It involves plotting all data points of the time series in a Cartesian coordinate system and comparing them with a diagonal straight 1:1 line [44,100]. The construction procedure for ITA is provided below:

(a) Divide the monthly, annual, or seasonal rainfall time series into two equal halves. Arrange each half series in ascending order.

(b) Place the first half of the time series on the X-axis and the second half on the Y-axis. Plot the rainfall series as a scatter diagram.
(c) Draw a straight 45° line diagonally in the scatter graph representing a 1:1 relationship. Divide the plot into upper and lower half triangles.
(d) If the scattered points align perfectly on the 45° line, the series does not exhibit a trend.
(e) When the scattered points lie within the upper half triangle, the rainfall series is said to have an increasing trend.
(f) When the scattered points lie within the lower half triangle, the rainfall series is said to have a decreasing trend.
(g) A time series is considered to have a monotonic trend if all scattered points lie above or below the upper or lower half triangles. Non-monotonic trends occur when some scattered points are located in the upper half triangle and others in the lower half triangles [8].
(h) The trend of low, medium, and high values can be observed in the scatter graph.
(i) To test the significance of the trend, a null hypothesis (Ho) is considered: there is no significant trend if the calculated slope value (S) is below the critical value (S_{cri}). The alternative hypothesis (Ha) states that there is a significant trend when $S > S_{cri}$.
(j) The slope (S) of the trend is computed using the formula [101]:

$$S = \frac{2(\overline{y}_2 - \overline{y}_1)}{n} \tag{2}$$

where, $\overline{y}_1$ and $\overline{y}_2$ represent the mean of the first and second halves of the rainfall time series, and n is the number of data points in the rainfall time series.
(k) The standard deviation (σ_s) of the trend slope is calculated using the formula [101]:

$$\sigma_S = \frac{2\sqrt{2}}{n\sqrt{n}}\sigma\sqrt{1 - \rho_{\overline{y}_1\overline{y}_2}} \tag{3}$$

where σ represents the standard deviation and $\rho_{\overline{y}_1\overline{y}_2}$ is the cross-correlation coefficient between the means of the two-half series.
(l) By utilizing the confidence limits (S_{cri}) for a standard normal probability density function, the confidence limits of the trend slope are calculated at a significance level of α using the formula [101]:

$$CL_{(1-\alpha)} = 0 \pm S_{cri}\sigma_S \tag{4}$$

If the slope value (S) falls beyond the lower or upper confidence limits, the null hypothesis of no significant trend is rejected at the α significance level.

2.12. Percentage Change in Magnitude of Trend from Mean Rainfall

The percentage change in magnitude of trend from mean rainfall of seasonal and annual time series can be calculated using the following expression [31,102,103]:

$$PC = \frac{n \times S}{\overline{x}} \times 100 \tag{5}$$

where PC represents the percentage change of rainfall (%), n is the length of time series in years, S is the magnitude of the trend slope calculated using methods such as Simple Linear Regression (SLR) coefficient, Sen's slope, or Innovative Trend Analysis (ITA) slope, and $\overline{x}$ is the mean value of the time series. This equation allows for quantifying the percentage of change in rainfall over the specified time period, relative to the mean value of the time series.

3. Results and Discussion

3.1. Rainfall Variability

The mean rainfall and coefficients of variation (CV) for the monthly, seasonal, and annual rainfall of 26 grid points in Vaippar during the period 1971–2019 is presented in

Tables A2 and A3. During the study period, the Vaippar basin experienced an annual average rainfall of 762.57 mm. The highest rainfall of 891.93 mm was observed in the G08 grid point, while the lowest rainfall of 570.2 mm occurred in the G18 grid point. Out of 26 grid points, 12 grid points located mostly on the western side of the basin experienced annual rainfall that exceeded the average annual rainfall. The NEM is the major rainy season in the study area, contributing approximately 54.7% of the annual rainfall. Within the NEM season, the majority of the rainfall, around 85%, occurred in two months: October and November. The month of December contributed to approximately 15% of the summer rainfall. Among the grid points, the highest NEM rainfall of 503.54 mm was recorded in the G01 grid point, while the lowest NEM rainfall of 361.39 mm was recorded in the G18 grid point. Apart from the NEM season, the remaining seasons, namely SWM, winter, and summer, contributed 19.5%, 5.4%, and 20.4% of the annual rainfall, respectively. The maximum monthly rainfall was observed in October, accounting for 24.1% of the annual rainfall. The month of November followed closely with 22.3% contribution, while September contributed 9.9%, and December contributed 8.3% of the annual rainfall.

In the Nagariyar sub basin (G01 grid point), the maximum rainfall was recorded during the months of February, March, April, November, December, and the winter season. The G08 grid point (part of Arjunanadhi) experienced high rainfall in January, September, and October, annually, and during the summer season. On the other hand, the G20 grid point (part of Kousiganadhi) recorded higher rainfall in June, July, August, and during the Southwest Monsoon (SWM) season. The grid points located in the lower part of the Sinkottaiyar sub basin (G25) and Sindapalli Uppodai sub basin (G14 and G18) registered the minimum rainfall. The G25 grid point recorded the minimum monthly rainfall from January to September, November, SWM, winter, and summer seasons. The G18 grid point experienced the minimum rainfall during October, while the G14 grid point recorded the minimum rainfall in December.

The coefficient of variation (CV) of monthly, seasonal, and annual rainfall calculated for each grid point is shown in Table A3. Rainfall variability as classified by Hare (2003) [82], shows the CV for annual mean rainfall ranges from 22.9% to 31.5%, indicating moderate to high variability in rainfall distribution across all grids. The G09 grid point recorded the highest CV, while the G06 grid point recorded the lowest CV for annual rainfall. Most of the grid points exhibited moderate variability in annual rainfall, except for G08, G09, and G14, which are located at the centre of the basin. Rainfall variability was found to be greater in seasonal rainfall compared to annual rainfall. However, it is worth noting that the NEM season experienced relatively lower variability compared to the other seasons. Among the months, October (51.47%) and November (61.33%) exhibited lower coefficients of variation (CV) compared to the other months.

3.2. Trends of Annual and Seasonal Rainfall Series

The temporal trends were identified using the SLR, MK/MMK, and ITA methods at different grid points of Vaippar basin for annual and seasonal rainfall series. The calculated values of the SLR t test, MK/MMK test statistic (Z test), and ITA (slope values) were spatially mapped for each grid point by IDW interpolation method using QGIS 3.30.2 software. The magnitude of the trends was identified and percentage changes in mean rainfall were also calculated. The comparison of number of grid points expressing the significant trends and correlation among the different trend methods were also attempted. The results are discussed in the subsequent sections.

3.3. Trends of Annual and Seasonal Rainfall Series by Simple Linear Regression

The spatial distributions of trends in annual and seasonal rainfall, detected by the SLR method at the 5% and 10% significance levels, are shown in Figure 3. The analysis of temporal trends using the simple linear regression for annual and seasonal rainfall series showed that approximately 73% of grid points exhibited non-significant upward trends in the annual rainfall series. Although the NEM rainfall is a major contributor to the annual rainfall, the trend pattern was not similar, with 50% of grid points showing a non-significant downward trend. The winter and summer rainfall series demonstrated that 57.7% and 76.9% of grid points, respectively, displayed non-significant upward trends.

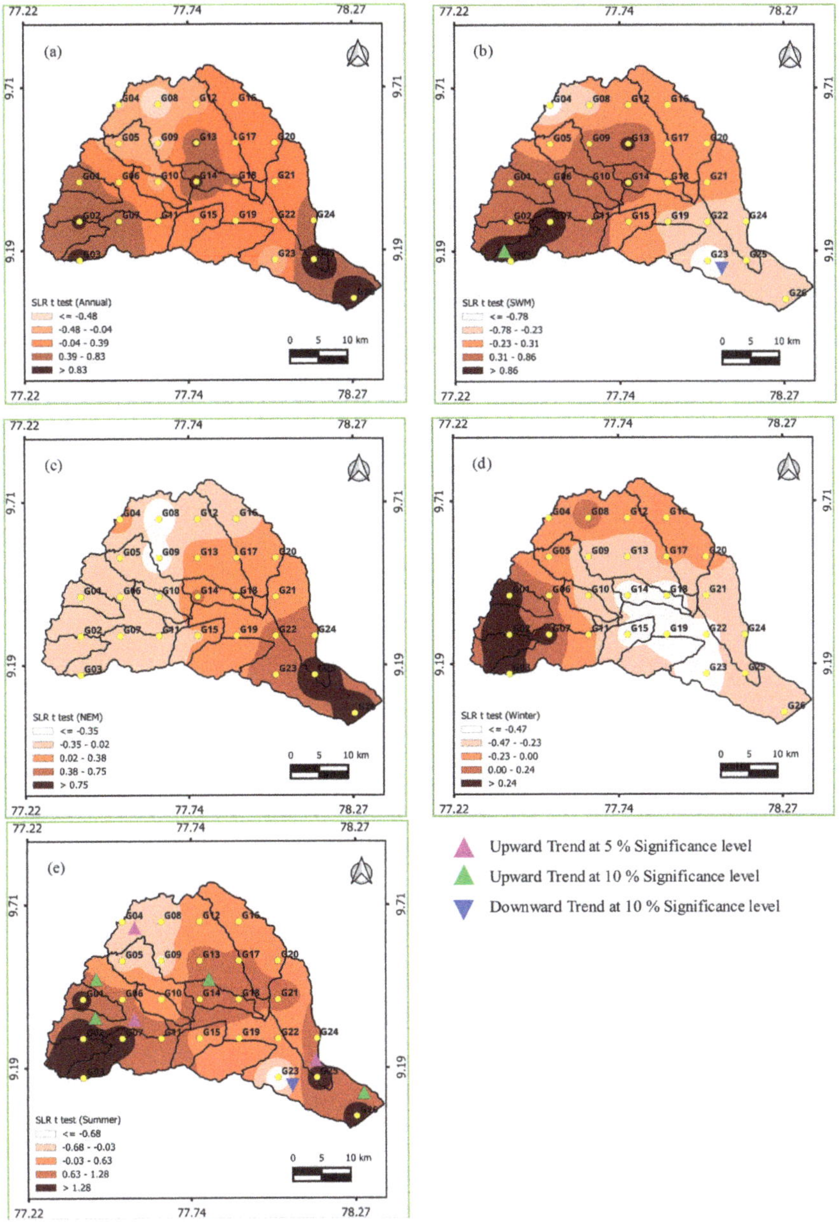

Figure 3. Test results of Simple Linear Regression (SLR) at 5 and 10% significance level for (**a**) annual (ANL), (**b**) southwest monsoon (SWM) (**c**) northeast monsoon (NEM), (**d**) winter (WIN), (**e**) summer (SUM) gridded rainfall series of Vaippar basin.

Among the five-rainfall series (annual and seasonal), only two series (SWM and summer) exhibited 7.69% of significant trends at the extreme ends of the basin. In the SWM series, an upward trend was observed at the G03 grid point and a downward trend at the G23 grid point, both at a 10% significance level. The summer rainfall series showed an upward trend at the G01, G02, G14, and G26 grid points at a 10% significance level, and at the G04, G07, and G25 grid points at a 5% significance level. A significant downward trend was noticed at the G23 grid point for the summer series. Consequently, the expected decrease in rainfall will not have a significant impact on water availability, as these seasons contribute very little to the annual rainfall.

3.4. Trends of Annual and Seasonal Rainfall Series by MK/MMK Test

The Mann–Kendall (MK) method was applied to the annual and seasonal rainfall series at different grid points of the Vaippar basin to identify significant trends at the 5% and 10% significance levels using the Z-test. The MK test was conducted for the annual and seasonal series, taking into account the auto-correlated non-significant series at lag-1. The modified MK (Z_c) test was performed only for statistically significant auto-correlated series.

Autocorrelation analysis was conducted to select the appropriate trend analysis method and evaluate the performance of both the original and normalized rainfall series. The autocorrelation coefficient for 26 grid points at lag-1 period for annual and seasonal rainfall series were worked out and the correlogram is presented in Figure 4. The upper and lower bound were decided by the 95% confidence interval to test the limits of the autocorrelation coefficient. The autocorrelation was considered as significant if it is greater than or lower than ±0.28. Since autocorrelation was found to be significant for five rainfall series viz ANL-G10, ANL-G12, SWM-G25, WIN-G23 and WIN-G25, the modified MK test was performed for these five series.

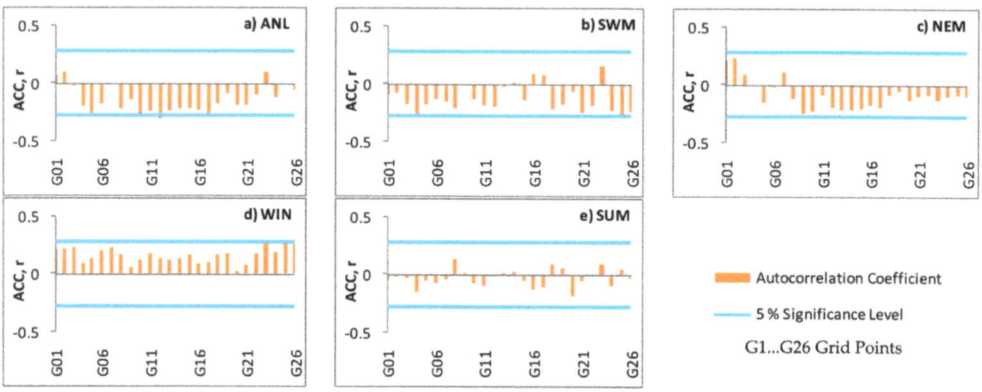

Figure 4. Autocorrelation coefficient at lag-1 period for annual and seasonal rainfall series of Vaippar basin. (ANL—annual, SWM—southwest monsoon, NEM—northeast monsoon, WIN—winter, SUM—summer).

The spatial pattern of trends in annual and seasonal rainfall, identified using the MK methods, is presented in Figure 5. Compared to the SLR test and ITA method, very few statistically significant trends were observed in the annual and seasonal rainfall series for both the MK and MMK tests. The temporal patterns of trend detected by MK test indicated that approximately 73% of the grid points for the annual series and 58% of the grid points for the NEM series showed non-significant upward trends. Furthermore, 54% of the grid points for the SWM series displayed a non-significant downward trend.

The MK test detected a downward trend at the 5% significance level in the G23 grid point for the summer series. A total of five rainfall series were tested using the modified MK method, which detected a significant downward trend at the G23 grid point for the SWM series, while the same grid point exhibited a significant upward trend for the winter season.

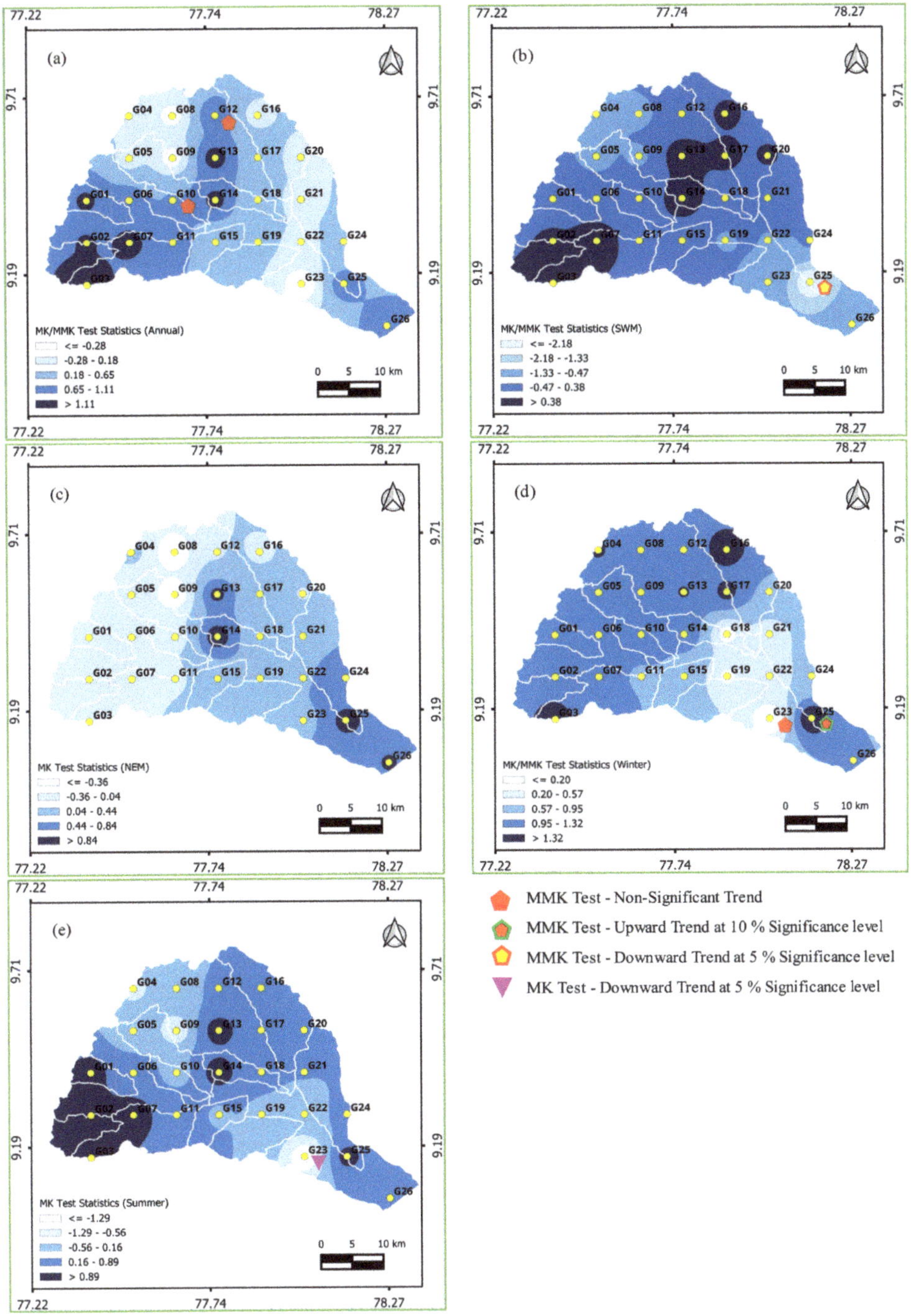

Figure 5. Mann–Kendall and Modified Mann–Kendall Z values at 5 and 10% significance level for (**a**) annual (ANL), (**b**) southwest monsoon (SWM) (**c**) northeast monsoon (NEM), (**d**) winter (WIN), (**e**) summer (SUM) gridded rainfall series of Vaippar basin.

3.5. Trends of Annual and Seasonal Rainfall Series by ITA Method

The grid-wise trend parameters for the annual and seasonal rainfall series of the Vaippar basin, as detected by the ITA method, are presented in Tables 2–4. These tables provide valuable insights into the trends observed in the rainfall patterns across different grids within the basin. As observed from the tables, the slope values of annual and seasonal rainfall series during the period 1971–2019 fall outside the lower and upper confidence limits (CL) for the particular grid point, suggesting existence of a significant trend in the rainfall pattern.

The slope values of the annual rainfall series are presented in Table 2. Among the 26 grid points analyzed, it is noteworthy that 15 grid points (57.69%) exhibited significant trends at both the 5% and 10% significance levels. Out of these significant trends, 11 grid points (42.3%) displayed a significant upward trend, while four grid points showed a significant downward trend.

Table 2. Grid-wise trend parameters detected by the ITA method for the annual and SWM rainfall series of the Vaippar basin.

Grid	Annual Series							SWM Series						
	S	σ	ρ	σs	LCL	UCL	Decision	S	σ	ρ	σs	LCL	UCL	Decision
G01	3.85	239.07	0.94	0.47	±0.77	±0.92	Ha **	0.13	59.89	0.98	0.06	±0.11	±0.13	Ha **
G02	4.47	241.49	0.95	0.47	±0.77	±0.91	Ha **	0.26	58.2	0.98	0.06	±0.10	±0.12	Ha **
G03	4.42	201.58	0.91	0.5	±0.82	±0.97	Ha **	0.64	48.23	0.96	0.08	±0.13	±0.16	Ha **
G04	−0.44	258.21	0.89	0.7	±1.16	±1.38	−Ha **	−1.39	85.92	0.97	0.13	±0.21	±0.25	Ho
G05	−1.11	206.69	0.96	0.36	±0.59	±0.71	Ho	−0.53	72.59	0.97	0.10	±0.16	±0.19	Ho
G06	1.41	193.6	0.92	0.44	±0.72	±0.86	Ha **	−0.01	60.38	0.95	0.11	±0.18	±0.21	−Ha **
G07	4.03	201.73	0.93	0.43	±0.72	±0.85	Ha **	0.5	50.63	0.97	0.07	±0.11	±0.13	Ha **
G08	−2.48	274.13	0.97	0.36	±0.59	±0.70	Ho	−0.5	104.81	0.97	0.16	±0.26	±0.31	Ho
G09	−4.24	264.56	0.92	0.6	±0.98	±1.17	Ho	−0.42	105.93	0.98	0.12	±0.20	±0.24	Ho
G10	−1.21	203.12	−0.27	1.89	±3.10	±3.70	−Ha **	−0.22	72.52	−0.19	0.65	±1.07	±1.28	−Ha **
G11	1.62	180.2	0.91	0.45	±0.74	±0.88	Ha **	0.23	53.13	0.94	0.11	±0.18	±0.21	Ha **
G12	−0.86	203.68	0.97	0.27	±0.44	±0.52	Ho	−0.41	69.96	0.97	0.1	±0.16	±0.19	Ho
G13	1.43	215.23	0.99	0.21	±0.35	±0.42	Ha **	0.19	71.67	0.97	0.11	±0.18	±0.21	Ha *
G14	2.05	221.59	0.98	0.23	±0.38	±0.46	Ha **	0.2	71.07	0.97	0.1	±0.17	±0.20	Ha **
G15	0.1	193.97	0.97	0.28	±0.46	±0.55	Ho	−0.37	61.36	0.96	0.11	±0.17	±0.21	Ho
G16	−0.77	200.28	0.99	0.2	±0.33	±0.39	Ho	−0.67	75.26	0.96	0.12	±0.19	±0.23	Ho
G17	−0.24	195.79	0.98	0.21	±0.34	±0.41	−Ha **	−0.51	71.02	0.96	0.12	±0.20	±0.24	Ho
G18	−0.72	188.77	0.98	0.23	±0.38	±0.45	Ho	−0.51	67.84	0.93	0.15	±0.25	±0.29	Ho
G19	−1.15	179.03	0.96	0.28	±0.47	±0.56	Ho	−0.72	62.15	0.95	0.11	±0.19	±0.22	Ho
G20	−0.47	211.24	0.96	0.37	±0.60	±0.72	−Ha **	−0.45	74.42	0.97	0.11	±0.19	±0.22	Ho
G21	−0.55	189.85	0.98	0.21	±0.34	±0.4	Ho	−0.47	61.99	0.96	0.1	±0.16	±0.19	Ho
G22	−0.94	172.76	0.97	0.25	±0.41	±0.48	Ho	−0.7	55.57	0.96	0.09	±0.15	±0.18	Ho
G23	−1.54	192.63	0.96	0.32	±0.52	±0.63	Ho	−0.76	54.19	0.96	0.09	±0.15	±0.18	Ho
G24	0.44	161.5	0.97	0.23	±0.37	±0.44	Ha *	−0.37	46.9	0.99	0.04	±0.07	±0.09	Ho
G25	2.35	160.77	0.97	0.23	±0.38	±0.46	Ha **	−0.13	42.57	0.98	0.05	±0.09	±0.11	Ho
G26	1.63	153.19	0.97	0.21	±0.35	±0.41	Ha **	−0.15	40.69	0.97	0.06	±0.10	±0.12	Ho

S—ITA slope; σ—standard deviation; ρ—correlation coefficient between the means of two half series; σs—standard deviation of trend slope; LCL/UCL—lower/upper confidence limits; Ho: no significant trend; Ha: there is a significant trend; **—significance level at 5 and 10%; *—significance level at 5%.

The ITA analysis of the seasonal rainfall series displayed in Tables 2–4 revealed that significant trends were observed in the NEM and summer rainfall series in 20 and 21 grids i.e., 76.92 and 80.77%, respectively. In contrast, the SWM and winter rainfall series detected significant trends in nine and eight grids i.e., 34.62 and 30.77%, respectively. For the NEM season, 53.85% of the grids showed a significant upward trend, while 23% of the grid points showed a significant downward trend. In the case of summer rainfall, a significant upward trend was observed in 73.1% of the grid points. SWM exhibited a significant upward trend in 26.9% of the grid points. Winter, on the other hand, experienced a significant downward trend in 19.2% of the grid points. These findings suggest that the NEM and summer seasons experienced more widespread and pronounced changes in rainfall patterns compared to the SWM and winter seasons.

Table 3. Grid-wise trend parameters detected by the ITA method for the NEM and winter rainfall series of the Vaippar basin.

Grid	NEM Series							Winter Series						
	S	σ	ρ	σs	LCL	UCL	Decision	S	σ	ρ	σs	LCL	UCL	Decision
G01	1.8	165.65	0.97	0.25	±0.41	±0.49	Ha **	0.42	74.07	0.95	0.14	±0.23	±0.27	Ha **
G02	2.03	170.53	0.97	0.23	±0.38	±0.45	Ha **	0.46	72.93	0.95	0.13	±0.22	±0.26	Ha **
G03	1.79	168.28	0.93	0.36	±0.60	±0.71	Ha **	0.21	52.93	0.9	0.14	±0.22	±0.27	Ho
G04	1.85	182.03	0.99	0.18	±0.30	±0.35	Ha **	−0.04	72.31	0.98	0.08	±0.13	±0.15	−Ha **
G05	0.01	146.05	0.97	0.21	±0.34	±0.40	Ho	−0.21	60.18	0.98	0.07	±0.11	±0.13	Ho
G06	0.71	140.19	0.98	0.16	±0.27	±0.32	Ha **	0.01	61.02	0.97	0.09	±0.15	±0.18	Ho
G07	1.69	161.06	0.95	0.28	±0.47	±0.55	Ha **	0.22	57.28	0.94	0.11	±0.19	±0.22	Ha *
G08	−1.61	200.77	0.97	0.29	±0.48	±0.57	Ho	0.09	65.16	0.93	0.14	±0.24	±0.28	Ho
G09	−1.89	153.9	0.96	0.25	±0.42	±0.50	Ho	−0.62	60.99	0.97	0.08	±0.13	±0.16	Ho
G10	−0.43	139.8	−0.15	1.24	±2.04	±2.43	−Ha **	−0.48	54.8	−0.12	0.48	±0.79	±0.94	−Ha **
G11	0.73	143.7	0.97	0.22	±0.35	±0.42	Ha **	−0.28	49.13	0.93	0.11	±0.18	±0.21	Ho
G12	−0.42	153.33	0.96	0.25	±0.41	±0.49	Ha #	−0.26	55.33	0.98	0.06	±0.11	±0.13	Ho
G13	0.61	158.35	0.98	0.2	±0.33	±0.40	Ha **	−0.54	53.49	0.90	0.14	±0.23	±0.27	Ho
G14	1.05	159.23	0.96	0.25	±0.41	±0.49	Ha **	−0.64	52.20	0.87	0.15	±0.25	±0.30	Ho
G15	0.31	146.56	0.97	0.19	±0.32	±0.38	Ho	−0.55	48.30	0.85	0.15	±0.25	±0.30	Ho
G16	−0.58	156.15	0.96	0.26	±0.43	±0.51	Ho	−0.24	57.62	0.78	0.22	±0.36	±0.43	−Ha **
G17	−0.32	153.28	0.97	0.23	±0.37	±0.45	−Ha **	−0.34	55.80	0.82	0.2	±0.32	±0.38	−Ha #
G18	−0.45	150.33	0.95	0.27	±0.44	±0.53	−Ha #	−0.6	46.23	0.92	0.11	±0.18	±0.22	Ho
G19	−0.36	146.27	0.96	0.24	±0.40	±0.47	−Ha **	−0.57	45.41	0.90	0.12	±0.20	±0.24	Ho
G20	−0.47	160.83	0.97	0.22	±0.36	±0.43	Ho	−0.3	57.04	0.93	0.12	±0.20	±0.24	Ho
G21	−0.25	149.28	0.97	0.2	±0.33	±0.40	−Ha **	−0.47	48.74	0.93	0.11	±0.18	±0.21	Ho
G22	−0.02	141.42	0.97	0.19	±0.32	±0.38	−Ha **	−0.5	45.57	0.91	0.11	±0.18	±0.22	Ho
G23	0.49	156.75	0.97	0.23	±0.37	±0.45	Ha **	−0.52	48.58	0.95	0.09	±0.15	±0.18	Ho
G24	0.54	130.83	0.98	0.16	±0.26	±0.32	Ha **	−0.37	44.09	0.91	0.11	±0.18	±0.21	Ho
G25	1.53	124.41	0.98	0.13	±0.22	±0.26	Ha **	−0.23	43.35	0.77	0.17	±0.28	±0.33	−Ha **
G26	1.11	122.91	0.96	0.19	±0.32	±0.38	Ha **	−0.28	42.42	0.88	0.12	±0.20	±0.23	Ho

S—ITA slope; σ—standard deviation; ρ—correlation coefficient between the means of two half series; σs—standard deviation of trend slope; LCL/UCL—lower/upper confidence limits; Ho: no significant trend; Ha: there is a significant trend; **—significance level at 5 and 10%; *—significance level at 5%; #—significance level at 10%.

Table 4. Grid-wise trend parameters detected by the ITA method for the summer rainfall series of the Vaippar basin.

Grid	S	σ	ρ	σs	LCL	UCL	Decision
G01	1.5	100.39	0.97	0.15	±0.25	±0.29	Ha **
G02	1.72	99.15	0.97	0.15	±0.24	±0.29	Ha **
G03	1.78	75.67	0.97	0.11	±0.19	±0.22	Ha **
G04	−0.85	109.55	0.88	0.31	±0.51	±0.61	Ho
G05	−0.38	88.4	0.85	0.29	±0.47	±0.56	−Ha **
G06	0.71	84.68	0.92	0.2	±0.33	±0.40	Ha **
G07	1.63	80.12	0.96	0.14	±0.23	±0.27	Ha **
G08	−0.46	104.04	0.94	0.21	±0.34	±0.40	Ho
G09	−1.32	109.89	0.80	0.41	±0.67	±0.80	Ho
G10	−0.08	80.80	0.85	0.26	±0.43	±0.51	−Ha **
G11	0.93	68.78	0.96	0.12	±0.20	±0.23	Ha **
G12	0.23	77.96	0.93	0.17	±0.28	±0.33	Ho
G13	1.16	75.63	0.99	0.07	±0.12	±0.15	Ha **
G14	1.43	75.76	0.99	0.04	±0.07	±0.09	Ha **
G15	0.7	69.62	0.97	0.1	±0.17	±0.20	Ha **
G16	0.73	74.36	0.98	0.09	±0.15	±0.18	Ha **
G17	0.93	71.56	0.97	0.1	±0.16	±0.20	Ha **
G18	0.84	65.38	0.96	0.11	±0.18	±0.22	Ha **
G19	0.5	62.65	0.98	0.07	±0.11	±0.14	Ha **
G20	0.75	67.65	0.97	0.09	±0.15	±0.18	Ha **
G21	0.65	60.68	0.97	0.08	±0.13	±0.16	Ha **
G22	0.27	57.60	0.98	0.06	±0.10	±0.12	Ha **
G23	−0.74	59.86	0.97	0.09	±0.14	±0.17	Ho
G24	0.64	53.79	0.96	0.09	±0.15	±0.18	Ha **
G25	1.19	59.00	0.94	0.12	±0.19	±0.23	Ha **
G26	0.96	53.79	0.96	0.09	±0.14	±0.17	Ha **

S—ITA slope; σ—standard deviation; ρ—correlation coefficient between the means of two half series; σs—standard deviation of trend slope; LCL/UCL—lower/upper confidence limits; Ho: no significant trend; Ha: there is a significant trend; **—significance level at 5 and 10%.

The spatial distribution of slope values obtained through the ITA method, along with their significance for seasonal and annual rainfall series, is illustrated in Figure 6. Grid points located in the southwestern and central parts (G01, G02, G03, G09, G10) of the basin exhibited a higher number of significant trends in annual rainfall. In terms of the NEM season, significant trends were observed across almost all parts of the basin. The southeastern, central, and extreme southern parts experienced a significant upward trend, while the eastern parts displayed a significant downward trend. Significant upward trends were identified in the southwestern parts for the SWM season. For the summer season, significant upward trends were observed in the western, southern, and eastern parts of the basin. Notably, the western parts exhibited a significant upward trend during the winter season. Among the grid points, G10 stood out with a notably higher number of significant downward trends in monthly, annual, and seasonal rainfall series. This indicates a consistent and significant decrease in rainfall at that specific location.

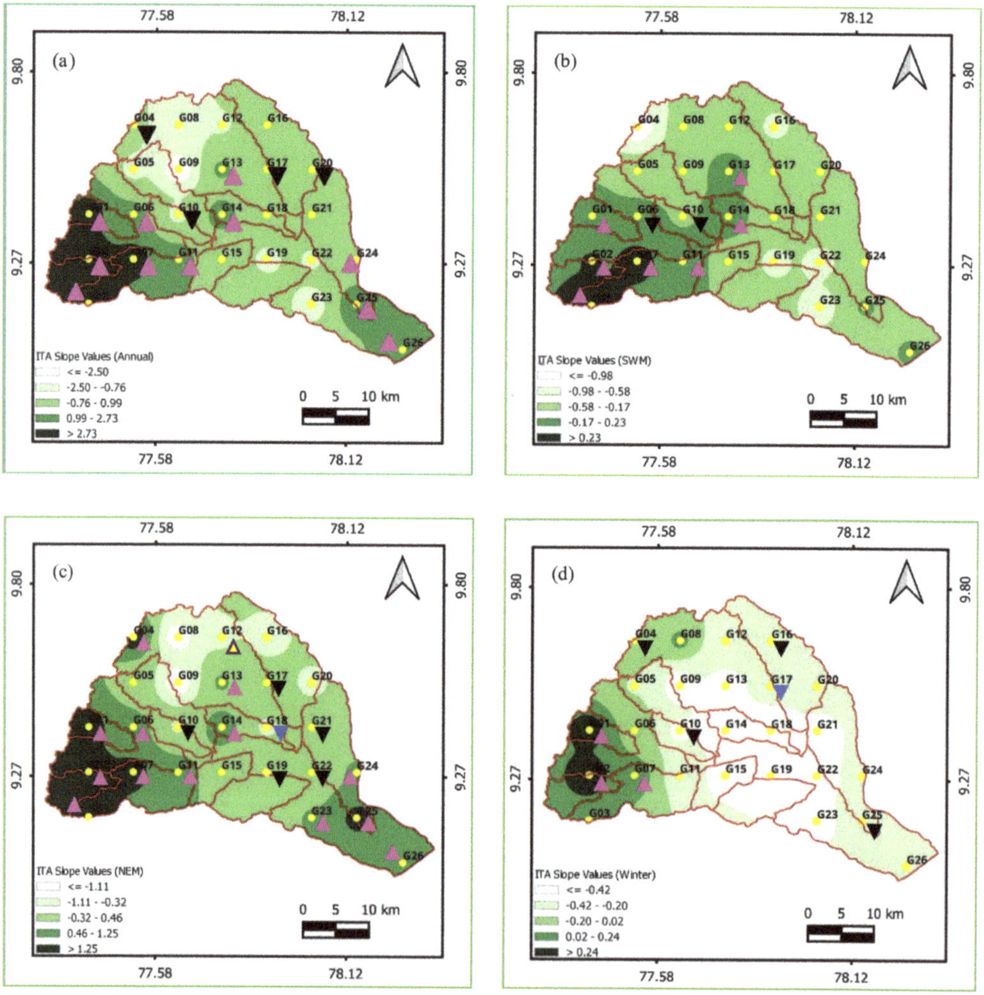

Figure 6. *Cont.*

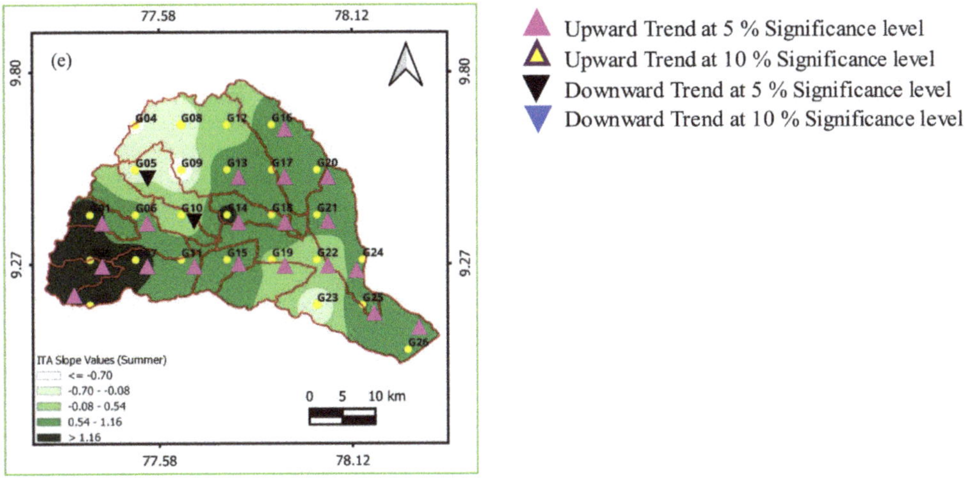

Figure 6. The ITA slope values at 5 and 10% significance level for (**a**) annual (ANL), (**b**) southwest monsoon (SWM) (**c**) northeast monsoon (NEM), (**d**) winter (WIN), (**e**) summer (SUM) gridded rainfall series of Vaippar basin.

From the spatial analysis of the Vaippar basin, it is evident that the southeastern, central, and extreme southern parts have exhibited a positive increase in annual rainfall. Moreover, when considering the different seasons, it is notable that the NEM season has displayed widespread positive trends across various parts of the basin. It is important to highlight that these positive trends in rainfall can have significant implications for water availability, agricultural productivity, and overall ecological balance within the Vaippar basin.

3.6. Identification and Nature of Subtrends by ITA Method

One of the most important features of the ITA method is its ability to identify the subtrend of a rainfall series [8]. Additionally, the scattered plot allows for the detection of both monotonic and non-monotonic upward or downward trends in a series. To conduct a comprehensive analysis for identifying subtrends within a rainfall series, scatter diagrams can be used, with data points plotted on 1:1 line graphs. These scatter diagrams are then divided into three segments based on rainfall depth: low, medium, and high rainfall segments. In the scenario where the annual series exhibits a combination of different trend patterns within the series, it is referred to as a non-monotonic trend. Conversely, if the series demonstrates a consistent upward or downward pattern, it is considered a monotonic trend. Additionally, it is also possible for the series to exhibit no trend if the scatter points align closely with the 1:1 line or if the slope values are zero.

The grid-wise subtrends and nature of trends observed in the annual and seasonal rainfall series for the Vaippar basin are displayed in Table 5. In the table, the subtrends of rainfall series, categorized as low, medium, and high rainfall segments, are indicated by upward or downward arrows. The nature of the trend is represented by whether it is monotonic or non-monotonic (upward/downward). Analyzing the annual and NEM series, it can be observed that five grid points (G01, G02, G03, G07 and G25) located at the southwestern parts of the basin exhibited monotonically upward trends. On the other hand, for the SWM, 10 grid points showed monotonically downward trends. In the case of the summer rainfall series, a monotonically upward trend was identified at 13 grid points.

For the annual rainfall series, upward trends were observed in the low rainfall segment for 21 grid points, in the medium rainfall segment for 14 grid points, and in the high rainfall segment for five grid points. As for the NEM rainfall, 24 grid points showed upward trends in the low rainfall segment, 15 grid points in the medium rainfall segment, and five grid points in the high rainfall segment. The high rainfall segments, as classified for annual and NEM rainfall, were found in the G01, G02, G03, G07, and G25 grid points, which are located in the southwestern parts of the basin. On the other hand, the SWM season exhibited the minimum number of grid points registering upward trends in the low, medium, and high rainfall segments. A higher number of grid points experienced upward trends in the low and medium rainfall segments for the winter series. Conversely, the summer rainfall series exhibited upward trends in the low, medium, and high rainfall segments for a larger number of grid points.

Table 5. Subtrends of annual and seasonal rainfall series for Vaippar basin.

Grid	RF	ANL	SWM	NEM	WIN	SUM	Grid	ANL	SWM	NEM	WIN	SUM
G01	Low	↑	↓	↑	↑	↑	G02	↑	↓	↑	↑	↑
	Medium	↑	↓	↑	↑	↑		↑	↓	↑	↑	↑
	High	↑	↑	↑	↓	↑		↑	↑	↑	↓	↑
	Nature	MU ↑	NMD ↓	MU ↑	NMU↑	MU↑		MU ↑	NMU ↑	MU ↑	NMU ↑	MU ↑
G03	Low	↑	↓	↑	↑	↑	G04	↓	↓	↑	↑	↓
	Medium	↑	↑	↑	↑	↑		↑	↓	↑	↑	↓
	High	↑	↑	↑	↓	↑		↓	↓	↑	↓	↓
	Nature	MU ↑	NMU ↑	MU ↑	NMD ↓	MU ↑		NMD ↓	MD ↓	MU ↑	NMD ↓	MD ↓
G05	Low	↑	↓	↑	↑	↓	G06	↑	↓	↑	↑	↑
	Medium	↑	↓	↓	↑	↓		↑	↓	↑	↑	↑
	High	↓	↓	↓	↓	↓		↓	↑	↓	↓	↓
	Nature	NMD ↓	MD ↓	NMD ↓	NMD ↓	MD ↓		NMU ↑	NMD ↓	NMU ↑	NMD ↓	NMU ↑
G07	Low	↑	↓	↑	↑	↑	G08	↓	↓	↓	↑	↓
	Medium	↑	↓	↑	↑	↑		↓	↓	↓	↓	↓
	High	↑	↑	↑	↓	↑		↓	↓	↓	↓	↓
	Nature	MU ↑	NMU ↑	MU ↑	NMD ↓	MU ↑		MD ↓	MD ↓	MD ↓	NMD ↓	MD ↓
G09	Low	↓	↓	↓	↑	↓	G10	↑	↑	↑	↑	↓
	Medium	↓	↓	↓	↓	↓		↓	↓	↑	↑	↓
	High	↓	↓	↓	↓	↓		↓	↓	↓	↓	↓
	Nature	MD ↓	MD ↓	MD ↓	NMD ↓	MD ↓		NMD ↓	NMD ↓	NMD ↓	NMD ↓	MD ↓
G11	Low	↑	↓	↑	↑	↑	G12	↑	↑	↑	↑	↓
	Medium	↑	↑	↑	↑	↑		↑	↓	↓	↑	↑
	High	↓	↑	↓	↓	↑		↓	↓	↓	↓	↓
	Nature	NMU ↑	NMU ↑	NMU ↑	NMD ↓	MU ↑		NMD ↓	NMD ↓	NMD ↓	NMD ↓	NMD ↓
G13	Low	↑	↑	↑	↑	↑	G14	↑	↑	↑	↑	↑
	Medium	↑	↑	↑	↑	↑		↑	↑	↑	↑	↑
	High	↓	↓	↓	↓	↑		↓	↓	↓	↓	↑
	Nature	NMU ↑	NMD ↓	NMU ↑	NMD ↓	MU ↑		NMU ↑	NMD ↓	NMU ↑	NMD ↓	MU ↑
G15	Low	↓	↓	↑	↑	↓	G16	↑	↑	↑	↑	↑
	Medium	↓	↓	↑	↑	↑		↓	↓	↓	↑	↑
	High	↓	↓	↓	↓	↑		↓	↓	↓	↓	↑
	Nature	MD ↓	MD ↓	NMD ↓	NMD ↓	NMU ↑		NMD ↓	NMD ↓	NMD ↓	NMD ↓	MU ↑
G17	Low	↑	↑	↑	↑	↑	G18	↑	↓	↑	↑	↑
	Medium	↓	↓	↓	↑	↑		↓	↓	↓	↓	↓
	High	↓	↓	↓	↓	↑		↓	↓	↓	↓	↑
	Nature	NMD ↓	NMD ↓	NMD ↓	NMD ↓	MU ↑		NMD ↓	MD ↓	NMD ↓	NMD ↓	NMU ↑
G19	Low	↑	↓	↑	↑	↑	G20	↑	↑	↑	↑	↑
	Medium	↓	↓	↓	↓	↓		↓	↓	↓	↑	↑
	High	↓	↓	↓	↓	↑		↓	↓	↓	↓	↑
	Nature	NMD ↓	MD ↓	NMD ↓	NMD ↓	NMU ↑		NMD ↓	NMD ↓	NMD ↓	NMD ↓	MU ↑
G21	Low	↑	↓	↑	↑	↑	G22	↑	↓	↑	↑	↓
	Medium	↓	↓	↓	↓	↑		↓	↓	↓	↑	↓
	High	↓	↓	↓	↓	↑		↓	↓	↓	↓	↑
	Nature	NMD ↓	MD ↓	NMD ↓	NMD ↓	MU ↑		NMD ↓	NMD ↓	NMD ↓	NMD ↓	NMD ↓
G23	Low	↓	↓	↑	↑	↓	G24	↑	↓	↑	↑	↑
	Medium	↓	↓	↑	↓	↓		↑	↓	↑	↑	↑
	High	↓	↓	↓	↓	↓		↓	↓	↓	↓	↑
	Nature	MD ↓	MD ↓	NMD ↓	NMD ↓	MD ↓		NMD ↓	MD ↓	NMU ↑	NMD ↓	MU ↑
G25	Low	↑	↓	↑	↑	↑	G26	↑	↓	↑	↑	↑
	Medium	↑	↓	↑	↑	↑		↑	↓	↑	↑	↑
	High	↑	↑	↓	↓	↑		↓	↑	↓	↓	↑
	Nature	MU ↑	NMD ↓	NMU ↑	NMD ↓	MU ↑		NMU ↑	NMD ↓	NMU ↑	NMD ↓	MU ↑

ANL—annual, SWM—southwest monsoon, NEM—northeast monsoon, WIN—winter, SUM—summer gridded rainfall series; MU ↑—Monotonic upward; NMU ↑—Non-monotonic upward; MD ↓—Monotonic downward; NMD ↓—Non-monotonic downward.

3.7. Magnitude of Trend in Rainfall Series

The magnitude of the trend (β), computed using the SLR method, and the Sen's slope (Q) are presented in Table 6. Considerable variability was observed in both the physical value and sign of the magnitude for different grid points. The SLR slopes indicate that the highest positive magnitude of 2.25 mm/year was recorded at the G02 grid point, while the lowest negative magnitude of −2.62 mm/year was observed at the G08 grid point for annual rainfall series. Furthermore, it is worth noting that these magnitudes indicate the absence of a statistically significant trend in the annual rainfall series. The upward trend was found to be more pronounced in the northern and southern parts of the basin. However, in the case of NEM rainfall, which contributes significantly to the annual rainfall, the trend magnitude did not follow a similar pattern as the annual rainfall. Grid point G25 (1.45 mm/year) and G08 (−1.52 mm/year) registered higher positive and negative magnitudes for the NEM series, respectively. Additionally, for the NEM series, the northern parts of the basin exhibited a negative magnitude, while the southern parts showed a positive magnitude.

Table 6. Magnitude of SLR slope (β) and Sen's Slope (Q) for Vaippar basin.

Grid	SLR Slope (β)					Sen's Slope (Q)				
	ANL	SWM	NEM	WIN	SUM	ANL	SWM	NEM	WIN	SUM
G01	1.94	0.43	−0.32	0.34	1.50	1.56	0.21	−0.14	0.47	1.17
G02	2.25	0.50	−0.30	0.37	1.67	2.22	0.28	−0.32	0.51	1.39
G03	1.96	0.71	−0.46	0.22	1.50	3.27	0.41	−0.19	0.34	1.19
G04	−1.37	−1.09	0.52	−0.05	−0.75	−0.35	−0.96	0.63	0.32	−0.73
G05	−0.83	0.01	−0.45	−0.09	−0.29	0.04	−0.53	−0.35	0.37	−0.43
G06	0.66	0.41	−0.52	0.07	0.71	1.78	−0.03	−0.25	0.40	0.42
G07	1.82	0.65	−0.47	0.21	1.43	2.81	0.48	−0.26	0.36	1.16
G08	−2.62	−0.51	−1.52	0.14	−0.73	−1.59	−0.72	−1.66	0.21	−0.63
G09	−1.68	0.68	−1.12	−0.31	−0.93	−1.49	−0.63	−1.32	0.16	−0.96
G10	−0.45	0.40	−0.55	−0.25	−0.07	0.80	−0.27	−0.25	0.31	−0.10
G11	0.58	0.46	−0.43	−0.12	0.67	1.95	0.18	−0.04	0.24	0.46
G12	−0.40	0	−0.35	−0.08	0.04	1.12	−0.22	0.01	0.34	0.13
G13	1.98	0.78	0.59	−0.27	0.88	3.76	0.78	1.69	0.24	1.10
G14	2.20	0.70	0.81	−0.35	1.04	3.89	0.83	1.74	0.22	1.09
G15	0.04	−0.08	0	−0.28	0.41	0.55	−0.06	0.38	0.13	0.06
G16	0.2	0.06	−0.17	−0.02	0.33	−0.09	0.30	−0.12	0.30	0.24
G17	0.74	0.21	0.1	−0.09	0.52	0.90	0.43	0.14	0.27	0.49
G18	0.81	0.04	0.51	−0.35	0.6	0.55	0.02	0.36	0	0.20
G19	0.28	−0.3	0.5	−0.28	0.35	0.44	−0.35	0.77	0.01	−0.03
G20	0.65	0.11	0.18	−0.06	0.43	−0.23	0.37	0.04	0.10	0.26
G21	0.68	−0.07	0.52	−0.23	0.45	−0.06	0.07	0.43	0.02	0.12
G22	0.19	−0.37	0.69	−0.26	0.13	0.19	−0.34	1.02	0.04	−0.18
G23	−0.87	−0.75	1.06	−0.36	−0.82	−1.60	−0.58	0.60	−0.01	−1.22
G24	0.98	−0.21	0.89	−0.18	0.48	0.60	−0.14	1.00	0.11	0.18
G25	2.12	−0.24	1.45	−0.11	1.02	1.42	−0.31	1.43	0.06	0.59
G26	1.66	−0.15	1.14	−0.13	0.8	1.52	−0.26	1.23	0.14	0.41

ANL—annual, SWM—southwest monsoon, NEM—northeast monsoon, WIN—winter, SUM—summer gridded rainfall series.

The magnitudes of Sen's slope (Q) calculated for annual and seasonal rainfall as presented in Table 6 showed higher positive magnitudes of the non-significant trends. For the annual rainfall series, the highest magnitude of 3.89 mm/year was observed at the G14 grid point, while the lowest magnitude of −1.6 mm/year was recorded at the G23 grid point for a non-significant trend. Positive magnitudes were noticed for more than 69.2% of the grids in the annual rainfall series. Higher positive magnitudes were observed in the southwestern and central parts of the basin for the annual series. In the case of NEM, the highest non-significant magnitude of 1.74 mm/year was observed at the G14 grid point, while the lowest magnitude of −1.66 mm/year was recorded at the G08 grid point.

Positive magnitudes were noticed for more than 57.7% of the grids in the NEM rainfall series. Negative magnitudes were noticed in the western and central parts of the basin for the NEM series, while the remaining part of the basin recorded positive magnitudes.

The slope values estimated using the ITA method for the annual and seasonal series are presented in Tables 2–4. The slope values of the annual series indicate that 42.3% of the grids exhibited positive magnitudes. Grid point G02 displayed a significantly upward trend with a slope of 4.47 mm/year, while G09 showed a non-significant downward trend with a slope value of −4.24 mm/year for the annual series. Regarding the seasonal rainfall, the NEM and summer series exhibited predominantly positive magnitudes, while the SWM and winter series showed negative magnitudes. Grid point G02 exhibited a significantly upward trend with a slope of 2.03 mm/year, whereas G09 showed a non-significant downward trend with a slope value of −1.89 mm/year for the NEM series. A similar pattern of positive and negative magnitudes, similar to the annual rainfall, was observed for the NEM series. Positive magnitudes were observed in the western, central, and eastern parts of the basin for both the annual and NEM series. Among the three methods considered, the ITA method consistently estimated a higher magnitude of trend for all the rainfall series compared to the other methods.

3.8. Percentage Change in Magnitude of Trend

The percentage change in magnitude of trend from the mean values over the study period calculated for the seasonal and annual rainfall series using the SLR slope (β), Sen's slope (Q), and ITA slope (S) are presented in Table 7.

Table 7. Percentage change in magnitude of trend from mean for SLR, Sen' Slope, and ITA slopes for Vaippar basin.

Grid	SLR Slope (β)					Sen's Slope (Q)					ITA Slope (S)				
	ANL	SWM	NEM	WIN	SUM	ANL	SWM	NEM	WIN	SUM	ANL	SWM	NEM	WIN	SUM
G01	12.47	14.18	−3.76	40.41	47.20	10.02	6.93	−1.64	55.86	36.82	24.74	4.29	21.15	49.92	47.2
G02	14.46	16.49	−3.52	43.97	52.55	14.26	9.24	−3.76	60.61	43.74	28.72	8.58	23.85	54.67	54.13
G03	12.59	23.42	−5.40	26.15	47.20	21.01	13.52	−2.23	40.41	37.45	28.4	21.11	21.03	24.96	56.01
G04	−8.80	−35.96	6.11	−5.94	−23.60	−2.25	−31.67	7.40	38.03	−22.97	−2.83	−45.85	21.73	−4.75	−26.75
G05	−5.33	0.33	−5.29	−10.70	−9.13	0.26	−17.48	−4.11	43.97	−13.53	−7.13	−17.48	0.12	−24.96	−11.96
G06	4.24	13.52	−6.11	8.32	22.34	11.44	−0.99	−2.94	47.54	13.22	9.06	−0.33	8.34	1.19	22.34
G07	11.69	21.44	−5.52	24.96	45.00	18.06	15.83	−3.05	42.78	36.5	25.9	16.49	19.85	26.15	51.29
G08	−16.84	−16.82	−17.86	16.64	−22.97	−10.22	−23.75	−19.5	24.96	−19.83	−15.94	−16.49	−18.91	10.7	−14.48
G09	−10.80	22.43	−13.16	−36.84	−29.27	−9.57	−20.78	−15.51	19.02	−30.21	−27.24	−13.85	−22.2	−73.68	−41.54
G10	−2.89	13.20	−6.46	−29.71	−2.20	5.14	−8.91	−2.94	36.84	−3.15	−7.78	−7.26	−5.05	−57.05	−2.52
G11	3.73	15.17	−5.05	−14.26	21.08	12.53	5.94	−0.47	28.52	14.48	10.41	7.59	8.58	−33.28	29.27
G12	−2.57	0.00	−4.11	−9.51	1.26	7.20	−7.26	0.12	40.41	4.09	−5.53	−13.52	−4.93	−30.9	7.24
G13	12.72	25.73	6.93	−32.09	27.69	24.16	25.73	19.85	28.52	34.62	9.19	6.27	7.17	−64.18	36.5
G14	14.14	23.09	9.52	−41.60	32.73	25.00	27.38	20.44	26.15	34.3	13.17	6.6	12.34	−76.06	45
G15	0.26	−2.64	0.00	−33.28	12.90	3.53	−1.98	4.46	15.45	1.89	0.64	−12.21	3.64	−65.37	22.03
G16	1.29	1.98	−2.00	−2.38	10.38	−0.58	9.90	−1.41	35.65	7.55	−4.95	−22.1	−6.81	−28.52	22.97
G17	4.75	6.93	1.17	−10.70	16.36	5.78	14.18	1.64	32.09	15.42	−1.54	−16.82	−3.76	−40.41	29.27
G18	5.20	1.32	5.99	−41.60	18.88	3.53	0.66	4.23	0	6.29	−4.63	−16.82	−5.29	−71.31	26.43
G19	1.80	−9.90	5.87	−33.28	11.01	2.83	−11.55	9.05	1.19	−0.94	−7.39	−23.75	−4.23	−67.74	15.73
G20	4.18	3.63	2.11	−7.13	13.53	−1.48	12.21	0.47	11.88	8.18	−3.02	−14.84	−5.52	−35.65	23.6
G21	4.37	−2.31	6.11	−27.33	14.16	−0.39	2.31	5.05	2.38	3.78	−3.53	−15.5	−2.94	−55.86	20.45
G22	1.22	−12.21	8.11	−30.90	4.09	1.22	−11.22	11.98	4.75	−5.66	−6.04	−23.09	−0.23	−59.42	8.5
G23	−5.59	−24.74	12.45	−42.78	−25.80	−10.28	−19.13	7.05	−1.19	−38.39	−9.9	−25.07	5.76	−61.8	−23.29
G24	6.30	−6.93	10.46	−21.39	15.11	3.86	−4.62	11.75	13.07	5.66	2.83	−12.21	6.34	−43.97	20.14
G25	13.62	−7.92	17.03	−13.07	32.10	9.12	−10.23	16.8	7.13	18.57	15.1	−4.29	17.97	−27.33	37.45
G26	10.67	−4.95	13.39	−15.45	25.18	9.77	−8.58	14.45	16.64	12.9	10.47	−4.95	13.04	−33.28	30.21

ANL—annual, SWM—southwest monsoon, NEM—northeast monsoon, WIN—winter, SUM—summer gridded rainfall series.

Among three methods, the ITA slope exhibited a higher percentage change in the magnitude of the trend compared to SLR and Sen's slope. Additionally, it was observed that the percentage changes in magnitude from mean rainfall were lower in the annual and NEM rainfall series compared to the SWM, winter, and summer series. Positive values of percentage change in rainfall were more prominent in the annual, NEM, and summer series across all three methods.

Specifically, in the annual series, the extreme western, central, and eastern parts of the basin showed positive magnitudes, while the northern side exhibited negative magnitudes for all three methods. Regarding the NEM series, the western parts displayed negative magnitudes, and the eastern parts showed positive magnitudes for SLR and Sen's slope methods. However, for the ITA slope method, this pattern was reversed, with positive magnitudes following a similar pattern to the annual series.

3.9. Comparison of Trend Methods

The number of grid points that exhibited a significant upward or downward trend in seasonal and annual rainfall series is displayed in Figure 7. A total of 130 rainfall series, including annual and four seasonal series, were analyzed for trends in 26 grid points using the SLR, MK/MMK, and ITA methods. The test results were compared based on the number of significant upward or downward trends identified.

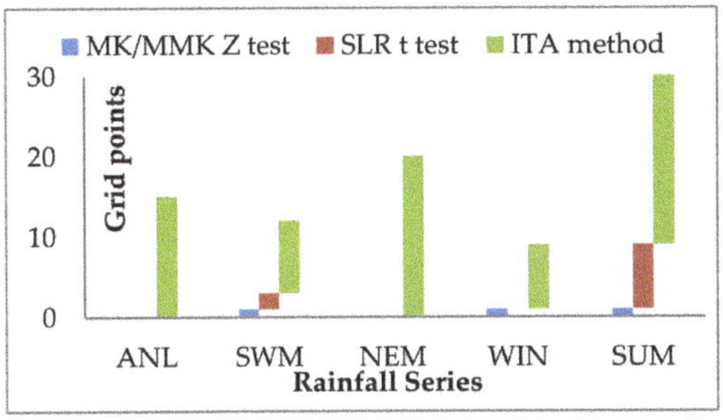

Figure 7. Grid points exhibited significant trend in Monthly, Seasonal and Annual rainfall series.

Among the three methods used, namely SLR, MK/MMK, and ITA, significant trends were observed in 2.3%, 7.7%, and 44.6% of the grid points, respectively. Notably, the SLR and MK/MMK methods detected significant trends only in the SWM, winter, and summer series, while no trends were identified in the annual and NEM series. It is noteworthy that SLR and MK/MMK methods only identified significant trends in the SWM, winter, and summer series, while they did not detect trends in the annual and NEM series. It is worth mentioning that all ten significant trends detected by the SLR test were also identified by the ITA method. However, the significant trend detected by the MK test did not align with the ITA method. One significant trend identified by the MMK method was also detected by the ITA method.

Table 8 presents the correlation between the test statistics of the MK and SLR methods with the slope values of the ITA method, regardless of the level of significance. It was observed that the MK test showed good correlation with the ITA method in the annual and summer series. Similarly, the correlation between the SLR test and the ITA method was found to be good for the annual, SWM, winter, and summer series.

Table 8. Correlation between the test statistics of MK and SLR with the slope values.

	ANL		SWM		NEM		WIN		SUM	
	MKT	**SLR**	**MKT**	**SLR**	**MKT**	**SLR**	**MKT**	**SLR**	**MKT**	**SLR**
ITA	0.896	0.828	0.692	0.851	0.359	0.335	0.461	0.969	0.937	0.958
MKT	1.00	0.841	1.00	0.803	1.00	0.887	1.00	0.592	1.00	0.948

The study findings suggest that the ITA method outperforms traditional trend detection methods. The ITA method proved beneficial in detecting many significant trends that could not be identified by the traditional methods in the annual and seasonal rainfall series. It effectively revealed hidden trends in rainfall series across the grid points. Many researchers have validated the Sen's (2012) [8] ITA methodology for different hydro-meteorological variables in various parts of the world [4,14,17,50,89,104–110]. Unlike classical methods, the ITA method does not require prewhitening prior to its application [8]. Trends of low, medium, and high data can be easily observed using this method [99,111]. One disadvantage of this test is that it must be applied to each recorded series individually [112]. The possibility of presenting results graphically in this method enables the easy observation of hidden subtrends and helps in identifying trends in extreme values [63,113].

The findings of the present study indicate that a majority of the grid points in the western and eastern parts of the Vaippar basin show a significant increasing trend in rainfall. This increasing trend is significant as it can contribute to an increase in runoff, which can be utilized for water management purposes, particularly for tapping and harnessing the runoff water in addition to the existing water conservation structures.

Conversely, it was also observed that some grid points exhibit a significant downward trend in rainfall. This decline in rainfall presents various challenges and implications for surface and groundwater management. One of the key challenges is the increased reliance on groundwater extraction for crop irrigation, leading to the depletion of groundwater reserves, drought-related issues, and diminished soil moisture [44].

Given these observations, it is crucial to carefully consider the changes in rainfall patterns and trends in long-term catchment-scale water management strategies. Adapting and planning for these changes can help mitigate the potential risks associated with declining precipitation, such as implementing measures to enhance water conservation, exploring alternative water sources, and promoting sustainable agricultural practices that optimize water usage. Long-term water management strategies should take into account the evolving rainfall patterns to ensure efficient and sustainable utilization of water resources in the Vaippar basin.

4. Conclusions

This study aimed to analyze the variation and trends of seasonal and annual rainfall series in the Vaippar basin using gridded rainfall data from 1971 to 2019. The SLR, MK/MMK, and ITA methods were employed to examine the trends, magnitudes, subtrends, and nature of the trend. In order to account for spatial variability, monthly rainfall data from 13 rain gauge stations within the basin were spatially interpolated using the inverse distance weighing (IDW) method under GIS environment. To further refine the spatial representation, the basin was divided into 26 grids, each covering approximately 205 km^2, and gridded rainfall data was generated from the interpolated gauge data. The key findings and conclusions of this study are summarized below.

The basin experienced moderate variability in annual rainfall, lower variability during the NEM season, and higher variability in other seasons. Among the three methods (SLR, MK/MMK, and ITA), significant trends were detected in 2.3%, 7.7%, and 44.6% of the grid points, respectively. The SLR and MK/MMK methods detected significant trends only in the SWM, winter, and summer series. The significant trend detected by the MK test did not align with the ITA method, but the significant trends detected by the SLR test were consistent with the ITA method. The ITA method indicated that 57.69%, 76.92%, and 80.77% of the grid points exhibited significant trends at 5% and 10% significance levels in annual, NEM, and summer rainfall, respectively. However, the SWM and winter series showed less than 35% significant trends. For the NEM season, 53.85% of the grids displayed a significant upward trend, which is a positive sign for improving water management. Grid points located in the southwestern and central parts of the basin showed a higher number of significant trends in annual rainfall. In terms of the NEM season, the southeastern, central, and extreme southern parts experienced a significant upward trend.

In the annual and NEM series, grid points located in the southwestern parts of the basin exhibited monotonically upward trends. Approximately 19.3% of the grid points in the southwestern parts of the basin showed upward trends in high rainfall segments, as classified for annual and NEM rainfall. Grid points in the western part of the basin exhibited a significantly upward trend with a slope of 2.03 mm/year, while the central part showed a non-significant downward trend with a slope value of −1.89 mm/year for the NEM series. Among the three methods considered, the ITA method consistently estimated a higher magnitude of trend for all the rainfall series compared to the other methods. Compared with traditional methods, the ITA method, which represents rainfall series graphically without making any assumptions, detected trends that were not identified by traditional methods. It facilitated the identification of monotonic or non-monotonic upward/downward trends and trends in low, medium, and high rainfall segments.

From the spatial analysis of the Vaippar basin, it is evident that the southeastern, central, and extreme southern parts have exhibited a positive increase in annual rainfall. Moreover, when considering different seasons, it is notable that the NEM season has displayed widespread positive trends across various parts of the basin. These positive rainfall trends have significant implications for water availability, agricultural productivity, and overall ecological balance within the Vaippar basin. The significant findings of this study will serve as a crucial scientific reference for policymakers, assisting in the preparation and management of extreme climate effects on land and water resources within and around the Vaipaar basin.

Author Contributions: M.M. and S.S. conceived and designed the study. M.M., R.A. and N.M. collected and processed the data and prepared mappings. M.M., S.S. and A.N. analysed the data, and M.M., A.N., S.S. and R.A. prepared the manuscript writings. All authors have read and agreed to the published version of the manuscript.

Funding: This research received no external funding.

Institutional Review Board Statement: Not applicable.

Informed Consent Statement: Not applicable.

Data Availability Statement: The data used in this study are available upon the request from the corresponding author.

Conflicts of Interest: The authors declare no conflict of interest.

Appendix A

Table A1. Space-time correlation analysis of the mean monthly rainfall data for Vaippar basin.

	Aruppukottai	Virudhunagar IB	Sattur	Sivakasi	Srivilliputhur	Watrap	Pilavakkal	Koilpatti (Rev)	Vilathikulam	Sankarankoil	Sivagiri	Vembakottai	Ettayapuram
Aruppukottai	1.00	0.99	0.98	0.98	0.97	0.95	0.95	0.98	0.95	0.88	0.88	0.95	0.95
Virudhunagar IB	0.99	1.00	0.97	0.97	0.95	0.93	0.93	0.96	0.93	0.84	0.84	0.95	0.94
Sattur	0.98	0.97	1.00	1.00	0.99	0.98	0.97	0.99	0.96	0.93	0.93	0.97	0.96
Sivakasi	0.98	0.97	1.00	1.00	0.99	0.99	0.98	0.99	0.96	0.93	0.93	0.97	0.97
Srivilliputhur	0.97	0.95	0.99	0.99	1.00	0.98	0.99	0.99	0.97	0.94	0.95	0.97	0.96
Watrap	0.95	0.93	0.98	0.99	0.98	1.00	0.99	0.99	0.97	0.96	0.97	0.97	0.96
Pilavakkal	0.95	0.93	0.97	0.98	0.99	0.99	1.00	0.98	0.98	0.97	0.97	0.96	0.97
Koilpatti (Rev)	0.98	0.96	0.99	0.99	0.99	0.99	0.98	1.00	0.99	0.94	0.94	0.97	0.98
Vilathikulam	0.95	0.93	0.96	0.96	0.97	0.97	0.98	0.99	1.00	0.95	0.95	0.97	0.99
Sankarankoil	0.88	0.84	0.93	0.93	0.94	0.96	0.97	0.94	0.95	1.00	1.00	0.92	0.92
Sivagiri	0.88	0.84	0.93	0.93	0.95	0.97	0.97	0.94	0.95	1.00	1.00	0.92	0.93
Vembakottai	0.95	0.95	0.97	0.97	0.97	0.97	0.96	0.97	0.97	0.92	0.92	1.00	0.98
Ettayapuram	0.95	0.94	0.96	0.97	0.96	0.96	0.97	0.98	0.99	0.92	0.93	0.98	1.00

Table A2. Mean rainfall of 26 grid points of Vaippar during the period 1971–2019.

Grid	JAN	FEB	MAR	APR	MAY	JUN	JUL	AUG	SEP	OCT	NOV	DEC	ANL	SWM	NEM	WIN	SUM
G01	22.18	43.25	50.90	81.96	55.67	14.96	19.69	27.93	66.92	199.76	224.74	79.04	886.99	129.49	503.54	65.43	188.53
G02	22.00	42.61	49.84	81.78	54.04	14.56	18.99	26.28	63.51	197.34	224.12	78.81	873.89	123.34	500.27	64.61	185.67
G03	18.89	29.84	40.00	70.40	51.94	12.36	16.95	24.75	55.81	181.46	197.44	70.52	770.37	109.87	449.42	48.73	162.34
G04	19.80	31.98	48.08	67.83	64.85	13.91	26.28	41.50	88.98	199.81	200.66	72.82	876.48	170.67	473.28	51.78	180.75
G05	19.05	29.51	45.49	68.58	67.14	14.65	22.12	38.34	83.47	196.20	191.58	70.00	846.13	158.58	457.78	48.56	181.21
G06	19.68	34.63	46.03	73.70	60.67	14.55	20.15	32.15	72.51	194.99	202.26	72.57	843.87	139.35	469.81	54.31	180.40
G07	19.55	33.11	42.82	73.35	53.58	13.07	17.72	26.07	59.56	186.79	203.54	72.29	801.44	116.41	462.62	52.66	169.74
G08	25.51	25.65	45.31	79.12	67.80	15.35	20.23	40.25	92.87	208.40	205.55	65.88	891.93	168.71	479.82	51.16	192.24
G09	15.81	27.84	47.92	66.23	76.73	15.46	19.72	38.90	86.84	196.78	178.65	69.87	840.75	160.92	445.30	43.65	190.88
G10	16.63	25.80	40.64	65.32	67.02	14.22	20.78	37.09	79.75	189.99	175.04	64.40	796.69	151.85	429.43	42.43	172.98
G11	17.06	24.46	36.05	64.55	57.29	12.57	19.64	32.02	67.87	185.25	177.31	63.62	757.69	132.10	426.19	41.51	157.89
G12	19.15	23.62	38.16	66.06	66.58	15.60	24.27	42.64	85.95	190.10	174.33	62.67	809.13	168.46	427.10	42.77	170.80
G13	15.57	22.31	34.67	61.23	64.34	14.63	21.28	40.27	82.27	177.51	158.50	56.86	749.44	158.45	392.87	37.88	160.24
G14	14.76	20.81	31.94	59.79	61.21	13.41	20.52	38.90	78.84	174.98	151.93	53.26	720.35	151.67	380.17	35.57	152.94
G15	16.19	19.62	31.28	59.98	58.42	12.26	22.37	36.56	75.28	186.68	159.39	56.34	734.37	146.48	402.41	35.80	149.68
G16	17.99	18.73	30.02	59.67	69.42	18.60	33.59	53.97	88.20	180.10	149.19	59.52	778.99	194.36	388.81	36.72	159.11
G17	16.95	18.40	29.32	58.66	68.04	17.71	31.64	51.50	86.27	176.19	146.39	57.78	758.84	187.12	380.36	35.35	156.01

Table A2. *Cont.*

Grid	JAN	FEB	MAR	APR	MAY	JUN	JUL	AUG	SEP	OCT	NOV	DEC	ANL	SWM	NEM	WIN	SUM
G18	14.38	17.26	29.01	57.64	63.38	12.47	24.38	39.20	77.10	163.33	142.84	55.22	696.21	153.15	361.39	31.64	150.03
G19	14.92	17.78	27.69	56.21	59.90	12.04	23.45	36.04	76.75	171.44	147.70	57.47	701.40	148.29	376.61	32.70	143.80
G20	16.52	16.79	26.12	52.61	70.60	21.63	37.44	55.51	87.49	173.36	141.51	57.78	757.36	202.07	372.64	33.31	149.33
G21	15.00	17.38	27.32	53.11	64.39	16.61	29.28	43.77	79.75	169.04	144.58	57.07	717.31	169.42	370.69	32.38	144.82
G22	14.61	17.85	25.99	52.24	56.03	12.49	23.86	36.39	74.35	172.80	147.23	58.18	692.02	147.09	378.21	32.46	134.27
G23	13.17	18.41	21.10	43.85	41.90	9.65	20.30	31.26	69.60	181.73	149.37	59.88	660.22	130.82	390.98	31.57	106.85
G24	14.25	17.40	23.90	47.20	49.98	12.29	22.99	34.23	68.88	173.48	146.12	59.21	669.93	138.40	378.81	31.65	121.08
G25	12.32	14.87	16.83	36.82	29.83	6.03	14.82	21.26	49.77	169.76	137.85	60.04	570.20	91.88	367.65	27.19	83.49
G26	13.51	16.78	21.11	42.77	39.40	8.80	18.23	27.03	59.17	173.16	144.70	60.26	624.92	113.22	378.12	30.29	103.28
Mean	17.13	24.10	34.91	61.56	59.24	13.84	22.72	36.68	75.30	183.48	170.10	63.51	762.57	148.54	417.09	41.23	155.71
Max	25.51	43.25	50.90	81.96	76.73	21.63	37.44	55.51	92.87	208.40	224.74	79.04	891.93	202.07	503.54	65.43	192.24
Min	12.32	14.87	16.83	36.82	29.83	6.03	14.82	21.26	49.77	163.33	137.85	53.26	570.20	91.88	361.39	27.19	83.49

Table A3. Coefficientof variation (CV) of gridded rainfall of Vaippar during the period 1971–2019.

Grid	JAN	FEB	MAR	APR	MAY	JUN	JUL	AUG	SEP	OCT	NOV	DEC	ANL	SWM	NEM	WIN	SUM
G01	154.18	140.78	140.79	69.21	91.50	102.63	80.10	91.22	71.29	53.74	49.67	101.42	26.95	46.25	32.90	113.22	53.25
G02	154.74	139.89	138.83	69.96	94.30	105.85	82.42	94.61	72.52	54.00	50.78	100.41	27.63	47.19	34.09	112.88	53.40
G03	160.41	137.16	140.52	63.02	75.10	112.95	100.21	89.63	67.49	46.10	55.65	95.75	26.17	43.89	37.44	108.60	46.61
G04	209.10	182.91	206.47	80.43	69.20	113.19	114.17	106.91	78.25	49.97	66.26	114.12	29.46	50.34	38.46	139.66	60.61
G05	191.89	157.09	178.35	62.38	56.22	95.67	95.11	80.86	72.20	47.33	54.61	98.78	24.43	45.78	31.90	123.93	48.79
G06	160.84	141.70	151.16	60.03	67.25	93.40	75.63	80.00	68.33	46.81	48.89	96.89	22.94	43.33	29.84	112.37	46.94
G07	155.48	137.10	140.31	62.92	76.39	106.01	86.15	87.91	67.81	46.82	52.82	95.45	25.17	43.49	34.81	108.77	47.20
G08	200.29	166.11	173.00	92.12	68.03	145.89	124.84	107.60	86.24	61.20	66.99	124.10	30.73	62.13	41.84	127.38	54.12
G09	243.12	168.77	192.18	67.97	67.11	119.19	114.92	101.25	99.80	57.18	62.06	102.65	31.47	65.83	34.56	139.72	57.57
G10	206.11	157.07	172.22	62.59	57.03	99.49	95.52	82.48	74.52	49.04	57.88	95.36	25.49	47.76	32.55	129.16	46.71
G11	177.12	145.75	151.89	62.93	57.59	97.41	90.44	82.70	62.11	44.84	55.16	91.14	23.78	40.22	33.72	118.37	43.56
G12	202.91	159.49	170.31	73.56	53.73	93.63	92.76	74.55	64.21	50.10	60.93	102.03	25.17	41.53	35.90	129.36	45.64
G13	219.95	173.14	171.89	78.27	63.15	105.55	111.57	90.21	63.59	52.38	71.66	101.21	28.72	45.23	40.31	141.20	47.20
G14	221.81	180.04	170.45	84.47	66.34	116.36	121.05	96.20	63.59	53.58	76.03	101.18	30.76	46.86	41.89	146.75	49.54
G15	207.47	159.99	162.01	73.34	52.34	104.70	103.38	87.00	60.13	49.81	61.49	92.45	26.41	41.89	36.42	134.90	46.51
G16	240.21	176.93	164.07	79.87	59.33	91.70	99.24	72.57	60.10	52.27	70.09	98.48	25.71	38.72	40.16	156.94	46.74
G17	243.80	178.31	163.36	78.11	59.17	91.64	100.12	73.89	57.23	51.61	71.25	97.91	25.80	37.95	40.30	157.86	45.87
G18	240.03	168.06	161.48	75.01	61.88	117.59	115.62	84.21	60.44	49.76	70.27	101.75	27.11	44.30	41.60	146.09	43.58
G19	232.13	155.85	159.91	72.59	61.30	114.08	109.24	83.74	58.03	46.93	65.70	94.24	25.52	41.91	38.84	138.89	43.57
G20	263.69	185.11	162.49	77.40	60.18	98.01	91.04	70.86	58.72	59.35	70.12	102.95	27.89	36.83	43.16	171.22	45.30
G21	244.70	164.01	160.37	71.41	57.94	101.21	90.37	70.06	55.71	52.92	65.67	100.62	26.47	36.59	40.27	150.53	41.90
G22	230.41	155.91	162.05	70.61	59.99	101.53	96.35	77.20	53.96	48.78	61.95	94.59	24.96	37.78	37.39	140.41	42.90
G23	239.88	175.30	180.07	81.78	77.29	113.59	111.86	98.10	54.59	59.61	60.99	96.21	29.18	41.43	40.09	153.86	56.02
G24	215.48	155.79	165.05	71.72	63.30	92.52	92.20	75.02	49.83	49.43	56.85	95.00	24.11	33.88	34.54	139.31	44.42
G25	199.82	191.86	189.23	114.12	117.77	155.16	143.20	111.09	65.75	54.41	56.78	105.52	28.19	46.33	33.84	159.45	70.67
G26	197.54	161.72	172.00	83.18	80.17	100.07	108.34	86.91	51.53	50.12	54.14	97.53	24.51	35.94	32.51	140.05	52.09

References

1. FAO. India at a Glance. *FAO in India*. Available online: https://www.fao.org/india/fao-in-india/india-at-a-glance/en/ (accessed on 22 July 2023).
2. TNENVIS. Agriculture. Available online: http://tnenvis.nic.in/Database/TN-ENVIS_792.aspx (accessed on 27 July 2023).
3. Kumaraperumal, R.; Pazhanivelan, S.; Ragunath, K.P.; Raman, M.G. Mapping of rainfed areas in Tamil Nadu using Remote Sensing Technology. *Madras Agric. J.* **2019**, *106*, 643.
4. Wang, Y.; Xu, Y.; Tabari, H.; Wangb, J.; Wangb, Q.; Songd, S.; Hu, Z. Innovative trend analysis of annual and seasonal rainfall in the Yangtze River Delta, eastern China. *Atmos. Res.* **2020**, *231*, 104673. [CrossRef]
5. IPCC (Intergovernmental Panel on Climate Change). Summary for policymakers. In *Global Warming of 1.5 °C—An IPCC Special Report on the Impacts of Global Warming of 1.5 °C above Pre-Industrial Levels and Related Global Greenhouse Gas Emission Pathways, in the Context of Strengthening the Global Response to World Meteorol*; Organ: Geneva, Switzerland, 2018; Volume 106, p. 32.
6. Mishra, A.K.; Singh, V.P. A review of drought concepts. *J. Hydrol.* **2010**, *391*, 202–216. [CrossRef]
7. Sun, F.; Roderick, M.L.; Farquhar, G.D. Rainfall statistics, stationarity, and climate change. *Proc. Natl. Acad. Sci. USA* **2018**, *115*, 2305–2310. [CrossRef]
8. Sen, Z. Innovative trend analysis methodology. *J. Hydrol. Eng.* **2012**, *17*, 1042–1046. [CrossRef]
9. Mondal, A.; Khare, D.; Kundu, S. Spatial and temporal analysis of rainfall and temperature trend of India. *Theor. Appl. Clim.* **2015**, *122*, 143–158. [CrossRef]
10. Ahmadi, F.; Tahroudi, M.N.; Mirabbasi, R.; Khalili, K.; Jhajharia, D. Spatiotemporal trend and abrupt change analysis of temperature in Iran. *Meteorol. Appl.* **2017**, *25*, 314–332. [CrossRef]
11. Machiwal, D.; Jha, M.K. *Hydrologic Time Series Analysis: Theory and Practice*, 1st ed.; Springer: Berlin/Heidelberg, Germany; Capital Publishing Company: New Delhi, India, 2012.
12. Machiwal, D.; Jha, M.K. Evaluating persistence and identifying trends and abrupt changes in monthly and annual rainfalls of a semi-arid region in Western India. *Theor. Appl. Clim.* **2017**, *128*, 689–708. [CrossRef]
13. Kumar, S.; Machiwal, D.; Dayal, D. Spatial modelling of rainfall trends using satellite datasets and geographic information system. *Hydrol. Sci. J.* **2017**, *62*, 1636–1653. [CrossRef]
14. Machiwal, D.; Gupta, A.; Jha, M.K.; Kamble, T. Analysis of trend in temperature and rainfall time series of an Indian arid region: Comparative evaluation of salient techniques. *Theor. Appl. Clim.* **2018**, *136*, 301–320. [CrossRef]
15. Malik, A.; Kumar, A. Spatio-temporal trend analysis of rainfall using parametric and non-parametric tests: Case study in Uttarakhand, India. *Theor. Appl. Clim.* **2020**, *140*, 183–207. [CrossRef]
16. Tabari, H.; Marofi, S.; Ahmadi, M. Long-term variations of waterquality parameters in the Maroon River Iran. *Environ. Monit. Assess.* **2010**, *177*, 273–287. [CrossRef]
17. Singh, R.; Sah, S.; Das, B.; Potekar, S.; Chaudhary, A.; Pathak, H. Innovative trend analysis of spatio-temporal variations of rainfall in India during 1901–2019. *Theor. Appl. Clim.* **2021**, *145*, 821–838. [CrossRef]
18. Hirsch, R.M.; Alexander, R.B.; Smith, R.A. Selection of methods for the detection and estimation of trends in water quality. *Water Resour. Res.* **1971**, *27*, 803–813. [CrossRef]
19. Ahmad, T.; Pandey, A.C.; Kumar, A. Long-term precipitation monitoring and its linkage with flood scenario in changing climate conditions in Kashmir valley. *Geocarto Int.* **2021**, *37*, 5497–5522. [CrossRef]
20. Singh, R.; Sah, S.; Das, B.; Vishnoi, L.; Pathak, H. Spatio-temporal trends and variability of rainfall in Maharashtra, India: Analysis of 118 years. *Theor. Appl. Clim.* **2020**, *143*, 883–900. [CrossRef]
21. Kumar, S.; Merwade, V.; Kam, J.; Thurner, K. Streamflow trends in Indiana: Effects of long term persistence, precipitation and subsurface drains. *J. Hydrol.* **2009**, *374*, 171–183. [CrossRef]
22. Zamani, R.; Mirabbasi, R.; Abdollahi, S.; Jhajharia, D. Streamflow trend analysis by considering autocorrelation structure, long-term persistence and Hurst coefficient in a semi-arid region of Iran. *Theor. Appl. Clim.* **2017**, *129*, 33–45. [CrossRef]
23. Dinpashoh, Y.; Mirabbasi, R.; Jhajharia, D.; Abianeh, H.Z.; Mostafaeipour, A. Effect of short-term and long-term persistence on identification of temporal trends. *J. Hydrol. Eng.* **2014**, *19*, 617–625. [CrossRef]
24. Sonali, P.; Kumar, D.N. Spatio-temporal variability of temperature and potential evapotranspiration over India. *J. Water Clim. Change* **2016**, *7*, 810–822. [CrossRef]
25. Amirataee, B.; Montaseri, M.; Sanikhani, H. The analysis of trend variations of reference evapotranspiration via eliminating the significance effect of all autocorrelation coefficients. *Theor. Appl. Clim.* **2016**, *126*, 131–139. [CrossRef]
26. Praveenkumar, C.; Jothiprakash, V. Spatio-temporal trend and homogeneity analysis of gridded and gauge precipitation in Indravati river basin, India. *J. Water Clim. Change* **2018**, *11*, 178–199. [CrossRef]
27. Kumar, M.S.; Geethalakshmi, V.; Ramanathan, S.; Senthil, A.; Senthilraja, K.; Bhuvaneswari, K.; Gowtham, R.; Kannan, B.; Priyanka, S. Rainfall Spatial-Temporal Variability and Trends in the Thamirabharani River Basin, India: Implications for Agricultural Planning and Water Management. *Sustainability* **2022**, *14*, 14948. [CrossRef]
28. Krishnakumar, K.N.; Rao, G.S.L.H.V.P.; Gopakumar, C.S. Rainfall trends in twentieth century over Kerala, India. *Atmos. Environ.* **2009**, *43*, 1940–1944. [CrossRef]
29. Kumar, V.; Jain, S. Rainfall trend in Ganga-Brahmputra-Meghna river basins of India (1951–2004). *J. Hydrol.* **2010**, *33*, 59–66.
30. Manikandan, M.; Tamilmani, D. Statistical Analysis of Spatial Pattern of Rainfall Trends in Parambikualam-Aliyar Sub Basin, Tamil Nadu. *J. Indian Water Resour. Soc.* **2012**, *32*, 40–49.

31. Suryavanshi, S.; Pandey, A.; Chaube, U.C.; Joshi, N. Long-term historic changes in climatic variables of Betwa Basin, India. *Theor. Appl. Clim.* **2014**, *117*, 403–418. [CrossRef]
32. Jhajharia, D.; Dinpashoh, Y.; Kahya, E. Trends in temperature over Godavari River basin in Southern Peninsular India. *Int. J. Clim.* **2014**, *34*, 1369–1384. [CrossRef]
33. Taxak, A.K.; Murumkar, A.R.; Arya, D.S. Long term spatial and temporal rainfall trends and homogeneity analysis in Wainganga basin, Central India. *Weather Clim. Extrem.* **2014**, *4*, 50–61. [CrossRef]
34. Gajbhiye, S.; Meshram, C.; Mirabbasi, R.; Sharma, S.K. Trend analysis of rainfall time series for Sindh river basin in India. *Theor. Appl. Clim.* **2015**, *125*, 593–608. [CrossRef]
35. Raju, B.C.K.; Nandagiri, L. Analysis of historical trends in hydrometeorological variables in the upper Cauvery Basin, Karnataka, India. *Curr. Sci.* **2017**, *112*, 577–587. [CrossRef]
36. Anand, B.; Karunanidhi, D. Long term spatial and temporal rainfall trend analysis using GIS and statistical methods in Lower Bhavani basin, Tamil Nadu, India. *Indian J. Geo-Mar. Sci.* **2020**, *49*, 419–427.
37. Bisht, D.S.; Chatterjee, C.; Raghuwanshi, N.S.; Sridhar, V. Spatiotemporal trends of rainfall across Indian river basins. *Theor. Appl. Clim.* **2018**, *132*, 419–436. [CrossRef]
38. Chandniha, S.K.; Meshram, S.G.; Adamowski, J.F.; Meshram, C. Trend analysis of precipitation in Jharkhand State, India. *Theor. Appl. Clim.* **2016**, *130*, 261–274. [CrossRef]
39. Sharma, S.; Singh, P.K. Long Term Spatiotemporal Variability in Rainfall Trends over the State of Jharkhand, India. *Climate* **2017**, *5*, 18. [CrossRef]
40. Meshram, S.G.; Singh, V.P.; Meshram, C. Long-term trend and variability of precipitation in Chhattisgarh State. India. *Theor. Appl. Clim.* **2016**, *129*, 729–744. [CrossRef]
41. Gajbhiye, S.; Meshram, C.; Singh, S.K.; Srivastava, P.K.; Islam, T. Precipitation trend analysis of Sindh River basin, India, from 102-year record (1901–2002). *Atmos. Sci. Lett.* **2016**, *17*, 71–77. [CrossRef]
42. Dhorde, A.G.; Korade, M.S.; Dhorde, A.A. Spatial distribution of temperature trends and extremes over Maharashtra and Karnataka states of India. *Theor. Appl. Clim.* **2017**, *130*, 191–204. [CrossRef]
43. Machiwal, D.; Kumar, S.; Dayal, D. Characterizing rainfall of hot arid region by using time-series modelling and sustainability approaches: A case study from Gujarat, India. *Theor. Appl. Clim.* **2016**, *124*, 593–607. [CrossRef]
44. Sanikhani, H.; Kisi, O.; Mirabbasi, R.; Meshram, S.G. Trend analysis of rainfall pattern over the Central India during 1901–2010. *Arab. J. Geosci.* **2018**, *11*, 437. [CrossRef]
45. Pingale, S.M.; Khare, D.; Jat, M.K.; Adamowski, J. Spatial and temporal trends of mean and extreme rainfall and temperature for the 33 urban centers of the arid and semi-arid state of Rajasthan, India. *Atmos. Res.* **2014**, *138*, 73–90. [CrossRef]
46. Pingale, S.M.; Khare, D.; Jat, M.K.; Adamowski, J. Trend analysis of climatic variables in an arid and semi-arid region of the Ajmer District, Rajasthan, India. *J. Water Land Dev.* **2016**, *28*, 3–18. [CrossRef]
47. Meena, H.M.; Machiwal, D.; Santra, P.; Moharana, P.C.; Singh, D.V. Trends and homogeneity of monthly, seasonal, and annual rainfall over arid region of Rajasthan, India. *Theor. Appl. Clim.* **2018**, *136*, 795–811. [CrossRef]
48. Panda, A.; Sahu, N. Trend analysis of seasonal rainfall and temperature pattern in Kalahandi, Bolangir and Koraput districts of Odisha, India. *Atmos. Sci. Lett.* **2019**, *20*, e932. [CrossRef]
49. Patakamuri, S.K.; Muthiah, K.; Sridhar, V. Long-term homogeneity, trend, and change-point analysis of rainfall in the arid district of Ananthapuramu, Andhra Pradesh State, India. *Water* **2020**, *12*, 211. [CrossRef]
50. Gujree, I.; Ahmad, I.; Zhang, F.; Arshad, A. Innovative Trend Analysis of High-Altitude Climatology of Kashmir Valley, North-West Himalayas. *Atmosphere* **2022**, *13*, 764. [CrossRef]
51. Wagesho, N.; Goel, N.K.; Jain, M.K. Temporal and spatial variability of annual and seasonal rainfall over Ethiopia. *Hydrol. Sci. J.* **2013**, *58*, 354–373. [CrossRef]
52. Ay, M.; Kisi, O. Investigation of trend analysis of monthly total precipitation by an innovative method. *Theor. Appl. Clim.* **2014**, *120*, 617–629. [CrossRef]
53. Manzanas, R.; Amekudzi, L.K.; Preko, K.; Herrera, S.; Guti'errez, J.M. Precipitation variability and trends in Ghana: An intercomparison of observational and reanalysis products. *Clim. Change* **2014**, *124*, 805–819. [CrossRef]
54. Daniels, E.E.; Lenderink, G.; Hutjes, R.W.A.; Holtslag, A.A.M. Spatial precipitation patterns and trends in The Netherlands during 1951–2009. *Int. J. Clim.* **2014**, *34*, 1773–1784. [CrossRef]
55. Farhangi, M.; Kholghi, M.; Chavoshian, S.A. Rainfall Trend Analysis of Hydrological Subbasins in Western Iran. *J. Irrig. Drain. Eng.* **2016**, *142*, 05016004. [CrossRef]
56. Minaei, M.; Irannezhad, M. Spatio-temporal trend analysis of precipitation, temperature, and river discharge in the northeast of Iran in recent decades. *Theor. Appl. Clim.* **2016**, *131*, 167–179. [CrossRef]
57. Lionello, P.; Abrantes, F.; Gacic, M.; Planton, S.; Trigo, R.; Ulbrich, U. The climate of the Mediterranean region: Research progress and climate change impacts. *Reg. Environ. Change* **2014**, *14*, 1679–1684. [CrossRef]
58. Valdes-Abellan, J.; Pardoand, M.A.; Tenza-Abril, A.J. Observed Precipitation Trend Changes in the Western Mediterranean Region. *Int. J. Clim.* **2017**, *37*, 1285–1296. [CrossRef]
59. Kabanda, T. Long-term rainfall trends over the Tanzania coast. *Atmosphere* **2018**, *9*, 155. [CrossRef]
60. Huang, J.; Sun, S.; Xue, Y.; Zhang, J. Spatial and temporal variability of precipitation indices during 1961–2010 in Hunan Province, central south China. *Theor. Appl. Clim.* **2014**, *118*, 581–595. [CrossRef]

61. Ye, J.S. Trend and variability of China's summer precipitation during 1955–2008. *Int. J. Clim.* **2014**, *34*, 59–566. [CrossRef]
62. Zhang, Y.; Xia, J.; She, D. Spatiotemporal variation and statistical characteristic of extreme precipitation in the middle reaches of the Yellow River Basin during 1960–2013. *Theor. Appl. Clim.* **2018**, *135*, 391–408. [CrossRef]
63. Wu, H.; Qian, H. Innovative trend analysis of annual and seasonal rainfall and extreme values in Shaanxi, China, since the 1950s *Int. J. Clim.* **2017**, *37*, 2582–2592. [CrossRef]
64. Qin, Z.; Peng, D.; Singh, V.P.; Chen, M. Spatio-temporal variations of precipitation extremes in Hanjiang River Basin, China during 1960–2015. *Theor. Appl. Clim.* **2019**, *138*, 1767–1783. [CrossRef]
65. Gado, T.A.; El-Hagrsy, R.M.; Rashwa, I.M.H. Spatial and Temporal Rainfall Changes in Egypt. Environ. *Sci. Pollut. Res.* **2019**, *26*, 28228–28242. [CrossRef]
66. Chattopadhyay, S.D.; Edwards, R.M. Long-term trend analysis of precipitation and air temperature for Kentucky, United States *Climate* **2016**, *4*, 10. [CrossRef]
67. Chakraborty, S.; Pandey, R.; Chaube, U.; Mishra, S. Trend and variability analysis of rainfall series at Seonath River Basin, Chhattisgarh (India). *Int. J. Appl. Sci. Eng.* **2013**, *2*, 425–434.
68. Chandrakar, A.; Khare, D.; Krishan, R. Assessment of Spatial and Temporal Trends of Long Term Precipitation over Kharun Watershed, Chhattisgarh, India. *Environ. Process.* **2017**, *4*, 959–974. [CrossRef]
69. Sen, Z. Trend identification simulation and application. *J. Hydrol. Eng.* **2014**, *19*, 635–642. [CrossRef]
70. Tosunoglu, F.; Kisi, O. Trend analysis of maximum hydrologicdrought variables using Mann–Kendall andŞen's innovative trend method. *River Res. Appl.* **2017**, *33*, 597–610. [CrossRef]
71. CGWB (Central Ground Water Board). Report on Aquifer Mapping and Management of Ground Water Resources for Viappar Aquifer System Tamil Nadu. Department of Water Resources, River Development and Ganga Rejuvenation, Ministry of Jal Shakti, Government of India. Available online: http://cgwb.gov.in/sites/default/files/2022-1/vaippar_aquifer_system_tn.pdf (accessed on 27 July 2023).
72. Season and Crop Report of Tamilnadu, 2020–21. Department of Economics and Statistics: Chennai, India. Available online: https://agritech.tnau.ac.in/pdf/season%20n%20crop%202021.pdf (accessed on 27 July 2023).
73. Kim, T.; Valdes, J.B.; Aparicio, J. Frequency and spatial characteristics of droughts in the Conchos River Basin, Mexico. *Water Int.* **2002**, *27*, 420–430. [CrossRef]
74. Edossa, D.C.; Babel, M.S.; Gupta, A.D. Drought analysis in the Awash River Basin, Ethiopia. *Water Resour. Manag.* **2010**, *24*, 1441–1460. [CrossRef]
75. Moradi, H.R.; Rajabi, M.; Faragzadeh, M. Investigation of meteorological drought characteristics in Fars province, Iran. *Catena* **2011**, *84*, 35–46. [CrossRef]
76. Mishra, A.K.; Desai, V.R. Drought forecasting using stochastic models. Stoch. *Environ. Res. Risk Assess.* **2005**, *19*, 326–339. [CrossRef]
77. Shepard, D. A Two-Dimensional Interpolation Function for Irregularly Data. In Proceedings of the 23rd ACM National Conference, Las Vegas, NV, USA, 27–29 August 1968; pp. 517–524.
78. Rase, W.D. Volume Preserving Interpolation of a Smooth Surface from Polygon Related Data. *J. Geogr. Syst.* **2001**, *3*, 199–213. [CrossRef]
79. Mishra, A.K.; Desai, V.R. Spatial and temporal drought analysis in the Kansabati River Basin, India. *Int. J. River Basin Manag.* **2005**, *3*, 31–41. [CrossRef]
80. Manikandan, M.; Tamilmani, D. Spatial and Temporal Variation of Meteorological Drought in the Parambikulam-Aliyar Basin, Tamil Nadu. *J. Inst. Eng. Ser. A* **2015**, *96*, 177–184. [CrossRef]
81. Asfaw, A.; Simane, B.; Hassen, A.; Bantider, A. Variability and time series trend analysis of rainfall and temperature in north central Ethiopia: A case study in Woleka sub-basin. *Weather Clim. Extrem.* **2018**, *19*, 29–41. [CrossRef]
82. Hare, W. *Assessment of Knowledge on Impacts of Climate Change, Contribution to the Specification of Art, 2 of the UNFCCC*; WBGU: Bowling Green, OH, USA, 2003.
83. Gadgil, A.; Dhorde, A. Temperature trends in twentieth century at Pune. India. *Atmos. Environ.* **2005**, *39*, 6550–6556. [CrossRef]
84. Hameed, T.; Marino, M.A.; DeVries, J.J.; Tracy, J.C. Method for trend detection in climatological variables. *J. Hydrol. Eng.* **1997**, *2*, 157–160. [CrossRef]
85. Jaiswal, R.K.; Lohani, A.K.; Tiwari, H.L. Statistical analysis for change detection and trend assessment in climatological parameters. *Environ. Process.* **2015**, *2*, 729–749. [CrossRef]
86. Mann, H.B. Non-parametric tests against trend. *Econometrica* **1945**, *13*, 245–259. [CrossRef]
87. Kendall, M.G. *Rank Correlation Methods*; Charles Griffin and Co., Ltd.: London, UK, 1975.
88. Sonali, P.; Kumar, D.N. Review of trend detection methods and their application to detect temperature changes in India. *J. Hydrol.* **2013**, *476*, 212–227. [CrossRef]
89. Malik, A.; Kumar, A.; Guhathakurta, P.; Kisi, O. Spatial-temporal trend analysis of seasonal and annual rainfall (1966–2015) using innovative trend analysis method with significance test. *Arab. J. Geosci.* **2019**, *12*, 10. [CrossRef]
90. Hamed, K.H.; Rao, A.R. A modified Mann-Kendall trend test for auto correlated data. *J. Hydrol.* **1998**, *204*, 182–196. [CrossRef]
91. Modarres, R.; Silva, V.P.R. Rainfall trends in arid and semi-arid regions of Iran. *J. Arid Environ.* **2007**, *70*, 344–355. [CrossRef]
92. Yue, S.; Pilon, P.; Phinney, B. Canadian streamflow trend detection: Impacts of serial and cross-correlation. *Hydrol. Sci. J.* **2003**, *48*, 51–63. [CrossRef]

93. Novotny, E.V.; Stefan, H.G. Stream flow in Minnesota: Indicator of Climate Change. *J. Hydrol.* **2007**, *334*, 319–333. [CrossRef]
94. Rai, R.K.; Upadhyay, A.; Ojha, C.S.P. Temporal variability of climatic parameters of Yamuna River Basin: Spatial analysis of persistence, trend and periodicity. *Open Hydrol. J.* **2010**, *4*, 184–210. [CrossRef]
95. Luo, Y.; Lio, S.; Sheng, L.; Fu, S.; Liu, J.; Wang, G.; Zhou, G. Trends of precipitation in Beijiang River Basin, Guangdong Province, China. *Hydrol. Process.* **2008**, *22*, 2377–2386. [CrossRef]
96. Karpouzos, D.K.; Kavalieratou, S.; Babajimopoulos, C. Trend analysis of precipitation data in Pieria region (Greece). *Eur. Water* **2010**, *30*, 31–40.
97. Anderson, R.L. Distribution of the serial correlation coefficient. *Ann. Math. Stat.* **1942**, *13*, 1–13. [CrossRef]
98. Sen, P.K. Estimates of the regression coefficient based on Kendall's tau. *J. Am. Stat. Assoc.* **1968**, *63*, 1379–1389. [CrossRef]
99. Kisi, O. An innovative method for trend analysis of monthly pan evaporations. *J. Hydrol.* **2015**, *527*, 1123–1129. [CrossRef]
100. Tabari, H.; Willems, P. Investigation of streamflow variation using an innovative trend analysis approach in Northwest Iran. In Proceedings of the 36th IAHR World Congress, The Hague, The Netherlands, 28 June–3 July 2015.
101. Sen, Z. Innovative trend significance test and applications. *Theor. Appl. Clim.* **2017**, *127*, 939–947. [CrossRef]
102. Yue, S.; Hashino, M. Long term trends of annual and monthly precipitation in Japan. *J. Am. Water Resour. Assoc.* **2003**, *39*, 587–596. [CrossRef]
103. Some'e, B.S.; Ezani, A.; Tabari, H. Spatiotemporal trends and change point of precipitation in Iran. *Atmos. Res.* **2012**, *113*, 1–12.
104. Caloiero, T.; Coscarelli, R.; Ferrari, E. Application of the Innovative Trend Analysis Method for the Trend Analysis of Rainfall Anomalies in Southern Italy. *Water Resour. Manag.* **2018**, *32*, 4971–4983. [CrossRef]
105. Güçlü, Y.S. Improved visualisation for trend analysis by comparing with classical Mann-Kendall test and ITA. *J. Hydrol.* **2020**, *584*, 124674. [CrossRef]
106. Marak, J.D.K.; Sarma, A.K.; Bhattacharjya, R.K. Innovative trend analysis of spatial and temporal rainfall variations in Umiam and Umtru watersheds in Meghalaya, India. *Theor. Appl. Clim.* **2020**, *142*, 1397–1412. [CrossRef]
107. Dong, Z.; Jia, W.; Sarukkalige, R.; Fu, G.; Meng, Q.; Wang, Q. Innovative Trend Analysis of Air Temperature and Precipitation in the Jinsha River Basin, China. *Water* **2020**, *12*, 3293. [CrossRef]
108. Ahmed, I.A.; Salam, R.; Naikoo, M.W.; Rahman, A.; Praveen, B.; Hoai, P.N.; Pham, Q.B.; DuongAnh, T.; Tri, D.Q.; Elkhrachy, I. Evaluating the variability in long-term rainfall over India with advanced statistical techniques. *Acta Geophys.* **2022**, *70*, 801–818. [CrossRef]
109. Chauhan, A.S.; Maurya, R.K.S.; Rani, A.; Malik, A.; Kisi, O.; Danodia, A. Rainfall dynamics observed over India during last century (1901–2020) using innovative trend methodology. *Water Supply* **2022**, *22*, 6909–6944. [CrossRef]
110. Alashan, S. Comparison of sub-series with different lengths using şen-innovative trend analysis. *Acta Geophys.* **2023**, *71*, 373–383. [CrossRef]
111. Katipoğlu, O.M. Analyzing the trend and change point in various meteorological variables in Bursa with various statistical and graphical methods. *Theor. Appl. Clim.* **2022**, *150*, 1295–1320. [CrossRef]
112. Haktanir, T.; Citakoglu, H. Trend, independence, stationarity, and homogeneity tests on maximum rainfall series of standard durations recorded in Turkey. *J. Hydrol. Eng.* **2014**, *19*, 9. [CrossRef]
113. Madane, D.A.; Waghaye, A.M. Spatio-temporal variations of rainfall using innovative trend analysis during 1951–2021 in Punjab State, India. *Theor. Appl. Clim.* **2023**, *153*, 923–945. [CrossRef]

Disclaimer/Publisher's Note: The statements, opinions and data contained in all publications are solely those of the individual author(s) and contributor(s) and not of MDPI and/or the editor(s). MDPI and/or the editor(s) disclaim responsibility for any injury to people or property resulting from any ideas, methods, instructions or products referred to in the content.

International Journal of
Geo-Information

MDPI

Article

Dry–Wet Changes in a Typical Agriculture and Pasture Ecotone in China between 1540 and 2019

Xiaodong Wang [1], Yujia Song [2], Yu An [3], Xiaohui Liu [3] and Xiaoqiang Li [1,*]

1 College of Geographical Sciences, Changchun Normal University, Changchun 130032, China; wangxd219@nenu.edu.cn
2 College of Tourism and Geographical Science, Jilin Normal University, Siping 136000, China; songyj224@nenu.edu.cn
3 Northeast Institute of Geography and Agroecology, Chinese Academy of Sciences, Changchun 130102, China; anyu@iga.ac.cn (Y.A.); liuxh@iga.ac.cn (X.L.)
* Correspondence: lixiaoqiang@ccsfu.edu.cn

Abstract: Exploring periodic dry–wet changes is an important topic in climate change research due to its impact on drought and flood disasters. The purpose of this research was to determine the occurrence law of dry–wet changes in China on a scale of several hundred years, using the example of transitional zones. In this study, we analyzed typical areas of the ecotone between agricultural land and pasture along the Great Wall of China. The ring width index of *Carya cathayensis* was fitted with the March–August Palmer drought severity index ($PDSI_{38}$). The $PDSI_{38}$ was divided into different periods using the stepwise function fitting method. The results indicated that there were two dry periods and one wet period in the region from 1543 to 2019. In each dry and wet period, there were also different temporal periods, including long (decades), intermediate (ten years), and short periods (several years). Drought represents a significant threat to agricultural production in China. In the first dry period (1543–1756), four periods with low $PDSI_{38}$ values (1633–1635, $PDSI_{38} = -1.71$; 1636–1939, $PDSI_{38} = -3.35$; 1640–1642, $PDSI_{38} = -4.68$; and 1643–1645, $PDSI_{38} = -2.92$) occurred, during which severe droughts ($PDSI_{38} < -4$) lasted for 13 years. The dry–wet change showed the characteristics of a 12-year or multiple 12-year cycle. The results can be used to prepare to effectively address extreme drought scenarios worldwide in the future.

Keywords: dry–wet change; agriculture and pasture ecotone; periodicity law; 12-year cycle; extreme drought

1. Introduction

Droughts and floods cause considerable losses around the world each year, and their occurrence is closely related to the periodicity of dry–wet changes [1,2]. The temporal variability of dry–wet events may correspond to the regularity of drought and flood disasters. Prolonged dry periods led to drought conditions in several parts of the world in the 1930s and 1950s [3]. The analysis of dry–wet changes showed that the frequency of droughts and floods was synchronous with the decadal timescale in the Loess Plateau within the catchment of the River Jing, China [4]. As the frequency of dry–wet changes accelerated, so too did the occurrence of droughts and floods in China. The periodicity of the dry–wet changes in the country was 4–5 years, which was highly consistent with the occurrence of drought–flood disasters in some parts of China [5,6]. Periodic dry–wet changes have a direct impact on the occurrence of droughts and floods in different regions worldwide [7,8]. The periodicity of dry–wet changes is the main basis for the accurate forecasting of droughts and floods [9]. Therefore, the analysis of dry–wet changes may aid in minimizing their associated risks. It is crucial to reveal the laws of such dry–wet changes to obtain a deeper understanding and early identification of the presenting features of droughts and floods.

Northern China, located in the mid-latitude, experiences profound negative impacts from dry–wet changes. The agricultural water supply, crop yields, and planting systems depend on the dry–wet state and climate change mode of the region [10,11]. In the ecologically fragile zone of Northern China, dry–wet changes play a significant role in the ecotone between agriculture and pasture, especially in the analysis of the occurrence of droughts and floods in the region. Recently, many studies have focused on the dry–wet changes in this region [12]. The results of a wavelet analysis of the precipitation and evapotranspiration indexes revealed that the dry–wet changes exhibited a 14-year periodic variation from 1961 to 2012 [13]. Jia and Zhang (2019) indicated that the average annual frequency of drought events was higher than that of flood events from 1960 to 2016 [14]. Previous studies have indicated that, as the wettability decreases, aridity increases, and the periodic fluctuation of the dry–wet changes becomes increasingly stronger; this inference was based on the moisture indexes for 1951–2000 extracted from 160 meteorological stations in China [15]. Notably, Ma et al. (2005) indicated a humidity trend of 100 years in this region, with arid periods in the 50th and 20th years [16]. Xiao et al. (2013) indicated that the large-scale emigration and social unrest in the region in the mid-17th century were driven by extreme droughts [17].

Tree growth is more sensitive to dry–wet changes in the ecotone between semi-arid and arid conditions. Therefore, tree-ring records can be utilized for the reconstruction of climate indexes. In most previous studies, the natural variability of the dry–wet cycles before the instrumental period was estimated using tree-ring-based reconstructions [18–20]. The March–August Palmer drought severity index (PDSI) for the past 424 years was reconstructed based on the tree-ring width chronology data of *Cedrus deodara* in Northern Pakistan. Additionally, the seasonal variability in the frequency and severity of the dry–wet cycle in the southwestern United States of America for 1663–2015 was reconstructed according to tree-ring records, using a total of 183 stem increment cores collected from seven *Pinus ponderosa* stands [21]. The standardized precipitation–evapotranspiration index (SPEI), extending back to 1773 and calculated using the ring width chronology of *Cedrus deodara*, revealed five droughts and pluvials (cold humid periods) during a 244-year period in the Western Himalayas, India [22].

These studies have addressed the characteristics of the dry–wet changes in Northern China, and have confirmed that these changes are closely related to droughts and floods. However, these studies have focused on a timescale of several decades, and there is little research on the occurrence of extreme droughts and floods based on the laws of dry–wet changes over a long timescale. Based on this, we studied the laws of dry–wet changes on a timescale of 480 years and used the periodic changes to infer the characteristics of extreme droughts and floods, proposing potential strategies to address them. We used the tree-ring chronology method to reconstruct the March–August Palmer drought severity index ($PDSI_{38}$) over 480 years and reveal the periodicity of the dry–wet changes within the region. The contributions of this work are summarized as follows: (1) We used the tree-ring chronology method to reconstruct the $PDSI_{38}$ for each year from 1540 to 2019; (2) we used the stepwise function method to reveal the periodic law of the dry–wet changes; (3) we offer an understanding of the periodic characteristics of the dry–wet changes in the short, intermediate, and long term to infer the characteristics of extreme droughts and floods and propose strategies to address them.

2. Materials and Methods

2.1. Study Area

Our study was located in Northern Hebei Province, adjacent to Liaoning (in the Inner Mongolia Provinces along the Great Wall) (Figure 1). The Great Wall was built along the Yanshan Mountains. The south of the Great Wall is an agricultural area, and the north is a pastoral area. Therefore, the study area is characterized as an agriculture and pasture ecotone that is typical of China (adjacent to the North China Plain) [23–25]. The annual temperatures range from 7.98 °C to 11.07 °C and the annual precipitation ranges from 388 mm to 1128 mm (Figure 2a). It is cold in winter and hot in summer, with the precipitation concentrated in the summer (Figure 2b). The North China Plain is a typical farming area and has a long history of farming [26]. The area is sensitive to dry–wet changes and is thus an ideal area for the study of the occurrence of droughts and floods resulting from dry–wet changes [27]. In the Yanshan Mountains, with a relatively small population, there are many ancient wild *Carya cathayensis* trees scattered across the area. The tree-ring width growth of *Carya cathayensis* is sensitive to climate change and can be used to obtain historical climate data [28].

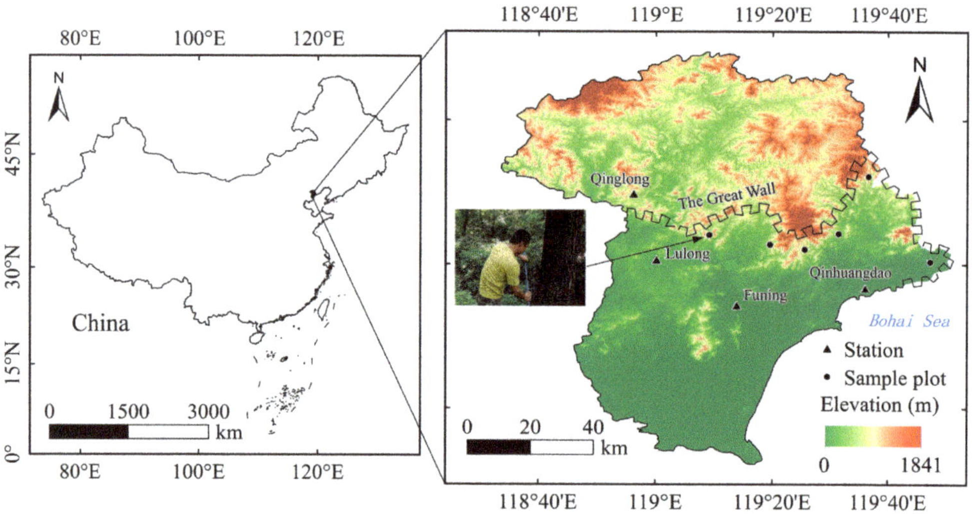

Figure 1. Map of the study area, sample plot, and field photos. The colored part in the bottom map of the right is a layered color topographic map, with green indicating low flat areas, yellow indicating slightly higher terrain, and brown yellow indicating the highest terrain. The elevation is shown in the legend.

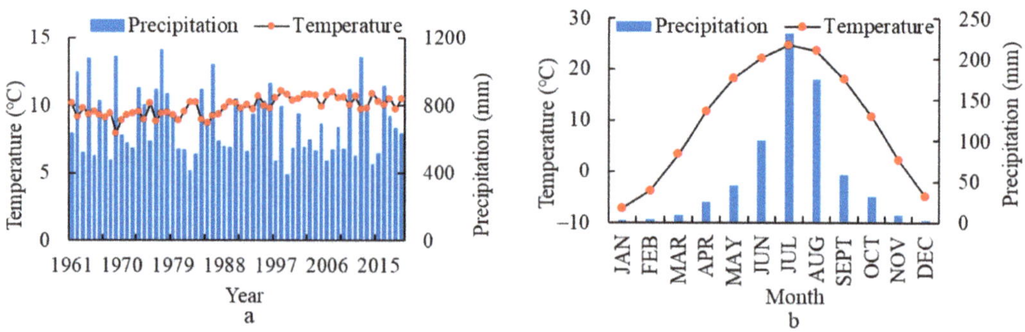

Figure 2. (**a**) Annual average temperature (black line with red squares) and yearly precipitation (blue bars) in the study area across 59 years from 1961 to 2019. (**b**) Monthly mean temperature (black line with red squares) and monthly precipitation (blue bars) for each year between 1961 and 2019.

2.2. Tree-Ring Sampling

From 2 July 2018 to 14 September 2018, we obtained 120 *Carya cathayensis* samples (cores) from six plots in the study area. These sampling points were distributed from west to east along the Yanshan Mountains (the Great Wall), with different elevations at different sampling points, forming a sampling system with several altitude gradients (Figure 1). For each tree, in all plots, two incremental tree-ring cores perpendicular to each other were extracted at a height of 1.2 m above the ground. The cores were sampled from 60 living trees with a diameter at breast height (DBH) > 25 cm, to determine the tree-ring chronologies (Table 1). Four cores (from the total of 120) that were damaged were not considered in our analysis. Firstly, the 116 remaining cores were dried and sanded (to enhance the visibility of the tree growth of the samples). The tree-ring widths were measured using the Velmex ring measurement system [29]. Below, we provide a flowchart to demonstrate the process of data gathering and method of analysis (Figure 3).

Table 1. Information about the plots considered in this study. C/T is the number of cores/trees, NCC is the number of cores for the chronology, and RP is the recording period for *Carya cathayensis*.

Plot	Longitude (E)	Latitude (N)	C/T	NCC	RP (Year)
1	119.162693	40.217760	18/9	18	1320–2018
2	119.446449	40.144764	20/10	18	1341–2018
3	119.530907	40.204247	20/10	20	1373–2018
4	119.561806	40.219650	24/12	22	1410–2018
5	119.620185	40.481356	16/8	16	1503–2018
6	119.742050	40.044175	22/11	22	1531–2018

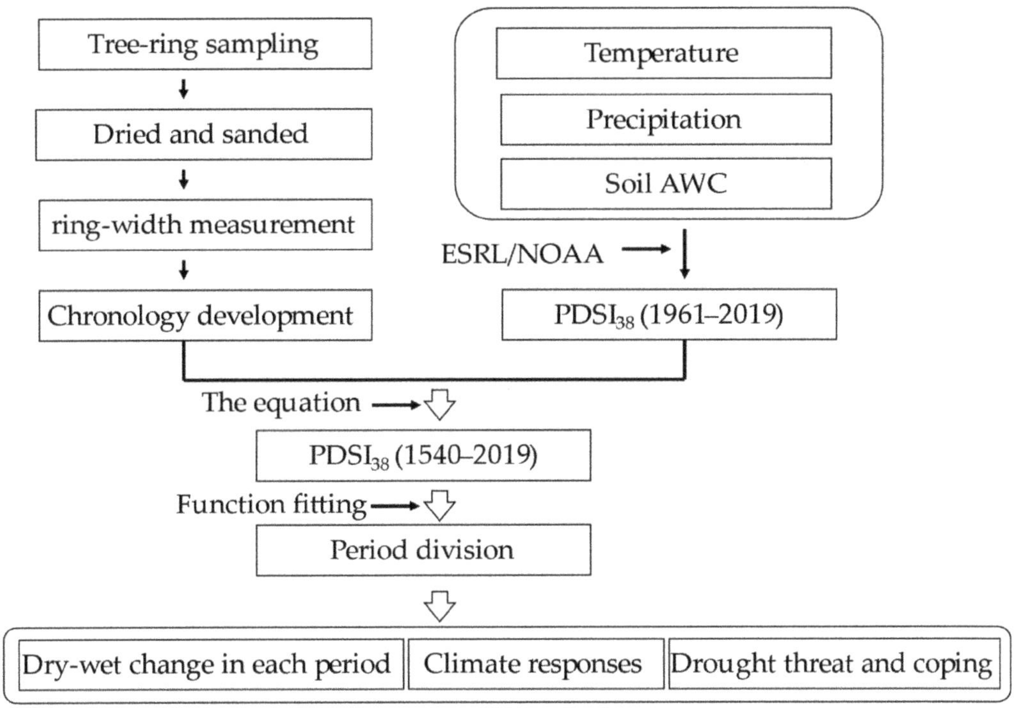

Figure 3. Flowchart of the relationship between the data and methods.

2.3. Chronology Development

All 116 cores were cross-dated by matching the patterns of relatively wide and narrow rings to account for the possibility of ring growth anomalies, such as missing or false rings or measurement errors [30]. The accuracy of the assigned dates regarding errors and measurements was further checked using the computer program COFECHA [31]. The tree-ring series that were poorly correlated with the master series were removed from the final dataset prior to chronology development. The remaining 91 cores were used to obtain the chronological order, using the ARSTAN software [32]. The ring width series were standardized using a smoothing spline with a 50% frequency–response cutoff of 67% of the length of each series. The detrended ring width index series were then pre-whitened by fitting an autoregressive model to remove any autocorrelation effects [33]. Finally, a standard deviation chronology (STD) was developed to reveal the relationship between the climate and tree growth. The mean index (MI) was greater than 0.95, the standard deviation (SD) was 0.03, and the variance of the 1st eigenvector (PCA1) was 43.28% of the STD; this could be used for the growth–climate analysis [34]. The mean series intercorrelation (0.36) and first-order autocorrelation (0.46) of the chronology's reliability indicated that the STD was sufficient for climate change analysis [35]. The expressed population signal (EPS) (threshold value: 0.85) was determined to reflect a valid time range for the recovery of climate data from a chronology [36]. Therefore, the time span of the STD from 1540 to 2018 was used to analyze the climate change. During this period, the EPS was greater than 0.85 and the sample depth was 25 (Figure 4).

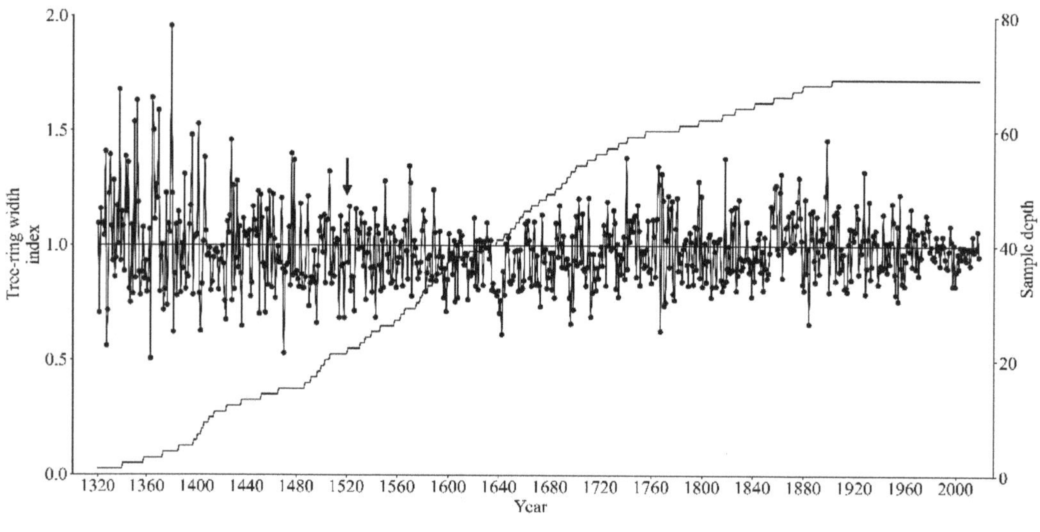

Figure 4. Standard deviation chronology (STD) for *Carya cathayensis* and the number of tree rings (sample depth). The numbers of tree rings are represented by solid lines; an arrow is used to mark the years with an expressed population signal (EPS) > 0.85.

2.4. Statistical Analysis

2.4.1. Reconstruction of $PDSI_{38}$

The temperature and precipitation data were acquired from the Qinglong (118.57° E, 40.25° N), Funing (119.23° E, 39.88° N), Qinhuangdao (119.60° E, 39.93° N), and Lulong stations (118.53° E, 39.53° N) near the Great Wall, from tree-ring sampling spots for 1961–2019. The $PDSI_{38}$ calculations included three data sets (temperature, precipitation, and the local available water content (AWC) of the soil) for each year (1961–2019) in the study area. Gridded precipitation and temperature data were extracted from the monthly 2.5° × 2.5° dataset compiled by the Climatic Research Unit from 1961 to 2019 [37], combined with the monthly precipitation and temperature data from five selected meteorological stations. The soil AWC data were extracted from the Global Soil Types 1-Degree Grid [38]. Calculations were performed on a 2.5° × 2.5° grid, chosen to achieve a trade-off between the spatial resolution and a reduced number of grid points for data processing. The $PDSI_{38}$ reconstruction was based on the monthly self-calibrated PDSI (scPDSI) grid data, with a resolution of 2.5° × 2.5° for 1961–2019, which were provided by the Earth System Research Laboratory/National Oceanic and Atmospheric Administration (ESRL/NOAA). The correlation coefficients (Rs) were computed to investigate the relationship between the chronology and climatic factors from 1961 to 2018. The climatic factors included the monthly average temperature and total precipitation from January to December of the current year. The R was also calculated between the $PDSI_{38}$ and tree-ring width. The relationship between the $PDSI_{38}$ value and the STD from 1961 to 2018 was revealed using a linear equation. The dependent variable was the $PDSI_{38}$ and the independent variable was the ring width index of the STD (Equation (1)). The equation passed the test of the leave-one-out model. The reduction of errors (RE) was 0.342, and the Durbin–Watson statistical coefficient (DW) was 1.720 (P (PMT and ST) < 0.05, where PMT represents the value derived using the product mean test and ST represents the sign test) (Table 2). The results indicated that the equation could be used to restore the values of the $PDSI_{38}$ [39]. Therefore, the $PDSI_{38}$ values between 1540 and 1960 were computed using Equation (1), based on the tree-ring width of the STD, as follows:

$$Y = -18.545 + 19.065 \cdot x \ (\mathrm{r} = 0.692, R^2 = 0.479, R^2\mathrm{adj} = 0.470, p < 0.01) \quad (1)$$

where the $PDSI_{38}$ is represented as Y, the tree-ring width index is represented as x and adjusted R^2 is represented as R^2_{adj}. The values of the $PDSI_{38}$ from 1540 to 1960 were computed using Equation (1). The $PDSI_{38}$ records were derived from the ESRL/NOAA data for 1961–2019. Then, we overlaid the $PDSI_{38}$ from 1540 to 1960 with the $PDSI_{38}$ from the ESRL/NOAA for 1961–2019. Finally, a 480-year (from 1540 to 2019) series for the $PDSI_{38}$ was reconstructed by combining the two sets of results (Figure 5).

Table 2. Calibration results of the leave-one-out (common period: 1961–2019) model. r is the correlation coefficient, R^2 is the explained variance, R^2_{adj} is the adjusted explained variance, RE is the reduction of error statistic, ST is the sign test, PMT is the product mean test, and DW is the Durbin–Watson test; ** $p < 0.01$, * $p < 0.05$.

	r	R^2	R^2_{adj}	RE	ST	PMT	DW
Calibration	0.69 **	0.479	0.470				
Verification	0.67 **	0.449	0.407	0.342	30+/29− *	3.534 *	1.720

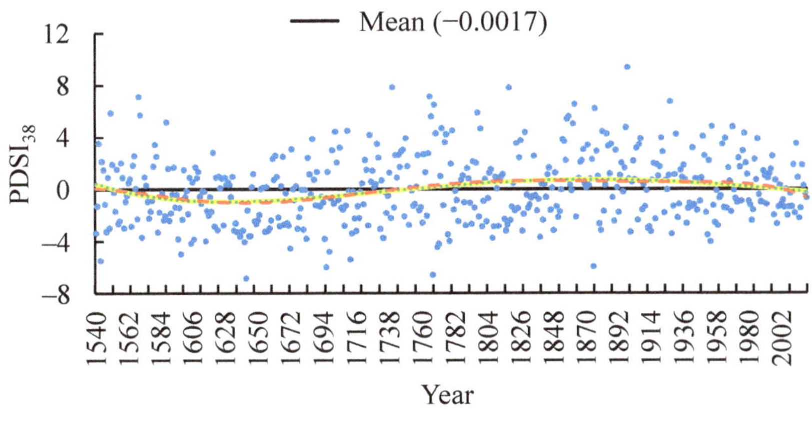

Figure 5. Changes in the reconstructed $PDSI_{38}$. The yellow solid line is the 4th function curve, the green dotted line is the 5th function curve, and the red dashed line is the 6th function curve.

2.4.2. Period Division of the $PDSI_{38}$ Related to Dry–Wet Changes

Division Using Function Fitting Methods

We used the year as the independent variable and the $PDSI_{38}$ of each year as the dependent variable for function fitting. If a function fitted well ($R^2 > 0.6$; $p < 0.05$), it indicated that the $PDSI_{38}$ change followed the law of the function curve. Based on the fluctuation characteristics of the curve, it could be divided into different dry–wet periods. However, in the 480-year period, the fitting effects of all functions from 1 to 30 times were not satisfactory ($R^2 < 0.2$; $p > 0.05$). Therefore, using a segmentation fitting method, the 480-year period was divided into different periods for function fitting, and good fitting functions ($R^2 > 0.6$; $p < 0.05$) were extracted for the different time periods. Based on the changes in the function curves, the variation law of the $PDSI_{38}$ was analyzed within the different periods. Well-fitting function curves can be used to divide the periodicity of the research object with the change in time. Therefore, the relationships may be used to divide the time series of the research object between the function curve and the mean line [40]. The periodic change in the $PDSI_{38}$ was applied using the stepwise regression method from the 4th to 6th functions. The functions were selected between the $PDSI_{38}$ (dependent variable) and the time series (independent variable) for 1540–2019. Firstly, the mean of the $PDSI_{38}$ was calculated for 1540–2019. Then, the different periods were divided using the intersections between the function curve and the mean line. The mean of three intersections between the three function curves and the mean line was calculated to divide the different periods, because the three function curves from the 4th to 6th functions were similar during the study period. The wet cycle was selected when the function curve was above the mean line. The dry cycle was represented by the curve below the mean line. Finally, the periodic change in the $PDSI_{38}$ values was divided completely in the first step. When the p values of the three functions were greater than 0.5 ($p > 0.05$), each period in the first step was divided again, using the same method as in the first step. The next step was carried out until the p value of the function was less than 0.05. The first step was considered as an example, to explain the periodic division of the $PDSI_{38}$.

Division Process of Function Fitting

The left-most intersection was the year of 1541 between the 4th function curve (from 1540 to 2019) and the mean line ($PDSI_{38} = -0.001$). The 5th function was the year of 1545 and the 6th function was the year of 1543. Then, the mean (1543 year) of the three years (1541, 1545, and 1543) was considered as the first point for periodic division. The second point was the year of 1757, and 1992 was the third point (for periodic division, using a similar method). The $PDSI_{38}$ values were divided into three cycles, from 1540 to 2019, according to the three intersections. The first dry cycle was from 1543 to 1756, because the three function curves (from 4th to 6th) were below the mean line (the three years from 1540 to 1542 were not considered in the study). The first wet cycle was from 1757 to 1992, because the three function curves (from 4th to 6th) were above the mean line during this period. The second dry cycle was from 1993 to 2019. The first step was concluded for the periodic division of the $PDSI_{38}$ change. The p values of all functions were greater than 0.5. Therefore, a second step was needed for periodic division. Finally, periodic division was performed successively until the 6th step in the first dry period (Figure 6a). The wet period was divided completely by the 5th step (Figure 6b), and by the 2nd step in the second dry period, as there were only 28 years in this period (Figure 6c).

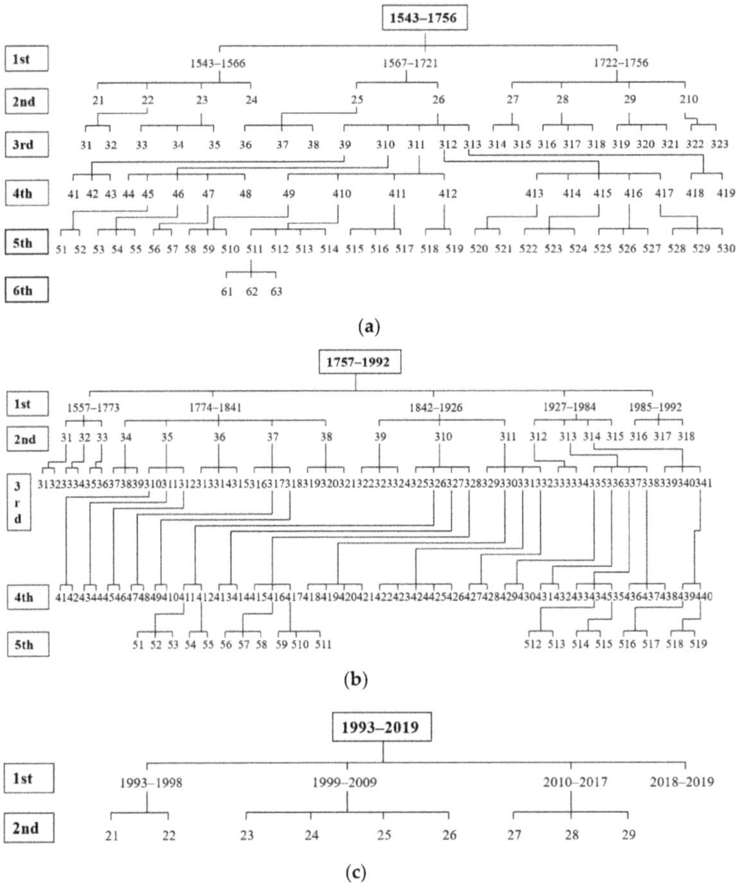

Figure 6. Stepwise regression of periodic change in $PDSI_{38}$ values for the (a) first dry (See Table S1), (b) first wet (See Table S2), and (c) second dry periods. The numbers represent the dry and wet periods (see Supplementary Materials). The means, standard deviations (SDs), and durations of the $PDSI_{38}$ values were calculated for each period of each step for the first dry, first wet, and second dry periods. The means of the $PDSI_{38}$ values were analyzed from the first to the last step, using the sixth function, to reveal the periodic characteristics of the dry–wet changes. The SDs of the $PDSI_{38}$ values were analyzed from the first to the last steps, using the sixth function, to reveal the periodic characteristics of the dry–wet changes. The durations of the periodic dry–wet periods were determined using the same method.

Analysis Methods

The means, STDs, and durations of the $PDSI_{38}$ were calculated for the average of each period within each of the first dry, the first wet, and the second dry periods, respectively. Regarding the results of the mean $PDSI_{38}$ for each step, the $PDSI_{38}$ change of each period was analyzed using the 6th function in order to reveal the periodic characteristics of the dry–wet change. Regarding the results of the STDs of the $PDSI_{38}$ for each period, the fluctuation law of the $PDSI_{38}$ was analyzed for each step using the 6th function in order to reveal the periodic characteristics of the dry–wet change. The duration of the periodic dry–wet change was understood in the same way.

3. Results

The chronology was strongly correlated with the temperature and precipitation from June to August of the current year (the year corresponding to climate factors was the same as the year of tree-ring width growth) ($p < 0.05$ and $p < 0.05$) (Figure 10a,b), indicating that the tree-ring growth responded very positively to changes in the temperature and precipitation in the active growth period. In addition, the R between the ring width chronology and $PDSI_{38}$ was 0.692 ($p < 0.01$). The tree growth was sensitive to dry–wet changes.

3.1. First Dry Period

3.1.1. Change in the Mean of the $PDSI_{38}$ in All Periods

The averages of the mean $PDSI_{38}$ values in all periods were −0.20 (first step), 0.12 (second step), −0.05 (third step), −0.35 (fourth step), −0.58 (fifth step), and −0.69 (sixth step) for all steps, respectively, for the long (the first step), intermediate (the second and third steps), and short (the fourth, fifth, and sixth steps) periods, indicating an initial trend of increasing followed by a decrease. The change range of the $PDSI_{38}$ was 0.81 [0.12 (maximum)-−0.69 (minimum)], with fluctuations above and below the mean line (−0.001). The results indicated that the regional climate can maintain the relative stability of the dry–wet period, which reflected the region's potential to become the first agricultural area in China. The SDs of the mean $PDSI_{38}$ values in all periods were 0.51, 0.85, 1.19, 1.23, 1.43, and 1.54 from the first to the sixth steps, respectively, indicating a gradual increase; thus, the differences between the periods increased gradually, and the fluctuations increased from the long to the short periods. All function curves exhibited a single concave shape; the concavity of the shape increased and the maximum values were found at the curve edges (at 1550–1581 and 1725–1749) from the long to the short periods (Figure 7A). The results indicated that the region had certain wet season characteristics when entering and exiting a dry period. The minimum values occurred in the middle (at 1636–1659) of the curve (Figure 7A), indicating that the strongest drought appeared in the middle of the period.

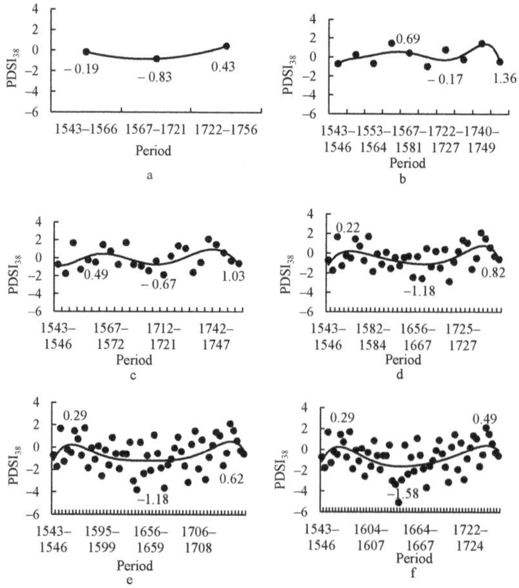

(A)

Figure 7. *Cont.*

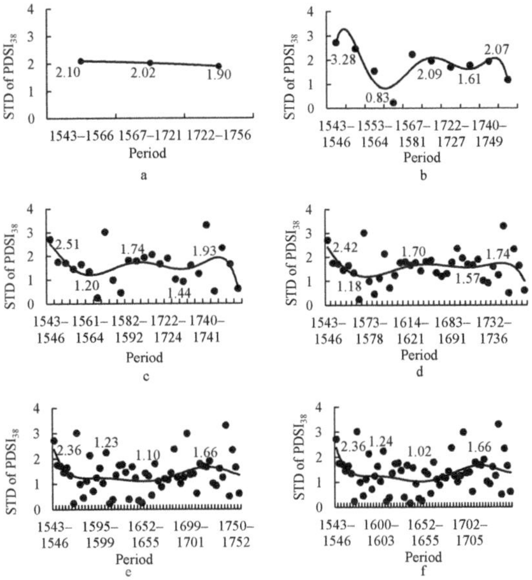

(B)

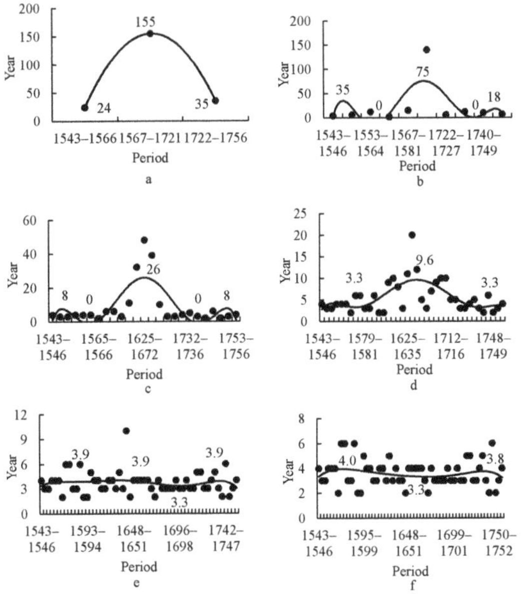

(C)

Figure 7. Sixth function curves of the (**A**) means, (**B**) standard deviations (SDs), and (**C**) durations of all periods of the $PDSI_{38}$ in the first dry period from the first step to the sixth step, where (**a**–**f**) denote the first to sixth steps. The numbers indicate the minimum and maximum values of all function curves. The solid lines are the function curves with the biggest R^2, and the dots are the index values of the vertical axis in the figures.

3.1.2. Change in the Standard Deviation (SD) of the $PDSI_{38}$ in All Periods

The averages of the SDs of the $PDSI_{38}$ values in all periods were 24, 23, 22, 22, 19, and 18 from the first to the sixth steps, respectively, indicating a gradual downward trend, and the duration of the dry period decreased from the long to the short periods. The SDs of the $PDSI_{38}$ values in all periods were 1, 9, 13, 11, 11, and 11 for the first to sixth steps, respectively, with a gradual upward trend, indicating that the dry difference among all periods increased gradually from the long to the short periods. The results indicated that the dry period relieves or even transforms into a wet period in the next period, to maintain the dry–wet balance, when the dry period within one cycle is long. The function curves presented a double concave shape in the intermediate and long periods (Figure 7B(b–d)), indicating that the fluctuations were small at the beginning and the end of the dry period, and the regional climate fluctuated relatively little in the early (approximately 1550–1581) and late periods (approximately 1725–1749) of the dry period. However, the dry fluctuation was the largest in the middle period (approximately 1636–1659) of the dry period, with the most severe droughts (Figure 7B(a–d)), which indicated that the regional climate tended to adjust itself as much as possible to overcome the drought in the long and intermediate periods. In the short periods, the function curves exhibited a single depression (Figure 7B(e,f)), indicating that the region could not change the severe drought conditions during 1633–1645.

3.1.3. Change in the Duration of the $PDSI_{38}$ in All Periods

The average durations of all periods were 71, 21, 9, 5, 3.7, and 3.6 years from the first to the sixth steps, respectively. The duration of the long period was 71, those of the intermediate periods were 21 and 9 years, and those of the short periods were 5, 3.7, and 3.6 years. The results were consistent with the historical data, i.e., during this period, disasters had periodic characteristics of 3–5, 20, and 70 years [41], which were influenced by the cyclical changes in the solar activity cycle and the cycle of the El Nino Southern Oscillation (ENSO) [42,43]. The SDs of the duration of the $PDSI_{38}$ in all periods decreased from 59, 40, 12, 3.6, 1.3, and 1.0 years (from the first step to the sixth step, respectively), indicating that the difference in the duration of each period in the long period was the largest; the difference was only one year for each period in the short period. The dry situation changed once every 3–5 years. The function curves showed a single peak in the intermediate and long periods (Figure 7C(a–d)), indicating that the durations of the different periods were long in the intermediate and long periods. Some short periods with high $PDSI_{38}$ values were interspersed to adjust the regional climate, so as to reduce the duration of a severe drought as far as possible. However, the function curves fluctuated very little in the short periods (Figure 7C(e,f)), indicating that the droughts were completely contained within the short periods.

3.2. *First Wet Cycle*

3.2.1. Changes in the Mean $PDSI_{38}$ in All Periods

The averages of the mean $PDSI_{38}$ values in all periods were 0.23, 0.13, 0.05, 0.12, and 0.17 from the first to the fifth steps, respectively. The difference in the $PDSI_{38}$ was only 0.18 [0.23 (maximum)–0.05 (minimum)] between the long (the first step), intermediate (the second step), and short (the third, fourth, and fifth steps) periods. Notably, all periods were relatively stable. Northern China is one of the most important grain-producing areas in China, and the planting structure is dominated by dry fields. The SDs of the mean $PDSI_{38}$ in all periods were 0.24, 0.64, 0.95, 1.13, and 1.25 from the first to the fifth steps, respectively, indicating a gradual increase, and the wet difference among all periods became larger as the period became shorter. The function curves were almost double concaves, progressing from the long to the intermediate and short periods (Figure 8A(a–e)), which indicated that the $PDSI_{38}$ was rather low at the beginning (1807–1839) and end (1938–1955) of the wet period. The regional climate tended to be dry in the early and late stages of the wet period. Meanwhile, the $PDSI_{38}$ was high in the middle of the period (approximately 1876–1904), which indicated that the flood with the wettest characteristics appeared during the wet period.

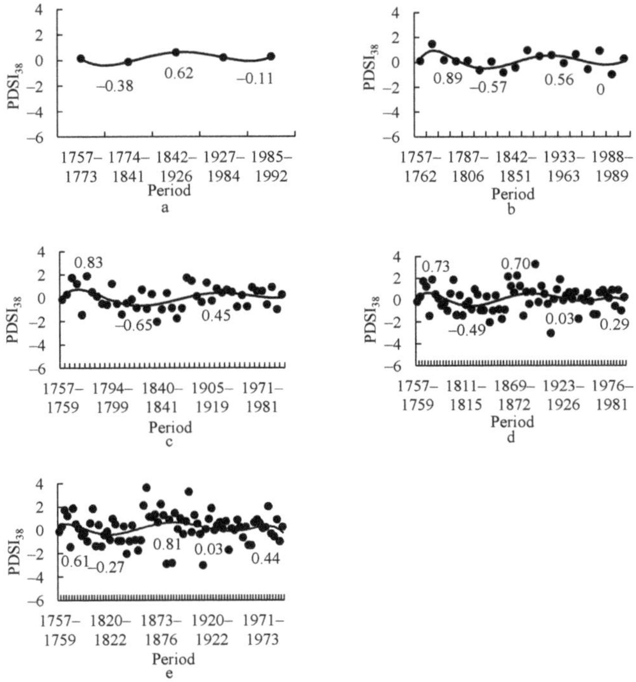

(A)

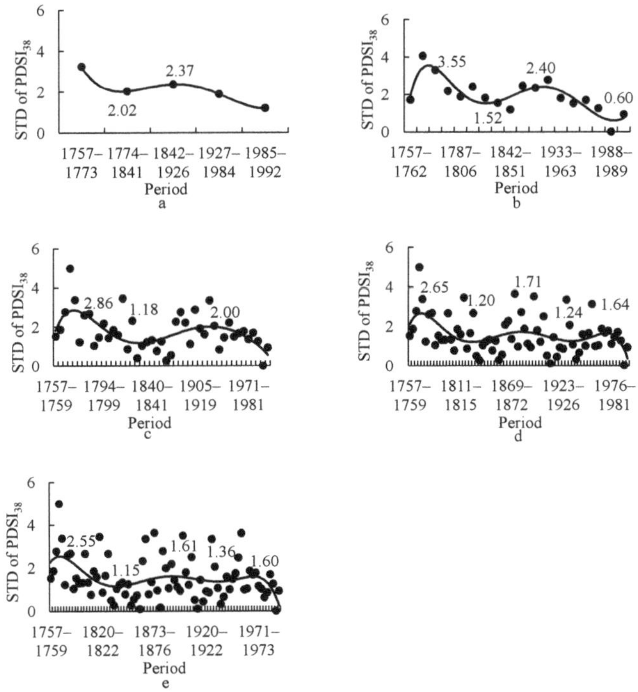

(B)

Figure 8. *Cont.*

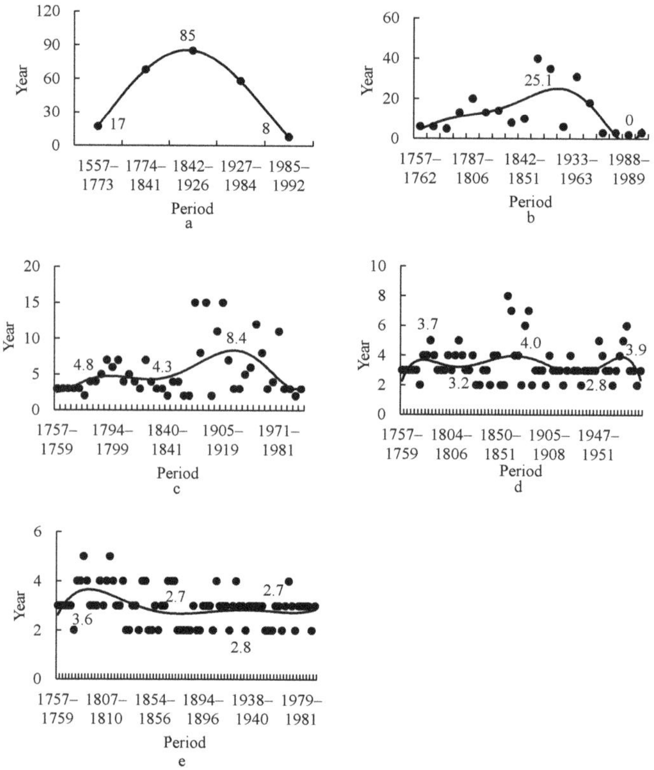

(C)

Figure 8. Sixth function curves of the (**A**) means, (**B**) standard deviations (SDs), and (C) durations of all periods in the first wet period from the first to the fifth steps, where (**a–e**) denote the first to fifth steps. The numbers on the graph indicate the minimum and maximum values of all function curves. The solid lines are the function curves with the biggest R^2, and the dots are the index values of the vertical axis in the figures.

3.2.2. Changes in the SD of the $PDSI_{38}$ in All Periods

The average SDs of the $PDSI_{38}$ values in all periods were 2.14, 1.94, 1.75, 1.58, and 1.53 from the first to the fifth steps, respectively, with a gradual downward trend, indicating that the wet period decreased from the long period to the short period. However, the SD values of all periods were 0.67, 0.89, 0.94, 0.96, and 0.98 (for the first to fifth steps, respectively), with a gradually increasing trend, indicating that the wet difference between the internal periods increased gradually. The function curves were double concaves from the long to the short periods, and the second concave was relatively flat (Figure 8B(a–e)), indicating that the $PDSI_{38}$ value fluctuated minimally and was relatively stable during the entire wet period, after a short period of adjustment at the beginning (1763–1768) of this period. The results indicated that the region maintained a relatively stable dry–wet state, which could provide suitable climatic conditions for agricultural production in wet periods.

3.2.3. Change in the Duration of the $PDSI_{38}$ in All Periods

The average durations of all periods were 47, 13, 5, 3, and 3 years from the first step to the fifth step, respectively. The duration of the long period was 47 years, that of the intermediate period was 13 years, and that of the short period was 3 years. The results were consistent with the changes in the drought and flood disasters in the Han River Basin over the past few hundred years. In addition, during this period, the number of floods (25 times) was greater than the number of droughts (13 times), indicating that the overall climate was relatively humid in this period [44]. The SDs of the duration of the $PDSI_{38}$ values of all periods decreased from 33, 12, 3.6, and 1.3 to 0.8 years, from the first step to the fifth step, respectively, indicating that the difference in duration was small in the short period. The function curves consisted of a single peak in the long periods (Figure 8C(a)), indicating that the duration was long in the middle stage (1842–1926). The curve had a small single peak at the back (Figure 8C(c)), indicating that the duration of each small period was not notably different in the short period. There was only one small single peak in the function curve in the initial stage of the short cycle, following which the shape was linear (Figure 8C(e)), indicating that the durations were relatively stable during the short cycles.

3.3. Second Dry Period

The function curves for the mean $PDSI_{38}$ values of all periods consisted of double concaves; however, the right concave was small, indicating an upward trend (Figure 9a,b), illustrating that the regional climate had a certain rectifying effect in the early part of the dry period, which caused the dry conditions to change to normal conditions. The function curve for the SDs of the $PDSI_{38}$ values of all periods was very complex (Figure 9d), indicating that the dry fluctuation increased at the beginning of the period. The function curve for the duration presented a double-concave shape (Figure 9f), indicating that the ability of the region to achieve climate adjustment and dry rectification was strong as the period began.

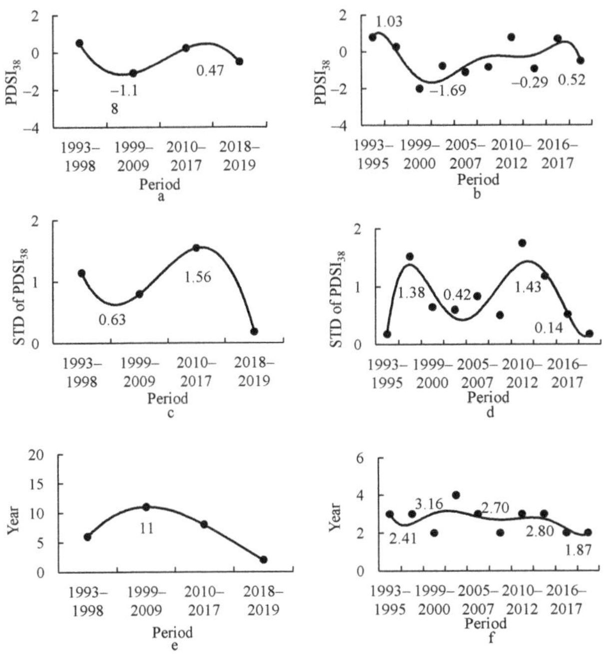

Figure 9. Sixth function curves of the means ((**a**): first step, (**b**): second step), standard deviations (STDs) ((**c**): first step, (**d**): second step), and durations ((**e**): first step, (**f**): second step) of all periods, indicating the $PDSI_{38}$ values in the second dry period for each step (from the first to the second steps). The numbers on the graphs indicate the minimum and maximum values of all function curves. The solid lines are the function curves with the biggest R^2, and the dots are the index values of the vertical axis in the figures.

4. Discussion

4.1. Climate Responses

The Rs between the monthly average temperature and STD from June to August were 0.35 (JUN, $p < 0.01$), 0.49 (JUL, $p < 0.01$), and 0.28 (AUG, $p < 0.05$), respectively (Figure 10a), indicating that, during the growing season, changes in temperature yielded a positive effect on tree growth [45]. From 1961 to 2019, the average monthly air temperature in the studied area was 22 °C in June, 25 °C in July, and 24 °C in August (Figure 10b). The temperatures were high and the conditions were favorable. The trees began their most active growth period, experiencing increased cell growth and radial expansion, indicating that the active growth period is crucial for tree-ring width growth. The STD was highly correlated with the precipitation from June to August of the current year ($p < 0.01$ and $p < 0.05$). The study by Stjepanovic et al. (2018) found that the growth of *Carya cathayensis* was affected by the precipitation, with significant differences observed when there was a lack of rainfall ($p < 0.05$) [46]. A drought during the growth period also contributed to reduced growth. The precipitation was highest in June, July, and August, with 100 mm, 231 mm, and 175 mm of rainfall, respectively (Figure 2b). The period was characterized by both high heat and high water availability in the region. Based on the findings of Li et al. (2016), trees grow better at higher temperatures when adequate water is available [45]. During the growth season, the tree rings widen in response to the temperature and rainfall. Moreover, there was a significant ($p < 0.05$) correlation between February's mean temperature and the ring width index. According to the lag characteristics of trees when adapting to cold conditions [47], the growth of the trees was related to the coldest temperature during January (−7.24 °C). To cope with low temperatures, the trees had to use their stored resources, which resulted in less growth. The Rs values between the STD and either the temperature or precipitation did not reach significant levels ($p > 0.05$), as shown in Figure 10a,b. Thus, the lack of extreme heat limited the growth of the trees. During the previous year's growing season, the trees accumulate storage reserves, utilizing carbon allocation to endure the harsh non-growing season [48]. Therefore, the cold temperatures (except during the coldest month) and low rainfall did not have a significant impact. The R value of 0.692 between the STD and $PDSI_{38}$ ($p < 0.01$) indicated that tree growth requires less water in agricultural and pastoral ecoregions [49], due to a combination of increased precipitation and the accelerated depletion of the water supply by warmer temperatures [48]. Thus, the water availability is influenced by these factors.

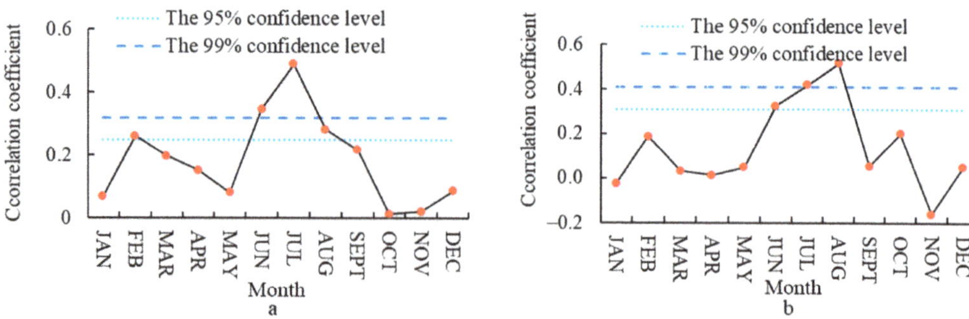

Figure 10. Correlation of the tree-ring chronology of *Carya cathayensis* and climatic factors (monthly average temperature (black line with red solid circles) (**a**) and total precipitation (black line with red solid circles) (**b**)) from the current January to December.

4.2. Drought Threat

There is one additional step in the first dry period (consisting of six steps) (Figure 6b) compared to the first wet period (consisting of five steps) (Figure 6c), and the number of cycles in each step of the first wet period was greater than that in the corresponding step of the first dry period (Figure 6b,c). The results indicated that droughts were relatively more concentrated and prominent, while floods were relatively scattered; thus, the floods were less severe than the droughts. The regional climate could not be effectively regulated in some periods, which could have had a serious impact on agriculture when severe droughts occurred. For example, the most severe drought in the first period occurred between 1640 and 1642, with the mean $PDSI_{38}$ value being only −4.68. Therefore, we could conclude that the region was experiencing an extreme drought at that time, when the value of the $PDSI_{38}$ was <−4 [50]. In addition, the mean $PDSI_{38}$ values of the two periods (1633–1635 and 1636–1639) were −1.71 and −3.35, respectively, which indicated that the droughts were mild and severe in these two periods [50]. According to historical records, this time period coincided with the reign of the late Ming Dynasty. During this period, successive droughts caused agricultural failures, resulting in a lack of agricultural production in Northern China [51]. In the first two drought periods (1633–1635 and 1636–1639), agricultural production was reduced significantly, and the population consumed all of the available food. At the end of the Ming Dynasty in 1644, the most severe drought (1640–1642) occurred in the region; there was no food to eat, and severe starvation and social unrest occurred [52]. Then, the Qing Dynasty gradually began its reign over China. Notably, there was no drought in China between 1646 and 1647 [53]. This was roughly consistent with the results of our study (1646–1647: $PDSI_{38}$ = 0.44, normal) [54]. During this period, i.e., around the middle of the 17th century, extreme droughts occurred in many parts of the world. For example, extreme droughts occurred in South Asia in 1615–1626 [55], South America in 1620–1630 [56], Europe in 1625–1632 [57,58], Australia in 1636–1645 [59], South Africa in 1640–1660 [60], and North America in 1655–1675 [61]. Therefore, many parts of the world experienced severe droughts in different years during this stage (1633–1642) (or near this stage).

4.3. Strategies to Address Severe Droughts

In the first dry period, from 1633 to 1645, four periods (1633–1635 ($PDSI_{38}$ = −1.71), 1636–1939 ($PDSI_{38}$ = −3.35), 1640–1642 ($PDSI_{38}$ = −4.68), and 1643–1645 ($PDSI_{38}$ = −2.92)) with low $PDSI_{38}$ values occurred successively, and severe droughts lasted 13 years. We speculate that there will be a severe drought in the early part of the 22nd century, based on the Chinese calendar of dry–wet changes that we have studied, where the duration of the first drought period was 214 years. In the middle of the period, the most severe drought occurred, with a half period of 108 years. The time point at which the second drought period began was 1992. According to this time point, it is speculated that 2100 will be the driest year in the future, occurring 108 years later. At the same time, according to the IPCC [62], at the end of the 21st century, sudden changes in climate factors such as temperature and precipitation may cause extreme droughts on a large scale throughout the world. Therefore, appropriate risk management measures should be planned accordingly. In addition, building water conservancy facilities and storing water and grain would provide active assistance in locations where severe droughts may occur in the future. The agricultural planting scheme should also be changed according to the actual dry–wet conditions in different regions. For example, according to historical records, there were 11 droughts in Sichuan during 1625–1645, with the average value of the $PDSI_{38}$ being −2.03 for all periods, based on the results of this study (the lowest of the research years). In order to survive this difficult period, the region was forced to transition from paddy fields to dry fields [63]. During 1646–1682, grain production increased gradually; finally, it led to large harvests for three consecutive years during 1680–1682 [63]. Therefore, some traditional rice-producing areas changed to dry land and planted food crops with higher yields. In the Yangtze River Basin, from 1633 to 1645, drought and wind disasters reached an average annual frequency of 2.6 times. Under the pressure of such drastic climate changes, millions of people moved from the Yangtze River Basin to the south of the Nanling Mountains (Pearl River Basin) [64], indicating that the impact of the drought on Guangdong and Guangxi was small. During this period, other tropical regions in the world were also insignificantly affected by extreme droughts [65–67]. Since China once purchased rice from the Indochina Peninsula and extreme droughts did not occur simultaneously worldwide, the time differences of the drought period could be used when purchasing different grains from different regions. For example, an extreme drought in South America (1620–1630) occurred earlier than that in North America (1655–1675). Owing to this asynchronism with North America, South America could have first purchased grains from South America; then, when North America entered the drought period, North America could have purchased grains from South America. Australia, South Asia, Europe, and East Asia could also have followed this method. In addition, the droughts in tropical areas are not serious; thus, the surplus produce from these regions could be exported to other parts of the world. Therefore, the threat of extreme drought can be minimized through global cooperation.

4.4. Period of Dry–Wet Change

The dry–wet change occurs in a cycle of 12 years or multiples of 12, and the first wet period (1757–1992) lasted a total of 236 years. By merging the period of 1985–1992 with the previous period of 1927–1984, we initially divided the first wet period into four periods (long cycle) (Figure 6c). The average length of the initial wet period's long intervals was 59 years, which was nearly a 60-year interval. The average duration (13 years) of all intermediate periods (the fourth step) in the initial wet period was comparable to the 12-year duration of the initial wet period. However, the duration of each interval in the dry period was slightly shorter than 12 years. The first dry period lasted 214 years (1543–1756), which is 26 years shorter than 240 years, and each period lasted 2 years less than two 12-year periods. The average length of the long periods (first step) during the first dry period was 71 years, one year less than the 72-year period of the sum of the 60- and 12-year periods. The average length of all intermediate periods (third step) during the first dry period was 9 years, or 3 years less than the 12-year average. These outcomes could reflect a 12-year change that is influenced by both the lowest point of the solar cycle and ENSO [68,69]. The dry–wet change over 60 years might be impacted by the Pacific Decadal Oscillation (PDO) [70], and the 120-year change could be linked to centennial-scale solar cycles [71]. Thus, the years of the dry–wet change can be roughly inferred using a 12-year or multiple 12-year cycle. The duration of the short period was 3–5 years in all dry and wet periods; this finding was in agreement with several previous studies. For example, the droughts in the region presented a 4-year periodic oscillation from 1961 to 2009 [72]. The main period of dry–wet changes in China lasted 4.4 years [73]. Even in Southern Africa, floods also occurred in one period with four years [74]. These periodic changes are all related to the El Nino and La Nina phenomena [75]. Therefore, that the shortest period of dry–wet changes in the region, and even in China, lasts 3–5 years indicates that China is deeply affected by these two phenomena.

5. Conclusions

The tree-ring data of *Carya cathayensis* were used to establish a chronology and restore the $PDSI_3$ of the ecotone between farming and animal husbandry along the Great Wall for the period of 1540–2019. The stepwise function method was used to study the periodicity of the dry–wet changes. There were two dry cycles and one wet cycle during the 477 years considered in this study. Each dry–wet period was divided into different levels, including long (decades), intermediate (10 years), and short periods (several years, less than 10 years). In the first dry period, the change range of the $PDSI_{38}$ value was not high for all periods (from the first to the fifth steps), indicating that the region maintained a stable dry–wet status. However, in the short period (the sixth step), there were periods of severe drought (1633–1645). In the first wet period, the fluctuations in the $PDSI_{38}$ values were very small in each individual period. The capacity for the regulation of regional floods was stronger than that of regional droughts, and the values of the $PDSI_{38}$ changed smoothly during this period; thus, the conditions were favorable for agricultural production. In the second drought period, only the initial stage of the drought was analyzed. During this period, the changes in the dry–wet period were corrected and, thus, the region maintained a relatively stable status. Compared with floods, droughts pose a greater threat, not only in China but throughout the entire world. Therefore, we should plan and implement effective measures to prepare for the severe droughts that may occur in the future. Notably, the durations of the long, intermediate, and short periods of the dry–wet periods were consistent with those indicated in a 12-year or multiple 12-year cycle; thus, the information in the sexagenary can be used to predict the occurrence of droughts and floods in China.

Supplementary Materials: The following supporting information can be downloaded at: https://www.mdpi.com/article/10.3390/ijgi13060191/s1, Table S1: The numbers in the Figure 6a indicate the years of the cycles in the first dry cycle (1540–1756); Table S2: The numbers in the Figure 6b indicate the years of the cycles in the first wet cycle (1756–1992) and the Figure 6c (the second dry cycle (1993–2019)). This study's supplementary materials can be available from the author upon.

Author Contributions: Conceptualization, methodology, and writing-original draft preparation, Xiaodong Wang; methodology, investigation, and writing-original draft preparation, Xiaoqiang Li; formal analysis, Yujia Song; investigation, Yu An; editing and project administration, Xiaohui Liu. All authors have read and agreed to the published version of the manuscript.

Funding: This research was funded by Joint Funds of the National Natural Science Foundation of China (U2243230); the National Natural Science Foundation of China (42230516); Technology Development Program of Jilin Province (20230203003SF); and Natural Science Foundation of Changchun Normal University (CSJJ2022008ZK).

Data Availability Statement: The data presented in this study are available from the author upon reasonable request.

Acknowledgments: The authors would like to thank the anonymous reviewers and handling editors for their constructive comments.

Conflicts of Interest: The authors declare no conflict of interest.

References

1. Hu, Z.Y.; Zhou, J.J.; Zhang, L.L.; Wei, W.; Cao, J.J. Climate dry-wet change and dry evolution characteristics of different dry-wet areas in northern China. *Acta Ecol. Sin.* **2018**, *38*, 1908–1919. (In Chinese with English Abstract)
2. Misra, V.; Solomon, S.; Mall, A.K.; Prajapati, C.P.; Hashem, A.; Allah, E.F.; Ansari, M.I. Morphological assessment of water stressed sugarcane: A comparison of waterlogged and dry affected crop. *Saudi J. Biol. Sci.* **2020**, *27*, 1228–1236. [CrossRef]
3. Wu, J.F.; Chen, X.H. Spatiotemporal trends of dryness/wetness duration and severity: The respective contribution of precipitation and temperature. *Atmos. Res.* **2019**, *216*, 176–185. [CrossRef]
4. Yu, X.F.; Wang, Y.; Yu, S.Y.; Kang, Z.H. Synchronous droughts and floods in the southern Chinese Loess Plateau since 1646 CE in phase with decadal solar activities. *Glob. Plant Change* **2019**, *183*, 103033. [CrossRef]
5. Huang, Q.Z.; Zhang, Q.; Singh, V.P.; Shi, P.J.; Zheng, Y.J. Variations of dryness/wetness across China: Changing properties, drought risks, and causes. *Glob. Planet Change* **2017**, *115*, 1–12. [CrossRef]
6. Peng, Y.; Long, S.F.; Ma, J.W.; Song, J.Y.; Liu, Z.W. Temporal-spatial variability in correlations of drought and flood during recent 500 years in Inner Mongolia, China. *Sci. Total Environ.* **2018**, *633*, 484–491. [CrossRef]
7. Cheng, Q.P.; Gao, L.; Zuo, X.A.; Zhong, F.L. Statistical analyses of spatial and temporal variabilities in total, daytime, and nighttime precipitation indices and of extreme dry/wet association with large-scale circulations of Southwest China, 1961–2016. *Atmos. Res.* **2019**, *219*, 166–182. [CrossRef]
8. Manoj, V.G.; Jain, K. Unravelling the teleconnections between ENSO and dry/wet conditions over India using nonlinear Granger causality. *Atmos. Res.* **2021**, *247*, 105168.
9. Melsen, L.A.; Teuling, A.J.; Torfs, P.J.J.F.; Zappa, M.; Mizukami, N.; Mendoza, P.A.; Clark, M.P.; Uijlenhoet, R. Subjective modeling decisions can significantly impact the simulation of flood and drought events. *J. Hydrol.* **2019**, *568*, 1093–1104. [CrossRef]
10. Hao, X.Y.; Wei, H.; Lam, S.K.; Li, P.; Zong, Y.Z.; Zhang, D.S.; Li, F.Y. Enhancement of no-tillage, crop straw return and manure application on field organic matter content overweigh the adverse effects of climate change in the arid and semi-arid Northwest China. *Agric. For. Meteorol.* **2020**, *295*, 108199. [CrossRef]
11. Xu, X.C.; Ge, Q.S.; Zheng, J.Y.; Liu, C.W. Reviews of agricultural drought risk assessment. *Agric. Res. Arid. Areas* **2010**, *6*, 263–270. (In Chinese with English Abstract)
12. Shi, W.J.; Liu, Y.T.; Shi, X.L. Contributions of climate change to the boundary shifts in the farming-pastoral ecotone in northern China since 1970. *Agric. Syst.* **2018**, *161*, 16–27. [CrossRef]

13. Wang, T.F.; Su, B.D.; Zhai, J.Q.; Jiang, T.; Wen, S.S. Variation of drys in Luanhe river basin, 1961–2012. *J. Arid. Land Resour. Environ.* **2015**, *29*, 180–185. (In Chinese with English Abstract)
14. Jia, Y.Q.; Zhang, B. Spatio-temporal changes of the extreme drought and wet events in Northern China from 1960 to 2016. *J. Nat. Resour.* **2019**, *34*, 1543–1554. (In Chinese with English Abstract)
15. Ma, Z.G.; Dan, L.; Hu, Y.W. The extreme dry/wet events in northern China during recent 100 years. *J. Geogr. Sci.* **2004**, *3*, 275–281
16. Ma, Z.G.; Huang, G.; Gan, W.Q.; Chen, M.L. Multi-Scale temporal characteristics of the dryness /wetness over northern China during the last century. *Chin. J. Atmos. Sci.* **2005**, *29*, 671–681. (In Chinese with English Abstract)
17. Xiao, L.B.; Fang, X.Q.; Ye, Y. Reclamation and revolt: Social responses in Eastern Inner Mongolia to flood/dry-induced refugees from the North China Plain 1644–1911. *J. Arid. Environ.* **2013**, *88*, 9–16. [CrossRef]
18. Gao, S.; Zhou, T.; Yi, C.X.; Shi, P.J.; Fang, W.; Liu, R.S.; Liang, E.Y.; Camarero, J.J. Asymmetric impacts of dryness and wetness on tree growth and forest coverage. *Agric. For. Meteorol.* **2020**, *288–289*, 107980. [CrossRef]
19. Klippel, L.; George, S.S.; Buntgen, U.; Krusic, P.J.; Esper, J. Differing pre-industrial cooling trends between tree rings and lower-resolution temperature proxies. *Clim. Past* **2020**, *16*, 729–742. [CrossRef]
20. Maxwell, J.T.; Harley, G.L.; Matheus, T.J.; Strange, B.M.; Van Aken, K.; Au, T.F.; Bregy, J.C. Sampling density and date along with species selection influence spatial representation of tree-ring reconstructions. *Clim. Past* **2020**, *5*, 1901–1916. [CrossRef]
21. Ziaco, E.; Miley, N.; Biondi, F. Reconstruction of seasonal and water-year precipitation anomalies from tree-ring records of the southwestern United States. *Palaeogeogr. Palaeoclimatol. Palaeoecol.* **2020**, *547*, 109689. [CrossRef]
22. Yadava, A.K.; Misra, K.G.; Singh, V.; Misra, S.; Sharma, Y.K.; Kotlia, B.S. 244-YEAR long tree-ring based drought records from Uttarakhand, western Himalaya, India. *Quat. Int.* **2021**, *599–600*, 128–137. [CrossRef]
23. Li, S.J.; Sun, Z.G.; Tan, M.H.; Guo, L.L.; Zhang, X.B. Changing patterns in farming pastoral ecotones in China between 1990 and 2010. *Ecol. Indic.* **2018**, *89*, 110–117. [CrossRef]
24. Lin, C.C.; Wang, K.; Sun, Y.M. Study on the change of temperature time series in medium section of agro-pastoral ectone of Northern China during the last 60 years. *Acta Agrestia Sin.* **2016**, *4*, 747–753. (In Chinese with English Abstract)
25. Zhou, Y.M.; Zhang, A.; Zhao, X.Y. Analysis of vulnerable characteristics in Chinese northern farming-pastoral region based on coordinated regional downscaling experiment. *Acta Sci. Nat. Univ. Pekin.* **2017**, *53*, 1099–1107. (In Chinese with English Abstract)
26. Shelach-Lavi, G.; Tu, D.D. Food, pots and socio-economic transformation: The beginning and intensification of pottery production in North China. *Archaeol. Res. Asia* **2017**, *12*, 1–10. [CrossRef]
27. Zhai, X.J.; Huang, D.; Tang, S.M.; Li, S.Y.; Guo, J.X.; Yang, Y.J.; Liu, H.F.; Li, J.S.; Wang, K. The emerge of metabolism in different ecosystems under the same environmental conditions in the agro-pastoral ecotone of northern China. *Ecol. Indic.* **2017**, *74*, 198–204. [CrossRef]
28. Stokes, M.; Harlan, T.; Harris, M.; Storey, J.B. Datability of pecan tree ring. *HortScience* **1995**, *30*, 523–524. [CrossRef]
29. Velmex, Inc. *The Velmex "TA" System for Research and Non-Contact Measurement Analysis*; Velmex Inc.: Bloomfield, NY, USA, 1992.
30. Fritts, H.C. *Tree Rings and Climate*; Academic Press: New York, NY, USA, 1976.
31. Holmes, R. Computer assisted quality control. *Tree-Ring Bull.* **1983**, *43*, 69–78.
32. Camarero, J.J.; Martin, E.; Gil-Pelegrin, E. The impact of a needleminer (*Epinotia subsequana*) outbreak on radial growth of silver fir (*Abies alba*) in the Aragon Pyrenees: A dendrochronological assessment. *Dendrochronologia* **2003**, *21*, 3–12. [CrossRef]
33. Cook, E.R.; Kairiukstis, L.A. *Methods of Dendrochronology: Applications in the Environmental Sciences*; Kluwer Academic Publishers: Dordrecht, The Netherlands, 1990.
34. Helama, S.; Begin, Y.; Vartiainen, M.; Peltola, H.; Kolstrom, T.; Merilainen, J. Quantifications of dendrochronological information from contrasting microdensitometric measuring circumstances of experimental wood samples. *Appl. Radiat. Isot.* **2012**, *70*, 1014–1023. [CrossRef]
35. Davi, N.K.; Jacoby, G.C.; Wiles, G.C. Boreal temperature variability inferred from maximum latewood density and tree-ring width data, Wrangell Mountain region, Alaska. *Quat. Res.* **2003**, *60*, 252–262. [CrossRef]
36. Wigley, T.; Briffa, K.R.; Jones, P.D. On the average value of correlated time series, with applications in dendroclimatology and hydrometeorology. *J. Appl. Meteorol. Climatol.* **1984**, *23*, 201–213. [CrossRef]
37. Dai, A.G. Drought under global warming: A review. Wiley Interdiscip. *Rev. Clim. Change* **2011**, *2*, 45–65. [CrossRef]
38. Zobler, L. *Global Soil Types, 1-Degree Grid*; Oak Ridge National Laboratory Distributed Active Archive Center: Auckland, New Zealand, 1999.
39. Fan, Z.X.; Brauning, A.; Tian, Q.H.; Yang, B.; Cao, K.F. Tree ring recorded May-August temperature variations since A.D. 1585 in the Gaoligong Mountains, southeastern Tibetan Plateau. *Palaeogeogr. Palaeoclimatol. Palaeoecol.* **2010**, *296*, 94–102. [CrossRef]
40. Cromer, T.L. A periodicity threshold theorem for a volterra integral equation. *Appl. Anal.* **1986**, *21*, 1–8. [CrossRef]
41. Yao, S.Y. The geographical distribution of floods and droughts in Chinese history, 206 B.C.-A.D. 1911. *Far East. Q.* **1943**, *4*, 357–378.
42. Sun, J.Q.; Wang, X.J.; Yin, Y.X.; Shahid, S. Analysis of historical drought and flood characteristics of Hengshui during the period 1649–2018: A typical city in North China. *Nat. Hazards* **2021**, *108*, 2081–2099. [CrossRef]
43. Liu, X.Y.; Li, D.L.; Wang, J.D. Spatiotemporal characteristics of dry over China during 1961–2009. *J. Desert Res.* **2012**, *32*, 473–483. (In Chinese with English Abstract)
44. Zhang, X.D.; Ren, G.Y.; Yang, Y.D.; Bing, H.; Hao, Z.X.; Zhang, P.F. Extreme historical droughts and floods in the Hanjiang River Basin, China, since 1426. *Clim. Past* **2022**, *18*, 1775–1796. [CrossRef]

45. Li, Y.J.; Wang, S.Y.; Niu, J.J.; Fang, K.Y.; Li, X.L.; Li, Y.; Bu, W.L.; Li, Y.H. Climate-adial growth relationship of Larix principis-upprechtii at different altitudes on Luya Mountain. *Acta Ecol. Sin.* **2016**, *36*, 1608–1618. (In Chinese with English Abstract)
46. Stjepanovic, S.; Matovic, B.; Stojanovic, D.; Lalic, B.; Levanic, T.; Orlovic, S.; Gutalj, M. The Impact of Adverse Weather and Climate on the Width of European Beech (*Fagus sylvatica* L.) Tree Rings in Southeastern Europe. *Atmosphere* **2018**, *9*, 451. [CrossRef]
47. Zeng, L.B.; Wang, X.P.; Cheng, J.F.; Lin, X.; Wu, Y.L.; Yin, W.L. Alpine timberline ecotone tree growth in relation to climatic variability for Picea crassifolia forests in the middle Qilian Mountains, northwestern China. *J. Beijing For. Univ.* **2012**, *34*, 50–56. (In Chinese with English Abstract)
48. Lockwood, B.R.; Leblanc, D.C. Radial growth-climate relationships of white ash (*Fraxinus americana* L. *Oleaceae*) in the eastern United States. *J. Torrey Bot Soc.* **2017**, *144*, 267–279.
49. Dai, J.H.; Wang, M.M.; Wang, H.J.; Bai, J.; Cui, H.T. Climate changes and their ecological impacts of the last 50 years in semihumid and semiarid transitional zones of the east part of north west China. *Quat. Sci.* **2009**, *29*, 920–930.
50. Gaire, N.P.; Dhakal, Y.R.; Shah, S.K.; Fan, Z.X.; Brauning, A.; Thapa, U.K.; Bhandari, S.; Aryal, S.; Bhuju, D.R. Drought (scPDSI) reconstruction of trans-Himalayan region of central Himalaya using *Pinus wallichiana* tree-rings. *Palaeogeogr. Palaeoclimatol. Palaeoecol.* **2019**, *514*, 251–264. [CrossRef]
51. Fang, X.Q.; Xiao, L.B.; Su, Y.; Zheng, J.Y.; Wei, Z.D.; Yin, J. Social impacts of climate change on the history of China. *J. Palaeogeogr.* **2017**, *19*, 729–736. (In Chinese with English Abstract)
52. Yang, Y. The influence of natural disasters on the extinction of the Ming Dynasty. *J. Suizhou Univ.* **2016**, *31*, 73–77. (In Chinese with English Abstract)
53. Jia, T.F.; Shi, W.Y.; Zheng, X.Y.; Zhang, W.G.; Yu, L.Z. The flood and drought disasters of the Chaohu lake basin in the past 600 years. *Sci. Geogr. Sin.* **2012**, *32*, 66–73. (In Chinese with English Abstract)
54. Lu, J.Y.; Sun, L.Y.; Wang, C.L.; Zhai, Z.H.; Wu, J.D. Analysis of drought characteristics in Guangdong in recent 40 years based on PDSI index. *J. Meteorol. Res. Appl.* **2022**, *43*, 36–40. (In Chinese with English Abstract)
55. Arsalani, M.; Grieinger, J.; Pourtahmasi, K.; Brauning, A. Multi-centennial reconstruction of drought events in South-Western Iran using tree rings of Mediterranean cypress (*Cupressus sempervirens* L). *Palaeogeogr. Palaeoclimatol. Palaeoecol.* **2021**, *567*, 110296. [CrossRef]
56. Hadad, M.A.; Gonzalez-Reyes, A.; Roig, F.A.; Matskovsky, V.; Cherubini, P. Tree-ring-based hydroclimatic reconstruction for the northwest Argentine Patagonia since 1055 CE and its teleconnection to large-scale atmospheric circulation. *Glob. Planet Change* **2021**, *202*, 103496. [CrossRef]
57. Buntgen, U.; Trouet, V.; Frank, D.; Leuschner, H.H.; Friedrichs, D.; Luterbacher, U.; Esper, J. Tree-ring indicators of German summer drought over the last millennium. *Quat. Sci. Rev.* **2010**, *29*, 1005–1016. [CrossRef]
58. Yue, R.P.H.; Lee, H.F. Drought-induced spatio-temporal synchrony of plague outbreak in Europe. *Sci. Total Environ.* **2020**, *698*, 134138. [CrossRef]
59. Karanitsch-Ackerl, S.; Mayer, K.; Gauster, T.; Laaha, G.; Holawe, F.; Wimmer, R.; Grabner, M. A 400-year reconstruction of spring–summer precipitation and summer low flow from regional tree-ring chronologies in North-Eastern Austria. *J. Hydrol.* **2019**, *577*, 123986. [CrossRef]
60. Stager, J.C.; Ryves, D.B.; King, C.; Madson, J.; Hazzard, M.; Neumann, F.H.; Maud, R. Late Holocene precipitation variability in the summer rainfall region of South Africa. *Quat. Sci. Rev.* **2013**, *67*, 105–120. [CrossRef]
61. Sutheimer, C.M.; Meunier, J.; Hotchkiss, S.C.; Rebitzke, E.; Radeloff, V.C. Historical fire regimes of North American hemiboreal peatlands. *For. Ecol. Manag.* **2021**, *498*, 119561. [CrossRef]
62. Zhou, T.J.; Chen, Z.M.; Chen, X.L.; Zou, M.; Jiang, J.; Hu, S. Interpreting IPCC AR6: Future global climate based on projection under scenarios and on near-term information. *Clim. Change Res.* **2021**, *17*, 652–663.
63. Huang, L.Y. Analysis on the change of crops structure in Chengdu plain in little ice age of Ming and Qing dynasties. *Agric. Archaeol.* **2018**, *4*, 66–71. (In Chinese with English Abstract)
64. Liu, T.Y. The she people's acculturation during the little ice age in the Ming and Qing dynasties: Environmental change and ethnic relations in the juncture areas of Fujian, Guangdong, Jiangxi and Hunan Provinces. *Guangxi Ethn. Stud.* **2019**, *146*, 96–102. (In Chinese with English Abstract)
65. Locosselli, G.M.; Brienen, R.J.W.; Martins, V.T.S.; Gloor, E.; Boom, A.; Camargo, E.P.; Saldiva, P.H.N.; Buckeridge, M.S. Intra-annual oxygen isotopes in the tree rings record precipitation extremes and water reservoir levels in the Metropolitan Area of Sao Paulo, Brazil. *Sci. Total Environ.* **2020**, *743*, 140798. [CrossRef]
66. Rigozon, N.R.; Prestes, A.; Nordemann, D.J.R.; Silva, H.E.; Echer, M.P.S.; Echer, E. Solar maximum epoch imprints in tree-ring width from Passo Fundo, Brazil (1741–2004). *J. Atmos. Solar-Terr. Phys* **2008**, *70*, 1025–1033. [CrossRef]
67. Yang, B.; Chen, X.H.; He, Y.H.; Wang, J.W.; Lai, C.G. Reconstruction of annual runoff since CE 1557 using tree-ring chronologies in the upper Lancang-Mekong River basin. *J. Hydrol.* **2019**, *569*, 771–781. [CrossRef]
68. Mikhailov, A.V.; Perrone, L.; Nusinov, A.A. Thermospheric parameters' long-term variations over the period including the 24/25 solar cycle minimum. Whether the CO_2 increase effects are seen? *J. Atmos. Solar-Terr. Phys* **2021**, *223*, 105736.
69. Wang, D.; Du, S.C.; Jia, W.J. Multiscale variability of China historical flood/drought index and precipitation teleconnections with ENSO using wavelet analyses. *Theor. Appl. Climatol.* **2022**, *149*, 1583–1597. [CrossRef]

70. Chen, Y.; Wang, H.; Liu, H.; Wang, F.; Li, F.M.; Cheng, H.; Ma, Z.B.; Cai, B.G.; Xiao, J.L. Summer monsoon precipitation variability during recent 700 years recorded in a high-resolution stalagmite from southwestern Yunnan, China. *Quat. Sci.* **2021**, *41*, 398–410, (In Chinese with English Abstract).
71. Gregory, B.R.B.; Patterson, R.T.; Galloway, J.M.; Reinhardt, E.G. The impact of cyclical, multi-decadal to centennial climate variability on arsenic sequestration in lacustrine sediments. *Palaeogeogr. Palaeoclimatol. Palaeoecol.* **2021**, *565*, 110189. [CrossRef]
72. Liu, Y.L.; Wen, Y.J.; Zhao, Y.Q.; Hu, H.N. Analysis of drought and flood variations on a 200-Year scale based on historical environmental information in western China. *Int. J. Environ. Res. Public Health* **2022**, *19*, 2771. [CrossRef]
73. Wang, Z.L.; Li, J.; Huang, Z.Q.; Zhong, R.D.; Chen, J.Y.; Qiu, Z.H. Spatiotemporal variations analysis of meteorological drought in China based on scPDSI. *Trans. Chin. Soc. Agric. Eng.* **2016**, *32*, 161–168, (In Chinese with English Abstract).
74. Franchi, F.; Ahad, J.M.E.; Geris, J.; Jhowa, G.; Petros, A.K.; Comte, J.C. Modern sediment records of hydroclimatic extremes and associated potential contaminant mobilization in semi-arid environments: Lessons learnt from recent flood-dry cycles in southern Botswana. *J. Soils Sediments* **2020**, *20*, 1632–1650. [CrossRef]
75. Stojanovic, M.; Liberato, M.L.R.; Sori, R.; Vazquez, M.; Phan-Van, T.; Hieu, D.; Hoang, C.T.; Nguyen, P.N.B.; Raquel, N.; Luis, G. Trends and extremes of dry episodes in vietnam sub-regions during 1980–2017 at different timescales. *Water* **2020**, *3*, 813. [CrossRef]

Disclaimer/Publisher's Note: The statements, opinions and data contained in all publications are solely those of the individual author(s) and contributor(s) and not of MDPI and/or the editor(s). MDPI and/or the editor(s) disclaim responsibility for any injury to people or property resulting from any ideas, methods, instructions or products referred to in the content.

climate

MDPI

Article

Historic Changes and Future Projections in Köppen–Geiger Climate Classifications in Major Wine Regions Worldwide

Cristina Andrade [1,2,3,*], André Fonseca [2,3], João A. Santos [2,3], Benjamin Bois [4] and Gregory V. Jones [5]

1 Polytechnic Institute of Tomar, Natural Hazards Research Center (NHRC.ipt), Quinta do Contador, Estrada da Serra, 2300-313 Tomar, Portugal
2 Centre for the Research and Technology of Agro-Environmental and Biological Sciences, CITAB, University of Trás-os-Montes e Alto Douro, UTAD, 5001-801 Vila Real, Portugal; jsantos@utad.pt (J.A.S.)
3 Institute for Innovation, Capacity Building and Sustainability of Agrifood Production, Inov4Agro, University of Trás-os-Montes e Alto Douro, UTAD, 5001-801 Vila Real, Portugal
4 Biogéosciences, UMR 6282 CNRS, Université de Bourgogne, 6 Boulevard Gabriel, 21000 Dijon, France; benjamin.bois@u-bourgogne.fr
5 Abacela Vineyards and Winery, 12500 Lookingglass Road, Roseburg, OR 97471, USA
* Correspondence: c.andrade@ipt.pt; Tel.: +351-249-328-100

Abstract: A valuable tool for comprehending and characterizing climate patterns on a global scale is the Köppen–Geiger climate classification system. When it comes to wine production, the climate of a region plays an essential role in determining whether specific grape varieties can be cultivated, largely determining the style of wine that can be made, and influencing the consistency of overall wine quality. In this study, the application of the Köppen–Geiger classification system to the latest Coupled Model Intercomparison Project (CMIP6) experiments has been explored. To establish a baseline for the historical period (1970–2000), the WorldClim dataset was used alongside a selection of an ensemble of 14 Global Climate Models. The evaluation of climate variability across winemaking regions is conducted by considering future climate projections from 2041 to 2060, which are based on different anthropogenic radiative forcing scenarios (Shared Socioeconomic Pathways, SSP2–4.5, and SSP5–8.5). The results are the most comprehensive documentation of both the historical climate classifications for most wine regions worldwide and the potential changes in these classifications in the future. General changes in climate types are projected to occur largely in a significant shift from a warm summer climate to a hot summer climate in temperate and dry zones worldwide (climate types C and B, respectively). This shift poses challenges for grape cultivation and wine production. The grape development process can be significantly affected by high temperatures, which could result in early ripening and changes in the grape berry's aromatic compounds. As regions transition and experience different climates, wine producers are required to adapt their vineyard management strategies by implementing suitable measures that can effectively counter the detrimental impacts of abiotic stresses on grape quality and vineyard health. These adaptation measures may include changes in canopy and soil management, using different variety-clone-rootstock combinations, adopting irrigation methods, or shifting into other microclimatic zones, among other effective techniques. To ensure long-term sustainability, wine producers must consider the climatic change projections that are specific to their region, allowing them to make more informed decisions about vineyard management practices, reducing risks, and ultimately making the wine industry more resilient and adaptive to the ongoing effects of climate change.

Keywords: climate classification systems; climate change; SSP projections; global viticulture; adaptation measures

1. Introduction

Climate classifications have been developed to help describe the conditions a region usually experiences over a year or longer temporal ranges [1]. Often, they include geographical names such as polar or tropical, biome names such as steppe or rainforest, and can be descriptive, such as desert scrub or dry tundra. Methods for classifying climates have included approaches using environmental indicators [2], weather patterns [3], water budgets [4], and life zones [5]. Numerous factors affect a region's climate, including the latitude, elevation, proximity to large bodies of water, mountains, and other surface features, ocean current circulation patterns, and global to regional patterns in atmospheric circulation. These factors control regional temperature, precipitation characteristics, and seasonality, which in turn control the ecology or types of native plants and animals in each region.

The most widely used climate classification was first developed by Wladimir Köppen in the late 1800s [2] and modified as more information about climate controls and ecology became available [6,7]. Further refinements were carried out by Geiger [8], resulting in what is known today as the Köppen–Geiger climate classification (KG). The KG uses seasonal temperatures and precipitation to divide climates into five main groups: tropical, arid, temperate, continental, and polar. These are further divided into 30 sub-types that are also based on the types of vegetation that are most common in each region, helping to define regional ecosystems and global biomes. The KG was further refined by Trewartha [9] and Trewartha and Horn [10] to better define the climate and vegetative zones in the mid-latitudes, especially in North America and Asia.

Given the importance of climate in growing grapes for wine production, the classification of climates suitable for viticulture has been important to the wine sector. Köppen [11] compared wine regions using an earlier version of his climate classification approach, and de Blij [12] (p. 112) noted that "viticulture is perhaps the most geographically expressive of all agricultural industries." Arising from the geography of the old-world wine regions of Europe and the use of the Köppen climate classification system, wine production has become synonymous with Mediterranean climates. Besides being found along the Mediterranean basin of Europe, these climate types are also located in the fynbos of South Africa, the Mallee of southern Australia, the matorral of Chile and Argentina, and the chaparral and coastal valleys of western North America. It is in these regions where a mild wet-winter, dry-summer climate is ideal for the cultivation of *Vitis vinifera* L. grapes. However, today, wine production occurs across a wide range of climate types [13]. Jones et al. [14] used an updated KG gridded dataset [15] and found that midlatitude climate types make up 55% of the world's wine regions. Within the midlatitude climate types, Mediterranean climates occur in only 15% of the surface area globally (e.g., Napa Valley, Coonawarra), with humid subtropical (24%; e.g., Barolo), maritime temperate (12%; e.g., Bordeaux), and maritime subarctic (4%; e.g., Tasmania) making up the rest. Other climate types that wine regions fall within include midlatitude dry (16%), subtropical dry (3%; e.g., Priorat and Mendoza), and humid continental climates (13%; e.g., Finger Lakes Region and Niagara).

Given that KG climate types describe observed biome distributions, they inherently have sharp boundaries due to plant sensitivity to various environmental factors [16]. As such, climate classifications have been used to examine how a changing climate has caused shifts in these boundaries [17,18] and could potentially alter entire biomes in the future [19]. For example, Chan and Wu [16] found that approximately 5.7% of the total global land area shifted to warmer and drier climate types during 1950–2010. Belda et al. [1], using both the KG and the Köppen-Trewartha climate classification, found observed changes in semi-arid, savanna, and tundra climate sub-types. Changes in tundra classes occurred across the 20th century but have accelerated in the last 40 years [20,21] and declined significantly in the western US [22]. Continental temperate climate types north of 50° have expanded over the last few decades as inland areas of the northern continents have warmed substantially more than most of the rest of the world [21]. Chen and Chen [17] found that polar climates shrank in size over the period 1901–2010, while there have been significant poleward shifts in temperate climates. Chan and Wu [16] found increasing average elevations in tropical and polar climates. Potential climate type changes in the future have been examined by Rubel and Kottek [19], finding that maximum changes are likely to be found in continental cold climates, with significant shifts to more temperate climate types. Additional changes were noted in low latitudes, where the tropical belt will likely widen, with rainforest and monsoon sub-types expanding in some regions while savanna sub-types expand in others. In Europe, de Castro et al. [23] found that 51–59% of land grid points would likely shift towards warmer or drier climate sub-types. They also noted that it is likely that ecosystems might be at risk for approximately 12% of the European land area, as these regions are projected to shift across two or more climate subtypes.

Climate change has been shown to affect wine quality and production globally [24,25] and impact the suitability of regions to continue growing the winegrape varieties they grow today [26,27]. In Europe, Santos et al. [28] detail how future climate change is expected to exacerbate issues of regional suitability for viticulture and wine production, requiring extensive short- and long-term adaptation strategies [29,30]. Moriondo et al. [31] also detail that climate change will likely reshape the geographical distribution of many of the famous wine regions in Europe. Given the strong connection between weather, climate, and environment in wine production, the purpose of this research is to examine the climate change projections for the KG classes in wine regions worldwide. Wine regions include those in most countries across Europe, in the Northern and Southern Hemispheres, as is detailed in the World Wine Production Outlook report [32]. The research examines the historic (1970–2000) and future (2041–2060) KG considering two Shared Socio-economic Pathways (SSP2–4.5 and SSP5–8.5) using the new state-of-the-art Coupled Model Intercomparison Projects models (CMIP6) downscaled for future climate projections [33].

2. Materials and Methods

2.1. Datasets

In this study, gridded datasets of monthly precipitation (mm), maximum, minimum, and mean temperatures (°C), with a 10-min spatial resolution, were retrieved from the WorldClim dataset [34]. For the historical period, monthly averages between 1970 and 2000 were considered, whereas monthly averages from 2041 to 2060 (20 years) were selected for the future climate. These datasets were generated by state-of-the-art Global Climate Models (GCM) within the framework of the Coupled Model Intercomparison Project (CMIP6). CMIP6 [33] is based on runs from around one hundred GCMs, produced across 49 different modeling groups, showing a higher sensitivity than the previous CMIP5 models [35]. Several Shared Socioeconomic Pathways (SSPs) have been outlined to drive climate models for CMIP6. As such, the previous Representative Concentration Pathways (RCPs) were reformulated in CMIP6. Two Shared Socioeconomic Pathways (SSPs) are considered herein: SSP2–4.5 and SSP5–8.5, or simply SSP2 and SSP5. The downscaling and bias correction of raw GCM data were carried out using WorldClim v2.1 as the baseline climate [33]. The calibration undertaken by the data providers assumes high spatial autocorrelation and was generated by computing the absolute or relative difference between the GCM outputs for the baseline periods and target periods (for this study, 2041–2060) with global cross-validation correlations of about 0.99 for temperature and about 0.86 for precipitation [34]. A 14-member ensemble of bias-corrected monthly datasets for 2041–2060 was chosen, comprising the following GCMs: ACCESS-CM2, BCC-CSM2-MR, CanESM5, CMCC-ESM2, CNRM-CM6-1, EC-Earth3-Veg, FIO-ESM-2-0, GISS-E2-1-G, HadGEM3-GC31-LL, INM-CM4-8, IPSL-CM6A-LR, MIROC-ES2L, MPI-ESM1-2-HR, UKESM1-0-LL. This bias-corrected ensemble was subsequently used to compute the KG climate classification system for the previously referred periods: 1970–2000 and 2041–2060 under both SSPs for each grid point worldwide. The methodology used to attain a worldwide KG climate classification is presented in the next section.

2.2. *Köppen's Climate Classification*

The KG climate classification system [2,6,7,11,36] is based on a subdivision of five major terrestrial climate types, represented by capital letters: A (tropical), B (dry), C (temperate), D (continental), and E (polar). The second letter denotes the seasonal precipitation type: s (dry summer), w (dry winter), and f (no dry season). Lastly, the third letter specifies the level of heat: a (hot summer), b (warm summer), c (cold summer), d (very cold winter), h (hot climate), and k (cold climate). It is worth mentioning that the A climate type can also be associated with a second letter, f (rainforest), or m, highlighting a well-defined monsoon regime. For the B type, W (arid) and S (semi-arid) are also used. Lastly, associated with the E type (polar), T or F are related to the tundra or ice cap, respectively.

This climate classification has been modified and updated over time ([8–10] and others). It is worth noting that the change in the classification may lead to different descriptions of the main group climates. In this study, the KG climate classification system denomination and calculation followed the methodology presented by Andrade and Contente [37] (Table A2; cf. full type description and definition criteria in their Tables 3–5), which in turn was based on Kottek et al. [38]. Therefore, a brief description of the 31 KG climate types is presented in Table A1. The color scheme for all figures presented herein was adopted by Peel et al. [15] and Beck et al. [39].

This study will assess changes in the worldwide KG climate classification system between the historical period (1970–2000) and the future (2041–2060), under SSP2 and SSP5.

2.3. Study Areas

This analysis utilizes both spatial data representing wine regions around the world and country-level listing of wine regions. The spatial data used to assess historic and future KG climate types is a database of major winegrowing regions worldwide developed by Bois et al. [40], called Vineyard Geodatabase (VGDB). The geodatabase consists of wine-producing regions delineated from various sources, including American Viticultural Areas in the USA, Geographic Indications in Australia and New Zealand, Wine Districts in South Africa, Protected Designation of Origin in Europe, and other region boundaries from the World Atlas of Wine. For Europe, vineyard limits were also adjusted to areas planted with grapevines using the Corine Land Cover database 2006 version [41]. VGDB has been used in various studies in the global analysis of wine region climate characteristics [42–44] and is continuously updated. This research utilized Version 1.2.3, which has the boundaries for 686 wine-producing regions worldwide. However, this number of wine-producing regions was restricted to match the country-level data in the World Wine Production Outlook report [32]. The list of countries included in this study can be found in Table 1, and their locations are shown in Figure 1. The final number of wine regions worldwide used in this analysis is 490, with 161 in the EU, 285 in the NH non-EU, and 44 in the Southern Hemisphere (SH).

Table 1. List of the major wine producers by geographical sector (based on [32]).

Geographical Sector	Countries (Acronym)
European Union (EU)	Italy (IT), France (FR), Spain (ES), Germany (DE), Portugal (PT), Romania (RO), Hungary (HU), Austria (AT), Greece (GR), Bulgaria (BG), Slovenia (SL), Czechia (CZ), Slovakia (SK), and Luxembourg (LU)
Outside the EU Northern Hemisphere (NH)	USA (US), China (CN), Georgia (GE), and Switzerland (CH)
Southern Hemisphere (SH)	Chile (CL), Australia (AU), Argentina (AR), South Africa (ZA), Brazil (BR), New Zealand (NZ), and Uruguay (UY)

Within the European Union (EU) countries, Cyprus and Malta were not considered due to the residual 5-year average wine production. Outside the EU, in the Northern Hemisphere (NH), Moldova and Russia were not included due to the lack of data. It is worth mentioning that the list of countries provided in Table 1 is ranked in decreasing order of wine production (data retrieved from Tables 1 to 3 in [32]).

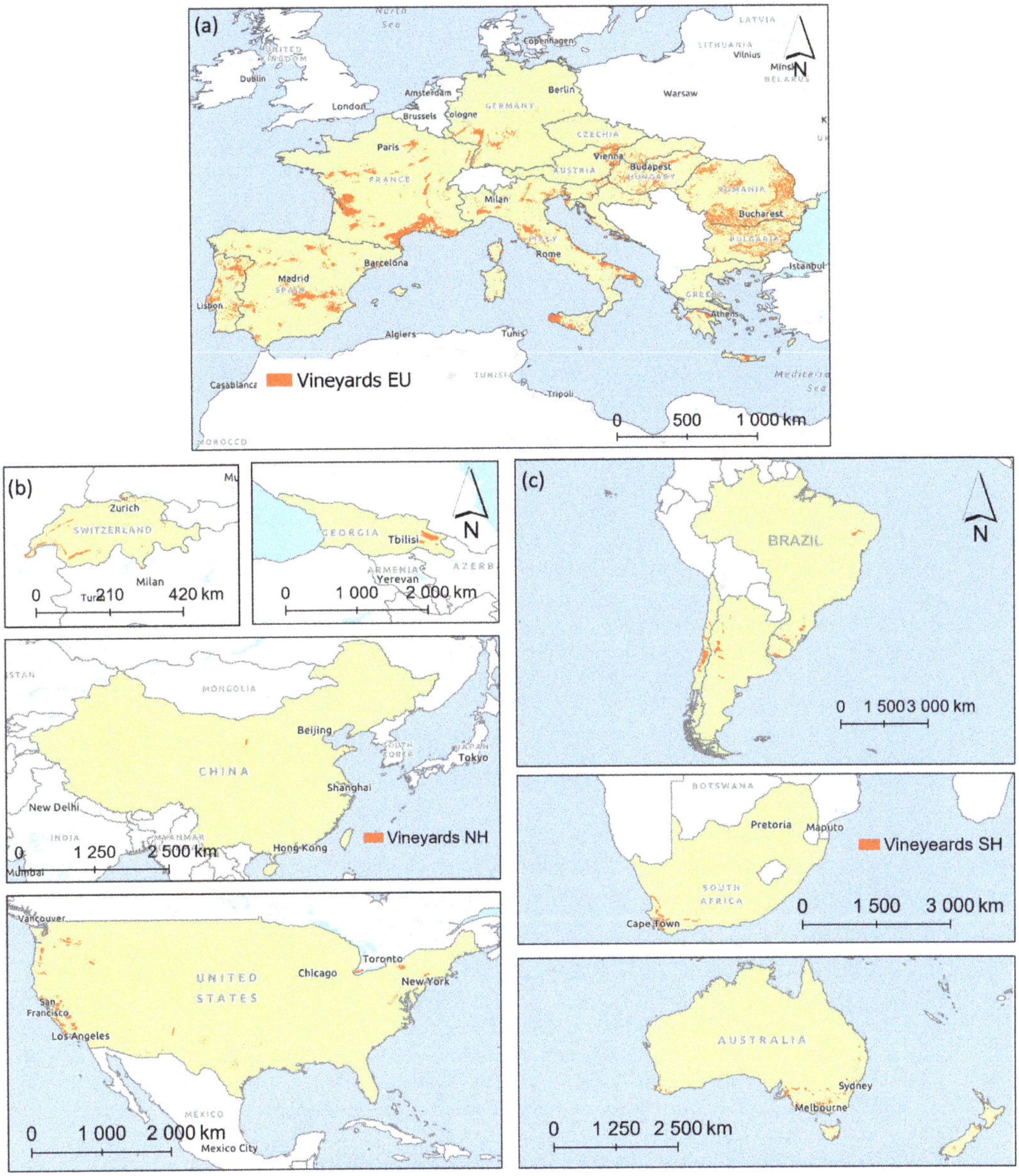

Figure 1. Study areas of the major wine producers divided by regions (**a**) European Union (EU) (**b**) outside the EU in the Northern Hemisphere (NH), and (**c**) Southern Hemisphere (SH) countries (list in Table 1).

3. Results

The results are presented following this framework. First, for the historical period (1970–2000), the wine regions and the KG climate types are analyzed and detailed. Second, the worldwide KG classification system map distributions for the historical and projections (2041–2060) under SSP2 and SSP5 are analyzed independently of the wine regions. In the third subsection, the relationships between climate change projections for the KG and wine regions are presented. Finally, following the work by Jones and Schultz [45], the impacts of climate change projections in emerging cool-climate wine regions (focused on the locations previously studied) are undertaken.

3.1. Wine Region KG Climate Types 1970–2000

During the historical period (1970–2000) for the climate data used in this research (Figure 2), wine regions were found to range across KG climate types, mostly in the mid-latitudes. Temperate C types represented 66.3% of the wine regions worldwide, while Dry B types were found in 23.1% and Continental D types in 7.2% (Table A3). Within the Temperate C types, temperate oceanic (Cfb) accounted for just over 30% of the wine regions, with humid subtropical (Cfa) representing close to 13%. Mediterranean climate types made up just under 20% of the worldwide wine regions during 1970–2000, with warm summer (Csb) accounting for 9.5% and hot summer (Csa) representing 10.3%. For the Dry climate types, semi-arid cold (BSk) was the most dominant, accounting for 16.6%. Within the Continental D climate types, warm summer humid continental (Dwb) made up 4.1% of the total worldwide.

Within the European sector wine regions examined in this research, Temperate C types represented 91.3% of the regions during 1970–2000, with temperate oceanic (Cfb) as the dominant climate type at 47.5%, followed by hot summer Mediterranean (Csa, 21.1%) and humid subtropical (Cfa, 14.9%) (Table A5). For other NH wine regions, Temperate C types were 47.8% of the total, with warm summer (Csb, 27.6%) and hot summer Mediterranean (Csa, 10.9%) as the most dominant climate classes. Other climate types in the NH regions accounting for significant percentages include semi-arid cold (BSk) at 17.2% of the total and warm summer humid continental (Dwb) at 12.8%. For the SH wine regions considered in the study, Temperate C types accounted for 60.6% of the wine regions. Within the C types, temperate oceanic (Cfb, 32.9%) and humid subtropical (Cfa, 16.9%) were the most common climate classes. Dry B types were also represented in the SH wine regions at 32.3% of the total area, with semi-arid cold (BSk, 23.2%) the most common type. Variations in climate types by country during the historic period (1970–2000) can be found in Tables A6–A8.

3.2. Climate Change Projections for the KG Climate Classification System

Figure 2 shows the global map series of the KG climate classification for the period 1970–2000, as well as under the selected future SSP scenarios (SSP2 and SSP5). The KG climate type map of eastern Europe for the historical period shows that only three main climate types are present: temperate (C), continental (D), and polar (E), in predominant order of climate type by land area. Under future scenarios, continental climate types depict a decrease in terms of area; the polar tundra (ET) climate completely disappears (previously in Iceland, the Alps, the Pyrenees, and the northern coast of Norway), while temperate climate types show a significant increase across eastern Europe. Across the Iberian Peninsula (IP), the temperate oceanic climate (Cfb) and the temperate warm-summer Mediterranean climate (Csb) shift to the temperate hot-summer Mediterranean climate (Csa), while in the southeast of the IP Csa areas change to dry semi-arid hot (BSh) and dry semi-arid cold (BSk) climates. Most of central and eastern Europe is predicted to change from a temperate oceanic climate (Cfb) to a temperate humid subtropical climate (Cfa).

In Asia, the biggest changes are observed in the Himalayan Mountain Range, where the polar tundra climate (ET) completely disappears, giving way to the continental warm-summer humid climate (Dfb) and continental subarctic climate (Dfc). Moreover, in the Mongolian region, the current Dfc climate is expected to change completely to a continental hot-summer humid climate (Dfa) and a continental warm-summer humid climate (Dfb). In northern Asia, monsoon-influenced summer humid continental and subarctic climate (Dwb and Dwc) areas contract due to the expansion of both monsoon-influenced hot summer humid continental (Dwa) and Mediterranean-influenced subarctic climate (Dsc) in the north-easternmost part of Asia (Figure 2).

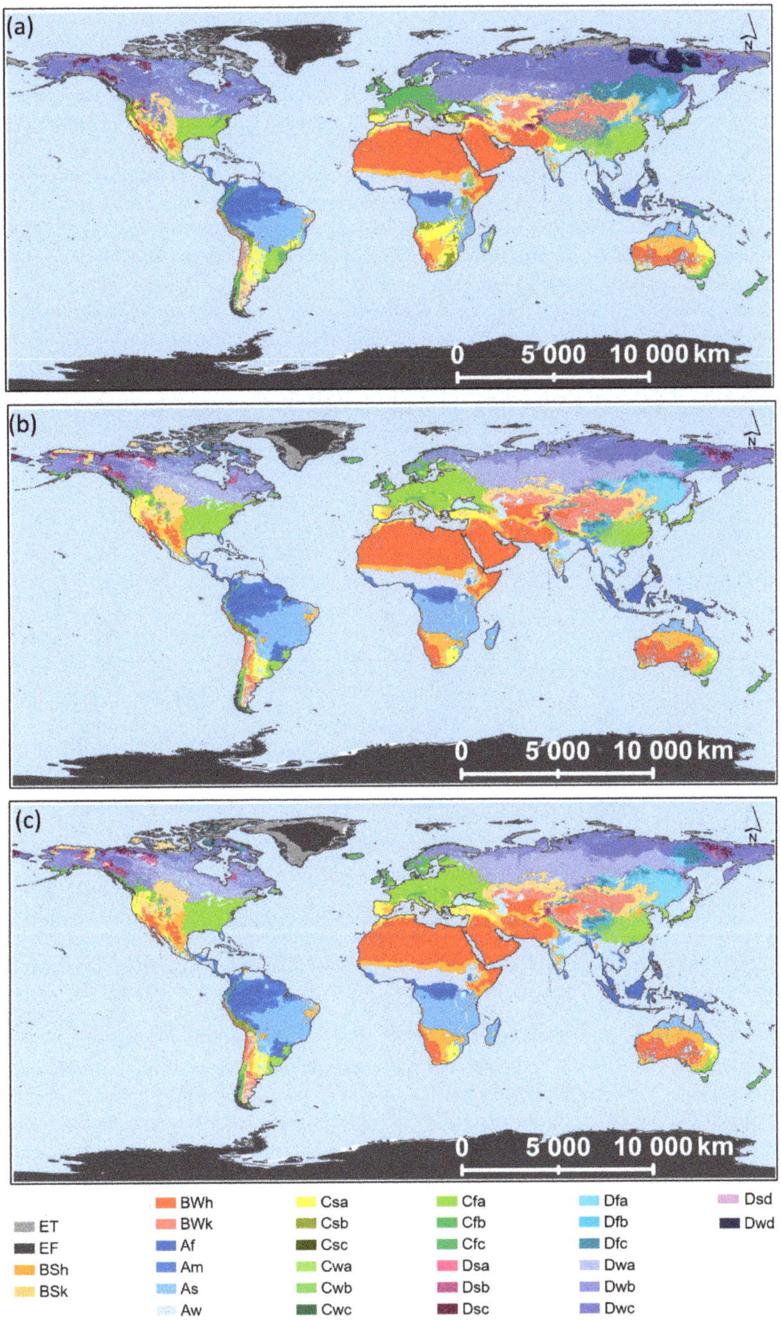

Figure 2. Worldwide Köppen–Geiger (KG) classification system between (**a**) 1970 and 2000, (**b**) from 2041 to 2060 under SSP2, and (**c**) from 2041 to 2060 under SSP5.

In North America, particularly in Canada, most of the changes are very similar to the northernmost part of Asia, where monsoon-influenced hot summer humid continental

(Dwa) climate areas expand and monsoon-influenced summer humid continental and subarctic climate (Dwb and Dwc, respectively) areas contract. In the United States of America (USA), there is an increase in the dry semi-arid hot (BSh), dry semi-arid cold (BSk), arid cold (BWk), and temperate humid subtropical climate (Cfa) climate areas. It is worth mentioning the decrease in polar icecap (EF) and the increase in polar tundra climate (ET) in Greenland (Figure 2).

Close to the equator, the KG classification areas remain very similar between historical and future periods under both scenarios. Some changes are projected southwards of the tropic of Capricorn (23.43° S), where temperate hot-summer Mediterranean climate (Csa) climate areas decrease due to the expansion of tropical savanna, tropical dry summer climate (As), and dry hot desert (BWh) areas.

Despite the shifts in the geographical distribution of the different KG climate type classes, the effects of climate change depict a significant impact only on a few KG climate types in terms of the global covered area (Figure 3; Table A4). Three KG climate types are projected to have a major decrease: continental subarctic (Dfc) climate (−7.3%), polar tundra (ET) climate (−4.6%), and polar icecap (EF) (−3.8%), while continental hot-summer humid (Dfa) and temperate humid subtropical (Cfa) climates are projected to increase by 5.2% and 2.8%, respectively.

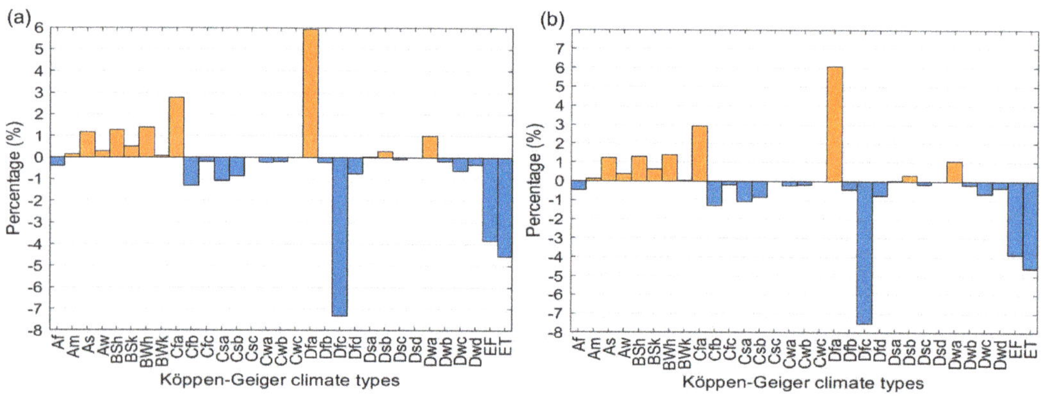

Figure 3. Percentage of the difference (in pixels) between the future 2041 to 2060 (**a**) under SSP2, and (**b**) under SSP5, and the historical (1970–2000) periods for each Köppen–Geiger (KG) climate type (values in Table A4).

3.3. *Relationship between Climate Change Projections for KG Classes and Wine Regions*

In global terms, within the selected world wine regions, the most important changes are projected to occur in temperate humid subtropical (Cfa) (from around 13% to just over 34%), dry semi-arid hot (BSh) (from around 2% to nearly 18%), and temperate hot-summer Mediterranean (Csa) (from around 10% to just over 17%) climate types when comparing the periods 1970–2000 and 2041–2060, under both SSPs (Table A3). The cold temperate climate types (D) in wine regions are projected to almost disappear, while the tropical climate types (A) are projected to increase in area. Furthermore, decreases in the temperate oceanic (Cfb) climate types (from around 30% to close to 6%), dry semi-arid cold (BSk) (from around 17% to just over 5%), and temperate warm-summer Mediterranean climate (Csb) (from over 9% to close to 1%) are also projected. It is worth mentioning that, due to the resolution of the gridded datasets used to compute the KG climate classification, it is not possible to resolve microclimatic conditions across wine regions. Therefore, some inconsistencies can be identified, such as the existence of vineyard land cover in areas classified as polar (ET) climates (Table A3), namely in some deep valleys in the Andes Mountain Range.

In most of the vineyard areas in the European Union (EU) (Figures 4a, 5a, 6a and 7a; Table A5), a general shift from a warmer summer climate (b type) to a hot summer climate (a type) is expected. In the north-westernmost part of the IP and along the Mediterranean coast, the temperate warm-summer Mediterranean (Csb) climate is projected to change entirely to the temperate hot-summer Mediterranean (Csa). Furthermore, the vineyard areas in the southeastern IP may experience a shift from the semi-arid cold steppe (BSk) to the semi-arid hot steppe (BSh) and even the arid hot desert climate (BWh), particularly in the Murcia region of Spain. Across the other major vineyard regions in Europe, the KG climate classification is expected to change from a temperate oceanic (Cfb) to a humid subtropical (Cfa) climate.

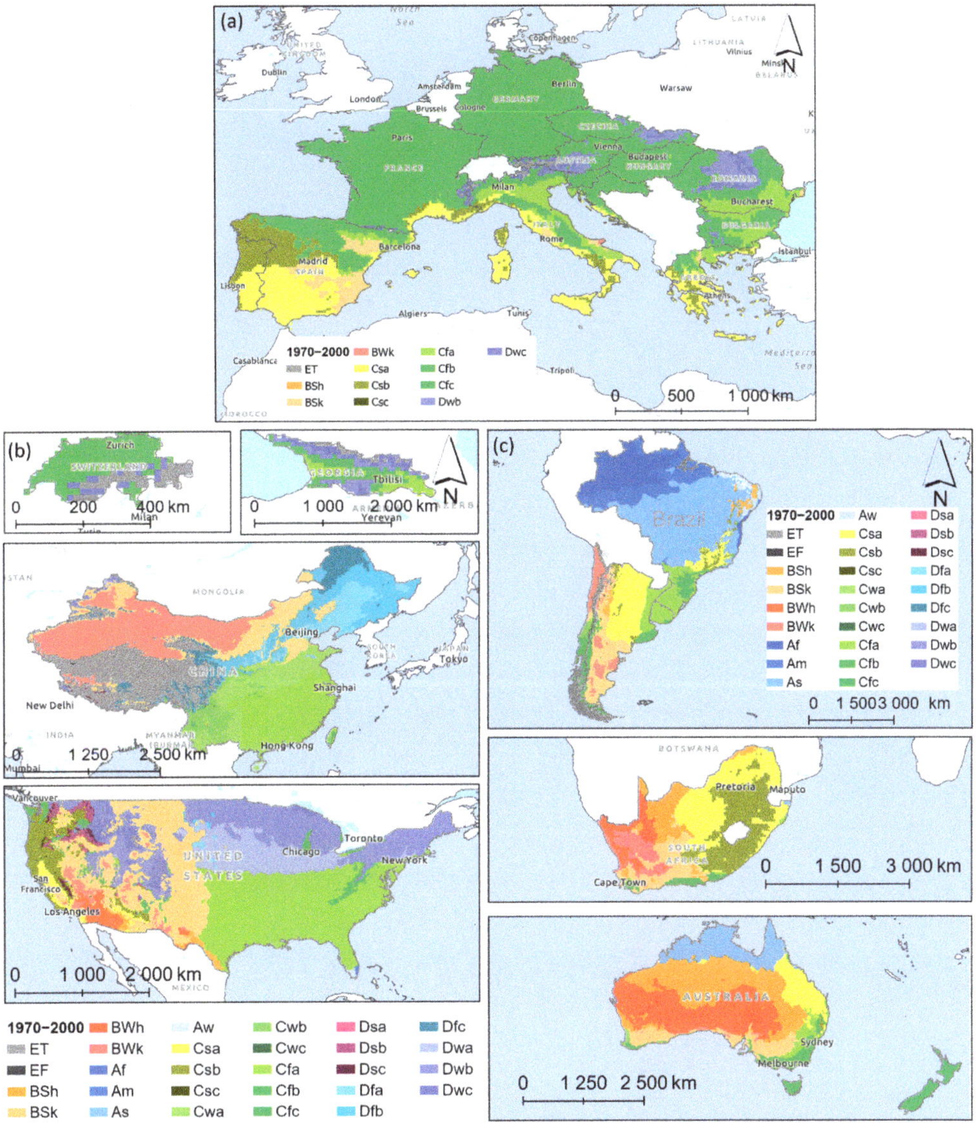

Figure 4. Worldwide Köppen–Geiger (KG) classification system of the major wine producers' countries for the (**a**) European Union (EU) countries; (**b**) Northern Hemisphere (NH) non-EU countries; and (**c**) Southern Hemisphere (SH) between 1970 and 2000.

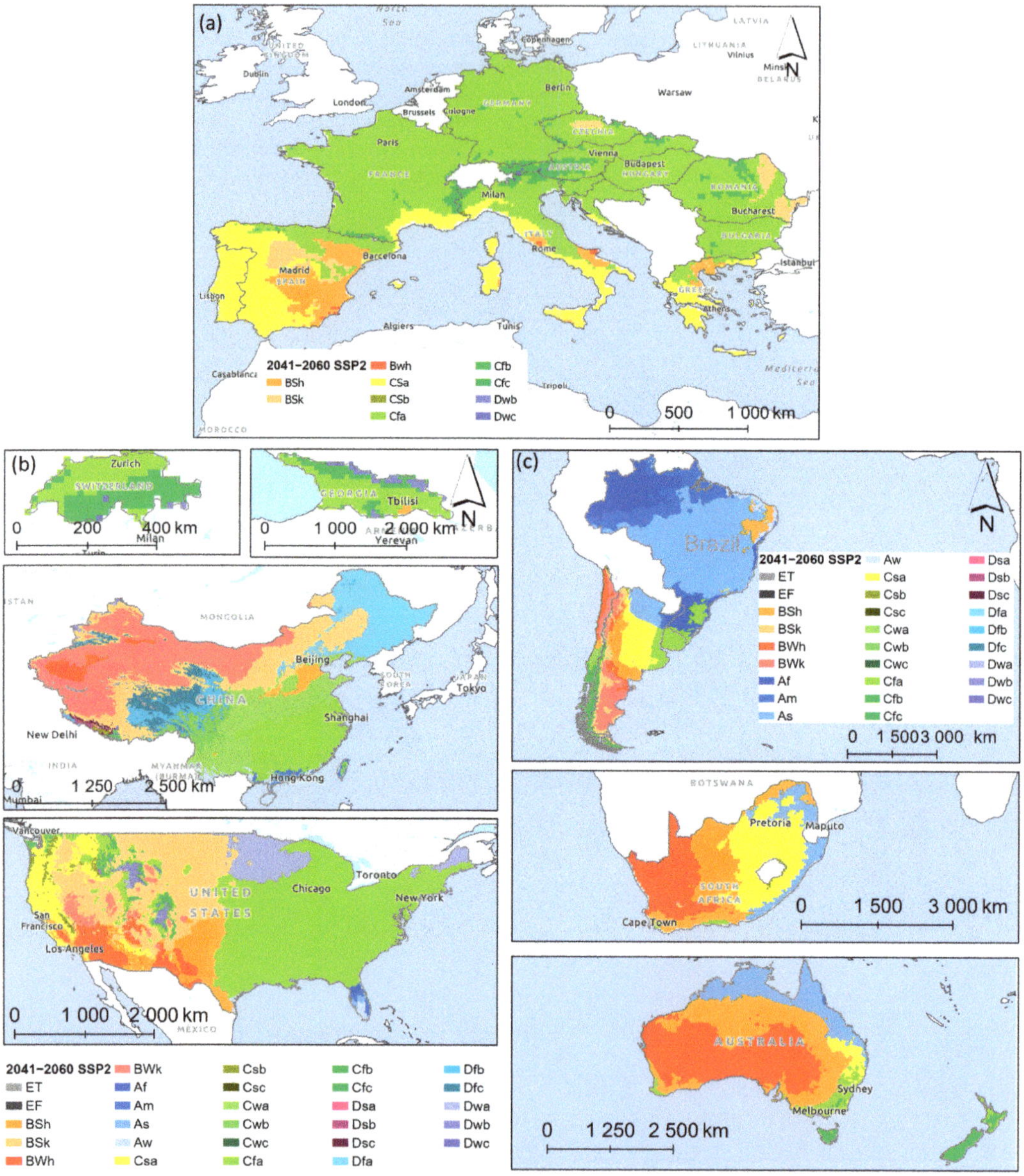

Figure 5. Worldwide Köppen–Geiger (KG) classification system of the major wine producers' countries for the (**a**) European Union (EU) countries; (**b**) Northern Hemisphere (NH) non-EU countries; and (**c**) Southern Hemisphere (SH) from 2041 to 2060 under SSP2.

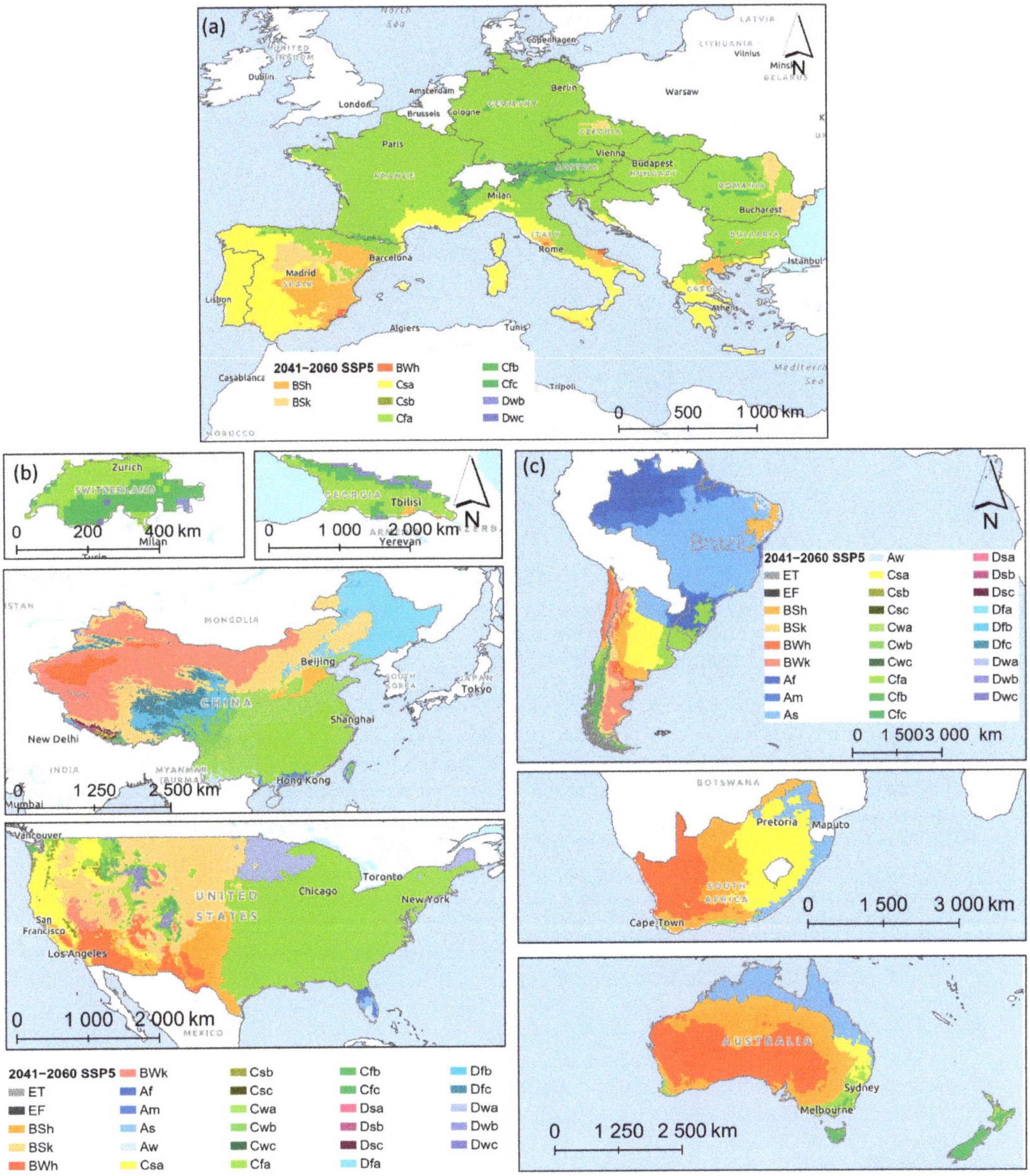

Figure 6. Worldwide Köppen–Geiger (KG) classification system for the major wine producers' countries for the (**a**) European Union (EU) countries; (**b**) Northern Hemisphere (NH) non-EU countries; and (**c**) Southern Hemisphere (SH) from 2041 to 2060 under SSP5.

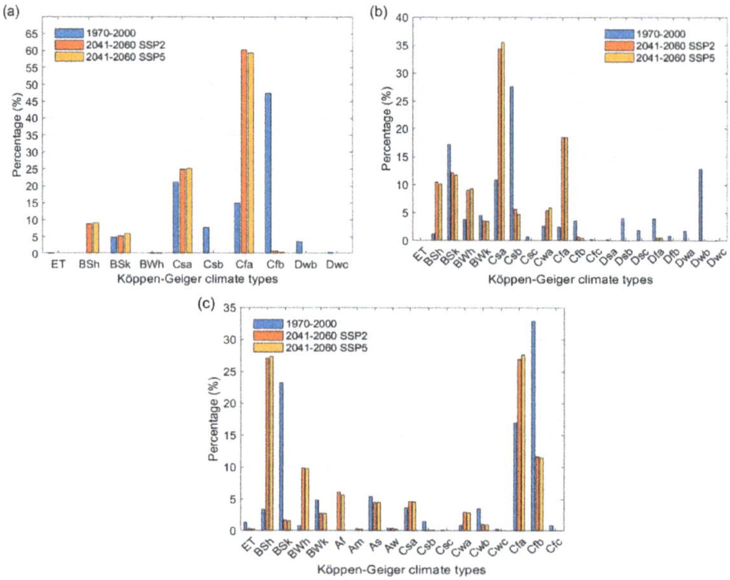

Figure 7. Köppen–Geiger climate type percentage of change within the study regions with vineyards (**a**) EU; (**b**) other NH countries; and (**c**) SH countries, for 1970–2000 and 2041–2060 under SSP2 and SSP5 for each climate type (values in Table A5).

For other areas across the NH (Figures 4b, 5b, 6b and 7; Table A5), wine regions examined in the research include areas in China, Switzerland, Georgia, and the USA, with the latter having significantly more areas on the west coast. For China, the KG classification shows that the climate in the vineyard areas will likely remain unchanged (Figure 8; Tables A6–A8).

Wine regions of the west coast of the USA are mostly projected to transition from a temperate warm-summer Mediterranean (Csb) to a temperate hot-summer Mediterranean (Csa) climate type, like what is projected in the IP, while in California, dry arid hot climate types (BWh-desert) are expected to increase, reflecting that the summers are expected to be much hotter. In contrast, the east coast of the USA is projected to experience a similar transformation as seen in much of central Europe, shifting from the temperate oceanic (Cfb) climate to the humid subtropical (Cfa) climate (Figure 8; Tables A6–A8).

In the SH, similar changes as projected for the NH are anticipated, with vineyard regions likely experiencing shifts from dry semi-arid cold (BSk) to dry semi-arid hot (BSh), namely in Argentina, South Africa, and Australia. In Chile, a transition to an arid hot desert (BWh) climate is projected since vineyard areas are already found in an arid cold (BWk) climate (Figure 8; Tables A6–A8).

3.4. Impacts of Climate Change Projections in Emerging Cool Climate Wine Regions

Jones and Schultz [45] assessed the impacts of climate change in emerging cool-climate wine regions. In these regions, the growing season average temperatures are between 13 and 15 °C, with a relatively short growing season (less than 7 months). These regions might also endure spring frost and fall frost depending on their proximity to the coast or other large bodies of water that buffer temperatures to some degree. In addition, higher latitudes along the coast usually have higher growing-season rainfall, which increases the risk of grapevine diseases. In this study, the area's growing grapevines located the furthest poleward in both hemispheres (Table A9) are revisited. For each location, a comparison between the KG climate classification system for the historical period and the future under both SSPs was performed. Results show that major changes are projected to occur in the

NH regions where greater warming rates are occurring at higher latitudes, while regions in the SH are less likely to see changes in climate types due to limited landmass at poleward latitudes. In the SH, cool climate regions in Argentina (arid cold, BWk), New Zealand (temperate oceanic, Cfb), Tasmania (temperate oceanic, Cfb), and Chile (temperate oceanic, Cfb) show that no changes in the KG are projected for 2041–2060 under both SSPs. Two regions in the NH are also not projected to change from their current climate type: Denmark (temperate oceanic, Cfb) and Britain (temperate oceanic, Cfb). Regions in the NH that are projected to see significant changes are located on the North American continent. In Canada and the USA, wine regions in the Sussex, the Leelanau Peninsula, and Annapolis areas that were classified as warm-summer humid continental (Dwb) during 1970–2000 are projected to shift to humid subtropical (Cfa). Other Canadian regions defined as warm-summer humid continental (Dwb) are projected to shift to semi-arid cold (BSk, Kamloops) or hot-summer humid continental (Dwa, Ontario, and Quebec). In Europe, Gothenburg (Sweden), Zilona Góra (Poland), and Maastricht (Netherlands) are projected to change from temperate oceanic (Cfb) to humid subtropical (Cfa) shortly and under both SSPs. Overall, these results highlight a shift towards warmer and milder climate types in most of the NH's coolest climate locations.

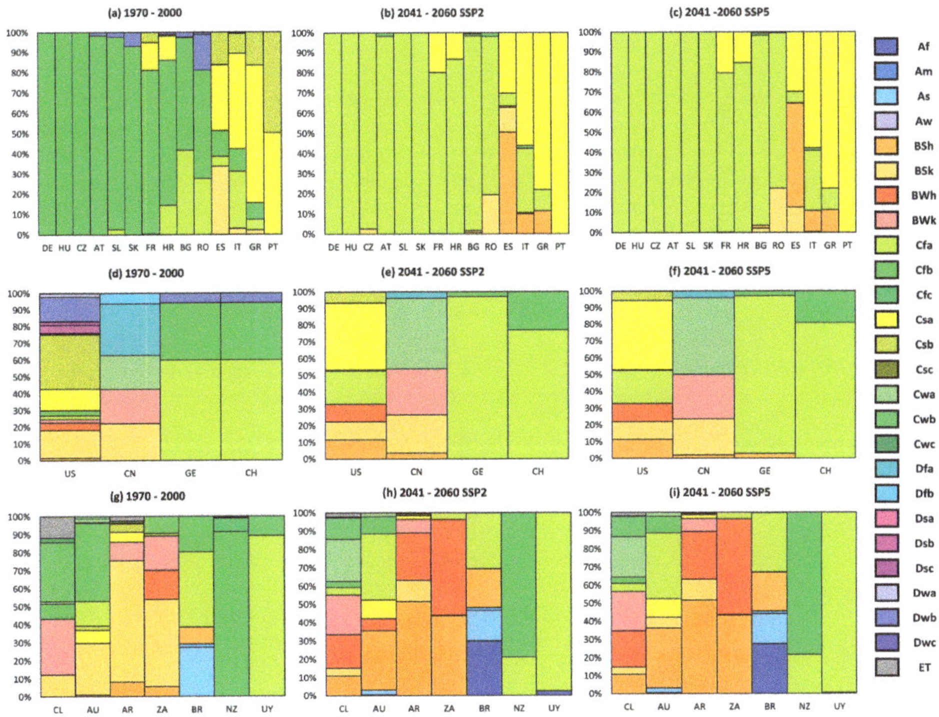

Figure 8. Köppen–Geiger (KG) climate type percentage of change per country within the study regions with vineyards (**a**–**c**) EU, (**d**–**f**) other countries in the NH, and (**g**–**i**) SH countries for 1970–2000 (left column) and 2041–2060 under SSP2 (middle column) and SSP5 (right column) for each climate type (values in Tables A6–A8).

4. Discussion

Climate and vegetation are intricately linked at both regional and global scales. The KG climate classification scheme has been used as an effective tool to describe and examine both natural and human-managed plant systems [46–48]. Compared with single-variable approaches (i.e., just temperature or precipitation), using the KG climate classification can add new dimensions to the description of climate variation and change and their impacts and potential changes to vegetation [17]. Spatial changes in climate zones can signal contraction or expansion of the influence of climatic conditions, which in turn implies risks for plant and animal species. While large-scale biome shifts are possible, edge effects may produce the most dramatic changes as species may not be able to adapt quickly enough to the changes in climate [49].

In terms of agriculture, the world map of zones suitable for various crops very closely correlates with the KG climate classification system [46–48,50]. Although the factors that influence agricultural suitability are many and complex, correlation is especially apparent in the temperate latitudes where a favorable climate and intensive higher-yield agriculture exist. Growing grapes for wine production is one of the most geographically expressive crops, having a relatively large footprint in the midlatitudes [12], which makes viticulture a model system with which to use the KG climate classification system. As such, wine production has become synonymous with the Mediterranean climate types found mostly in southern European wine regions. Previous efforts to depict the climatic environments for growing winegrapes worldwide have used bioclimatic models and indices designed for grapevine aptitude to grow and produce quality grapes, e.g., [42,44,51–53]. By using KG, the most common reference climate classification system in use today, our analysis makes it possible to compare grape production climate characteristics to those of other crops and biomes. The results allow for both assessing the current and future potential to grow grapevines for wine production rather than other crops and estimating which type of ecosystem might replace viticulture if removed from a given area for conservation purposes.

Recent research has shown that climate change is already affecting viticulture and wine production through changes in grapevine phenological development [14], harvest dates [54], and grape acidity and sugar content [55,56]. However, establishing relationships between environmental changes and other components, like polyphenols or aroma compounds, is more complex [57,58]. This research has added to the baseline knowledge of the potential changes of climate on viticulture and wine production by examining global and regional-scale shifts in the KG climate types for wine regions. The two main climatic shifts identified in this study are: (1) regions shifting to a warmer and drier climate (temperate warm-summer Mediterranean (Csb) to temperate hot-summer Mediterranean (Csa); dry semi-arid cold (BSk) to dry semi-arid hot (BSh) and arid hot desert (BWh) climates); and (2) other regions shifting to a warmer and moister climate (temperate oceanic (Cfb) to humid subtropical (Cfa)).

These changes in climate conditions are believed to be causing a gradual shift in grapevine varieties being grown in temperate climate wine regions, with an increasing loss of suitability for white wine varieties towards suitability for red wine varieties [14,59,60]. In more semi-arid regions, where water for viticulture is already scarce, warmer and drier conditions may induce greater risk for the sustainability of vineyards since water consumption for viticulture is usually higher than total annual precipitation in these regions [61]. To address the increasing intensity of drought stress and to ensure more dependable and predictable yields, the use of water through irrigation in historically rainfed regions will likely be required as a necessary means to adapt to warmer and drier climates [28,62,63]. Additionally, shifts from more temperate to more subtropical or more continental climate types are likely to produce a wider annual temperature range and more extreme temperatures [64]. In regions such as southern Europe, Australia, and California, where arid, hot climate types are expected to become even drier and hotter in the summer, the chance of wildfire occurrences will likely increase as well [65].

At the cool limits of viticulture, warming and shifting to more suitable climate types identified in this research will likely increase the potential ripening of the main varieties being grown today and expand the range to other varieties. The most common varieties currently grown in these emerging cool climate regions are early or very early Vitis vinifera cv. Müller-Thurgau, Chardonnay, Pinot noir, Riesling, Pinot gris, Sauvignon blanc, Madeleine

Angevine, and Siegerrebe, as well as cold and disease-resistant hybrids such as Rondo, Vidal, and Regent [45].

The potential shifts in KG climate types identified in this research will likely produce different challenges depending on the region, and as such, the strategies to ensure a sustainable product must be adapted accordingly. In general, cool-climate wine regions have already seen a change to intermediate climates, with many regions even expected to become warm- or hot-climate regions. However, while cool to intermediate to warm regions have experienced generally more consistent ripening due to warming, shifts to earlier bud break have left grapevines more prone to spring frosts, which still occur in many wine regions [66,67]. In addition, heat stress has already increased in many regions, and the results of this study indicate that it will most likely continue to increase in the future, bringing greater challenges to grape quality and yield in the warmest and driest wine regions [68,69]. Furthermore, Wang et al. [70] find that shifting climate zones due to climate change are impacting steep-slope agricultural areas much more than more broadacre global agricultural lands. The researchers found that the effect is more prevalent with the expansion of arid zones, which would have a major impact on many areas of southern Europe where steep slope viticulture is prevalent.

The potential future shifts in climate types in wine regions worldwide are likely to be non-uniform over both space and time. Sgubin et al. [66] find that for Europe, future warming is likely to produce a non-linear loss of suitable land in traditional wine regions from 4% at 1 °C to 17% at 2 °C of additional warming. The research finds that there is some offset of the loss in traditional regions by the gain of newly suitable regions due to warming. In addition, the research suggests that adopting grapevine varieties better suited to warmer and drier conditions will provide a viable long-term response to a changing climate, up to a point. Therefore, additional adaptation measures are needed to increase the resilience of the winemaking sector and can be divided into long-term [29] and short-term strategies [30]. Some examples of adaptive measures include appropriate canopy management and changes in training systems, soil management, more effective pest and disease control, the selection of scion-rootstock combinations, or vineyard relocation [28].

5. Conclusions

Global bioclimatic conditions have been documented via climate classifications that help depict the spatial characteristics of ecosystems and biomes. The Köppen–Geiger climate classification has been the most widely used vegetation-based empirical system. Using temperature and precipitation, the classification system subdivides terrestrial climates into five major types (tropical, arid, temperate, continental, and polar), which are further defined by the seasonal precipitation characteristics and the level of heat. Tied to the plant life of a given region, the system is useful in predicting future changes in ecosystems at the global and regional levels.

This research utilizes a global-scale climate data set to categorize most wine regions worldwide into recent and future KG climate types. This research has provided a comprehensive look at the range of climate types that winegrapes have been grown in worldwide. The results show in mid-latitudes that temperate C types represented most wine regions worldwide (66.3%), with Dry B types (23.1%) and Continental D types (7.2%) also prominent. Within the Temperate C types, temperate oceanic (Cfb) made up roughly 30% of the wine regions, while humid subtropical (Cfa) climate types were close to 13%. Contrary to popular belief, Mediterranean climate types were found in a lower percentage of wine regions, making up just under 20% of the worldwide wine regions (warm summer, Csb 9.5%, and hot summer, Csa 10.3%). Additionally, the Dry climate types (semi-arid cold, BSk) were also dominant, accounting for 16.6%, while Continental D climate types such as warm summer humid continental (Dwb) types made up 4.1% of the total worldwide. The results show that wine regions are more diverse than the commonly held view that Mediterranean climates are where most wine is produced.

Wine regions in the recent past have been largely located in temperate oceanic (Cfb, around 30%), Mediterranean (Csa,b,c, nearly 20%), cold dry semi-arid (BSk, around 17%), and humid subtropical (Cfa, around 13%) KG climate types. Observations from numerous studies have shown that many of these regions have already warmed, and some have become drier. These changes in KG climate types reflect accelerated warming in the Earth's climate system, with polar climates shrinking and arid regions expanding. Research examining the potential changes in geographical distributions of climate zones in the future points to shifts into warmer and drier climates in this century. Tropical and arid climates will likely expand into the midlatitudes, while continental hot summer and humid climates are projected to expand, and high-latitude subarctic and polar climates will likely continue to shrink in area. Projected changes in broadscale climatic conditions such as these are likely to lead to the reorganization of ecosystems and the redistribution of species diversity.

Future climate projections used in this research show that significant shifts from warm summer climates to hot summer climates in the temperate zones where many wine regions are located are likely. These shifts are a further indication of the need for adaptive measures in the wine sector to address potential challenges to grape cultivation and wine production in the future. As such, this study adds to our knowledge of how climate change is likely to alter wine region climates around the world and provides additional scope to the framework by which adaptation and mitigation measures can be considered.

Author Contributions: Conceptualization, C.A. and J.A.S.; methodology, C.A.; software, C.A.; validation, C.A., A.F., J.A.S., G.V.J. and B.B.; formal analysis, C.A.; investigation, C.A.; data curation, C.A.; writing—original draft preparation, C.A., A.F. and G.V.J.; writing—review and editing, C.A., A.F., J.A.S., G.V.J. and B.B.; visualization, C.A.; project administration, J.A.S.; funding acquisition, J.A.S. All authors have read and agreed to the published version of the manuscript.

Funding: This research was funded by National Funds by FCT—Portuguese Foundation for Science and Technology, under the projects UIDB/04033/2020 and LA/P/0126/2020. Vine & Wine Portugal—Driving Sustainable Growth Through Smart Innovation, PRR—Plano de Recuperação e Resiliência e pelos Fundos Europeus Next Generation EU, no âmbito das Agendas Mobilizadoras para a Reindustrialização, Projeto n.º C644866286-011.

Data Availability Statement: The original contributions presented in the study are included in the article, further inquiries can be directed to the corresponding author.

Acknowledgments: The authors acknowledge the use of WorldClim version 2.1 climate datasets, which used CRU-TS 4.06 [71] downscaled with WorldClim 2 [34].

Conflicts of Interest: The authors declare no conflicts of interest.

Appendix A

Table A1. Köppen–Geiger (KG) climate classification system.

Climate Type	Description
A Tropical	Af Tropical rainforest Am Tropical monsoon Aw Tropical savanna with dry winter As Tropical savanna with a dry summer
B Dry	BWh Arid hot (Desert) BWk Arid cold (Desert) BSh Semi-arid hot (Steppe) BSk Semi-arid cold (Steppe)

Table A1. *Cont.*

Climate Type	Description
C Temperate	Cfa Humid subtropical Cfb Temperate oceanic Cfc Subpolar oceanic Cwa Monsoon-influenced humid subtropical Cwb Subtropical highland climate or monsoon-influenced temperate oceanic Cwc Cold subtropical highland climate or monsoon-influenced subpolar oceanic climate Csa Hot-summer Mediterranean climate Csb Warm-summer Mediterranean climate Csc Cold-summer Mediterranean climate
D Continental	Dfa Hot-summer humid continental Dfb Warm-summer humid continental climate Dfc Subarctic climate Dfd Extremely cold subarctic Dwa Monsoon-influenced hot-summer humid continental Dwb Monsoon-influenced warm-summer humid continental Dwc Monsoon-influenced subarctic climate Dwd Monsoon-influenced extremely cold subarctic Dsa Mediterranean-influenced hot-summer humid continental Dsb Mediterranean-influenced warm-summer humid continental climate Dsc Mediterranean-influenced subarctic climate Dsd Mediterranean-influenced extremely cold subarctic climate
E Polar	ET Tundra EF Icecap

Table A2. Criteria to calculate the Köppen–Geiger classification for the main climates and subsequent precipitation conditions for the first two letters. Criteria to compute the third level temperature classification (h and k) for dry climates (B) and (a to d) for the warm temperate (C) and continental (D) climates following the Köppen's climate classification. For the warm summer (b) type, a threshold temperature value of +10 °C must occur for at least four months (adapted from [36]).

Climate	Description	Criteria
A Tropical or Equatorial T_{min} [1] $\geq +18$ °C	**Af** Tropical rainforest, fully humid **Am** Tropical monsoon **As** Equatorial savannah with a dry summer **Aw** Equatorial savannah with dry winter	$P_{min} \geq 60$ mm $P_a \geq 25(100 - P_{min})$ $P_{min} < 60$ mm in summer $P_{min} < 60$ mm in winter
B Dry or Arid $P_a < 10\ P^{th}$	**BW** Desert **BS** Semi-arid or steppe	$P_a > 5\ P^{th}$ $P_a \leq 5\ P^{th}$
C Temperate or Warm temperate -3 °C $< T_{min} < +18$ °C	**Cs** Warm temperate (or Mediterranean) with dry summer **Cw** Warm temperate (or Mediterranean) with dry winter **Cf** Warm temperate (or Mediterranean) fully humid	$P_{smin} < P_{wmin}$, $P_{wmax} > 3\ P_{smin}$ and $P_{smin} < 40$ mm $P_{wmin} < P_{smin}$ and $P_{smax} > 10\ P_{wmin}$ neither Cs nor Cw
D Continental or Cold $T_{min} \leq -3$ °C	**Ds** Continental with dry summer **Dw** Continental with dry winter **Df** Continental, fully humid	$P_{smin} < P_{wmin}$, $P_{wmax} > 3\ P_{smin}$ and $P_{smin} < 40$ mm $P_{wmin} < P_{smin}$ and $P_{smax} > 10\ P_{wmin}$ neither Ds nor Dw
E Polar $T_{max} < +10$ °C	**ET** Tundra **EF** Ice cap or frost	0 °C $\leq T_{max} < +10$ °C $T_{max} < 0$ °C
h	hot steppe or desert	$T_a \geq +18$ °C
k	Cold steppe or desert	$T_a < +18$ °C
a	Hot summer	$T_{max} \geq +22$ °C
b	Warm summer	Not a and at least 4 $T_{min} \geq +10$ °C
c	Cold summer and cold winter	Not b and $T_{min} > -38$ °C
d	Extremely continental	Like c but $T_{min} \leq -38$ °C

[1] T_a—Annual mean temperature (°C); T_{max}—Monthly mean temperature of the warmest month (°C); T_{min}—Monthly mean temperature of the coldest month (°C); P_a—Accumulated annual precipitation (mm); P_{min}—Precipitation of the driest month (mm); P_{smin}, P_{smax}, P_{wmin}, P_{wmax}—Lowest and highest monthly precipitation values for the summer and winter half-years (April to September and October to March), respectively (°C). For the polar climates (E) no precipitation differentiations are given, only temperature conditions are defined, consequently, the polar climates (E) must be determined first, followed by the arid climates (B) and subsequent differentiations into the equatorial climates (A) and the warm temperate and snow climates (C) and (D), respectively.

Table A3. KG climate type percentage of wine regions worldwide in the historical period 1970–2000 and the future period 2041–2060 under SSP2 and SSP5.

Vineyards	ET	BSh	BSk	BWh	BWk	Af	Am	As	Aw	Csa	Csb	Csc	Cwa	Cwb	Cwc	Cfa	Cfb	Cfc	Dsa	Dsb	Dsc	Dfa	Dfb	Dwa	Dwb	Dwc
70–00	0.68	1.9	16.6	1.3	3.4			2.6	0.17	10.3	9.47	0.19	1.07	1.7	0.10	12.9	30.08	0.47	0.05	0.95	0.44	0.93	0.2	0.4	4.05	0.14
41–60 SSP2	0.12	18.0	5.2	6.9	2.2	2.9	0.12	2.1	0.14	17.4	1.39	0.01	2.72	0.5	0.02	34.2	5.95	0.01				0.12				
41–60 SSP5	0.11	18.1	5.3	6.9	2.2	2.6	0.12	2.2	0.13	17.8	1.18	0.01	2.78	0.48	0.01	34.2	5.76					0.12				

Table A4. Differences in the land area percentages of each Köppen–Geiger (KG) climate type between future (2041–2060), under SSP2 (upper rows, Δ1) or SSP5 (lower rows, Δ2), and the historical period (1970–2000) (Table linked to Figure 3).

KG (%)	ET	EF	BSh	BSk	BWh	BWk	Af	Am	As	Aw	Csa	Csb	Csc	Cwa	Cwb	Cwc
Δ1	−4.55	−3.82	1.27	0.52	1.40	0.07	−0.41	0.16	1.19	0.31	−1.09	−0.86	−0.01	−0.23	−0.20	−0.01
Δ2	−4.58	−3.86	1.29	0.63	1.40	0.03	−0.46	0.14	1.24	0.38	−1.07	−0.86	−0.01	−0.21	−0.20	−0.01
KG (%)	**Cfa**	**Cfb**	**Cfc**	**Dsa**	**Dsb**	**Dsc**	**Dsd**	**Dfa**	**Dfb**	**Dfc**	**Dfd**	**Dwa**	**Dwb**	**Dwc**	**Dwd**	
Δ1	2.78	−1.30	−0.20	0.03	0.30	−0.10	0.00	5.92	−0.23	−7.34	−0.74	1.02	−0.18	−0.62	−0.34	
Δ2	2.91	−1.29	−0.20	0.04	0.32	−0.14	0.00	6.06	−0.44	−7.52	−0.74	1.06	−0.17	−0.67	−0.34	

Table A5. KG climate type percentage of change within the study regions with vineyards (EU, NH non-EU countries, and SH) for 1970–2000 and 2041–2060 under SSP2 and SSP5 (Table linked to Figure 7).

Vineyards EU	70–00	41–60 SSP2	41–60 SSP5	Vineyards NH	70–00	41–60 SSP2	41–60 SSP5	Vineyards SH	70–00	41–60 SSP2	41–60 SSP5
ET	0.02%			ET	0.03%			ET	1.37%	0.25%	0.23%
BSh		8.82%	9.09%	BSh	1.17%	10.43%	10.09%	BSh	3.39%	27.07%	27.30%
BSk	4.85%	5.25%	5.89%	BSk	17.22%	12.07%	11.73%	BSk	23.24%	1.70%	1.65%
BWh		0.12%	0.10%	BWh	3.80%	9.01%	9.21%	BWh	0.79%	9.82%	9.77%
BWk				BWk	4.50%	3.56%	3.45%	BWk	4.83%	2.76%	2.74%
Af				Af				Af		6.08%	5.57%
Am				Am				Am		0.24%	0.25%
As				As				As	5.40%	4.46%	4.50%
Aw				Aw				Aw	0.36%	0.30%	0.27%
Csa	21.11%	24.87%	25.22%	Csa	10.85%	34.30%	35.55%	Csa	3.64%	4.63%	4.58%
Csb	7.77%			Csb	27.56%	5.59%	4.69%	Csb	1.53%	0.10%	0.10%
Csc				Csc	0.67%	0.06%	0.06%	Csc	0.07%		
Cwa				Cwa	2.60%	5.37%	5.82%	Cwa	0.96%	2.94%	2.87%
Cwb				Cwb				Cwb	3.50%	1.05%	0.99%
Cwc				Cwc				Cwc	0.21%	0.04%	0.03%
Cfa	14.93%	60.28%	59.33%	Cfa	2.43%	18.45%	18.45%	Cfa	16.92%	26.90%	27.62%
Cfb	47.46%	0.66%	0.37%	Cfb	3.52%	0.68%	0.45%	Cfb	32.89%	11.65%	11.52%
Cfc				Cfc	0.17%			Cfc	0.90%	0.01%	0.01%
Dsa				Dsa	0.17%			Dsa			
Dsb				Dsb	4.00%			Dsb			
Dsc				Dsc	1.87%			Dsc			
Dfa				Dfa	3.94%	0.48%	0.50%	Dfa			
Dfb				Dfb	0.81%			Dfb			
Dwa				Dwa	1.73%			Dwa			
Dwb	3.54%			Dwb	12.83%			Dwb			
Dwc	0.32%			Dwc	0.13%			Dwc			

Table A6. KG climate type percentage of change per country within the study regions with vineyards EU (EU, NH non-EU countries, and SH) for 1970–2000 and 2041–2060 under SSP2 and SSP5 (Tables linked to Figure 8).

Vineyards EU										
70–00	**ET**	**BSh**	**BSk**	**BWh**	**Csa**	**Csb**	**Cfa**	**Cfb**	**Dwb**	**Dwc**
AT								98.33%	1.67%	
BU					0.27%		41.67%	55.65%	2.41%	
CR					12.12%	0.61%	14.55%	71.52%		1.20%
CZ								100.00%		
DE								100.00%		
ES			33.77%		32.46%	16.23%	4.85%	12.69%		
FR					13.74%	5.19%	0.50%	80.57%		
GR			2.21%		68.01%	16.54%	5.15%	8.09%		
HR			0.00%					100.00%		
IT	0.14%		2.85%		47.08%	10.04%	28.36%	10.99%	0.14%	0.40%
PT					50.26%	49.74%				
RO							27.59%	53.58%	17.64%	1.19%
SK								92.93%	7.07%	
SL							2.50%	95.00%	2.50%	
41–60 SSP2	**ET**	**BSh**	**BSk**	**BWh**	**Csa**	**Csb**	**Cfa**	**Cfb**	**Dwb**	**Dwc**
AT							98.33%	1.67%		
BU		0.82%	0.82%		0.54%		96.73%	1.09%		
CR					13.29%		86.71%			
CZ			2.56%				97.44%			
DE							100.00%			
ES		50.47%	12.24%	0.75%	30.14%		6.40%			
FR		0.17%			19.83%		80.00%			
GR		11.43%			78.10%		10.47%			
HR							100.00%			
IT		9.89%		0.62%	56.16%		31.96%	1.37%		
PT					100.00%					
RO			19.41%				78.86%	1.73%		
SK							100.00%			
SL							100.00%			
41–60 SSP5	**ET**	**BSh**	**BSk**	**BWh**	**Csa**	**Csb**	**Cfa**	**Cfb**	**Dwb**	**Dwc**
AT			0.000%				100.00%			
BU		1.36%	2.17%		0.54%		94.84%	1.09%		
CR					15.38%		84.62%			
CZ							100.00%			
DE							100.00%			
ES		51.43%	12.84%	0.19%	29.68%		5.86%			
FR		0.17%			20.51%		79.32%			
GR		11.42%			78.10%		10.48%			
HR							100.00%			
IT		10.50%	0.46%		57.84%		29.98%	1.22%		
PT					100.00%					
RO			22.07%				77.53%	0.40%		
SK							100.00%			
SL							100.00%			

Table A7. KG climate type percentage of change per country within the study regions with vineyards NH non-EU countries for 1970–2000 and 2041–2060 under SSP2 and SSP5 (Tables linked to Figure 8).

Vineyards NH																				
70–00	**ET**	**BSh**	**BSk**	**BWh**	**BWk**	**Csa**	**Csb**	**Csc**	**Cwa**	**Cfa**	**Cfb**	**Cfc**	**Dsa**	**Dsb**	**Dsc**	**Dsd**	**Dfa**	**Dfb**	**Dwb**	**Dwc**
CH											96.15%									3.85%
GE										60.00%	34.29%								5.71%	
US	0.03%	1.37%	16.81%	4.45%	2.19%	12.69%	32.25%	0.79%		2.16%	2.91%	0.20%	0.20%	4.68%	2.19%	2.03%	14.95%	0.10%		
CN			22.22%		20.48%				20.26%								30.72%	6.32%		
41–60 SSP2	**ET**	**BSh**	**BSk**	**BWh**	**BWk**	**Csa**	**Csb**	**Csc**	**Cwa**	**Cfa**	**Cfb**	**Cfc**	**Dsa**	**Dsb**	**Dsc**	**Dsd**	**Dfa**	**Dfb**	**Dwb**	**Dwc**
CH										76.92%	23.08%									
GE										97.14%	2.86%									
US		11.66%	10.70%	10.54%	0.07%	40.08%	6.54%	0.07%		19.78%	0.56%									
CN		3.56%	22.89%		27.56%				42.21%								3.78%			
41–60 SSP5	**ET**	**BSh**	**BSk**	**BWh**	**BWk**	**Csa**	**Csb**	**Csc**	**Cwa**	**Cfa**	**Cfb**	**Cfc**	**Dsa**	**Dsb**	**Dsc**	**Dsd**	**Dfa**	**Dfb**	**Dwb**	**Dwc**
CH										80.77%	19.23%									
GE		2.86%								94.28%	2.86%									
US		11.46%	10.50%	10.77%	0.07%	41.55%	5.48%	0.07%		19.78%	0.32%									
CN		2.00%	21.55%		26.67%				45.78%								4.00%			

Table A8. KG climate type percentage of change per country within the study regions with vineyards SH for 1970–2000 and 2041–2060 under SSP2 and SSP5 (Tables linked to Figure 8).

Vineyards SH																			
70–00	**ET**	**BSh**	**BSk**	**BWh**	**BWk**	**Af**	**Am**	**As**	**Aw**	**Csa**	**Csb**	**Csc**	**Cwa**	**Cwb**	**Cwc**	**Cfa**	**Cfb**	**Cfc**	**Dwc**
AR	2.43%	8.18%	67.39%		10.23%					5.50%	4.60%	0.64%					0.90%	0.13%	
AU		0.89%	28.83%		0.18%					7.03%	2.36%		1.98%	1.34%		13.71%	43.30%	0.38%	
BR		9.35%						27.41%	1.83%							41.81%	19.60%		
CL	11.89%		12.36%		30.83%								1.10%	33.02%	2.35%		8.45%		
NZ	0.59%																91.36%	7.61%	0.44%
ZA		5.68%	48.30%	16.19%	19.03%											1.42%	9.38%		
UY																89.30%	10.70%		
41–60 SSP2	**ET**	**BSh**	**BSk**	**BWh**	**BWk**	**Af**	**Am**	**As**	**Aw**	**Csa**	**Csb**	**Csc**	**Cwa**	**Cwb**	**Cwc**	**Cfa**	**Cfb**	**Cfc**	**Dwc**
AR	0.51%	51.41%	11.76%	25.96%	7.42%					1.92%	0.90%							0.12%	
AU		32.61%		6.57%			0.53%	2.56%		10.25%			2.07%			36.06%	9.35%		
BR		21.34%				29.93%	0.07%	16.69%	1.48%							30.35%	0.14%		
CL	2.21%	11.04%	4.26%	18.30%	21.61%								22.87%	11.67%	0.47%	4.26%	3.31%		
NZ																21.22%	78.78%		
ZA		43.71%	0.30%	52.40%												3.59%			
UY						2.50%										97.50%			
41–60 SSP5	**ET**	**BSh**	**BSk**	**BWh**	**BWk**	**Af**	**Am**	**As**	**Aw**	**Csa**	**Csb**	**Csc**	**Cwa**	**Cwb**	**Cwc**	**Cfa**	**Cfb**	**Cfc**	**Dwc**
AR	0.51%	51.66%	11.64%	26.21%	7.16%					1.92%	0.90%								
AU		32.97%	6.04%				0.56%	2.73%		10.15%			2.07%			36.35%	9.13%		
BR		21.62%				27.68%	0.07%	16.55%	1.34%							32.60%	0.14%		
CL	1.89%	11.04%	3.94%	19.87%	21.77%								22.08%	11.05%	0.32%	4.57%	3.47%		
NZ																21.83%	78.17%		
ZA		43.42%	0.30%	52.69%												3.59%			
UY						0.50%										99.50%			

Table A9. KG climate types for cool climate wine regions. Geographical coordinates, location, and climate type are listed, ranked by decreasing latitude in each hemisphere (EU, NH non-EU countries, and SH) for 1970–2000 and 2041–2060 under SSP2 and SSP5.

Latitude	Longitude	Location	70–00	41–60 SSP2	41–60 SSP5
57.7° N	12.0° E	Gothenburg, Sweden	Cfb	Cfa	Cfa
51.6° N	15.5° E	Zilona Góra, Poland	Cfb	Cfa	Cfa
57.1° N	9.9° E	Aalborg, Denmark	Cfb	Cfb	Cfb
54.0° N	1.1° W	York, Britain	Cfb	Cfb	Cfb
50.9° N	5.7° E	Maastricht, Netherlands	Cfb	Cfa	Cfa
50.7° N	120.4° W	Kamloops, British Colombia, Canada	Dwb	BSk	BSk
46.4° N	79.9° W	Lake Te, Timagami, Ontario, Canada	Dwb	Dwa	Dwa
46.4° N	71.4° W	Quebec City, Quebec, Canada	Dwb	Dwa	Dwa
45.4° N	65.5° W	Sussex, New Brunswick, Canada	Dwb	Cfa	Cfa
45.2° N	85.5° W	Leelanau Peninsula, Michigan, USA	Dwb	Cfa	Cfa
45.2° N	65.2° W	Annapolis Valley, Nova Scotia, Canada	Dwb	Cfa	Cfa
45.6° S	69.1° W	Sarmiento, Argentina	Bwk	Bwk	Bwk
45.3° S	169.4° E	Alexandra, New Zealand	Cfb	Cfb	Cfb
43.3° S	147.3° E	Bruny Island, Tasmania, Australia	Cfb	Cfb	Cfb
42.7° S	73.8° W	Chiloe Island, Chile	Cfb	Cfb	Cfb

References

1. Belda, B.; Holtanová, E.; Halenka, T.; Kalvová, J. Climate classification revisited: From Köppen to Trewartha. *Clim. Res.* **2014**, *59*, 1–13. [CrossRef]
2. Köppen, W. Die Wärmezonen der Erde, nach der Dauer der heissen, gemässigten und kalten Zeit und nach der Wirkung der Wärme auf die organische Welt betrachtet. *Meteorol. Z.* **2011**, *20*, 351–360. [CrossRef] [PubMed]
3. Bryson, R.A. Air Masses, Streamlines, and the Boreal Forest. *Geogr. Bull.* **1966**, *8*, 228–269.
4. Thornwaite, C.W. An Approach toward a Rational Classification of Climate. *Geogr. Rev.* **1948**, *38*, 55–94. [CrossRef]
5. Holdridge, L.R. Determination of world plant formations from simple climatic data. *Science* **1947**, *105*, 367–368. [CrossRef] [PubMed]
6. Köppen, W. Klassifikation der Klimate nach Temperatur, Niederschlag und Jahreslauf. *Petermanns Geogr. Mitteilungen* **1918**, *64*, 193–203, 243–248.
7. Köppen, W. Das geographische System der Klimate. In *Handbuch der Klimatologie*; Borntraeger: Berlin, Germany, 1936; Volume 1.
8. Geiger, R.F. *Klassifikation der Klimate nach W. Köppen*; Landolt-Börnstein—Zahlenwerte und Funktionen aus Physik, Chemie, Astronomie, Geophysik und Technik, alte Serie; Springer: Berlin/Heidelberg, Germany, 1954; Volume 3, pp. 603–607.
9. Trewartha, G.T. *An Introduction to Climate*; McGraw-Hill: New York, NY, USA, 1968.
10. Trewartha, G.T.; Horn, L.H. *Introduction to Climate*, 5th ed.; McGraw Hill: New York, NY, USA, 1980.
11. Köppen, W. *Grundiss der Klimakunde*, 2nd ed.; Walter de Gruyter & Co.: Berlin, Germany, 1931.
12. de Blij, H.J. Geography of viticulture: Rationale and resource. *J. Geogr.* **1983**, *82*, 112–121. [CrossRef]
13. Anderson, K.; Nelgen, S. *Which Winegrape Varieties Are Grown Where?, Revised ed.*; University of Adelaide Press: Adelaide, Australia, 2020; p. 962. ISBN 978-1925261868.
14. Jones, G.V.; Reid, R.; Vilks, A. Climate, Grapes, and Wine: Structure and Suitability in a Variable and Changing Climate. In *The Geography of Wine: Regions, Terroir, and Techniques*; Dougherty, P., Ed.; Springer Press: Dordrecht, The Netherlands, 2012; pp. 109–133.
15. Peel, M.C.; Finlayson, B.L.; McMahon, T.A. Updated world map of the Köppen–Geiger climate classification. *Hydrol. Earth Syst. Sci.* **2007**, *11*, 1633–1644. [CrossRef]
16. Chan, D.; Wu, Q. Significant anthropogenic-induced changes of climate classes since 1950. *Sci. Rep. Nat. Res.* **2015**, *5*, 13487. [CrossRef]
17. Chen, D.; Chen, H.W. Using the Köppen classification to quantify climate variation and change: An example for 1901–2010. *Environ. Dev.* **2013**, *6*, 69–79. [CrossRef]
18. Mahlstein, I.; Daniel, J.S.; Solomon, S. Pace of shifts in climate regions increases with global temperature. *Nat. Clim. Chang.* **2013**, *3*, 739–743. [CrossRef]
19. Rubel, F.; Kottek, M. Observed and projected climate shifts 1901–2100 depicted by world maps of the Köppen-Geiger climate classification. *Meteorol. Z.* **2010**, *19*, 135–141. [CrossRef]
20. Wang, M.; Overland, J.E. Detecting Arctic climate change using Köppen climate classification. *Clim. Chang.* **2004**, *67*, 43–62. [CrossRef]
21. Feng, S.; Ho, C.H.; Hu, Q.; Oglesby, R.J.; Jeong, S.J.; Kim, B.M. Evaluating observed and projected future climate changes for the Artic using the Köppen-Trewartha climate classification. *Clim. Dyn.* **2012**, *38*, 1359–1373. Available online: https://www.jstor.org/stable/e24897565 (accessed on 21 January 2023). [CrossRef]

22. Diaz, H.F.; Eischeid, J.K. Disappearing "alpine tundra" Köppen climatic type in the western United States. *Geophys. Res. Lett.* **2007**, *34*, L18707. [CrossRef]
23. de Castro, M.; Gallardo, C.; Jylha, K.; Tuomenvirta, H. The use of a climate-type classification for assessing climate change effects in Europe from an ensemble of nine regional climate models. *Clim. Chang.* **2007**, *81*, 329–341. [CrossRef]
24. Jones, G.V.; White, M.A.; Cooper, O.R.; Storchmann, K. Climate Change and Global Wine Quality. *Clim. Chang.* **2005**, *73*, 319–343. [CrossRef]
25. Gambetta, G.A.; Kurtural, S.K. Global warming and wine quality: Are we close to the tipping point? *OENO One* **2021**, *55*, 353–361. [CrossRef]
26. Hannah, L.; Roehrdanz, P.R.; Ikegami, M.; Hijmans, R.J. Climate change, wine, and conservation. *Biol. Sci.* **2013**, *110*, 6907–6912. [CrossRef] [PubMed]
27. Sgubin, G.; Swingedouw, D.; Mignot, J.; Gambetta, G.A.; Bois, B.; Loukos, H.; Noël, T.; Pieri, P.; García de Cortázar-Atauri, I.; Ollat, N.; et al. Non-linear loss of suitable wine regions over Europe in response to increasing global warming. *Glob. Chang. Biol.* **2023**, *29*, 808–826. [CrossRef]
28. Santos, J.A.; Fraga, H.; Malheiro, A.C.; Moutinho-Pereira, J.; Dinis, L.T.; Correia, C.; Moriondo, M.; Leolini, L.; Dibari, C.; Costafreda-Aumedes, S.; et al. A review of the potential climate change impacts and adaptation options for European viticulture. *Appl. Sci.* **2020**, *10*, 3092. [CrossRef]
29. Santos, J.A.; Yang, C.; Fraga, H.; Malheiro, A.C.; Moutinho-Pereira, J.M.; Dinis, L.; Correia, C.M.; Moriondo, M.; Bindi, M.; Leolini, L.; et al. Long-term adaptation of European viticulture to climate change: An overview from the H2020 Clim4Vitis action. *IVES Tech. Rev. Vine Wine* **2021**. [CrossRef]
30. Santos, J.A.; Yang, C.; Fraga, H.; Malheiro, A.C.; Moutinho-Pereira, J.M.; Dinis, L.; Correia, C.M.; Moriondo, M.; Bindi, M.; Leolini, L.; et al. Short-term adaptation of European viticulture to climate change: An overview from the H2020 Clim4Vitis action. *IVES Tech. Rev. Vine Wine* **2021**. [CrossRef]
31. Moriondo, M.; Jones, G.V.; Bois, B.; Dibari, C.; Ferrise, R.; Trombi, G.; Bindi, M. Projected shifts of wine regions in response to climate change. *Clim. Chang.* **2013**, *119*, 825–839. [CrossRef]
32. World Wine Production Outlook. *Wine Production—OIV First Estimates*; International Organization of Vine and Wine: Dijon, France, 2022. Available online: https://www.google.com/url?sa=t&source=web&rct=j&opi=89978449&url=https://www.oiv.int/sites/default/files/documents/OIV_World_Wine_Production_Outlook_2023.pdf&ved=2ahUKEwiNh4ac5fuGAxU-AtsEHe4GD20QFnoECBQQAQ&usg=AOvVaw0BxkIobYX0LNebJLer5E2T (accessed on 21 January 2023).
33. Swart, N.M.; Cole, J.N.S.; Kharin, V.V.; Lazare, M.; Scinocca, J.F.; Gillett, N.P.; Anstey, J.; Arora, V.; Christian, J.R.; Jiao, Y.; et al. *CCCma CanESM5 Model Output Prepared for CMIP6 ScenarioMIP*; Earth System Grid Federation, 2019. Available online: https://esgf.llnl.gov/ (accessed on 21 January 2023). [CrossRef]
34. Fick, S.E.; Hijmans, R.J. WorldClim 2: New 1km spatial resolution climate surfaces for global land areas. *Int. J. Climatol.* **2017**, *37*, 4302–4315. [CrossRef]
35. Intergovernmental Panel on Climate Change. Sixth Assessment Report (IPCC, AR6), AR6 Synthesis Report: Climate Change. 2023. Available online: https://www.ipcc.ch/assessment-report/ar6/ (accessed on 21 January 2023).
36. Köppen, W.; Geiger, R. *Handbuch der Klimatologie*; Gebrüder Bornträger: Berlin, Germany, 1930.
37. Andrade, C.; Contente, J. Köppen's climate classification projections for the Iberian Peninsula. *Clim. Res.* **2020**, *81*, 71–80. [CrossRef]
38. Kottek, M.J.; Grieser, C.; Beck, B.; Rudolf, R.F. World map of the Köppen–Geiger climate classification updated. *Meteorol. Z.* **2006**, *15*, 259–263. [CrossRef]
39. Beck, H.; Zimmermann, N.; McVicar, T.; Vergopolan, N.; Berg, A.; Wood, E. Present and future Köppen-Geiger climate classification maps at 1-km resolution. *Sci. Data* **2018**, *5*, 180214. [CrossRef]
40. Bois, B.; Blais, A.; Moriondo, M.; Jones, G.V. High resolution climate spatial analysis of European winegrowing regions. In Proceedings of the 9th International Terroir Congress, Reims, France, 25–29 June 2012; Volume 1, pp. 217–220.
41. European Environment Agency. *CLC2006 Technical Guidelines*; EEA Technical Report 17/2007; Office for Official Publications of the European Communities: Luxembourg, 2007.
42. Bois, B.; Gavrilescu, C.; Moriondo, M.; Jones, G.V. Wine growing region global climate analysis. In Proceedings of the 11th International Terroir Congress 2016, McMinnville, OR, USA, 10–14 July 2016; pp. 9–14.
43. Gavrilescu, C.; Bois, B. Chardonnay wines climate plasticity: A worldwide geographical approach. *BIO Web Conf.* **2016**, *7*, 01013. [CrossRef]
44. Bois, B.; Zito, S.; Calonnec, A. Climate vs. grapevine pests and diseases worldwide: The first results of a global survey. *OENO One* **2017**, *51*, 133–139. [CrossRef]
45. Jones, G.V.; Schultz, H.R. Climate change and emerging cool climate wine regions. *Wine Vitic. J.* **2016**, *31*, 51–53.
46. Cramer, W.P.; Leemans, R. Assessing impacts of climate change on vegetation using climate classification systems. In *Vegetation Dynamics and Global Change*; Solomon, A.M., Shugart, H.H., Eds.; Springer: Boston, MA, USA, 1993; pp. 190–217.
47. Rohli, R.V.; Joyner, T.A.; Reynolds, S.J.; Ballinger, T.J. Overlap of global Köppen–Geiger climates, biomes, and soil orders. *Phys. Geogr.* **2015**, *36*, 158–175. [CrossRef]
48. Sohoulande, C.D. Vegetation and water resource variability within the Köppen-Geiger global climate classification scheme: A probabilistic interpretation. *Theor. Appl. Climatol.* **2024**, *155*, 1081–1092. [CrossRef]

49. Chen, I.C.; Hill, J.K.; Ohlemüller, R.; Roy, D.B.; Thomas, C.D. Rapid range shifts of species associated with high levels of climate warming. *Science* **2011**, *333*, 1024–1026. [CrossRef] [PubMed]
50. Müller, M.S.; Dourado-Neto, D.; Timm, L.C.; Reichardt, K.; Sartori, F.F.; Felisberto, G. Climate analysis for agricultural improvement of the Economic Community of West African States according to Köppen and Thornthwaite. *Afr. J. Agric. Res.* **2018**, *13*, 1198–1212. [CrossRef]
51. Wolkovich, E.M.; García de Cortázar-Atauri, I.; Morales-Castilla, I.; Nicholas, K.A.; Lacombe, T. From Pinot to Xinomavro in the world's future wine-growing regions. *Nat. Clim. Chang.* **2018**, *8*, 29–37. [CrossRef]
52. Morales-Castilla, I.; García de Cortázar-Atauri, I.; Cook, B.I.; Lacombe, T.; Parker, A.; van Leeuwen, C.; Nicholas, K.A.; Wolkovich, E.M. Diversity buffers winegrowing regions from climate change losses. *Proc. Natl. Acad. Sci. USA* **2020**, *117*, 2864–2869. [CrossRef]
53. Puga, G.; Anderson, K.; Jones, G.; Tchatoka, F.D.; Umberger, W. A climatic classification of the world's wine regions. *OENO One* **2022**, *56*, 165–177. [CrossRef]
54. Labbé, T.; Pfister, C.; Brönnimann, S.; Rousseau, D.; Franke, J.; Bois, B. The longest homogeneous series of grape harvest dates, Beaune 1354–2018, and its significance for the understanding of past and present climate. *Clim. Past* **2019**, *15*, 1485–1501. [CrossRef]
55. Koufos, G.C.; Mavromatis, T.; Koundouras, S.; Jones, G.V. Adaptive capacity of winegrape varieties cultivated in Greece to climate change: Current trends and future projections. *OENO One* **2020**, *54*, 1201–1219. [CrossRef]
56. Bécart, V.; Lacroix, R.; Puech, C.; García de Cortázar-Atauri, I. Assessment of changes in Grenache grapevine maturity in a Mediterranean context over the last half-century. *OENO One* **2022**, *56*, 53–72. [CrossRef]
57. Schultz, H.R. Global climate change, sustainability, and some challenges for grape and wine production. *J. Wine Econ.* **2016**, *11*, 181–200. [CrossRef]
58. Dinis, L.T.; Bernardo, S.; Yang, C.; Fraga, H.; Malheiro, A.C.; Moutinho-Pereira, J.; Santos, J.A. Mediterranean viticulture in the context of climate change. *Ciência Téc. Vitiv.* **2022**, *37*, 139–158. [CrossRef]
59. Kenny, G.J.; Harrison, P.A. The effects of climate variability and change on grape suitability in Europe. *J. Wine Res.* **1992**, *3*, 163–183. [CrossRef]
60. Stock, M.; Gerstengarbe, F.W.; Kartschall, T.; Werner, P.C. Reliability of climate change impact assessments for viticulture. In Proceedings of the VII International Symposium on Grapevine Physiology and Biotechnology, Davis, CA, USA, 21–25 July 2004; Volume 689, pp. 29–40. [CrossRef]
61. Medrano, H.; Tomás, M.; Martorell, S.; Escalona, J.M.; Pou, A.; Fuentes, S.; Flexas, J.; Bota, J. Improving water use efficiency of vineyards in semi-arid regions. A review. *Agron. Sustain. Dev.* **2015**, *35*, 499–517. [CrossRef]
62. Chaves, M.M.; Zarrouk, O.; Francisco, R.; Costa, J.M.; Santaos, T.; Regalado, A.P.; Rodriges, M.L.; Lopes, C.M. Grapevine under deficit irrigation: Hints from physiological and molecular data. *Ann. Bot.* **2010**, *105*, 661–676. [CrossRef]
63. Flexas, J.; Galmés, J.; Gallé, A.; Gulías, J.; Pou, A.; Ribas-Carbo, M.; Tomás, M.; Medrano, H. Improving water use efficiency in grapevines: Potential physiological targets for biotechnological improvement. *Aust. J. Grape Wine Res.* **2010**, *16*, 106–121. [CrossRef]
64. Skalák, P.; Farda, A.; Zahradníček, P.; Trnka, M.; Hlásny, T.; Štěpánek, P. Projected shift of Köppen–Geiger zones in the central Europe: A first insight into the implications for ecosystems and the society. *Int. J. Climatol.* **2018**, *38*, 3595–3606. [CrossRef]
65. Williams, A.P.; Abatzoglou, J.T.; Gershunov, A.; Guzman-Morales, J.; Bishop, D.A.; Balch, J.K.; Lettenmaier, D.P. Observed impacts of anthropogenic climate change on wildfire in California. *Earth's Futur.* **2019**, *7*, 892–910. [CrossRef]
66. Sgubin, G.; Swingedouw, D.; Dayon, G.; García de Cortázar-Atauri, I.; Ollat, N.; Pagé, C.; van Leeuwen, C. The risk of tardive frost damage in French vineyards in a changing climate. *Agric. For. Meteorol.* **2018**, *250–251*, 226–242. [CrossRef]
67. Lamichhane, J.R. Rising risks of late-spring frosts in a changing climate. *Nat. Clim. Chang.* **2021**, *11*, 554–555. [CrossRef]
68. Mira de Orduña, R. Climate change associated effects on grape and wine quality and production. *Food Res. Int.* **2010**, *43*, 1844–1855. [CrossRef]
69. White, M.A.; Diffenbaugh, N.S.; Jones, G.V.; Pal, J.S.; Giorgi, F. Extreme heat reduces and shifts United States premium wine production in the 21st century. *Proc. Natl. Acad. Sci. USA* **2006**, *103*, 11217–11222. [CrossRef] [PubMed]
70. Wang, W.; Pijl, A.; Tarolli, P. Future climate-zone shifts are threatening steep-slope agriculture. *Nat. Food* **2022**, *3*, 193–196. [CrossRef]
71. Harris, I.; Osborn, T.J.; Jones, P.; Lister, D. Version 4 of the CRU TS monthly high-resolution gridded multivariate climate dataset. *Sci. Data* **2020**, *7*, 109. [CrossRef] [PubMed]

Disclaimer/Publisher's Note: The statements, opinions and data contained in all publications are solely those of the individual author(s) and contributor(s) and not of MDPI and/or the editor(s). MDPI and/or the editor(s) disclaim responsibility for any injury to people or property resulting from any ideas, methods, instructions or products referred to in the content.

remote sensing

MDPI

Article

Causes of Increased Compound Temperature and Precipitation Extreme Events in the Arid Region of Northwest China from 1961 to 2100

Huihui Niu [1], Weijun Sun [1], Baojuan Huai [1], Yuzhe Wang [1], Rensheng Chen [2], Chuntan Han [2], Yingshan Wang [1], Jiaying Zhou [1] and Lei Wang [1,2,*]

[1] College of Geography and Environment, Shandong Normal University, Jinan 250358, China; niuhuihui@stu.sdnu.edu.cn (H.N.); sunweijun@sdnu.edu.cn (W.S.); 616070@sdnu.edu.cn (B.H.); wangyuzhe@sdnu.edu.cn (Y.W.); 2022010109@stu.sdnu.edu.cn (Y.W.); zhoujiaying@stu.sdnu.edu.cn (J.Z.)

[2] Qilian Alpine Ecology and Hydrology Research Station, Northwest Institute of Eco-Environment and Resources, Chinese Academy of Sciences, Lanzhou 730000, China; crs2008@lzb.ac.cn (R.C.); hancht@lzb.ac.cn (C.H.)

* Correspondence: leiwang@sdnu.edu.cn

Abstract: Compound extreme events pose more grave threats to human health, the natural environment, and socioeconomic systems than do individual extreme events. However, the drivers and spatiotemporal change characteristics of compound extreme events under climate transition remain poorly understood, especially in the arid region of Northwest China. This study examined the spatiotemporal change characteristics and driving mechanisms of extreme temperature and precipitation compound events in Northwest China based on data from 86 national meteorological stations and 11 climate models of the Coupled Model Intercomparison Project, Phase 6. The results indicated that (1) the frequency values of heat extremity–dry (1.60/10a) and heat extremity–heavy precipitation (0.60/10a) events increased from 1961 to 2020, and showed a faster uptrend after 1990 than before. (2) Under four shared socioeconomic pathway scenarios, there is also the likelihood of an upward trend in heat extremity–dry and heat extremity–heavy precipitation events in Northwest China by the end of 21 century, especially under SSP585, with probability values of 1.70/10a and 1.00/10a, respectively. (3) A soil moisture deficit leads to decreased evaporation and increased sensible heat by reduction in the soil–atmosphere exchange; the non-adiabatic heating process leads to a higher frequency of hot days. This land–air interaction feedback mechanism is a significant driver of heat extremity–dry events in Northwest China. (4) In the Northwest China region, the warmer trend surpasses the wetter trend, contributing to increased specific humidity, and the vapor pressure deficit may lead to an increasing frequency of extreme precipitation, consequently increasing heat extremity–heavy precipitation events. These results provide new insights for the understanding of compound extreme events, in order to cope with their risks.

Keywords: compound extreme events; spatiotemporal change; projections; land–atmosphere interaction

1. Introduction

The alterations in the severity and recurrence of climatic extremes typically correlate with the magnitude of global warming, which will likely persist, even when the global temperature rise reaches 1.5 °C [1]. This especially holds for temperature extremes, worsening droughts in arid regions, and intensification of heavy precipitation [2,3]. These exceptional extreme events may occur concurrently in various regions, resulting in significant and irreparable effects on both human and natural systems [4,5]. The recurrence of CEEs, defined as the concurrent or sequential incidence of two or more extreme events in the same locality or neighboring areas, has been on the rise in recent years. The damage caused by CEEs to the economy, society, and the natural environment are substantially greater than those caused by individual extreme events [6]. Additionally, conducting separate analyses on univariate extreme events might significantly underrate the detriments inflicted by multivariate CEEs [7]. Consequently, robust event attribution is important for quantifying the current risk of such events and can support decisions, such as how to rebuild after a disaster [8]. However, CEEs can arise randomly and may be either independent of each other or induced by an identical underlying mechanism or meteorological system [9,10]. Accordingly, exploration of the relevant spatiotemporal characteristics and potential harms to human society attributable to CEEs are hot topics in the field of climate change.

The CEEs are typically correlated with the combined influences of precipitation and temperature. In recent years, considerable effort has been directed towards elucidating the occurrence patterns and explaining the mechanisms of individual/compound temperature–precipitation events. It is generally acknowledged that the atmospheric water vapor content could increase by 7% as the temperature rises by 1 °C [11–13]. Accordingly, the future occurrence of temperature–precipitation CEEs on land may be subject to mean precipitation patterns and warming trends [9,10]. Hoell et al. [14] found that the increased risks of extreme temperature are principally due to anthropogenic effects, while the increased risks of extreme low precipitation are mainly due to La Niña. High-temperature stress combined with water stress has caused drought–heat CEEs in California-Nevada to increase more than sixfold. Yin et al. [15] conducted ensemble simulations and estimated that the global incidence of drought–heatwave CEEs will increase tenfold under the highest emission scenario. In China, Peng et al. [7] demonstrated that warm–wet CEEs are commonly observed in the Tibetan Plateau and NWC, whereas warm–dry CEEs predominantly cluster in the southwest and northern regions, particularly post-1996, aligning with the warming and wetting trends in the NWC. Yu et al. [3] discovered a pronounced rise in the occurrence of drought–heat CEEs in cities located in arid regions, with the population affected by compound drought–heat extreme events doubling in eastern China since the late 1990s. Likewise, there has been a marked escalation in the number of individuals subjected to simultaneous daytime and nocturnal heatwaves [16]. Many research works have indicated that the number of CEEs has been increasing worldwide in different regions and periods. Despite the escalating attention to the spatial–temporal characteristics of CEEs in numerous investigations, and the partial elucidation of their changing characteristics, comprehension of the underlying formation mechanisms of CEEs remains inadequate. Currently, historical records are commonly used to estimate the probability of a prospective occurrence via a technique known as frequency analysis. These methods are suitable when the climatic factors driving these events remain constant. However, frequency analysis is no longer used to estimate the probability of future events under global warming associated with significant climate change. Thus, we need to understand the natural reasons for the changes that affect extreme events more explicitly [4].

NWC is approximately 2.5×10^7 km^2 and is located in the Eurasian hinterland; it has a dry climate, low levels of precipitation, and an extremely vulnerable ecological environment. Water vapor cannot enter NWC because of mountains which block it, and this results in a climate typified by scorching summers, frigid winters, and little precipitation. Nevertheless, NWC has undergone a remarkable climate transition from warm and dry to warm and wet. The precipitation and temperature in NWC show significant trends of increase, at 4.98 mm/10a and 0.37 °C/10a, respectively, which are higher than the average rates for China (−1.6 mm/10a and 0.25 °C/10a) for the same period. The NWC is one of the regions most vulnerable to climate change, as its fragile ecosystems are particularly susceptible to damage from both extreme weather events and compounded extreme events. The frequent droughts and extreme precipitation events have resulted in severe vegetation degradation and an accelerated retreat of glaciers, thereby posing a significant threat to both the ecosystems and water security in NWC. Therefore, it is essential to urgently understand the changing future characteristics of compound extreme events in NWC, a project which could help to improve prevention and preparedness measures. Therefore, the limited in situ data in NWC are not sufficient to explain and project the changing characteristics of CEEs under climate change, as the region changes from warm–dry to warm–wet.

This study aims to examine the spatial–temporal patterns of CEEs based on in situ data from 1961 to 2020 to project and explain the changing future characteristics, using the mechanism of CMIP6 models. The findings not only provide a scientific basis for the formulation of CEE-based disaster prevention measures and mitigation tactics for NWC, but also could help to expand readers' understanding of CEEs in other arid regions of the world and help policymakers keenly aware of the urgency of climate-based governance.

2. Materials and Methods

2.1. Study Area

The scope of NWC in this study refers to the NWC desert region within the seven natural regions of China's comprehensive natural geographic zoning, considering the multi-year average precipitation, humidity, and temperature. The specific geographic location is 74.42–103.25°, north latitude 36.25–49.42°, encompassing roughly one-fifth of China's total land area.

NWC is located deep in Eurasia, far from the ocean, and is mainly characterized by semi-arid and arid climates, which are further subdivided into temperate and warm-temperate climates, with scanty precipitation, high summer temperatures, and dry and cold winters. The multiannual mean temperature in the study area from 1961 to 2020 was 7.85 °C, within which the highest mean temperature was reached in 2007, at 9.10 °C. The mean temperature in NWC over the last 60a exhibits a rising pattern, with a tendency of rising 0.37 °C/10a. The mean precipitation in NWC is 287.57 mm, with summer precipitation accounting for 48.15% of the total. The precipitation exhibits a tendency to rise at a rate of 4.98 mm/10a, with high precipitation variability, extremely uneven spatial distribution, and high evaporation capacity, making it among the driest areas globally situated at a similar latitude. The climate of NWC is primarily controlled by westerly circulation, with year-round water vapor being transported mainly by the mid-latitude westerly belt. In addition, the summer months are influenced by the southeast monsoon, the local recirculation of water vapor, and the southwest monsoon.

The allocation of water resources exhibits significant imbalance, as most of the land is within the endorheic zone. An increase in population density has occurred in the endorheic river and lake areas, where oasis agriculture has developed. Being the driest region at a similar latitude globally, NWC is a typical ecologically fragile and climate-change-sensitive area. The impact of climate change on NWC is mainly characterized by changes in warming and precipitation patterns, which not only affect the water cycle process in the region but also precipitate surges in both the recurrence and the severity of extreme meteorological occurrences, including droughts and floods. Thus, NWC, which is highly dependent on climatic conditions and water resources, faces more severe challenges in terms of ecosystem stability, economic and industrial development, and the survival of tangible cultural heritage sites.

2.2. Data

2.2.1. Measured Data

Ground-based observations of precipitation and temperature at 86 national meteorological stations were used to determine the CEEs. The China Meteorological Administration (CMA) provided daily precipitation and air temperature data spanning the years 1961 to 2020. Although ground-based observations are susceptible to inaccuracies and indeterminacies, all datasets underwent rigorous quality assurance and consistency evaluation according to the standard protocols of the CMA [17]. Figure 1 illustrates the geographic distribution of meteorological stations.

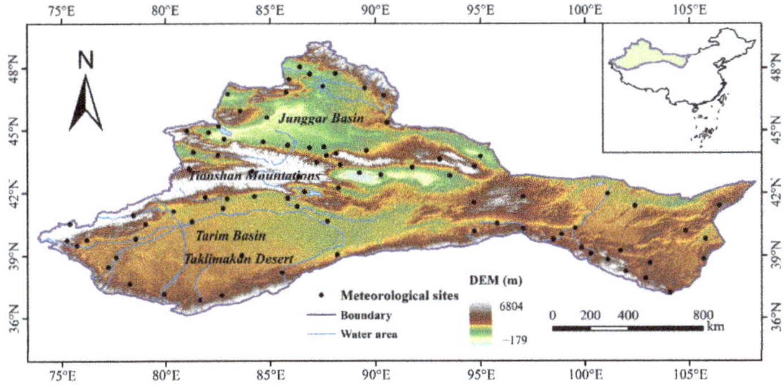

Figure 1. The topography of NWT and distribution of meteorological sites.

2.2.2. CMIP6 Data

The observational data are employed to evaluate the historical products of CMIP6 models based on several statistical methods such as model bias and pattern correlation coefficient (PCC). The daily air temperature data was chosen for analysis, using the period from 1961 to 2015; this was employed to simulate monthly mean air temperature. To quantify the differences among models, we further calculated the PCC and normalized root-mean-square error (NRMSE) of each model, as compared with observations (Figure S1). Both PCC and NRMSE need to be considered when choosing high-performance models. The best 11 models with high PCC (greater than 0.5) and low NRMSE (smaller than the mean) were selected (upper left quadrant in Figure S1). The MME (multi-model ensemble) of the best 11 models ('Best 11 MME') is better than most of the other models in capturing the observed pattern.

Daily variations in near-surface maximum temperature (tasmax) and precipitation (pr) were analyzed using 11 CMIP6 models (Table 1). Based on the anomalies in the daily precipitation and near-surface maximum temperature, extreme events were identified using thresholds of 25% and 75%, respectively. Experimental data from four scenarios were selected for this study and the timescales were standardized to 2021–2100. To ensure consistency, most CMIP6 climate simulations at lower resolutions were remapped onto a standard $0.1° \times 0.1°$ grid using bilinear interpolation [18]. To enhance accuracy, the MME was derived through the averaging of outcomes from each model. The MME mitigates the constraints inherent to a solitary model and has been shown to exhibit superior precision relative to a singular model, as evidenced by previous studies [19].

Table 1. Basic information about the CMIP6 model used in this study.

No.	CMIP6 Model Name	Country	Atmospheric Resolution (Lon × Lat in Deg)	Variant Label
1	ACCESS-CM2	Australia	1.9° × 1.3°	r1i1p1f1
2	AWI-CM-1-1-MR	Germany	0.9° × 0.9°	r1i1p1f1
3	BCC-CSM2-MR	China	1.1° × 1.1°	r1i1p1f1
4	CNRM-CM6-1	France	1.4° × 1.4°	r1i1p1f2
5	CNRM-ESM2-1	France	1.4° × 1.4°	r1i1p1f2
6	EC-Earth3-Veg-LR	Europe	1.1° × 1.1°	r1i1p1f1
7	GFDL-ESM4	USA	1.3° × 1°	r1i1p1f1
8	INM-CM4-8	Russia	2° × 1.5°	r1i1p1f1
9	INM-CM5-0	Russia	2° × 1.5°	r1i1p1f1
10	MPI-ESM1-2-LR	Germany	1.9° × 1.9°	r1i1p1f1
11	NorESM2-LM	Norway	2.5° × 1.9°	r1i1p1f1

2.3. Definition of Compound Extreme Temperature and Extreme Precipitation Events

The delineation and computational approaches for CEEs in this study followed the methodology set forth by Beniston et al. [20] and Hao et al. [21]. In this research, the 25th and 75th percentile quantiles of precipitation and temperature over 30 years at a standard climate calculation time were employed as the threshold criteria for delineating the conjoined extremes. The calculations of the 25th/75th percentiles excluded the value of zero. Compound hot extremes and heavy precipitation (HE-HP) events, characterized by daily maximum temperatures exceeding the 75th percentile and daily precipitation higher than the 75th percentile, were delineated in this study. Compound heat extremity and dry (HE-D) events are characterized by daily maximum temperatures exceeding the 75th percentile and a lack of rainfall for a continuous seven-day period. In this study, 'extreme' refers to a moderate deviation from the mean that exceeds the 75th percentile or falls below the 25th percentile.

3. Results

3.1. Observed Changes in CEEs

3.1.1. Annual Variations in Compound Events

The annual frequency and changing trends of CEEs at each weather station in NWC over the past 60 years were analyzed based on in situ datasets. As shown in Figure 2c, both the HE-D and HE-HP events showed a significant trend of increase from 1961 to 2020. The average annual HE-D events increased by 1.6/10a, whereas the average annual HE-HP events increased by 0.60/10a. Notably, the average annual HE-D events increased significantly more than the HE-HP events. The average annual HE-D events were considerably more frequent than HE-HP events, with annual HE-HP events occurring on average 5–10 times and annual HE-D events occurring on average 10–20 times.

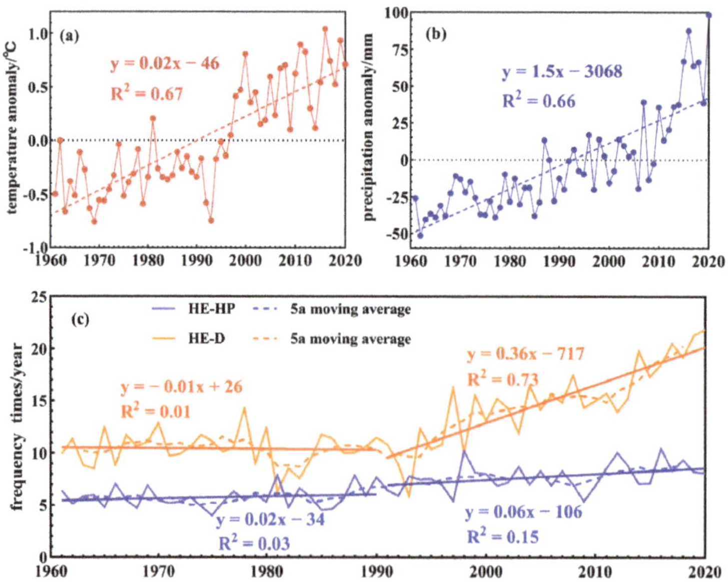

Figure 2. Temporal changes in temperature anomalies (**a**), precipitation anomalies (**b**), and CEES (**c**) in NWC during 1961–2020. Orange in (c) indicates HE-D events and blue indicates HE-HP events.

Between 1961 and 2020, the year with the most HE-D events in NWC was 2020. According to the 2020 China Climate Bulletin, the summer temperature in 2020 showed a 0.5 °C positive anomaly based on the average summer temperature from 1961 to 2020. During this season, several regions experienced multiple high-temperature heat waves and prolonged periods of extreme heat. The number of days with high temperatures exceeded the multi-year average by 1.7 days. In northwestern Inner Mongolia and southern and eastern Xinjiang, high-temperature days amounted to 20–30 days, with eastern Xinjiang surpassing 30 days. The year 2020 was a dry year, with the precipitation in NWC less than 20–50% of the annual average [22]. Positive summer temperature anomalies contribute to increased evapotranspiration, which leads to moisture deficits. Furthermore, when the annual precipitation is significantly lower than the multi-year average, these factors collectively result in a substantially higher probability of drought in NWC. Additionally, the frequent occurrence of high-temperature heat waves significantly increases the probability of HE-D events. The aforementioned factors may account for the high frequency of HE-D events observed in the NWC during 2020.

3.1.2. Spatial Distribution of Compound Events

As shown in Figure 3, the probability of HE-HP events at almost all stations was lower than that in NWC, except for several stations located in the Tian Shan Mountains. The HE-HP events at most stations have an average annual frequency of 5–10 and only the stations located in the western Central and Southern Tienshan Mountains and western Junggar Basin had more than 15 HE-HP events on average annually. However, most stations located on the southern margin of the Tarim Basin had a lower HE-HP frequency, of <5 (Figure 3a). Almost all stations had more than 10 HE-D events, the number was especially high in eastern NWC and the southern margin of the Taklamakan Desert, with more than 20 events (Figure 3b).

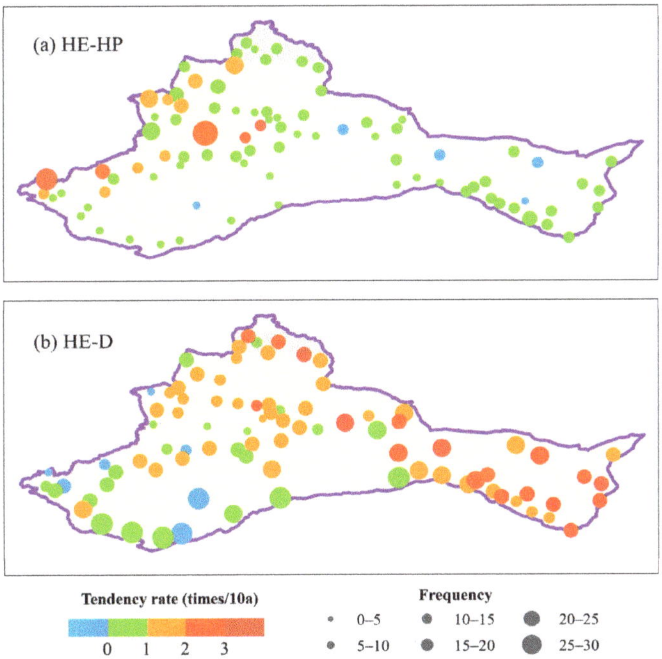

Figure 3. Spatial distribution of the average annual frequency and changing trends of HE-HP events (**a**) and HE-D events (**b**) at all stations within the studied region, from 1961 to 2020.

On the spatial scale, different stations exhibited different patterns of change. However, HE-HP and HE-D events showed trends of increase at almost all stations from 1961 to 2020. For HE-HP events, only five stations, located in eastern NWC and the central Tarim Basin, showed a trend to decrease. Although HE-HP events at the other stations exhibited a rising pattern, the increases at most sites were not significant (approximately 1/10a). As for HE-HP events, only several stations located in western NWC showed a significant trend of increase (>2/10a). The greatest rates of increase were observed at the Animal Husbandry Meteorological Experiment Station of Urumqi and the Turgart station, which are located in the central Tian Shan Mountains and Pamir, respectively (Figure 3a). For HE-D events (Figure 3b), all stations exhibited a rising pattern, except for six stations in western NWC; however, the rising trend was not significant in the surrounding regions of the Taklamakan Desert (<1/10a). Additionally, the rising trend of HE-D events in eastern NWC (almost all stations > 2/10a) was greater than that found in western NWC (almost all stations < 2/10a).

3.2. Projections of Compound Weather Extreme Events

3.2.1. Annual Variations in Compound Events

Under the SSP126, SSP245, SSP370, and SSP585 scenarios, the average annual frequency of HE-D events showed a trend to increase between 2020 and 2100 (Figure 4a), which is consistent with the findings of the observational data in the previous section. Under SSP126, the mean annual frequency of HE-D events decreased, whereas under SSP245, SSP370, and SSP585, the mean annual frequency of HE-D events increased. Under SSP126, the average annual frequency of HE-D events increased by 0.60/10a from 2015 to 2100, compared to 0.70/10a under SSP245, 1.50/10a under SSP370, and 1.70/10a under SSP585.

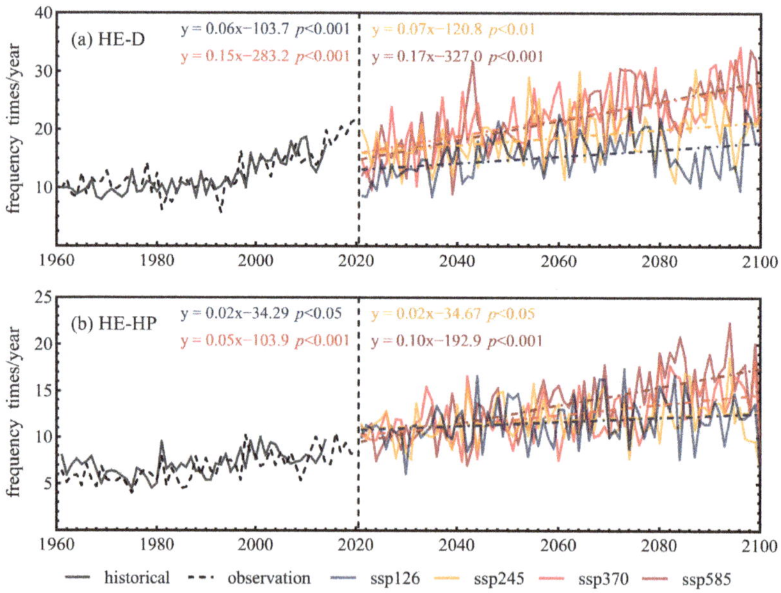

Figure 4. Multi-model ensemble (MME) average annual frequency of HE-D events (**a**) and HE-HP events (**b**), 1961–2100. The black dashed line represents changes in CMIP6 historical data, while the gray, blue, orange, red, and burgundy solid lines represent changes in measured data under the ssp126 scenario, the ssp245 scenario, the ssp370 scenario, and the ssp585 scenario, respectively. The colored dotted lines are trend lines.

Under the four scenarios, there was a trend of increase in the yearly mean occurrence rate of HE-HP events between 2015 and 2100 (Figure 4b). This agrees with the observational data presented in the previous section. The trend in the mean annual frequency of HE-HP events was smaller under SSP126 and SSP245 than under SSP370 and SSP585. Under SSP126, the average annual growth rate of the average annual frequency of HE-HP events between 2015 and 2100 was 0.22/10a, compared to 0.23/10a under SSP245, 0.50/10a under SSP370, and 1.00/10a under SSP585.

3.2.2. Spatial Distribution of Compound Events

The study analyzed the spatial distribution characteristics of future CEEs in NWC using three time frames: near future (2021–2050), mid-future (2051–2080), and far future (2081–2100).

In terms of spatial distribution, HE-D events were more distributed in the northern NWC (Figure 5). A gradual increase in CEEs occurred in the central NWC from 2021 to 2100.

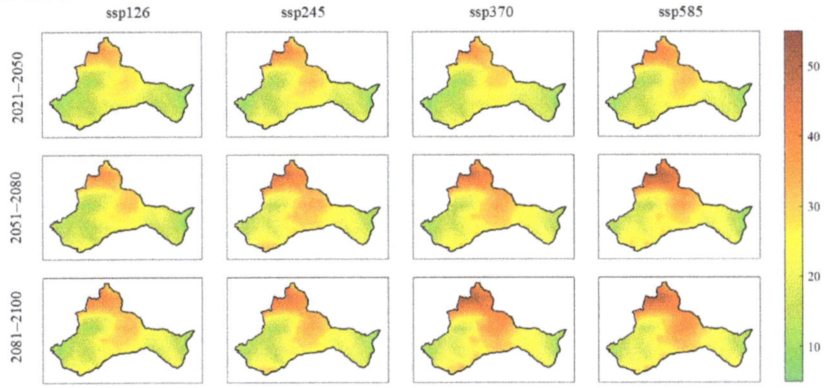

Figure 5. Spatial distribution of annual mean frequency of HE-D events under four scenarios.

The CMIP6 data exhibited a visible rise in the occurrence of HE-D events from 2051 to 2080, compared with from 2021 to 2050, under SSP126, SSP245, SSP370, and SSP585. However, there is no significant increase in the frequency of HE-D events between 2081 and 2100 compared to 2051–2080, indicating a steady state under the SSP126 and SSP245 scenarios. Under SSP370 and SSP585, there was a small increase in the frequency of HE-D events from 2081 to 2100 compared to from 2051 to 2080. This finding concurs with previous research indicating that episodes of extreme heat and heavy precipitation have grown more frequent since the year 1950 [23].

The frequency of HE-D events has an insignificant change between 2021 and 2050 under four scenarios, and there is only a small increase observed from SSP126 to SSP245 and from SSP370 to SSP585. From 2051 to 2080, and 2081 to 2100, there was a more pronounced increase in the recurrence of HE-D events when comparing SSP126 to SSP245 and SSP370 to SSP585.

In terms of spatial distribution, HE-HP events were more prevalent in the eastern and western areas of NWC (Figure 6). The CMIP6 data indicate that there will be a modest increase in HE-HP events from 2021 to 2100 under SSP126 and SSP245, but a significant increase in HE-HP events under SSP370 and SSP585.

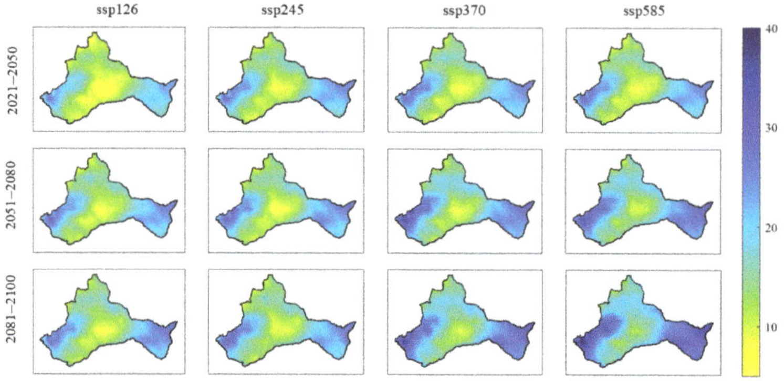

Figure 6. Spatial distribution of the annual mean frequency of HE-HP events under four scenarios.

From 2021 to 2050, the frequency of HE-HP events increased more significantly under SSP245 than under SSP126. However, the frequency of HE-HP events tended to be steady under SSP245, SSP370, and SSP585. From 2051 to 2080, the frequency of HE-HP events became more stable under SSP126 and SSP245, as it was under SSP370 and SSP585 and the rise in the recurrence of HE-HP events was more pronounced under SSP370 and SSP585 than under SSP126 and SSP245. Across SSP126, SSP245, SSP370 and SSP585, the average annual frequency of HE-HP events increases more significantly throughout NWC between 2081 and 2100, with the most notable rises in the east and west. HE-HP events spread from east and west to the center.

The frequency of HE-HP events may be increased in the periods 2021–2050 and 2051–2080 under the SSP126 and SSP370 scenarios. Under SSP245, the frequency of HE-HP events remains steady. After 2050, the frequency of HE-HP events under SSP126 and SSP370 will stabilize. This finding aligns with the outcomes documented by Sun et al. [24]. Under SSP585, the frequency of HE-HP events continues to increase with time.

4. Discussion

4.1. Influence Factor for the Warmer and Wetter Trend in the NWC

Several studies have confirmed that global warming plays a crucial role in the generation and development of extreme events and CEEs [9,25,26]. It is generally acknowledged that high levels of greenhouse gas emissions are a vital cause of global warming, which is also the case in NWC [27]. In addition, the temperatures in NWC have been influenced by atmospheric circulation anomalies. The intensity, weakness, and fluctuations of the subtropical high-pressure system restrain the mid-latitude climate. The placement and strength of the subtropical high-pressure system can affect precipitation patterns, whereas the intensity of the polar vortex is related to global temperature changes. For example, extreme temperature changes in NWC are closely related to the Arctic Oscillation [28], which can cause unusual easterly winds [29], and the Pacific Decadal Oscillation has an important implication for humidity in NWC [30]. The emergence of EI Niño results in deviations in global climate, serving as a significant driver behind droughts and floods in China [30]. Accordingly, the present research effort computed correlations to examine the implications of each circulation factor for temperature and precipitation patterns. Second, the various factors affecting CEEs have only been qualitatively analyzed; a quantitative assessment of the magnitude of influence which each determinant exerts on the recurrence of such events is lacking, particularly as to the influence of human activities (Table S1).

Table 2 shows a notable difference in the correlations between the circulation factors and the maximum temperature. The correlations of the NAS and the NH with the maximum temperature were greater than that with NANA and NHA. The maximum temperatures exhibited the strongest response to NH influences, which was followed by NAS, and finally NANA and NHA. The associated series plots in Figure 7c,d exhibit a robust negative correlation between the NH/ NHA and the maximum temperature in NWC. The correlation coefficient between the NH and the maximum temperature in NWC was −0.87 and that between the NHA and the maximum temperature in NWC was −0.84, both passing the 99% confidence test. The associated series plots in Figure 7a,b exhibit a robust positive correlations of the NAS and NANA with the maximum temperature in NWC. The correlation coefficient between the NAS and the maximum temperature in NWC was 0.86 and that between the NANA and the maximum temperature in NWC was 0.84, both passing the 99% confidence test. In NWC, the maximum temperatures show a trend to decrease with an increase in NH/NHA and a trend to increase with an increase in NAS/NANA.

Table 2 shows significant differences in the correlations between circulation factors and precipitation. The correlations of ESA and ES with precipitation were found to be stronger than those with NAM and PS. The corresponding series plots in Figure 8 show strong positive correlations of ESA (Figure 8a), ES (Figure 8b), NAM (Figure 8c), and PS (Figure 8d) with precipitation in the NWC; precipitation shows the strongest response to the ESA and ES influences, followed by the NAM and PS influences, at 0.97, 0.97, 0.96, and 0.96, all passing the 99% confidence test. In NWC, precipitation exhibits an increasing pattern with rising ESA/ES/NAM/PS values.

Table 2. The correlation between temperature or precipitation and the Atmospheric Circulation Index.

Atmospheric Circulation Index	R^2	
	Temperature	Precipitation
North Atlantic Subtropical High Intensity Index (NAS)	0.74 **	0.72 **
North American-North Atlantic Subtropical High Intensity Index (NANA)	0.71 **	0.86 **
Northern Hemisphere Polar Vortex Area Index (NHA)	0.71 **	0.74 **
Northern Hemisphere Polar Vortex Intensity Index (NH)	0.76 **	0.68 **
Eastern Pacific Subtropical High Area Index (ESA)	0.64 **	0.94 **
Eastern Pacific Subtropical High Intensity Index (ES)	0.61 **	0.94 **
North American Subtropical High Intensity Index (NAM)	0.65 **	0.91 **
Pacific Subtropical High Intensity Index (PS)	0.64 **	0.93 **

** represents a statistically significant result at the 0.99 confidence level.

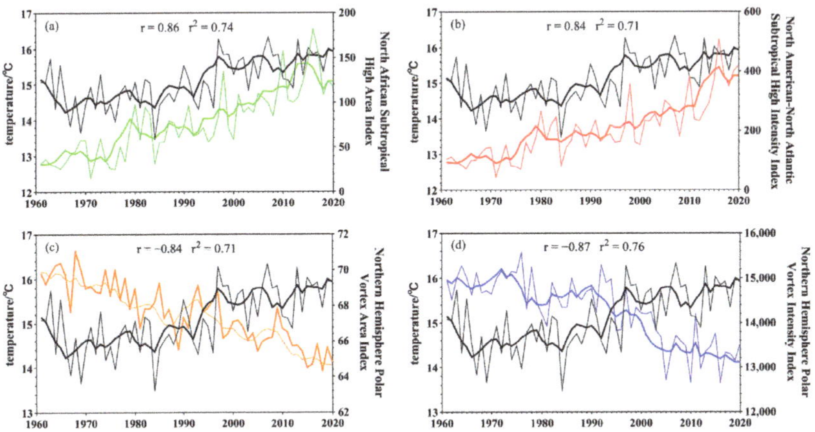

Figure 7. Relationship between temperature and the four Atmospheric Circulation Index. The black (or colored) in (**a**–**d**) indicates temperature (or the Atmospheric Circulation Index) and its five-year moving average.

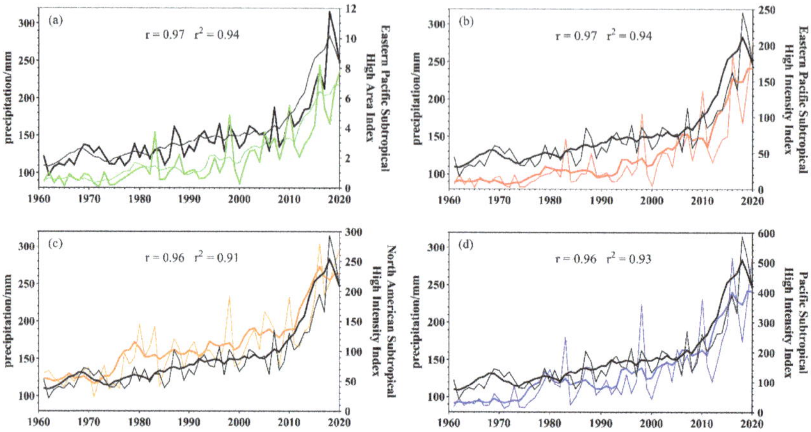

Figure 8. Relationship between precipitation and the four Atmospheric Circulation Index. The black (or colored) in (**a–d**) indicates precipitation (or the Atmospheric Circulation Index) and its five-year moving average.

4.2. Driving Mechanism for HE-D Events

Under global warming, the probability of HE-D events in NWC has substantially increased because of the influence of atmospheric circulation anomalies and land–air coupling processes [31]. Byrne et al. [32] confirmed the mechanism of "the drier the hotter", and the decreased soil water content leads to the increased probability of warming. Increased temperatures and net radiation accelerate soil evaporation and plant transpiration. This reduces soil water content, which, in turn, inhibits soil evaporation and plant evapotranspiration. Soil moisture decreased significantly from 1980 to 2010 and has increased since 2010 due to increased precipitation and irrigation (Figures 2b and 9) [33,34]. Increased soil–air latent heat interactions lead to decreased soil moisture. However, decreased soil moisture leads to increased soil–air sensible-heat interactions. In Figures 5 and 10, regardless of the scenario, the trend of the latent heat change is similar to that of the HE-D event change. In regions where HE-D events are increasing, latent heat is also increasing, especially in the northern region of NWC. The region of change for latent heat is distinct from that of the HE-HP events, whereas the trend of sensible-heat change mirrors that of the HE-HP events. Latent heat exhibits a clear upward trend in the eastern and western regions of the NWC, whereas the upward trend is less pronounced in the central and northern regions of the NWC. (Figure 11). Therefore, the soil–atmosphere latent heat exchange increases, and this leads to a further decrease in soil moisture [35,36]. Moreover, increased soil–atmosphere sensible-heat exchange results in a warmer atmospheric environment and a thickened atmospheric boundary layer [37–39]. The water vapor pressure deficit in the atmospheric boundary layer increases, which leads to a further soil moisture deficit [40]. This procedure continues until the evening. The land–atmosphere interaction feedback mechanism under global warming serves as a significant driver of HE-D events in the NWC (Figure 12).

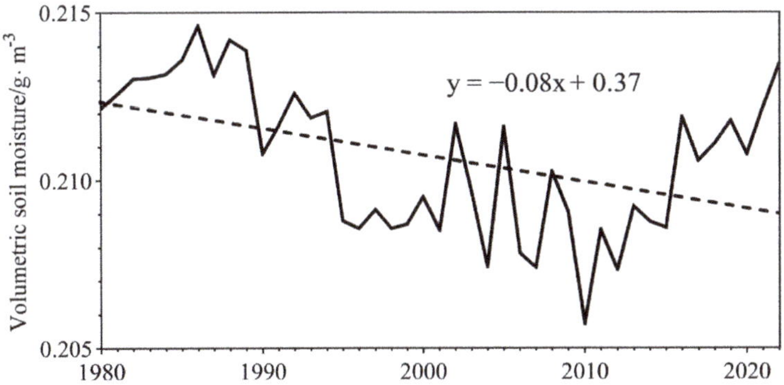

Figure 9. Soil moisture variation in NWC from 1980 to 2022, based on ERA5 reanalysis data.

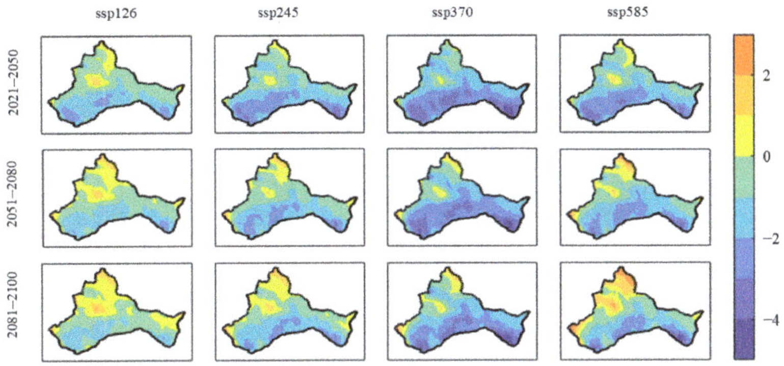

Figure 10. Differences in surface upward latent heat flux (unit: W·m^{-2}) between 2021 and 2100, and 1961 and 1990, across four scenarios.

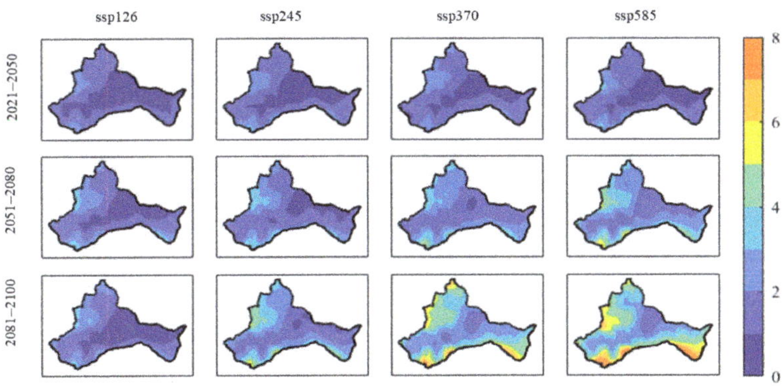

Figure 11. Difference in surface upward sensible-heat flux (unit: W·m^{-2}) between 2021 and 2100, and 1961 and 1990, across four scenarios.

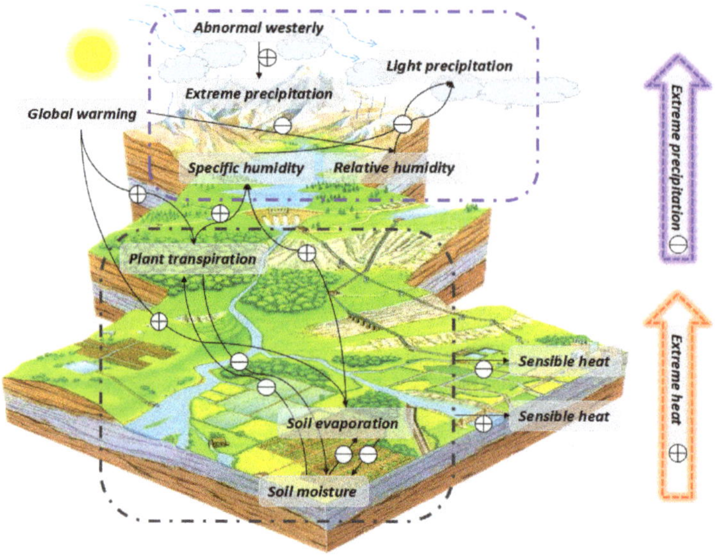

Figure 12. Schematic diagram of the driving mechanism of HE-D and HE-HP events in NWC under global warming (Adapted from: https://mp.weixin.qq.com/s/gcidw2ysZkcx1QeIGgLMlw, accessed on 27 June 2024).

4.3. Driving Mechanism of HE-HP Events

Located at mid-to-high latitudes, NWC experiences atmospheric water vapor transport that is impacted by both the circulation of westerly winds and the southwest monsoon circulation, as described by [41]. The water vapor transport within the westerly wind belt furnishes the fundamental source of moisture for the majority of NWC; alterations in the westerly winds directly influence the interannual variability of water vapor transport, and high-altitude jets shift northward and eastward [42], which has shown a trend of increase in recent years.

Multiple studies have confirmed trends of increasing temperature and precipitation in NWC; however, further research indicates that the warmer trend is more pronounced than the wetter trend. In other words, the increased temperature was more significant than the increased precipitation, which was confirmed by the changing trends of specific humidity and relative humidity in the future (as shown in Figures 13 and 14). According to Figure 13, the specific humidity change trends in NWC show increases from 2021 to 2100 under all scenarios, showing a more pronounced rate of growth in the western and eastern regions of NWC compared to the central areas. However, according to Figure 14, the projection of the relative humidity change trends of SSP126–585 are generally indicative of decrease in NWC. However, different spatial patterns were observed in different scenarios. Under SSP126 and SSP370, the central region of NWC will exhibit a persistent trend of increasing specific humidity until approximately 2050. In contrast, other areas continue to experience a significant drying trend until 2100, particularly in the northern and western areas. Under SSP585, the central region exhibits an insignificant wetting trend, whereas the western and northern regions display pronounced drying trends. This drying trend does not contradict the rise in extreme precipitation. Although the relative humidity of the air decreases, the moisture content of the air increases. This leads to an increase in the vapor pressure deficit of air. In addition, an increase in the vapor pressure deficit of air leads to a situation in which water vapor is less likely to condense and precipitate. Moreover, the increased specific humidity and the rise in the vapor pressure deficit are unfavorable for the occurrence of light precipitation events [43–45].

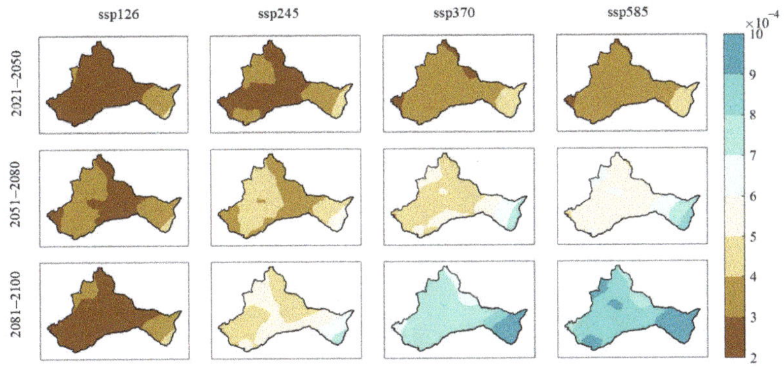

Figure 13. Differences in specific humidity (unit: $g \cdot kg^{-1}$) between 2020 and 2100 (divided into three time periods: 2021–2050, 2051–2080, and 2081–2100), and 1961–1990, across four scenarios.

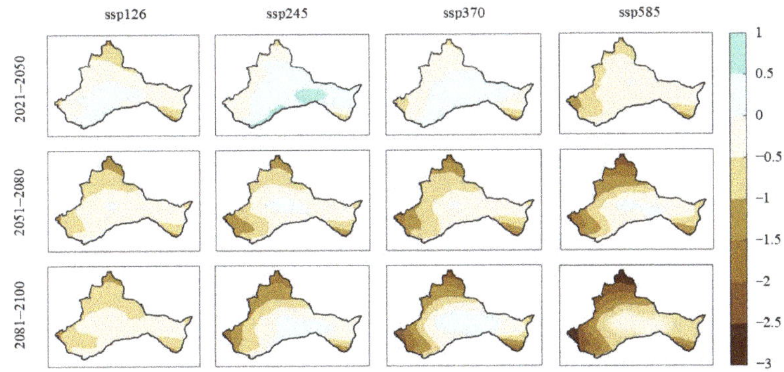

Figure 14. Differences in relative humidity between 2020 and 2100 (divided into three time periods: 2021–2050, 2051–2080, and 2081–2100), and 1961–1990, across four scenarios.

Since the 1980s, the NWC climate has transitioned from warm and dry to warm and wet [46,47]. There is growing evidence that the interdecadal increase in westerly circulation and the longitudinal shift of subtropical westerly jets have resulted in increased transportation of water vapor, which likely explains the increased precipitation [48].

However, these studies merely illuminate the causative factors behind the rise in precipitation levels within NWC, without elucidating the underlying reasons for the escalation in instances of extreme precipitation. The observational data shows that, compared with the average from 1961 to 1990, the precipitation in NWC generally showed a trend of increase, especially in July, August, and September, contributing approximately 47.71% of the total precipitation increment (Figure 15). In addition, current research findings indicate that the quantity of precipitation days and instances of light precipitation in NWC exhibit trends of decrease [49–51]. The increase in the vapor pressure deficit, which results in water vapor that is less likely to condense, may be one of the reasons for this (Figures 13 and 14). Furthermore, the significant negative phase of the Silk Road teleconnection wave train induces a longitudinal displacement of the Asian jet stream. Consequently, this has led to an interdecadal anomaly of summer precipitation in the eastern part of NWC since 1998 [52]. Figure 16 presents the anomaly maps of the 500 hpa atmospheric circulation patterns between 1991 and 2020, and 1961 and 1990. Wind anomalies in the low troposphere exhibited significant cyclonic circulation on both sides of NWC. The anomalous cyclone on the eastern side facilitated the transport of water vapor from the western North Pacific to NWC. Meanwhile, the instability between the Mongolian High and Siberian Low induces anomalous cyclones over NWC, promoting the anomalous intensification of westerly and easterly airflows and leading to the generation of extreme precipitation [53]. As a consequence, there was a significant rise in the recurrence of HE-HP events (Figure 12).

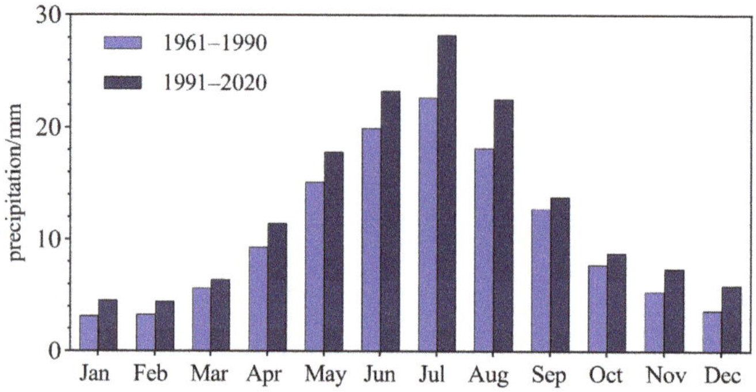

Figure 15. Mean monthly precipitation in NWC during different periods.

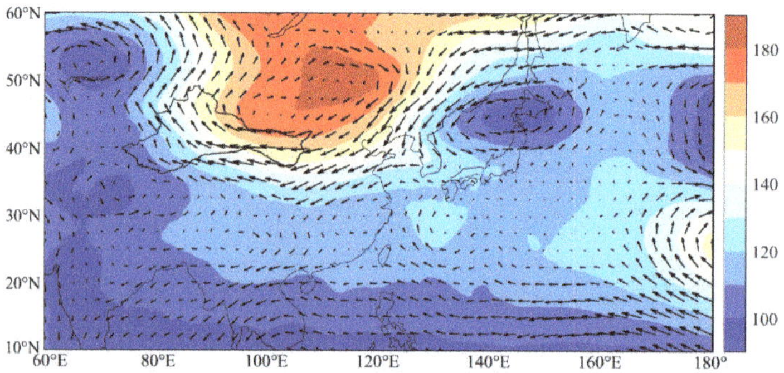

Figure 16. Overlay of 500-hPa geopotential height (unit: gpm) and winds between 1991 and 2020, and 1961 and 1990.

4.4. Limitations and Outlooks

First, the definitions based on the 25th/75th percentile threshold were slightly subjective. The relative threshold method is a widely used approach that employs a specified extreme index and 90th/95th percentile for individual extreme occurrences [54]. Nevertheless, there is disagreement concerning the determination of thresholds for composite events, with some scholars retaining the 90th/95th percentile from an individual extreme index [55,56] and others opting for the 25th/75th percentile [31]. Nevertheless, Hao et al. [21] discovered that combined precipitation–temperature statistical metrics exhibit relative insensitivity to the selection of quantile thresholds. Their research revealed consistent trends in extreme events, even with an alternative 10th/90th percentile threshold. Although different threshold combinations yielded essentially the same trends in the results, the degree of detail varied [57]. An important future direction for CEEs is to find a way to define them accurately. There is also a need to strengthen research on the linkages between extreme events to assess future trends in the regularity and attributes of extreme compound events [10].

Secondly, while the multitude of elements impacting compound extreme occurrences have undergone qualitative examination, there lacks a quantitative assessment of the magnitude of influence that each determinant exerts on the recurrence of such events, notably the impact of anthropogenic actions. Human activities are a significant driver of climate change. Determining the precise impact of anthropogenic activities on regional climate change is a priority for future research [58,59].

Third, the results have large uncertainties in the study area because of the complex terrain and climatic conditions, sparse and non-uniform distribution of meteorological stations, poor representativeness, and poor resolution of the model. In addition, inter-model uncertainties remain high in precipitation and temperature projections [60,61]. CMIP6 Models can overestimate annual precipitation in drylands by approximately 33%, compared with observations [62]. CMIP6 projections overestimate warming in China [63]. This leads to uncertainty in the projection of compound temperature and precipitation events.

Fourth, water scarcity is a primary factor contributing to the vulnerability of the arid ecosystem. The CEEs have resulted in an increasingly imbalanced distribution of water resources under climate change, which has had profound impacts on arid ecosystems [64–66]. In the future, as the frequency of CEEs continues to rise, their impacts are expected to become even more pronounced. Additionally, the limited recovery capacity is a key characteristic of the arid ecosystem, and it is difficult to regain its prior state once damaged. Therefore, enhancing the accuracy of CEE prediction and, specifically, their likely effects on arid ecosystems, based on new technologies and methodologies, may represent a promising research field in the future.

Ultimately, these aspects pose certain challenges for disaster prediction over the next 30–50 years. Future studies should focus on a mechanistic study of climate warming and wetting in the NWC, quantitatively analyze the impacts of various influencing factors, and develop appropriate risk response strategies. In addition, by establishing more meteorological stations, refining observation methodologies, and harnessing sophisticated modeling approaches, the precision of future forecasts for complex extreme weather phenomena can be increased.

5. Conclusions

We used different threshold combinations to construct two distinct sets of event networks in the northwest region. The principal findings are summarized as follows.

Over the past 60 years, the frequency of CEEs in NWC has increased. Nevertheless, the linear trend of HE-D events is larger than that of HE-HP events, reaching 1.60 times/decade, while the HE-HP events occur at the rate of 0.60 times/decade. The CEEs in NWC varied abruptly around 1990, which aligns fundamentally with the hypothesis that the climate of NWC underwent a transformation commencing from 1987. Spatial differences between HE-D and HE-HP events in NWC have been prominent, particularly in eastern NWC, specifically as to the sites of stations recording the peak and nadir frequency levels of HE-D and HE-HP events. However, CEEs exhibited a rising tendency across nearly all stations. A pronounced trend of increase in CEEs was observed from 2021 to 2100. The linear trend of SSP585 was the largest, reaching 1.70 times/10a (HE-D events) and 1.00 times/10a (HE-HP events), followed by SSP370. SSP126 and SSP245 did not show a significant linear trend. From 2020 to 2100, the spatial distribution of HE-D events showed a trend of "higher in the north and lower in the south". The space allocation pattern of HE-HP events showed "more in the east and west, less in the middle".

Warming increases soil evaporation and plant transpiration, resulting in decreased soil moisture content. Simultaneously, soil moisture deficiency suppresses soil evaporation and plant transpiration, resulting in a reduction in latent heat and an augmentation in sensible heat, thereby enhancing the heating efficiency of the underlying surface relative to the atmosphere. Moreover, soil moisture evaporates quickly at higher temperatures. These feedback processes contribute to the sustained intensification of HE-D events. The rise in temperature in NWC led to an augmentation in evapotranspiration, leading to a reduction in soil moisture content and a reduction in atmospheric-soil latent heat exchange. This soil warming contributes to the frequent occurrence of HE-D events. Anomalies in atmospheric circulation and the land–atmosphere feedback mechanism under the background of climate warming are significant drivers of HE-D events in NWC.

Commencing in the 1980s, the climate of NWC has undergone a transformation from a warm and dry state to a warm and wet state. However, the warmer trend was stronger than the wetter trend. The warmer trend being greater than that of the wetter trend reduced the frequency of light rain events. Concurrently, the anomalous intensification of westerly and easterly winds has further concentrated the precipitation, leading to the generation of extreme precipitation, which triggers HE-HP events.

Supplementary Materials: The following supporting information can be downloaded at: https://www.mdpi.com/article/10.3390/rs16173111/s1, Table S1: Relationship between the temperature, precipitation and the Atmospheric Circulation Index. Table S2: Symbols and Abbreviations Used in the Manuscript. Figure S1: Performance of the 22 CMIP6 models and their MME (solid black dot) on simulating the observed monthly mean air temperature over the NWC during 1961–2015. The vertical and horizontal coordinates in Figure S1 are PCC and NRMSE, respectively. The horizontal gray line in Figure S1 indicates the PCC greater than 0.5. The vertical gray line in Figure S1 indicates the mean value of the NRMSE of 22 models. The hollow black square in Figure S1 depicts the MME of the se-lected best 11 models enclosed in the upper left quadrant.

Author Contributions: Conceptualization, W.S., R.C. and C.H.; methodology, H.N.; software, Y.W. (Yuzhe Wang); validation, L.W.; formal analysis, B.H.; investigation, J.Z.; resources, R.C. and W.S.; data curation, H.N.; writing—original draft preparation, H.N.; writing—review and editing, L.W. and B.H.; visualization, H.N.; supervision, Y.W. (Yuzhe Wang); project administration, Y.W. (Yingshan Wang); funding acquisition, L.W., B.H. and W.S. All authors have read and agreed to the published version of the manuscript.

Funding: The research was funded by the National Natural Sciences Foundation of China, Nos. 42101120, 42171121, and 42271145; the Support Plan on Science and Technology for Youth Innovation of Universities in Shandong Province, No. 2023KJ195; and the National Natural Science Foundation of Shandong Province, No. ZR2021QD138; and the Taishan Scholars Program of Shandong Province, No. tsqn202312158.

Data Availability Statement: The original contributions presented in the study are included in the article/Supplementary Material; further inquiries can be directed to the corresponding author/s.

Conflicts of Interest: The authors declare no conflicts of interest.

References

1. IPCC. AR6 Climate Change 2021: The Physical Science Basis. 2021. Available online: https://www.ipcc.ch (accessed on 8 May 2024).
2. Martius, O.; Pfahl, S.; Chevalier, C. A Global Quantification of Compound Precipitation and Wind Extremes. *Geophys. Res. Lett.* **2016**, *43*, 7709–7717. [CrossRef]
3. Yu, R.; Zhai, P. Changes in Compound Drought and Hot Extreme Events in Summer over Populated Eastern China. *Weather Clim. Extrem.* **2020**, *30*, 100295. [CrossRef]
4. Zscheischler, J.; Westra, S.; Van Den Hurk, B.J.J.M.; Seneviratne, S.I.; Ward, P.J.; Pitman, A.; AghaKouchak, A.; Bresch, D.N.; Leonard, M.; Wahl, T.; et al. Future Climate Risk from Compound Events. *Nat. Clim. Chang.* **2018**, *8*, 469–477. [CrossRef]
5. Zscheischler, J.; Martius, O.; Westra, S.; Bevacqua, E.; Raymond, C.; Horton, R.M.; Van Den Hurk, B.; AghaKouchak, A.; Jézéquel, A.; Mahecha, M.D.; et al. A Typology of Compound Weather and Climate Events. *Nat. Rev. Earth Environ.* **2020**, *1*, 333–347. [CrossRef]
6. Yu, R.; Zhai, P. Advances in scientific understanding on compound extreme events. *Trans. Atmos. Sci.* **2021**, *44*, 645–649. [CrossRef]
7. Peng, T.; Zhao, L.; Zhang, L.; Shen, X.; Ding, Y.; Wang, J.; Li, Q.; Liu, Y.; Hu, Y.; Ling, J.; et al. Changes in Temperature-Precipitation Compound Extreme Events in China during the Past 119 Years. *Earth Space Sci.* **2023**, *10*, e2022EA002777. [CrossRef]
8. Stott, P. How Climate Change Affects Extreme Weather Events. *Science* **2016**, *352*, 1517–1518. [CrossRef] [PubMed]
9. Alizadeh, M.R.; Adamowski, J.; Nikoo, M.R.; AghaKouchak, A.; Dennison, P.; Sadegh, M. A Century of Observations Reveals Increasing Likelihood of Continental-Scale Compound Dry-Hot Extremes. *Sci. Adv.* **2020**, *6*, eaaz4571. [CrossRef] [PubMed]
10. Bevacqua, E.; Zappa, G.; Lehner, F.; Zscheischler, J. Precipitation Trends Determine Future Occurrences of Compound Hot–Dry Events. *Nat. Clim. Chang.* **2022**, *12*, 350–355. [CrossRef]
11. Allan, R.P.; Willett, K.M.; John, V.O.; Trent, T. Global Changes in Water Vapor 1979–2020. *JGR Atmos.* **2022**, *127*, e2022JD036728. [CrossRef]
12. Ren, D.; Wang, Y.; Wang, G.; Liu, L. Rising Trends of Global Precipitable Water Vapor and Its Correlation with Flood Frequency. *Geod. Geodyn.* **2023**, *14*, 355–367. [CrossRef]
13. Zhang, J.; Zhao, T.; Dai, A.; Zhang, W. Detection and Attribution of Atmospheric Precipitable Water Changes since the 1970s over China. *Sci. Rep.* **2019**, *9*, 17609. [CrossRef]

14. Hoell, A.; Quan, X.-W.; Hoerling, M.; Diaz, H.F.; Fu, R.; He, C.; Lisonbee, J.R.; Mankin, J.S.; Seager, R.; Sheffield, A.; et al. Water Year 2021 Compound Precipitation and Temperature Extremes in California and Nevada. *Bull. Am. Meteorol. Soc.* **2022**, *103*, E2905–E2911. [CrossRef]
15. Yin, J.; Gentine, P.; Slater, L.; Gu, L.; Pokhrel, Y.; Hanasaki, N.; Guo, S.; Xiong, L.; Schlenker, W. Future Socio-Ecosystem Productivity Threatened by Compound Drought–Heatwave Events. *Nat. Sustain.* **2023**, *6*, 259–272. [CrossRef]
16. Zhang, Y.; Mao, G.; Chen, C.; Lu, Z.; Luo, Z.; Zhou, W. Population Exposure to Concurrent Daytime and Nighttime Heatwaves in Huai River Basin, China. *Sustain. Cities Soc.* **2020**, *61*, 102309. [CrossRef]
17. Xing, Z.; Yu, Z.; Wei, J.; Zhang, X.; Ma, M.; Yi, P.; Ju, Q.; Wang, J.; Laux, P.; Kunstmann, H. Lagged Influence of ENSO Regimes on Droughts over the Poyang Lake Basin, China. *Atmos. Res.* **2022**, *275*, 106218. [CrossRef]
18. Wang, X.; Li, Y.; Chen, Y.; Li, Y.; Wang, C.; Kaldybayev, A.; Gou, R.; Luo, M.; Duan, W. Intensification of Heatwaves in Central Asia from 1981 to 2020—Role of Soil Moisture Reduction. *J. Hydrol.* **2023**, *627*, 130395. [CrossRef]
19. Ma, Z.; Sun, P.; Zhang, Q.; Zou, Y.; Lv, Y.; Li, H.; Chen, D. The Characteristics and Evaluation of Future Droughts across China through the CMIP6 Multi-Model Ensemble. *Remote Sens.* **2022**, *14*, 1097. [CrossRef]
20. Beniston, M. Trends in Joint Quantiles of Temperature and Precipitation in Europe since 1901 and Projected for 2100: Joint heat and moisture extremes. *Geophys. Res. Lett.* **2009**, *36*, L07707. [CrossRef]
21. Hao, Z.; AghaKouchak, A.; Phillips, T.J. Changes in Concurrent Monthly Precipitation and Temperature Extremes. *Environ. Res. Lett.* **2013**, *8*, 034014. [CrossRef]
22. Ma, Q.; Li, Z.; Chen, Z.; Su, T.; Wu, Y.; Feng, G. Moisture Changes with Increasing Summer Precipitation in Qilian and Tienshan Mountainous Areas. *Atmos. Sci. Lett.* **2023**, *24*, e1154. [CrossRef]
23. Easterling, D.R.; Meehl, G.A.; Parmesan, C.; Changnon, S.A.; Karl, T.R.; Mearns, L.O. Climate Extremes: Observations, Modeling, and Impacts. *Science* **2000**, *289*, 2068–2074. [CrossRef]
24. Sun, P.; Zou, Y.; Yao, R.; Ma, Z.; Bian, Y.; Ge, C.; Lv, Y. Compound and Successive Events of Extreme Precipitation and Extreme Runoff under Heatwaves Based on CMIP6 Models. *Sci. Total Environ.* **2023**, *878*, 162980. [CrossRef] [PubMed]
25. Xu, X.; Huang, A.; Huang, D.; Zhang, Y.; Gu, C.; Cai, S.; Tang, Y.; Zhao, Z.; Zeng, J. What Are the Dominant Synoptic Patterns Leading to the Summer Regional Hourly Extreme Precipitation Events over Central-Eastern Tibetan Plateau and Sichuan Basin? *Geophys. Res. Lett.* **2023**, *50*, e2022GL102342. [CrossRef]
26. Zhang, X.; Chen, Y.; Fang, G.; Xia, Z.; Li, Z.; Duan, W.; Xia, Q. Increasing Multiscale Variability in Extreme Precipitation under Global Warming in the Tienshan Mountains, Central Asia. *J. Hydrol.* **2023**, *626*, 130158. [CrossRef]
27. Ham, Y.-G.; Kim, J.-H.; Min, S.-K.; Kim, D.; Li, T.; Timmermann, A.; Stuecker, M.F. Anthropogenic Fingerprints in Daily Precipitation Revealed by Deep Learning. *Nature* **2023**, *622*, 301–307. [CrossRef] [PubMed]
28. Liu, X.; Xu, Z. Spatial and Temporal Pattern of Extreme Temperature during 1961–2018 in China. *J. Water Clim. Chang.* **2020**, *11*, 1633–1644. [CrossRef]
29. Zhang, J.; Li, D.; Li, L.; Deng, W. Decadal Variability of Droughts and Floods in the Yellow River Basin during the Last Five Centuries and Relations with the North Atlantic SST. *Int. J. Climatol.* **2013**, *33*, 3217–3228. [CrossRef]
30. Ouyang, R.; Liu, W.; Fu, G.; Liu, C.; Hu, L.; Wang, H. Linkages between ENSO/PDO Signals and Precipitation, Streamflow in China during the Last 100 Years. *Hydrol. Earth Syst. Sci.* **2014**, *18*, 3651–3661. [CrossRef]
31. Hao, Z.; Hao, F.; Xia, Y.; Feng, S.; Sun, C.; Zhang, X.; Fu, Y.; Hao, Y.; Zhang, Y.; Meng, Y. Compound Droughts and Hot Extremes: Characteristics, Drivers, Changes, and Impacts. *Earth-Sci. Rev.* **2022**, *235*, 104241. [CrossRef]
32. Byrne, M.P. Amplified Warming of Extreme Temperatures over Tropical Land. *Nat. Geosci.* **2021**, *14*, 837–841. [CrossRef]
33. Zhang, J.; Ma, P.; Wu, S.; Li, S.; Li, M.; Zhu, Y.; Xing, Y. Study on Characteristics of Soil Water Change and Law of Crop Water Consumption under Shallow Buried Drip Irrigation. *J. Soil Water Conserv.* **2013**, *37*, 111–118. [CrossRef]
34. Wang, B.; Yan, C.; Wang, Y.; Li, C.; Zhang, Q.; Qiu, G. Water Budget Characteristics of Over-Irrigated Oasis in Arid Region of Northwest China. *Acta Sci. Nat. Univ. Pekin* **2020**, *56*, 1122–1128. [CrossRef]
35. Jiang, K.; Pan, Z.; Pan, F.; Wang, J.; Han, G.; Song, Y.; Zhang, Z.; Huang, N.; Ma, S.; Chen, X. Influence Patterns of Soil Moisture Change on Surface-Air Temperature Difference under Different Climatic Background. *Sci. Total Environ.* **2022**, *822*, 153607. [CrossRef] [PubMed]
36. Qiao, L.; Zuo, Z.; Zhang, R.; Piao, S.; Xiao, D.; Zhang, K. Soil Moisture–Atmosphere Coupling Accelerates Global Warming. *Nat. Commun.* **2023**, *14*, 4908. [CrossRef] [PubMed]
37. García-García, A.; Cuesta-Valero, F.J.; Miralles, D.G.; Mahecha, M.D.; Quaas, J.; Reichstein, M.; Zscheischler, J.; Peng, J. Soil Heat Extremes Can Outpace Air Temperature Extremes. *Nat. Clim. Chang.* **2023**, *13*, 1237–1241. [CrossRef]
38. Liu, W.; Zhang, Q.; Li, C.; Xu, L.; Xiao, W. The Influence of Soil Moisture on Convective Activity: A Review. *Theor. Appl. Clim.* **2022**, *149*, 221–232. [CrossRef]
39. Xu, Z.; Chen, H.; Guo, J.; Zhang, W. Contrasting Effect of Soil Moisture on the Daytime Boundary Layer under Different Thermodynamic Conditions in Summer Over China. *Geophys. Res. Lett.* **2021**, *48*, e2020GL090989. [CrossRef]
40. Li, F.; Xiao, J.; Chen, J.; Ballantyne, A.; Jin, K.; Li, B.; Abraha, M.; John, R. Global Water Use Efficiency Saturation Due to Increased Vapor Pressure Deficit. *Science* **2023**, *381*, 672–677. [CrossRef]
41. Liu, Y.; Wu, C.; Jia, R.; Huang, J. An overview of the influence of atmospheric circulation on the climate in arid and semi-arid region of Central and East Asia. *Sci. China Earth Sci.* **2018**, *61*, 1183–1194. [CrossRef]

42. Xu, G.; Liu, X.; Trouet, V.; Treydte, K.; Wu, G.; Chen, T.; Sun, W.; An, W.; Wang, W.; Zeng, X.; et al. Regional Drought Shifts (1710–2010) in East Central Asia and Linkages with Atmospheric Circulation Recorded in Tree-Ring δ18O. *Clim. Dyn.* **2019**, *52*, 713–727. [CrossRef]
43. Dai, A.; Rasmussen, R.M.; Liu, C.; Ikeda, K.; Prein, A.F. A New Mechanism for Warm-Season Precipitation Response to Global Warming Based on Convection-Permitting Simulations. *Clim. Dyn.* **2020**, *55*, 343–368. [CrossRef]
44. Giorgi, F.; Im, E.-S.; Coppola, E.; Diffenbaugh, N.S.; Gao, X.J.; Mariotti, L.; Shi, Y. Higher Hydroclimatic Intensity with Global Warming. *J. Clim.* **2011**, *24*, 5309–5324. [CrossRef]
45. Yu, X.; Gu, X.; Kong, D.; Zhang, Q.; Cao, Q.; Slater, L.J.; Ren, G.; Luo, M.; Li, J.; Liu, J.; et al. Asymmetrical Shift toward Less Light and More Heavy Precipitation in an Urban Agglomeration of East China: Intensification by Urbanization. *Geophys. Res. Lett.* **2022**, *49*, e2021GL097046. [CrossRef]
46. Shi, Y.; Shen, Y.; Hu, R. Preliminary study on signal, impact and foreground of climatic shift from warm-dry to warm-humid in Northwest China. *J. Glaciol. Geocryol.* **2022**, *24*, 219–226. [CrossRef]
47. Shi, Y.; Shen, Y.; Li, D.; Zhang, G.; Ding, Y.; Hu, R.; Kang, E. Discussion on the present climate change from warm-dry to warm-wet in Northwest China. *Quat. Sci.* **2003**, *23*, 152–164. [CrossRef]
48. Zhang, Q.; Zhu, B.; Yang, J.; Ma, P.; Liu, X.; Lu, G.; Wang, Y.; Yu, H.; Liu, W.; Wang, D. New characteristics about the climate humidification trendin Northwest China. *Chin. Sci. Bull.* **2021**, *66*, 3757–3771. [CrossRef]
49. Wen, G.; Huang, G.; Hu, K.; Qu, X.; Tao, W.; Gong, H. Changes in the Characteristics of Precipitation over Northern Eurasia. *Theor. Appl. Clim.* **2015**, *119*, 653–665. [CrossRef]
50. Liu, W.; Zhang, Q.; Fu, Z. Variation characteristics of precipitation and its affecting factors in Northwest China over the past 55 Years. *Plateau Meteorol.* **2017**, *36*, 1533–1545. [CrossRef]
51. Wang, L.; Chen, R.; Song, Y. Study of statistical characteristics of wet season hourly rainfall at Hulu watershed with Γ function in Qilian Mountains. *Adv. Earth Sci.* **2016**, *31*, 840–848. [CrossRef]
52. Shang, S.; Zhu, G.; Li, R.; Xu, J.; Gu, J.; Chen, H.; Liu, X.; Han, T. Decadal Change in Summer Precipitation over the East of Northwest China and Its Associations with Atmospheric Circulations and Sea Surface Temperatures. *Int. J. Climatol.* **2020**, *40*, 3731–3747. [CrossRef]
53. Zhang, Q.; Yang, J.; Wang, P.; Yu, H.; Yue, P.; Liu, X.; Lin, J.; Duan, X.; Zhu, B.; Yan, X. Progress and prospect on climate warming and humidification in Northwest China. *Chin. Sci. Bull.* **2023**, *68*, 1814–1828. (In Chinese) [CrossRef]
54. Findell, K.L.; Berg, A.; Gentine, P.; Krasting, J.P.; Lintner, B.R.; Malyshev, S.; Santanello, J.A.; Shevliakova, E. The Impact of Anthropogenic Land Use and Land Cover Change on Regional Climate Extremes. *Nat. Commun.* **2017**, *8*, 989. [CrossRef] [PubMed]
55. Li, C.; Min, R.; Gu, X.; Gulakhmadov, A.; Luo, S.; Liu, R.; Slater, L.J.; Xie, F.; Kong, D.; Liu, J.; et al. Substantial Increase in Heavy Precipitation Events Preceded by Moist Heatwaves Over China during 1961–2019. *Front. Environ. Sci.* **2022**, *10*, 951392. [CrossRef]
56. Zhang, G.; Wang, H.; Gan, T.Y.; Zhang, S.; Shi, L.; Zhao, J.; Su, X.; Song, S. Climate Change Determines Future Population Exposure to Summertime Compound Dry and Hot Events. *Earth's Future* **2022**, *10*, e2022EF003015. [CrossRef]
57. Meng, Y.; Hao, Z.; Feng, S.; Zhang, X.; Hao, F. Increase in Compound Dry-Warm and Wet-Warm Events under Global Warming in CMIP6 Models. *Glob. Planet. Chang.* **2022**, *210*, 103773. [CrossRef]
58. Duan, J.; Zhu, H.; Dan, L.; Tang, Q. Recent Progress in Studies on the Influences of Human Activity on Regional Climate over China. *Adv. Atmos. Sci.* **2023**, *40*, 1362–1378. [CrossRef]
59. Yin, H.; Sun, Y. Anthropogenic Influence on Temperature Change in China over the Period 1901–2018. *J. Clim.* **2023**, *36*, 2131–2146. [CrossRef]
60. Yang, X.; Zhou, B.; Xu, Y.; Han, Z. CMIP6 Evaluation and Projection of Temperature and Precipitation over China. *Adv. Atmos. Sci.* **2021**, *38*, 817–830. [CrossRef]
61. Zhang, S.; Hu, Y.; Zhang, J.; Xia, Y. Attribution of Biases of Interhemispheric Temperature Contrast in CMIP6 Models. *Adv. Atmos. Sci.* **2024**, *41*, 325–340. [CrossRef]
62. Yu, X.; Zhang, L.; Zhou, T.; Zheng, J. Assessing the Performance of CMIP6 Models in Simulating Droughts across Global Drylands. *Adv. Atmos. Sci.* **2024**, *41*, 193–208. [CrossRef]
63. Chen, Z.; Zhou, T.; Chen, X.; Zhang, W.; Zuo, M.; Man, W.; Qian, Y. Emergent Constrained Projections of Mean and Extreme Warming in China. *Geophys. Res. Lett.* **2023**, *50*, e2022GL102124. [CrossRef]
64. Cao, S.; He, Y.; Zhang, L.; Sun, Q.; Zhang, Y.; Li, H.; Wei, X.; Liu, Y. Spatiotemporal Dynamics of Vegetation Net Ecosystem Productivity and Its Response to Drought in Northwest China. *GIScience Remote Sens.* **2023**, *60*, 2194597. [CrossRef]
65. Yin, J.; Slater, L. Understanding Heatwave-Drought Compound Hazards and Impacts on Socio-Ecosystems. *Innov. Geosci.* **2023**, *1*, 100042. [CrossRef]
66. Lu, J.; Qin, T.; Yan, D.; Lv, X.; Yuan, Z.; Wen, J.; Xu, S.; Yang, Y.; Feng, J.; Li, W. Response of Vegetation to Drought in the Source Region of the Yangtze and Yellow Rivers Based on Causal Analysis. *Remote Sens.* **2024**, *16*, 630. [CrossRef]

Disclaimer/Publisher's Note: The statements, opinions and data contained in all publications are solely those of the individual author(s) and contributor(s) and not of MDPI and/or the editor(s). MDPI and/or the editor(s) disclaim responsibility for any injury to people or property resulting from any ideas, methods, instructions or products referred to in the content.

forests MDPI

Article

Estimation of the Short-Term Impact of Climate-Change-Related Factors on Wood Supply in Poland in 2023–2025

Jan Kotlarz [1,2,*] and Sylwester Bejger [2]

1 Łukasiewicz Research Network—Institute of Aviation, al. Krakowska 110/114, 02-256 Warsaw, Poland
2 Faculty of Economic Sciences and Management, Nicolaus Copernicus University, Gagarina 13a, 87-100 Toruń, Poland
* Correspondence: jan.kotlarz@ilot.lukasiewicz.gov.pl

Abstract: In this study, we analyzed in situ data from the years 2018–2022 encompassing entire forest plantations in Poland. Based on data regarding stand density and the occurrence of fungal, water-related, climate-related, fire, and insect factors that may intensify with climate changes, we determined the correlation between their occurrence and the decline in wood increments for six tree species: pine, birch, oak, spruce, beech, and alder. Subsequently, we identified age intervals in which the species–factor interaction exhibited statistically significant effects. Next, we developed neural network models for short-term wood increment predictions. Utilizing these models, we estimated a reduction in wood supply harvested in accordance with the plans for the years 2023–2025 assuming a tenfold greater intensity of factors than in 2022. Findings indicate: birch: water-related factors may reduce wood production by 0.1%–0.2%. This aligns with previous research linking drought to birch wood decline, highlighting its sensitivity to water-related issues. Oak: fungal and insect factors could decrease wood production by up to 0.1%. Prior studies emphasize the significant influence of fungal diseases on oak health and regeneration, as well as the impact of insect infestation on wood production. Alder: water-related factors may lead to a slight reduction in wood production, approximately 0.02%. The impact is significant within specific age ranges, indicating potential effects on harvesting. Pine: water- and climate-related factors may result in up to a 0.05% reduction in wood production. Pine, a key forest-forming species in Poland, is notably sensitive to these factors, especially as it nears harvesting age. Spruce: insects, fungi, and climate-related factors could lead to a reduction in wood production of up to 0.2%–0.3%. Analyses demonstrate sensitivity, resulting in a noticeable growth differential compared to the typical rate. Short-term predictions based on neural networks were developed, acknowledging their suitability for short-term forecasts due to uncertainties regarding long-term factor impacts. Additionally, our study discussed modeling wood increments in divisions well below the harvesting time, emphasizing that the influence of current and 2023–2025 factors on wood increments and supply may only manifest several decades from now. These results imply important indications for the economic and financial performance of the wood industry.

Keywords: climate change; wood industry economics; forest sector; wood supply; machine learning; neural networks

1. Introduction

Forests are increasingly vulnerable to the impacts of climate change, with rising temperatures, altered precipitation patterns [1,2], and more frequent extreme weather events posing significant challenges to their health and stability [3]. Forests also play a crucial role in mitigating climate change. They act as carbon sinks, absorbing substantial amounts of carbon dioxide from the atmosphere and storing it in trees and soil [4].

Climate change also has significant implications for forests in Poland [5,6]. The primary causes potentially related to climate change observed in Polish forests include fungal diseases [7,8] and insect pests, such as the bark beetle [9]. Over the past approximately 15 years, there has been a trend towards milder winters, cooler summers, and reduced rainfall during the growing seasons [10]. These changes have led to increased mortality among various tree species [5]. Alterations in soil moisture levels have contributed to a heightened impact of fungal diseases, both in terms of root pathogens—*Phytophthora* spp., *Armillaria* spp., *Heterobasidion* spp. [11–13]—and above-ground pathogens like *Hymenoscyphus* sp. [14], which is responsible for crown dieback. Furthermore, shifts in rainfall patterns have resulted in fluctuations in groundwater levels, further complicating the challenges faced by Polish forests [15]. The increasing intensity of climate change effects, characterized by a heightened impact of factors such as insect pests, fungal pathogens, water-related variables (such as declining groundwater levels), and wildfires, may lead to a decline in the growth rates of key tree species cultivated in Poland. As a consequence, this could result in a reduced supply of wood in the forestry sector. All of these factors collectively pose significant challenges to the sustained growth and supply of important timber species in Poland's forestry industry. The Polish wood sector as a whole generates approximately 2.5 percent of national GDP, provides budget revenues exceeding PLN 30 billion annually, and directly employs approximately 350,000 employees. Wood harvesting is an extremely important element of this sector. Each change in the wood supply has a significant impact on the economic result of the entire wood sector.

Forests in Poland, primarily managed by the State Forests National Forest Holding, cover a vast expanse of approximately 6,980,000 hectares (according to a holding report in 2022). These forests boast a total wood volume of around 2070 million cubic meters. The average stand volume across the forested area stands at 297 cubic meters per hectare. The forests exhibit robust growth, with an annual increment in wood volume reaching 63.6 million cubic meters, so an average wood production is approximately 9 cubic meters per hectare. In 2022, the forestry sector harvested a total of 42.3 million cubic meters of wood. The sale of timber from these forests reached a substantial 41.7 million cubic meters [16].

Even a marginal 1% reduction in the rate of wood increment in Poland would have significant consequences on future timber supply, resulting in a decline of nearly 0.5 million cubic meters per year. However, it is important to note that this decline will not appear immediately in wood supply. The forest management guidelines specify the optimal harvesting age for individual species, and the impact of current factors would reduce harvested wood volume several, if not tens, of years down the line. Thus, the consequences of these influences may only become fully evident over an extended period, emphasizing the long-term and complex nature of forest management and its connection to future timber supply.

Machine-learning methods were used previously for predicting tree diameter increment, particularly focusing on mixed and uneven-aged forests. In Bayat et al. (2022) [17], the authors employ multilayer perceptron (MLP) artificial neural networks and linear mixed-effect models to assess tree diameter increment under current and anticipated climate change scenarios. Despite the modest impact of predicted climate change, the MLP model outperforms the linear mixed-effect model, emphasizing the effectiveness of neural networks in capturing complex, nonlinear relationships within forest ecosystems. In Salehnasab et al. (2022) [18], the study investigates the application of MLP and an adaptive neuro-fuzzy inference system (ANFIS) for developing diameter increment models in the Hyrcanian forests. Evaluating four distinct species groups, the ANFIS technique demonstrated superior performance for beech and chestnut-leaved oak, while the ANN excelled with hornbeam and other species. The authors emphasize the deep relationship between modeling techniques and the nature of tree species, providing valuable insights for future studies in diverse forests worldwide. The choice of neural networks in these studies is justified by their ability to model complex, nonlinear relationships within the datasets. The flexible nature of neural networks allows them to capture intricate patterns influenced by various factors affecting tree growth. The use of MLPs in these articles reflects a recognition of the capacity of artificial neural networks to provide accurate predictions in the context of complex ecological systems, reaffirming their efficacy in forestry-related applications. Importantly, we did not want to assume the nature of the influence of individual factors; hence, we opted for the neural network version allowing for the modeling of both linear and nonlinear influences.

In this article, we have used forest assessment data gathered by the Polish national forestry holding between 2018 and 2022. These comprehensive datasets include the entirety of the forests managed by the holding. Based on these data, wood increments were computed for each forest division and each tree species. These wood increments were then linked to data on the occurrence of various factors, potentially related to climate change, including wildfires, fungal pathogens, insect pests, climatic conditions (snow, hurricane winds, windfalls etc.), and hydrological factors. Subsequently, for the primary tree species, specific age periods were determined in which the wood increments were statistically ($p < 0.05$) sensitive to the occurrence of these factors.

To build growth models for trees, convolutional neural networks (CNNs) were developed separately for forest divisions unaffected by external factors and those affected by each of the factors. Using a sample of 100,000 forest divisions, the growth models were developed. These models were then applied to the current (2022) wood volumes, taking into account two scenarios: (1) the absence of any external factor and (2) the occurrence of factors in each of the years from 2023 to 2025. Subsequently, the wood volumes harvested for both scenarios, based on the harvesting ages for all tree species, were compared. This analysis allowed us to estimate the potential maximum change in wood supply for the years 2023–2025 caused by occurrence of potentially climatic-related factors. Our results help to predict the economic performance of the wood industry and encourage sustainable and efficient management of forestry.

2. Methods

To investigate the impact of climate-related factors on wood increment values, we used forest inventory data from the Forest Data Bank. A representative sample of 100,000 forest divisions was selected from a set of 3,494,871 forest divisions. From this subset, we selected forest divisions affected by factors such as wildfires, climatic factors, water-related factors, insects, and fungal pathogens. Comparisons were calculated between wood increment rates of dominant tree species in stands unaffected by any factor versus those influenced. Sensitivity ranges of tree age to specific factors were estimated. Using these sensitivity ranges, convolutional neural network (CNN) models were developed to predict growth, with the presence or absence of each factor as one of the input parameters. Utilizing these CNN models, forecasts for wood supply were formulated for scenarios representing minimum and maximum occurrences of the factors from the years 2023 to 2025.

2.1. Data Sample

The data used in this analysis were sourced from the Forest Data Bank, which is regularly updated on an annual basis through in situ forest inventory conducted by forestry authorities [19]. Each record within the dataset describes an individual forest stand, defined in forestry as a homogenous forest area characterized by economically significant features, necessitating uniform management practices. The typical area of these individual forest stands ranges from one to several hectares. The database for the year 2022 covered a total of 3,494,871 records. For further analysis, specific categories of stand parameters were taken into account, including: (1) forest habitat type, (2) soil type on which cultivation occurs, (3) habitat degradation level, (4) moisture regime, (5) economic and protective functions of the stands, and (6) species composition and age structure of the stands (see: Figure 1).

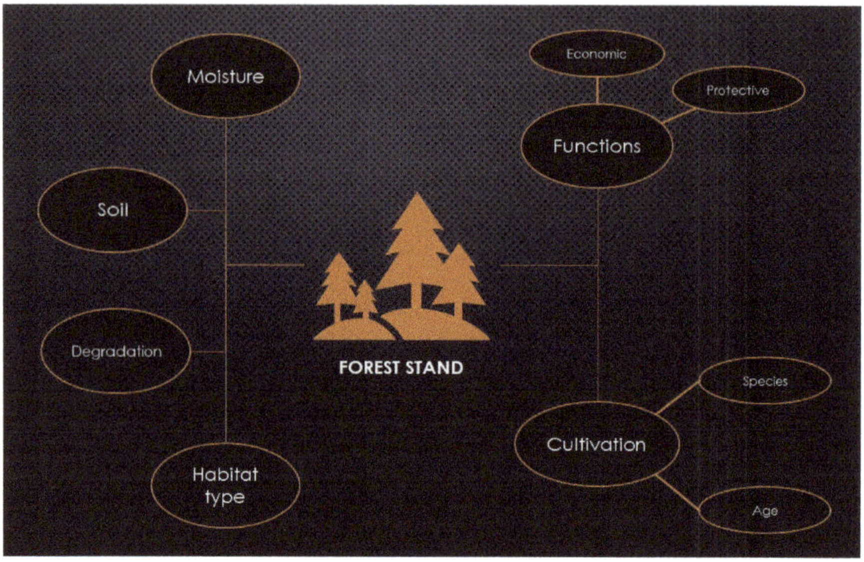

Figure 1. The structure of data describing forest stands in Poland used in the analysis. Each record contains data describing: habitat type, soil type and subtype, moisture, degradation, functions, and trees' species and current age.

Each record of the forest division contains 398 parameters from the aforenamed categories. To create a representative sample of all forest stands in Poland, a supervised classification was performed using the ISOCLASS algorithm [20] in the 398-dimensional space. This resulted in the formation of 9 classes, collectively encompassing over 98% of all forest stands. Subsequently, a random sample of 100,000 stands was randomly selected for further analyses, representing each of the derived classes.

Data from the years 2018–2022 were downloaded from the database and stored locally on disk. Data processing and analysis were carried out using Python (version 3.0) with *pandas* and *numpy* libraries. The ISOCLASS algorithm, responsible for data classification, was applied within the code, with a condition added to ensure that the maximum proportion of records outside the classes was less than 5%. The classification process was performed using data from 2022. The resulting classes are presented in Table 1.

Table 1. Classification of forest cultivation in Poland using the ISOCLASS algorithm based on habitat data, soil, environmental moisture, tree species, cultivation functions, and degradation.

Class	Habitat	Soil	Species	Moisture Variant	Degradation	Functions
1	fresh coniferous forest	mainly rusty podzol, additionally rusty, ferrous	pine with a mixture of birch	fresh	natural	mainly economic forest
2	deciduous or mixed forest, fresh, upland or mountainous	brown, podzol, river alluvial	mainly: beech, oak, spruce, pine, additionally: birch, Douglas fir, hornbeam	fresh	natural and lightly degraded	mainly protective
3	mixed forest	rusty	pine, oak	fresh	lightly degraded, natural	economic forest

Table 1. *Cont.*

Class	Habitat	Soil	Species	Moisture Variant	Degradation	Functions
4	deciduous forest	rusty	pine	fresh	lightly degraded	economic forest
5	mixed moist conifer or deciduous forest, alder	gleyic podzol, gley, precipitation-dependent	pine, birch, spruce	moist, very moist, marshy, floodplain	mainly natural	protective
6	fresh forests	podzols, ferric	pine, additionally spruce	very moist	lightly degraded, natural	economic forest
7	alder, ash	peaty soils of low peat bogs, peat—marsh soils	birch, ash, spruce, pine	marshy or floodplain	natural	protective
8	mixed fresh coniferous forest	podzols, ferric	pine, additionally spruce	moist	varied	protective
9	mixed forest—fresh, mountain or upland	brown, acidic or leached	pine, fir, spruce	very moist	lightly degraded	protective

One hundred thousand randomly selected forest divisions represented all derived classes. The sample of forest divisions was collected as follows:

- For each class i, the number n_i of divisions belonging to that class was determined, and the ratio $k_i = n_i/N$ was calculated, where N = 3,494,871 is the total number of forest divisions in Poland.
- For all divisions in class i, the average position $\boldsymbol{r}$ in the 398-dimensional space and the standard deviation dr of the Cartesian distance d between the position of divisions in the class and the average position were computed.
- Among the divisions in class i, $k_i \times 100{,}000$ divisions meeting the condition $d < dr$ were randomly selected. These selected divisions were added to the sample.

This process resulted in the selection of the most representative divisions for forest cultivation in Poland.

2.2. Historical Impact Analysis

In the initial stage of data analysis, historical data for a selected sample of forest stands were examined to identify age periods for the main tree species (pine, oak, alder, birch, spruce, and beech), in which wood volume increment in stands affected by specific factors was statistically different (p-value < 0.05) from increments in stands unaffected by any factor. Due to the limited scope of data reflecting wood increment for only a few consecutive vegetative seasons in individual forest plantations, we opted to employ tree age as the temporal axis on our timeline. This decision allowed us to construct growth curves for the dominant tree species over an extended period, spanning from 20 to 120 years of age. By aggregating data from 100,000 forest plantations across five vegetative seasons, we were able to model growth curves for both reference scenarios and forest plantations affected by external factors. This approach not only facilitated the representation of baseline growth curves but also provided insights into the growth patterns of forest stands impacted by various environmental and external influences. We analyzed only volume increments in the vegetation season immediately following the occurrence of the factor. The statistics were performed separately for each tree species and age with a temporal resolution of 1 year.

The procedure involved selecting forest stands for each tree species s and each age s_a between 20 and 120 years, where the species covered at least 10% of the stand's area.

Subsequently, the dataset of forest stands was divided into two subsets: those with the presence of the factor and those where none of the examined factors occurred. This step was repeated for each of the studied factors. The mean wood volume increment was calculated for both subsets, and a *t*-Student test was conducted to determine whether the observed difference in means was statistically significant. In cases where the data distribution did not follow a normal distribution, the obtained result was excluded from further analysis, considering it statistically insignificant. This algorithm is illustrated in Figure 2.

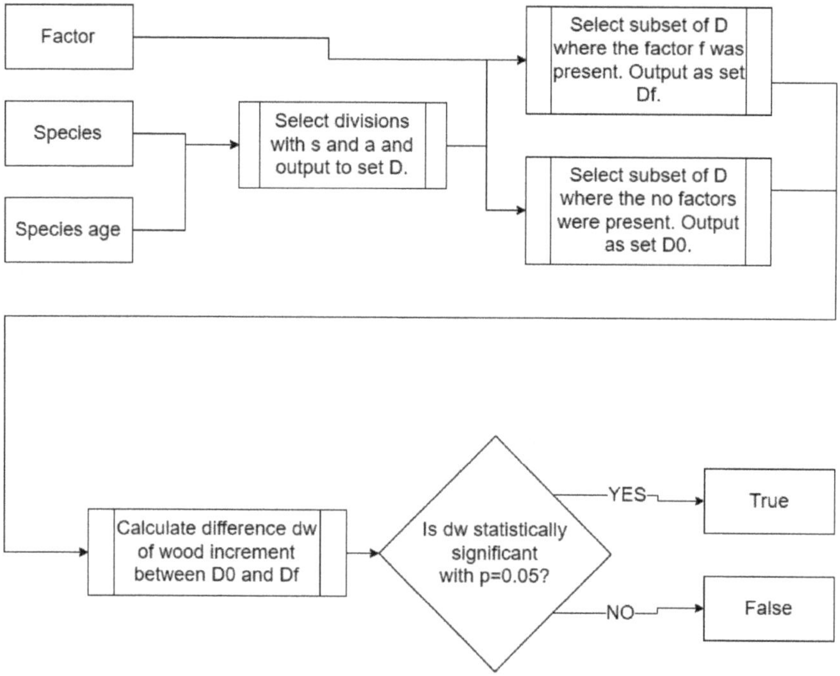

Figure 2. The algorithm determines the age range in which the impact of factor *f* on the cultivation of species *s* is statistically significant. The same algorithm was repeated for all considered species, for each factor, and for ages between 20 and 120 years.

This process yielded, for each dominant tree species (pine, oak, birch, spruce, alder, beech), age periods during which they were sensitive to the studied factors: insect pests, fungal pathogens, wildfires, climatic factors, and water-related factors. Subsequently, it was assumed that only during age periods where the difference was statistically significant did a given factor have an impact on the wood increment of that species.

The applied simplifications in the analysis were as follows:

(a) Only information about the occurrence of the factor was extracted from the database, without considering its intensity or severity.
(b) It was assumed that the factor had an impact on the subsequent vegetation season following its occurrence.
(c) It was assumed that the factor only affected the wood increment in the specific forest stand where it occurred, without influencing the growth of neighboring stands.

These simplifications were made to streamline the analysis and focus on the most essential aspects of the data, although they do represent certain limitations in terms of the depth of analysis and the real-world complexity of ecological systems.

The division of a tree's lifespan into periods where the statistically significant impact of external factors on wood increment is observed and those where data did not indicate statistically significant influence does not imply the absence of such an impact. Several reasons could account for the lack of observed negative or positive effects on wood increment. For instance, the absence of a sufficient quantity of data illustrating the increment of a specific species under the influence of a particular factor at a precise age may be a contributing factor. Other influential environmental parameters, such as soil conditions, economic or protective management practices, variances in validation, and similar factors, may overshadow the impact of external factors. The outlined division in this publication merely reflects the statistical analysis of a selectively chosen sample of 100,000 delineated forest plantations. It is essential to recognize that the absence of statistically significant influence during certain periods does not negate the potential influence of external factors, emphasizing the complexities involved in understanding the nuanced relationship between tree growth and environmental variables.

2.3. Neural Network Wood Increment Prediction Model

Neural networks are currently used for predicting wood growth increments [21,22]. We developed and used neural network models in simulations. For each species s, the following procedure was performed. From a sample database containing 100,000 records, forest divisions containing species s were selected. This way, the sample database was limited to samples for species s. Then, in order to create a neural network model that predicts increments for divisions without external factors, divisions were selected where either no external factor was present or the external factor was present but did not have a statistically significant impact on the growth of that species at a given age based on historical studies. Using this constructed database, reference models were developed. Next, for each pair of species s–external factor f from the sample for species s, a subset where the factor f occurred was selected. Models were then developed for each s–f pair. This scheme is illustrated in Figure 3.

Each neural network model has the following data as input in its first layer: habitat type, soil, moisture variant, degradation, forest stand function, tree stand age, and the current volume of the studied species. The output of the model had a single parameter: the wood volume increment of the studied species (see Figure 4).

To assess the maximum potential impact of a given factor on forest cultivations, two models of wood volume increments were developed. In the first model, it was assumed that the factor occurred in each year of the time period 2023–2025 with probability analogue as in 2022. In the second model, it was assumed that the factor occurred in each year: 2023, 2024, and 2025 in every forest stand with probability 10 times higher. In the year 2022, within the scrutinized sample of 100,000 delineated forest plantations reported by forestry services, the occurrence rates of individual factors were carefully selected. We focused on factors reported by forestry services within the studied sample of 100,000 forest plantations, specifically choosing those that occurred in at least 500 plantations. An exception to this criterion was made for forest fires, which, although less frequent during the study period, were included for comprehensive analysis. Additionally, only factors affecting at least 20% of the forest plantation area were considered. The selected factors and their occurrence rates in 2022 are as follows: insect pests accounted for 1.18%, fungal pathogens for 3.46%, wildfires for 0.04%, climatic factors for 2.05%, and water-related factors for 0.54%. This selection process aimed to capture the most impactful and prevalent factors affecting forest plantations, providing a robust basis for our analysis.

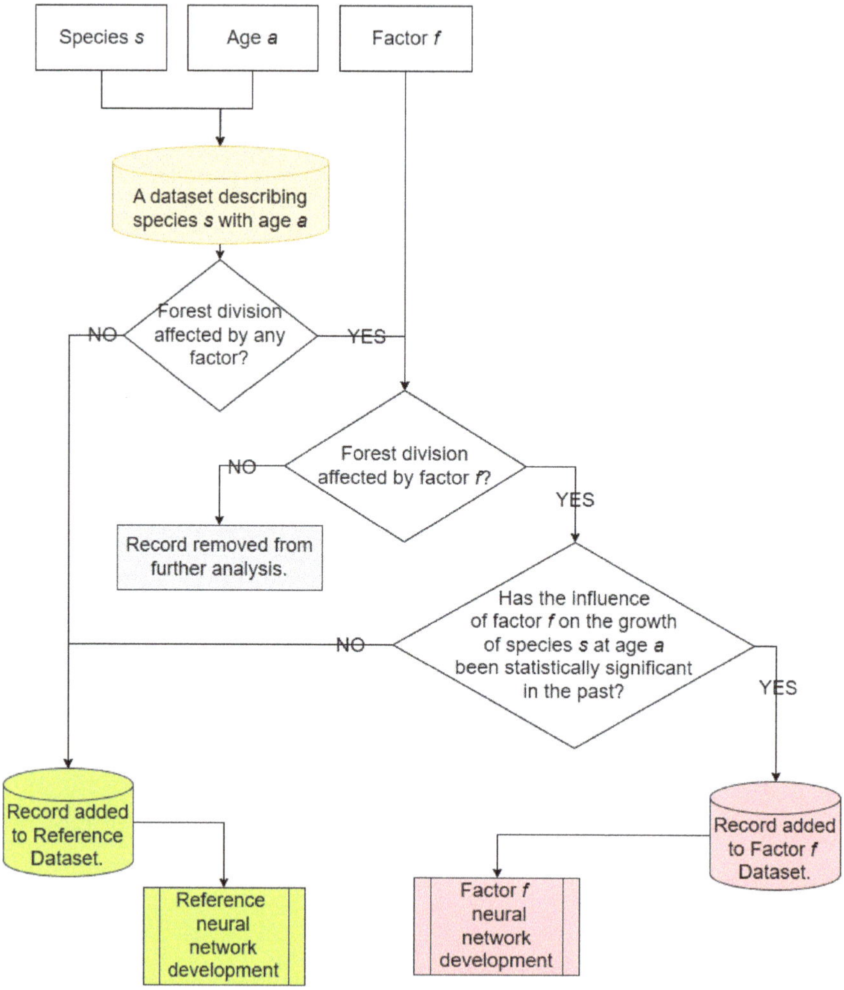

Figure 3. Development process of wood increment models using neural networks. Each tree species—external factor pair has its own neural network. Reference outputs marked by green and affected by external factors marked by red.

This resulted in wood volume increment values in these two scenarios: (a) current reference scenario and (b) tenfold increase in the frequency of all factors. To estimate the impact on the volume of harvested wood in the years 2023–2025, it was assumed that all trees are harvested according to their harvesting time specified in the instructions.

Finally, the harvested wood volumes in both scenarios were compared, and the percentage decrease in the scenario where the factor occurred in each year was calculated. This analysis provided insights into the potential impact of the factor on wood production in the studied forest cultivations.

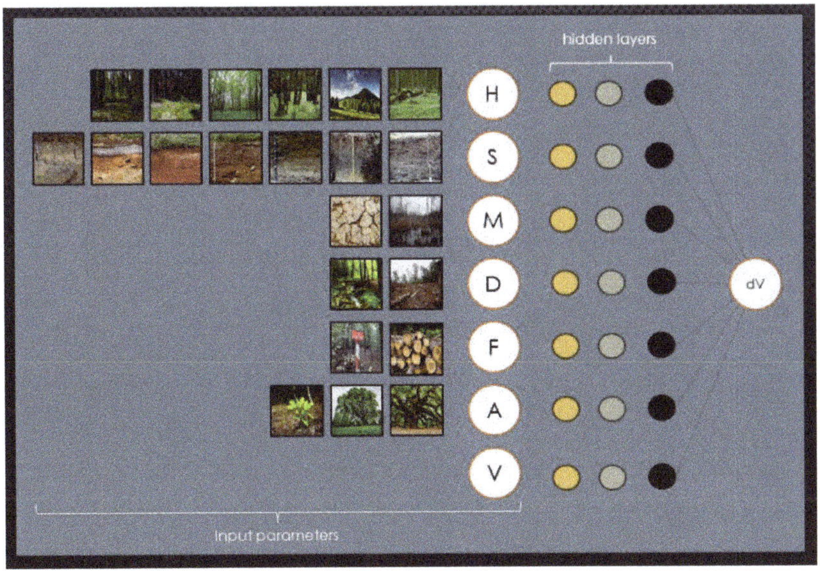

Figure 4. Structure of a neural network model for an individual tree species and for an individual external factor or the absence of such a factor. The input of the neural networks includes information: habitat type (H), soil (S), moisture (M), habitat degradation (D), cultivation functions (F), tree stand age (A), current volume of the species (V). At the output, we obtain the value dV, which represents the volume increment. Input data are derived from forest stands where the studied species is present and where cultivation occurs at an age sensitive to the occurrence of a given factor (or the absence of any factor).

3. Results

The sample size of 100,000 records allowed for obtaining statistically significant insights into the differences in wood increment between conditions with and without external factors for the following tree species: pine, oak (both pedunculate and sessile oak, as there were insufficient data for other oak species in the sample), alder, birch, spruce, and beech. The spatial distribution of the sample forest stands is presented in Appendix A (Figure A1).

3.1. The Influence of External Factors on Historical Wood Growth Increments

In Table 2 we present species–factor pairs for which significant impacts on wood increment were observed.

Table 2. The age ranges of tree species for which the annual wood volume increment significantly depended on the occurrence of external factors. The statistical analysis was conducted based on a sample of 100,000 forest stands from across Poland, using assessment data. The lack of influence of wildfires on the increments is due to insufficient data to achieve statistical significance ($p < 0.05$).

	Pine	Oak	Alder	Birch	Spruce	Beech
Insects	51–63 106–109	40–42 113–117	64, 67	66, 67	65–118	75–85
Fungi	37–56 67–76 106–117	23–28 112–120	82–83	-	36–38 87–120	46
Fires	-	-	-	-	-	-
Climate	50–97 109–115	88, 91	25–31	70–75 98–101	87–120	107–114
Water	20–35 55–77 85–97	20–25 92	20–67 75–90	20–85	-	-

The charts indicating the average increment for a given species and its change in the presence of an external factor are presented below.

3.1.1. Pine

Figure 5 show the average annual increments of pine wood volume in a sample of 100,000 forest stands in Poland, as well as the average increments in forest stands affected by external factors: water, climate, fungi and insects. The age of trees, for which the growth differences are statistically significant, has been highlighted.

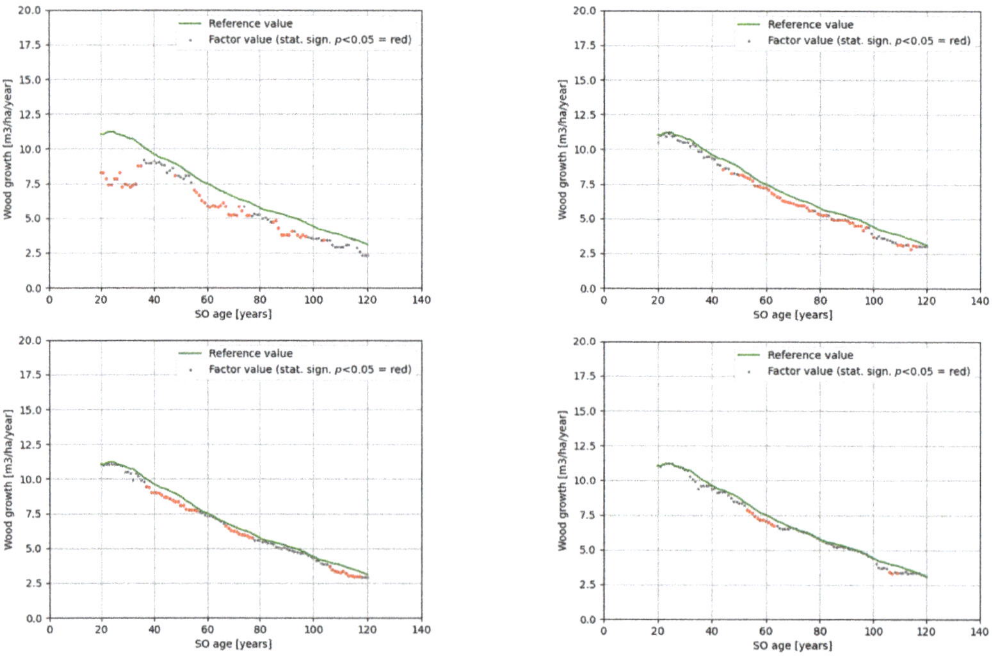

Figure 5. External factors: water (**top left**), climate (**top right**), fungi (**bottom left**), insects (**bottom right**) influence on the average annual increments of pine wood volume.

3.1.2. Oak

Figure 6 show the average annual increments of oak wood volume in a sample of 100,000 forest stands in Poland, as well as the average increments in forest stands affected by external factors: water, climate, fungi and insects. The age of trees, for which the growth differences are statistically significant, has been highlighted.

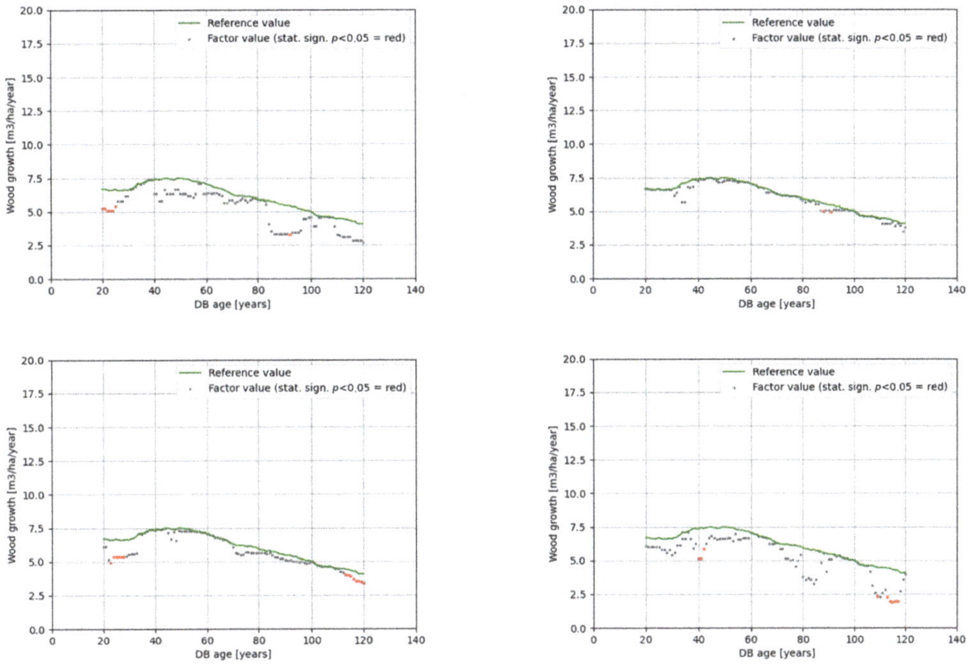

Figure 6. External factors: water (**top left**), climate (**top right**), fungi (**bottom left**), insects (**bottom right**) influence on the average annual increments of oak wood volume.

3.1.3. Alder

Figure 7 show depict the average annual increments of alder wood volume in a sample of 100,000 forest stands in Poland, as well as the average increments in forest stands affected by external factors: water and climate. The age of trees, for which the growth differences are statistically significant, has been highlighted.

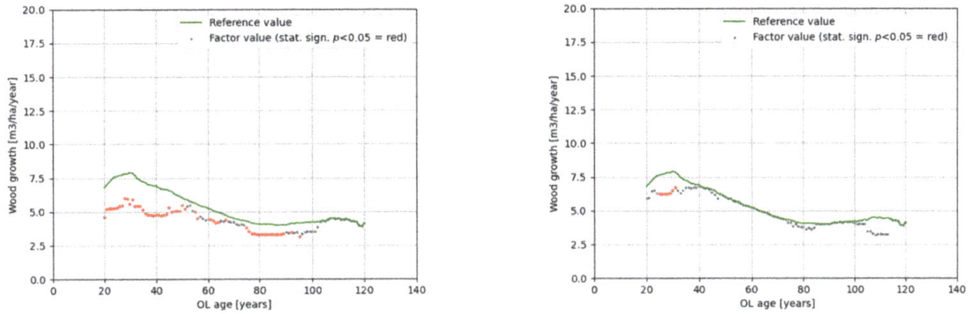

Figure 7. External factors: water (**left**), climate (**right**) influence on the average annual increments of alder wood volume.

3.1.4. Birch

Figure 8 show below depict the average annual increments of birch wood volume in a sample of 100,000 forest stands in Poland, as well as the average increments in forest stands affected by external factors: water and climate. The age of trees, for which the growth differences are statistically significant, has been highlighted.

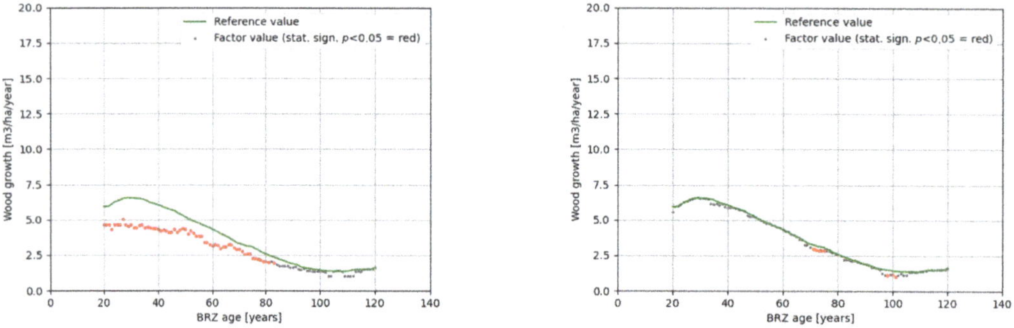

Figure 8. External factors: water (**left**), climate (**right**) influence on the average annual increments of birch wood volume.

3.1.5. Spruce

Figure 9 show depict the average annual increments of spruce wood volume in a sample of 100,000 forest stands in Poland, as well as the average increments in forest stands affected by external factors: insects, fungi and climate. The age of trees, for which the growth differences are statistically significant, has been highlighted.

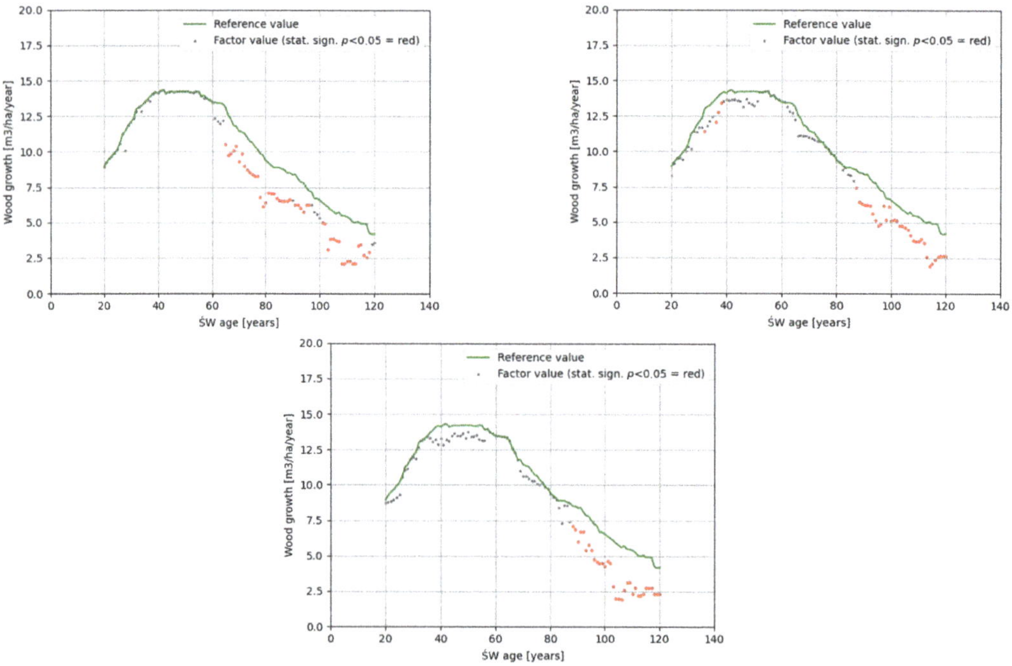

Figure 9. External factors: insects (**top left**), fungi (**top right**), climate (**bottom**) influence on the average annual increments of spruce wood volume.

3.1.6. Beech

Figure 10 show depict the average annual increments of beech wood volume in a sample of 100,000 forest stands in Poland, as well as the average increments in forest stands affected by external factors: insects and climate. The age of trees, for which the growth differences are statistically significant, has been highlighted.

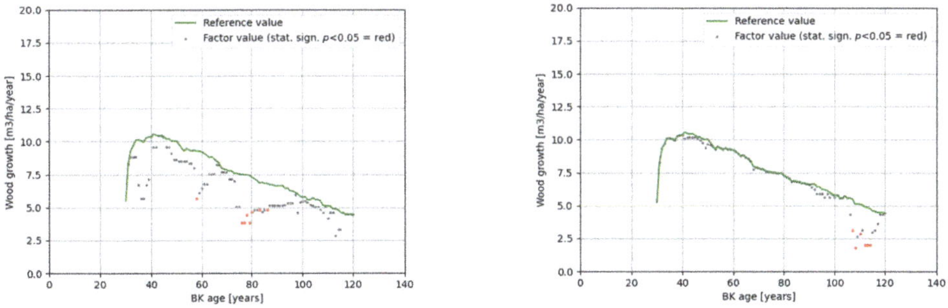

Figure 10. External factors: insects (**left**), climate (**right**) influence on the average annual increments of beech wood volume.

3.2. Estimation of the Potential Impact of External Factors on Wood Supply in the Years 2023–2025 Using Neural Networks

The variations in growth increments between the scenarios with the occurrence of a factor in each growing season (2023, 2024, 2025) for individual species are illustrated in: Figure 11—birch, Figure 12—oak, Figure 13—alder, Figure 14—pine and Figure 15—spruce. For each year, the percentage difference in the growth increments of trees at harvest age is indicated. Only the results of models for which the estimated cumulative three-year impact of a single factor was higher than 0.1% are presented. The charts also include the estimated impact for the year 2022.

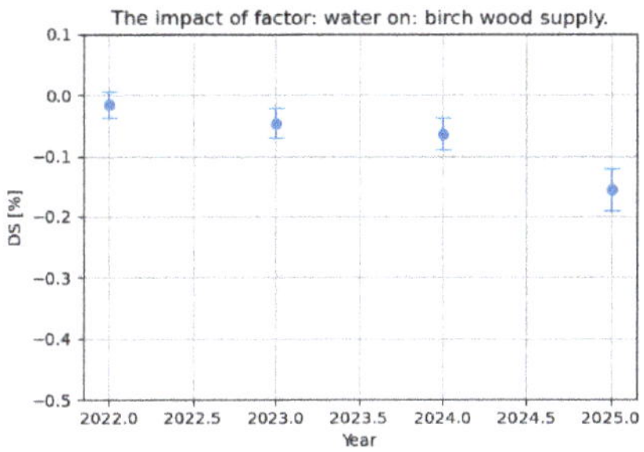

Figure 11. The impact of external factors on birch wood supply from 2022–2025.

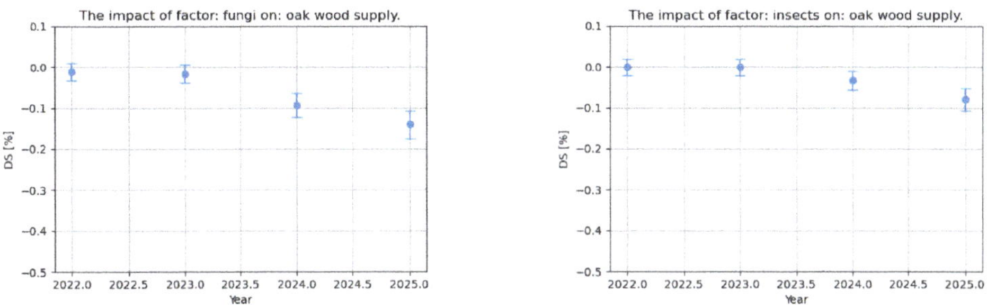

Figure 12. The impact of external factors on oak wood supply from 2022–2025.

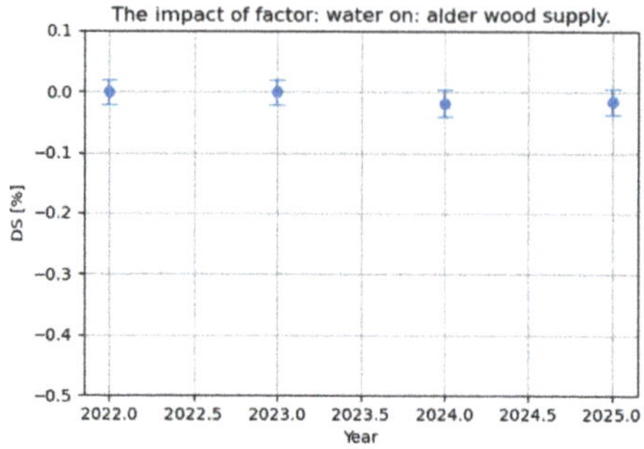

Figure 13. The impact of external factors on alder wood supply from 2022–2025.

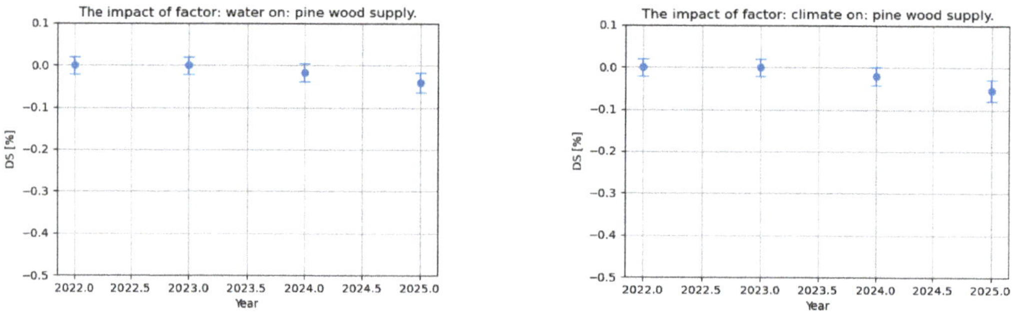

Figure 14. The impact of external factors on pine wood supply from 2022–2025.

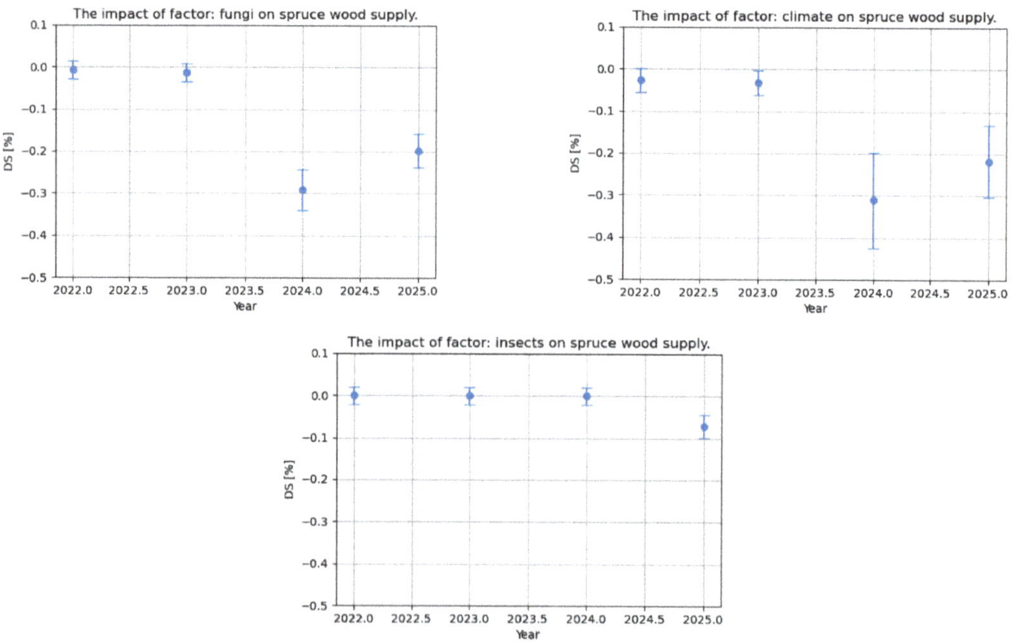

Figure 15. The impact of external factors on spruce wood supply from 2022–2025.

3.2.1. Birch

Below, we present the impact of water relations disturbance on birch harvesting volume. It is assumed that the intensity of the factor increased tenfold compared to 2022. Figure 11 presents the results based on a neural network model: during the period of 2023–2025, the harvesting volume diminishes, reaching a decline of nearly 0.2% in 2025 compared to the scenario without the occurrence of this factor.

3.2.2. Oak

Below, we present the impact of fungi and insects on oak harvesting volume. It is assumed that the intensity of these factors increased tenfold compared to 2022. Figure 12 presents the results based on a neural network model: during the period of 2023–2025, fungi caused a decline in harvesting volume, especially in 2024–2025, with a decrease exceeding 0.1%. In contrast, insects led to a decrease in harvesting volume by less than 0.1% in 2024–2025 compared to the scenario without the occurrence of these factors.

3.2.3. Alder

The impact of water-related factors on alder harvesting volume is less significant. Figure 13 presents the results based on a neural network model: during the period of 2023–2025, water-related factors caused a decrease in harvesting volume, with the decline staying below 0.05% for each year compared to the scenario without the occurrence of these factors.

3.2.4. Pine

Below, we present the impact of water-related factors and climate-related factors (snow icing, hurricane winds, droughts, and frosts) on pine harvesting volume. It is assumed that the intensity of these factors increased tenfold compared to 2022. Figure 14 presents the results based on a neural network model: during the period of 2023–2025, both water-related and climate-related factors contribute to a decrease in harvesting volume, particularly in 2024–2025, with the decline remaining below 0.1% compared to the scenario without the occurrence of these factors.

3.2.5. Spruce

The impact of fungi, climate, and insects on the harvesting volume of spruce is the most significant. It is assumed that the intensity of these factors increased tenfold compared to 2022. Figure 15 presents the results based on a neural network model: during the period of 2023–2025, both fungi and climate-related factors contribute to the highest decrease in harvesting volume among the studied species, reaching above 0.2%. In 2024, it amounts to 0.3% for both fungi and climate-related factors. The influence of insects on harvesting volume is most pronounced in 2025, with a decrease exceeding 0.05%.

4. Discussion

Below, we present a discussion of the results for each species separately.

4.1. Birch: Reduction in Wood Production Due to Water-Related Factors up to 0.1%–0.2%

In previous studies, the impact of water-related factors, such as drought, on the growth of birch trees has been closely examined, and these investigations have often linked these influences to climate change. Kharuk et al. (2013) [23] examined the phenomenon of declining birch wood production in Trans-Baikal using forest inventory and satellite data. In the first decade of the 21st century, a decrease and mortality of birch trees were observed in approximately 5% of the total area within their 1250 km^2 study area. The decline and mortality of birch forests primarily occurred on the outskirts of stands, within the forest-steppe ecotone, on sun-exposed slopes. During the first decade of the 21st century, summer precipitation (June–August) was approximately 25% below average. Soil moisture measurements revealed the lowest values in areas where trees had died, the highest in healthy forests, and intermediate levels in damaged areas. Rojo et al. (2021) [24] examined the sensitivity of birch trees to climate change using data from 28 research stations in Bavaria. Their study revealed that the birch population is declining in lowland areas, and there is a migration of this species to higher elevations. The research spanned a 30-year period and primarily focused on the impact of summer droughts. Our research confirms the relationship between droughts and birch wood production observed in the literature. Dox et al. (2022) [25] observed the impact of drought on accelerating the cessation of wood growth (CWG) by approximately 5–6 weeks. Interestingly, despite the prolonged retention of green canopies in trees, this phenomenon led to a reduction in wood production. Their findings underscore the significant influence of dry conditions and warm summers on growth patterns in both the current and subsequent years.

We have demonstrated that, in Poland, birch is sensitive to water-related factors throughout its entire growth period, from 20 to 80 years of age (Figure 8, left). The age of birch harvesting in Poland is precisely 80 years, and therefore, all factors associated with insufficient water supply will have an impact on wood growth and birch wood supply. The upper estimation limit of birch wood production decline is near 0.2% in 2025 (Figure 4). Drought and other water-related factors can be considered as the primary factors that may potentially cause this decrease in production in the near future, as well as in the long term.

4.2. Oak: Reduction in Wood Production Due to Fungal and Insect Factors of up to 0.1%

Relations between oak wood production and the presence of fungal and insect factors were described in the literature. Marçais and Desprez-Loustau (2014) [26] have described how fungal diseases can significantly impact the natural regeneration of oak trees and play a significant role in reducing the health of mature trees. Climate change can influence the severity of the disease primarily by altering the phenological synchrony between the host plant and the pathogen. The high impact of the disease was often associated with modified growth patterns, either due to environmental factors (such as insects or frost). Also, wide research on fungal impact on oak decline and wood production was performed in Poland. Oszako et al. (2005) [27] emphasized the detrimental impact of fungal diseases, particularly *Phytophthora* sp. on oak wood production. The indirect influence of fungal diseases on the wood growth of already weakened trees was also indicated in the doctoral thesis by G. Attocchi [28]. It is also indicated that damage to oak forest stands in typical cultivation or felling works is the reason for the formation of input ports in which fungal decay may cause further wood depreciation (devaluation) or even lower increments [29]. Marcais and Desprez-Loustau (2014) [26] showed that fungal factors can have important impacts in natural oak regenerations and a significant role in the decline of mature trees.

A similar relationship was noted in the literature between the occurrence of insect pests and the production of oak wood. The potential impact on wood production by insects was already noted by Rafes (1973) [30]. He noted that the consumption of wood tissue itself reduces wood production, however, wood consumption usually affects weakened plants, and the damage to the entire stand caused by xylophages is not as great as with the consumption of leaves. The opposite result was obtained by Grigri et al. (2020) [31]. The authors, in their research on oaks ($n > 700$), indicated that simulated insect infestation did not affect wood production but did lead to a reduction in leaf area at the end of the growing season. The authors hypothesized that wood production would be disrupted in subsequent growing seasons.

In our study, we found that the growth of oak wood was influenced by fungal diseases, primarily below 30 years and above 110 years of age (Figure 6, left) and by insects at around 40 years and also above 110 years (Figure 6, right). Trees within the age range of timber harvesting may experience reduced growth in the presence of fungal and insect diseases, as evident in our forecasts for the years 2024–2025, indicating a potential decline in oak wood supply not higher than 0.1% (Figure 12).

4.3. Alder: Reduction in Wood Production Due to Water-Related Factors up to 0.02%

The age of alder felling in Poland is around 70. According to Figure 7, this is the age between two periods in which the impact of water-related factors is statistically significant, meaning—theoretically—that the impact of water-related factors would be lower than for birch harvesting. The obtained result indicates a 0.02% lower harvest of this species in 2024 and 2025. This result does not reflect the impact on the harvesting of alder wood in general, which may manifest itself in subsequent years, because the impact of water-related factors—according to our result—on younger individuals is of the order of 1–2.5 m^3/ha for normal growth of 4–8 m^3/ha.

Nossov et al. (2010) [32] demonstrated that increased precipitation and river discharge, particularly in June and August, positively impact alder growth, while dry conditions and low hyporheic flow during these months can limit its growth, underscoring the susceptibility of alder to early-season moisture deficits. Similar results are presented in various papers [33–35]. However, it is important to note that our study only partially confirmed these influences, highlighting the need for further investigations and long-term simulations. It should be noted that in the studies, younger alder trees were primarily examined, which were younger than the typical harvesting age in Poland investigated in our research (e.g., average age = 39 years—see Table 1 in [34]).

4.4. Pine: Reduction in Wood Production Due to Water- and Climate-Related Factors up to 0.05%

Pine covers 67% of Poland's forested land area and stands as the primary forest-forming species in the country, thriving in a wide range of environmental conditions. While the study identified the influence of four factors on pine growth (insects, fungi, water-related factors, and climatic factors), only water-related and climatic factors exert direct effects on pine growth just before the harvesting age of pine (100 years). The remaining factors are projected to exert their impact in the subsequent years. These findings are further corroborated by the supply reduction simulations derived from machine-learning models.

Eilmann et al. (2011) [36] examined the impact of water-related factors on mature pine growth in the Inner Alpine region. They found that trees without irrigation had significantly shorter wood formation periods and lower growth. The significant reduction in the growth period of control trees suggests that the actual wood formation period can be much shorter under drought conditions than the "potential" period, which corresponds to the phenological growth period. This finding is relevant to our study, which also includes mature pine trees. Seidling et al. (2012) [37] observed that the decline in pine stem growth was particularly noticeable in relation to water deficiency during early springs. Matveev et al. (2017) [38] examined the correlation between pine circumference increments and climatic factors spanning from the 20th to the 21st century. They uncovered trends, variability in climate indices, and corresponding radial tree growth in regional Scots pine stands. Our result aligns with the findings of the latest publication. Although the prediction of a decrease in wood supply due to water-related factors stands at only 0.05%, it should still be considered, given the significant presence of pine in forestry practices in Poland.

4.5. Spruce: Reduction in Wood Production Due to Insects, Fungi, and Climate-Related Factors up to 0.2%–0.3%

Our analyses have revealed that spruce trees were sensitive to three factors: climatic conditions, insects, and fungal pathogens. In the case of insects, the sensitivity was observed in trees above the age of 60, while for climatic factors and fungal pathogens, it was prominent in trees exceeding 90 years of age. Across all three scenarios, there was a growth differential ranging from 1.5 to 2.5 m^3/ha, compared to the typical growth rate of 5.0–7.5 m^3/ha. Results of machine-learning-based model simulations have indicated that under scenarios involving climatic factors and fungal pathogens, there could be a decrease in wood supply of approximately 0.2%–0.3% in the years 2024 and 2025. In the case of insect-related scenarios, the decline in wood supply remained below 0.1% in the year 2025. It should be emphasized that the supply simulation for insect-related factors included a model version that did not exhibit statistically significant influence on wood production during the last two years prior to the harvesting age, i.e., at 120 years. This could potentially explain the lower impact of this factor on supply in the years 2023 and 2024.

Mäkinen et al. (2007) [39] have previously observed that mechanical damage to spruce trees can lead to wood decay. The source of the disease could originate from both root and stem damage. Hrib et al. (1983) [40] observed the influence of wood-inhabiting fungi from the *Amalaria* sp. family on wood increments. It should be emphasized here that, similar to many other species, fungal pathogens attack trees weakened by various factors such as drought, floods, mechanical damage, and a range of other factors. Such damage can result from the direct influence of climatic factors, for example, in windbreaks or during droughts. This convergence in predicting a reduction in spruce wood supply between climatic factors and fungal pathogens is observed as a result of machine-learning-based simulations.

4.6. The Short-Term Predictions Based on Neural Networks

During the design of our experiment, we assumed that, at the initial stage of the study, we should not use developed neural networks models for long-term predictions. This was due to uncertainties regarding the following issues and open questions:

(1) In estimating the impact of factors on the growth of individual tree species, we assumed during the development process that this impact occurs over the course of one year, i.e., a single growing season. We could speculate that the use of machine-learning models in long-term forecasting would not account for overlooked dependencies. Hence, the forecast based on the prepared models covers a period comparable to the time frame of the reference data.

(2) Due to the limited observation period in the reference data, spanning only a few years, we are uncertain about how the occurrence of multiple factors simultaneously or even one factor for two or more consecutive years may interfere with each other. Our simulation assumed that a single factor occurs for three consecutive years (2023, 2024, 2025). In this way, we wanted to estimate the maximum impact of a given factor on wood supply. Here, we assumed that this impact is additive. Naturally, this approach limits the feasibility of using it for long-term forecasting.

The obtained forecasts actually depict changes in the stands and those already at the harvesting age. These results serve as a good starting point for developing long-term forecasts, which would also encompass trees currently in the maintenance stage.

A similar issue arises in modeling wood increments in divisions significantly below the harvesting age. These stands, which will determine wood supply in 10, 20, or 50 years, can be treated as a "stock state." As shown in Figure 5, the impact of factors is often significant in a period prior to harvesting (e.g., water factors' impact on birch and alder, or insects' impact on pine). It can be assumed that the main influence on wood increments and wood supply of factors occurring currently and in the years 2023–2025 will only be observed several decades from now.

5. Summary

Our research analyzes the individual impacts of various factors (climate, water-related, fungi, insects, and fires) on wood production for the main tree species in Poland (pine, oak, spruce, birch, alder, and beech). The following summary pertains to the impact on short-term wood supply in the years 2023–2025. The research underlines usability of our results for assessing economic and financial implications of wood supply changes.

Birch: Water-related factors may lead to a reduction in wood production of up to 0.1%–0.2%. Previous studies have established a connection between drought and declining birch wood production. Our research supports this, showing a sensitivity to water-related factors throughout the growth period.

Oak: Fungal and insect factors can reduce wood production by up to 0.1%. Previous research highlights the significant impact of fungal diseases on oak regeneration and mature tree health. Similarly, insect infestation can affect wood production, especially in weakened plants.

Alder: Water-related factors may lead to a slight reduction in wood production, around 0.02%. The impact of these factors is statistically significant between specific age ranges, suggesting a potential impact on harvesting.

Pine: Water- and climate-related factors can reduce wood production by up to 0.05%. Pine, being the primary forest-forming species in Poland, is sensitive to these factors, particularly as it approaches its harvesting age.

Spruce: Insects, fungi, and climate-related factors may lead to a reduction in wood production of up to 0.2%–0.3%. Our analyses indicate sensitivity to these factors, with a noticeable growth differential compared to the typical growth rate.

Additionally, short-term predictions based on neural networks were discussed. It was noted that these models were designed for short-term forecasts due to uncertainties regarding the long-term impact of various factors.

Furthermore, modeling wood increments in divisions significantly below the harvesting age was discussed. These stands were referred to as a "stock state," and it was acknowledged that the impact of current and 2023–2025 factors on wood increments and supply may only become apparent several decades from now.

6. Conclusions

In conclusion, our study utilized in situ data spanning 2018–2022 to analyze entire forest plantations in Poland, focusing on factors affecting wood increments for six tree species. We identified age intervals exhibiting statistically significant effects of species-factor interactions and developed neural network models for short-term wood increment predictions. Estimating a potential reduction in wood supply for 2023–2025 due to intensified factors, our findings highlight specific impacts on different tree species. For birch: water-related factors may decrease wood production by 0.1%–0.2%, aligning with prior research on birch sensitivity to drought. For oak: fungal and insect factors could lead to a 0.1% reduction in wood production, emphasizing the significant influence of fungal diseases and insect infestation on oak health. For alder: water-related factors may slightly reduce wood production (approximately 0.02%), with a statistically significant impact within specific age ranges, hinting at potential effects on harvesting. For pine: water- and climate-related factors may result in a 0.05% reduction in wood production, underscoring the sensitivity of this key forest-forming species in Poland. For spruce: insects, fungi, and climate-related factors may cause a 0.2%–0.3% reduction in wood production, showcasing noticeable growth differentials compared to the typical rate.

Our short-term predictions based on neural networks acknowledge their suitability in light of uncertainties regarding long-term factor impacts. Furthermore, our study discussed the modeling of wood increments in divisions well below harvesting time, emphasizing that the influence of current and 2023–2025 factors on wood increments and supply may only manifest several decades from now. These results carry crucial implications for the economic and financial performance of the wood industry, providing insights into the potential impacts of various factors on short-term wood supply in the coming years.

Author Contributions: Conceptualization, J.K. and S.B.; methodology, J.K and S.B.; software, J.K.; validation, S.B.; formal analysis, S.B.; investigation, J.K.; writing—original draft preparation, J.K.; writing—review and editing, J.K. and S.B.; supervision, S.B.; All authors have read and agreed to the published version of the manuscript.

Funding: This research received no external funding.

Data Availability Statement: Wood volume data for forests is accessible through the Forest Data Bank: www.bdl.lasy.gov.pl. Neural network models are available upon request in the form of MD5 files after contacting Jan Kotlarz.

Conflicts of Interest: The authors declare no conflict of interest.

Appendix A

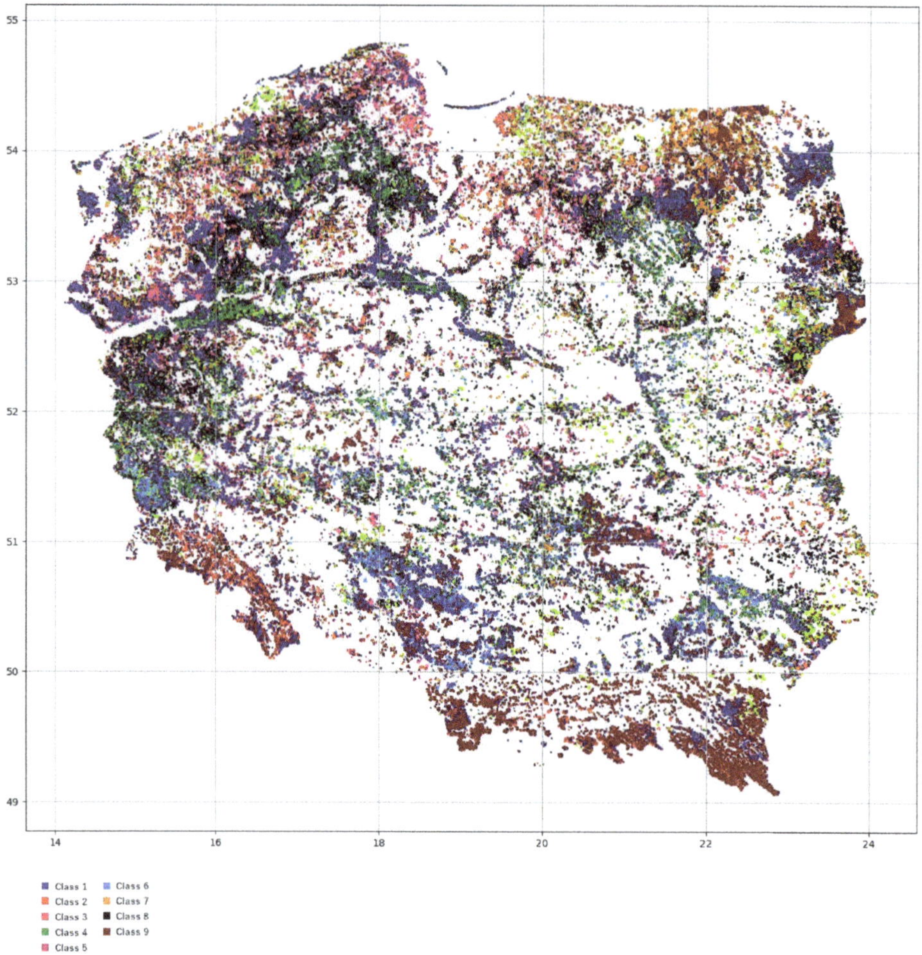

Figure A1. The spatial distribution of 100,000 forest plantations and their classification into classes—the description of classes is provided in Table 1.

References

1. Kirilenko, A.P.; Sedjo, R.A. Climate change impacts on forestry. *Proc. Natl. Acad. Sci. USA* **2007**, *104*, 19697–19702. [CrossRef] [PubMed]
2. De Frenne, P.; Lenoir, J.; Luoto, M.; Scheffers, B.R.; Zellweger, F.; Aalto, J.; Ashcroft, M.B.; Christiansen, D.M.; Decocq, G.; De Pauw, K.; et al. Forest microclimates and climate change: Importance, drivers and future research agenda. *Glob. Change Biol.* **2021**, *27*, 2279–2297. [CrossRef] [PubMed]
3. De Rigo, D.; Libertà, G.; Durrant, T.H.; Vivancos, T.A.; San-Miguel-Ayanz, J. Forest fire danger extremes in Europe under climate change: Variability and uncertainty. Ph.D. Thesis, Publications Office of the European Union, Luxembourg, 2017.
4. Luyssaert, S.; Marie, G.; Valade, A.; Chen, Y.Y.; Njakou Djomo, S.; Ryder, J.; Otto, J.; Naudts, K.; Lansø, A.S.; Ghattas, J.; et al. Trade-offs in using European forests to meet climate objectives. *Nature* **2018**, *562*, 259–262. [CrossRef] [PubMed]
5. Sierota, Z.; Grodzki, W.; Szczepkowski, A. Abiotic and biotic disturbances affecting forest health in Poland over the past 30 years: Impacts of climate and forest management. *Forests* **2019**, *10*, 75. [CrossRef]
6. Wypych, A.; Sulikowska, A.; Ustrnul, Z.; Czekierda, D. Variability of growing degree days in Poland in response to ongoing climate changes in Europe. *Int. J. Biometeorol.* **2017**, *61*, 49–59. [CrossRef] [PubMed]

7. Skrzecz, I.; Perlinska, A. Current problems and tasks of forest protection in Poland. *Folia For. Polonica Ser. A For.* **2018**, *60*, 161–172. [CrossRef]
8. Kubiak, K.; Żółciak, A.; Damszel, M.; Lech, P.; Sierota, Z. Armillaria pathogenesis under climate changes. *Forests* **2017**, *8*, 100. [CrossRef]
9. Nowakowska, J.A.; Hsiang, T.; Patynek, P.; Stereńczak, K.; Olejarski, I.; Oszako, T. Health assessment and genetic structure of monumental Norway spruce trees during a bark beetle (*Ips typographus* L.) outbreak in the Białowieża Forest District, Poland. *Forests* **2020**, *11*, 647. [CrossRef]
10. Sierota, Z.; Starzyk, J.R.; Grodzki, W. Ochrona lasu i ochrona drzewostanów–spojrzenie retrospektywne. *Sylwan* **2020**, *164*, 1011–1026.
11. Jung, T.; Orlikowski, L.; Henricot, B.; Abad-Campos, P.; Aday, A.G.; Aguín Casal, O.; Bakonyi, J.; Cacciola, S.O.; Cech, T.; Chavarriaga, D.; et al. Widespread Phytophthora infestations in European nurseries put forest, semi-natural and horticultural ecosystems at high risk of Phytophthora diseases. *For. Pathol.* **2016**, *46*, 134–163. [CrossRef]
12. Kotlarz, J.; Nasiłowska, S.A.; Rotchimmel, K.; Kubiak, K.; Kacprzak, M. Species diversity of oak stands and its significance for drought resistance. *Forests* **2018**, *9*, 126. [CrossRef]
13. Nasiłowska, S.A.; Kotlarz, J.; Kacprzak, M.; Rynkiewicz, A.; Rotchimmel, K.; Kubiak, K. The impact of drought in 2015 on the health forest condition determined using Landsat-8 OLI images. *Leśne Pr. Badaw.* **2019**, *80*, 55–68. [CrossRef]
14. Landolt, J.; Gross, A.; Holdenrieder, O.; Pautasso, M. Ash dieback due to Hymenoscyphus fraxineus: What can be learnt from evolutionary ecology? *Plant Pathol.* **2016**, *65*, 1056–1070. [CrossRef]
15. Sojka, M.; Kozłowski, M.; Kęsicka, B.; Wróżyński, R.; Stasik, R.; Napierała, M.; Jaskuła, J.; Liberacki, D. The effect of climate change on controlled drainage effectiveness in the context of groundwater dynamics, surface, and drainage outflows. Central-western Poland case study. *Agronomy* **2020**, *10*, 625. [CrossRef]
16. Lasy Polskie. Sprawozdanie Finansowo-Gospodarcze PGL LP za Rok 2022. 2022. Available online: https://www.lasy.gov.pl/pl/informacje/publikacje/informacje-statystyczne-i-raporty/sprawozdanie-finansowo-gospodarcze-pgl-lp/sprawozdanie-finansowo-gospodarcze-pgl-lp-za-rok-2022.pdf/view (accessed on 1 September 2023).
17. Bayat, M.; Knoke, T.; Heidari, S.; Hamidi, S.K.; Burkhart, H.; Jaafari, A. Modeling Tree Growth Responses to Climate Change: A Case Study in Natural Deciduous Mountain Forests. *Forests* **2022**, *13*, 1816. [CrossRef]
18. Salehnasab, A.; Bayat, M.; Namiranian, M.; Khaleghi, B.; Omid, M.; Masood Awan, H.U.; Al-Ansari, N.; Jaafari, A. Machine learning for the estimation of diameter increment in mixed and uneven-aged forests. *Sustainability* **2022**, *14*, 3386. [CrossRef]
19. Talarczyk, A. Bank Danych o Lasach w Polsce na tle baz danych i systemów udostępniania informacji z zakresu leśnictwa w innych krajach. *Probl. World Agric./Probl. Rol. Swiat.* **2015**, *15*, 150–158. [CrossRef]
20. Kotlarz, J.; Kacprzak, M. Algorytm automatycznego oszacowania zróżnicowania gatunkowego drzewostanu z wykorzystaniem zdjęć RGB koron drzew. *Pomiary Autom. Robot.* **2017**, *21*, 63–70. [CrossRef]
21. De Freitas, E.C.S.; de Paiva, H.N.; Neves, J.C.L.; Marcatti, G.E.; Leite, H.G. Modeling of eucalyptus productivity with artificial neural networks. *Ind. Crops Prod.* **2020**, *146*, 112149. [CrossRef]
22. Bayat, M.; Bettinger, P.; Hassani, M.; Heidari, S. Ten-year estimation of Oriental beech (*Fagus orientalis* Lipsky) volume increment in natural forests: A comparison of an artificial neural networks model, multiple linear regression and actual increment. *For. Int. J. For. Res.* **2021**, *94*, 598–609. [CrossRef]
23. Kharuk, V.I.; Ranson, K.J.; Oskorbin, P.A.; Im, S.T.; Dvinskaya, M.L. Climate induced birch mortality in Trans-Baikal lake region, Siberia. *For. Ecol. Manag.* **2013**, *289*, 385–392. [CrossRef]
24. Rojo, J.; Oteros, J.; Picornell, A.; Maya-Manzano, J.M.; Damialis, A.; Zink, K.; Werchan, M.; Werchan, B.; Smith, M.; Menzel, A.; et al. Effects of future climate change on birch abundance and their pollen load. *Glob. Change Biol.* **2021**, *27*, 5934–5949. [CrossRef] [PubMed]
25. Dox, I.; Skrøppa, T.; Decoster, M.; Prislan, P.; Gascó, A.; Gričar, J.; Lange, H.; Campioli, M. Severe drought can delay autumn senescence of silver birch in the current year but advance it in the next year. *Agric. For. Meteorol.* **2022**, *316*, 108879. [CrossRef]
26. Marçais, B.; Desprez-Loustau, M.L. European oak powdery mildew: Impact on trees, effects of environmental factors, and potential effects of climate change. *Ann. For. Sci.* **2014**, *71*, 633–642. [CrossRef]
27. Oszako, T.; Orlikowski, L.B. First data on the occurrence of Phytophthora cinnamomi on pedonculate oak in Poland. *Sylwan* **2005**, *149*, 47–53.
28. Attocchi, G. Silviculture of oak for high-quality wood production. *Acta Univ. Agric. Sueciae* **2015**, *39*, 1–86.
29. Tsioras, P.A.; Liamas, D.K. Hauling damages in a mixed beech oak stand. In Proceedings of the FORMEC 2010, Padova, Italy, 11–14 July 2010; pp. 1–8.
30. Rafes, P.M. Estimation of the effects of phytophagous insects on forest production. In *Analysis of Temperate Forest Ecosystems*; Springer: Berlin/Heidelberg, Germany, 1973; pp. 100–106.
31. Grigri, M.S.; Atkins, J.W.; Vogel, C.; Bond-Lamberty, B.; Gough, C.M. Aboveground wood production is sustained in the first growing season after phloem-disrupting disturbance. *Forests* **2020**, *11*, 1306. [CrossRef]
32. Nossov, D.R.; Ruess, R.W.; Hollingsworth, T.N. Climate sensitivity of thinleaf alder growth on an interior Alaskan floodplain. *Ecoscience* **2010**, *17*, 312–320. [CrossRef]
33. Elferts, D.; Dauškane, I.; Ūsele, G.; Treimane, A. Effect of water level and climatic factors on the radial growth of black alder. *Proc. Latv. Acad. Sci. Sect. B Nat. Exact Appl. Sci.* **2011**, *65*, 164–169. [CrossRef]

34. Rodríguez-González, P.M.; Campelo, F.; Albuquerque, A.; Rivaes, R.; Ferreira, T.; Pereira, J.S. Sensitivity of black alder (*Alnus glutinosa* [L.] Gaertn.) growth to hydrological changes in wetland forests at the rear edge of the species distribution. *Plant Ecol.* **2014**, *215*, 233–245. [CrossRef]
35. Zhu, L.; Liao, Z.; Liu, L.; Du, B. Effects of Extreme Drought and Heat Events on Leaf Metabolome of Black Alder (*Alnus glutinosa* L.) Growing at Neighboring Sites with Different Water Availability. *Forests* **2023**, *14*, 151. [CrossRef]
36. Eilmann, B.; Zweifel, R.; Buchmann, N.; Graf Pannatier, E.; Rigling, A. Drought alters timing, quantity, and quality of wood formation in Scots pine. *J. Exp. Bot.* **2011**, *62*, 2763–2771. [CrossRef] [PubMed]
37. Seidling, W.; Ziche, D.; Beck, W. Climate responses and interrelations of stem increment and crown transparency in Norway spruce, Scots pine, and common beech. *For. Ecol. Manag.* **2012**, *284*, 196–204. [CrossRef]
38. Matveev, S.M.; Chendev, Y.G.; Lupo, A.R.; Hubbart, J.A.; Timashchuk, D.A. Climatic changes in the East-European forest-steppe and effects on Scots pine productivity. *Pure Appl. Geophys.* **2017**, *174*, 427–443. [CrossRef]
39. Mäkinen, H.; Hallaksela, A.M.; Isomäki, A. Increment and decay in Norway spruce and Scots pine after artificial logging damage. *Can. J. For. Res.* **2007**, *37*, 2130–2141. [CrossRef]
40. Hřib, J.; Kyncl, J.; ČErný, A. A tree-ring study of Norway spruce infected with the wood-destroying fungus Armillaria mellea. *Eur. J. For. Pathol.* **1983**, *13*, 160–165. [CrossRef]

Disclaimer/Publisher's Note: The statements, opinions and data contained in all publications are solely those of the individual author(s) and contributor(s) and not of MDPI and/or the editor(s). MDPI and/or the editor(s) disclaim responsibility for any injury to people or property resulting from any ideas, methods, instructions or products referred to in the content.

remote sensing

MDPI

Article

Intra-Annual Cumulative Effects and Mechanisms of Climatic Factors on Global Vegetation Biomes' Growth

Guoming Du [1], Shouhong Yan [1,2], Hang Chen [3], Jian Yang [2,4] and Youyue Wen [2,4,*]

[1] School of Geography and Planning, Sun Yat-sen University, Guangzhou 510275, China; eesdgm@mail.sysu.edu.cn (G.D.); yanshh6@mail2.sysu.edu.cn (S.Y.)
[2] South China Institute of Environmental Science, Ministry of Ecology and Environment, Guangzhou 510535, China; yangjian@scies.org
[3] Guangzhou Planning and Natural Resources Automation Center, Guangzhou 510055, China; chenhang@gz.gov.cn
[4] National Key Laboratory of Urban Ecological Environment Simulation and Protection, Guangzhou 510535, China
* Correspondence: wenyouyue@scies.org; Tel.: +86-180-2241-8972

Abstract: Previous studies have shown that climate change has significant cumulative effects on vegetation growth. However, there remains a gap in understanding the characteristics of cumulative climatic effects on different vegetation types and the underlying driving mechanisms. In this study, using the normalized difference vegetation index data from 1982 to 2015, along with accumulated temperature, precipitation, and solar radiation data, we quantitatively investigated the intra-annual cumulative effects of climatic factors on global vegetation biomes across climatic zones. We also explored the underlying driving mechanisms. The results indicate that precipitation has a longer intra-annual cumulative effect on vegetation, with effects lasting up to 12 months for large percentages of most vegetation biomes. The cumulative effect of solar radiation is mostly concentrated within 0–6 months. Temperature has a shorter cumulative effect, with no significant cumulative effect of temperature on large percentages of tree-type vegetation. Compared to other vegetation types, evergreen broadleaf forests, close shrublands, open shrublands, savannas, and woody savannas exhibit more complex cumulative climatic effects. Each vegetation type shows a weak-to-moderate correlation with accumulated precipitation while exhibiting strong-to-extremely-strong positive correlations with accumulated temperature and accumulated solar radiation. The climate-induced regulations of water, heat, and nutrient, as well as the intrinsic mechanisms of vegetation's tolerance, resistance, and adaptation to climate change, account for the significant heterogeneity of cumulative climatic effects across vegetation biomes in different climatic zones. This study contributes to enriching the theoretical understanding of the relationship between vegetation growth and climate change. It also offers crucial theoretical support for developing climate change adaptation strategies and improving future "vegetation-climate" models.

Keywords: cumulative effects; NDVI; vegetation biomes; temperature; precipitation; solar radiation

1. Introduction

Climate change poses one of the greatest challenges to humanity, with intricate mechanisms and far-reaching impacts. As a vital nexus between the soil, atmosphere, and water, green vegetation serves as a sensitive barometer of climate change's effects on terrestrial ecosystems. Through photosynthesis, green vegetation absorbs atmospheric carbon dioxide and sequesters it in biomass and the soil, while releasing oxygen. This process is critical for achieving carbon neutrality in land ecosystems and will be pivotal in humanity's efforts to mitigate global climate change.

Extensive research has examined climate change's impacts on vegetation growth. For example, studies have shown that warming induces earlier onset and extended duration of the growing season, altering vegetation's phenological traits. Additionally, warming enables the expansion of habitats suitable for vegetation growth towards higher latitudes and elevations [1–3]. Warming can also boost vegetation's carbon sequestration capacity [4,5]. However, increases in extreme weather like heatwaves and droughts can indirectly constrain vegetation growth, transforming some carbon sinks into sources [6–8].

Climate change is typified by fluctuations in factors like temperature, solar radiation, and precipitation—the three most widely used climatic indices [4,9,10]. These factors shape plant growth and distribution by influencing processes including photosynthesis, respiration, and soil carbon decomposition. As key determinants of vegetation growth, temperature enhances productivity in colder regions [4,11], while solar radiation is the prime driver of tropical rainforest productivity [4,8,12]. In arid areas, precipitation most strongly governs vegetation productivity changes [13,14]. These factors are critical and frequently used to characterize vegetation growth shifts. Research suggests that over half of vegetation change can be attributed to variations in these three factors [4,9].

Climate change does not immediately impact vegetation. Instead, a noticeable time-lag exists. For instance, vegetation growth at middle and low latitudes exhibits a 1–3 month lag in responding to temperature changes, while middle- and high-latitude areas show a 1–3 month lag in responding to solar radiation [15]. Precipitation also induces a 1 month lag in arid and semi-arid regions [15]. Across China, a 0–1 month lag is typical in vegetation responses to temperature and precipitation changes [16]. In the Amazon basin, lags of 0–6 months (temperature), 0–9 months (radiation), and 0–6 months (precipitation) have been observed [17]. At the global scale, radiation, soil moisture, and temperature drive 0–3 month lags in vegetation growth [18]. These studies indicate that historical climatic conditions often have the strongest impact on current growth. This lag provides vegetation time to respond to changes, preventing drastic effects from abrupt anomalies.

The lag duration is primarily determined by changes in vegetation water content [19]. The timing of nutrient and water storage, transport, and excretion within the vegetation also play a role [20]. In addition, the long-term balancing and synergistic mechanisms between vegetation growth and environmental factors like soil moisture and carbon/nitrogen [21–23], along with intrinsic mechanisms conferring climate resilience like drought tolerance [24] and the thermal adaptation of photosynthesis/respiration [25–27], are also important driving factors behind the lagged effects.

Furthermore, recent research shows that climate change has a long-term, cumulative impact on vegetation growth [28]. This cumulative effect is widely observed across climates. For instance, temperature has a negligible cumulative effect in most of the Arctic and Northern Temperate Zones but a substantial 2–3 months effect in arid/semi-arid areas. Solar radiation induces a noticeable 3-month cumulative effect in the Arctic and Northern Temperate Zones, while it is shorter in other regions. Precipitation shows complex spatial cumulative effects. This research highlights the prolonged, compounding nature of climate change impacts on vegetation growth and indicates that considering the cumulative effect of climate, the relationship between "climate change-vegetation growth" can be better characterized. On a global scale, the average fitting capability is improved by 12.94%, with a maximum improvement of up to 76.85%. The fitting capability of climatic factors to vegetation growth can be effectively enhanced in approximately 77% of vegetation regions globally, especially in regions with strong interaction between vegetation and climatic factors, such as arid and semi-arid areas. These research findings have been widely confirmed by subsequent studies.

Based on the concept of cumulative effects [28], many scholars have further revealed that the intensity of cumulative climatic effects on vegetation growth varies over different growing seasons [29–34]. Specifically, a study conducted in Siberia [29] found that climatic factors exhibit an average cumulative effect of 3.6 months on vegetation growth throughout the entire growing season, with more complex ecosystems showing longer cumulative effects. In a study conducted in Inner Mongolia [35], it was discovered that temperature and precipitation have a cumulative effect on vegetation growth for up to two months, but the contribution of climatic factors to the current vegetation growth varies depending on the month, with the highest contribution coming from the current month. It has been found that cumulative drought effects impact the photosynthesis process of 52.11% of global vegetation, with an average cumulative duration of 1–4 months [32]. Furthermore, more than three-quarters (78.55%) of grasslands globally are affected by cumulative drought effects, lasting 8–10 months in arid regions of the northern hemisphere (40°N–55°N) [31] and 5–10 months in the Qinghai–Tibet Plateau region of China [34]. Meanwhile, some researchers have attempted to couple cumulative climatic effects into the "climate change-vegetation growth" relationship model in order to better characterize the response of vegetation growth to climate change [30,33,35].

Based on previous research, it can be summarized that climate change has significant cumulative effects on vegetation growth, exhibiting heterogeneity across different climate regions and vegetation types. However, the existing studies mentioned before did not systematically disentangled the effects of cumulative climatic effects on different vegetation types at global a scale, and the underlying driving mechanisms remain poorly understood. Specifically, as climatic conditions transition throughout the year, vegetation growth undergoes a distinct cyclic pattern of changes, these studies have not effectively captured the crucial intra-annual impacts of climatic variability on vegetation growth cycles, which hinders a comprehensive understanding of the relationship between climate change and vegetation growth. Therefore, this study utilized the satellite-derived Normalized Difference Vegetation Index (NDVI) and corresponding climate data (temperature, precipitation, solar radiation) from 1982 to 2015 to investigate the differential responses of major global vegetation types in different climatic zones to the cumulative effects of climatic factors. The present study also attempted to unravel the underlying driving mechanisms. Specifically, this study first extracted the spatial distribution of unchanged vegetation types during the study period. Then, through partial correlation and time-lag analyses, we examined the magnitude and duration of the cumulative effects of temperature, precipitation, and radiation on vegetation growth in these unchanged regions. We analyzed how these effects differ across vegetation types and climatic zones. Finally, we explored the potential mechanisms driving the observed cumulative climatic effects. Overall, this study provides new insights into differential vegetation responses to cumulative climatic effects over time. Our findings bridge knowledge gaps regarding vegetation–climate interactions and support climate-change-mitigation policies.

2. Data

2.1. NDVI Data

This study utilized the Global Inventory Monitoring and Modelling Studies Group (GIMMS)'s third generation NDVI dataset (GIMMS NDVI3g v1.0) to characterize vegetation growth conditions. The NDVI3g dataset has a spatial resolution of 8 km, has a temporal resolution of 15 days, and spans July 1981 to December 2015 for global land areas outside of Antarctica. We processed the data into monthly values using the Maximum Value Composite (MVC) method to minimize contamination from non-vegetation signals. The NDVI3g dataset provides a long time series, extensive spatial coverage, and high accuracy for monitoring and evaluating large-scale vegetation dynamics [15,36–38]. Moreover, the statistical analyses in this study were based on the vegetation growing season, which was defined as those months with monthly average air temperature > 0 °C and NDVI > 0.2 [15,39]. We used the subset of January 1982 to December 2015 in this study.

2.2. Climate Data

The climate data include monthly mean temperature (TEM), total monthly precipitation (PRE), and total monthly shortwave solar radiation (SOLAR) at 0.5° spatial resolution from January 1982 to December 2015. TEM and PRE are from the Climatic Research Unit (CRU) Time-Series Version 3.24 dataset, which incorporates additional station data and corrects known errors relative to earlier versions [40]. SOLAR is from the Princeton Global

Meteorological Forcing Dataset version 2, which implements improved downscaling and bias correction methods [41]. These datasets have been widely used in global vegetation–climate studies [15,28,36,37,42–44]. The climatic datasets were interpolated into 8 km using kriging interpolation.

2.3. Vegetation Type Data

The vegetation classification was based on the 1 km University of Maryland (UMD) land cover product derived from 14 years (1981–1994) of Advanced Very High Resolution Radiometer (AVHRR) data. It provides 14 land use/cover types with a relatively high overall classification accuracy ranging from 81.4% to 90.3% [45]. Furthermore, this study obtained the 500 m land use/cover product (MCD12Q1) from The Moderate Resolution Imaging Spectroradiometer (MODIS) satellite for the year of 2015. The UMD scheme with 11 vegetation types (Figure 1) was adopted, with non-vegetated areas masked out. Using the UMD and MCD12Q1 land use/cover data, the unchanged vegetation pixels for each vegetation biome were identified during the study period. Combined with the distribution of four climatic zones (Figure 1), we extracted the average values of NDVI and climatic factors for each vegetation biome in each climatic zone and then analyzed the climatic cumulative effects on different types of vegetation in different climatic zones. In addition, to conduct a more detailed analysis, we utilized the climatic zone data from IPCC (Figure A8) and the global distribution data of the four major crops (maize, rice, soybean, and wheat) provided by the Chinese Academy of Agricultural Sciences.

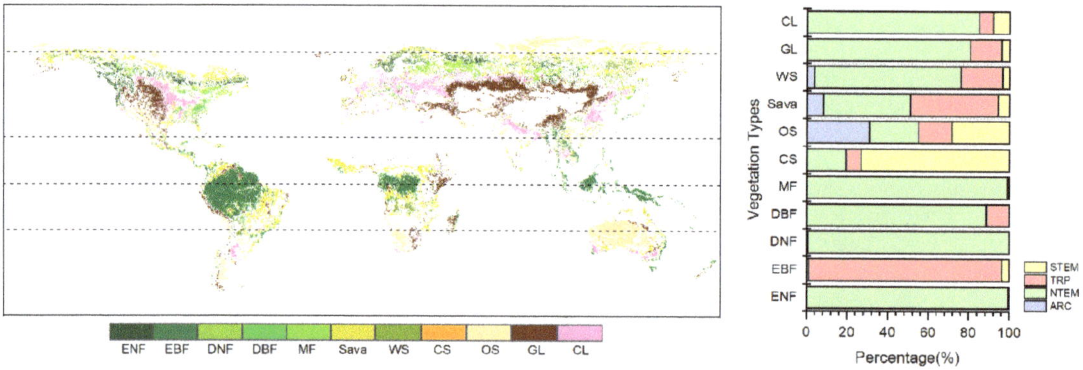

(a) (b)

Figure 1. (a) Spatial distribution map of unchanged vegetation types worldwide from 1982 to 2015; (b) a stacked bar chart illustrating the percentage distribution of different climatic zones. Evergreen Needleleaf Forest (ENF), Evergreen Broadleaf Forest (EBF), Deciduous Needleleaf Forest (DNF), Deciduous Broadleaf Forest (DBF), Mixed Forest (MF), Closed Shrubland (CS), Open Shrubland (OS), Savannas (Sava), Woody Savannas (WS), Grassland (GL), and Cropland (CL). Four climatic zones were divided [28]: the Arctic Zone (ARC, 66.5°N–90.0°N), the Northern Temperate Zone (NTEM, 23.5°N–66.5°N), the Tropic Zone (TRP, 23.5°S–23.5°N), and the Southern Temperate Zone (STEM, 23.5°S–66.5°S).

The data processing procedure and research methods are shown in Figure 2. In the figure, the blue boxes represent data processed using ArcGIS 10.8 software, and red box represents data processed using MATLAB. Please note that, to minimize the loss of data information, the post-processed datasets were resampled to 500 m to match the spatial resolution of MCD12Q1. Our analyses focused on areas where vegetation types remained unchanged during the study period.

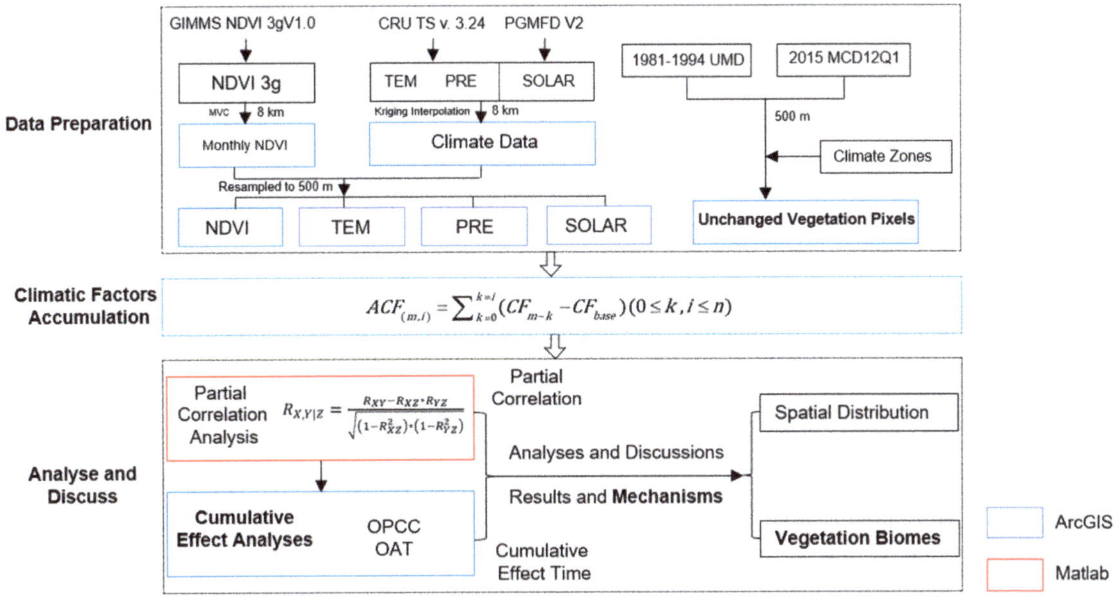

Figure 2. Technology roadmap.

3. Methods

3.1. Accumulated Climatic Factors

We defined accumulated climatic factors, including accumulated temperature (ATEM), accumulated solar radiation (ASOLAR), and accumulated precipitation (APRE), to represent the historical magnitude of climate change. These cumulative factors were calculated by summing the monthly climate data exceeding specified thresholds over a given time period, as follows:

$$ACF_{(m,i)} = \sum\nolimits_{k=0}^{k=i} (CF_{m-k} - CF_{base})(0 \leq k, i \leq n) \quad (1)$$

where ACF is the accumulated climatic factors, namely, ATEM, ASOLAR, and APRE. CF is the time series data of the original climatic factors corresponding to ACF. CF_{base} is the threshold corresponding to the cumulative climate factors. Based on previous studies, the thresholds for ATEM, ASOLAR, and APRE were set at 0 °C, 0 W/m^2, and 0 mm, respectively [28]. m is the monthly time series, ranging from January 1982 to December 2015. i is the cumulative time period. k is the difference between the current month and the cumulative months. When monthly data were below the thresholds, the cumulative values remained unchanged from the prior month. We set the maximum accumulation period (n) to 12 months, consistent with typical annual vegetation cycles [46,47].

For example, the cumulative temperature data for December 2000 with an accumulative time of 3 months was obtained by summing the temperature values for October, November, and December of 2000. If the temperature for a month was below 0 °C, 0 was added instead.

3.2. Partial Correlation Analyses and Cumulative Effect Analyses

Partial correlation analysis was used to examine relationships between NDVI and one typical accumulated climatic factor while controlling others [28]. The partial correlation coefficient takes values [−1, 1], representing complete negative correlation to complete positive correlation in statistics. The calculation formula is as follows:

$$R_{X,Y|Z} = \frac{R_{XY} - R_{XZ} \times R_{YZ}}{\sqrt{(1 - R_{XZ}^2) \times (1 - R_{YZ}^2)}} \quad (2)$$

where X, Y and Z are the independent variable, dependent variable, and control variable, respectively. $R_{X,Y|Z}$ represents the partial correlation coefficient between X and Y after removing the influence of the Z variable. R_{XY}, R_{XZ}, and R_{YZ} are the Pearson correlation coefficients between X and Y, X and Z, and Y and Z, respectively. When the absolute value of the partial correlation coefficient is between 0.8 and 1.0, it indicates a very strong correlation. A correlation between 0.6 and 0.8 is considered strong, 0.4 to 0.6 indicates a moderate degree of correlation, 0.2 to 0.4 suggests a weak correlation, and 0.0 to 0.2 indicates a very weak correlation or no correlation.

In this study, the accumulated duration ranges from 0 to 12 months. For each assumed accumulated duration, a partial correlation coefficient can be calculated. When a partial correlation coefficient reaches the maximum coefficient of determination, the corresponding accumulated duration is considered the optimal accumulated time (OAT). The partial correlation coefficient at this duration is also considered the optimal partial correlation coefficient ($OPCC$). The specific formula is as follows:

$$OPCC = PCC_k, when\ PCC_k^2 = Maximum\left\{PCC_0^2, PCC_1^2, \ldots, PCC_i^2, \ldots, PCC_{12}^2\right\}(0 \le k, i \le 12) \quad (3)$$

$$OAT = k, when\ PCC_k^2 = Maximum\left\{PCC_0^2, PCC_1^2, \ldots, PCC_i^2, \ldots, PCC_{12}^2\right\}(0 \le k, i \le 12) \quad (4)$$

where $OPCC$ is the optimal partial correlation coefficient. i is the assumed accumulated duration, with values ranging from 0 to 12. k is the cumulative duration associated with the maximum deterministic coefficient. PCC_k is the partial correlation coefficient at a certain assumed cumulative duration. PCC_k^2 is the coefficient of determination for that correlation coefficient. OAT is the optimal accumulated time.

For example, we first calculate the partial correlation coefficients between the NDVI of a certain pixel and 13 cumulative temperature time series data from 0 to 12 months, with the corresponding cumulative precipitation and cumulative solar radiation data as control variables. The cumulative time corresponding to the largest squared value of these 13 partial correlation coefficients is the optimal cumulative time for the temperature of this pixel.

4. Results

4.1. Global Distribution of Unchanged Vegetation Types

According to Figure 1, it can be observed that the unchanged vegetation types were widely distributed in the northern temperate (NTEM) and tropical (TRP) climatic zones during the study period. Most of the ENFs, DNFs, DBFs, MFs, WS, GLs, and CLs were located in NTEM, accounting for 99.20%, 99.13%, 88.65%, 99.10%, 72.40%, 80.32%, and 84.94%, respectively. EBFs were predominantly in the TRP at 95.24%. CS were mainly in the southern temperate (STEM) zone at 73.31%. OSs were more evenly distributed across climatic zones. Sava were mostly found in NTEM and TRP (85.94% combined).

Comparing with the global climate regions defined by the Intergovernmental Panel on Climate Change (IPCC), the following were found (Figure 3): ENFs are mainly distributed in the Cool Temperate Moist (42%) and Boreal Moist (48%) climate regions. EBFs are primarily found in the Tropical Wet (60%) and Tropical Moist (32%) climate regions. DNFs are mainly in the Boreal Moist (64%) and Boreal Dry (36%) climate regions. DBFs are primarily in the Cool Temperate Moist (42%), Warm Temperate Moist (39%), and Tropical Moist (10%) climate regions. MFs are predominantly located in the Cool Temperate Moist (67%) and Boreal Moist (25%) climate regions. CSs are largely distributed in the Tropical

Dry (69%) and Warm Temperate Dry (10%) zones. OS are mostly in the Tropical Dry (38%), Polar Moist (19%), Boreal Moist (18%), and Warm Temperate Dry (13%) climate regions. Sava are mainly distributed in the Tropical Moist (30%) and Boreal Moist (28%) climate regions. WS are mostly located in the Boreal Moist (40%), Cool Temperate Moist (13%), and Tropical Moist (10%) climate regions. GLs are primarily found in the Cool Temperate Dry (50%) and Warm Temperate Dry (13%) climate regions. CLs are mainly distributed in the Cool Temperate Dry (30%), Warm Temperate Dry (25%), and Cool Temperate Moist (21%) climate regions.

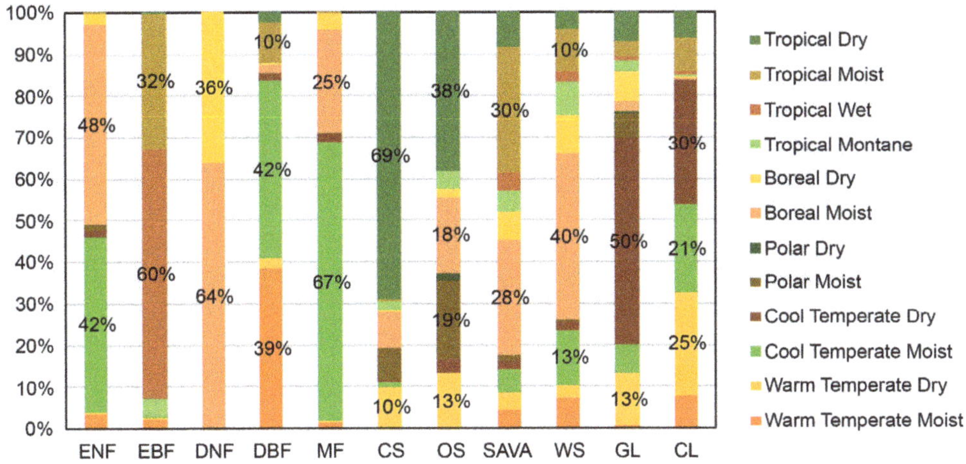

Figure 3. Distribution proportion of different types of vegetation in dry and wet climate zones. Here, the climate zones were defined by IPCC (https://esdac.jrc.ec.europa.eu/projects/RenewableEnergy/, accessed on 2 September 2023).

4.2. Accumulated Climatic Effect Durations

The cumulative effects of precipitation on vegetation growth showed complex patterns across climatic zones and vegetation types (Figures 4 and 5). Different optimal accumulation durations were observed, varying significantly by region and vegetation type. Globally, ENFs exhibited cumulative precipitation effects mainly at 6–7 months (31.15%) and 12 months (32.79%). EBFs responded primarily at 0–3 months (40.49%). DNFs showed a major response at 2–3 months (58.07%). DBF effects accumulated over 12 months (43.33%) and 0–3 months (31.55%). MFs were mainly distributed in NTEM, and 59.00% of MFs in this climate zone exhibited a cumulative response to precipitation for up to 12 months, resulting in a 12-month cumulative effect on 58.59% of global MF growth. CSs in NTEM accumulated effects over 7–12 months (78.71%) versus 3–10 months in STEM (94.32%), showing a mixed global pattern. OSs, Sava, and WS demonstrated similar heterogeneity as CSs. However, it is worth noting that for Sava and WS in NTEM, 12-month effects were notable at 30.36% and 22.28%. GLs and CLs exhibited primary global responses at 12 months (38.62% and 40.52%) and 0–3 months (41.35% and 32.52%).

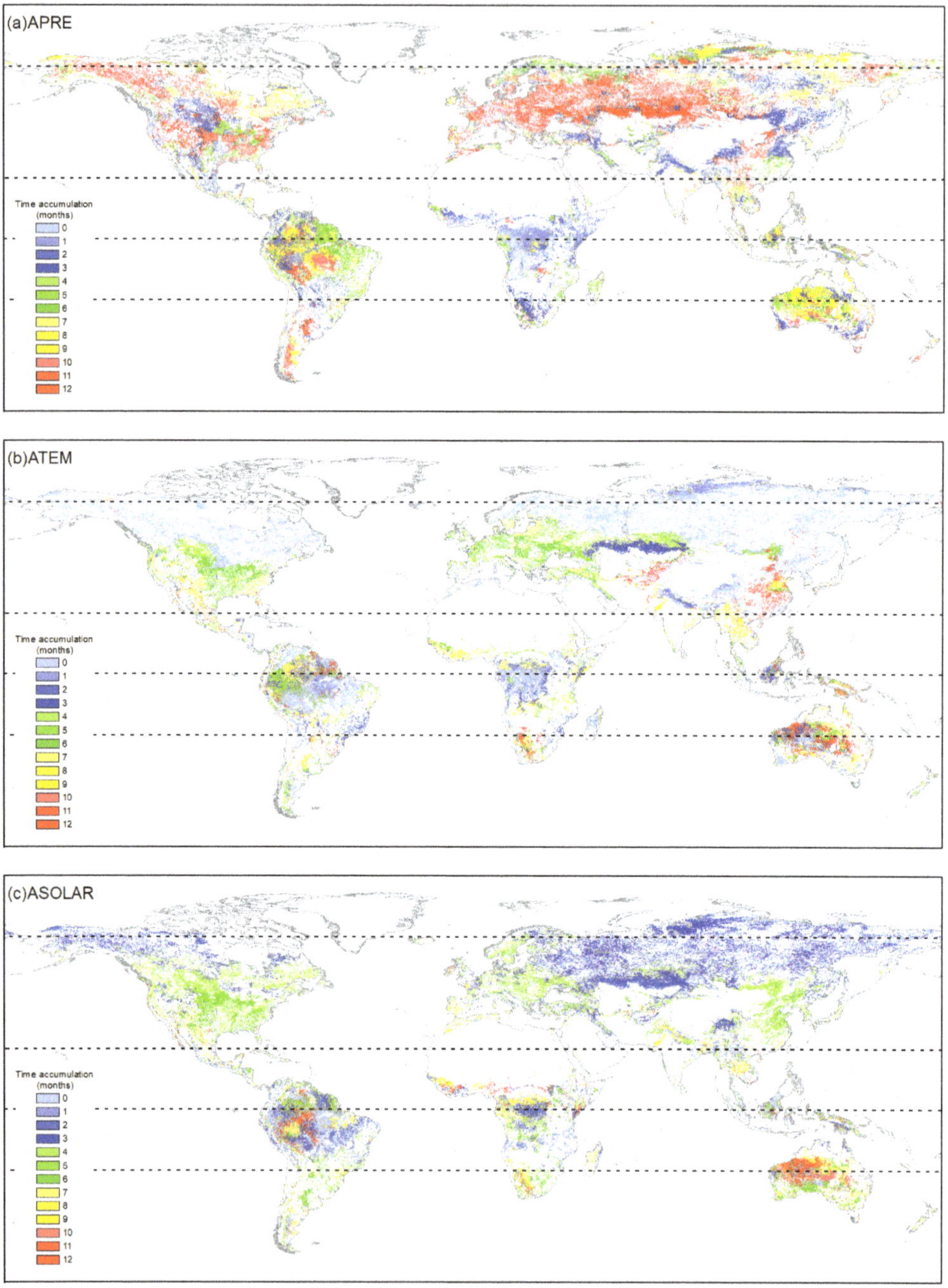

Figure 4. The distribution of accumulation effects durations of climatic factors (**a**) APRE, (**b**) ATEM, and (**c**) ASOLAR on vegetation growth from 1982 to 2015. The white areas represent regions with a *p*-value greater than 0.05, areas with vegetation type transitions, or non-vegetated areas.

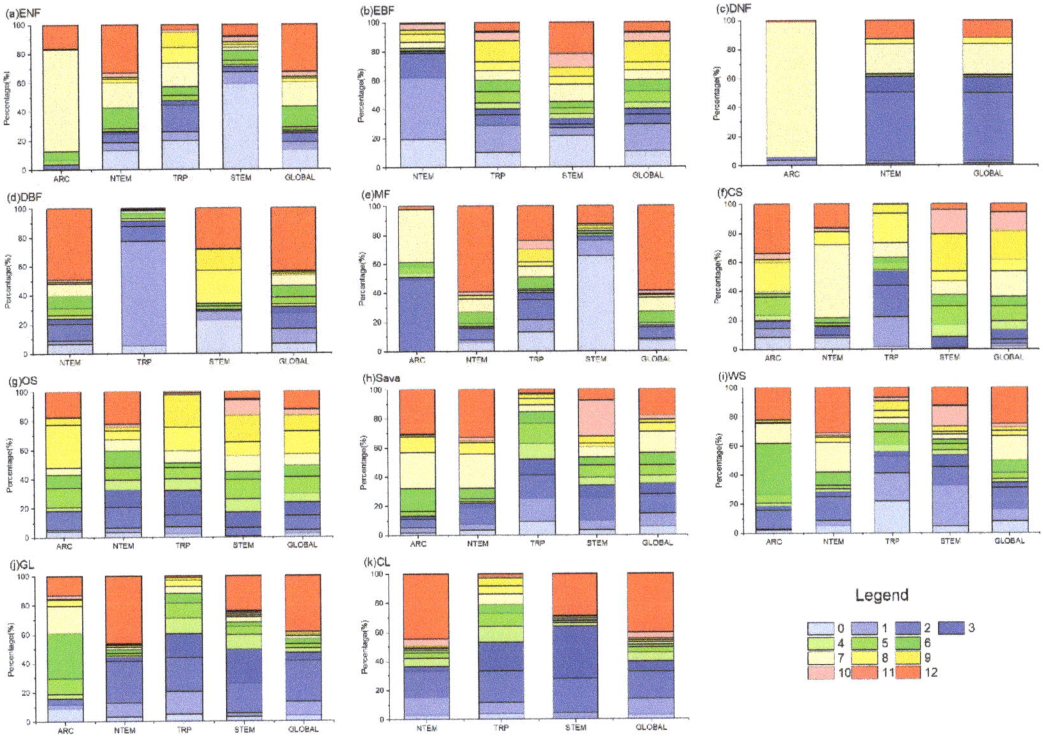

Figure 5. The proportion of OAT of PRE on global vegetation biomes. In the figure, (**a**) ENF, (**b**) EBF, (**c**) DNF, (**d**) DBF, (**e**) MF, (**f**) CS, (**g**) OS, (**h**) Sava, (**i**) WS, (**j**) GL, and (**k**) CL are shown.

According to Figures 4 and 6, the cumulative effects of solar radiation on vegetation growth were mainly concentrated within 0–6 months across climate zones and vegetation types. However, some differences existed in the dominant response durations. Globally, ENFs showed primary cumulative solar effects within 3–4 months (70.42%), driven by the response in NTEM. EBFs exhibited a 0–3-month response (57.55%) worldwide, with balanced effects in other months. Over 97% of global DNFs responded within 1 month consistently across all climatic zones. Regarding DBFs, NTEM exhibited cumulative solar effects accumulated over 3–5 months (93.36%), versus 1–2 months (48.32%) and 5–6 months (29.83%) in TRP, and 6–7 months (84.08%) in STEM. Considering all zones, the global cumulative effects of solar radiation on DBF mainly occurred within 3–5 months (84.31%). MFs responded predominantly within 3–4 months (91.69%). CS showed shorter cumulative solar effects in NTEM (0–4 months, 96.13%) than in STEM (4–7 months, 60.41%; 10–12 months, 24.80%). The cumulative effects of solar radiation on OS, Sava, WS, and GLs shared similar characteristics with CSs. However, unlike CSs, a significant proportion of OSs (31.39%) exhibited a response to accumulated solar radiation for up to 12 months in TRP, and a considerable proportion of GLs (41.18%) showed an instantaneous response in ARC. At a global scale, solar radiation had a dominant cumulative effect on CLs within 3–6 months (79.55%).

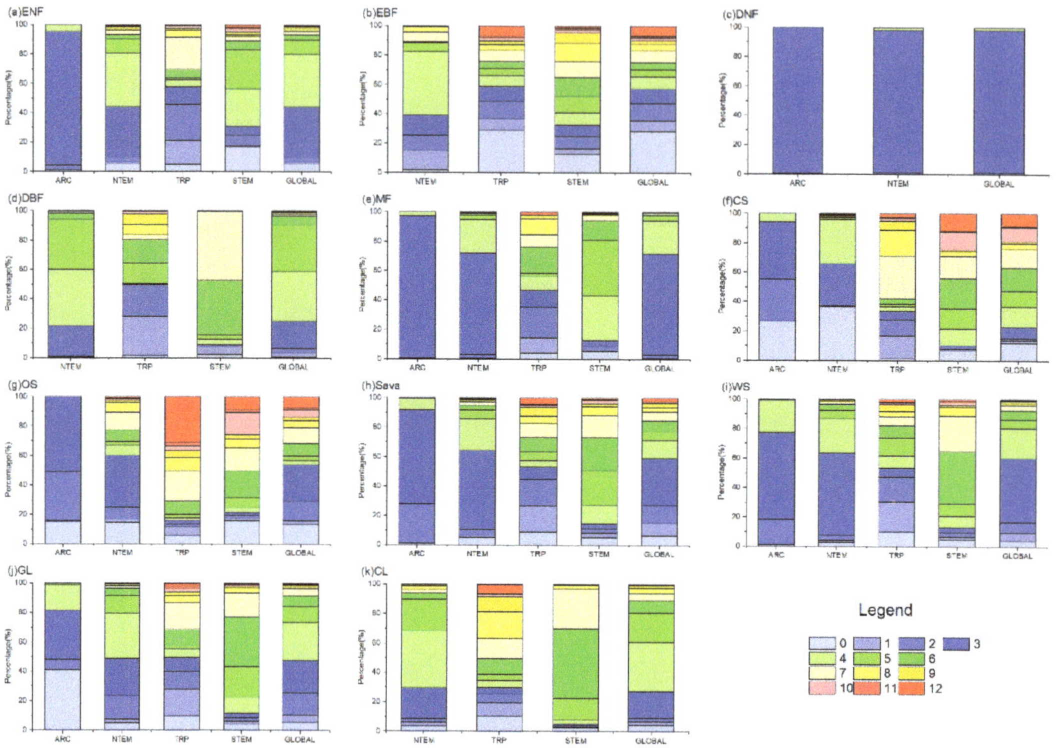

Figure 6. The proportion of OAT of SOLAR on different vegetation types globally. In the figure, (**a**) ENF, (**b**) EBF, (**c**) DNF, (**d**) DBF, (**e**) MF, (**f**) CS, (**g**) OS, (**h**) Sava, (**i**) WS, (**j**) GL, and (**k**) CL are shown.

Figures 4 and 7 show that in ARC, the cumulative temperature effect on almost all vegetation types was within 0–1 month, indicating an urgent temperature requirement for vegetation growth in these regions. In NTEM, a large portion of ENFs (80.38%), DNFs (98.63%), MFs (87.09%), CSs (80.71%), Sava (76.64%), and WS (79.58%) exhibited a cumulative temperature response of 0 months. EBFs (67.65%) responded predominantly over 8–9 months, while DBFs showed 0 months (39.86%) and 6–7 months (54.87%). GL (87.41%) and CLs (85.94%) responded mostly within 0–6 months. In TRP, the cumulative effect of temperature on vegetation biomes was more complex, with no clear dominance. However, ENFs, MFs, and OSs showed notable 12-month responses compared to other zones. In STEM, there were significant cumulative temperature effects of 5–7 months for ENFs (56.85%), 1–2 months for DBFs (81.17%), and 6–7 months for MFs (65.36%). CSs and OSs both demonstrated 0-month (37.18%, 30.83%) and 12-month (21.96%, 21.52%) effects. CLs showed 0-month (44.36%) and 7-month (28.17%) effects. Other types had relatively even temperature responses.

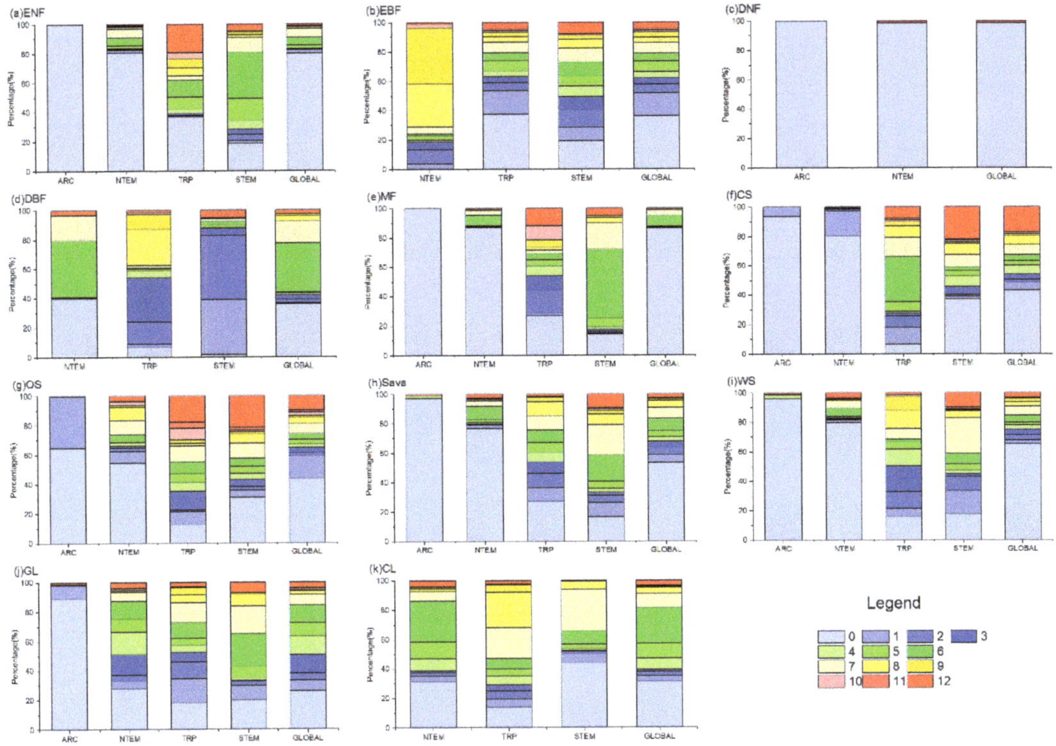

Figure 7. The proportion of OAT of TEM on different vegetation types globally. In the figure, (**a**) ENF, (**b**) EBF, (**c**) DNF, (**d**) DBF, (**e**) MF, (**f**) CS, (**g**) OS, (**h**) Sava, (**i**) WS, (**j**) GL, and (**k**) CL are shown.

4.3. Correlations between Accumulated Climatic Factors and Vegetation Growth

Regarding APRE correlation (Figures A1 and 8), vegetation types generally exhibited a weak correlation. In ARC, on average, over 70% of pixels for each vegetation type showed a weak or very weak negative correlation, especially ENFs, DNFs, and GLs (81.41%, 93.02%, 88.52%). For MFs, 52.52% of the pixels showed a weak positive correlation. In NTEM and STEM, over 75% of the pixels for DNFs, DBFs, OSs, Sava, WS, GLs, and CLs were primarily weakly positively correlated. Notably, OSs and GLs had a significant proportion of pixels showing a strong positive correlation in STEM (averaging around 15%). ENFs was mainly weakly correlated in NTEM (43.71% positive, 54.11% negative), but weakly negatively correlated in STEM (76.84%). EBFs exhibited a weak or very weak negative correlation in NTEM (73.05%) but a weak or very weak positive correlation (72.94%) in STEM. The situation was reversed for MFs compared to EBFs. CSs was weakly negatively correlated in NTEM (75.90%) but moderately, strongly positively correlated in STEM (86.21%). In TRP, there was a relatively strong correlation between APRE and vegetation growth. Specifically, on average, around 65% of the pixels for each vegetation type exhibited a weak-to-moderate positive correlation. CSs, OSs, Sava, WS, GLs, and CLs had significant strong positive proportions (67.63%, 23.82%, 21.57%, 32.85%, 23.57%, respectively).

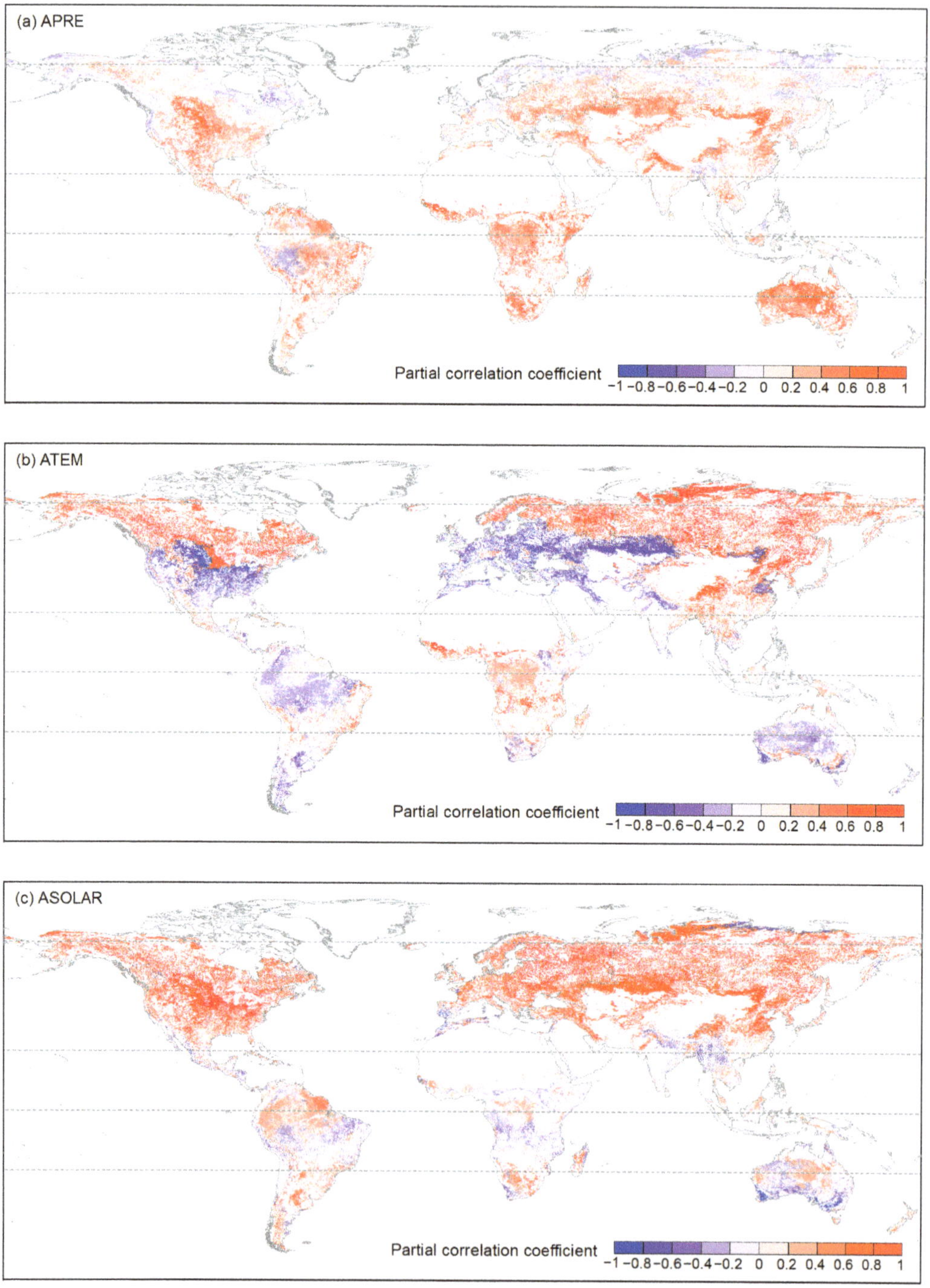

Figure 8. The distribution of OPCC between global vegetation growth and accumulated climate factors (**a**) APRE, (**b**) ATEM, and (**c**) ASOLAR from 1982 to 2015. The white areas represent regions with a *p*-value greater than 0.05, areas with vegetation type transitions, or bare areas.

For the ASOLAR correlation (Figures A2 and 8), vegetation types showed significant heterogeneity across climate zones. In ARC, all vegetation biomes exhibited strong to very strong positive correlations (>98% for ENFs, DNFs, MFs, Sava, WS). In NTEM, over 90% of DNFs, DBFs, MFs, WS, and CLs showed strong to very strong positive correlations. ENF, Sava, and GL showed moderate and above-positive correlations (~90%). A total of 86.82% of EBFs exhibited weak-to-strong negative correlations, while CSs correlated strongly positively (53.19%) and negatively (24.08%). In TRP, all vegetation types showed weak or lower correlations with roughly equal positive/negative proportions. In STEM, the correlation patterns were more complex. ENFs, DBFs, and MFs had weakly strongly positive correlations (77.79%, 90.48%, 87.51%), while over half of EBFs and WS had a weakly moderately positive correlation (58.42%, 55.42%). A total of 48.62% of CSs exhibited a weak-to-moderate negative correlation. A total of 81.63% of OSs displayed a weak or lower correlation (with positive correlation accounting for 49.69%). Sava and WS showed predominantly weak-to-moderate correlations (79.60% 78.85%) with positive proportions of 43.12% and 55.42%, respectively. A total of 64.61% of GLs exhibited moderate-to-strong positive correlations, while 51.89% of CLs showed strong and above-negative correlations. Lastly, 38.20% of CLs also exhibited moderate-to-strong positive correlations.

Regarding the correlation with ATEM (Figures A3 and 8), vegetation types differed significantly in strength. In ARC, most ENFs, DNFs, MFs, CSs, OSs, Sava, WS, and GLs exhibited strong-to-very strong positive correlations (>95% pixels). In NTEM, the majority of ENFs, DNFs, MFs, CSs, Sava, and WS showed strong-to-very strong positive correlations (>72%). In TRP, most vegetation biomes displayed weak or lower correlations. In STEM, 80.44% of ENFs showed negative weak-to-strong correlations, and 78.75% of EBF exhibited a weak or lower correlation. A total of 89.78% of DBF showed weak-to-moderate positive correlations. A total of 79.92% of MF showed moderate-to-strong negative correlations. A total of 70.04% of CS exhibited weak-to-moderate negative correlations, a total of 83.03% of OS showed moderate and below-negative correlations. Sava and WS exhibited weak-to-moderate correlations (80.65%, 78.55%) with positive and negative areas being roughly equal. A total of 78.98% of GL exhibited moderate-to-strong correlations (with negative correlations accounting for 47.06%), and 75.43% of CL showed negative moderate-to-strong correlations.

5. Discussion

This study found that a large proportion of ENFs, MFs, and DBFs had a cumulative effect on precipitation for up to 12 months (weakly positive correlation). These vegetation types are mainly distributed in the Cool Temperate Moist and Boreal Moist climate regions, accounting for over 85% of the total. This may be due to the deep root systems of these vegetation types, which allow them to effectively utilize moisture from the deeper soil layers to sustain growth while reducing dependency on current precipitation. However, the long-term accumulation of precipitation replenishes moisture in deeper soils, promoting the growth of these forests [48]. Additionally, the complex ecosystem composition of MFs leads to niche partitioning and growth-promotion mechanisms. MFs exhibit strong adaptation mechanisms, including tolerance, avoidance, and recovery, to cope with drought [49–52]. The higher evapotranspiration, lower ecological connectivity, and increased water demand during the growing season in DBFs likely contributes to its long-term cumulative precipitation effect [48,53,54]. DNFs show a shorter cumulative effect of precipitation compared to DBFs, consistent with previous findings [15,48]. The study also indicates that a significant proportion of DNF pixels exhibit a short response (2 months) to APRE, mainly in the Boreal Moist climate, suggesting a significant and urgent demand for precipitation by DNFs in this climate zone. EBFs primarily grow in hot and humid tropical and subtropical regions. This study found that EBFs in these two regions mainly exhibited weak or very weak negative correlations with APRE, potentially due to increased evapotranspiration, reduced radiation, and unfavorable temperatures for EBFs' growth [15].

GANG C et al. [55] found that the NPP of GLs is positively correlated with annual mean precipitation, while its correlation with annual average temperature varies. This aligns with the current study's findings. The regions with a high GL distribution in NTEM, TRP, and STEM all showed a positive correlation of over 90% with APRE and a varying proportion of positive and negative correlations with ATEM. GLs are mainly distributed in the Cool Temperate Dry climate regions of North America and Eurasia in NTEM, and the Qinghai–Tibet Plateau (Cool Temperate Moist climate region). In the Cool Temperate Dry regions, limited water resources due to scarce precipitation and high evapotranspiration become the most important growth-limiting factors, resulting a moderate positive correlation between grassland and precipitation in NTEM as found by this study. The shallow root systems of GLs enable a quick response to alleviated water limitations from precipitation inputs. Due to the urgent precipitation demand, any rainfall during a growing season significantly promotes growth [56,57]. Therefore, previous research [15] discovered a short lag effect (1 month) of precipitation on GLs, while this study found a longer cumulative effect of precipitation (12 months). On the other hand, GLs in the Qinghai–Tibet Plateau (Cool Temperate Moist climate region) can mitigate water limitations through snow-cover melting. As a result, the demand for precipitation by GLs is not strong, leading to a significantly smaller cumulative effect of precipitation (3 months). CLs are primarily distributed in the Cool Temperate Dry, Cool Temperate Moist, and Warm Temperate Dry climate regions. In these areas, approximately half of the CLs exhibited a cumulative effect of precipitation lasting up to 12 months, indicating a long-term influence of precipitation.

Vegetation growing in high-latitude and high-altitude regions is primarily temperature-dependent [15]. Temperature influences vegetation growth by increasing photosynthetic enzyme activity [58] and reducing autumn foliage degradation rates [59], promoting vegetation growth. This rapid response [15,48] has led to urgent temperature response mechanisms in these vegetation types. This may be an important reason for the strong positive temperature correlation and short cumulative temperature effects (0–1 month) for Cool Temperate climate vegetation types such as ENFs, DNFs, DBFs, MFs, WS, Sava, and, specifically, GLs on the Qinghai–Tibet Plateau. Furthermore, we found that the GLs growing in the North American Great Plains and the Eurasian continent west of Lake Balkhash to the Caspian Sea were significantly negatively correlated with temperature, and the cumulative response to temperature was relatively long (more than 4 months), which may be because the two regions belong to a temperate humid climate, and the increase in temperature significantly reduces soil moisture, which is unfavorable for local vegetation growth. In contrast, vegetation in arid regions has evolved mechanisms to withstand drought, allowing them to tolerate prolonged environmental anomalies [60]. The heterogeneous response durations of EBFs, OSs, CSs, Sava, WS, GLs, and CLs suggest complex temperature response mechanisms.

Increasing growing-season solar radiation enhances photosynthetic enzyme activity [58], providing more photosynthetically active radiation (PAR) to improve plant photosynthetic capacity and carbon sequestration [61,62]. Sunlight also delays leaf abscisic acid accumulation, slowing down aging and extending the growing season [63,64]. Wen et al. (2019) [28] documented that solar radiation promotes snow melting and indirectly increases surrounding temperatures, thus facilitating vegetation growth. However, this energy conversion process requires a certain accumulation period. It is likely why this study found that ENFs, DNFs, MFs, Sava, and WS, which are widely distributed in ARC and NTEM, had a strong or above-positive correlation with ATEM and ASOLAR, and they mainly exhibited a 0–1 month cumulative effect of temperature, while the cumulative effect of solar radiation mainly occurred within 2–3 months. This result further validates previous findings [28,48] that vegetation growth in ARC and NTEM can respond rapidly to temperature changes but requires a certain amount of accumulated solar radiation. This study found varying cumulative effects of solar radiation on vegetation growth in low-latitude arid and semi-arid regions, confirming previous research findings [15,48]. This relationship is closely related to the complex vegetation composition in these regions. The growth of vegetation in these regions shows a negative correlation with ASOLAR. This could be because an increase in long-term solar radiation promotes evapotranspiration and leads to water loss, which is unfavorable for vegetation growth [24]. However, in the EBF of the Amazon rainforest, there is a high positive correlation with solar radiation, consistent with previous study [65], which indicates that solar radiation is an important driving factor for leaf shedding and new leaf growth in the Amazon rainforest. Additionally, this study's results suggest that the cumulative response of grasslands in the North American Great Plains to temperature and solar radiation is longer than that of grasslands in the Eurasian continent west of Lake Balkhash to the Caspian Sea. This indicates that grasslands in the Great Plains have developed stronger mechanisms to tolerate drought and light conditions. We believe this may be related to the widespread presence of C4 grasslands in North America, which have strong characteristics of adaptation to high temperatures, intense light, and arid environments [60].

This study found that, globally, the average OAT for vegetation types is generally ranked as precipitation > solar radiation > temperature, aligning with the research conducted by DING et al. [48]. Previous studies have indicated that green vegetation can adapt to seasonal variations in water, heat, and solar radiation resources through leaf shedding [66]. For tropical rain forests, the peak of leaf litter occurs during the dry spring or winter seasons and is predominantly regulated by precipitation and radiation, while for boreal forests, the peak of leaf litter occurs in different seasons and is mostly regulated by radiation and temperature [66]. Furthermore, studies have shown that temperature drives the distribution of deciduous and evergreen forests through its impact on nitrogen mineralization rates [67]. The effect of temperature on aboveground biomass accumulation in DBFs transitions from initial inhibition to subsequent promotion during the maturation stage [67]. Additionally, precipitation contributes to biomass accumulation in maturing ENFs but inhibits biomass accumulation in maturing EBFs, DNFs, and DBFs [67]. This indicates that different vegetation types exhibit complex and not absolute response mechanisms to climatic factors, which is also a major cause of spatiotemporal heterogeneity in cumulative climatic effects on global vegetation.

This study found stronger cumulative solar radiation effects, followed by temperature, and then precipitation, for widely distributed Arctic (ARC) vegetation types, including OSs, Sava, WS, and GLs, congruent with the results of SHI et al. [29]. DBFs, Sava, and WS exhibited prolonged intra-annual response durations, which is consistent with the findings of DING et al. [48]. This study indicates that the global distributions of the cumulative effects of climate on GLs and CLs, Sava and WS, and CSs and OSs share a similarity pattern, aligning with previous research [15,48]. This may be related to the similarity in community types, intrinsic characteristics, and growth environments shared by these vegetation types [10,68]. For Sava and WS, the cumulative effects of each climate factor in NTEM and STEM are more similar to those of OSs, but in TRP, they are more similar to those of GLs. This may be because in TRP, the NDVI information captured by satellites is mainly derived from grasslands, while in NTEM and STEM, it is more derived from trees [69]. CSs are predominantly distributed in STEM, while OSs are ubiquitous across climatic zones. This study reveals that there is no clear absolute advantage in the cumulative effects of precipitation on OSs and CSs, which is true also for solar radiation on CSs, indicating complex water and radiation demand for OSs' and CSs' growth [15,48]. Additionally, this study reveals that solar radiation mainly exhibits cumulative effects on OSs within 0–3 months, while temperature exhibits cumulative effects of 0 months on OSs and CSs, illustrating the urgent demand of different shrub categories for light and heat energy [15,48]. However, the specific reasons behind these effects need further investigation.

Combining the global distribution of four major crops as mapped by the Chinese Academy of Agricultural Sciences, it can be observed that (Figures A5–A7) different crops exhibit varying cumulative effects on temperature, precipitation, and solar radiation. For example, a larger proportion of corn, soybean, and wheat show cumulative effects of precipitation lasting up to 12 months, while rice primarily shows 2–3 months. Rice exhibits cumulative effects of solar radiation mainly within 4–8 months, while corn and soybean show mainly within 4–6 months, and wheat exhibits shorter times (around 3–4 months). Additionally, four major crops exhibit cumulative effects on temperature mainly within 0 and around 6 months. It is worth noting that CLs are a vegetation type that is heavily influenced by human activities [70]. People cultivate different crops based on factors such as seasons, climate, soil conditions, and the water requirements of these crops, which can be met through irrigation [71]. The growth of CLs is also influenced by local weather conditions, pests and diseases, fertilizer application, and other factors [72–74]. These factors contribute to the complexity of the duration and relationships of climatic factors' cumulative effects on CLs, which may be more intricate than what is revealed in this study.

In summary, the intrinsic tolerance, resistance, and adaptation mechanisms evolved in diverse vegetation types over time, alongside shifts in water, thermal, and nutrient limitations imposed by changing climatic factors, represent critical determinants of the spatiotemporal heterogeneities in cumulative climatic effects across vegetation biomes. The present findings enrich the theoretical comprehension of vegetation–climate interactions, elucidating the intricate processes governing vegetative responses to climate change. These insights contribute to formulating strategies for addressing and increasing adaptation to climate change. However, it is important to note the complex, multifaceted nature of climate change impacts on vegetation. This study relied solely on linear modeling to characterize vegetation–climate linkages, potentially oversimplifying true interactions. Furthermore, quantitative assessments of anthropogenic effects, extreme events, and other disturbances were not conducted here. Future work could explore nonlinear techniques to further disentangle cumulative climatic effect traits across vegetation types. Examining additional climatic and non-climatic parameters (e.g., human activity, natural disturbances) is also needed to elucidate specific factors eliciting pronounced heterogeneous responses among tropical/Australian vegetation biomes, EBFs, OSs, CSs, Sava, WS, and CLs. This study spans over 30 years of data, and through comparative analyses of land use at the beginning and end of the study period, efforts were made to eliminate regions with significant changes in vegetation types during the research period. However, few pixels might still have exhibited land use changes that were not captured, which could have introduced a certain degree of error into the research results. Future research could attempt to further eliminate these pixels. In addition, global climate change is a complex phenomenon, with cumulative effects influenced by both natural and anthropogenic factors. This study only employed three main climatic factors to investigate the cumulative effects of climate change on vegetation growth, and, thus, our conclusions explain the main cumulative effects reflected by these climatic factors. Future research could incorporate more influencing factors, thereby more comprehensively characterizing this phenomenon. Furthermore, consistent with previous studies [28], heterogeneities exist in the desert of central Australia, as well as the African and the Amazon rainforests. The reason might be that NDVI is sensitive to soil background, which makes it difficult for the vegetation index based on red and near-infrared reflectance to distinguish perennial vegetation in arid and semi-arid ecosystems from the red soil, as these plants often lack contrast between red and near-infrared reflectance [75,76]. NDVI tends to saturate over regions like the Amazon rainforest, making it more susceptible to residual cloud or aerosol noise than other vegetation indices [77,78]. We argue and strongly recommend that field-based studies be conducted to reduce this confusion.

6. Conclusions

This study utilized accumulated temperature, precipitation, and solar radiation data, as well as GIMMS NDVI3g data, to examine the intra-annual cumulative climatic effects on global vegetation biomes from 1982 to 2015 quantitatively. The main conclusions are as follows.

Precipitation exerted the longest cumulative effects on vegetation growth worldwide with significant intra-annual heterogeneity. Specifically, precipitation primarily displayed

6–7 and 12-month cumulative effects on ENFs, 0–3- and 12-month effects on GLs and CLs, 2–3-month effects on DNFs, and up to 12-month effects on over 40% of DBFs and MFs. Solar radiation's cumulative effects on vegetation growth were concentrated within 0–6 months, with 3–5-month effects dominating in ENFs and DBFs and 3-month effects in DNFs and MFs. Temperature exhibited the shortest cumulative effects, with no significant impacts observed for many tree types. EBFs, CSs, OSs, Sava, and WS showed the most complex cumulative effects without a clear duration dominance. Vegetation types demonstrated weak-to-moderate correlations with APRE, highlighting the complex role of precipitation. Apart from EBFs, CSs, OSs, and Sava, other vegetation types showed strong-to-extremely-strong positive ATEM and ASOLAR correlations, indicating temperature and radiation dependency during growth cycles. This research indicates that the intrinsic tolerance, resistance, and adaptation mechanisms evolved in vegetation types over time, alongside shifting water, thermal, and nutrient limitations imposed by climate changes, are critical determinants of the spatiotemporal heterogeneities in intra-annual cumulative climatic effects. The results of this study provide key theoretical underpinnings to construct more accurate vegetation–climate relational models moving forward.

Author Contributions: G.D.: Data curation, Formal analysis, Software, Supervision, Writing—original draft, Writing—review and editing, Validation, Visualization. S.Y.: Data curation, Formal analysis, Software, Writing—original draft, Writing—review and editing, Validation, Visualization. H.C.: Resources, Supervision, Writing review and editing. J.Y.: Methodology, Resources, Supervision, Writing review and editing. Y.W.: Conceptualization, Methodology, Funding acquisition, Resources, Supervision, Writing original draft, Writing—review and editing. All authors have read and agreed to the published version of the manuscript.

Funding: This work was funded by the National Science Foundation for Young Scientists of China (No. 42007406), the Guangzhou Science and Technology Plan Project (No. 202102020666), the Guangdong Natural Science Foundation-General Program (No. 2022A1515010632), and the Fundamental Research Funds for the Central Public Welfare Research Institutes (No. PM-zx097-202305-214).

Data Availability Statement: Data are contained within the article.

Acknowledgments: The authors would like to thank the researchers who provided the opensource algorithms, which were extremely helpful to the research reported in this paper.

Conflicts of Interest: The authors declare no conflicts of interest.

Appendix A

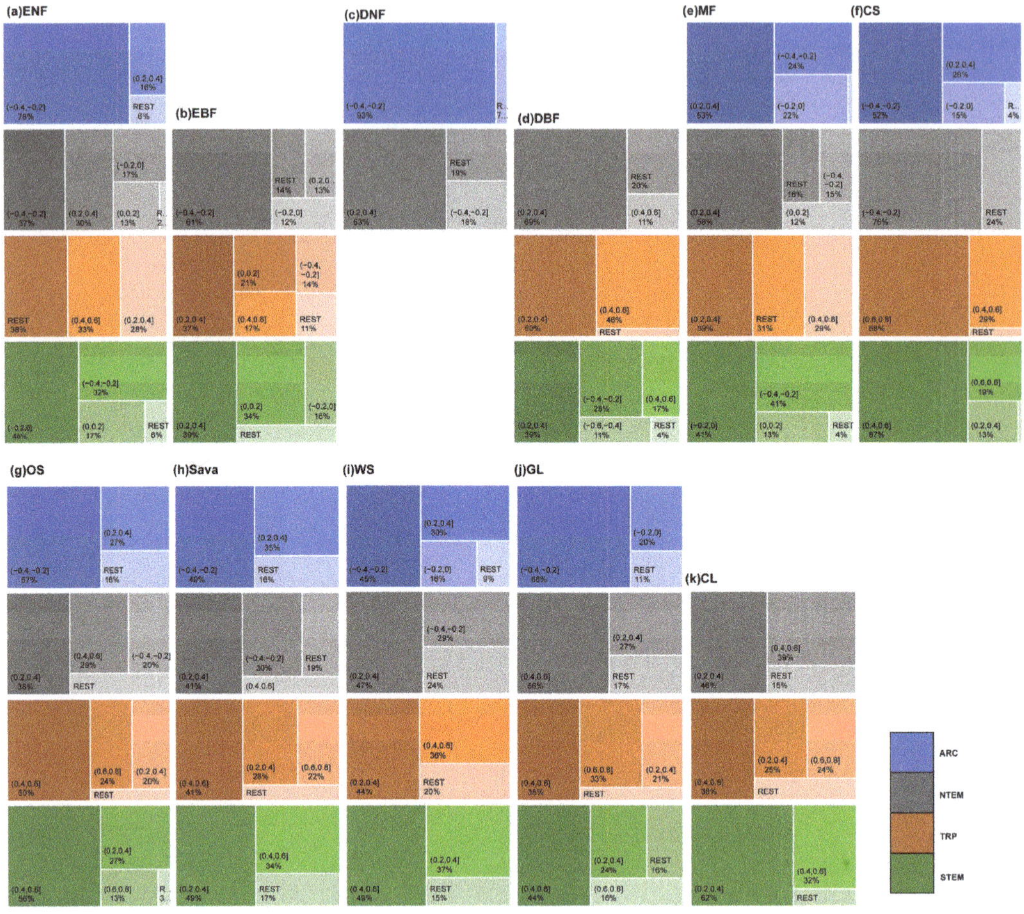

Figure A1. The rectangular tree map illustrates the proportion of the partial correlation strength between the growth of different vegetation types in various climatic zones and APRE. Different color schemes represent different climatic zones, and each vegetation type is labeled in the top left corner of the corresponding box. The map displays the large proportion of partial correlations and merges the smaller proportions into the category "REST".

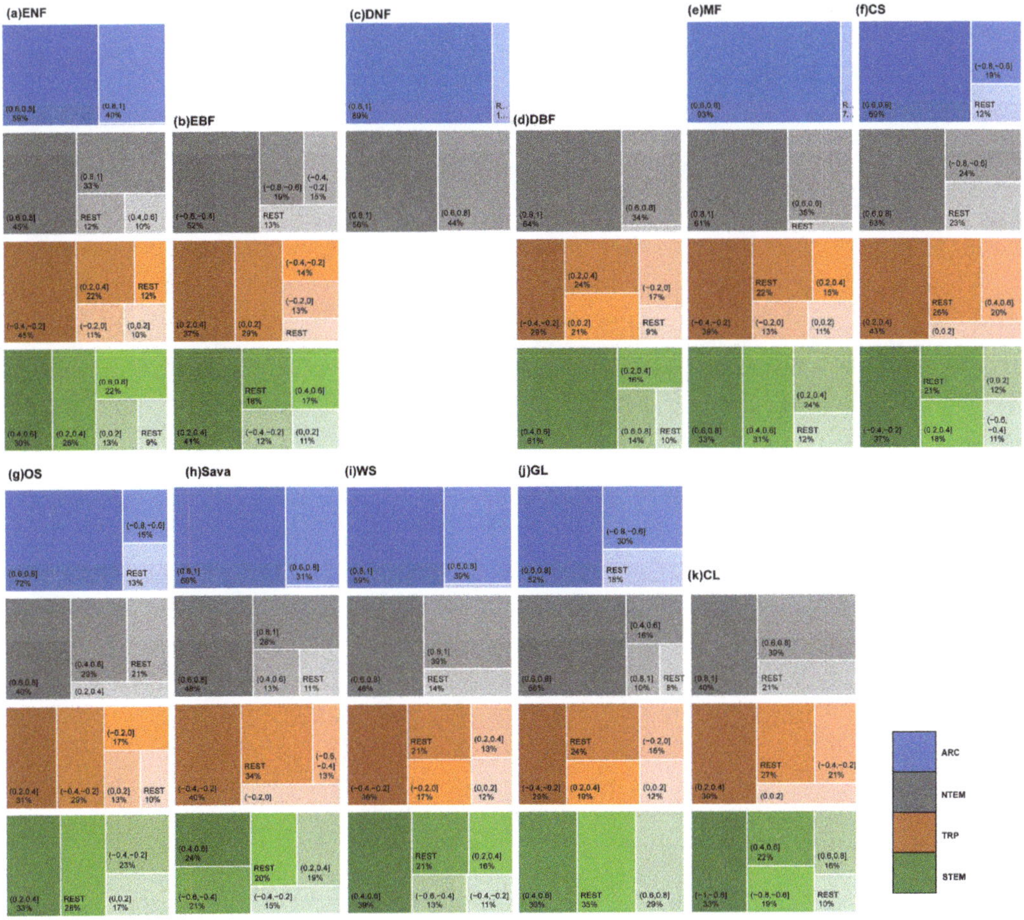

Figure A2. The rectangular tree map illustrates the proportion of the partial correlation strength between the growth of different vegetation types in various climatic zones and ASOLAR. Different color schemes represent different climatic zones, and each vegetation type is labeled in the top left corner of the corresponding box. The map displays the large proportion of partial correlations and merges the smaller proportions into the category "REST".

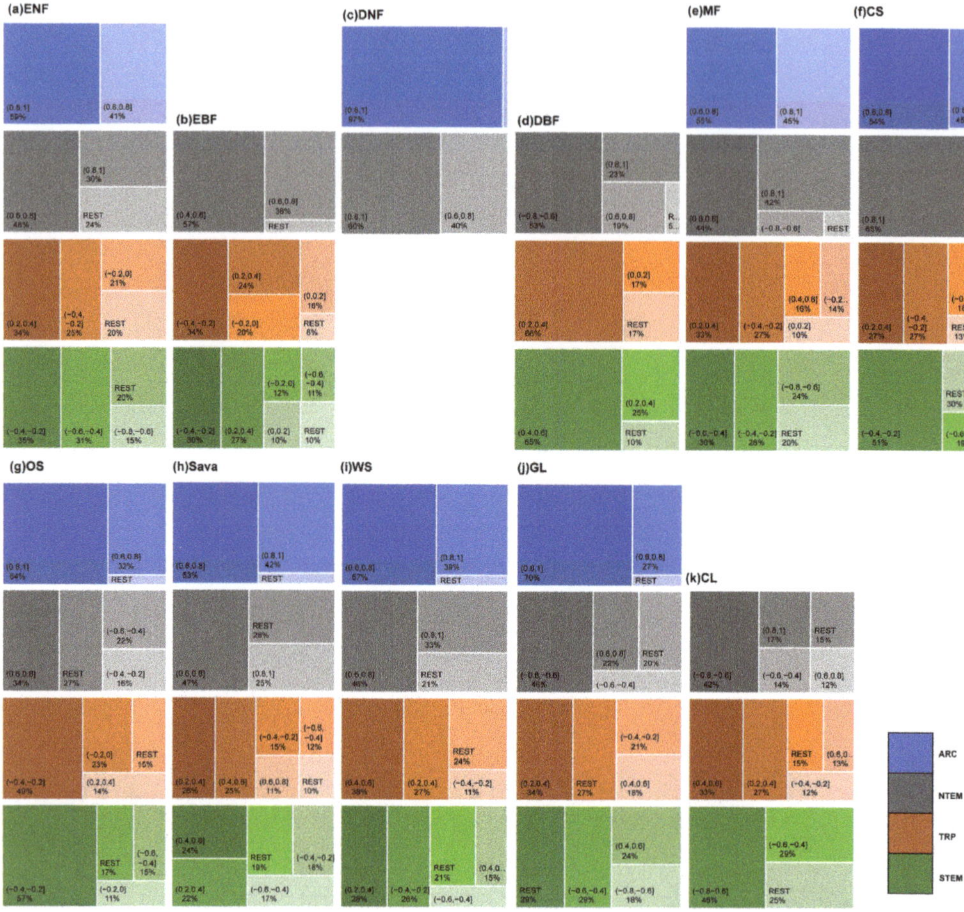

Figure A3. The rectangular tree map illustrates the proportion of the partial correlation strength between the growth of different vegetation types in various climatic zones and ATEM. Different color schemes represent different climatic zones, and each vegetation type is labeled in the top left corner of the corresponding box. The map displays the large proportion of partial correlations and merges the smaller proportions into the category "REST".

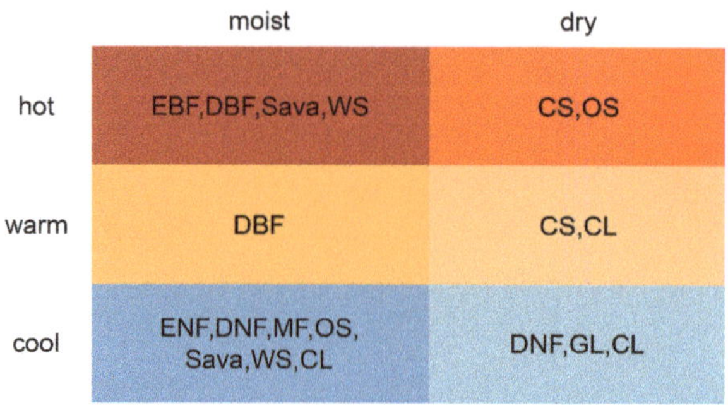

Figure A4. The climatic conditions of vegetation growth regions.

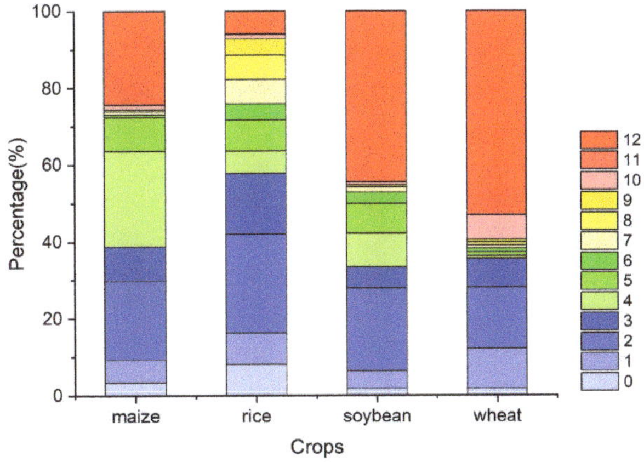

Figure A5. The percentage of the OAT of accumulation effects of precipitation on four major crops globally (maize, rice, soybean, and wheat).

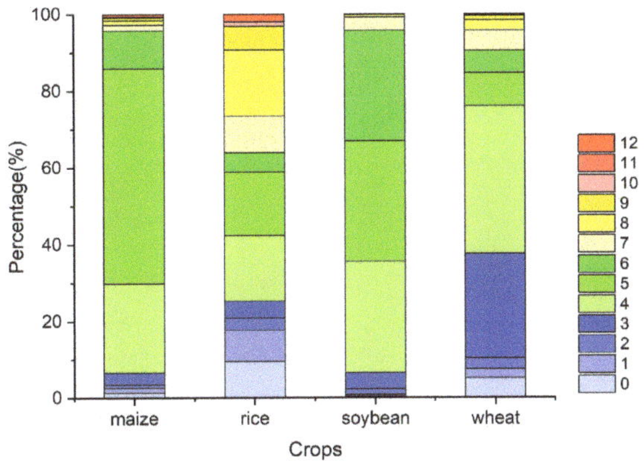

Figure A6. The percentage of the OAT of accumulation effects of solar radiation on four major crops globally (maize, rice, soybean, and wheat).

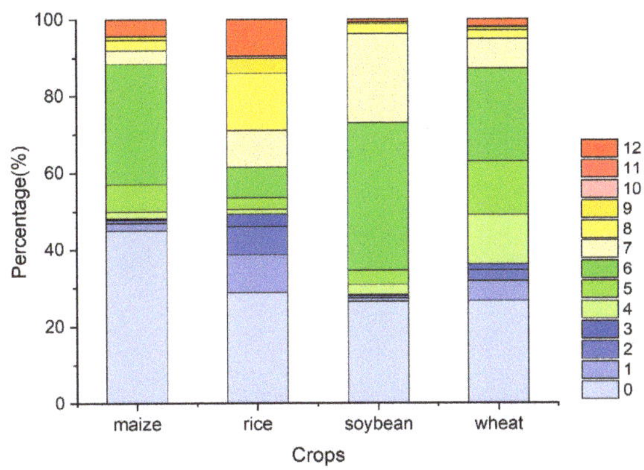

Figure A7. The percentage of the OAT of accumulation effects of temperature on four major crops globally (maize, rice, soybean, and wheat).

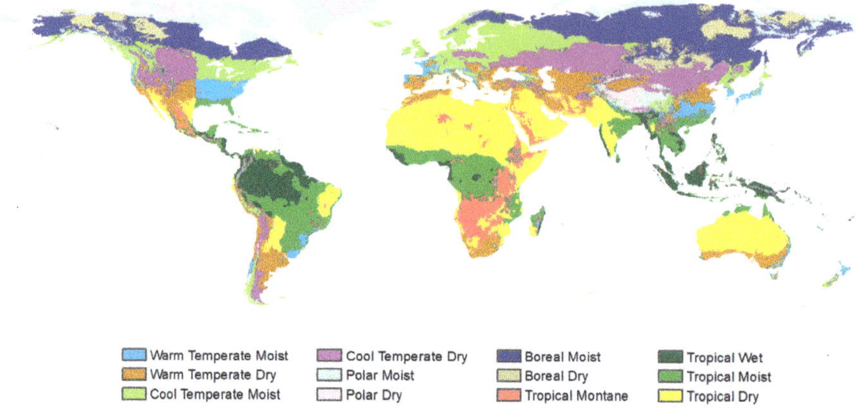

Figure A8. The climate zones defined by IPCC (https://esdac.jrc.ec.europa.eu/projects/RenewableEnergy/, accessed on 2 September 2023).

Table A1. Percentage of unchanged vegetation in different climate regions.

Climate Regions	ENF	EBF	DNF	DBF	MF	CS	OS	SAVA	WS	GL
Warm Temperate Moist	3.58%	2.51%	0.00%	38.55%	1.51%	0.02%	0.00%	4.36%	7.40%	0.64%
Warm Temperate Dry	0.42%	0.43%	0.00%	2.58%	0.27%	10.08%	13.33%	4.22%	2.86%	12.62%
Cool Temperate Moist	42.00%	0.41%	0.01%	42.37%	67.10%	0.99%	0.12%	5.74%	13.21%	6.92%
Cool Temperate Dry	1.89%	0.00%	0.00%	1.72%	2.13%	0.61%	3.12%	2.00%	1.86%	49.60%
Polar Moist	1.12%	0.00%	0.05%	0.02%	0.04%	7.69%	18.93%	1.16%	0.57%	5.80%
Polar Dry	0.00%	0.00%	0.00%	0.00%	0.00%	0.00%	1.90%	0.14%	0.01%	0.57%
Boreal Moist	48.18%	0.00%	63.71%	2.36%	24.73%	8.87%	17.91%	27.86%	40.07%	2.58%
Boreal Dry	2.61%	0.00%	36.23%	0.13%	4.08%	0.04%	2.19%	6.63%	9.46%	7.11%
Tropical Montane	0.03%	4.07%	0.00%	0.21%	0.01%	2.32%	4.32%	4.99%	7.83%	2.56%
Tropical Wet	0.00%	59.91%	0.00%	0.17%	0.00%	0.00%	0.00%	4.14%	2.66%	0.70%
Tropical Moist	0.16%	32.46%	0.00%	9.58%	0.12%	0.20%	0.01%	30.42%	10.13%	3.71%
Tropical Dry	0.00%	0.21%	0.00%	2.31%	0.00%	69.19%	38.17%	8.34%	3.95%	7.19%

Table A2. The proportion of the top ten optimal accumulation durations for APRE, ATEM, and ASOLAR across all climatic zones for different vegetation types.

Biomes	APRE			ATEM			ASOLAR		
	Time Accumulation	Climate Zones	Percentage	Time Accumulation	Climate Zones	Percentage	Time Accumulation	Climate Zones	Percentage
ENF	12	7	18.58%	0	7	46.48%	3	7	24.55%
	7	7	11.51%	0	3	27.52%	4	7	17.83%
	12	3	11.09%	7	3	5.00%	4	3	15.23%
	0	3	8.86%	6	3	4.31%	5	3	8.17%
	6	7	8.24%	0	8	2.43%	3	3	6.99%
	6	3	5.52%	1	3	1.19%	0	3	2.97%
	7	3	4.55%	5	3	1.12%	2	7	2.83%
	1	3	3.32%	0	4	1.06%	6	3	2.18%
	2	7	2.86%	0	0	0.96%	7	3	2.14%
	2	3	2.26%	0	5	0.89%	3	8	1.72%

Table A2. *Cont.*

Biomes	APRE			ATEM			ASOLAR		
	Time Accumulation	Climate Zones	Percentage	Time Accumulation	Climate Zones	Percentage	Time Accumulation	Climate Zones	Percentage
EBF	9	10	10.86%	0	10	19.09%	0	10	19.95%
	1	11	10.41%	0	11	14.67%	3	10	6.15%
	1	10	7.01%	1	10	8.01%	2	11	5.71%
	6	10	5.41%	1	11	6.90%	0	11	5.69%
	0	10	5.07%	5	10	5.58%	2	10	5.57%
	5	10	4.83%	6	10	4.37%	4	10	5.42%
	10	10	4.71%	7	10	4.06%	12	10	5.01%
	8	10	4.32%	2	10	3.62%	1	10	3.86%
	0	11	4.10%	12	10	3.33%	7	11	3.80%
	2	10	4.09%	4	10	2.88%	7	10	3.48%
DNF	2	7	35.10%	0	7	62.40%	3	7	61.57%
	2	8	12.12%	0	8	36.17%	3	8	35.50%
	7	8	11.87%	12	7	1.23%	4	7	1.13%
	12	7	10.54%	0	5	0.05%	2	7	0.69%
	7	7	9.36%	12	8	0.05%	4	8	0.49%
	3	8	6.84%	5	7	0.03%	0	7	0.22%
	3	7	3.97%	4	7	0.03%	2	8	0.21%
	8	7	2.08%	0	0	0.02%	9	7	0.10%
	12	8	1.92%	1	7	0.02%	3	5	0.05%
	8	8	1.54%	0	3	0.01%	3	0	0.01%
DBF	12	1	22.93%	0	3	29.25%	5	1	21.41%
	12	3	15.98%	6	1	26.05%	4	3	17.19%
	2	3	7.49%	7	1	8.93%	3	3	15.17%
	1	11	6.52%	6	3	5.53%	4	1	13.67%
	6	1	6.37%	7	3	5.36%	5	3	7.20%
	0	3	5.01%	8	11	2.62%	1	11	2.64%
	7	3	4.68%	0	7	2.18%	6	1	2.39%
	5	1	3.32%	3	11	2.17%	2	11	2.13%
	3	3	3.18%	0	1	1.66%	3	7	1.29%
	12	2	1.72%	0	0	1.31%	5	2	1.24%
MF	12	3	41.80%	0	3	55.77%	3	3	44.82%
	12	7	11.61%	0	7	24.14%	3	7	18.89%
	7	3	6.04%	6	3	6.56%	4	3	16.63%
	6	3	5.40%	0	8	4.00%	4	7	4.09%
	0	3	5.09%	7	3	2.99%	3	8	2.85%
	2	7	4.37%	0	4	1.83%	5	3	2.61%
	7	7	2.81%	0	0	0.50%	3	4	1.51%
	2	3	2.79%	6	1	0.46%	2	7	1.25%
	12	8	2.46%	5	3	0.43%	4	8	0.84%
	10	3	1.77%	7	1	0.33%	6	3	0.62%
CS	9	12	19.17%	0	12	25.29%	6	12	13.12%
	10	12	9.83%	12	12	16.26%	5	12	9.33%
	5	12	9.19%	0	7	7.22%	10	12	9.01%
	7	12	7.39%	0	5	6.53%	7	12	8.75%
	8	12	5.88%	4	12	5.36%	12	12	8.57%
	7	7	5.78%	8	12	4.64%	4	12	7.81%
	6	12	5.37%	7	12	4.27%	0	7	4.43%
	4	12	5.04%	3	12	3.68%	0	12	3.78%
	7	5	2.84%	0	2	2.62%	4	5	3.29%
	3	12	2.55%	7	2	2.31%	7	2	3.07%

Table A2. *Cont.*

Biomes	APRE			ATEM			ASOLAR		
	Time Accumulation	Climate Zones	Percentage	Time Accumulation	Climate Zones	Percentage	Time Accumulation	Climate Zones	Percentage
OS	9	12	8.11%	0	7	14.68%	3	7	12.24%
	8	5	5.13%	0	5	10.35%	3	5	9.07%
	12	7	4.84%	0	12	8.55%	12	12	7.53%
	8	12	4.82%	1	5	8.27%	2	5	7.04%
	5	12	4.80%	12	12	7.86%	7	12	6.02%
	3	12	4.37%	0	2	5.00%	6	12	5.06%
	7	12	4.23%	7	12	3.49%	0	12	4.66%
	8	7	3.59%	3	12	3.47%	10	12	4.51%
	4	12	3.57%	1	7	3.06%	2	7	3.01%
	10	12	2.72%	7	2	2.31%	8	12	2.78%
Sava	7	7	10.09%	0	7	27.11%	3	7	19.42%
	12	7	6.67%	0	3	2.88%	2	11	5.72%
	1	11	5.56%	0	8	6.38%	4	7	5.23%
	2	11	4.80%	6	1	1.47%	3	8	4.87%
	5	11	4.17%	3	9	0.75%	1	11	4.46%
	12	8	3.36%	0	11	7.79%	6	11	3.31%
	2	7	3.28%	7	1	0.41%	7	11	2.92%
	4	11	2.97%	2	11	2.75%	3	11	2.57%
	6	11	2.69%	12	1	0.75%	0	11	2.32%
	0	11	2.67%	4	9	0.58%	9	11	2.20%
WS	7	7	11.53%	0	7	38.95%	3	7	28.37%
	12	7	9.28%	0	3	9.85%	4	7	8.88%
	2	7	8.01%	0	8	9.07%	3	8	7.36%
	12	3	4.99%	6	1	2.12%	3	3	5.31%
	6	7	4.81%	3	9	1.93%	4	3	4.92%
	12	8	4.14%	0	11	1.70%	1	9	2.25%
	0	9	3.10%	7	1	1.63%	6	1	1.88%
	12	1	2.53%	2	11	1.46%	5	1	1.80%
	1	11	2.21%	12	1	1.40%	4	1	1.62%
	2	8	1.85%	4	9	1.39%	2	7	1.55%
GL	12	4	27.98%	3	4	10.99%	4	4	13.73%
	2	4	12.70%	4	4	10.97%	3	4	11.99%
	1	4	5.54%	0	4	7.34%	2	4	11.38%
	12	8	3.40%	6	4	5.73%	5	4	6.47%
	2	8	3.30%	5	4	5.23%	4	8	4.37%
	2	5	3.17%	0	8	4.91%	3	5	2.93%
	12	2	3.14%	0	5	3.70%	6	2	2.85%
	2	2	2.76%	6	2	3.53%	5	2	2.31%
	2	12	2.40%	2	4	2.82%	6	4	2.07%
	1	12	1.87%	0	3	2.51%	4	3	2.06%
CL	12	4	16.54%	0	4	12.77%	4	4	15.87%
	12	3	11.54%	6	3	8.76%	3	3	7.85%
	12	2	8.19%	0	3	7.58%	3	4	7.23%
	2	2	5.99%	6	4	5.47%	4	3	6.51%
	1	4	5.09%	0	2	5.46%	4	2	6.42%
	2	4	4.87%	4	4	5.10%	5	2	5.67%
	3	2	3.38%	5	4	4.59%	5	4	4.95%
	12	1	3.07%	6	2	4.54%	5	3	4.82%
	2	12	2.69%	6	1	4.44%	6	2	4.09%
	1	2	2.67%	5	2	3.57%	5	1	2.93%

Note: Climate regions: 1: Tropical Montane; 2: Tropical Wet; 3: Tropical Moist; 4: Tropical Dry; 5: Warm Temperate Moist; 6: Warm Temperate Dry; 7: Cool Temperate Moist; 8: Cool Temperate Dry; 9: Boreal Moist; 10: Boreal Dry; 11: Polar Moist; 12: Polar Dry.

References

1. Buitenwerf, R.; Rose, L.; Higgins, S.I. Three decades of multi-dimensional change in global leaf phenology. *Nat. Clim. Chang.* **2015**, *5*, 364–368. [CrossRef]
2. Peñuelas, J.; Filella, I. Responses to a Warming World. *Science* **2001**, *294*, 793–795. [CrossRef] [PubMed]
3. Zhang, G.; Zhang, Y.; Dong, J.; Xiao, X. Green-up dates in the Tibetan Plateau have continuously advanced from 1982 to 2011. *Proc. Natl. Acad. Sci. USA* **2013**, *110*, 4309–4314. [CrossRef] [PubMed]
4. Nemani, R.R.; Keeling, C.D.; Hashimoto, H.; Jolly, W.M.; Piper, S.C.; Tucker, C.J.; Myneni, R.B.; Running, S.W. Climate-Driven Increases in Global Terrestrial Net Primary Production from 1982 to 1999. *Science* **2003**, *300*, 1560–1563. [CrossRef]
5. Ostberg, S.; Lucht, W.; Schaphoff, S.; Gerten, D. Critical impacts of global warming on land ecosystems. *Earth Syst. Dynam.* **2013**, *4*, 347–357. [CrossRef]
6. Ciais, P.; Reichstein, M.; Viovy, N.; Granier, A.; Ogée, J.; Allard, V.; Aubinet, M.; Buchmann, N.; Bernhofer, C.; Carrara, A.; et al. Europe-wide reduction in primary productivity caused by the heat and drought in 2003. *Nature* **2005**, *437*, 529–533. [CrossRef]
7. Gatti, L.V.; Gloor, M.; Miller, J.B.; Doughty, C.E.; Malhi, Y.; Domingues, L.G.; Basso, L.S.; Martinewski, A.; Correia, C.S.C.; Borges, V.F.; et al. Drought sensitivity of Amazonian carbon balance revealed by atmospheric measurements. *Nature* **2014**, *506*, 76–80. [CrossRef]
8. Zhao, M.; Running, S.W. Drought-Induced Reduction in Global Terrestrial Net Primary Production from 2000 Through 2009. *Science* **2010**, *329*, 940–943. [CrossRef]
9. de Jong, R.; Schaepman, M.E.; Furrer, R.; de Bruin, S.; Verburg, P.H. Spatial relationship between climatologies and changes in global vegetation activity. *Glob. Chang. Biol.* **2013**, *19*, 1953–1964. [CrossRef]
10. Woodward, F.I.; Lomas, M.R.; Kelly, C.K. Global climate and the distribution of plant biomes. *Philos. Trans. R. Soc. London Ser. B Biol. Sci.* **2004**, *359*, 1465–1476. [CrossRef]
11. Fu, G.; Sun, W.; Li, S.; Zhang, J.; Yu, C.; Shen, Z. Modeling Aboveground Biomass Using MODIS Images and Climatic Data in Grasslands on the Tibetan Plateau. *J. Resour. Ecol.* **2017**, *8*, 42–49. [CrossRef]
12. Garbulsky, M.F.; Peñuelas, J.; Papale, D.; Ardö, J.; Goulden, M.L.; Kiely, G.; Richardson, A.D.; Rotenberg, E.; Veenendaal, E.M.; Filella, I. Patterns and controls of the variability of radiation use efficiency and primary productivity across terrestrial ecosystems. *Glob. Ecol. Biogeogr.* **2010**, *19*, 253–267. [CrossRef]
13. Musau, J.; Patil, S.; Sheffield, J.; Marshall, M. Spatio-temporal vegetation dynamics and relationship with climate over East Africa. *Hydrol. Earth Syst. Sci. Discuss.* **2016**, *2016*, 1–30. [CrossRef]
14. Zhu, L.; Southworth, J. Disentangling the Relationships between Net Primary Production and Precipitation in Southern Africa Savannas Using Satellite Observations from 1982 to 2010. *Remote Sens.* **2013**, *5*, 3803–3825. [CrossRef]
15. Wu, D.; Zhao, X.; Liang, S.; Zhou, T.; Huang, K.; Tang, B.; Zhao, W. Time-lag effects of global vegetation responses to climate change. *Glob. Chang. Biol.* **2015**, *21*, 3520–3531. [CrossRef]
16. Xu, G.; Zhang, H.; Chen, B.; Zhang, H.; Innes, J.L.; Wang, G.; Yan, J.; Zheng, Y.; Zhu, Z.; Myneni, R.B. Changes in Vegetation Growth Dynamics and Relations with Climate over China's Landmass from 1982 to 2011. *Remote Sens.* **2014**, *6*, 3263–3283. [CrossRef]
17. Zhao, W.; Zhao, X.; Zhou, T.; Wu, D.; Tang, B.; Wei, H. Climatic factors driving vegetation declines in the 2005 and 2010 Amazon droughts. *PLoS ONE* **2017**, *12*, e0175379. [CrossRef]
18. Papagiannopoulou, C.; Miralles, D.G.; Dorigo, W.A.; Verhoest, N.E.C.; Depoorter, M.; Waegeman, W. Vegetation anomalies caused by antecedent precipitation in most of the world. *Environ. Res. Lett.* **2017**, *12*, 074016. [CrossRef]
19. Sherry, R.A.; Weng, E.; Arnone Iii, J.A.; Johnson, D.W.; Schimel, D.S.; Verburg, P.S.; Wallace, L.L.; Luo, Y. Lagged effects of experimental warming and doubled precipitation on annual and seasonal aboveground biomass production in a tallgrass prairie. *Glob. Chang. Biol.* **2008**, *14*, 2923–2936. [CrossRef]
20. Kuzyakov, Y.; Gavrichkova, O. REVIEW: Time lag between photosynthesis and carbon dioxide efflux from soil: A review of mechanisms and controls. *Glob. Chang. Biol.* **2010**, *16*, 3386–3406. [CrossRef]
21. Braswell, B.H.; Schimel, D.S.; Linder, E.; Moore, B. The Response of Global Terrestrial Ecosystems to Interannual Temperature Variability. *Science* **1997**, *278*, 870–873. [CrossRef]
22. VukiĆEviĆ, T.; Braswell, B.H.; Schimel, D. A diagnostic study of temperature controls on global terrestrial carbon exchange. *Tellus B* **2001**, *53*, 150–170. [CrossRef]
23. Zhang, T.; Yang, S.; Guo, R.; Guo, J. Correction: Warming and Nitrogen Addition Alter Photosynthetic Pigments, Sugars and Nutrients in a Temperate Meadow Ecosystem. *PLoS ONE* **2016**, *11*, e0158249. [CrossRef]
24. Vicente-Serrano, S.M.; Gouveia, C.; Camarero, J.J.; Beguería, S.; Trigo, R.; López-Moreno, J.I.; Azorín-Molina, C.; Pasho, E.; Lorenzo-Lacruz, J.; Revuelto, J.; et al. Response of vegetation to drought time-scales across global land biomes. *Proc. Natl. Acad. Sci. USA* **2013**, *110*, 52–57. [CrossRef]
25. Gunderson, C.A.; Norby, R.J.; Wullschleger, S.D. Acclimation of photosynthesis and respiration to simulated climatic warming in northern and southern populations of Acer saccharum: Laboratory and field evidence. *Tree Physiol.* **2000**, *20*, 87–96. [CrossRef]
26. Slot, M.; Kitajima, K. General patterns of acclimation of leaf respiration to elevated temperatures across biomes and plant types. *Oecologia* **2015**, *177*, 885–900. [CrossRef]
27. Yamori, W.; Hikosaka, K.; Way, D.A. Temperature response of photosynthesis in C3, C4, and CAM plants: Temperature acclimation and temperature adaptation. *Photosynth. Res.* **2014**, *119*, 101–117. [CrossRef] [PubMed]

28. Wen, Y.; Liu, X.; Xin, Q.; Wu, J.; Xu, X.; Pei, F.; Li, X.; Du, G.; Cai, Y.; Lin, K.; et al. Cumulative Effects of Climatic Factors on Terrestrial Vegetation Growth. *J. Geophys. Res. Biogeosci.* **2019**, *124*, 789–806. [CrossRef]
29. Shi, S.; Wang, P.; Zhang, Y.; Yu, J. Cumulative and time-lag effects of the main climate factors on natural vegetation across Siberia. *Ecol. Indic.* **2021**, *133*, 108446. [CrossRef]
30. Sun, Q.; Liu, C.; Chen, T.; Zhang, A. A Weighted-Time-Lag Method to Detect Lag Vegetation Response to Climate Variation: A Case Study in Loess Plateau, China, 1982–2013. *Remote Sens.* **2021**, *13*, 923. [CrossRef]
31. Wei, X.; He, W.; Zhou, Y.; Ju, W.; Xiao, J.; Li, X.; Liu, Y.; Xu, S.; Bi, W.; Zhang, X.; et al. Global assessment of lagged and cumulative effects of drought on grassland gross primary production. *Ecol. Indic.* **2022**, *136*, 108646. [CrossRef]
32. Wu, C.; Wang, T. Evaluating Cumulative Drought Effect on Global Vegetation Photosynthesis Using Numerous GPP Products. *Front. Environ. Sci.* **2022**, *10*, 908875. [CrossRef]
33. Yuan, Y.; Bao, A.; Jiapaer, G.; Jiang, L.; De Maeyer, P. Phenology-based seasonal terrestrial vegetation growth response to climate variability with consideration of cumulative effect and biological carryover. *Sci. Total Environ.* **2022**, *817*, 152805. [CrossRef]
34. Zhao, A.; Yu, Q.; Feng, L.; Zhang, A.; Pei, T. Evaluating the cumulative and time-lag effects of drought on grassland vegetation: A case study in the Chinese Loess Plateau. *J. Environ. Manag.* **2020**, *261*, 110214. [CrossRef]
35. Liu, H.; Zhang, A.; Liu, C.; Zhao, Y.; Zhao, A.; Wang, D. Analysis of the time-lag effects of climate factors on grassland productivity in Inner Mongolia. *Glob. Ecol. Conserv.* **2021**, *30*, e01751. [CrossRef]
36. Peng, S.; Piao, S.; Ciais, P.; Myneni, R.B.; Chen, A.; Chevallier, F.; Dolman, A.J.; Janssens, I.A.; Peñuelas, J.; Zhang, G.; et al. Asymmetric effects of daytime and night-time warming on Northern Hemisphere vegetation. *Nature* **2013**, *501*, 88–92. [CrossRef]
37. Piao, S.; Tan, J.; Chen, A.; Fu, Y.H.; Ciais, P.; Liu, Q.; Janssens, I.A.; Vicca, S.; Zeng, Z.; Jeong, S.-J.; et al. Leaf onset in the northern hemisphere triggered by daytime temperature. *Nat. Commun.* **2015**, *6*, 6911. [CrossRef]
38. Tan, J.; Piao, S.; Chen, A.; Zeng, Z.; Ciais, P.; Janssens, I.A.; Mao, J.; Myneni, R.B.; Peng, S.; Peñuelas, J.; et al. Seasonally different response of photosynthetic activity to daytime and night-time warming in the Northern Hemisphere. *Glob. Chang. Biol.* **2015**, *21*, 377–387. [CrossRef]
39. Piao, S.; Friedlingstein, P.; Ciais, P.; Viovy, N.; Demarty, J. Growing season extension and its impact on terrestrial carbon cycle in the Northern Hemisphere over the past 2 decades. *Glob. Biogeochem. Cycles* **2007**, *21*, GB3018. [CrossRef]
40. Harris, I.; Jones, P.D.; Osborn, T.J.; Lister, D.H. Updated high-resolution grids of monthly climatic observations—The CRU TS3.10 Dataset. *Int. J. Climatol.* **2014**, *34*, 623–642. [CrossRef]
41. Stackhouse Jr, P.; Gupta, S.; Cox, S.; Mikovitz, J.C.; Zhang, T.; Chiacchio, M. 12-year surface radiation budget data set. *GEWEX News* **2004**, *14*, 10–12.
42. Müller Schmied, H.; Müller, R.; Sanchez-Lorenzo, A.; Ahrens, B.; Wild, M. Evaluation of Radiation Components in a Global Freshwater Model with Station-Based Observations. *Water* **2016**, *8*, 450. [CrossRef]
43. Sheffield, J.; Wood, E.F.; Roderick, M.L. Little change in global drought over the past 60 years. *Nature* **2012**, *491*, 435–438. [CrossRef]
44. Wu, X.; Liu, H.; Li, X.; Liang, E.; Beck, P.S.A.; Huang, Y. Seasonal divergence in the interannual responses of Northern Hemisphere vegetation activity to variations in diurnal climate. *Sci. Rep.* **2016**, *6*, 19000. [CrossRef] [PubMed]
45. De Fries, R.S.; Hansen, M.; Townshend, J.R.G.; Sohlberg, R. Global land cover classifications at 8 km spatial resolution: The use of training data derived from Landsat imagery in decision tree classifiers. *Int. J. Remote Sens.* **1998**, *19*, 3141–3168. [CrossRef]
46. Barnes, M.L.; Moran, M.S.; Scott, R.L.; Kolb, T.E.; Ponce-Campos, G.E.; Moore, D.J.P.; Ross, M.A.; Mitra, B.; Dore, S. Vegetation productivity responds to sub-annual climate conditions across semiarid biomes. *Ecosphere* **2016**, *7*, e01339. [CrossRef]
47. Wen, Y.; Liu, X.; Du, G. Nonuniform Time-Lag Effects of Asymmetric Warming on Net Primary Productivity across Global Terrestrial Biomes. *Earth Interact.* **2018**, *22*, 1–26. [CrossRef]
48. Ding, Y.; Li, Z.; Peng, S. Global analysis of time-lag and -accumulation effects of climate on vegetation growth. *Int. J. Appl. Earth Obs. Geoinf.* **2020**, *92*, 102179. [CrossRef]
49. Aubin, I.; Munson, A.D.; Cardou, F.; Burton, P.J.; Isabel, N.; Pedlar, J.H.; Paquette, A.; Taylor, A.R.; Delagrange, S.; Kebli, H.; et al. Traits to stay, traits to move: A review of functional traits to assess sensitivity and adaptive capacity of temperate and boreal trees to climate change. *Environ. Rev.* **2016**, *24*, 164–186. [CrossRef]
50. Anderegg, L.D.L.; HilleRisLambers, J. Drought stress limits the geographic ranges of two tree species via different physiological mechanisms. *Glob. Chang. Biol.* **2016**, *22*, 1029–1045. [CrossRef]
51. Mayoral, C.; Pardos, M.; Sánchez-González, M.; Brendel, O.; Pita, P. Ecological implications of different water use strategies in three coexisting mediterranean tree species. *For. Ecol. Manag.* **2016**, *382*, 76–87. [CrossRef]
52. Pardos, M.; del Río, M.; Pretzsch, H.; Jactel, H.; Bielak, K.; Bravo, F.; Brazaitis, G.; Defossez, E.; Engel, M.; Godvod, K.; et al. The greater resilience of mixed forests to drought mainly depends on their composition: Analysis along a climate gradient across Europe. *For. Ecol. Manag.* **2021**, *481*, 118687. [CrossRef]
53. Liu, Y.; Liu, R.; Pisek, J.; Chen, J.M. Separating overstory and understory leaf area indices for global needleleaf and deciduous broadleaf forests by fusion of MODIS and MISR data. *Biogeosciences* **2017**, *14*, 1093–1110. [CrossRef]
54. Ershadi, A.; McCabe, M.F.; Evans, J.P.; Wood, E.F. Impact of model structure and parameterization on Penman–Monteith type evaporation models. *J. Hydrol.* **2015**, *525*, 521–535. [CrossRef]

55. Gang, C.; Zhou, W.; Wang, Z.; Chen, Y.; Li, J.; Chen, J.; Qi, J.; Odeh, I.; Groisman, P.Y. Comparative Assessment of Grassland NPP Dynamics in Response to Climate Change in China, North America, Europe and Australia from 1981 to 2010. *J. Agron. Crop Sci.* **2015**, *201*, 57–68. [CrossRef]
56. Aouissi, J.; Benabdallah, S.; Lili Chabaâne, Z.; Cudennec, C. Evaluation of potential evapotranspiration assessment methods for hydrological modelling with SWAT—Application in data-scarce rural Tunisia. *Agric. Water Manag.* **2016**, *174*, 39–51. [CrossRef]
57. Liu, N.; Shi, Y.; Ding, Y.; Liu, L.; Peng, S. Temporal effects of climatic factors on vegetation phenology on the Loess Plateau, China. *J. Plant Ecol.* **2023**, *16*, rtac063. [CrossRef]
58. Shi, C.; Sun, G.; Zhang, H.; Xiao, B.; Ze, B.; Zhang, N.; Wu, N. Effects of Warming on Chlorophyll Degradation and Carbohydrate Accumulation of Alpine Herbaceous Species during Plant Senescence on the Tibetan Plateau. *PLoS ONE* **2014**, *9*, e107874. [CrossRef]
59. Fracheboud, Y.; Luquez, V.; Björkén, L.; Sjödin, A.; Tuominen, H.; Jansson, S. The Control of Autumn Senescence in European Aspen. *Plant Physiol.* **2009**, *149*, 1982–1991. [CrossRef]
60. Ke, Y.; Leung, L.R.; Huang, M.; Coleman, A.M.; Li, H.; Wigmosta, M.S. Development of high resolution land surface parameters for the Community Land Model. *Geosci. Model Dev.* **2012**, *5*, 1341–1362. [CrossRef]
61. Kim, J.-H.; Moon, Y.R.; Wi, S.G.; Kim, J.-S.; Lee, M.H.; Chung, B.Y. Differential Radiation Sensitivities of Arabidopsis Plants at Various Developmental Stages. In *Photosynthesis. Energy from the Sun: Proceedings of the 14th International Congress on Photosynthesis, Glasgow, UK, 22–27 July 2007*; Springer: Dordrecht, The Netherlands, 2008; pp. 1491–1495.
62. He, P.; Osaki, M.; Takebe, M.; Shinano, T.; Wasaki, J. Endogenous hormones and expression of senescence-related genes in different senescent types of maize. *J. Exp. Bot.* **2005**, *56*, 1117–1128. [CrossRef] [PubMed]
63. Liu, Q.; Fu, Y.H.; Zeng, Z.; Huang, M.; Li, X.; Piao, S. Temperature, precipitation, and insolation effects on autumn vegetation phenology in temperate China. *Glob. Chang. Biol.* **2016**, *22*, 644–655. [CrossRef] [PubMed]
64. Gepstein, S.; Thimann, K.V. Changes in the abscisic acid content of oat leaves during senescence. *Proc. Natl. Acad. Sci. USA* **1980**, *77*, 2050–2053. [CrossRef] [PubMed]
65. Chen, X.; Ciais, P.; Maignan, F.; Zhang, Y.; Bastos, A.; Liu, L.; Bacour, C.; Fan, L.; Gentine, P.; Goll, D.; et al. Vapor Pressure Deficit and Sunlight Explain Seasonality of Leaf Phenology and Photosynthesis Across Amazonian Evergreen Broadleaved Forest. *Glob. Biogeochem. Cycles* **2021**, *35*, e2020GB006893. [CrossRef]
66. Zhang, H.; Yuan, W.; Dong, W.; Liu, S. Seasonal patterns of litterfall in forest ecosystem worldwide. *Ecol. Complex.* **2014**, *20*, 240–247. [CrossRef]
67. Weng, E.; Farrior, C.E.; Dybzinski, R.; Pacala, S.W. Predicting vegetation type through physiological and environmental interactions with leaf traits: Evergreen and deciduous forests in an earth system modeling framework. *Glob. Chang. Biol.* **2017**, *23*, 2482–2498. [CrossRef]
68. Loidi, J.; Navarro-Sánchez, G.; Vynokurov, D. Climatic definitions of the world's terrestrial biomes. *Veg. Classif. Surv.* **2022**, *3*, 231–271. [CrossRef]
69. Lehmann, C.E.R.; Archibald, S.A.; Hoffmann, W.A.; Bond, W.J. Deciphering the distribution of the savanna biome. *New Phytol.* **2011**, *191*, 197–209. [CrossRef]
70. Shi, X.; Wang, W.; Shi, W. Progress on quantitative assessment of the impacts of climate change and human activities on cropland change. *J. Geogr. Sci.* **2016**, *26*, 339–354. [CrossRef]
71. Frisvold, G.B.; Konyar, K. Climate Change Mitigation Policies: Implications for Agriculture and Water Resources. *J. Contemp. Water Res. Educ.* **2013**, *151*, 27–42. [CrossRef]
72. Chen, Q.; Yang, L.E.; Luo, J.; Liu, F.; Zhang, Y.; Zhou, Q.; Guo, R.; Gu, X. The 300 years cropland changes reflecting climate impacts and social resilience at the Yellow River–Huangshui River Valley, China. *Environ. Res. Lett.* **2021**, *16*, 065006. [CrossRef]
73. Mueller, N.D.; Butler, E.E.; McKinnon, K.A.; Rhines, A.; Tingley, M.; Holbrook, N.M.; Huybers, P. Cooling of US Midwest summer temperature extremes from cropland intensification. *Nat. Clim. Chang.* **2016**, *6*, 317–322. [CrossRef]
74. Reinermann, S.; Gessner, U.; Asam, S.; Kuenzer, C.; Dech, S. The Effect of Droughts on Vegetation Condition in Germany: An Analysis Based on Two Decades of Satellite Earth Observation Time Series and Crop Yield Statistics. *Remote Sens.* **2019**, *11*, 1783. [CrossRef]
75. Jafari, R.; Lewis, M.M.; Ostendorf, B. Evaluation of vegetation indices for assessing vegetation cover in southern arid lands in South Australia. *Rangel. J.* **2007**, *29*, 39–49. [CrossRef]
76. Leprieur, C.; Kerr, Y.H.; Mastorchio, S.; Meunier, J.C. Monitoring vegetation cover across semi-arid regions: Comparison of remote observations from various scales. *Int. J. Remote Sens.* **2000**, *21*, 281–300. [CrossRef]
77. Brantley, S.T.; Zinnert, J.C.; Young, D.R. Application of hyperspectral vegetation indices to detect variations in high leaf area index temperate shrub thicket canopies. *Remote Sens. Environ.* **2011**, *115*, 514–523. [CrossRef]
78. Zhang, Y.; Tian, Y.; Knyazikhin, Y.; Martonchik, J.V.; Diner, D.J.; Leroy, M.; Myneni, R.B.; Sensing, R. Prototyping of MISR LAI and FPAR algorithm with POLDER data over Africa. *IEEE Trans. Geosci. Remote Sens.* **2000**, *38*, 2402–2418. [CrossRef]

Disclaimer/Publisher's Note: The statements, opinions and data contained in all publications are solely those of the individual author(s) and contributor(s) and not of MDPI and/or the editor(s). MDPI and/or the editor(s) disclaim responsibility for any injury to people or property resulting from any ideas, methods, instructions or products referred to in the content.

agronomy

MDPI

Article

Predicting the Potential Geographic Distribution of Invasive Freshwater Apple Snail *Pomacea canaliculate* (Lamarck, 1819) under Climate Change Based on Biomod2

Tao Wang [1], Tingjia Zhang [1], Weibin An [2], Zailing Wang [1,*] and Chuanren Li [1,*]

[1] Hubei Engineering Research Center for Pest Forewarning and Management, Institute of Entomology, College of Agriculture, Yangtze University, Jingzhou 434025, China; wt852434057@163.com (T.W.); z15248891080@163.com (T.Z.)
[2] Agriculture and Rural Affairs Bureau of Zengdu, Suizhou 441300, China; szzdzjz@163.com
* Correspondence: 13554596516@163.com (Z.W.); 13986706558@163.com (C.L.)

Abstract: *Pomacea canaliculata* is widely distributed in the Chinese provinces south of the Yangtze River, causing serious damage to aquatic ecosystems, rice cultivation, and human health. Predicting the potential geographic distributions (PGDs) of *P. canaliculata* under current and future climate conditions in China is crucial for developing effective early warning measures and facilitating long-term monitoring. In this study, we screened various species distribution models (SDMs), including CTA, GBM, GAM, RF, and XGBOOST, to construct an ensemble model (EM) and then predict suitable habitats for *P. canaliculata* under current and future climate scenarios (SSP1-26, SSP2-45, SSP3-70, SSP5-85). The EM (AUC = 0.99, TSS = 0.96) yielded predictions that were more precise than those from the individual models. The Annual Mean Temperature (Bio1) and Precipitation of the Warmest Quarter (Bio18) are the most significant environmental variables affecting the PGDs of *P. canaliculata*. Under current climate conditions, the highly suitable habitats for *P. canaliculata* are primarily located south of the Yangtze River, collectively accounting for 17.66% of the nation's total area. Unsuitable habitats predominate in higher-latitude regions, collectively covering 66.79% of China's total land area. In future climate scenarios, the total number of suitable habitats for *P. canaliculata* is projected to expand into higher latitude regions, especially under SSP3-70 and SSP5-85 climate conditions. The 4.1 °C contour of Bio1 and the 366 mm contour of Bio18 determine the northernmost geographical distribution of *P. canaliculata*. Climate change is likely to increase the risk of *P. canaliculata* expanding into higher latitudes.

Keywords: Biomod2; climate change; *Pomacea canaliculata*; potential geographic distributions

1. Introduction

Biological invasion refers to the process by which organisms encroach upon a new environment, either naturally or artificially, resulting in economic losses or ecological disasters for biodiversity, agriculture, forestry, animal husbandry, fishery production, and human health in the invaded area [1,2]. The majority of invasive species exhibit pronounced reproductive capacities and high tolerance, enabling them to surpass native species in the competition for resources and habitat [3]. Consequently, once an invasive alien species (IAS) has established a population in a new habitat, eradicating it becomes extremely challenging [4,5]. With the escalation of economic and trade interactions, as well as international exchanges between China and other nations, the number of IAS has been increasing in China at an unprecedented rate, representing a significant threat to China's agricultural development, ecological environment, and public health [6]. As of 2016, China had confirmed 560 IAS, of which 125 species are pests, including 92 that adversely impact agricultural ecosystems. IAS are estimated to incur economic losses exceeding $18.9 billion annually [7].

Pomacea canaliculata (Lamarck, 1819), Ampullariidae, Architaenioglossa, and Gastropoda, originating from the Amazon Basin in South America, was listed by the International Union for Conservation of Nature (IUCN) in 2000 as one of the 100 most destructive invasive species in the world [8]. In the 1980s and early 1990s, *P. canaliculata* initiated their invasion of numerous countries in Southeast Asia and East Asia [9,10]. As of the latest data, in Southeast Asian countries like Thailand, the Philippines, Malaysia, and Vietnam, *P. canaliculata* has emerged as one of the most significant rice pests, resulting in over $1 million expended annually for their control [11–14]. Concurrently, *P. canaliculata* exerts significant detrimental effects on aquatic vegetation [15]. Its high population density resulted in the near-complete eradication of aquatic vegetation and led to the eutrophication of water bodies and increased phytoplankton biomass [15,16]. This culminates in water turbidity and the disturbance of ecosystem function [17]. Additionally, *P. canaliculata* serves as the intermediate host for *Angiostrongylus cantonensis*, with up to 6000 parasites detectable in a single snail [18]. Consumption of raw or undercooked snail meat can readily lead to eosinophilic meningitis, as well as myelitis and spinal meningitis [19]. Moreover, the empty shells of deceased snails represent a health hazard, as they may cause injuries to the feet of individuals involved in planting, harvesting, or managing crops [20,21].

Pomacea canaliculata demonstrates characteristics of amphibiousness, euryphagy, remarkable adaptability, and high fecundity, enabling it to rapidly establish substantial populations in invaded areas [22]. After its introduction into Zhongshan City, Guangdong Province, in 1981 as a valuable species for freshwater fisheries, *P. canaliculata* subsequently expanded its range to include Hubei, Zhejiang, Fujian, Jiangsu, Shanghai, Jiangxi, and other provinces through artificial introduction [23]. In 2003, *P. canaliculata* was included in the inaugural "List of invasive alien species in China". It then disseminated over long distances in China primarily by attaching its eggs to ships or serving as prey for carnivorous fish (herring) or leeches [24,25]. Over the past two decades, Chinese government authorities have engaged in extensive efforts to prevent and control *P. canaliculata* on a large scale. Although there has been some reduction in the damage and spread of *P. canaliculata*, the overall impact remains relatively limited. By 2023, *P. canaliculata* had spread as far north as Chaoyang City, Liaoning Province, and as far west as Lanzhou City, Gansu Province [26]. Although low temperatures may limit *P. canaliculata*'s spread to higher latitudes, global climate warming aids its survival further north in the future [27–29]. Consequently, given the condition of global climate warming, predicting the potential geographic distributions (PGDs) of *P. canaliculata* could be a more effective strategy for preventing and controlling its invasion in non-endemic areas.

Species distribution models (SDMs) utilize known distribution locations of species and related environmental data to ascertain the ecological needs of species and then forecast the potential distribution of species through mathematical-statistical methods or machine learning theory [30]. SDMs are extensively employed in biodiversity conservation research, species invasion studies, evolutionary biology, and the prediction of species' suitable habitats [31,32]. However, various SDMs exhibit differences in data requirements and algorithms, which leads to variations in the impact of environmental factors on species distribution, thus introducing uncertainty in the simulation results [33]. For instance, Classification Tree Analysis (CTA) [34], Random Forest (RF) [35], and eXtreme Gradient Boosting (XGBoost) [36] can process nonlinear and complex data, offering significant advantages like no need for data preprocessing, automatic feature value selection, and the ability to handle missing values. However, these models have complex structures, complicating the explanation of the relationship between variables and predictions, and are prone to overfitting. Conversely, the Generalized Linear Model (GLM) [37] and Generalized Additive Model (GAM) [38] excel in flexibly handling various data types, are well-suited for regression and classification problems, and offer more interpretable results. However, these models are sensitive to outliers and limited in processing high-dimensional data. MaxEnt, widely utilized in species distribution prediction, excels in modeling uncertainty and managing small sample data. However, it is highly sensitive to the quality of training data and requires a complex parameter adjustment process [39]. Therefore, leveraging the strengths of different models while mitigating their shortcomings has become a focal point in research. Ensemble models (EM) preserve the authentic data information captured in each individual model and subsequently minimize the errors inherent in the model, thus enhancing the reliability of the modeling results [40–42]. BIOMOD2 (BIOdiversity MODeling) has the capability to select superior SDMs for constructing EM, thereby enabling a more precise simulation of PGDs and permitting the deduction of the primary environmental factors influencing species distribution [43]. Due to BIOMOD2's ability to compensate for the limitations of individual models, it has rapidly emerged as the predominant model for predicting species distribution in recent years [44–46].

Since the introduction into China of *P. canaliculata* in 1981, numerous researchers have employed various SDMs such as Maxent [47], CLIMEX [48], and GARP [49] to predict the suitable habitat of *P. canaliculata* in China. However, thus far, no researchers have employed Biomod2 to predict the suitable habitat of *P. canaliculata* in China under the conditions of future climate change. Therefore, in this study, we hypothesize that Biomod2 could provide more precise prediction of PGDs for *P. canaliculata* in China than traditional SDMs and that the PGDs of *P. canaliculata* may continue to spread toward higher latitudes in northern China under future warming climate conditions of *P. canaliculata*. In order to identify the important environmental variables that affect the spatial distribution of *P. canaliculata* and predict the distribution pattern and change law of *P. canaliculata* in China under future climate change conditions, We (1) selected five optimal SDMs to combine with an EM; (2) analyzed key environmental variables influencing PGDs of *P. canaliculata*; (3) predicted PGDs of *P. canaliculata* under current climate and climate change conditions in the 2021–2040, 2041–2060s, 2061–2080 and 2081–2100 by EM model; and (4) analyzed centroid shifts of *P. canaliculata* under current climate conditions versus climate change conditions.

2. Materials and Methods

2.1. Occurrence Records of *P. canaliculata*

The global distribution data of *P. canaliculata* are obtained from the Global Biodiversity Information Facility (GBIF; https://www.gbif.org/, accessed on 1 December 2023), Mediterranean Plant Protection Organization Global database (EPPO; https://gd.eppo.int/, accessed on 1 December 2023), CABI Compendium Invasive Species (https://www.cabidigitallibrary.org/product/qi, accessed on 1 December 2023). The distribution data of snails in China were obtained by searching the China National Knowledge Infrastructure (CNKI; https://www.cnki.net/, accessed on 15 December 2023) database. A total of 107 articles, ranging from 1980 to the present, focusing on the birthplace of the *P. canaliculata*, were identified. A total of 5001 original records were obtained, of which 3001 records were selected through screening and deduplication using R for subsequent data processing and analysis (Supplementary S1). Due to the lack of data on actual absence points, we followed the methodologies outlined by Zhao (2022) [50] and Crego (2022) [51], selecting 1000 coordinate points at random from the global geographical scope to serve as pseudo-absence points [52] (Supplementary S2).

2.2. Environmental Variables Related to *P. canaliculata*

Given that *P. canaliculata* are aquatic organisms, environmental factors like soil conditions and wind speed do not influence their PGDs. Consequently, to enhance the efficiency of the model's operation, 19 bioclimatic variables (downloaded from the WorldClim Database (www.worldclim.org/data/worldclim21.html, accessed on 18 December 2023)) [53], along with three terrain factors (elevation, aspect, and slope), downloaded from the Geospatial Data Cloud (www.gscloud.org, accessed on 18 December 2023), and water bodies coverage (WBG), downloaded from the ASTER Global Water Bodies Database (https://search.earthdata.nasa.gov/search, accessed on 1 December 2023), were identified as the primary environmental factors. A total of 19 future bioclimate variables were obtained from BCC-CSM2-MR, a climate model from the Coupled Model Intercomparison Project Phase 6 (CMIP6). These variables are derived from the latest shared socioeconomic pathways (SSPs) and representative concentration pathways (RCPs) to reflect various greenhouse gas emission scenarios. These scenarios encompass SSP1-26 (low forcing), SSP2-45 (moderate forcing), SSP3-70 (moderate to high forcing), and SSP5-85 (high forcing) for the periods of the 2021–2040, 2041–2060, 2061–2080, and 2081–2100. Correlation analysis was conducted using R to address the potential issue of multicollinearity among 19 bioclimate variables, which may result in model overfitting and affect prediction accuracy. Whenever the correlation coefficient between environmental variables exceeded 0.8, variables demonstrating a higher contribution rate to the distribution of *P. canaliculata* were prioritized for inclusion. Finally, seven bioclimate indicators (Bio1, Bio4, Bio7, Bio12, Bio17, Bio18, Bio19), one terrain indicator (elevation), and Water Bodies Coverage (WBC) were employed as the environmental variables. Additionally, the study assumed that elevation and WBC would remain constant over time to guarantee model comparability.

2.3. Construction and Evaluation of SDMs

Biomod2 is an integrated platform that establishes the relationship between species and environmental variables using statistical and machine learning algorithms. Biomod2 (version 4.2.1) includes 11 SDMs, such as the Generalized Linear Model (GLM, Generalized Boosting Model (GBM), Generalized Additive Model (GAM), Multivariate Adaptive Regression Splines (MARS), Classification Tree Analysis (CTA), Artificial Neural Networks (ANN), Surface Range Envelope (SRE, commonly known as BIOCLIM), Flexible Discriminant Analysis (FDA), Random Forest (RF), Maximum Entropy (MaxEnt) and eXtreme Gradient Boosting (XGBOOST). In each process, the distribution data of *P. canaliculata* were randomly divided into two subsets. One subset, comprising 80% of all distribution data, was used to train the model, while the other, containing 20% of the data, was used to evaluate the model's performance. The Area under the ROC Curve (AUC) and True Skill Statistic (TSS) were utilized to assess the model's accuracy. As the AUC value is unaffected by the diagnostic threshold and remains insensitive to species occurrence distribution rate, it enables a comprehensive comparison of the accuracy of diagnostic tests [54]. AUC is currently recognized as one of the most effective evaluation indicators for SDMs [55]. TSS was proposed by Allouche and Kadmon in 2006. TSS not only inherits the advantages of Kappa statistics but also addresses the limitation of Kappa statistics related to unimodal curve responses to species incidence [56]. In this study, PGM models with an AUC > 0.95 and a TSS > 0.85 were selected to integrate the model via the weighted average method. The weight of each model is determined by the ratio of its AUC value to the sum of the AUC values of all models. Ultimately, the EM model was applied to identify the PGDs for *P. canaliculata* under current and future climate scenarios.

2.4. Division of the PGDs and Calculation of Centroids

The suitable habitat of *P. canaliculata* is classified into four categories using ArcGIS 10.6: unsuitable habitats (0–0.25), low suitability habitats (0.25–0.50), moderate suitability habitats (0.50–0.75), and high suitability habitats (0.75–1.00). Simultaneously, the study compares the total suitable area of *P. canaliculata* between the current period and the 2021–2040, 2041–2060, 2061–2080 and 2081–2100, evaluating changes in the habitat (increases, decreases, or no change). Calculate the direction and distance of the spatial movement of the centroids based on the changes in the coordinates of the centroids, and then predict the overall changes in the future distribution trend of *P. canaliculata*. The direction and distance of the spatial movement of the centroids are calculated based on changes in their coordinates, followed by predictions of the overall changes in the future distribution trend of *P. canaliculata*.

3. Results

3.1. Evaluation of SDMs

Utilizing the global and China-specific distribution data of *P. canaliculata*, along with environmental variables, the TSS and AUC values of 11 models were obtained after conducting ten runs. XGBOOST (TSS = 1, AUC = 1), RF (TSS = 0.99, AUC = 1), and CTA (TSS = 0.915, AUC = 0.97) exhibited the highest average TSS and AUC values, demonstrating the best accuracy and stability. GAM (TSS = 0.86, AUC = 0.97) and GBM (TSS = 0.88, AUC = 0.98) ranked second in overall performance. However, SER, ANN, FDA, GLM, MAXENT, and MARS exhibited lower TSS and AUC values, indicating poorer accuracy and stability. Models with an AUC value > 0.95 and a TSS value > 0.85 (CTA, GBM, GAM, RF, and XGBOOST) were screened to derive the EM (AUC = 0.99, TSS = 0.96) (Figure 1).

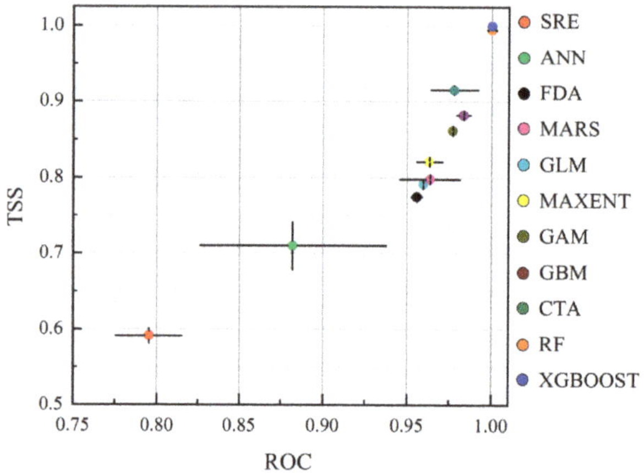

Figure 1. True skill statistics (TSS) and area under the receiver operating characteristic curve (AUC) of 11 SDMs.

3.2. Evaluation of Environmental Variables

The weights assigned to 9 environmental variables differ across various SDMs. Bio1 and Bio18 have the highest weight proportions and are the most significant environmental variables affecting the PGDs of *P. canaliculata*, followed by Water Bodies Coverage (WBC), Bio4, and Bio19. Bio7, Bio12, Bio17, and elevation each contribute less than 10 to the PGDs of *P. canaliculata*. In the XGBOOST, RF, and EM models, Bio1 has the highest weight, with respective values of 32.61, 29.21, and 25.13, while in the GBM, GAM, and CTA models, Bio18 has the highest weight, with values of 33.24, 33.23, and 32.96 respectively. (Figure 2). There is a non-linear relationship between the habitat adaptability of *P. canaliculata* and various environmental variables. The habitat adaptability of *P. canaliculata* increases with the rise in Bio1 and Bio18 values. As WBC, Bio4, and Bio19 have minor contributions to the PGDs, the probability of occurrence, as indicated by the response curves of these environmental variables, is relatively stable, leading to a lower impact on the suitable distribution of *P. canaliculata*. Concurrently, variations are observed across different models in how the changing intervals of environmental variables affect the suitability of *P. canaliculata* habitats. In models such as EM, CTA, GBM, GAM, XGBOOST, and RF, the critical Bio1 thresholds for *P. canaliculata* survival are 4.1 °C, 10 °C, 4.6 °C, 6.9 °C, 4.6 °C, and 8.6 °C, respectively (Figure 3). However, across the various models, there is a consensus regarding the critical Bio18 threshold for *P. canaliculata*, identified as 366 mm.

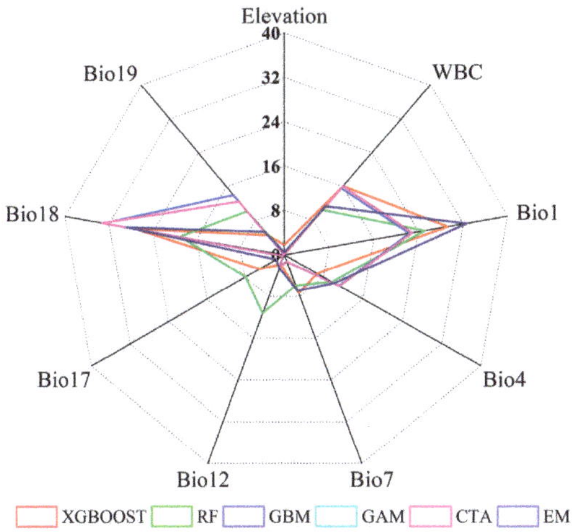

Figure 2. Contribution values of the environmental variables on six SDMs.

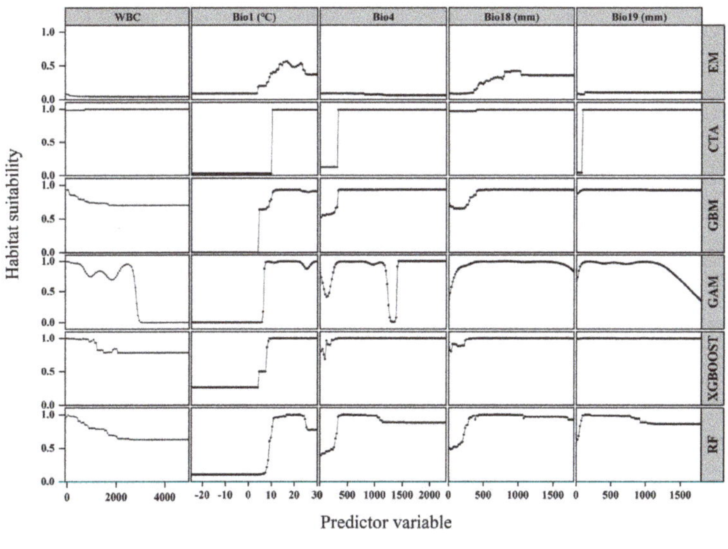

Figure 3. Response curves of predicted occurrence probability of *P. canaliculata* under five key environmental variables using six SDMs.

3.3. Occurrence Records and PGDs of P. canaliculata in China under the Current Climate Conditions

Since the initial discovery of *P. canaliculata* in China in 1981, its distribution has gradually extended from the southern regions towards the north. From 1981 to 1990, *P. canaliculata* was primarily found in the coastal areas of Guangdong Province and was introduced to various counties in Hubei, Sichuan, Anhui, Chongqing, Zhejiang, and Yunnan provinces through aquaculture. Between 1991 and 2000, *P. canaliculata* expanded its range from the coastal to inland areas of Guangdong Province. Concurrently, its presence steadily increased in Guangxi Province. The period from 2001 to 2010 witnessed *P. canaliculata*'s continued northward expansion, with notable invasions in Hunan, Chongqing, Zhejiang, and Sichuan. This expansion resulted in *P. canaliculata* occurring in most counties of Zhejiang and Chongqing. From 2011 to 2020, the distribution of *P. canaliculata* further expanded into Fujian, Hubei, and Sichuan provinces. Additionally, *P. canaliculata* invasions were recorded in southern regions, including Shanxi, Gansu, and Beijing. Since 2020, *P. canaliculata* has progressively spread to northern Jiangsu provinces, with invasions also reported in Henan and Shandong provinces (Figure 4A). Under the current climate, high-suitability habitats for *P. canaliculata* in China are mainly located in Guangdong, Guangxi, Hunan, Hubei, Jiangxi, Anhui, Zhejiang, Fujian, Chongqing, eastern Sichuan, and northern Yunnan provinces, together accounting for 17.66 of the nation's total area. Moderate suitability habitats for *P. canaliculata* are mainly found in Yunnan, central Guizhou, central Henan, central Shandong, and parts of Beijing and Hebei, covering 8.73 of China's total area. Low suitability habitats are primarily situated in select areas of western Yunnan, northern Henan and Hebei, and southern Inner Mongolia, encompassing 6.81 of the nation's area. Unsuitable habitats are predominant in Inner Mongolia, Liaoning, Gansu, Shanxi, Shaanxi, Qinghai, Xinjiang, and other higher-latitude regions, collectively occupying 66.79 of China's total land area (Figure 4B).

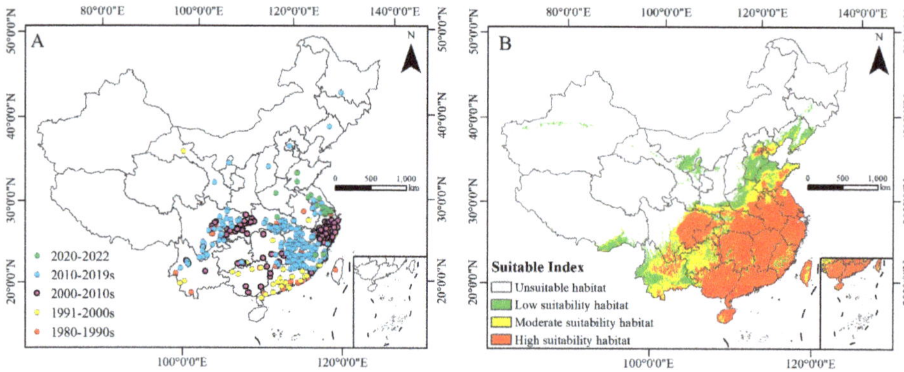

Figure 4. Historical spread dynamics of *P. canaliculata* in China (**A**). Potential habitat suitability of *P. canaliculata* in China based on EM (**B**).

3.4. PGDs of P. canaliculata under Future Climate Conditions

Compared to its current climate suitability habitats, the total suitable habitats for *P. canaliculata* show an increasing trend under the four future climate scenarios (SSP1-26, SSP2-45, SSP3-70, SSP5-85) (Figure 5). Under the SSP1-26 climate scenario, there is a notable expansion of the total suitable area for *P. canaliculata*, particularly in 2021–2040. The proportion of highly suitable habitats for *P. canaliculata* significantly grows, reaching 28.39%. Regions including northern provinces like Shandong, southern Henan, eastern Hebei, and parts of Beijing emerge as highly suitable habitats for *P. canaliculata*. Subsequently, from 2041 to 2100, the proportion of highly suitable habitats gradually decreases while moderately suitable habitats increase. In the SSP2-45 climate scenario, the total suitable habitat for *P. canaliculata* increases, primarily due to the expansion of highly and moderately suitable habitats, with a decrease in low suitable habitats. Both highly and moderately suitable habitats extend northward, reaching the southern part of Liaoning Province. Under SSP3-70 climate conditions, the total suitable habitats for *P. canaliculata* gradually increase, with an expansion in both highly and moderately suitable areas. However, from 2021 to 2040, there will be a temporary decrease in the total suitable area for *P. canaliculata*, mainly due to a reduction in low-suitability habitats. Similarly, under SSP5-85 climate conditions, there is a gradual increase in the total, highly, moderately, and low suitable areas for *P. canaliculata*. The habitat for moderately suitable areas extends to the central region of Heilongjiang (Figures 5 and 6).

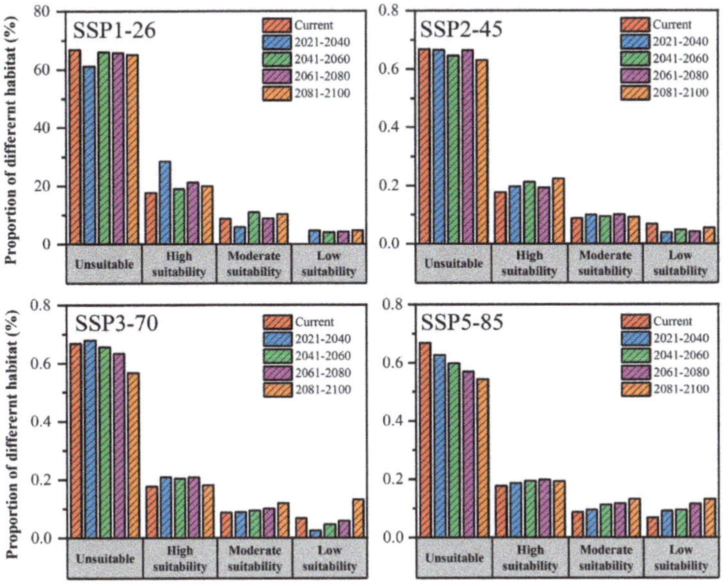

Figure 5. Proportion of different habitats of *P. canaliculata* for current and future climate conditions.

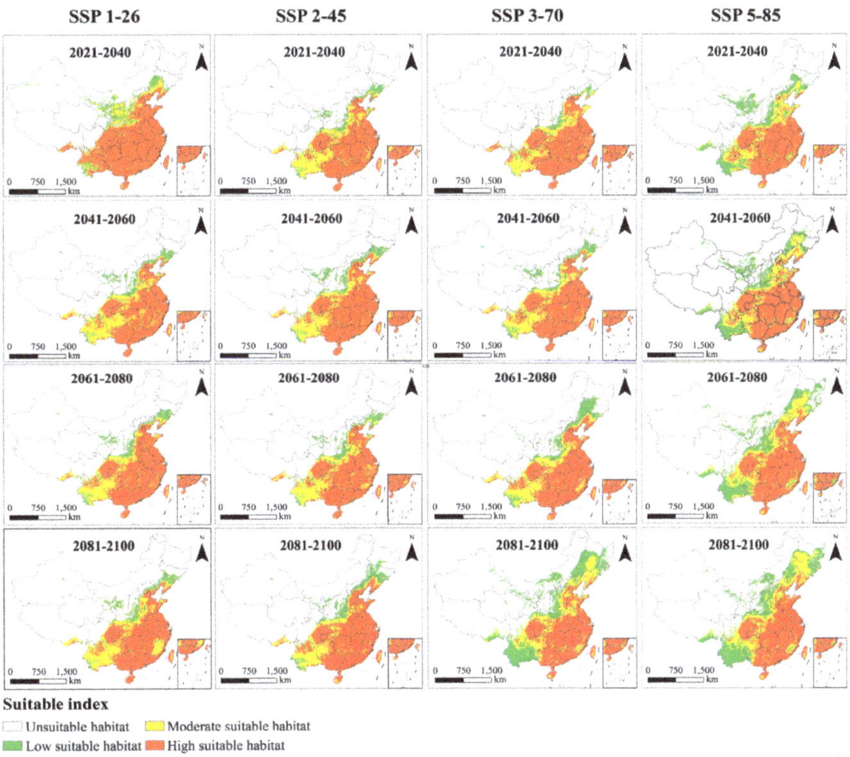

Figure 6. PGDs of *P. canaliculata* in China under four shared socioeconomic pathways (SSP) based on EM.

3.5. Centroids Shifts of P. canaliculata under Future Climate Change Conditions

The centroid of total suitable habitats for *P. canaliculata* under various climatic scenarios is depicted in Figure 7. Under current climate conditions, the centroid of suitable habitats is located in the central region of Chongqing. In the SSP1-26 climate scenario, the centroid of suitable habitats initially shifts northward before gradually moving southward over time. Under SSP2-45 climate conditions, the centroid of suitable habitats exhibits oscillatory lateral movements. Under SSP3-70 conditions, the centroid of suitable habitats initially shifts southward, then gradually migrates northward, ultimately settling in Shaanxi province. Lastly, under SSP5-85 conditions, the centroid of suitable habitats steadily migrates northward over time.

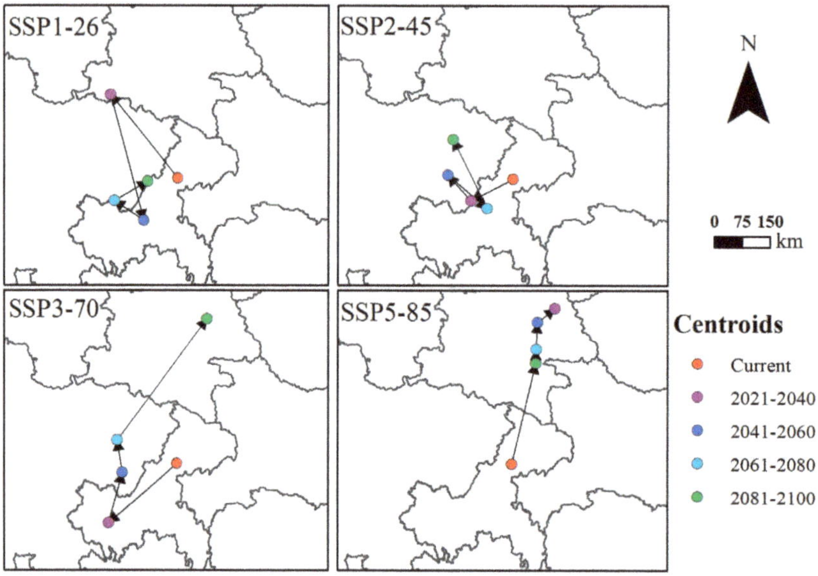

Figure 7. Changes in the centroid shifts of *P. canaliculata* in China under four shared socioeconomic pathways (SSP) based on EM.

4. Discussion

4.1. Model Evaluation

The precision of the PGDs for IAS depends on various factors, such as the availability of species distribution data, the choice of SDMs, and the types of environmental variables. Predictions incorporating both true presence and absence data generally outperform those based only on true presence data in forecasting PGDs [57]. Additionally, MacLeod (2008) [58] suggests that in cases where a species has a broad distribution range, and the true presence data is authentic, its potential distribution range can be theoretically predicted with accuracy, although sampling bias may affect the analysis results. Therefore, this study has employed a method of gathering global distribution data for *P. canaliculata* and incorporating data of *P. canaliculata* invasion in China to ensure the accuracy of predicting its PGDs. Due to algorithmic variations among different SDMs, various environmental factors have diverse impacts on species distribution, resulting in uncertainties in predictions from various SDMs. Elith (2006) [59] evaluated the predictive outcomes of 16 models for 226 species, classifying their performance into three categories: Group 1 consists of the most effective models, mainly using machine learning techniques such as MARS community and Boosted Regression Trees (BRT); Group 2 includes models with the second-best predictive performance, primarily based on standard regression methods like Generalized Additive Models (GAM) and Generalized Linear Models (GLM); Conversely, group 3 encompasses models with the lowest predictive performance, primarily using only presence data without inferred absences, such as BIOCLIM, LIVES, and DOMAIN. Similarly, our findings are largely in agreement with those of Elith (2006) [59]. In this study, machine learning methods such as XGBOOST, GTA, and RF were found to have the highest average TSS and AUC values, indicating superior accuracy and stability. Meanwhile, standard regression models such as GAM and GBM ranked second in overall performance. However, models such as SRE (commonly known as BIOCLIM) exhibited lower TSS and AUC values, placing them in the third group. While machine learning models are capable of achieving high predictive accuracy, Breiman (2001) [35], Loh (2011) [34], and Chen (2016) [36] note that this model often lacks explicit statistical principles and is prone to risks like overfitting. The Ensemble model, which combines multiple models, enhances predictive accuracy, especially for rare species with low occurrence rates. This approach effectively mitigates overfitting concerns while maintaining prediction accuracy [60]. Marmion (2009) [61] and Crimmins (2013) [62] used data proportional segmentation for verification evaluation and found that the Ensemble model outperformed individual models, achieving a notable prediction accuracy of 0.97. Predictions from the Ensemble model align with the accuracy of those from machine learning methods. Therefore, this study integrates five different SDMs (XGBOOST, RF, GBM, GAM, and CTA) to enhance the accuracy of the Ensemble model and overcome the limitations of single-model approaches, yielding a more accurate representation of the geographic distribution of *P. canaliculata*.

4.2. PGDs of *P. canaliculata*

Although certain regions in Yunnan, Guizhou, and Chongqing are moderately suitable habitats and low suitable habitats for *P. canaliculata*, it is clear that *P. canaliculata* can survive in all areas south of the Yangtze River (Figure 4B). Therefore, our study focuses more on the distribution range of *P. canaliculata* in suitable and unsuitable habitats north of the Yangtze River or at higher latitudes. Under current climate conditions, unsuitable habitats for *P. canaliculata* are primarily found in Shaanxi, Shanxi, northern Sichuan, Tibet, Qinghai, Xinjiang, and other northern regions. This finding contradicts Yin's (2022) [47] Maxent predictions, which identified Yunnan, Guizhou, and Chongqing as unsuitable habitats and the western regions of Hunan and Hubei as moderately suitable for *P. canaliculata*. This discrepancy may stem from Yin's (2022) [47] reliance solely on global distribution data (GBIF data) and the neglect of population distribution data of *P. canaliculata* in China. As mentioned in my discussion earlier, predictions of a species' potential geographic distribution are influenced by data on its actual presence. In Figure 4A, *P. canaliculata* is found in Chongqing, Yunnan, and Guizhou, damaging rice fields across up to 50,000 hectares [63]. This also explains why our prediction model identifies Chongqing, Yunnan, and Guizhou as suitable areas for *P. canaliculata*. At the same time, the prediction results align closely with the current occurrence areas of *P. canaliculata* in China, further validating the accuracy of our integrated model. Furthermore, compared to Zhang (2016) [48] Maxent-based predictions, which indicated that suitable habitats for *P. canaliculata* were mainly within the Yangtze River Basin and its southern regions, our predictions show a continued northward expansion of suitable habitats for *P. canaliculata*, the northernmost reaches Jilin Province. This expansion is attributed to the rapid proliferation of *P. canaliculata* in recent years, particularly in areas of northern Hubei and northern Jiangsu, north of the Yangtze River (Figure 4A). Concurrently, sporadic damage caused by *P. canaliculata* has been reported in higher latitude areas such as Henan, Shandong, and Beijing within China. The distribution data of *P. canaliculata* in China significantly influences the prediction of their suitable habitat within the country.

4.3. Environmental Variables

Environmental variables Bio1 and Bio18 have the highest weight, influencing the water and heat requirements crucial for the northward expansion of *P. canaliculata*. These findings are consistent with the research of Lv (2011) [64], Yang (2022) [26], Yin (2023) [47],

Seuffert (2021, 2024) [28,29], and Qin (2023) [65], all highlighting the primary environmental variables—water and heat—as influencers of the suitable habitat for *P. canaliculata*. *P. canaliculata* predominantly thrives in warm and humid environments, while dry conditions and cold winters act as limiting factors for its range expansion. In the EM, CTA GBM, GAM, XGBOOS, and RF, the survival thresholds of *P. canaliculata* on annual mean temperature (Bio1) are 4.1 °C, 10 °C, 4.6 °C, 6.9 °C, 4.6 °C, and 8.6 °C, respectively. This means that different models lead to differing estimates of suitable habitat areas for *P. canaliculata* in China. For example, the CTA model, which has the highest Bio1 threshold (10 °C), predicts a smaller suitable habitat area for *P. canaliculata* compared to the EM, GBM, GAM, and XGBOOST models. The tolerance of *P. canaliculata* to extremely low-temperature conditions in the invaded areas is a critical factor in determining whether it can successfully establish new populations there. Yoshida (2009) [66] has demonstrated that although *P. canaliculata* eggs cannot overwinter in extremely low-temperature conditions, they can endure cold winters and resume reproduction when temperatures rise in the following year. Furthermore, *P. canaliculata* can increase its energy supply and tolerance capacity, thereby enhancing its cold resistance [67]. For example, Wada and Matsukura (2007) [27] discovered that *P. canaliculata* in Ibaraki Prefecture, Japan, not only survived cold winters but also exhibited greater cold tolerance compared to *P. canaliculata* in southern Japan. This phenomenon helps explain why, despite *P. canaliculata* being introduced as food in Tianmen City, Hubei, and Nanjing City, Jiangsu, in 1988, a large-scale outbreak occurred in Hubei only after 2015 and in the Jiangsu region after 2020, potentially due to the gradual increase in *P. canaliculata* cold tolerance.

4.4. PGDs under Climate Change

Environmental variables Bio1 and Bio18 play a crucial role in influencing the geographic distribution of *P. canaliculata*. Consequently, changes in Bio1 and Bio18 under different future climate scenarios will dictate the PGDs of this species. To elucidate the relationship between changes in *P. canaliculata*'s suitable habitats and Bio1 and Bio18, we visualized raster data of Bio1 and Bio18 under current and future climates and marked the lowest thresholds of Bio1 (4.1 °C) and Bio18 (366 mm) for *P. canaliculata*'s survival, as well as the optimal thresholds of Bio1 (16.7 °C) and Bio18 (897 mm) (Figures S1–S3). Under SSP1-26 climate conditions, the 4.1 °C contour does not exhibit significant northward movement. However, under the SSP2-45, SSP3-70, and SSP5-85 climate scenarios, both the 4.1 °C and 16.7 °C contours exhibit significant northward movement over time (Figure S2). Similarly, under SSP1-26 and SSP2-45 climate conditions, the 366 mm contour has not extended northward. However, under SSP3-70 and SSP5-85 climate scenarios, the 366 mm contour for *P. canaliculata* survival has moved northward over time (Figure S3). These observations suggest that due to climate change, currently inhospitable dry and cold regions may become warm and humid, providing suitable environments for *P. canaliculata*'s survival. In our study, under SSP1-26 and SSP2-45 climate scenarios, the total suitable habitats for *P. canaliculata* remain stable. In contrast, under SSP3-70 and SSP5-85 climate scenarios, the total suitable habitats for *P. canaliculata* increase. The northward expansion of the PGDs for *P. canaliculata* under various climate conditions aligns closely with the northward shift of the 4.1 °C and 366 mm contours. Therefore, limiting high fossil fuel consumption not only protects the environment but also helps curtail the colonization and spread of invasive *P. canaliculata*.

5. Conclusions

In this study, we reconstructed the invasive history and analyzed the PGDs of *P. canaliculata* in China. We selected five SDMs—CTA, GBM, GAM, RF, and XGBOOST—to develop an ensemble model (EM) that predicts suitable habitats for *P. canaliculata* under current and future climate scenarios (SSP1-2.6, SSP2-4.5, SSP3-7.0, SSP5-8.5). The EM, exhibiting an AUC of 0.99 and a TSS of 0.96, produced more precise predictions compared to the individual models. Under current climate conditions, the highly suitable habitats for *P. canaliculata* are primarily located south of the Yangtze River. The 4.1 °C contour of Bio1 and the 366 mm contour of Bio18 determine the northernmost geographical distribution of *P. canaliculata*. Sufficient water and heat conditions provide a favorable living environment for *P. canaliculata*. Under future climate scenarios, the total suitable habitats for *P. canaliculata* are expected to expand into higher latitude regions, particularly under SSP3-7.0 and SSP5-8.5 conditions. Future efforts should include field surveys in high-risk areas, such as Shandong, Henan, and Hebei provinces, and increased focus on preventing the further spread of *P. canaliculata* in northern China.

Supplementary Materials: The following supporting information can be downloaded at: https://www.mdpi.com/article/10.3390/agronomy14040650/s1, Figure S1: 4.1 °C (red line) and 16.7 °C (blue line) contours of Bio1 (A) and 366 mm (red line) and 897 mm (blue line) contours of Bio18 (B) on current climate conditions; Figure S2: 4.1 °C (red line) and 16.7 °C (blue line) contours of Bio1 on future climate conditions; Figure S3: 366 mm (red line) and 897 mm (blue line) contours of Bio18 on future climate conditions; Supplementary S1: True Presence Points; Supplementary S2: Pseudo-Absence Points.

Author Contributions: T.W., writing—original draft preparation; T.Z. and W.A., data collection; Z.W., analyze data and review and editing; C.L., editing, supervision. All authors have read and agreed to the published version of the manuscript.

Funding: This research received no external funding.

Data Availability Statement: All data generated or analyzed during this study are included in this published article.

Conflicts of Interest: The authors declare no conflicts of interest.

References

1. Wan, F.; Guo, J.; Wang, D. Alien invasive species in China: Their damages and management strategies. *Biodivers. Sci.* **2002**, *10*, 119–125.
2. Biondi, A.; Guedes, R.N.C.; Wan, F.H.; Desneux, N. Ecology, worldwide spread, and management of the invasive South American tomato pinworm, *Tuta absoluta*: Past, present, and future. *Annu. Rev. Entomol.* **2018**, *63*, 239–258. [CrossRef]
3. Kolar, C.S.; Lodge, D.M. Progress in invasion biology: Predicting invaders. *Trends Ecol. Evol.* **2001**, *16*, 199–204. [CrossRef]
4. Gonçalves da Silva, A.; Kolokotronis, S.O.; Wharton, D. Modeling the eradication of invasive mammals using the sterile male technique. *Biol. Invasions* **2010**, *12*, 751–759. [CrossRef]
5. Simberloff, D. How much information on population biology is needed to manage introduced species? *Conserv. Biol.* **2003**, *17*, 83–92. [CrossRef]
6. Liu, X.; Blackburn, T.M.; Song, T.; Li, X.; Huang, C.; Li, Y. Risks of biological invasion on the belt and road. *Curr. Biol.* **2019**, *29*, 499–505. [CrossRef] [PubMed]
7. Wan, F.H.; Yang, N.W. Invasion and management of agricultural alien insects in China. *Annu. Rev. Entomol.* **2016**, *61*, 77–98. [CrossRef]
8. Luque, G.M.; Bellard, C.; Bertelsmeier, C.; Bonnaud, E.; Genovesi, P.; Simberloff, D.; Courchamp, F. The 100th of the world's worst invasive alien species. *Biol. Invasions* **2014**, *16*, 981–985. [CrossRef]
9. Mochida, O. Spread of freshwater *Pomacea* snails (Pilidae, Mollusca) from Argentina to Asia. *Micronesica* **1991**, *3*, 51–62.
10. Naylor, R. Invasions in agriculture: Assessing the cost of the golden apple snail in Asia. *Ambio* **1996**, *25*, 443–448.
11. Adalla, C.; Magsino, E. Understanding the golden apple snail (*Pomacea canaliculata*): Biology and early initiatives to control the pest in the Philippines. In *Global Advances in Ecology and Management of Golden Apple Snails*; Philippine Rice Research Institute (PhilRice): Los Baños, Philippines, 2006; pp. 199–213.
12. Sawangproh, W.; Poonswad, P. Population dynamics of *Pomacea canaliculata* (Lamarck, 1822) in relation to rice cultivation practice and seasons in Nakhon Pathom, Thailand. *KKU Sci. J.* **2010**, *38*, 228–238.
13. Yahaya, H.; Nordin, M.; Muhamad, H.; Sivapragasam, A. Golden apple snails in Malaysia. In *Global Advances in Ecology and Management of Golden Apple Snails*; Philippine Rice Research Institute (PhilRice): Los Baños, Philippines, 2006; pp. 215–230.
14. Duong, N.C. The golden apple snail in Vietnam. In *Global Advances in Ecology and Management of Golden Apple Snails*; Philippine Rice Research Institute (PhilRice): Los Baños, Philippines, 2006; pp. 243–254.
15. Carlsson, N.O.; Lacoursiere, J.O. Herbivory on aquatic vascular plants by the introduced golden apple snail (*Pomacea canaliculata*) in Lao PDR. *Biol. Invasions* **2005**, *7*, 233–241. [CrossRef]
16. Carlsson, N.O.; Brönmark, C. Size-dependent effects of an invasive herbivorous snail (*Pomacea canaliculata*) on macrophytes and periphyton in Asian wetlands. *Freshw. Biol.* **2006**, *51*, 695–704. [CrossRef]

17. Fang, L.; Wong, P.K.; Lin, L.; Lan, C.; Qiu, J. Impact of invasive apple snails in Hong Kong on wetland macrophytes, nutrients, phytoplankton and filamentous algae. *Freshw. Biol.* **2010**, *55*, 1191–1204. [CrossRef]
18. Cowie, R.H. Angiostrongylus cantonensis: Agent of a sometimes fatal globally emerging infectious disease (rat lungworm disease). *ACS Chem. Neurosci.* **2017**, *8*, 2102–2104. [CrossRef] [PubMed]
19. Hochberg, N.S.; Bhadelia, N. Infections associated with exotic cuisine: The dangers of delicacies. *Microbiol. Spectr.* **2015**, *3*, 355–374. [CrossRef] [PubMed]
20. Cowie, R.H. Apple snails (Ampullariidae) as agricultural pests: Their biology, impacts and management. In *Molluscs as Crop Pests*; CABI Publishing: Wallingford, UK, 2002; p. 145192.
21. Marwoto, R.M.; Isnaningsih, N.R.; Joshi, R.C. The invasive apple snail (*Pomacea* spp.) in Indonesia. *Agric. Dev.* **2018**, *35*, 41–48.
22. Salleh, N.H.M.; Arbain, D.; Daud, M.Z.M.; Pilus, N.; Nawi, R. Distribution and management of *Pomacea canaliculata* in the northern region of Malaysia: Mini review. *APCBEE Procedia* **2012**, *2*, 129–134. [CrossRef]
23. Yang, Y.; Hu, Y.; Li, X.; Wang, X.; Mu, X.; Song, H.; Wang, P.; Liu, C.; Liu, C.; Luo, J. Historical invasion, expansion process and harm investigation of *Pomacea canaliculata* in China. *Sci. Agric. Sic. Bull.* **2010**, *26*, 245–250.
24. Baker, R.; Candresse, T.; sné Simon, E.; Gilioli, G.; Grégoire, J.; Jeger, M.; Karadjova, O.; Lövei, G.; Makowski, D.; Manceau, C. Statement on the identity of apple snails. *EFSA J.* **2012**, *10*, 1–11.
25. Lucero, J.M. *Regional Expansion and Evaluation of Potential Chemical Control for Invasive Apple Snails (Pomacea maculata) in Southwest Louisiana*; Louisiana State University and Agricultural Mechanical College: Baton Rouge, LA, USA, 2021.
26. Yang, Q.Q.; Liu, S.W.; He, C.; Yu, X.P. Distribution and the origin of invasive apple snails, *Pomacea canaliculata* and *P. maculata* (Gastropoda: Ampullariidae) in China. *Sci. Rep.* **2018**, *8*, 1185. [CrossRef] [PubMed]
27. Wada, T.; Matsukura, K. Seasonal changes in cold hardiness of the invasive freshwater apple snail, *Pomacea canaliculata* (Lamarck)(Gastropoda: Ampullariidae). *Malacologia* **2007**, *49*, 383–392. [CrossRef]
28. Seuffert, M.E.; Martín, P.R. Influence of temperature, size and sex on aerial respiration of *Pomacea canaliculata* (Gastropoda: Ampullariidae) from southern Pampas, Argentina. *Malacologia* **2009**, *51*, 191–200. [CrossRef]
29. Seuffert, M.E.; Burela, S.; Martín, P.R. Influence of water temperature on the activity of the freshwater snail *Pomacea canaliculata* (Caenogastropoda: Ampullariidae) at its southernmost limit (Southern Pampas, Argentina). *J. Therm. Biol.* **2010**, *35*, 77–84. [CrossRef] [PubMed]
30. Lith, J.; Leathwick, J.R. Species distribution models: Ecological explanation and prediction across space and time. *Annu. Rev. Ecol. Evol. Syst.* **2009**, *40*, 677–697.
31. Cayuela, L.; Golicher, D.; Newton, A.; Kolb, M.; De Alburquerque, F.; Arets, E.; Alkemade, J.; Pérez, A. Species distribution modeling in the tropics: Problems, potentialities, and the role of biological data for effective species conservation. *Trop. Conserv. Sci.* **2009**, *2*, 319–352. [CrossRef]
32. Barbet-Massin, M.; Rome, Q.; Villemant, C.; Courchamp, F. Can species distribution models really predict the expansion of invasive species? *PLoS ONE* **2018**, *13*, e0193085. [CrossRef]
33. Guo, C.; Lek, S.; Ye, S.; Li, W.; Liu, J.; Li, Z. Uncertainty in ensemble modelling of large-scale species distribution: Effects from species characteristics and model techniques. *Ecol. Model.* **2015**, *306*, 67–75. [CrossRef]
34. Loh, W. Classification and regression trees. *Wires. Data. Min. Knowl.* **2011**, *1*, 14–23. [CrossRef]
35. Breiman, L. Random forests. *Mach. Learn.* **2001**, *45*, 5–32. [CrossRef]
36. Chen, T.; Guestrin, C. Xgboost: A scalable tree boosting system. In Proceedings of the 22nd Acm Sigkdd International Conference on Knowledge Discovery and Data Mining, San Francisco, CA, USA, 13–17 August 2016; pp. 785–794.
37. Nelder, J.A.; Wedderburn, R.W. Generalized linear models. *J. R. Stat. Soc. Ser. A Stat. Soc.* **1972**, *135*, 370–384. [CrossRef]
38. Hastie, T.J. Generalized additive models. In *Statistical Models in S*; Routledge: Oxfordshire, UK, 2017; pp. 249–307.
39. Berger, A.; Della Pietra, S.A.; Della Pietra, V.J. A maximum entropy approach to natural language processing. *Comput. Linguist.* **1996**, *22*, 39–71.
40. Lomba, A.; Pellissier, L.; Randin, C.; Vicente, J.; Moreira, F.; Honrado, J.; Guisan, A. Overcoming the rare species modelling paradox: A novel hierarchical framework applied to an Iberian endemic plant. *Biol. Conserv.* **2010**, *143*, 2647–2657. [CrossRef]
41. Jones-Farrand, D.T.; Fearer, T.M.; Thogmartin, W.E.; Thompson, F.R., III; Nelson, M.D.; Tirpak, J.M. Comparison of statistical and theoretical habitat models for conservation planning: The benefit of ensemble prediction. *Ecol. Appl.* **2011**, *21*, 2269–2282. [CrossRef] [PubMed]
42. Dormann, C.F.; Calabrese, J.M.; Guillera-Arroita, G.; Matechou, E.; Bahn, V.; Bartoń, K.; Beale, C.M.; Ciuti, S.; Elith, J.; Gerstner, K. Model averaging in ecology: A review of Bayesian, information-theoretic, and tactical approaches for predictive inference. *Ecol. Monogr.* **2018**, *88*, 485–504. [CrossRef]
43. Hao, T.; Elith, J.; Guillera-Arroita, G.; Lahoz-Monfort, J.J. A review of evidence about use and performance of species distribution modelling ensembles like BIOMOD. *Divers. Distrib.* **2019**, *25*, 839–852. [CrossRef]
44. Park, C.; Byeon, J.; Park, J.; Lee, S.; Park, B.; Kim, J.; Woo, H. Impact of climate change on species distribution of *Picea jezoensis* Carrière in Baekdudaegan using ensemble modeling. *Sens. Mater.* **2022**, *34*, 4629–4637. [CrossRef]
45. Huang, D.; An, Q.; Huang, S.; Tan, G.; Quan, H.; Chen, Y.; Zhou, J.; Liao, H. Biomod2 modeling for predicting the potentialecological distribution of three Fritillaria species under climate change. *Sci. Rep.* **2023**, *13*, 18801. [CrossRef] [PubMed]
46. Zhao, G.; Cui, X.; Sun, J.; Li, T.; Wang, Q.; Ye, X.; Fan, B. Analysis of the distribution pattern of Chinese *Ziziphus jujuba* under climate change based on optimized biomod2 and MaxEnt models. *Ecol. Indic.* **2021**, *132*, 108256. [CrossRef]

47. Yin, Y.; He, Q.; Pan, X.; Liu, Q.; Wu, Y.; Li, X. Historical invasion, expansion process and harm investigation of *Pomacea canaliculata* in China. *Biology* **2022**, *11*, 110.
48. Zhang, H.; Luo, D.; Mu, X.; Xu, M.; Wei, H.; Luo, J.; Zhang, J.; Hu, Y. Predicting the potential suitable distribution area of the apple snail *Pomacea canaliculata* in China based on multiple ecological niche models. *Chin. J. Appl. Ecol.* **2016**, *27*, 1277–1284.
49. Wang, Y.; Zheng, P.; Pan, W. Predicting the potential suitable distribution area of *Pomacea canliculate* in China based on the GARP ecological niche modeling. *J. Fujian Agric. For. Univ.* **2018**, *47*, 21–25.
50. Zhao, Z.; Xiao, N.; Shen, M.; Li, J. Comparison between optimized MaxEnt and random forest modeling in predicting potential distribution: A case study with *Quasipaa boulengeri* in China. *Sci. Total Environ.* **2022**, *842*, 156867. [CrossRef] [PubMed]
51. Crego, R.D.; Stabach, J.A.; Connette, G. Implementation of species distribution models in Google Earth Engine. *Divers. Distrib.* **2022**, *28*, 904–916. [CrossRef]
52. Barbet-Massin, M.; Jiguet, F.; Albert, C.H.; Thuiller, W. Selecting pseudo-absences for species distribution models: How, where and how many? *Methods Ecol. Evol.* **2012**, *3*, 327–338. [CrossRef]
53. Rashid, W.; Shi, J.; Rahim, I.; Qasim, M.; Baloch, M.N.; Bohnett, E.; Yang, F.; Khan, I.; Ahmad, B. Modelling potential distribution of snow Leopards in pamir, northern Pakistan: Implications for human–snow leopard conflicts. *Sustainability* **2021**, *13*, 13229. [CrossRef]
54. Fawcett, T. An introduction to ROC analysis. *Pattern Recogn. Lett.* **2006**, *27*, 861–874. [CrossRef]
55. Huang, J.; Ling, C.X. Using AUC and accuracy in evaluating learning algorithms. *IEEE Trans. Knowl. Data Eng.* **2005**, *17*, 299–310. [CrossRef]
56. Allouche, O.; Tsoar, A.; Kadmon, R. Assessing the accuracy of species distribution models: Prevalence, kappa and the true skill statistic (TSS). *J. Appl. Ecol.* **2006**, *43*, 1223–1232. [CrossRef]
57. Brotons, L.; Thuiller, W.; Araújo, M.B.; Hirzel, A.H. Presence-absence versus presence-only modelling methods for predicting bird habitat suitability. *Ecography* **2004**, *27*, 437–448. [CrossRef]
58. MacLeod, C.D.; Mandleberg, L.; Schweder, C.; Bannon, S.M.; Pierce, G.J. A comparison of approaches for modelling the occurrence of marine animals. In *Essential Fish Habitat Mapping in the Mediterranean*; Springer Science & Business Media: Berlin/Heidelberg, Germany, 2008; pp. 21–32.
59. Elith, J.; Graham, C.H.; Anderson, R.P.; Dudík, M.; Ferrier, S.; Guisan, A.; Hijmans, R.J.; Huettmann, F.; Leathwick, J.R.; Lehmann, A. Novel methods improve prediction of species' distributions from occurrence data. *Ecography* **2006**, *29*, 129–151. [CrossRef]
60. Breiner, F.T.; Guisan, A.; Bergamini, A.; Nobis, M.P. Overcoming limitations of modelling rare species by using ensembles of small models. *Methods Ecol. Evol.* **2015**, *6*, 1210–1218. [CrossRef]
61. Marmion, M.; Parviainen, M.; Luoto, M.; Heikkinen, R.K.; Thuiller, W. Evaluation of consensus methods in predictive species distribution modelling. *Divers. Distrib.* **2009**, *15*, 59–69. [CrossRef]
62. Crimmins, S.M.; Dobrowski, S.Z.; Mynsberge, A.R. Evaluating ensemble forecasts of plant species distributions under climate change. *Ecol. Model.* **2013**, *266*, 126–130. [CrossRef]
63. Deng, X.; Liu, D.; Guo, X.; Tian, Z.; Yang, L.; Gu, R.; Huang, Z.; Li, Y.; Liu, S.; Xiao, W. Temporal and spatial distribution of *Pomacea canaliculata* in Er'hai Lake catchment. *Asian J. Ecotox.* **2018**, *13*, 134–142.
64. Lv, S.; Zhang, Y.; Steinmann, P.; Yang, G.; Yang, K.; Zhou, X.; Utzinger, J. The emergence of angiostrongyliasis in the People's Republic of China: The interplay between invasive snails, climate change and transmission dynamics. *Freshw. Biol.* **2011**, *56*, 717–734. [CrossRef]
65. Qin, Z.; Zhang, J.; Yao, F.; Liu, J.; Shi, Z.; Zhao, B.; Guo, J. Niche conservatism and geographical range expansion of *Pomacea canaliculata* and *Pomacea maculata* in non-native United States and China. *Biol. Invasions* **2023**, *25*, 3391–3405. [CrossRef]
66. Yoshida, K.; Hoshikawa, K.; Wada, T.; Yusa, Y. Life cycle of the apple snail *Pomacea canaliculata* (Caenogastropoda: Ampullariidae) inhabiting Japanese paddy fields. *Appl. Entomol. Zool.* **2009**, *44*, 465–474. [CrossRef]
67. Liu, J.; Sun, Z.; Wang, Z.; Peng, Y. A comparative transcriptomics approach to analyzing the differences in cold resistance in *Pomacea canaliculata* between Guangdong and Hunan. *J. Immunol. Res.* **2020**, *2020*, 8025140. [CrossRef]

Disclaimer/Publisher's Note: The statements, opinions and data contained in all publications are solely those of the individual author(s) and contributor(s) and not of MDPI and/or the editor(s). MDPI and/or the editor(s) disclaim responsibility for any injury to people or property resulting from any ideas, methods, instructions or products referred to in the content.

remote sensing

MDPI

Article

Exploring the Spatiotemporal Alterations in China's GPP Based on the DTEC Model

Jie Peng [1,2], Yayong Xue [1,2,*], Naiqing Pan [3], Yuan Zhang [4], Haibin Liang [5,6] and Fei Zhang [7]

[1] College of Geography and Remote Sensing Sciences, Xinjiang University, Urumqi 830046, China; pengjie@stu.xju.edu.cn
[2] Xinjiang Key Laboratory of Oasis Ecology, Xinjiang University, Urumqi 830046, China
[3] Schiller Institute for Integrated Science and Society, Boston College, Chestnut Hill, MA 02467, USA
[4] College of Environmental and Resource Sciences, Shanxi University, Taiyuan 030006, China
[5] Institute of Geographical Science, Taiyuan Normal University, Jinzhong 030619, China
[6] Shanxi Key Laboratory of Earth Surface Processes and Resource Ecological Security in Fenhe River Basin, Taiyuan Normal University, Jinzhong 030619, China
[7] College of Geography and Environmental Sciences, Zhejiang Normal University, Jinhua 321004, China; zhangfei3s@zjnu.edu.cn
* Correspondence: xueyy@xju.edu.cn

Abstract: Gross primary productivity (GPP) is a reliable measure of the carbon sink potential of terrestrial ecosystems and is an essential element of terrestrial carbon cycle research. This study employs the diffuse fraction-based two-leaf light-use efficiency (DTEC) model to imitate China's monthly GPP from 2001 to 2020. We studied the trend of GPP, investigated its relationship with climatic factors, and separated the contributions of climate change and human activities. The findings showed that the DTEC model was widely applicable in China. During the study period, China's average GPP increased significantly, by 9.77 g C m^{-2} yr^{-1} ($p < 0.001$). The detrimental effect of aerosol optical depth (AOD) on GPP was more widespread than that of total precipitation, temperature, and solar radiation. Areas that benefited from AOD, such as Northwest China, experienced significant increases in GPP. Climate change and human activities had a primary and positive influence on GPP during the study period, accounting for 28% and 72% of the increase, respectively. Human activities, particularly ecological restoration projects and the adoption of advanced agricultural technologies, played a significant role in China's GPP growth. China's afforestation plan was particularly notable, with the GPP increasing in afforestation areas at a rate greater than 10 g C m^{-2} yr^{-1}. This research provides a theoretical foundation for the long-term management of China's terrestrial ecosystems and helps develop adaptive ecological restoration tactics.

Keywords: diffuse fraction-based two-leaf light-use efficiency model; gross primary productivity; carbon cycle; climate change; ecological restoration projects

1. Introduction

Global carbon sequestration is an essential step in the terrestrial carbon cycle [1]. For determining the state and dynamics of the terrestrial carbon cycle, Gross Primary Productivity (GPP) is an essential indicator that describes the overall amount of atmospheric carbon dioxide that plants absorb via photosynthesis [2]. Therefore, accurate modelling of GPP on a zonal or worldwide scale will contribute to understanding the stability of terrestrial ecosystems and assist in decision making in response to more climatic changes [3].

Currently, two main methods are used to estimate GPP: ground-based observations and model simulations [4]. Ground-based observational methods generally tend to use eddy correlation techniques [5]. However, as the number of flux sites is few and sparsely distributed, ground-based observations are commonly used for validation analyses of GPP models [6]. Ecosystem modelling, which primarily consists of process-based models and is based on vegetation photosynthesis light-use efficiency (LUE) models, is a useful technique for predicting GPP on a zonal or global scale. Due to its simpler structure and fewer parameter inputs, the LUE model is widely applied on a larger scale than process-based models that require many parameters and data inputs.

However, there are large differences in the variety of LUE models. For example, the MODIS GPP product was based on the big-leaf LUE model, but the fixed parameters of the model led to an underestimation of simulated GPP values, especially in farmland ecosystems, with an underestimation of approximately 30% [7]. Additionally, GPP is underestimated by the LUE model because its algorithms do not distinguish between the maximum LUE of C3 and C4 vegetation [8]. To compensate for this deficiency, Yan [9] proposed a terrestrial ecosystem carbon flux model (TEC) to differentiate the LUEs of C3 and C4 vegetation. C3 and C4 plants are two fundamental plant functional types (PFTs) with different responses during photosynthesis. Nonetheless, the model encountered limitations, as the LUE of scattered radiation from vegetation canopies was greater than that of direct radiation. He [10] proposed the two-leaf light-use efficiency model (TL-LUE), which distinguishes between shaded and sunlit leaves. However, the two-leaf model does not consider the influence of soil moisture stress and the diffuse fraction on vegetative photosynthesis. To address this limitation, Yan [11] proposed a diffuse fraction-based two-leaf light-use efficiency model (DTEC) based on previous models. The novel model adds a diffuse fraction to recalculate the solar radiation received and the LUE by the sunny and shady leaves, allowing the model to better capture seasonal variations in the GPP [12].

The effects of climate change on the dynamics of GPP have been a hot topic in global change research [13]. Surplus solar radiation may cause an increase in soil moisture, leading to anaerobic conditions in the soil, thus reducing vegetation productivity on the Tibetan Plateau [14]. Furthermore, aerosol particles have a scattering effect on solar radiation, increasing the proportion of scattered radiation to total solar radiation, which in turn affects vegetation productivity by influencing photosynthesis [15]. In particular, climate extremes can alter the ecosystem's structure and capacity to function; this phenomenon may negatively impact the terrestrial carbon cycle [16]. Another aspect is human activity, notably grazing [17], the use of ecological restoration initiatives [18], land management strategies [19], and urbanization [20].

China has become the most important and sensitive region in global research on carbon and water cycles in terrestrial ecosystems [21]. Moreover, China, the world's largest carbon emitter, has pledged to achieve carbon neutrality by 2060 [22]. To achieve this goal, an in-depth understanding of the status of the terrestrial carbon cycle in China is needed. Since the conclusion of the 20th century, China has initiated a range of environmental recovery and conservation measures, including China's afforestation program (CAP) and the Conversion of Cropland to Forest Program (CCFP). The main objective of these initiatives is to restore the natural environment through a variety of human endeavours, such as afforestation and the protection of natural forests [23]. Therefore, understanding how anthropogenic influences and climate change affect carbon sinks in detail is essential for understanding China's terrestrial ecosystems [24].

In this research, the GPP of Chinese areas from 2001 to 2020 was calculated using the DTEC model, and we analysed the spatial-temporal patterns and their influencing factors. The main research objectives were as follows: (1) to evaluate the applicability of the DTEC model in the Chinese region and create a new GPP dataset, (2) to analyse the spatial and temporal distribution characteristics of the new GPP data, (3) to explore the correlations with different climatic variables, and (4) to evaluate the relative impacts and contributions of anthropogenic activities and the climate to GPP at the regional and national levels.

2. Materials and Methods

2.1. Study Area

In this study, nine zones were created within the study area (Figure 1) according to the geo-ecological factors. These regions included Northeast China (NEC), Inner Mongolia

(IM), Northwest China (NWC), North China (NC), Central China (CC), the Tibetan Plateau (TP), Southeast China (SEC), South China (SC), and Southwest China (SWC) [25].

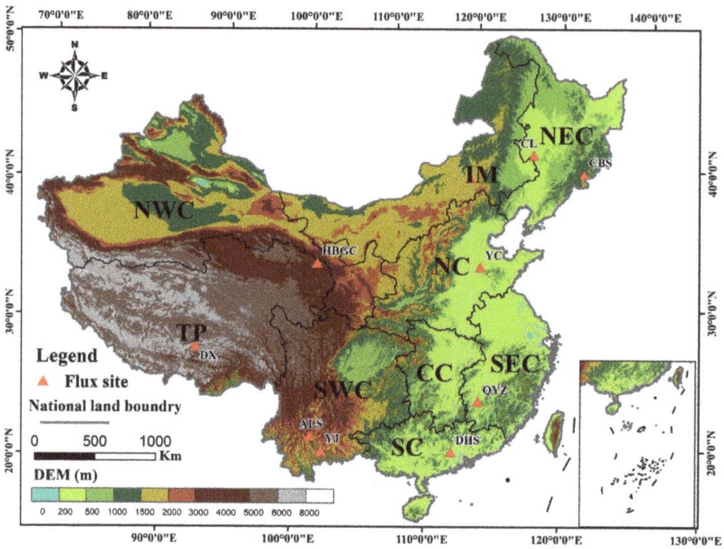

Figure 1. Elevation, flux tower location distribution and ecological regions in the study area.

2.2. Data Sources

The flux observation data used in this paper were obtained from the China Terrestrial Ecosystem Flux Observation Research Network (http://chinaflux.org, accessed on 11 April 2022) and the FLUXENT2015 dataset (https://fluxnet.org, accessed on 20 April 2022). This study selected 10 flux towers in China, and detailed information about each site is shown in Table 1.

Table 1. Flux site details for 10 flux towers in China.

Flux Sites	Latitude/°N	Longitude/°E	Vegetation Type	Period/Year
Changbaishan (CBS)	42.40	128.10	Deciduous Broadleaf Forests	2004–2010
Dinghushan (DHS)	23.17	112.53	Evergreen Broadleaf Forests	2004–2010
Dangxiong (DX)	30.50	91.07	Grassland	2004–2010
Haibei (HB)	37.62	101.32	Shrubland	2004–2010
Inner Mongolia (NMG)	43.55	116.67	Grassland	2004–2010
Qianyanzhou (QYZ)	26.74	115.06	Evergreen Broadleaf Forests	2004–2010
Yucheng (YC)	36.83	116.57	Cropland	2004–2010
Changling (CL)	44.59	123.51	Grassland	2007–2010
Ailaoshan (ALS)	24.54	101.29	Mixed Forest	2009–2012
Yuanjiang (YJ)	23.47	102.18	Savannas	2014–2015

The two types of GPP product data used in this paper were MODIS17A2H.006 (MOD17) and the global dataset of sun-induced chlorophyll fluorescence (GO-SIF). The GO-SIF data were obtained from the high spatial and temporal resolution (0.05°, 8 days) dataset provided by the Global Ecology Team (https://globalecology.unh.edu, accessed on 15 April 2022). The MOD17 data were obtained from the MOD17A2H.006 product on the Google Earth Engine (GEE) platform, which provides spatial and temporal resolution (500 m, 8 days). The performance of the DTEC model was evaluated in this paper with these two products and flux data. The C4 vegetation distribution percentage data were obtained from the International-Satellite-Land-Surface-Climatology-Project-II (https://doi.org/10.3334/ORNLDAAC/932, accessed on 15 April 2022), which provides a 1° spatial resolution static map of the globe. The aerosol optical depth (AOD) data were obtained from https://doi.org/10.5281/zenodo.5652257 (accessed on 15 April 2022), which was produced by the Bai team and had a spatial and temporal resolution of 1000 m/day.

The following data were obtained from the GEE platform. Meteorological driving data such as temperature, precipitation, and solar radiation were obtained from the ERA5-Land Monthly Averaged-ECMWF Climate Reanalysis data, with a spatiotemporal resolution of 11,132 m/1 month. The digital elevation model (DEM) data were taken from the NASANASADEM_HGT.001 data product with a spatial resolution of 30 m. Evapotranspiration and potential evapotranspiration data were obtained from the MOD16A2.006 product, solar zenith angle data were obtained from the MOD09A1.061 product, the leaf area index (LAI) was obtained from the MOD15A2H.061 product, and the temporal and spatial resolutions were both 500 m, 8 days. The land cover type data were taken from the IGBP classification data of the MCD12Q1.006 data, with a spatial resolution of 500 m. The land cover types were reclassified into 10 categories: evergreen needleleaf forest (ENF), evergreen broadleaf forest (EBF), deciduous needleleaf forest (DNF), deciduous broadleaf forest (DBF), mixed forest (MF), shrublands, savannas, grasslands, croplands, and non-vegetated areas (NOA). Considering the consistency of the model runs with the cycles of each dataset, all the data were resampled to 500 m, and the period was chosen to be 2001–2020 month by month.

2.3. Methods

2.3.1. DTEC Model

DTEC is a two-leaf light-use efficiency model based on the diffusion fraction. The model considers the effects of scattered radiation, diffuse fraction and soil water stress on GPP [11]. The main structure of the model is as follows:

$$\mathrm{GPP} = (\varepsilon_{\mathrm{msu}} \times \mathrm{APAR}_{\mathrm{sun}} + \varepsilon_{\mathrm{msh}} \times \mathrm{APAR}_{\mathrm{shd}}) \times W_{\varepsilon} \times T_{\varepsilon} \quad (1)$$

$$\mathrm{APAR}_{\mathrm{sun}} = \left[\mathrm{PAR}_{\mathrm{dir}} \times \frac{\cos(\beta)}{\cos(\theta)} + \frac{\mathrm{PAR}_{\mathrm{dif}} - \mathrm{PAR}_{\mathrm{dif,u}}}{\mathrm{LAI}} + \mathrm{C}\right] \times \mathrm{LAI}_{\mathrm{sun}} \quad (2)$$

$$\mathrm{APAR}_{\mathrm{shd}} = \left[\frac{\mathrm{PAR}_{\mathrm{dif}} - \mathrm{PAR}_{\mathrm{dif,\,u}}}{\mathrm{LAI}} + \mathrm{C}\right] \times \mathrm{LAI}_{\mathrm{shd}} \quad (3)$$

$$\mathrm{LAI}_{\mathrm{sum}} = 2 \times \cos(\theta) \times \left[1 - \exp\left(-0.5 \times \Omega \times \frac{LAI}{\cos(\theta)}\right)\right] \quad (4)$$

$$\mathrm{LAI}_{\mathrm{shd}} = \mathrm{LAI} - \mathrm{LAI}_{\mathrm{sum}} \quad (5)$$

$$\mathrm{PAR}_{\mathrm{dif}} = \mathrm{PAR} \times D_f \quad (6)$$

$$D_f = 0.7527 + 3.8453SI - 16.316SI^2 + 18.962SI^3 - 7.0802SI^4 \quad (7)$$

$$SI = \frac{PAR}{[0.48S_0\cos(\theta)]} \tag{8}$$

$$W_\varepsilon = E/E_{PT} \tag{9}$$

$$T_\varepsilon = \frac{(T_a - T_{min})(T_a - T_{max})}{(T_a - T_{min})(T_a - T_{max}) - (T_a - T_{opt})^2} \tag{10}$$

where ε_{msu} and ε_{msh} are the maximum LUEs of the sun and shade leaves, respectively. For C3 plants: $\varepsilon_{msh} = 3.78 \times D_f^{1.8}$, $\varepsilon_{msu} = 1.67$ g C/MJ, and for C4 plants: $\varepsilon_{msh} = 5.78 \times D_f^{1.8}$, $\varepsilon_{msu} = 2.56$ g C/MJ. $APAR_{sun}$ and $APAR_{shd}$ are the photosynthetically active radiation absorbed by the sun and shade leaves, respectively. W_ε is the water stress factor, in which E is the actual evapotranspiration and E_{PT} is the potential evapotranspiration. T_ε is the temperature stress factor, in which T_a is the air temperature and T_{min}, T_{max}, and T_{opt} are the minimum, maximum, and optimal temperatures for photosynthesis in biomes, respectively. C is the contribution of direct radiation to the scattered radiation after multiple scattering, and the formula is $0.07\Omega \times PAR_{dir} \times (1.1 - 0.1LAI) \times e^{(-\cos(\theta))}$. Ω is the vegetation type index. β is the mean leaf solar angle, which is 60°, and θ is the solar zenith angle. PAR_{dir}, PAR_{dif} and $PAR_{dif,u}$ are the direct photosynthetically active radiation, scattered photosynthetically active radiation and scattered radiation under the canopy, respectively, in which $PAR_{dir} = PAR - PAR_{dif}$ and $PAR_{dif,u} = PAR_{dif} \times e^{(-0.5\times\Omega\times LAI)/(0.537+0.025LAI)}$. LAI_{sun} and LAI_{shd} are the LAIs for sunny and shady leaves, respectively, which were obtained by separating the LAI products. PAR is the photosynthetically active radiation, calculated as $0.48R_g$, where R_g is the solar radiation. D_f is the scattering ratio, SI is the clear sky index, and S_0 is the solar constant (1367 W/m^{-2}).

2.3.2. Analytical Methods

The accuracy of the simulated values was determined using three metrics: coefficient of determination (R^2), root mean square error (RMSE), and bias. The Theil–Sen median trend was used for the GPP time series trend analysis [26]. Partial correlation analysis was used to quantify complex connections between variables with multiple correlations [27]. Residual analysis was used to evaluate the contributions of climate change and human activities to GPP changes [28]. To more accurately assess how human activities and climate conditions affect GPP, the contribution of climatic factors (GPP_{CC}) and contribution of human activities (GPP_{HA}) were classified into 7 levels based on their linear trend values [29], as shown in Table S1. The classification of the main drivers of change in China and the calculations of the relative contributions to GPP [30] are shown in Table S2.

Figure 2 shows the steps and sequences adopted in the present study's data acquisition and analysis. The flow charts were divided into four parts: data preparation, model accuracy verification, GPP change analysis, and determination of the contributions of anthropogenic and climate factors to GPP.

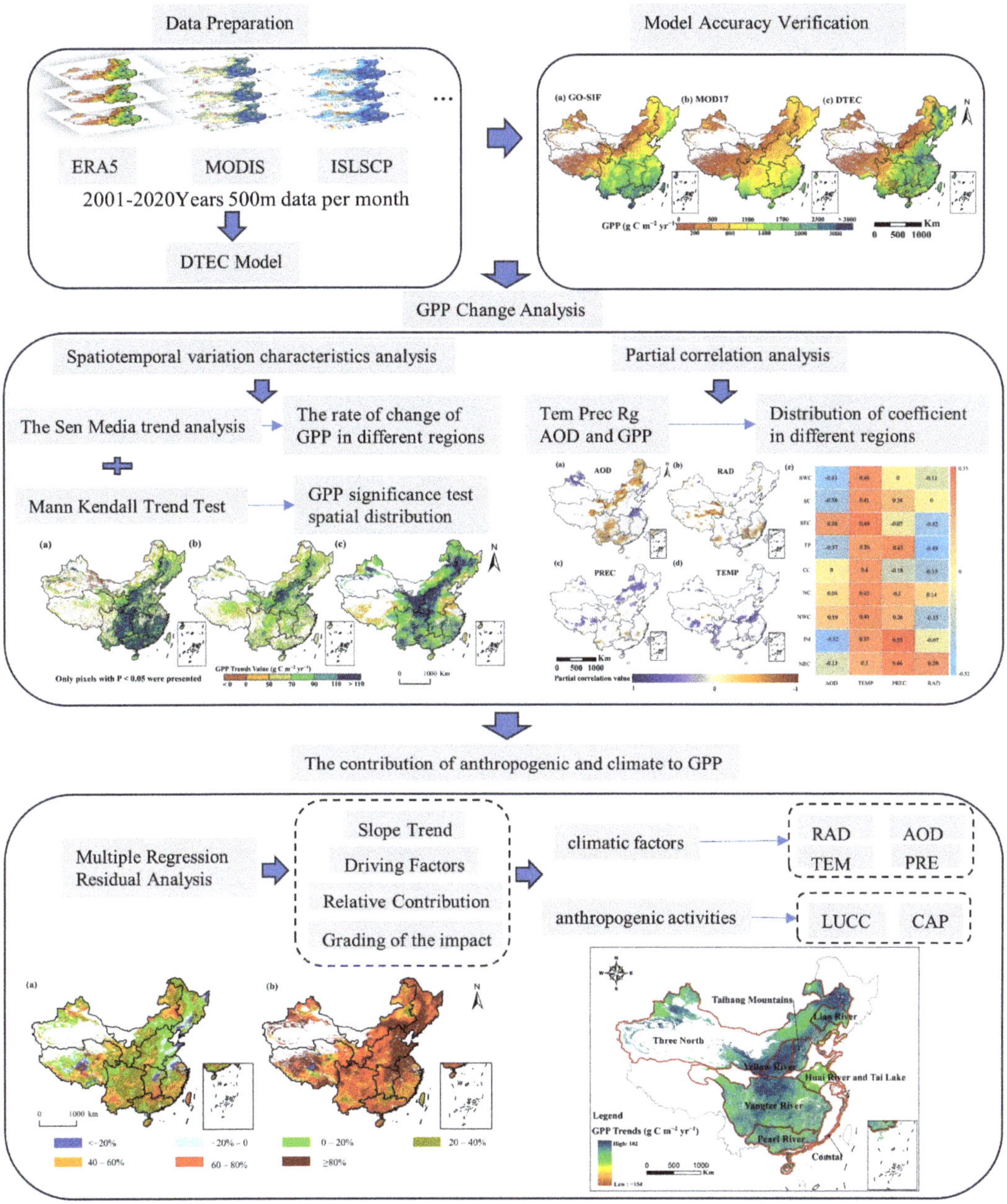

Figure 2. Flow charts show the steps and sequences adopted in the present study's data acquisition and analysis. According to the order of the arrows in the figure, the pictures displayed were: spatial distribution of the multi-year mean value of three GPP models; three model significance test results; DTEC model partial correlation analysis results; residual analysis contribution results; and GPP growth trend in China's afforestation areas.

3. Results

3.1. Verification of GPP Simulation Accuracy

The monthly GPP values simulated by the DTEC model (GPP_{DT}) with GO-SIF GPP (GPP_{SIF}) and MOD17 GPP (GPP_{MOD}) were compared at the site scale with flux measurement data (GPP_{EC}). These accuracy validation results (Figure 3) showed that the three models simulated and captured the dynamics of the GPP well, but the simulation accuracy changed with flux site and ecosystem type. At the cropland site (YC), GPP_{DT} (R^2 = 0.84, RMSE = 68.37 g C m^{-2} $month^{-1}$, bias = 0.52 g C m^{-2} $month^{-1}$) was more accurate than GPP_{SIF} (R^2 = 0.90, RMSE = 124.33 g C m^{-2} $month^{-1}$, bias = 79.98 g C m^{-2} $month^{-1}$) and GPP_{MOD} (R^2 = 0.81, RMSE = 164.09 g C m^{-2} $month^{-1}$, bias = 109.20 g C m^{-2} $month^{-1}$) (Figure 3b). Among the QYZ, CBS and DHS forest sites, GPP_{DT} achieved the best simulation, and the CBS site had the highest accuracy (R^2 = 0.95, RMSE = 27.52 g C m^{-2} $month^{-1}$, bias = 0.37 g C m^{-2} $month^{-1}$) (Figure 3f). The GPP_{DT} values were more accurate than the GPP_{MOD} and GPP_{SIF} values at the mixed forest (ALS) (Figure 3g) and savanna (YJ) (Figure 3c) sites. In general, GPP_{DT} (R^2 = 0.85, RMSE = 41.12 g C m^{-2} $month^{-1}$, bias = 5.98 g C m^{-2} $month^{-1}$) was better than GPP_{SIF} (R^2 = 0.65, RMSE = 64.04 g C m^{-2} $month^{-1}$, bias = −2.94 g C m^{-2} $month^{-1}$) and GPP_{MOD} (R^2 = 0.56, RMSE = 74.00 g C m^{-2} $month^{-1}$, bias = 23.33 g C m^{-2} $month^{-1}$) (Figure 3a).

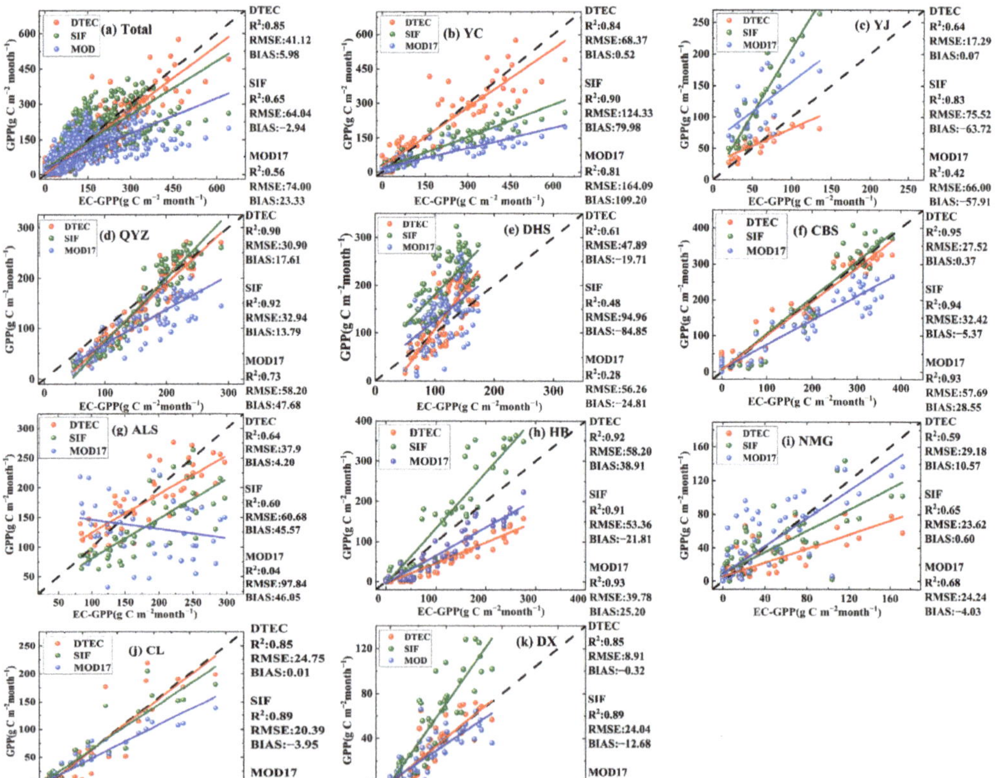

Figure 3. The accuracy validation results of the DTEC, GO-SIF, and MOD17 models at 10 flux sites in China, where the x-axis represents the measured data at the flux sites, the y-axis represents the simulated values of the three models, and R^2, RMSE, and BIAS were used as the coefficients to evaluate the accuracy of the models.

This paper further compared GPP_{DT} data with GPP_{SIF} and GPP_{MOD} data to evaluate the precision and accuracy of the DTEC model from three aspects: multiyear average (Figure S1), spatial distribution (Figure S2), and GPP trend significance test (Figure S3). Overall, the DTEC model had the highest accuracy, could better reflect the status of GPP in China, and could provide more accurate data for subsequent analyses.

3.2. Spatiotemporal Variation Characteristics of Chinese Terrestrial Ecosystem GPP

The DTEC model showed a significant increasing trend from 2001 to 2020 (Figure 4j), but there were fluctuations between years. The minimum value occurred in 2001, and the maximum value occurred in 2018. The average annual increase is 9.77 g C m^{-2} yr^{-1} ($p < 0.001$). The GPP growth trends of the nine regions passed the significance test, and all showed a very significant increase (Figure 4). Among them, the NC region had the largest growth rate of 20.65 g C m^{-2} yr^{-1}, and the lowest growth trend was in the TP region, at 2.03 g C m^{-2} yr^{-1}.

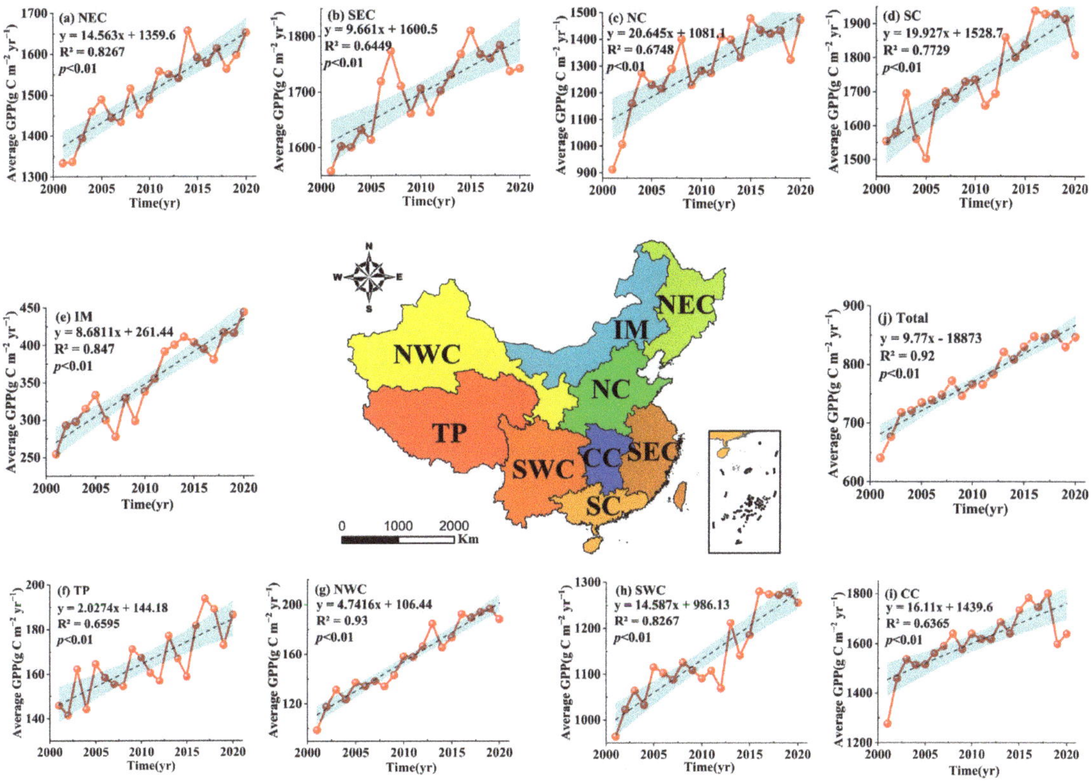

Figure 4. Annual average trends of GPP in China as a whole and in each ecological region from 2001 to 2020, with different colours representing different ecological regions. Among them, NEC (**a**) represents Northeast China, IM (**e**) represents Inner Mongolia, NWC (**g**) represents Northwest China, NC (**c**) represents North China, CC (**i**) represents Central China, TP (**f**) represents the Tibetan Plateau, SEC (**b**) represents Southeast China, SC (**d**) represents South China, SWC (**h**) represents Southwest China, and Total (**j**) represents the overall trend of China.

The spatial distribution of the multiyear average GPP in China from 2001 to 2020 is shown in Figure 5a. According to the colour of the spatial distribution of the average GPP value from 2001 to 2020, the high values were mainly distributed in the eastern region, and the low values were mainly distributed in the western region. The overall trend decreased from the southeast to the northwest and from the coast to the interior. The areas with increasing GPP trends in China from 2001 to 2020 were 67.29%, approximately 6.46 million km^2, and the areas with decreasing trends were 32.71%, approximately 3.14 million km^2. The results revealed that the trend of GPP change differed among the different regions, and in general, the area with an increasing trend of GPP in China was larger than the area with a decreasing trend. Among them, the decreasing regions were mainly in NWC, IM, TP, and SEC, and the increasing regions were mainly in SC, CC, NC, and SWC. According to the significance test results, 43.08% of China's regional GPP changes were nonsignificant, and 56.92% of the area passed the significance test; within that range, the percentages of the areas with very significant increases, significant increases, very significant decreases, and significant decreases were 47.88%, 8.23%, 0.24%, and 0.57%, respectively (Figure 5b).

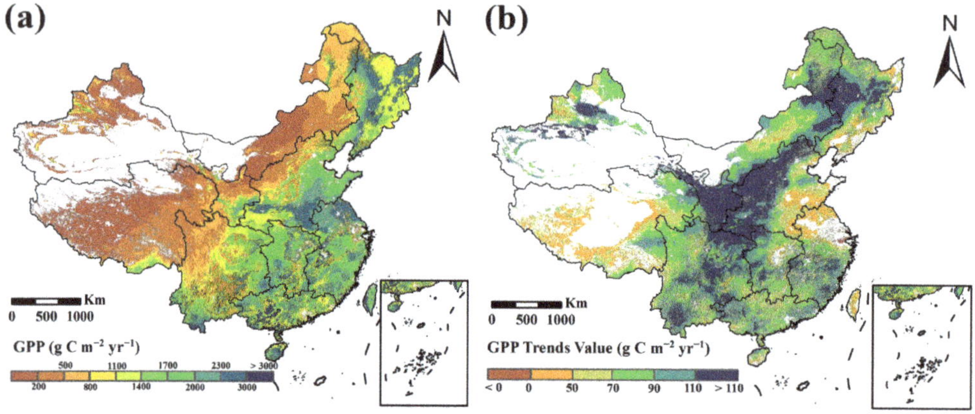

Figure 5. The spatial distribution of the multiyear average of China's GPP from 2001 to 2020 (**a**) and the spatial distribution of the GPP trend that passed the significance test (**b**).

3.3. *The Effects of Climate Factors on GPP*

By counting the partial correlation coefficients between GPP and various climate factors that passed the significance test, we found that the average partial correlation coefficients between China's GPP and temperature, precipitation, solar radiation, and AOD were 0.41, 0.34, −0.29, and −0.22, respectively (Figure 6). The percent areas with a positive correlation between GPP and temperature and precipitation were larger than the areas with a negative correlation (9.5% and 9.8% vs. 0.9% and 0.2%, respectively), and the overall correlation was positive. The percent area of negative correlation between GPP and solar radiation and AOD was larger than the area of positive correlation (8.8% and 15.3% vs. 2.5% and 6.2%, respectively), and an overall negative correlation was present. At the national scale, the order of the response of GPP to climate factors in China was temperature > precipitation > solar radiation > AOD.

However, at the regional scale, the response of GPP to climate factors exhibited significant spatial heterogeneity (Figure 6e). Precipitation had the greatest effect on GPP in the NEC and IM regions, with partial correlation coefficients of 0.46 and 0.55, respectively. In the NWC, NC, CC, SEC, and SC regions, temperature had the greatest influence on GPP, with partial correlation coefficients of 0.41, 0.43, 0.4, 0.49, and 0.41, respectively. The greatest influence on GPP in the TP region was from solar radiation, with a partial correlation coefficient of −0.49. Temperature (0.44) and solar radiation (−0.43) had the greatest impacts on GPP in the SWC region.

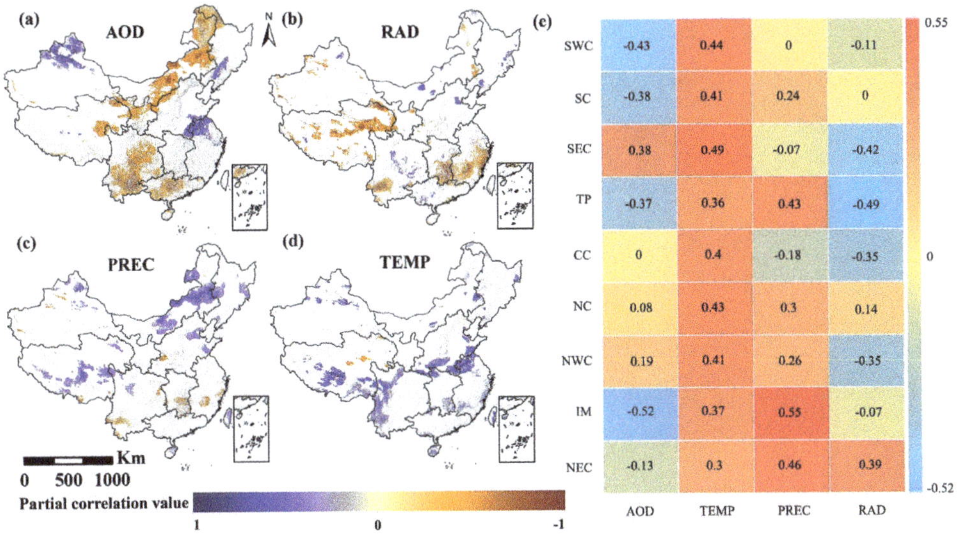

	AOD	TEMP	PREC	RAD
SWC	-0.43	0.44	0	-0.11
SC	-0.38	0.41	0.24	0
SEC	0.38	0.49	-0.07	-0.42
TP	-0.37	0.36	0.43	-0.49
CC	0	0.4	-0.18	-0.35
NC	0.08	0.43	0.3	0.14
NWC	0.19	0.41	0.26	-0.35
IM	-0.52	0.37	0.55	-0.07
NEC	-0.13	0.3	0.46	0.39

Figure 6. Spatial distribution of partial correlation coefficients between GPP and aerosol optical depth (**a**), solar radiation (**b**), precipitation (**c**), and temperature (**d**) in China from 2001 to 2020 and partial correlations of nine ecological regions to four climate factor coefficients (**e**); only pixels that passed the significance test are shown in the figure.

3.4. Drivers of China's GPP Change and Their Relative Contributions

The regions where climate change contributed (C_{CC}) to GPP were equal to an area of approximately 50.2% (Figure 7a). The regions where it contributed moderately and significantly were equal to an area of approximately 23.3%, mainly in the junction regions of NWC, NC, SWC, CC, IM, and NEC. The area where climate change suppressed GPP was approximately 8.6%. The area where GPP moderately and significantly decreased was approximately 4.7%, mainly in the central region of CC, the eastern coastal region of NC, the southern region of NEC, and the northeastern region of SEC. The areas where human activity contributed (C_{HA}) to GPP accounted for approximately 61.1% of the total area (Figure 7b). The area of moderate and significant contribution of human activities to GPP was larger than that of climate change (approximately 43.3%). The regions where human activities suppressed GPP accounted for approximately 6.1%, and these regions were mainly in the northern part of the NC and the northeastern part of the SEC.

Figure 8a shows that approximately 66.4% of the regional C_{CC} had a positive effect on GPP. The area with a contribution between 0% and 60% was larger, accounting for 59.4% of the total area, and the area with a contribution greater than 80% accounted for approximately 2.7% of the total area, which was mainly concentrated in the central part of the TP and the eastern coast of the SEC. The regions with negative C_{CC} to GPP accounted for approximately 8.1% of the total area, which was mainly distributed in the southern part of the NEC, the eastern part of the NC, the northern part of the SEC, the TP, and the central part of the CC. Figure 8b shows that approximately 72.6% of the regional C_{HA} had a positive effect on GPP. Within this range, the areas with contributions in the range of 40% to 80% and greater than 80% are larger, and these areas account for approximately 67.4% of the total area. The areas with a contribution rate of more than 80% are mainly concentrated in NEC, NC, the TP, and NWC, with an area share of 23.3%. The regions with negative C_{HA} to GPP accounted for approximately 1.8% of the total area and were mainly concentrated in the central part of the TP and the northeastern coastal area of the SEC. In comparison, the contribution of HA was greater than that of CC in most regions. The contributions of climate change and human activities to changes in the gross primary productivity of vegetation in China were approximately 28% and 72%, respectively.

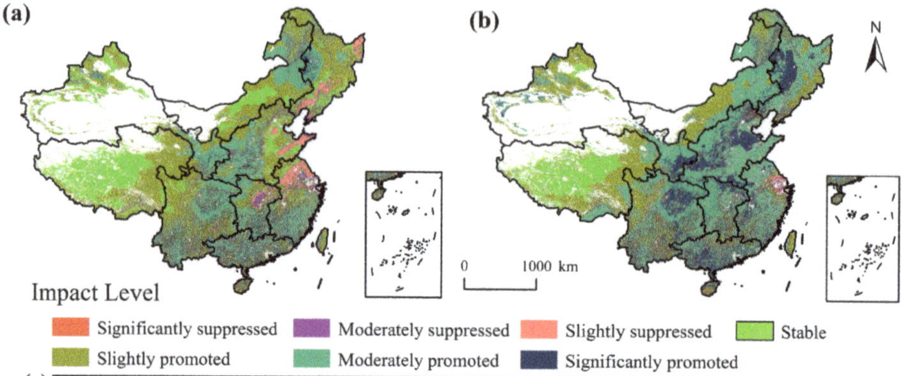

(c)

Impact on GPP	Area Ratio (%) Significantly suppressed	Moderately suppressed	Slightly suppressed	Stable	Slightly promoted	Moderately promoted	Significantly promoted
Climate Change	1.4	3.2	3.9	15.5	26.9	20.6	2.7
Human Activity	2.4	2.5	1.2	7.2	17.7	34.1	9.2

Figure 7. Spatial distribution of the impacts of climate change (**a**) and human activities (**b**) on GPP in China from 2001 to 2020 and the percentage of area affected by each impact level (**c**).

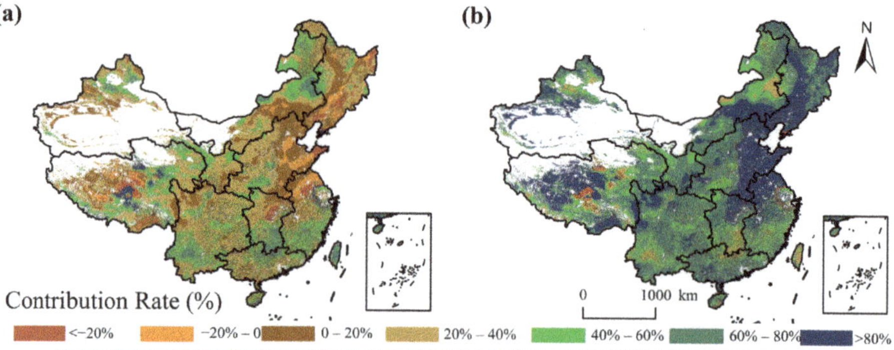

(c)

Contribution Rate (%)	Area Ratio (%) < −20%	20% – 0	0 – 20%	20% – 40%	40% – 60%	60% – 80%	> 80%
Climate Change	3.2	4.8	15.3	26.3	17.7	4.3	2.7
Human Activity	1.3	0.4	1	4.3	17.7	26.3	23.3

Figure 8. Spatial distribution of the contribution rates of climate change (**a**) and human activities (**b**) to the GPP in China from 2001 to 2020 and the area percentage of each contribution rate (**c**).

At the regional scale, the C_{CC} ranged from 18% to 37%, as shown in Figure 9. Among them, the C_{CC} in the SWC, TP, IM, and SC regions were greater than 30%; the highest contribution was in the IM and SWC regions, both with 37%, and the lowest was in the NC region, with 18%. The C_{HA} ranged from 63% to 82%; statistically, the C_{HA} exceeded 60% in all nine ecoregions; the highest was 82% in NC, and the lowest was 63% in IM.

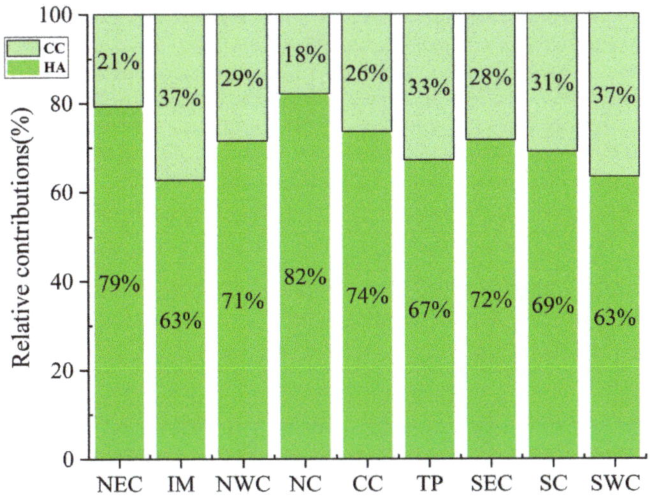

Figure 9. Relative contributions of climate change and human activities to GPP in nine eco-regions of China, 2001–2020.

4. Discussion

4.1. The Influence of Climate Change on China's Dynamic Changes in GPP

The relationships between GPP and climate factors in China show significant spatial heterogeneity. At higher elevations and latitudes (e.g., the TP) (Figure 6d), temperature was positively correlated with GPP. Because the temperature in these areas is low, it affects the activity of vegetation photosynthesis-related enzymes, making temperature the main factor limiting the photosynthetic rate of vegetation [31]. In contrast, solar radiation had an inhibitory effect on vegetation in the TP region [32] (Figure 6b). This result may be due to sufficient solar radiation causing snow and permafrost to melt, increasing the water content of vegetation roots and keeping the soil in an anaerobic state, thus reducing vegetation productivity [33]. In the southern part of the humid region (SWC, CC, and SEC) (Figure 6c), an increase in precipitation reduced temperature and radiation, which led to precipitation suppression of vegetation growth, making GPP significantly negatively correlated with precipitation [34]. In arid and semiarid regions (IM and NWC), GPP was more sensitive to precipitation and had a significant positive correlation with precipitation; precipitation was the main factor affecting GPP in northern China [35]. This was because with the increase in global warming, the water demand of vegetation in arid regions has increased significantly compared to that in non-arid regions [36]. Jiao [37] also noted that vegetation water constraints are associated with greening trends, which leads to vegetation becoming increasingly sensitive to precipitation.

In recent years, it has been concluded that aerosol optical depth (AOD) can enhance scattered radiation by weakening direct radiation and total radiation, thereby promoting or inhibiting vegetation photosynthesis [38]. This has attracted increasing attention from scholars to the impact of AOD on GPP [39]. At the regional scale, GPP was significantly positively correlated with AOD in most of the regions. These regions were significantly negatively correlated with solar radiation and vice versa (Figure 6a,e). This phenomenon indicates that when vegetation is undergoing photosynthesis, the high canopy of vegetation can absorb both direct and scattered radiation and easily reach light saturation, after which photosynthesis is inhibited with increasing solar radiation [40,41]. In contrast, as AOD increases, direct radiation is attenuated, and the proportion of scattered radiation increases, resulting in an increase in radiation absorbed by shady lower canopy leaves that have not yet reached light saturation, thereby improving productivity [42]. In the northwestern region of NWC, AOD had a significant positive correlation with GPP, which may be related to the vegetation types of grassland, ENF and desert in this region [43]. In the SWC and SC regions, GPP had a large and significant negative correlation with AOD. This phenomenon was due to the reduction in total radiation, which decreased the productivity of more complex ecosystems, such as forests, at a level greater than the compensatory effect of scattered radiation, thus reducing overall productivity [43]. GPP was also significantly negatively correlated with AOD in IM and its surrounding areas. As the vegetation type in the zone was mostly grassland, where AOD leads to a reduction in direct radiation, the canopy structure of grassland is simpler, so it cannot sufficiently absorb scattered radiation [44].

4.2. The Influence of Human Activities on China's GPP Dynamic Changes

The effects of climate change and anthropogenic activities on increasing GPP trends were 1.39 and 5.67 g C m^{-2} yr^{-1}, respectively. Anthropogenic activities had a greater positive impact on GPP than did climate change, showing that anthropogenic activities are the dominant factor in China's GPP growth. Zhang [45] also showed that anthropogenic activities had a stronger effect on vegetation changes than did climate change. The overall trend of change was positive, showing that China's ecological construction and protection projects have been effective [46]. Zhang [47] reported that China's forest area increased by 13.48% from 2001 to 2015. By 2020, the forest area had increased to 16%, with the increase obtained mainly by the conversion of shrubland and savanna types (Figure S4). According to the eighth consecutive inventory of forest resources in China, compared with the second inventory, China's plantation forest preservation area increased by 2.13 times, and the plantation forest stock increased by approximately five times [48]. This conclusion is in agreement with the regions covered in this paper where the GPP growth trend was greater than 10 g C m^{-2} yr^{-1}. Against the background of the continuous reduction in global forest area, the average annual growth rate of China's forest coverage was the highest in the world, especially in terms of the growth of artificial forest area, accounting for approximately 41% of the global afforestation rate [49]. China's forest ecosystems have had a significant carbon sink function in the last two decades [50]. Notably, land use changes caused by human activities are also among the key factors affecting vegetation GPP [51]. Among them, the promotion of agricultural intensification techniques, such as replanting [52] and various irrigation techniques [53], has greatly contributed to the increase in GPP in farmlands in China. Zhong [54] noted that the grain output of farmland in China has increased by more than 35% since 2000. In this paper, by comparing the land use data from 2001 to 2020, we found that the area of cultivated land in China increased by 3%; this increase was mainly obtained by the conversion of grassland and savanna vegetation types to crops through the ecological reclamation activities of adjusting land use [55] (Figure S4). The spatial distribution of cropland coincided with regions with greater positive anthropogenic contributions (Figure 8b). However, crops are seasonal [56], and after crops and straw are harvested and consumed during the nongrowing season, carbon is no longer stored [57]. With the respiration of microorganisms in the soil and the decomposition of soil residual fertilizers, the carbon exchange process between farmland ecosystems and other systems intensifies [58]. Therefore, many scholars generally believe that the carbon sink of crops is zero [59]. The DTEC model was driven by remote sensing data to assess the productivity of terrestrial ecosystems. Therefore, this model only focuses on the carbon sequestration effect of vegetation. For issues such as carbon exchange in nongrowing seasons and soil carbon consumption in farmland ecosystems, this model's mechanism cannot be reflected at present, so we will not discuss it too much here.

There were significant regional variations in the effects of anthropogenic activities on GPP. A comparison of the trends of human activity in the nine regions revealed that the difference between the extremes was 15.66 g C m^{-2} yr^{-1}. The NC area was the most effective area for implementing vegetation construction projects [60], and the relative contribution of human activities was also the highest (82%). The regions with growth trends greater than 10 g C m^{-2} yr^{-1} (NEC, NC, SC, SWC, and CC) all corresponded to the major areas in China's afforestation program (Figure 10); this finding was consistent with that of Cai [61], and in the abovementioned areas, the contribution of human activities exceeded 62%. This result highlights the positive effect of ecological engineering on China's vegetation GPP [62]. However, there are also studies showing the negative impacts of afforestation [63]. For example, in the Loess Plateau region, afforestation has led to increasing negative impacts such as deep soil drying, erosion, and water scarcity [64]. Cai [65] suggested that nonnative forests are not suitable for controlling erosion in the Loess Plateau region unless there is a protective ground cover of herbaceous plants. In arid and semiarid grassland areas, artificial herbs consume more water than natural herbs, which can easily lead to soil degradation and further environmental degradation in ecologically fragile areas [66]. Therefore, in the future, we should further strengthen the construction of ecological engineering in ecologically fragile areas and degraded vegetation areas to realize the coupling between China's urbanization and development and the quality of the natural environment.

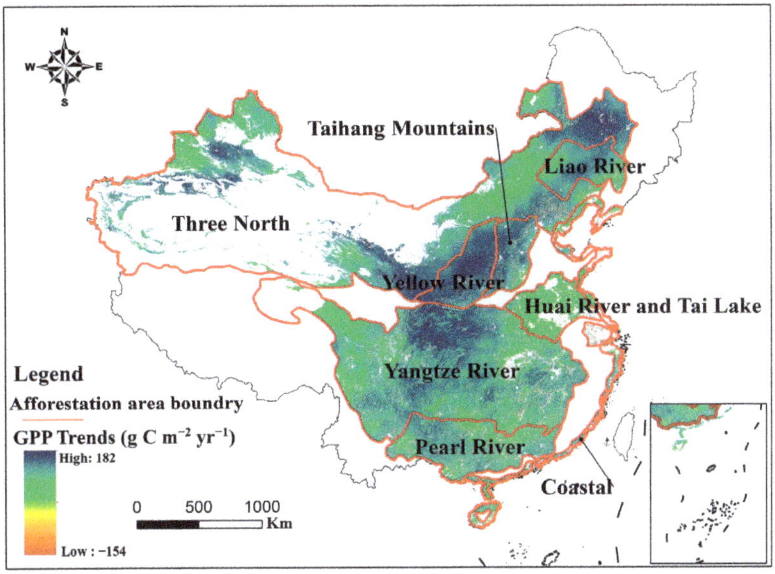

Figure 10. The distribution of GPP changes in China's major afforestation program from 2001–2020.

4.3. DTEC Model Evaluation

The results of this study show that the accuracy and applicability of the DTEC model in China were greater than those of the GO-SIF and MOD17 products. The model parameters in this paper do not use the parameters given in previous studies but were adjusted and modified for the Chinese region on the basis of the original parameters to conform to the characteristics of vegetation in China, which is one of the reasons why the accuracy of the DTEC model is higher than that of the other two models in China. Second, it distinguishes the LUE of C4 and C3 vegetation, which can be seen in the accuracy validation of the farmland site. The simulation accuracy of the DTEC model is significantly better than that of the GO-SIF and MOD17 products. As farmland ecosystems are the most active carbon reservoirs in terrestrial ecosystems, the high-accuracy estimation of the GPP of farmland ecosystems will be effective in reducing the error in the simulation of the total amount of China's terrestrial GPP. Finally, the effect of scattered radiation on vegetation photosynthesis was considered. In the past, the LUE of low-canopy vegetation leaves was often underestimated [67], which made the simulation error larger in areas with complex vegetation structures and dense canopies, such as tropical rainforests, while the DTEC model considered this problem and effectively improved the simulation accuracy in these areas. In summary, the DTEC model has good applicability in China and can meet the needs of subsequent studies.

4.4. Uncertainties and Limitations

The DTEC model distinguishes the different light-use efficiencies of C3 and C4 plants in GPP calculations; however, due to the lack of large-scale, long-term, and high-resolution C3 and C4 vegetation type data, we can use only a static map of C3 and C4 vegetation for GPP estimation, which affects the estimation accuracy of GPP. The atmospheric CO_2 concentration is an important factor affecting photosynthetic efficiency because vegetation adjusts leaf chemistry in response to elevated CO_2 concentrations [68], leading to changes in photosynthetic capacity [69]. Accounting for the CO_2 fertilization effect (CFE) in models can improve GPP estimates to reproduce long-term changes [70], and studies [71] have also shown that accounting for the CFE can better capture long-term trends in elevated GPP in the tropics; the CFE can also better explain and quantify the causes of global vegetation greening [72]. CFE should be incorporated into models in future studies.

5. Conclusions

In this study, the diffuse fraction-based two-leaf light-use efficiency (DTEC) model was used to calculate China's gross primary productivity (GPP) from 2001 to 2020, and the contributions of climate change (temperature, precipitation, solar radiation, and aerosol optical depth) and human activities to China's GPP were further investigated. The results showed that the DTEC model has good applicability in China compared with the 10 flux sites, and its overall accuracy coefficients were $R^2 = 0.85$, RMSE = 41.12 g C m^{-2} $month^{-1}$, and bias = 5.98 g C m^{-2} $month^{-1}$, which achieved great accuracy for different vegetation types. From 2001 to 2020, 67.29% of China's regions showed an increasing trend in GPP, with an average increasing trend of 9.77 g C m^{-2} yr^{-1} ($p < 0.01$), and within different ecological regions, the increasing trend in GPP ranged from 20.65 to 2.03 g C m^{-2} yr^{-1} ($p < 0.01$). Temperature has a greater impact on GPP than precipitation, solar radiation, and aerosol optical depth under climate change, especially on the Qinghai–Tibet Plateau. In addition, multiple regression residual methods show that there is great spatial heterogeneity in the impacts of climate change and human activities on China's GPP. The impacts on GPP growth were 1.39 and 5.67 g C m^{-2} yr^{-1}, respectively, and the relative contributions were 28% and 72%, respectively. Our conclusions show that human activities, mainly ecological restoration projects and agricultural intensification technologies, have been the main driving force behind China's GPP growth over the past 20 years.

Supplementary Materials: The following supporting information can be downloaded at: https://www.mdpi.com/article/10.3390/rs16081361/s1, Figure S1: Inter-annual changes in China's GPP from 2001 to 2020, where DTEC stands for the simulated value of the DTEC model, SIF stands for the simulated value of the GO-SIF model, and MOD17 stands for the simulated value of the MOD17 model. Figure S2: Spatial distribution of the annual mean values of GPP from 2001 to 2020: GO-SIF (a), MOD17 (b), and DTEC (c). Among them, the darker the blue color, the higher the value of GPP in the region. Figure S3: Spatial distribution of GPP passing the significance test for 2001–2020: GO-SIF (a), MOD17 (b), and DTEC (c). Among them, the darker the blue color, the higher the trend of GPP growth in the region. Figure S4: Spatial distribution of land use types in China from 2001 (a) to 2020 (b) and conversion between different land use types from 2001 to 2020 (c). The land use types in China were resampled into six: forest, shrubland, savanna, grassland, crop, and non-vegetated area (NOA). Table S1: Grading of the impact of GPP_{CC} and GPP_{HA}. Table S2: Determination Criteria for Drivers of China's GPP Changes and Calculation Method of Contribution Rate.

Author Contributions: Methodology, writing original draft, J.P.; conceptualization, review and editing, Y.X.; methodology and resources, N.P., Y.Z. and H.L.; review and editing, F.Z. All authors contributed to the editing of the manuscript. All authors have read and agreed to the published version of the manuscript.

Funding: This study was financially supported by the Key Research and Development Program of the Xinjiang Uygur Autonomous Region (Grant Number: 2022B01012), the National Nature Science Foundation of China (Grant Number: 42301127), the Ph.D. Start-up Fund in Xinjiang University (Grant Number: 620321046), and the Tianchi Doctor Program of the Xinjiang Uygur Autonomous Region (Grant Number: TCBS202140).

Data Availability Statement: Data sharing is not applicable to this article.

Acknowledgments: We are sincerely grateful to the reviewers and editors for their constructive comments towards the improvement of the manuscript.

Conflicts of Interest: The authors declare no conflicts of interest.

References

1. Friedlingstein, P.; Jones, M.W.; O'Sullivan, M.; Andrew, R.M.; Hauck, J.; Peters, G.P.; Peters, W.; Pongratz, J.; Sitch, S.; Le Quéré, C.; et al. Global Carbon Budget 2019. *Earth Syst. Sci. Data* **2019**, *11*, 1783–1838. [CrossRef]
2. Chen, C.; Park, T.; Wang, X.; Piao, S.; Xu, B.; Chaturvedi, R.K.; Fuchs, R.; Brovkin, V.; Ciais, P.; Fensholt, R.; et al. China and India Lead in Greening of the World through Land-Use Management. *Nat. Sustain.* **2019**, *2*, 122–129. [CrossRef] [PubMed]
3. Wang, S.; Zhang, Y.; Ju, W.; Chen, J.M.; Ciais, P.; Cescatti, A.; Sardans, J.; Janssens, I.A.; Wu, M.; Berry, J.A.; et al. Recent Global Decline of CO_2 Fertilization Effects on Vegetation Photosynthesis. *Science* **2020**, *370*, 1295–1300. [CrossRef] [PubMed]
4. Xiao, J.; Chevallier, F.; Gomez, C.; Guanter, L.; Hicke, J.A.; Huete, A.R.; Ichii, K.; Ni, W.; Pang, Y.; Rahman, A.F. Remote Sensing of the Terrestrial Carbon Cycle: A Review of Advances over 50 Years. *Remote Sens. Environ.* **2019**, *233*, 111383. [CrossRef]
5. Wofsy, S.C.; Goulden, M.L.; Munger, J.W.; Fan, S.-M.; Bakwin, P.S.; Daube, B.C.; Bassow, S.L.; Bazzaz, F.A. Net Exchange of CO_2 in a Mid-Latitude Forest. *Science* **1993**, *260*, 1314–1317. [CrossRef] [PubMed]
6. Wang, X.; Ma, M.; Li, X.; Song, Y.; Tan, J.; Huang, G.; Zhang, Z.; Zhao, T.; Feng, J.; Ma, Z.; et al. Validation of MODIS-GPP Product at 10 Flux Sites in Northern China. *Int. J. Remote Sens.* **2013**, *34*, 587–599. [CrossRef]
7. Zhang, F.; Chen, J.M.; Chen, J.; Gough, C.M.; Martin, T.A.; Dragoni, D. Evaluating Spatial and Temporal Patterns of MODIS GPP over the Conterminous U.S. against Flux Measurements and a Process Model. *Remote Sens. Environ.* **2012**, *124*, 717–729. [CrossRef]
8. Zhang, K.; Kimball, J.S.; Running, S.W. A Review of Remote Sensing Based Actual Evapotranspiration Estimation. *WIREs Water* **2016**, *3*, 834–853. [CrossRef]
9. Yan, H.; Wang, S.; Billesbach, D.; Oechel, W.; Bohrer, G.; Meyers, T.; Martin, T.A.; Matamala, R.; Phillips, R.P.; Rahman, F.; et al. Improved Global Simulations of Gross Primary Product Based on a New Definition of Water Stress Factor and a Separate Treatment of C3 and C4 Plants. *Ecol. Model.* **2015**, *297*, 42–59. [CrossRef]
10. He, M.; Ju, W.; Zhou, Y.; Chen, J.; He, H.; Wang, S.; Wang, H.; Guan, D.; Yan, J.; Li, Y.; et al. Development of a Two-Leaf Light Use Efficiency Model for Improving the Calculation of Terrestrial Gross Primary Productivity. *Agric. For. Meteorol.* **2013**, *173*, 28–39. [CrossRef]
11. Yan, H.; Wang, S.-Q.; Yu, K.-L.; Wang, B.; Yu, Q.; Bohrer, G.; Billesbach, D.; Bracho, R.; Rahman, F.; Shugart, H.H. A Novel Diffuse Fraction-Based Two-Leaf Light Use Efficiency Model: An Application Quantifying Photosynthetic Seasonality across 20 AmeriFlux Flux Tower Sites. *J. Adv. Model. Earth Syst.* **2017**, *9*, 2317–2332. [CrossRef]
12. Kanniah, K.D.; Beringer, J.; Hutley, L. Exploring the Link between Clouds, Radiation, and Canopy Productivity of Tropical Savannas. *Agric. For. Meteorol.* **2013**, *182*, 304–313. [CrossRef]
13. Cao, D.; Zhang, J.; Yan, H.; Xun, L.; Yang, S.; Bai, Y.; Zhang, S.; Yao, F.; Zhou, W. Regional Assessment of Climate Potential Productivity of Terrestrial Ecosystems and Its Responses to Climate Change Over China From 1980–2018. *IEEE Access* **2020**, *8*, 11138–11151. [CrossRef]
14. Yang, J.; Zhang, X.C.; Luo, Z.H.; Yu, X.J. Nonlinear Variations of Net Primary Productivity and Its Relationship with Climate and Vegetation Phenology, China. *Forests* **2017**, *8*, 361. [CrossRef]
15. Wild, M.; Trüssel, B.; Ohmura, A.; Long, C.N.; König-Langlo, G.; Dutton, E.G.; Tsvetkov, A. Global Dimming and Brightening: An Update beyond 2000. *J. Geophys. Res. Atmos.* **2009**, *114*. [CrossRef]
16. von Buttlar, J.; Zscheischler, J.; Rammig, A.; Sippel, S.; Reichstein, M.; Knohl, A.; Jung, M.; Menzer, O.; Arain, M.A.; Buchmann, N.; et al. Impacts of Droughts and Extreme-Temperature Events on Gross Primary Production and Ecosystem Respiration: A Systematic Assessment across Ecosystems and Climate Zones. *Biogeosciences* **2018**, *15*, 1293–1318. [CrossRef]
17. Batunacun; Wieland, R.; Lakes, T.; Nendel, C. Using Shapley Additive Explanations to Interpret Extreme Gradient Boosting Predictions of Grassland Degradation in Xilingol, China. *Geosci. Model Dev.* **2021**, *14*, 1493–1510. [CrossRef]
18. Xue, Y.; Liang, H.; Zhang, B.; He, C. Vegetation Restoration Dominated the Variation of Water Use Efficiency in China. *J. Hydrol.* **2022**, *612*, 128257. [CrossRef]
19. Xue, Y.; Liang, H.; Ma, Y.; Xue, G.; He, J. The Impacts of Climate and Human Activities on Grassland Productivity Variation in China. *Remote Sens.* **2023**, *15*, 3864. [CrossRef]
20. Yang, K.; Sun, W.; Luo, Y.; Zhao, L. Impact of Urban Expansion on Vegetation: The Case of China (2000–2018). *J. Environ. Manag.* **2021**, *291*, 112598. [CrossRef] [PubMed]

21. Piao, S.; Ciais, P.; Huang, Y.; Shen, Z.; Peng, S.; Li, J.; Zhou, L.; Liu, H.; Ma, Y.; Ding, Y.; et al. The Impacts of Climate Change on Water Resources and Agriculture in China. *Nature* **2010**, *467*, 43–51. [CrossRef] [PubMed]
22. Mallapaty, S. How China Could Be Carbon Neutral by Mid-Century. *Nature* **2020**, *586*, 482–483. [CrossRef]
23. Doelman, J.C.; Stehfest, E.; van Vuuren, D.P.; Tabeau, A.; Hof, A.F.; Braakhekke, M.C.; Gernaat, D.E.H.J.; van den Berg, M.; van Zeist, W.-J.; Daioglou, V.; et al. Afforestation for Climate Change Mitigation: Potentials, Risks and Trade-Offs. *Glob. Chang. Biol.* **2020**, *26*, 1576–1591. [CrossRef] [PubMed]
24. Chen, Y.; Feng, X.; Tian, H.; Wu, X.; Gao, Z.; Feng, Y.; Piao, S.; Lv, N.; Pan, N.; Fu, B. Accelerated Increase in Vegetation Carbon Sequestration in China after 2010: A Turning Point Resulting from Climate and Human Interaction. *Glob. Chang. Biol.* **2021**, *27*, 5848–5864. [CrossRef] [PubMed]
25. Piao, S.; Fang, J.; Ciais, P.; Peylin, P.; Huang, Y.; Sitch, S.; Wang, T. The Carbon Balance of Terrestrial Ecosystems in China. *Nature* **2009**, *458*, 1009–1013. [CrossRef] [PubMed]
26. Jiang, W.; Yuan, L.; Wang, W.; Cao, R.; Zhang, Y.; Shen, W. Spatio-Temporal Analysis of Vegetation Variation in the Yellow River Basin. *Ecol. Indic.* **2015**, *51*, 117–126. [CrossRef]
27. Qu, S.; Wang, L.; Lin, A.; Yu, D.; Yuan, M.; Li, C. Distinguishing the Impacts of Climate Change and Anthropogenic Factors on Vegetation Dynamics in the Yangtze River Basin, China. *Ecol. Indic.* **2020**, *108*, 105724. [CrossRef]
28. Evans, J.; Geerken, R. Discrimination between Climate and Human-Induced Dryland Degradation. *J. Arid Environ.* **2004**, *57*, 535–554. [CrossRef]
29. Badgley, G.; Anderegg, L.D.L.; Berry, J.A.; Field, C.B. Terrestrial Gross Primary Production: Using NIRV to Scale from Site to Globe. *Glob. Chang. Biol.* **2019**, *25*, 3731–3740. [CrossRef] [PubMed]
30. Jin, K.; Wang, F.; Zong, Q.; Qin, P.; Liu, C.; Wang, S. Spatiotemporal Differences in Climate Change Impacts on Vegetation Cover in China from 1982 to 2015. *Environ. Sci. Pollut. Res.* **2022**, *29*, 10263–10276. [CrossRef] [PubMed]
31. Keenan, T.F.; Riley, W.J. Greening of the Land Surface in the World's Cold Regions Consistent with Recent Warming. *Nat. Clim. Chang.* **2018**, *8*, 825–828. [CrossRef] [PubMed]
32. Luo, L.; Ma, W.; Zhuang, Y.; Zhang, Y.; Yi, S.; Xu, J.; Long, Y.; Ma, D.; Zhang, Z. The Impacts of Climate Change and Human Activities on Alpine Vegetation and Permafrost in the Qinghai-Tibet Engineering Corridor. *Ecol. Indic.* **2018**, *93*, 24–35. [CrossRef]
33. Yang, Y.; Guan, H.; Shen, M.; Liang, W.; Jiang, L. Changes in Autumn Vegetation Dormancy Onset Date and the Climate Controls across Temperate Ecosystems in China from 1982 to 2010. *Glob. Chang. Biol.* **2015**, *21*, 652–665. [CrossRef] [PubMed]
34. Nemani, R.R.; Keeling, C.D.; Hashimoto, H.; Jolly, W.M.; Piper, S.C.; Tucker, C.J.; Myneni, R.B.; Running, S.W. Climate-Driven Increases in Global Terrestrial Net Primary Production from 1982 to 1999. *Science* **2003**, *300*, 1560–1563. [CrossRef] [PubMed]
35. Shi, S.; Yu, J.; Wang, F.; Wang, P.; Zhang, Y.; Jin, K. Quantitative Contributions of Climate Change and Human Activities to Vegetation Changes over Multiple Time Scales on the Loess Plateau. *Sci. Total Environ.* **2021**, *755*, 142419. [CrossRef] [PubMed]
36. Lai, W.; Wang, M.; Wei, J.; Zhang, J.; Song, J.; Zhou, H.; Chou, S.; Wang, Y. Separating the Impact of Climate Changes and Human Activities on Vegetation Growth Based on the NDVI in China. *Adv. Meteorol.* **2022**, *2022*, e6294029. [CrossRef]
37. Jiao, W.; Wang, L.; Smith, W.K.; Chang, Q.; Wang, H.; D'Odorico, P. Observed Increasing Water Constraint on Vegetation Growth over the Last Three Decades. *Nat. Commun.* **2021**, *12*, 3777. [CrossRef] [PubMed]
38. Shu, Y.; Liu, S.; Wang, Z.; Xiao, J.; Shi, Y.; Peng, X.; Gao, H.; Wang, Y.; Yuan, W.; Yan, W.; et al. Effects of Aerosols on Gross Primary Production from Ecosystems to the Globe. *Remote Sens.* **2022**, *14*, 2759. [CrossRef]
39. Ma, W.; Ding, J.; Wang, J.; Zhang, J. Effects of Aerosol on Terrestrial Gross Primary Productivity in Central Asia. *Atmos. Environ.* **2022**, *288*, 119294. [CrossRef]
40. Kanniah, K.D.; Beringer, J.; North, P.; Hutley, L. Control of Atmospheric Particles on Diffuse Radiation and Terrestrial Plant Productivity: A Review. *Prog. Phys. Geogr. Earth Environ.* **2012**, *36*, 209–237. [CrossRef]
41. Urban, O.; Janouš, D.; Acosta, M.; Czerný, R.; Marková, I.; Navrátil, M.; Pavelka, M.; Pokorný, R.; Šprtová, M.; Zhang, R.; et al. Ecophysiological Controls over the Net Ecosystem Exchange of Mountain Spruce Stand. Comparison of the Response in Direct vs. Diffuse Solar Radiation. *Glob. Chang. Biol.* **2007**, *13*, 157–168. [CrossRef]
42. Zan, M.; Zhou, Y.; Ju, W.; Zhang, Y.; Zhang, L.; Liu, Y. Performance of a Two-Leaf Light Use Efficiency Model for Mapping Gross Primary Productivity against Remotely Sensed Sun-Induced Chlorophyll Fluorescence Data. *Sci. Total Environ.* **2018**, *613–614*, 977–989. [CrossRef] [PubMed]
43. Cirino, G.G.; Souza, R.A.F.; Adams, D.K.; Artaxo, P. The Effect of Atmospheric Aerosol Particles and Clouds on Net Ecosystem Exchange in the Amazon. *Atmos. Chem. Phys.* **2014**, *14*, 6523–6543. [CrossRef]
44. Niyogi, D.; Chang, H.-I.; Saxena, V.K.; Holt, T.; Alapaty, K.; Booker, F.; Chen, F.; Davis, K.J.; Holben, B.; Matsui, T. Direct Observations of the Effects of Aerosol Loading on Net Ecosystem CO_2 Exchanges over Different Landscapes. *Geophys. Res. Lett.* **2004**, *31*. [CrossRef]
45. Zhang, X.; Huang, X. Human Disturbance Caused Stronger Influences on Global Vegetation Change than Climate Change. *PeerJ* **2019**, *7*, e7763. [CrossRef] [PubMed]
46. Chen, J.M.; Ju, W.; Ciais, P.; Viovy, N.; Liu, R.; Liu, Y.; Lu, X. Vegetation Structural Change since 1981 Significantly Enhanced the Terrestrial Carbon Sink. *Nat. Commun.* **2019**, *10*, 4259. [CrossRef] [PubMed]
47. Zhang, Y.; Xiao, X.; Jin, C.; Dong, J.; Zhou, S.; Wagle, P.; Joiner, J.; Guanter, L.; Zhang, Y.; Zhang, G.; et al. Consistency between Sun-Induced Chlorophyll Fluorescence and Gross Primary Production of Vegetation in North America. *Remote Sens. Environ.* **2016**, *183*, 154–169. [CrossRef]

48. Yu, Z.; You, W.; Agathokleous, E.; Zhou, G.; Liu, S. Forest Management Required for Consistent Carbon Sink in China's Forest Plantations. *For. Ecosyst.* **2021**, *8*, 54. [CrossRef]
49. Zhang, L.; Sun, P.; Huettmann, F.; Liu, S. Where Should China Practice Forestry in a Warming World? *Glob. Chang. Biol.* **2022**, *28*, 2461–2475. [CrossRef] [PubMed]
50. Qiu, Z.; Feng, Z.; Song, Y.; Li, M.; Zhang, P. Carbon Sequestration Potential of Forest Vegetation in China from 2003 to 2050: Predicting Forest Vegetation Growth Based on Climate and the Environment. *J. Clean. Prod.* **2020**, *252*, 119715. [CrossRef]
51. Zhou, Q.; Li, B.; Chen, Y. Remote Sensing Change Detection and Process Analysis of Long-Term Land Use Change and Human Impacts. *AMBIO* **2011**, *40*, 807–818. [CrossRef] [PubMed]
52. Zhang, C.; Zhang, H.; Tian, S. Phenology-Assisted Supervised Paddy Rice Mapping with the Landsat Imagery on Google Earth Engine: Experiments in Heilongjiang Province of China from 1990 to 2020. *Comput. Electron. Agric.* **2023**, *212*, 108105. [CrossRef]
53. Banerjee, S.; Walder, F.; Büchi, L.; Meyer, M.; Held, A.Y.; Gattinger, A.; Keller, T.; Charles, R.; van der Heijden, M.G.A. Agricultural Intensification Reduces Microbial Network Complexity and the Abundance of Keystone Taxa in Roots. *ISME J.* **2019**, *13*, 1722–1736. [CrossRef] [PubMed]
54. Zhong, T.; Mitchell, B.; Scott, S.; Huang, X.; Li, Y.; Lu, X. Growing Centralization in China's Farmland Protection Policy in Response to Policy Failure and Related Upward-Extending Unwillingness to Protect Farmland since 1978. *Environ. Plan. C Polit. Space* **2017**, *35*, 1075–1097. [CrossRef]
55. Gerber, S.; Hedin, L.O.; Keel, S.G.; Pacala, S.W.; Shevliakova, E. Land Use Change and Nitrogen Feedbacks Constrain the Trajectory of the Land Carbon Sink. *Geophys. Res. Lett.* **2013**, *40*, 5218–5222. [CrossRef]
56. Huang, M.; Tian, A.; Zhou, X.; Gao, W.; Li, Z.; Chen, G.; Li, Z.; Chen, Y.; Liu, L.; Yin, X.; et al. Yield Performance of Machine-Transplanted Double-Season Rice Grown Following Oilseed Rape. *Sci. Rep.* **2019**, *9*, 6818. [CrossRef] [PubMed]
57. Lu, X.; Lu, X. Tillage and Crop Residue Effects on the Energy Consumption, Input–Output Costs and Greenhouse Gas Emissions of Maize Crops. *Nutr. Cycl. Agroecosyst.* **2017**, *108*, 323–337. [CrossRef]
58. Almusaed, A. *Landscape Ecology: The Influences of Land Use and Anthropogenic Impacts of Landscape Creation*; BoD—Books on Demand: Norderstedt, Germany, 2016; ISBN 978-953-51-2513-6.
59. Li, M.; Peng, J.; Lu, Z.; Zhu, P. Research Progress on Carbon Sources and Sinks of Farmland Ecosystems. *Resour. Environ. Sustain.* **2023**, *11*, 100099. [CrossRef]
60. Wang, S.; Fu, B.; Piao, S.; Lü, Y.; Ciais, P.; Feng, X.; Wang, Y. Reduced Sediment Transport in the Yellow River Due to Anthropogenic Changes. *Nat. Geosci.* **2016**, *9*, 38–41. [CrossRef]
61. Cai, Y.; Zhang, F.; Duan, P.; Yung Jim, C.; Weng Chan, N.; Shi, J.; Liu, C.; Wang, J.; Bahtebay, J.; Ma, X. Vegetation Cover Changes in China Induced by Ecological Restoration-Protection Projects and Land-Use Changes from 2000 to 2020. *Catena* **2022**, *217*, 106530. [CrossRef]
62. Chen, P.; Liu, H.; Wang, Z.; Mao, D.; Liang, C.; Wen, L.; Li, Z.; Zhang, J.; Liu, D.; Zhuo, Y. Vegetation Dynamic Assessment by NDVI and Field Observations for Sustainability of China's Wulagai River Basin. *Int. J. Environ. Res. Public Health* **2021**, *18*, 2528. [CrossRef] [PubMed]
63. Niu, Y.; Squires, V.; Jentsch, A. Risks of China's Increased Forest Area. *Science* **2023**, *379*, 447–448. [CrossRef] [PubMed]
64. Wang, Z.; Liu, B.; Liu, G.; Zhang, Y. Soil Water Depletion Depth by Planted Vegetation on the Loess Plateau. *Sci. China Ser. Earth Sci.* **2009**, *52*, 835–842. [CrossRef]
65. Jiang, W.; Yang, S.; Yang, X.; Gu, N. Negative Impacts of Afforestation and Economic Forestry on the Chinese Loess Plateau and Proposed Solutions. *Quat. Int.* **2016**, *399*, 165–173. [CrossRef]
66. Li, W.; Yan, M.; Qingfeng, Z.; Zhikaun, J. Effects of Vegetation Restoration on Soil Physical Properties in the Wind–Water Erosion Region of the Northern Loess Plateau of China. *CLEAN—Soil Air Water* **2012**, *40*, 7–15. [CrossRef]
67. Cheng, S.J.; Bohrer, G.; Steiner, A.L.; Hollinger, D.Y.; Suyker, A.; Phillips, R.P.; Nadelhoffer, K.J. Variations in the Influence of Diffuse Light on Gross Primary Productivity in Temperate Ecosystems. *Agric. For. Meteorol.* **2015**, *201*, 98–110. [CrossRef]
68. Smith, N.G.; Keenan, T.F. Mechanisms Underlying Leaf Photosynthetic Acclimation to Warming and Elevated CO_2 as Inferred from Least-Cost Optimality Theory. *Glob. Chang. Biol.* **2020**, *26*, 5202–5216. [CrossRef] [PubMed]
69. Wu, Q.; Wang, X.; Chen, S.; Wang, L.; Jiang, J. Land Surface Greening and CO_2 Fertilization More than Offset the Gross Carbon Sequestration Decline Caused by Land Cover Change and the Enhanced Vapour Pressure Deficit in Europe. *Remote Sens.* **2023**, *15*, 1372. [CrossRef]
70. Zheng, Y.; Shen, R.; Wang, Y.; Li, X.; Liu, S.; Liang, S.; Chen, J.M.; Ju, W.; Zhang, L.; Yuan, W. Improved Estimate of Global Gross Primary Production for Reproducing Its Long-Term Variation, 1982–2017. *Earth Syst. Sci. Data* **2020**, *12*, 2725–2746. [CrossRef]
71. Kang, Y.; Gaber, M.; Bassiouni, M.; Lu, X.; Keenan, T. CEDAR-GPP: Spatiotemporally Upscaled Estimates of Gross Primary Productivity Incorporating CO_2 Fertilization. *Earth Syst. Sci. Data Discuss.* **2023**, 1–51. [CrossRef]
72. Chen, C.; Riley, W.J.; Prentice, I.C.; Keenan, T.F. CO_2 Fertilization of Terrestrial Photosynthesis Inferred from Site to Global Scales. *Proc. Natl. Acad. Sci. USA* **2022**, *119*, e2115627119. [CrossRef]

Disclaimer/Publisher's Note: The statements, opinions and data contained in all publications are solely those of the individual author(s) and contributor(s) and not of MDPI and/or the editor(s). MDPI and/or the editor(s) disclaim responsibility for any injury to people or property resulting from any ideas, methods, instructions or products referred to in the content.

agronomy

MDPI

Article

Spatial Heterogeneity in the Response of Winter Wheat Yield to Meteorological Dryness/Wetness Variations in Henan Province, China

Cheng Li [1,*], Yuli Gu [1], Hui Xu [2,*], Jin Huang [3], Bo Liu [4], Kwok Pan Chun [5] and Thanti Octavianti [5]

1 Department of Ecology, School of Plant Protection, Joint International Research Laboratory of Agriculture and Agri-Product Safety of the Ministry of Education, Yangzhou University, Yangzhou 225009, China; mz120231474@stu.yzu.edu.cn
2 Nanjing Municipal Academy of Ecological and Environmental Protection Science, Nanjing 210013, China
3 School of Ecology and Applied Meteorology, Nanjing University of Information Science and Technology, Nanjing 210044, China; hjtyforlove@aliyun.com
4 College of Hydraulic Science and Engineering, Yangzhou University, Yangzhou 225009, China; boliu@yzu.edu.cn
5 CATE School of Architecture and Environment, University of the West of England, Bristol BS16 1QY, UK; kwok.chun@uwe.ac.uk (K.P.C.); thanti.octavianti@uwe.ac.uk (T.O.)
* Correspondence: licheng@yzu.edu.cn (C.L.); xuhuisafe@sina.com (H.X.)

Abstract: Knowledge of the responses of winter wheat yield to meteorological dryness/wetness variations is crucial for reducing yield losses in Henan province, China's largest winter wheat production region, under the background of climate change. Data on climate, yield and atmospheric circulation indices were collected from 1987 to 2017, and monthly self-calibrating Palmer drought severity index (sc-PDSI) values were calculated during the winter wheat growing season. The main results were as follows: (1) Henan could be partitioned into four sub-regions, namely, western, central-western, central-northern and eastern regions, based on the evolution characteristics of the time series of winter wheat yield in 17 cities during the period of 1988–2017. Among them, winter wheat yield was high and stable in the central-northern and eastern regions, with a remarkable increasing trend ($p < 0.05$). (2) The sc-PDSI in February had significantly positive impacts on climate-driven winter wheat yield in the western and central-western regions ($p < 0.05$), while the sc-PDSI in December and the sc-PDSI in May had significantly negative impacts on climate-driven winter wheat yield in the central-northern and eastern regions, respectively ($p < 0.05$). (3) There were time-lag relationships between the sc-PDSI for a specific month and the atmospheric circulation indices in the four sub-regions. Furthermore, we constructed multifactorial models based on selected atmospheric circulation indices, and they had the ability to simulate the sc-PDSI for a specific month in the four sub-regions. These findings will provide scientific references for meteorological dryness/wetness monitoring and risk assessments of winter wheat production.

Keywords: winter wheat yield; meteorological dryness/wetness features; spatial differences; Henan province

1. Introduction

Global warming has led to an ever-intensifying water cycle, and in particular, increasing rates of evaporation and precipitation, which have also changed the occurrence and severity of meteorological dryness/wetness events in different regions of the world [1–4]. Crop farming has been the backbone of human society, providing food and other basic needs for people [5]. With the rapid development of urbanization and industrialization, a large amount of cultivated land has been occupied, so improving crop yield per unit area has proven to be an important approach to boost food production [6,7]. However, the weight of observational evidence indicates that unfavorable meteorological dryness/wetness events during the crop growing season can lead to reduced crop yields and even threaten food security [8–10]. Therefore, analyzing the spatial heterogeneity of yield variability and its response to meteorological dryness/wetness conditions is of great importance in evaluating regional yield vulnerability and proposing adaptation strategies under global warming.

With the increasing importance of long time-series data, time series clustering methods have become widespread in the field of hydrometeorology [11,12]; they divide a large amount of data into smaller groups according to their similarity so that the data belonging to the same group are as similar as possible. Traditional clustering methods designed for static data are susceptible to noise when dealing with the time series clustering problem due to the high dimensionality of the time series [13]. Under these circumstances, the characteristic-based time series clustering method was proposed [14], which clusters based on structural characteristics extracted from the time series rather than using a distance measure to cluster point values. The characteristic measures used to reflect the time series are derived from statistical operations, such as statistical, time-domain and frequency-domain characteristics, and they are then input into arbitrary clustering algorithms. For example, Garcia-Gutierrez et al. (2022) [15] used a K-means clustering algorithm to regionalize solar radiation in Spain based on statistical parameters computed from clear-sky index series. As a typical example of time series data, the regionalization of crop yield for a given region not only analyzes spatial variations in yield but also provides fine-grained strategies to cope with unfavorable meteorological dryness/wetness conditions [16]. Despite the widespread interest in the mean values of and trends in crop yield [17], the statistical, time-domain and frequency-domain characteristics of yield data have rarely been systematically examined in previous studies.

To understand the response of crop yield to meteorological dryness/wetness variations, several meteorological dryness/wetness indices have been proposed in recent decades [3,18,19], such as the standardized precipitation index (SPI), standardized precipitation evapotranspiration index (SPEI) and self-calibrating Palmer drought severity index (sc-PDSI). Both SPI and SPEI have similar multi-timescale characteristics, with the former considering only precipitation and the latter considering precipitation and potential evapotranspiration [20]. In contrast to the SPI and SPEI, the sc-PDSI is more physical, as it takes into account the water balance using a two-layer bucket model and uses self-calibrating processes that can adjust the original PDSI to the local climate [21]. Recent studies have suggested that the sc-PDSI demonstrates good performance in monitoring meteorological dryness/wetness variations in China, and studies have also been carried out to show that climate–yield relationships vary between regions and crop types [22,23]. Taking China as an example, the sc-PDSI showed positive relationships with maize yield anomalies in Northeast China and wheat in North, South and Northwest China, but some regions (rice in East and Southeast China) had negative correlations [23]. These existing studies have focused mainly on relationships by considering the entire crop-growing season as a whole [24,25]. Few studies have examined seasonal differences in yield responses to meteorological dryness/wetness conditions, particularly the key month affecting yield and the corresponding key meteorological drought/wetness index (KMDWI). Considering that atmospheric circulations can be regarded as important premonitory influencing signals in the analysis of meteorological dryness/wetness variations [26,27], it is necessary to investigate the relationships between the KMDWI and atmospheric circulation indices in order to construct an appropriate simulation model, which may be a valuable approach to sustainable crop production and agricultural water management.

Wheat is an ancient and important grain crop. China is the world's largest winter wheat producer, accounting for nearly 20% of the global wheat yield [28]. As the largest winter wheat production region in China, Henan province accounts for nearly 25% of the winter wheat yield in China [29]. However, an uneven spatio-temporal distribution of precipitation has severely affected sustainable winter wheat production in Henan [30]. Although drought and flood events frequently occur in Henan [31], the spatial heterogeneity of yield variability and its response to meteorological dryness/wetness conditions are still unclear in this region. Therefore, the main research objectives are (1) to understand the spatial difference in winter wheat yield using a characteristic-based time series clustering analysis and (2) to identify the KMDWIs that affect yield fluctuations in different sub-regions and evaluate the potential impacts of atmospheric circulation indices. The detailed results are of great significance for meteorological dryness/wetness monitoring and risk assessments of winter wheat production.

2. Materials and Methods

2.1. Study Region

Located in central China (Figure 1), Henan province covers an area of 167,000 km^2. It consists of 18 cities, i.e., Zhengzhou, Kaifeng, Luoyang, Pingdingshan, Anyang, Hebi, Xinxiang, Jiaozuo, Puyang, Xuchang, Luohe, Sanmenxia, Shangqiu, Zhoukou, Zhumadian, Nanyang, Xinyang and Jiyuan, and it has a wide altitude ranging from 16 to 2272 m. Henan is located mainly in the warm temperate zone, with four distinct seasons and frequent meteorological disasters [31]. The annual average temperature ranges from 10.5 °C in the north to 16.7 °C in the south, and the annual precipitation ranges from 407.7 mm to 1295.8 mm, mainly concentrated between June and August [16]. As a major grain-producing province in China, the output of major agricultural products such as grain, cotton and oil in Henan ranks at the forefront of the country. Among them, winter wheat is the first grain crop in Henan, with 1/16 of the nation's arable land producing 1/4 of the winter wheat in China [29].

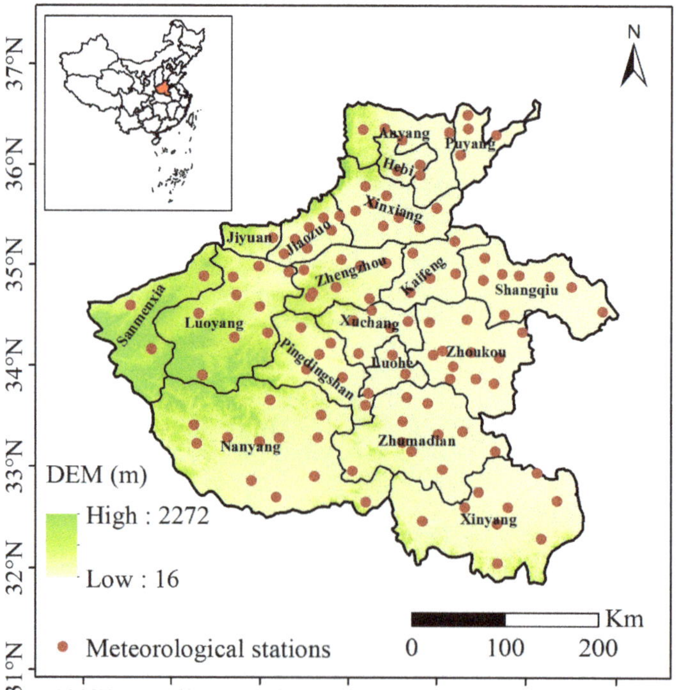

Figure 1. Geographical location of the study region and distribution of meteorological stations.

2.2. Datasets

The homogenized datasets of monthly temperature, precipitation, atmospheric pressure, relative humidity, sunshine and wind speed from 116 stations in Henan during the period of 1987–2017 were obtained from the National Climate Center of China. Data on annual winter wheat yield (AWWY, unit: kg/hm^2) in 17 cities in Henan, with the exception of Jiyuan, during the period of 1987–2017 were obtained from the Henan Province Bureau of Statistics. Monthly atmospheric circulation indices that may drive the regional climate in Henan were used to analyze the premonitory influencing signals of meteorological dryness/wetness variations, which included Niño 1+2, Niño 3, Niño 3.4, Niño 4, North Atlantic Oscillation (NAO), Arctic Oscillation (AO) and Pacific Decadal Oscillation (PDO), and they were obtained from http://www.cpc.ncep.noaa.gov/products/precip/CWlink/ (accessed on 5 November 2023). All data used in this study were obtained from official sources and were subjected to rigorous quality controls.

2.3. Methods

2.3.1. Characteristic-Based Time Series Clustering Analysis

In this study, AWWY was selected for a given indicator, and a series of characteristic-based indices were constructed and are presented in Table 1. As shown in Table 1, we calculated the statistical characteristics of the AWWY time series, such as the minimum value (MIN), maximum value (MAX), mean value (AVE), range value (RNG), standard deviation (STD) and coefficient of variation (CV). Furthermore, a trend analysis is one of the most important measurements in studying time series data, and it can reflect their time-domain characteristics. In this study, the Z statistic (Z_{MK}), Sen's slope (MAG) and the Hurst index (H) were used to describe the trends and persistent patterns of the AWWY time series using the Mann–Kendall test, Sen's slope estimator and the Hurst method, respectively [32]. In addition to a trend analysis, multi-time scale characteristics are also important in studying time series data. For example, the AWWY time series can be decomposed into three periodic oscillation intrinsic mode functions (IMFs) and a trend component, and this was achieved in this study using ensemble empirical mode decomposition (EEMD) [33]. The variance contribution of each IMF (i.e., CIMF1, CIMF2 and CIMF3) and the trend component (CTrend) to AWWY was selected to reflect the frequency-domain characteristics of the AWWY time series. More details about the Mann–Kendall test, Sen's slope estimator, the Hurst method and EEMD are shown in the Supplementary Materials [34–41].

Table 1. Definitions of characteristic-based indices of annual winter wheat yield (AWWY).

No.	Types	Indices	Definition
1	Statistical characteristics	MIN	The minimum value of the AWWY time series
2		MAX	The maximum value of the AWWY time series
3		AVE	The mean value of the AWWY time series
4		RNG	The range value of the AWWY time series
5		STD	The standard deviation of the AWWY time series
6		CV	The coefficient of variation in the AWWY time series
7	Time-domain characteristics	Z_{MK}	The Z statistic of the AWWY time series
8		MAG	The magnitude of the AWWY time series
9		H	The persistent pattern of the AWWY time series
10	Frequency-domain characteristics	CIMF1	The variance contribution of a short-period oscillation component to AWWY
11		CIMF2	The variance contribution of a medium-period oscillation component to AWWY
12		CIMF3	The variance contribution of a long-period oscillation component to AWWY
13		CTrend	The variance contribution of a trend component to AWWY

Note: MIN, MAX, AVE, RNG, STD, CV, Z_{MK}, MAG, H, CIMF1, CIMF2, CIMF3 and CTrend are abbreviations for the minimum value, maximum value, mean value, range value, standard deviation, coefficient of variation, Z statistic, Sen's slope, Hurst index and variance contribution of each IMF and a trend component, respectively.

Taking into account the potential collinearity and redundancy among these indices, a principal component analysis (PCA) and a K-means clustering analysis were used in our study to partition the 17 cities into k clusters, and the clustering result was evaluated using silhouette coefficients (SCs) [11,42]. The highest SC value determined the optimum number of clusters. More details about the PCA and K-means clustering analysis are shown in the Supplementary Materials [43–50].

2.3.2. Calculation of sc-PDSI

The sc-PDSI is a typical meteorological dryness/wetness index based on a physical water balance model, calculated from precipitation and potential evapotranspiration, together with parameters related to the characteristics of the soil/surface at each location [51]. In the calculation of the sc-PDSI, potential evapotranspiration is based on the Penman–Monteith equation, which takes into account energy balance and aerodynamic theory, rather than the Thornthwaite method [52]. The sc-PDSI values range from −4 to +4, with a negative (positive) value indicating a dry (wet) period. In this study, we calculated the monthly sc-PDSI during the winter wheat growing season (from October of the previous year to May of the current year) in Henan using climate data on temperature, precipitation, relative humidity, sunshine and wind speed to reflect meteorological dryness/wetness conditions, and this was achieved using an open-access R package (https://github.com/Sibada/scPDSI, accessed on 5 November 2023).

2.3.3. Calculations of Climate-Driven Winter Wheat Yield (CDWWY)

Crop yield is affected by a variety of factors such as crop varieties, agricultural technology and meteorological conditions, which is a combination of trend yield and climate-driven yield [53]. Extracting the climate-driven yield is a crucial step in understanding the yield impacts of meteorological dryness/wetness variations [9]. In this study, four statistical methods, namely, EEMD, linear regression (LR), logistic function (LF) and the Hodrick–Prescott (HP) filter, were used to extract CDWWY [54].

In this study, we mainly focused on the relationships between time series variables (i.e., CDWWY and sc-PDSI) during the period of 1987–2017. Therefore, we used a Pearson correlation analysis to examine the relationships between the CDWWY time series and sc-PDSI time series during the winter wheat growing season, which allowed us to select the key month that affected the yield and the corresponding KMDWI in Henan. Furthermore, a regression equation was constructed between the KMDWI and CDWWY using the following formula [42]:

$$\text{std (CDWWY)} = a \times \text{std (KMDWI)} + b \quad (1)$$

where std (CDWWY) and std (KMDWI) are the standardized values of the KMDWI and CDWWY, respectively. Coefficient a is the sensitivity of the yield to the KMDWI, and it is multiplied by the trend of the KMDWI for a given period to determine the impact of the KMDWI trend on yield [42].

2.3.4. Relationships between the KMDWI and Atmospheric Circulation Indices

Atmospheric circulation indices are considered important premonitory influencing signals in the analysis of meteorological dryness/wetness variations [26,27]. A time-lag correlation analysis was performed to examine the relationships between the KMDWI and seven atmospheric circulation indices [55,56], and a time lag of 0–11 months was selected in this study. According to the results of the correlation analysis, we then selected the atmospheric circulation indices with the highest correlation with the corresponding KMDWI for each month as the premonitory influencing signals. Finally, we used the multiple linear regression method to construct an empirical KMDWI simulation model based on the atmospheric circulation indices selected in this study. Two indicators were used to evaluate the performance of the multiple linear regression model: the determination coefficients (R^2) and root-mean-square error (RMSE) [57]. More details about the time-lag correlation analysis and multiple linear regression method are shown in the Supplementary Materials.

3. Results

3.1. AWWY-Based Regionalization in Henan

As described in Section 2.3.1, the values of 13 indices in 17 cities in Henan were first calculated. Then, the principal components of 13 columns (13 indices) × 17 rows (17 cities) were analyzed using the PCA method (Table 2). To reduce the dimensionality of the data, we selected the first three principal components (PCs) because their eigenvalues were greater than 1, and their cumulative variances exceeded 89%. As shown in Table 2, PC1 had higher loadings on five indices, namely, Z_{MK}, H, CIMF1, CIMF3 and Ctrend, representing the persistent trend of the AWWY time series. PC2 had higher loadings on four indices, namely, RNG, STD, CV and MAG, representing the magnitude of the change in AWWY. PC3 had higher loadings on three indices, namely, MIN, MAX and AVE, representing the high (or low) levels of AWWY. According to the first three PCs, we obtained their corresponding score sequences. A matrix of 3 columns (3 score sequences) × 17 rows (17 cities) was clustered using the K-means method.

Table 2. The results of the application of the principal component analysis (PCA) method to a matrix composed of 13 characteristic-based indices of AWWY in 17 cities in Henan.

	Eigenvalue	Variance (%)	Cumulative (%)
PC1	6.64	51.11	51.11
PC2	3.61	27.80	78.90
PC3	1.42	10.92	89.83

The loadings of PC1, PC2 and PC3 on 13 indices

	MIN	MAX	AVE	RNG	STD	CV	Z_{MK}	MAG	H	CIMF1	CIMF2	CIMF3	Ctrend
PC1	0.51	0.51	0.52	0.14	0.23	−0.14	0.93	0.51	0.86	−0.90	−0.57	−0.84	0.98
PC2	−0.33	0.59	0.37	0.96	0.98	0.85	0.36	0.92	0.32	−0.07	−0.49	0.08	0.17
PC3	0.82	0.90	0.98	0.30	0.35	−0.32	0.66	0.46	0.71	−0.63	−0.21	−0.29	0.60

Note: PC1, PC2 and PC3 are abbreviations for the first three principal components. MIN, MAX, AVE, RNG, STD, CV, Z_{MK}, MAG, H, CIMF1, CIMF2, CIMF3 and CTrend are abbreviations for the minimum value, maximum value, mean value, range value, standard deviation, coefficient of variation, Z statistic, Sen's slope, Hurst index and variance contribution of each IMF and a trend component, respectively.

As shown in Figure 2, the SC values varied for different numbers of clusters. The highest SC value occurred when the number of clusters reached four. Therefore, 17 cities in Henan could be partitioned into four sub-regions characterized by different AWWY variations, i.e., Sanmenxia and Luoyang in the western region (Region I); Zhengzhou, Pingdingshan and Nanyang in the central-western region (Region II); Xuchang, Kaifeng, Jiaozuo, Xinxiang, Hebi, Puyang and Anyang in the central-northern region (Region III); and Luohe, Shangqiu, Zhoukou, Zhumadian and Xinyang in the eastern region (Region IV) (Table S1).

3.2. Spatial Differences in AWWY in Henan

The statistical values of 13 indices characterizing the AWWY time series in the four sub-regions are shown in Table 3. As shown in Table 3, the AVE of AWWY was the highest in Region III (5770.9 kg/hm^2), followed by Region IV (5091.3 kg/hm^2) and it was the lowest in Region I (3536.3 kg/hm^2). Compared with the AVE, the RNG of AWWY was the highest in Region IV (4736 kg/hm^2), followed by Region III (3561.2 kg/hm^2) and it was the lowest in Region II (2543.8 kg/hm^2). Furthermore, the Z_{MK}, MAG and CTrend were higher in Regions III and IV than in Regions II and I. These results indicate that in Henan, Regions III and IV had a remarkable increasing trend in AWWY, as well as a high and stable yield.

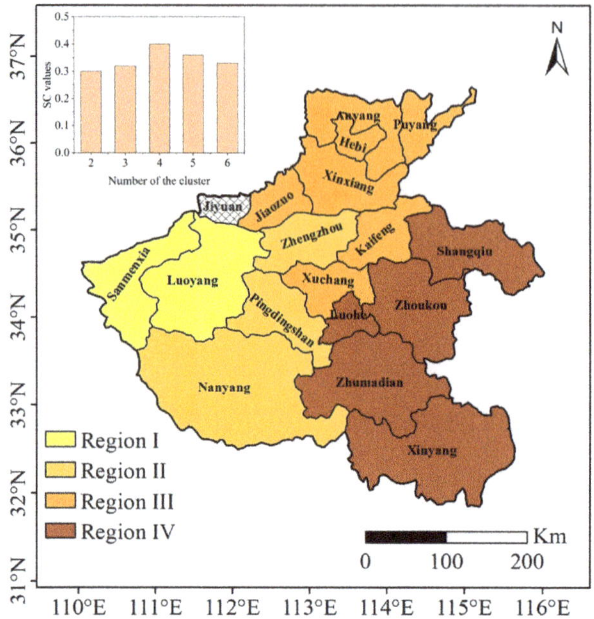

Figure 2. The clustering results of AWWY in Henan.

Table 3. Statistical values of characteristic-based indices of AWWY for four sub-regions in Henan.

No.	Indices	Region I	Region II	Region III	Region IV
1	MIN (kg/hm^2)	1861.0	2554.3	3592.8	2118.3
2	MAX (kg/hm^2)	4812.0	5098.1	7154.0	6855.2
3	AVE (kg/hm^2)	3536.3	4023.1	5770.9	5091.3
4	RNG (kg/hm^2)	2951.0	2543.8	3561.2	4736.9
5	STD	796.0	710.1	1048.8	1416.1
6	CV (%)	22.5	17.8	18.3	27.8
7	Z_{MK}	4.2	5.2	6.9	6.5
8	MAG (kg/(hm^2·a))	62.2	63.3	111.5	138.1
9	H	0.7	0.8	0.8	0.8
10	CIMF1 (%)	26.9	16.8	2.3	11.2
11	CIMF2 (%)	5.8	4.4	2.6	2.6
12	CIMF3 (%)	3.4	2.4	1.1	2.5
13	CTrend (%)	63.9	76.4	94.0	83.8

Note: MIN, MAX, AVE, RNG, STD, CV, Z_{MK}, MAG, H, CIMF1, CIMF2, CIMF3 and CTrend are abbreviations for the minimum value, maximum value, mean value, range value, standard deviation, coefficient of variation, Z statistic, Sen's slope, Hurst index and variance contribution of each IMF and a trend component, respectively.

3.3. Relationships between CDWWY and sc-PDSIs during the Wheat Growing Season in Henan

In this study, we employed four commonly used methods to extract CDWWY (Figure 3). Compared with the LR, LF and HP filter, the CDWWY calculated using EEMD showed a higher correlation with the sc-PDSIs during the winter wheat growing season for the four sub-regions in Henan. Taking the CDWWY calculated using EEMD as an example, it was found that the sc-PDSIs from January to February of the current year had a significant positive impact on CDWWY in Region I ($p < 0.05$), and the sc-PDSI in February had the largest impact on CDWWY, which was considered the KMDWI in this sub-region. Similarly, the sc-PDSI in February also had the largest impact on CDWWY in Region II ($p < 0.05$), and a wet February was more favorable for winter wheat production in this sub-region. In contrast to Regions I and II, the sc-PDSI in December of the previous year and the sc-PDSI in May of the current year had significant negative impacts on CDWWY in Regions III and IV, respectively ($p < 0.05$), which were considered the KMDWI in both sub-regions, respectively. These results indicate that meteorological dryness/wetness conditions in different key months limited winter wheat production in Regions III and IV.

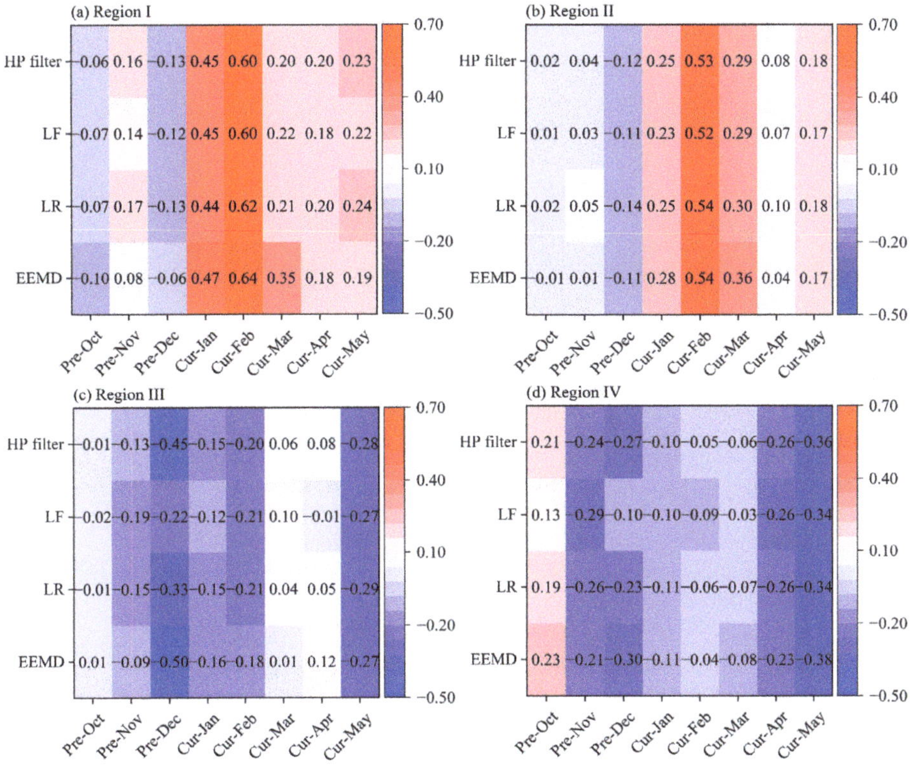

Figure 3. Relationships between the climate-driven winter wheat yield (CDWWY) and the self-calibrating Palmer drought severity index (sc-PDSI) during the wheat growing season in Henan. Note: Pre- and Cur- denote the previous year and current year, respectively. A correlation coefficient value of 0.36 corresponds to the 5% significance level.

The responses of CDWWY to KMDWI changes in different sub-regions and periods are shown in Figure 4. If the KMDWI increased by 1% during the study period, CDWWY increased by 0.66% and 0.46% in Regions I and II, respectively, and it decreased by 0.35% and 0.30% in Regions III and IV, respectively (Figure 4a). However, limited impacts of the KMDWI trends on CDWWY were found for the four sub-regions, ranging from −0.06% to 0.05%, due to a weak KMDWI trend in each sub-region from 1988 to 2017 (Figure 4b,c). Furthermore, we classified the study period into three sub-periods, i.e., 1988–1999, 2000–2009 and 2010–2017, and we then found that the responses of CDWWY to KMDWI changes showed obvious interdecadal differences in Regions I and II. The sensitivities of CDWWY to KMDWI changes were the highest in 1988–1999, followed by 2000–2009 and they were the lowest in 2010–2017 in Regions I and II. Consequently, the impacts of the KMDWI trends on CDWWY gradually weakened from 1988–1999 to 2010–2017 in both sub-regions. Compared with Regions I and II, the impacts of the KMDWI trends on CDWWY were obviously weaker in Regions III and IV during the three sub-periods.

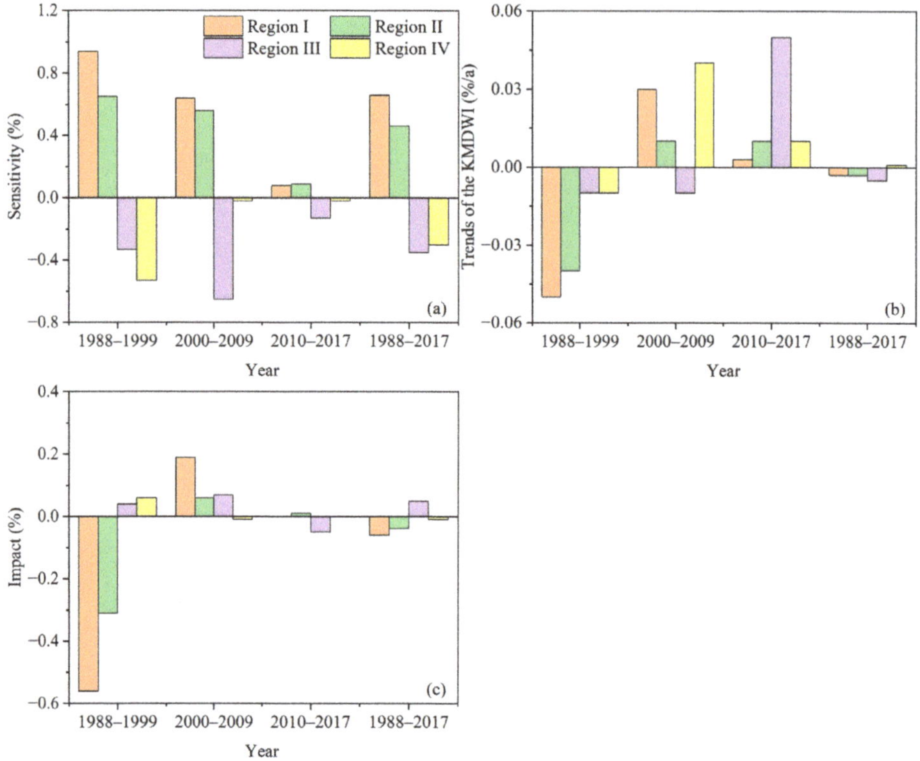

Figure 4. Responses of CDWWY to key meteorological drought/wetness index (KMDWI) changes in Henan. (**a**) Sensitivity of the yield to the KMDWI. (**b**) Trends of the KMDWI. (**c**) Impact of the KMDWI trend on yield.

3.4. Relationships between the KMDWI and Atmospheric Circulation Indices in Henan

We used a Pearson correlation analysis to examine the relationship between the KMDWI of the key month and seven atmospheric circulation indices of the 12 months previous to that key month (Figure 5). As shown in Figure 5, the most significant correlation was found in Region IV, with the highest absolute value of 0.49. Compared with this, the highest correlation coefficient in Region III was less than 0.40. Furthermore, the atmospheric circulation indices had time-lag effects on the KMDWIs in the different sub-regions. For example, NAO in December, PDO in November and AO in April of the previous year had a significant impact on the KMDWI ($p < 0.05$) in Region I (Figure 5a), while NAO in June and December and AO in April of the previous year had a significant impact on the KMDWI ($p < 0.05$) in Region II (Figure 5b). As for Regions III and IV, NAO in February and AO in August of the previous year had a significant impact on the KMDWI ($p < 0.05$) in Region III (Figure 5c), while the atmospheric circulation indices that affected the KMDWI in Region IV ($p < 0.05$) mainly included Niño 1+2, Niño 3 and Niño 3.4 from June to November of the previous year, as well as NAO in June of the previous year (Figure 5d).

Given the results of the correlation analysis described above, we selected the atmospheric circulation indices with the highest correlation with the corresponding KMDWI for each month as the premonitory influencing signals in different sub-regions. Then, we constructed regression models between the KMDWI and selected atmospheric circulation indices in each sub-region. As shown in Table 4, the multifactorial models that we constructed had the ability to simulate the KMDWIs for the four sub-regions. Their R^2 values were in the range of 0.49 to 0.65, and the corresponding RMSE values were in the range of 0.58 to 0.72 in the different sub-regions.

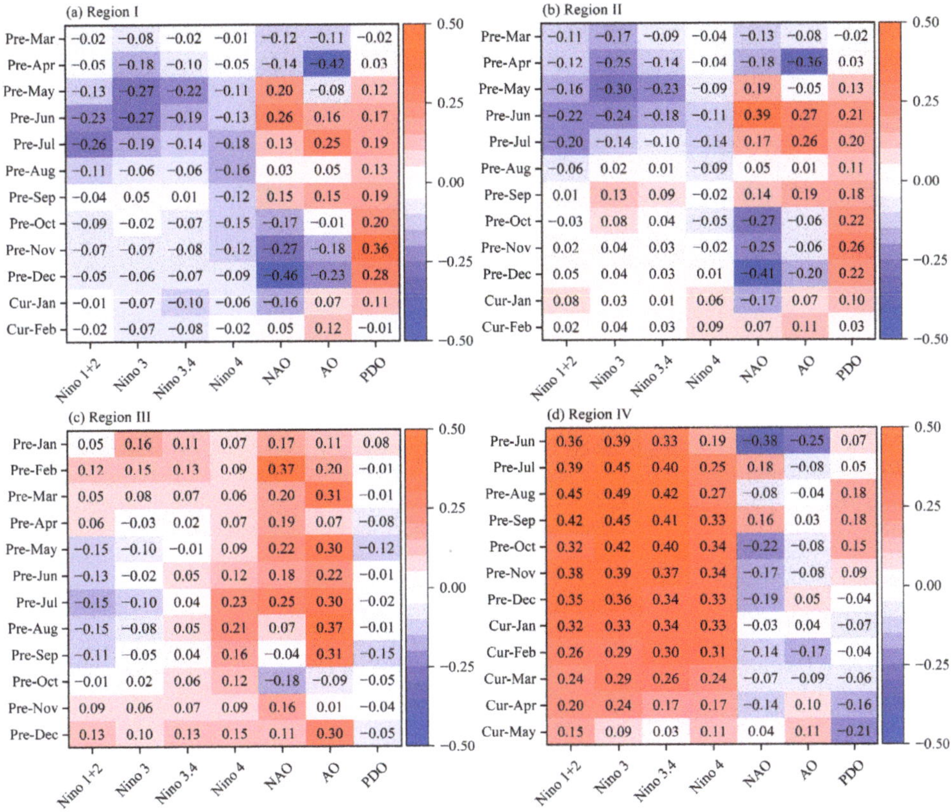

Figure 5. Relationships between KMDWIs and atmospheric circulation indices in Henan. Note: Pre- and Cur- denote the previous year and current year, respectively. A correlation coefficient value of 0.36 corresponds to the 5% significance level.

Table 4. Regression results between KMDWIs and selected atmospheric circulation indices in Henan.

Lag Time	Region I		Region II		Region III		Region IV	
	Indices	Coefficients	Indices	Coefficients	Indices	Coefficients	Indices	Coefficients
0	AO	−0.08	AO	0.02	AO	0.06	PDO	−0.45
1	NAO	0.19	NAO	0.10	NAO	0.35	Niño 3	0.27
2	NAO	−0.30	NAO	−0.39	NAO	−0.10	Niño 3	−0.11
3	PDO	0.30	PDO	0.27	AO	0.73	Niño 4	0.92
4	PDO	0.13	NAO	−0.11	AO	1.07	Niño 3.4	−0.62
5	PDO	0.08	AO	0.35	AO	0.44	Niño 3	−0.50
6	Niño 4	−0.57	PDO	−0.05	AO	0.25	Niño 3	1.19
7	Niño 1+2	−0.18	AO	−0.05	AO	0.17	Niño 3	−0.97
8	Niño 3	0.29	NAO	0.38	NAO	−0.04	Niño 3	−0.11
9	Niño 3	−0.53	Niño 3	−0.04	AO	0.19	Niño 3	1.90
10	AO	−0.08	AO	−0.13	NAO	0.08	Niño 3	−0.99
11	NAO	−0.10	Niño 3	−0.14	NAO	−0.29	Niño 3	0.24

Table 4. *Cont.*

Lag Time	Region I		Region II		Region III		Region IV	
	Indices	Coefficients	Indices	Coefficients	Indices	Coefficients	Indices	Coefficients
Intercept	27.29		5.16		−0.02		−32.42	
R^2	0.50		0.55		0.65		0.49	
RMSE	0.66		0.64		0.58		0.72	

Note: Niño 1+2, Niño 3, Niño 3.4 and Niño 4 reflect the Niño regions, which are areas of the tropical Pacific Ocean that are used to monitor and measure the El Niño–Southern Oscillation climate pattern, and each Niño region covers a different area of the Pacific Ocean. AO, NAO and PDO are abbreviations for North Atlantic Oscillation, Arctic Oscillation and Pacific Decadal Oscillation.

4. Discussion

According to the AWWY time series from 17 cities, Henan province could be partitioned into four sub-regions using a characteristic-based time series clustering analysis (Figure 2). The complex geographical environment could be one of the important factors responsible for the spatial differences in AWWY. Regions I and II are surrounded by the Taihang Mountains and the Funiu Mountains (Figure 1), and the high mountain terrain restricts the use of agricultural machinery [58]. Furthermore, the complex climate, poor quality of arable land and low level of irrigation in the mountainous areas have a combined effect on the production of winter wheat in both sub-regions [30,58–60]. However, our finding is similar to that of a risk assessment of an agrometeorological disaster in Henan, which was the result of a combination of four main factors (hazard, exposure, vulnerability and restorability) [59]. Therefore, a characteristic-based time series clustering analysis could also provide a new perspective for the regionalization of crop production risk under the background of climate change due to its simple data requirements and reliable results [14].

The sc-PDSIs in February, May and December were the KMDWIs that influenced yield in the four sub-regions (Figure 3). Among them, the sc-PDSI in February had significantly positive impacts on yield in Regions I and II ($p < 0.05$), while the sc-PDSI in December and the sc-PDSI in May had significantly negative impacts on yield in Regions III and IV, respectively ($p < 0.05$). The uneven distribution of precipitation during the winter wheat growing season led to a severe water shortage in the greening stage, and the drought risk gradually decreased from western to eastern Henan [31,61]. Thus, the higher the sc-PDSI in February of the current year, the higher the yields in Regions I and II. The impacts of drought risk on yield could be alleviated by maintaining soil moisture through appropriate tillage methods, improving irrigation efficiency and using drought-resistant varieties of wheat [62,63]. Spring waterlogging disasters occurred mainly in the eastern parts of Henan, and a wetter climate in the maturity stage was unfavorable to spike development and grain formation [64,65], resulting in lower yields in Region IV with a higher sc-PDSI in May of the current year. Therefore, conducting farmland drainage measures in a timely manner is critical for achieving drainage and water logging, as well as coordinated regulation [66,67]. Additionally, the occurrence of wheat lodging caused by heavy rainfall and high winds should also be noted [68]. Chill damage in the overwintering stage was the main unfavorable factor for the growth of winter wheat in the northern parts of Henan, and more rainfall exacerbated low temperatures and low sunshine [69,70], which led to lower yields in Region III with a higher sc-PDSI in December of the previous year. Therefore, shifting the wheat planting calendar, spraying foliar fertilizers and using cold-resistant winter wheat varieties could reduce the negative effects of chill damage in the overwintering stage [71–73].

In addition to identifying the KMDWIs that affect yield fluctuations, we also found a time-lag relationship between the KMDWIs and atmospheric circulation indices. Although similar results have been reported in previous studies [9,55,56], our study indicates that the time-lag relationship exhibited spatial differences in the different sub-regions (Figure 5). In particular, the time lag of the same KMDWI response was different for the different atmospheric circulation indices in the four sub-regions. Furthermore, the multifactorial models used to simulate the KMDWIs for the four sub-regions were constructed by selecting suitable atmospheric circulation indices (Table 4). Compared with previous studies conducted at a larger spatial scale [74,75], the accuracy of our empirical models, together with R^2 and RMSE, could give a reliable output and be applied at the provincial scale. Furthermore, atmospheric circulation indices can be forecasted about 6 to 9 months in advance by the National Climate Center of China [76], suggesting that our empirical models obtain KMDWIs for the four sub-regions at an early stage, and they are of great significance for meteorological dryness/wetness monitoring for winter wheat production and providing some coping strategies in each sub-region [77,78].

The limitations of our study are based on two main aspects. First, we mainly focused on the response of winter wheat yield to meteorological dryness/wetness variations in Henan using statistical data from the perspective of the agricultural climate, without fully considering the effects of other yield-related factors, e.g., plant cultivation, soil category and treatments [79,80]. Second, we constructed multifactorial models capable of simulating KMDWIs in four sub-regions. However, the mechanisms of the natural and anthropogenic factors, including atmospheric circulation, underlying the KMDWIs are more complex and still need to be studied in depth [81]. Therefore, we will collect more data on climate, soil and crop management to understand climate–yield relationships and to improve the simulation performance of unfavorable meteorological dryness/wetness conditions by incorporating machine learning methods [82–84].

5. Conclusions

Using long-term data on climate, yield and atmospheric circulation indices, this study analyzed the response of winter wheat yield to sc-PDSI variations in Henan from 1987 to 2017. The main results were as follows:

(1) A PCA and a K-means clustering analysis were used to partition the AWWY time series from 17 cities into four sub-regions. Although AWWY exhibited an increasing trend in all of the four sub-regions ($p < 0.05$), it was high and stable in Regions III and IV.

(2) The sc-PDSI for a specific month could be considered the KMDWI affecting CDWWY in each sub-region, for example, the sc-PDSI in February of the current year for Regions I and II, the sc-PDSI in December of the previous year for Region III and the sc-PDSI in May of the current year for Region IV.

(3) The atmospheric circulation indices had time-lag effects on the KMDWIs, and empirical KMDWI simulation models were constructed based on selected atmospheric circulation indices in the four sub-regions. Their R^2 values were in the range of 0.49 to 0.65, and the corresponding RMSE values were in the range of 0.58 to 0.72.

Supplementary Materials: The following supporting information can be downloaded at: https://www.mdpi.com/article/10.3390/agronomy14040817/s1, Table S1: statistical values of climate elements for four sub-regions in Henan. Section S1: detailed explanation of statistical methods used in our study.

Author Contributions: Conceptualization, C.L. and H.X.; methodology, Y.G. and J.H.; software, C.L.; formal analysis, H.X. and J.H.; data curation, B.L.; writing—original draft preparation, C.L. and Y.G.; writing—review and editing, H.X., J.H., B.L., K.P.C. and T.O.; funding acquisition, C.L. and K.P.C. All authors have read and agreed to the published version of the manuscript.

Funding: This research was funded by the National Natural Science Foundation of China (41801013), the Royal Society International Exchanges 2022 (IEC\NSFC\223132), the Young Scientific and Technological Talents Support Project of Jiangsu Association for Science and Technology (TJ-2023-032), the Humanities and Social Sciences Foundation of Yangzhou University (xjj2023-05) and the Qinglan Project of Yangzhou University.

Data Availability Statement: The data presented in this study are available on request from the corresponding author.

Conflicts of Interest: The authors declare no conflicts of interest.

References

1. Konapala, G.; Mishra, A.K.; Wada, Y.; Mann, M.E. Climate change will affect global water availability through compounding changes in seasonal precipitation and evaporation. *Nat. Commun.* **2020**, *11*, 3044. [CrossRef] [PubMed]
2. Tabari, H. Climate change impact on flood and extreme precipitation increases with water availability. *Sci. Rep.* **2020**, *10*, 13768. [CrossRef]
3. Huang, Q.; Zhang, Q.; Singh, V.P.; Shi, P.; Zheng, Y. Variations of dryness/wetness across China: Changing properties, drought risks, and causes. *Global Planet. Chang.* **2017**, *155*, 1–12. [CrossRef]
4. Wei, W.; Wang, J.; Ma, L.; Wang, X.; Xie, B.; Zhou, J.; Zhang, H. Global Drought-Wetness Conditions Monitoring Based on Multi-Source Remote Sensing Data. *Land* **2024**, *13*, 95. [CrossRef]
5. Chen, S.; Gong, B. Response and adaptation of agriculture to climate change: Evidence from China. *J. Dev. Econ.* **2021**, *148*, 102557. [CrossRef]
6. Kang, L.; Zhao, R.; Wu, K.; Huang, Q.; Zhang, S. Impacts of Farming Layer Constructions on Cultivated Land Quality under the Cultivated Land Balance Policy. *Agronomy* **2021**, *11*, 2403. [CrossRef]
7. Tian, X.; Engel, B.A.; Qian, H.; Hua, E.; Sun, S.; Wang, Y. Will reaching the maximum achievable yield potential meet future global food demand? *J. Clean. Prod.* **2021**, *294*, 126285. [CrossRef]
8. Hasegawa, T.; Sakurai, G.; Fujimori, S.; Takahashi, K.; Hijioka, Y.; Masui, T. Extreme climate events increase risk of global food insecurity and adaptation needs. *Nat. Food* **2021**, *2*, 587–595. [CrossRef] [PubMed]
9. Li, C.; Wang, R.; Xu, J.; Luo, Y.; Tan, M.L.; Jiang, Y. Analysis of meteorological dryness/wetness features for spring wheat production in the Ili River basin, China. *Int. J. Biometeorol.* **2018**, *62*, 2197–2204. [CrossRef]
10. Wang, X.; Qian, L.; Dong, C.; Tang, R. Evaluating the Impacts of Waterlogging Disasters on Wheat and Maize Yields in the Middle and Lower Yangtze River Region, China, by an Agrometeorological Index. *Agronomy* **2023**, *13*, 2590. [CrossRef]
11. Li, C.; Chen, J.; Wang, R.; Huang, J.; Qian, Z.; Xu, Y. Multi-indices analysis of heavy precipitation changes in Anhui Province, China. *Meteorol. Atmos. Phys.* **2021**, *133*, 1317–1325. [CrossRef]
12. Huang, J.; Zhang, F.; Zhou, L.; Hu, Z.; Li, Y. Regional changes of climate extremes and its effect on rice yield in Jiangsu province, southeast China. *Environ. Earth. Sci.* **2018**, *77*, 106. [CrossRef]
13. Song, C.; Pei, T. Research Progress in Time Series Clustering Methods Based on Characteristics. *Prog. Geogr.* **2012**, *10*, 1307–1317, (In Chinese with English abstract).
14. Wang, X.; Smith, K.; Hyndman, R. Characteristic-Based Clustering for Time Series Data. *Data Min. Knowl. Discov.* **2006**, *13*, 335–364. [CrossRef]
15. Garcia-Gutierrez, L.; Voyant, C.; Notton, G.; Almorox, J. Evaluation and Comparison of Spatial Clustering for Solar Irradiance Time Series. *Appl. Sci.* **2022**, *12*, 8529. [CrossRef]
16. Huang, J.; Zhou, L.; Zhang, F.; Hu, Z.; Tian, H. Responses of yield variability of summer maize in Henan province, north China, to large-scale atmospheric circulation anomalies. *Theor. Appl. Climatol.* **2021**, *143*, 1655–1665. [CrossRef]
17. Iizumi, T.; Sakai, T. The global dataset of historical yields for major crops 1981–2016. *Sci. Data* **2020**, *7*, 97. [CrossRef] [PubMed]
18. Li, L.; She, D.; Zheng, H.; Lin, P.; Yang, Z. Elucidating diverse drought characteristics from two meteorological drought indices (SPI and SPEI) in China. *J. Hydrometeorol.* **2020**, *21*, 1513–1530. [CrossRef]
19. Liu, Q.; Zhang, J.; Zhang, H.; Yao, F.; Bai, Y.; Zhang, S.; Meng, X.; Liu, Q. Evaluating the performance of eight drought indices for capturing soil moisture dynamics in various vegetation regions over China. *Sci. Total Environ.* **2021**, *789*, 147803. [CrossRef]
20. Saharwardi, M.S.; Kumar, P. Future drought changes and associated uncertainty over the homogenous regions of India: A multimodel approach. *Int. J. Climatol.* **2022**, *42*, 652–670. [CrossRef]
21. Wang, H.; Chen, Y.; Pan, Y.; Li, W. Spatial and temporal variability of drought in the arid region of China and its relationships to teleconnection indices. *J. Hydrometeorol.* **2015**, *523*, 283–296. [CrossRef]
22. Zhao, H.; Gao, G.; An, W.; Zou, X.; Li, H.; Hou, M. Timescale differences between sc-PDSI and SPEI for drought monitoring in China. *Phys. Chem. Earth* **2017**, *102*, 48–58. [CrossRef]
23. Faiz, M.A.; Zhang, Y.; Tian, X.; Tian, J.; Zhang, X.; Ma, N.; Aryal, S. Drought index revisited to assess its response to vegetation in different agro-climatic zones. *J. Hydrometeorol.* **2022**, *614*, 128543.
24. Ray, D.; Gerber, J.; MacDonald, G.; West, P.C. Climate variation explains a third of global crop yield variability. *Nat. Commun.* **2015**, *6*, 5989. [CrossRef] [PubMed]
25. Shi, W.; Tao, F.; Zhang, Z. A review on statistical models for identifying climate contributions to crop yields. *J. Geogr. Sci.* **2013**, *23*, 567–576. [CrossRef]
26. Nikraftar, Z.; Mostafaie, A.; Sadegh, M.; Afkueieh, J.H.; Pradhan, B. Multi-type assessment of global droughts and teleconnections. *Weather Clim. Extrem.* **2021**, *34*, 100402. [CrossRef]
27. Kretschmer, M.; Adams, S.V.; Arribas, A.; Prudden, R.; Robinson, N.; Saggioro, E.; Shepherd, T.G. Quantifying causal pathways of teleconnections. *Bull. Am. Meteorol. Soc.* **2021**, *102*, E2247–E2263. [CrossRef]
28. Shiferaw, B.; Smale, M.; Braun, H.J.; Duveiller, E.; Reynolds, M.; Muricho, G. Crops that feed the world 10. Past successes and future challenges to the role played by wheat in global food security. *Food Secur.* **2013**, *5*, 291–317. [CrossRef]
29. Wang, T.; Zhang, L.; Zhou, S.; Zhang, T.; Zhai, S.; Yang, Z.; Wang, D.; Song, H. Effects of ground-level ozone pollution on yield and economic losses of winter wheat in Henan, China. *Atmos. Environ.* **2021**, *262*, 118654. [CrossRef]

30. Zheng, H.; Huang, J.; Chen, J. Climate-Induced Yield Losses for Winter Wheat in Henan Province, North China and Their Relationship with Circulation Anomalies. *Water* **2021**, *13*, 3341. [CrossRef]
31. Shi, B.; Zhu, X.; Hu, Y.; Yang, Y. Drought characteristics of Henan province in 1961-2013 based on Standardized Precipitation Evapotranspiration Index. *J. Geogr. Sci.* **2017**, *27*, 311–325. [CrossRef]
32. Liu, X.; Cao, K.; Li, M. Assessing the impact of meteorological and agricultural drought on maize yields to optimize irrigation in Heilongjiang Province, China. *J. Clean. Prod.* **2024**, *434*, 139897. [CrossRef]
33. Ghose, B.; Islam, A.R.M.T.; Islam, H.M.T.; Hasanuzzaman, M.; Huang, J.; Hu, Z.; Moniruzzaman, M.; Gustave, W.; Karim, M.; Ibrahim, S.M. Rain-fed rice yield fluctuation to climatic anomalies in Bangladesh. *Int. J. Plant Prod.* **2021**, *15*, 183–201. [CrossRef]
34. Gocic, M.; Trajkovic, S. Analysis of changes in meteorological variables using Mann-Kendall and Sen's slope estimator statistical tests in Serbia. *Global Planet. Chang.* **2013**, *100*, 172–182. [CrossRef]
35. Wang, W.; Chen, X.; Shi, P.; van Gelder, P.H.A.J.M. Detecting changes in extreme precipitation and extreme streamflow in the Dongjiang River basin in southern China. *Hydrol. Earth Syst. Sci.* **2008**, *12*, 207–221. [CrossRef]
36. Sen, P.K. Estimates of the regression coefficient based on Kendall's tau. *J. Am. Stat. Assoc.* **1968**, *63*, 1379–1389. [CrossRef]
37. Mandelbrot, B.; Willis, J.R. Robustness of the rescaled range R/S in the measurement of noncyclic long-run statistical dependence. *Water Resour. Res.* **1969**, *5*, 967–988. [CrossRef]
38. Peng, J.; Liu, Z.; Liu, Y.; Wu, J.; Han, Y. Trend analysis of vegetation dynamics in Qinghai-Tibet Plateau using Hurst Exponent. *Ecol. Indic.* **2012**, *14*, 28–39. [CrossRef]
39. Kim, T.; Shin, J.Y.; Kim, S.; Heo, J.H. Identification of relationships between climate indices and long-term precipitation in South Korea using ensemble empirical mode decomposition. *J. Hydrol.* **2018**, *557*, 726–739. [CrossRef]
40. Wu, Z.; Huang, N.E. Ensemble empirical mode decomposition: A noise-assisted data analysis method. *Adv. Adapt. Data Anal.* **2009**, *1*, 1–41. [CrossRef]
41. Lei, Y.; He, Z.; Zi, Y. Application of the EEMD method to rotor fault diagnosis of rotating machinery. *Mech. Syst. Signal Process.* **2009**, *23*, 1327–1338. [CrossRef]
42. Li, C.; Li, Z.; Xu, H.; Huang, J.; Zhang, F.; Qian, Z. Fluctuation Characteristics of Wheat Yield and Their Relationships with Precipitation Anomalies in Anhui Province, China. *Int. J. Plant Prod.* **2022**, *16*, 483–494. [CrossRef]
43. Santos, J.F.; Pulido-Calvo, I.; Portela, M.M. Spatial and temporal variability of droughts in Portugal. *Water Resour Res.* **2010**, *46*, W03503. [CrossRef]
44. Bonaccorso, B.; Bordi, I.; Cancelliere, A.; Rossi, G.; Sutera, A. Spatial variability of drought: An analysis of the SPI in Sicily. *Water Resour. Manag.* **2003**, *17*, 273–296. [CrossRef]
45. Bordi, I.; Sutera, A. Fifty years of precipitation: Some spatially remote teleconnections. *Water Resour. Manag.* **2001**, *15*, 247–280. [CrossRef]
46. Jain, A.K. Data clustering: 50 years beyond K-means. *Pattern Recogn. Lett.* **2010**, *31*, 651–666. [CrossRef]
47. Liu, H.; Zhan, Q.; Yang, C.; Wang, J. Characterizing the Spatio-temporal pattern of land surface temperature through time series clustering: Based on the latent pattern and morphology. *Remote Sens.* **2018**, *10*, 654. [CrossRef]
48. Kazor, K.; Hering, A.S. Assessing the performance of model-based clustering methods in multivariate time series with application to identifying regional wind regimes. *Agric. Biol. Environ. Stat.* **2015**, *20*, 192–217. [CrossRef]
49. Zhe, M.; Zhang, X. Time-lag effects of NDVI responses to climate change in the Yamzhog Yumco Basin, South Tibet. *Ecol. Indic.* **2021**, *124*, 107431. [CrossRef]
50. Li, Z.; Wu, Y.; Wang, R.; Liu, B.; Qian, Z.; Li, C. Assessment of Climatic Impact on Vegetation Spring Phenology in Northern China. *Atmosphere* **2023**, *14*, 117. [CrossRef]
51. Wells, N.; Goddard, S.; Hayes, M.J. A self-calibrating Palmer drought severity index. *J. Clim.* **2004**, *17*, 2335–2351. [CrossRef]
52. Han, J.; Singh, V.P. A review of widely used drought indices and the challenges of drought assessment under climate change. *Environ. Monit. Assess.* **2023**, *195*, 1438. [CrossRef]
53. Najafi, E.; Devineni, N.; Khanbilvardi, R.M.; Kogan, F. Understanding the changes in global crop yields through changes in climate and technology. *Earth's Future* **2018**, *6*, 410–427. [CrossRef]
54. Huang, J.; Chen, X.; Zhou, L.; Xue, Y.; Lin, J. Statistical analysis of the relationship between climate-induced maize yield and rainy-season precipitation across Inner Mongolia, North China. *Theor. Appl. Climatol.* **2017**, *129*, 1145–1156. [CrossRef]
55. Fan, L.; Wang, Y.; Cao, C.; Chen, W. Teleconnections of Atmospheric Circulations to Meteorological Drought in the Lancang-Mekong River Basin. *Atmosphere* **2024**, *15*, 89. [CrossRef]
56. Yue, Y.; Shen, S.; Wang, Q. Trend and Variability in Droughts in Northeast China Based on the Reconnaissance Drought Index. *Water* **2018**, *10*, 318. [CrossRef]
57. Zhou, W.; Guan, K.; Peng, B.; Shi, J.; Jiang, C.; Wardlow, B.; Pan, M.; Kimball, J.S.; Franz, T.E.; Gentine, P.; et al. Connections between the hydrological cycle and crop yield in the rainfed US Corn Belt. *J. Hydrometeorol.* **2020**, *590*, 125398.
58. Liu, Y.; Wu, K.; Li, X.; Li, X.; Cao, H. Adaptive Management of Cultivated Land Use Zoning Based on Land Types Classification: A Case Study of Henan Province. *Land* **2022**, *11*, 346. [CrossRef]
59. Hu, Z.; Wu, Z.; Islam, A.R.M.T.; You, X.; Liu, C.; Li, Q.; Zhang, S. Spatiotemporal characteristics and risk assessment of agricultural drought disasters during the winter wheat-growing season on the Huang-Huai-Hai Plain, China. *Theor. Appl. Climatol.* **2021**, *143*, 1393–1407. [CrossRef]

60. Guan, X.; Wang, X.; Chen, W. Risk Assessment and Regulation Strategy of Farmland Marginalization: A Case Study of Mengjin County, Henan Province. *Front. Environ. Sci.* **2022**, *10*, 892665. [CrossRef]
61. Zeng, R.; Yao, F.; Zhang, S.; Yang, S.; Bai, Y.; Zhang, J.; Wang, J.; Wang, X. Assessing the effects of precipitation and irrigation on winter wheat yield and water productivity in North China Plain. *Agric. Water Manag.* **2021**, *256*, 107063. [CrossRef]
62. Wang, C.; Zhao, J.; Feng, Y.; Shang, M.; Bo, X.; Gao, Z.; Chen, F.; Chu, Q. Optimizing tillage method and irrigation schedule for greenhouse gas mitigation, yield improvement, and water conservation in wheat–maize cropping systems. *Agric. Water Manag.* **2021**, *248*, 106762. [CrossRef]
63. Mwadzingeni, L.; Shimelis, H.; Dube, E.; Laing, M.D.; Tsilo, T.J. Breeding wheat for drought tolerance: Progress and technologies. *J. Integr. Agric.* **2016**, *15*, 935–943. [CrossRef]
64. Chen, C.; Lu, W.; Sun, X.; Yu, H. Regional differences of winter wheat phenophase and grain yields response to global warming in the Huang-Huai-Hai plain in China since 1980s. *Int. J. Plant Prod.* **2018**, *12*, 33–41. [CrossRef]
65. Song, Y.; Linderholm, H.W.; Wang, C.; Tian, J.; Huo, Z.; Gao, P.; Song, Y.; Guo, A. The influence of excess precipitation on winter wheat under climate change in China from 1961 to 2017. *Sci. Total Environ.* **2019**, *690*, 189–196. [CrossRef]
66. Schultz, B.; Wrachien, D.D. Irrigation and drainage systems research and development in the 21st century. *Irrig. Drain.* **2002**, *51*, 311–327. [CrossRef]
67. Tang, R.; Wang, X.; Han, X.; Yan, Y.; Huang, S.; Huang, J.; Shen, T.; Wang, Y.; Liu, J. Effects of Combined Main Ditch and Field Ditch Control Measures on Crop Yield and Drainage Discharge in the Northern Huaihe River Plain, Anhui Province, China. *Agriculture* **2022**, *12*, 1167. [CrossRef]
68. Martinez-Vazquez, P. Crop Lodging Induced by Wind and Rain. *Agric. For. Meteorol.* **2016**, *228*, 265–275. [CrossRef]
69. Zheng, D.; Yang, X.; Mínguez, M.I.; Mu, C.; He, Q.; Wu, X. Effect of freezing temperature and duration on winter survival and grain yield of winter wheat. *Agric. For. Meteorol.* **2018**, *260*, 1–8. [CrossRef]
70. Hao, Z.; Geng, X.; Wang, F.; Zheng, J. Impacts of climate change on agrometeorological indices at winter wheat overwintering stage in Northern China during 2021–2050. *Int. J. Climatol.* **2018**, *38*, 5576–5588. [CrossRef]
71. Nouri, M.; Homaee, M.; Bannayan, M.; Hoogenboom, G. Towards shifting planting date as an adaptation practice for rainfed wheat response to climate change. *Agric. Water Manag.* **2017**, *186*, 108–109. [CrossRef]
72. Rafiullah; Khan, M.J.; Muhammad, D.; Fahad, S.; Adnan, M.; Wahid, F.; Alamri, S.; Khan, F.; Dawar, K.M.; Irshad, I.; et al. Phosphorus Nutrient Management through Synchronization of Application Methods and Rates in Wheat and Maize Crops. *Plants* **2020**, *9*, 1389. [CrossRef] [PubMed]
73. Dong, Y.; Wei, B.; Wang, L.; Zhang, Y.; Zhang, H.; Zhang, Y. Performance of Winter-Seeded Spring Wheat in Inner Mongolia. *Agronomy* **2019**, *9*, 507. [CrossRef]
74. Sigaroodi, S.K.; Chen, Q.; Ebrahimi, S.; Nazari, A.; Choobin, B. Long-term precipitation forecast for drought relief using atmospheric circulation factors: A study on the Maharloo Basin in Iran. *Hydrol. Earth Syst. Sci.* **2014**, *18*, 1995–2006. [CrossRef]
75. Tian, Y.; Xu, Y.P.; Wang, G. Agricultural drought prediction using climate indices based on Support Vector Regression in Xiangjiang River basin. *Sci. Total Environ.* **2018**, *622–623*, 710–720. [CrossRef] [PubMed]
76. Han, Y.G.; Kim, J.H.; Luo, J.J. Deep learning for multi-year ENSO forecasts. *Nature* **2019**, *573*, 568–572.
77. Schillerberg, T.A.; Tian, D.; Miao, R. Spatiotemporal patterns of maize and winter wheat yields in the United States: Predictability and impact from climate oscillations. *Agric. For. Meteorol.* **2019**, *275*, 208–222. [CrossRef]
78. Feng, P.; Wang, B.; Luo, J.J.; Liu, D.L.; Waters, C.; Ji, F.; Ruan, H.; Xiao, D.; Shi, L.; Yu, Q. Using large-scale climate drivers to forecast meteorological drought condition in growing season across the Australian wheatbelt. *Sci. Total Environ.* **2020**, *724*, 138162. [CrossRef] [PubMed]
79. Iwańska, M.; Paderewski, J.; Stępień, M.; Rodrigues, P.C. Adaptation of Winter Wheat Cultivars to Different Environments: A Case Study in Poland. *Agronomy* **2020**, *10*, 632. [CrossRef]
80. Ljubičić, N.; Popović, V.; Ćirić, V.; Kostić, M.; Ivošević, B.; Popović, D.; Pandžić, M.; El Musafah, S.; Janković, S. Multivariate Interaction Analysis of Winter Wheat Grown in Environment of Limited Soil Conditions. *Plants* **2021**, *10*, 604. [CrossRef]
81. Allen, R.J.; Sherwood, S.C. The impact of natural versus anthropogenic aerosols on atmospheric circulation in the Community Atmosphere Model. *Clim. Dynam.* **2011**, *36*, 1959–1978. [CrossRef]
82. Arshad, S.; Kazmi, J.H.; Prodhan, F.A.; Mohammed, S. Exploring dynamic response of agrometeorological droughts towards winter wheat yield loss risk using machine learning approach at a regional scale in Pakistan. *Field Crop Res.* **2023**, *302*, 109057. [CrossRef]
83. Farooq, A.; Farooq, N.; Akbar, H.; Hassan, Z.U.; Gheewala, S.H. A Critical Review of Climate Change Impact at a Global Scale on Cereal Crop Production. *Agronomy* **2023**, *13*, 162. [CrossRef]
84. Li, C.; Li, Z.Z.; Zhang, F.M.; Lu, Y.Y.; Duan, C.F.; Xu, Y. Seasonal dynamics of carbon dioxide and water fluxes in a rice-wheat rotation system in the Yangtze-Huaihe region of China. *Agric. Water Manag.* **2023**, *275*, 107992. [CrossRef]

Disclaimer/Publisher's Note: The statements, opinions and data contained in all publications are solely those of the individual author(s) and contributor(s) and not of MDPI and/or the editor(s). MDPI and/or the editor(s) disclaim responsibility for any injury to people or property resulting from any ideas, methods, instructions or products referred to in the content.

sustainability

MDPI

Article

Climate Change Impact on the Distribution of Forest Species in the Brazilian Amazon

Ingrid Lana Lima de Morais [1], Alexandra Amaro de Lima [2], Ivinne Nara Lobato dos Santos [1], Carlos Meneses [3], Rogério Freire da Silva [3], Ricardo Lopes [4], Santiago Linorio Ferreyra Ramos [5], Ananda Virginia de Aguiar [6], Marcos Silveira Wrege [6] and Maria Teresa Gomes Lopes [1,*]

[1] Faculty of Agricultural Sciences, Federal University of Amazonas, Avenue Rodrigo Otávio, 3000, Manaus 69060-000, AM, Brazil; ingridmorais24@gmail.com (I.L.L.d.M.); ivinne.lobato@gmail.com (I.N.L.d.S.)
[2] Institute of Technology and Education Galileo of Amazon, Avenue Joaquim Nabuco, 1950, Manaus 69020-030, AM, Brazil; xanduca@gmail.com
[3] Graduate Program in Agricultural Sciences, Department of Biology, Center for Biological and Health Sciences, State University of Paraíba, Campina Grande 58429-500, PB, Brazil; carlos.meneses@servidor.uepb.edu.br (C.M.); rogerio.freire.silva@aluno.uepb.edu.br (R.F.d.S.)
[4] Embrapa Western Amazon, Route AM 10, Km 29, s/n, C.P. 319, Manaus 69010-970, AM, Brazil; ricardo.lopes@embrapa.br
[5] Institute of Exact Sciences and Technology, Federal University of Amazonas, Nossa Senhora do Rosário, 3863, Itacoatiara 69103-128, AM, Brazil; slfr@ufam.edu.br
[6] Pollen Laboratory, Embrapa Florestas, Km 111, BR 476, CP. 319, Colombo 83411-000, PR, Brazil; ananda.aguiar@embrapa.br (A.V.d.A.); marcos.wrege@embrapa.br (M.S.W.)
* Correspondence: mtglopes@ufam.edu.br; Tel.: +55-92-98121-0021

Abstract: Studies using ecological niche models highlight the vulnerability of forest species to climate change. This work aimed to analyze the distribution of timber species *Aspidosperma desmanthum*, *Cariniana micranta*, *Clarisia racemosa*, *Couratari oblongifolia*, and *Vouchysia guianensis*, which are targets of deforestation, to predict the impacts of climate change and identify areas for their conservation in the Amazon. For this purpose, 37 environmental variables were used, including climatic and edaphic factors. The models were fitted using five algorithms, and their performance was evaluated by the metrics Area Under the Curve (AUC), True Skill Statistic, and Sorensen Index. The deforestation analysis was conducted using data accumulated over a period of 14 years. The study indicated that under the most pessimistic predictions, considering continued high emissions of greenhouse gases (GHGs) from the use of fossil fuels, SSP5–8.5, potential habitat loss for the studied species was more significant. Analyses of the species show that the Western Amazon has a greater climatic suitability area for the conservation of its genetic resources. Further study of the accumulated deforestation over 14 years showed a reduction in area for all species. Therefore, in situ conservation policies and deforestation reduction are recommended for the perpetuation of the analyzed forest species.

Keywords: conservation; niche modeling; impacts of climate change

1. Introduction

The Amazonian ecosystem has an important role in the conservation of biodiversity and the regulation of the global climate because it is home to the largest rainforest, which has been intensively devastated [1,2]. The maintenance and conservation of biodiversity have been the biggest challenges in the last few decades [3–5]. The loss and fragmentation of biodiversity are related to factors such as deforestation, fires, the spread of invasive and/or exotic species, and anthropic activities, especially those influencing climate change [6,7].

Deforestation in the Amazon promotes environmental degradation and consequently the reduction in species and natural resources in the region [8]. During 2022, the increase in deforestation in the Brazilian Amazon was about 12,480 km^2, presenting an increase of 2.3% compared to 2021 (12,188 km^2) [9]. The rainforest is highly heterogeneous and considered a complex system, which makes it difficult to conduct species monitoring and to obtain more information about mortality, growing and natural regeneration, and climate change impacts [10–12].

Several studies highlight the vulnerability of forest species to climate change [13], including changes in the region's rainfall regime leading to much more intense and longer dry and rainy seasons, which can cause changes in the distribution of these species within an ecosystem [14–16], causing changes in its natural habitat. The vulnerability of these areas to land use and climate change has significant implications for their conservation and management, as well as the impact of climate change on the natural ecosystem comprehension [17,18]. The tree species have genetically specific characteristics that may limit their capacity to adapt quickly to environmental changes, from the long cycle of vegetative growth until the period of reproduction and seed formation [19].

Another way of monitoring these changes is through ecological niche models (ENMs). These models allow monitoring on temporal and spatial scales to predict the distribution of species and habitats in future scenarios [20–22]. We were able to identify the presence areas of and deduce the relevant holes in economics, culture, and forest ecosystems [23].

To predict the geographic distribution of bats in Europe and Asia, a study performed in the ENMTML package used climatic and landscape relationships. The results showed that the effects of climate, land use, and cover changes made the presence of the studied species more fragmented [24]. ENM application is an important tool for evaluating the threat status of species.

The objective of this study was to analyze the potential distribution of Amazonian timber forest species, *Aspidosperma desmanthum* Benth. ex Müll. Arg., *Cariniana micranta* Ducke, *Clarisia racemosa* Ruiz & Pav., *Couratari oblongifolia* Ducke & Knuth, and *Vouchysia guianensis* Aubl.; present the effect of global climate changes in the studied area under the current period and future scenarios; and identify the areas with more climatic suitability for species conservation.

2. Materials and Methods

2.1. Study Area

In this study, occurrence and distribution analyses of the species *A. desmanthum*, *C. micranta*, *C. racemosa*, *C. oblongifolia*, and *V. guianensis* were carried out. The tree species data were obtained from the Environmental Information Reference Center (CRIA) database via the *SpeciesLink* network (https://specieslink.net/; accessed on 1 March 2023), the Global Biodiversity Information Facility database (GBIF, https://www.gbif.org/; accessed on 1 March 2023), and the Botanical Information and Ecology Network database (BIEN) version 1.2.6, available in R software version 4.2. (https://bien.nceas.ucsb.edu/bien/biendata; accessed on 1 March 2023) [25].

For modeling, this study used data from South America; however, the results were limited to the Brazilian Amazon. Furthermore, the scientific name verification of all the species was carried out through the platform Tropicos (https://www.tropicos.org/name/Search?name=; accessed on 1 March 2023). The concentration of these occurrence points was analyzed using kernel density estimation (KDE), obtained through the extension of the interpolation "Heat Map" Epanechnikov and a 1° radius (~111 km) in QGIS version 3.22.8.

2.2. Environmental Variables

Ecological niche model (ENM) development was carried out using 37 environmental variables, of which 19 corresponded to bioclimatic variables and 9 to edaphic ones (Table S1). The climate data were obtained from WorldClim version 2.1, covering a temporal interval of 30 years (1970–2000) and using 9000 to 60,000 meteorological stations [26]. Bioclimatic variables combined include minimum, mean, and maximum temperatures and precipitation, available on the *WorldClim* dataset (https://www.worldclim.org/; accessed on 1 March 2023) at a temporal resolution of 2.5 min or ~0.041° (~4 km^2 per pixel).

Climate data predictions utilized averages from the projections of the sixth Intergovernmental Panel on Climate Changes (IPCC) report of the atmospheric circulating models HadGEM-GC31-LL [27], IPSL-CM6Ä-LR [27], and MIROC6 [28] for two periods: 2021–2040 and 2041–2060. In addition, two Shared Socioeconomic Pathways (SSPs) were considered: SSP2–4.5, which predicts a temperature rise of 2.5 °C, and SSP5–8.5, with a prediction of a global temperature rise of 3 °C by 2060 [26,29].

The effect of edaphic variables was represented using two soil depths (0–20 cm, 20–40 cm) for a total of 18 variables considering soil chemical and physical properties (Table S1). The data used were from the Harmonized World Soil Database at a spatial resolution of 1 km^2 (30 s) (version 2.0; [30]) (https://www.fao.org/documents/card/en/c/cc3823en; accessed on 1 March 2023). Edaphic variables were used to improve the modeling process for ENMs since soil attributes may be prevalent as determinants in modeling plant species [31–34]. Furthermore, the spatial resolution of these variables was reduced to 2.5 min, or ~0041° (~4 km^2 per pixel), the same resolution as climatic data.

2.3. Creation and Predictor Modeling Analysis

All the simulations were carried out in the ENMTML Ecological Niche Models at the MetaLand EcologyLab package [20] from RStudio software version 4.3.2, an integrated tool of R software (version 4.2), and when utilizing the tool, it was necessary to extend the Rtools extent (version 4.2). The ENMTML package provides models with different methodologies developed in the ENM field, which are divided into preprocessing, processing, and postprocessing, and it offers different kinds of algorithms to create individual and combined ecological niche models (Table S1).

In the first stage, known as preprocessing, georeferenced occurrence points were checked, along with the environmental variables. Despite the rigorous verification of data presence, these data often showed sampling bias due to sampling concentration in easily accessible locations. Thus, to decrease spatial data autocorrelation, the extension from an occurrence site was reduced to 5 km, through the argument "thin_occ" [20] using the "spThin" [35] package. In order to avoid sampling bias, the data were partitioned into 4 folds through the K-fold, and the validation was carried out according to the total number of folds [36].

The climate variable data present a high collinearity, which is undesirable during the modeling process [37]. To reduce this collinearity, principal component analysis (PCA) was carried out, and fourteen principal components were selected, namely those with more contribution to the analysis, such as the environmental layer, which explains >95% of variation from the original data [20,34,37].

Due to the lack of absence occurrence of the studied species, the geographical and environmental combination methodology [38] was used to allocate pseudo-absence and backgrounds. To relocate the backgrounds, they were delimited in a buffer of 50 km around the presence points, and all the sites without similarity with the presence points were extracted as possible backgrounds to select pseudo-absence. These sites, without similarity, were grouped by K-mean and used to select a representative sample [39]. Additionally, in order to maintain homogeneity in the distribution of pseudo-absences, we determined that the number of pseudo-absences would be the same as the presence number [40].

In the processing stage, thirteen algorithms available in the ENMTML package were tested, and the analysis of the average metric values of AUC, TSS, and Sorensen for each algorithm was performed, considering the five studied species. Then, five best-performing algorithms were selected. Thus, the fit of the consensus model was carried out through the algorithms: Random Forest (RDF), Boosted Regression Trees (BRT), Support Vector Machine (SVM), Bayesian Gaussian Process (GAU), and Maximum Entropy Default (MXD). The consensus model was based on the PCA method, which uses the first principal component to generate the final map [41].

In the postprocessing stage, a predictive performance evaluation of the consensus model was carried out using the following metrics: Area Under the Curve (AUC) [36], True

Skill Statistic (TSS) [42], and Sorensen Index [43]. The suitability range establishment was performed by the threshold value TSS [42]. Cell values equal to 1 indicate the suitability area; on the other hand, 0 indicates the unsuitability area for the studied species. Models that presented metric values above 0.7 were considered satisfactory [41,42].

2.4. Loss Analysis of the Environmental Suitability Areas Due to Deforestation

The area loss due to accumulated deforestation for 14 years (2008–2022) was quantified as potential occurrence areas for the species *A. desmanthum, C. micrantha, C. oblongifolia, C. racemosa,* and *V. guianensis* in the Brazilian Amazon, with the intention of subtracting these areas due to the probability of the occurrence of these species decreasing in deforested areas. To do so, accumulated deforestation data were based on [9] from the Monitoring Project of Legal Amazon Deforestation by Satellite (PRODES) (terrabrasilis.dpi.inpe.br/downloads/#cat_5; accessed on 1 March 2023).

3. Results

3.1. Distribution Analysis of Species Occurrence

After the species occurrence data were manually cleaned, there were 249 occurrence points for *A. desmanthum*, 120 points for *C. micranta*, 852 points for *C. racemosa*, 110 points for *C. oblongifolia*, and 115 occurrence points for *V. guianensis* (Table 1). These points are distributed in countries that are part of the International Amazon; however, it was observed that the concentration is located in the Brazilian Amazon (Figure 1).

Table 1. Quantity of georeferenced occurrence points after the checking of the following timber wood species: *A. desmanthum, C. micranta, C. racemosa, C. oblongifolia,* and *V. guianensis*, plus the quantity of occurrence points after spatial reduction to 5 km using the SpThin package.

Species	Number of Points	
	Without Spatial Reduction	With Spatial Reduction
A. desmanthum	249	203
C. micrantha	120	58
C. racemosa	852	455
C. oblongifolia	110	71
V. guianensis	115	75

Despite the notorious reduction in occurrence points sampled after the manual cleaning of data, it was observed that some of these occurrence points were concentrated in easy locations, which is undesirable. The concentration of records may generate sampling bias, which influences the model performance, creating an underestimation or overestimation in modeling and, consequently, an increase in high spatial autocorrelation [35,44,45].

After the application of a spatial reduction of 5 km through the *spThin* packages to the species, it was observed that there was a spatial reduction among occurrence points, which decreased their autocorrelation. It was possible to identify the species *A. desmanthum, C. micrantha, C. racemosa, C. oblongifolia,* and *V. guianensis*, which presented a reduction of 18%, 52%, 47%, 35%, and 35%, respectively, in the occurrence points (Table 1). This reduction was observed near roads, research institutions, and regions that comprise the International Amazon. The spatial reduction did not exclude the sites that presented new conditions but decreased point agglomeration in the site. Likewise, a reduction can be observed in Figure 1, where the KDE is essentially based on color density; the orange color indicates high density, and the green color indicates low density of occurrence points within a 111 km radius.

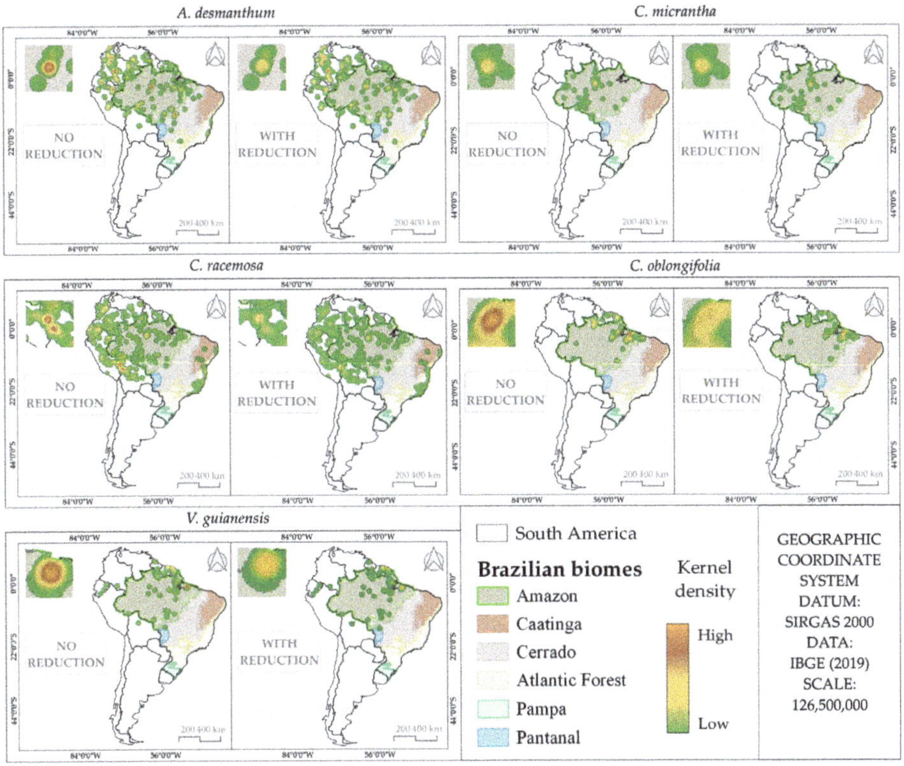

Figure 1. Geographical distribution and kernel density estimation (KDE) of *A. desmanthum*, *C. micranta*, *C. racemosa*, *C. oblongifolia*, and *V. guianensis* in South America, without a spatial reduction and with spatial reduction of 5 km. South America: Brazilian biomes: Amazon, Caatinga, Cerrado, Atlantic Forest, Pampa, Pantanal; kernel density, high to low; also shown are the geographic coordinate system, cartographic base, and scale. Points in orange indicate higher Kernel density and those in green indicate lower Kernel density, from points of occurrence within a radius of 111 km.

3.2. Environmental Variable Analysis

A principal component analysis (PCA) was carried out using the available dataset and generated 37 new environmental variables [46]. Among the principal components (PCs) generated, the first fourteen were selected, representing 97% of the entire variability data. The first and second PCs were those that best explained the variation, representing 32% and 14%, respectively, of the entire variability dataset (Figure 2).

In an analysis of the variation in the environmental data used and the eigenvectors that make up PC1, temperature seasonality was the most significant variable among the environmental variables, with the highest eigenvector (eigenvector = 0.2432). In PC2, the most significant eigenvector was the sand content for depth intervals of 0–20 m below the surface, presenting values around 0.3383. Among the most expressive eigenvector values for the fourteen PCs are those according to edaphic features, precipitation, and temperature.

Temperature seasonality refers to regular variations and predicted temperatures throughout the year, with increase or decrease periods of analysis. This variable plays an important role in the occurrence and distribution of forest species, as it influences plant growth, phenology, dormancy, and survival [47]. Concerning edaphic features, sand contents are an important element in the ability of rainwater to percolate; consequently, they affect water retention for plants as well as the availability of nutrients in the soil [48].

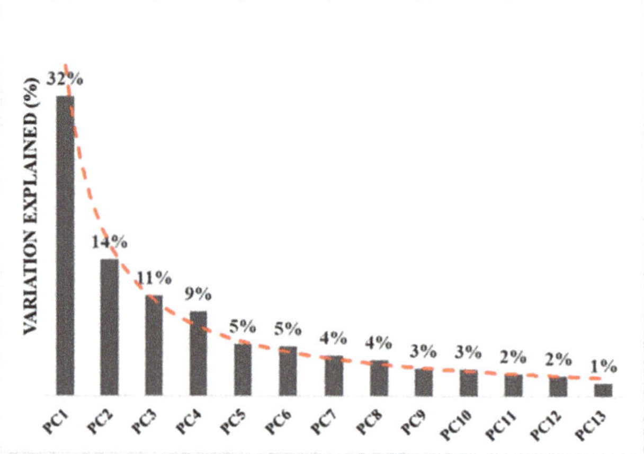

Figure 2. The variance plot, the fourteen principal components (PCs) that best represent the variation in bioclimatic and edaphic data together. The gray bars represent the PCs, while the red line represents the explained variable.

3.3. Areas of Environmental Suitability for the Reference Period

The potential areas for the occurrence of the species *A. desmanthum*, *C. micranta*, *C. racemosa*, *C. oblongifolia*, and *V. guianensis* generated using the ENMTML package are shown in Figure 3. The results showed that in base scenarios, all the species presented an environmental suitability area in the Brazilian Amazon.

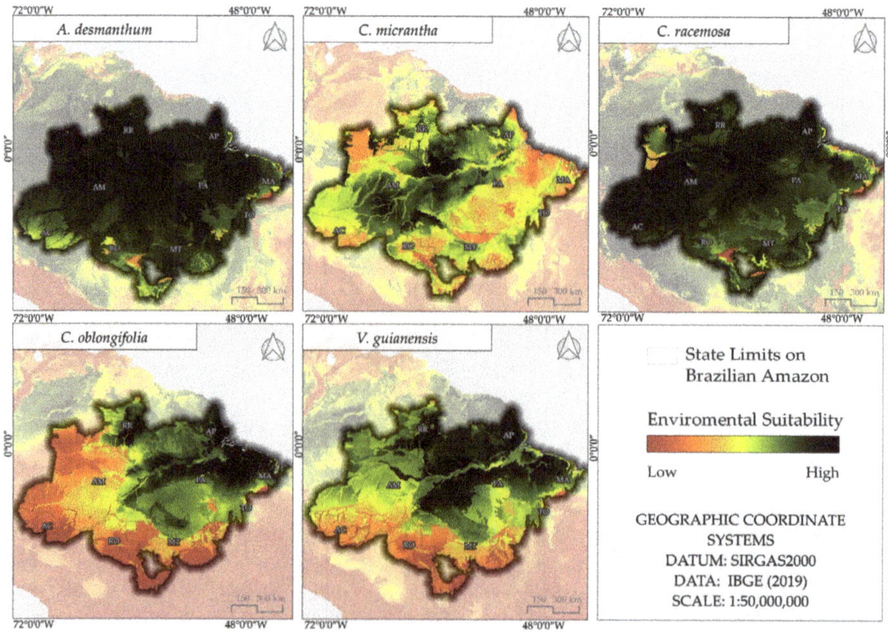

Figure 3. Environmental suitability for the base scenario (1970–2000). State boundaries in the Brazilian Amazon; environmental suitability from low to high; also shown are the geographic coordinate system, cartographic base, and scale. AC: Acre, AM: Amazonas; RO: Rondônia; PA: Pará; MT: Mato Grosso; TO: Tocantins; MA: Maranhão; AP: Amapá; and RR: Roraima.

The species *A. desmanthum* and *C. racemosa* showed the highest areas of environmental suitability, approximately 4,136,188 km^2 and 4,093,816 km^2, respectively, representing 98% and 97% of the total area of the Amazon. The species *C. micranta* presented an environmental suitability area of approximately 2,465,926 km^2, of which 30% is in the Amazonas State, while the area of the species *V. guianensis* was 2,999,466 km^2. *C. oblongifolia* was the species that showed the lowest area of environmental suitability in the Brazilian Amazon, covering approximately 2,224,906 km^2 (Table 2).

Table 2. Area with environmental suitability for the occurrence of the timber forest species *A. desmanthum, C. micranta, C. racemosa, C. oblongifolia,* and *V. guianensis* according to two climate scenarios (SSP2–4.5 and SSP5–8.5) for two periods 2021–2040 and 2041–2060 in km^2 and the reduction in these areas in relation to the base period in %.

Species	**Base Period**	**SSP2–4.5**			
		2021–2040	**%**	**2041–2060**	**%**
A. desmanthum	4,136,188	4,102,344	−1	4,067,493	−2
C. micranta	2,465,926	2,628,901	7	2,448,709	−1
C. racemosa	4,093,816	3,544,000	−13	2,889,697	−29
C. oblongifolia	2,224,906	2,484,873	12	2,472,128	11
V. guianensis	2,999,466	2,892,412	−4	2,596,902	−13
Species	Base Period	SSP5–8.5			
		2021–2040	%	2041–2060	%
A. desmanthum	4,136,188	4,087,326	−1	2,450,481	−41
C. micranta	2,465,926	2,599,843	5	2,393,141	−3
C. racemosa	4,093,816	3,454,992	−16	1,548,902	−62
C. oblongifolia	2,224,906	2,431,419	9	1,803,522	−19
V. guianensis	2,999,466	2,796,509	−7	1,886,705	−37

3.4. Area of Environmental Suitability for the Future

Table 2 shows the area of environmental suitability of species occurrence in the Brazilian Amazon for the periods 2021–2040 and 2041–2060. In relation to the base period, it is also possible to observe that for the scenarios SSP2–4.5 and SSP5–8.5, *A. desmanthum*, despite suffering losses in the estimation of future scenarios, still presented the highest total area of suitability for the studied periods when compared to the other species (*C. micranta, C. racemosa, C. oblongifolia,* and *V. guianensis*).

The estimated area of the species was 4,102,344 km^2 for the intermediate scenario, SSP2–4.5, for the period of 2021–2040; however, the species has presented a reduction of 33,844 km^2, which is equivalent to 1% of the niche area in the base period. It continued to have the largest area in comparison to the other species. For the period of 2041–2060, the reduced area was 2%. The estimates for the scenario SSP5–8.5 infer that the area losses of the species were more significant for the worst-case scenario, it was observed that there was a reduction of 48,862 km^2 in the period 2021–2040, approximately 1% of the estimated area for the base period, and 2,446,345 km^2 for the period 2041–2060, approximately 41% (Table 2; Figure 3).

The niche areas of the species *C. micrantha* and *C. oblongifolia* tend to increase in both scenarios for the period 2021–2040 in the Brazilian Amazon. The species *C. micrantha* presented an increase of 7% for the scenario SSP2–4.5 and 5% for the scenario SSP5–8.5, according to the base scenario, although, for the period 2041–2060, it was estimated that the species suffered a loss of 1% from its area under the scenario SSP2–4.5 and 3% in the scenario SSP5–8.5. These results revealed that species C. micrantha showed the lowest area variation for the studied scenarios of emission of GHGs, demonstrating that it was the most resilient among the species analyzed. *C. micranta* is known in the Amazon due to the fact that its seeds easily germinate and survive in open areas.

Although the species *C. oblongifolia* presents an increase of 12% and 9% in the scenarios SSP2–4.5 and SSP5–8.5, respectively, for the period 2021–2040, it presents a loss of 19% from the environmental suitability area for the period 2041–2060 under the scenario SSP5–8.5. This is a significant difference compared to the estimated area to the scenario SSP2–4.5, where an expansion of 11% under the area for the period 2041–2060 was foreseen (Figure 4).

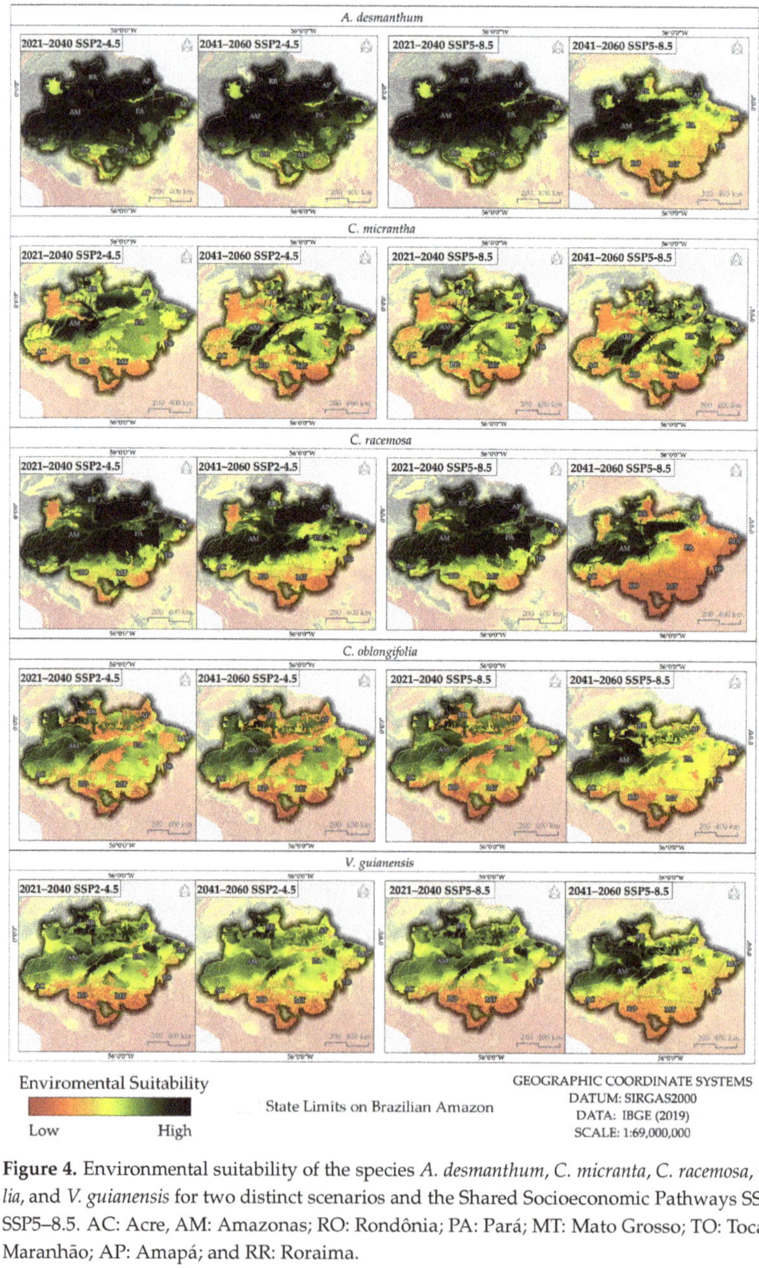

Figure 4. Environmental suitability of the species *A. desmanthum*, *C. micranta*, *C. racemosa*, *C. oblongifolia*, and *V. guianensis* for two distinct scenarios and the Shared Socioeconomic Pathways SSP2–4.5 and SSP5–8.5. AC: Acre, AM: Amazonas; RO: Rondônia; PA: Pará; MT: Mato Grosso; TO: Tocantins; MA: Maranhão; AP: Amapá; and RR: Roraima.

The estimated niche of C. racemosa in the Brazilian Amazon under the intermediate scenario (SSP2–4.5) predicts a loss of approximately 549,816 km^2 by 2040, the equivalent of 13% of the estimated area in the base period. By 2060, the area lost is expected to be around 1,204,119 km^2 (29%), more than double the area previously predicted. In the worst-case scenario, SSP5–8.5 predicts a reduction of 16% in the suitability area for the species occurrence from 2021 to 2060 and 62% between 2041 and 2060. The species C. *racemosa* was demonstrated to be more sensitive to environmental conditions under different GHG emission scenarios; the prediction indicates that the loss for the scenario SSP5–8.5 was 33% higher than the prediction for intermediate scenario for the period 2041–2060 (Figure 4).

We estimated the loss of niche area for the species *V. guianensis* for the intermediate scenario (SSP2–4.5) and worst-case scenario (SSP5–8.5) during both periods analyzed. Under the pessimistic scenario, the reduction in the niche area was more significant than the estimated areas in the intermediate scenario. *V. guianensis* species showed 7% and 37% loss for the periods 2021–2040 and 2041–2060, respectively. For the intermediate scenario, the reduction was estimated at 4% and 13% for the same period, respectively (Figure 4).

The SSP5–8.5 scenario showed that species suffered more expressive losses than SSP2–4 (Figure 4). The SSP2–4.5 scenario predicts a moderate increase in GHG emissions, even after adopting climate change mitigation policies. For the SSP5–8.5 scenario, the emissions were considered higher and less sustainable when compared to other scenarios. The world is heading towards the SSP5–8.5 scenario, which predicts an increase in temperature of 1.9 °C to 3.0 °C by 2060 [33] due to the continuous increase in CO_2 emissions caused mainly by the burning of fossil fuels, deforestation, and industrial processes. In the study of *Urena lobata* L., the results of ecological modeling showed the extinction of the species by 2070 for the most pessimistic scenario (SSP5–8.5) [49].

There is variability in emissions between the Eastern and Western Amazon regions. In the case of the Eastern region, there is a significant reduction in the occurrence area of the species *A. desmanthum*, *C. racemosa*, *C. oblongifolia*, and *V. guianensis* (Figure 4). This is an area that coincides with the "area of deforestation", where there has also been an expansion of agriculture and livestock farming. Higher rates of GHG emissions contribute to global warming and its associated effects, such as rising temperatures, changes in rainfall patterns, and extreme climate events [50].

The migration of forest species from the Eastern Amazon to other regions has been observed as a response to climate change, mainly due to increasing temperatures and changes in rainfall patterns [49]. These changes could lead to a decrease in water availability in some areas, affecting the survival of plants, animals, and insects that migrate and are part of the maintenance ecosystem. The loss of Amazon species caused by climate change and deforestation by 2050 has been reported at 30% and 47%, considering SSP1–2.6 and SSP5–8.5, respectively, and the Eastern Amazon is reported as the most affected in the analysis [51]. The most vulnerable areas to environmental losses are those located in Pará (Brazil), where there is a high level of deforestation, an increase in temperature, and a trend for precipitation reduction.

Although forest species are known as resilient, in recent years, there has been a significant increase in the occurrence of extreme climatic events in shorter periods, such as long-term droughts and extreme rainfall, making it necessary for these species to adapt quickly in order to survive. Some individuals have specific genetic characteristics that may limit their ability to adapt quickly to changes. Considering natural selection within natural plant populations, climate change may act as a selective agent, reducing their genetic diversity, gene flow, and ability to adapt to extreme events [52].

After the 2005 drought, some forests in the Amazon decreased, and this reduction was observed mainly in areas that presented the highest water deficits, as registered in the south of the Amazon [53]. When adaptive responses have not been noticed, such as phenological changes, the species may become locally extinct or migrate to other regions looking for the most favorable environment to develop. This pattern can be observed in the niche distribution of the species *A. desmanthum*, *C. racemosa*, *C. oblongifolia*, and *V. guianensis*.

The Western Amazon has relatively more preserved forest cover and less deforestation than the Eastern Amazon. GHG emissions tend to be relatively lower in this region, increasing the possibility of other species migrating in order to find an ecosystem in favor of their perpetuation. The areas of the Western Amazon are recommended for in situ conservation, through conservation units, because they are suitable habitats for the perpetuation of the species and because in situ conservation is more cost-effective than ex situ conservation.

Climate changes may cause a reduction in the species distribution of this study, plus a decrease of genetic variability of populations, and consequently may increase the inbreeding within each species. The effects of inbreeding and inadequate effective population size hinder the adaptation of forest species [16], making them more vulnerable to extreme climate changes and habitat fragmentation [54].

In addition to the studies carried out that address bioclimatic variables with the geospatial distribution of plants, experimental studies can be conducted under controlled conditions to clarify how rising temperatures can affect the optimal conditions necessary for the growth and survival of specific plant species. Such research should include monitoring, at different temperatures, for the same species, pollen production, pollination, fruiting, growth, and mortality of plant populations, as well as the evaluation of physiological responses, such as photosynthesis and transpiration rates [49].

Based on the prediction of reduction areas of several forest species in the Amazon for future scenarios, it is relevant to the development of genetic conservation to carry out studies that may simulate the extreme conditions of climate change, namely germination tests using higher temperatures, plantings in dry conditions, and mutation experiments in the lab in order to study genetic variability, aiming to find resilient mutants in response to extreme conditions [16]. Field planting of selected individuals for adverse conditions, anticipating climate change, and introducing the most resilient individuals will reduce the risk of extinction and allow continuity in the species evolution process.

Considering the economic and ecological importance of the species *A. desmanthum*, *C. micranta*, *C. racemosa*, *C. oblongifolia*, and *V. guianensis*, this study suggests the characterization of their genetic resources in order to make management and conservation strategies more efficient. Molecular marker studies, which intend to evaluate the magnitude and distribution of the genetic variability of populations in different areas over a long period, will allow us to infer how climate change will affect this distribution.

The species populations in the Eastern Amazon will be the most affected by climate change. In order not to lose genetic variability, current conservation should prioritize the conservation of the most affected population in the region, particularly those in the arc of deforestation, which are most vulnerable. The recommendation for the conservation of forest species is that it should be carried out in situ, once it comes to gene flow, which is crucial to the continuity of genetic variability [55]. Moreover, due to the fact that the Eastern Amazon presents areas of difficult climatic suitability in the future, it would be prudent to collect seeds and pollen from species in these sites of low climatic suitability in order to also carry out ex situ conservation of the genetic resources of threatened populations. This study suggests adopting cryopreservation strategies to initiate the seed conservation of these forest species.

Addressing the impacts of climate change on the distribution of forest species in the Brazilian Amazon requires an approach that involves multidisciplinary studies and actions, including robust conservation measures, sustainable land management, and global actions to mitigate climate change.

3.5. Accumulated Deforestation in Potential Areas of Species Occurrence

Analyses of the accumulated deforestation up to the year 2022, over 14 years, showed a reduction of 112,977 km^2, 56,478 km^2, 114,396 km^2, 65,961 km^2, and 68,564 km for *A. desmanthum*, *C. micranta*, *C. racemosa*, *C. oblongifolia*, and *V. guianensis*, respectively, according to the base period, from PRODES data (Figure 5). The species are considered vulnerable to exploitation and commercialization due to the good quality of the wood, and there are no records of commercial plantation areas. They are slow-growing and large species, for which reproduction is difficult; seed production is low in quantity or quality, and the matrices are in places that are difficult to access for collecting seeds [11].

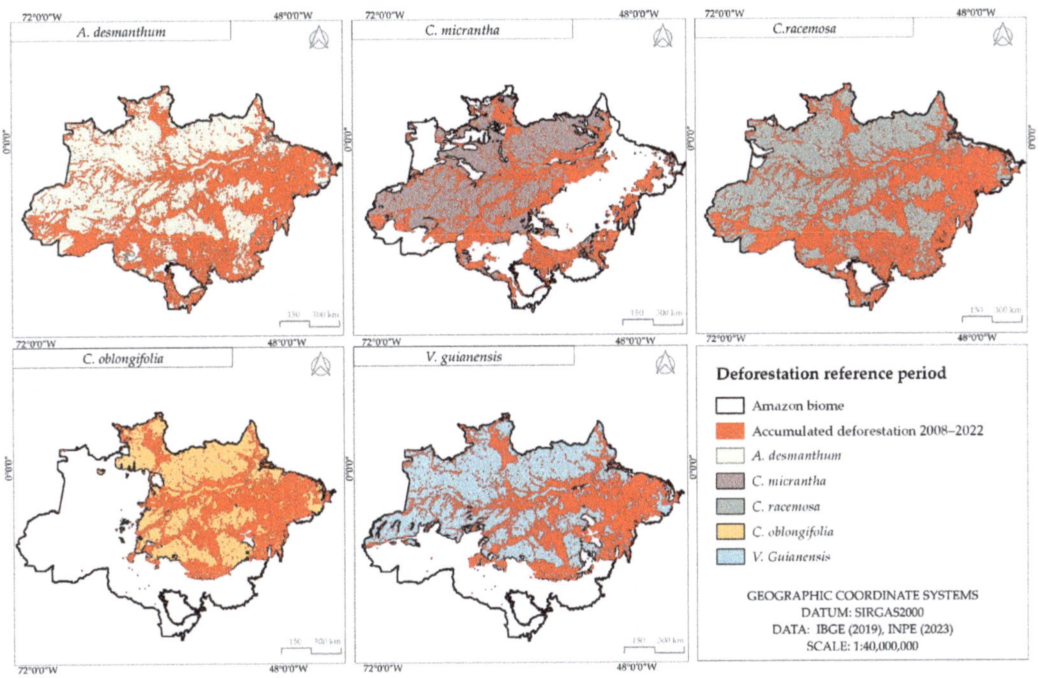

Figure 5. Map of accumulated deforestation considering the projected areas of occurrence of the species *A. desmanthum*, *C. micrantha*, *C. racemosa*, *C. oblongifolia*, and *V. guianensis* for the base period (1970–2000).

Deforestation is the principal cause of the temperature increases. When the forests are cleared, part of the carbon fixation in their biomass is released into the atmosphere, which increases their concentration and contributes to the temperature increase. Furthermore, due to climatic factors, deforestation promotes forest fragmentation, leading to populational isolation and genetic diversity reduction [52,56].

The deforestation registered in 2013 was responsible for a decline of 7% in estimated occurrence areas of Amazon tree species, and it will achieve a rise of 19–33% by 2050, considering the projected deforestation analyses [49]. It is recommended to adopt conservation strategies to reduce deforestation, recover degraded areas, promote the proper management of natural resources, and protect hotspots for biodiversity. In addition, environmental awareness and education programs are important, with the aim of involving local communities in the conservation and valuation of ecosystem services offered by the Amazon rainforest.

3.6. Model Evaluation

The values of the used metrics were AUC > 0.96, TSS > 0.84, and Sorensen > 0.92 (Table 3). The developed consensus models, combined with five algorithms, were considered as fitted metrics in their evaluation: AUC = 0.995 ± 0.001, TSS = 0.961 ± 0.016, Sorensen = 0.980 ± 0.008 for the species *A. desmanthum*; AUC = 0.965 ± 0.019, TSS = 0.847 ± 0.053, Sorensen = 0.927 ± 0.024 for the species *C. micrantha*; AUC = 0.989 ± 0.007, TSS = 0.932 ± 0.029,

Sorensen = 0.966 ± 0.014 for the species *C. racemosa*; AUC = 0.986 ± 0.009, TSS = 0.916 ± 0.055, Sorensen = 0.956 ± 0.031 for the species *C. oblongifolia*; and AUC = 0.981 ± 0.018, TSS = 0.906 ± 0.090, Sorensen = 0.951 ± 0.050 for the species *V. guianensis*.

Table 3. Results of the metrics Area Under the Curve (AUC), True Skill Statistic (TSS), and Sorensen Index generated from the use of the consensus model using the algorithms Random Forest (RDF), Boosted Regression Trees (BRT), Support Vector Machine (SVM), Bayesian Gaussian Process (GAU), and Maximum Entropy Default (MXD), based on the analysis of the algorithms available in the package ENMTML, for the species *A. desmanthum*, *C. micranta*, *C. racemosa*, *C. oblongifolia*, and *V. guianensis*.

Species	AUC	TSS	Sorensen
A. desmanthum	0.995 ± 0.001	0.961 ± 0.016	0.980 ± 0.008
C. micranta	0.965 ± 0.019	0.847 ± 0.053	0.927 ± 0.024
C. racemosa	0.989 ± 0.007	0.932 ± 0.029	0.966 ± 0.014
C. oblongifolia	0.986 ± 0.009	0.916 ± 0.055	0.956 ± 0.031
V. guianensis	0.981 ± 0.018	0.906 ± 0.090	0.951 ± 0.050

From the results obtained, it could be seen that all the models presented had favorable fit results with values over 0.7 for all the evaluation metrics [41,42], and this can be confirmed from the observed standard deviation values.

4. Conclusions

The occurrence points of the species *A. desmanthum*, *C. micranta*, *C. racemosa*, *C. oblongifolia*, and *V. guianensis* are in countries that compose the International Amazon, but their widely concentrated distribution is in the phytogeographic domain of the Brazilian Amazon.

The spatial reduction of the occurrence points of the species *A. desmanthum*, *C. micranta*, *C. racemosa*, *C. oblongifolia*, and *V. guianensis* recommended for data analysis aimed to avoid model bias by reducing the aggregation of points in easy access without exclusion of areas representative of the natural occurrence of these species, thus showing the efficiency of reducing points with overlap or very close proximity.

C. micrantha reveals itself to be more resilient to climate change, presenting less variation in area in the different scenarios of GHG emissions in comparison to the species *A. desmanthum*, *C. micranta*, *C. racemosa*, *C. oblongifolia*, and *V. guianensis*, while *C. racemosa* showed itself to be the most sensitive to the most significant niche area losses.

The habitat reduction for the species *A. desmanthum*, *C. micranta*, *C. racemosa*, *C. oblongifolia*, and *V. guianensis* occurs mainly in the Eastern Amazon, in the region named the "arc of deforestation", considering the periods of 2021–2040 and 2041–2060 based on the scenarios SSP2–4.5 and SSP5–8.5. For the scenario SSP5–8.5 (2041–2060), it has been observed that the suitability areas extend toward Eastern Amazonia, even for the resilient *C. micranta*, which has also been maintained in the Eastern Amazon area.

Analysis of the species shows that the Western Amazon has a greater climatic suitability area for the conservation of *A. desmanthum*, *C. micrantha*, *C. racemosa*, *C. oblongifolia*, and *V. guianensis*.

The losses in areas with environmental suitability for species occurrence in the SSP5–8.5 scenario, considering high rates of greenhouse gas (GHG) emissions, are more significant when compared to the losses in other scenarios.

Supplementary Materials: The following supporting information can be downloaded at https://www.mdpi.com/article/10.3390/su16083458/s1: Table S1: Coefficients of the principal components (PCs) selected by the PCAs for bioclimatic and edaphic variables together.

Author Contributions: Conceptualization, I.L.L.d.M., S.L.F.R., I.N.L.d.S., C.M., R.L. and M.T.G.L.; methodology, I.L.L.d.M., M.S.W., S.L.F.R., R.L., A.V.d.A., M.T.G.L. and A.A.d.L.; software, M.S.W., S.L.F.R. and A.A.d.L.; validation, R.L., I.N.L.d.S., M.S.W., R.F.d.S., A.V.d.A. and S.L.F.R.; formal analysis, M.S.W., S.L.F.R. and M.T.G.L.; resources, M.T.G.L., A.V.d.A., C.M., A.A.d.L. and S.L.F.R.;

writing—original draft preparation, I.L.L.d.M., R.F.d.S., S.L.F.R., R.L., M.T.G.L. and C.M.; writing—review and editing, R.F.d.S., M.T.G.L., R.L. and C.M.; funding acquisition, M.T.G.L. and C.M. All authors have read and agreed to the published version of the manuscript.

Funding: This research has the financial support of the National Council for Scientific and Technological (CNPq) process No. 442914/2020-2. This study was financed in part by Paraiba State University, PRPGP grant #01/2024, and by the Coordination of Superior Level Staff Improvement—Brazil (CAPES)—Edital No. 16/2020—PROCAD-SPCF.

Institutional Review Board Statement: Not applicable.

Informed Consent Statement: Not applicable.

Data Availability Statement: Data are contained within the article.

Acknowledgments: The authors wish to thank the Forest and Environmental Sciences Graduate Program of the Federal University of Amazonas (PPGCIFA/UFAM). Carlos Henrique Salvino Gadelha Meneses (Process No. 313075/2021-2), Santiago Linorio Ferreyra Ramos (Process No. 305280/2022-8), Ricardo Lopes (Process No. 308815/2023-8) and Maria Teresa Gomes Lopes (Process No. 310307/2018-0) were supported by fellowships from CNPq. Ingrid Lana Lima de Morais was supported by fellowships from FAPEAM—Amazonas State Research Support Foundation.

Conflicts of Interest: Author Ricardo Lopes was employed by the company Embrapa Western Amazon; Authors Ananda Virginia de Aguiar and Marcos Silveira Wrege were employed by the company Embrapa Florestas; The remaining authors declare that the research was conducted in the absence of any commercial or financial relationships that could be construed as a potential conflict of interest.

References

1. Artaxo, P.; Hansson, H.C.; Machado, L.A.T.; Rizzo, L.V. Tropical forests are crucial in regulating on Earth. *PLoS Clim.* **2022**, *1*, e0000054. [CrossRef]
2. Val, A.L.; Wood, C.M. Global changes and physiological challenges for fish of the Amazon today and in the near future. *J. Exp. Biol.* **2022**, *225*, jeb216440.
3. Qin, Y.; Xiao, X.; Liu, F.; de Sa e Silva, F.; Shimabukuro, Y.; Arai, E.; Fearnside, P.M. Forest Conservation in Indigenous Territories and Protected Areas in the Brazilian Amazon. *Nat. Sustain.* **2023**, *6*, 295–305. [CrossRef]
4. Green, J.M.H.; Croft, S.A.; Durán, A.P.; West, C.D. Linking global drivers of agricultural trade to on-the-ground impacts on biodiversity. *Proc. Natl. Acad. Sci. USA* **2019**, *116*, 23202–23208. [CrossRef]
5. Linares, L.M.D. The awkward question: What baseline should be used to measure biodiversity loss? The role of history, biology and politics in setting up an objective and fair baseline for the international biodiversity regime. *Environ. Sci. Policy* **2022**, *135*, 137–146. [CrossRef]
6. Sintayehu, D. Impact of climate change on biodiversity and associated key ecosystem services in Africa: A systematic review. *Ecosyst. Health Sustain.* **2018**, *4*, 225–239. [CrossRef]
7. Habibullah, M.S.; Din, B.H.; Tan, S.H.; Zahid, H. Impact of climate change on biodiversity loss: Global evidence. *Environ. Sci. Pollut. Res. Int.* **2022**, *29*, 1073–1086. [CrossRef] [PubMed]
8. Silva, L.B.; Oliveira, G.; Frederico, R.G.; Loyola, R.; Zacarias, D.; Ribeiro, B.R.; Mendes-Oliveira, A.C. How future climate change and deforestation can drastically affect the species of monkeys endemic to the eastern Amazon, and priorities for conservation. *Biodivers. Conserv.* **2022**, *31*, 971–988. [CrossRef]
9. Inpe—Instituto Nacional de Pesquisas Espaciais. Coordenação Geral de Observação da Terra. Programa de Monitoramento da Amazônia e Demais Biomas. Desmatamento—Amazonia. Available online: http://terrabrasilis.dpi.inpe.br/app/dashboard/deforestation/biomes/legal_amazon/rates (accessed on 30 July 2023).
10. Jardim, F.C.S. Natural regeneration in tropical forests. *Amaz. J. Agric. Environ. Sci.* **2015**, *58*, 105–113.
11. Diosinio, L.F.; Schwartz, G.; Lopes, J.C.; Oliveira, F.A. Growth, mortality, and recruitment of tree species in a Amazonian rainforest over 13 years of reduced impact logging. *For. Ecol. Manag.* **2018**, *430*, 150–156.
12. Dionisio, L.F.S. Mid-term effects of selective logging on the growth, mortality, and recruitment of *Manilkara huberi* (Ducke) A. Chev. in an Amazonian rainforest. *Sci. For.* **2020**, *48*, e3154.
13. Jump, A.S.; Ruiz-Benito, P.; Greenwood, S.; Allen, C.D.; Kitzbeger, T.; Fensham, R.; Martinez-Vilalta, J.; Lloret, F. Structural overshoot of tree growth with climate variability and the global spectrum of drought-induced forest dieback. *Glob. Chang. Biol.* **2017**, *23*, 3742–3757. [CrossRef] [PubMed]
14. Pilotto, I.L.; Rodriguez, D.A.; Chou, S.C.; Garofolo, L.; Gomes, J.L. Impacts of the land use and land-cover changes on local Hydroclimate in Southwestern Amazon. *Clim. Dynam.* **2023**, *61*, 5597–5612. [CrossRef]

15. Sierra, J.P.; Espinoza, J.C.; Junquas, C.; Wongchuing, S.; Polcher, J.; Morons, V.; Fitas, L.; Arias, P.A.; Schrapffers, A.; Pennel, R. Impacts of land-surface heterogeneities and Amazonian deforestation on the wet season onset in southern Amazon. *Clim. Dyn.* **2023**, *61*, 4867–4898. [CrossRef]
16. Tomaz, J.S.; Bezerra, C.S.; De Aguiar, A.V.; Wrege, M.S.; Lopes, M.T. Prediction of the natural distribution, habitat and conservation of *Stryphnodendron pulcherrimum* (Willd) Hochr. In response to global climate change. *Pesq. Agropec. Trop.* **2022**, *52*, e72422.
17. Metzger, M.J.; Rounsevell, M.D.A.; Acosta-Michlik, R.; Leemans, R.; Schroter, D. The vulnerability of ecosystem services to land use change. *Agric. Ecosyst. Environ.* **2006**, *114*, 69–85. [CrossRef]
18. Chishugi, D.U.; Sonwa, D.J.; Kahindo, J.M.; Itunda, D.; Chishugi, J.B.; Felix, F.L.; Sahani, M. How climate change and land use/land cover change affect domestic water vulnerability in Yangambi Watersheds. *Land* **2021**, *10*, 165. [CrossRef]
19. Bellard, C.; Bertelsmeier, C.; Leadley, P.; Thuiller, W.; Courchamp, F. Impactos of climate change on the future of biodiversity. *Ecol. Lett.* **2012**, *15*, 365–377. [CrossRef] [PubMed]
20. Andrade, A.F.; Velazco, S.J.E.; Marco Junior, P. ENMTML: An R package for a straightforward construction of complex ecological niche models. *Environ. Model. Softw.* **2020**, *125*, 104615. [CrossRef]
21. Elith, J.; Leathwick, J.R. Species Distribution Models: Ecological Explanation and Prediction Across Space and Tim. *Annu. Rev. Ecol. Evol. Syst.* **2009**, *40*, 677–697. [CrossRef]
22. Ma, Y.; You, X. A sustainable conservation strategy of wildlife in urban ecosystems: Case of *Gallinua chloropus* in Beijing-Tianjin-Hebei region. *Ecol. Inform.* **2022**, *68*, 101571. [CrossRef]
23. Aksoy, Ö.K. Predicting the Potential Distribution Area of the *Platanus orientalis* L. in Turkey Today and in the Future. *Sustainability* **2022**, *14*, 11706. [CrossRef]
24. Muylaert, R.; Kingston, T.; Luo, J.; Vancine, M.H.; Galli, K.; Carlson, C.J.; John, R.S.; Rulli, M.C.; Hayman, D.T.S. Present and future distribution of bat hosts of sarbecoviruses: Implications for conservation and public health. *Proc. R. Soc. Lond. Ser. B. Biol. Sci.* **2022**, *289*, 20220397. [CrossRef]
25. Maitner, B.S.; Boyle, B.; Casler, N.; Condit, R.; Donoghue, J.; Durán, S.M.; Guaderrama, D.; Hinchliff, C.E.; Jorgensen, P.M.; Kraft, N.J.B.; et al. The Bien R package: A tool to access the Botanical Information and Ecology Network (BIEN) database. *Methods Ecol. Evol.* **2018**, *9*, 373–379. [CrossRef]
26. Fick, S.E.; Hijmans, R.J. WordClim 2: New 1-km spatial resolution climate surfaces for global land areas. *Int. J. Climatol.* **2017**, *37*, 4302–4315. [CrossRef]
27. Firpo, M.A.F.; Guimarães, B.S.; Dantas, L.G.; Da Silva, M.G.B.; Alvez, L.M.; Chadwick, R.; Llopart, M.P.; De Oliveira, G.S. Assessment of CPIP6 models performance in simulating present day climate in Brazil. *Front. Clim.* **2022**, *4*, 1–27. [CrossRef]
28. Monteverde, C.; De Sales, F.; Jones, C. Evaluation of the CMIP6 Performance in Simulating Precipitation in the Amazon River Basin. *Climate* **2022**, *10*, 122. [CrossRef]
29. IPCC. Climate Change 2021: The Physical Science Basis. In *Contribution of Working Group I to the Sixth Assessment Report of the Intergovernmental Panel on Climate Change*; Masson-Delmotte, V.P., Zhai, A., Pirani, S.L., Connors, C., Péan, S., Berger, N., Caud, Y., Chen, L., Goldfarb, M.I., Gomis, M., et al., Eds.; Cambridge University Press: Cambridge, UK; New York, NY, USA, 2021; *In press*.
30. FAO; IIASA. Harmonized World Soil Database Version 2.0. Rome and Laxenburg. 2023, 69p. Available online: https://www.fao.org/documents/card/en/c/cc3823en (accessed on 30 March 2023).
31. Alvarez, F.; Morandi, P.S.; Marimon-Junior, B.H.; Exavier, R.; Araújo, I.; Mariano, L.H.; Muller, A.O.; Felpausch, T.R.; Marimon, B.S. Climate defined but not soil-restricted: The distribution of a Neotropical tree through space and time. *Plant Soil* **2022**, *471*, 175–191. [CrossRef]
32. Collevatti, R.; Terribile, L.C.; Lima-Ribeiro, M.S.; Nabout, J.C.; Oliveira, G.; Rangel, T.F.; Rabelo, S.G.; Diniz-Filho, J.F. A coupled phylogeographical and species distribution modelling approach recovers the demographical history of Neotropical seasonally dry forest tree species. *Mol. Ecol.* **2012**, *21*, 5845–5863. [CrossRef]
33. Oliveira, G.; Rangel, T.F.; Lima-Ribeiro, M.S.; Terribile, L.C.; Diniz-Filho, A.F. Evaluating, partitioning, and mapping the spatial autocorrelation component in ecological niche modeling: A new approach based on environmentally equidistant records. *Ecography* **2014**, *37*, 637–647. [CrossRef]
34. Velazco, S.J.E.; Galvão, F.; Villalobos, F.; Marco-Junior, P. Using worldwide edaphic data to model plant species niches: Na assessment at a continental extent. *PLoS ONE* **2017**, *12*, e0186025. [CrossRef] [PubMed]
35. Aiello-Lammens, M.; Boria, R.A.; Radosavljevic, A.; Vilela, B.; Anderson, R. SpThin: Na the R package for spatial thinning of species occurrence records for use in ecological niche models. *Ecography* **2015**, *38*, 541–545. [CrossRef]
36. Fielding, A.H.; Bell, J.F. A review of methods for the assessment of prediction errors in conservation presence/absence model. *Environ. Conserv.* **1997**, *24*, 38–49. [CrossRef]
37. De Marco Junior, P.; Villen, S.; Mendes, P.; Nobrega, C.; Cortes, L.; Castro, T.; Souza, R. Vulnerability of Cerrado threatened mammals: Na the integrative landscape and climate modeling approach. *Biodivers. Conserv.* **2020**, *29*, 1637–1658. [CrossRef]
38. Lobo, J.; Jiménez-Valverde, A.; Hortal, J. The uncertain nature of absences and their importance in species distribution modelling. *Ecography* **2010**, *33*, 103–114. [CrossRef]
39. Senay, S.D.; Worner, S.P.; Ikeda, T. Novel Three-Step Pseudo-Adsence Selection Technique for Improved Species Distribution Modelling. *PLoS ONE* **2013**, *8*, e71218. [CrossRef] [PubMed]
40. Zaniewski, A.E.; Lehmann, A.; Overton, J.M. Predicting species spatial distributions using presence-only data: A case study of native New Zealand ferns. *Ecol. Modell.* **2002**, *157*, 261–280. [CrossRef]

41. Allouche, O.; Tsoar, A.; Kadmon, R. Assuming the accuracy of species distribution models: Prevalence, kappa and the true skill statistic (TSS). *J. Appl. Ecol.* **2006**, *43*, 1223–1232. [CrossRef]
42. Thuiller, W.; Guéguen, M.; Renaud, J.; Karge, D.N.; Zimmermann, N.E. Uncertainty in ensembles of global biodiversity scenarios. *Nat. Commun.* **2019**, *10*, 1446. [CrossRef]
43. Leroy, B.; Delsol, R.; Hugueny, B.; Meynard, C.N.; Barhoumi, C.; Massin, M.B.; Bellard, C. Without quality presence-absence data, discrimination metrics such as TSS can be misleading measures of model performance. *J. Biogeogr.* **2018**, *45*, 1994–2002.
44. Boria, R.; Olson, L.; Goodman, S.M.; Anderson, R.P. Spatial filtering to reduce sampling bias can improve the performance of ecological niche models. *Ecol. Modell.* **2014**, *275*, 73–77. [CrossRef]
45. Sillero, N.; Barbosa, A.M. Common mistakes in ecological niche models. *Int. J. Geogr. Inf. Sci.* **2020**, *35*, 213–226. [CrossRef]
46. Abdi, H.; Williams, L.J. Principal component analysis. *Wiley Interdiscip. Rev.* **2010**, *2*, 433–459. [CrossRef]
47. Mcmeans, B.C.; Mccann, K.S.; Guzzo, M.M.; Vartley, T.J.; Bieg, C.; Blanchfield, P.J.; Fernandes, T.; Giacomini, H.C.; Middel, T.; Rennie, M.D.; et al. Winter in water: Differential responses and the maintenance of biodiversity. *Ecol. Lett.* **2020**, *23*, 922–938. [CrossRef] [PubMed]
48. Bordoloi, R.; Das, B.; Yam, G.; Pandey, P.K.; Tripathi, O.P. Modeling of water holding capacity using readily available soil characteristics. *Agric. Res.* **2019**, *8*, 347–355. [CrossRef]
49. Gomes, L.M.; Bezerra, C.S.; Aguiar, A.V.; Wrege, M.S.; Lopes, M.T.G. Prediction of the natural distribution and conservation of *Urena lobata* L. in Brazil. *Pesq. Agropec. Trop.* **2022**, *52*, e72594. [CrossRef]
50. Brandão, D.O.; Barata, L.E.S.; Nobre, C.A. The effects of environmental changes on plant species and forest dependent communities in the Amazon region. *Forests* **2022**, *13*, 466. [CrossRef]
51. Gomes, V.H.F.; Vieira, I.C.G.; Salomão, R.P. Amazonian tree species threatened by deforestation an climate change. *Nat. Clim. Change* **2019**, *9*, 547–553. [CrossRef]
52. Naranjo, A.; Melton, A.E.; Soltis, D.E.; Soltis, P.S. Endemism, projected climate change, and identifying species of critical concern in the Scrub Mint clade (Lamiaceae). *Conserv. Sci. Pract.* **2022**, *4*, e621. [CrossRef]
53. Yang, Y.; Saatchi, S.S.; Xu, L.; Choi, S.; Philips, N.; Kennedy, R.; Keller, M.; Knyazikhin, Y.; Myneni, R.B. Post-drought decline of the Amazon carbon sink. *Nat. Commun.* **2018**, *9*, 3172–3181. [CrossRef]
54. Kramer, A.T.; Ison, J.L.; Ashley, M.V.; Howe, H.F. The genetic paradox of forest fragmentation. *Conserv. Biol.* **2008**, *22*, 878–885. [CrossRef] [PubMed]
55. Van Dyck, H.; Baguette, M. Dispersal behaviour in fragmented landscape: Routine or special movements? *Basic Appl. Ecol.* **2005**, *5*, 535–545. [CrossRef]
56. Jump, A.S.; Peñuelas, J. Running to stand still: Adaptation and the response of plants to rapid climate change. *Ecol. Lett.* **2005**, *8*, 1010–1020. [CrossRef] [PubMed]

Disclaimer/Publisher's Note: The statements, opinions and data contained in all publications are solely those of the individual author(s) and contributor(s) and not of MDPI and/or the editor(s). MDPI and/or the editor(s) disclaim responsibility for any injury to people or property resulting from any ideas, methods, instructions or products referred to in the content.

forests

MDPI

Article

Climate Change-Induced Growth Decline in Planted Forests of *Quercus variabilis* Blume near Beijing, China

Ayjamal Keram [1,2], Puyuan Liu [1], Guolei Li [1,*], Wen Liu [3] and Ümüt Halik [2,*]

1 Key Laboratory for Silviculture and Conservation, Ministry of Education, College of Forestry, Beijing Forestry University, Beijing 100083, China; ayjamal0995@bjfu.edu.cn (A.K.); puyuan622531@bjfu.edu.cn (P.L.)
2 Ministry of Education Key Laboratory of Oasis Ecology, College of Ecology and Environment, Xinjiang University, Ürümqi 830046, China
3 Beijing Xishan Experimental Forest Farm, Beijing 100041, China; liuwen10397146@gmail.com
* Correspondence: glli226@163.com (G.L.); halik@xju.edu.cn (Ü.H.)

Abstract: A progressive decline in tree growth may be induced by global warming, affecting tree health and eventually resulting in death, particularly for trees growing in rocky mountainous regions where seasonal droughts have become more pronounced. However, tree growth dynamics in areas experiencing pronounced climate change have received little attention. In this study, a total of 100 (10 m × 10 m) grid plots were investigated in planted forests of Chinese cork oak (*Quercus variabilis* Blume), which were established in the 1960s and 1970s in a rocky mountainous area near Beijing, northern China. Furthermore, the radial growth of *Q. variabilis* (a total of 843 trees sampled from the dominant [D], co-dominant [CD], and suppressed [S] crown classes) was analyzed using retrospective dendrochronology and generalized additive models. The effects of meteorological changes between 1962 and 2020 on radial growth across the three crown classes were examined using correlation analysis. The results indicated that the growth of *Q. variabilis* initially increased and then decreased after 2010 at the regional level, and these trends varied according to crown class. The radial growth of the D trees responded more positively to an increase in temperature and drought severity index (<0 for dry and >0 for wet conditions) compared with that of the CD and S trees. The growth of the D and CD trees continuously increased under higher temperatures during the rainy seasons (June to September); however, the increases were higher for D than for CD trees. In contrast, the radial growth of S trees declined. We confirmed the historical effects of rising temperatures on tree growth and health, which are linked to water availability. Our data suggested that *Q. variabilis* trees will be considerably affected by intensified droughts. This study furthers our knowledge regarding the impact of climate change on tree and forest growth and provides management strategies for afforestation projects in rocky mountainous areas that are facing climate change.

Keywords: climatic variables; radial growth; deciduous tree; forest plantation; rocky mountainous area

1. Introduction

Understanding the responses of forest ecosystems to global warming is an emerging and serious concern. This is particularly true for forest ecosystems in rocky mountainous areas, where seasonal droughts may be more evident. In addition, global temperatures have increased over the last 100 years, particularly since the 1980s. Planted forests are more abundant than other forest types in China [1]. As essential constituents of regional and global climate systems, planted forests contribute to carbon balance and the water cycle [2,3] and provide a wide range of ecosystem services to local populations. Most planted forests are widely distributed in mountainous areas because of their limited geographical conditions. However, large-scale forests planted in rocky mountainous areas often exhibit low growth rates, diminished tree stability and resilience, and poor quality of the dominant trees [4,5] because of frequent seasonal droughts. Therefore, foresters, ecologists, and land managers have increasingly focused on strategies to improve the quality of planted forests in rocky mountainous regions because of climate change.

Tree growth and dynamics are mainly governed by meteorological factors (temperature, precipitation, and drought severity). Recent studies have focused primarily on the impact of climate variables on physiological responses versus the characteristics of tree species [6], soil properties [7–9], and community structures (i.e., density, composition, and species competition) [10,11]. For example, Luo et al. (2017) noted that rising air temperatures can increase the decomposition rate of soil organic matter and stimulate the absorption of essential nutrients such as C, N, P, and K [12]. Moreover, warmer air temperatures versus rising atmospheric CO_2 concentrations increase plant photosynthesis and water-holding capacity, further increasing the accumulation of above- and below-ground biomass [13]. Meanwhile, these impacts may alter species composition and distribution. For instance, rising temperatures promote population growth at high altitudes; in contrast, forest ecosystems at lower elevations exhibit a decline in growth because of frequent seasonal droughts, resulting in range contraction [14–16]. A rocky mountainous area at low altitude is often characterized by thin soils and a high gravel content with strong seasonality [17,18], which results in insufficient water availability for tree growth during the growing season. The frequency of seasonal droughts in rocky mountainous areas may be one of the main factors affecting the growth and development of planted trees under climate warming. However, there remains a notable gap in the research on the adaptation of pure-planted forests to climate change in seasonal drought areas. Thus, to explore the climatic driving forces of the decline in planted forests, systematic research must be performed to evaluate the effects of meteorological factors on the radial growth of trees in this region.

The Chinese cork oak (*Quercus variabilis* Blume), a fast-growing deciduous species, is widely distributed throughout Eastern Asia (24°–42° N, 97°–140° E) and commonly grows on sunny slopes at low altitudes (<800 m a.s.l.). Considering its adaptability, rapid growth, and drought tolerance, *Q. variabilis* is commonly used for afforestation in mountainous areas of China, benefiting the ecology, economy, and culture of local populations. However, the distribution range of planted forests of *Q. variabilis* has diminished, and their growth may have declined under recent progressive seasonal droughts [19–21]; nevertheless, the radial growth of *Q. variabilis* in this region, which is sensitive to meteorological factors, remains unclear. Therefore, understanding the impacts of climate on the growth and quality of *Q. variabilis* plantations is essential for designing appropriate forest management strategies for rocky mountainous areas.

To date, existing studies on the response of oak trees to climate change have been conducted on healthy individuals. We hypothesized that trees with various signs of vigor may exhibit distinct sensitivity to climate variables. This study aimed to determine whether climate change considerably affects tree growth in planted forests and which meteorological factors are considered the main driving forces of tree decline in seasonal drought areas. Thus, we investigated the radial growth of *Q. variabilis* in three vigor classes (dominant [D], co-dominant [CD], and suppressed [S] trees) and explored how the classes were related to climate change using correlation coefficient analysis. The main objectives of this study were to (i) analyze the temporal radial growth of *Q. variabilis* in the three crown classes; (ii) determine whether the growth of planted forests of *Q. variabilis* is affected by changing climatic conditions, including rising temperatures and seasonal droughts; and (iii) propose implications for improving the negative effects of global warming on planted forests in rocky mountainous areas. This study will help us identify the mechanisms by which meteorological factors affect the growth of the *Q. variabilis* plantations in seasonal drought areas.

2. Materials and Methods

2.1. Site Description

The research was conducted in the Xishan National Forest Park (39°58′18.17″ N, 116°11′51.20″ E), which is situated in a mountainous area in the north of China (Figure 1).

The study site is located 30 km away from Beijing at an average elevation of 450 m a.s.l., and it experiences a hot, rainy summer and a cold, dry winter. The climate of the area is warm temperate with continental monsoons and strong seasonality. Approximately 90% of the 465 mm precipitation occurs during the growing season, with the majority concentrated in July and August. In this region, the growing season generally continues for approximately 240 days, from April to the end of November, whereas the non-growing season lasts for 180 days. The site is relatively flat, and the soil is shallow (with a depth of 0.3–0.8 m), rocky, and rich in clay. The soil has a high soil humus content in the upper layers and high gravel density in the lower layers. As a typical lithoid mountainous area, it has pure artificial forests of *Q. variabilis*, which were cultivated via seedling planting in the 1960s and 1970s. In the study area, *Grewia biloba*, *Rhamnus davurica*, *Vitex negundo* var *heterophylla*, and other shrubs are below the canopy, with a forest cover of approximately 90%. The radial growth of this species usually peaks in the summer (June–September).

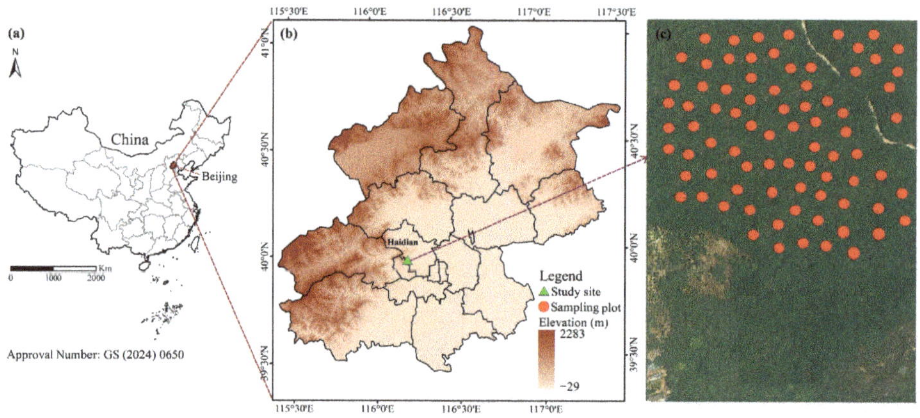

Figure 1. Study site (**a**) of *Quercus variabilis* plantations near Beijing, northern China. Note: Sampling plots (**c**) were located at the study site (**b**) (red symbol).

2.2. Climate Data

Meteorological data of the research site covering the period 1962–2020, including annual maximum temperature (T_{max}), mean temperature (T_{mean}), minimum temperature (T_{min}), precipitation (PPT), Standardized Precipitation Evapotranspiration Index in 12 months (SPEI 12 MONTHS), and Palmer Drought Severity Index in 12 months (PDSI 12 MONTHS), were derived from the World Meteorological Organization (http://climexp.knmi.nl/), URL (accessed on 5 October 2022). SPEI 12 MONTHS and PDSI 12 MONTHS are good indicators of drought severity that consider the effects of other climatic factors, such as temperature and wind [22,23]. Therefore, long-term SPEI 12 MONTHS and PDSI 12 MONTHS are used as alternatives to water availability in the study area. Positive PDSI 12 MONTHS and SPEI 12 MONTHS values are often regarded to be indicative of wet climates, whereas negative values indicate dry climates (<0 for dry and >0 for wet conditions).

According to climate data from 1962 to 2020, the temperatures (T_{max}, T_{mean}, and T_{min}) showed a consistent increase trend with annual variation, which was particularly evident after 1980. Conversely, SPEI 12 MONTHS and PDSI 12 MONTHS decreased considerably over time, and a weak downtrend in PPT was observed ($p < 0.001$). Overall, climate warming and drought severity drastically increased with time at the study site (Figure 2). The annual evapotranspiration is approximately 1100 mm. The area has a mean annual air temperature of 11.6 °C with 2662 h of annual sunshine and a mean monthly air temperature of −3.4 °C in December and 26.7 °C in July. It often experiences long-term droughts because of the uneven rainfall distribution, and <15 mm PPT was recorded in the non-growing season (December of the previous year to March of the current year), with lower SPEI 12 MONTHS and PDSI 12 MONTHS values (Figure 3).

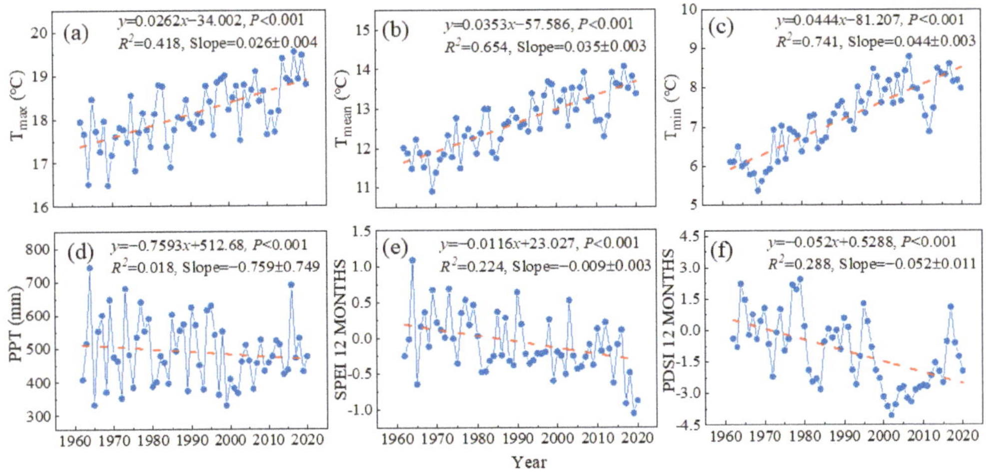

Figure 2. Annual variations in meteorological factors at the study site during 1962–2020 period. T_{max}—maximum temperature (**a**); T_{mean}—mean temperature (**b**); T_{min}—minimum temperature (**c**); PPT—precipitation (**d**); SPEI 12 MONTHS—Standardized Precipitation Evapotranspiration Index in 12 months (**e**); and PDSI 12 MONTHS—Palmer Drought Severity Index in 12 months (**f**).

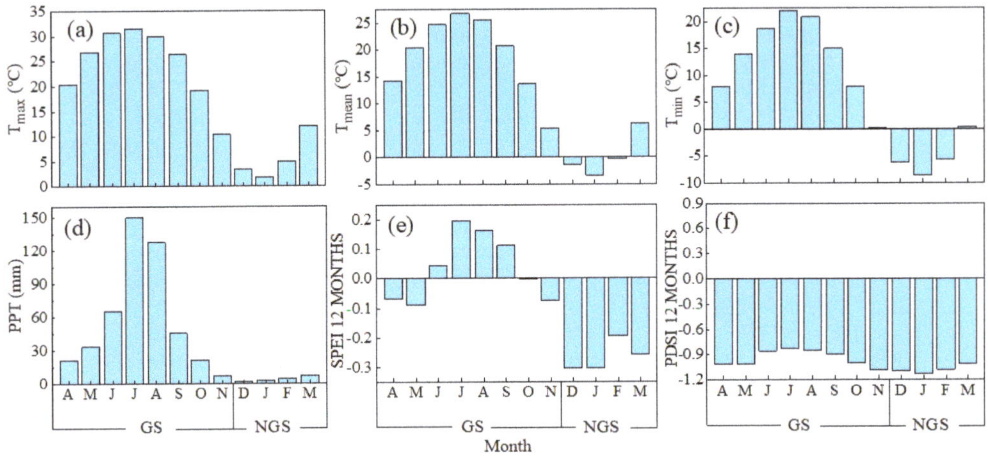

Figure 3. Monthly variations in meteorological factors at the study site during 1962–2020. A—April; M—May; J—June; J—July; A—August; S—September; O—October; N—November; D—December; J—January; F—February; M—March; GS—growing season; NGS—non-growing season; T_{max}—maximum temperature (**a**); T_{mean}—mean temperature (**b**); T_{min}—minimum temperature (**c**); PPT—precipitation (**d**); SPEI 12 MONTHS—Standardized Precipitation Evapotranspiration Index in 12 months (**e**); and PDSI 12 MONTHS—Palmer Drought Severity Index in 12 months (**f**).

We calculated climate anomalies to evaluate the meteorological factors that drive temporal changes in tree growth. These anomalies were identified as a departure of the mean between two sequential measurements from the long-term climate mean [24]. Six sets of climate anomalies were calculated: annual T_{max}, T_{mean}, T_{min}, PPT, SPEI 12 MONTHS, and PDSI 12 MONTHS.

2.3. Forest Inventory Data

Fieldwork was conducted from July to September 2021. At the study site, a total of 100 sampling plots (10 m × 10 m) were established, with a minimum distance of 40 m between each plot (Figure 1). The criteria for plot selection were as follows: (a) the plots were situated at least 100 m away from openings to avoid edge effects; (b) the stand conditions were relatively homogeneous in structure and composition with no anthropogenic disturbances (i.e., wood cutting, thinning, pruning, camping, or grazing); and (c) all plots were located in areas with the same topography, aspect, and slope.

At each plot, trees with a diameter at breast height (DBH) ≥ 4 cm were identified, measured, and recorded, including species, size (DBH), location (x–y coordinates), height, and crown class. The crown classes were categorized into three groups based on tree vitality (D, CD, and S trees). In this study, D trees had crowns located above the main forest layers, CD trees had crowns equal to those in the main forest layers, and S trees had crowns located below the main forest layers [25,26]. At each plot, 3–4 trees were cut and felled to collect tree ring disks. In total, 864 trees were sampled for dendroecological analyses. Each sampled tree was stem-sectioned at 0 m (stem base), at a height of 1.3 m, and at intervals of 2 m above the DBH along the remaining length of the stem. In this study, disks sectioned at the stem base were used. The stem disks were air-dried, progressively sanded, and polished using fine sandpaper until the tree rings were clearly visible, after which they were scanned at high resolution (1200 dpi) [27,28]. Furthermore, the tree ring widths (0.001 mm), consisting of the earlywood (EW) and latewood (LW) widths of each stem disk, were gauged along two (south and north) or four (south, north, east, and west) lines from the core to the pith using the software WinDENDRO STD 4800 (Regent Instruments Inc., Québec City, QC, Canada, 2009). We identified the interannual ring as the boundary between light-colored earlywood and dark-colored latewood. All the ring width series were calibrated via cross-dating using the program COFECHA v6.06P [29]. Samples of poor quality (low COFECHA correlation coefficients) were removed from the subsequent analyses. Finally, we obtained an extensive dataset comprising 100 plots with 843 *Q. variabilis* trees categorized into the D (212 trees), CD (315 trees), and S (316 trees) classes, providing a rich source of data for examining the interaction between tree radial growth and climate variables. We acknowledge the significant impact of the data collection on the environment and regret the necessity of cutting down trees to obtain the ring width data. Specific sample information is presented in Table 1. The earlywood width (EW) and latewood width (LW) of the corresponding years in two or four directions of each stem disk were averaged, and the radial growth of each tree was expressed as the basal area increment (BAI, annual increment in tree ring area). The age of the trees in the core sample ranged from 36 to 58 years.

Table 1. Statistical information for dominant (D), co-dominant (CD), and suppressed (S) trees in the sampled *Quercus variabilis*. No.—number; Ave.—average; DBH—diameter at breast height; TH—tree height; CD—crown diameter; BAI—basal area increment; EW—earlywood width; LW—latewood width.

Crown Classes	Period	No. of Trees (Disks)	Ave. Age (a)	Ave. DBH (cm)	Ave. TH (m)	Ave. CD (m)	Ave. BAI (cm^2/year)	Ave. EW (cm/year)	Ave. LW (cm/year)
D	1962–2020	213/212	47	20.2	15.8	5.23	17.05	0.361	1.230
CD	1970–2020	320/315	45	17.9	13.7	3.52	15.71	0.355	1.156
S	1979–2020	331/316	38	14.7	10.3	2.23	12.15	0.341	1.116

2.4. Data Processing

2.4.1. Hegyi Competition Index

The Hegyi competition index [30] was modified to capture the impact of intraspecific competition on tree growth.

$$H_{ijk} = \left[\exp\left(\frac{maxDBH_{jk} - DBH_{ijk}}{maxDBH_{jk} - minDBH_{jk}}\right)\right]^{\alpha} \times \sum_{n \neq i} \frac{DBH_{njk}}{DBH_{ijk} \times [(Dis\tan ce)_{in}]_{jk}{}^{\beta}} \tag{1}$$

where i, n, j, and k are the ith subject tree, nth neighboring tree, jth census, and kth plot, respectively. Here, N is the number of trees per plot. $maxDBH_{jk}$ and $minDBH_{jk}$ are the maximum and minimum DBH at the jth census in plot k, respectively. DBH_{ijk} and DBH_{njk} are the DBH of the focal and neighboring trees, respectively. $(Distance_{in})_{jk}$ is the distance between trees i and n at the jth census in the kth plot. α and β are the cross-stand asymmetric coefficient and the crowdedness coefficient, respectively.

2.4.2. Tree Growth

The radiating ring was derived from the mean ring widths measured in four distinct directions on each disk sample, and the annual BAI was computed based on the following model:

$$BAI = \pi \times \left(R_n^2 - R_{n-1}^2\right) \tag{2}$$

where R_n and R_{n-1} denote the tree radius at breast height in the current year (n) and the previous year (n − 1) of ring formation, respectively [31].

To detect temporal trends in BAI, EW, and LW, we developed Year models using generalized additive models (GAMs) based on calendar year, ontogeny, competition, and their two-way interactions on tree growth. Because tree age (A), tree size (DBH), and Hegyi competition (H) strongly affected tree growth, they were considered to account for the effects of the exogenous and endogenous processes associated with forest development on the growth of various crown classes (i.e., D, CD, and S). To examine the long-term temporal trends in the tree growth of the various crown classes, we fitted the GAMs to include the main effects of Year, A, DBH, and H and their two-way interactions (Equation (3)). The first model was developed for all sampled trees and included three hierarchical layers in the dataset, i.e., observations nested in trees, trees nested in plots, and plots nested in class groups. Here, the temporal trends of the annual BAI, EW, and LW in the three crown classes were compared to assess the classes promoting the decline in forest growth. The trends in BAI, EW, and LW were modeled as smoothing functions of the temporal variations, expressed as years. A, DBH, and H were included as random effects in the model to account for intra-annual variations. In general, the GAMs were used to assess the relationship of the calendar year with BAI, EW, and LW in different vigor classes (i.e., D, CD, and S), which accounted for A, DBH, and H, to remove the effects of endogenous and exogenous processes on tree growth. The final model is expressed as follows:

$$\begin{aligned}(BAI, EW, LW)_{ijkg} = {} & \beta_0 + \beta_1 Year_{ijkg} + \delta_1 f(A) + \delta_2 f(DBH) + \delta_3 f(H) \\ & + \delta_4 Year_{ijkg} \times f(A) + \delta_5 Year_{ijkg} \times f(DBH) \\ & + \delta_6 Year_{ijkg} \times f(H) + \delta_7 f(A) \times f(DBH) \\ & + \delta_8 f(A) \times f(H) + \delta_9 f(DBH) \times f(H) + \varepsilon_i\end{aligned} \tag{3}$$

where i, j, and k are the ith census period observation for jth tree in the kth plot of the gth class group; β_0 is the smoothing function; β_1 is the intercept of the model; δ_n is the fixed coefficient to be estimated; and *f* is a spline function fitted to each individual time series (average BAI, EW, and LW of all D, CD, and S trees). In this study, the two-way interactions of A, DBH, and H with tree growth were modeled simultaneously. To examine the temporal growth trends, analyses were extended to the D, CD, and S trees. Based on these models, the class group was not specified in Equation (3).

We also fitted nonlinear mixed-effect models to explore the temporal trends in radial growth (BAI in cm^2/year and EW and LW in cm/year) and comparatively examine the growth patterns of *Q. variabilis* in the three vigor classes. For climate–growth interactions, Spearman's correlation coefficients were calculated between the climate anomalies and radial growth (BAI, EW, and LW) to comparatively analyze the responses of the D, CD, and S trees to climate drivers (statistical significance was set at $p < 0.05$, and a higher significance was set at $p < 0.01$). The data analysis was performed in R 4.2.2 (R Core Team, 2022) using the packages "data table" and "dplyr", and the GAMs were fitted using the "mgcv" package. The location map of the sampling plots was generated using ArcGIS 10.9 (Esri, Redlands, CA, USA, 2021), and the other figures were produced using OriginPro v2022 9.9.0.225 (OriginLab Corp., Northampton, MA, USA, 2022).

3. Results

3.1. Temporal Changes in Tree Growth

In this study, the temporal changes in BAI, EW, and LW were analyzed using GAMs (Figure 4). Over 60 years, the BAI of *Q. variabilis* forest increased drastically but stagnated after 2010; however, these trends varied among the three crown classes (Figure 4a). The BAI of the D trees showed a continuously increasing trend during 1962–2020, whereas that of the CD trees started to decrease after 2010. In contrast, the BAI of the S trees has been constrained since the early 2000s, and their growth status is consistently lower than that of the other two classes.

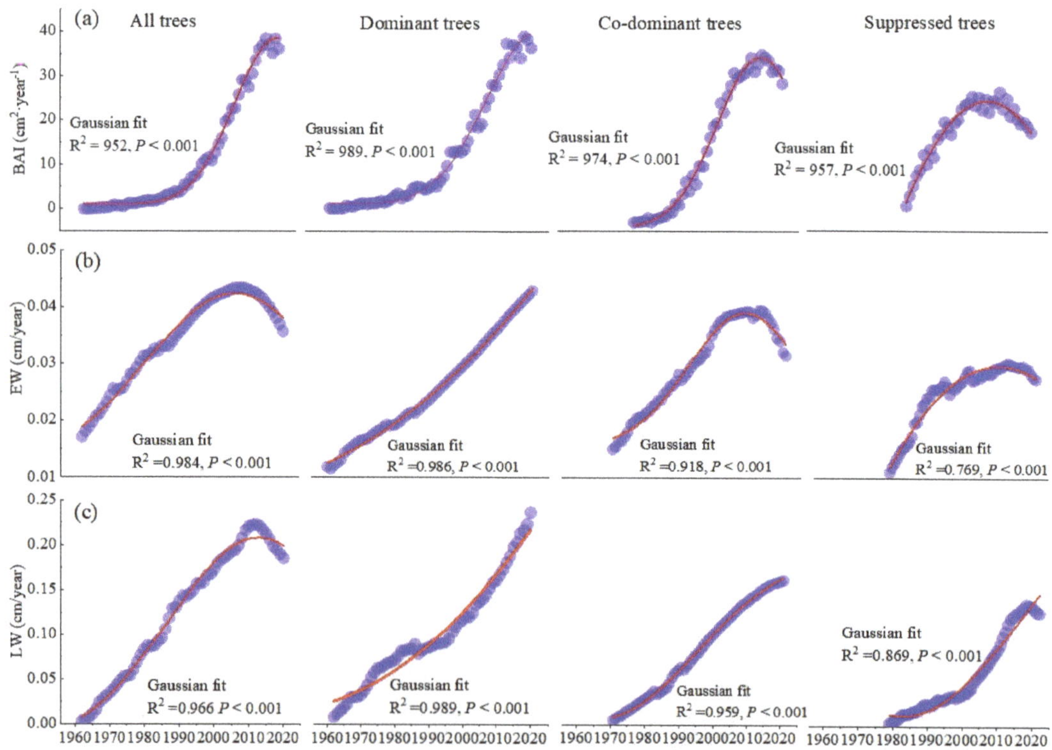

Figure 4. Predicted temporal trends in (**a**) basal area increment (BAI), (**b**) earlywood width (EW), and (**c**) latewood width (LW). Blue dots represent the temporal trends in BAI, EW, and LW during 1962–2020; the red curves represent the fitted linear line and 95% confidence intervals summarized from GAMs (detailed in the Materials and Methods section).

Q. variabilis growing in this area demonstrated a narrow EW and a wide LW because it was affected by seasonal climatic factors. To quantify the annual evolution of radial growth in the three crown classes, the EW and LW were plotted against time (Figure 4b,c). For 60 years, the EW and LW of all the trees initially increased and then decreased after 2010; however, these trends varied among the three crown classes (Figure 4). Similar to BAI, the EW in the D trees continuously increased from 1962 to 2020, whereas that in the CD trees decreased after 2010 and decreased in the S trees after the early 1990s (Figure 4b). However, the LW in the D and CD trees sharply increased, whereas in the S trees, it decreased after 2015 (Figure 4c). These results indicate that the temporal change in EW contributed more to the decline in growth of *Q. variabilis* than the temporal change in LW in this region.

3.2. *Effects of Meteorological Factors on Tree Growth*

3.2.1. Effects of Climate Variables on BAI

For the three crown classes of *Q. variabilis*, the responses of BAI growth to climate change were investigated ($p < 0.05$) (Figure 5). The BAIs of the three crown classes were strongly and positively correlated with T_{mean} and T_{min} in almost every month. However, different results were obtained for the effect of T_{max} on BAI for the three crown classes. The BAI in the D and CD trees exhibited a significantly positive correlation with T_{max} during the growing season, whereas the BAI in the S trees exhibited a significantly negative correlation with high temperature in the late non-growing (March) and early growing seasons (April and May). Additionally, the positive correlation coefficients were higher for the D trees than for the other classes, particularly from July to September. Significantly positive correlations were found between the BAI and PPT during the growing season, and the D tree responses were relatively significant in July and September. Similarly, the responses of BAI to SPEI 12 MONTHS and PDSI 12 MONTHS demonstrated significant differences among the three crown classes. SPEI 12 MONTHS was positively correlated with BAI in the D and CD trees in the hot summer (June to August), whereas no significant effects were found for the S trees in any of the months. Although PDSI 12 MONTHS positively affected the radial growth of *Q. variabilis*, some variations were observed across the different crown classes; specifically, PDSI 12 MONTHS had an obvious positive effect on BAI in the D and CD trees, and the D trees were more strongly positive than the CD trees. Similar to SPEI 12 MONTHS, weak correlations were observed between BAI in S trees and the PDSI during the whole year ($p < 0.05$) (Figure 5). We hypothesized that, in general, higher temperatures during drought seasons are the most dominant factors driving the differentiation of tree growth and quality in planted forests.

3.2.2. Effects of Climate Variables on EW and LW

Correlation analysis revealed that the growth of EW and LW in *Q. variabilis* was significantly affected by climate variables (Figure 6). For the three crown classes, similar patterns of radial growth (EW and LW) according to T_{mean} and T_{min} were identified. For instance, strong and significantly positive growth–climate correlations were observed for almost every month. However, significant variations in the growth response to T_{max} were observed among the three crown classes (Figure 6). The EW and LW in the D trees were positively correlated with T_{max} during the growing season (June to August), and the correlation coefficients were higher for LW ($p < 0.01$) than for EW ($p < 0.05$) during July and August. For the CD trees, LW was significantly positively correlated with T_{max} during the growing season (July–September), but the correlation coefficients were lower than those for the D trees. In contrast to the D and CD trees, significantly negative patterns were found in the EW of the S trees from February to June (late non-growing and early growing seasons) ($p < 0.05$ and $p < 0.01$). Moreover, the EW of the S trees was more strongly restricted by T_{max} during the early growing season than during the late non-growing season.

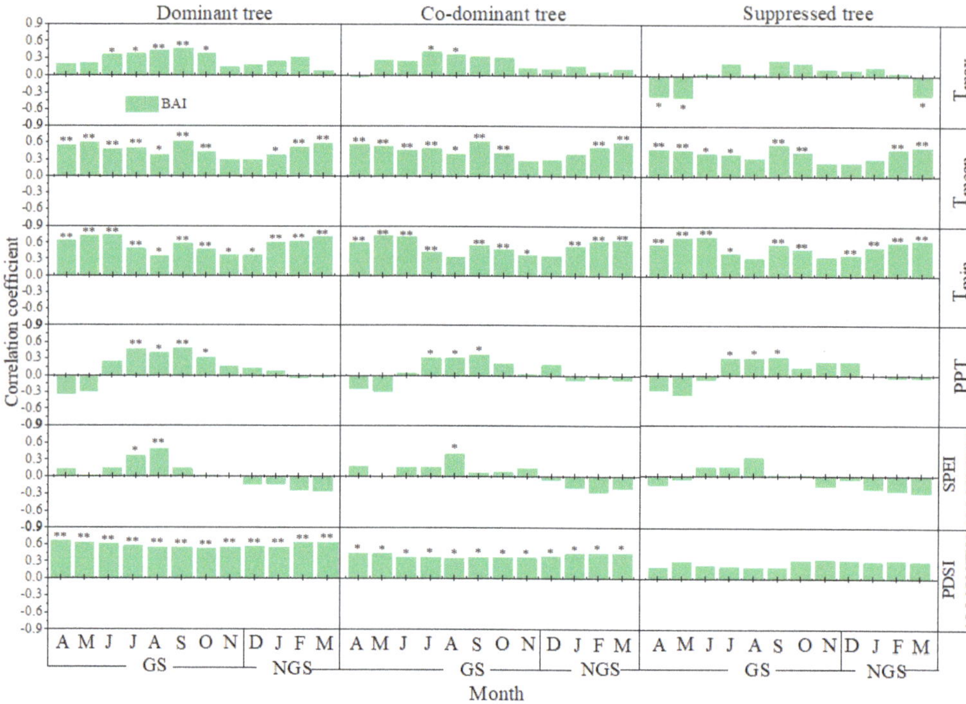

Figure 5. Correlation coefficients between the BAI and climate anomalies. BAI—basal area increment; A—April; M—May; J—June; J—July; A—August; S—September; O—October; N—November; D—December; J—January; F—February; M—March; GS—growing season; NGS—non-growing season; T_{max}—maximum temperature; T_{mean}—mean temperature; T_{min}—minimum temperature; PPT—precipitation; SPEI 12 MONTHS—Standardized Precipitation Evapotranspiration Index in 12 months; PDSI 12 MONTHS—Palmer Drought Severity Index in 12 months. * = correlation is significant at the 0.05 level, and ** = correlation is significant at the 0.01 level.

Furthermore, the responses of EW and LW in the three crown classes to PPT, SPEI 12 MONTHS, and PDSI 12 MONTHS were analyzed ($p < 0.05$) (Figure 6). There were significant patterns in the EW and LW responses to PPT during the growing seasons, and the D tree responses were more significant than those of the other classes ($p < 0.01$). For the D and CD trees, SPEI 12 MONTHS was significantly positively correlated with EW and LW in the growing season (July–September), and the correlation coefficients were higher in D than in CD trees; in particular, SPEI 12 MONTHS in August strongly accelerated the growth of the LW in the D and CD trees ($p < 0.01$). However, weak correlations were observed between SPEI 12 MONTHS and the radial growth (EW and LW) of the S trees throughout the year. The EW and LW in the three crown classes were positively sensitive to PDSI 12 MONTHS for the entire year, but the correlation coefficients exhibited some differences across the three classes ($p < 0.05$). The EW and LW in the D trees were more strongly correlated with PDSI 12 MONTHS every month ($p < 0.01$), whereas only the LW in the CD trees was significantly sensitive to PDSI 12 MONTHS ($p < 0.05$). Nevertheless, low correlation coefficients were found between the radial growth (EW and LW) of the S trees and PDSI 12 MONTHS throughout the year. The results indicated that high temperatures under drought stress were more obviously impactful on the EW than on the LW of *Q. variabilis* in the lithoid mountainous area; moreover, double rings (EW and LW) of the D and CD trees continuously grew under high water availability, whereas the S trees exhibited no response.

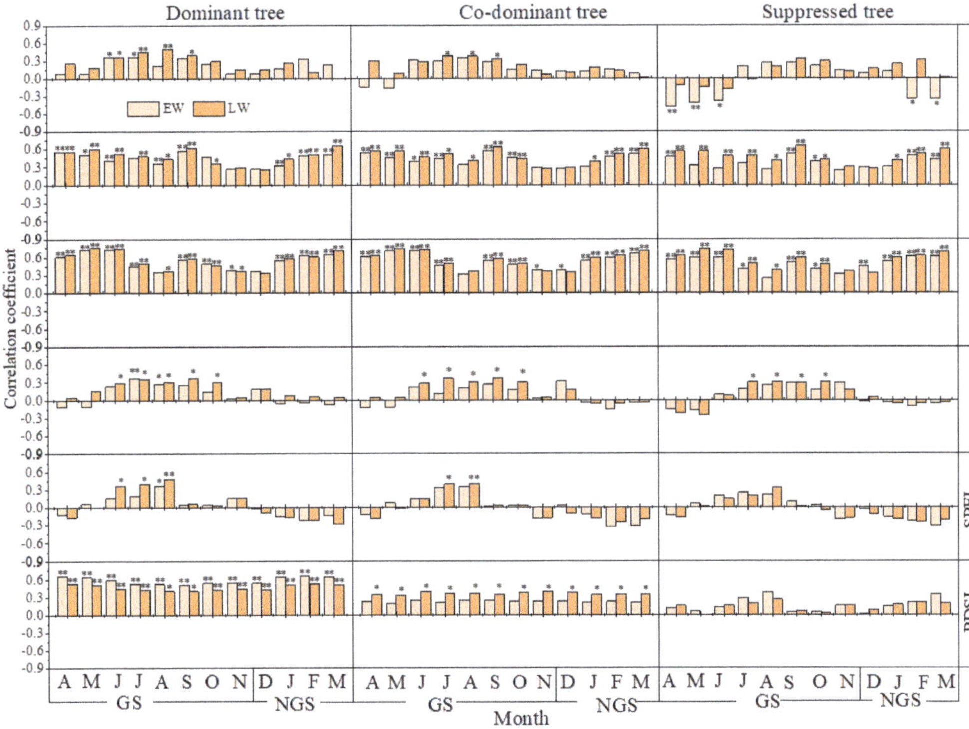

Figure 6. Relationship between the ring widths (EW and LW) and climate anomalies. EW—earlywood width; LW—latewood width; A—April; M—May; J—June; J—July; A—August; S—September; O—October; N—November; D—December; J—January; F—February; M—March; GS—growing season; NGS—non-growing season; T_{max}—maximum temperature; T_{mean}—mean temperature; T_{min}—minimum temperature; PPT—precipitation; SPEI 12 MONTHS—Standardized Precipitation Evapotranspiration Index in 12 months; and PDSI 12 MONTHS—Palmer Drought Severity Index in 12 months. * = correlation is significant at the 0.05 level, and ** = correlation is significant at the 0.01 level.

4. Discussion

4.1. Variations in Meteorological Conditions

Climatic variations are major natural factors influencing plant growth and survival in seasonal drought regions [32]. The study area, which is situated at low elevations in the lithoid mountainous area near Beijing, is a typical seasonal drought region in northern China [33]. Our investigation showed that the temperatures in this area rose by 0.035 °C/year from 1962 to 2020, which is relatively higher than the overall rate in China (0.025 °C/year) and worldwide (0.013 °C/year) [34]. An irregular rise in air temperature can enhance soil evapotranspiration and water shortages [35]. The study area had high summer precipitation and low winter precipitation during the cold winter ($p < 0.01$). These results imply that the climate was in contrast to that in previous studies focusing on Mediterranean regions [36,37] and northwest Iberia [38]. Further, we observed that the vigorous growth of this forest can span 6–7 months each year, and the legacy effects of low precipitation in the non-growing seasons (24.5 mm) strengthened the drought severity in the early growing season (April and May). Therefore, the study area has often experienced long-term seasonal droughts. The long-term insufficient soil water availability caused by climate warming may be a critical factor that restricts tree growth because limited water input influences the physiological activities of plants (such as leaf photosynthesis, water potential, and hydrologic conductivity). Consequently, several trees with developed root systems can grow depending on limited water availability, whereas vulnerable trees face severe decline because of insufficient water absorption [33]. Hence, climate change may drastically affect the growth and quality of the planted forests in this area.

4.2. Radial Growth of *Q. variabilis*

Growth slowing was observed in most trees before the decline, which was illustrated by an increase in annual autocorrelation and a change in the variables of the time series [39–41]. Based on our observations, *Q. variabilis* germinates at the beginning of April, loses its leaves at the end of November, and grows vigorously in summer (June to September). Moreover, we found that *Q. variabilis* comprises wide LW and narrow EW vessels. This finding is contrary to that of oak trees in northern and western Spain, which are known to foster wide EW and narrow LW vessels [38,42,43]. Consequently, the distinction between EW and LW is straightforward, and the tree ring compartment appears to undergo various climate-associated changes [44–46]. In this study, yearly variations in the BAI, EW, and LW of *Q. variabilis* were revealed, initially increasing rapidly with time and then decreasing after 2010, and the changing trend in EW was consistent with that in BAI. Moreover, EW decreased earlier than LW. This result was consistent with the radial growth of *Q. variabilis* in Ji Yuan, northern China [26]. EW formation contributed more to the declining growth of *Q. variabilis*. In addition, various trends in radial growth were observed across the three vigor classes. The growth of the D trees was continuous, whereas the growth of the CD and S trees decreased after 1990. In particular, the S trees exhibited a decline earlier than the CD trees. One hypothesis may be that the limiting factors substantially varied across the three crown classes, and the study area covered rainy, hot summers and dry, cold winters, which significantly affected the formation of ring wood in the trees. Previous studies have revealed that a temperature rise can stimulate xylem formation by extending the growing seasons or causing them to begin earlier [47]. However, the impact of high temperatures may weaken during seasonal droughts because drought severity increases to a threshold that negates the positive effects of increasing temperature. Moreover, wood formation was not completed at the same time each year, and the duration of formation varied [48]. Based on previous studies, the radial growth response of oak trees can record climatic changes, particularly its response to rainfall in the summer [49]. Similarly, our study revealed that high precipitation during the hot summer drastically stimulates LW formation, whereas low water availability during cold winter had negative effects on EW formation. Therefore, climate change may be critical in restricting the growth of *Q. variabilis*.

4.3. Climate–Growth Relationships

In addition to geophysical and edaphic factors, climate change influences radial growth and development in planted forests. In this study, the temperatures exhibited a significant effect on the radial growth of *Q. variabilis*, where the radial growth increased sharply at mean and minimum temperatures. Similar to our findings, previous studies on boreal tree species, such as *Pinus sylvestris* and *Pinus nigra* in east–central Spain and *Picea mariana* and *Pinus banksiana* in eastern Canada, observed a positive correlation between growth and temperatures [13,50,51]. In contrast, Gao et al. (2018) reported that the growth of *Q. variabilis* in the subtropical humid monsoon climate zone was negatively sensitive to temperature [20]. Moreover, Tardif et al. (2010) illustrated that the growth of *Q. alba* and *Q. rubra* at the northern limit of their distribution was confined to low temperatures [52]. These results indicate that limiting factors differ across various climate zones and not just by tree species. The radial growth of *Q. variabilis* was also significantly sensitive to PPT, SPEI 12 MONTHS, and PDSI 12 MONTHS. Higher precipitation and water availability increased the growth of *Q. variabilis*. Similarly, a study demonstrated a positive correlation between tree growth and wet weather at the northern and southern boundaries of the distribution of *Pilgerodendron uviferum* [53]. A recent study revealed that rising temperatures restricted tree growth at low elevations, particularly through impacts on soil evaporation and water use [54]. For *Q. variabilis*, 79% of the primary root system is strongly developed and homogeneously distributed in various soil layers [33], making it suitable for many water sources. Planted forests of *Q. variabilis* in this region may opportunistically increase their growth when water availability is sufficient, but there were variations in climate sensitivity among different crown classes.

Many studies have indicated the impact of climate change on several species in the same vigor class but have not performed comparison analyses on the sensitivity of various vigor classes to climate change. In this study, we found that climate change affected D, CD, and S trees. The radial growth (BAI, EW, and LW) of the D and CD trees continuously increased under high temperatures, whereas the S trees showed a reduction from February of the previous year to May of the current year. The growth of *Q. variabilis* remained significantly sensitive to precipitation during the growing season, and the responses of the D trees were more significant than those of the CD and S trees. Moreover, our study showed a higher positive sensitivity of D trees to SPEI 12 MONTHS and PDSI 12 MONTHS than CD trees, but S trees were not significantly influenced by these drought severity parameters throughout the whole year. Jia et al. (2022) found a typical pattern for *Q. variabilis* subject to frequent drought events, where there was a higher percentage of stored nonstructural carbohydrates consumed during drought events. This resulted in a significantly positive relationship between resistance and intrinsic efficient water use [26], which further limited tree growth following drought [55,56]. Previous studies in Mediterranean regions were characterized by less rainfall and high air temperatures in summer [57,58], which were contrary to our study parameters, but similar tree growth responses to PPT. Similar to other studies on oak tree species [59,60], our results demonstrated the overriding importance of meteorological factors for the radial growth of EW and LW in planted forests of *Q. variabilis*. Based on our results, we observed that the radial growth of *Q. variabilis* usually peaks in the rainy season, and it fosters wide LW vessels, whereas EW mostly comprises narrow vessels that are fostered during the spring. The distinction between EW and LW was evident because their tree ring compartments appeared to undergo various climate-associated changes. Considerable research has also demonstrated that trees may accumulate a larger proportion of nonstructural carbohydrates through photosynthesis under drought conditions, which are used to maintain respiration in the dormant season and the growth of EW in the following year, further leading to the legacy of tree growth [61]. To further clarify which climatic variables induced the growth decline in *Q. variabilis* in the rocky mountainous regions, we compared and analyzed the responses of EW and LW to climate changes across the three crown classes. We observed that the EW and LW of the S trees were limited by high temperatures in the late non-growing and early growing seasons, whereas the D and CD trees continuously grew. Moreover, the EW and LW of the S trees were not sensitive to high water availability for the whole year, and the growth of LW was less restricted by climate changes than EW for all classes.

The study sites were situated at low elevations with strong seasonality (~90% of the 465 mm of precipitation in the growing season, and it is concentrated in July and August). In general, *Q. variabilis* primarily uptakes water from the deep soil layers (32.8%) and natural springs (43.3%) during the seasonally dry periods, and it draws up as much water as possible if there are excessive precipitation incidents [34]. Liu et al. (2019) reported that *Q. variabilis* used water mostly from deep soil layers and natural springs to ensure a continuous water supply because the water availability in the upper soil layers was insufficient [34]. Wang et al. (2017) and Peddinti et al. (2020) found that some vegetation that has developed roots can draw from deep soil water during the growing season [62,63], thereby ensuring a stable water supply. However, there may be a historical effect of short-term seasonal drought, and young trees with fewer root systems will be restricted again when seasonal drought recurs or lasts for a long time [64]. Thus, many D and CD trees obtain water from groundwater using deeper root systems if the upper soil water supply is limited, but S trees are restricted because of their undeveloped root systems. In contrast, the energy required to draw water from the deeper soil layers is higher than that from the shallow layers. The S trees with a lower root density may not have had sufficient energy and time to absorb water from the deep soil water or natural springs during the drought seasons. In addition, the S trees were often located below the forest layer, resulting in less exposure to the atmosphere, and they may not have benefitted from better environmental conditions (i.e., light intensity, ventilation, and temperature) during the vigorous growing period compared with the other classes. Consequently, there was significant growth divergence of *Q. variabilis* in the D, CD, and S tree classes because they suffered long-term and severe seasonal droughts differently. This result is consistent with that of existing studies, which illustrated that the growth of *Q. variabilis* at this northern boundary was substantially affected by autumn water shortages [32]. In general, our findings indicate that climate change does not often restrict the growth of dominant trees, but it induces tree divergence and forest decline if warm and dry climate conditions continue beyond the adaptive capacity of tree species.

5. Conclusions

Our study site was a typical strong seasonality region, and it experiences a prolonged dry season (4–6 months). Hence, the effects of meteorological factors on the growth of planted forests in this region were a necessary concern for indication to forest managers seeking to proactively address climate-related global challenges. The novelty of our research lies in determining the impact of climatic factors on the radial growth and quality of planted forests of *Q. variabilis* in the lithoid mountainous area near Beijing, northern China. Our study revealed various responses of different crown classes to climatic changes. During drought seasons, S trees exhibit slower growth than D and CD trees under high temperatures; subsequently, they recover slowly [26]. Rising temperature under seasonal drought conditions was a critical meteorological driver that accelerated the growth decline in *Q. variabilis* plantations. Furthermore, it induced young trees to diverge significantly and early from the main canopy layers in response to continued climate change. As an extension of the planted forest density in mountainous areas, some young trees with undeveloped root systems experienced decreased water availability, low light density, and poor ventilation. Thus, restricting tree growth may also be induced by other factors, i.e., abiotic factors (light, soil, insects, and pathogens) or stand structures (tree density, total basal area, and species competition), accompanied by climatic change [20]. Indeed, the impact of climate change on radial growth and tree quality may be alleviated by adjusting and improving the stand characteristics. Our study emphasized that the resource utilization (water, light, and nutrients) of each crown class could be improved using forest thinning treatments and stand structure regulation. In this way, declining growth of planted forests under future changes in climate can be reduced and the ecosystem services that they provide maintained. To maximize the long-term forest resilience against rising aridity, we suggest the following methods based on previous studies, in addition to our findings:

1. Implementing forest thinning treatments

As a pure artificial forest ecosystem, the study area has a canopy cover of approximately 90% and a tree density of 1105 trees/ha^2. Consequently, excessive tree density in mountainous areas may reduce resource utilization because of enhanced species competition. Apart from climatic variables, stand structures, light density, and soil properties were also the major factors affecting the growth of planted forests. In addition, species competition for resources can gradually increase when environmental conditions improve [65]; for instance, competition among neighboring trees has gradually increased with global warming [58]. Meanwhile, tree density and total basal area reflect the crowdedness of a forest and resource efficiency levels, which can be considered competitive parameters [66]. Most studies have underscored that the growth of trees is significantly correlated with stand conditions (density and total basal area) and their interaction with the climate. For example, oak trees had higher growth in lower-density stands than in higher-density stands under drought conditions, because reducing wood density (i.e., thinning effect) mitigates species competition and increases the moisture availability for each tree class [67–69]. Therefore, reducing stand density through thinning treatments may be considered a potential adaptive strategy for the planted forests of *Q. variabilis* in seasonal drought areas; it promotes tree vitality by increasing environmental resources (i.e., space, light, nutrients, and water), thereby increasing tree vitality [68]. Furthermore, this strategy enhances the adaptive capacity of planted forests and prevents tree mortality because of impending global warming.

2. Improving species composition

Planted forests often suffer from symmetrical competition because of monocultures or unreasonable population structures. Consequently, limited water availability during drought seasons may not satisfy the growth needs of all trees simultaneously. However, growth sensitivity to changes in resource competition stress is generally diverse among plant species. Therefore, apart from reducing stand density, forest managers may improve the resistance of forest stands to future adverse climate warming by favoring species composition. As an important silviculture strategy, diversifying tree species has many advantages, one of which is good tree growth, likely because of the complementarity effect. For example, resource partitioning or positive interactions promote resource availability and the higher growth of individuals [70]. Indeed, nutrient, light, or water availability could be more sufficient in mixture stands compared with monoculture [71,72]. Moreover, most studies indicate higher productivity for tree mixtures than monoculture [73]. An optimal proportion of the other species should mitigate the competition effect caused by complementarity. For example, mixed species could share soil water and nutrients via their different root prospecting volumes [74]. Because *Q. variabilis* grows with other species, it could prospect deeper soil layers through its root distribution, and this strategy decreases competition for water resources. In addition, if niche complementarity occurred in the above-ground component, less shade-tolerant species could occupy the upper canopy, whereas more shade-tolerant species could populate the understory. In this way, each species could use resources to increase their vitality [75]. These strategies mitigate competition for resources among different tree species and promote symbiotic relationships across community species.

In general, the planted forests of *Q. variabilis* are too dense. To avoid increased mortality and the possibility of fire becoming a natural density control factor, forests need to be thinned. Determining the most ecologically sustainable methods for maintaining ecosystem goods and services requires further investigation. Considering the increase in global warming, close-to-nature principles appear to be a suitable management option for the cultivation of more adapted forests [76,77]. However, the present study focused on a single tree species that is sensitive to climatic changes. Further research is necessary to comprehensively understand which species should be selected to improve the growth and quality of planted forests and to determine how these forests use resources during seasonal droughts in rocky mountainous areas. Studies considering several tree species will provide scientific information on silviculture and management treatments for forest plantations in rocky mountainous regions under global warming conditions.

Author Contributions: A.K. conceptualized, designed, and performed the experiments, contributed to the data production and formal analyses, prepared the figures and tables, and wrote the original draft. G.L. supervised the study experiments and approved the final draft. P.L. and W.L. participated in the fieldwork and contributed to the investigation. Ü.H. discussed the research framework and provided useful advice and revised the final manuscript. All the authors have checked and approved the submitted version. All authors have read and agreed to the published version of the manuscript.

Funding: The research was financially supported by the Beijing Municipal Bureau of Landscape and Greening Project (Grant No. 20210618).

Data Availability Statement: The dataset generated for the study area is available from the corresponding author on reasonable request.

Acknowledgments: We cordially thank the Xishan Forest Farm of Beijing for its assistance with the fieldwork. We also thank the master's students (Jiatong Yu, Minghao Ge, Rui Li, Jialu Kang, and Yadong Liu) from the Beijing Forestry University for their contributions during the field investigation. The authors are grateful to the anonymous editors and reviewers for their constructive comments, which greatly helped us improve the quality of this manuscript.

Conflicts of Interest: The authors declare that they have no competing interests.

References

1. Cheng, S.; Xu, Z.; Su, Y.; Zhen, L. Spatial and temporal flows of China's forest resources: Development of a framework for evaluating resource efficiency. *Ecol. Econ.* **2010**, *69*, 1405–1415. [CrossRef]
2. Jia, G.; Liu, Z.; Chen, L.; Yu, X. Distinguish water utilization strategies of trees growing on earth-rocky mountainous area with transpiration and water isotopes. *Ecol. Evol.* **2017**, *7*, 10640–10651. [CrossRef] [PubMed]
3. Hu, J.X.; Pradit, S.; Loh, P.S.; Wang, J.J. Storage and dynamics of soil organic carbon in allochthonous-dominated and nitrogen-limited natural and planted mangrove forests in southern Thailand. *Mar. Pollut. Bull.* **2024**, *200*, 116064. [CrossRef] [PubMed]
4. Sun, S.J.; He, C.; Qiu, L.; Li, C.; Zhang, J.S.; Meng, P. Stable isotope analysis reveals prolonged drought stress in poplar plantation mortality of the three-north shelter forest in Northern China. *Agric. For. Meteorol.* **2018**, *252*, 39–48. [CrossRef]
5. Zhao, P.; Xu, C.; Zhou, M.; Zhang, B.; Ge, P.; Zeng, N. Rapid regeneration offsets losses from warming-induced tree mortality in an aspen dominated broad-leaved forest in Northern China. *PLoS ONE* **2018**, *13*, e0195630. [CrossRef] [PubMed]
6. Trouvé, R.; Bontemps, J.D.; Collet, C.; Seynave, I.; Lebourgeois, F. Growth partitioning in forest stands is affected by stand density and summer drought in Sessile oak and Douglas-fir. *For. Ecol. Manag.* **2014**, *334*, 358–368. [CrossRef]
7. Russo, S.E.; Davies, S.J.; King, D.A.; Tan, S. Soil-related performance variation and distributions of tree species in a Bornean rain forest. *J. Ecol.* **2005**, *93*, 879–889. [CrossRef]
8. Coudun, C.; Gegout, J.; Piedallu, C.; Rameau, J. Soil nutritional factors improve models of plant species distribution: An illustration with *Acer campestre* (L.) in France. *J. Biogeogr.* **2006**, *33*, 1750–1763. [CrossRef]
9. Lévesque, M.; Walthert, L.; Weber, P. Soil nutrients influence growth response of temperate tree species to drought. *J. Ecol.* **2016**, *104*, 377–387. [CrossRef]
10. Seynave, I.; Gégout, J.C.; Hervé, J.C.; Dhôte, J.F. Is the spatial distribution of European beech (*Fagus sylvatica* L.) limited by its potential height growth? *J. Biogeogr.* **2008**, *35*, 1851–1862. [CrossRef]
11. Foster, J.R.; Finley, A.O.; D'Amato, A.W.; Bradford, J.B.; Banerjee, S. Predicting tree biomass growth in the temperate-boreal ecotone: Is tree size, age, competition, or climate response most important? *Glob. Chang. Biol.* **2016**, *22*, 2138–2151. [CrossRef]
12. Luo, D.W.; Huang, J.G.; Jiang, X.Y.; Ma, Q.Q.; Liang, H.X.; Guo, X.L.; Zhang, S.K. Effect of climate and competition on radial growth of *Pinus massoniana* and *Schima superba* in China's subtropical monsoon mixed forest. *Dendrochronologia* **2017**, *46*, 24–34. [CrossRef]
13. Huang, J.; Tardif, J.C.; Bergeron, Y.; Denneler, B.; Berninger, F.; Girardin, M.P. Radial growth response of four dominant boreal tree species to climate along a latitudinal gradient in the eastern Canadian boreal forest. *Glob. Chang. Biol.* **2010**, *16*, 711–731. [CrossRef]
14. Jump, A.S.; Hunt, J.M.; Penuelas, J. Rapid climate change-related growth decline at the southern range edge of *Fagus sylvatica*. *Glob. Chang. Biol.* **2006**, *12*, 2163–2174. [CrossRef]
15. Babst, F.; Poulter, B.; Trouet, V.; Tan, K.; Neuwirth, B.; Wilson, R.; Carrer, M.; Grabner, M.; Tegel, W.; Levanic, T. Site- and species-specific responses of forest growth to climate across the European continent. *Glob. Ecol. Biogeogr.* **2013**, *22*, 706–717. [CrossRef]
16. Matias, L.; Linares, J.C.; Sanchez-Miranda, A.; Jump, A.S. Contrasting growth forecasts across the geographical range of Scots pine due to altitudinal and latitudinal differences in climatic sensitivity. *Plant Biotechnol.* **2017**, *33*, 4106–4116. [CrossRef]
17. Andrade, J.L.; Meinzer, F.C.; Goldstein, G.; Schnitzer, S.A. Water uptake and transport in Lianas and co-occurring trees of a seasonally dry tropical forest. *Trees Struct. Funct.* **2005**, *19*, 282–289. [CrossRef]
18. Nie, Y.P.; Chen, H.S.; Wang, K.L.; Tant, W.; Deng, P.Y.; Yang, J. Seasonal water use patterns of woody species growing on the continuous dolostone outcrops and nearby thin soils in subtropical China. *Plant Soil* **2011**, *341*, 399–412. [CrossRef]
19. Gao, W.Q.; Ni, Y.Y.; Xue, Z.M.; Wang, X.F.; Kang, F.F.; Hu, J.; Gao, Z.H.; Jiang, Z.P.; Liu, J.F. Population structure and regeneration dynamics of *Quercus variabilis* along latitudinal and longitudinal gradients. *Ecosphere* **2017**, *8*, e01737. [CrossRef]
20. Gao, W.Q.; Liu, J.F.; Xuea, Z.M.; Zhang, Y.T.; Gao, Z.H.; Ni, Y.Y.; Wang, X.F.; Jiang, Z.P. Geographical patterns and drivers of growth dynamics of *Quercus variabilis*. *For. Ecol. Manag.* **2018**, *429*, 256–266. [CrossRef]
21. Zhang, X.P.; Xu, X.L.; Su, W.; Zhao, X.P.; Xu, X.L. Spring precipitation effects on formation of first row of earlywood vessels in *Quercus variabilis* at Qinling Mountain (China). *Trees* **2019**, *33*, 457–468. [CrossRef]
22. Santiago, B.; Sergio, M.; Vicente, S.; Reig, F.; Latorre, B. Standardized precipitation evapotranspiration index (SPEI) revisited: Parameter fitting, evapotranspiration models, tools, datasets and drought monitoring. *Int. J. Climatol.* **2014**, *34*, 3001–3023. [CrossRef]
23. Ram, S. Tree growth–climate relationships of conifer trees and reconstruction of summer season Palmer drought severity index (PDSI) at Pahalgam in Srinagar, India. *Quatern. Int.* **2012**, *254*, 152–158. [CrossRef]

24. D'Orangeville, L.; Duchesne, L.; Houle, D.; Kneeshaw, D.; Côté, B.; Pederson, N. Northeastern North America as a potential refugium for boreal forests in a warming climate. *Science* **2016**, *352*, 1452–1455. [CrossRef]
25. Clark, J.S.; Bell, D.M.; Hersh, M.H.; Nichols, L. Climate change vulnerability of forest biodiversity: Climate and competition tracking of demographic rates. *Glob. Chang. Biol.* **2011**, *17*, 1834–1849. [CrossRef]
26. Seidel, D.; Leuschner, C.; Müller, A.; Krause, B. Crown plasticity in mixed forests–quantifying asymmetry as a measure of competition using terrestrial laser scanning. *For. Ecol. Manag.* **2011**, *261*, 2123–2132. [CrossRef]
27. Jia, H.; Guan, C.; Zhang, J.; He, C.; Yin, C.; Meng, P. Drought effects on tree growth, water use efficiency, vulnerability and canopy health of *Quercus variabilis-Robinia pseudoacacia* mixed plantation. *Front. Plant Sci.* **2022**, *13*, 1018405–1018417. [CrossRef]
28. Stokes, M.A.; Smiley, T.L. *An Introduction to Tree-Ring Dating*; The University of Arizona Press: Tucson, Arizona, 1996; 73p. [CrossRef]
29. Arsalani, M.; Pourtahamsi, K.; Azizi, G.; Mohammadi, H. Tree-ring based December-February precipitation reconstruction in the southern Zagros Mountains, Iran. *Dendrochronologia* **2018**, *49*, 45–56. [CrossRef]
30. Holmes, R.L. Computer-assisted quality control in tree-ring dating and measurement. *Tree-Ring Bull.* **1983**, *44*, 69–75. [CrossRef]
31. Hegyi, F. *A Simulation Model for Managing Jack-Pine Stands*; Department of Forest Yield Research, Royal College of Forestry: Stockholm, Sweden, 1974.
32. Bai, X.P.; Zhang, X.L.; Li, J.X.; Duan, X.Y.; Jin, Y.T.; Chen, Z.J. Altitudinal disparity in growth of Dahurian larch (*Larix gmelinii* Rupr.) in response to recent climate change in northeast China. *Sci. Total Environ.* **2019**, *670*, 466–477. [CrossRef]
33. Aminzadeh, M.; Or, D.; Stevens, B.; AghaKouchak, A.; Shokri, N. Upper bounds of maximum land surface temperatures in a warming climate and limits to plant growth. *Earth's Future* **2023**, *11*, e2023EF003755. [CrossRef]
34. Liu, Z.Q.; Yu, X.X.; Jia, G.D. Water uptake by coniferous and broad-leaved forest in a rocky mountainous area of northern China. *Agric. For. Meteorol.* **2019**, *265*, 381–389. [CrossRef]
35. Gao, W.Q.; Lei, D.X.; Fu, L.Y.; Duan, G.S.; Zhou, M.L.; Cao, J. Radial growth response of two oaks to climate at their disparate distribution limits in semiarid areas, Beijing, China. *Ecosphere* **2020**, *11*, e03062. [CrossRef]
36. Intergovernmental Panel on Climate Change (IPCC). *Climate Change 2007: The Physical Science Basis*; Solomon, S., Ed.; Contribution of Working Group 1 to the Cambridge University Press: Cambridge, UK, 2007.
37. Lebourgeois, F. Climatic signal in annual growth variation of silver fir (*Abies alba* Mill.) and spruce (*Picea abies* Karst.) from the French Permanent Plot Network (RENECOFOR). *Ann. For. Sci.* **2007**, *64*, 333–343. [CrossRef]
38. Bogino, S.; Fernández Nieto, M.J.; Bravo, F. Climate effect on radial growth of *Pinus sylvestris* at its southern and western distribution limits. *Silva Fenn.* **2009**, *43*, 609–623. [CrossRef]
39. Manuel, S.H.; Vicente, R.I.; García, G. Chronologies of earlywood vessels and latewood width disentangle climate drivers of oak growth in a mild oceanic region. *Dendrochronologia* **2018**, *51*, 40–53. [CrossRef]
40. Dakos, V.; Carpenter, S.R.; Brock, W.A.; Ellison, A.M.; Guttal, V.; Ives, A.R.; Scheffer, M. Methods for detecting early warnings of critical transitions in time series illustrated using simulated ecological data. *PLoS ONE* **2012**, *7*, e41010. [CrossRef]
41. de Luis, M.; Čufar, K.; Di Filippo, A.; Novak, K.; Papadopoulos, A.; Piovesan, G.; Rathgeber, C.B.K.; Raventós, J.; Saz, M.A.; Smith, K.T. Plasticity in dendroclimatic response across the distribution range of Aleppo pine (*Pinus halepensis*). *PLoS ONE* **2013**, *8*, e83550. [CrossRef]
42. Rozas, V.; Camarero, J.J.; Sangüesa-Barreda, G.; Souto, M.; García-González, I. Summer drought and ENSO-related cloudiness distinctly drive *Fagus sylvatica* growth near the species rear-edge in Northern Spain. *Agric. For. Meteorol.* **2015**, *201*, 153–164. [CrossRef]
43. Rozas, V.; Lamas, S.; García, G.I. Differential tree-growth responses to local and large-scale climatic variation in two *Pinus* and two *Quercus* species in Northwest Spain. *Ecoscience* **2009**, *16*, 299–310. [CrossRef]
44. Aloni, R. Ecophysiological implications of vascular differentiation and plant evolution. *Trees* **2015**, *29*, 1–16. [CrossRef]
45. García-González, I.; Eckstein, D. Climatic signal of earlywood vessels of oak on a maritime site. *Tree Physiol.* **2003**, *23*, 497–504. [CrossRef]
46. García-González, I.; Souto-Herrero, M.; Campelo, F. Ring-porosity and earlywood vessels: A review on extracting environmental information through time. *IAWA J.* **2016**, *37*, 295–314. [CrossRef]
47. Fonti, P.; von Arx, G.; García-González, I.; Eilmann, B.; Sass-Klaassen, U.; Gärtner, H.; Eckstein, D. Studying global change through investigation of the plastic responses of xylem anatomy in tree rings. *New Phytol.* **2010**, *185*, 42–53. [CrossRef]
48. King, G.; Fonti, P.; Nievergelt, D.; Buntgen, U.; Frank, D. Climatic drivers of hourly to yearly tree radius variations along a 6 °C natural warming gradient. *Agric. For. Meteorol.* **2013**, *168*, 36–46. [CrossRef]
49. Rossi, S.; Deslauriers, A.; Griçar, J. Critical temperatures for xylogenesis in conifers of cold climates. *Glob. Ecol. Biogeogr.* **2008**, *17*, 696–707. [CrossRef]
50. Bowman, D.M.; Williamson, G.J.; Keenan, R.; Prior, L.D. A warmer world will reduce tree growth in evergreen broadleaf forests: Evidence from Australian temperate and subtropical eucalypt forests. *Glob. Ecol. Biogeogr.* **2014**, *23*, 925–934. [CrossRef]
51. Martin, B.D.; Anchukaitis, K.J.; Evans, M.N.; Del, R.M.; Beeckman, H.; 'Cañellas, I. Effects of drought on xylem anatomy and water-use efficiency of two co-occurring pine species. *Forests* **2017**, *8*, 332. [CrossRef]
52. Holz, A.; Hart, S.J.; Williamson, G.J.; Veblen, T.T.; Aravena, J.C. Radial growth response to climate change along the latitudinal range of the world's southernmost conifer in southern South America. *J. Biogeogr.* **2018**, *45*, 1140–1152. [CrossRef]

53. Goldblum, D. The geography of white Oak's (*Quercus alba* L.) response to climatic variables in North America and speculation on its sensitivity to climate change across its range. *Dendrochronologia* **2010**, *28*, 73–83. [CrossRef]
54. Camarero, J.J.; Linares, J.C.; García-Cervigón, A.I.; Batllori, E.; Martínez, I.; Gutiérrez, E. Back to the future: The responses of alpine treelines to climate warming are constrained by the current ecotone structure. *Ecosystems* **2017**, *20*, 683–700. [CrossRef]
55. Tardif, J.C.; Conciatori, F.; Gagnon, P.N.; Gagnon, D. Radial growth and climate responses of white Oak (*Quercus alba*) and Northern red oak (*Quercus rubra*) at the Northern distribution limit of white Oak in *Quebec*, Canada. *J. Biogeogr.* **2010**, *33*, 1657–1669. [CrossRef]
56. Galiano, L.; Martínez-Vilalta, J.; Lloret, F. Drought-induced multifactor decline of Scots pine in the Pyrenees and potential vegetation change by the expansion of co-occurring Oak Species. *Ecosystems* **2010**, *13*, 978–991. [CrossRef]
57. Lyu, L.X.; Suvanto, S.; Njd, P.; Henttonen, H.M.; Zhang, Q.B. Tree growth and its climate signal along latitudinal and altitudinal gradients: Comparison of tree rings between Finland and the Tibetan Plateau. *Biogeosciences* **2017**, *14*, 3083–3095. [CrossRef]
58. Olsen, S.L.; Töpper, J.P.; Skarpaas, O.; Vandvik, V.; Klanderud, K. From facilitation to competition: Temperature-driven shift in dominant plant interactions affects population dynamics in semi natural grasslands. *Glob. Chang. Biol.* **2016**, *22*, 1915–1926. [CrossRef]
59. Rozas, V.; Garcia-Gonzalez, I. Non-stationary influence of El Nino-Southern Oscillation and winter temperature on Oak latewood growth in NW Iberian Peninsula. *Int. J. Biometeorol.* **2012**, *56*, 787–800. [CrossRef]
60. Trouvé, R.; Bontemps, J.D.; Collet, C.; Seynave, I.; Lebourgeois, F. Radial growth resilience of sessile oak after drought is affected by site water status, stand density, and social status. *Trees* **2017**, *31*, 517–529. [CrossRef]
61. Liu, Z.; Yu, X.; Jia, G.; Jia, J.; Lou, Y.; Zhang, K. Water use characteristics of *Platycladus orientalis* and *Quercus variabilis* in Beijing mountain area. *Sci. Silv. Sin.* **2016**, *52*, 22–30. [CrossRef]
62. Wang, J.; Fu, B.; Lu, N.; Zhang, L. Seasonal variation in water uptake patterns of three plant species based on stable isotopes in the semi-arid Loess Plateau. *Sci. Total Environ.* **2017**, *609*, 27–37. [CrossRef] [PubMed]
63. Peddinti, S.R.; Kambhammettua, B.V.N.P.; Lad, R.S.; Šimůnek, J.; Gade, R.M.; Adinarayana, J. A macroscopic soil-water transport model to simulate root water uptake in the presence of water and disease stress. *J. Hydrol.* **2020**, *587*, 124940. [CrossRef]
64. Richburg, J.A. Timing Treatments to the Phenology of Root Carbohydrate Reserves to Control Woody Invasive Plants. Ph.D. Thesis, University of Massachusetts Amherst, America, MA, USA, 2005.
65. Maestre, F.T.; Callaway, R.M.; Valladares, F.; Lortie, C.J. Refining the stress-gradient hypothesis for competition and facilitation in plant communities. *J. Ecol.* **2009**, *97*, 199–205. [CrossRef]
66. Mantgem, P.J.V.; Stephenson, N.L.; Byrne, J.C.; Daniels, L.D.; Franklin, J.F.; Fulé, P.Z.; Harmon, M.E.; Larson, A.J.; Smith, J.M.; Taylor, A.H. Widespread increase of tree mortality rates in the Western United States. *Science* **2009**, *323*, 521–524. [CrossRef] [PubMed]
67. Jordan, B.; Patrick, V.; Thomas, P.; Philippe, B.; Vincent, S.; Sandrine, P.; Camille, C.; Nathalie, K. How do mixing tree species and stand density affect seasonal radial growth during drought events? *For. Ecol. Manag.* **2019**, *432*, 436–445. [CrossRef]
68. Manrique-Alba, A.; Beguería, S.; Tomas-Burguera, M.; Camarero, J.J. Increased post-drought growth after thinning in *Pinus nigra* plantations. *Forests* **2021**, *12*, 985. [CrossRef]
69. Bréda, N.; Granier, A.; Aussenac, G. Effects of thinning on soil and tree water relations, transpiration and growth in an Oak forest (*Quercus petrea* (Matt.) Liebl.). *Tree Physiol.* **1995**, *15*, 295–306. [CrossRef] [PubMed]
70. Loreau, M.; Hector, A. Partitioning selection and complementarity in biodiversity experiments. *Nature* **2001**, *412*, 72–76. [CrossRef] [PubMed]
71. Forrester, D.I. Transpiration and water-use efficiency in mixed-species forests versus monocultures: Effects of tree size, stand density and season. *Tree Physiol.* **2015**, *35*, 289–304. [CrossRef] [PubMed]
72. Liang, J.; Crowther, T.W.; Picard, N.; Wiser, S.; Zhou, M.; Alberti, G.; Schulze, E.D.; McGuire, A.D.; Bozzato, F.; Pretzsch, H.; et al. Positive biodiversity-productivity relationship predominant in global forests. *Science* **2016**, *354*, aaf8957. [CrossRef]
73. Tian, D.Y.; He, P.; Jiang, L.C.; Gaire, D. Developing crown width model for mixed forests using soil, climate and stand factors. *J. Ecol.* **2024**, *112*, 427–442. [CrossRef]
74. Richards, A.E.; Forrester, D.I.; Bauhus, J.; Scherer, L.M. The influence of mixed tree plantations on the nutrition of individual species: A review. *Tree Physiol.* **2010**, *30*, 1192–1208. [CrossRef]
75. Niinemets, Ü. A review of light interception in plant stands from leaf to canopy in different plant functional types and in species with varying shade tolerance. *Ecol. Res.* **2010**, *25*, 693–714. [CrossRef]
76. Eduardo, C.; Míriam, P.; Jaime, C.; Javier, D.D.G.; Carla, F.; Lluís, C. Close-to-nature management effects on tree growth and soil moisture in Mediterranean mixed forests. *For. Ecol. Manag.* **2023**, *549*, 121457–121466. [CrossRef]
77. Merlin, M.; Perot, T.; Perret, S.; Korboulewsky, N.; Vallet, P. Effects of stand composition and tree size on resistance and resilience to drought in Sessile oak and Scots pine. *For. Ecol. Manag.* **2015**, *339*, 22–33. [CrossRef]

Disclaimer/Publisher's Note: The statements, opinions and data contained in all publications are solely those of the individual author(s) and contributor(s) and not of MDPI and/or the editor(s). MDPI and/or the editor(s) disclaim responsibility for any injury to people or property resulting from any ideas, methods, instructions or products referred to in the content.

agronomy

Article

Effects of Warming and Increased Precipitation on Root Production and Turnover of *Stipa breviflora* Community in Desert Steppe

Qi Li [1,2], Jianying Guo [1], Zhanyi Wang [2,*], Chengjie Wang [2], Pengbo Liu [3], Guangyi Lv [2], Zhenqi Yang [1], Chunjie Wang [4] and Xiao Qiu [5]

1 Yinshanbeilu Grassland Eco-Hydrology National Observation and Research Station, China Institute of Water Resources and Hydropower Research, Beijing 100038, China; liqi2024@emails.imau.edu.cn (Q.L.); guojy@iwhr.com (J.G.); yangzq@iwhr.com (Z.Y.)
2 Key Laboratory of Grassland Resources of the Ministry of Education, College of Grassland, Resources and Environment, Inner Mongolia Agricultural University, Hohhot 010018, China; nmgcjwang3@imau.edu.cn (C.W.); lgy1995@emails.imau.edu.cn (G.L.)
3 Arong Banner Forestry and Grassland Bureau, Arong Banner 162750, China; pbliu@emails.imau.edu.cn
4 Research Center for Eco-Environmental Sciences, Chinese Academy of Sciences, Beijing 100085, China; cjwang@rcees.ac.cn
5 Inner Mongolia Academy of Agricultural and Animal Husbandry Sciences, Hohhot 010031, China; qiuxiao@imaaahs.ac.cn
* Correspondence: wangzy@imau.edu.cn

Abstract: Organic carbon in grassland mainly exists in the soil, and root production and turnover play important roles in carbon input into the soil. However, the effects of climate change on plant root dynamics in desert steppe are unknown. We conducted an experiment in a desert steppe, which included ambient temperature (T0); temperature increased by 2 °C (T1); temperature increased by 4 °C (T2); natural precipitation (P0); precipitation increased by 25% (P1); precipitation increased by 50% (P2); and the interaction between warming and increased precipitation. Plant community aboveground characteristics; root production; and root turnover were measured. We found that the root length production of the T0P2; T1P1; T2P0; and T2P1 treatments were significantly higher than that of the T0P0 treatment, with an increment of 98.70%, 11.72%, 163.03%, and 85.14%, respectively. Three treatments with temperature increased by 2 °C (T1P0; T1P1; and T1P2) and significantly increased root turnover rate compared to the T0P0 treatment, with increases of 62.53%, 42.57%, and 35.55%, respectively. The interaction between warming and increased precipitation significantly affected the root production of the community ($p < 0.01$), but this interaction was non-additive. Future climate warming will benefit the accumulation of root-derived carbon in desert steppe communities.

Keywords: climatic change; root standing crop; *Stipa breviflora* community; aboveground net primary productivity; root turnover rate

1. Introduction

Due to human activities emitting large amounts of greenhouse gases since the Industrial Revolution, global and regional climate change has been exacerbated, mainly climate warming and extreme precipitation. It is estimated that the global surface temperature will rise by 1.0–3.7 °C by the end of the 21st century [1]. Global warming is accelerating hydrological cycles, leading to significant changes in precipitation patterns in terrestrial ecosystems, particularly in high-latitude regions [2]. China has been continuously warming with a temperature of 0.39 °C/decade, with significant warming occurring especially in northwestern China, including Inner Mongolia [3]. Annual precipitation in the northwest region of China has significantly increased at a rate of 0.55 mm/year [4]. Temperature and moisture determine most of the processes involved in photosynthetic carbon sequestration, including photosynthesis, carbon distribution in the aboveground and belowground parts, transpiration, and root growth during plant growth [5].

The Inner Mongolia desert steppe is an indispensable part of the temperate steppe of Central Asia [6,7]. Desert steppes belong to the transitional zone between desert and typical steppes. As a result of harsh ecological environments and climate fluctuation, the desert steppe ecosystem is very fragile, which not only has an impact on the production and life of Inner Mongolia but also has a crucial influence on the carbon cycle in China and even the world [8]. Fine roots play an extremely important role in the carbon exchange process that connects the aboveground parts and the soil [9]. The root turnover rate is often defined as the number of times per year that fine root biomass is replaced [10]. The rate of root turnover is closely related to biochemical cycling in terrestrial ecosystems [11]. Plants can produce and store carbohydrates synthesized by photosynthesis through their roots and then transfer carbon from dead roots to the soil through root turnover [12]. The net primary productivity (NPP) of ecosystems includes aboveground net primary productivity (ANPP) and belowground net primary productivity (BNPP) [13]. The distribution of NPP between aboveground and belowground components is an important process in the carbon cycle and energy flow of ecosystems and is also one of the main components of the global carbon budget [14,15]. Climate change directly or indirectly affects NPP, so understanding NPP dynamics is essential to enhance our understanding of the impact of grassland belowground carbon allocation and carbon storage. Because of this, a great number of climate change experiments have been carried out, such as studies showing how plant root production will increase [16], decrease [17], or remain unchanged [18] due to climate warming. Changes in precipitation also strongly influence changes in root production, with reduced precipitation increasing root production in alpine steppes and increased precipitation increasing root production in typical steppes. Differences between the results of different studies may be due to their different regional environments due to factors such as climatic conditions, vegetation types, and altitude.

Climatic conditions are considered to be the main drivers of root growth and turnover [19], and a meta-analysis of the world's grasslands found that temperature and precipitation had a stronger impact on root turnover than soil physico-chemical properties [20]. Climate warming increases soil net mineralization rate and reduces soil water availability, which may affect root productivity and turnover [21,22]. Temperature and precipitation are necessary for plant growth and development, but the effects are still unclear. Higher temperatures lead to lower soil water content [23], which exacerbates drought stress, so plants allocate more carbon to root production to obtain an equal amount of water, thus promoting root growth [24]. In the semi-arid temperate steppe of northern China, it was also concluded that warming promoted root production and turnover [25]. Therefore, our first hypothesis is that warming will increase root production and turnover of *Stipa breviflora* communities in the desert steppe examined in this study. Increased precipitation will increase soil moisture content, so our second hypothesis is that increased precipitation within a certain range will reduce the root production and turnover of *Stipa breviflora* communities in the desert steppe. Climate change experiments on the temperate steppe in northern China have shown that warming and increased precipitation have a significant interactive effect on root production and mortality [25]. Thus, our third hypothesis is that there is an interaction between warming and increased precipitation on root production and turnover in the desert steppe. The objective of this study is to characterize the root production and turnover of plant communities in the desert steppe and to evaluate the effects of warming and increased precipitation on plant community aboveground characteristics, root production, and root turnover.

2. Materials and Methods

2.1. Study Site

The field experiment site is located in Siziwang Banner (41°47′19″ N, 111°53′45″ E, 1456 m elevation). This region has a typical temperate continental monsoon climate, with cold and long winters, hot summers, and a great temperature difference between day and night. The mean annual precipitation is 280 mm, with precipitation mostly occurring from May to September. The mean annual temperature is 7.2 °C. The soil in the region is light castanozems with a lack of phosphorus, as well as less nitrogen, more potassium, and low organic matter content (the total phosphorus content was 0.31 g/kg, the total nitrogen content was 1.27 g/kg, the total potassium content was 44.07 mg/kg, and the soil organic matter content was 19.27 g/kg) [26]. The vegetation type in the region is that of the desert steppe, and the constructive species is *Stipa breviflora*; dominant species include *Cleistogenes songorica* and *Artemisia frigida*, and associated species include *Convolvulus ammannii* and *Kochia prostrata* and other plants.

2.2. Experimental Design

The experiment commenced in 2014, and an open-top chamber (OTC) was used to increase temperature. The OTC used was a hexagonal open-top chamber with a side length of 1.5 m and two kinds of height, 1 m and 2.3 m (1 m and 2.3 m can increase temperature by 1–2 °C and 2–4 °C, respectively), with the top design being a fully or partial opening. The OTC was set up with a 3 m diameter, and a 1 m × 1 m quadrat placed in the center [27]. Precipitation increase was achieved by using rainwater interceptors with 25% and 50% of the basal area of the OTC, respectively. After each rainfall, the rainwater collected in the interceptors was poured into the corresponding plots by the workers (Figure 1a). The experiment treatments included three temperature gradients, ambient temperature (T0), temperature increased by 1–2 °C (T1), and temperature increased by 2–4 °C (T2), as well as three precipitation gradients, natural precipitation (P0), precipitation increased by 25% (P1), and precipitation increased by 50% (P2), with a total of nine treatments (i.e., T0P0, T0P1, T0P2, T1P0, T1P1, T1P2, T2P0, T2P1, and T2P2) and four replicates for each treatment, with a total of 36 plots (Figure 1b).

(a) (b)

Figure 1. Diagram of OTC warming device and precipitation collection device (**a**). Schematic diagram of the plot. T0, T1, T2, P0, P1, and P2 are ambient temperature, temperature increased by 1–2 °C, temperature increased by 2–4 °C, natural precipitation, precipitation increased by 25%, and precipitation increased by 50% (**b**).

Stipa breviflora was selected as the research object in the quadrat, and a 20 cm deep soil hole was drilled vertically next to it. An organic glass tube with a diameter of 7.5 cm was buried. The tube was marked at 10 cm and 20 cm below the ground with a black line. The exposed part of the tube was painted black and covered with an opaque lid.

2.3. Data Collection and Calculation

Image collection was carried out from May to October 2019 using a root scanner (CI-600, CID Bio-Science Inc., Camas, WA, USA), scanning every 15 days with a resolution of 11.6 pixels per mm. Rooyfly software (Version 2.0.2, Rootfly Development Team, Clemson, SC, USA) was used to process the scanned map for root growth. A total of 360 images were collected during the growth season. In total, 3673 roots were tracked and marked. Dead and living roots were identified by the color of the root, all black and disappearing roots were recorded as dead, and light brown and white roots were recorded as alive.

The calculations of root production, root standing crop, and root turnover mainly follow Bai and Huo et al. [25,28]. Root production was calculated by the sum of the length of all new roots and the extension length of previously existing roots in each sampling period. Annual mean root standing crop were calculated by the average length of live roots at each observing date of a year. Root turnover rate was calculated by annual root length production, dividing it by annual mean root standing crop

Plant community surveys were conducted in August 2019, with one permanent 1 m × 1 m quadrat in each experimental replicate, where the height, cover, density, and frequency of all species within the quadrats were recorded. Since the experimental restrictions did not allow for the destruction of the plants within the quadrats, 5–10 plants of equal size were selected around the quadrats to be mowed with the ground based on the results of the recordings and were brought back to the laboratory to be dried and weighed before calculating the aboveground production. We used Equation (1) to calculate the importance values of plant species in the quadrats [29].

$$\mathrm{IV} = (\mathrm{Hr} + \mathrm{Cr} + \mathrm{Dr} + \mathrm{Fr} + \mathrm{Br})/5 \quad (1)$$

Hr, Cr, Dr, Fr, and Br are the relative height, cover, density, frequency, and biomass, respectively.

Meteorological data were collected by a meteorological station (Dynamet, Dynamax Inc., CA, USA) in the experiment site.

Soil temperature and humidity were measured in the topsoil (0–10 cm), using a soil temperature and humidity meter (HH2 Moisture Meter, Delta-T Devices Ltd., Cambridge, UK), at noon on the same day as root observation, once in each plot.

2.4. Data Analysis

The Shapiro–Wilk test and Levene test were used to test the normality hypothesis and the homoscedasticity hypothesis of the residuals. The differences in soil temperature, soil moisture, ANPP, root production, root standing crop, and root turnover rate among the treatments were tested using both LSD and post hoc Duncan tests for multiple comparisons, with a significance level of 0.05. Two-way ANOVA was employed to compare the effects of warming, increased precipitation, and their interaction on the ANPP, root production, and root turnover rate (Duncan, $p < 0.05$). A structural equation model was executed in AMOS 21.0. First, possible explanatory variables were included in the path analysis by reviewing the relevant literature on plant root production and turnover. Secondly, we used a maximum likelihood estimation technique to obtain path coefficients and then eliminated the insignificant paths until we obtained the final model. Finally, we used a Chi-square test and the root mean square error of approximation to evaluate the fit of the model. All statistical analyses were conducted with IBM SPSS Statistics 26. Figures were elaborated using Origin 2021, and tables were elaborated using Microsoft Excel 2019.

3. Result

3.1. Seasonal Variations in the Root Length Production and Root Standing Crop of Plant Communities

Root length production was greatest in August during the experimental period (Figure 2). The root length production of the four treatments with T2P0, T2P1, T2P2, and T1P2 showed a trend of decreasing first in July and then increasing with a peak in production in August.

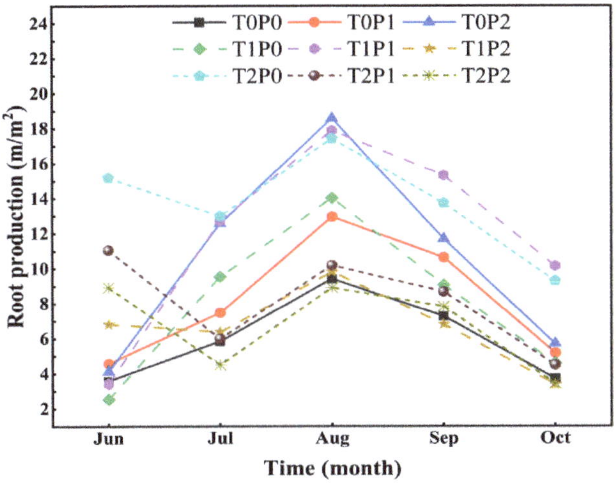

Figure 2. Seasonal variation in root production of plant community with different treatments in 0–20 cm soil. T0, T1, T2, P0, P1, and P2 are ambient temperature, temperature increased by 1–2 °C, temperature increased by 2–4 °C, natural precipitation, precipitation increased by 25%, and precipitation increased by 50%.

The peaks of the root standing crop appeared from the 22nd of July to the 8th of August, and the peaks of the T1P0 treatment, the T2P0 treatment, and the T2P2 treatment were in early July; the peak of the T0P2 treatment was in early August. The root standing crop of all the treatments showed a trend of first increasing and then decreasing throughout the experimental period, and the three treatments with precipitation increased by 50% (T0P2, T1P2, and T2P2), showing an increasing trend in the second half of August (Figure 3).

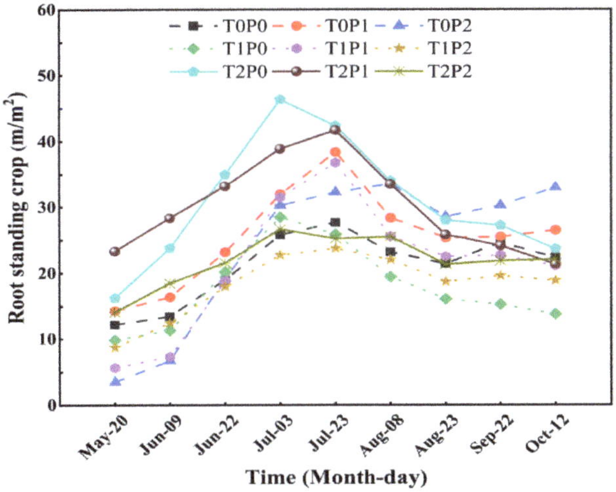

Figure 3. Seasonal variation in root standing crop with different treatments in 0–20 cm soil. T0, T1, T2, P0, P1, and P2 are ambient temperature, temperature increased by 1–2 °C, temperature increased by 2–4 °C, natural precipitation, precipitation increased by 25%, and precipitation increased by 50%.

3.2. *Effects of Different Treatments on ANPP and Root Length Production*

Warming, increased precipitation, and their interaction significantly affected ANPP ($p < 0.001$, Table 1). ANPP in all treatments was 132.34 ± 6.57 g/m^2 (Figure 4). The T1P0 and T2P2 treatment significantly increased community ANPP ($p < 0.05$) in comparison to the T0P0 treatment, with an increase of 49.20% and 79.14%, respectively.

Table 1. Two-way ANOVA of ANPP, root production, and root turnover of warming and increased precipitation.

Factor	df	ANPP		Root Production		Root Turnover	
		F	*p*	F	*p*	F	*p*
Warming	2	7.909	<0.001	2.792	0.088	28.588	<0.001
Increased precipitation	2	8.64	<0.001	2.363	0.123	7.583	<0.01
Warming × Increased precipitation	4	6.161	<0.001	6.755	<0.01	1.686	0.197

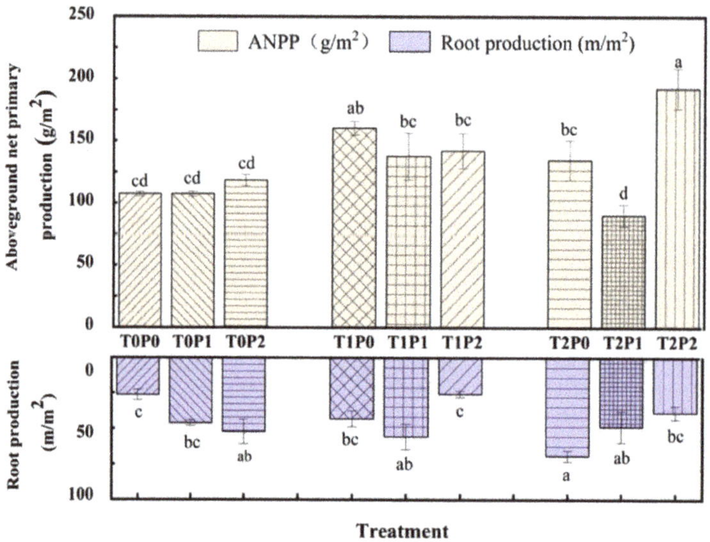

Figure 4. Effects of different treatments on ANPP and root production in *Stipa breviflora* community in 0–20 cm soil. ANPP: aboveground net primary production. Same lowercase letters indicate no significant differences between treatments ($p > 0.05$). T0, T1, T2, P0, P1, and P2 are ambient temperature, temperature increased by 1–2 °C, temperature increased by 2–4 °C, natural precipitation, precipitation increased by 25%, and precipitation increased by 50%.

The interaction of warming and increased precipitation significantly affected the root length production of the community in the growing season ($p < 0.01$, Table 1). In the 0–20 cm soil, the root length production of all the treatments was 45.14 ± 3.11 m/m^2 (Figure 4). Under natural temperature conditions, with the increase in precipitation, root length production showed an increasing trend, and root length production of the T0P2 treatment was significantly higher than that of the T0P0 treatment ($p < 0.05$), with an increase of 98.70%. The root length production of T1P1 was significantly higher than that of the T1P2 treatment and the T0P0 treatment ($p < 0.05$), with an increment of 116.15% and 111.72%, respectively. Under the condition of warming by 4 °C, with the increase in precipitation, root length production showed a decreasing trend, and the root length production of T2P2 treatment was significantly lower than that in the T2P0 treatment ($p < 0.05$), with a decrease of 78.45%. The T2P0 treatment and T2P1 treatment significantly increased root length production ($p < 0.05$) in comparison to the T0P0 treatment, with an increase of 163.03% and 85.14%, respectively.

3.3. Effects of Different Treatments on Root Turnover Rate of Plant Community

Root turnover rate tended to decrease with increasing precipitation under the same warming conditions (Figure 5). Both warming and increased precipitation significantly affected root turnover ($p < 0.01$, Table 1). In the 0–20 cm soil, the root turnover rate of all the treatments was 1.98 ± 0.11 year^{-1}. Under the condition of warming by 2 °C, the three precipitation treatments (T1P0, T1P1, and T1P2) significantly increased the root turnover rate ($p < 0.05$) in comparison to the T0P0 treatment, with increases of 62.53%, 42.57%, and 35.55%, respectively. Under natural precipitation conditions, warming resulted in an increasing trend in root turnover, which diminished with increasing temperatures. Root turnover showed a decreasing trend with increasing precipitation under warming conditions, and this decreasing trend was more pronounced at higher temperatures.

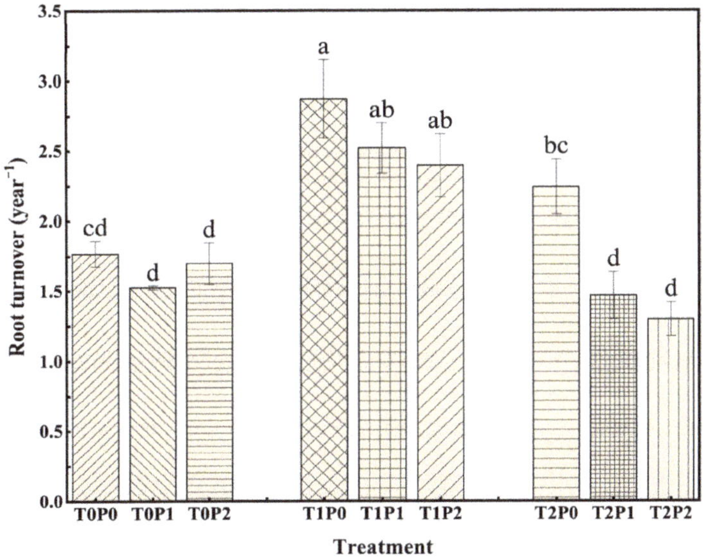

Figure 5. Effects of different treatments on root turnover in *Stipa breviflora* community in 0–20 cm soil. Same lowercase letters indicate no significant differences between treatments ($p > 0.05$). T0, T1, T2, P0, P1, and P2 are ambient temperature, temperature increased by 1–2 °C, temperature increased by 2–4 °C, natural precipitation, precipitation increased by 25%, and precipitation increased by 50%.

Through the structural equation model, it can be seen that the correlation between root turnover and environmental factors (temperature and precipitation) was high (Figure 6). Warming had a significant positive effect on the root turnover rate of the *Stipa breviflora* community (effect value of 0.332); however, increased precipitation had a significant negative effect on the root turnover of the *Stipa breviflora* community and important value of *Stipa breviflora* (effect values of −0.678 and −0.404). The effect of precipitation increase on the root turnover rate of the *Stipa breviflora* community was greater than that of warming.

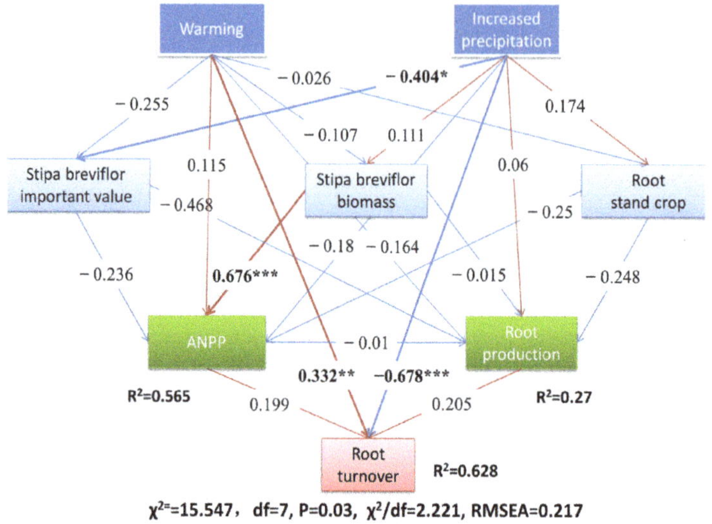

Figure 6. Structural equation model of warming and increased precipitation with *Stipa breviflora* and aboveground and belowground characteristics of community. The red and blue lines indicate significant positive and negative impacts, respectively. Numbers are standardized path coefficients. Goodness-of-fit index: χ^2 is Chi-square, df is degrees of freedom, and RMSEA is the root mean square error of approximation. *, **, and *** indicate $p < 0.05$, $p < 0.01$, and $p < 0.001$, respectively.

4. Discussion

4.1. Effects of Warming and Increased Precipitation on the Root Length Production of the *Stipa breviflora* Community

The results of this study only partially support our first and second hypotheses. The positive effects of warming on root length production and turnover rate were observed under natural precipitation treatments, which is consistent with other results obtained by warming experiments [19,25]. However, climate warming has an inhibitory effect on these variables under the conditions of increased precipitation. We found a significant interaction between warming and increased precipitation on root production and standing crop, which is partly consistent with the third hypothesis. The structural equation model showed that warming had a negative correlation with root production, and increased precipitation had a positive correlation with root production. The strong interaction between warming and increased precipitation implied that warming and increased precipitation in the desert steppe had a non-additive relationship with root production in the *Stipa breviflora* community. Because the increased temperatures can accelerate the rate of soil moisture evaporation [23], plants grow in order to absorb more water, which in turn promotes the growth of the primary root to deeper soil [9,24]. In environments with relatively abundant precipitation, the positive impact of climate change may be stronger than the negative impact, thereby promoting root production [30]. However, in arid or semi-arid ecosystems, moisture is the primary limiting factor, which limits the root production in desert steppe ecosystems, and the negative impact of climate warming may be stronger than the positive impact [25]. This study indicated that warming by 1.5–2 °C had a negative effect on root production and standing crop in alpine meadows [18], and similar conclusions were also drawn in typical temperate steppes [25]. In addition, temperature and precipitation had a significant interaction and effect on root production, but precipitation had different effects on root production under different temperature conditions. Because different magnitudes of warming have different effects on soil moisture content, different magnitudes of precipitation increase can improve soil moisture limitations [25].

This study found that warming, increased precipitation, or their interactions could promote root production, but the effect on the root standing crop was not significant. Because plants constantly adjust their nutrient distribution patterns according to different temperatures and precipitation environments [31], such as increased precipitation in desert steppes, more species will compete for this limited resource. Due to the increase in species, the height, coverage, and density of the community will change [32]. Plants with a better environment need to compete for more resources, such as sunlight, CO_2, moisture, and so on. Therefore, in order to better survive, plants will have different aboveground and belowground growth patterns and nutrient distribution patterns at different times. The difference in root production is the performance of plant roots in adapting to different environments, while the lack of difference in standing crop may be due to the long-term adaptation of plants to desert steppes or the fact that the difference has not yet been shown due to the short treatment time. Because this study only observed the roots in 0–20 cm soil, it may also affect the results of root production and standing crop. However, a study of temperate steppes in northern China (including meadow steppes, typical steppes, and desert steppes) showed that most of the root lengths of the fibrous-rooted plant species were within 0–20 cm, and only a few taproot plant species had a root length longer than 20 cm [33], but these did not belong to the dominant species of desert steppes. Therefore, the root observation in 0–20 cm soil in this study conforms to the majority of root growth ranges.

This study found that increasing precipitation by 50% under warming conditions did not affect root production but increased ANPP. Because substantial warming leads to the evaporation of soil moisture, increased precipitation reduces soil moisture dissipation, lowers soil temperatures, increases plant cover, and promotes aboveground plant growth [34]. Small increases in temperature have been shown to increase plant ANPP in American short-grass prairie [35], which is the same as the results of this study. Perhaps it is because slight warming enhances the aboveground productivity of plants by promoting photosynthesis and prolonging their phenology [36]. Significant warming promotes the early re-greening of plants, and the appropriate temperature in the early growing season can promote the accumulation of nutrients in plants [36]. However, when the temperature is high in the late growing season, significant warming increases the severity of soil drought and slows down plant growth.

4.2. Effects of Warming and Increased Precipitation on the Root Turnover of the Stipa breviflora Community

In this study, the root turnover rate was positively correlated with warming and negatively correlated with precipitation increase, which was consistent with the first and second hypotheses. This study also found that warming by 2 °C promoted root turnover. Warming accelerated nutrient mineralization rate and root respiration rate, resulting in shortened optimal root lifespan [37,38], Decayed roots promote the decomposition of roots by microorganisms and return nutrients from the roots to the soil for plant reabsorption [39,40], thereby accelerating nutrient cycling. One study found that the fine root turnover of steppes increases exponentially with global warming [36], which is similar to some of the results of this study. Studies have shown that shrubs have short-lived roots under drought conditions [41], and studies have shown that desert plants grow fine roots quickly when it rains and fall off when it is dry [42], which may be a strategy for plants to cope with drought stress.

4.3. Prediction of Root Production and Root Turnover in the Desert Steppe in the Future, Based on Historical Climate Change

The changes in the root production and turnover of plant communities in the desert steppe in the future depend on the precipitation variation in the region. From 1960 to 2013, the annual average temperature and rainfall in most regions of northwest China showed an increasing trend [4]. According to this trend, it is predicted that the root productivity of communities in the desert steppe in northwest China will increase, but this is also regulated by precipitation; a small increase in temperature and precipitation (a 2 °C and 25% increase in precipitation in this study) is beneficial to the turnover of community roots, but a greater increase in temperature and precipitation (a 4 °C and 50% increase in precipitation in this study) does not affect the turnover of community roots. The experimental site in this study is located in the central part of Inner Mongolia. Based on the meteorological data of Inner Mongolia from 1961 to 2012, the annual rainfall in the central part of Inner Mongolia has not changed after 1975 (except for several wet years), while the temperature has increased by 0.37 °C/decade, which is significantly higher than the global warming rate (0.14 °C/decade) [43].

5. Conclusions

In the desert steppe of northern China, the main effect of experimental warming on root turnover was positive, and the main effect of increased precipitation on root turnover was negative. This study found that warming and increased precipitation had a significant interaction and effect on root production and standing crop, and the interaction was non-additive. Based on this trend, it is predicted that future climate change will be beneficial to the root production and turnover of desert steppe communities in Inner Mongolia. These results will provide theoretical assistance for predicting the response of temperate desert steppes to global climate change.

Supplementary Materials: The following supporting information can be downloaded at https://www.mdpi.com/article/10.3390/agronomy14071521/s1, Figure S1: Effects of different treatments on mean root standing crop in *Stipa breviflora* community in 0–20 cm soil; Table S1: Precipitation and temperature in the growing season of the experimental site in 2019; Table S2: Effects of different treatments on soil temperature and moisture (August).

Author Contributions: Conceptualization, J.G., Q.L. and Z.W.; Methodology, J.G., Q.L. and Z.W.; Experimentation, Q.L. and P.L.; Analysis, Q.L., P.L., C.W. (Chengjie Wang), X.Q. and G.L.; Writing—original draft preparation, J.G. and Q.L.; Writing—review and editing, Z.W., C.W. (Chunjie Wang), G.L., Z.Y. and X.Q.; Fund acquisition, Z.W., J.G. and Z.Y. All authors have read and agreed to the published version of the manuscript.

Funding: This study was funded by the Center Leading Local Science and Technology Development Plan (2021ZY0020), the National Science Foundation of China (32160331), the IWHR Research & Development Support Program (MK0145B022021), and the Open Research Fund of Yinshanbeilu Grassland Eco-Hydrology National Observation and Research Station (YSS202109), and the Natural Science Foundation of Inner Mongolia Autonomous Region of China (2024MS03005).

Data Availability Statement: The original contributions presented in the study are included in the article/Supplementary Materials, further inquiries can be directed to the corresponding author.

Conflicts of Interest: The authors declare no conflicts of interest.

References

1. Boucher, O.; Randall, D.; Artaxo, P.; Bretherton, C.; Feingold, G.; Forster, P.; Kerminen, V.-M.; Kondo, Y.; Liao, H.; Lohmann, U.; et al. Contribution of Working Group to the Fifth Assessment Report of the Intergovernmental Panel on Climate Change. Clouds and Aerosols. In *The Physical Science Basis*; Cambridge University Press: Cambridge, UK, 2013; Volume 2013, pp. 616–617.
2. Buytaert, W.; Cuesta-Camacho, F.; Tobón, C. Potential Impacts of Climate Change on the Environmental Services of Humid Tropical Alpine Regions. *Glob. Ecol. Biogeogr.* **2011**, *20*, 19–33. [CrossRef]
3. You, Q.L.; Cai, Z.Y.; Wu, F.Y.; Jiang, Z.H.; Pepin, N.; Shen, S.S.P. Temperature Dataset of Cmip6 Models over China: Evaluation, Trend and Uncertainty. *Clim. Dyn.* **2021**, *57*, 17–35. [CrossRef]
4. Yang, P.; Xia, J.; Zhang, Y.Y.; Hong, S. Temporal and Spatial Variations of Precipitation in Northwest China during 1960–2013. *Atmos. Res.* **2017**, *183*, 283–295. [CrossRef]
5. Liu, H.Y.; Lin, L.; Wang, H.; Zhang, Z.H.; Shangguan, Z.J.; Feng, X.J.; He, J.S. Simulating Warmer and Drier Climate Increases Root Production but Decreases Root Decomposition in an Alpine Grassland on the Tibetan Plateau. *Plant Soil* **2021**, *458*, 59–73. [CrossRef]
6. Zhang, R.Y.; Wang, Z.W.; Han, G.D.; Schellenberg, M.P.; Wu, Q.; Gu, C. Grazing Induced Changes in Plant Diversity Is a Critical Factor Controlling Grassland Productivity in the Desert Steppe, Northern China. *Agric. Ecosyst. Environ.* **2018**, *265*, 73–83. [CrossRef]

7. Tariq, A.; Graciano, C.; Sardans, J.; Ullah, A.; Zeng, F.J.; Ullah, I.; Ahmed, Z.; Ali, S.; Al-Bakre, D.A.; Zhang, Z.H.; et al. Decade-Long Unsustainable Vegetation Management Practices Increase Macronutrient Losses from the Plant-Soil System in the Taklamakan Desert. *Ecol. Indic.* **2022**, *145*, 109653. [CrossRef]
8. Zhang, Y.N.; Wang, Z.Y.; Liu, P.B.; Wang, C.J. Mixed Cattle and Sheep Grazing Reduces the Root Lifespan of the Community in a Desert Steppe. *Ecol. Indic.* **2022**, *143*, 109422. [CrossRef]
9. Wang, C.T.; Zhao, X.Q.; Zi, H.B.; Hu, L.; Ade, L.; Wang, G.X.; Lerdau, M. The Effect of Simulated Warming on Root Dynamics and Soil Microbial Community in an Alpine Meadow of the Qinghai-Tibet Plateau. *Appl. Soil Ecol.* **2017**, *116*, 30–41. [CrossRef]
10. Lukac, M. Fine Root Turnover. In *Measuring Roots: An Updated Approach*; Mancuso, S., Ed.; Springer: Berlin/Heidelberg, Germany, 2012; pp. 363–373.
11. McCormack, M.L.; Guo, D.L. Impacts of Environmental Factors on Fine Root Lifespan. *Front. Plant Sci.* **2014**, *5*, 205. [CrossRef]
12. McCormack, M.L.; Adams, T.S.; Smithwick, E.A.H.; Eissenstat, D.M. Variability in Root Production, Phenology, and Turnover Rate among 12 Temperate Tree Species. *Ecology* **2014**, *95*, 2224–2235. [CrossRef]
13. Gao, Y.Z.; Chen, Q.; Lin, S.; Giese, M.; Brueck, H. Resource Manipulation Effects on Net Primary Production, Biomass Allocation and Rain-Use Efficiency of Two Semiarid Grassland Sites in Inner Mongolia, China. *Oecologia* **2011**, *165*, 855–864. [CrossRef]
14. Niu, B.; Zeng, C.X.; Zhang, X.; He, Y.T.; Shi, P.L.; Tian, Y.; Feng, Y.F.; Li, M.; Wang, Z.P.; Wang, X.T.; et al. High Below-Ground Productivity Allocation of Alpine Grasslands on the Northern Tibet. *Plants* **2019**, *8*, 535. [CrossRef]
15. Diabate, B.; Wang, X.Y.; Gao, Y.Z.; Yu, P.J.; Wu, Z.F.; Zhou, D.W.; Yang, H.J. Tillage and Haymaking Practices Speed up Belowground Net Productivity Restoration in the Degraded Songnen Grassland. *Soil Tillage Res.* **2018**, *175*, 62–70. [CrossRef]
16. Yang, Y.; Klein, J.A.; Winkler, D.E.; Peng, A.H.; Lazarus, B.E.; Germino, M.J.; Suding, K.N.; Smith, J.G.; Kueppers, L.M. Warming of Alpine Tundra Enhances Belowground Production and Shifts Community towards Resource Acquisition Traits. *Ecosphere* **2020**, *11*, e03270. [CrossRef]
17. Yan, C.; Yuan, Z.Y.; Liu, Z.C.; Zhang, J.J.; Liu, K.; Shi, X.R.; Lock, T.R.; Kallenbach, R.L. Aridity Stimulates Responses of Root Production and Turnover to Warming but Suppresses the Responses to Nitrogen Addition in Temperate Grasslands of Northern China. *Sci. Total Environ.* **2021**, *753*, 142018. [CrossRef] [PubMed]
18. Wu, Y.B.; Zhu, B.; Eissenstat, D.M.; Wang, S.P.; Tang, Y.H.; Cui, X.Y. Warming and Grazing Interact to Affect Root Dynamics in an Alpine Meadow. *Plant Soil* **2021**, *459*, 109–124. [CrossRef]
19. Yuan, Z.Y.; Chen, H.Y.H. Fine Root Biomass, Production, Turnover Rates, and Nutrient Contents in Boreal Forest Ecosystems in Relation to Species, Climate, Fertility, and Stand Age: Literature Review and Meta-Analyses. *Crit. Rev. Plant Sci.* **2010**, *29*, 204–221. [CrossRef]
20. Wang, J.S.; Sun, J.; Yu, Z.; Li, Y.; Tian, D.S.; Wang, B.X.; Li, Z.L.; Niu, S.L. Vegetation Type Controls Root Turnover in Global Grasslands. *Glob. Ecol. Biogeogr.* **2019**, *28*, 442–455. [CrossRef]
21. Yin, H.J.; Li, Y.F.; Xiao, J.; Xu, Z.F.; Cheng, X.Y.; Liu, Q. Enhanced Root Exudation Stimulates Soil Nitrogen Transformations in a Subalpine Coniferous Forest under Experimental Warming. *Glob. Chang. Biol.* **2013**, *19*, 2158–2167. [CrossRef]
22. Hertel, D.; Strecker, T.; Müller-Haubold, H.; Leuschner, C. Fine Root Biomass and Dynamics in Beech Forests across a Precipitation Gradient—Is Optimal Resource Partitioning Theory Applicable to Water-Limited Mature Trees? *J. Ecol.* **2013**, *101*, 1183–1200. [CrossRef]
23. Gimbel, K.F.; Felsmann, K.; Baudis, M.; Puhlmann, H.; Gessler, A.; Bruelheide, H.; Kayler, Z.; Ellerbrock, R.H.; Ulrich, A.; Welk, E.; et al. Drought in Forest Understory Ecosystems—A Novel Rainfall Reduction Experiment. *Biogeosciences* **2015**, *12*, 961–975. [CrossRef]
24. Joslin, J.D.; Wolfe, M.H.; Hanson, P.J. Effects of Altered Water Regimes on Forest Root Systems. *New Phytol.* **2000**, *147*, 117–129. [CrossRef]
25. Bai, W.M.; Wan, S.Q.; Niu, S.L.; Liu, W.X.; Chen, Q.S.; Wang, Q.B.; Zhang, W.H.; Han, X.G.; Li, L.H. Increased Temperature and Precipitation Interact to Affect Root Production, Mortality, and Turnover in a Temperate Steppe: Implications for Ecosystem C Cycling. *Glob. Chang. Biol.* **2010**, *16*, 1306–1316. [CrossRef]
26. Zhang, G.G.; Sui, X.; Li, Y.; Jia, M.Q.; Wang, Z.W.; Han, G.D.; Wang, L.C. The Response of Soil Nematode Fauna to Climate Drying and Warming in *Stipa breviflora* Desert Steppe in Inner Mongolia, China. *J. Soils Sediments* **2020**, *20*, 2166–2180. [CrossRef]
27. Lv, G.; He, M.; Wang, C.; Wang, Z. The Stability of Perennial Grasses Mediates the Negative Impacts of Long-Term Warming and Increasing Precipitation on Community Stability in a Desert Steppe. *Front. Plant Sci.* **2023**, *14*, 1235510. [CrossRef]
28. Huo, C.F.; Gu, J.C.; Yu, L.Z.; Wang, P.; Cheng, W.X. Temporal Dynamics of Fine Root Production, Mortality and Turnover Deviate across Branch Orders in a Larch Stand. *Oecologia* **2022**, *199*, 699–709. [CrossRef] [PubMed]
29. Curtis, J.T.; McIntosh, R.P. An Upland Forest Continuum in the Prairie-Forest Border Region of Wisconsin. *Ecology* **1951**, *32*, 476–496. [CrossRef]
30. Reynolds, J.F.; Virginia, R.A.; Kemp, P.R.; de Soyza, A.G.; Tremmel, D.C. Impact of Drought on Desert Shrubs: Effects of Seasonality and Degree of Resource Island Development. *Ecol. Monogr.* **1999**, *69*, 69–106. [CrossRef]
31. Domisch, T.; Finér, L.; Lehto, T. Growth, Carbohydrate and Nutrient Allocation of Scots Pine Seedlings after Exposure to Simulated Low Soil Temperature in Spring. *Plant Soil* **2002**, *246*, 75–86. [CrossRef]
32. Hu, Y.; Li, X.Y.; Guo, A.X.; Yue, P.; Guo, X.X.; Lv, P.; Zhao, S.L.; Zuo, X.A. Species Diversity Is a Strong Predictor of Ecosystem Multifunctionality under Altered Precipitation in Desert Steppes. *Ecol. Indic.* **2022**, *137*, 108762. [CrossRef]

33. Jiang, L.B.; Liu, H.Y.; Peng, Z.Y.; Dai, J.Y.; Zhao, F.J.; Chen, Z.T. Root System Plays an Important Role in Responses of Plant to Drought in the Steppe of China. *Land Degrad. Dev.* **2021**, *32*, 3498–3506. [CrossRef]
34. Li, G.Y.; Han, H.Y.; Du, Y.; Hui, D.F.; Xia, J.Y.; Niu, S.L.; Li, X.N.; Wan, S.Q. Effects of Warming and Increased Precipitation on Net Ecosystem Productivity: A Long-Term Manipulative Experiment in a Semiarid Grassland. *Agric. For. Meteorol.* **2017**, *232*, 359–366. [CrossRef]
35. Morgan, J.A.; LeCain, D.R.; Mosier, A.R.; Milchunas, D.G. Elevated CO_2 Enhances Water Relations and Productivity and Affects Gas Exchange in C3 and C4 Grasses of the Colorado Shortgrass Steppe. *Glob. Chang. Biol.* **2001**, *7*, 451–466. [CrossRef]
36. Dorji, T.; Totland, O.; Moe, S.R.; Hopping, K.A.; Pan, J.B.; Klein, J.A. Plant Functional Traits Mediate Reproductive Phenology and Success in Response to Experimental Warming and Snow Addition in Tibet. *Glob. Chang. Biol.* **2013**, *19*, 459–472. [CrossRef]
37. Gill, R.A.; Jackson, R.B. Global Patterns of Root Turnover for Terrestrial Ecosystems. *New Phytol.* **2000**, *147*, 13–31. [CrossRef]
38. Burton, A.J.; Pregitzer, K.S.; Hendrick, R.L. Relationships between Fine Root Dynamics and Nitrogen Availability in Michigan Northern Hardwood Forests. *Oecologia* **2000**, *125*, 389–399. [CrossRef]
39. Bartsch, N. Responses of Root Systems of Young Pinus-Sylvestris and Picea-Abies Plants to Water Deficits and Soil Acidity. *Can. J. For. Res.* **1987**, *17*, 805–812. [CrossRef]
40. Eissenstat, D.M.; Wells, C.E.; Yanai, R.D.; Whitbeck, J.L. Building Roots in a Changing Environment: Implications for Root Longevity. *New Phytol.* **2000**, *147*, 33–42. [CrossRef]
41. Stewart, A.M.; Frank, D.A. Short Sampling Intervals Reveal Very Rapid Root Turnover in a Temperate Grassland. *Oecologia* **2008**, *157*, 453–458. [CrossRef]
42. Jackson, R.B.; Caldwell, M.M. The Timing and Degree of Root Proliferation in Fertile-Soil Microsites for 3 Cold-Desert Perennials. *Oecologia* **1989**, *81*, 149–153. [CrossRef]
43. Hu, Q.; Pan, F.F.; Pan, X.B.; Zhang, D.; Li, Q.Y.; Pan, Z.H.; Wei, Y.R. Spatial Analysis of Climate Change in Inner Mongolia during 1961–2012, China. *Appl. Geogr.* **2015**, *60*, 254–260. [CrossRef]

Disclaimer/Publisher's Note: The statements, opinions and data contained in all publications are solely those of the individual author(s) and contributor(s) and not of MDPI and/or the editor(s). MDPI and/or the editor(s) disclaim responsibility for any injury to people or property resulting from any ideas, methods, instructions or products referred to in the content.

forests

MDPI

Article

Repeated Mild Summer Drought in *Crataegus monogyna* Jacq. Provokes Compensation Growth in the Following Year

Kristine Vander Mijnsbrugge [1,*], Stefaan Moreels [1], Laura Decorte [1], Marie Stessens [1] and Eduardo Notivol Paino [2]

[1] Department of Forest Ecology and Management, Research Institute for Nature and Forest, 9500 Geraardsbergen, Belgium; stefaan.moreels@inbo.be (S.M.); laura.decorte@kuleuven.be (L.D.); m.e.m.stessens@rug.nl (M.S.)

[2] Department for Environment, Agricultural and Forest Systems, Agri-Food Research and Technology Centre of Aragon (CITA), 50059 Zaragoza, Spain; enotivol@cita-aragon.es

* Correspondence: kristine.vandermijnsbrugge@inbo.be

Abstract: Water limitations will pose significant challenges to forest ecosystems across Europe. To gain a deeper understanding of the potential impacts, we investigated the response of the common shrub *Crataegus monogyna* to two summer droughts, each followed by rewatering. The experimental design consisted of a common garden with potted saplings from a local Belgian (n = 48), a Swedish (n = 47), and a Spanish-Pyrenean provenance (n = 48). We quantified the effects on growth and leaf phenology, focusing on the legacies in the year following the droughts. Responses were influenced by the severity of the drought and by its timing. Most strikingly, height increment was enhanced by 24% ($p = 0.046$) in comparison to the controls in the year following the droughts in the group of plants that endured the two drought treatments, each time without developing visible stress symptoms. Only one such mild drought, whether early or late summer, did not lead to this response, suggesting stress memory acting as a growth promoter. A late summer drought that resulted in visible drought symptoms led to a reduced diameter increment in the year following the droughts, independent of the preceding treatment (severe, mild, or no drought), whereas this was not the case for a similar drought in early summer. Minor leaf phenological responses were detected in the year following the droughts. Finally, the non-local provenances did not respond in a deviating way to the droughts compared to the local provenance. Our findings contribute to the prediction of carbon sequestration in forests and other woody vegetations in the temperate regions of Europe.

Keywords: stress memory; drought stress; provenance trial; increment growth; common hawthorn; recurrent drought

1. Introduction

Forests play a fundamental role in hydrological and biogeochemical cycles, offering a wide array of ecosystem services crucial for biodiversity conservation and maintenance [1,2]. The escalation of extreme weather events such as heatwaves, droughts, and storms due to climate change can heighten tree mortality rates, potentially resulting in a net release of atmospheric CO_2. Among these events, drought and its associated disturbances exert the most substantial global impact on forests [3,4]. It is hypothesized that trees may not readily adapt to abrupt aridity increases through evolutionary mechanisms, given their long reproductive cycles and limited ability to migrate away from stressful conditions. The looming threat of prolonged and intensified droughts due to global warming justifies the study of the effects of drought on woody plants [4,5].

During drought conditions, plants eventually experience desiccation, which leads to a decrease in cell turgor pressure, resulting in the closure of stomatal pores on leaf surfaces. This closure significantly slows down the dehydration process [6]. Studies indicate that in trees, stomata typically close before significant cavitation in the hydraulic system occurs, at the expense of negative impacts of stomatal closure, such as reduced carbon assimilation [7].

However, additional water loss can still occur through cuticular conductance, stomatal leakiness, and other tissues like bark [7].

Plant hydraulic strategies are finely tuned to their environmental conditions, aiming to maximize carbon assimilation while avoiding hydraulic failure during drought stress. Leaf shedding is a common adaptive strategy during drought conditions, considered vital for plant survival [8]. Typically, leaf shedding occurs after stomatal closure [7], with xylem embolism in the leaves serving as a primary driver of leaf mortality during drought [9–11]. This shedding reduces the evaporative leaf surface area, helping woody perennials delay cavitation initiation in stem conductive tissues [12,13]. Consequently, it eases water stress on the remaining foliage, slowing desiccation rates [14,15]. However, shedding leaves without complete nutrient resorption leads to net nutrient losses, affecting long-term tree performance [16]. Rebuilding a damaged crown after severe drought requires additional carbon investment, either from non-structural carbohydrate reserves or from assimilation of remaining or newly grown leaves after the drought stress [17].

While there has been significant attention on understanding the physiological factors leading to tree mortality during drought, it is equally important to grasp the mechanisms involved in the recovery of drought [18]. Drought resilience can be measured by assessing both the impact of the disturbance and the rate of recovery following it [19]. If little or no cavitation has occurred, recovery after rainfall is rapid, with stomata reopening to assimilate new carbon as plants rehydrate. However, when cavitation thresholds are exceeded, photosynthetic recovery is considerably slower [20]. In summary, summer drought significantly impacts the growth, gas exchange, and photosynthetic performance of both shrubs and trees, including differences in drought tolerance strategies among co-occurring species [21,22].

Drought and heat pose a threat to ecosystem carbon uptake, which currently helps in mitigating the rising atmospheric CO_2 levels by offsetting human-caused fossil fuel emissions [23]. Reduced forest carbon uptake during drought and heat comes from stress-related declines in photosynthesis [24]. While the connection between drought and reduced carbon uptake is well-documented, several key questions remain [23]. These include the impact of recurrent droughts, the strength of seasonal and regional compensation effects, land–atmosphere feedbacks that can intensify heatwaves, and forest management strategies in a changing climate.

In this study, we conducted a controlled recurrent drought experiment using three provenances of *Crataegus monogyna* Jacq. in a common garden setting. *C. monogyna*, commonly known as hawthorn, is a shrub species traditionally and widely planted in hedgerows in the temperate regions of Europe. Despite being a common shrub, hawthorn has been relatively neglected both in scientific research on woody species, as it holds no economic value in the forestry sector, and in conservation efforts, which typically prioritize rare species [25]. Apart from in situ measurements of physiological traits [26] or secondary metabolites in *Crataegus* spp. [27] in relation to drought, this study describes an experimental drought in controlled conditions, focusing on growth responses.

Potted saplings were exposed to water deprivation during early and/or late summer within the same growing season, followed by rewatering. This experimental setup allowed us to investigate the potential impact of prior drought exposure. We hypothesized that the response to the imposed drought would be influenced by (i) the severity of the drought and (ii) whether saplings had previously experienced a drought. In addition, we hypothesized that (i) drought not only leads to above-ground growth reduction but can also act as an above-ground growth stimulus, and that (ii) different provenances may react in a different way to imposed drought regimes. Our main objective was to understand the post-drought recovery process, focusing specifically on the effects on growth traits and leaf phenology in the year following the drought. The common garden setting provided a controlled environment for assessing variability among the different provenances.

2. Materials and Methods

2.1. Plant Material

Our study species is *Crataegus monogyna*. It is widespread across nearly all of Europe, thriving on soils with moderate pH, nutrient levels, and water supply [28]. In May and June, the shrubs bloom abundantly, offering pollen and nectar to various insects. By August, the shrubs bear numerous berries that are a food source for many birds [28]. Hawthorn is extensively planted in hedgerows, wooded banks, and the edges and understory of forest stands in Belgium to enhance species diversity, restore historical landscapes, and support wildlife [25,29].

We utilized 143 *C. monogyna* plants belonging to three provenances (48 plants from a local Belgian provenance, 48 plants from a Spanish-Pyrenean provenance, and 47 plants from a south-Swedish provenance). The seed collection process is already outlined in [30], as *C. monogyna* was collected in the same growth environments as described for *Prunus spinosa*. Seeds were gathered in 2016 and germinated in 2018 (*C. monogyna* seeds need a relative long stratification time). In the first growing season, the seedlings remained in forestry trays (54.5 × 31 cm with 28 cells) filled with standard nursery potting soil composed of 1.5 kg/m^3 NPK 12 + 14 + 24, 20% organic matter, pH levels ranging from 5.0 to 6.5, an electrical conductivity (E.C.) of 450 µS/cm, and dry matter content of 25%. No additional fertilizer was added to the potting soil. In the winter of 2018, the seedlings were transferred to 1 L pots, using the same standard potting soil, without additional fertilizer. A common garden of potted seedlings was established in 2019 on an outdoor container field situated at the Research Institute for Agriculture and Fisheries in Melle, Belgium, where the pots were watered using automated sprinklers that were programmed and monitored by experienced technical staff. The three provenances were individually mingled (single tree plot design).

2.2. Drought Treatments

In mid-June 2020, we transferred all potted plants from the container field to a greenhouse equipped with an automatic grey shade net to shield the plants from intense solar radiation. Monthly mean temperatures in the greenhouse for July till September were 21.8, 24.3, and 19.7 °C, whereas the mean relative humidity was 54.8, 55.3, and 60%, respectively. We conducted controlled drought treatments during early and late summer of 2020 following a full-factorial design (Figure 1). The early summer treatment was conducted from 1 July to 23 July, while the late summer treatment ran from 6 August to 21 August. To avoid mortality, both treatments were ended when several plants displayed (nearly) total leaf desiccation. At the beginning and at the end of each treatment, we ensured all plants, both control and drought treated, were fully hydrated by immersing the pots overnight in a water basin, with the water level maintained at 5 cm above the pot base. Subsequently, excess water was allowed to drain naturally. In this way, field capacity was approached.

For the initial early summer treatment, half of the young plants (designated as the drought group D, n = 72, each provenance with 24 plants) were subjected to water withholding, while the other half (the control group C, n = 71, each provenance with 24 or 23 plants) received regular watering by skilled personnel to maintain optimal moisture levels. In the subsequent late summer treatment, we divided both the control and drought groups from the early summer treatment into halves, with each half then assigned to either the control or the drought condition for the late summer treatment (Figure 1). The four experimental groups resulting from this design are referred to as control-control (C-C), control-drought (C-D), drought-control (D-C), and drought-drought (D-D). The allocation of the three provenances was evenly distributed among the control and drought groups for both the early and late summer treatment.

After the two treatments, the plants were kept in the greenhouse until October 2020, ensuring they were consistently maintained under well-watered conditions as monitored by experienced technical staff. In October 2020, they were transferred back to the container field outside. All treatment groups were individually intermingled.

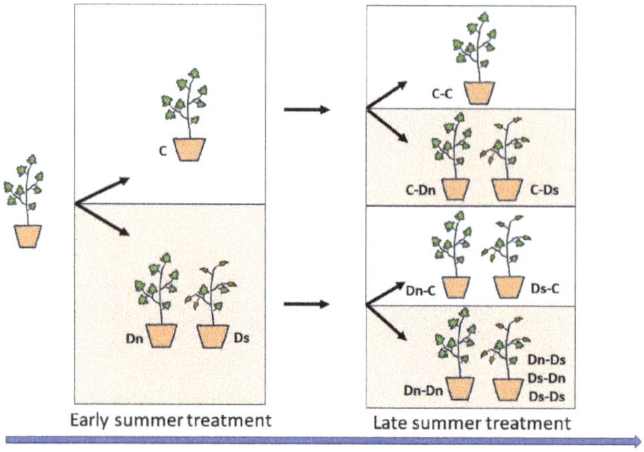

Figure 1. Factorial design for the two consecutive drought treatments during the 2020 growing season with four treatment groups: C-C (control-control), C-D (control-drought), D-C (drought-control), and D-D (drought-drought). For each treatment, plants subjected to drought conditions were additionally categorized based on whether they exhibited visible drought symptoms (Ds) or remained symptom-free (Dn).

2.3. Measurements and Observations

We conducted several measurements and observations throughout the experiment. We weighed all pots at the beginning of each drought treatment, after drainage of excess water following the overnight soaking in water (proxy for field capacity), and more or less weekly thereafter during the treatments. The reduction in pot weight served as an indicator of water scarcity (Figure 2).

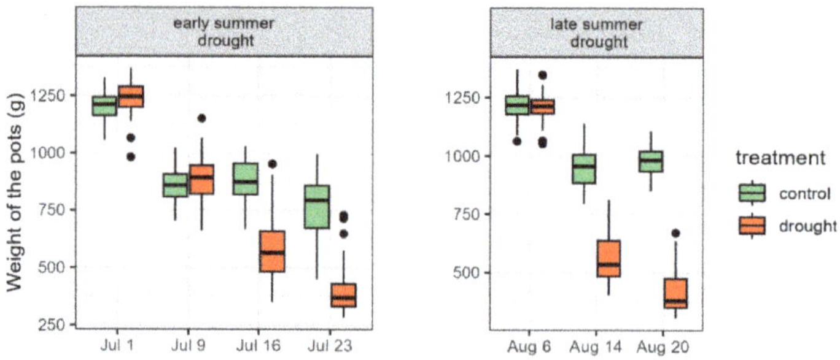

Figure 2. Weights of the pots during the two treatments.

Height and diameter measurements were taken at the start of each treatment (1 July and 6 August) and at the end of 2020 (20 December for diameter) or at the beginning of the growing season in 2021 (25 March for height) (Figure 3). Height was measured up to the still-living part of the plant. Height measurements in December 2020 were considered less reliable as it was visually less clear if top parts of stems had died off. This became unmistakable at the spring bud burst in 2021. We measured the stem diameter using a measuring rod positioned 2 cm above the soil level. At the end of the growing season of 2021, height and diameter were measured once more. We calculated the increment height and diameter by subtracting the height and diameter on 1 July from the ones on 6 August for the early summer treatment, by subtracting the height and diameter on August 6 from the ones in the winter between 2020 and 2021 for the late summer treatment, and by subtracting the height and diameter in the winter between 2020 and 2021 from the ones at the end of 2021 for the following growing season.

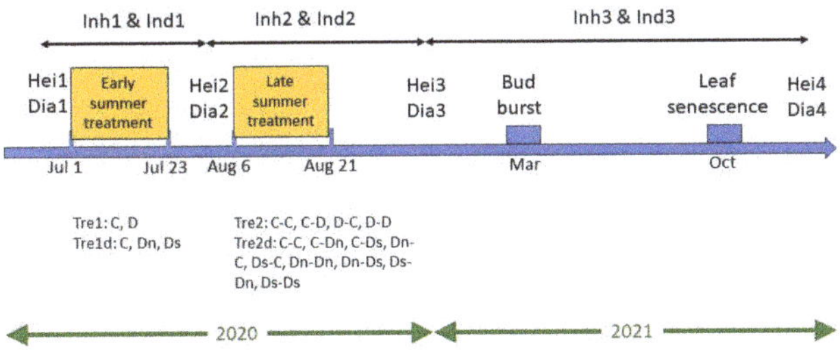

Figure 3. Timeline with indication of treatments and measurements. Included are heights (Hei) and diameters (Dia) with their respective increments (Inh and Ind). Treatments are shown (Tre1 and Tre2) with control (C), drought (D), control-control (C-C), control-drought (C-D), drought-control (D-C), and drought-drought (D-D). Plants subjected to drought conditions were additionally categorized (Tre1d and Tre2d) based on whether they exhibited visible drought symptoms (Ds) or remained symptom-free (Dn).

We assessed the presence/absence of visual drought symptoms on the plants, i.e., the desiccation of leaves, about one week after the end of each treatment (29 July and 28 August). At this time, affected leaves would not regain turgor and revive anymore. Based on this assessment, the groups of plants in both drought treatments were divided in subgroups: whether or not they were displaying visual drought symptoms (Ds and Dn, respectively). This resulted in the following number of plants in each treatment group: 36 in C-C, 13 in C-Dn, 23 in C-Ds, 25 in Dn-C, 11 in Ds-C, 17 in Dn-Dn, 7 in Dn-Ds, 8 in Ds-Dn, and 3 in Ds-Ds. Because of the low number, results from the Ds-Ds group are shown for completeness but are not discussed.

Bud burst in the spring of 2021 was evaluated following a scoring protocol with 1: buds in rest, 2: buds swelling and changing colour from brown/red to white and green, 3: buds opening and first leaves protruding but not yet unfolding, 4: leaves unfolding, and 5: leaves unfolded and enlarged [25]. Bud burst was scored on 5, 15, and 26 March and 6 April. Leaf senescence in the autumn of 2021 was observed using a scoring protocol with 1: green leaves, 2: leaves light green, 3: less than half of the leaves becoming brown, 4: more than half of the leaves becoming brown, and 5: all leaves brown and starting to fall off [31]. This phenophase was scored on 19 September and 18 October. For both phenological traits, the whole plant (i.e., all buds or all leaves) was evaluated and a mean score was given.

2.4. Statistical Analysis

We utilized the open-source statistical software R [32] for all data analyses. We employed linear models for height and diameter or the change in height and diameter (increments) and we processed the phenological observations, which are ordinal data, using cumulative logistic regression in the package ordinal [33]. Figures were generated using ggplot2 [34]. For the phenological data, we included a unique plant identifier as a random effect in the mixed-effect models to address repeated observations on the same plants.

To evaluate the effects of the droughts on the growth of the plants, we modelled the change in height and diameter over different time spans (increment response variables), each time taking into account the initial height and diameter. For each response variable, we first looked at the influence of the treatment, distinguishing the effects of mild (having no visual drought stress symptoms at the end of the treatment) or more severe (having visual drought stress symptoms) drought. The treatment variables (Tre1d or Tre2d) consisted of three (C, Dn, and Ds for the first treatment) or nine levels (C-C, C-Dn, C-Ds, Dn-C, Ds-C, Dn-Dn, Dn-Ds, Ds-Dn, and Ds-Ds for the first and second treatment). Secondly, we questioned whether the change in height or diameter due to the treatments differed between the provenances (Pro). In other words, we looked at significant interaction terms between provenance and treatment in the models. The amount of plants in the experiment was not large enough to distinguish in these provenance models between mild and more severe drought. Therefore, the treatment variables in these provenance models (Tre1 or Tre2) consisted of two (C and D for the first treatment) or four levels (C-C, C-D, D-C, and D-D for the second treatment).

Height (Hei1) and diameter (Dia1) at the start of the experiment were modelled to look for initial growth differences between the provenances:

$$\text{Hei1} = \beta_0 + \beta_1\text{Pro}$$

$$\text{Dia1} = \beta_0 + \beta_1\text{Pro}$$

The change in height (Inh1) and diameter (Ind1) during the early summer treatment was modelled:

$$\text{Inh1} = \beta_0 + \beta_1\text{Tre1d} + \beta_2\text{Hei1}$$

$$\text{Ind1} = \beta_0 + \beta_1\text{Tre1d} + \beta_2\text{Dia1}$$

Influence of provenance:

$$\text{Inh1}_\text{P} = \beta_0 + \beta_1\text{Tre1} \times \beta_2\text{Pro} + \beta_3\text{Hei1}$$

$$\text{Ind1}_\text{P} = \beta_0 + \beta_1\text{Tre1} \times \beta_2\text{Pro} + \beta_3\text{Dia1}$$

The change in height (Inh2) and diameter (Ind2) during and after the late summer treatment was modelled:

$$\text{Inh2} = \beta_0 + \beta_1\text{Tre2d} + \beta_2\text{Hei2}$$

$$\text{Ind2} = \beta_0 + \beta_1\text{Tre2d} + \beta_2\text{Dia2}$$

Influence of provenance:

$$\text{Inh2}_\text{P} = \beta_0 + \beta_1\text{Tre2} \times \beta_2\text{Prov} + \beta_3\text{Hei2}$$

$$\text{Ind2}_\text{P} = \beta_0 + \beta_1\text{Tre2} \times \beta_2\text{Prov} + \beta_3\text{Dia2}$$

The change in height (Inh3) and diameter (Ind3) during the growing season of 2021 was modelled:

$$\text{Inh3} = \beta_0 + \beta_1\text{Tre2d} + \beta_2\text{Hei3}$$

$$\text{Ind3} = \beta_0 + \beta_1\text{Tre2d} + \beta_2\text{Dia3}$$

Influence of provenance:

$$\text{Inh3}_\text{P} = \beta_0 + \beta_1\text{Tre2} \times \beta_2\text{Prov} + \beta_3\text{Hei3}$$

$$\text{Ind3}_\text{P} = \beta_0 + \beta_1\text{Tre2} \times \beta_2\text{Prov} + \beta_3\text{Dia3}$$

We looked for carry-over effects of the treatments on the leaf phenological traits bud burst (Bud) and leaf senescence (Sen) in the year after the treatments. We modelled the chance (p_{bud}) that a plant on a given day had already reached a given bud burst score, or a score higher than this. Similarly, we modelled the chance (p_{sen}) that the leaves of a plant on a given day were still in a given senescence score level or in a level lower than this.

$$(p_{bud}/1\text{-}p_{bud}) = \beta_0 - \beta_1\text{Day} - \beta_2\text{Tre2d} - \beta_3\text{Hei3}$$

$$(p_{sen}/1\text{-}p_{sen}) = \beta_0 - \beta_1 Day - \beta_2 Tre2d - \beta_3 Hei4$$

Influence of provenance:

$$(p_{budP}/1\text{-}p_{budP}) = \beta_0 - \beta_1 Day - \beta_2 Tre2 \times \beta_3 Prov - \beta_4 Hei3$$

$$(p_{senP}/1\text{-}p_{senP}) = \beta_0 - \beta_1 Day - \beta_2 Tre2 \times \beta_3 Prov - \beta_4 Hei4$$

3. Results

3.1. Growth Traits in the Year of the Drought Treatments

At the start of the early summer drought, the Belgian provenance was higher than the Spanish one but shorter than the Swedish one (Table 1, Figure 4a). The diameter of this local provenance was still smaller than the diameter of the Swedish provenance, but there was no significant difference between the diameters of the local and the Spanish provenance (Table 1, Figure 4b).

Table 1. Test statistics for the height (Hei1) and diameter (Dia1) at the start of the first treatment. The Belgian provenance is the standard to which the Spanish-Pyrenean (Sp) and the Swedish (Sw) provenances are compared.

Response Variable	Variable	Estimate	St. Error	t-Value	*p*-Value
Hei1	(intercept)	40.40	1.84	21.90	<0.001 ***
	Sp	−5.44	2.61	−2.09	0.039 *
	Sw	8.88	2.62	3.39	<0.001 ***
Dia1	(intercept)	4.32	0.13	34.40	<0.001 ***
	Sp	−0.19	0.18	−1.08	0.283
	Sw	0.43	0.18	2.41	0.017 *

*** $p < 0.001$; * $p < 0.05$.

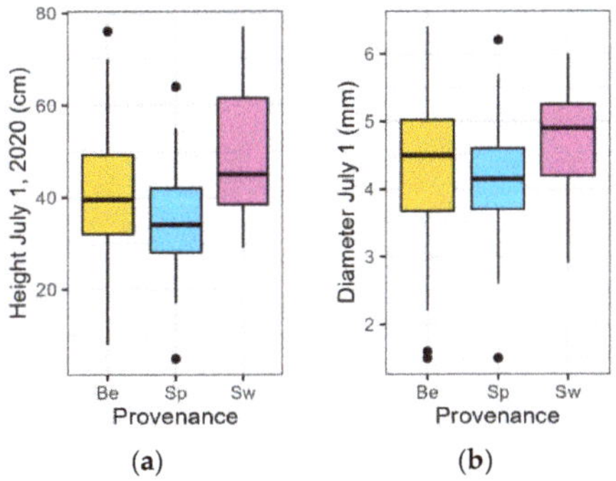

Figure 4. Boxplots displaying the initial height (**a**) and diameter (**b**) of the plants at the start of the first treatment, according to their provenance.

When considering plants of equal height (height is a co-factor in the model), the early summer drought caused a smaller height increment between 1 July and 6 August in the group of saplings that displayed leaf desiccation symptoms due to the drought (significant treatment group Ds in Table 2, Figure 5a). The Spanish-Pyrenean and Swedish provenances did not show a deviating response to this early summer drought compared to the Belgian one (no significant interaction term between provenance and treatment in Table 2). For the diameter increment between 1 July and 6 August, no influence of the treatment was present (no significant treatment groups in Table 2) and the provenances also reacted in a similar manner to the treatment (no significant interaction term between provenance and treatment in Table 2).

Table 2. Test statistics for the height (Inh1 and Inh1p) and diameter increment (Ind1 and Ind1p) during the first treatment, between July 1 and August 6. The control group is the standard to which the drought treated group is compared. For the models without provenance (Inh1 and Ind1), the drought-treated group (D) is further divided according to the absence (Dn) or presence (Ds) of visual drought symptoms. This division is not retained in the provenance models (Inh1p and Ind1p). Here, the Belgian provenance is the standard to which the Spanish-Pyrenean (Sp) and the Swedish (Sw) provenances are compared.

Response Variable	Variable	Estimate	St. Error	t-Value	*p*-Value
Inh1	(intercept)	−2.35	1.73	−1.35	0.178
	Dn	−1.67	1.14	−1.47	0.145
	Ds	−3.96	1.49	−2.66	0.009 **
	Hei1	0.22	0.04	5.94	<0.001 ***
Ind1	(intercept)	0.52	0.20	2.63	0.010 **
	Dn	0.07	0.08	0.82	0.411
	Ds	0.03	0.11	0.29	0.775
	Dia1	−0.03	0.04	−0.73	0.465
Inh1p	(intercept)	0.72	2.06	0.35	0.726
	Sp	−4.70	1.74	−2.70	0.008 **
	Sw	−3.78	1.75	−2.16	0.033 *
	D	−3.26	1.72	−1.89	0.060
	Hei1	0.22	0.04	5.44	<0.001 ***
	Sp:D	2.24	2.43	0.92	0.358
	Sw:D	0.30	2.44	0.12	0.903
Ind1p	(intercept)	0.68	0.20	3.34	0.001 **
	Sp	−0.16	0.12	−1.35	0.179
	Sw	0.33	0.12	2.74	0.007 **
	D	0.09	0.12	0.75	0.457
	Dia1	−0.08	0.04	−1.93	0.056
	Sp:D	0.16	0.17	0.91	0.365
	Sw:D	−0.29	0.17	−1.71	0.091

*** $p < 0.001$; ** $p < 0.01$; * $p < 0.05$.

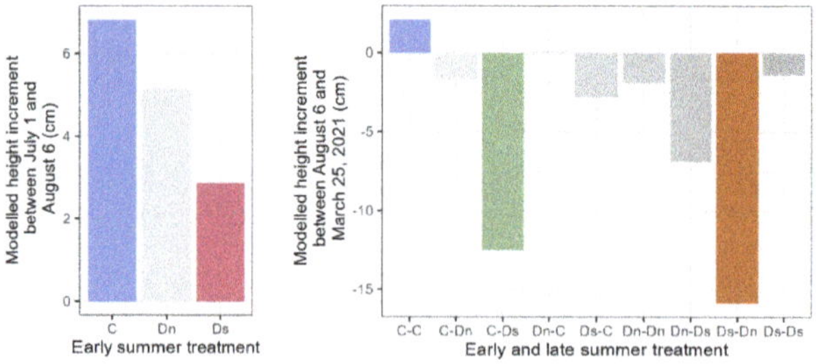

Figure 5. Modelled change in height during the first treatment, between 1 July and 6 August (**a**), and during and after the second treatment, between 6 August and 25 March in the following year (**b**). Treatment groups with control (C), drought (D), control-control (C-C), control-drought (C-D), drought-control (D-C), and drought-drought (D-D), and with a further division of droughted groups according to the absence (Dn) or presence (Ds) of visible drought symptoms. Groups not differing significantly from the control (C and C-C) are in grey.

Due to the late summer drought, a reduction in height increment, as measured between 6 August and 25 March of 2021, was observed in the C-Ds and Ds-Dn groups (significant treatment groups in Table 3, Figure 5b) because of top parts dying off. Similar to the early summer drought treatment, the Spanish-Pyrenean and Swedish provenances did not respond differently to the late summer drought compared to the Belgian one (no significant interaction term between provenance and treatment groups in Table 3). For the diameter increment between 6 August and 20 December, no influence of the treatment was present (no significant treatment groups in Table 3) and the provenances also reacted in a similar manner to the treatments (no significant interaction terms between provenance and treatment in Table 3).

Table 3. Test statistics for the height (Inh2 and Inh2$_P$) and diameter increment (Ind2 and Ind2$_P$) during and after the second treatment, between August 6 and March 25 of 2021 for height and between August 6 and December 20 for diameter. The control-control group is the standard to which the other groups are compared. Treatment groups with control-control (C-C), control-drought (C-D), drought-control (D-C), and drought-drought (D-D). For the models without provenance (Inh2 and Ind2), the drought-treated groups are further divided according to the absence (Dn) or presence (Ds) of visual drought symptoms. This division is not retained in the provenance models (Inh2$_P$ and Ind2$_P$). Here, the Belgian provenance is the standard to which the Spanish-Pyrenean (Sp) and the Swedish (Sw) provenances are compared.

Response Variable	Variable	Estimate	St. Error	t-Value	*p*-Value
Inh2	(intercept)	8.18	4.11	1.99	0.048 *
	C-Dn	−3.82	4.50	−0.85	0.397
	C-Ds	−14.65	3.59	−4.08	0.000 ***
	Dn-C	−2.02	3.48	−0.58	0.563
	Ds-C	−4.96	4.58	−1.08	0.280
	Dn-Dn	−4.05	4.02	−1.01	0.315
	Dn-Ds	−9.02	5.49	−1.64	0.103
	Ds-Dn	−18.02	5.19	−3.47	0.001 ***
	Ds-Ds	−3.56	8.01	−0.45	0.657
	Hei2	−0.14	0.07	−2.04	0.043 *
Ind2	(intercept)	13.14	5.26	2.50	0.014 *
	Sp	−4.68	5.65	−0.83	0.409
	Sw	1.19	5.72	0.21	0.835
	C-D	−11.27	5.66	−1.99	0.048 *
	D-C	−4.60	5.63	−0.82	0.416
	D-D	−11.48	5.63	−2.04	0.044 *
	Hei2	−0.22	0.07	−3.01	0.003 **
	Sp: C-D	6.37	8.02	0.80	0.428
	Sw: C-D	−5.17	8.03	−0.64	0.521
	Sp: D-C	5.80	7.96	0.73	0.467
	Sw: D-C	−1.63	7.97	−0.21	0.838
	Sp: D-D	9.32	7.96	1.17	0.244
	Sw: D-D	−1.19	8.08	−0.15	0.883
Inh2$_P$	(intercept)	0.70	0.24	2.92	0.004 **
	C-Dn	−0.13	0.16	−0.80	0.427
	C-Ds	−0.13	0.12	−1.11	0.271
	Dn-C	−0.23	0.12	−1.88	0.062
	Ds-C	−0.14	0.16	−0.88	0.378
	Dn-Dn	−0.27	0.14	−1.94	0.055
	Dn-Ds	−0.13	0.19	−0.71	0.476
	Ds-Dn	−0.06	0.18	−0.34	0.736
	Ds-Ds	0.47	0.27	1.70	0.091
	Dia2	−0.07	0.05	−1.50	0.137
Ind2$_P$	(intercept)	0.86	0.25	3.40	0.001 ***
	Sp	−0.38	0.19	−2.00	0.047 *
	Sw	−0.10	0.19	−0.51	0.614
	C-D	−0.27	0.19	−1.42	0.157
	D-C	−0.42	0.19	−2.25	0.026 *
	D-D	−0.32	0.19	−1.71	0.090
	Dia2	−0.07	0.04	−1.52	0.131
	Sp: C-D	0.39	0.27	1.46	0.148
	Sw: C-D	0.01	0.26	0.05	0.957
	Sp: D-C	0.50	0.26	1.90	0.060
	Sw: D-C	0.17	0.27	0.63	0.533
	Sp: D-D	0.26	0.26	1.00	0.322
	Sw: D-D	0.32	0.27	1.21	0.230

*** $p < 0.001$; ** $p < 0.01$; * $p < 0.05$.

3.2. Growth Traits in the Year Following the Drought Treatments

When plants of the same height were compared (height is a co-factor in the models), the treatment group C-Ds exhibited a reduced height increment in 2021 compared to the control group (C-C), while, surprisingly, Dn-Dn showed a greater increment (significant treatment groups in Table 4, Figure 6a). The change in height increment in 2021 due to the drought treatments in 2020 did not differ among the three studied provenances (no significant interaction terms between provenance and treatment in Table 4). For the diameter increment in the year after the treatments, the groups experiencing a drought treatment in the late summer and displaying visual drought symptoms (C-Ds, Dn-Ds, and Ds-Ds) had a significant smaller increment compared to the control (C-C) (significant treatment groups in Table 4, Figure 6b). The Swedish provenance displayed a significantly smaller diameter increment than the Belgian provenance for the treatment group C-D (significant interaction term between provenance and treatment group in Table 4, Figure 7).

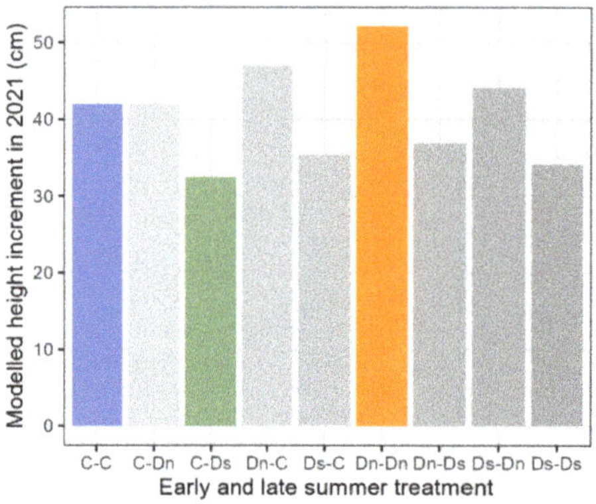

(a)

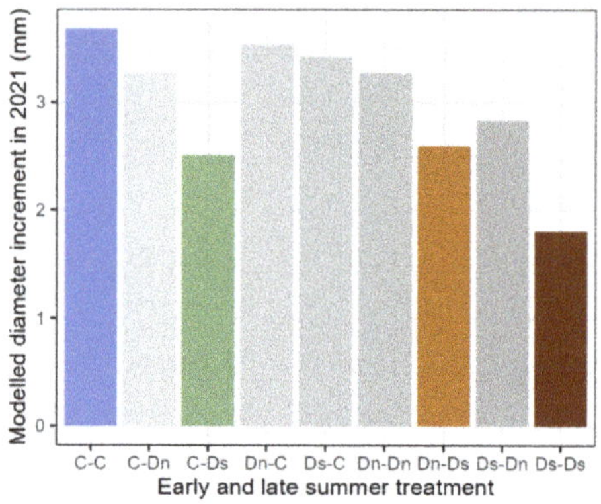

(b)

Figure 6. Modelled change in height (**a**) and diameter (**b**) during the year after the drought treatments. Treatment groups with control-control (C-C), control-drought (C-D), drought-control (D-C), and drought-drought (D-D), and with a further division of droughted groups according to the absence (Dn) or presence (Ds) of visible drought symptoms. Groups not differing significantly from the control (C-C) are in grey.

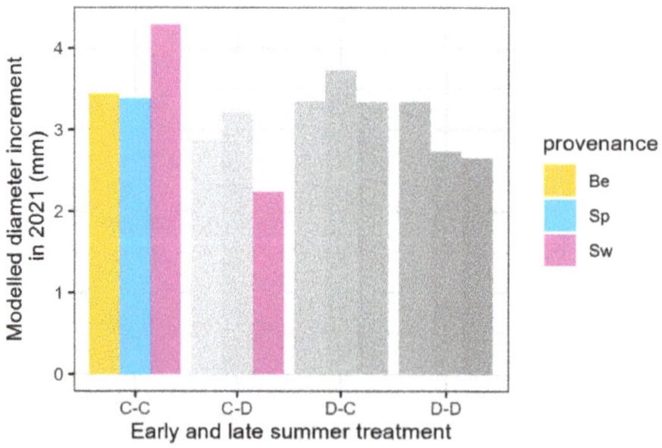

Figure 7. Modelled diameter increment during the year after the drought treatments, according to the provenance. Treatment groups with control-control (C-C), control-drought (C-D), drought-control (D-C), and drought-drought (D-D). Be: Belgian provenance, Sp: Spanish-Pyrenean provenance, Sw: Swedish provenance. Groups not differing significantly from the control (C-C) are in grey.

Table 4. Test statistics for the height (Inh3 and Inh3p) and diameter increment (Ind3 and Ind3p) in the year following the treatments. The control-control group is the standard to which the other groups are compared. Treatment groups are control-control (C-C), control-drought (C-D), drought-control (D-C), and drought-drought (D-D). For the models without provenance (Inh3 and Ind3), the drought-treated groups are further divided according to the absence (Dn) or presence (Ds) of visual drought symptoms. This division is not retained in the provenance models (Inh3p and Ind3p). Here, the Belgian provenance is the standard to which the Spanish-Pyrenean (Sp) and the Swedish (Sw) provenances are compared.

Response Variable	Variable	Estimate	St. Error	t-Value	*p*-Value
Inh3	(intercept)	48.92	4.69	10.44	<0.001 ***
	C-Dn	0.00	5.58	0.00	1.000
	C-Ds	−9.48	4.47	−2.12	0.036 *
	Dn-C	5.02	4.36	1.15	0.252
	Ds-C	−6.61	5.73	−1.15	0.251
	Dn-Dn	10.12	5.03	2.01	0.046 *
	Dn-Ds	−5.08	6.88	−0.74	0.462
	Ds-Dn	2.17	6.62	0.33	0.743
	Ds-Ds	−7.81	10.02	−0.78	0.437
	Hei3	−0.16	0.07	−2.19	0.030 *
Ind3	(intercept)	2.98	0.58	5.13	<0.001 ***
	C-Dn	−0.40	0.38	−1.07	0.288
	C-Ds	−1.17	0.29	−4.00	<0.001 ***
	Dn-C	−0.14	0.29	−0.49	0.625
	Ds-C	−0.25	0.38	−0.67	0.502
	Dn-Dn	−0.41	0.33	−1.22	0.226
	Dn-Ds	−1.08	0.45	−2.39	0.018 *
	Ds-Dn	−0.84	0.45	−1.86	0.066
	Ds-Ds	−1.87	0.66	−2.85	0.005 **
	Dia3	0.14	0.10	1.34	0.182
Inh3p	(intercept)	47.62	6.28	7.58	<0.001 ***
	Sp	4.00	6.97	0.57	0.567
	Sw	6.32	6.99	0.90	0.368
	C-D	−3.61	6.93	−0.52	0.603
	D-C	−1.39	6.94	−0.20	0.841
	D-D	0.84	7.01	0.12	0.905
	Hei3	−0.20	0.08	−2.60	0.010 *
	Sp: C-D	2.57	9.79	0.26	0.793
	Sw: C-D	−11.44	9.87	−1.16	0.249
	Sp: D-C	0.85	9.80	0.09	0.931
	Sw: D-C	6.96	9.79	0.71	0.478
	Sp: D-D	5.12	9.81	0.52	0.602
	Sw: D-D	1.98	9.92	0.20	0.842
Ind3p	(intercept)	3.19	0.62	5.17	<0.001 ***
	Sp	−0.05	0.44	−0.12	0.903
	Sw	0.85	0.45	1.90	0.060
	C-D	−0.56	0.44	−1.28	0.205
	D-C	−0.09	0.45	−0.19	0.846
	D-D	−0.09	0.45	−0.20	0.838
	Dia3	0.05	0.10	0.49	0.623
	Sp: C-D	0.40	0.63	0.63	0.527
	Sw: C-D	−1.49	0.62	−2.38	0.019 *
	Sp: D-C	0.43	0.63	0.69	0.493
	Sw: D-C	−0.86	0.63	−1.37	0.173
	Sp: D-D	−0.56	0.63	−0.89	0.376
	Sw: D-D	−1.54	0.64	−2.41	0.017 *

*** $p < 0.001$; ** $p < 0.01$; * $p < 0.05$.

3.3. Leaf Phenological Traits in the Year Following the Drought Treatments

In the year following the treatments, only the group Ds-C displayed an advanced bud burst (significant treatment group in Table 5, Figure 8a). Regarding leaf senescence in autumn, only the group C-Ds displayed a delayed leaf senescence (significant treatment group in Table 5, Figure 8b). When looking at the different provenances, only the Spanish-Pyrenean provenance differed from the Belgian provenance for both bud burst (later) and leaf senescence (later) (significant provenances in Table 5, Figure 9a,b). The phenological responses of the Spanish-Pyrenean and Swedish provenances in 2021 to the different treatments in 2020 did not deviate from the Belgian provenance (no significant interaction terms between provenance and treatment groups in Table 5).

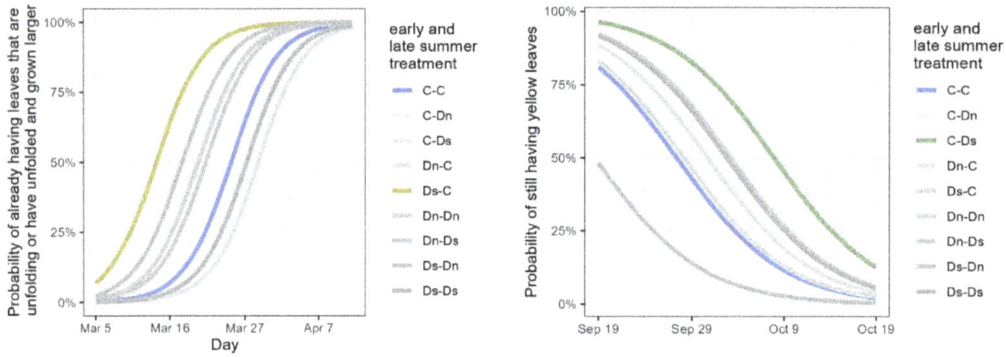

(**a**) (**b**)

Figure 8. Modelled bud burst (**a**) and leaf senescence (**b**) in the year after the treatments, according to the treatments in the year before. Treatment groups with control-control (C-C), control-drought (C-D), drought-control (D-C), and drought-drought (D-D), and with the drought treated groups further divided according to the absence (Dn) or presence (Ds) of visual drought symptoms. Groups not differing significantly from the control (C-C) are in grey.

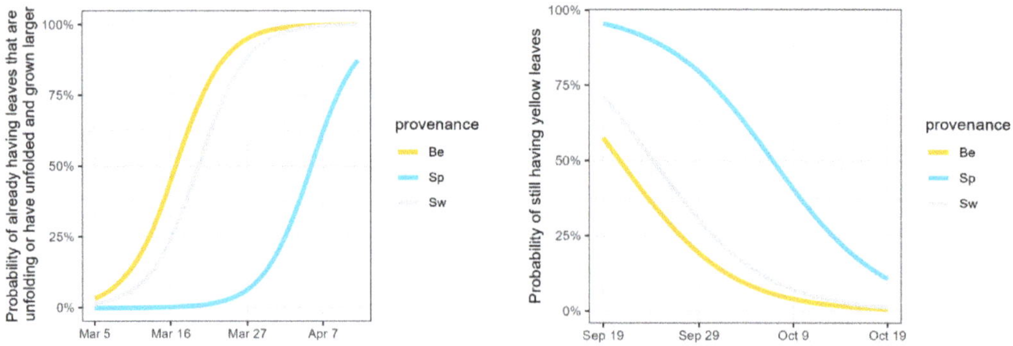

(**a**) (**b**)

Figure 9. Modelled bud burst (**a**) and leaf senescence (**b**) in the year after the drought treatments according to the provenance. Treatment groups with control-control (C-C), control-drought (C-D), drought-control (D-C), and drought-drought (D-D). Be: Belgian provenance, Sp: Spanish-Pyrenean provenance, Sw: Swedish provenance. Groups not differing significantly from the control are in grey (i.e., Swedish provenance).

Table 5. Test statistics for bud burst (Bud) and leaf senescence (Sen) in the year following the treatments. The control-control group is the standard to which the other groups are compared. Treatment groups are control-control (C-C), control-drought (C-D), drought-control (D-C), and drought-drought (D-D). For the phenological models (Bud and Sen) without provenance, the drought-treated groups are further divided according to the absence (Dn) or presence (Ds) of visual drought symptoms. This division is not retained in the provenance models (Bud_P and Sen_P). Here, the Belgian provenance is the standard to which the Spanish-Pyrenean (Sp) and the Swedish (Sw) provenances are compared.

Response Variable	Variable	Estimate	St. Error	z-Value	p-Value
Bud	day	−0.29	0.02	−14.66	<0.001 ***
	C-Dn	0.64	1.18	0.55	0.585
	C-Ds	−1.39	0.99	−1.41	0.160
	Dn-C	1.19	0.91	1.31	0.192
	Ds-C	−3.14	1.22	−2.56	0.010 *
	Dn-Dn	−1.37	1.07	−1.29	0.199
	Dn-Ds	−1.05	1.54	−0.69	0.493
	Ds-Dn	−2.04	1.46	−1.40	0.162
	Ds-Ds	0.69	2.10	0.33	0.743
	Hei3	0.01	0.02	0.53	0.597
Sen	day	0.17	0.03	6.80	<0.001 ***
	C-Dn	−1.91	1.03	−1.85	0.064
	C-Ds	−1.84	0.87	−2.11	0.035 *
	Dn-C	−0.59	0.81	−0.73	0.468
	Ds-C	−0.14	1.07	−0.13	0.895
	Dn-Dn	−1.03	0.92	−1.12	0.262
	Dn-Ds	−0.96	1.27	−0.75	0.452
	Ds-Dn	1.53	1.24	1.23	0.217
	Ds-Ds	−0.95	1.84	−0.52	0.606
	Hei4	0.06	0.01	4.19	<0.001 ***
Bud_P	day	−0.29	0.02	−14.64	<0.001 ***
	Sp	5.62	1.27	4.44	<0.001 ***
	Sw	0.93	1.22	0.76	0.445
	C-D	0.40	1.23	0.33	0.743
	D-C	−0.50	1.21	−0.42	0.678
	D-D	−0.92	1.24	−0.74	0.461
	Hei3	0.03	0.01	1.95	0.052
	Sp: C-D	−1.38	1.75	−0.79	0.428
	Sw: C-D	−1.04	1.78	−0.58	0.560
	Sp: D-C	0.62	1.71	0.36	0.716
	Sw: D-C	0.92	1.71	0.54	0.591
	Sp: D-D	−1.73	1.73	−1.00	0.317
	Sw: D-D	1.30	1.77	0.74	0.462
Sen_P	day	0.17	0.03	6.95	<0.001 ***
	Sp	−2.79	1.20	−2.32	0.020 *
	Sw	−0.60	1.19	−0.51	0.611
	C-D	−2.78	1.20	−2.32	0.021 *
	D-C	−0.97	1.17	−0.83	0.408
	D-D	0.60	1.17	0.51	0.611
	Hei4	0.05	0.01	4.28	0.000
	Sp: C-D	1.06	1.63	0.65	0.515
	Sw: C-D	1.65	1.68	0.98	0.327
	Sp: D-C	0.90	1.64	0.55	0.582
	Sw: D-C	0.66	1.66	0.40	0.691
	Sp: D-D	−1.08	1.65	−0.66	0.511
	Sw: D-D	−2.10	1.68	−1.25	0.210

*** $p < 0.001$; * $p < 0.05$.

4. Discussion

In our experiment, we observed not only a reduced radial growth in the year following a repeated summer drought treatment, a well-known reaction, but also an enhanced height growth. The responses were influenced by the severity of the drought and by its timing within the growing season.

4.1. Compensation Growth

Height increment in the year following the droughts was enhanced in the group of plants that experienced both of the two drought treatments without developing visible drought stress symptoms (Dn-Dn). Although rather counter-intuitive, enhanced growth during the recovery phase following adverse growing conditions is a known phenomenon and is called compensation growth [35,36]. The term is used among plant and animal systems [37]. It is an adaptive strategy employed by organisms facing challenging environmental conditions, serving as a mechanism that contributes to ecosystem stability [38]. Accelerated growth has been studied as a response to herbivory, both in herbaceous species [39] and in woody perennials [40,41], and also upon other disturbances such as arthropod attack, disease, fire, lightning strikes, floods, drought, and blowdown (reviewed by [reviewed by 37]). In grassland ecosystems responding to drought, two distinct biotic mechanisms have been proposed to lead to compensation growth [42]. The first occurs at the population level and involves an increase in the abundance of drought-tolerant plants, which functionally compensate for the decline in drought-intolerant species [43]. The second mechanism, at the individual level, is termed physiological compensation [44] and is suggested to take place when all individuals within the community recover following a drought event. Physiological compensation typically occurs in response to short-term or moderate drought conditions, whereas shifts in species abundances are more likely to manifest under long-term or extreme drought [36].

Our result of enhanced height increment in the year following two successive mild summer droughts fits in the description of physiological compensation growth, specifically as environmental conditions for all individual plants were similar in our common garden set-up, with exception of the water supply during the treatments. This contrasts with the description of compensation growth in forests by Li et al. [37], where enhanced growth of individual trees is associated with improved access to resources such as nutrients, light, water, etc. resulting from a disturbance, which can be natural (fire, disease, etc.) or man-made (e.g., thinning). Thus, the question arises of which resources plants use for physiological compensation growth when access to external resources is not improved. The answer can possibly be found in the distinction between accumulated versus stored reserves [45]. The accumulated fraction of the total non-structural carbohydrates in trees is characterised by a rapid turnover rate that mitigates negative carbon balances in the recurrent annual growth cycle. Conversely, the stored fraction is marked by a gradual turnover rate that may meet the demand during stressful conditions. In this sense, physiological compensation growth can be achieved by using stored non-structural carbohydrates which are only released upon disturbances.

Compensation growth on an individual level following rehydration after drought stress has already been described in oak [46,47], Douglas-fir [48], Scotch pine [49], and Thuja occidentalis [50]. Oak and Douglas fir are tree species characterized by a cyclic growth pattern whereby drought-treated plants displayed a higher chance of an extra growth cycle after post-drought rewatering. An interesting observation in our experiment is the occurrence of compensation growth only after two mild drought treatments, not after only one, suggesting stress memory. A plant can adjust to environmental pressures, improving a plant's subsequent response to future stressors [51,52]. Drought stress can act as a priming event that initiates stress memory [17,18]. Our experiment exemplifies the triggering of compensation growth due to stress memory.

4.2. Timing of the Droughts and Legacies in the Timing of Leaf Phenological Traits

Woody vegetation experiences negative legacy effects after droughts, including lower growth than expected in the years following the drought [53]. Even more, reduced radial growth can occur together with post-drought upregulation of photosynthesis [54]. It can be expected that the seasonal growth stages have typical windows of vulnerability to droughts [55]. Our experiment corroborates this. A late summer drought that caused visible stress symptoms (C-Ds and Dn-Ds) resulted in a reduced radial increment in the subsequent year, whereas this was not the case for a severe early summer drought (Ds-C and Ds-Dn), suggesting that post-drought photosynthesis during late summer was able to help the recovery of the early summer droughted plants. This finding may be related to the enhanced post-drought photosynthesis observed in beech [56]. When considering the impact of the severe late summer drought on the height increments (C-Ds and Dn-Ds), only the C-Ds group displayed a reduced height increment in the following year, possibly (partly) attributable to stress memory in the Dn-Ds group [18].

Minor effects were observed on the phenological traits in the year after the treatments. Only the treatment group Ds-C displayed an advanced bud burst. This is possibly a rather random effect. Regarding leaf senescence in autumn, only the treatment group C-Ds displayed a delayed leaf senescence, while this group also demonstrated a reduced growth increment in both height and diameter in 2021, suggesting that more severe droughts may influence the timing of leaf senescence. Although drought often advances autumnal leaf senescence, it may also delay it [57–59]. Finally, whereas bud burst in general is more genetically determined, the timing of leaf senescence may be more sensitive to environmental triggers such as drought [60].

4.3. *Influence of Provenance*

For all the growth traits that we studied, none displayed a deviating response to the drought among the different provenances, suggesting a general drought-response strategy in *C. monogyna*. In the timing of the leaf phenological traits, we found the Spanish-Pyrenean provenance to deviate from the local Belgian provenance with a later bud burst and a later leaf senescence. Although the Swedish provenance did not display a later or earlier bud burst and leaf senescence, the C-D treatment reduced the diameter increment in the year following the drought treatment in this provenance more than in the Belgian provenance. This effect may have been caused by the fact that initial plants in the Swedish provenance were higher.

5. Conclusions

As has similarly been observed in agricultural crops and herbaceous plants, mild drought can act as a growth stimulus. This was observed for the shrub *Crataegus monogyna* after repeated mild drought within one growing season, each time followed by plentiful rewatering. Although these conditions may occur rarely in situ, it still may influence the outcome of carbon sequestration modelling in woody vegetations. Research is needed to look for similar responses in major forest tree species.

Author Contributions: Conceptualization, K.V.M., E.N.P. and S.M.; methodology, K.V.M., E.N.P. and S.M.; investigation, M.S., L.D. and S.M.; validation, K.V.M.; formal analysis, K.V.M., L.D. and M.S.; data curation, K.V.M., L.D. and M.S.; writing—original draft preparation, K.V.M.; writing—review and editing, K.V.M., M.S., L.D. and E.N.P.; supervision, K.V.M. and E.N.P. All authors have read and agreed to the published version of the manuscript.

Funding: This research received no external funding.

Data Availability Statement: Data are available at 10.5281/zenodo.12205688.

Acknowledgments: We thank Yorrick Aguas Guerreiro, Cédric Van Dun, and Matthieu Gallin for contributing to the data acquisition. Marc Schouppe and Nico De Regge helped in taking care of the plants.

Conflicts of Interest: The authors declare no conflicts of interest.

References

1. Bonan, G.B. Forests and climate change: Forcings, feedbacks, and the climate benefits of forests. *Science* **2008**, *320*, 1444–1449. [CrossRef] [PubMed]
2. Pan, Y.; Birdsey, R.A.; Fang, J.; Houghton, R.; Kauppi, P.E.; Kurz, W.A.; Phillips, O.L.; Shvidenko, A.; Lewis, S.L.; Canadell, J.G.; et al. A Large and Persistent Carbon Sink in the World's Forests. *Science* **2011**, *333*, 988–993. [CrossRef] [PubMed]
3. Reichstein, M.; Bahn, M.; Ciais, P.; Frank, D.; Mahecha, M.D.; Seneviratne, S.I.; Zscheischler, J.; Beer, C.; Buchmann, N.; Frank, D.C.; et al. Climate extremes and the carbon cycle. *Nature* **2013**, *500*, 287–295. [CrossRef] [PubMed]
4. Choat, B.; Brodribb, T.J.; Brodersen, C.R.; Duursma, R.A.; López, R.; Medlyn, B.E. Triggers of tree mortality under drought. *Nature* **2018**, *558*, 531–539. [CrossRef] [PubMed]
5. Trenberth, K.E.; Dai, A.; van der Schrier, G.; Jones, P.D.; Barichivich, J.; Briffa, K.R.; Sheffield, J. Global warming and changes in drought. *Nat. Clim. Chang.* **2014**, *4*, 17–22. [CrossRef]
6. Monteith, J.L. A reinterpretation of stomatal responses to humidity. *Plant Cell Environ.* **1995**, *18*, 357–364. [CrossRef]
7. Martin-StPaul, N.; Delzon, S.; Cochard, H. Plant resistance to drought depends on timely stomatal closure. *Ecol. Lett.* **2017**, *20*, 1437–1447. [CrossRef]
8. Munne-Bosch, S.; Alegre, L. Die and let live: Leaf senescence contributes to plant survival under drought stress. *Funct. Plant Biol.* **2004**, *31*, 203–216. [CrossRef] [PubMed]
9. Blackman, C.J.; Brodribb, T.J.; Jordan, G.J. Leaf hydraulics and drought stress: Response, recovery and survivorship in four woody temperate plant species. *Plant Cell Environ.* **2009**, *32*, 1584–1595. [CrossRef]
10. Blackman, C.J.; Creek, D.; Maier, C.; Aspinwall, M.J.; Drake, J.E.; Pfautsch, S.; O'Grady, A.; Delzon, S.; Medlyn, B.E.; Tissue, D.T.; et al. Drought response strategies and hydraulic traits contribute to mechanistic understanding of plant dry-down to hydraulic failure. *Tree Physiol.* **2019**, *39*, 910–924. [CrossRef]
11. Cardoso, A.A.; Batz, T.A.; McAdam, S.A.M. Xylem Embolism Resistance Determines Leaf Mortality during Drought in Persea americana. *Plant Physiol.* **2020**, *182*, 547–554. [CrossRef] [PubMed]
12. Wolfe, B.T.; Sperry, J.S.; Kursar, T.A. Does leaf shedding protect stems from cavitation during seasonal droughts? A test of the hydraulic fuse hypothesis. *New Phytol.* **2016**, *212*, 1007–1018. [CrossRef] [PubMed]
13. Hochberg, U.; Windt, C.W.; Ponomarenko, A.; Zhang, Y.-J.; Gersony, J.; Rockwell, F.E.; Holbrook, N.M. Stomatal Closure, Basal Leaf Embolism, and Shedding Protect the Hydraulic Integrity of Grape Stems. *Plant Physiol.* **2017**, *174*, 764–775. [CrossRef] [PubMed]
14. Vilagrosa, A.; Bellot, J.; Vallejo, V.R.; Gil-Pelegrin, E. Cavitation, stomatal conductance, and leaf dieback in seedlings of two co-occurring Mediterranean shrubs during an intense drought. *J. Exp. Bot.* **2003**, *54*, 2015–2024. [CrossRef] [PubMed]
15. Schuldt, B.; Buras, A.; Arend, M.; Vitasse, Y.; Beierkuhnlein, C.; Damm, A.; Gharun, M.; Grams, T.E.E.; Hauck, M.; Hajek, P.; et al. A first assessment of the impact of the extreme 2018 summer drought on Central European forests. *Basic Appl. Ecol.* **2020**, *45*, 86–103. [CrossRef]
16. Marchin, R.; Zeng, H.; Hoffmann, W. Drought-deciduous behavior reduces nutrient losses from temperate deciduous trees under severe drought. *Oecologia* **2010**, *163*, 845–854. [CrossRef]
17. Ruehr, N.K.; Grote, R.; Mayr, S.; Arneth, A. Beyond the extreme: Recovery of carbon and water relations in woody plants following heat and drought stress. *Tree Physiol* **2019**, *39*, 1285–1299. [CrossRef] [PubMed]
18. Gessler, A.; Bottero, A.; Marshall, J.; Arend, M. The way back: Recovery of trees from drought and its implication for acclimation. *New Phytol.* **2020**, *228*, 1704–1709. [CrossRef] [PubMed]
19. Ingrisch, J.; Bahn, M. Towards a Comparable Quantification of Resilience. *Trends Ecol. Evol.* **2018**, *33*, 251–259. [CrossRef]
20. Zeppel, M.J.B.; Harrison, S.P.; Adams, H.D.; Kelley, D.I.; Li, G.; Tissue, D.T.; Dawson, T.E.; Fensham, R.; Medlyn, B.E.; Palmer, A.; et al. Drought and resprouting plants. *New Phytol.* **2015**, *206*, 583–589. [CrossRef]
21. Bryant, K.; Fredericksen, B.; Hudiburg, T.; Rosenthal, D. Physiological strategies for handling summer water stress differ among co-existing species and between juvenile and mature trees. *Front. For. Glob. Chang.* **2023**, *5*, 1018789. [CrossRef]
22. Weemstra, M.; Eilmann, B.; Sass-Klaassen, U.G.W.; Sterck, F.J. Summer droughts limit tree growth across 10 temperate species on a productive forest site. *For. Ecol. Manag.* **2013**, *306*, 142–149. [CrossRef]
23. Wolf, S.; Paul-Limoges, E. Drought and heat reduce forest carbon uptake. *Nat. Commun.* **2023**, *14*, 6217. [CrossRef] [PubMed]
24. Ciais, P.; Reichstein, M.; Viovy, N.; Granier, A.; Ogée, J.; Allard, V.; Aubinet, M.; Buchmann, N.; Bernhofer, C.; Carrara, A.; et al. Europe-wide reduction in primary productivity caused by the heat and drought in 2003. *Nature* **2005**, *437*, 529–533. [CrossRef] [PubMed]
25. Vander Mijnsbrugge, K.; Onkelinx, T.; De Cuyper, B. Variation in bud burst and flower opening responses of local versus non-local provenances of hawthorn (*Crataegus monogyna* Jacq.) in Belgium. *Plant Syst. Evol.* **2014**, *301*, 1171–1179. [CrossRef]
26. Urban, J.; Matoušková, M.; Robb, W.; Jelínek, B.; Úradníček, L. Effect of Drought on Photosynthesis of Trees and Shrubs in Habitat Corridors. *Forests* **2023**, *14*, 1521. [CrossRef]
27. Kirakosyan, A.; Seymour, E.; Kaufman, P.B.; Warber, S.; Bolling, S.; Chang, S.C. Antioxidant Capacity of Polyphenolic Extracts from Leaves of Crataegus laevigata and Crataegus monogyna (Hawthorn) Subjected to Drought and Cold Stress. *J. Agric. Food Chem.* **2003**, *51*, 3973–3976. [CrossRef]
28. Fichtner, A.; Wissemann, V. Biological Flora of the British Isles: Crataegus monogyna. *J. Ecol.* **2021**, *109*, 541–571. [CrossRef]

29. Uyttenbroeck, R.; De Vos, B.; Vander Mijnsbrugge, K. *Verspreiding en Standplaats van Inheemse Bomen en Struiken in Vlaanderen. INBO.R.2014.1388322*; Research Institute of Nature and Forest: Brussels, Belgium, 2014.
30. Vander Mijnsbrugge, K.; Malanguis, J.M.; Moreels, S.; Turcsán, A.; Van der Schueren, N.; Notivol Mijnsbrugge, E. Direct Phenological Responses but Later Growth Stimulation upon Spring and Summer/Autumn Warming of *Prunus spinosa* L. in a Common Garden Environment. *Forests* **2022**, *13*, 23. [CrossRef]
31. Vander Mijnsbrugge, K.; Janssens, A. Differentiation and Non-Linear Responses in Temporal Phenotypic Plasticity of Seasonal Phenophases in a Common Garden of *Crataegus monogyna* Jacq. *Forests* **2019**, *10*, 293. [CrossRef]
32. R_Core_Team. *R: A Language and Environment for Statistical Computing. R Foundation for Statistical Computing, version 4.3.0*; R_Core_Team: Vienna, Austria, 2023.
33. Christensen, R.H.B. Ordinal: Regression Models for Ordinal data. R Package Version 2015.6-28. Available online: http://www.cran.r-project.org/package=ordinal/ (accessed on 11 July 2024).
34. Wickham, H. *ggplot2: Elegant Graphics for data Analysis*; Springer: New York, NY, USA, 2016.
35. Belsky, A.J. Does Herbivory Benefit Plants? A Review of the Evidence. *Am. Nat.* **1986**, *127*, 870–892. [CrossRef]
36. Zhou, H.; Hou, L.; Lv, X.; Yang, G.; Wang, Y.; Wang, X. Compensatory growth as a response to post-drought in grassland. *Front. Plant Sci.* **2022**, *13*, 1004553. [CrossRef] [PubMed]
37. Li, C.; Barclay, H.; Roitberg, B.; Lalonde, R. Ecology and Prediction of Compensatory Growth: From Theory to Application in Forestry. *Front Plant Sci.* **2021**, *12*, 655417. [CrossRef] [PubMed]
38. Gonzalez, A.; Loreau, M. The Causes and Consequences of Compensatory Dynamics in Ecological Communities. *Annu. Rev. Ecol. Evol. Syst.* **2009**, *40*, 393. [CrossRef]
39. Maschinski, J.; Whitham, T.G. The Continuum of Plant Responses to Herbivory: The Influence of Plant Association, Nutrient Availability, and Timing. *Am. Nat.* **1989**, *134*, 1–19. [CrossRef]
40. Aldous, S.E. Deer Browse Clipping Study in the Lake States Region. *J. Wildl. Manag.* **1952**, *16*, 401–409. [CrossRef]
41. Garrison, G.A. Effects of Clipping on Some Range Shrubs. *J. Range Manag.* **1953**, *6*, 309–317. [CrossRef]
42. Wilcox, K.R.; Koerner, S.E.; Hoover, D.L.; Borkenhagen, A.K.; Burkepile, D.E.; Collins, S.L.; Hoffman, A.M.; Kirkman, K.P.; Knapp, A.K.; Strydom, T.; et al. Rapid recovery of ecosystem function following extreme drought in a South African savanna grassland. *Ecology* **2020**, *101*, e02983. [CrossRef]
43. Hoover, D.L.; Knapp, A.K.; Smith, M.D. Resistance and resilience of a grassland ecosystem to climate extremes. *Ecology* **2014**, *95*, 2646–2656. [CrossRef]
44. Connell, S.D.; Ghedini, G. Resisting regime-shifts: The stabilising effect of compensatory processes. *Trends Ecol. Evol.* **2015**, *30*, 513–515. [CrossRef]
45. El Omari, B. Accumulation versus storage of total non-structural carbohydrates in woody plants. *Trees* **2022**, *36*, 869–881. [CrossRef]
46. Turcsán, A.; Steppe, K.; Sárközi, E.; Erdélyi, É.; Missoorten, M.; Mees, G.; Vander Mijnsbrugge, K. Early summer drought stress during the first growing year stimulates extra shoot growth in oak seedlings (*Quercus petraea*). *Front. Plant Sci.* **2016**, *7*, 193. [CrossRef] [PubMed]
47. Spiess, N.; Oufir, M.; Matusikova, I.; Stierschneider, M.; Kopecky, D.; Homolka, A.; Burg, K.; Fluch, S.; Hausman, J.F.; Wilhelm, E. Ecophysiological and transcriptomic responses of oak (*Quercus robur*) to long-term drought exposure and rewatering. *Environ. Exp. Bot.* **2012**, *77*, 117–126. [CrossRef]
48. Kaya, Z.; Adams, W.T.; Campbell, R.K. Adaptive significance of intermittent shoot growth in Douglas-fir seedlings. *Tree Physiol* **1994**, *14*, 1277–1289. [CrossRef] [PubMed]
49. Seidel, H.; Matiu, M.; Menzel, A. Compensatory Growth of Scots Pine Seedlings Mitigates Impacts of Multiple Droughts within and Across Years. *Front. Plant Sci.* **2019**, *10*, 519. [CrossRef]
50. van Kampen, R.; Fisichelli, N.; Zhang, Y.-J.; Wason, J. Drought timing and species growth phenology determine intra-annual recovery of tree height and diameter growth. *AoB Plants* **2022**, *14*, plac012. [CrossRef]
51. Walter, J.; Jentsch, A.; Beierkuhnlein, C.; Kreyling, J. Ecological stress memory and cross stress tolerance in plants in the face of climate extremes. *Environ. Exp. Bot.* **2013**, *94*, 3–8. [CrossRef]
52. Hilker, M.; Schmülling, T. Stress priming, memory, and signalling in plants. *Plant Cell Environ.* **2019**, *42*, 753–761. [CrossRef] [PubMed]
53. Anderegg, W.R.L.; Schwalm, C.; Biondi, F.; Camarero, J.J.; Koch, G.; Litvak, M.; Ogle, K.; Shaw, J.D.; Shevliakova, E.; Williams, A.P.; et al. Pervasive drought legacies in forest ecosystems and their implications for carbon cycle models. *Science* **2015**, *349*, 528–532. [CrossRef]
54. Kannenberg, S.A.; Maxwell, J.T.; Pederson, N.; D'Orangeville, L.; Ficklin, D.L.; Phillips, R.P. Drought legacies are dependent on water table depth, wood anatomy and drought timing across the eastern US. *Ecol. Lett.* **2019**, *22*, 119–127. [CrossRef]
55. Anderegg, L.D.L.; Anderegg, W.R.L.; Berry, J.A. Not all droughts are created equal: Translating meteorological drought into woody plant mortality. *Tree Physiol.* **2013**, *33*, 672–683. [CrossRef] [PubMed]
56. Pflug, E.E.; Buchmann, N.; Siegwolf, R.T.W.; Schaub, M.; Rigling, A.; Arend, M. Resilient Leaf Physiological Response of European Beech (*Fagus sylvatica* L.) to Summer Drought and Drought Release. *Front. Plant Sci.* **2018**, *9*, 187. [CrossRef]

57. Vander Mijnsbrugge, K.; Vandepitte, J.; Moreels, S.; Mihaila, V.-V.; De Ligne, L.; Notivol, E.; Van Acker, J.; Van den Bulcke, J. Timing of autumnal leaf senescence in a common shrub species depends on the level of preceding summer drought symptoms. *Environ. Exp. Bot.* **2023**, *216*, 105539. [CrossRef]
58. Vander Mijnsbrugge, K.; Turcsan, A.; Maes, J.; Duchene, N.; Meeus, S.; Steppe, K.; Steenackers, M. Repeated Summer Drought and Re-watering during the First Growing Year of Oak (*Quercus petraea*) Delay Autumn Senescence and Bud Burst in the Following Spring. *Front. Plant Sci.* **2016**, *7*, 419. [CrossRef]
59. Mariën, B.; Dox, I.; De Boeck, H.J.; Willems, P.; Leys, S.; Papadimitriou, D.; Campioli, M. Does drought advance the onset of autumn leaf senescence in temperate deciduous forest trees? *Biogeosciences* **2021**, *18*, 3309–3330. [CrossRef]
60. Vander Mijnsbrugge, K.; Moreels, S. Varying Levels of Genetic Control and Phenotypic Plasticity in Timing of Bud Burst, Flower Opening, Leaf Senescence and Leaf Fall in Two Common Gardens of *Prunus padus* L. *Forests* **2020**, *11*, 1070. [CrossRef]

Disclaimer/Publisher's Note: The statements, opinions and data contained in all publications are solely those of the individual author(s) and contributor(s) and not of MDPI and/or the editor(s). MDPI and/or the editor(s) disclaim responsibility for any injury to people or property resulting from any ideas, methods, instructions or products referred to in the content.

International Journal of *Geo-Information*

MDPI

Article

Challenges to Viticulture in Montenegro under Climate Change

António Fernandes [1,*], Nataša Kovač [2], Hélder Fraga [1], André Fonseca [1], Sanja Šućur Radonjić [3], Marko Simeunović [2], Kruna Ratković [2], Christoph Menz [4], Sergi Costafreda-Aumedes [5] and João A. Santos [1]

1 Centre for the Research and Technology of Agro-Environmental and Biological Sciences (CITAB), Institute for Innovation, Capacity Building, and Sustainability of Agri-Food Production (Inov4Agro), University of Trás-os-Montes e Alto Douro (UTAD), P.O. Box 1013, 5000-801 Vila Real, Portugal; hfraga@utad.pt (H.F.); andre.fonseca@utad.pt (A.F.); jsantos@utad.pt (J.A.S.)

2 Faculty of Applied Sciences, University of Donja Gorica, Oktoih 1, 81000 Podgorica, Montenegro; natasa.kovac@udg.edu.me (N.K.); marko.simeunovic@udg.edu.me (M.S.); kruna.ratkovic@udg.edu.me (K.R.)

3 Faculty for Food Technology, Food Safety and Ecology, University of Donja Gorica, Oktoih 1, 81000 Podgorica, Montenegro; sanja.radonjic@plantaze.com

4 Potsdam Institute for Climate Impact Research (PIK) e. V., Telegrafenberg A 31, 14473 Potsdam, Germany; christoph.menz@pik-potsdam.de

5 Institute of BioEconomy, Italian National Research Council (IBE-CNR), 50019 Sesto Fiorentino, Italy; sergi.costafredaaumedes@cnr.it

* Correspondence: acpf91@utad.pt

Abstract: The Montenegrin climate is characterised as very heterogeneous due to its complex topography. The viticultural heritage, dating back to before the Roman empire, is settled in a Mediterranean climate region, located south of the capital Podgorica, where climate conditions favour red wine production. However, an overall increase in warmer and drier periods affects traditional viticulture. The present study aims to discuss climate change impacts on Montenegrin viticulture. Bioclimatic indices, ensembled from five climate models, were analysed for both historical (1981–2010) and future (2041–2070) periods upon three socio-economic pathways: SSP1-2.6, SSP3-7.0 and SSP5-8.5. CHELSA (≈1 km) was the selected dataset for this analysis. Obtained results for all scenarios have shown the suppression of baseline conditions for viticulture. The average summer temperature might reach around 29.5 °C, and the growing season average temperature could become higher than 23.5 °C, advancing phenological events. The Winkler index is estimated to range from 2900 °C up to 3100 °C, which is too hot for viticulture. Montenegrin viticulture requires the application of adaptation measures focused on reducing temperature-increase impacts. The implementation of adaptation measures shall start in the coming years, to assure the lasting productivity and sustainability of viticulture.

Keywords: grapevine; winegrowing; bioclimatic indices; Shared Socioeconomic Pathways; projections; climate impacts; adaptation; Montenegrin viticulture; Western Balkans; CHELSA

1. Introduction

Climate change involves an alteration of the mean and the variability of a wide range of climatic variables, associated with shifts in the Earth's atmosphere and subsequent effects on the other Earth system components [1,2]. In 1896, Savante Arrhenius had already anticipated anthropogenic climate change [3] by quantifying the contribution of CO_2 to the greenhouse effect. The first major conference on this topic, human activity affecting the world's climate, was the World Climate Conference in February 1979 in Geneva [4]. Thereafter, climate change communication to the general public and awareness thereof have witnessed a steep rise [5,6]. According to IPCC [7], the global temperature has faced an abrupt increase since 1950, mostly from anthropogenic-induced changes in climate [8]. The most pronounced climate change effects are the rise in temperature, increased occurrence of extreme weather events, and sea level rise [9–12]. Impacts of climate change affect all social and economic sectors to a higher frequency, and intensity of floods [13], severe droughts, and heatwaves, among other extreme weather and climate events [11]. Multiple laws, programs, and objectives have been settled to guide society to offset this escalating trend, but it is uncertain which pathway society will follow [14,15], and so, the application of mitigation measures is intrinsically vulnerable to political change and socio-economic development [16,17]. Under this uncertainty, future climate projections shall be analysed through an overarching overview encompassing different anthropogenic forcing scenarios, based on the Representative Concentration Pathways (RCPs) and the Shared Socio-economic Pathways (SSPs) [18,19]. Data from climate projections are essential inputs for simulation models to quantify future environmental and/or socio-economic consequences. For example, in the scope of agriculture, using future climate data as an input for crop models will enable the estimation of future crop responses [20].

Specifically, for viticulture, there are indeed multiple consequences of climate change, starting with phenology [21,22]. With increased temperatures, the natural dormancy of grapevines is disturbed [23], and phenological events are advanced [24]. As a result, grape berry composition and wine quality parameters change [25]. When climate conditions are not suitable, wine quality may become compromised [26]. With warm temperatures, the trend is to produce unbalanced wines; grape berries accumulate more sugar and, subsequently, higher wine alcohol content and fewer flavours [27,28]. In the worst case, climate change makes areas unsuitable for viticulture, as conditions become too warm and dry [29–31]. Relative to other world regions, the Mediterranean is highly vulnerable to climate change, as it is anticipated to experience a faster temperature increase, with longer warmer and drier periods [32], and a decline in precipitation [33]. Accordingly, the viticulture sector direly needs strategies to guarantee water availability, achieve a balanced phenology alignment, maximise the yield, and achieve the necessary berry composition properties for wine production [21,34,35]. The review from Naulleau et al. [36] revealed that canopy management is effective for adapting phenology in the short term. In the long term, relocation, plant material changes, and changes to vineyard design are strategies with proven results. The selection of plant material, irrigation, and soil management are the most crucial aspects to achieve sustainable yields [36]. In connection with viticulture, several studies have been conducted that provide a global analysis concerning large researched areas (such as continents). Other studies, such as Jones and Alves' [37], portray smaller regions, depicting more detailed information. Nonetheless, there is still a lack of climate change impact studies focusing on viticulture in specific parts of the world. This is the case for Montenegro, where viticulture plays a significant role in the country's economy [38].

Montenegro's climate is characterised by multiple studies [39–49]. Resorting to observed data, temperatures in Montenegro have increased in recent decades [40,49]. The country is also facing a decrease in precipitation during the summer, and an increase in the autumn, but overall, the total annual precipitation is slightly decreasing [39]. However, future projections of precipitation do not reveal significant changes until 2100 [42], while average annual temperatures will increase from 2 °C up to 5 °C [50,51]. Among the multiple consequences, it is estimated that there will be decreases in rainfed potato yields above 30% [52] and for olives there will be a decrease of approximately 15% [53]. In the scope of climate change impacts on viticulture, there are no studies for Montenegro, raising the necessity of exploring this. For this reason, the present study was accomplished to fulfil three main objectives: (1) to estimate future climate changes in Montenegro; (2) to assess potential impacts on local viticulture; and (3) to delineate adaptation strategies, to ensure the resilience of grape production in Montenegro.

2. Materials and Methods

2.1. Study Area

Montenegro is located in Southeastern Europe on the Balkan Peninsula (Figure 1a). According to CORINE Land Cover [54], the land comprises 79% forested areas, 16% agricultural areas, 3% wetlands and water bodies and 2% artificial surfaces (Figure 1b).

Montenegro's landscape is characterised by its mountainous terrain (Figure 1c), over two-thirds of which features diverse macro relief units, such as mountain plains, valleys, basins, and mountains, leading to a rich variety of natural resorts [55]. This regional complex topography [56] results in the presence of multiple climate types, subtypes, and varieties across the country [43,45], from a Mediterranean climate on the coast [57] to alpine conditions in the inner mountains [58]. According to weather station data, the total annual precipitation ranges from 768 mm in the northern zone up to 4700 mm in the mountains [45,59]. Mean annual temperatures vary from 4 °C in the northern mountains to 17 °C near the Adriatic coast [45]. Montenegro's flora and fauna are remarkably diverse due to the complex interplay between topography, climate, geology, and soil complexity, making it a biodiversity hotspot [55,60].

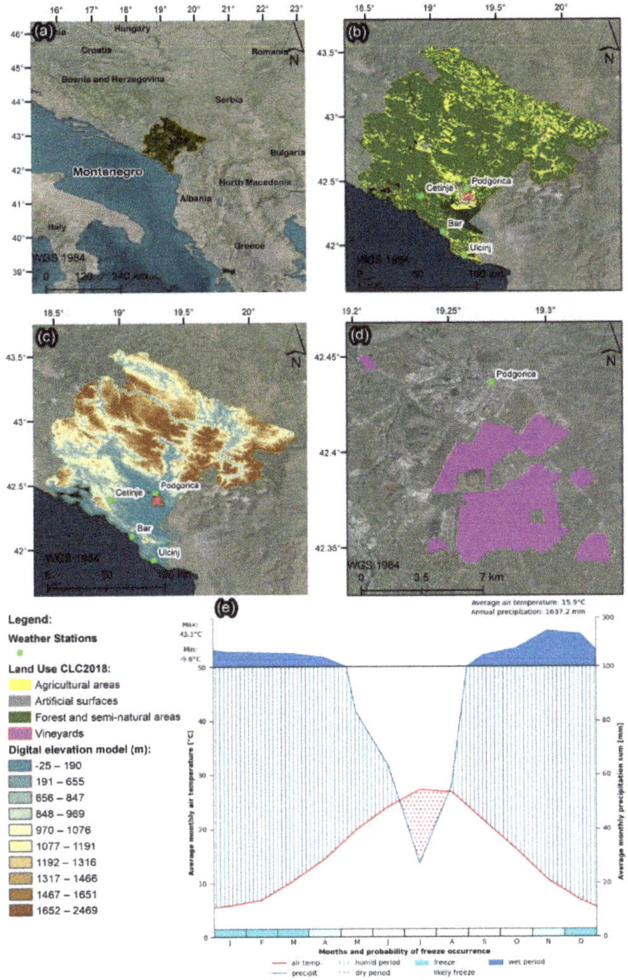

Figure 1. Representation of the Study Area. (**a**) is the location of Montenegro and contiguous countries. (**b**) is the location of Montenegro vineyards and other land uses, according to CLC 2018, and the location of weather stations. (**c**) portrays the elevation of Montenegro, in 10 quantiles. In (**d**), the vineyards of Montenegro near Podgorica's weather station are shown. In (**e**), an ombrothermic diagram of Podgorica weather station data (cf. legend for details) during the Historical period (1981–2010) is depicted.

The viticulture sector in Montenegro is particularly relevant due to the quality of grapes and wine, with lower-elevation wines being distinctive for their superior quality [61]. Montenegro is located within the global vine belt at a similar latitude to famous wine regions like Tuscany and Umbria in Italy and La Rioja in Spain [62]. Considering this geographical position and the influence of the Mediterranean climate (Figure 1e), the long tradition of grape growing and winemaking legacy dates back to the pre-roman period [63]. Viticulture has a rich heritage in grapevine growing and winemaking, contributing to its economic success [64]. Montenegro is among the top 50 wine-producing countries; however, exportation has been decreasing in the last decade from USD 17M down to USD 15M in 2023 [65], with an average production of 38 thousand tons per year [66]. Still, viticulture practice is increasing in Montenegro, with currently more than 500 grape producers [67] across the four defined viticulture regions [68]. The biggest one is the Montenegrin basin of Lake Skadar, followed by the Montenegrin coast, the Montenegrin north, and the smaller viticulture region of Nudo. In terms of soil types, the basin of Lake Skadar contains mostly red clays, eutric cambisols, and alluvial soil; the coast also contains red clays and eutric cambisols, but also includes rankers [61,69,70]. The region of Nudo is dominated by rendzina, while, in the North Region, there are brown cambisols and also rendzina [69,70].

The core of viticulture, including the public and private sectors (with 97% of producers), is located between Podgorica and Lake Skadar, where ideal conditions for the growth and development of authentic grape varieties can be found. Plantaže, a state-owned company, owns around 2000 ha of vineyards (Figure 1d). The assortment of grape varieties is in line with warmer climatic conditions and the tradition of red wine production, considering the vineyard areas that are primarily based on black grape varieties. Still, other vineyards are also dedicated to white wine varieties, and the least acreage is allocated to table grape varieties intended for direct consumption [68]. The main part of the vineyard area is situated at elevations from 0 to 100 metres, and most of the registered vineyards are within this elevation range. Most vineyards are planted with Montenegrin autochthonous grape varieties, Vranac variety (over 97% of the vineyards), and a Kratošija variety (50.5% of the vineyards). Kratošija was the main grape variety in Montenegro until the onset of the phylloxera crisis caused the removal of many ageing vines, prompting their substitution with fresh planting material [71]. During this period, the replacement of withered Kratošija vines was done with grafted Vranac plants, so favoured by grape growers for their production of deeply dark red ruby wines [72]. Consequently, the cultivated area of Kratošija decreased, paving the way for Vranac to emerge as the prevailing and emblematic cultivar in Montenegro throughout the twentieth century [73]. Alongside these prominent varieties, lesser-known indigenous cultivars such as Žižak, Krstač, Čubrica, and Bioka can be identified in Montenegro, with unique simple sequence repeats [73–76].

In Montenegrin vineyards, agro-technical operations such as winter and green pruning, protection, cluster reduction and defoliation are applied. The space between rows varies from 0.6 to 1.2 m, with double and single Guyots and cordons being used. Montenegrin viticulture is moving towards the application of precision viticulture, with the installation of soil moisture and meteorological sensors, to support smart irrigation [77–79], and also the usage of digital pheromone traps. Towards the adaptation to climate change, rootstocks resistant to higher temperatures are being used and research is being performed to find autochthonous grape varieties which are resistant to climate change [80].

2.2. Datasets

Montenegro's future climate was analysed using the CHELSA dataset [81,82] of observations and climate model projections downscaled to a resolution of 30 arc-sec (approximately 1 km). The CHELSA dataset provides multiple sub-datasets, with different time steps, from daily to climate normal (30-year averages); also, the dataset is already bias-adjusted [83] using a trend-preserving bias correction [84]. Our analysis focuses on the historical period from 1981 to 2010 and the future period from 2041 to 2070. Future climate scenarios are derived from the Coupled Model Intercomparison Project Phase 6 scenarios (CMIP6) [85], more specifically the Shared Socio-Economic Pathways [86]: SSP1-2.6; SSP3-7.0 and SSP5-8.5. For each pathway, simulations of 5 global climatic models (GCMs) were available [87], as presented in Table 1. For each future projection, data from the 5 GCMs were ensembled by median.

To further test the robustness of the CHELSA dataset, historical period data was compared with observation-based gridded climate data (with a spatial resolution of 0.1° latitude × longitude) from ERA5-Land [88] and E-OBS [89], version 27.0e. Additionally, the 3 datasets were compared with observed data from 4 weather stations (WSs), provided by the Institute of Hydrometeorology and Seismology [90]. These contain daily temperature and precipitation data in the cities of Podgorica, Bar, Cetinje, and the Ulcinj Municipality [90], Figure 1.

Table 1. Used climate models.

Abbreviation	Model	Source
GFDL-ESM4	Geophysical Fluid Dynamics Laboratory Earth System Model Version 4.1	Dunne et al. [91]
UKESM1-0-LL	UK Earth System Model 1.0 Low Resolution	Sellar et al. [92]
MPI-ESM1-2-HR	Max Planck Institute Earth System Model Version 1.2—High Resolution	Gutjahr et al. [93]
IPSL-CM6A-LR	Institut Pierre-Simon Laplace Earth System Model—Coupled Model version 6A—Low Resolution	Lurton et al. [94]
MRI-ESM2-0	Meteorological Research Institute Earth System Model Version 2.0	Yukimoto et al. [95]

2.3. *Methods*

2.3.1. Validation of CHELSA Dataset

The first step was to validate the CHELSA historical observation dataset, by comparing it with observation-based data from E-OBS and ERA5-Land. Here, we focused on climatological means of temperature and precipitation within the grapevine growing season, from 1 April to 31 October [21,96]. This comparison was performed to validate if the spatial pattern of precipitation and temperature would be similar between the 3 datasets.

Furthermore, the three observational datasets were compared with the observed data from the 4 WSs, to confirm the feasibility of CHELSA. Temperature and precipitation climatologies of each dataset (CHELSA, E-OBS, ERA5-Land) were compared for the historical period using 6 performance metrics: Mean Absolute Error (MAE); Root Mean Square Error (RMSE), Mean Absolute Percentage Error (MAPE), Percent Bias (PBIAS), Nash–Sutcliffe Efficiency (NSE), and Pearson Correlation.

2.3.2. Evaluation of Future Climate Changes

The next step was to estimate the climate for Montenegro's Viticulture, calculating bioclimatic indices for the historical and future periods, using the 5 GCMs for future SSP scenarios. Projections for each scenario were ensembled from the GCMs, to capture the range of uncertainty in regional climate projections [97], providing better results than individual models [98]. Between the ensemble, through average or median, median was selected to minimize the effect of outliers on the reported projections for Montenegro [99].

Six bioclimatic indices were selected: mean temperature from April to October; mean temperature of the warmest quarter; temperature seasonality; growing degree days heat sum above 10 °C (GDD10); Winkler index (WI); and growing degree days (NGD10) with average temperature above 10 °C. The average temperature from April to October is particularly important since it marks the beginning of flowering until leaf fall. The mean temperature of the warmest quarter is relevant, not only for its influence on grape growth and development but also because high temperatures during the warmest quarter might lead to heat stress, and in consequence, impact the grape development and subsequently the wine quality. Temperature seasonality, the standard deviation of monthly average temperature, is a relevant indicator since it assesses seasonal temperature transitions.

Growing degree days (GDD10) and Winkler index were selected for this study since both are characteristic bioclimatic indices in the scope of viticulture. Both are the sum of temperature degrees above the growth threshold (10 °C, for grapevines), with the difference that GDD10 is calculated for the full year, while WI is specific for the growing season. The growing degree days with an average temperature above 10 °C (NGD10) is a similar parameter, representing the number of days with suitable growing conditions. Furthermore, we also considered three additional bioclimatic indices for this study, but these are only reported in the Supplementary Material: mean annual temperature; precipitation amount from April to October, and annual precipitation amount.

The vineyards of Montenegro (henceforth MVAs—Montenegro Vineyard Areas) were identified in CORINE Land Cover 2018 (Figure 1b–d). For the MVAs, data was analysed for the historical and future periods, but the values of all models for each SSP scenario were analysed without ensembling, to understand the differences among GCMs, while for the whole country, projections were ensembled by the median.

The characterisation of the climate in the MVAs was performed using the data from Podgorica's WS, which is located in the vicinity of the vineyards (Figure 1d). To compare the temporal trend of the bioclimatic indexes in vineyards during the historical period, the annual average of each of the selected indices was calculated, and then Sen's slope was calculated with a statistical significance of 5% using the Scipy Python library.

3. Results and Discussion

3.1. Validation of CHELSA Dataset

Climate change in Montenegro and MVAs was assessed through the CHELSA dataset. To understand if CHELSA is a viable dataset, the CHELSA observations (for the historical period) were compared with the data from E-OBS and ERA5-Land (Figures 2 and 3).

A clear difference in resolution among datasets is visible, and in terms of temperature range, the dataset differences are not perceptible (Figure 2). Precipitation (Figure 3) patterns and values are highly different between the three datasets, with CHELSA showing a much more heterogeneous distribution and values up to 1700 mm during the growing season. E-OBS and ERA5-Land show a considerably lower peak precipitation with 600 mm and 1200 mm, respectively. In the Supplementary Material a scatterplot of the grid boxes of each dataset is provided, converted into the grid distribution and grid size of E-OBS, Figure S1.

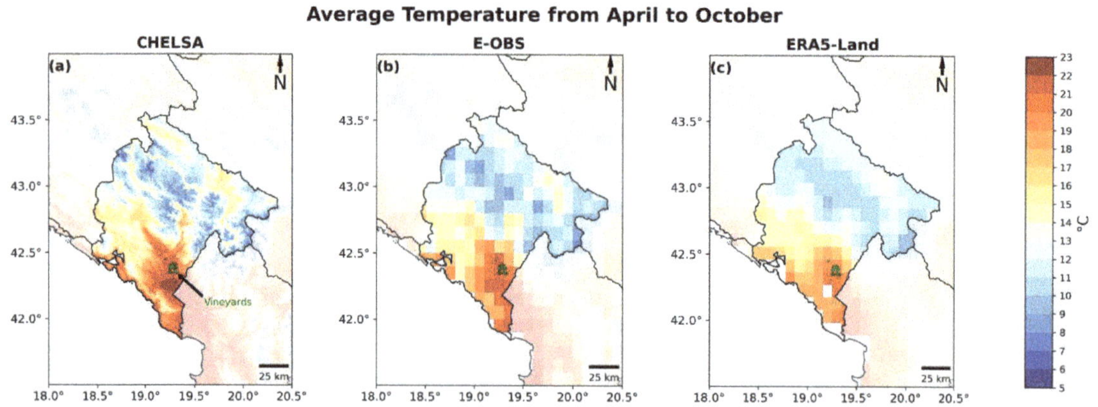

Figure 2. Representation of the climate average of daily average temperature during the growing season (from April to October), for the Historical period (1981–2010), for CHELSA (**a**) with a resolution of 0.008°, E-OBS (**b**) with a resolution of 0.1°, and ERA5-Land (**c**) with a resolution of 0.1°.

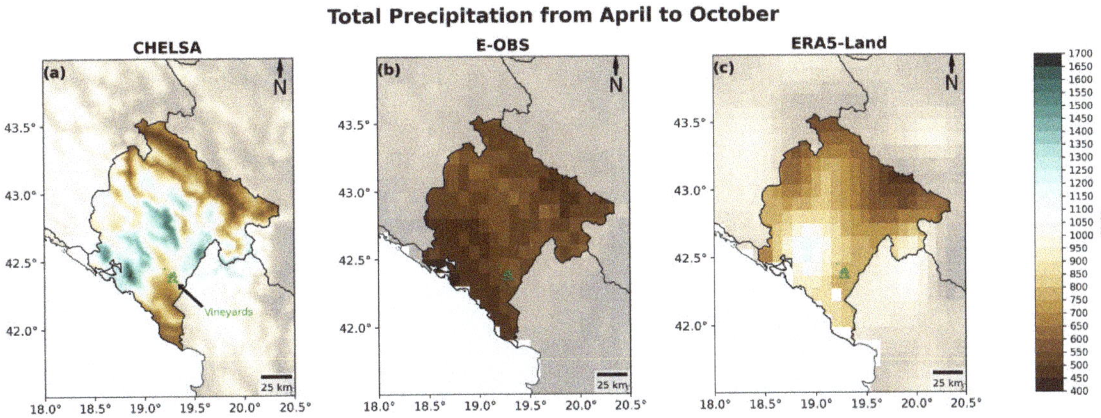

Figure 3. Representation of the climate annual precipitation during the growing season (from April to October), for the Historical period (1981–2010), for CHELSA (**a**) with a resolution of 0.008°, E-OBS (**b**) with a resolution of 0.1°, and ERA5-Land (**c**) with a resolution of 0.1°.

Temperature distribution from CHELSA is closer to E-OBS, while ERA5-Land features a more scattered pattern (Figure S1a–c). Precipitation reveals apparent differences among all datasets (Figure S1a–c). E-OBS precipitation is far more homogeneous and lower compared to the other two datasets, with values ranging between 400 to 600 mm in all of Montenegro. CHELSA and ERA5-Land are more similar in terms of precipitation intensities but also differ considerably in patterns. In general, CHELSA temperatures and precipitation resemble the orographic features of Montenegro. The dataset differences, especially for precipitation, raise the question of if CHELSA is a reliable dataset, portraying a realistic climate of Montenegro. Therefore, various statistics for temperature and precipitation (Table 2) were calculated for each dataset to compare them with the data from four weather stations in lowland Montenegro. The ERA5-Land grid contains data for Ulcinj weather station. CHELSA and E-OBS show a higher agreement across the temperature statistics with the WSs, except for the Cetinje station. E-OBS is most accurate for Podgorica's WS, which is the closest WS to the MVAs, with an MAE and RMSE of 0.3 °C. However, CHELSA MAE and RMSE are only slightly higher (0.5 °C). Despite ERA5-Land showing a more homogeneous MAPE among the weather stations (13.3% to 23.7%), the error is higher. In summary, CHELSA is considered the most suitable database to portray Montenegro's temperature, since it reaches the best scores across all statistics for most WSs.

When it comes to precipitation, all datasets have a lower accuracy when compared to temperature. The CHELSA dataset overestimates precipitation for 3 weather stations, ERA5-Land overestimates two stations, and EOBS underestimates all WSs. E-OBS seems to be an accurate database (MAPE ranging from 12.9% to 18%); however, for the Cetinje weather station, the MAPE is a much higher 62.4%. ERA5-Land is not as accurate as E-OBS, as MAPE values are 14.5%, 25.2%, and 49.1%. For CHELSA, the error is more homogeneous among weather stations. However, for CHELSA, the monthly precipitation in Podgorica is overestimated by 49 mm, while in E-OBS, it is overestimated by 19 mm, and in ERA5-Land, it is overestimated by 20 mm. In this sense, the precipitation data raises questions, possibly due to the complex topography, so precipitation data is hard to validate. Therefore, for precipitation, CHELSA seems to be the best dataset for Montenegro, but with the disadvantage of having a higher error for Podgorica WS, and overestimating values comparatively relative to the other databases.

Table 2. Comparison of monthly average of daily temperature and monthly precipitation among stations and datasets. The comparison is made between the observed data (4 weather stations) and data from different databases for the period 1981 to 2010.

Average Temperature							
Station	Dataset	MAE (°C)	RMSE (°C)	MAPE (%)	PBIAS (%)	NSE	Pearson Correlation
Bar_(B)	CHELSA	0.8	0.9	4.6	−5.0	0.97	1.00
Cetinje_(C)		2.3	2.4	66.7	−22.6	0.88	1.00
Podgorica_(P)		0.5	0.5	3.4	−0.6	0.99	1.00
Ulcinj_(U)		0.7	0.8	6.4	−4.3	0.98	1.00
Bar_(B)	E-OBS	2.9	3.0	22.8	17.7	0.72	1.00
Cetinje_(C)		2.0	2.1	57.6	−19.6	0.91	1.00
Podgorica_(P)		0.3	0.3	2.3	2.0	1.00	1.00
Ulcinj_(U)		0.6	0.7	5.0	1.5	0.99	1.00
Bar_(B)	ERA5-Land	2.2	2.2	16.8	13.4	0.85	1.00
Cetinje_(C)		0.8	0.9	23.7	−5.3	0.98	0.99
Podgorica_(P)		1.6	1.7	13.3	10.2	0.95	1.00
Ulcinj_(U)		-	-	-	-	-	-
Precipitation							
Station	Dataset	MAE (mm)	RMSE (mm)	MAPE (%)	PBIAS (%)	NSE	Pearson Correlation
Bar_(B)	CHELSA	11	14	8.9	−10.2	0.90	1.00
Cetinje_(C)		44	53	18.5	5.5	0.88	0.97
Podgorica_(P)		43	49	30.4	−31.4	0.42	1.00
Ulcinj_(U)		12	14	12.4	−12.1	0.88	0.99
Bar_(B)	E-OBS	19	21	18.0	16.9	0.77	0.97
Cetinje_(C)		184	219	62.4	67.3	−1.04	0.95
Podgorica_(P)		18	19	13.0	12.9	0.91	1.00
Ulcinj_(U)		9	10	12.9	6.7	0.94	0.99
Bar_(B)	ERA5-Land	59	69	49.1	−53.3	−1.36	0.98
Cetinje_(C)		85	116	25.2	28.9	0.43	0.92
Podgorica_(P)		18	20	14.5	−12.3	0.90	0.99
Ulcinj_(U)		-	-	-	-	-	-

3.2. Montenegro Climate Change

To assess the impact of climate change on viticulture, we evaluated various bioclimatic indices from CHELSA climate simulations (Figures 4–9 and S2–S4). These figures contain the data for the historical period (Figures 4a–9a), and the results for the future climate (2041–2070) for different SSP scenarios in (Figures 4b–d–9b–d). The Kernel density estimations (KDEs) are also displayed (Figures 4e–9e).

The spatial pattern of average growing season temperature (Figure 4) is similar among the historical period and future projections, as well as for the warmest quarter (Figure 5) and annual mean temperature (Figure S2). When looking at the KDEs (Figures 4e, 5e and S2e) a clear bimodal distribution is visible, resulting from the complex reliefs of Montenegro. The first peak corresponds to the "cold" temperatures of the mountain regions. Warmer regions are found in the valleys between the northeast mountains, and also on the coast, corresponding to the second peak of the KDEs. The second peak also renders the temperatures in the MVAs, which represent the valleys near Podgorica and Bjelopavlici Valley located in the south of Podgorica city, the warmest zones of Montenegro. Additionally, the terrain in this area is predominantly flat (Figure 1c), favouring agricultural practices. In terms of SSP scenarios, the results reveal that the SSP3-7.0 and SSP5-8.5 temperature increases are similar, while the SSP1-2.6 temperature is closer to that of the historical period (Figures 4e, 5e and S2).

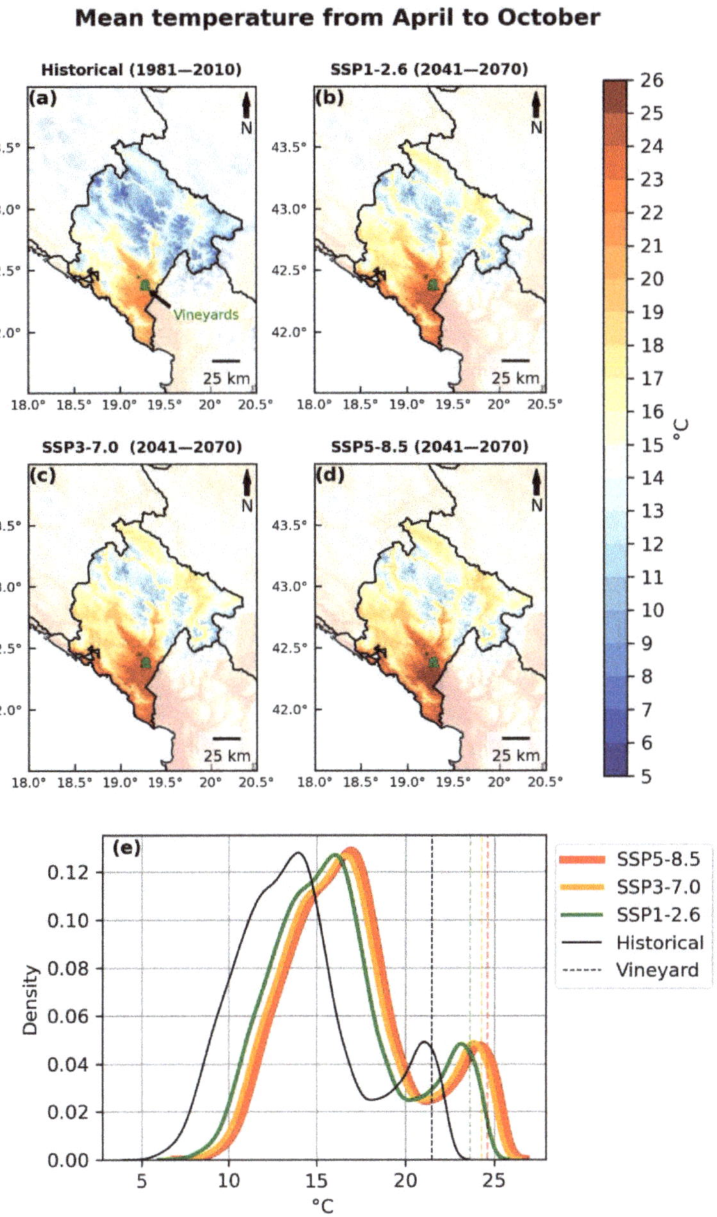

Figure 4. Climate mean daily average temperature during the vineyard growing season from April to October, for Montenegro, according to the CHELSA dataset (**a**) over the historical period (1981–2010) and the future period (2041–2070) under (**b**) SSP1-2.6, (**c**) SSP3-7.0, and (**d**) SSP5-8.5. (**e**) Corresponding Kernel density plot of the spatial variability in (**a**–**d**). The median value in the vineyard area is shown as a vertical dashed line. Projections were ensembled from 5 climate models through the median.

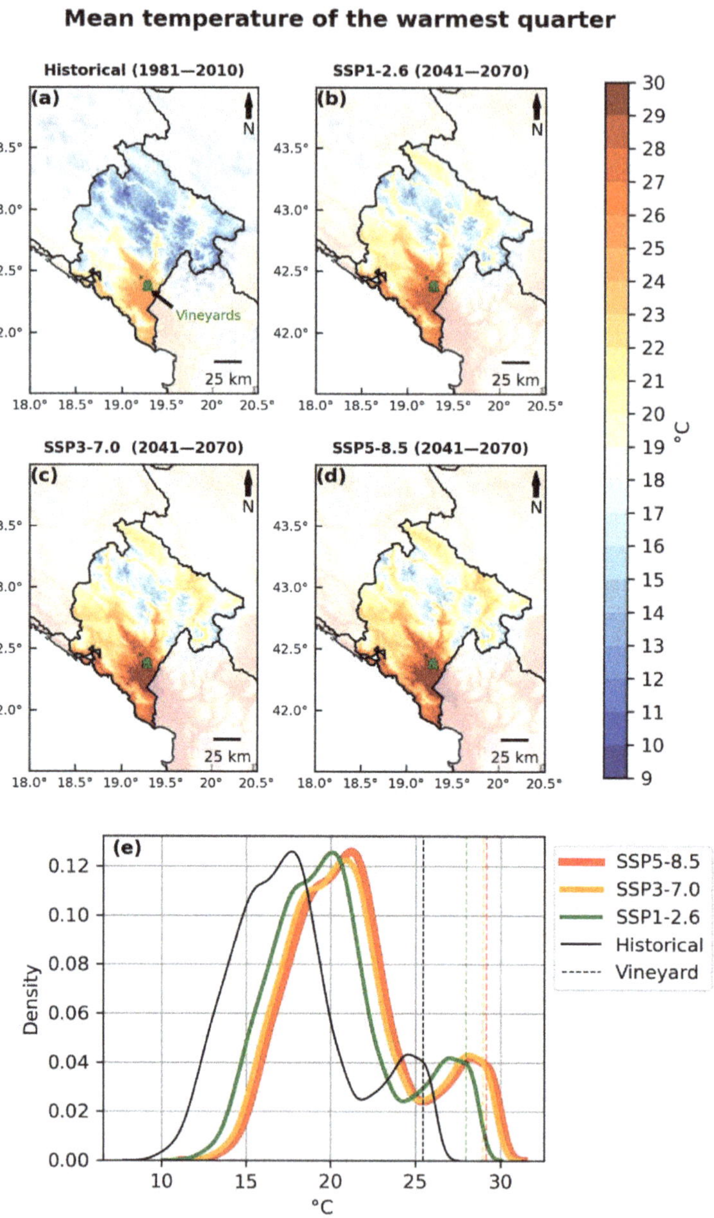

Figure 5. Climate mean daily average temperature during the warmest quarter for Montenegro according to the CHELSA dataset (**a**) over the historical period (1981–2010) and the future period (2041–2070) under (**b**) SSP1-2.6, (**c**) SSP3-7.0, and (**d**) SSP5-8.5. (**e**) Corresponding Kernel density plot of the spatial variability in (**a**–**d**). The median value in the vineyard area is shown as a vertical dashed line. Projections were estimated by the ensemble median from 5 climate models.

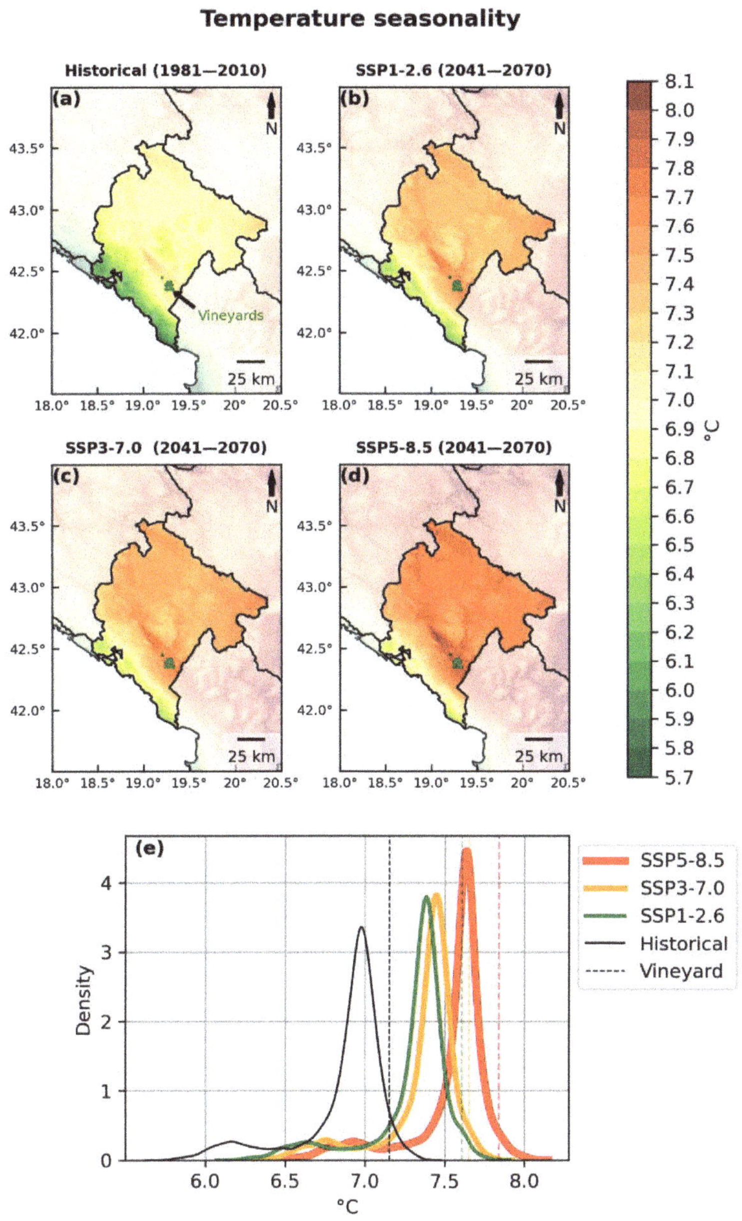

Figure 6. Climate mean of temperature seasonality; the standard deviation of the monthly means for Montenegro according to the CHELSA dataset (**a**) over the historical period (1981–2010) and the future period (2041–2070) under (**b**) SSP1-2.6, (**c**) SSP3-7.0 and (**d**) SSP5-8.5. (**e**) Corresponding Kernel density plot of the spatial variability in (**a–d**). The median value in the vineyard area is shown as a vertical dashed line. Projections were estimated by the ensemble median from 5 climate models.

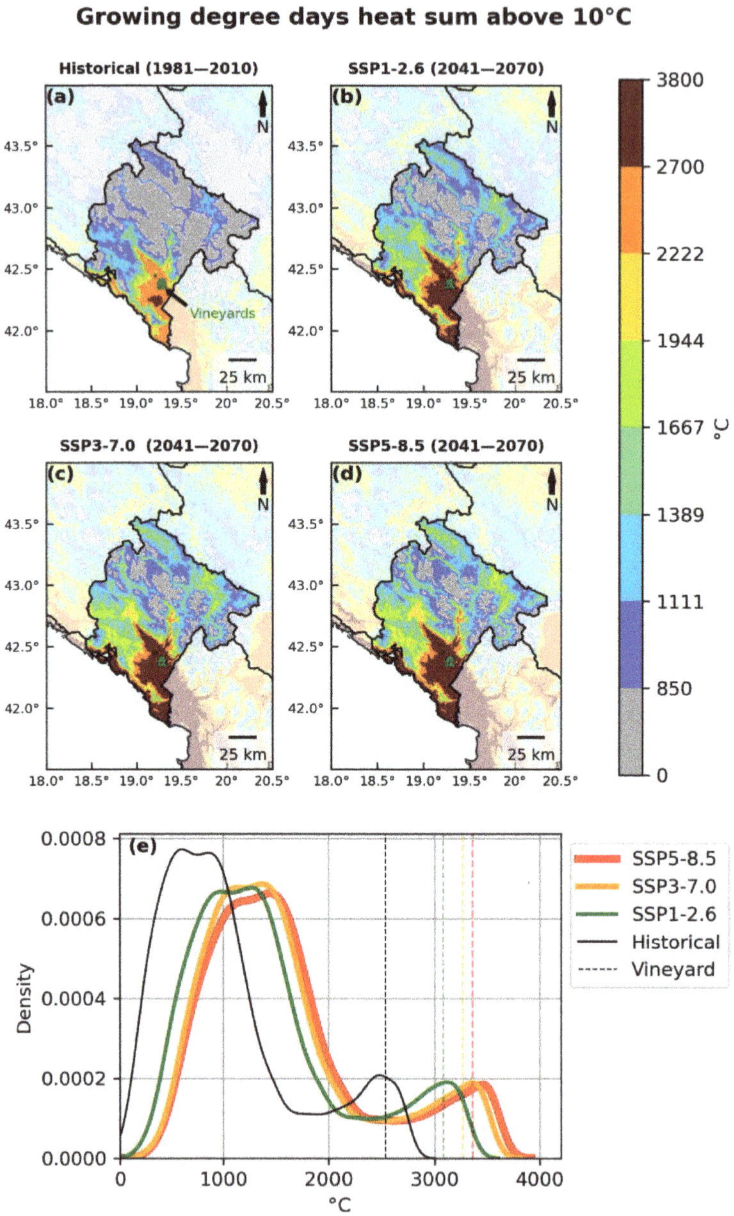

Figure 7. Climate mean growing degree days heat sum above 10 °C for Montenegro according to the CHELSA dataset (**a**) over the historical period (1981–2010) and the future period (2041–2070) under (**b**) SSP1-2.6, (**c**) SSP3-7.0 and (**d**) SSP5-8.5. (**e**) Corresponding Kernel density plot of the spatial variability in (**a–d**). The median value in the vineyard area is shown as a vertical dashed line. Projections were estimated by the ensemble median from 5 climate models.

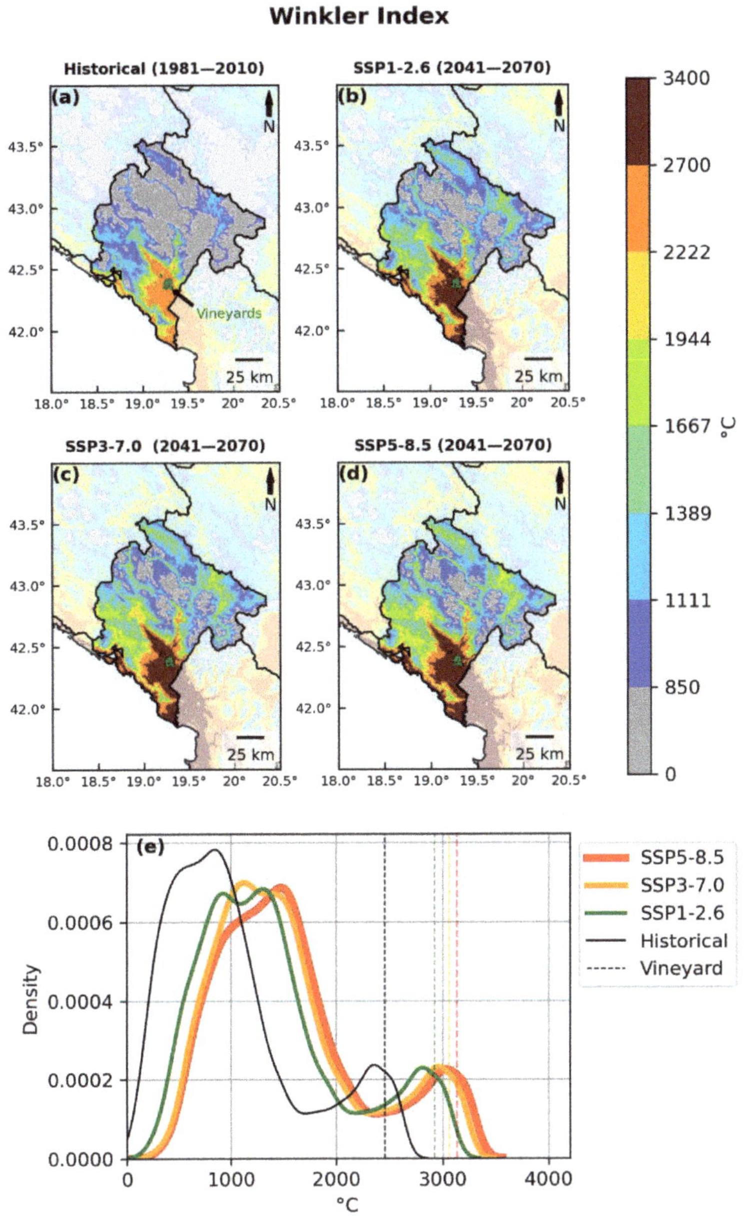

Figure 8. Climate annual mean of Winkler index for Montenegro according to the CHELSA dataset (**a**) over the historical period (1981–2010) and the future period (2041–2070) under (**b**) SSP1-2.6, (**c**) SSP3-7.0 and (**d**) SSP5-8.5. (**e**) Corresponding Kernel density plot of the spatial variability in (**a–d**). The median value in the vineyard area is shown as a vertical dashed line. Projections were estimated by the ensemble median from 5 climate models.

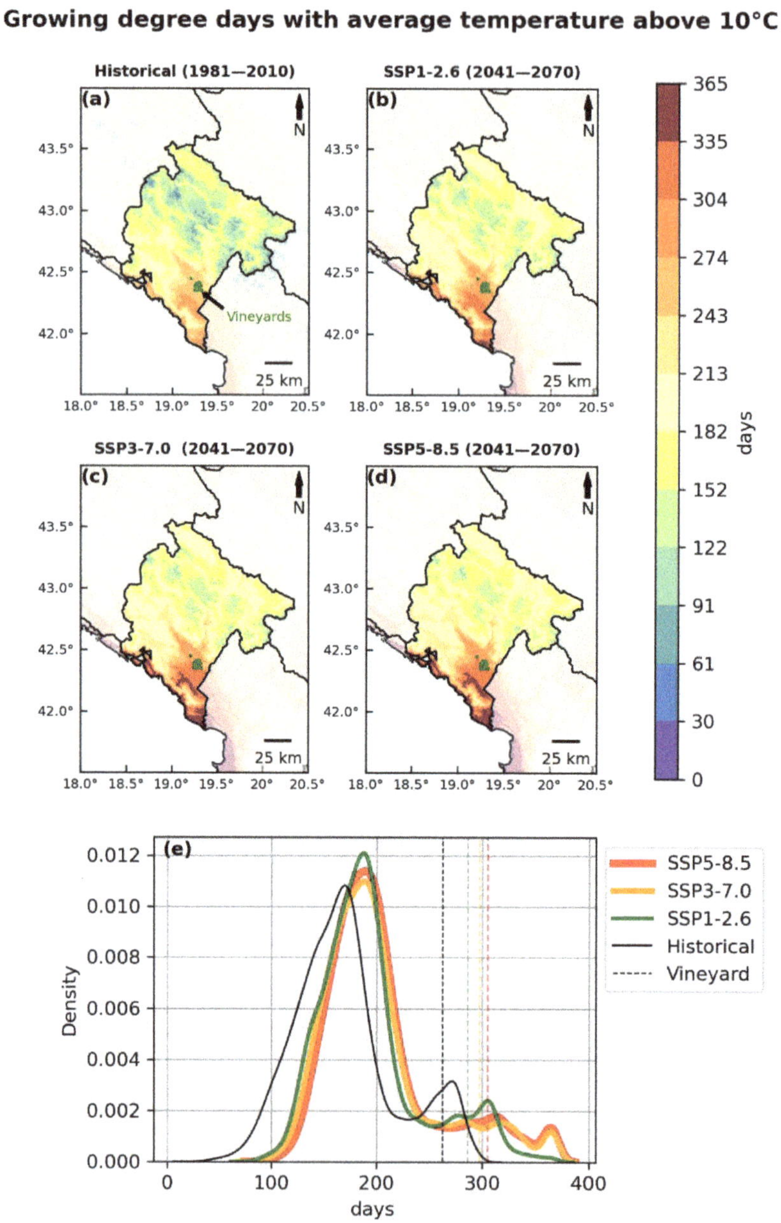

Figure 9. Climate mean of growing degree days with an average daily temperature higher than 10 °C for Montenegro, according to the CHELSA dataset (**a**) over the historical period (1981–2010) and the future period (2041–2070), under (**b**) SSP1-2.6, (**c**) SSP3-7.0 and (**d**) SSP5-8.5. (**e**) Corresponding Kernel density plot of the spatial variability in (**a**–**d**). The median value in the vineyard area is shown as a vertical dashed line. Projections were estimated by the ensemble median from 5 climate models.

In Montenegro, the average temperature from April to October ranges from 5.4 °C up to 22.1 °C (Figure 4). For the future period (Figure 4b–d) it is expected to shift to 7.5 °C and 25.6 °C, respectively, depending on the future scenario. In general, a temperature increase above 2 °C is estimated across all scenarios. Hence, MVA temperature is expected to increase by a similar amount. The warmest quarter is typically dated from June to August [100] as illustrated in Podgorica's weather data (Figure 1e). During these months, in the historical period, the temperature ranges from 9.2 °C up to 26.1 °C. Regardless of the SSP scenario, the warmest quarter temperatures across Montenegro are expected to increase to a range from 11.6 °C up to 29.9 °C (Figure 5).

According to Figure 5e, MVAs warmest quarter temperatures may increase by at least 2.5 °C. In contrast, median annual temperatures of the MVAs (Figure S2e) are expected to increase by a lesser amount of at least 1.7 °C. Hence, the temperature increase will be higher during the summer period and lower for the colder months. This inhomogeneous temperature increase leads to a stronger temperature seasonality (Figure 6). From 1981 to 2010, temperature seasonality was lower in the coastal region, down to 5.7 °C, and it increases with the distance to the shore of the Adriatic Sea, reaching 7 °C. Bjelopavlici Valley [101] faces the highest temperature seasonality, of up to 7.4 °C, also due to the hot summers, as in Podgorica [45]. Like for the other temperature indices, the spatial pattern of temperature seasonality (Figure 6) is similar across the historical period and future projections. In contrast to temperature (Figures 4e, 5e and S2), temperature seasonality for SSP3-7.0 is closer to SSP1-2.6, while SSP5-8.5 reaches higher seasonality (Figure 6e).

The temperature seasonality for the period 2041–2070 is not of major concern, but the increase in temperatures brings challenges to Montenegro's crops [102] by advancing phenological development [35], and decreasing crop productivity and also quality [103]. For a better understanding of the temperature effects, specific bioclimatic indices suitable for viticulture, GDD10, WI, and NGD10 for a base temperature of 10 °C [104] were also analysed. Both the annual (GDD10, Figure 7) and growing season (WI, Figure 8) heat sums significantly increase under the selected SSPs. In Montenegro, GDD10 during the historical period ranges from 6 °C to 2723 °C, and the range is anticipated to increase to between 142 °C and 3601 °C. Particularly for the MVAs, a GDD10 increase of at least 500 °C is expected. During the growing season (Figure 8), the MVAs heat sum is 2457 °C, slightly lower than the annual, but is anticipated to reach values above 3000 °C under climate change. In Montenegro, NGD10 goes from 26 days in the high-altitude areas up to 298 days on the coast. In the MVAs, the median number of days is 262, and this is projected to extend by over 30 days (Figure 6e). The results reveal that there are no distinct differences between SSP3-7.0 and SSP5-8.5 regarding GDD10, WI, and NGD10 (Figure 7e, Figure 8e, and Figure 9e, respectively).

According to Burić et al. [59], annual precipitation across Montenegro ranges from 768 mm up to 4650 mm, while according to the CHESLA dataset, it ranges from 831 mm to 3827 mm. According to the future projections, it is expected that there will be no distinct changes in annual precipitation (Figure S3) since precipitation in future scenarios (all) is close to precipitation in the historical period. During the growing season, in Figure S4, precipitation decreases in all projections. The projected precipitations for SSP3-7.0 and SSP5-8.5 are quite the same, and the SSP1-2.6 scenario is closer to the precipitation for the historical period. Consequently, precipitation will increase in the colder months.

3.3. *Climate Change in Vineyards*

The present section reports the bioclimatic indices for the historical period in Podgorica's WS, Figure 10, and also the analysis of the climate change in the vineyards: Figure 11 for temperature bios, and Figure S6 for precipitation. Here, the climate change data is accessed through all model data pooled together instead of considering ensemble metrics. Median values can be found in Table S1.

Podgorica WS is crucial for this study, as it is the closest WS to the main vineyard area, thus reliably representing its mesoclimate. When comparing the vineyard's median temperature-based indices during the historical period (Figures 10 and 11) with the Podgorica WS data (Figure 10), the thermal conditions are very similar. In Podgorica WS, the mean growing season temperature is 21.5 ± 0.8 °C (average and standard deviation), with temperatures in the warmest quarter at 26.0 ± 1.1 °C and the annual mean at 15.9 ± 0.6 °C; while, in the vineyards, the median according to the CHELSA dataset is 21.5 °C, 25.4 °C, and 16.2 °C, respectively. When it comes to the temperature seasonality in Podgorica WS, it is 7.7 ± 0.4 °C, which is close to the seasonality in vineyards, at 7.2 °C. The GDD10 in Podgorica is 2574 ± 181 °C, and so it is close to the vineyards' 2530 °C. For the WI, it is also clear that Podgorica has the same heat sum during the growing season as the vineyards, as in Podgorica it is 2457 ± 168 °C, and in the vineyards, it is 2457 °C. Moreover, the NGD10 in Podgorica (257 ± 12) is close to the number of days in the vineyards, which is 262 days.

According to Figure S5, for the WSs, the annual precipitation is 1637 ± 308 mm, while it is 2047 mm in the vineyards according to CHELSA (Figure S3). Precipitation during the growing season is 688 ± 175 mm in Podgorica (Figure S5), and in the vineyards (according to the CHELSA dataset) it is 872 mm (Figure S4). Therefore, the weather station precipitation data is fairly close to the CHELSA observational precipitation in the vineyards. The precipitation patterns (Figure S6) do not change over the years. As for projections (Figures S3 and S4), our results reveal that future precipitation may not change substantially and may slightly decrease during the growing season. Other authors also did not perceive significant changes in Podgorica [42,52]. At this point, it is worth noting that precipitation levels might not undergo any significant changes, though the hydric safety of vineyards can become at-risk due to the increasing temperature, which will lead to higher evapotranspiration [105], and so the water demand will increase [31]. In this sense, it is necessary to develop hydrological modelling studies and compile future scenarios, to perceive if water availability will be a threat to viticulture.

From Figure 11, temperatures are projected to increase in the MVAs. These results are in agreement with the increasing trend in Podgorica (Figure 10). Mean temperature during the growing season has increased in the last 30 years (Figure 10a), at a rate of 0.5 °C per decade, which is equivalent to the increase in the annual mean (Figure 10c). However, the increase is higher in the warmest quarter, i.e., 0.8 °C per decade (Figure 11b). Bačević et al. [106] have already mentioned the temperature increase between 1947 and 2018 (72 years), revealing an increase of 1.4 °C in average temperature. When analysing the increase in summer air temperature from 1951 to 2010, Doderović and Burić [44] estimated an increase of 0.27 °C per decade, which is discordant with Figure 10c. Possibly, this is due to the average temperature decreasing trend from 1961 to 1980, and increasing trend during 1981–2010, as the results from Bačević et al. [106] show, which was also noted for Serbia [107]. All the results reveal an increase in temperature; thus, the city's climate is becoming more arid and extreme [42]. For the vineyards, the results show that the mean annual temperature could increase up to 4.9 °C, according to the UKESM1-0-LL model (Figure 11, Table S1). These estimations seem to deviate from other models, resulting in the outliers shown in Figure 11a–c,e–g. As Sellar et al. [92] have mentioned, this model portrays high climate sensitivity, compared to other models. In this regard, the median value from the 5 climate models is considered by the authors as the representative future value. In this order, is expected that (in MVAs), the mean annual temperature will increase from 16.2 °C, up to a temperature between 18 °C and 19 °C. At first sight, the increase in mean annual temperature is not severe, since according to Schultz and Jones [108] a mean annual temperature between 10 °C and 20 °C is suitable for viticulture. During the growing season, the mean temperature of 21.5 °C is estimated to increase up to a value between 23.6 °C and 24.6 °C, which is high for successful grape growing, as the suitable mean temperature during the growing season ranges from 12 to 22 °C [108]. In Montenegro's vineyards, the growing season temperature is already at the limit of the threshold, and the increasing trend might exceed it by at least 2.6 °C. For the warmest quarter, the temperature (25.4 °C) could increase to 28.0 °C or, in the worst-case scenario, up to 29.2 °C. Future consequences seem to compromise viticulture suitability, which is a shared trend in the southern Mediterranean regions of Europe [109,110].

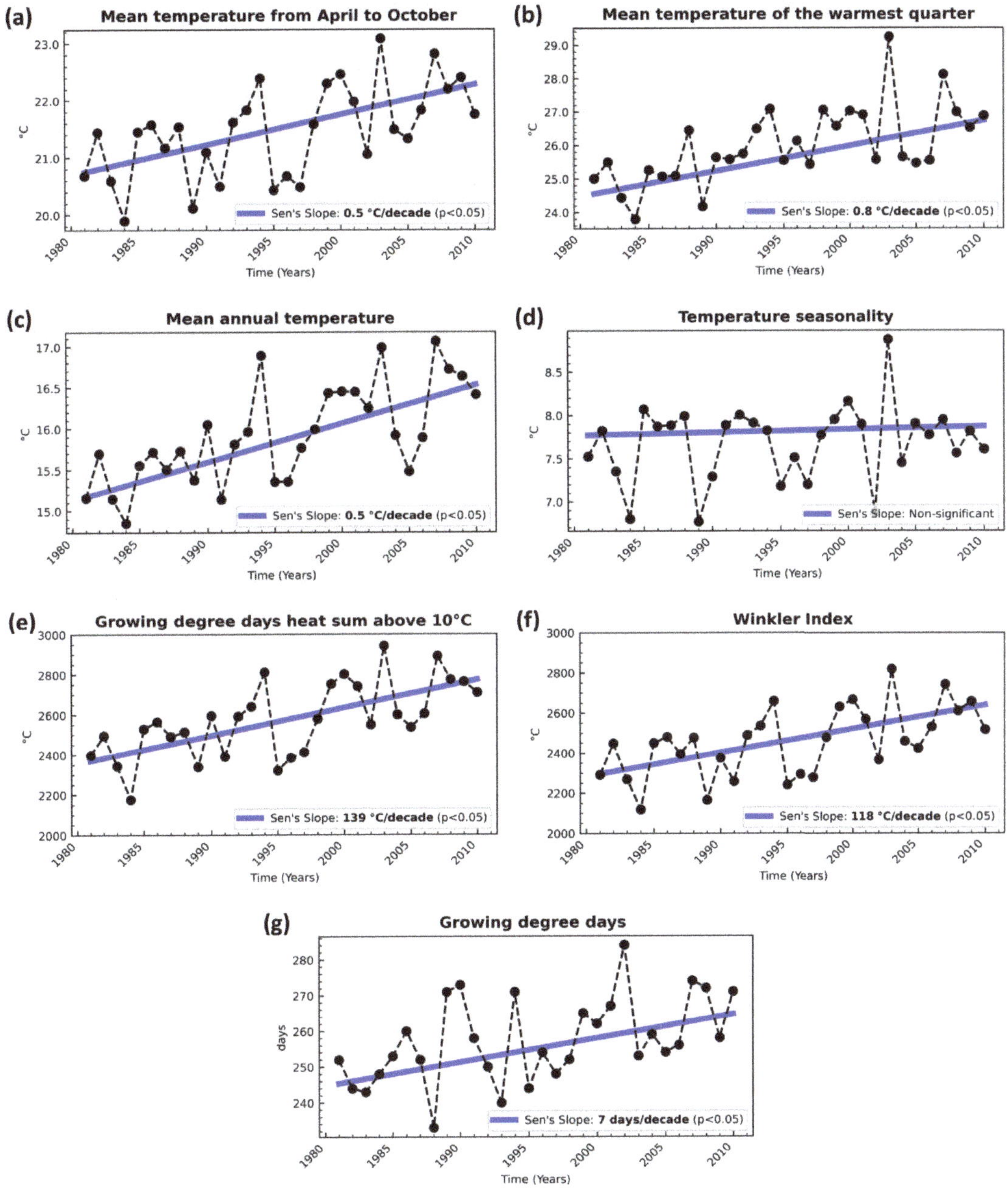

Figure 10. Chronograms of the annual averages of the outlined indices for Podgorica weather station during the historical period (1981–2010). (**a**) Daily mean temperature during the vineyard growing season, from April to October. (**b**) Daily mean temperature during the warmest quarter. (**c**) Daily average temperature. (**d**) Temperature seasonality, the standard deviation of the monthly means. (**e**) Growing degree days heat sum above 10 °C. (**f**) Growing degree days heat sum above 10 °C during the growing season. (**g**) Climate mean of growing degree days with an average daily temperature higher than 10 °C.

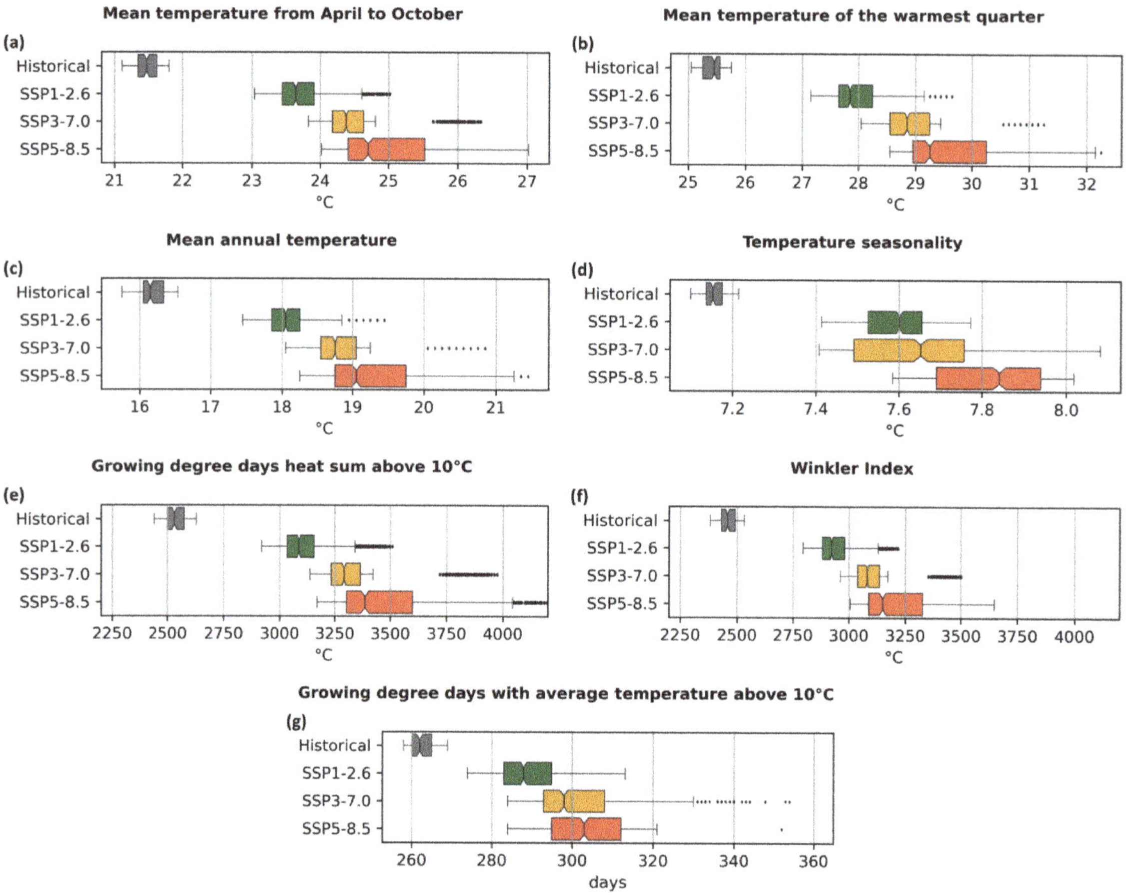

Figure 11. Notch boxplots of the outlined indices from the CHELSA dataset in the vineyard areas of Montenegro, for the historical period (1971–2010) and for the future period (2041–2070) models under SSP1-2.6, SSP3-7.0 and SSP5-8.5. (**a**) Climate means daily average temperature during the vineyard growing season, from April to October. (**b**) Climate means daily average temperature during the warmest quarter. (**c**) Climate annual means daily average temperature. (**d**) Climate means of temperature seasonality, the standard deviation of the monthly means. (**e**) Climate means of GDD10. (**f**) Climate means of WI. (**g**) Climate means of NGD10.

The future increase in temperature during the growing season leads to a higher temperature seasonality. Seasonality tends to remain constant in Podgorica (7.7 ± 0.4 °C, Figure 10). The vineyard's future seasonality could range from 7.5 to 8.0 °C, which are not record-breaking values, even for the worst-case scenario. In Podgorica, the average seasonality is close to the vineyard's future seasonality (Figure 11d). Still, the trend of increase is relatively weak and must be estimated beyond 2070.

The heat sum (Figure 10e,f), and NGD10 (Figure 10g), reveal an increasing trend, in agreement with the future projections (Figure 11e–g). According to the WI scale [111], Podgorica is classified as region V, with a WI of 2457 ± 168 °C, which is equal to that of the vineyards. The WI tends to increase by 118 °C per decade. This trend encompasses the results from (Figure 11f and Table S1), showing that the WI might reach values from 2919 °C up to 3130 °C in vineyards, which are classified as "too hot" [112]. According to historical averages of the WI, there are no regions in Europe with a value higher than

2700 °C, above "Region V", for the historical period [113–115]. When it comes to future projections, Cardell et al. [113] have already shown that multiple regions will surpass the 2700 °C threshold. This heat sum would extend out of the growing season, as the annual heat sum could increase from 2530 °C to up to 3352 °C (Figure 11e). For the historical period, both indexes have similar values. This means that daily average temperatures above 10 °C are mostly found during the growing season. Future scenarios show that the difference between GDD10 and the WI is likely to increase (Figure 11e,f). For SSP1—2.6 the difference is 157 °C, for SSP3-7.0 it is 208 °C and for SSP5-8.5 it is 222 °C; thus, the growing season can start earlier. According to Podgorica WS data Figure 1e, the growing season (>10 °C) starts in the middle of March and extends to the middle of November, 257 ± 12 days. However, is estimated to increase to between 286 and 305 days (Figure 11g), reaching approximately 10 months, and so starting during February and ending in December. For a clear understanding of the possible consequences, it is necessary to apply future projections in phenological models [31,116,117]. Still, the results from Fraga et al. [24], when using a lower resolution dataset, predicted that the budburst, flowering, veraison, and harvest will be anticipated by 20 to 30 days, according to RCP scenarios. Furthermore, the anticipation of phenological events will deeply require management adaptations to hold sustainable yields.

4. Adaptation Strategies

Crops are commonly settled in regions where the climate and soil conditions favour their development. Due to climate change, it is clear for most wine regions that baseline conditions will not prevail in the upcoming decades, thus challenging their viticultural suitability and implying geographical shifts in the optimal/adequate conditions [118]. In the worst-case scenario, the selection of new crops, better adapted to the anticipated conditions, is an answer to ensure socio-economic sustainability. Still, when the option of keeping the same crop is chosen, there are two options left: relocate or adapt on-site.

According to our projections, it is likely that new vineyards will emerge. Savić and Vukotić [61] performed a climate zoning for Montenegro, using climate data from 1950 to 2005. By combining their results with the present study's climate projections, it is possible to identify areas that will have suitable climate conditions for vineyards in Montenegro, keeping the already grown grapevine cultivars [80]. However, to plant new vineyards, soil properties must be considered as well, as they largely determine local terroirs and wine attributes [119]. Montenegro's complex orography challenges the settlement of new vineyards, due to the lack of flat land areas. Viticulture is, however, possible in steep slope landscapes, such as the Douro Wine Region [120] in Portugal, or the Aosta Valley [121] in Italy. New climate conditions will emerge in the future (Figure 8), as well as new suitable regions for viticulture, with a WI mostly ranging from 850 °C to 2000 °C, which are classified as regions Ib, II, III [111]. Region "V", currently located in the Lake Skadar basin, will only prevail on the hillsides of rivers. Upon the SSP1-2.6 scenario, the vicinity of Nikšić city will become region III, but "IV" under SSP3-7.0 or SSP5-8.5, with a growing season average temperature from 16 to 20 °C. Nikšić has already been suggested as a potential future zone for viticulture [68,122]. As such, it is important to perform viticulture zoning under future climates in Montenegro, considering soil and landscape features, and specifically regarding soil erosion in steep slopes [123] and ecological impacts due to land use change [124].

Changing the crop and relocation are the most extreme adaptation measures in the long-term, which should be adopted in the worst-case scenario [62]. However, long- and short-term adaptation measures on-site should be considered [34]. Adaptation starts with awareness, which is the first step to delineate successful strategies. For Montenegro, the results from Ćeranić et al. [125] suggested that it is necessary to raise climate change awareness among Montenegrin citizens. This is the first and most important adaptation strategy, to foster a proactive stance from winemakers and decision-makers to reverse the negative effects of climate change on wine production [126]. Adaptation to the necessity is not a good practice, as this reduces the range of options for short-term measures. This is a common choice for winegrowers, while long-term measures are taken as a last resort [127]. Nevertheless, to guarantee future viticultural sustainability and profitability, the key is to combine short- and long-term strategies [36]. To maintain viticulture on the same site, short-term strategies focus on measures like irrigation, soil management, canopy management, pest and disease control, and harvest management, whereas long-term strategies comprise developing adequate farm strategies, selection of plant material, or adjusting the vineyard design [31,34,36,62].

For the studied vineyards, the water supply for irrigation comes from groundwater extraction from intergranular aquifers [128]. Water supply is not a current concern for MVA as there are no reported concerns about water availability in the vineyards, and also because precipitation patterns will not change significantly, However, water availability might become a future problem, since evapotranspiration will increase and, with it, the water demand. From this perspective, the improvement and optimization of irrigation systems is an unavoidable necessity in Montenegro's vineyards, which might need to be supported by the construction of water storage structures. In the scope of precision viticulture, sensors in Montenegrin vineyards measuring soil moisture at different depths are already being implemented [77–79].

Based on the received data obtained through the installed equipment, soil moisture content is determined and complemented with meteorological data, thus supporting smart irrigation [77–79]. Due to the inherent uncertainties of climate change and weather irregularity, sensors are indeed critical for precision viticulture, enhancing short-term predictions (e.g., based on weather forecasting) to optimise canopy and harvest management. For pest control, digital pheromone traps are being used in Montenegro for monitoring the presence and abundance of pests to ensure the optimal use of protective agents based on the presence and number of pests.

In Montenegro, autochthonous and locally adapted grapevine varieties have survived because age-old natural selection represents a source of genetic variety and diversity, and this enables them to have a greater ability to adapt to climate change. Enviable results were achieved in the work on the genetics and genomics of autochthonous and domestic varieties [80], as well as in the clonal selection of the Vranac variety [79], the dominant grapevine variety for the production of red wines in Montenegro. Montenegrin wine is mainly produced with Vranac, which is a variety adapted to warm conditions [129,130]. In this sense, Podgorica winegrowers are already implementing the usage of late-ripening varieties, which is a key long-term adaptation strategy [131]. However, vineyards are already placed in the warmest regions, which will become "too-hot" regions according to the Winkler scale. Moreover, as Vranac is a late-ripening grape variety, it ripens in the middle of September, according to the agroecological conditions of Montenegro [132]. Therefore, later-ripening varieties are mandatory in this location, as well as, possibly, varieties even more resistant to climate change.

From a long-term perspective, research is being performed to find old grape native neglected varieties, which are likely more resistant to climate change [80]. This brings the need for another adaptation measure, i.e., the adaptation of the consumers and markets [133]. Changing varieties will bring a new challenge, as new flavours might not be easily accepted. As the typical vineyard lifespan is around 25–30 years [134], the replacement of old varieties with new ones should be undertaken gradually, enabling a smooth consumer adaptation. Apart from Podgorica's vineyards, there is a small section of vineyards scattered across Montenegro that produce foreign white wine varieties, and also red wine varieties. As the temperature increase across Montenegro is inevitable, these small vineyards might need to change their farm strategy, increasing wine diversification and, e.g., planting red wine varieties. Overall, warmer climates tend to be better for red varieties [135,136].

When it comes to rootstocks, in the newer vineyards the dominant rootstocks are 1103 P, R110, R140 and 41B, while in the older plantings, the rootstocks Kober 5BB and SO4 were used more often. Vine rootstock Teleki 5C and 161-49 are also used in smaller areas. According to the metanalysis from Santesteban et al. [137] 161-49 C and 41-B MGt are suitable for climate change, allowing grapes to attain a moderate sugar accumulation and pH, while Ruggeri 140 and Kober 5BB provide high yields, but higher acidity and sugar content. Among the used ones, Paulsen 1103 and Richter 110, the most used are highly resistant to water stress [31,138]. Overall, the most used rootstocks in Montenegro, 1103 Paulsen, 110 Richter, and 140-Ruggeri, are suitable for climate change, as they are among the best options for hot climates [139,140], and so this is already an implemented long-term strategy.

The state-of-the-art of Montenegrin viticulture in climate change adaptation requires further research incorporating more local data for a better selection and the fine-tuning of adaptation measures. Here, the biggest challenge is to implement long-term adaptation measures. For short-term adaptation, soil and canopy management require further research. Moreover, collecting data on the perception and adoption of measures by viticulturists and winemakers is essential to foster an avant-garde stance, break away from older practices, and embracing new suitable practices.

Further studies should be devoted to the quantification of the adaptative potential of these strategies in response to climate change and risk reduction, to select the best to adopt, knowing that the usage of a single adaptation measure might not be enough, as the results from Fraga et al. [141] suggest. It is also important to downscale available climate datasets to a spatial resolution of <1 km in some targeted vineyard sites over complex orography [142,143]. The impacts of increasing temperature can thereby be better quantified by modelling, e.g., phenology, yield, or wine quality. Future research should also address the impacts of climate change on soil for viticulture. Soil properties are essential for any cultivation and certainly soil quality is vulnerable to climate change [144]. Increasing temperatures accelerate biological processes in soil, altering the nutrient cycles and also the soil microbiome [145]. With increased evapotranspiration, soil moisture is expected to decrease and aridity to increase. This leads to the loss of soil's carbon content, an increase in erosion, and a decrease in water and nutrient holding capacities, and so soils become less fertile [146]. Consequently, grass cover cropping in vineyards becomes harder to maintain, which is essential to limit erosion, stimulate microbial activity, and, among other factors, promote soil health and carbon sequestration as a mitigation measure [147].

The present study outcomes provide a foundation for the improvement and adaptation of Montenegrin viticulture amidst the challenges posed by climate change. Climate change projections suggest significant consequences for Montenegrin viticulture, regardless of the SSP scenario. Even in the best-case scenario, besides adaptation, the implementation of mitigation measures should be fostered, as the effectiveness of the adaptation measures strongly depends on the intensity of climate change impacts. Adaptation measures have a higher chance of success under SSP1-2.6, for which baseline conditions are slightly exceeded. Under SSP5-8.5, the exceedance is superior, and it is uncertain that most adaptation measures would preserve Montenegrin viticulture heritage.

5. Conclusions

The future climate of Montenegrin viticulture was accessed in the present study, using the CHELSA dataset and a representative weather station located in the vicinity of the vineyards, in Podgorica. Three SSPs were used, and the results showed that regardless of the scenario, Montenegrin viticulture will be challenged by climate change impacts driven by increases in temperature.

Our results revealed that precipitation might decrease, but not below a concerning point. Nevertheless, the optimisation of the irrigation system will have to be conducted for vineyards to adequately cope with higher evapotranspiration rates in the future, and guarantee water availability.

In Podgorica's vineyards, the mean annual temperature of 16.2 °C is estimated to increase up to 19.0 °C. During the growing season, the average temperature could reach between 23.6 to 24.6 °C. Increasing temperatures will be also felt in the colder season, at a point where the growing season might expand to over 1 month, and so the advancement of phenological development is expected, along with other consequences.

Montenegro is already applying adaptation measures, including the application of precision viticulture, usage of late-ripening varieties, and rootstocks tolerant to higher temperatures. Furthermore, additional adaptation measures must be taken for the expected climate changes from 2041 to 2070 to maintain the viticultural heritage.

Our results also revealed the suitability of new locations for the settlement of vineyards, which opens up new opportunities for the development of viticulture, bringing new wines to the market.

Supplementary Materials: The following supporting information can be downloaded at: https://www.mdpi.com/article/10.3390/ijgi13080270/s1, Figures S1–S6 and Table S1.

Author Contributions: Conceptualization, António Fernandes, Nataša Kovač, Marko Simeunović, Kruna Ratković, Christoph Menz, Sergi Costafreda-Aumedes and João A. Santos; Data curation, António Fernandes, Nataša Kovač and André Fonseca; Formal analysis, António Fernandes, Nataša Kovač, Hélder Fraga, André Fonseca and João A. Santos; Funding acquisition, Nataša Kovač, Marko Simeunović and Kruna Ratković; Investigation, António Fernandes, Nataša Kovač, Hélder Fraga, André Fonseca, Sanja Šućur Radonjić, Christoph Menz and João A. Santos; Methodology, António Fernandes, Nataša Kovač, Marko Simeunović, Kruna Ratković, Christoph Menz, Sergi Costafreda-Aumedes and João A. Santos; Project administration, Nataša Kovač, Marko Simeunović and Kruna Ratković; Resources, Kruna Ratković and João A. Santos; Software, António Fernandes, Nataša Kovač, Hélder Fraga and André Fonseca; Supervision, Nataša Kovač, Marko Simeunović, Kruna Ratković and João A. Santos; Validation, António Fernandes, Nataša Kovač, Hélder Fraga, André Fonseca, Christoph Menz, Sergi Costafreda-Aumedes and João A. Santos; Visualization, António Fernandes, Hélder Fraga, André Fonseca, Christoph Menz, Sergi Costafreda-Aumedes and João A. Santos; Writing—original draft, António Fernandes, Nataša Kovač, Hélder Fraga, André Fonseca, Sanja Šućur Radonjić, Marko Simeunović, Kruna Ratković, Christoph Menz, Sergi Costafreda-Aumedes and João A. Santos; Writing—review & editing, António Fernandes, Nataša Kovač, Hélder Fraga, André Fonseca, Sanja Šućur Radonjić, Marko Simeunović, Kruna Ratković, Christoph Menz, Sergi Costafreda-Aumedes and João A. Santos. All authors have read and agreed to the published version of the manuscript.

Funding: This research was funded by the MONTEVITIS project "Integrating a comprehensive European approach for climate change mitigation and adaptation in Montenegro viticulture", funded by the European Union's Horizon Europe—the Framework Programme for Research and Innovation (2021–2027)—under grant agreement n° 101059461 (https://doi.org/10.3030/101059461).

Data Availability Statement: Weather station observations data belong to State Institute for Hydrometeorology and Seismology of Montenegro (IHMSM), available on IHMSM request. The original data from E-OBS is available at: https://www.ecad.eu/download/ensembles/download.php, accessed on 10 July 2023. The original data from ERA5-Land is available at: https://cds.climate.copernicus.eu/cdsapp#!/dataset/reanalysis-era5-land?tab=form, accessed on 10 July 2023. The original data from ERA5-Land is available at https://chelsa-climate.org/cmip6/, accessed on 14 August 2023.

Acknowledgments: Authors thank the Portuguese Foundation for Science and Technology, under the project UIDB/04033/2020 (https://doi.org/10.54499/UIDB/04033/2020) and LA/P/0126/2020 (https://doi.org/10.54499/LA/P/0126/2020).

Conflicts of Interest: The authors declare no conflicts of interest.

References

1. Aleixandre-Benavent, R.; Aleixandre-Tudó, J.L.; Castelló-Cogollos, L.; Aleixandre, J.L. Trends in Scientific Research on Climate Change in Agriculture and Forestry Subject Areas (2005–2014). *J. Clean. Prod.* **2017**, *147*, 406–418. [CrossRef]
2. Wu, X.; Lu, Y.; Zhou, S.; Chen, L.; Xu, B. Impact of Climate Change on Human Infectious Diseases: Empirical Evidence and Human Adaptation. *Environ. Int.* **2016**, *86*, 14–23. [CrossRef]
3. Arrhenius, S. On the Influence of Carbonic Acid in the Air upon the Temperature of the Ground. *Philos. Mag. J. Sci.* **1896**, *41*, 237–276. [CrossRef]

4. Chesney, M.; Gheyssens, J.; Pana, A.C.; Taschini, L. International Efforts to Tackle Climate Change. In *Environmental Finance and Investments*; Chesney, M., Gheyssens, J., Pana, A.C., Taschini, L., Eds.; Springer: Berlin/Heidelberg, Germany, 2016; pp. 17–48. ISBN 978-3-662-48175-2.
5. Moser, S.C. Communicating Climate Change: History, Challenges, Process and Future Directions. *Wiley Interdiscip. Rev. Clim. Change* **2010**, *1*, 31–53. [CrossRef]
6. Pasqui, M.; Di Giuseppe, E. Climate Change, Future Warming, and Adaptation in Europe. *Anim. Front.* **2019**, *9*, 6–11. [CrossRef]
7. IPCC. *Climate Change 2021—The Physical Science Basis*; IPCC: Geneva, Switzerland, 2023.
8. Chan, D.; Wu, Q. Significant Anthropogenic-Induced Changes of Climate Classes since 1950. *Sci. Rep.* **2015**, *5*, 13487. [CrossRef]
9. Cruz, J.; Belo-Pereira, M.; Fonseca, A.; Santos, J.A. Dynamic and Thermodynamic Drivers of Severe Sub-Hourly Precipitation Events in Mainland Portugal. *Atmosphere* **2023**, *14*, 1443. [CrossRef]
10. Malhi, G.S.; Kaur, M.; Kaushik, P. Impact of Climate Change on Agriculture and Its Mitigation Strategies: A Review. *Sustainability* **2021**, *13*, 1318. [CrossRef]
11. Beniston, M.; Stephenson, D.B.; Christensen, O.B.; Ferro, C.A.T.; Frei, C.; Goyette, S.; Halsnaes, K.; Holt, T.; Jylhä, K.; Koffi, B.; et al. Future Extreme Events in European Climate: An Exploration of Regional Climate Model Projections. *Clim. Change* **2007**, *81*, 71–95. [CrossRef]
12. Vaidya, H.N.; Breininger, R.D.; Madrid, M.; Lazarus, S.; Kachouie, N.N. Generalized Additive Models for Predicting Sea Level Rise in Coastal Florida. *Geosciences* **2023**, *13*, 310. [CrossRef]
13. Eccles, R.; Zhang, H.; Hamilton, D. A Review of the Effects of Climate Change on Riverine Flooding in Subtropical and Tropical Regions. *J. Water Clim. Change* **2019**, *10*, 687–707. [CrossRef]
14. Volchenko, N.; Zhmakin, S.; Udovenko, R.; Soldatkin, S.; Soldatkin, I. Combating Climate Change through the International Law Perspective: The Role of the EU in Environmental Diplomacy. *Eur. Energy Environ. Law Rev.* **2023**, *32*, 257–266. [CrossRef]
15. Debebe, Y.; Merine, M.; Argaw, M. An Overview of Climate Change Mitigation, Mitigation Strategies, and Technologies to Reduce Atmospheric Greenhouse Gas Concentrations: A Review Review Article. *Environ. Sci. Res. Rev.* **2023**, *6*, 595–603.
16. Jütersonke, S.; Groß, M. The Effect of Social Recognition on Support for Climate Change Mitigation Measures. *Sustainability* **2023**, *15*, 16486. [CrossRef]
17. Kolodko, G.W. Political System and Socio-Economic Development. In *Political Economy of New Pragmatism: Implications of Irreversible Globalization*; Kolodko, G.W., Ed.; Springer International Publishing: Cham, Switzerland, 2022; pp. 53–108, ISBN 978-3-031-12263-7.
18. Kawase, H.; Nagashima, T.; Sudo, K.; Nozawa, T. Future Changes in Tropospheric Ozone under Representative Concentration Pathways (RCPs). *Geophys. Res. Lett.* **2011**, *38*, L05801. [CrossRef]
19. Riahi, K.; van Vuuren, D.P.; Kriegler, E.; Edmonds, J.; O'Neill, B.C.; Fujimori, S.; Bauer, N.; Calvin, K.; Dellink, R.; Fricko, O.; et al. The Shared Socioeconomic Pathways and Their Energy, Land Use, and Greenhouse Gas Emissions Implications: An Overview. *Glob. Environ. Change* **2017**, *42*, 153–168. [CrossRef]
20. Wimalasiri, E.M.; Sirishantha, D.; Karunadhipathi, U.L.; Ampitiyawatta, A.D.; Muttil, N.; Rathnayake, U. Climate Change and Soil Dynamics: A Crop Modelling Approach. *Soil. Syst.* **2023**, *7*, 82. [CrossRef]
21. Droulia, F.; Charalampopoulos, I. Future Climate Change Impacts on European Viticulture: A Review on Recent Scientific Advances. *Atmosphere* **2021**, *12*, 495. [CrossRef]
22. Leolini, L.; Moriondo, M.; Fila, G.; Costafreda-Aumedes, S.; Ferrise, R.; Bindi, M. Late Spring Frost Impacts on Future Grapevine Distribution in Europe. *Field Crops Res.* **2018**, *222*, 197–208. [CrossRef]
23. Dinu, D.G.; Ricciardi, V.; Demarco, C.; Zingarofalo, G.; De Lorenzis, G.; Buccolieri, R.; Cola, G.; Rustioni, L. Climate Change Impacts on Plant Phenology: Grapevine (*Vitis vinifera*) Bud Break in Wintertime in Southern Italy. *Foods* **2021**, *10*, 2769. [CrossRef] [PubMed]
24. Fraga, H.; Garcia de Cortazar-Atauri, I.; Malheiro, A.; Santos, J. Modelling Climate Change Impacts on Viticultural Yield, Phenology and Stress Conditions in Europe. *Glob. Change Biol.* **2016**, *22*, 3774–3788. [CrossRef]
25. Jones, G.; Davis, R. Climate Influences on Grapevine Phenology, Grape Composition, and Wine Production and Quality for Bordeaux, France. *Am. J. Enol. Vitic.* **2000**, *51*, 249–261. [CrossRef]
26. Van Leeuwen, C.; Darriet, P. The Impact of Climate Change on Viticulture and Wine Quality. *J. Wine Econ.* **2016**, *11*, 150–167. [CrossRef]
27. Xynas, B.; Barnes, C. Yeast or Water: Producing Wine with Lower Alcohol Levels in a Warming Climate: A Review. *J. Sci. Food Agric.* **2022**, *103*, 3249–3260. [CrossRef]
28. Leolini, L.; Moriondo, M.; Romboli, Y.; Gardiman, M.; Costafreda-Aumedes, S.; García De Cortázar-Atauri, I.; Bindi, M.; Granchi, L.; Brilli, L. Modelling Sugar and Acid Content in Sangiovese Grapes under Future Climates: An Italian Case Study. *Clim. Res.* **2019**, *78*, 211–224. [CrossRef]
29. Bonfante, A.; Monaco, E.; Langella, G.; Mercogliano, P.; Bucchignani, E.; Manna, P.; Terribile, F. A Dynamic Viticultural Zoning to Explore the Resilience of Terroir Concept under Climate Change. *Sci. Total Environ.* **2018**, *624*, 294–308. [CrossRef]
30. Sgubin, G.; Swingedouw, D.; Mignot, J.; Gambetta, G.A.; Bois, B.; Loukos, H.; Noël, T.; Pieri, P.; García de Cortázar-Atauri, I.; Ollat, N.; et al. Non-Linear Loss of Suitable Wine Regions over Europe in Response to Increasing Global Warming. *Glob. Change Biol.* **2023**, *29*, 808–826. [CrossRef]

31. Van Leeuwen, C.; Destrac-Irvine, A.; Dubernet, M.; Duchêne, E.; Gowdy, M.; Marguerit, E.; Pieri, P.; Parker, A.; De Rességuier, L.; Ollat, N. An Update on the Impact of Climate Change in Viticulture and Potential Adaptations. *Agronomy* **2019**, *9*, 514. [CrossRef]
32. Lionello, P.; Scarascia, L. The Relation between Climate Change in the Mediterranean Region and Global Warming. *Reg. Environ. Change* **2018**, *18*, 1481–1493. [CrossRef]
33. Tuel, A.; Eltahir, E. Why Is the Mediterranean a Climate Change Hot Spot? *J. Clim.* **2020**, *33*, 5829–5843. [CrossRef]
34. Santos, J.A.; Fraga, H.; Malheiro, A.C.; Moutinho-Pereira, J.; Dinis, L.T.; Correia, C.; Moriondo, M.; Leolini, L.; Dibari, C.; Costafreda-Aumedes, S.; et al. A Review of the Potential Climate Change Impacts and Adaptation Options for European Viticulture. *Appl. Sci.* **2020**, *10*, 3092. [CrossRef]
35. Droulia, F.; Charalampopoulos, I. A Review on the Observed Climate Change in Europe and Its Impacts on Viticulture. *Atmosphere* **2022**, *13*, 837. [CrossRef]
36. Naulleau, A.; Gary, C.; Prévot, L.; Hossard, L. Evaluating Strategies for Adaptation to Climate Change in Grapevine Production—A Systematic Review. *Front. Plant Sci.* **2021**, *11*, 607859. [CrossRef]
37. Jones, G.; Alves, F. Impact of Climate Change on Wine Production: A Global Overview and Regional Assessment in the Douro Valley of Portugal. *Int. J. Glob. Warm.* **2012**, *4*, 383–406. [CrossRef]
38. Usmonova, D. The Use of Marketing Strategies in Increasing the Export Potential of Enterprises of the Viticultural Industry. *Econ. Educ.* **2023**, *24*, 268–273. [CrossRef]
39. Ćulafić, G.; Popov, T.; Gnjato, S.; Bajić, D.; Trbić, G.; Mitrović, L. Spatial and Temporal Patterns of Precipitation in Montenegro. *Időjárás* **2020**, *124*, 499–519. [CrossRef]
40. Burić, D.; Doderović, M. Trend of Percentile Climate Indices in Montenegro in the Period 1961–2020. *Sustainability* **2022**, *14*, 12519. [CrossRef]
41. Burić, D.; Doderović, M. Precipitation, Humidity and Cloudiness in Podgorica (Montenegro) during the Period 1951–2018. *Geogr. Pannonica* **2019**, *23*, 233–244. [CrossRef]
42. Burić, D.; Doderović, M. Changes in Temperature and Precipitation in the Instrumental Period (1951–2018) and Projections up to 2100 in Podgorica (Montenegro). *Int. J. Climatol.* **2021**, *41*, E133–E149. [CrossRef]
43. Burić, D.; Vladan, D.; Mihajlovic, J. The Climate of Montenegro: Modificators and Types—Part One. *Glas. Srp. Geogr. Drus.* **2013**, *93*, 83–102. [CrossRef]
44. Doderović, M.M.; Burić, D.B. Atlantic Multi-Decadal Oscillation and Changes of Summer Air Temperature in Montenegro. *Therm. Sci.* **2015**, *19*, 405–414. [CrossRef]
45. Burić, D.; Ducic, V.; Mihajlovic, J. The Climate of Montenegro: Modificators and Types—Part Two. *Glas. Srp. Geogr. Drus.* **2014**, *94*, 73–90. [CrossRef]
46. Lukovic, J.; Buric, D.; Mihajlovic, J.; Pejovic, M. Spatial and Temporal Variations of Aridity-Humidity Indices in Montenegro. *Theor. Appl. Climatol.* **2024**, *155*, 4553–4566. [CrossRef]
47. Burić, D.; Banjak, D.; Doderović, M.; Marčev, A. Example of the Importance of Early Warning of Extreme Weather Events in Montenegro in the Context of Recent Climate Change. *Zb. Rad.-Geogr. Fak. Univ. Beogr.* **2022**, *2022*, 57–72. [CrossRef]
48. Doderović, M.; Burić, D.; Ducić, V.; Mijanović, I. Recent and Future Air Temperature and Precipitation Changes in the Mountainous North of Montenegro. *J. Geogr. Inst. Jovan Cvijic SASA* **2020**, *70*, 189–201. [CrossRef]
49. Burić, D.; Dragojlović, J.; Penjišević-Sočanac, I.; Luković, J.; Doderović, M. Relationship between Atmospheric Circulation and Temperature Extremes in Montenegro in the Period 1951–2010. In *Climate Change Adaptation in Eastern Europe: Managing Risks and Building Resilience to Climate Change*; Leal Filho, W., Trbic, G., Filipovic, D., Eds.; Springer International Publishing: Cham, Switerland, 2019; pp. 29–42. ISBN 978-3-030-03383-5.
50. Buric, D. Detected and Projected Temperature Changes in the Area of Mediterranean Montenegro. *Geogr. J.* **2024**, e12580. [CrossRef]
51. Burić, D.; Doderović, M. Projected Temperature Changes in Kolašin (Montenegro) up to 2100 According to EBU-POM and ALADIN Regional Climate Models. *Idojaras* **2020**, *124*, 427–445. [CrossRef]
52. Knežević, M.; Zivotić, L.; Čereković, N.; Topalović, A.; Koković, N.; Todorovic, M. Impact of Climate Change on Water Requirements and Growth of Potato in Different Climatic Zones of Montenegro. *J. Water Clim. Change* **2018**, *9*, 657–671. [CrossRef]
53. Knezević, M.; Zivotić, L.; Perović, V.; Topalović, A.; Todorović, M. Impact of Climate Change on Olive Growth Suitability, Water Requirements and Yield in Montenegro. *Ital. J. Agrometeorol.* **2017**, *22*, 39–52. [CrossRef]
54. CLC CORINE Land Cover 2018 (Vector). Available online: https://sdi.eea.europa.eu/catalogue/copernicus/api/records/71c95a07-e296-44fc-b22b-415f42acfdf0?language=all (accessed on 5 January 2024).
55. Djurović, P.; Djurović, M. Physical Geographic Characteristics and Sustainable Development of the Mountain Area in Montenegro. In *Sustainable Development in Mountain Regions: Southeastern Europe*; Springer: Cham, Switerland, 2016; ISBN 978-3-319-20109-2.
56. Del Bianco, F.; Gasperini, L.; Angeletti, L.; Giglio, F.; Bortoluzzi, G.; Montagna, P.; Ravaioli, M.; Kljajic, Z. Stratigraphic Architecture of the Montenegro/N. Albania Continental Margin (Adriatic Sea—Central Mediterranean). *Mar. Geol.* **2014**, *359*, 61–74. [CrossRef]
57. Pajek, L.; Košir, M. Overheating Vulnerability Assessment of Energy Retrofit Actions in a Multi-Apartment Building in Podgorica, Montenegro. In *E3S Web of Conferences*; EDP Sciences: Les Ulis, France, 2023; Volume 396.
58. Körner, C.; Hiltbrunner, E. Why Is the Alpine Flora Comparatively Robust against Climatic Warming? *Diversity* **2021**, *13*, 383. [CrossRef]

59. Burić, D.; Luković, J.; Bajat, B.; Kilibarda, M.; Živković, N. Recent Trends in Daily Rainfall Extremes over Montenegro (1951–2010). *Nat. Hazards Earth Syst. Sci.* **2015**, *15*, 2069–2077. [CrossRef]
60. Ljubisavljević, K.; Tomović, L.; Urošević, A.; Gvozdenović Nikolić, S.; Vuk, I.; Vernes, Z.; Labus, N. Species Diversity and Distribution of Lizards in Montenegro. *Acta Herpetol.* **2018**, *13*, 3–11. [CrossRef]
61. Savić, S.; Vukotić, M. Viticulture Zoning in Montenegro. *Bull. Univ. Agric. Sci. Vet. Med. Cluj-Napoca Hortic.* **2018**, *75*, 73. [CrossRef]
62. Fraga, H.; Freitas, T.R.; Fonseca, A.; Fernandes, A.; Santos, J.A. Climate Change Implications on the Viticulture Geography. In *Advances in Botanical Research*; Academic Press: Cambridge, MA, USA, 2024; ISBN 0065-2296.
63. Maraš, V.; Popović, T.; Gazivoda, A.; Raičević, J.; Kodžulović, V.; Mugoša, M.; Radonjić, S. Origin and Characterization of Montenegrin Grapevine Varieties. *Vitis—J. Grapevine Res.* **2015**, *54*, 135–137.
64. Maraš, V. Ampelographic and Genetic Characterization of Montenegrin Grapevine Varieties. In *Advances in Grape and Wine Biotechnology*; BoD—Books on Demand: Norderstedt, Germany, 2019; ISBN 978-1-78984-612-6.
65. TrendEconomy. Annual International Trade Statistics by Country (HS). Available online: https://trendeconomy.com/data/h2/Montenegro/2204 (accessed on 13 July 2024).
66. Zejak, D.; Dudic, B.; Bartáková, G.P.; Gubíniová, K. Contribution to the Knowledge of Grapevine Production in Southeastern Europe—Case Study of Montenegro. In *New Technologies, Development and Application VI. NT 2023*; Lecture Notes in Networks and Systems; Springer: Cham, Switerland, 2023; Volume 707 LNNS.
67. Jakšić, D.; Basha, E.; Blesić, M.; Kuçi, Y.; Maraš, V.; Beleski, K. *Report for the Viticulture and Wine Sector in the Western Balkans*; Regional Rural Development Standing Working Group in SEE (SWG): Skopje, North Macedonia, 2023.
68. CRSFA. *Studija o Rejonizaciji Vinogradarskih Geografskih Proizvodnih Područja Crne Gore (A Study on Regionization of Wine Growing Geographic Production Areas of Montenegro)*; CRSFA: Bari, Italy, 2017.
69. Burić, D.; Mihajlović, J.; Ducić, V. Climatic Regionalization of Montenegro by Applying Different Methods of Cluster Analysis. *Geogr. Pannonica* **2023**, *27*, 119–131. [CrossRef]
70. Spalevic, V. Pedological Characteristics of Montenegro. In *Speleology of Montenegro*; Barovic, G., Ed.; Springer International Publishing: Cham, Switerland, 2024; pp. 85–97, ISBN 978-3-031-49375-1.
71. Tello, J.; Mammerler, R.; Čajić, M.; Forneck, A. Major Outbreaks in the Nineteenth Century Shaped Grape Phylloxera Contemporary Genetic Structure in Europe. *Sci. Rep.* **2019**, *9*, 17540. [CrossRef]
72. Pajović-Šćepanović, R.; Wendelin, S.; Eder, R. Phenolic Composition and Varietal Discrimination of Montenegrin Red Wines (*Vitis vinifera* var. Vranac, Kratošija, and Cabernet Sauvignon). *Eur. Food Res. Technol.* **2018**, *244*, 2243–2254. [CrossRef]
73. Maraš, V.; Bozovic, V.; Giannetto, S.; Crespan, M. SSR Molecular Marker Analysis of the Grapevine Germplasm of Montenegro. *J. Int. Sci. Vigne Vin* **2014**, *48*, 87–97. [CrossRef]
74. Štajner, N.; Tomić, L.; Ivanišević, D.; Korać, N.; Cvetković-Jovanović, T.; Beleski, K.; Angelova, E.; Maraš, V.; Javornik, B. Microsatellite Inferred Genetic Diversity and Structure of Western Balkan Grapevines (*Vitis vinifera* L.). *Tree Genet. Genomes* **2014**, *10*, 127–140. [CrossRef]
75. Mihaljević, M.Ž.; Anhalt, U.C.M.; Rühl, E.; Mugoša, M.T.; Maraš, V.; Forneck, A.; Zdunić, G.; Preiner, D.; Pejić, I. Cultivar Identity, Intravarietal Variation, and Health Status of Native Grapevine Varieties in Croatia and Montenegro. *Am. J. Enol. Vitic.* **2015**, *66*, 531–541. [CrossRef]
76. Madžgalj, V.; Petrović, A.; Cakar, U.; Maras, V.; Sofrenic, I.; Tešević, V. The Influence of Different Enzymatic Preparations and Skin Contact Time on Aromatic Profile of Wines Produced from Autochthonous Grape Varieties Krstac and Zizak. *J. Serbian Chem. Soc.* **2023**, *88*, 11–23. [CrossRef]
77. Maraš, V.; Popovic, T.; Gajinov, S.; Mugosa, M.; Popovic, V.; Savovic, A.; Pavicevic, K.; Mirovic, V. Optimal Irrigation as a Tool of Precision Agriculture. In Proceedings of the 2019 8th Mediterranean Conference on Embedded Computing (MECO), Budva, Montenegro, 10–14 June 2019.
78. Stojanovic, R.; Maraš, V.; Radonjić, S.; Martic, A.; Đurković, J.; Pavicevic, K.; Mirovic, V.; Cvetkovic, M. A Feasible IoT-Based System for Precision Agriculture. In Proceedings of the 2021 10th Mediterranean Conference on Embedded Computing (MECO), Budva, Montenegro, 7–10 June 2021.
79. Maraš, V.; Kodžulović, V.; Mugoša, M.; Raičević, J.; Gazivoda, A.; Šućur, S.; Perišić, M. Clonal Selection of Autochthonous Grape Variety Vranac in Montenegro. In Proceedings of the CMBEBIH 2017, Sarajevo, Bosnia and Herzegovina, 16–18 March 2017; Badnjevic, A., Ed.; Springer: Singapore, 2017; pp. 787–790.
80. Maraš, V.; Tello, J.; Gazivoda, A.; Mugoša, M.; Perišić, M.; Raičević, J.; Stajner, N.; Ocete, R.; Bozovic, V.; Popović, T.; et al. Population Genetic Analysis in Old Montenegrin Vineyards Reveals Ancient Ways Currently Active to Generate Diversity in *Vitis vinifera*. *Sci. Rep.* **2020**, *10*, 15000. [CrossRef]
81. Karger, D.N.; Conrad, O.; Böhner, J.; Kawohl, T.; Kreft, H.; Soria-Auza, R.W.; Zimmermann, N.E.; Linder, H.P.; Kessler, M. Climatologies at High Resolution for the Earth's Land Surface Areas. *Sci. Data* **2017**, *4*, 170122. [CrossRef]
82. Brun, P.; Zimmermann, N.E.; Hari, C.; Pellissier, L.; Karger, D.N. Global Climate-Related Predictors at Kilometer Resolution for the Past and Future. *Earth Syst. Sci. Data* **2022**, *14*, 5573–5603. [CrossRef]
83. Beck, H.E.; Wood, E.F.; McVicar, T.R.; Zambrano-Bigiarini, M.; Alvarez-Garreton, C.; Baez-Villanueva, O.M.; Sheffield, J.; Karger, D.N. Bias Correction of Global High-Resolution Precipitation Climatologies Using Streamflow Observations from 9372 Catchments. *J. Clim.* **2020**, *33*, 1299–1315. [CrossRef]

84. Lange, S. Trend-Preserving Bias Adjustment and Statistical Downscaling with ISIMIP3BASD (v1.0). *Geosci. Model. Dev.* **2019**, *12*, 3055–3070. [CrossRef]
85. O'Neill, B.C.; Tebaldi, C.; van Vuuren, D.P.; Eyring, V.; Friedlingstein, P.; Hurtt, G.; Knutti, R.; Kriegler, E.; Lamarque, J.-F.; Lowe, J.; et al. The Scenario Model Intercomparison Project (ScenarioMIP) for CMIP6. *Geosci. Model. Dev.* **2016**, *9*, 3461–3482. [CrossRef]
86. Meinshausen, M.; Nicholls, Z.R.J.; Lewis, J.; Gidden, M.J.; Vogel, E.; Freund, M.; Beyerle, U.; Gessner, C.; Nauels, A.; Bauer, N.; et al. The Shared Socio-Economic Pathway (SSP) Greenhouse Gas Concentrations and Their Extensions to 2500. *Geosci. Model. Dev.* **2020**, *13*, 3571–3605. [CrossRef]
87. Karger, D.N.; Brun, P.; Zimmermann, N. *Climatologies at High Resolution for the Earth Land Areas, CHELSA V2.1: Technical Specification*; Nature Publishing Groups: London, UK, 2021; Volume 4.
88. Muñoz Sabater, J. ERA5-Land Hourly Data from 1950 to Present. Copernicus Climate Change Service (C3S) Climate Data Store (CDS). Available online: https://cds.climate.copernicus.eu/cdsapp#!/dataset/10.24381/cds.e2161bac?tab=overview (accessed on 1 August 2023).
89. Cornes, R.C.; van der Schrier, G.; van den Besselaar, E.J.M.; Jones, P.D. An Ensemble Version of the E-OBS Temperature and Precipitation Data Sets. *J. Geophys. Res. Atmos.* **2018**, *123*, 9391–9409. [CrossRef]
90. IHMS Institute of Hydrometeorology and Seismology. Available online: https://www.meteo.co.me/ (accessed on 17 July 2023).
91. Dunne, J.P.; Horowitz, L.W.; Adcroft, A.J.; Ginoux, P.; Held, I.M.; John, J.G.; Krasting, J.P.; Malyshev, S.; Naik, V.; Paulot, F.; et al. The GFDL Earth System Model Version 4.1 (GFDL-ESM 4.1): Overall Coupled Model Description and Simulation Characteristics. *J. Adv. Model Earth Syst.* **2020**, *12*, e2019MS002015. [CrossRef]
92. Sellar, A.A.; Jones, C.G.; Mulcahy, J.P.; Tang, Y.; Yool, A.; Wiltshire, A.; O'Connor, F.M.; Stringer, M.; Hill, R.; Palmieri, J.; et al. UKESM1: Description and Evaluation of the U.K. Earth System Model. *J. Adv. Model Earth Syst.* **2019**, *11*, 4513–4558. [CrossRef]
93. Gutjahr, O.; Putrasahan, D.; Lohmann, K.; Jungclaus, J.H.; von Storch, J.-S.; Brüggemann, N.; Haak, H.; Stössel, A. Max Planck Institute Earth System Model (MPI-ESM1.2) for the High-Resolution Model Intercomparison Project (HighResMIP). *Geosci. Model Dev.* **2019**, *12*, 3241–3281. [CrossRef]
94. Lurton, T.; Balkanski, Y.; Bastrikov, V.; Bekki, S.; Bopp, L.; Braconnot, P.; Brockmann, P.; Cadule, P.; Contoux, C.; Cozic, A.; et al. Implementation of the CMIP6 Forcing Data in the IPSL-CM6A-LR Model. *J. Adv. Model Earth Syst.* **2020**, *12*, e2019MS001940. [CrossRef]
95. Yukimoto, S.; Kawai, H.; Koshiro, T.; Oshima, N.; Yoshida, K.; Urakawa, S.; Tsujino, H.; Deushi, M.; Tanaka, T.; Hosaka, M.; et al. The Meteorological Research Institute Earth System Model Version 2.0, MRI-ESM2.0: Description and Basic Evaluation of the Physical Component. *J. Meteorol. Soc. Jpn. Ser. II* **2019**, *97*, 931–965. [CrossRef]
96. Comte, V.; Schneider, L.; Calanca, P.; Rebetez, M. Effects of Climate Change on Bioclimatic Indices in Vineyards along Lake Neuchatel, Switzerland. *Theor. Appl. Climatol.* **2022**, *147*, 423–436. [CrossRef]
97. Christensen, J.H.; Christensen, O.B. A Summary of the PRUDENCE Model Projections of Changes in European Climate by the End of This Century. *Clim. Change* **2007**, *81*, 7–30. [CrossRef]
98. Jacob, D.; Bärring, L.; Christensen, O.B.; Christensen, J.H.; De Castro, M.; Déqué, M.; Giorgi, F.; Hagemann, S.; Hirschi, M.; Jones, R.; et al. An Inter-Comparison of Regional Climate Models for Europe: Model Performance in Present-Day Climate. *Clim. Change* **2007**, *81*, 31–52. [CrossRef]
99. Wallach, D.; Mearns, L.; Ruane, A.; Rötter, R.P.; Asseng, S. Lessons from Climate Modeling on the Design and Use of Ensembles for Crop Modeling. *Clim. Change* **2016**, *139*, 551–564. [CrossRef]
100. UNDP. *Montenegro Third National Communication on Climate Change*; UNDP: New York, NY, USA, 2020.
101. Živković, K.; Radulović, M.; Lojen, S.; Pucarević, M. Overview of the Chemical and Isotopic Investigations of the Mareza Springs and the Zeta River in Montenegro. *Water* **2020**, *12*, 957. [CrossRef]
102. Cataldo, E.; Eichmeier, A.; Mattii, G.B. Effects of Global Warming on Grapevine Berries Phenolic Compounds—A Review. *Agronomy* **2023**, *13*, 2192. [CrossRef]
103. Gordo, O.; Sanz, J. Impact of Climate Change on Plant Phenology in Mediterranean Ecosystems. *Glob. Change Biol.* **2010**, *16*, 1082–1106. [CrossRef]
104. Grillakis, M.G.; Doupis, G.; Kapetanakis, E.; Goumenaki, E. Future Shifts in the Phenology of Table Grapes on Crete under a Warming Climate. *Agric. For. Meteorol.* **2022**, *318*, 108915. [CrossRef]
105. Williams, L.E. Potential Vineyard Evapotranspiration (ET) Due to Global Warming: Comparison of Vineyard et at Three Locations in California Differing in Mean Seasonal Temperatures. *Acta Hortic.* **2012**, *931*, 221–226. [CrossRef]
106. Bačević, N.; Valjarević, A.; Kićović, D.; Milentijević, N.; Ivanović, M.; Mujević, M. Analysis of Air Temperature Trends: City of Podgorica (Montenegro). *Univ. Thought—Publ. Nat. Sci.* **2020**, *10*, 31–36. [CrossRef]
107. Ruml, M.; Gregorić, E.; Vujadinović, M.; Radovanović, S.; Matović, G.; Vuković, A.; Počuča, V.; Stojičić, D. Observed Changes of Temperature Extremes in Serbia over the Period 1961–2010. *Atmos. Res.* **2017**, *183*, 26–41. [CrossRef]
108. Schultz, H.R.; Jones, G.V. Climate Induced Historic and Future Changes in Viticulture. *J. Wine Res.* **2010**, *21*, 137–145. [CrossRef]
109. Fraga, H.; Santos, J.; Malheiro, A.; Oliveira, A.; Moutinho Pereira, J.; Jones, G. Climatic Suitability of Portuguese Grapevine Varieties and Climate Change Adaptation. *Int. J. Climatol.* **2015**, *36*, 1–12. [CrossRef]
110. Moriondo, M.; Jones, G.V.; Bois, B.; Dibari, C.; Ferrise, R.; Trombi, G.; Bindi, M. Projected Shifts of Wine Regions in Response to Climate Change. *Clim. Change* **2013**, *119*, 825–839. [CrossRef]
111. Winkler, A.J.; Cook, J.A.; Kliewer, W.M.; Lider, L.A. General Viticulture. *Soil. Sci.* **1975**, *120*, 462. [CrossRef]

112. Alba, V.; Gentilesco, G.; Tarricone, L. Climate Change in a Typical Apulian Region for Table Grape Production: Spatialisation of Bioclimatic Indices, Classification and Future Scenarios. *OENO One* **2021**, *55*, 1–20. [CrossRef]
113. Cardell, M.; Amengual, A.; Romero, R. Future Effects of Climate Change on the Suitability of Wine Grape Production across Europe. *Reg. Environ. Change* **2019**, *19*, 2299–2310. [CrossRef]
114. Spinoni, J.; Vogt, J.; Barbosa, P. European Degree-Day Climatologies and Trends for the Period 1951–2011. *Int. J. Climatol.* **2015**, *35*, 25–36. [CrossRef]
115. Santos, J.; Malheiro, A.; Pinto, J.; Jones, G. Macroclimate and Viticultural Zoning in Europe: Observed Trends and Atmospheric Forcing. *Clim. Res.* **2012**, *51*, 89–103. [CrossRef]
116. Tello, J.; Ibáñez, J. Review: Status and Prospects of Association Mapping in Grapevine. *Plant Sci.* **2023**, *327*, 111539. [CrossRef]
117. Yang, C.; Menz, C.; Fraga, H.; Reis, S.; Machado, N.; Malheiro, A.C.; Santos, J.A. Simultaneous Calibration of Grapevine Phenology and Yield with a Soil–Plant–Atmosphere System Model Using the Frequentist Method. *Agronomy* **2021**, *11*, 1659. [CrossRef]
118. Ollat, N.; Touzard, J.-M.; Van Leeuwen, C. Climate Change Impacts and Adaptations: New Challenges for the Wine Industry. *J. Wine Econ.* **2016**, *11*, 139–149. [CrossRef]
119. De Santis, D.; Frangipane, M.T.; Brunori, E.; Cirigliano, P.; Biasi, R. Biochemical Markers for Enological Potentiality in a Grapevine Aromatic Variety under Different Soil Types. *Am. J. Enol. Vitic.* **2017**, *68*, 100–111. [CrossRef]
120. Oliveira, C.; Ferreira, A.C.; Costa, P.; Guerra, J.; De Pinho, P.G. Effect of Some Viticultural Parameters on the Grape Carotenoid Profile. *J. Agric. Food Chem.* **2004**, *52*, 4178–4184. [CrossRef]
121. Biddoccu, M.; Zecca, O.; Audisio, C.; Godone, F.; Barmaz, A.; Cavallo, E. Assessment of Long-Term Soil Erosion in a Mountain Vineyard, Aosta Valley (NW Italy). *Land. Degrad. Dev.* **2018**, *29*, 617–629. [CrossRef]
122. Radonjić, S.; Krstić, O.; Cvrković, T.; Hrnčić, S.; Marinković, S.; Mitrović, M.; Toševski, I.; Jović, J. The First Report on the Occurrence of Flavescence Dorée Phytoplasma Affecting Grapevine in Vineyards of Montenegro and an Overview of Epidemic Genotypes in Natural Plant Reservoirs. *J. Plant Pathol.* **2023**, *105*, 419–427. [CrossRef]
123. Rodrigo Comino, J.; Ruiz Sinoga, J.D.; Senciales González, J.M.; Guerra-Merchán, A.; Seeger, M.; Ries, J.B. High Variability of Soil Erosion and Hydrological Processes in Mediterranean Hillslope Vineyards (Montes de Málaga, Spain). *Catena* **2016**, *145*, 274–284. [CrossRef]
124. Paiola, A.; Assandri, G.; Brambilla, M.; Zottini, M.; Pedrini, P.; Nascimbene, J. Exploring the Potential of Vineyards for Biodiversity Conservation and Delivery of Biodiversity-Mediated Ecosystem Services: A Global-Scale Systematic Review. *Sci. Total Environ.* **2020**, *706*, 135839. [CrossRef] [PubMed]
125. Ćeranić, G.; Krivokapić, N.; Šarović, R.; Živković, P. Perception of Climate Change and Assessment of the Importance of Sustainable Behavior for Their Mitigation: The Example of Montenegro. *Sustainability* **2023**, *15*, 10165. [CrossRef]
126. Jones, G.V.; Webb, L.B. Climate Change, Viticulture, and Wine: Challenges and Opportunities. *J. Wine Res.* **2010**, *21*, 103–106. [CrossRef]
127. Neethling, E.; Petitjean, T.; Quénol, H.; Barbeau, G. Assessing Local Climate Vulnerability and Winegrowers' Adaptive Processes in the Context of Climate Change. *Mitig. Adapt. Strateg. Glob. Change* **2017**, *22*, 777–803. https://doi.org/10.1007/s11027-015-9698-0. Erratum in *Mitig. Adapt. Strateg. Glob. Change* **2020**, *25*, 737–738. [CrossRef]
128. Sekulić, G.; Radulovic, M. The Hydrology and Hydrogeology of Montenegro. In *The Rivers of Montenegro*; Springer: Cham, Switerland, 2019; ISBN 978-3-030-55711-9.
129. Maraš, V.; Bogicevic, M.; Tomic, M.; Kodžulović, V.; Radonjić, S.; Čizmović, M.; Raicevic, D. Genetic and Sanitary Evaluation of CV. Vranac. *Bull. UASVM Hortic.* **2011**, *68*, 155–162.
130. Pajovic Scepanovic, R.; Raicevic, D.; Popovic, T.; Sivilotti, P.; Lisjak, K.; Vanzo, A. Polyphenolic Characterisation of Vranac, Kratosija and Cabernet Sauvignon (*Vitis vinifera* L. Cv.) Grapes and Wines from Different Vineyard Locations in Montenegro. *S. Afr. J. Enol. Vitic.* **2014**, *35*, 139–148. [CrossRef]
131. Rodrigues, P.; Pedroso, V.; Reis, S.; Yang, C.; Santos, J.A. Climate Change Impacts on Phenology and Ripening of Cv. Touriga Nacional in the Dão Wine Region, Portugal. *Int. J. Climatol.* **2022**, *42*, 7117–7132. [CrossRef]
132. Milosavljević, M. *Biotehnika Vinove Loze*; Samostalna izdavačka agencija; NIK-PRESS: Beograd, Srbija, 2012.
133. Ollat, N.; Quénol, H.; Barbeau, G.; van Leeuwen, C.; Darriet, P.; Garcia de Cortazar-Atauri, I.; Bois, B.; Ojeda, H.; Duchêne, E.; Lebon, E.; et al. Adaptation to Climate Change of the French Wine Industry: A Systemic Approach—Main Outcomes of the Project LACCAVE. *E3S Web Conf.* **2018**, *50*, 01020. [CrossRef]
134. Keller, M. Managing Grapevines to Optimise Fruit Development in a Challenging Environment: A Climate Change Primer for Viticulturists. *Aust. J. Grape Wine Res.* **2010**, *16*, 56–69. [CrossRef]
135. Adelsheim, D.; Busch, C.; Catena, L.; Champy, B.; Coetzee, J.; Coia, L.; Croser, B.; Draper, P.; Durbourdieu, D.; Frank, F.; et al. Climate Change: Field Reports from Leading Winemakers. *J. Wine Econ.* **2016**, *11*, 5–47. [CrossRef]
136. Kemp, B.; Pedneault, K.; Pickering, G.; Usher, K.; Willwerth, J. Red Winemaking in Cool Climates. *Red Wine Technol.* **2019**, 341–356. [CrossRef]
137. Santesteban, L.G.; Rekarte, I.; Torres, N.; Galar, M.; Villa-Llop, A.; Visconti, F.; Intrigliolo, D.S.; Escalona, J.M.; de Herralde, F.; Miranda, C. The Role of Rootstocks for Grape Growing Adaptation to Climate Change. Meta-Analysis of the Research Conducted in Spanish Viticulture. *OENO One* **2023**, *57*, 283–290. [CrossRef]

138. Ollat, N.; Peccoux, A.; Papura, D.; Esmenjaud, D.; Marguerit, E.; Tandonnet, J.-P.; Bordenave, L.; Cookson, S.; Barrieu, F.; Rossdeutsch, L.; et al. Rootstocks as a Component of Adaptation to Environment. In *Grapevine in a Changing Environment: A Molecular and Ecophysiological Perspective*; Wiley: Hoboken, NJ, USA, 2015.
139. Morales-Henríquez, T.; Gutiérrez-Gamboa, G.; Zheng, W.; Martínez de Toda, F. Principles of Vineyard Establishment and Strategies to Delay Ripening under a Warming Climate. *IVES Tech. Rev. Vine Wine* **2022**. [CrossRef]
140. Corso, M.; Bonghi, C. Grapevine Rootstock Effects on Abiotic Stress Tolerance. *Plant Sci. Today* **2014**, *1*, 108–113. [CrossRef]
141. Fraga, H.; García de Cortázar Atauri, I.; Santos, J.A. Viticultural Irrigation Demands under Climate Change Scenarios in Portugal. *Agric. Water Manag.* **2018**, *196*, 66–74. [CrossRef]
142. De Cáceres, M.; Martin-StPaul, N.; Turco, M.; Cabon, A.; Granda, V. Estimating Daily Meteorological Data and Downscaling Climate Models over Landscapes. *Environ. Model. Softw.* **2018**, *108*, 186–196. [CrossRef]
143. Fonseca, A.; Cruz, J.; Fraga, H.; Andrade, C.; Valente, J.; Alves, F.; Neto, A.; Flores, R.; Santos, J. Vineyard Microclimatic Zoning as a Tool to Promote Sustainable Viticulture under Climate Change. *Sustainability* **2024**, *16*, 3477. [CrossRef]
144. Rounsevell, M.D.A.; Evans, S.P.; Bullock, P. Climate Change and Agricultural Soils: Impacts and Adaptation. *Clim. Change* **1999**, *43*, 683–709. [CrossRef]
145. Dai, Z.; Yu, M.; Chen, H.; Zhao, H.; Huang, Y.; Su, W.; Xia, F.; Chang, S.X.; Brookes, P.C.; Dahlgren, R.A.; et al. Elevated Temperature Shifts Soil N Cycling from Microbial Immobilization to Enhanced Mineralization, Nitrification and Denitrification across Global Terrestrial Ecosystems. *Glob. Change Biol.* **2020**, *26*, 5267–5276. [CrossRef] [PubMed]
146. Costantini, E.A.C.; Castaldini, M.; Diago, M.P.; Giffard, B.; Lagomarsino, A.; Schroers, H.J.; Priori, S.; Valboa, G.; Agnelli, A.E.; Akça, E.; et al. Effects of Soil Erosion on Agro-Ecosystem Services and Soil Functions: A Multidisciplinary Study in Nineteen Organically Farmed European and Turkish Vineyards. *J. Environ. Manag.* **2018**, *223*, 614–624. [CrossRef] [PubMed]
147. Abad, J.; Hermoso de Mendoza, I.; Marín, D.; Orcaray, L.; Santesteban, L.G. Cover Crops in Viticulture. A Systematic Review (1): Implications on Soil Characteristics and Biodiversity in Vineyard. *OENO One* **2021**, *55*, 295–312. [CrossRef]

Disclaimer/Publisher's Note: The statements, opinions and data contained in all publications are solely those of the individual author(s) and contributor(s) and not of MDPI and/or the editor(s). MDPI and/or the editor(s) disclaim responsibility for any injury to people or property resulting from any ideas, methods, instructions or products referred to in the content.

climate

MDPI

Review

Net Zero Dairy Farming—Advancing Climate Goals with Big Data and Artificial Intelligence

Suresh Neethirajan

Faculty of Agriculture and Computer Science, Dalhousie University, Halifax, NS B3H 1W5, Canada; sureshraja@dal.ca

Abstract: This paper explores the transformative potential of Big Data and Artificial Intelligence (AI) in propelling the dairy industry toward net zero emissions, a critical objective in the global fight against climate change. Employing the Canadian dairy sector as a case study, the study extrapolates its findings to demonstrate the global applicability of these technologies in enhancing environmental sustainability across the agricultural spectrum. We begin by delineating the environmental challenges confronting the dairy industry worldwide, with an emphasis on greenhouse gas (GHG) emissions, including methane from enteric fermentation and nitrous oxide from manure management. The pressing need for innovative approaches in light of the accelerating climate crisis forms the crux of our argument. Our analysis delves into the role of Big Data and AI in revolutionizing emission management in dairy farming. This includes applications in optimizing feed efficiency, refining manure management, and improving energy utilization. Technological solutions such as predictive analytics for feed optimization, AI in herd health management, and sensor networks for real-time monitoring are thoroughly examined. Crucially, the paper addresses the wider implications of integrating these technologies in dairy farming. We discuss the development of benchmarking standards for emissions, the importance of data privacy, and the essential role of policy in promoting sustainable practices. These aspects are vital in supporting the adoption of technology, ensuring ethical use, and aligning with international climate commitments. Concluding, our comprehensive study not only suggests a pathway for the dairy industry towards environmental sustainability but also provides insights into the role of digital technologies in broader agricultural practices, aligning with global environmental sustainability efforts.

Keywords: climate change; net zero emissions; dairy farming; Big Data; Artificial Intelligence (AI); greenhouse gas emissions; sustainable agriculture; technological innovation; policy framework; environmental sustainability

1. Climate Change and the Drive for Net Zero Emissions

1.1. Global Climate Crisis and its Impacts

The global climate crisis, driven by anthropogenic activities, has precipitated a series of environmental upheavals with dire consequences. The Intergovernmental Panel on Climate Change (IPCC) reports indicate a substantial increase in global temperatures, attributable to the amplified concentration of greenhouse gases (GHGs) such as carbon dioxide (CO_2), methane (CH_4), and nitrous oxide (N_2O) in the atmosphere [1,2]. The primary sources of these emissions include fossil fuel combustion, deforestation, and industrial processes. The resulting global warming is not a distant threat but a present reality, manifesting in melting polar ice caps, rising sea levels, and increases in the frequency and severity of extreme weather events. These environmental changes have a cascading effect on ecosystems and biodiversity. The alterations in temperature and precipitation patterns disrupt natural habitats, leading to species migration and extinction [3]. Additionally, the acidification of oceans due to increased CO_2 levels poses a significant threat to marine life [4]. For humanity, these changes entail dire consequences, ranging from health risks due to heatwaves and pollution to economic losses in agriculture, fisheries, and other sectors crucial for livelihoods.

1.2. The Concept and Importance of Achieving Net Zero Emissions

Net zero emissions represent a state where the amount of GHGs emitted into the atmosphere is balanced by an equivalent amount removed [5]. Achieving this balance is crucial for stabilizing global temperatures. The concept hinges on the principle of carbon neutrality, where every ton of anthropogenic GHG emitted is counterbalanced by a ton removed or sequestered from the atmosphere [6].

Achieving net zero emissions necessitates a twofold strategy: reducing the existing emissions and enhancing carbon sinks. The former involves transitioning to renewable energy sources, boosting energy efficiency, and altering land use practices [7]. Simultaneously, enhancing natural carbon sinks like forests and oceans, alongside technological solutions such as carbon capture and storage (CCS), is imperative. CCS, for instance, involves capturing CO_2 emissions at their source (like power plants) and sequestering them underground or using them in various applications [8]. This balance is not just an environmental imperative but also a socio-economic necessity. The transition to a low-carbon economy promises innovation, job creation, and energy security, fostering sustainable development [9]. However, achieving net zero is a monumental task that requires concerted global action and significant investment in technology and infrastructure.

1.3. The Role of International Commitments in Climate Change Mitigation

International commitments are pivotal in the fight against climate change. The Paris Agreement, a landmark international treaty adopted in 2015, exemplifies global efforts to combat climate change [10]. The agreement's central aim is to limit global warming to well below 2 degrees Celsius, preferably to 1.5 degrees Celsius, compared to pre-industrial levels [11]. This goal is grounded in scientific evidence suggesting that surpassing this threshold could lead to catastrophic climate impacts.

Under the Paris Agreement, countries have committed to nationally determined contributions (NDCs), which are plans to outline each country's efforts to reduce national emissions and adapt to the impacts of climate change. The effectiveness of the Paris Agreement lies in its structure, which allows for periodic review and enhancement of these NDCs, fostering a progressive increase in global ambition. However, the challenge extends beyond mere commitment. The implementation of these NDCs requires a radical overhaul of national policies and economies. It demands a shift from fossil fuel dependence to renewable energy sources, a transformation in agricultural and industrial practices, and significant changes in consumption patterns. Furthermore, international cooperation is vital in this endeavor. Developed countries are expected to provide financial and technical support to developing nations, acknowledging the principle of common but differentiated responsibilities. This support is crucial as developing countries often lack the resources to implement substantial climate action.

The role of international commitments extends to fostering a global culture of sustainability. By setting clear targets and establishing a framework for collaboration, these commitments provide a roadmap for nations to collectively address the climate crisis. They also send a strong signal to businesses and investors, shifting financial flows towards more sustainable projects and technologies.

The drive for net zero emissions is not just an environmental goal but a comprehensive strategy for sustainable development. It requires a paradigm shift in how economies operate and societies function [12]. Achieving net zero emissions is pivotal in mitigating the global climate crisis, a task that demands unwavering commitment, innovative solutions, and robust international cooperation. As we stand at the crossroads of a climatic emergency, the path we choose today will determine the future of our planet and the generations to come.

2. Environmental Footprint of the Canadian Dairy Industry

The Canadian dairy industry, a significant contributor to the national economy, also plays a pivotal role in the environmental landscape, particularly concerning greenhouse gas (GHG) emissions. The sector's environmental footprint is predominantly characterized by emissions arising from enteric fermentation in dairy cattle and manure management [13,14]. These emissions are critical factors in the broader context of climate change and environmental sustainability.

2.1. The Economic and Environmental Significance of Dairy Farming in Canada

The dairy sector in Canada is a substantial segment of the nation's agricultural framework, contributing to both the economy and food security. According to Agriculture and Agri-Food Canada, the industry provides significant economic benefits, encompassing over 9000 dairy farms and a considerable contribution to the national GDP [15,16]. This economic importance, however, is juxtaposed against the sector's environmental impacts, chiefly its contribution to GHG emissions.

The environmental footprint of dairy farming extends beyond GHG emissions. It includes land use for feed production, water consumption, and the impact on biodiversity [17–19]. Nevertheless, the emission of GHGs remains the most pressing environmental issue associated with dairy farming, directly influencing climate change.

2.2. Main Sources of GHG Emissions in Dairy Farming: Enteric Fermentation and Manure Management

2.2.1. Enteric Fermentation

Enteric fermentation is a natural digestive process in ruminants, where microbes in the stomach break down food, producing methane (CH_4) as a byproduct [20,21]. This methane, a potent greenhouse gas, is then released into the atmosphere, predominantly through belching. The global warming potential of methane is approximately 28 times greater than that of CO_2 over a 100-year period, as reported by the IPCC [22–24]. In the context of the Canadian dairy industry, enteric fermentation represents a significant portion of its GHG emissions. Strategies to mitigate these emissions include dietary modifications to reduce fermentable substrates in the rumen and the use of feed additives like lipids or nitrates that can decrease methanogenesis [25,26].

2.2.2. Manure Management

The management of manure on dairy farms is another major source of GHG emissions, particularly methane and nitrous oxide (N_2O) [27]. Methane emissions occur during the storage and handling of manure under anaerobic conditions. Nitrous oxide, with a global warming potential 265 times that of CO_2, is emitted primarily during the storage and application of manure to fields [28,29]. N_2O emissions are linked to the nitrification and denitrification processes in soil, exacerbated by the application of nitrogen-rich manure. The mitigation of these emissions can be achieved through improved manure management practices, such as the adoption of aerobic composting techniques, and the utilization of anaerobic digesters that convert manure into biogas, a renewable energy source [30].

2.2.3. Synthetic Fertilizers

The use of synthetic fertilizers in feed crop production for dairy cattle is a notable source of N_2O emissions. When applied to soil, these fertilizers undergo microbial transformations, leading to N_2O release. The extent of these emissions is influenced by factors such as soil type, climate, and application method. Precision agriculture techniques, which optimize fertilizer application based on soil needs and environmental conditions, are effective in reducing N_2O emissions from synthetic fertilizers.

2.2.4. Fossil Fuel Use

Dairy farming operations, including machinery for fieldwork, transportation, and dairy processing, contribute to CO_2 emissions through the combustion of fossil fuels [31,32]. The energy-intensive nature of these activities necessitates a shift towards more sustainable energy sources and the adoption of energy-efficient technologies.

2.3. Environmental Impact of Dairy Farming Emissions

The environmental impact of emissions from dairy farming is multifaceted. Methane and nitrous oxide, due to their high global warming potentials, significantly contribute to the greenhouse effect and climate change. These emissions are a concern not only for their contribution to global warming but also for their role in altering weather patterns, impacting ecosystems and biodiversity. Furthermore, the emission of N_2O presents an additional environmental challenge due to its role in ozone layer depletion [33,34]. This aspect underscores the need for effective strategies to manage N_2O emissions from both manure and fertilizer applications.

In addressing these emissions, the dairy industry faces several challenges. Technological barriers include the development and implementation of effective mitigation strategies that are economically viable and scalable [35]. Economic constraints, particularly the cost of adopting new technologies and practices, pose significant hurdles for many dairy farmers. Additionally, policy and regulatory challenges can impact the pace and extent of emission reduction efforts.

In the realm of dairy farming, quantifying and understanding emissions is vital for crafting sustainable practices. This sector, integral to global agriculture, faces the dual challenge of maintaining productivity while mitigating its environmental footprint [36]. Key emission sources include enteric fermentation, manure management, synthetic fertilizers, and fossil fuel use. Each of these sources contributes uniquely to the overall environmental impact of dairy farming.

The environmental footprint of dairy farming, notably its GHG emissions, poses significant challenges to sustainability goals. Enteric fermentation and manure management are the main contributors to these emissions, necessitating targeted mitigation strategies. The Canadian dairy industry's approach to addressing its environmental impact involves a multifaceted strategy, incorporating technological innovations, farm management practices, and policy interventions.

Technological innovations, such as precision agriculture, can optimize feed efficiency and manure management, thereby reducing emissions. Precision agriculture utilizes data-driven insights to tailor feed formulations and application rates of manure, minimizing excess nutrient application and optimizing digestion processes [37]. In addition to technological solutions, farm management practices play a crucial role. Strategies such as altering feed composition to reduce enteric fermentation, improving manure storage and treatment facilities, and adopting nutrient management plans are integral to reducing the dairy industry's environmental footprint.

Policy interventions and incentives are also critical. Regulations and guidelines that promote sustainable practices, alongside financial incentives for farmers who adopt emission-reducing technologies, can significantly contribute to mitigating the environmental impact of dairy farming. The environmental footprint of the Canadian dairy industry, particularly its contribution to GHG emissions, is a complex issue that requires a holistic approach. The mitigation of emissions from enteric fermentation and manure management necessitates a combination of technological innovations, effective farm management practices, and supportive policy frameworks. As the dairy industry continues to evolve, the integration of these strategies will be key in ensuring its sustainability and alignment with broader environmental objectives.

3. Sustainability Challenges and Industry Efforts in Dairy Farming

The dairy industry, a vital component of global agriculture, faces significant sustainability challenges, particularly in mitigating its environmental impact. Central to these challenges are the greenhouse gas (GHG) emissions from dairy farms, primarily methane emissions from enteric fermentation, the environmental implications of manure management, and energy use in dairy processing. These factors significantly contribute to climate change, necessitating urgent and innovative responses from the industry.

3.1. *Addressing Methane Emissions from Enteric Fermentation*

The dairy industry has been exploring various strategies to mitigate these emissions. One approach is dietary manipulation, where the feed composition is adjusted to reduce fermentable substrates in the rumen that contribute to methane production. This includes incorporating lipids, nitrates, or certain enzymes that can decrease methanogenesis. Another method is the use of feed additives like 3-nitrooxypropanol (3-NOP), which directly inhibits the enzyme (methyl coenzyme M reductase) responsible for methane production in the rumen [38,39].

3.2. *Innovations in Manure Management*

Innovative manure management practices are being developed to mitigate these emissions. These include the adoption of aerobic composting techniques, which reduce methane production by promoting aerobic conditions. Anaerobic digesters are another solution, converting manure into biogas, a renewable energy source, while significantly reducing methane emissions [40]. Additionally, precision agriculture tools are being utilized to optimize the application rates of manure, ensuring nutrients are used efficiently and minimizing excess nitrogen application, which contributes to nitrous oxide emissions.

3.3. *Energy Use in Dairy Processing and its Environmental Implications*

Dairy processing, encompassing pasteurizing, homogenizing, and packaging milk, is an energy-intensive process. The reliance on fossil fuels in dairy processing contributes significantly to GHG emissions [41,42]. Reducing the energy footprint of dairy processing is crucial for the overall sustainability of the industry.

Efforts to mitigate energy use in dairy processing include transitioning to renewable energy sources, such as solar or wind power [43]. Energy efficiency measures, such as the use of energy-efficient equipment and the optimization of processing operations, are also being implemented. These measures not only reduce GHG emissions but also decrease operational costs. Another innovative approach is the recovery and utilization of waste heat generated during processing. Heat recovery systems can repurpose this waste heat for other processing steps or for heating purposes within the facility, thereby improving overall energy efficiency.

Continuous research and development, coupled with industry-wide adoption of these practices, are essential for achieving substantial reductions in GHG emissions. The integration of these strategies, supported by policy frameworks and incentives, will be key in ensuring the dairy industry's progress towards sustainability. As the industry evolves, the continuous refinement and implementation of these approaches will play a critical role in mitigating its environmental footprint and contributing to global efforts to combat climate change.

4. The Transformative Impact of Big Data and AI in Dairy Farming

The advent of Big Data and Artificial Intelligence (AI) has revolutionized dairy farming, introducing enhanced efficiency, improved sustainability, and innovative management practices. These technologies address key challenges in dairy farming, such as optimizing resource usage, improving herd health, and benchmarking for sustainability [44].

Big Data in dairy farming involves collecting and analyzing extensive data from various sources, like sensor networks and animal health monitoring systems. This comprehensive approach enables farmers to make data-driven decisions, optimizing productivity and sustainability. Big Data analytics play a crucial role in resource management, particularly in optimizing feed and water usage, and in analyzing a farm's climate impact, especially concerning greenhouse gas emissions [45].

AI technologies bring sophistication to herd management and resource optimization. Predictive analytics in herd health, powered by AI, revolutionize animal care by anticipating health issues, thereby enhancing preventive care [46]. AI also optimizes feed composition and energy management, integrating renewable energy sources like solar or biogas into operations, reducing reliance on fossil fuels.

4.1. Benchmarking for Performance and Sustainability

Benchmarking is a critical process in dairy farming, enabling the comparison of farm performance against industry standards [47]. It identifies areas for improvement in resource utilization, animal health management [Figure 1], and environmental sustainability. Benchmarking also helps in the adoption of best practices, tracking progress over time, and ensuring compliance with environmental standards [48]. The process of benchmarking emissions in the dairy industry is comprehensive, involving identification and quantification of emissions sources such as enteric fermentation, manure management, and fossil fuel usage. It entails comparing emission levels with industry standards and best practices, setting reduction targets, and regular monitoring and reporting.

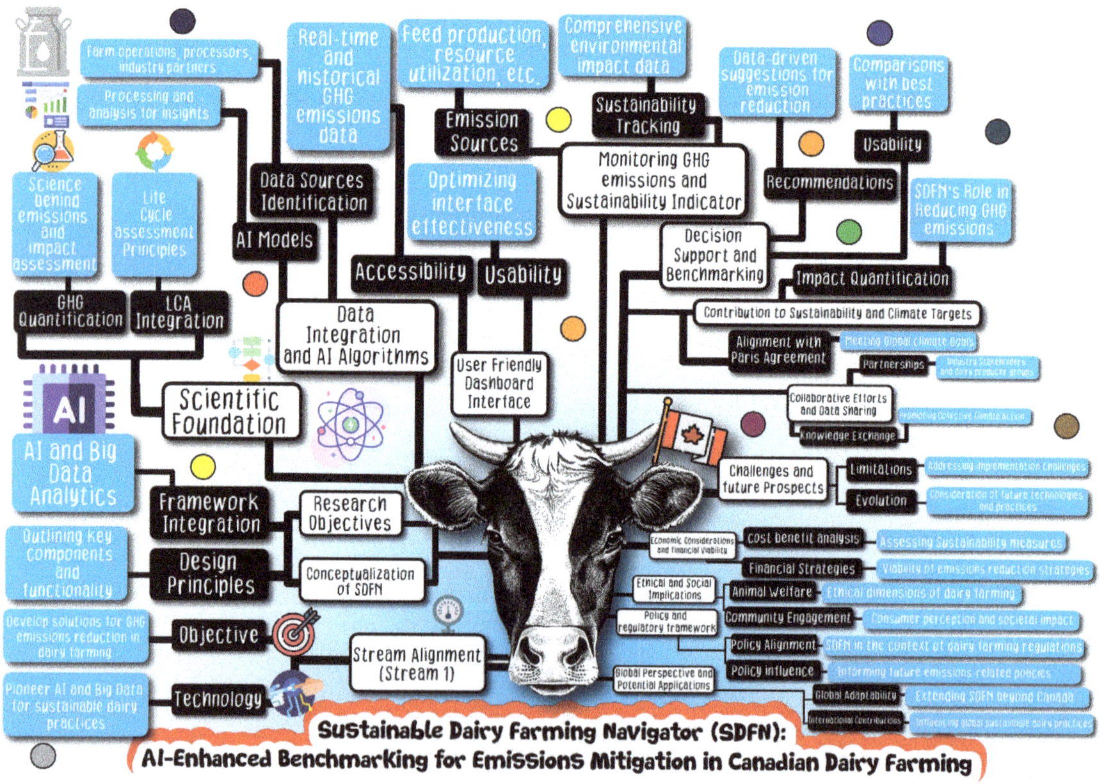

Figure 1. AI-enhanced benchmarking for emissions in Canadian dairy farming.

4.2. Leveraging Cross-Industry Insights for Dairy Farming

The dairy industry can learn from methodologies and sustainability practices from sectors like food processing, biomedicine, and industrial manufacturing. Techniques from industrial manufacturing for process optimization, energy management systems from other industries, and waste-to-resource models from food processing can be adapted to enhance sustainability in dairy farming. The implementation of Big Data, AI, and benchmarking in dairy farming represents a transformative leap towards sustainability and efficiency [49]. By leveraging these technologies, the dairy industry can achieve significant gains in environmental stewardship, resource optimization, and overall farm management. The critical role of benchmarking, particularly in emissions reduction, provides a roadmap (Table 1) for the dairy industry to align with global sustainability goals and environmental regulations. The integration of these technologies, coupled with lessons learned from other industries, paves the way for a more sustainable, efficient, and innovative future in dairy farming.

Table 1. Benchmarking of emissions in dairy industry.

Aspect of Benchmarking	Enteric Fermentation	Manure Management	Energy Use	Feed Efficiency	Water Usage
Emission Source	Methane from digestion	Methane and nitrous oxide from storage	CO_2 from fossil fuel use	Methane from digestion	Water management related energy use
Quantification Method	CH_4 measurements, LCA	Gas capture and analysis, LCA	Energy audits, carbon footprinting	Feed analysis, LCA	Water usage metering, energy audits
Industry Average	X kg CH_4/cow/year	Y kg N_2O/ton manure	Z kWh/liter of milk	A kg CH_4/ton feed	B liters/milk liter
Best Practice Standard	Low-methane feed	Aerobic composting, digesters	Renewable energy use	High-efficiency feed	Water recycling systems
Improvement Strategy	Feed additives, diet optimization	Manure treatment, efficient application	Solar, biogas energy	Precision feeding	Irrigation management, leak repairs
Monitoring Frequency	Bi-annually	Annually	Quarterly	Annually	Bi-annually
Compliance Requirement	Voluntary industry standards	Environmental regulations	Energy conservation laws	Sustainable farming certifications	Water management regulations
Technology Used	Rumen sensors, data analytics	Emission capturing systems	Smart energy meters	AI-driven feeding systems	Automated irrigation
Training Required	Feed management	Manure handling, system operation	Energy management	Nutritional management	Water conservation techniques
Cost Implication	Moderate investment	High investment	High investment for renewables	Moderate investment	Moderate investment
Potential Emission Reduction	Significant CH_4 reduction	CH_4 and N_2O reduction	Significant CO_2 reduction	CH_4 reduction	Energy-related emission reduction

In our comprehensive benchmarking analysis of emissions in the dairy industry, as detailed in Table 1, we systematically evaluate various aspects such as enteric fermentation, manure management, and energy use. This includes a thorough examination of emission sources, ranging from methane from digestion to CO_2 from fossil fuel use, and strategies for their quantification, like CH_4 measurements and energy audits. The industry averages for these emissions provide a baseline for setting best practice standards, which include low-methane feed and renewable energy use. Our study also outlines improvement strategies, such as diet optimization and efficient manure application, and underscores the importance of regular monitoring, compliance with environmental regulations, and the use of advanced technologies like AI-driven feeding systems for effective emission reduction.

5. The Economic and Structural Landscape of Canadian Dairy Industry

The Canadian dairy industry, integral to the country's agriculture, epitomizes a complex interplay of economic impact, diverse farm structures, and a stringent regulatory framework, all of which align to create a unique economic and structural landscape.

5.1. Industry Size and Economic Impact

As of recent data, Canada's dairy sector is a substantial component of the agricultural economy. It is not only a significant contributor to the national GDP but also a pivotal player in the rural economy. The industry is characterized by its extensive production capacity, involving thousands of dairy farms and processing facilities across the country. These farms collectively contribute billions to the national economy, underscoring the sector's economic significance. The dairy industry's impact extends beyond direct production. It encompasses a wide range of associated industries, including feed production, veterinary services, equipment supply, and transportation. This extensive value chain [50,51] amplifies the sector's economic influence, providing employment and stimulating local economies in rural and urban areas alike.

5.2. Global Market Presence

Despite its robust domestic market, the Canadian dairy industry has a relatively modest footprint in the global dairy trade. This is primarily due to the country's supply management system, which focuses on balancing domestic supply with demand, thus limiting export volumes [52,53]. However, the industry still contributes significantly to the global dairy market through specialized products and niche markets. Its reputation for quality and safety standards positions Canadian dairy products favorably in the international arena, where they command a premium.

5.3. Diversity in Farm Structures and Operations

Canadian dairy farms exhibit a diverse range of structures and operational scales. This diversity is reflective of the country's vast geographical and climatic variations. From small family-owned farms to large corporate-run facilities, the industry encompasses a wide spectrum of farming models. This structural diversity is underpinned by a common commitment to high-quality milk production, animal welfare, and sustainable farming practices. Technological integration varies across these farms, with some employing cutting-edge technologies for milking, feeding, and herd management, while others adhere to more traditional farming methods. This variation is indicative of the sector's adaptability and its ability to cater to different market needs and preferences.

5.4. Regulatory Framework Governing Canadian Dairy Farming

The Canadian dairy sector operates under a unique regulatory framework known as supply management [54]. This system, characterized by production quotas, price setting, and import controls, ensures stable income for producers and processors. It balances supply with domestic demand, protecting the industry from market volatilities and ensuring consistent quality of dairy products. Additionally, the industry adheres to strict standards regarding animal welfare and environmental sustainability. These regulations, coupled with initiatives like the ProAction program, underscore the sector's commitment to responsible farming practices. The ProAction program, in particular, sets benchmarks [55,56] in areas such as milk quality, animal health, and environmental stewardship.

5.5. Economic and Policy Challenges

Despite its strengths, the Canadian dairy industry faces several economic and policy challenges. These include adapting to changing market demands, both domestically and globally, and responding to pressures from international trade agreements. The sector also grapples with the need for continuous innovation and sustainability improvements, particularly in the face of global environmental concerns. Emerging trends, such as the growing demand for plant-based alternatives and the increasing importance of sustainable practices, are set to shape the industry's future. Adapting to these trends while maintaining its core values and economic viability will be crucial for the sector's continued success.

6. Strategies for GHG Emission Reduction in Dairy Farming

In the pursuit of reducing greenhouse gas (GHG) emissions in dairy farming, a combination of innovative practices, technological advancements, and supportive policy initiatives plays a pivotal role. Addressing the emissions from this sector is crucial for mitigating climate change impacts given the significant contribution of dairy farming to global GHG emissions, particularly methane (CH_4) and nitrous oxide (N_2O).

6.1. Innovative Farming Practices for Emission Reduction

6.1.1. Optimized Feed Efficiency

One of the primary strategies in emission reduction is optimizing feed efficiency. Research has shown that improved feed quality and dietary adjustments can significantly reduce methane emissions from enteric fermentation. Feed additives, such as lipids, tannins, and certain enzymes, have been studied for their potential to reduce methane production during digestion [57–59]. Furthermore, precision feeding techniques, ensuring each animal receives a diet tailored to its specific needs, not only improve the feed conversion efficiency but also reduce the overall carbon footprint of the dairy operation.

6.1.2. Advanced Manure Management

Manure management represents a substantial opportunity for emission reduction. Traditional manure storage and treatment methods often lead to significant methane and nitrous oxide emissions. Implementing advanced techniques, such as anaerobic digestion, can convert manure into biogas, a renewable energy source, while significantly reducing emissions [60,61]. Moreover, innovative practices like composting and improved storage and application methods can minimize emissions from manure handling.

6.1.3. Pasture-Based Farming

Transitioning to pasture-based systems can also contribute to lowering GHG emissions. Grazing systems promote better manure distribution and can enhance soil carbon sequestration, reducing the overall carbon footprint compared to conventional confinement systems [62,63].

6.1.4. Precision Agriculture

The application of precision agriculture technologies is a game-changer in reducing emissions. Using tools such as GPS, sensors, and data analytics, dairy farmers can optimize resource use, thereby reducing emissions [64,65]. For instance, sensor-based technologies can monitor soil health and moisture levels, enabling more efficient use of fertilizers and irrigation, thereby reducing nitrous oxide emissions from synthetic fertilizers.

6.1.5. Biogas Systems

The adoption of biogas systems for energy production from dairy waste not only provides a sustainable energy source but also plays a critical role in methane emission reduction. These systems capture methane from manure and convert it into bioenergy, effectively reducing the methane emissions associated with manure management practices. The biogas produced can be used for heating, electricity generation, or even as a vehicle fuel, contributing to the overall energy sustainability of the dairy farm.

6.1.6. Carbon Capture and Storage (CCS) Technologies

Emerging CCS technologies offer potential in mitigating the climatic impact of dairy operations [66]. While still in development, these technologies aim to capture carbon emissions directly from various sources within dairy farms, including manure management systems and energy consumption processes. Implementing CCS could play a significant role in reducing the carbon footprint of dairy farming.

6.2. Role of Policy Initiatives and Incentives in Emission Reduction

6.2.1. Federal and Provincial Incentives

Government incentives are crucial in supporting emission reduction efforts in the dairy sector [67]. These incentives can take various forms, such as funding for adopting sustainable practices, tax breaks for implementing green technologies, or technical support for transitioning to low-emission practices. For example, programs like the Sustainable Canadian Agricultural Partnership (CAP) offer financial assistance for projects aimed at improving environmental sustainability [68].

6.2.2. Carbon Pricing and Trading Systems

Implementing carbon pricing mechanisms, like carbon taxes or cap-and-trade systems, encourages emission reductions by attributing a cost to carbon emissions [69,70]. Such mechanisms can incentivize dairy farmers to adopt sustainable practices by making emission-intensive operations more costly. This economic approach can drive significant changes in the industry towards more sustainable and low-emission practices.

6.2.3. Research and Development Support

Governmental support for research and development is essential in advancing new technologies and practices for emission reduction. This includes funding for research into low-emission animal diets, manure management technologies, and precision farming tools. Partnerships between the government, academia, and the dairy industry are vital for fostering innovation and developing practical, scalable solutions for emission reduction.

6.2.4. Extension Services and Education

Providing extension services and education about sustainable practices and technologies is critical for enabling dairy farmers to effectively implement emission reduction strategies. These services can offer technical advice, training, and support, helping farmers to understand the benefits of sustainable practices and how to integrate them into their operations.

Reducing GHG emissions in dairy farming requires a multifaceted approach that incorporates innovative farming practices, technological advancements, and supportive policy frameworks [71,72]. The adoption of these strategies not only contributes to climate change mitigation but also enhances the overall sustainability and resilience of the dairy sector. Collaboration between farmers, industry stakeholders, researchers, and policymakers is essential in driving these changes and achieving substantial emission reductions in dairy farming.

7. The Future of Dairy Farming with Big Data

The integration of Big Data into dairy farming is not merely an incremental change but represents a fundamental shift in agricultural practices. By harnessing the power of vast datasets, dairy farmers can optimize various aspects of their operations, from herd management to resource utilization, ushering in a new era of efficiency and sustainability.

7.1. Emerging Technologies in Big Data for Dairy Farming

Big Data in dairy farming involves the collection, analysis, and utilization of large and complex datasets from various sources, including sensor networks, farm equipment, and environmental data inputs. These extensive data encompass information ranging from milk yield and quality metrics to animal feeding patterns, health indicators, and environmental conditions.

7.1.1. Advanced Sensors and Monitoring Systems

The deployment of advanced sensor technologies in dairy farming is reshaping the management of herd health and environmental monitoring. Sensors that track animal health indicators, such as body temperature and activity levels, along with environmental sensors measuring factors like air quality and humidity, are generating vast amounts of data. These sensors provide critical insights into the well-being of the herd and the optimization of the farm environment.

7.1.2. Data Analytics for Enhanced Decision Making

The analysis of collected data enables farmers to gain comprehensive insights into their operations. This involves using sophisticated algorithms and machine learning techniques to process and interpret complex datasets, leading to informed decision making. For instance, data-driven insights can guide feed composition adjustments, resulting in improved feed efficiency and reduced waste.

7.1.3. Predictive Modeling for Proactive Management

Utilizing predictive analytics, farmers can forecast future scenarios based on historical data trends. This proactive approach allows for early intervention in potential issues such as health problems in the herd or environmental stressors, minimizing risks and enhancing productivity.

7.2. Integration with Other Technologies: IoT and Robotics

The convergence of Big Data with the Internet of Things (IoT) and robotics is driving significant advancements in dairy farming. IoT in Dairy Farming: Integrating Big Data with IoT devices facilitates a more connected and automated farm environment [73,74]. Connected devices such as milking robots, feed dispensers, and climate control systems can generate real-time data, enhancing precision in farming operations. The integration of these devices through IoT platforms allows for seamless data flow and analytics, leading to more efficient and sustainable farm management.

Robotics and Automation

The incorporation of robotics in dairy farming, coupled with Big Data, is revolutionizing traditional practices. Automated systems for milking, feeding, and cleaning not only improve efficiency but also reduce the physical labor required. The data generated by these robotic systems contribute to a deeper understanding of farm operations, enabling continuous improvement and innovation.

7.3. Policy and Industry Implications of Technological Advancements

The advancements in Big Data and its integration with other technologies have significant implications for both policy and the dairy industry. Sustainability and Environmental Impact—Big Data technologies contribute to the environmental sustainability of dairy farms by optimizing resource use and reducing waste. This aligns with global efforts to minimize agriculture's environmental footprint, particularly concerning greenhouse gas emissions. Policies encouraging the adoption of such technologies can support the industry's transition towards more sustainable practices.

7.3.1. Regulatory Frameworks for Data Use and Privacy

The use of Big Data raises important considerations regarding data privacy and security. Regulatory frameworks need to be established to protect sensitive farm data, ensure their ethical use, and clarify issues around data ownership and control [75,76]. Policymakers must balance the need for data protection with the potential benefits of data sharing and utilization in the industry.

7.3.2. Industry Adaptation and Skill Development

As dairy farming becomes increasingly data-driven, there is a growing need for skills development in this area. Training and educational programs focusing on data analytics, IoT, and robotics are essential for preparing the current and future workforce. The industry must adapt to these technological changes by fostering a culture of continuous learning and innovation.

7.3.3. Economic Considerations and Market Adaptation

The economic implications of adopting Big Data technologies in dairy farming cannot be overlooked. While the initial investment may be substantial, the long-term benefits in terms of efficiency, productivity, and sustainability can offer a significant return on investment. The market readiness to accept and integrate these technologies is also a crucial factor, influencing their widespread adoption and economic viability.

7.4. Digital Technology's Role in Enhancing Dairy Farm Emission Efficiency

In their 2023 study, Liu et al. [77] provide a comprehensive analysis of how digital technology applications enhance carbon emission efficiency in dairy farms. Utilizing a robust methodology involving the Undesirable Outputs-SBM model, Tobit model, propensity score matching, quantile regression model, and an instrumental variable approach, the study examines 136 Chinese dairy farms. The focus is on the impact of digital technology, particularly precision feeding, on carbon emission efficiency, and how environmental regulations modulate this relationship.

The key findings reveal that digital technology significantly improves carbon emission efficiency, with the most substantial impact observed in farms with lower initial efficiency levels. Precision feeding technology emerges as the most effective, followed by manure treatment, environmental monitoring, and cow monitoring technologies. The study also highlights the critical role of environmental regulations in enhancing this effect; more stringent regulations amplify the positive impact of digital technology on emission efficiency.

Further, the study underscores the influence of factors such as farmers' educational background and technical training in adopting digital technologies. It notes that implementation of digital technology could potentially increase carbon emission efficiency by 11.61% in dairy farms. This finding emphasizes the need for enhanced educational and policy initiatives to promote digital technology use in dairy farming, aligning with low-carbon and efficient production goals.

The research by Liu et al., (2023) [77] provides empirical evidence on the pivotal role of digital technologies in reducing carbon emissions in the dairy industry, highlighting the need for supportive educational and regulatory environments to maximize these technologies' benefits.

8. Artificial Intelligence in Emission Management and Energy Efficiency in Dairy Farming

Artificial Intelligence (AI) has emerged as a pivotal tool in the agricultural sector, particularly in dairy farming, where it offers substantial potential for improving emission management and enhancing energy efficiency. The deployment of AI in this context addresses two critical areas: predictive analytics for optimizing feed and manure management, and monitoring of and reduction in energy use. AI's role in emission management and energy efficiency in dairy farming is multidimensional and highly impactful. As AI technology advances, its integration into dairy farming practices promises to drive significant improvements in environmental sustainability, operational efficiency, and economic viability. The key to realizing these benefits lies in overcoming the current challenges and continuing innovation and development in AI applications specific to the dairy industry. This evolution of AI in dairy farming not only contributes to reducing the environmental footprint of the sector but also aligns with the broader goals of sustainable agriculture and climate change mitigation.

The application of Artificial Intelligence (AI) in feed optimization for dairy farming involves the development and implementation of algorithms capable of analyzing extensive datasets. These algorithms are designed to predict the most efficient feed compositions and schedules, taking into account individual animal health, milk production, and varying environmental conditions [78,79]. This data analysis is critical in understanding how different feeds impact milk yield, quality, and the overall health of the herd.

A significant focus of feed optimization using AI is the reduction in methane emissions produced during enteric fermentation. AI models are equipped to suggest dietary adjustments that minimize methane production, considering breed-specific responses and the diverse nutritional needs of the herd. Optimizing feed through AI not only contributes to reducing methane emissions but also enhances feed efficiency. This leads to lower production costs and a reduced environmental impact while simultaneously improving animal health and productivity.

In the realm of manure management, another critical area for reducing greenhouse gas emissions in dairy farming, AI systems offer significant benefits. They can predict optimal times and methods for manure application, storage, and processing, taking into account external factors such as weather patterns and soil conditions. This guidance helps farmers make decisions that not only reduce emissions but also enhance soil quality.

AI's role extends to optimizing anaerobic digestion processes, where it can improve the conversion of manure into biogas, thereby reducing methane emissions. By determining the optimal mix of manure and other organic wastes, AI models ensure maximum biogas production with minimum greenhouse gas emissions. This optimization is a testament to AI's growing significance in enhancing environmental sustainability in dairy farming practices.

AI in Monitoring and Reducing Energy Use

Energy management is a critical aspect where AI can make a significant impact. The application of AI in energy use within dairy farms entails:

AI-driven smart energy systems are increasingly pivotal in managing and optimizing energy consumption on dairy farms. These sophisticated systems utilize AI algorithms to forecast energy needs and effectively integrate renewable energy sources. For example, AI can optimize the use of solar panels or biogas, a renewable energy source generated from manure, enhancing the energy efficiency of dairy farming operations.

In the realm of energy consumption, AI algorithms excel in monitoring and analyzing patterns across various dairy farm operations, including milking, cooling, and processing [80]. This monitoring is instrumental in identifying areas where energy consumption can be reduced and more efficient practices can be implemented. By pinpointing these areas, dairy farms can significantly cut down on energy usage, contributing to both economic and environmental sustainability.

Another crucial aspect of AI in dairy farming is predictive maintenance. AI can accurately predict when machinery or equipment requires maintenance, thereby avoiding energy wastage due to inefficient operation. This predictive maintenance ensures that farm equipment operates at optimal energy efficiency, leading to a reduction in unnecessary energy consumption and associated costs.

The integration of AI with renewable energy sources on dairy farms offers a promising pathway toward reducing dependence on fossil fuels. AI can skillfully balance the use of renewable energy sources with traditional energy sources. This balance is essential for maintaining a consistent energy supply and minimizing the carbon footprint of dairy farming operations. Additionally, AI systems are adept at forecasting weather conditions and energy requirements. This capability aids in planning and optimizing the use of renewable energy sources, such as solar and wind energy, on dairy farms. This integration not only enhances energy efficiency but also supports the broader goal of sustainable and environmentally responsible dairy farming.

9. Future Pathways—Implementing AI and Big Data in Dairy Farming for Climate Change Mitigation and Emissions Reduction

In envisioning a roadmap for future research and application of AI and Big Data in dairy farming (Figure 2), particularly in addressing climate change challenges and emissions reduction, a strategic and methodical approach is required. This roadmap should encompass both immediate and long-term strategies, integrating the latest advancements in technology with practical on-the-ground farming practices.

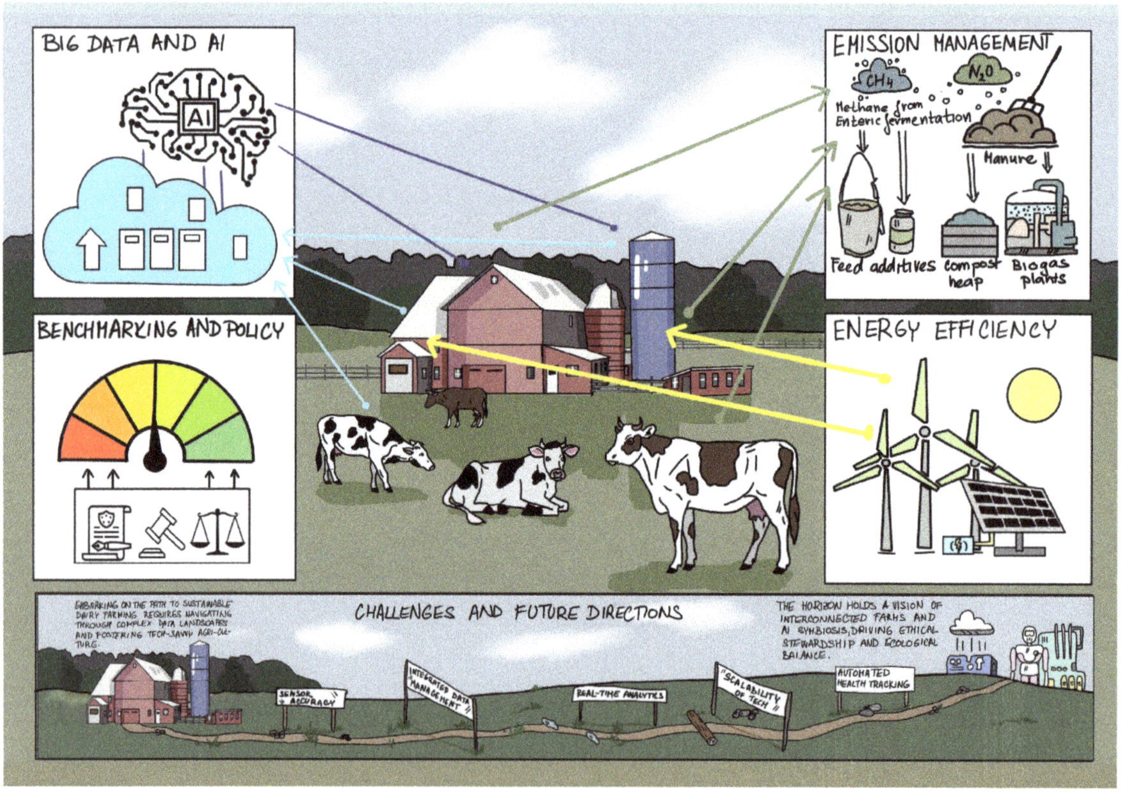

Figure 2. Harnessing Big Data and AI for sustainable dairy farming—a visual guide to emission reduction and benchmarking.

9.1. Immediate Strategies

9.1.1. Enhanced Data Collection and Analysis

- Invest in sensor technologies and IoT devices to collect comprehensive data on various aspects of dairy farming, including feed intake, animal health, and manure management.
- Utilize advanced data analytics and machine learning algorithms to process these data, identifying patterns and correlations that can inform more sustainable farming practices.

9.1.2. Predictive Analytics for Resource Optimization

- Develop AI models that can predict optimal feeding strategies, reducing waste and improving the efficiency of resource use.
- Implement systems that can forecast environmental impacts and provide recommendations for minimizing carbon footprints.

9.2. Long-Term Strategies

9.2.1. Integration of Climate Models with Farm Management Systems

- Collaborate with climate scientists to integrate global and regional climate models into AI-driven farm management systems.
- Use these integrated models to anticipate and mitigate the impacts of climate variability on dairy farming operations.

9.2.2. Development of AI-Driven Sustainable Practices

- Focus on AI innovations that promote sustainable practices, such as precision agriculture techniques to optimize the use of water, fertilizers, and energy.
- Explore AI-based solutions for managing waste and manure in environmentally friendly ways, such as converting waste into renewable energy sources.

9.3. Cross-Disciplinary Collaboration and Knowledge Sharing

Fostering Industry–Academia Partnerships

- Encourage partnerships between dairy farmers, technology companies, and academic institutions to drive innovation and research in AI and Big Data applications.
- Facilitate knowledge sharing and technology transfer through workshops, conferences, and joint research projects.

9.4. Policy Advocacy and Support for Technological Adoption

- Engage with policymakers to advocate for support and funding for the adoption of AI and Big Data technologies in dairy farming.
- Work towards the development of regulatory frameworks that encourage innovation while ensuring ethical and sustainable use of technology.

9.5. Adaptive and Evolving Methodologies

Continuous Improvement and Adaptation

- Establish a culture of continuous improvement, where technologies and methodologies are regularly reviewed and updated based on new research and changing climate conditions.
- Embrace adaptive management approaches that allow for flexibility and responsiveness to new challenges and opportunities.

9.6. Monitoring and Evaluation

- Implement robust monitoring and evaluation systems to assess the effectiveness of AI and Big Data applications in reducing emissions and addressing climate change.
- Use these assessments to refine and improve strategies over time, ensuring that the dairy farming industry can effectively contribute to global climate change mitigation efforts.

By following this roadmap, the dairy farming industry can not only significantly reduce its environmental impact but also enhance productivity and resilience in the face of climate change. This comprehensive approach, blending cutting-edge technology with practical farming practices and policy support, paves the way for a more sustainable and environmentally responsible future in dairy farming.

10. Challenges and Future Directions

The integration of Artificial Intelligence (AI) in dairy farming, while holding immense potential, faces several challenges that need to be addressed for wider adoption and effective utilization. One of the primary challenges is the complexity of integrating AI with existing farming systems. This integration is not just a matter of software compatibility; it requires the systems to effectively process and analyze diverse data types, which is essential for the AI to function optimally.

Another critical issue involves the quality and availability of data, which are foundational for the effectiveness of AI. The performance of AI systems can be significantly hindered by inconsistent data recording practices and the limited availability of comprehensive data. This challenge underscores the need for standardized, high-quality data collection practices in the dairy farming sector.

Moreover, the successful implementation of AI in dairy farming hinges on the training and adaptation of farmers. It is imperative for farmers to receive adequate training to understand and manage AI systems effectively. This aspect involves overcoming potential resistance to new technologies and facilitating a smooth transition from traditional farming practices to more technologically advanced methods.

Looking towards future advancements, several key developments are anticipated to enhance the role of AI in dairy farming. One such advancement is the continuous improvement in the accuracy of AI algorithms, which will lead to more efficient and effective farm management practices. Additionally, the integration of AI with Internet of Things (IoT) devices is expected to play a significant role. This integration will provide more detailed and granular data, further optimizing farm operations and energy usage.

Furthermore, the development of customized AI solutions tailored to the specific needs of different farming operations will greatly enhance the applicability and effectiveness of AI in the dairy industry. These customized solutions will address the unique challenges and requirements of various farming setups, thereby maximizing the benefits of AI in dairy farming.

While the integration of AI in dairy farming presents challenges, the future holds promising advancements that will likely overcome these hurdles. Continuous improvements in AI technology, coupled with better integration and customization, are poised to transform dairy farming into a more efficient, sustainable, and economically viable sector.

Addressing Challenges and Enhancing Policies for Sustainable Transformation

Addressing the multifaceted challenges and enhancing policies for a sustainable transformation in dairy farming involves navigating through technological, economic, and policy barriers. These barriers need to be methodically addressed to facilitate the transition to more sustainable practices. Simultaneously, it is crucial to ensure that technological advancements are harmonized with ethical farming practices.

Table 2 presents a comprehensive policy framework essential for sustainable transformation in the dairy farming industry. It details various policy focuses, from establishing benchmarking standards to promoting Big Data utilization and integrating AI technologies, which collectively aim to enhance farm efficiency and achieve significant reductions in greenhouse gas emissions. The table also highlights the key components of each policy, the stakeholders involved, the challenges faced, and the potential outcomes. This framework serves as a roadmap for stakeholders at various levels, including farmers, industry regulators, tech companies, and government bodies, guiding the industry towards a sustainable and environmentally responsible future while balancing economic viability.

Table 2. Policy framework for sustainable transformation in dairy farming.

Policy Focus	Objective	Key Components	Stakeholders Involved	Challenges	Potential Outcomes
Benchmarking Standards	Establish industry-wide benchmarking standards for emissions and efficiency	Comparative analysis, Best practices, Performance metrics	Farmers, Industry regulators, Environmental agencies	Standardization across diverse farm operations	Consistent quality and sustainability metrics across the industry

Table 2. *Cont.*

Policy Focus	Objective	Key Components	Stakeholders Involved	Challenges	Potential Outcomes
Big Data Utilization	Promote the use of Big Data for farm management optimization	Data collection, Analysis tools, Real-time monitoring	Farmers, Tech companies, Data analysts	Managing large datasets, Interpreting complex information.	Optimized farm operations, Enhanced decision making
Emission Reduction Targets	Set clear and achievable emission reduction targets	Methane and N_2O reduction, Carbon footprint assessment	Government, Environmental groups, Dairy producers	Balancing economic viability with environmental goals.	Significant reduction in greenhouse gas emissions
AI Integration in Farming	Facilitate the integration of AI technologies in dairy farming	Predictive analytics, Herd management, Resource optimization	Tech developers, Farmers, AI experts	Adapting to new technologies, Overcoming resistance	Increased farm efficiency and reduced labor costs
Data Privacy and Security	Ensure the security and privacy of farm data collected via digital means.	Data encryption, User consent, Ethical use guidelines	Farmers, Data protection agencies, Legal experts	Risk of data breaches, Maintaining farmer trust	Protected farmer data, Ethical technology usage
Renewable Energy Incentives	Encourage the use of renewable energy sources through incentives.	Subsidies, Tax breaks, Green energy solutions	Energy companies, Farmers, Government bodies	Initial investment costs, Long-term sustainability	Reduced carbon footprint, Energy self-sufficiency
Education and Training	Enhance farmer knowledge and skills in advanced technologies	Workshops, Online courses, Technical assistance	Educational institutions, Dairy farmers, Industry experts	Varied technological proficiency among farmers	Improved adoption of advanced farming technologies
Public-Private Partnerships	Foster collaboration between government, industry, and academia	Joint funding, Knowledge exchange, Innovation incubators	Government, Corporates, Research institutions	Aligning goals and resources of different entities	Synergy in innovation, Accelerated technology transfer
Environmental Regulation	Implement stringent environmental regulations to control emissions.	Compliance standards, Monitoring, Penalties for non-compliance	Regulatory bodies, Dairy farmers, Environmentalists	Ensuring regulations are effective yet feasible	Enhanced environmental protection, Sustainable dairy practices.
Innovation and R&D Support	Support research and development in sustainable dairy farming technologies.	Innovative feed solutions, Waste management, Energy-efficient practices	Scientists, Dairy industry, Government funders	Translating research into practical, scalable solutions	Breakthroughs in sustainable farming techniques

11. Overcoming Technological, Economic, and Policy Barriers

Technological barriers in dairy farming often stem from the complexity of integrating new technologies with existing systems. This includes ensuring compatibility with current hardware and software and effectively processing diverse data types. To overcome these challenges, there is a need for technologies that are adaptable, user-friendly, and can seamlessly integrate with various farming operations. Additionally, addressing data quality and availability is crucial. Inconsistent data recording practices and limited data availability can impede AI systems' performance. Establishing standardized data collection methods and ensuring the availability of high-quality, comprehensive datasets are imperative.

Economic barriers, primarily the high costs associated with adopting new technologies, pose a significant hurdle. This is particularly challenging for small- and medium-sized farms. Addressing these barriers requires the implementation of financial incentives such as subsidies, grants, or tax breaks to make investments more feasible for farmers. Moreover, demonstrating the long-term economic benefits of these technologies through case studies and pilot projects can encourage broader adoption.

Policy barriers often involve existing regulations that are not adapted to new technologies, hindering their implementation. To address this, regulatory frameworks need regular review and reform to keep pace with technological advancements. This includes relaxing certain regulatory barriers and developing standards that encourage innovation while ensuring safety and sustainability.

11.1. Enhancing Policies for Emission Reduction and Sustainable Practices

Enhancing policies to support emission reduction and sustainable practices in dairy farming is essential. This can be achieved by developing and implementing policies that support the transition to sustainable dairy farming, providing a regulatory environment that encourages the adoption of sustainable practices. Governments can play a pivotal role by offering financial incentives and investing in research and development of sustainable dairy farming technologies.

Creating sustainability guidelines is also critical. Policymakers should establish guidelines that promote sustainability and environmental responsibility in dairy farming, ensuring that the industry's growth aligns with environmental conservation goals. This could involve setting clear, measurable emission reduction targets specific to the dairy sector, aligning with broader sustainability goals.

11.2. Balancing Technological Advancement with Ethical Farming Practices

The integration of technology in dairy farming should complement, not replace, traditional farming practices. It should respect the cultural heritage of farming communities and adapt to fit within their existing cultural framework. This includes developing technologies that enhance rather than overhaul existing practices and ensuring that technological development involves farmers in the design process.

Ensuring equity and inclusivity in technological advancement is crucial. Technologies must be accessible to all farmers, including smallholders and family-owned farms, and training and support must be provided to ensure that farmers have the knowledge and skills to effectively utilize these tools.

Regulatory and policy frameworks should establish clear guidelines for the ethical use of technology in dairy farming. These guidelines could cover aspects such as data privacy, animal welfare, and labor rights. Effective regulatory frameworks should also include mechanisms for monitoring the implementation of technology and enforcing compliance with ethical standards.

Addressing challenges and enhancing policies for sustainable transformation in dairy farming requires a multifaceted approach. By overcoming technological, economic, and policy barriers, enhancing policies for emission reduction and sustainable practices, and ensuring that technological advancements are balanced with ethical farming practices, the dairy industry can move towards a more sustainable and environmentally responsible future. This transition not only aligns with environmental imperatives but also ensures the economic viability and cultural sustainability of the dairy farming sector.

12. Summary and Conclusions

Dairy processing, an energy-intensive process, significantly contributes to GHG emissions. Mitigating these emissions involves transitioning to renewable energy sources and implementing energy efficiency measures. Innovations like waste heat recovery and the use of energy-efficient equipment can reduce the energy footprint of dairy processing, decreasing operational costs and GHG emissions.

The advent of Big Data and AI has revolutionized dairy farming, offering enhanced efficiency, improved sustainability, and innovative management practices. These technologies address challenges in resource optimization and herd health management. Big Data analytics, involving collecting and analyzing extensive data, enables farmers to make informed decisions for productivity and sustainability. AI brings sophistication to herd management and resource optimization, with applications in predictive analytics for feed optimization and manure management. Benchmarking for performance and sustainability, involving comparing farm performance against industry standards, helps in identifying improvement areas and ensuring compliance with environmental standards.

The implementation of Big Data and AI in dairy farming marks a transformative leap towards sustainability and efficiency. By leveraging these technologies, the dairy industry can make significant gains in environmental stewardship, resource optimization, and overall farm management. However, this transformation also brings challenges, including integrating AI with existing farming systems, ensuring data quality, and farmer training. Addressing these challenges involves developing adaptable, user-friendly technologies, standardized data collection methods, and comprehensive farmer training programs.

Policies play a crucial role in supporting this transformation. Governments can offer financial incentives, invest in research and development, and implement sustainability guidelines. Creating a regulatory environment that encourages the adoption of sustainable practices and aligns industry growth with environmental conservation goals is essential. The study concludes that a multifaceted approach is needed to address technological, economic, and policy barriers and enhance policies for emission reduction and sustainable practices. The integration of technology in dairy farming should complement traditional practices and respect the cultural heritage of farming communities. Ensuring equity and inclusivity in technological advancement is crucial, as is establishing guidelines for the ethical use of technology in dairy farming.

By overcoming these challenges and enhancing policies for sustainable transformation, the dairy industry can progress towards a more sustainable and environmentally responsible future. This transition not only aligns with environmental imperatives but also ensures the economic viability and cultural sustainability of the dairy farming sector. The integration of Big Data and AI emerges as a crucial factor in this journey towards sustainability, offering innovative solutions to reduce GHG emissions and optimize resource use, thereby contributing significantly to global efforts in environmental sustainability.

Funding: This research received no external funding.

Conflicts of Interest: The author declares no conflicts of interest.

References

1. Easterling, D.R.; Kunkel, K.E. Climate Change in the Earth System. In *Climate Change and Estuaries*; CRC Press: Boca Raton, FL, USA, 2023; pp. 23–42.
2. Sudo, K. Atmospheric Mixing Ratios of Ozone and Radiative Forcing. In *Handbook of Air Quality and Climate Change*; Springer Nature Singapore: Singapore, 2023; pp. 997–1029.
3. Pouresmaeily, M. Ecological Responses to Climate Change. In *Climate Change: The Social and Scientific Construct*; Springer International Publishing: Cham, Switzerland, 2022; pp. 133–149.
4. Falkenberg, L.J.; Bellerby, R.G.; Connell, S.D.; Fleming, L.E.; Maycock, B.; Russell, B.D.; Sullivan, F.J.; Dupont, S. Ocean acidification and human health. *Int. J. Environ. Res. Public Health* **2020**, *17*, 4563. [CrossRef]
5. Rogelj, J.; Geden, O.; Cowie, A.; Reisinger, A. Three ways to improve net-zero emissions targets. *Nature* **2021**, *591*, 365–368. [CrossRef]
6. McLaren, D.P.; Tyfield, D.P.; Willis, R.; Szerszynski, B.; Markusson, N.O. Beyond "net-zero": A case for separate targets for emissions reduction and negative emissions. *Front. Clim.* **2019**, *1*, 4. [CrossRef]
7. Huang, M.T.; Zhai, P.M. Achieving Paris Agreement temperature goals requires carbon neutrality by middle century with far-reaching transitions in the whole society. *Adv. Clim. Change Res.* **2021**, *12*, 281–286. [CrossRef]
8. Wilberforce, T.; Olabi, A.G.; Sayed, E.T.; Elsaid, K.; Abdelkareem, M.A. Progress in carbon capture technologies. *Sci. Total Environ.* **2021**, *761*, 143203. [CrossRef]
9. Mentes, M. Sustainable development economy and the development of green economy in the European Union. *Energy Sustain. Soc.* **2023**, *13*, 32. [CrossRef]
10. Petri, F.; Biedenkopf, K. "United we stand, divided we fall": The effects of US contestation on EU foreign climate policy ambition. *Glob. Aff.* **2020**, *6*, 381–397. [CrossRef]

11. Hughes, L. Warming of 1.5 °C What we do now matters more than ever. *ReNew Technol. Sustain. Future* **2019**, *146*, 24–27.
12. Stern, N.; Valero, A. Innovation, growth and the transition to net-zero emissions. *Res. Policy* **2021**, *50*, 104293. [CrossRef]
13. Wattiaux, M.A.; Uddin, M.E.; Letelier, P.; Jackson, R.D.; Larson, R.A. Invited Review: Emission and mitigation of greenhouse gases from dairy farms: The cow, the manure, and the field. *Appl. Anim. Sci.* **2019**, *35*, 238–254. [CrossRef]
14. Gavrilova, O.; Leip, A.; Dong, H.; MacDonald, J.D.; Gomez Bravo, C.A.; Amon, B.; Barahona Rosales, R.; Prado, A.D.; de Lima, M.A.; Oyhantcabal, W.; et al. *Emissions from Livestock and Manure Management*; IPCC: Geneva, Switzerland, 2019.
15. Windfeld, E.; Lhermie, G. The value of Canadian agriculture: Direct, indirect, and induced economic impacts. *Front. Sustain. Food Syst.* **2022**, *6*, 940968. [CrossRef]
16. Biden, S.; Ker, A.P.; Duff, S. Impacts of trade liberalization in Canada's supply managed dairy industry. *Agric. Econ.* **2020**, *51*, 535–552. [CrossRef]
17. Berton, M.; Bovolenta, S.; Corazzin, M.; Gallo, L.; Pinterits, S.; Ramanzin, M.; Ressi, W.; Spigarelli, C.; Zuliani, A.; Sturaro, E. Environmental impacts of milk production and processing in the Eastern Alps: A "cradle-to-dairy gate" LCA approach. *J. Clean. Prod.* **2021**, *303*, 127056. [CrossRef]
18. Clay, N.; Garnett, T.; Lorimer, J. Dairy intensification: Drivers, impacts and alternatives. *Ambio* **2020**, *49*, 35–48. [CrossRef]
19. Biagetti, E.; Gislon, G.; Martella, A.; Zucali, M.; Bava, L.; Franco, S.; Sandrucci, A. Comparison of the use of life cycle assessment and ecological footprint methods for evaluating environmental performances in dairy production. *Sci. Total Environ.* **2023**, *905*, 166845. [CrossRef]
20. Cholewińska, P.; Czyż, K.; Nowakowski, P.; Wyrostek, A. The microbiome of the digestive system of ruminants—A review. *Anim. Health Res. Rev.* **2020**, *21*, 3–14. [CrossRef]
21. Ungerfeld, E.M.; Cancino-Padilla, N.; Vera-Aguilera, N. Fermentation in the rumen. In *Microbial Fermentations in Nature and as Designed Processes*; Hurst, C.J., Ed.; Wiley: Hoboken, NJ, USA, 2023; pp. 133–165.
22. Liu, S.; Proudman, J.; Mitloehner, F.M. Rethinking methane from animal agriculture. *CABI Agric. Biosci.* **2021**, *2*, 1–13. [CrossRef]
23. Dobson, S.; Goodday, V.; Winter, J. If it matters, measure it: A review of methane sources and mitigation policy in Canada. *Int. Rev. Environ. Resour. Econ.* **2023**, *16*, 309–429. [CrossRef]
24. Parker, K. "Cow-nting down": Regulatory measures to reduce New Zealand's biogenic methane emissions. *N. Z. J. Environ. Law* **2021**, *25*, 191–215.
25. Beauchemin, K.A.; Ungerfeld, E.M.; Eckard, R.J.; Wang, M. Fifty years of research on rumen methanogenesis: Lessons learned and future challenges for mitigation. *Animal* **2020**, *14*, s2–s16. [CrossRef]
26. Honan, M.; Feng, X.; Tricarico, J.M.; Kebreab, E. Feed additives as a strategic approach to reduce enteric methane production in cattle: Modes of action, effectiveness and safety. *Anim. Prod. Sci.* **2022**, *62*, 1303–1317. [CrossRef]
27. Rivera, J.E.; Chará, J. CH4 and N2O emissions from cattle excreta: A review of main drivers and mitigation strategies in grazing systems. *Front. Sustain. Food Syst.* **2021**, *5*, 657936. [CrossRef]
28. Costa, C.; Wironen, M.; Racette, K.; Wollenberg, E.K. *Global Warming Potential*(GWP*): Understanding the Implications for Mitigating Methane Emissions in Agriculture*; CGIAR Research Program on Climate Change, Agriculture and Food Security (CCAFS): Wageningen, The Netherlands, 2021.
29. Rotz, A.; Stout, R.; Leytem, A.; Feyereisen, G.; Waldrip, H.; Thoma, G.; Holly, M.; Bjorneberg, D.; Baker, J.; Vadas, P.; et al. Environmental assessment of United States dairy farms. *J. Clean. Prod.* **2021**, *315*, 128153. [CrossRef]
30. Khoshnevisan, B.; Duan, N.; Tsapekos, P.; Awasthi, M.K.; Liu, Z.; Mohammadi, A.; Angelidaki, I.; Tsang, D.C.; Zhang, Z.; Pan, J.; et al. A critical review on livestock manure biorefinery technologies: Sustainability, challenges, and future perspectives. *Renew. Sustain. Energy Rev.* **2021**, *135*, 110033. [CrossRef]
31. O'Brien, P.L.; Hatfield, J.L. Dairy Manure and Synthetic Fertilizer: A Meta-Analysis of Crop Production and Environmental Quality. *Agrosyst. Geosci. Environ.* **2019**, *2*, 1–12. [CrossRef]
32. Woolery, S.; Osei, E.; Yu, M.; Guney, S.; Lovell, A.; Jafri, H. The Carbon Footprint of a 5000-Milking-Head Dairy Operation in Central Texas. *Agriculture* **2023**, *13*, 2109. [CrossRef]
33. Arvidsson Segerkvist, K.; Hansson, H.; Sonesson, U.; Gunnarsson, S. Research on environmental, economic, and social sustainability in dairy farming: A systematic mapping of current literature. *Sustainability* **2020**, *12*, 5502. [CrossRef]
34. Ravishankara, A.R.; Daniel, J.S.; Portmann, R.W. Nitrous oxide (N_2O): The dominant ozone-depleting substance emitted in the 21st century. *Science* **2009**, *326*, 123–125. [CrossRef]
35. Neethirajan, S. Artificial Intelligence and Sensor Technologies in Dairy Livestock Export: Charting a Digital Transformation. *Sensors* **2023**, *23*, 7045. [CrossRef]
36. Neethirajan, S. The role of sensors, big data and machine learning in modern animal farming. *Sens. Bio-Sens. Res.* **2020**, *29*, 100367. [CrossRef]
37. Neethirajan, S. AI-Driven Climate Neutrality in Dairy Farming: Benchmarking Emissions for Sustainable Transformation. 2023. Available online: https://osf.io/preprints/osf/4znq5 (accessed on 15 December 2023).
38. Pitta, D.W.; Indugu, N.; Melgar, A.; Hristov, A.; Challa, K.; Vecchiarelli, B.; Hennessy, M.; Narayan, K.; Duval, S.; Kindermann, M.; et al. The effect of 3-nitrooxypropanol, a potent methane inhibitor, on ruminal microbial gene expression profiles in dairy cows. *Microbiome* **2022**, *10*, 1–21. [CrossRef]
39. Fouts, J.Q.; Honan, M.C.; Roque, B.M.; Tricarico, J.M.; Kebreab, E. Enteric methane mitigation interventions. *Transl. Anim. Sci.* **2022**, *6*, txac041. [CrossRef]

40. Obileke, K.; Nwokolo, N.; Makaka, G.; Mukumba, P.; Onyeaka, H. Anaerobic digestion: Technology for biogas production as a source of renewable energy—A review. *Energy Environ.* **2021**, *32*, 191–225. [CrossRef]
41. Malliaroudaki, M.I.; Watson, N.J.; Ferrari, R.; Nchari, L.N.; Gomes, R.L. Energy management for a net zero dairy supply chain under climate change. *Trends Food Sci. Technol.* **2022**, *126*, 153–167.
42. Malliaroudaki, M.I.; Watson, N.J.; Glover, Z.J.; Nchari, L.N.; Gomes, R.L. Net zero roadmap modelling for sustainable dairy manufacturing and distribution. *Chem. Eng. J.* **2023**, *475*, 145734. [CrossRef]
43. Paris, B.; Vandorou, F.; Tyris, D.; Balafoutis, A.T.; Vaiopoulos, K.; Kyriakarakos, G.; Manolakos, D.; Papadakis, G. Energy use in the EU livestock sector: A review recommending energy efficiency measures and renewable energy sources adoption. *Appl. Sci.* **2022**, *12*, 2142. [CrossRef]
44. Bertoglio, R.; Corbo, C.; Renga, F.M.; Matteucci, M. The digital agricultural revolution: A bibliometric analysis literature review. *IEEE Access* **2021**, *9*, 134762–134782. [CrossRef]
45. Neethirajan, S.; Kemp, B. Digital livestock farming. *Sens. Bio-Sens. Res.* **2021**, *32*, 100408. [CrossRef]
46. Neethirajan, S. SOLARIA-SensOr-driven resiLient and adaptive monitoRIng of farm Animals. *Agriculture* **2023**, *13*, 436. [CrossRef]
47. Kofler, J.; Suntinger, M.; Mayerhofer, M.; Linke, K.; Maurer, L.; Hund, A.; Fiedler, A.; Duda, J.; Egger-Danner, C. Benchmarking based on regularly recorded claw health data of Austrian dairy cattle for implementation in the Cattle Data Network (RDV). *Animals* **2022**, *12*, 808. [CrossRef]
48. Poulopoulou, I.; Zanon, T.; Alrhmoun, M.; Katzenberger, K.; Holighaus, L.; Gauly, M. Development of a benchmarking tool to assess the welfare of dairy cattle on small-scale farms. *J. Dairy Sci.* **2023**, *106*, 6464–6475. [CrossRef]
49. Kakani, V.; Nguyen, V.H.; Kumar, B.P.; Kim, H.; Pasupuleti, V.R. A critical review on computer vision and artificial intelligence in food industry. *J. Agric. Food Res.* **2020**, *2*, 100033. [CrossRef]
50. Grandsir, C.; Falagán, N.; Alamar, M.C. Application of novel technologies to reach net-zero greenhouse gas emissions in the fresh pasteurised milk supply chain: A review. *Int. J. Dairy Technol.* **2023**, *76*, 38–50. [CrossRef]
51. Snorek, J.; Cummings, W.; Hryniewicz, E.; Stevens, K.; Iannuzzi, R. Diversification strategies for the resilience of small New England dairies. *J. Agric. Food Syst. Community Dev.* **2023**, *12*, 1–21. [CrossRef]
52. Mazzetto, A.M.; Falconer, S.; Ledgard, S. Mapping the carbon footprint of milk production from cattle: A systematic review. *J. Dairy Sci.* **2022**, *105*, 9713–9725. [CrossRef]
53. Auclair, O.; Burgos, S.A. Carbon footprint of Canadian self-selected diets: Comparing intake of foods, nutrients, and diet quality between low-and high-greenhouse gas emission diets. *J. Clean. Prod.* **2021**, *316*, 128245. [CrossRef]
54. Peta, C. Canada's Supply Management System and the Dairy Industry in the Era of Trade Liberalization: A Cultural Commodity? *Am. Rev. Can. Stud.* **2019**, *49*, 547–562. [CrossRef]
55. Ritter, C.; Mills, K.E.; Weary, D.M.; von Keyserlingk, M.A. Perspectives of western Canadian dairy farmers on the future of farming. *J. Dairy Sci.* **2020**, *103*, 10273–10282. [CrossRef]
56. Denis-Robichaud, J.; Kelton, D.F.; Bauman, C.A.; Barkema, H.W.; Keefe, G.P.; Dubuc, J. Biosecurity and herd health management practices on Canadian dairy farms. *J. Dairy Sci.* **2019**, *102*, 9536–9547. [CrossRef]
57. Beauchemin, K.A.; Ungerfeld, E.M.; Abdalla, A.L.; Alvarez, C.; Arndt, C.; Becquet, P.; Benchaar, C.; Berndt, A.; Mauricio, R.M.; McAllister, T.A.; et al. Invited review: Current enteric methane mitigation options. *J. Dairy Sci.* **2022**, *105*, 9297–9326. [CrossRef]
58. Arndt, C.; Hristov, A.N.; Price, W.J.; McClelland, S.C.; Pelaez, A.M.; Cueva, S.F.; Oh, J.; Bannink, A.; Bayat, A.R.; Crompton, L.A.; et al. Strategies to mitigate enteric methane emissions by ruminants-a way to approach the 2.0 °C target. *AgriRxiv* **2021**, AgriRxiv: 20210085288. [CrossRef]
59. Min, B.R.; Solaiman, S.; Waldrip, H.M.; Parker, D.; Todd, R.W.; Brauer, D. Dietary mitigation of enteric methane emissions from ruminants: A review of plant tannin mitigation options. *Anim. Nutr.* **2020**, *6*, 231–246. [CrossRef]
60. Bhatnagar, N.; Ryan, D.; Murphy, R.; Enright, A.M. A comprehensive review of green policy, anaerobic digestion of animal manure and chicken litter feedstock potential–Global and Irish perspective. *Renew. Sustain. Energy Rev.* **2022**, *154*, 111884. [CrossRef]
61. Yao, Y.; Huang, G.; An, C.; Chen, X.; Zhang, P.; Xin, X.; Shen, J.; Agnew, J. Anaerobic digestion of livestock manure in cold regions: Technological advancements and global impacts. *Renew. Sustain. Energy Rev.* **2020**, *119*, 109494. [CrossRef]
62. Franzluebbers, A.J. Cattle Grazing Effects on the Environment: Greenhouse Gas Emissions and Carbon Footprint. In *Management Strategies for Sustainable Cattle Production in Southern Pastures*; Academic Press: San Diego, CA, USA, 2020; pp. 11–34.
63. Lal, R. Reducing carbon footprints of agriculture and food systems. *Carbon Footpr.* **2022**, *1*, 3. [CrossRef]
64. Niloofar, P.; Francis, D.P.; Lazarova-Molnar, S.; Vulpe, A.; Vochin, M.C.; Suciu, G.; Balanescu, M.; Anestis, V.; Bartzanas, T. Data-driven decision support in livestock farming for improved animal health, welfare and greenhouse gas emissions: Overview and challenges. *Comput. Electron. Agric.* **2021**, *190*, 106406. [CrossRef]
65. Lovarelli, D.; Bacenetti, J.; Guarino, M. A review on dairy cattle farming: Is precision livestock farming the compromise for an environmental, economic and social sustainable production? *J. Clean. Prod.* **2020**, *262*, 121409. [CrossRef]
66. Vitillo, J.G.; Eisaman, M.D.; Aradóttir, E.S.; Passarini, F.; Wang, T.; Sheehan, S.W. The role of carbon capture, utilization, and storage for economic pathways that limit global warming to below 1.5 °C. *iScience* **2022**, *25*, 5. [CrossRef]
67. Winsten, J.R.; Gorman, E.; Gravitz, A. Coordinating a "basket of incentives" to facilitate resilience in the dairy sector. *J. Soil Water Conserv.* **2020**, *75*, 144A–149A. [CrossRef]

68. Cameron, G.; Rosado, F.R.P.; Mederos, D.D.D. Agricultural co-operatives in Canada and Cuba: Trends, prospects and ways forward. *Environ. Dev. Sustain.* **2020**, *22*, 643–660. [CrossRef]
69. Stavins, R.N. *Carbon Taxes vs. Cap and Trade: Theory and Practice*; Harvard Project on Climate Agreements: Cambridge, MA, USA, 2019.
70. Green, J.F. Does carbon pricing reduce emissions? A review of ex-post analyses. *Environ. Res. Lett.* **2021**, *16*, 043004. [CrossRef]
71. Henchion, M.M.; Regan, Á.; Beecher, M.; MackenWalsh, Á. Developing 'Smart' Dairy Farming Responsive to Farmers and Consumer-Citizens: A Review. *Animals* **2022**, *12*, 360. [CrossRef]
72. Steenwerth, K.L.; Hodson, A.K.; Bloom, A.J.; Carter, M.R.; Cattaneo, A.; Chartres, C.J.; Hatfield, J.L.; Henry, K.; Hopmans, J.W.; Horwath, W.R.; et al. Climate-smart agriculture global research agenda: Scientific basis for action. *Agric. Food Secur.* **2014**, *3*, 1–39. [CrossRef]
73. van Hilten, M.; Wolfert, S. 5G in agri-food-A review on current status, opportunities and challenges. *Comput. Electron. Agric.* **2022**, *201*, 107291. [CrossRef]
74. Tian, F.; Wang, X.; Yu, S.; Wang, R.; Song, Z.; Yan, Y.; Li, F.; Wang, Z.; Yu, Z. Research on Navigation Path Extraction and Obstacle Avoidance Strategy for Pusher Robot in Dairy Farm. *Agriculture* **2022**, *12*, 1008. [CrossRef]
75. Duzha, A.; Alexakis, E.; Kyriazis, D.; Sahi, L.F.; Kandi, M.A. From Data Governance by design to Data Governance as a Service: A transformative human-centric data governance framework. In Proceedings of the 2023 7th International Conference on Cloud and Big Data Computing, Manchester, UK, 17–19 August 2023; pp. 10–20.
76. Bibri, S.E.; Krogstie, J. A novel model for data-driven smart sustainable cities of the future: A strategic roadmap to transformational change in the era of big data. *Future Cities Environ.* **2021**, *7*, 3. [CrossRef]
77. Liu, C.; Wang, X.; Bai, Z.; Wang, H.; Li, C. Does Digital Technology Application Promote Carbon Emission Efficiency in Dairy Farms? Evidence from China. *Agriculture* **2023**, *13*, 904. [CrossRef]
78. Ji, B.; Banhazi, T.; Phillips, C.J.; Wang, C.; Li, B. A machine learning framework to predict the next month's daily milk yield, milk composition and milking frequency for cows in a robotic dairy farm. *Biosyst. Eng.* **2022**, *216*, 186–197. [CrossRef]
79. Cockburn, M. Application and prospective discussion of machine learning for the management of dairy farms. *Animals* **2020**, *10*, 1690. [CrossRef]
80. Krupitzer, C.; Stein, A. Unleashing the Potential of Digitalization in the Agri-Food Chain for Integrated Food Systems. *Annu. Rev. Food Sci. Technol.* **2023**, *15*, 1–22. [CrossRef]

Disclaimer/Publisher's Note: The statements, opinions and data contained in all publications are solely those of the individual author(s) and contributor(s) and not of MDPI and/or the editor(s). MDPI and/or the editor(s) disclaim responsibility for any injury to people or property resulting from any ideas, methods, instructions or products referred to in the content.

forests

MDPI

Article

Adaptation of Tree Species in the Greater Khingan Range under Climate Change: Ecological Strategy Differences between *Larix gmelinii* and *Quercus mongolica*

Bingyun Du [1], Zeqiang Wang [1], Xiangyou Li [1], Xi Zhang [1], Xuetong Wang [1] and Dongyou Zhang [1,2,*]

[1] Heilongjiang Province Key Laboratory of Geographical Environment Monitoring and Spatial Information Service in Cold Regions, Harbin Normal University, Harbin 150025, China; duby@stu.hrbnu.edu.cn (B.D.); wangzeqiang@stu.hrbnu.edu.cn (Z.W.); hsdlxy1999@gmail.com (X.L.); zhangxi@stu.hrbnu.edu.com (X.Z.); 18845725592@163.com (X.W.)

[2] Heilongjiang Wuyiling Wetland Ecosystem National Observation and Research Station, Yichun 153000, China

* Correspondence: zhangdy@hrbnu.edu.cn

Abstract: Global warming significantly affects forest ecosystems in the Northern Hemisphere's mid-to-high latitudes, altering tree growth, productivity, and spatial distribution. Additionally, spatial and temporal heterogeneity exists in the responses of different tree species to climate change. This research focuses on two key species in China's Greater Khingan Range: *Larix gmelinii* (Rupr.) Kuzen. (Pinaceae) and *Quercus mongolica* Fisch. ex Ledeb. (Fagaceae). We utilized a Maxent model optimized by the kuenm R package to predict the species' potential habitats under various future climate scenarios (2050s and 2070s) considering three distinct Shared Socioeconomic Pathways: SSP1-2.6, SSP2-4.5, and SSP5-8.5. We analyzed 313 distribution records and 15 environmental variables and employed geospatial analysis to assess habitat requirements and migration strategies. The Maxent model demonstrated high predictive accuracy, with Area Under the Curve (AUC) values of 0.921 for *Quercus mongolica* and 0.985 for *Larix gmelinii*. The high accuracy was achieved by adjusting the regularization multipliers and feature combinations. Key factors influencing the habitat of *Larix gmelinii* included the mean temperature of the coldest season (BIO11), mean temperature of the warmest season (BIO10), and precipitation of the driest quarter (BIO17). Conversely, *Quercus mongolica's* habitat suitability was largely affected by annual mean temperature (BIO1), elevation, and annual precipitation (BIO12). These results indicate divergent adaptive responses to climate change. *Quercus mongolica's* habitable area generally increased in all scenarios, especially under SSP5-8.5, whereas *Larix gmelinii* experienced more complex habitat changes. Both species' distribution centroids are expected to shift northwestward. Our study provides insights into the divergent responses of coniferous and broadleaf species in the Greater Khingan Range to climate change, contributing scientific information vital to conserving and managing the area's forest ecosystems.

Keywords: global warming; Maxent; GIS; potential habitats; parameter optimization

1. Introduction

Forest ecosystems are sensitive to climate variations and are crucial for ecological services, notably for carbon sequestration [1,2]. Of the estimated 2.19×10^{18} kg of carbon in the total biosphere carbon pool, a significant portion, approximately 1.0×10^{18} kg, is sequestered in forest ecosystems, which hold about 50% more carbon than the current atmospheric pool [3]. Predictions suggest a mean global temperature rise of 0.3–1.7 °C to 2.6–4.8 °C by 2100, posing significant challenges to these ecosystems, including potential productivity decline, biodiversity loss, and forest degradation [4–7]. Trees mitigate varied responses to global warming [8–10]. Although studies have shown that warming may limit habitats for species like *Nothofagus pumilio* (Poepp.& Endl.) Krasser (Nothofagaceae) [8] and *Picea mariana* (Mill.) Britton, Sterns & Poggenb. (Pinaceae) [9], it could benefit tree growth in temperate and boreal regions [10]. This variability in tree responses to climate change highlights the importance of adaptive management and conservation in forest ecosystems.

Researchers have utilized statistical models, species distribution data, and environmental variables to simulate potential future distributions of forest ecosystems under climate change [11–13]. The Maximum Entropy Model (Maxent), integrating machine learning with maximum entropy principles, effectively predicts species' potential distributions by utilizing known geographical data and environmental factors [14]. Elith et al. [15] compared Maxent's accuracy and consistency against 16 species distribution models across 226 species and found it superior in performance and stability across varied sample sizes. Maxent's low bias, high data tolerance, and effectiveness with small samples make it a preferred choice for predicting species spatial distribution, especially for tree species [16]. For instance, Carlos et al. [11] demonstrated Maxent's high predictive accuracy and stability for 17 economically significant Mexican pine species, where it outperformed nine other models, which highlights its potential for modeling tree species distribution.

However, Maxent has several limitations, including the need for high-quality data, appropriate environmental variables, optimal parameters, and uncertainty assessment [17,18]. Using the default or a priori parameters and including all environmental variables can complicate the model and limit its application and accuracy, resulting in a decrease in the capacity of species translocation [19,20]. Moreover, the Intergovernmental Panel on Climate Change (IPCC) has updated the future climate scenarios, requiring more in-depth simulation analysis [4]. The WorldClim database (http://worldclim.org, in its 2020 update to version 2.1, extended the period of historical climate data from the 1960s–1990s to include the 1970s–2000s [21]. This update marked a significant shift from the Representative Concentration Pathways (RCPs) used for the IPCC's Fifth Coupled Model Intercomparison Project (CMIP5) to the Shared Socioeconomic Pathways (SSPs) of the CMIP6 framework. The SSPs in version 2.1 provided granular future climate projections: SSP1-2.6 (Sustainability) targeted a warming scenario substantially below 1.5 °C by 2100; SSP2-4.5 (Middle of the Road) projected a radiative force of approximately 4.5 W/m^2 by 2100; SSP3-7.0 (Regional Rivalry) indicated scenarios of high social fragility coupled with a significant anthropogenic radiative force; and SSP5-8.5 (Fossil-fueled Development) forecast the most extreme scenario, with radiative force reaching 8.5 W/m^2 by 2100 [4,22,23]. These SSPs enhanced our understanding of the interplay between socioeconomic development and climate change, resulting in more precise and realistic climate simulations [24]. Hence, the meticulous choice of environmental factors and model parameters is crucial for effective Maxent applications.

Research on tree habitats often focuses on single species, whereas few comprehensive studies have been conducted on different species, such as conifer and broadleaf trees. Therefore, there is limited understanding of their growth suitability and response to climate change, which will affect forest management. Climate warming has affected the geographic distribution and growth patterns of coniferous and broadleaf trees in northeast China [25–27]. Research by Lyu et al. [28] on *Pinus koraiensis* Siebold & Zucc. (Pinaceae) and *Quercus mongolica* Fisch. ex Ledeb. (Fagaceae) in this region revealed that the radial growth of *Quercus mongolica* is predominantly precipitation-dependent during the growing season, while *Pinus koraiensis* growth is more temperature-sensitive. Located in the Greater Khingan Range, the forests in the northern region of northeast China not only represent 29.9% of the country's natural forests but also encompass one-third of the carbon stock within China's forest ecosystems [29,30]. Notably, since 1961, the Greater Khingan Range has exhibited a significant warming trend, with its annual average temperature increase reaching 0.88 °C, far exceeding the national average rate of 0.02 °C [31]. This pronounced climate change has made the Greater Khingan Range an ideal location for studying the responses and adaptive mechanisms of trees to global warming.

Against this backdrop, two key species in the Greater Khingan Range forests, *Larix gmelinii* (Rupr.) Kuzen. (Pinaceae) and *Quercus mongolica* Fisch. ex Ledeb. (Fagaceae) have become focal points of study. These species are critical indicators of climate change and play vital roles in their respective forest ecosystems [27]. They exhibit a complex coexistence with other species, such as *Pinus koraiensis* Siebold & Zucc. (Pinaceae) and *Betula platyphylla* Sukaczev (Betulaceae), forming diverse inter-species relationships [32–34]. *Larix gmelinii*, known for its cold tolerance and drought resistance, is one of the dominant vegetation species in the northern part of China's cold temperate coniferous forest zone [35]. It accounts for over 70% of the forest coverage in the Greater Khingan Range [36], and its carbon storage represents approximately 8.5% of the total forest carbon reserves in China [37]. This species plays a crucial role in maintaining the ecological balance in the northeast region of the country [38]. *Quercus mongolica*, as one of the most important deciduous tree species in northeast China, is classified as a secondary precious tree species. It is characterized by rapid growth and high-quality timber [39] and exhibits significant biomass accumulation and productivity in temperate zones. The ecosystem biocycling capacity of *Quercus mongolica* forests not only surpasses that of subtropical *Cunninghamia lanceolata* (Lamb.) Hook. (Cupressaceae) mature forests but also exceeds that of common pioneer tree species in temperate regions, such as *Betula platyphylla* [39–44].

To address this research gap, we employed a Maxent model optimized with the kuenm R package (https://github.com/marlonecobos/kuenm, accessed on 2 January 2023)) to predict the future distributions of *Larix gmelinii* and *Quercus mongolica* in the Greater Khingan Range. Leveraging updated IPCC data [21], we aimed to assess the impact of future climate change on these species in detail. Our study focuses on (1) defining the species' current and future distributions and suitabilities under various climate scenarios and (2) comparing the distributions, trends, and mechanisms of these species. We hypothesize a northward shift under future climate change and expect *Larix gmelinii* to be more sensitive to these changes than *Quercus mongolica*.

This study utilized a Maxent model optimized with R language to investigate the diverse responses and adaptation strategies of coniferous and broadleaf species to climate change, thereby contributing to a deeper understanding of forest–climate interactions. This research lays a scientific foundation for the conservation of the Greater Khingan Range's forest ecosystem. Additionally, the integration of Geographic Information System (GIS) methods with the Maxent model demonstrates the practical application value of this methodological combination in ecological protection and management.

2. Materials and Method

2.1. Collection of Species Occurrence Data

The occurrence data of *Larix gmelinii* and *Quercus mongolica* were obtained from two primary sources: (1) extensive field surveys by our research team in the Greater Khingan Range between 2019 and 2023 to map the species distribution comprehensively and (2) digitized historical specimen data from the Global Biodiversity Information Facility (GBIF; http://www.gbif.org) and the National Plant Specimen Resource Center (NSII; http://www.cvh.ac.cn), which provided essential historical insights into the species' distributions.

To ensure data accuracy and minimize spatial autocorrelation, we carefully screened records for precise latitude and longitude information, retaining only one record within a 1 km radius to avoid redundancy [45,46]. This strategy resulted in a robust dataset of 186 and 127 unique, spatially distinct occurrence records for *Larix gmelinii* and *Quercus mongolica*, respectively, serving as the basis for niche modeling and predicting potential future distributions under various climate scenarios (Figure 1).

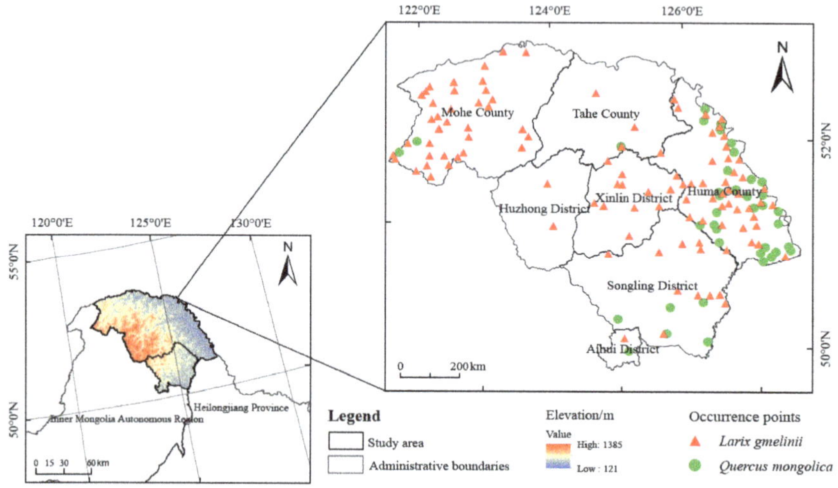

Figure 1. Administrative borders and research zone (**left**) alongside Greater Khingan Prefecture (areas surveyed) featuring the presence data of *Larix gmelinii* and *Quercus mongolica* (**right**).

2.2. Environmental Data Acquisition and Processing

We integrated topographical, soil, and 19 bioclimatic variables (Tables 1 and A1). The bioclimatic variables were derived from the WorldClim database (http://worldclim.org), and Shared Socioeconomic Pathways (SSPs) from the Sixth Coupled Model Intercomparison Project (CMIP6) were used to project future scenarios (2050s and 2070s) under three climate scenarios (SSP1-2.6, SSP2-4.5, and SSP5-8.5) at a 30-arc-second resolution [47,48]. These climate data, generated by the Beijing Climate Center's Medium-Resolution Climate System Model (BCC.CSM2.MR), are noted for their accuracy at simulating China's precipitation and temperature patterns [49,50]. Soil data came from the World Soil Database (http://www.fao.org), and topographic data (high-resolution (90 m) elevation data) were sourced from the Chinese Academy of Sciences' Resource and Environmental Science Data Center (http://www.resdc.cn/). The vector zoning map of the study area was sourced from the National Administration of Surveying, Mapping and Geoinformation's standard map service website (http://bzdt.ch.mnr.gov.cn/).

The Maxent model necessitates that the environmental data inputs must rigorously maintain uniform spatial reference coordinates and resolution. Accordingly, the vector map of the study area is utilized as a boundary, and the ArcGIS (Version 10.4, ESRI, Redland, CA, USA) mask tool is employed for batch extraction of the pertinent environmental variable layers. Additionally, the resampling tool under raster processing is utilized to standardize the spatial resolution: that is, the pixel size (X, Y) of the environmental layers. Subsequently, utilizing the 'From Raster to ASCII' functionality in ArcGIS, the datasets containing environmental variables were converted into ASCII format.

2.3. Model

The complexity of the model constitutes a critical aspect in the investigation of species distribution models. An effective model must strike a balance between accurately fitting the data and constraining overly complex functional relationships [51]. Maxent, as an algorithmic model, is inclined to form complex functional relationships to align with observational data. This tendency significantly contributes to the model's enhanced predictive capabilities: a notable strength of the maximum entropy approach [52].

The complexity of the Maxent model is primarily influenced by three factors: the number of environmental variables included, the types of functional modes (or 'features') used, and the regularization multiplier [53,54].

The choice of environmental variables is pivotal; an excess can lead to computational inefficiencies and challenges in interpreting the results, especially when variables display high collinearity [19,51]. Therefore, a balanced approach for selecting these variables is crucial for both computational efficiency and clarity of interpretation.

Regularization plays a vital role in Maxent by constraining variable weights, ensuring a balance between model fit and extrapolation. This prevents overfitting, allowing the model to maintain a margin of error [55]. Maxent's versatility is further illustrated by its six functional modes: linear (L), quadratic (Q), product (P), threshold (T), hinge (H), and category (C), each catering to different variable interactions and types [56,57]. Maxent dynamically tests combinations of these modes, ultimately selecting the one that achieves the highest Area Under the Curve (AUC) for prediction.

Model performance is critically evaluated using the AUC, a metric derived from the Receiver Operating Characteristic (ROC) curve [14]. The ROC curve is a plot of the false positive rate against the true positive rate at various thresholds [14]. The AUC, measuring the area under this curve, provides a robust measure of model performance and is applicable across various thresholds [58]. Its range from 0 to 1, with values nearer to 1 reflecting greater predictive accuracy, classifies performance into four categories [59,60]: poor (0.6–0.7), fair (0.7–0.8), good (0.8–0.9), and excellent (0.9–1).

2.3.1. Environmental Variable Selection

Considering the potentially high correlations among 19 bioclimate variables, such multicollinearity could lead to data overfitting when these factors are directly used in model construction using the Maxent software (v3.4.4, Biodiversity Informatics, Cambridge, MA, USA), thereby affecting the simulation results [61]. Pearson correlation analysis was conducted by the software ENMTools v1.3 [62,63] to refine our variable selection; we focused on factors with a correlation coefficient above 0.8 (Figure A1). Utilizing Maxent 3.4.4, ten calculations were performed on the selected environmental variables to establish an initial model. Based on the average results of these ten computations, environmental factors with a contribution rate of less than 1% were excluded. This selection process was guided by a comprehensive consideration of the species' environmental needs and physiological traits, resulting in the identification of key variables for modeling (Table 1).

Table 1. Environmental variables considered in the Maxent models of *Larix gmelinii* and *Quercus mongolica* in the Greater Khingan Range.

Variable Types	**Code and Unit**	**Species**	
		Larix gmelinii	*Quercus mongolica*
Annual Mean Temperature	BIO1 (°C)	–	✓
Isothermality × 100	BIO3 (°C)	✓	✓
Mean Temperature of Warmest Quarter	BIO10 (°C)	✓	✓
Mean Temperature of Coldest Quarter	BIO11 (°C)	✓	–
Annual Precipitation	BIO12 (mm)	✓	–
Precipitation Seasonality (Co-efficient of Variation)	BIO15 (mm)	✓	–
Precipitation of Driest Quarter	BIO17 (mm)	✓	–
Soil Total Organic Carbon Content	T_CACO3 (%)	✓	✓
Total Organic Carbon Content	T_OC (%)	–	✓
Soil Base Saturation	S_BS (%)	✓	–
Elevation	Elevation (m)	✓	✓
Slope	Slope (degree)	–	✓
Aspect	Aspect (degree)	–	✓

The variables marked with "✓" are used for building the model.

2.3.2. Model Calibration

Model calibration is a process in which the aim is to determine which combination of parameters best represents the phenomenon of interest by finding the best fit with the data [64]. Performed manually, detailed model calibration and final model creation is quite time-consuming (e.g., a week or more). Hence, automating the process is essential for increasing the robustness of Ecological Niche Models (ENMs) [54]. Here, we employed kuenm, an R package that automates important calibration and evaluation steps in ENM. In its current version, this package uses Maxent as the modeling algorithm and automates model calibration and the creation of final models [54]. For each species, we created 1240 candidate models by combining 3 sets of environmental variables, 40 values of regularization multipliers (RMs, 0.1–4 at intervals of 0.1), and all 31 possible combinations of 5 feature classes (FCs). The R software (Version 4.1.1, R Foundation for Statistical Computing, Vienna, Austria) evaluates candidate model performance and initially selects statistically significant models with omission rates $\leqslant$ 5% [54]. Subsequently, models with a Delta-corrected Akaike Information Criterion (AICc) of $\leqslant$2 are chosen [17,51,54].

2.3.3. Threshold Determination

After determining the best configuration of the FC and RM, we reserved 25% of the samples for testing. The significance of the environmental variables was evaluated through jackknife tests, which were conducted ten times to ensure robustness. The mean values of the iterations were used as predictive results. Utilizing the ArcGIS software (Version 10.4, ESRI, Redland, CA, USA) (specifically, the ArcToolbox: Conversion Tools), the species distribution maps generated by the Maxent model (.asc files) were converted into raster data (.tif files), resulting in suitability maps that represented the probability of species occurrence ($0 < p \leqslant 1$).

We adopted the maximum test sensitivity plus specificity (MTSPS) threshold as a criterion to delineate suitable and unsuitable areas to effectively consider both omission and commission errors [65–67]. Suitable habitat was further categorized into three levels: low (MTSPS $< p \leqslant 0.4$), moderate ($0.4 < p \leqslant 0.6$), and high suitability ($p > 0.6$) [68].

2.4. *Habitat Change Analysis and Centroid Shifts*

To ensure the temporal comparability of Maxent, we maintained the consistency of topographical and soil variables while focusing exclusively on changes to bioclimatic variables to predict future spatial patterns of species distribution [69,70]. Additionally, we cited a hypothesis that species possess unrestricted migration capacities when evaluating the impact of future climatic changes on species distribution [71]. This implies that in response to climatic alterations, species may either adapt and persist in their current habitats, potentially relocate to newly viable regions, or face in situ extinction. Based on this hypothesis, we categorized the changes in the species' potential habitats into four types (expansion, unsuitable, no change, and contraction) based on future area changes relative to the current suitable habitats. In the habitat change matrix, transitions from 0 to 1 indicated expansion, 1 to 0 contraction, 1 to 1 persistence, and 0 to 0 consistent unsuitability. Finally, to compare the responses of the geographical suitability areas of *Larix gmelinii* and *Quercus mongolica* under various future climate scenarios, we implemented the following steps: initially, the raster data were reclassified; subsequently, a raster calculator was utilized to determine the expansion or contraction of the suitable habitat ranges for both tree species under different future climate scenarios.

Due to the irregular boundaries of habitats, the centroids of the species records in the entire geographic range were determined to assess the changes between current and future conditions. We used SDMtoolbox 2.4 to assess the trends of the centroid shifts between adjacent habitat stages and calculated the centroid migration distances [46].

3. Results

3.1. *Model Performance*

Automatic analysis using the R program indicated that all 1240 candidate models were statistically significant. One recommended model was selected for each species (Figure 2); both had a DeltaAICc value of 0. This result indicated optimal model transferability from known to predicted distribution areas and the avoidance of overfitting; thus, these were the best models. The FC was Q, P, and T for *Larix gmelinii* and H for *Quercus mongolica*. The RM values were 1.7 for *Larix gmelinii* and 1 for *Quercus mongolica*. For these optimal FC and RM settings, the AUC values of the Maxent prediction results were 0.921 for *Larix gmelinii* and 0.985 for *Quercus mongolica* (Figure 3).

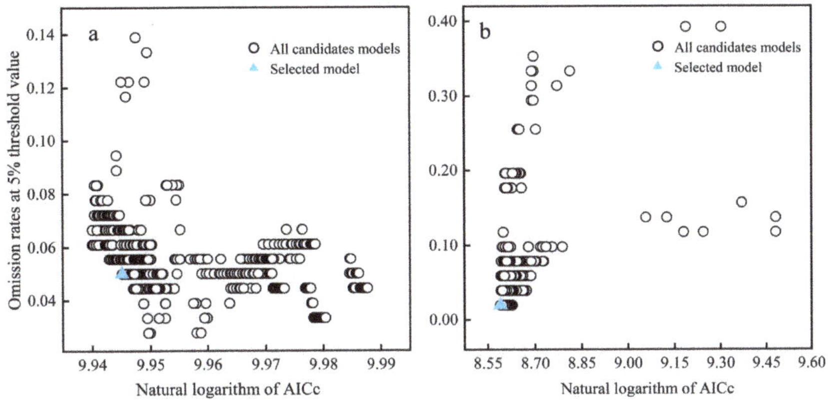

Figure 2. Omission rates and AICc values for *Larix gmelinii* (**a**) and *Quercus mongolica* (**b**), including all candidate models and the chosen 'best' models.

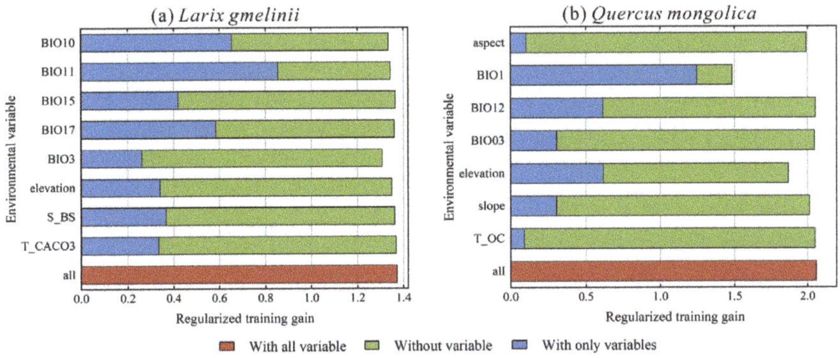

Figure 3. The outcomes of the jackknife test elucidate the significance of various variables for the habitat distribution modeling of (**a**) *Larix gmelinii* and (**b**) *Quercus mongolica*. The regularized training gain quantifies the enhancement in the Maxent model's fit to the existing data, as opposed to a uniform distribution. Dark blue bars represent the gain achieved by employing each variable independently for each species, light blue bars reflect the reduction in gain consequent to the exclusion of a single variable from the comprehensive model for each species, and the red bar denotes the gain attained with the incorporation of all variables for each species.

3.2. Variable Importance

Environmental variables crucial for *Larix gmelinii* and *Quercus mongolica* were identified through contribution rate analysis (Figure 4 and Table 2). For *Larix gmelinii*, BIO11, BIO10, and BIO17 accounted for 71.6% of the model's predictive power, underlining their critical role in determining its potential geographic distribution. In contrast, for *Quercus mongolica*, BIO1, elevation, and BIO12 contributed 90%, dominating the model and shaping the species' current potentially suitable habitat.

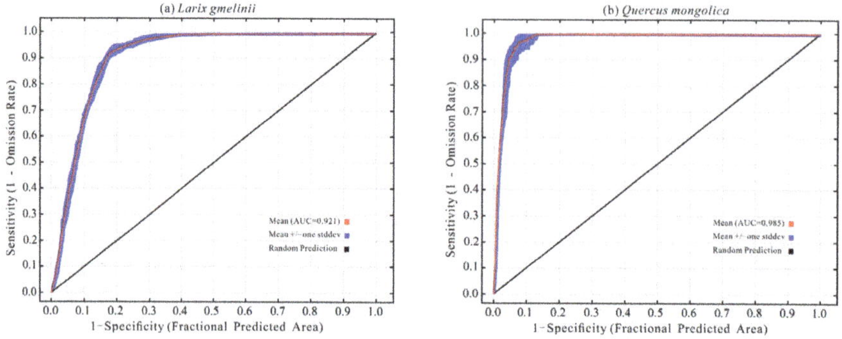

Figure 4. AUC curve representation: (**a**) *Larix gmelinii* and (**b**) *Quercus mongolica* distribution model development. The red curve represents the training data, indicating model alignment with the training dataset; the blue curve represents the testing data, reflecting the model's predictive performance in practical tests.

Table 2. Contribution rate of environmental variable importance obtained by jackknife test.

Species	Variable	Percent Contribution (%)
Larix gmelinii	BIO11	37.1
	BIO10	19.5
	BIO17	15
	T_CACO3	11.4
	BIO3	5.9
	S_BS	5.3
	BIO15	3.3
	elevation	2.6
Quercus mongolica	BIO1	64.4
	elevation	21.1
	BIO12	4.8
	slope	4.2
	aspect	3.1
	BIO3	1.1
	T_OC	1

The red lines in Figure 5 display the influence of individual environmental factors on the predicted probability of occurrence. Details on other environmental factors can be found in Figures A2 and A4. Specifically, Figure 5a illustrates that the occurrence probability for *Larix gmelinii* increases as BIO11 and BIO10 decrease: peaking before stabilizing and forming a response crest. This finding reveals that *Larix gmelinii* consistently reacts to temperature variations, implying a decline in its suitable habitat as temperatures increase. Conversely, BIO17 shows a positive correlation with the distribution probability. As BIO17 increases, the logistical probability value rises and remains consistently high. For *Quercus mongolica*, elevation and BIO1 stabilize at high probabilities at fixed thresholds but decrease with increasing variable values and then stabilize. The relationship between BIO12 and the distribution probability differs slightly, with a minor increase in probability as annual precipitation rises, followed by stabilization. Most precipitation values correspond to high probability levels, indicating a comparatively lower contribution and importance of annual precipitation to the distribution probability prediction for *Quercus mongolica* relative to the other two dominant factors.

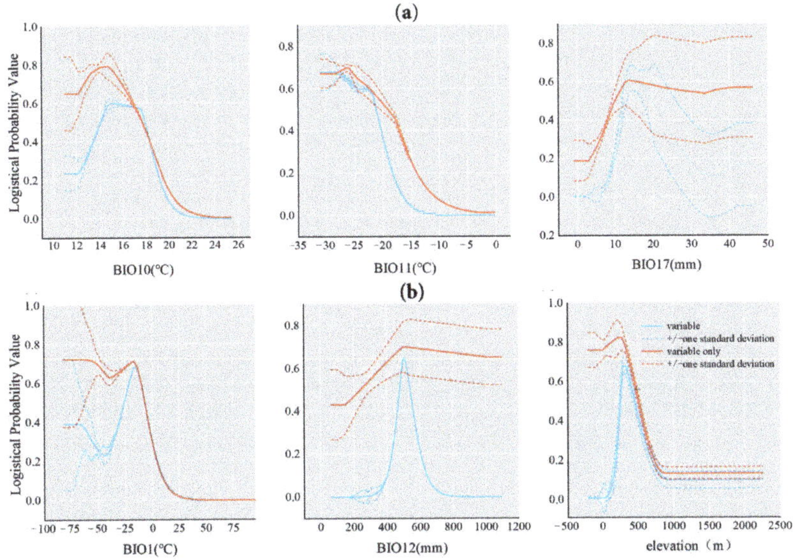

Figure 5. Response curves characterizing how dominant variables affected the Maxent predictions for *Larix gmelinii* (**a**) and *Quercus mongolica* (**b**). Red lines represent logistical predictions varying with each climatic variable while holding other variables at their means. Based on individual variables, the Maxent models generate blue lines. Solid lines show the average of ten Maxent runs, and dashed lines denote one standard deviation from this mean.

The blue lines in Figure 5 display the distinct effects of the key variables in the optimal model on forecasting the likelihood of favorable conditions, as demonstrated by a Maxent model that incorporates each variable independently. Details on other environmental factors can be found in Figures A3 and A5. *Larix gmelinii's* occurrence probability approaches zero with BIO11 above −10 °C, increases sharply as temperatures decrease, and peaks below −28 °C. This result suggests a stringent upper limit but no lower limit for BIO11. The suitable distribution for *Larix gmelinii* (probability ⩾ 0.24) requires that BIO11 is below −19 °C. In contrast to BIO11, BIO10 and BIO17 have upper and lower limits. The probability approaches zero for BIO17 below 7 mm, rapidly increases to its highest level at 14 mm, and sharply declines above this value. Similarly, the probability of BIO10 stabilizes around 0.23 when temperatures are below 12 °C and increases with rising temperatures, maintaining high levels between 15 °C and 17 °C, followed by a downward trend. The suitable distribution area for *Larix gmelinii* requires BIO10 and BIO17 ranges of 12–19 °C and 10–28 mm, respectively. The occurrence probability of *Quercus mongolica* approaches zero when BIO01 exceeds 18 °C, with the optimal BIO1 range of −20 °C to −10 °C. Unlike temperature, precipitation and elevation require stringent thresholds. The occurrence probability approaches zero when the elevation is below 140 m and BIO12 is below 400 mm. This probability surges with increasing values, peaking at 340 m for elevation and 505 mm for BIO12 before sharply decreasing beyond these points. The suitable distribution for *Quercus mongolica* (probability ⩾ 0.15) necessitates an elevation range of 166–776 m and annual precipitation of 415–620 mm.

3.3. Potential Current and Future Habitat Distributions

3.3.1. *Larix gmelinii*

Under the current climate, the unsuitable habitat area for *Larix gmelinii* accounts for only 2.06% of the research zone, whereas the suitable habitat encompasses a significant 98.02%, demonstrating its high adaptability to current climate conditions (Figure 6 and

Table 3). Within the suitable habitat, marginally suitable, moderately suitable, and highly suitable habitats comprise 14.17%, 37.33%, and 46.52%, respectively, indicating that most areas offer ideal conditions for *Larix gmelinii*. Future climate shifts, however, are expected to alter these habitats (Figure 7 and Table 3). By the 2050s, various climate scenarios predict an expansion of unsuitable habitats: notably, to 29.02% under the SSP5-8.5 scenario. However, the conditions will differ by the 2070s. Under SSP1-2.6, suitable habitats will revert to near-current levels at 98.43%, whereas under the SSP2-4.5 and SSP5-8.5 scenarios, unsuitable areas could expand to 17.65% and 54.06%, respectively. In particular, a substantial increase in unsuitable habitats is observed in the SSP5-8.5 scenario, potentially covering over half of the study area. These results highlight the notable influence of varying climate scenarios on the suitable habitat of *Larix gmelinii* and the looming threat of climate change to its distribution. Additionally, the shift in suitability zones is noteworthy. Under most climate scenarios, moderately and highly suitable habitats for *Larix gmelinii* decrease, whereas marginally suitable areas increase, hinting at a potential overall decline in future growth conditions. These effects are especially pronounced in the SSP5-8.5 scenario, where highly suitable habitats will shrink to 3.66% by the 2070s—much lower than in other scenarios—suggesting a scarcity of optimal growth environments under extreme climate conditions.

Table 3. Forecast areas unsuitable and suitable for *Larix gmelinii* under current and future climate scenarios.

SSPs (Year)	Unsuitable Habitats	Total Suitable Habitats	Minimally Suitable Habitats	Moderately Suitable Habitats	Highly Suitable Habitats
	Area /km^2 (%)	Area/km^2 (%)	Area /km^2 (%)	Area /km^2 (%)	Area /km^2 (%)
Current	1701.8 2.06	81,104.7 98.02	11,724.9 14.17	30,885.3 37.33	38,494.5 46.52
SSP1-2.6 (2050s)	6142.0 7.42	76,687.9 92.68	7138.4 8.63	24,326.6 29.40	45,222.9 54.65
SSP2-4.5 (2050s)	12,183.3 14.72	70,646.6 85.38	9965.7 12.04	18,502.4 22.36	42,178.5 50.97
SSP5-8.5 (2050s)	24012.2 29.02	58817.7 71.08	11,972.8 14.47	42,373.2 51.21	4471.7 5.40
SSP1-2.6 (2070s)	1360.0 1.64	81,446.6 98.43	3656.6 4.42	22,681.3 27.41	55,108.7 66.60
SSP2-4.5 (2070s)	14,602.1 17.65	68,227.8 82.46	12,050.2 14.56	34,774.7 42.03	21,402.8 25.87
SSP5-8.5 (2070s)	44,732.1 54.06	38,097.7 46.04	24,590.3 29.72	10,480.6 12.67	3026.9 3.66

SSPs: Shared Socioeconomic Pathways; SSP1-2.6: Sustainable Development Pathway; SSP2-4.5: Moderate Development Pathway; SSP5-8.5: Conventional Development Pathway Dominated by Fossil Fuels. The proportion of suitable habitat area under different future climate scenarios is the average for the periods 2021–2040, 2041–2060, and 2061–2080.

In summary, future climate change will significantly affect the habitats of *Larix gmelinii*, with diverse impacts under different climate scenarios. These findings not only underline potential risks to this species but also emphasize the diverse responses for different levels of suitable habitats.

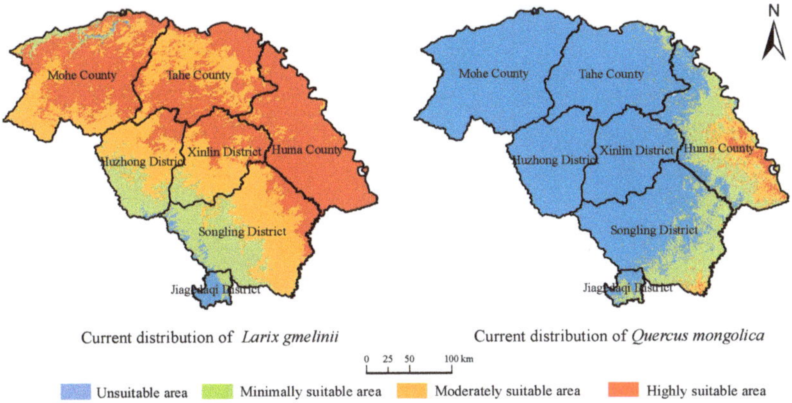

Figure 6. Suitable habitat distribution of *Larix gmelinii* (**left**) and *Quercus mongolica* (**right**) in the Greater Khingan Range under current climatic conditions.

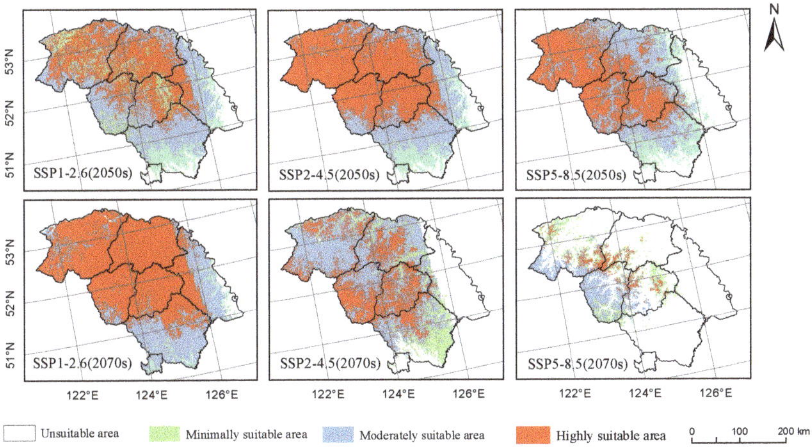

Figure 7. Future distribution of *Larix gmelinii* as a reaction to future climate change scenarios during 2050s and 2070s. SSPs: Shared Socioeconomic Pathways; SSP1-2.6: Sustainable Development Pathway; SSP2-4.5: Moderate Development Pathway; SSP5-8.5: Conventional Development Pathway Dominated by Fossil Fuels.

3.3.2. *Quercus mongolica*

Currently, 79.64% of the study area is unsuitable for *Quercus mongolica*; only 20.36% is suitable (Figure 6 and Table 4). Future climate scenarios predict a significant increase in suitable habitat (Figure 8 and Table 4). Suitable habitat will comprise the majority by the 2050s under the SSP1-2.6, SSP2-4.5, and SSP5-8.5 scenarios (66.28%, 78.39%, and 89.09%, respectively). This expansion intensifies by the 2070s, particularly under SSP5-8.5, with suitable habitats nearly encompassing the whole area (99.99%). A key trend in this shift is the marked rise in marginally suitable habitats; they will account for 67.84% under SSP5-8.5 by the 2050s, increasing to 95.42% by the 2070s. Meanwhile, moderately suitable habitats will exhibit relatively minor changes.

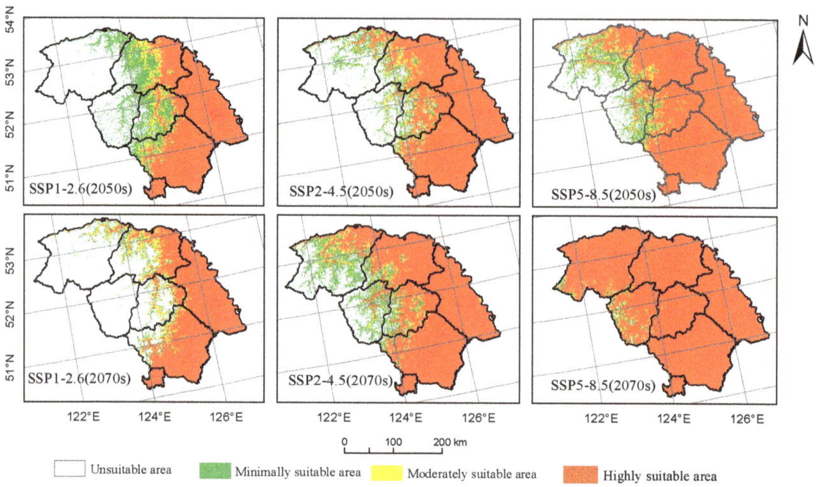

Figure 8. Future distribution of *Quercus mongolica* as a reaction to future climate change scenarios during 2050s and 2070s. SSPs: Shared Socioeconomic Pathways; SSP1-2.6: Sustainable Development Pathway; SSP2-4.5: Moderate Development Pathway; SSP5-8.5: Conventional Development Pathway Dominated by Fossil Fuels.

Table 4. Forecast areas unsuitable and suitable for *Quercus mongolica* under current and future climate scenarios.

SSPs (Year)	Unsuitable Habitats	Total Suitable Habitats	Minimally Suitable Habitats	Moderately Suitable Habitats	Highly Suitable Habitats
	Area /km^2 (%)	Area/km^2 (%)	Area /km^2 (%)	Area /km^2 (%)	Area /km^2 (%)
Current	65,896.4 79.64	16,848.4 20.36	11,967.0 14.46	3745.5 4.53	1135.9 1.37
SSP1-2.6 (2050s)	27,897.9 33.72	54,846.9 66.28	12,473.4 15.07	6610.1 7.99	35,763.4 43.22
SSP2-4.5 (2050s)	17,958.6 21.70	64,866.1 78.39	12,156.8 14.69	6658.5 8.05	46,050.8 55.65
SSP5-8.5 (2050s)	9111.2 11.01	73,713.5 89.09	10,446.4 12.62	7137.0 8.63	56,130.1 67.84
SSP1-2.6 (2070s)	29,908.1 36.14	52,916.6 63.95	12,642.7 15.28	6660.7 8.05	33,613.3 40.62
SSP2-4.5 (2070s)	8058.2 9.74	74,766.6 90.36	15,597.5 18.85	7757.7 9.38	51,411.4 62.13
SSP5-8.5 (2070s)	86.5 0.10	82,738.2 99.99	1220.2 1.47	2564.9 3.10	78,953.2 95.42

SSPs: Shared Socioeconomic Pathways; SSP1-2.6: Sustainable Development Pathway; SSP2-4.5: Moderate Development Pathway; SSP5-8.5: Conventional Development Pathway Dominated by Fossil Fuels; The proportion of suitable habitat area under different future climate scenarios is the average for the periods 2021–2040, 2041–2060, and 2061–2080.

These data reveal that *Quercus mongolica's* habitat will change significantly under different future climate scenarios. Generally, its suitable habitat will expand, especially under extreme scenarios, potentially covering the entire study area. However, the majority of this expanded habitat will fall into the low suitability category, suggesting that the quality of its habitat could diminish although the species may extend its range.

3.4. *Differences in the Distribution of Current and Future Habitats*

3.4.1. *Larix gmelinii*

The results from the Maxent model showed a shift in the habitat distributions of *Larix gmelinii* over time (Figure 9 and Table 5). The largest no-change area regarding suitability is 75,766.29 km^2 by the 2050s under the SSP1-2.6 scenario. In contrast, the SSP5-8.5 scenario predicts the most extensive contraction of suitable areas (23,202.01 km^2). The ability of the species to adapt will be more complex by the 2070s. The large, stable area of 79,760.56 km^2 expands by 1686.04 km^2 under SSP1-2.6, indicating resilience. In contrast, the suitable area sharply declines to 43,317.24 km^2 under the more extreme SSP5-8.5 scenario, nearly doubling the reduction observed in the 2050s and suggesting survival challenges. Interestingly, the stability of the unsuitable area at 687.06 km^2 under the 70s SSP2-4.5 scenario marks a significant shift from its decrease in the 2050s, indicating evolving habitat dynamics for *Larix gmelinii* over time, even for the same climate scenario.

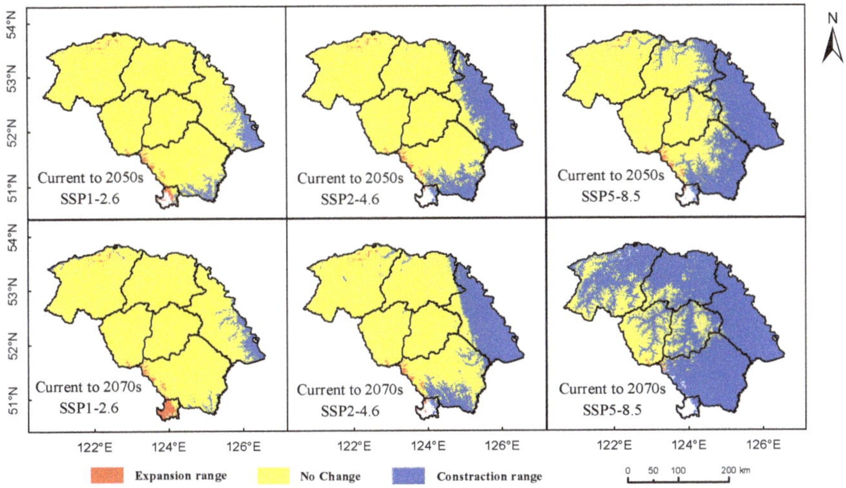

Figure 9. Future distribution of *Larix gmelinii* as a reaction to future climate change scenarios during 2050s and 2070s. SSPs: Shared Socioeconomic Pathways; SSP1-2.6: Sustainable Development Pathway; SSP2-4.5: Moderate Development Pathway; SSP5-8.5: Conventional Development Pathway Dominated by Fossil Fuels.

Table 5. Changes in the distribution of *Larix gmelinii's* habitats comparing current to future climate change projections.

Current to SSPs (Year)	Expansion Range		Unsuitable		No Change		Contraction Range	
	Area (km^2)	%	Area (km^2)	%	Area (km^2)	%	Area (km^2)	%
SSP1-2.6 (2050s)	901.7	1.09	800.2	0.97	75,766.3	91.57	5338.4	6.45
SSP2-4.5 (2050s)	901.7	1.09	800.2	0.97	75,766.3	91.57	5338.4	6.45
SSP5-8.5 (2050s)	908.3	1.10	793.5	0.96	57,902.7	69.98	23,202.0	28.0
SSP1-2.6 (2070s)	1686.0	2.04	15.8	0.02	79,760.6	96.39	1344.2	1.62
SSP2-4.5 (2070s)	1014.8	1.23	687.1	0.83	67,194.7	81.21	13,910.1	16.81
SSP5-8.5 (2090s)	306.1	0.37	1395.7	1.69	37,787.5	45.67	43,317.2	52.35

SSPs: Shared Socioeconomic Pathways; SSP1-2.6: Sustainable Development Pathway; SSP2-4.5: Moderate Development Pathway; SSP5-8.5: Conventional Development Pathway Dominated by Fossil Fuels.

Overall, these findings indicate that *Larix gmelinii's* suitable habitat in the Greater Khingan Range is generally stable under future climate scenarios, but the species faces varying degrees of change and contraction risks.

3.4.2. *Quercus mongolica*

The evolving patterns of *Quercus mongolica's* habitat distributions exhibits notable characteristics (Figure 10 and Table 6). First, a stable suitable area of 16,916.15888 km^2 is consistent in all scenarios, indicating habitat stability. However, the degree of expansion differs. The areas of expansion are 47,941.181 km^2 and 56,754.02 km^2 in the 2050s under SSP2-4.5 and SSP5-8.5, respectively, surpassing the 37,933.05 m^2 under SSP1-2.6. A similar trend of larger expansions under SSP2-4.5 and SSP5-8.5 is observed in the 2070s. Second, the stable unsuitable areas show significant differences between the scenarios, especially under the 2070s SSP5-8.5 scenario, where it reduces to just 89.00 km^2. This finding suggests that *Quercus mongolica's* unsuitable areas might contract more under certain scenarios.

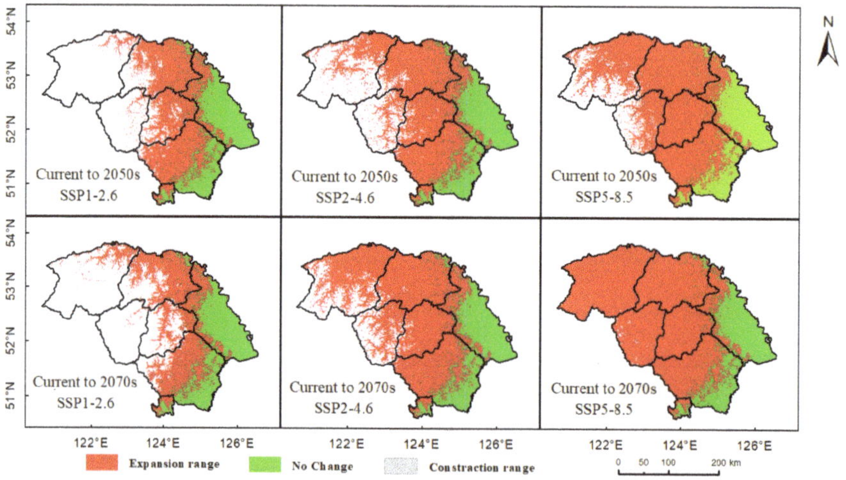

Figure 10. Changes in the distribution of *Quercus mongolica's* habitats comparing current to future climate change projections during 2050s and 2070s. SSPs: Shared Socioeconomic Pathways; SSP1-2.6: Sustainable Development Pathway; SSP2-4.5: Moderate Development Pathway; SSP5-8.5: Conventional Development Pathway Dominated by Fossil Fuels.

In summary, *Quercus mongolica's* future suitable habitat in the Greater Khingan Range is somewhat stable. However, the expansion and stability of the suitable and unsuitable areas differ markedly for different climate scenarios.

Table 6. Changes in the distribution of *Quercus mongolica's* habitats comparing current to future climate change projections.

Current to SSPs (year)	Expansion Range		Unsuitable		No Change		Contraction Range	
	Area (km^2)	%	Area (km^2)	%	Area (km^2)	%	Area (km^2)	%
SSP1-2.6 (2050s)	37,933.1	45.84	27,895.0	33.71	16,916.2	20.44	0.0	0.00
SSP2-4.5 (2050s)	47,941.2	57.94	17,886.9	21.62	16,916.2	20.44	0.0	0.00
SSP5-8.5 (2050s)	56,754.0	68.59	90,74.0	10.97	16,916.2	20.44	0.0	0.0
SSP1-2.6 (2070s)	35,924.3	43.42	29,903.8	36.14	16,916.2	20.44	0.0	0.00
SSP2-4.5 (2070s)	57,833.7	69.89	7994.4	9.66	16,916.2	20.44	0.0	0.00
SSP5-8.5 (2090s)	65,739.0	79.45	89.0	0.11	16,916.2	20.44	0.0	0.00

SSPs: Shared Socioeconomic Pathways; SSP1-2.6: Sustainable Development Pathway; SSP2-4.5: Moderate Development Pathway; SSP5-8.5: Conventional Development Pathway Dominated by Fossil Fuels.

3.5. *Scope and Intensity of Distribution Shifts*

3.5.1. *Larix gmelinii*

The current geographic center for *Larix gmelinii* is located at 52.03° N, 124.33° E in Xinlin's northwest (Figure 11 and Table 7). It will migrate to the northwest by the 2050s under different climate scenarios. Specifically, it migrates 8.45 km to 52.08° N and 124.24° E under SSP1-2.6. The shifts are more pronounced for SSP2-4.5 and SSP5-8.5 (30.3 km to 52.18° N, 123.97° E and 51.9 km to 52.27° N, 123.97° E, respectively).

Table 7. Predicted migration distances for *Larix gmelinii* and *Quercus mongolica* under different SSP scenarios.

Species	Initial Location (Lat, Long)	Scenario	Shifted Location (Lat, Long)	Migration Distance (km)
Larix gmelinii	52.03° N, 124.33° E	SSP1-2.6 (2050s)	52.08° N, 124.24° E	8.45
		SSP1-2.6 (2070s)	52.04° N, 124.28° E	5.67
		SSP2-4.5 (2050s)	52.18° N, 123.97° E	30.3
		SSP2-4.5 (2070s)	52.03° N, 124.32° E	1.19
		SSP5-8.5 (2050s)	52.27° N, 123.97° E	51.9
		SSP5-8.5 (2070s)	52.20° N, 123.35° E	25.3
Quercus mongolica	51.28° N, 125.83° E	SSP1-2.6 (2050s)	51.69° N, 125.18° E	64.2
		SSP1-2.6 (2070s)	51.73° N, 125.24° E	5.95
		SSP2-4.5 (2050s)	51.85° N, 124.87° E	92.24
		SSP2-4.5 (2070s)	51.95° N, 124.64° E	19.75
		SSP5-8.5 (2050s)	51.94° N, 124.64° E	110.6
		SSP5-8.5 (2070s)	52.01° N, 124.34° E	22.1

SSPs: Shared Socioeconomic Pathways; SSP1-2.6: Sustainable Development Pathway; SSP2-4.5: Moderate Development Pathway; SSP5-8.5: Conventional Development Pathway Dominated by Fossil Fuels.

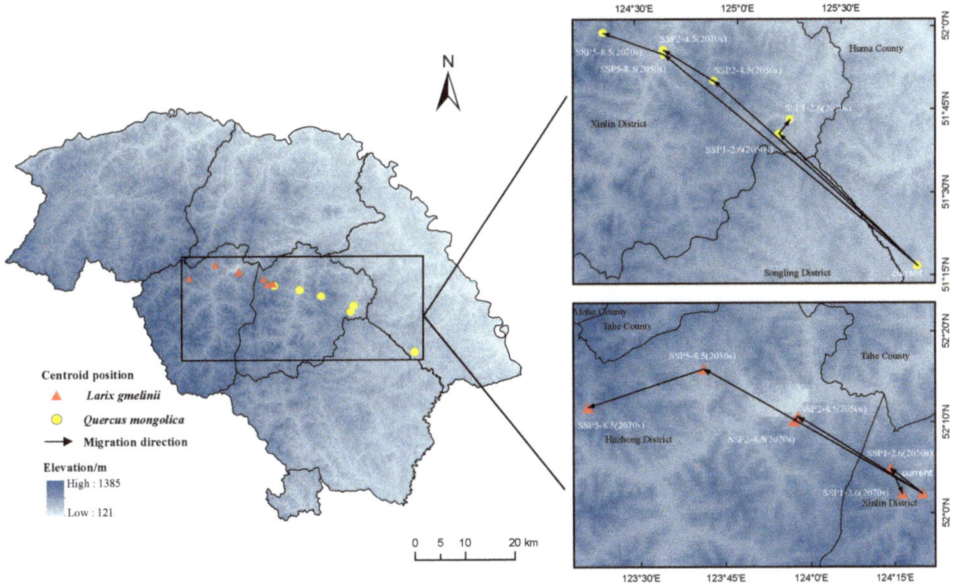

Figure 11. Location of the center of mass in the fitness zone of two tree species under different climate scenarios during 2050s and 2070s. SSPs: Shared Socioeconomic Pathways; SSP1-2.6: Sustainable Development Pathway; SSP2-4.5: Moderate Development Pathway; SSP5-8.5: Conventional Development Pathway Dominated by Fossil Fuels..

Larix gmelinii's centroid migrates southeast to 52.04° N, 124.28° E in the Xinlin District under SSP1-2.6, covering 5.67 km from the 2050s to the 2070s. It shifts a short distance southwest by 1.19 km under SSP2-4.5, indicating stability. In contrast, the shift is substantial under SSP5-8.5, and the centroid moves to 52.20° N, 123.35° E in the northwest Huzhong District, a distance of 25.3 km. These centroid movements reflect the nuanced climate impacts and adaptability challenges for *Larix gmelinii* in all scenarios.

3.5.2. *Quercus mongolica*

Currently, *Quercus mongolica's* centroid is southwest of Huma County at 51.28° N, 125.83° E (Figure 11). Under SSP126, it migrates to the western Xinlin District by the 2050s (51.69° N, 125.18° E; 64.2 km) and shifts northeast by the 2070s (51.73° N, 125.24° E; 5.95 km). Under SSP245, the centroid migrates northwest to the northern Xinlin District by the 2050s (51.85° N, 124.87° E; 92.24 km) and continues northwest in the 2070s (51.95° N, 124.64° E; 19.75 km). It relocates northwest to the north-central Xinlin District by the 2050s under SSP5-8.5 (51.94° N, 124.64° E; 110.60 km) and then to the western part by the 2070s (52.01° N, 124.34° E; 22.1 km).

This result reveals that *Quercus mongolica* predominantly migrates toward the northwest and to higher altitudes in different periods and climate scenarios, with the exception of a northeast shift in the 2070s under SSP1-2.6. The migration distance increases with higher greenhouse gas emissions over the same period, and the distance migrated in the 2070s is consistently shorter than that in the 2050s in each climate scenario.

4. Discussion

The niche model is based on the statistical correlation between species distribution and environmental variables and primarily focuses on significant environmental influences on species distribution but tends to overlook interspecies interactions, evolutionary processes, extreme disturbances, and species dissemination [72,73]. While the existing relationships between species and their environmental factors, particularly climatic conditions, may not consistently provide precise forecasts for the distribution of species in the future, niche modeling remains the predominant approach for analyzing the effects of climate change on species distributions within a region [19]. It is crucial to understand that these models predict potential suitable habitats for species, not their actual distribution. The interaction between species and their migration capabilities restricts the presence of specific species in certain areas, meaning that the actual distribution is only a part of the potential distribution [74]. Future research should more comprehensively consider factors such as vegetation competition and succession, botanical functional processes, and insect infestation impacts in order to develop and improve the integration techniques between regional climate models and dynamic ecosystem models.

4.1. *Potential Current and Future Habitat Distributions*

The Greater Khingan Range area is characterized by a typical cold temperate continental monsoon climate [75,76]. During winter, it is dominated by the Mongolian High, leading to temperature inversion and stable snow cover, which contribute to the region's significant permafrost heterogeneity [77]. The extreme low temperatures (reaching −52.3 °C), prolonged winter snow cover (approximately five months), and the presence of permafrost pose substantial challenges to plant growth [78,79]. *Larix gmelinii*, as a principal constituent tree species of the cold temperate coniferous forests, exhibits high wood density and possesses notable frost resistance, enabling it to maintain elevated levels of water transport capacity in cold environments [80]. Furthermore, its well-developed root system can penetrate the permafrost to access water and nutrients, concurrently enhancing soil permeability and stability. These characteristics empower it to thrive in harsh environments. In contrast, *Quercus mongolica*, a secondary forest species, predominantly emerges in environments following the disturbance of primary forests. Its adaptability to environmental conditions may be comparatively lower [81].

Under the majority of future climatic scenarios, the suitable habitat of *Larix gmelinii* is anticipated to contract, with a trend of the distribution centroid shifting northwestward. This aligns with the conclusions derived by Zhang et al. [82] using the Maxent model, which predict a significant reduction in the suitable habitat of *Larix gmelinii* forests in northeast China during the 2050s and 2070s. Shi et al. [83] further corroborate this perspective, noting that under the effects of climate change, the suitability of *Larix gmelinii* in northeast China will decline, particularly in the eastern foothills of the Greater Khingan Range, the Lesser Khingan Range, and the Changbai Mountains. In these areas, the decrease in suitable habitat area is especially pronounced, with potential distribution areas expected to move northward and contract towards central regions. While many studies [82–85] have shown variability in the specific figures regarding the rate of habitat loss for *Larix gmelinii*, they collectively highlight a significant trend: under the overarching context of global warming and increased precipitation, the distribution area of *Larix gmelinii* is expected to contract, indicating its sensitivity and vulnerability to future climate change. The variations in these figures likely arise from differences in study areas, environmental factors, models, and parameter settings, which in turn limit the comparability of the results [86].

Conversely, *Quercus mongolica* shows considerable adaptability to future climatic changes, with all predictive scenarios indicating an expansion in its suitable habitat area. These findings are consistent with previous studies on climate change impacts in northeastern forests. For instance, Chen et al. [87] employed the ECHAM5-OM and HadCM3 atmospheric circulation models for simulations and revealed that climate warming over the next century (2002–2102) will favor birches and *Quercus mongolica* in northeastern forest communities. Similarly, Yan et al. [88] conducted simulations using the NEWCOP model under GFDL climate change scenarios for forests in the Greater Khingan Range and observed a gradual increase in the proportion of *Quercus mongolica*. Other studies, such as those referenced [89–91], consistently report that regional warming benefits the growth of temperate broadleaf forests.The expansion of *Quercus mongolica* forests is primarily attributed to its significant physiological traits. Firstly, the stability of this species is underpinned by its remarkable adaptability, resistance, and resilience, leading to a longer lifespan for *Quercus mongolica* and its ability to dominate habitats once it establishes a community [40,92,93]. Secondly, *Quercus mongolica* is among the most drought-tolerant species owing to its well-developed root system. Additionally, *Quercus mongolica* possesses drought-resistant leaf structures with strong stomatal closure capabilities, low water potential, and flexible regulatory mechanisms [94,95]. Among broadleaf trees, it also exhibits high stem water flow—attributable to its unique leaf morphology—high branching angle of branches, and a higher leaf area index, enabling it to effectively absorb more water and nutrients [96]. These physiological structures enable *Quercus mongolica* to maintain a high rate of photosynthesis even under future warming conditions. Compared to other species, it shows greater survivability in poor soils [95,97]. Moreover, projections indicate that the future distribution centroid of *Quercus mongolica* is expected to shift predominantly towards the northwest, a continuation of the migration trend observed in past research, which is highlighted by a study showing that from 1896 to 1986, the northern boundary of the broadleaf forests in Heilongjiang Province extended approximately 290 km towards the northwest [98].

4.2. Environmental Variables Affecting the Distribution of Larix gmelinii and Quercus mongolica

Larix gmelinii, owing to its preference for sunlight, cold tolerance, and high adaptability to soil and moisture [99], is significantly more influenced by temperature (contributing 62.2%) than by precipitation (contributing 18.3%). This finding is corroborated by previous research; for instance, Shi [83] discovered through probability of presence analysis that spring maximum temperature (contributing over 50%) and annual mean temperature are key factors influencing its distribution rather than spring precipitation. Similarly, Yang et al. [84] identified that the primary climatic factors affecting its geographic distribution are the mean temperature of the coldest month (contributing over 50%), annual temperature range, annual radiation, and accumulated temperature of ≥5 °C. Although different studies vary in their data sources and methodologies, and the main climatic factors differ, a consistent conclusion is the predominance of temperature over precipitation.

Annual temperature, elevation, and precipitation are crucial factors affecting *Quercus mongolica's* distribution, with temperature being the most significant at 64.7%. This finding is consistent with studies identifying similar influencing factors. For example, Yin et al. [100] found that annual precipitation (330–910 mm) and annual temperature differences greater than 29 °C were dominant influencing factors, and Jia et al. [101] emphasized the impact of rainfall and altitude. Zhou et al. [102] focused on the influence of low temperatures in May on natural regeneration. Despite differences in factor contributions, all factors were related to temperature, precipitation, and altitude. Our study's logistical response curves show a distinct threshold for annual precipitation for *Quercus mongolica's* suitable areas and its adaptability to lower temperatures, aligning with its drought resistance and cold tolerance. The elevation range of *Quercus mongolica* identified in this study aligns with the data presented in *Flora of China* (Volume 22, p. 236) [79], noting that *Quercus mongolica* commonly grows at elevations below 600 m in northeast China. Complementary literature [103] indicates that this species predominantly occupies an elevation range of 250 to 400 m in the Greater and Lesser Khingan Ranges and is typically found on low mountain summits, ridges, and variously inclined slopes, which supports these findings.

4.3. *Implications for Conservation*

Larix gmelinii and *Quercus mongolica*, key species in the Greater Khingan Range, face various challenges and opportunities due to climate change. Therefore, we provide the following conservation recommendations:

Larix gmelinii has high sensitivity and vulnerability to climate shifts; thus, it is crucial to enhance monitoring and research. This involves mitigating climate impacts, improving pest control, and carefully planning human activities to protect its population and ecological services. For example, the projected 54.06% habitat reduction by the 2070s under the SSP5-8.5 scenario suggests the need for stronger conservation efforts, such as expanding reserves, promoting tree planting, and controlling pests.

Although *Quercus mongolica* is adaptable, it will face challenges, particularly as future suitable areas are predicted to be of lower quality. Conservation planning should account for future climate impacts and include strategies to improve current and future suitable habitats. The projected 20.35% habitat increase by the 2070s under SSP5-8.5 is promising, but the potential impact on other species and its adaptability to less suitable areas must be considered. Thus, conservation efforts should aim for a balance: ensuring species diversity and the species' ability to thrive in varied conditions.

5. Conclusions

We used the Maxent model to predict the current and future habitat distribution of *Larix gmelinii* and *Quercus mongolica* in the Greater Khingan Range under varying climate scenarios. Significant differences were observed in the species' responses to climate change. *Larix gmelinii* is projected to experience a reduction in its suitable habitat and a likely shift towards the northwest in future scenarios. In contrast, *Quercus mongolica* is expected to expand its habitat in all scenarios, mostly shifting northwest, demonstrating stronger adaptability.

Our findings provide a baseline for further research into the region's forest ecosystems and biodiversity conservation. The combination of GIS methods and Maxent modeling proved effective for developing conservation strategies. Future research should broaden niche modeling to include more tree species and use diverse climate models and environmental factors to enhance the distribution models to deepen our understanding of the Greater Khingan Range's forest response to climate change.

Author Contributions: Conceptualization, B.D. and D.Z.; methodology, B.D.; software, B.D.; validation, D.Z.; formal analysis, B.D.; investigation, X.L.; resources, B.D., X.L., X.W., X.Z. and X.W.; data curation, B.D. and X.L.; writing—original draft preparation, B.D.; writing—review and editing, B.D., D.Z. and Z.W.; project administration, B.D. All authors have read and agreed to the published version of the manuscript.

Funding: This research was funded by the National Natural Science Foundation of China (No. 41671064) and the Natural Science Foundation of Heilongjiang Province (No. LH2021D012).

Data Availability Statement: The data that support the findings of this study are not openly available but are available from the corresponding author upon reasonable request.

Acknowledgments: We acknowledge with great appreciation the technical support provided by Xi'an Keshuo Co., Ltd., Xi'an, China.

Conflicts of Interest: The authors declare no conflicts of interest.

Appendix A

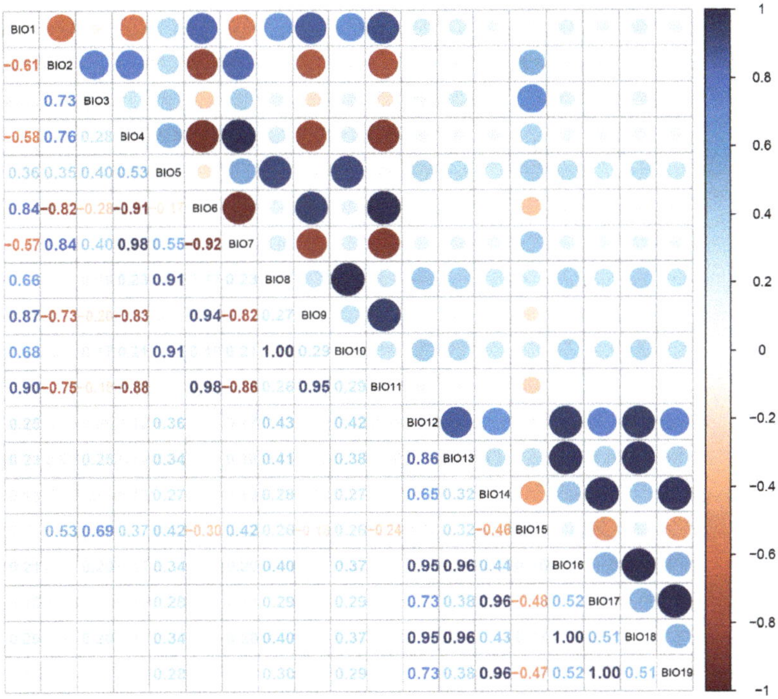

Figure A1. Assessing interrelations of 19 bioclimatic variables: Pearson's (r) correlation analysis via ENM tools: $|r| \geqslant 0.8$ is displayed in bold.

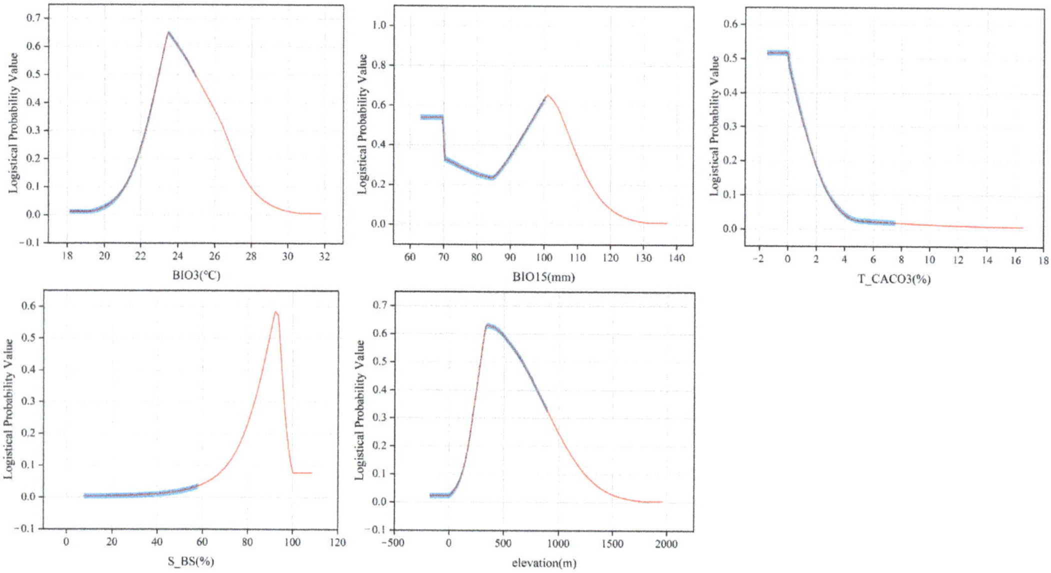

Figure A2. Response curves characterizing how variables affected the Maxent predictions for *Larix gmelinii*. Red lines represent logistical predictions varying with each climatic variable while holding other variables at their means. Shaded regions denote one standard deviation from this mean.

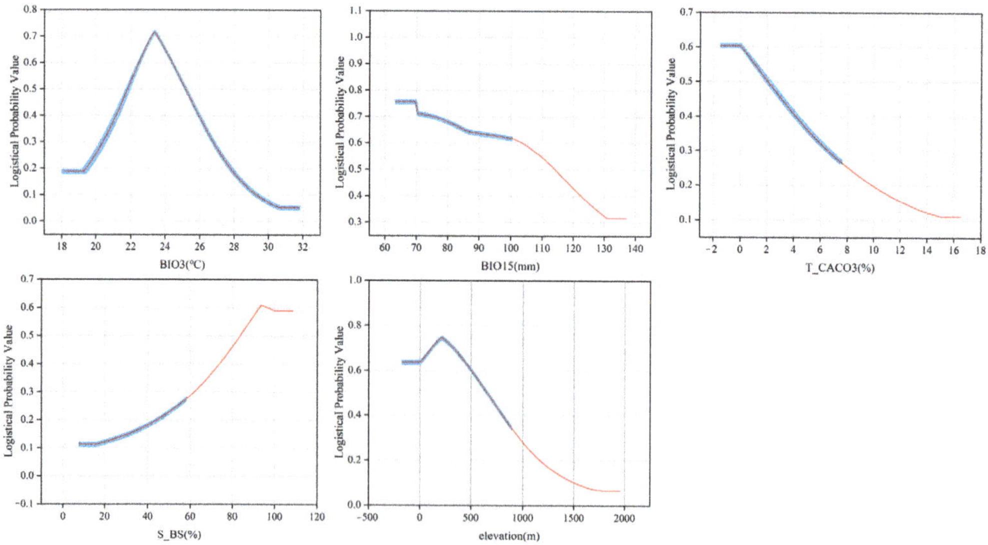

Figure A3. Response curves characterizing how variables affected the Maxent predictions for *Larix gmelinii*. Based on individual variables, the Maxent models generate red lines. Shaded regions denote one standard deviation from this mean.

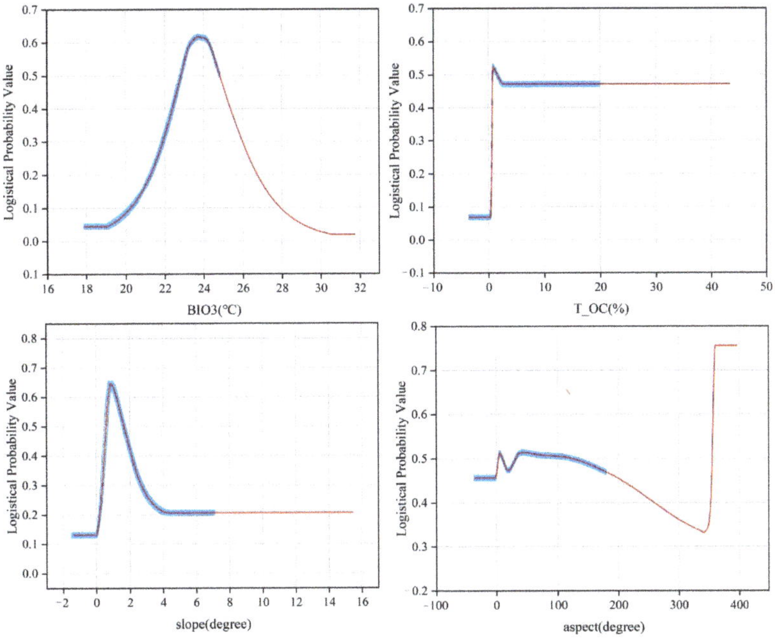

Figure A4. Response curves characterizing how variables affected the Maxent predictions for *Quercus mongolica*. Red lines represent logistical predictions varying with each climatic variable while holding other variables at their means. Shaded regions denote one standard deviation from this mean.

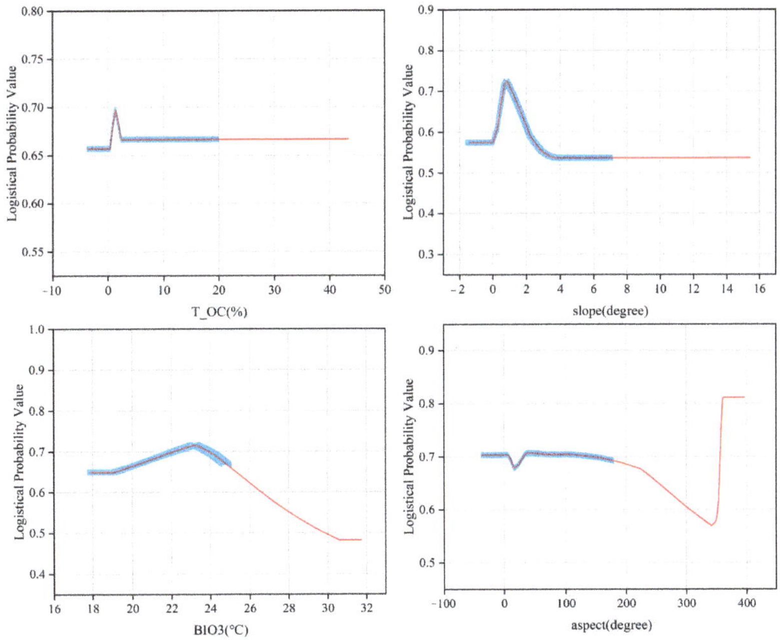

Figure A5. Response curves characterizing how variables affected the Maxent predictions for *Quercus mongolica*. Based on individual variables, the Maxent models generate red lines. Shaded regions denote one standard deviation from this mean.

Table A1. WorldClim version 2.1: standard 19 bioclimatic variables. The monthly values are averages over 20-year periods: 2021–2040, 2041–2060 and 2061–2080.

Variables	Description	Units
BIO1	Annual Mean Temperature	°C
BIO2	Mean Diurnal Range (mean of monthly (max temp − min temp))	°C
BIO3	Isothermality (BIO2/BIO7) ×100	/
BIO4	Temperature Seasonality (standard deviation ×100)	/
BIO5	Max Temperature of Warmest Month	°C
BIO6	Min Temperature for Coldest month	°C
BIO7	Temperature Annual Range (BIO5−BIO6)	°C
BIO8	Mean Temperature of Wettest Quarter	°C
BIO9	Mean Temperature of Driest Quarter	°C
BIO10	Mean Temperature of Warmest Quarter	°C
BIO11	Mean Temperature of Coldest Quarter	°C
BIO12	Annual Precipitation	mm
BIO13	Precipitation of Wettest Month	mm
BIO14	Precipitation of Driest Month	mm
BIO15	Precipitation Seasonality (Coefficient of Variation)	/
BIO16	Precipitation of Wettest Quarter	mm
BIO17	Precipitation of Driest Quarter	mm
BIO18	Precipitation of Warmest Quarter	mm
BIO19	Precipitation of Coldest Quarter	mm

References

1. Hui, D.; Deng, Q.; Tian, H.; Luo, Y. Climate Change and Carbon Sequestration in Forest Ecosystems. *Handb. Clim. Chang. Mitig. Adapt.* **2017**, *555*, 555–594. [CrossRef]
2. Ninan, K.N.; Inoue, M. Valuing forest ecosystem services: What we know and what we don't. *Ecol. Econ.* **2013**, *93*, 137–149. [CrossRef]
3. Yu, P.; Richard, A.B.; Oliver, L.P.; Robert, B.J. The Structure, Distribution, and biomass of the World's Forests. *Annu. Rev. Ecol. Evol. Syst.* **2013**, *44*, 593–622. [CrossRef]
4. IPCC. Summary for Policymakers. In *Climate Change 2021—The Physical Science Basis: Working Group I Contribution to the Sixth Assessment Report of the Intergovernmental Panel on Climate Change*; Intergovernmental Panel on Climate Change (IPCC), Ed.; Cambridge University Press: Cambridge, UK, 2023; pp. 3–32.
5. Prăvălie, R. Major perturbations in the Earth's forest ecosystems. Possible implications for global warming. *Earth-Sci. Rev.* **2018**, *185*, 544–571. [CrossRef]
6. Martinez del Castillo, E.; Zang, C.S.; Buras, A.; Hacket-Pain, A.; Esper, J.; Serrano-Notivoli, R.; Hartl, C.; Weigel, R.; Klesse, S.; Resco de Dios, V.; et al. Climate-change-driven growth decline of European beech forests. *Commun. Biol.* **2022**, *5*, 163. [CrossRef]
7. Mishra, S.N.; Gupta, H.S.; Kulkarni, N. Impact of climate change on the distribution of Sal species. *Ecol. Inform.* **2021**, *61*, 101244. [CrossRef]
8. Paritsis, J.; Veblen, T.T. Dendroecological analysis of defoliator outbreaks on *Nothofagus pumilio* and their relation to climate variability in the Patagonian Andes. *Glob. Chang. Biol.* **2011**, *17*, 239–253. [CrossRef]
9. Girardin, M.P.; Hogg, E.H.; Bernier, P.Y.; Kurz, W.A.; Guo, X.J.; Cyr, G. Negative impacts of high temperatures on growth of black spruce forests intensify with the anticipated climate warming. *Glob. Chang. Biol.* **2016**, *22*, 627–643. [CrossRef] [PubMed]
10. Wang, J.; Taylor, A.R.; D'Orangeville, L. Warming-induced tree growth may help offset increasing disturbance across the Canadian boreal forest. *Proc. Natl. Acad. Sci. USA* **2023**, *120*, e2212780120. [CrossRef]
11. Montoya-Jimenez, J.C.; Valdez-Lazalde, J.R.; Angeles-Perez, G.; de los Santos-posadas, H.M.; Cruz-Cardenas, G. Predictive capacity of nine algorithms and an ensemble model to determine the geographic distribution of tree species. *iForest-Biogeosci. For.* **2022**, *15*, 363–371. [CrossRef]
12. Hebbar, K.B.; Abhin, P.S.; Sanjo Jose, V.; Neethu, P.; Santhosh, A.; Shil, S.; Prasad, P.V.V. Predicting the Potential Suitable Climate for Coconut (*Cocos nucifera* L.) Cultivation in India under Climate Change Scenarios Using the Maxent Model. *Plants* **2022**, *11*, 731. [CrossRef] [PubMed]
13. Yousaf, A.; Hadi, R.; Khan, N.; Ibrahim, F.; Moin, H.; Rahim, S.; Hussain, M. Identification of suitable habitat for Taxus wallichiana and Abies pindrow in moist temperate forest using Maxent modelling technique. *Saudi J. Biol. Sci.* **2022**, *29*, 103459. [CrossRef] [PubMed]
14. Phillips, S.J.; Dudík, M. Modeling of species distributions with Maxent: New extensions and a comprehensive evaluation. *Ecography* **2008**, *31*, 161–175. [CrossRef]
15. Elith, J.; Kearney, M.; Phillips, S. The art of modelling range-shifting species. *Methods Ecol. Evol.* **2010**, *1*, 330–342. [CrossRef]

16. Chen, B.; Zou, H.; Meng, X.; Wang, L.; Hao, X.; Kang, X.; Wang, C.; Zhang, X. Distribution Patterns and Change Prediction of Suitable Habitats for Chaihu (*Radix Bupleuri*) and Narrow-Leaved Chaihu (*Radix Bupleuri Chinensis*) in China under Climate Change. *Acta Ecol. Sin.* **2022**, *20*, 1–13.
17. Warren, D.L.; Wright, A.N.; Seifert, S.N.; Shaffer, H.B. Incorporating model complexity and spatial sampling bias into ecological niche models of climate change risks faced by 90 California vertebrate species of concern. *Divers. Distrib.* **2014**, *20*, 334–343. [CrossRef]
18. Schymanski, S.J.; Dormann, C.F.; Cabral, J.; Chuine, I.; Graham, C.H.; Hartig, F.; Kearney, M.; Morin, X.; Römermann, C.; Schröder, B.; et al. Process, correlation and parameter fitting in species distribution models: A response to Kriticos et al. *J. Biogeogr.* **2013**, *40*, 612–613. [CrossRef]
19. Li, Y.; Li, M.; Li, C.; Liu, Z. Optimized Maxent Model Predictions of Climate Change Impacts on the Suitable Distribution of *Cunninghamia lanceolata* in China. *Forests* **2020**, *11*, 302. [CrossRef]
20. Dakhil, M.A.; Xiong, Q.; Farahat, E.A.; Zhang, L.; Pan, K.; Pey, B.; Olatunji, O.A.; Tariq, A.; Wu, X.; Zhang, A.; et al. Past and future climatic indicators for distribution patterns and conservation planning of temperate coniferous forests in southwestern China. *Ecol. Indic.* **2019**, *107*, 105559. [CrossRef]
21. Fick, S.E.; Hijmans, R.J. WorldClim 2: New 1-km Spatial Resolution Climate Surfaces for Global Land Areas. *Int. J. Climatol.* **2017**, *37*, 4302–4315. [CrossRef]
22. Harris, R.M.B.; Grose, M.R.; Lee, G.; Bindoff, N.L.; Porfirio, L.L.; Fox-Hughes, P. Climate Projections for Ecologists. *Wiley Interdiscip. Rev. Clim. Chang.* **2014**, *5*, 621–637. [CrossRef]
23. Moss, R.H.; Edmonds, J.A.; Hibbard, K.A.; Manning, M.R.; Rose, S.K.; van Vuuren, D.P.; Carter, T.R.; Emori, S.; Kainuma, M.; Kram, T.; et al. The Next Generation of Scenarios for Climate Change Research and Assessment. *Nature* **2010**, *463*, 747–756. [CrossRef] [PubMed]
24. Fan, X.W.; Miao, C.Y.; Duan, Q.Y.; Shen, C.W.; Wu, Y. The Performance of CMIP6 Versus CMIP5 in Simulating Temperature Extremes over the Global Land Surface. *J. Geophys. Res. Atmos.* **2020**, *125*, e2020JD033031. [CrossRef]
25. Du, Q.; Wei, C.; Liang, C.; Yu, J.; Wang, H.; Wang, W. Maxent Model Analysis of the Response of 12 Pioneer Tree Species to Climate Change in Northeast China. *Acta Ecol. Sin.* **2022**, *42*, 9712–9725. [CrossRef]
26. Han, J.; Zhao, H.; Zhu, L.; Zhang, Y.; Li, Z.; Wang, X. A Comparative Study on the Response of Radial Growth of *Quercus mongolica* and *Pinus koraiensis* to Climate Change in the Xiaoxing'an Mountains. *Chin. J. Appl. Ecol.* **2019**, *30*, 2218–2230. [CrossRef]
27. Li, W.; Jiang, Y.; Dong, M.; Du, E.; Wu, F.; Zhao, S.; Xu, H. Species-specific growth-climate responses of Dahurian larch (*Larix gmelinii*) and Mongolian pine (*Pinus sylvestris* var. mongolica) in the Greater Khingan Range, northeast China. *Dendrochronologia* **2021**, *65*, 125803. [CrossRef]
28. Lyu, S.; Wang, X.; Zhang, Y.; Li, Z. Different responses of Korean pine (*Pinus koraiensis*) and Mongolia oak (*Quercus mongolica*) growth to recent climate warming in northeast China. *Dendrochronologia* **2017**, *45*, 113–122. [CrossRef]
29. Wang, C.; Gower, S.T.; Wang, Y.; Zhao, H.; Yan, P.; Bond-Lamberty, B.P. The Influence of Fire on Carbon Distribution and Net Primary Production of Boreal *Larix gmelinii* Forests in North-Eastern China. *Glob. Chang. Biol.* **2001**, *7*, 719–730. [CrossRef]
30. Huang, C.; Liang, Y.; He, H.S.; Wu, M.M.; Liu, B.; Ma, T. Sensitivity of Aboveground Biomass and Species Composition to Climate Change in Boreal Forests of Northeastern China. *Ecol. Model.* **2021**, *445*, 109472. [CrossRef]
31. Zhang, K.; Li, T.; Qu, Y.; Gao, F.; Lin, J.; Jiang, L. Analysis of Precipitation and Temperature Trends in the Great Xing'an Mountains Region. *Forests* **2018**, *34*, 8–14. [CrossRef]
32. Zhou, L.; Li, W.; Wang, W.; Zhang, X.; Li, W.; Li, X.; Li, Y.; Wang, Z.; Zhang, G.; Zhang, J. *Vegetation of the Daxinganling Region in China*; Science Press: Beijing, China, 1991.
33. Dong, L.; Liu, Z.; Li, F.; Jiang, L. Spatial Structure and Optimal Tree Species Composition of Major Forest Types in the Daxinganling Mountains. *For. Sci. Res.* **2014**, *27*, 6. [CrossRef]
34. Bai, X. Analysis of Point Pattern Distribution of Several Major Plant Species in Typical Forest Areas of the Daxinganling Mountains. *China High Technol. Enterp.* **2012**, 106 –107. [CrossRef]
35. Zhou, Y.; Dong, S.; Nie, S. *Heilongjiang Tree Record*; Heilongjiang Science Press: Harbin, China, 1986.
36. Xu, H. *Forests of the Greater Khingan Mountains, China*; Science Press: Beijing, China, 1998.
37. Zhou, Y.; Yu, Z.; Zhao, S. Carbon Storage and Carbon Balance in Major Forest Ecosystems in China. *Chin. J. Plant Ecol.* **2000**, *24*, 518–522.
38. Gao, J.; Ao, W.; Liu, D.; Xiang, L.; Li, J.; Cheng, X. Origin and Biological Characteristics of Xing'an Larch in the Greater Khingan Mountains. *Inner Mong. Sci. Technol. Econ.* **2003**, *10*, 99–100.
39. Xu, Z.-Q.; Li, W.-H.; Liu, W.-Z.; Wu, X.-B. Study on the Biomass and Productivity of Mongolian Oak Forests in Northeast Region of China. *Chin. J. Eco-Agric.* **2006**, *14*, 21–24.
40. Wei, X.H. Comprehensive Study on Mongolian Oak Ecosystem. Ph.D. Thesis, Northeast Forestry University, Harbin, China, 1989.
41. Xu, Z.B. Study on the Root Distribution Patterns of Main Tree Species in the Broadleaf Korean Pine Forests of Changbai Mountain. *J. Ecol.* **1992**, *41*, 1924.
42. Wen, S.Y.; Yang, S.H.; Yin, Z.F.; Gao, H.Y.; Li, N.N.; Li, J.H.; Xia, J.; Luo, H.H.; Zhang, Q.X.; Bai, J.Y. Study on Canopy Structure, Light Distribution, and Leaf Growth in Oak Silkworm Forests. *J. Appl. Ecol.* **1991**, *2*, 286–291.
43. Wen, Z.D.; Yang, S.H.; Jiang, B.; Gao, H.Y.; Li, N.N.; Li, J.H.; Xia, J.; Luo, H.H.; Zhang, Q.X.; Bai, J.Y. Study on Biological Productivity and Substance Transformation of Oak Silkworm Forests. *J. Ecol.* **1993**, *12*, 5–10.

44. Chen, D.K.; Zhou, X.F. Structure, Function, and Succession of Four Types of Natural Secondary Forests. *J. Northeast For. Univ.* **1982**, *10*, 1–20.
45. Veloz, S.D. Spatially autocorrelated sampling falsely inflates measures of accuracy for presence-only niche models. *J. Biogeogr.* **2009**, *36*, 2290–2299. [CrossRef]
46. Brown, J.L. SDMtoolbox: A python-based GIS toolkit for landscape genetic, Biogeographic and species distribution model analyses. *Methods Ecol. Evol.* **2014**, *5*, 694–700. [CrossRef]
47. Tebaldi, C.; Debeire, K.; Eyring, V.; Fischer, E.; Fyfe, J.; Friedlingstein, P.; Knutti, R.; Lowe, J.; O'Neill, B.; Sanderson, B.; et al. Climate model projections from the Scenario Model Intercomparison Project (ScenarioMIP) of CMIP6. *Earth Syst. Dyn.* **2020**, *12*, 253–293. [CrossRef]
48. Fan, X.; Miao, C.; Duan, Q.; Shen, C.; Wu, Y. Future Climate Change Hotspots Under Different 21st Century Warming Scenarios. *Earths Future* **2021**, *9*, e2021EF002027. [CrossRef]
49. Xiao-Ge, X.; Tong-Wen, W.; Jiang-Long, L.; Zai-Zhi, W.; Wei-Ping, L.; Fang-Hua, W. How Well does BCC_CSM1.1 Reproduce the 20th Century Climate Change over China? *Atmos. Ocean. Sci. Lett.* **2013**, *6*, 21–26. [CrossRef]
50. Wu, T.; Li, W.; Ji, J.; Xin, X.; Li, L.; Wang, Z.; Zhang, Y.; Li, J.; Zhang, F.; Wei, M.; et al. Global carbon budgets simulated by the Beijing Climate Center Climate System Model for the last century. *J. Geophys. Res. Atmos.* **2013**, *118*, 4326–4347. [CrossRef]
51. Warren, D.L.; Seifert, S.N. Ecological Niche Modeling in MaxEnt: The Importance of Model Complexity and the Performance of Model Selection Criteria. *Ecol. Appl.* **2011**, *21*, 335–342. [CrossRef] [PubMed]
52. Jiménez-Valverde, A.; Lobo, J.M.; Hortal, J. Not as Good as They Seem: The Importance of Concepts in Species Distribution Modeling. *Divers. Distrib.* **2008**, *14*, 885–890. [CrossRef]
53. Radosavljevic, A.; Anderson, R.P. Making better MAXENT models of species distributions: Complexity, overfitting and evaluation. *J. Biogeogr.* **2014**, *41*, 629–643. [CrossRef]
54. Cobos, M.E.; Peterson, A.T.; Barve, N.; Osorio-Olvera, L. Kuenm: An R package for detailed development of ecological niche models using Maxent. *PeerJ* **2019**, *7*, e6281. [CrossRef]
55. Wintle, B.A.; McCarthy, M.A.; Parris, K.M.; Bugman, M.A. Precision and Bias of Methods for Estimating Point Survey Detection Probabilities. *Ecol. Appl.* **2004**, *14*, 703–712. [CrossRef]
56. Elith, J.; Phillips, S.J.; Hastie, T.; Dudík, M.; Chee, Y.E.; Yates, C.J. A Statistical Explanation of MaxEnt for Ecologists. *Divers. Distrib.* **2011**, *17*, 43–57. [CrossRef]
57. Zhao, H.; Xian, X.; Zhao, Z.; Zhang, G.; Liu, W.; Wan, F. Climate Change Increases the Expansion Risk of *Helicoverpa zea* in China According to Potential Geographical Distribution Estimation. *Insects* **2022**, *13*, 79. [CrossRef] [PubMed]
58. Hanley, J.A.; McNeil, B.J. The Meaning and Use of the Area Under a Receiver Operating Characteristic (ROC) Curve. *Radiology* **1982**, *143*, 29–36. [CrossRef]
59. Fielding, A.H.; Bell, J.F. A review of methods for the assessment of prediction errors in conservation presence/absence models. *Environ. Conserv.* **1997**, *24*, 38–49. [CrossRef]
60. Swets, J.A. Measuring the Accuracy of Diagnostic Systems. *Science* **1988**, *240*, 1285–1293. [CrossRef] [PubMed]
61. Li, L.H.; Liu, H.Y.; Lin, Z.S.; Jia, J.H.; Liu, X. Identifying Priority Areas for Monitoring the Invasion of *Solidago canadensis* Based on MAXENT and ZONATION. *Sci. Technol. Rev.* **2017**, *37*, 3124–3132. [CrossRef]
62. Warren, D.L.; Matzke, N.J.; Cardillo, M.; Baumgartner, J.B.; Beaumont, L.J.; Turelli, M.; Glor, R.E.; Huron, N.A.; Simøes, M.; Iglesias, R.E.; et al. ENMTools 1.0: An R package for comparative ecological biogeography. *Ecography* **2021**, *44*, 504–511. [CrossRef]
63. Warren, D.L.; Glor, R.E.; Turelli, M. ENMTools: A toolbox for comparative studies of environmental niche models. *Ecography* **2010**. [CrossRef]
64. Steele, K.; Werndl, C. Climate Models, Calibration, and Confirmation. *Br. J. Philos. Sci.* **2013**, *64*, 609–635. [CrossRef]
65. Jiménez-Valverde, A.; Lobo, J.M. Threshold criteria for conversion of probability of species presence to either-or presence-absence. *Acta Oecol.* **2007**, *31*, 361–369. [CrossRef]
66. Xue, Y.; Lin, C.; Wang, Y.; Liu, W.; Wan, F.; Zhang, Y.; Ji, L. Predicting Climate Change Effects on the Potential Distribution of Two Invasive Cryptic Species of the *Bemisia tabaci* Species Complex in China. *Insects* **2022**, *13*, 1081. [CrossRef]
67. Liu, C.; Berry, P.M.; Dawson, T.P.; Pearson, R.G. Selecting Thresholds of Occurrence in the Prediction of Species Distributions. *Ecography* **2005**, *28*, 385–393. [CrossRef]
68. Jiao, X.; Long, M.; Li, J.; Yang, Q.; Liu, Z. Reconstructing the Invasive History and Potential Distribution Prediction of *Amaranthus palmeri* in China. *Agronomy* **2023**, *13*, 2498. [CrossRef]
69. Li, W.Q.; Xu, Z.F.; Shi, M.M.; Chen, J. Prediction of the Potential Geographic Distribution Pattern Changes of *Salix tetrasperma* under Different Climate Scenarios. *Acta Ecol. Sin.* **2019**, *39*, 3224–3234. [CrossRef]
70. Liao, J.F.; Yi, Z.L.; Li, S.C.; Xiao, L. Study on the Potential Distribution of *Dichanthium annulatum* Based on the Maxent Model in Different Periods. *Acta Ecol. Sin.* **2020**, *40*, 8297–8305. [CrossRef]
71. Zhang, Y.B.; Liu, Y.L.; Qin, H.; Meng, Q.X. Spatial Migration Prediction of Suitable Distribution Areas for *Elaeagnus mollis* under Climate Change Conditions. *Chin. J. Appl. Ecol.* **2019**, *30*, 496–502. [CrossRef]
72. Zhang, L. Estimation and Uncertainty Analysis of the Impact of Climate Change on the Geographic Distribution of Major Afforestation Tree Species/Natural Vegetation in China. Ph.D. Thesis, Chinese Academy of Forestry, Beijing, China, 2011.

73. Zhao, H.; Xian, X.; Liang, T.; Wan, F.; Shi, J.; Liu, W. Constructing an Ensemble Model and Niche Comparison for the Management Planning of Eucalyptus Longhorned Borer *Phoracantha semipunctata* under Climate Change. *Insects* **2023**, *14*, 84. [CrossRef]
74. Wang, W.; Gao, S.; Wang, S. Prediction of Potential Invasion Areas of Four Toxic Weeds in Grassland in Gansu. *Acta Ecol. Sin.* **2019**, *39*, 5301–5307.
75. Duan, B.; Man, X.; Cai, T.; Xiao, R.; Ge, Z. Increasing Soil Organic Carbon and Nitrogen Stocks along with Secondary Forest Succession in Permafrost Region of the Daxing'an Mountains, Northeast China. *Glob. Ecol. Conserv.* **2020**, *24*, e01258. [CrossRef]
76. Duan, L.; Cai, T. Changes in Magnitude and Timing of High Flows in Large Rain-Dominated Watersheds in the Cold Region of North-Eastern China. *Water* **2018**, *10*, 1658. [CrossRef]
77. Chang, X.; Jin, H.; He, R.; Zhang, Y.; Li, X.; Jin, X.; Li, G. Permafrost changes in the northwestern Da Xing'anling Mountains, Northeast China, in the past decade. *Earth Syst. Sci. Data* **2022**, *14*, 3947–3959. [CrossRef]
78. Cui, Y. Response of Larix decidua Radial Growth to Climate Change in the Da Xing'anling Region. Master's Thesis, Harbin Normal University, Harbin, China, 2020.
79. Editorial Committee of Flora of China, Chinese Academy of Sciences. *Flora of China*; Science Press: Beijing, China, 1998; Volume 7, pp. 187–237.
80. Zhang, X.; Bai, X.; Hou, M.; Chen, Z.; Manzanedo, R.D. Warmer Winter Ground Temperatures Trigger Rapid Growth of Dahurian Larch in the Permafrost Forests of Northeast China. *JGR Biogeosci.* **2019**, *124*, 1088–1097. [CrossRef]
81. Li, X.; Liu, W.S. Analysis of Secondary Forest Community Structure and Dominant Species Population Patterns in Mongolian Oak Forests. *Plant Res.* **2020**, *40*, 830. [CrossRef]
82. Zhang, X.; Chen, C.; Gao, F.; Yuan, S.; Han, S.; Ni, Z.; Yu, J. Spatial Distribution of *Larix gmelinii* Forests in Northeastern China and Their Response to Climate Change. *Ecol. J.* **2022**, *41*, 1041–1049. [CrossRef]
83. Shi, W. The Impact of Climate Change on the Distribution of *Larix gmelinii* in Northeastern China. Master's Thesis, Beijing Forestry University, Beijing, China, 2013.
84. Yang, Z.; Zhou, G.; Yin, X.; Jia, B. Geographic Distribution and Climatic Suitability of Natural Larix gmelinii Forests in China. *Ecol. J.* **2014**, *33*, 1429–1436. [CrossRef]
85. Li, F.; Zhou, G.S.; Cao, M.C. Simulation of the Geographic Distribution Response of Xing'an Larch to Climate Change. *J. Appl. Ecol.* **2006**, *17*, 2255–2260. Available online: https://kns.cnki.net/KXReader/Detail?invoice=Qly4Q6b9ZAP3LZBcrJUA9zbIekmd%2BSFeh9W5joB2lJd4SIJTWGqERCZFIG3HSLOYkNyzXj3JhUwsfM%2FyCtusi%2FHKk7ZnHdCPUkVYBLU%2Flc2GEN%2FRteE%2BmlZi0yytyueiFXJ%2BRtWtFw1%2FlK60T6j9FOW1GpT29DEU6PWOqgBF4lc%3D&DBCODE=CJFQ&FileName=YYSB200612004&TABLEName=cjfd2006&nonce=4943124702934F38AD474B51F15D23AE&TIMESTAMP=1705626680615&uid= (accessed on 2 January 2024).
86. Zhang, L.; Liu, S.-R.; Sun, P.-S.; Wang, T.-L. Comparative evaluation of multiple models of the effects of climate change on the potential distribution of *Pinus massoniana*. *Chin. J. Plant Ecol.* **2011**, *35*, 1091–1105. [CrossRef]
87. Cheng, X.X.; Yan, X.D. Effects of Climate Change on Typical Forest in the Northeast of China. *Acta Ecol. Sin.* **2008**, *28*, 534–543.
88. Yan, X.D.; Zhao, S.H.D.; Yu, Z.H.L. Modeling Growth and Succession of Northeastern China Forests and Its Applications in Global Change Studies. *Acta Phytoecol. Sin.* **2000**, *24*, 1–8.
89. Zhao, S.D. Advance on the Study of Potential Impacts of Climate Change on Northeastern China Forest. *Acta Ecol. Sin.* **1995**, *15* (Supp. B), 1–9.
90. Liu, S.H.E. *Selected Works of Liu Shene*; Science Press: Beijing, China, 1985; pp. 7, 190, 195.
91. Chen, D.K.; Zhou, X.F.; Zhu, N. *Natural Secondary Forest—Structure, Function, Dynamics and Management*; Northeast Forestry University Press: Harbin, China, 1994; p. 237.
92. Yu, S.-L.; Ma, K.-P.; Chen, L.-Z. Preliminary Discussion on the Origin of *Quercus mongolica* Forest in North China. *Guihaia* **2000**, *20*, 131–137.
93. Xu, Z.Q.; Wang, Y.H. Progress in Research on *Quercus mongolica*. *Hebei J. For. Orchard Res.* **2002**, *17*, 365–370, ISSN 1007-4961.
94. Zheng, X.W.; Zhao, R.H.; Song, X.J. Research on Drought Resistance of Main Afforestation Tree Species in Western Liaoning Region. *For. Sci.* **1990**, *26*, 353–358.
95. Abrams, D.M.; Franzk, S. Does the Absence of Sediment Charcoal Provide Substantial Evidence against the Fire and Oak Hypothesis. *J. Ecol.* **1997**, *85*, 373–377. [CrossRef]
96. Wei, X.H.; Zhou, X.F. Study on Runoff of Three Types of Broadleaved Secondary Forests. *Acta Ecol. Sin.* **1989**, *9*, 325–332.
97. Abrams, D.M. Fire and the Development of Oak Forest. *Bioscience* **1992**, *42*, 346–353. [CrossRef]
98. Chen, X.-W. Characteristic Change of Several Forest Landscapes between 1896 and 1986 in Heilongjiang Province. *Acta Bot. Sin.* **2000**, *42*, 979–984.
99. Li, Z. *Natural Geography of the Northeast Region*; Higher Education Press: Beijing, China, 1993.
100. Yin, X.; Zhou, G.; Sui, X.; He, Q.; Li, R. Dominant Climatic Factors and Their Thresholds for the Geographic Distribution of Mongolian Oak. *Acta Ecol. Sin.* **2013**, *33*, 103–109.
101. Jia, X.; Ma, F.; Zhou, W.; Zhou, L.; Yu, D.; Qin, J.; Dai, L. The Impact of Climate Change on the Potential Geographic Distribution of Korean Pine Broad-Leaved Forests. *Acta Ecol. Sin.* **2017**, *37*, 464–473.

102. Zhou, X.; Zhang, Y.; Sun, H.; Chai, Y.; Wang, Y. The Impact of Climate Change on the Population Dynamics of Mongolian Oak in Northern Greater Khingan Mountains. *Acta Ecol. Sin.* **2002**, *7*, 1035–1040.
103. Li, X.; Liu, W.-S.; Zhou, W.; Chen, F.-Y.; Mu, L.-Q. Analysis on Community Structure and Dominant Population Point Pattern of Secondary Forest of *Quercus mongolica*. *Plant Res.* **2020**, *40*, 830–838. [CrossRef]

Disclaimer/Publisher's Note: The statements, opinions and data contained in all publications are solely those of the individual author(s) and contributor(s) and not of MDPI and/or the editor(s). MDPI and/or the editor(s) disclaim responsibility for any injury to people or property resulting from any ideas, methods, instructions or products referred to in the content.

climate

MDPI

Article

Addressing the Climate Change Adaptation Gap: Key Themes and Future Directions

Ishfaq Hussain Malik [1,*] and James D. Ford [1,2]

1 School of Geography, University of Leeds, Leeds LS2 9JT, UK; j.ford2@leeds.ac.uk
2 Priestley Centre for Climate Futures, University of Leeds, Leeds LS2 9JT, UK
* Correspondence: i.h.malik@leeds.ac.uk

Abstract: Climate change adaptation is a critical response to the challenges posed by climate change and is important for building resilience. Progress in adaptation efforts has been made globally, nationally, and locally through international agreements, national plans, and community-based initiatives. However, significant gaps exist in knowledge, capacity, and finance. The Adaptation Gap Report 2023, published by the United Nations Environment Programme (UNEP), examines the status of climate change adaptation efforts globally. The report highlights the widening adaptation finance gap and the deepening climate crisis. We analyse the key themes of the report and incorporate an analysis of the wider literature and insights from COP28 to substantiate key points and identify gaps where more work is needed to develop an understanding of climate change adaptation. This paper focuses on the underfinanced and underprepared state of global climate change adaptation efforts, the widening adaptation finance gap, slow progress in adaptation, gender equality and social inclusion issues, and challenges in addressing loss and damage. We provide a way forward for climate change adaptation and offer recommendations for future actions.

Keywords: adaptation gap; climate change; equity; inclusion; COP28

1. Introduction

The United Nations Environment Programme Adaptation Gap Report (AGR) series provides an annual science-based assessment of global progress in adaptation planning, financing, and implementation [1]. Since its inception in 2014, UNEP has consistently produced the AGR with the objective of providing insights to inform climate negotiations among United Nations Member States, and the AGR 2023 is the ninth report in the series. The Adaptation Gap Report 2023, titled "Underfinanced. Underprepared. Inadequate investment and planning on climate adaptation leaves world exposed", was published on 2 November 2023 by the United Nations Environment Programme [1]. It highlights the "underfinanced" resources allocated to climate change adaptation and emphasises the "underprepared" state on a global scale. The report emphasises the significant implications of the failure to effectively address the severe consequences of climate change, particularly for the most vulnerable populations [1]. The discourse surrounding climate change adaptation is experiencing a surge in attention as the draft proposal for the Global Goal on Adaptation at the 28th Conference of the Parties (COP28) proposes to prioritise the development of climate-resilient food and agricultural systems, enhance resilience against health impacts arising from climate change, ensure the provision of climate-resilient health services, and mitigate climate-related morbidity [2]. The UAE Consensus at COP28 advocates for nations to demonstrate their commitment towards a just, orderly, and equitable shift away from fossil fuel reliance within their energy systems, with a particular emphasis on expediting efforts during the present pivotal decade. The ultimate objective is to attain a state of net zero greenhouse gas emissions by the year 2050, aligning with scientific findings and recommendations. It concluded with a historic agreement that, for the first time in three decades, incorporates provisions related to oil and gas [2].

The effects of climatic change are becoming increasingly evident in various regions across the world, making communities worldwide face increasing exposure to alterations and disruptions in weather patterns, wildlife and vegetation dynamics, and the quality, accessibility, and availability of water and food resources [3]. The increasing urgency of the climate crisis highlights the need for effective climate change adaptation measures and policies to address climate impacts. However, the global tracking of adaptation progress faces challenges, including a vast and diverse evidence base, reliance on proxies, and the ongoing debate on defining adaptation success [4–6]. Environmental justice plays a pivotal role in the context of climate change, as evidenced by the experiences of distributional, procedural, and recognition (in)justice observed at various levels [7]. The focus on equity and justice in climate change adaptation is unequal across geographical regions, topical sectors, and marginalised communities [8]. Certain aspects of justice, such as distributive and procedural concerns, receive more attention than others, and there is a need to incorporate ethics and justice theories into adaptation and resilience planning to address the vulnerabilities of those most affected by climate change and advance progress in social justice [8,9]. The injustices range from the global scale, where historically low-emitting countries bear a disproportionate burden of climate change impacts, to the local scale, where the most vulnerable members of society often endure the gravest impacts [10].

This paper aims to analyse the key themes covered in the Adaptation Gap Report 2023 and situate them in the broader literature. It provides an overview of the underfinanced and underprepared state of climate change adaptation efforts globally, emphasising the implications for vulnerable populations and the need for urgent action. It explores the widening adaptation finance gap, slow progress in adaptation, gender equality and social inclusion issues, and challenges in addressing loss and damage caused by climate change. This paper aims to provide insights and recommendations for addressing these challenges and calls for an international consensus and innovative funding sources to bridge the adaptation gap.

2. Methodology

The method employed in this study involved a systematic approach to analysing the key themes of the Adaptation Gap Report 2023, supplementing it with insights from a comprehensive literature review and discussions from COP28. The goal was to present a thorough overview of the global state of climate change adaptation, including issues of underfinancing, slow progress, gender equality, and challenges in addressing climate-induced losses. The methodology included steps such as a literature review, a detailed analysis of the Adaptation Gap Report 2023, the incorporation of COP28 insights, the identification of knowledge gaps, and outlining future research directions. A comprehensive review of the existing literature on climate change adaptation, equity, gender equality, and loss and damage was conducted. This review included academic papers, reports, and other relevant sources. The literature review aimed to gather information on the key themes covered in the Adaptation Gap Report 2023 and identify gaps in the existing knowledge and understanding of climate change adaptation. The Adaptation Gap Report 2023 was thoroughly analysed to identify the key themes and findings. The analysis focused on extracting relevant data, statistics, and case studies from the report to support the arguments made in this paper. Insights from COP28 were identified to contextualise the key themes discussed in this paper and provide additional evidence and examples to support the arguments. The paper is structured to integrate findings, statistics, and case studies to substantiate arguments and concludes with recommendations for enhancing climate change adaptation efforts. The combined methodology ensures a comprehensive understanding of the current situation and provides a roadmap for future actions in climate change adaptation.

3. Results and Discussion

3.1. Key Themes

3.1.1. Widening Adaptation Finance Gap

The Adaptation Gap Report (AGR) 2023 highlights that the adaptation finance gap is growing and estimates it to be between USD 194 billion and USD 366 billion per year, which indicates a deepening climate crisis [1]. Climate change adaptation finance is essential for supporting vulnerable communities and ecosystems in adapting to the impacts of global environmental change [11]. However, there is growing concern about the widening climate change adaptation finance gap. This gap is caused by the inadequate allocation of funds by governments and international institutions, as well as a misalignment between adaptation and mitigation funding [12–14]. The complex nature of climate change and its impacts also presents challenges in accurately estimating financial needs for adaptation [15]. The insufficiency of adaptation finance in adequately addressing the adaptation priorities of developing nations can be attributed primarily to the inherent design of guidance [16,17]. Overcoming this gap requires prioritizing adaptation funding, increasing financial commitments, improving access to funding, and developing innovative financial instruments [18,19].

Climate change adaptation is an urgent global challenge for which actions are urgently needed, necessitating collaborative endeavours across multiple disciplines and stakeholders to achieve effective and sustainable results [20,21]. However, a significant gap persists between the numerous expectations linked to these discussions and the current state of adaptation finance [22]. The climate change adaptation finance gap was between 5 and 10 times larger in 2022 [23], and it is now estimated to be 10 to 18 times greater than current international funding [1]. This estimation of the adaptation finance gap holds significance in the discourse surrounding the nature and magnitude of the newly established collective, quantified goal for climate finance [24]. This objective plays a crucial role in bridging the adaptation finance gap, particularly for countries that are highly vulnerable to climate change impacts, such as the Least Developed Countries (LDCs) and Small Island Developing States (SIDS). This estimation holds relevance to the decision (Decision CMA.3, COP26) [25] made at the 26th Conference of the Parties (COP 26) in Glasgow, which urged developed nations to increase their financial support for adaptation in developing countries to at least twice the amount provided in 2019 by the year 2025 [26].

Climate finance has consistently emerged as a crucial concern during United Nations climate negotiations, and the provision of assistance to developing nations in their efforts to mitigate climate change, such as by promoting the adoption of renewable energy sources or low-carbon transportation and aiding in their adaptation efforts, such as enhancing the resilience of agricultural and water sectors to physical climate effects, plays a crucial role in establishing and upholding trust among participating nations [27–29]. However, there are power struggles surrounding international climate adaptation finance and conflicts related to key questions such as the amount of finance, providers, delivery channels, allocation, prioritisation, compensation, and climate justice [30,31]. Limited progress has been made in integrating Indigenous Peoples into international climate change governance, and the material constraints and the designation of Indigenous Peoples as nonstate observers pose challenges for their meaningful participation [32]. Ethnic and racial minorities, migrants, and people with disabilities are seldom taken into consideration, and it is crucial to pay attention to how equity is integrated into adaptation research and practice to ensure fair and just adaptation [8]. The emergence of neoliberal characteristics in the present-day climate regime poses unique challenges in the pursuit of justice pertaining to adaptation finance [31]. The ambiguous definitions of climate finance, the shortcomings of the current USD 100 billion per year goal, allowing for multiple interpretations and reducing transparency and trust between countries, and the current objectives of doubling international finance flows to developing countries by 2025 are insufficient to effectively address the existing finance gap, which necessitates a need for a new collective, quantified goal for climate finance [33,34].

The increasing frequency of climatic anomalies, such as prolonged droughts in East Africa, floods in China and Europe, and the emergence of extreme temperatures and wildfires in the United States and Canada and increasing hailstorms and extreme weather events in different parts of the world, underscores the urgency of addressing the disparity in adaptation finance [35,36]. This is particularly significant due to the substantial benefits that investments in adaptation can offer in terms of mitigating climate-related risks and promoting equity and climate justice [1]. The adaptation gap is widening and requires addressing both quantitative and qualitative factors, including access to finance and equity [31].

3.1.2. Slow Progress in Adaptation

Global progress in adaptation is slowing down instead of accelerating, despite the increasing climate risks and impacts worldwide [1]. There exist notable impediments affecting successful adaptation to climate risks, like the dearth of policies pertaining to climate change adaptation, as well as a deficiency in comprehensive climate vulnerability and risk assessments [37,38]. The state of progress towards climate adaptation is uncertain. Studies suggest that global adaptation is only halfway towards its full potential, with urban areas generally scoring higher than rural areas, and there is an imbalance in adaptation efforts across different dimensions and a limited focus on long-term strategies [11,39]. Assessing global progress in human adaptation to climate change is an urgent priority. Adaptations are mostly fragmented, local, and incremental, with limited evidence of transformational adaptation and risk reduction outcomes [40]. The current comprehension of policy instruments in adaptation remains limited as most studies have failed to establish a connection between the impacts and risks posed by climate change and the efficacy of policy instruments, thus rendering the evaluation of adaptation effectiveness a challenging task [41,42]. Public multilateral and bilateral adaptation finance flows to developing countries decreased by 15% in 2021. This lack of funding has implications for vulnerable populations and could lead to significant impacts [1].

The rate of adaptation progress is unequal, with lower-income population groups experiencing the greatest adaptation gaps [3]. Slowing progress in adaptation implementation is due to factors such as fluctuations driven by non-climate-related events, limited access to finance, a lack of knowledge about adaptation policies, institutional factors like the adoption of national climate change adaptation strategies, socioeconomic factors like population density, the GDP, unemployment rates, environmental factors like flood risk, and political factors like ineffective leadership [3,43,44]. Constraints and limits to adaptation are important to understand human and natural systems' responses to climate change [45]. Little is known about the extent to which national climate change adaptation plans are implemented [46]. Central and South America, along with the Small Islands, encounter significant constraints and limitations to adaptation, particularly in technological, infrastructural, and ecosystem-based approaches, while economic, socio-cultural, and governance constraints are prevalent globally [47].

Current climate action is inadequate to meet the goals of the Paris Agreement. The projected global temperature increase is already exceeding 1.1 °C above pre-industrial levels, and current plans put us on a path towards 2.4–2.6 °C by the end of the century [1]. Global greenhouse gas emissions and temperature have reached new highs, primarily due to fossil fuel combustion and industrial processes [48]. There is an uneven distribution of emissions among countries, with high-income nations needing to speed up their own emission reductions and assist low- and middle-income countries, and failure to reduce emissions and reliance on carbon dioxide removal (CDR) technologies will make it impossible to limit warming to 1.5 °C or 2 °C [49]. Growing evidence indicates that equity and justice are not incorporated much in climate adaptation responses, and these responses may exacerbate inequality and increase vulnerability [8].

Climate change is increasingly driving adaptation efforts, but there is limited focus on potential opportunities and benefits [50]. It poses significant risks, and adaptation measures are necessary to mitigate these risks. However, the effectiveness and feasibility of adaptation in a changing world are still uncertain. Adaptation can be beneficial in reducing risks, but there are limits to adaptation, and even ambitious adaptation efforts may not be enough to offset the failure to mitigate greenhouse gas emissions [51,52]. The successful implementation of climate change adaptation depends on the capabilities of societies to create and mobilise the adaptive capacities of their socio-ecological systems [53].

Tracking global progress in adaptation poses various challenges, including a lack of comprehensive and coherent data sources and inconsistent reporting guidelines, necessitating a need for consistency, comparability, comprehensiveness, and coherency in adaptation tracking [54]. Further efforts are needed to integrate process and output indicators into tracking frameworks and to collect better data on adaptation [55]. There is a need to address the inclusion of vulnerable populations in adaptation policies and the durability of adaptation policies in the face of political transitions [55,56]. Global adaptation research needs to prioritise assessing the effectiveness of adaptation responses, understanding limits to adaptation, enabling individuals and civil society to adapt, and improving methods for synthesising different forms of evidence [40]. There is a need to refine adaptation targets, prioritise actions beyond development levels, and develop comprehensive and context-specific adaptation strategies [39]. The present estimation of adaptation finance needs indicates an increased necessity for the enhanced mobilisation and accessibility of financial resources to effectively implement adaptation measures and reduce adaptation gaps [10].

Current adaptation measures focus on physical infrastructure and short-term economic goals, neglecting the social aspects of vulnerability [57,58]. It is anticipated that key constraints will persist until well past 2050, impeding effective adaptation efforts, especially in vulnerable communities [59]. To enhance global adaptation efforts, it is important to evaluate the effectiveness of adaptation responses, deepen our understanding of the limitations of adaptation, empower individuals and civil society to adapt, and address overlooked areas [40].

3.1.3. Gender Equality and Social Inclusion

The AGR 2023 highlights the inadequate inclusion of gender equality and social inclusion in adaptation finance needs and flows. Only a small percentage of national plans have dedicated budgets for gender and social inclusion activities. The report reveals that only 20% of costed Nationally Determined Contributions (NDCs) and National Adaptation Plans (NAPs) allocate a specific budget for gender equality and social inclusion, with an average allocation of 2%. In terms of international public finance for adaptation, only 2% is considered gender-responsive, while an additional 24% is considered gender-specific or integrative [1]. Although previous gap reports mentioned gender [23], this is the first instance in a gap report in which an examination of gender equality and social inclusion has been conducted.

Climate change significantly contributes to the exacerbation of inequality across various dimensions, such as gender, socioeconomics, the loss of traditional knowledge and culture, and the perpetuation of colonial stigmas through the depletion of resources and the disruption of livelihoods. Conversely, the escalation of inequality also plays a pivotal role in intensifying climate-related risks [3]. Structurally disadvantaged people, who have historically faced discrimination, marginalisation, or disenfranchisement due to factors such as gender, age, ethnicity, class, language, ability, and/or sexual orientation are disproportionately susceptible to the adverse consequences of climate change [60,61]. Extreme weather events worldwide highlight the unequal effects of climate change on different populations due to a combination of social, economic, historical, and political factors [62] exacerbating the existing differential vulnerability across axes of social difference such as race, class, ethnicity, and gender [63]. Though adapting to climate change is important, understanding why certain communities face greater vulnerability to and consequences from climate hazards is crucial [62]. Understanding power and justice considerations is important in making locally led adaptation effective [64]. Retrofitting adaptation into existing development agendas may lead to maladaptation, and to address these challenges, it is crucial to involve marginalised groups to enhance the use of climate finance [65].

Gender inequality and discrimination serve as significant obstacles to adaptation. A reduction in gender disparities can have a transformative impact on the pursuit of climate justice [66]. The gendered implications of climate change, especially in the Global South, are particularly poignant, as patriarchal norms, inequities, and inequalities often place women and men in differentiated positions in their abilities to respond to and cope with social and environmental changes and foreground the complex ways in which social power relations operate in communal responses to adaptation strategies [67]. Garcia et al. [68] elucidate the socio-political drivers behind gendered inequalities that generate discriminatory conditions for adaptation. Employing an intersectional subjectivities framework, they explore how entrenched power dynamics and social norms pertaining to gender give rise to barriers to adaptation, such as limited resources and agency. The analysis reveals a stark dichotomy as women bear the brunt of these obstacles and face a persistent power imbalance that positions them as less capable of adapting compared to men. There is a need for gender-responsive climate finance and adaptation funding, social inclusion, and improvements in integrating gender equality and women's empowerment in multilateral climate funds to support more equitable and effective adaptation [69]. The issue is gaining importance as Gender Day—Technical Dialogue: Financing for gender-responsive just transitions and climate action, held on 4 December 2023 during COP 28 in Dubai, stressed the gendered impacts of climate change and the need for gender-responsive financing to advance gender equality and women's empowerment in climate action. The technical dialogue focused on a common understanding of opportunities and gaps in financing gender-responsive just transitions and climate action. The COP28 Presidency announced the establishment of the COP28 Gender-Responsive Just Transitions and Climate Action Partnership, which received endorsement from 68 parties. The partnership involves various commitments, such as addressing data, finance, and equal opportunities. A review of the implementation will take place during a second meeting at COP31 [2].

3.1.4. Issues in Loss and Damage (L&D)

The AGR 2023 highlights the need to address loss and damage caused by climate change, particularly in vulnerable developing countries. It supports the establishment of a loss and damage fund and funding arrangements to assist these countries [1], and it is the first time that a chapter on loss and damage has been included in an AGR. The environmental justice lens emphasises that loss and damage are not solely results of climate hazards but are also influenced by differential vulnerabilities to climate change which are frequently shaped by a range of socio-political processes, including racism and histories of colonialism and exploitation [70]. Vulnerable people are currently experiencing loss and damage to their fundamental human rights, particularly in relation to the right to a healthy environment and the ability to own, use, develop, and control land [71]. This has led to significant impacts on property rights, communal assets, standards of living, and family and social cohesion, affecting interconnected rights such as customs, Indigenous knowledge, family, agency, and identity, which can transcend across generations [71,72]. The COP27 of the UN Framework Convention on Climate Change (UNFCCC) established a Transitional Committee to operationalise funding for loss and damage. However, tactical opposition is obstructing loss and damage finance, and the fund is unlikely to involve direct reparations from developed countries to the most vulnerable nations. Instead, it builds upon existing financing mechanisms within the UNFCCC, including loans, grants, and private investments [73,74]. The creation of a loss and damage fund has been seen as a positive development for Global South negotiators. However, there are concerns about how funding will be allocated to vulnerable developing countries [75]. The use of objective means to determine funding allocation may cause division and delay as vulnerability indicators are complex and politically influenced, and there is a need for transformation in adaptation, extension, and liberation as a response to loss and damage for long-term climate change adaptation [75,76].

The concept of loss and damage (L&D) had prominence as a policy domain since 1991, when the Alliance of Small Island States (AOSIS) put forth a proposition to establish a fund and insurance mechanism to address the adverse effects of climate-induced sea-level rise [77]. However, its advancement within international frameworks has been impeded by conflicting viewpoints and varying interpretations. This has resulted in limited action in addressing losses and damages [78]. It has been a contentious topic in international climate policy for many years. While formal mechanisms have been established and conceptually, loss and damage considers the impacts that occur beyond the limits of adaptation [79], there is a lack of conceptual clarity and the language used is unclear, and this ambiguity is important politically [80]. It is important to provide adequate funding for loss and damage to support vulnerable nations that are unable to adapt to the impacts of climate change. Various innovative funding approaches like the Financial Transaction Tax (FTT), International Airline Passenger Levy (IAPAL), Solidarity Levy, Bunker Fuels Levy, Fossil Fuel Majors Carbon Levy, and Global Carbon Tax are important for funding loss and damage [81].

Noneconomic losses and damages (NELDs) encompass a wide spectrum of impacts that are not easily assigned a monetary value, such as the loss of life, health, or mobility; the loss of territory, cultural heritage, or Indigenous or local knowledge; and the loss of biodiversity [82]. It is important to consider people's lived experiences with climate-related loss, and many intangible aspects, such as culture and identity, are often overlooked and omitted in discussions about climate change that matter most to people [83]. Understanding ecological grief is important for understanding the mental health implications of climate change and identifying coping strategies, and there are three categories of ecological grief: grief associated with physical ecological losses, grief associated with the loss of environmental knowledge and identity, and grief associated with anticipated future losses. Confronting ecological grief is challenging but necessary to understand and address the emotional and psychological impacts of climate change [84,85].

The COP28 has achieved a major milestone by operationalizing the Loss and Damage Fund to assist developing countries vulnerable to the effects of climate change. The fund was agreed upon during COP27 and became operational following five transitional committee meetings. The UAE has committed USD 100 million to the fund, with other countries such as Germany, the UK, Japan, and the USA also making notable contributions. This action regarding loss and damage will allow the parties to focus on the Global Stocktake, which evaluates progress towards the goals of the Paris Agreement [2].

4. Where Do We Need to Go?

We argue that the discourse surrounding climate change adaptation is experiencing a surge in attention; the focus on equity and justice in climate change adaptation is unequal across geographical regions, topical sectors, and marginalised communities; the state of progress towards climate adaptation is uncertain; and little is known about the extent to which national climate change adaptation plans are implemented.

Understanding the factors that enable or limit adaptation is crucial for both improving adaptability and conducting risk assessments in the face of future climate change, and the impacts of future climate change not only rely on changing climate hazards and land use changes but also heavily depend on the level of adaptation achieved [86–88]. The AGR 2023 provides insights into where we need to go to enhance climate change adaptation ambition. Currently, only 25% of countries have legal instruments in place that require national governments to prepare national adaptation planning instruments [1]. It is important for more countries to adopt these instruments to prioritise adaptation planning and ensure regular updates. However, increasing international public adaptation finance alone is unlikely to bridge the finance gap. The information provided in national adaptation plans is diverse, making it challenging to estimate the finance needs of developing countries.

Financing mechanisms should be tailored to meet the needs of small- and medium-sized enterprises (SMEs) as they have the potential to unlock climate adaptation solutions [73,89]. Standardised reporting on climate resilience integration in investment decisions can help monitor progress. Identifying climate-related risks could negatively impact developing economies in the short term. Regional cooperation is needed to address loss and damage, and alternative models of finance disbursement should be developed to ensure funds reach affected communities promptly and effectively. Indigenous knowledge needs to drive climate change adaptation strategies to enable communities to cultivate sustainable practices that mitigate environmental consequences and uphold ecological balance while also integrating Indigenous knowledge and institutions into resource management. The report does not address adaptation efforts in highly susceptible regions like the Arctic, Himalayas, and Antarctic. Immediate action, increased financial support, and a focus on equity and inclusion are necessary to bridge the adaptation gap and mitigate the consequences of climate change, particularly in the most vulnerable regions and communities.

There is a need for conceptual clarity on loss and damage, an international consensus on key concepts, and increased coordination across global frameworks to strengthen loss and damage management approaches. The uncertainties in financial needs for addressing loss and damage call for innovative funding sources, institutional arrangements, and governance structures. The provision of access to the Loss and Damage Fund should be extended to encompass a broader range of communities, ensuring that Indigenous Peoples are granted access regardless of their geographical location. The Loss and Damage Fund for climate change should be spent on compensating direct losses, supporting adaptation measures, and investing in insurance mechanisms. The funds should be allocated for capacity building, research, and community-based initiatives, emphasising transparency, equity, and international cooperation. Regular monitoring and evaluation are crucial to ensure effective utilisation in addressing the specific needs of vulnerable communities. Bridging policy, research, and practice to support discussions on adaptation is important. Addressing the ethical connections between climate impacts and finance mechanisms is important, and developed nations should do more to assist developing nations in climate change adaptation through technology transfer and capital.

Future research directions include exploring the potential for transformational change; understanding the values and engagement of people with loss and grief and lived experiences; incorporating equity, gender equality, and social inclusion; incorporating the perspectives of vulnerable groups and Indigenous Peoples in decision making; incorporating equity and justice; and conducting policy-relevant research and critical analyses of climate change in the context of global capitalism and neoliberal development. To successfully monitor adaptation, methods must rapidly handle large datasets, recognise fine-grained distinctions, and consider context-dependent meanings in the face of the evolving climate crisis.

Climate change adaptation necessitates a comprehensive evaluation of the efficacy of nature-based solutions, the incorporation of socio-economic considerations into adaptation planning, the reinforcement of climate resilience in urban settings, the assessment of technological innovations, the exploration of Indigenous knowledge and traditional practices, the appraisal of financial mechanisms for adaptation effectiveness, the examination of governance and policy frameworks, the comprehensive analysis of the impacts of climate change on vulnerable populations, and the promotion of knowledge exchange and capacity sharing. It is important to note that the term "capacity building" may inadvertently perpetuate power dynamics and hierarchical structures; therefore, the emphasis should be placed on knowledge exchange and capacity sharing to foster equitable relationships among stakeholders.

Author Contributions: I.H.M. and J.D.F. conceptualised the ideas and contributed to the writing of the article. All authors have read and agreed to the published version of the manuscript.

Funding: We acknowledge the financial support provided by UK Research and Innovation (IMAGINE Project, reference NE/X003868/1).

Data Availability Statement: Data are contained within the article.

Conflicts of Interest: The authors declare no conflicts of interest.

References

1. United Nations Environment Programme. Adaptation Gap Report 2023: Underfinanced. Underprepared. Inadequate Investment and Planning on Climate Adaptation Leaves World Exposed. Nairobi. 2023. Available online: https://doi.org/10.59117/20.500.11822/43796 (accessed on 10 November 2023).
2. United Nations Climate Change (UNCC). COP28 2023. Available online: https://www.cop28.com/en/ (accessed on 15 December 2023).
3. IPCC. *Climate Change 2022: Impacts, Adaptation and Vulnerability. Contribution of Working Group II to the Sixth Assessment Report of the Intergovernmental Panel on Climate Change*; Pörtner, H.-O., Roberts, D.C., Tignor, M., Poloczanska, E.S., Mintenbeck, K., Alegría, A., Craig, M., Langsdorf, S., Löschke, S., Möller, V., et al., Eds.; Cambridge University Press: Cambridge, UK; New York, NY, USA, 2022; p. 3056.
4. Sietsma, A.J.; Ford, J.D.; Minx, J.C. The next generation of machine learning for tracking adaptation texts. *Nat. Clim. Chang.* **2023**, *14*, 31–39. [CrossRef]
5. Wani, S.N.; Malik, I.H. Disaster vulnerability and flood management policy framework in the Himalayas. *Arab. J. Geosci.* **2023**, *16*, 609. [CrossRef]
6. Malik, I.H.; Hashmi, S.N.I. Ethnographic account of flooding in North-Western Himalayas: A study of Kashmir Valley. *GeoJournal* **2022**, *87*, 1265–1283. [CrossRef]
7. Thomas, A.; Benjamin, L. Non-economic loss and damage: Lessons from displacement in the Caribbean. *Clim. Policy* **2021**, *20*, 715–728. [CrossRef]
8. Araos, M.; Jagannathan, K.; Shukla, R.; Ajibade, I.; de Perez, E.C.; Davis, K.; Ford, J.D.; Galappaththi, E.K.; Grady, C.; Hudson, A.; et al. Equity in human adaptation-related responses: A systematic global review. *One Earth* **2021**, *4*, 1454–1467. [CrossRef]
9. Coggins, S.; Berrang-Ford, L.; Hyams, K.; Satyal, P.; Ford, J.; Paavola, J.; Arotoma-Rojas, I.; Harper, S. Empirical assessment of equity and justice in climate adaptation literature: A systematic map. *Environ. Res. Lett.* **2021**, *16*, 073003. [CrossRef]
10. IPCC. *Climate Change 2023: Synthesis Report. Contribution of Working Groups I, II and III to the Sixth Assessment Report of the Intergovernmental Panel on Climate Change*; Core Writing Team, Lee, H., Romero, J., Eds.; IPCC: Geneva, Switzerland, 2023; p. 184. [CrossRef]
11. Owen, G. What makes climate change adaptation effective? A systematic review of the literature. *Glob. Environ. Chang.* **2020**, *62*, 102071. [CrossRef]
12. Malik, I.H. Anthropogenic causes of recent floods in Kashmir Valley: A study of 2014 flood. *SN Soc. Sci.* **2022**, *2*, 1–18. [CrossRef]
13. Bracking, S.; Leffel, B. Climate finance governance: Fit for purpose? *Wiley Interdiscip. Rev. Clim. Chang.* **2021**, *12*, e709. [CrossRef]
14. Savvidou, G.; Atteridge, A.; Omari-Motsumi, K.; Trisos, C.H. Quantifying international public finance for climate change adaptation in Africa. *Clim. Policy* **2021**, *21*, 1020–1036. [CrossRef]
15. Fankhauser, S. Adaptation to climate change. *Annu. Rev. Resour. Econ.* **2017**, *9*, 209–230. [CrossRef]
16. Malik, I.H. Flood risk assessment and analysis of Kashmir Valley Floor. In *Re-Envisioning Advances in Remote Sensing*; CRC Press: Boca Raton, FL, USA, 2022; pp. 133–141.
17. Conevska, A.; Ford, J.; Lesnikowski, A. Assessing the adaptation fund's responsiveness to developing country's needs. *Clim. Dev.* **2020**, *12*, 436–447. [CrossRef]
18. Mitra, P.; Ribeiro, M.P.; Melina, G.; Belianska, A.; Bohme, N.; Cai, K.; Diallo, Y.; Jain, S.; Zerbo, S. *Climate Change and Select Financial Instruments: An Overview of Opportunities and Challenges for Sub-Saharan Africa*; International Monetary Fund: Washington, DC, USA, 2022.
19. McDonald, J.; McCormack, P. C. Rethinking the role of law in adapting to climate change. *Wiley Interdiscip. Rev.: Clim. Change* **2021**, *12*, e726. [CrossRef]
20. Arteaga, E.; Nalau, J.; Biesbroek, R.; Howes, M. Unpacking the theory-practice gap in climate adaptation. *Clim. Risk Manag.* **2023**, *42*, 100567. [CrossRef]
21. Biswas, R.R.; Rahman, A. Adaptation to climate change: A study on regional climate change adaptation policy and practice framework. *J. Environ. Manag.* **2023**, *336*, 117666. [CrossRef]
22. Weikmans, R. *The Normative Foundations of International Climate Adaptation Finance*; Cambridge University Press (CUP): Cambridge, UK, 2023.
23. United Nations Environment Programme. Adaptation Gap Report 2022: Too Little, Too Slow—Climate Adaptation Failure Puts World at Risk. Nairobi. 2022. Available online: https://www.unep.org/adaptation-gap-report-2022 (accessed on 10 November 2023).
24. United Nations Framework Convention on Climate Change. Compilation and Synthesis of Inputs on the Sixth Technical Expert Dialogue under the Ad Hoc Work Programme on the New Collective Quantified Goal on Climate Finance. 1 June. NCQG/2023/TED6/C&S/3. 2023. Available online: https://unfccc.int/sites/default/files/resource/TED6_Compilation_Synthesis_final.pdf (accessed on 30 December 2023).

25. Decision CMA.3, COP26. In Proceedings of the United Nations Climate Change Conference, Glasgow, Scotland, UK. 2021. Available online: https://unfccc.int/event/cma-3#decisions_reports (accessed on 15 December 2023).
26. United Nations Framework Convention on Climate Change. Report of the Conference of the Parties Serving as the Meeting of the Parties to the Paris Agreement on Its Third Session, Held in Glasgow from 31 October to 13 November 2021. 8 March. FCCC/PA/CMA/2021/10/Add.3. 2021. Available online: https://unfccc.int/documents/460952 (accessed on 5 November 2023).
27. Munira, S.; Bashar, R.; Easher, T.H.; Khan, M.R. Climate Finance in the UNFCCC Negotiations: Bridging Gaps with Lessons Learnt. In *Climate Change in Bangladesh: A Cross-Disciplinary Framework*; Springer: Berlin/Heidelberg, Germany, 2021; pp. 1–24.
28. Pickering, J.; Betzold, C.; Skovgaard, J. Managing fragmentation and complexity in the emerging system of international climate finance. *Int. Environ. Agreem. Politics Law Econ.* **2017**, *17*, 1–16. [CrossRef]
29. Ciplet, D.; Roberts, J.T.; Khan, M.R. *Power in a Warming World: The New Global Politics of Climate Change and the Remaking of Environmental Inequality*; MIT Press: Cambridge, MA, USA, 2015.
30. Méndez, M. *Climate Change from the Streets: How Conflict and Collaboration Strengthen the Environmental Justice Movement*; Yale University Press: New Haven, CT, USA, 2020.
31. Khan, M.; Robinson, S.-A.; Weikmans, R.; Ciplet, D.; Roberts, J.T. Twenty-five years of adaptation finance through a climate justice lens. *Clim. Chang.* **2020**, *161*, 251–269. [CrossRef]
32. Belfer, E.; Ford, J.D.; Maillet, M.; Araos, M.; Flynn, M. Pursuing an indigenous platform: Exploring opportunities and constraints for indigenous participation in the UNFCCC. *Glob. Environ. Politics* **2019**, *19*, 12–33. [CrossRef]
33. United Nations Conference on Trade and Development (UNCTAD). Considerations for a New Collective Quantified Goal: Bringing Accountability, Trust and Developing Country Needs to Climate Finance. 2023. Available online: https://unctad.org/system/files/official-document/gds2023d7_en.pdf (accessed on 30 December 2023).
34. Pauw, W.P.; Moslener, U.; Zamarioli, L.H.; Amerasinghe, N.; Atela, J.; Affana, J.P.B.; Buchner, B.; Klein, R.J.T.; Mbeva, K.L.; Puri, J.; et al. Post-2025 climate finance target: How much more and how much better? *Clim. Policy* **2022**, *22*, 1241–1251. [CrossRef]
35. Trisos, C.H.; Adelekan, I.O.; Totin, E.; Ayanlade, A.; Efitre, J.; Gemeda, A.; Kalaba, K.; Lennard, C.; Masao, C.; Mgaya, Y.; et al. Africa. In *Climate Change 2022: Impacts, Adaptation and Vulnerability. Contribution of Working Group II to the Sixth Assessment Report of the Intergovernmental Panel on Climate Change*; Pörtner, H.-O., Roberts, D.C., Tignor, M., Poloczanska, E.S., Mintenbeck, K., Alegría, A., Craig, M., Langsdorf, S., Löschke, S., Möller, V., Eds.; Cambridge University Press: Cambridge, UK; New York, NY, USA, 2022; pp. 1285–1455. [CrossRef]
36. Raupach, T.H.; Martius, O.; Allen, J.T.; Kunz, M.; Lasher-Trapp, S.; Mohr, S.; Rasmussen, K.L.; Trapp, R.J.; Zhang, Q. The effects of climate change on hailstorms. *Nat. Rev. Earth Environ.* **2021**, *2*, 213–226. [CrossRef]
37. Fatorić, S.; Biesbroek, R. Adapting cultural heritage to climate change impacts in the Netherlands: Barriers, interdependencies, and strategies for overcoming them. *Clim. Chang.* **2020**, *162*, 301–320. [CrossRef]
38. Biesbroek, R.; Berrang-Ford, L.; Ford, J.D.; Tanabe, A.; Austin, S.E.; Lesnikowski, A. Data, concepts and methods for large-n comparative climate change adaptation policy research: A systematic literature review. *Wiley Interdiscip. Rev. Clim. Chang.* **2018**, *9*, e548. [CrossRef]
39. Magnan, A.K.; Bell, R.; Duvat, V.K.E.; Ford, J.D.; Garschagen, M.; Haasnoot, M.; Lacambra, C.; Losada, I.J.; Mach, K.J.; Noblet, M.; et al. Status of global coastal adaptation. *Nat. Clim. Chang.* **2023**, *13*, 1213–1221. [CrossRef]
40. Berrang-Ford, L.; Siders, A.R.; Lesnikowski, A.; Fischer, A.P.; Callaghan, M.W.; Haddaway, N.R.; Mach, K.J.; Araos, M.; Shah, M.A.R.; Wannewitz, M.; et al. A systematic global stocktake of evidence on human adaptation to climate change. *Nat. Clim. Chang.* **2021**, *11*, 989–1000. [CrossRef]
41. Biesbroek, R. Policy integration and climate change adaptation. *Curr. Opin. Environ. Sustain.* **2021**, *52*, 75–81. [CrossRef]
42. Biesbroek, R.; Delaney, A. Mapping the evidence of climate change adaptation policy instruments in Europe. *Environ. Res. Lett.* **2020**, *15*, 083005. [CrossRef]
43. Milordis, A.C.; Butler, W.H.; Holmes, T.J. What is slowing progress on climate change adaptation? Evaluating barriers to planning for sea level rise in Florida. *Mitig. Adapt. Strat. Glob. Chang.* **2023**, *28*, 42. [CrossRef]
44. Reckien, D.; Flacke, J.; Olazabal, M.; Heidrich, O. The influence of drivers and barriers on urban adaptation and mitigation plans—An empirical analysis of European cities. *PLoS ONE* **2015**, *10*, e0135597. [CrossRef]
45. Mechler, R.; Singh, C.; Ebi, K.; Djalante, R.; Thomas, A.; James, R.; Tschakert, P.; Wewerinke-Singh, M.; Schinko, T.; Ley, D.; et al. Loss and Damage and limits to adaptation: Recent IPCC insights and implications for climate science and policy. *Sustain. Sci.* **2020**, *15*, 1245–1251. [CrossRef]
46. Leiter, T. Do governments track the implementation of national climate change adaptation plans? An evidence-based global stocktake of monitoring and evaluation systems. *Environ. Sci. Policy* **2021**, *125*, 179–188. [CrossRef]
47. Thomas, A.; Theokritoff, E.; Lesnikowski, A.; Reckien, D.; Jagannathan, K.; Cremades, R.; Campbell, D.; Joe, E.T.; Sitati, A.; Singh, C.; et al. Global evidence of constraints and limits to human adaptation. *Reg. Environ. Chang.* **2021**, *21*, 85. [CrossRef]
48. Yoro, K.O.; Daramola, M.O. CO_2 emission sources, greenhouse gases, and the global warming effect. In *Advances in Carbon Capture*; Woodhead Publishing: Sawston, UK, 2020; pp. 3–28.
49. United Nations Environment Programme. *Emissions Gap Report 2023: Broken Record–Temperatures Hit New Highs, Yet World Fails to Cut Emissions (Again)*; United Nations Environment Programme: Nairobi, Kenya, 2023. [CrossRef]

50. Canosa, I.V.; Ford, J.D.; McDowell, G.; Jones, J.; Pearce, T. Progress in climate change adaptation in the Arctic. *Environ. Res. Lett.* **2020**, *15*, 093009. [CrossRef]
51. Magnan, A.K.; Oppenheimer, M.; Garschagen, M.; Buchanan, M.K.; Duvat, V.K.E.; Forbes, D.L.; Ford, J.D.; Lambert, E.; Petzold, J.; Renaud, F.G.; et al. Sea level rise risks and societal adaptation benefits in low-lying coastal areas. *Sci. Rep.* **2022**, *12*, 10677. [CrossRef] [PubMed]
52. Tamura, M.; Kumano, N.; Yotsukuri, M.; Yokoki, H. Global assessment of the effectiveness of adaptation in coastal areas based on RCP/SSP scenarios. *Clim. Chang.* **2019**, *152*, 363–377. [CrossRef]
53. Phuong, L.T.H.; Biesbroek, G.R.; Wals, A.E. The interplay between social learning and adaptive capacity in climate change adaptation: A systematic review. *NJAS—Wagening J. Life Sci.* **2017**, *82*, 1–9. [CrossRef]
54. Ford, J.D.; Berrang-Ford, L. The 4Cs of adaptation tracking: Consistency, comparability, comprehensiveness, coherency. *Mitig. Adapt. Strateg. Glob. Chang.* **2016**, *21*, 839–859. [CrossRef] [PubMed]
55. Lesnikowski, A.; Ford, J.; Biesbroek, R.; Berrang-Ford, L.; Heymann, S.J. National-level progress on adaptation. *Nat. Clim. Chang.* **2016**, *6*, 261–264. [CrossRef]
56. Atteridge, A.; Remling, E. Is adaptation reducing vulnerability or redistributing it? *WIREs Clim. Chang.* **2018**, *9*, e500. [CrossRef]
57. Kehler, S.; Birchall, S.J. Climate change adaptation: How short-term political priorities trample public well-being. *Environ. Sci. Policy* **2023**, *146*, 144–150. [CrossRef]
58. Malik, I.H.; Hashmi, S.N.I. The great flood and its aftermath in Kashmir Valley: Impact, consequences and vulnerability assessment. *J. Geol. Soc. India* **2021**, *97*, 661–669. [CrossRef]
59. Theokritoff, E.; van Maanen, N.; Andrijevic, M.; Thomas, A.; Lissner, T.; Schleussner, C.-F. Adaptation constraints in scenarios of socio-economic development. *Sci. Rep.* **2023**, *13*, 19604. [CrossRef] [PubMed]
60. Otto, F.; Allen, M.; Stott, P.; van Oldenborgh, G.; Eden, J.; Karoly, D. Framing the question of attribution of extreme weather events. *Nat. Clim. Chang.* **2016**, *6*, 1–10.
61. Kaijser, A.; Kronsell, A. Climate change through the lens of intersectionality. *Environ. Politics* **2014**, *23*, 417–433. [CrossRef]
62. Thomas, K.; Hardy, R.D.; Lazrus, H.; Mendez, M.; Orlove, B.; Rivera-Collazo, I.; Roberts, J.T.; Rockman, M.; Warner, B.P.; Winthrop, R. Explaining differential vulnerability to climate change: A social science review. *Wiley Interdiscip. Rev. Clim. Chang.* **2019**, *10*, e565. [CrossRef] [PubMed]
63. Shepherd, M.; Kc, B. Climate change and African Americans in the USA. *Geogr. Compass* **2015**, *9*, 579–591. [CrossRef]
64. Rahman, M.F.; Falzon, D.; Robinson, S.A.; Kuhl, L.; Westoby, R.; Omukuti, J.; Schipper, E.L.; McNamara, K.E.; Resurrección, B.P.; Mfitumukiza, D.; et al. Locally led adaptation: Promise, pitfalls, and possibilities. *Ambio* **2023**, *52*, 1543–1557. [CrossRef] [PubMed]
65. Eriksen, S.; Schipper, E.L.F.; Scoville-Simonds, M.; Vincent, K.; Adam, H.N.; Brooks, N.; Harding, B.; Khatri, D.; Lenaerts, L.; Liverman, D.; et al. Adaptation interventions and their effect on vulnerability in developing countries: Help, hindrance or irrelevance? *World Dev.* **2021**, *141*, 105383. [CrossRef]
66. Birkmann, J.; Liwenga, E.; Pandey, R.; Boyd, E.; Djalante, R.; Gemenne, F.; Leal Filho, W.; Pinho, P.; Stringer, L.; Wrathall, D. Poverty, livelihoods and sustainable development. In *Climate Change 2022: Impacts, Adaptation and Vulnerability. Contribution of Working Group II to the Sixth Assessment Report of the Intergovernmental Panel on Climate Change*; Pörtner, H.-O., Roberts, D.C., Tignor, M., Poloczanska, E.S., Mintenbeck, K., Alegría, A., et al., Eds.; Cambridge University Press: Cambridge, UK; New York, NY, USA, 2022; pp. 1171–1284.
67. Sultana, F. Gendering climate change: Geographical insights. *Prof. Geographer.* **2014**, *66*, 372–381. [CrossRef]
68. Garcia, A.; Tschakert, P.; Karikari, N.A. 'Less able': How gendered subjectivities warp climate change adaptation in Ghana's Central Region. *Gend. Place Cult.* **2020**, *27*, 1602–1627. [CrossRef]
69. Kreibiehl, S.; Jung, T.Y.; Battiston, S.; Carvajal, P.E.; Clapp, C.; Dasgupta, D.; Dube, N.; Jachnik, R.; Morita, K.; Samargandi, N.; et al. Investment and finance. In *IPCC, 2022: Climate Change 2022: Mitigation of Climate Change. Contribution of Working Group III to the Sixth Assessment Report of the Intergovernmental Panel on Climate Change*; Shukla, P.R., Skea, J., Slade, R., Al Khourdajie, A., van Diemen, R., McCollum, D., Pathak, M., Some, S., Vyas, P., Fradera, R., Eds.; Cambridge University Press: Cambridge, UK; New York, NY, USA, 2022. [CrossRef]
70. Falzon, D.; Batur, P. Lost and damaged: Environmental racism, climate justice, and conflict in the Pacific. In *Handbook of the Sociology of Racial and Ethnic Relations*; Springer: Berlin/Heidelberg, Germany, 2018; pp. 401–412.
71. McNamara, K.E.; Clissold, R.; Westoby, R.; Stephens, S.; Koran, G.; Missack, W.; Bartlett, C.Y. Using a human rights lens to understand and address loss and damage. *Nat. Clim. Chang.* **2023**, *13*, 1334–1339. [CrossRef]
72. Knox, J.H.; Morgera, E. *Human Rights and the Environment: The Interdependence of Human Rights and a Healthy Environment in the Context of National Legislation on Natural Resources*; Food & Agriculture Organization: Rome, Italy, 2022; Volume 109.
73. Naylor, A.W.; Ford, J. Vulnerability and loss and damage following the COP27 of the UN Framework Convention on Climate Change. *Reg. Environ. Chang.* **2023**, *23*, 38. [CrossRef]
74. Falzon, D.; Shaia, F.; Roberts, J.T.; Hossain, F.; Robinson, S.-A.; Khan, M.R.; Ciplet, D. Tactical opposition: Obstructing loss and damage finance in the United Nations climate negotiations. *Glob. Environ. Politics* **2023**, *23*, 95–119. [CrossRef]
75. Robinson, S.A.; Roberts, J.T.; Weikmans, R.; Falzon, D. Vulnerability-based allocations in loss and damage finance. *Nat. Clim. Chang.* **2023**, *13*, 1055–1062. [CrossRef]

76. Roberts, E.; Pelling, M. Loss and damage: An opportunity for transformation. In *The Third Pillar of International Climate Change Policy;* Routledge: Oxfordshire, UK, 2021; pp. 105–118.
77. Roberts, E.; Huq, S. Coming full circle: The history of loss and damage under the UNFCCC. *Int. J. Glob. Warm.* **2015**, *8*, 141–157. [CrossRef]
78. Lindegaard, L.S.; White, H.; Shawoo, Z. Making Headway on Loss and Damage: Bridging Policy, Research and Practice (No 2022: 07). DIIS Report. 2022. Available online: https://www.econstor.eu/bitstream/10419/272806/1/1820491692.pdf (accessed on 10 November 2023).
79. McNamara, K.E.; Jackson, G. Loss and damage: A review of the literature and directions for future research. *WIREs Clim. Chang.* **2019**, *10*, e564. [CrossRef]
80. Boyd, E.; James, R.A.; Jones, R.G.; Young, H.R.; Otto, F.E.L. A typology of loss and damage perspectives. *Nat. Clim. Chang.* **2017**, *7*, 723–729. [CrossRef]
81. Roberts, J.T.; Natson, S.; Hoffmeister, V.; Durand, A.; Weikmans, R.; Gewirtzman, J.; Huq, S. How will we pay for loss and damage? *Ethics Policy Environ.* **2017**, *20*, 208–226. [CrossRef]
82. Serdeczny, O.; Waters, E.; Chan, S. *Non-Economic Loss and Damage in the Context of Climate Change: Understanding the Challenges;* (No. 3/2016) Discussion Paper; Deutsches Institut für Entwicklungspolitik (DIE): Bonn, Germany, 2016.
83. Tschakert, P.; Ellis, N.; Anderson, C.; Kelly, A.; Obeng, J. One thousand ways to experience loss: A systematic analysis of climate-related intangible harm from around the world. *Glob. Environ. Chang.* **2019**, *55*, 58–72. [CrossRef]
84. Cunsolo, A.; Harper, S.L.; Minor, K.; Hayes, K.; Williams, K.G.; Howard, C. Ecological grief and anxiety: The start of a healthy response to climate change? *Lancet Planetary Health* **2020**, *4*, e261–e263. [CrossRef]
85. Cunsolo, A.; Ellis, N.R. Ecological grief as a mental health response to climate change-related loss. *Nat. Clim. Chang.* **2018**, *8*, 275–281. [CrossRef]
86. O'Neill, B.C.; van Aalst, M.; Zaiton Ibrahim, Z.; Berrang-Ford, L.; Bhadwal, S.; Buhaug, H.; Diaz, D.; Frieler, K.; Garschagen, M.; Magnan, A.K.; et al. Key Risks Across Sectors and Regions. In *Climate Change 2022: Impacts, Adaptation and Vulnerability. Contribution of Working Group II to the Sixth Assessment Report of the Intergovernmental Panel on Climate Change;* Cambridge University Press: Cambridge, UK; New York, NY, USA, 2022.
87. Malik, I.H. Spatial dimension of impact, relief, and rescue of the 2014 flood in Kashmir Valley. *Nat. Hazards* **2022**, *110*, 1911–1929. [CrossRef] [PubMed]
88. Jamal, S.; Malik, I.H.; Ahmad, W.S. Dynamics of urban land use and its impact on land surface temperature (LST) in Aligarh City, Uttar Pradesh. In *Re-Envisioning Advances in Remote Sensing;* CRC Press: Boca Raton, FL, USA, 2022; pp. 25–40.
89. Meath, C.; Linnenluecke, M.; Griffiths, A. Barriers and motivators to the adoption of energy savings measures for small- and medium-sized enterprises (SMEs): The case of the ClimateSmart Business Cluster program. *J. Clean. Prod.* **2016**, *112*, 3597–3604. [CrossRef]

Disclaimer/Publisher's Note: The statements, opinions and data contained in all publications are solely those of the individual author(s) and contributor(s) and not of MDPI and/or the editor(s). MDPI and/or the editor(s) disclaim responsibility for any injury to people or property resulting from any ideas, methods, instructions or products referred to in the content.

applied sciences

MDPI

Article

Visualising the Relevance of Climate Change for Spatial Planning by the Example of Serbia

Marijana Pantić *, Tamara Maričić and Saša Milijić

Institute of Architecture and Urban & Spatial Planning of Serbia, Bulevar Kralja Aleksandra 73/II, 11000 Belgrade, Serbia; tamara@iaus.ac.rs (T.M.); sasam@iaus.ac.rs (S.M.)
* Correspondence: marijana.d.pantic@gmail.com; Tel.: +381-11-337-0091; Fax: +381-11-337-0203

Featured Application: As the urgency of climate change intensifies, the need for effective spatial planning becomes paramount. This article explores techniques in visualization that empower planners to tackle climate-related challenges and estimate whether balance between state of the art, mitigation measures and adaptation measures is achieved.

Abstract: After decades of rising awareness and undertaken actions, climate change is still one of several focal global challenges. Additionally, the latest report by researchers at the International Panel for Climate Change indicates that the crisis has deepened. With its comprehensive nature, spatial planning is one of the management tools responsible for dealing with climate change and combating its effects. Land use definition is the foundation on which we build mitigation and adaptation systems. It is a complex process that involves (or should involve) a range of stakeholders—experts, politicians, the civil sector, and citizens—in which the clear transmission of messages to stakeholders regarding the state of the art and planned actions is significant. The use of visualisation tools is one of the important ways to achieve this. This research aims to present a set of visualisation tools, applying them in analysis and decision making in the field of spatial planning with regard to climate change. We combined content analysis, colour-graded classification, and the spider method applied to the example of Serbia. The results showed that application of the suggested visualisation methods in combination with regular planning tools (maps) facilitates an understanding of the problem and its presentation to other stakeholders. In the case of Serbia, visualisation tools have shown that adaptation measures prevail over mitigation measures and that the effects of climate change addressed in spatial-planning documents do not significantly match the most challenging effects as perceived from the citizens' perspective. These are aspects that should be corrected in the next generation of planning documents. The suggested visualisation tools are replicable, with slight adjustments to a specific case, to any other region in the world.

Keywords: climate change; spatial planning; content analysis; mitigation measures; adaptation measures; visualisation tools; Serbia

1. Introduction

Long-term shifts in temperature and weather patterns, recognized as *climate change*, have a strong impact around the globe. Scientists first acknowledged rising global temperatures [1], subsequently accompanied by extreme weather events (floods, heat waves, droughts, cyclones, and wildfires), which can have huge negative impacts on human health and lives, animals, plants, ecosystems, and built infrastructure [2].

There are several theories about the causes of global warming. The most commonly accepted theory by scientists and international organisations/institutions is called anthropogenic (man-made) global warming, which is reflected in Article 1 of the Framework Convention on Climate Change (UNFCCC) as part of the Intergovernmental Panel on Climate Change (IPCC) Glossary: "a change of climate which is attributed directly or indirectly to human activity that alters the composition of the global atmosphere and which is in addition to natural climate variability observed over comparable time periods" [3]. Reflections on the connection between human activities and climate change started at the end of the 19th century, when Swedish researcher Svante Arrhenius developed the idea that the use of fossil fuels adds CO_2 to the atmosphere, thus altering some parameters of climate conditions [1,4]. This important hypothesis was rather neglected till the 1950s, when researchers started to test it with technically improved methods [5]. The results indicated that the hypothesis was correct, and measurements in the following decades showed a constant rise in temperature and CO_2 concentration in the atmosphere [6–8]. However, the World Meteorological Organization (WMO) had more serious concerns at the end of the 1970s, warning that an increase in CO_2 in the atmosphere could be the cause of global warming [2].

It took another decade before WMO and the United Nations Environment Programme (UNEP) managed to establish the IPCC in 1988 as a United Nations (UN) body to advance knowledge on climate change. The result was the creation of responsible and suitable climate policies and constant international negotiations on climate change mitigation and adaptation [9]. Following IPCC efforts, the UN proclaimed taking "urgent action to combat climate change and its impacts" [8] as one of the sustainable development goals in 2015 [10]. Showing their awareness about climate change issues and in order to put solutions in practice, 196 parties ratified the Paris Agreement—a legally binding international treaty on climate change, which was adopted at the Conference of Parties (COP21) in Paris in 2015 and entered into force in 2016 [8].

The global temperature in Europe is rising. Serbia belongs to the group of countries where the number of cold days and nights has decreased, while the number of warm days and nights has increased [11]. Central European countries greatly contribute to the net short-term warming caused by emissions [12]. According to the latest scientific data (ibid.), observed trends for Central Europe show significant increases in the intensity and frequency of hot extremes and significant decreases in the intensity and frequency of cold extremes. There is significant intensification of heavy precipitation, days with thunderstorms have an increasing trend along with an increase in river floods, while drought trends remain unchanged. It is very likely that these man-influenced trends will continue. The frequency of heat waves will likely increase in all European regions, and heat stress due to both high temperatures and humidity will affect morbidity, mortality, and labour capacity (ibid.).

Data presented in the Second National Communication of the Republic of Serbia under the UN Framework Convention on Climate Change [13] show that there has been a significant increase in daily mean temperatures since 1960, as well as in daily maximum and daily minimum temperatures (with an average 0.3 °C increase per decade). The Republic of Serbia has experienced an increased number of episodes with heavy precipitation. The major changes have been in rising temperature trends, coupled with extremely high temperatures and extended warm periods. Climate change scenarios for the time horizon 2040–2070–2100 predict a plausible temperature rise and precipitation decrease, along with a reduction in frost days and an increase in summer days (ibid.). If they are realised, these estimates will have numerous negative effects on many sectors, especially hydrology and water resources, forestry, agriculture, and healthcare.

Spatial planning, owing to its multidisciplinary nature, is recognized as a key element in dealing with climate change, referring to both mitigation and adaptation [14,15]. It is considered to be a framework that plays a crucial role in linking mitigation and adaptation to climate change [16,17], as well as reconciling various elements of built-up space and environment in which interventions are required [18]. The latest prognoses of climate change trends indicate that it will have serious negative impacts on humanity and consequently create challenges in the field of spatial planning, particularly in aspects of land use regulation, urban design, planning engineering measures, and adaptation/mitigation infrastructure which play a crucial role in coping with future challenges [11,19,20].

At the city level, it has been proved that engaged spatial planning reduces the severity of the heat island effect [18], while on a more general scale, it will be the key to solving challenges such as population displacement caused by climate change impact [11]. Although the history of spatial planning is very long, decision-makers in Europe have only recently started to consider it a tool in combating the negative effects of climate change [21], but still not as much as they should or could [22,23]. Some case studies have been conducted to analyse and assess the impact of spatial planning on climate resilience (e.g., [21,22]). There are also a few international initiatives that emphasize the role of spatial planning instruments in coping with climate change, such as the CLISP—Climate Change Adaptation by Spatial Planning in the Alpine Space Project [24] or OECD Climate Change in the European Alps—Adapting Winter Tourism and Natural Hazards Management publication [25].

The results of the literature search on the topic of visualization methods in spatial planning related to climate change did not indicate that researchers have addressed this issue. From experience, as well as a review of existing plans analysed in this work, it has been shown that, apart from maps, other visualization tools for representing or analysing climate change issues are not in use.

Maps, as a means of visualisation, have always been part of spatial planning. Geographically represented visualisation has been greatly improved and popularized with the introduction of Geographical Informational Systems (GIS) [26]. But, maps are not the only tool for visual communication with stakeholders. Namely, the development and popularization of technology have brought many new opportunities and tools to visualise urban and spatial planning measures and concepts [27,28]. Currently, the use of virtual reality and augmented reality are new trends with growing relevance for planning and public participation processes [29,30]. All these methods that are closely related to the use of technology and visualisation became indispensable during the COVID-19 pandemic because it was the only way to include stakeholders in the planning process [31].

This research aims to suggest a set of visualisation tools in analysis and decision making in the spatial planning process with regard to climate change. Spatial planning (spatial plans) is understood here to be a management tool to deal with the global challenge of climate change effects and awareness. The tools were applied to the example of Serbia, i.e., its spatial-planning documents on the national and regional level. The suggested tools have an internationally applicable value since, with slight adjustments, they can be applied in any other territorial context. The Data and Methods chapter will present more details on the chosen tools and research process, followed by the Results chapter, where the presented tools will be applied to the example of Serbia's spatial-planning documents. A broader context for understanding the suggested tools and results will be given in the Discussion chapter. Finally, the Conclusions chapter will give a quick overview of the main messages, the applicability of the suggested visualisation methods, and potential improvements for future research.

2. Data and Methods

This research is built around two main pillars: one addresses spatial planning as a tool to deal with climate change, and the other addresses the methodology of visual analysis and presents the role of spatial planning in dealing with climate change effects. In this regard, this research combines several methods in order to visualise relations between climate change and spatial planning. We use Serbia only as an example because the suggested visualisation methodology can be applied in the context of any country or even at the international, regional and local level.

2.1. Data Collection Methods

The Law on Planning and Construction (2023) [32] defines four types of spatial plans in Serbia: (1) spatial plans for the Republic of Serbia, (2) regional spatial plans, (3) spatial plans for special-purpose areas, and (4) spatial plans for local self-governing units. The spatial plan for the Republic of Serbia, as an overarching strategic document, provides strategic guidelines for regional and other spatial plans at lower planning levels. The whole territory of Serbia is encompassed by regional plans as well as local plans. The relationship between climate change and spatial planning is analysed here through the current set of national and regional documents, i.e., the national spatial plan, all nine regional plans, and some additional planning documents that accompany them such as strategic environmental assessment (SEA) and implementation plans: the Spatial Plan of the Republic of Serbia (2010) [33] and its SEA (2010) [34], and Implementation Program (2011) [35]; the Regional Spatial Plan for Južno Pomoravlje (2010) [36]; the Regional Spatial Plan for the Administrative Area of the City of Belgrade (2011) [37]; the Regional Spatial Plan for the Autonomous Region of Vojvodina (2011) [38], with its Implementation Program (2013) [39]; the Regional Spatial Plan for Timočka Krajina (2011) [40] and its Implementation Program (2015) [41]; the Regional Spatial Plan for Nišavski, Toplički, and Pirotski Districts (2013) [42]; the Regional Spatial Plan for Zlatiborski and Moravički Districts (2013) [43]; the Regional Spatial Plan for Šumadijski, Pomoravski, Raški, and Rasinski Districts (2014) [44]; the Regional Spatial Plan for Podunavski and Braničevski Districts (2015) [45]; and the Regional Spatial Plan for Kolubarski and Mačvanski Districts (2015) [46].

2.2. Data Analysis and Visualisation Methods

Content analysis is a tool for determining the presence of certain words, themes, or concepts within qualitative data (i.e., spatial-planning documents) [47,48]; Content analysis enables researchers to analyse and quantify the presence and frequency of selected terms, thus interpreting their relevance. It is commonly used in social sciences and the context of climate change e.g., [49,50]., although not in the field of spatial planning, which is the intention of this paper. We used content analysis with the aim of developing a visual method to present it.

Therefore, we used information visualisation methods in order to communicate in a graphical, interactive, and understandable way, since 90% of the information transmitted to the brain is visual [51,52]. Until recently, research in social sciences posed fundamental challenges to the application of visualisation techniques [53]. However, in the field of spatial planning, it has been acknowledged since the very beginning that visual communication, featuring significantly more graphic constituents than text, has an indispensable role in providing the necessary information. A review analysis by Shen et al. [54] showed that most common visualisation methods are based on lines, plots, maps, parallel coordinates, and radial coordinates. Based on a comparative assessment of some of the most popular visualisation methods (presented in Table 1), we further elaborated our content analysis findings through a *colour-graded classification* and the *spider* method as the simplest, most readable, affordable, and fastest to produce.

Charts (line charts, bar charts, pie charts, and histograms) are one of the most common visualisation methods that are simple and fast to produce and which do not require basic software and skills, but are rather easy to read and understand. Diverse types of charts correspond to different needs. A line chart displays data points connected by straight lines, illustrating trends and variations over a continuous range. It is effective for showing changes in data over time or across categories. A bar chart represents data using rectangular bars of varying lengths that correspond to the value it represents, making it easy to compare values among different categories. A histogram represents the distribution of a dataset. It consists of bars where the width represents intervals and the height represents the frequency or count of data points within each interval. It is commonly used in statistical analysis to show the frequency distribution of continuous data. Finally, a pie chart represents data in a circular graph, divided into slices to illustrate the proportion of each category relative to the whole. It is effective for displaying the composition of a whole and showing the relationship of parts to the whole.

Table 1. Comparison of selected information visualisation methods.

Visualisation Method	Simplicity and Speed of Production	Affordability (Software, Skilled Workforce)	Clarity and Readability	Comment
Line chart, bar chart, histogram, pie chart	++	+	+	Simple and fast production, requires basic software and skills, rather easy to read and understand.
Colour-graded classification	++	++	++	Very simple and fast to produce; does not require special software nor skills; easy to read.
Cartographic representations/maps	−−	−−	++	Requires time, software and skills to produce. Easy to read, but not all data can be spatially related; too much information or too many colours can make reading difficult.
Radar chart/spider	+	+	++	Simple and fast to produce; semi-skilled workforce (can be used to visualise the relative strengths and weaknesses of the selected case studies or different scenarios for various chosen factors).
Word cloud	+	+	+	Simple to produce but requires software. Limited clarity/readability, for text summarisation (and analysis).

Source: [55–57]. Note: the symbols "+" and "−" and their multiplication signify the intensity of positive and negative aspects associated with each specific visualisation method.

Maps as cartographic representations are inextricably linked with spatial planning. They are graphical representations of geographic information, providing a visual depiction of spatial relationships and features on the Earth's surface. They use symbols, colours, and scale to convey information about locations, landscapes, and various spatial attributes. However, their production requires time, costly software and experts with specific skills that require time to develop. They are easy to read, but cannot be applied in the representation of all aspects (measures) of climate change since they do not all have spatial connotations nor are they convenient to present on a map.

A word cloud is mainly used for text summarisation and the analysis of qualitative data (in business and academia); it is simple to produce but requires software and has limited clarity/readability. A word cloud is a visual representation of text data where words are arranged in a graphical manner, and the size of each word is determined by its frequency or importance in the given text. Commonly used words appear larger and more prominent. It is often employed to quickly identify and communicate key themes or topics within a set of textual information. According to Table 1, it is clear that simplicity and speed of production are often opposite to costs (i.e., simple methods are usually not very expensive), while correlation with readability varies.

Colour-graded classification is a simple visual tool developed for this research. Colour-graded classification is a method of categorizing or representing data using a colour spectrum. In this technique, different colours are assigned to different categories or values within a dataset. The variation in colour intensity or hue indicates variations in the underlying data, providing a quick visual understanding of patterns or differences. It is commonly used in maps, charts, and visualizations to enhance the interpretation of information and highlight distinctions within a dataset. In this case, it is a visual representation, resembling a table, where selected measures related to climate change are distributed in four categories according to frequency: (1) dark blue—measures present in all analysed documents (100%); (2) light blue—measures present in the majority of the analysed documents (more than 50%, less than 100%); (3) light red—measures occurring only in some of the analysed documents (above 0%, less than or equal to 50%); and (4) dark red—measures not occurring in the analysed documents (Figure 1). This colour-graded classification lists all the measures tracked via content analysis by order of their occurrence in the analysed spatial plans. The frequency of their occurrence was measured by the number of planning documents addressing each particular measure and by the share of planning documents addressing each particular measure.

No.	Keyword (Measure)	Planning documents addressing keyword in a context of a measure(s)	
		No.	%
1	Landslide "Klizište"	14	100.0
2	Afforestation "Pošumljavanje"	14	100.0
3	Biomass "Biomasa"	14	100.0
4	Landfill "Deponije"	14	100.0
5	Waste water "Otpadne vode"	14	100.0
6	Irrigation "Navodnjavanje", "irigacija"	13	92.9
7	Education "Edukacija"	13	92.9
8	Dam "Brana"	12	85.7
9	Water reservoir "(Vodo)akumulacija"	12	85.7
10	River bed "Rečno korito"	12	85.7
11	Etablishing new prot. areas "Proglašenje novih zašt. područja"	12	85.7
12	Notification "Obaveštavanje"	12	85.7
13	Embankment "Nasip"	11	78.6
14	Drainage "Odvodnjavanje", "drenaža"	11	78.6
15	Channel "Kanal"	11	78.6
16	Plan (combating drought, weather disasters, fires, climate change, etc.) "Plan"	11	78.6
17	Credit "Kredit"	11	78.6
18	Farmers cooperative "Zemljoradnička zadruga"	11	78.6
19	Monitoring "Monitoring"	11	78.6
20	Climate change "Klimatske promene"	11	78.6
21	Greenhouse "Plastenik"	9	64.3
22	Strategy, strategic "Strategija", "strateški"	8	57.1
23	Civil protection "Civilna zaštita"	8	57.1
24	Bank revetments "Obaloutvrda"	7	50.0
25	Retention "Retenzija"	6	42.9
26	Rehabilitation of buildings "Sanacija objekata"	6	42.9
27	Insurance "Osiguranje"	6	42.9
28	Meteorological station "Meteorološka stanica"	6	42.9
29	Warning system "Uzbunjivanje"	4	28.6
30	Electric drive, Electric power "Električni pogon"	3	21.4
31	Drop by drop "Kap po kap"	3	21.4
32	Financial support "Finansijska podrška"	3	21.4
33	Hail "Grad"	2	14.3
34	Anti-hail "Protivgradni"	1	7.1
35	Sprinkling (system) "Orošavanje"	0	0
36	Correcting the course of the river "Ispravljanje rečnog toka"	0	0
37	Sand bag dyke "Zečji nasip"	0	0
38	Terrain slope "Nagib terena"	0	0

Figure 1. Colour-graded classification of keywords (addressed measures) by frequency. (dark blue—the highest frequency, dark red—the lowest frequency).

The *spider method*, based on radial coordinates and radar charts, or the practice of displaying data in a circular pattern, is an increasingly common technique in information visualisation research. It is a graphical method of displaying multivariate data in the form of a two-dimensional chart with three or more quantitative variables represented on axes radiating from a central point. It uses connected lines to illustrate data points, helping to visualize patterns and comparisons across multiple variables simultaneously. Each axis represents a different variable, and data points are plotted where these axes intersect. Radar charts are useful for displaying and comparing data distribution patterns. This visualisation tool is characteristic for comparative scenario studies in decision-making processes [58]. It has already been applied by researchers in the field of spatial planning in Serbia e.g., [59–61], although not in the visualisation of the relationship between climate change and spatial planning. The essence of the method is to transform unrelated variables (e.g., population density, employment rate) into standardized spider values, and scale the axes from zero (interior) to ten (outer edge) so that a higher score represents better performance. Our research included only related variables (all presenting the occurrence frequency of the measures); hence, absolute values were used instead of standardized spider values. As in the spider tool with standardized values, occurrence frequency in the analysed spatial-planning documents rises in the graph with distance from the graph interior.

2.3. Research Design

Visualisation methods have been part of extensive research because their application "clarifies the complexity of the models [and] ... creates an overview and transparency" ([62] p. 1). Some researchers chose visualisation to increase the expressiveness and attractiveness of overall environmental issues [63,64], urban planning, and carbon emissions [65], etc. In urban and spatial planning, visualisation has gone beyond its role as a tool for researchers and experts and has become a requirement in decision-making processes when addressing citizens and other stakeholders [31,66] Therefore, the research design presented in this paper can have relevant applicability beyond this paper and the example of Serbia.

For content analysis, the focus of this research dictated that one of the concepts and keywords to look for was "climate change". Based on this, we selected 38 keywords that were traced and observed in the context of climate change, i.e., in the form of a measure proclaimed by the analysed planning document. The keywords were selected based on climate change measures and concepts recognized in the previously conducted study by Pantić et al. [61]. In that study, citizens were asked to enumerate all the climate change measures they were familiar with. Since the nature of some enumerated measures does not pertain to the field of spatial planning (e.g., the selection of crop type, changes of field work schedules), we chose not to consider those measures in the selection of keywords and concepts. The list of 38 keywords (measures) is presented in Figure 1. To trace the selected keywords in the documents, we used the command "find" in digital versions of the analysed planning documents (Word or PDF files) and subsequently checked the context in which the keyword was used. If the context referred to climate change and represented a proclaimed measure, we considered it a positive result and used it in the next step of the analysis.

In the next step, we used binary analysis and frequency grading. A matrix was created where spatial-planning documents intersect keywords. If the keyword was used in the document in the context of climate change measures, this was indicated with a "1"; if the keyword was not found, this was indicated with a "0". In our document analysis, the keywords we sought served as indicators, guiding us to specific sections within the documents. After identifying each keyword, we meticulously examined the context in which it was employed, assessing whether it pertained to climate change measures. This process not only allowed us to confirm climate-change-related content but also to scrutinize different sections of the text and potentially identify new keywords that enriched our analytical framework. Part of the matrix is shown in Table 2 as an example. Table 2 also illustrates that "TOTAL" represents the frequency of keywords throughout the analysed documents, not in terms of the frequency of occurrence in individual documents but the number of documents in which they appear. Figure 1 is the outcome and the first visual representation in the analysis. It shows the relevance of certain measures to spatial planning in Serbia in four categories, based on the number/percentage of plans in which measures occur in the climate change context.

Table 2. Measures addressing climate change—example of the matrix for binary analysis and frequency grading.

		Anti-Hail "Protivgradni"	Hail "Grad"	Bank Revetments "Obaloutvrda"	...
1	Spatial Plan of the Republic of Serbia 2010–2020	1	1	1	...
2	Strategic Environmental Impact Assessment for the SPRS 2010–2020	0	0	0	...
3	Implementation Program for the SPRS 2010–2020	0	0	0	...
	...	...	...	...	...
	TOTAL	1	2	7	...

The next task was to simplify represent the results. To achieve that, we distributed the 38 keywords into categories, with each corresponding to a different climate change effect. The climate change effects we selected based on the relevant literature review e.g., [31,67–71], typical for the territory of Serbia, are the following: drought, floods, hail/storms, heat waves, landslides, and premature snow melting. The distribution of keywords, hence, measures, is displayed in Table 3. Some keywords/measures were considered in several categories (e.g., 2—afforestation, 11—establishing new protection areas) since they influence multiple aspects, i.e., climate change effects.

Table 3. Classification of keywords (measures) by climate change effect.

Climate Change Effect	Keywords (Measures) (Ordinal Numbers in Figure 1)
Drought	2, 5, 6, 7, 8, 9, 11, 12, 15, 16, 17, 18, 19, 20, 22, 23, 25, 27, 28, 29, 31, 32, 35
Floods	1, 2, 7, 8, 9, 10, 11, 12, 14, 13, 15, 16, 17, 18, 19, 20, 22, 23, 24, 26, 27, 28, 29, 32, 36, 37, 38
Hail/Storms	7, 11, 12, 16, 17, 18, 19, 20, 21, 22, 23, 26, 27, 28, 29, 32, 33, 34
Heat waves	2, 3, 6, 7, 8, 9, 12, 11, 15, 16, 17, 18, 19, 20, 21, 22, 23, 25, 28, 29, 30, 31, 32, 35
Landslides	1, 2, 7, 8, 9, 10, 11, 12, 13, 16, 19, 20, 22, 23, 24, 26, 27, 28, 29, 32, 38
Premature snow melting	2, 3, 5, 7, 9, 11, 16, 19, 20, 22, 25, 28, 30, 38

Only then was it possible to proceed with the quantification of the appearance of each effect for each planning document (Figure 2). This was performed by summing up the number of keywords addressed in each plan per each climate change effect, as shown in Figure 2. The result is a matrix that shows how many aspects of dealing with climate change were addressed in each analysed plan in relation to each climate change effect separately. Additionally, the matrix allows for comparisons between plans and between extents to which climate change effects are addressed ("Total per plan" and "Total per effect" in Figure 2). Parallel to the matrix, we applied the spider visualisation method that targets the presentation of climate change effects in spatial planning, i.e., the spatial planning of Serbia (Figure 2).

Count	Planning document	Droughts	Floods	Hail/Storms	Heat waves	Landslides	Premature snow melting	TOTAL PER PLAN
1	Spatial Plan of the Republic of Serbia 2010-2020 (SPRS)	20	22	15	21	18	13	109
2	Strategic Environmental Impact Assessment for SPRS	16	16	10	16	14	12	84
3	Implementation Program for SPRS	12	14	8	12	12	9	67
4	Regional Spatial Plan for Južno Pomoravlje	20	19	15	22	17	12	105
5	Regional Spatial Plan for the Administrative Area of the City of Belgrade	17	19	12	20	15	11	94
6	Regional Spatial Plan for the Autonomous Region of Vojvodina (RSPARV)	16	17	10	19	14	13	89
7	Implementation Program for RSPARV	12	13	10	13	10	9	67
8	Regional Spatial Plan for Timočka Krajina (RSPTK)	16	19	11	18	15	9	88
9	Implementation Program for RSPTK	15	18	10	16	14	9	82
10	Regional Spatial Plan for Nišavski, Toplički, and Pirotski Districts	14	18	10	16	15	9	82
11	Regional Spatial Plan for Zlatiborski and Moravički Districts	21	24	16	22	20	13	116
12	Regional Spatial Plan for Šumadijski, Pomoravski, Raški, and Rasinski Districts	14	15	10	15	14	8	76
13	Regional Spatial Plan for Podunavski and Braničevski Districts	14	16	10	16	12	9	77
14	Regional Spatial Plan for Kolubarski and Mačvanski Districts	19	21	12	20	17	13	102
	TOTAL PER EFFECT	226	251	159	246	207	149	

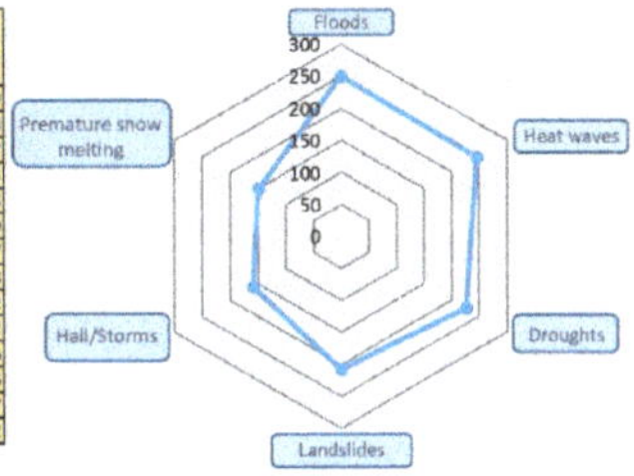

Figure 2. Addressing climate change effects in spatial-planning documents.

As maps are a consistent part of spatial plans where measures related to climate change can be potentially presented, we added a map analysis to the analysis of spatial plans. The focus was on infrastructural, conceptual, and territorial elements (river dams, errosian endangered areas, protection zones, etc.). In addition to the fact that such an analysis indicates the involvment of specific plans in climate change, these examples contribute to the general interperetation of mapping as a standard visualisation method in spatial planning.

Therefore, this study proposes a methodology for the visual evaluation of the extent to which climate change is tackled in spatial-planning documents using the example of Serbia, while noting its applicability in any other international, national, regional, or local context.

3. Results

This section presents the results of the planning document analysis, i.e., the presence of climate change themes and climate change measures in the spatial planning of Serbia. The results are interpreted descriptively and visually using combined visual methods.

The Spatial Plan of the Republic of Serbia 2010–2020, as the overarching planning act, included the topic of climate change in Serbia's spatial planning with the introduction of an independent section titled "Effects of Climate Change". The Regional Spatial Plan for the Administrative Area of the City of Belgrade (2011) [37] followed this example identically. Other analysed plans, however, did not treat climate change issues through a separate chapter; rather, climate change effects and climate change-related measures were addressed in sections about agricultural development, environmental protection, energy consumption, and others.

Some keywords (i.e., measures) in Figure 1 can be interpreted as measures directly triggered by climate change concerns, such as education and increased awareness, calls for specific plans to combat drought, weather disasters, fires, etc., the use of biomass, and the introduction of infrastructure for electric vehicles. According to Figure 1, only a few measures are integrated in all analysed documents: landslide regulation, afforestation, the intensification of biomass use, the treatment of landfills in a sanitary manner (modernisation), and wastewater treatment. On the other hand, there are measures that none of the spatial-planning documents prescribed, such as sprinkling systems in agriculture or for urban green infrastructure maintenance, the correction of the course of rivers, sandbag dykes, or the correction of terrain slopes to mitigate erosion and other effects of heavy precipitation, storms, and floods. The same colour-graded classification of measures indicates that the number of vastly prescribed measures is slightly larger than those omitted in spatial-planning documents. The majority of measures (light blue) are contained in most of the analysed documents and there are measures (light red) whose presence should be improved in the new generation of spatial plans in Serbia.

Figure 2 shows how the most relevant climate change effects are addressed in the analysed plans in Serbia. The final row (total per effect) and the accompanying spider graph show that the most addressed climate change effect is flooding, followed by heat waves and droughts to a significant extent, and less significantly by landslides, hail storms, and premature snow melting. The final column (total per plan) shows how the extent of the reaction towards climate change effects is distributed over the analysed documents, i.e., that the Regional Spatial Plan for Zlatiborski and Moravički Districts has the best coverage and that the implementation programs are not as engaged with climate change effects as the plans themselves. Besides the mentioned regional spatial plan, the Spatial Plan of the Republic of Serbia has good coverage of all climate change effects. Various types of documents serve distinct purposes, and as a result, it may seem unnecessary for all of them to uniformly address climate change. Nevertheless, identifying omissions in the table can signal aspects that warrant potential consideration.

Out of ten analysed spatial plans, eight include maps on which some climate change-related measures are presented. It appears that their relation to climate change issues is more auxiliary than explicit, which can also be concluded from the map titles, e.g., a map of hazards, a map of tourism and environmental protection, a map of infrastructure, etc. Also, on their maps, the Regional Plan for Južno Pomoravlje, the Regional Spatial Plan for Timočka Krajina, and the Regional Spatial Plan for Šumadijski, Pomoravski, Raški, and Rasinski Districts present only zones of living environment according to four quality categories, which might also be related to climate change effects, but this cannot be clearly read from the maps. The elements related to climate change on the maps of the remaining spatial plans are most often water-management infrastructure—dams, the regulation of river beds, and channels; highlighting environmental impairment categories; and locations/areas exposed or at risk of exposure to erosion processes and landslides. The most prominent example would be the Regional Spatial Plan of Nišavski, Toplički and Pirotski Districts (2013), with an accompanying map of tourism and environmental protection, which shows erosion, potential flooding areas, riverbed regulation activities, and an infrastructure map, which shows water-management infrastructure and planned dams (Figure 3).

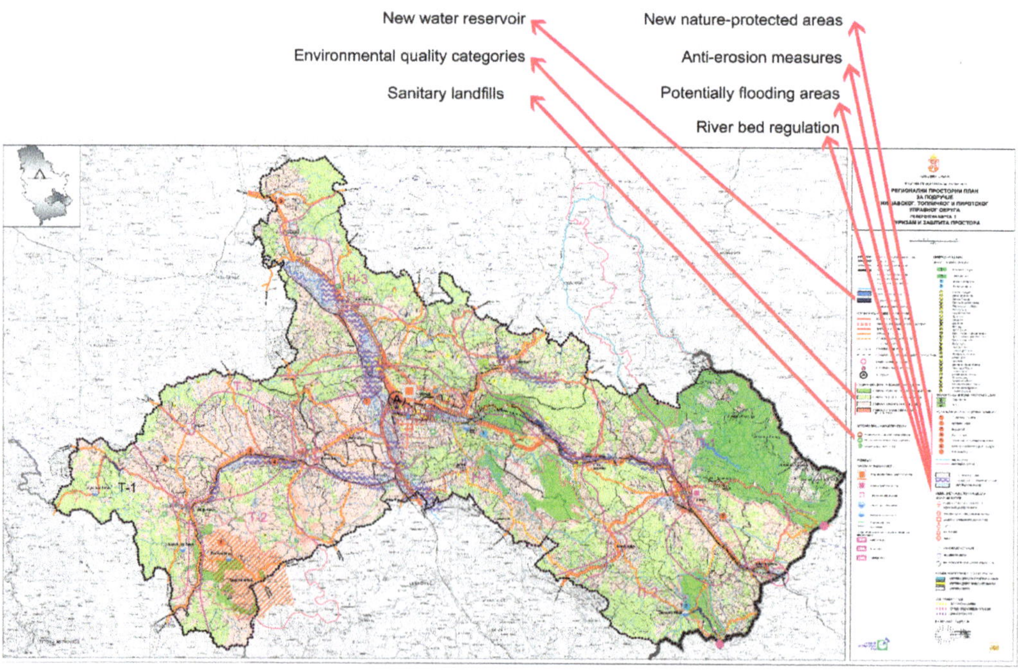

Figure 3. Map as an example of presenting climate change measures. Source: prepared based on the internal IAUS database, ©JP Zavod za urbanizam Niš & IAUS.

4. Discussion

Our analysis shows that most of the measures to combat climate change effects in Serbia have been circumstantial so far. The regulation of riverbeds, care of channels, irrigation-drainage systems, etc. were part of the spatial planning setup before climate change took the global spotlight. But, over time, their significance increased with the occurrence of new environmental problems. Döll [72] stresses the increasing significance of irrigation systems from the perspective of global climate change trends, Van Binh et al. [73] indicate the close relationship between climate change and changes in riverbeds, and Doelman et al. [74] investigate and prove the high importance of afforestation in climate change mitigation. All these are "traditional" measures in spatial planning in Serbia, which are beneficial in the context of climate change.

Climate change measures are differentiated as mitigation and adaptation measures, where the former are in charge of reducing and stabilizing the levels of heat-trapping greenhouse gases in the atmosphere, prevailingly CO_2, and the latter are adjustments in response to climate change effects [75,76]. By observing the list of measures in Figure 1, we can conclude where the adjustment is regarding the analysed plans. The reduction of CO_2 is one of the globally accepted aspects in the combat against climate change [2]; however, adaptation measures (preventing the intensity of climate change impact: irrigation system, dams, water reservoirs, etc.) in Serbia's spatial plans are more frequent than mitigation measures (preventing the intensity of climate change itself: afforestation, biomass as an energy production resource, education, etc.). This is in contrast to the 28 cases of EU cities analysed by Reckien et al. [77]: almost 70% of the analysed cities have mitigation plans and less than 30% have adaptation plans. Serbia's spatial-planning documents more often prescribe traditional anti-pollution measures (afforestation, sanitary landfills, proclaiming protected areas, etc.), but the analysis has showed that they have also introduced new categories in land use—reserving land for power supply stations for e-vehicles, wastewater facilities and energy production from biomass. New measures that are not related to specific locations or land but still refer to climate change include educating the population on this topic.

Colour-graded categorisation allows the beneficiaries an easy and quick overview of the measures that are missing and the actions that should be considered in the creation of future spatial planning and other types of documents. Garrigos-Simon et al. [78], who applied science mapping and the visualisation of similarities to visualise results regarding content analyses on the topic of sustainability, share the same attitude. In their view, visualisation helps professionals and researchers understand the examined problem. The tabulation and application of the spider tool visually facilitates comprehension of the potential gap between climate change effects and selected measures. The analysis of the table behind the spider (Table 3) also has value in understanding which measures have multiple impacts, which is convenient in the processes of prioritizing measures: those that impact several and the most crucial climate change effects are to be given priority over others that have only a limited impact.

With regard to their readability, the tools suggested in this paper use up to four colours, i.e., a maximum of two colours in two shades each. This approach eases readability for the viewers because, as Mitchell [79] advocates, using more than seven colours (on a map) can make it difficult to recognize subjects with closely aligned values. Maps are also a relevant means of communicating messages. Even though this research only shows the role of maps in spatial planning in Serbia, it indicates that maps are one of the visualisation tools that help professionals and stakeholders involved in decision making to realise which aspects of combating climate change have been included in a set of solutions and which have been omitted. The rapid increase in geographical information systems (GIS) is one of the signs that mapping different topics improves planning [80–82], implementation, and monitoring processes and that the interactive perspective of GIS is the next step that mapping climate change measures should take in Serbia and elsewhere. Since synthesis maps with an abundance of subjects are the most common characteristic of maps in spatial planning, the application of the suggested methods tends to increase readability, which is relevant for experts' understanding and even more for non-professional stakeholders [83].

What we see from the cross-check of Table 3 and Figure 2 is that measures that address educating the population and other stakeholders about climate change, establish new protection areas, deal with wastewater treatment, create planning and strategic documents for extreme weather events, and monitor climate change-related parameters by introducing new meteorological stations and instruments would have an impact on all climate change effects—floods, droughts, hail storms, heat waves, landslides and premature snow melting. Five types of interventions—afforestation, building dams, establishing notification and warning systems, organizing civil protection, and financial support to farmers for the introduction of protection systems (e.g., anti-hail nets, drainage systems) or reimbursement in the case of damage due to extreme weather events—tackle five out of six climate change effects. They also have diverse impacts on different fields of climate change effects. Drainage systems, correcting river courses, sandbag dykes, and alterations of terrain slopes are measures influencing only one climate change effect—flooding. However, selecting priority measures should be made with caution because, for example, floods are the most common effect of climate change in Serbia [61] (Pantić et al. 2022); hence, it would not be wise to disregard them.

Survey results collected by Pantić et al. [61] (2022) indicate that citizens in the rural areas of Serbia suffer the most from (1) drought, (2) flooding, and (3) hail, and less from (4) heat waves, (5) landslides, and (6) premature snow melting. This gradation does not completely coincide with the gradation of climate change-effect measures incorporated in the analysed spatial plans, and the differences indicate aspects that are changing. According to Figure 2, floods are considered more comprehensively in spatial-planning documents than drought, and hail is only considered after heat waves and landslides. If we compare the Spatial Plan of the Republic of Serbia for 2010–2020 with the new one for the period of 2021–2035 [84], which is in the adoption process, we will see that the presence of climate change measures is strengthened in the new document (Figure 4), although their prioritization per effect remains the same. The least considered effect in the plans of both generations is the premature melting of snow, which is in contrast to the efforts of Alpine countries to socio-economically adapt to climate change—looking for a solution to prevent avalanches or prolong the skiing season [25] (Agrawala 2007). The reason behind this might be the fact that the majority of households in Serbia still depend prevailingly on agriculture instead of tourism. Therefore, premature snow melting is not considered to be a particular threat to the economy or health.

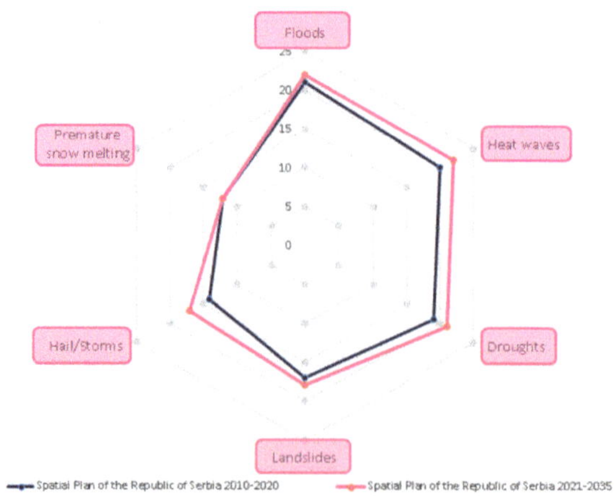

Figure 4. Comparison of the spatial plans of the Republic of Serbia.

Furthermore, applying the recommended visualisation methods to Serbia's spatial-planning documents has revealed a noticeable increase in awareness regarding climate change mitigation over the analysed period (2010 to 2023). A review of the keywords and measures addressing climate change in these planning documents indicates significant progress. Most of the measures (as shown in Figure 1) have been part of the spatial-planning discipline in Serbia for some time, including afforestation, dam and reservoir construction, retentions, landfills, and waste collection. While these measures predate the significant emergence of climate change concerns, they are now receiving heightened attention in spatial-planning documents. Notably, only four measures are consistently present in all analysed spatial-planning documents, highlighting the potential for introducing new measures to further improve these documents.

According to the Law on Planning and Construction (2023) [32], there is no obligation to address climate change in Serbia's spatial planning system. Therefore, their integration depends greatly on the awareness of experts and decision-makers, as well as initiatives coming from citizens in the early public viewing of a plan. According to the analysed spatial-planning documents, the extent to which climate change was addressed depended on the spatial-plan coordinators, since the plans with independent chapters on climate change were led by the same spatial-plan coordinators. This is why raising awareness of this topic is relevant, not only for professionals, but also for teachers [85], the media [86], citizens [87], and all other stakeholders [88]. Sending a clear message is relevant in communication, and the visualisation suggested here can play an important role in communications between professionals and final users as a non-technical, easily understandable tool.

It is important to stress that public opinion is not always based on facts and can be influenced by the media and politicians, including climate change issues [89]. Therefore, the role of visualising results is to facilitate an understanding of the state of the art and the future we want in all phases and by all stakeholders [31]. The role of visualisation techniques and methods became more significant during the COVID-19 pandemic when communication and participation were possible only online, i.e., using technology [26,90]. The visualisation methods suggested in this research are applicable not only to screen groups of spatial-planning documents but also at the level of each plan individually. Most important, they can be applied already in the drafting phase, so that corrections in defining final measures can be conducted on time. Their implementation does not have to be limited to the field of spatial planning either because they are also suitable for other decision-making opportunities.

The novelty of this paper is the introduction of new visualisation tools in spatial planning and evaluating the appropriateness of responses to climate change defined in spatial plans. These suggested visualisation tools implicitly demonstrate their effectiveness in enhancing stakeholder communication.

5. Conclusions

The research presented in this paper contributes significantly to the existing body of international empirical evidence concerning the capacity of spatial planning to facilitate climate change through adequate measures. Along with the analysis of climate change measures proposed in current spatial-planning documents in Serbia, this research has employed two information visualisation techniques to enhance our understanding of these measures from multiple perspectives.

The findings of this study underscore the potential role of visualisation as a tool in evaluation and decision-making processes related to climate change and its adverse impacts. These visualisation techniques serve to simplify complex information, fostering effective communication with all stakeholders involved in the planning process. The visualisations utilized here enable the recognition of climate change measures, assess the extent to which they are incorporated into the existing spatial planning framework, and identify which climate change impacts are receiving the most attention. As a result, the emphasis on specific climate change effects can be compared with and adjusted in response to the main climate change aspects and issues (e.g., floods, hail, and drought).

The visualisation methods discussed in this research, such as colour-graded classification, spider diagrams, and maps, can be readily adapted for application in various national and international contexts. The simplicity of colour-graded classification and spider diagrams facilitates clear and accessible communication between professionals and the general public in planning processes, including public participation, which increases the feasibility of the plans and the appropriateness of measures to the real problems which citizens face. Once the climate change measures are identified, they can be easily categorized as adaptation, mitigation, or a combination of both, offering insight into whether there is a need to strike a better balance between these two approaches. It is important to note that when applying the suggested visualisation methods in other cases, the selection of keywords for content analysis and the climate change effects considered should be adapted to suit the ever-evolving measures addressing climate change and the region-specific effects. The advantage of implementing the suggested visualisation methods is that they do not require substantial resources in terms of technology, expertise, and time. They are also not limited in terms of generalizability because they can be easily adapted to different regions or contexts with distinct climate challenges. The findings might not be universally applicable, especially for regions facing unique climate challenges, thus emphasizing the need for context-specific approaches and adaptations.

A limitation of this approach lies in the fact that analysed measures cannot always be compared to real local issues. This would require additional data collection, which might be demanding should it involve a non-standardized step in the creation of a spatial plan. In addition, it is important to acknowledge that certain limitations may arise regarding the potential omission of key words in the content analysis. Therefore, there is a possibility that we omitted some climate change measures in the analysed documents, believing that they did not affect the results. In addition, it is important to acknowledge that certain limitations may arise regarding the potential omission of key words in the content analysis. Therefore, there is a possibility that we omitted some climate change measures in the analysed documents, believing that they did not affect the results. By addressing these limitations and following these recommendations, future studies can further enhance the use of visualisation techniques in climate change planning and adapt them to meet the unique challenges faced by different regions.

In conclusion, we strongly recommend the integration of visualisation techniques in the decision-making process, as they have demonstrated their effectiveness in this study. While the combination of methods presented here is a valuable starting point, it is important to recognize that the field of visualisation offers a wide array of tools and methods that can contribute to a more comprehensive understanding of various aspects related to climate change planning. As we continue to address the pressing issue of climate change, the dynamic nature of this field should encourage ongoing innovation and the exploration of visualisation techniques to support effective spatial planning for climate change mitigation and adaptation. Our recommendations for the authors of future studies are to conduct further research to validate the effectiveness of visualisation techniques in different regions and under various climate change scenarios. This will help establish the generalizability of the approach. An interdisciplinary collaboration, i.e., the encouragement of collaboration between experts in climate science, spatial planning, and data visualisation, would ensure that the chosen visualisation methods are tailored to address the specific needs of each stakeholder. Practitioners are encouraged to apply, test, and explore similar methods to involve the public and stakeholders in the design and interpretation of visualisations. This can enhance transparency and help mitigate subjectivity concerns.

Author Contributions: All authors contributed to the study conception and design. Material preparation, data collection, and analysis were performed by M.P., T.M. and S.M. The first draft of the manuscript was written by M.P., and all authors commented on previous versions of the manuscript. Conceptualization: M.P. and T.M.; Methodology: M.P.; Formal analysis and investigation: M.P., T.M.

and S.M.; Investigation: M.P., T.M. and S.M.; Writing—original draft preparation: M.P.; Visualisation: M.P. and T.M.; Writing—review and editing: M.P., T.M. and S.M.; Funding acquisition: S.M.; Resources: M.P., T.M. and S.M.; Supervision: M.P.; Data curation: M.P.; Software: M.P. and T.M. All authors have read and agreed to the published version of the manuscript.

Funding: This research was supported by the Ministry of Education, Science and Technological Development of the Republic of Serbia (Grant No. 451-03-66/2024-03/200006).

Institutional Review Board Statement: Not applicable.

Informed Consent Statement: Not applicable.

Data Availability Statement: Publicly available datasets were analyzed in this study. This data can be found here: http://www.pravno-informacioni-sistem.rs/ (accessed on 14 January 2024).

Conflicts of Interest: The authors declare no conflicts of interest.

References

1. Crawford, E. *Arrhenius: From Ionic Theory to the Greenhouse Effect*; Watson Publishing—Science History: Canton, MA, USA, 1996.
2. Cherni, J.; Daley, B.; Dorward, A.; Guendel, S.; Macartney, J.; Nelson, V. *Climate Change and Development*; Centre for Development, Environment and Policy, SOAS University of London: London, UK, 2020.
3. IPCC. 2018: Annex I: Glossary. In *Global Warming of 1.5 °C*; An IPCC Special Report on the impacts of global warming of 1.5 °C above pre-industrial levels and related global greenhouse gas emission pathways, in the context of strengthening the global response to the threat of climate change, sustainable development, and efforts to eradicate poverty; Masson-Delmotte, V., Zhai, P., Pörtner, H.-O., Roberts, D., Skea, J., Shukla, P.R., Pirani, A., Moufouma-Okia, W., Péan, C., Pidcock, R., et al., Eds.; Cambridge University Press: Cambridge, UK; New York, NY, USA; pp. 541–562. [CrossRef]
4. Henning, R.; Charlson, R. (Eds.) *The Legacy of Svante Arrhenius. Understanding the Greenhouse Effect*; Royal Swedish Academy of Sciences: Stockholm, Sweden, 1998.
5. Fleming, J.R. *The Callendar Effect. The Life and Work of Guy Stewart Callendar (1898–1964), the Scientist Who Established the Carbon Dioxide Theory of Climate Change*; American Meteorological Society: Boston, MA, USA, 2007.
6. Philander, G.S. *Is the Temperature Rising—The Uncertain Science of Global Warming*; Princeton University Press: Princeton, NJ, USA, 1998.
7. Ekwurzel, B.; Boneham, J.; Dalton, M.W.; Heede, R.; Mera, R.J.; Allen, M.R.; Frumhoff, P.C. The Rise in Global Atmospheric CO_2, Surface Temperature, and Sea Level from Emissions Traced to Major Carbon Producers. *Clim. Chang.* **2017**, *144*, 579–590. [CrossRef]
8. United Nations. Sustainable Development Goals—Goal 13: Take Urgent Action to Combat Climate Change and Its Impacts. 2020. Available online: www.un.org/sustainabledevelopment/climate-change/ (accessed on 3 March 2020).
9. Intergovernmental Panel on Climate Change (IPCC). About the IPCC. 2020. Available online: www.ipcc.ch/about/ (accessed on 3 March 2020).
10. United Nations. Sustainable Development Goals—The Sustainable Development Agenda. 2020. Available online: www.un.org/sustainabledevelopment/development-agenda/ (accessed on 3 March 2020).
11. Intergovernmental Panel on Climate Change (IPCC). *Climate Change 2014: Synthesis Report. Contribution of Working Groups, I.; II and III to the Fifth Assessment Report of the Intergovernmental Panel on Climate Change*; IPCC: Geneva, Switzerland, 2015.
12. Intergovernmental Panel on Climate Change (IPCC 2021). Climate Change 2021: The Physical Science Basis. Contribution of Working Group, I. to the Sixth Assessment Report of the Intergovernmental Panel on Climate Change. Available online: www.ipcc.ch/report/ar6/wg1/downloads/report/IPCC_AR6_WGI_FrontMatter.pdf (accessed on 28 December 2022).
13. Ministry of Environmental Protection. *Second National Communication of the Republic of Serbia under the United Nations Framework Convention on Climate Change (SNC)*; Ministry of Environmental Protection: Belgrade, Serbia, 2017.
14. Hurlimann, A.C.; March, A.P. The Role of Spatial Planning in Adapting to Climate Change. *Wiley Interdiscip. Rev. Clim. Chang.* **2012**, *3*, 477–488. [CrossRef]
15. Davoudi, S. Climate Change and the Role of Spatial Planning in England. In *Climate Change Governance. Climate Change Management*; Knieling, J., Leal Filho, W., Eds.; Springer: Berlin/Heidelberg, Germany, 2013; pp. 153–169.
16. Wilson, E.; Piper, J. *Spatial Planning and Climate Change*; Routledge: London, UK, 2010.
17. Kruse, S.; Pütz, M. Adaptive capacities of spatial planning in the context of climate change in the European Alps. *Eur. Plan. Stud.* **2014**, *22*, 2620–2638. [CrossRef]
18. Biesbroek, G.R.; Swart, R.J.; van der Knapp, W.G.M. The Mitigation-Adaptation Dichotomy and the Role of Spatial Planning. *Habitat Int.* **2009**, *33*, 230–237. [CrossRef]
19. Shalaby, H.; Somaya, A. Climate Change Impacts on Urban Planning in the Cities. In Proceedings of the 1st International Conference on Towards a Better Quality of Life, El Gouna, Egypt, 24–26 November 2017. [CrossRef]
20. Yiannakou, A.; Salata, K.D. Adaptation to Climate Change through Spatial Planning in Compact Urban Areas: A Case Study in the City of Thessaloniki. *Sustainability* **2017**, *9*, 271. [CrossRef]

21. Swart, R.; Sedee, A.G.J.; de Pater, F.; Goosen, H.; Pijnappels, M.; Vellinga, P. Climate-Proofing Spatial Planning and Water Management Projects: An Analysis of 100 Local and Regional Projects in the Netherlands. *J. Environ. Policy Plan.* **2014**, *16*, 55–74. [CrossRef]
22. Macintosh, A.; Foerster, A.; McDonald, J. Policy Design, Spatial Planning and Climate Change Adaptation: A Case Study from Australia. *J. Environ. Plan. Manag.* **2015**, *58*, 1432–1453. [CrossRef]
23. Dhar, T.K.; Khirfan, L. Climate Change Adaptation in the Urban Planning and Design Research: Missing Links and Research Agenda. *J. Environ. Plan. Manag.* **2017**, *60*, 602–627. [CrossRef]
24. Climate ADAPT. CLISP: Climate Change Adaptation by Spatial Planning in the Alpine Space. 2015. Available online: https://climate-adapt.eea.europa.eu/metadata/projects/climate-change-adaptation-by-spatial-planning-in-the-alpine-space (accessed on 27 December 2022).
25. Agrawala, S. *Climate Change in the European Alps—Adapting Winter Tourism and Natural Hazards Management*; OECD Publishing: Paris, France, 2007.
26. Rajhans, V.; Rege, S.; Memon, U.; Shinde, A. Adopting a modified Delphi technique for revisiting the curriculum: A useful approach during the COVID-19 pandemic. *Qual. Res. J.* **2020**, *20*, 373–382. [CrossRef]
27. Stanyer, J. The British Public and Political Attitude Expression: The Emergence of a Self-expressive Political Culture? *Contemp. Politics* **2005**, *11*, 19–32. [CrossRef]
28. Vromen, A. Building virtual spaces: Young people, participation and the Internet. *Aust. J. Political Sci.* **2008**, *43*, 79–97. [CrossRef]
29. Jamei, E.; Mortimer, M.; Seyedmahmoudian, M.; Horan, B.; Stojcevski, A. Investigating the Role of Virtual Reality in Planning for Sustainable Smart Cities. *Sustainability* **2017**, *9*, 2006. [CrossRef]
30. Olszewski, R.; Gnat, M.; Trojanowska, H.; Turek, A.; Wieladek, A. Towards social fuzzy geoparticipation stimulated by gamification and augmented reality. In Proceedings of the 13th International Conference on Natural Computation, Fuzzy Systems and Knowledge Discovery (ICNC-FSKD), Guilin, China, 29–31 July 2017; pp. 1363–1370.
31. Pantić, M.; Cilliers, J.; Cimadomo, G.; Montaño, F.; Olufemi, O.; Torres Mallma, S.; van den Berg, J. Challenges and Opportunities for Public Participation in Urban and Regional Planning during the COVID-19 Pandemic—Lessons Learned for the Future. *Land* **2021**, *10*, 1379. [CrossRef]
32. Gazette of the Republic of Serbia. *Law on Planning and Construction (2023) "Official Gazette of the Republic of Serbia", No. 72/2009, 81/2009—Corr.; 64/2010—decision, C.C.; 24/2011, 121/2012, 42/2013—Decision, C.C.; 50/2013—Decision, C.C.; 98/2013—Decision, C.C.; 132/2014, 145/2014, 83/2018, 31/2019, 37/2019—Other law, 9/2020, 52/2021 and 62/2023*; Gazette of the Republic of Serbia: Belgrade, Serbia, 2023.
33. Official Gazette of the Republic of Serbia. *Spatial Plan of the Republic of Serbia 2010–2020*; Official Gazette of the Republic of Serbia 88/10: 2010; Official Gazette of the Republic of Serbia: Belgrade, Serbia, 2010. (In Serbian)
34. Official Gazette of the Republic of Serbia. *Strategic Environmental Assessment of the Spatial Plan of the Republic of Serbia from 2010–2020*; Official Gazette of the Republic of Serbia 88/10: 2010; Official Gazette of the Republic of Serbia: Belgrade, Serbia, 2020. (In Serbian)
35. Implementation Program for Spatial Plan of the Republic of Serbia from 2010–2020 for the Period from 2011–2015 (In Serbian). 2011. Available online: www.putevi-srbije.rs/images/pdf/strategija/PI_PPRS_2011-2015.pdf (accessed on 7 May 2023).
36. Official Gazette of the Republic of Serbia. *Regional Spatial Plan for the Municipalities in Južno Pomoravlje 2006–2021*; Official Gazette of the Republic of Serbia 350/10: 2010; Official Gazette of the Republic of Serbia: Belgrade, Serbia, 2010. (In Serbian)
37. Official Gazette of the Republic of Serbia. *Regional Spatial Plan for the Administrative Area of the City of Belgrade*; Official Gazette of the City of Belgrade 31/11; Official Gazette of the Republic of Serbia: Belgrade, Serbia, 2011. (In Serbian)
38. Official Gazette of the Autonomous Region of Vojvodina. *Regional Spatial Plan for the Autonomous Region of Vojvodina*; Official Gazette of the Autonomous Region of Vojvodina 22/11; Official Gazette of the Autonomous Region of Vojvodina: Novi Sad, Serbia, 2011. (In Serbian)
39. Official Gazette of the Autonomous Region of Vojvodina. *Implementation Program for the Autonomous Region of Vojvodina's Regional Spatial Plan from 2013–2017*; Official Gazette of the Autonomous Region of Vojvodina 22/13; Official Gazette of the Autonomous Region of Vojvodina: Novi Sad, Serbia, 2013. (In Serbian)
40. Official Gazette of the Republic of Serbia. *Regional Spatial Plan for Timočka Krajina*; Official Gazette of the Republic of Serbia 51-11; Official Gazette of the Republic of Serbia: Belgrade, Serbia, 2011. (In Serbian)
41. Official Gazette of the Republic of Serbia. *Implementation Program for Timočka Krajina District until 2017*; Official Gazette of the Republic of Serbia 17/15; Official Gazette of the Republic of Serbia: Belgrade, Serbia, 2015. (In Serbian)
42. Official Gazette of the Republic of Serbia. *Regional Spatial Plan for Nišavski, Toplički, and Pirotski Districts*; Official Gazette of the Republic of Serbia 1/13; Official Gazette of the Republic of Serbia: Belgrade, Serbia, 2013. (In Serbian)
43. Official Gazette of the Republic of Serbia. *Regional Spatial Plan for Zlatiborski and Moravički Districts*; Official Gazette of the Republic of Serbia 1/13; Official Gazette of the Republic of Serbia: Belgrade, Serbia, 2013. (In Serbian)
44. Official Gazette of the Republic of Serbia. *Regional Spatial Plan for Šumadijski, Pomoravski, Raški, and Rasinski Districts*; Official Gazette of the Republic of Serbia 39/14; Official Gazette of the Republic of Serbia: Belgrade, Serbia, 2014. (In Serbian)
45. Official Gazette of the Republic of Serbia. *Regional Spatial Plan for Podunavski and Braničevski Districts*; Official Gazette of the Republic of Serbia 8/15; Official Gazette of the Republic of Serbia: Belgrade, Serbia, 2015. (In Serbian)

46. Official Gazette of the Republic of Serbia. *Regional Spatial Plan for Kolubarski and Mačvanski Districts*; Official Gazette of the Republic of Serbia 11/15; Official Gazette of the Republic of Serbia: Belgrade, Serbia, 2015. (In Serbian)
47. Cole, F.L. Content Analysis: Process and Application. *Clin. Nurse Spec.* **1988**, *2*, 53–57. [CrossRef] [PubMed]
48. Downe-Wamboldt, B. Content analysis: Method, applications, and issues. *Health Care Women Int.* **1992**, *13*, 313–321. [CrossRef] [PubMed]
49. Stecula, D.; Merkley, E. Framing Climate Change: Economics, Ideology, and Uncertainty in American News Media Content From 1988 to 2014. *Front. Commun.* **2019**, *4*, 6. [CrossRef]
50. Walter, S.; Lörcher, I.; Brüggemann, M. Scientific networks on Twitter: Analyzing scientists' interactions in the climate change debate. *Public Underst. Sci.* **2019**, *28*, 696–712. [CrossRef] [PubMed]
51. Draper, G.M.; Livnat, Y.; Riesenfeld, R.F. A Survey of Radial Methods for Information Visualization. *IEEE Trans. Vis. Comput. Graph.* **2009**, *15*, 759–776. [CrossRef]
52. Calzon, B. 17 Essential Data Visualization Techniques, Concepts & Methods to Improve Your Business—Fast. 2023. Available online: www.datapine.com/blog/data-visualization-techniques-concepts-and-methods/ (accessed on 12 September 2023).
53. Osinska, V.; Osinski, G. *Information Visualization Techniques in the Social Sciences and Humanities*; IGI Global: Hershey, PA, USA, 2018.
54. Shen, H.; Bednarz, T.; Nguyen, H.; Feng, F.; Wyeld, T.; Hoek, P.J.; Lo, E.H. Information visualisation methods and techniques: State-of-the-art and future directions. *J. Ind. Inf. Integr.* **2019**, *16*, 100102. [CrossRef]
55. Baycan-Levent, T.; Van Leeuwen, E.; Rodenburg, C.; Nijkamp, P. Development and management of green spaces in European cities: A comparative analysis. *Res. Memo.* **2002**, *2002*, 25.
56. Dühr, S. *The Visual Language of Spatial Planning: Exploring Cartographic Representations for Spatial Planning in Europe*; Routledge: New York, NY, USA, 2007.
57. Heimerl, F.; Lohmann, S.; Lange, S.; Ertl, T. Word cloud explorer: Text analytics based on word clouds. In Proceedings of the 47th Hawaii International Conference on System Sciences, Waikoloa, HI, USA, 6–9 January 2014; pp. 1833–1842.
58. Baycan-Levent, T.; Brunisma, F.; Nijkamp, P. Urban Spiders: A Comparative Framework for Evaluation and Scenario Analysis. In *Sustainable Urban Development Volume 2, The Environmental Assessment Methods*; Deakin, M., Mitchell, G., Nijkamp, P., Vreeker, R., Eds.; Taylor & Francis: New York, NY, USA, 2007; pp. 236–253.
59. Vujošević, M.; Zeković, S.; Maričić, T. *The Post-Socialist Transition in Serbia and the Territorial Capital of Serbia—The State of the Art, Future Prospects and Scenarios*; Institute of Architecture and Urban & Spatial Planning of Serbia: Belgrade, Serbia, 2010. (In Serbian)
60. Zeković, S.; Vujošević, M.; Maričić, T. A Preliminary Analysis of Sustainable Development in the Belgrade Metropolitan Area. *Int. J. Soc. Behav. Educ. Econ. Manag. Eng.* **2015**, *9*, 839–846.
61. Pantić, M.; Maričić, T.; Milijić, S. Citizen Participation Offers Lessons to Climate Change Management. In Proceedings of the 29th International Conference Ecological Truth and Environmental Research, Sokobanja, Serbia, 21–24 June 2022; pp. 489–494.
62. Berding, V.; Schwartz, S.; Matthies, M. Visualisation of the complexity of EUSES. *Environ. Sci. Pollut. Res.* **1999**, *6*, 37–43. [CrossRef] [PubMed]
63. Cserny, A.; Kovács, Z.; Domokos, E.; Rédey, Á. Environmental information system for visualizing environmental impact assessment information. *Environ. Sci. Pollut. Res.* **2009**, *16*, 36–41. [CrossRef] [PubMed]
64. Koondhar, M.A.; Shahbaz, M.; Memon, K.A.; Ozturk, I.; Kong, R. A visualization review analysis of the last two decades for environmental Kuznets curve "EKC" based on co-citation analysis theory and pathfinder network scaling algorithms. *Environ. Sci. Pollut. Res.* **2021**, *28*, 16690–16706. [CrossRef]
65. Zhu, E.; Qi, Q.; Sha, M. Identify the effects of urbanization on carbon emissions (EUCE): A global scientometric visualization analysis from 1992 to 2018. *Environ. Sci. Pollut. Res.* **2021**, *28*, 31358–31369. [CrossRef] [PubMed]
66. Pietsch, S.M. Computer Visualisation in the Design Control of Urban Environments: A Literature Review. *Environ. Plan. B Plan. Des.* **2000**, *27*, 521–536. [CrossRef]
67. Intergovernmental Panel on Climate Change (IPCC). AR6 Synthesis Report: Climate Change 2022. Available online: www.ipcc.ch/report/ar6/wg2/ (accessed on 27 December 2022).
68. Djordjevic, D.S.; Secerov, V.; Filipovic, D.; Lukic, B.; Jeftic, M.R. The impact of climate change on the planning of mountain tourism development in Serbia: Case studies of Kopaonik and Zlatibor. *Fresenius Environ. Bull.* **2016**, *25*, 5027–5034.
69. Božanić, D.; Mitrović, Đ. *Study on the Socio-Economic Aspects of Climate Change in the Republic of Serbia*; United Nations Development Programme: Belgrade, Serbia, 2019.
70. Popovicki, T. *Study on Nature-Based Climate Solutions in Serbia*; United Nations Development Programme: Belgrade, Serbia, 2019.
71. Stričević, R.; Prodanović, S.; Đurović, N.; Petrović Obradović, O.; Đurović, D. *Climate Change Impacts on Serbian Agriculture*; United Nations Development Programme: Belgrade, Serbia, 2019.
72. Döll, P. Impact of Climate Change and Variability on Irrigation Requirements: A Global Perspective. *Clim. Chang.* **2002**, *54*, 269–293. [CrossRef]
73. Van Binh, D.; Kantoush, A.S.; Sumi, T.; Phuong Mai, N.; Anh Ngoc, T.; Vinh Trung, L.; Dang An, T. Effects of riverbed incision on the hydrology of the Vietnamese Mekong Delta. *Hydrol. Process.* **2021**, *35*, e14030. [CrossRef]
74. Doelman, J.C.; Stehfest, E.; van Vuuren, D.P.; Tabeau, A.; Hof, A.F.; Braakhekke, M.C.; Gernaat, D.E.; van den Berg, M.; van Zeist, W.J.; Daioglou, V.; et al. Afforestation for climate change mitigation: Potentials, risks and trade-offs. *Glob. Chang. Biol.* **2020**, *26*, 1576–1591. [CrossRef] [PubMed]

75. NASA. Responding to Climate Change. 2022. Available online: https://climate.nasa.gov/solutions/adaptation-mitigation/ (accessed on 28 December 2022).
76. United Nations Climate Change. Introduction. Adaptation and Resilience. 2022. Available online: https://unfccc.int/topics/adaptation-and-resilience/the-big-picture/introduction#:~:text=Adaptation%20refers%20to%20adjustments%20in,opportunities%20associated%20with%20climate%20change (accessed on 28 December 2022).
77. Reckien, D.; Salvia, M.; Heidrich, O.; Church, J.M.; Pietrapertosa, F.; De Gregorio-Hurtado, S.; D'Alonzo, V.; Foley, A.; Simoes, S.G.; Krkoška Lorencová, E.; et al. How are cities planning to respond to climate change? Assessment of local climate plans from 885 cities in the EU-28. *J. Clean. Prod.* **2018**, *191*, 207–219. [CrossRef]
78. Garrigos-Simon, F.J.; Narangajavana-Kaosiri, Y.; Lengua-Lengua, I. Tourism and Sustainability: A Bibliometric and Visualization Analysis. *Sustainability* **2018**, *10*, 1976. [CrossRef]
79. Mitchell, A. *The ESRI Guide to GIS Analysis, Vol. 1: Geographic Patterns & Relationships*; ESRI Press: Redlands, CA, USA, 1999.
80. Carver, S. Developing Web-based GIS/MCE: Improving Access to Data and Spatial Decision Support Tools. In *Spatial Multicriteria Decision Making and Analysis*, 1st ed.; Jean-Claude, T., Ed.; Routledge: Oxfordshire, UK, 2019.
81. Marzouk, M.; Othman, A. Planning utility infrastructure requirements for smart cities using the integration between BIM and GIS. *Sustain. Cities Soc.* **2020**, *57*, 102120. [CrossRef]
82. Wang, F. Why public health needs GIS: A methodological overview. *Ann. GIS* **2020**, *26*, 1–12. [CrossRef]
83. Butt, M.A.; Li, S. Usability evaluation of collaborative PPGIS-GeoCWMI for supporting public participation during municipal planning and management services. *Appl. Geomat.* **2015**, *7*, 139–161. [CrossRef]
84. *Spatial Plan of the Republic of Serbia 2021–2035*; An Internal Database of the Institute of Architecture and Urban & Spatial Planning of Serbia: Belgrade, Serbia, 2023. (In Serbian)
85. Dorji, Y.; Rai, C.M.; Nidup, T. Climate Change Awareness among the Teachers of Higher Secondary Schools. *Asian J. Soc. Sci.* **2021**, *15*, 12–22. [CrossRef]
86. Elia, E.F. Journalists' Awareness and Understanding of Climate Change in Tanzania. *Int. J. Commun.* **2021**, *15*, 22–40.
87. Gärtner, L.; Schoen, H. Experiencing climate change: Revisiting the role of local weather in affecting climate change awareness and related policy preferences. *Clim. Chang.* **2021**, *167*, 31. [CrossRef]
88. Iturriza, M.; Labaka, L.; Hernantes, J.; Abdelgawad, A. Shifting to climate change aware cities to facilitate the city resilience implementation. *Cities* **2020**, *101*, 102688. [CrossRef]
89. Freestone, J. To Get Conservative Climate Change Contrarians to Really Listen, Try Speaking Their Language, The Conversation. Available online: http://theconversation.com/to-get-conservative-climate-contrarians-to-really-listen-try-speaking-their-language-94296 (accessed on 3 March 2020).
90. Bricout, J.; Baker, M.A.P.; Moon, W.N.; Sharma, B. Exploring the Smart Future of Participation: Community, Inclusivity, and People with Disabilities. *Int. J. E-Plan. Res.* **2021**, *10*, 94–108. [CrossRef]

Disclaimer/Publisher's Note: The statements, opinions and data contained in all publications are solely those of the individual author(s) and contributor(s) and not of MDPI and/or the editor(s). MDPI and/or the editor(s) disclaim responsibility for any injury to people or property resulting from any ideas, methods, instructions or products referred to in the content.

sustainability

MDPI

Article

Environmental Management of Ecuador's Business Sector in the Fight against Climate Change

Miguel Aizaga *, Marcelo Ramírez, María Carmen Colmenárez Mujica and Renato M. Toasa *

Faculty of Administrative Sciences, Central Campus, Universidad Tecnológica Israel, Quito 170516, Ecuador; mramirez@uisrael.edu.ec (M.R.); mcolmenarez@uisrael.edu.ec (M.C.C.M.)

* Correspondence: maizaga@uisrael.edu.ec (M.A.); rtoasa@uisrael.edu.ec (R.M.T.)

Abstract: The private sector is part of the United Nations Global Compact, which promotes the voluntary participation of organizations to implement environmental care strategies. The purpose of this article is to examine the performance of Ecuadorian companies in regard to environmental management, especially in the fight against climate change, considering the economic sectors (manufacturing, mining, commerce, construction and services). Figures from the National Institute of Statistics and the Census of Ecuador (2020) are used, including descriptive statistics and cross-tabulations with Cramer's V index. The results show that approximately 5% of companies had the ISO 14001:2015 certification. In the seven actions against climate change considered, the proportion of companies that did not consider them to be current expenses predominated. Cramer's V index, for associating the economic sector and the environmental spend, revealed that certain economic sectors (manufacturing and mining) are contributing significantly to environmental management spending in the areas of air, soil, wastewater and waste treatment, while no economic sector dominates the others in areas such as radiation treatment, the use of mineral or energy resources and water resources.

Keywords: companies; economic sectors; combating climate change; environmental expenditure

1. Introduction

Companies worldwide must contribute to socio-economic development, beyond what they contribute by creating jobs and paying taxes, this implies so-called corporate social responsibility, which involves company practices that seek to avoid harm, while promoting the welfare of stakeholders by abiding by existing regulations and standards, and voluntarily going beyond what is required [1]. This responsibility encompasses not only their internal and external customers, but the community in general and the environment. In regard to this last aspect, today more than ever, they are committed to joining the fight against climate change.

In this sense, the United Nations Global Compact "is a non-binding agreement that seeks to promote sustainable development and compliance with human rights and labour standards, as well as to fight corruption, in business activities worldwide". It was established in 2000, with the aim of creating a common framework for developing corporate sustainability practices, and two decades after its creation, 10,000 companies have signed up to it [2].

This pact is based on 10 principles that have human rights, labor standards, the environment and anti-corruption as fundamental pillars, establishing that signatory companies commit to implement and report annually about progress in these areas, indicating the actions and policies implemented [3]. Among the principles, those that are directly linked to environmental care and that are aligned with the business commitment to fight climate change are Principle 7 "Businesses should maintain a precautionary approach to environmental challenges", Principle 8 "Businesses should encourage initiatives that promote greater environmental responsibility", and Principle 9 "Businesses should encourage the development and diffusion of environmentally friendly technologies" [4].

In addition to this, globally there is the ISO 14001:2015 standard [5]. "This International Standard specifies requirements for an environmental management system that an organization can use to improve its environmental performance" and can be implemented by any organization wishing to manage its environmental responsibilities towards sustainability [5].

In Ecuador, within the framework of the Global Compact and with the support of Ørsted and Natura & Co and sponsored by En+ Group, there is a program called the Climate Ambition Accelerator, which aims to "scale up credible climate action in companies of all sizes, sectors and regions, enabling them to meet meaningful commitments to reduce emissions at scale". This program equips companies with the knowledge and skills needed to accelerate progress towards setting science-based emissions targets aligned with the 1.5 °C pathway, whose participation requires that they are committed to a local Global Compact network in their region or willing to join, that they are committed to the Paris Agreement and that they are willing to appoint two representatives to participate in the program's activities and events and a representative person at a senior level to follow program developments, provide support and participate in high-level events [6].

The latest results from the Global Compact Ecuador 2020 show that 34 recognized companies in the country participated in the Climate Ambition Accelerator program [7]. In addition, there is the Ecuador Verde program, whose purpose is to provide training on the reduction of carbon emissions.

Climate change is a consequence of an infinite number of factors, which are not only related to direct air pollution, but also to the use and treatment of soil, waste, wastewater, energy and radioactivity, among others. Wastewater treatment generates greenhouse gases, inadequate solid waste management also has an impact, as organic waste in landfills produces methane, a potent greenhouse gas. On the soil side, deforestation and unsustainable agricultural practices can result in the release of large amounts of carbon stored in the soil.

Fossil fuels, such as coal, oil and gas, "are the major causes of global climate change, as they are responsible for more than 75% of total global greenhouse gas emissions and about 90% of all carbon dioxide emissions" [8]. Also, the handling of radioactive waste can contribute to climate change if greenhouse gases (GHGs) are emitted during the treatment, transportation, storage or waste disposal process. Hence, the importance of the transition to renewable energies, produced by the sun, wind, water, waste or the Earth's own heat, which are renewed by nature itself and emit few (or none) pollutants or greenhouse gases into the air.

In this sense, it is important to analyze corporate environmental management in order to visualize the efforts made by private companies to promote environmentally sustainable development. This is the case in the research by Ochoa and other authors, which considers "estimates of current expenditure and investment in environmental management of the private sector versus GDP during the period 2010–2015 in Ecuador, supported by the finding of the environmental Kuznets curve (CKA)". The study uses linear regression to visualize the relationship between economic growth (GDP) and environmental degradation (CO_2 emissions), incorporating the current expenditure and environmental investment related to the variables made by private companies.

The main conclusion in this study is that "based on estimates, spending on environmental management is at 0.34% of GDP on average during the study period and environmental performance is in an increasing phase of the Kuznets curve", i.e., the higher the country's economic growth, the greater the environmental deterioration. It is ratified that Ecuador, "being a developing country, has not yet reached the inflection point of this curve, this being possible when it reaches a level of per capita income that transforms environmental degradation into a decreasing trend". In this study, indicators obtained from the Central Bank of Ecuador, such as GDP, and from the INEC, related to the survey of economic environmental information in private companies, were used [9].

Also, it is relevant to mention the analysis by Duarte, who categorizes the possible postures of companies facing the environmental problem, through an adaptation of Robbins and Coulter in 2010. Duarte, points out four postures [10]:

1. The reactive posture. Companies that face environmental problems only when they arrive, indicating zero sensitivity to environmental issues.
2. The accommodating posture. Companies that make decisions to comply with legal regulations related to environmental issues.
3. The stakeholder posture. Companies that focus on satisfying the environmental demands of internal and external stakeholders.
4. The proactive stance. Companies that anticipate or plan for environmental problems that could occur in their internal and external environment, through the implementation of actions to prevent their occurrence.

In addition, Duarte presents a general scheme of environmental innovations in private enterprise, as shown in Figure 1.

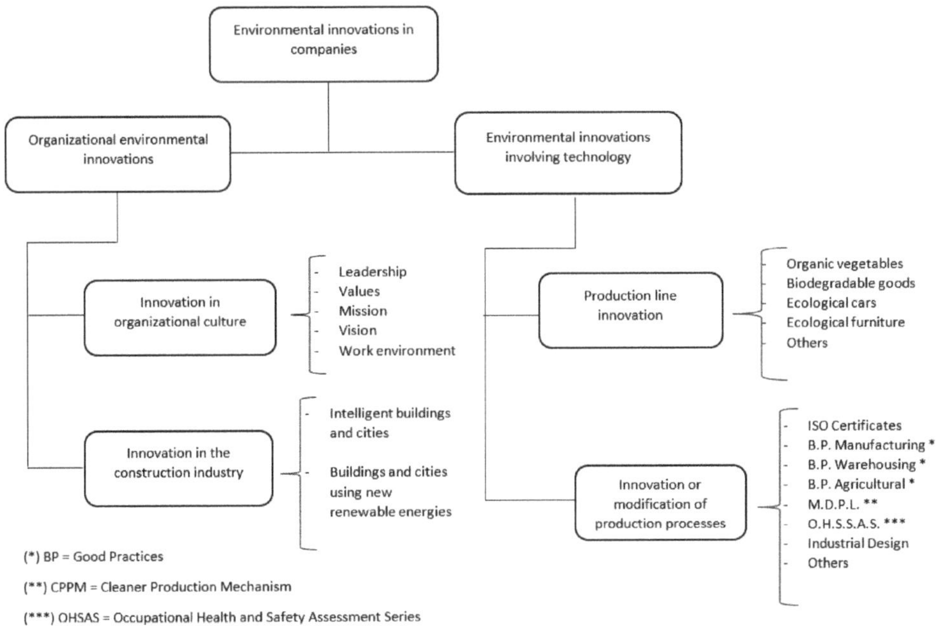

Figure 1. General outline of environmental innovations [10].

On the other hand, the analysis of corporate social responsibility (CSR) conducted by [11] aimed to determine the measures for sustainable environmental management in the face of the effects caused by productive processes.

This study "identifies and analyzes the implementation of corporate social responsibility towards environmental practices in Ecuadorian companies during the period 2017 to 2020, through the systematization of information from 13 articles from journals indexed in databases". In this sense, it extracts articles from Scopus, ScienceDirect, Doaj, Dialnet and Google Scholar, in the context of: knowledge and compliance with corporate social responsibility, dimensions of corporate social responsibility towards environmental management, and good environmental practices. "The main results show that there is a lack of knowledge about the application of corporate social responsibility, and companies tend to implement strategies that favor economic and social practices leaving aside environmental practices" [11].

Identifying which sector affects climate change the most can be complicated, as they all contribute in different ways to greenhouse gas (GHG) emissions, and the magnitude of each sector's impact may vary by region and specific practices within each sector; however, some general observations can be made. The manufacturing sector, due to its manufacturing processes, uses chemicals and a lot of energy, generally generated from fossil fuels. The mining sector involves deforestation, soil disturbance and the release of methane during the extraction of coal. This sector also consumes large amounts of energy. The construction sector involves the production of materials, such as cement and steel, as well as the generation of construction and demolition waste. The commerce sector has an impact, especially through the long-distance transportation of goods, and the services sector, although smaller, has an impact mainly through energy consumption by commercial buildings, transportation and public services.

In this research, the main objective is to analyze the performance of each of the five business sectors established in Ecuador, namely manufacturing, commerce, mining, construction and services, to determine their contribution in regard to environmental management actions, especially those that impact the fight against climate change. Hence the research question is: What is the contribution of Ecuador's business sectors in the fight against climate change, considering actions in specific areas, such as the current expenditure?

2. Methodology

In Ecuador, the National Institute of Statistics and Census (INEC) conducts surveys of companies in the national territory to determine progress in setting science-based emissions targets, in order to assess an organization's transition to net zero. The results of these surveys are recorded in the Environmental Economic Information Module for Companies [12].

This study is a documentary, descriptive study, whose main source is the Environmental Economic Information Survey of Companies (EIAEE) [12] conducted by the INEC in 2020, involving a total of 3777 companies throughout the country, whose purpose is to generate statistics on the expenditure and investment made by companies to prevent, reduce or eliminate pollution resulting from the development of their activities (emissions of air pollutants, noise pollution, treatment of wastewater and solid waste generated, etc.), defined by the United Nations' Classification of Environmental Protection Activities.

In this case, the classification by sector is considered, which has five categories, namely manufacturing, commerce, mining, construction and services, examining, using the EIAEE, the environmental expenditure in general and the expenditure for each of the actions linked to the fight against climate change, considered as the current expenditure of the company. To analyze these variables, descriptive statistics are considered, with the main measures being the mean, maximum, minimum and percentiles. These statistics are calculated both for the total environmental management expenditure and for each of the seven variables related to the reduction, prevention and minimization factors focused on the fight against climate change.

In addition, the classifications by economic sector of the companies are crossed with the type of action linked to the fight against climate change, which the company has considered as its current expenditure; since, in the private sector, current expenditure is the set of expenses that are considered essential for the company's ordinary activities and, therefore, are not superfluous [13]. The variables are analyzed using quantitative and qualitative measurements (Table 1).

Table 1. Study variables.

Variable	Quantitative Measurement	Qualitative Measurement
ISO 14001:2015	-	Obtaining certification YES/NO
Total expenditure on environmental management	Total amount and by economic sector	-
Reduction of emissions or pollutant concentrations in the air	Total amount	Inclusion as current expense by economic sector YES/NO
Prevention of surface water pollution	Total amount	Inclusion as current expense by economic sector YES/NO
Prevention of waste generation	Total amount	Inclusion as current expense by economic sector YES/NO
Prevention of infiltration of soil and groundwater pollutants	Total amount	Inclusion as current expense by economic sector YES/NO
Reduction or elimination of the negative consequences of radiation emitted by any source	Total amount	Inclusion as current expense by economic sector YES/NO
Minimization of the use of mineral and energy resources	Total amount	Inclusion as current expense by economic sector YES/NO
Minimization of water resource extraction	Total amount	Inclusion as current expense by economic sector YES/NO

Qualitative and quantitative variables.

Using cross-tables with Cramer's V index, the hypotheses are established to determine the association between the variables mentioned. Cramer's V test is an index that measures the relationship between two categorical variables. It is based on the chi-square coefficient, but is corrected to obtain a value between 0 and 1. A value of 0 indicates that there is no association between the variables, and a value of 1 indicates a perfect association. A weak association is considered to exist if the value is between 0 and 0.2, a moderate association if it is between 0.2 and 0.6, and a strong association if it is between 0.6 and 1.0 [14].

Cramer's V index is calculated from the chi-square statistic and the total number of observations in the contingency table. The formula is (1):

$$Cramer's\ V = \sqrt{\frac{x^2}{n \cdot \min(r-1, c-1)}} \tag{1}$$

where:

- x^2 is the value of the chi-square statistic;
- n is the total number of observations in the table;
- c is the number of columns in the table;
- r is the number of rows in the table.

The expression "min(r − 1, c − 1)" means that the number of rows and columns is reduced by 1 and the smaller of the two values is used in the equation. The Statistical Package for the Social Sciences (SPSS) program is used for data processing [15]. In this case, Cramer's V coefficient is a measure of the strength of association between two categorical variables in a contingency table, the structure of which is, on the one hand, the rows made up of the five economic sectors and, on the other, the columns formed by the YES/NO responses to each of the seven actions represented in the survey items. For data processing, the Statistical Package for the Social Sciences (SPSS), Version 25, program is used, which allows, through the analyze menu, the calculation of descriptive statistics, including frequencies for measures of the central tendency, position and dispersion, and also in crosstabs, Cramer's V index.

3. Results

The distribution of the surveyed companies by economic sector shows that the sector with the highest proportion is commerce, while the lowest is the mining sector. Of the total 3777 companies, only 199 have the ISO 14001:2015 certification, which represents only a little more than 5%. The companies belonging to the manufacturing and mining sectors have the highest proportion of certifications, with 11.38 and 23.08%, respectively. Although, in absolute numbers, it is the manufacturing and services sectors that have the most companies with this certification (Table 2).

Table 2. Distribution of companies by economic sector and their proportion with respect to ISO 14001:2015 certification.

Was the Company ISO 14001:2015 Certified?—Yes/No				Total
		Yes	No	
Economic Sector Code covering the main economic activity of the company 2020	Manufacturing	80	623	703
	Mining	24	80	104
	Trade	28	1634	1662
	Construction	16	165	181
	Services	51	1076	1127
Total		199	3578	3777

As for the main statistics on environmental spending reported by the companies, it can be clearly seen that, of the companies surveyed, the average is a little over USD 40,000. When examining the percentiles, it is observed that 25% of the companies do not spend anything on the environment, 50% spend below USD 1200 and 75% spend below USD 15,600. The amazing thing is that the maximum value is a little over USD 16,000,000, which implies that there is a large gap in the annual spending of more than a thousand times, between the 25% of companies that expressed the highest volume of expenditure and the other 75% of companies that spend the least on the environment.

The action to prevent the generation of waste or residues stands out in terms of the largest volume of companies participating in such activities with more than 14,000, with the maximum expenditure of almost USD 4,000,000 and the maximum total exceeding USD 28,000,000. In relation to minimizing the use of mineral and energy resources, the highest average expenditure stands out at a little more than USD 165,000, and in regard to the highest expenditure figures, considering the 25th, 50th and 75th percentiles, the latter percentile reaches an amount of more than USD 94,000.

These statistics provide key information on environmental spending, from the average amount invested to the variability and distribution of that investment. These figures help researchers, policymakers and stakeholders better understand how resources are being allocated to address environmental issues and identify areas where adjustments or different approaches may be needed.

When looking at environmental spending overall, the mining sector has the highest environmental spending, averaging USD 382,636 (Figure 2). This makes perfect sense, as the impact of mining on the environment is significant and worrisome. Large-scale mining causes irreparable physical, biological and socio-economic damage, including water and air pollution, deforestation, loss of biodiversity and land degradation [16].

Of the total environmental expenditure, considering each of the provinces of Ecuador, as shown in the illustration, it can be seen that the province in which the companies reported the highest amount is the province of Zamora Chinchipe, with an average of USD 918,283, which represents a very marked contrast with the other provinces in the country (Figure 3). In Ecuador, the province with the highest mining activity is Zamora Chinchipe, where the mines Fruta del Norte and Mirador are located, both involved in the mining production phase [17].

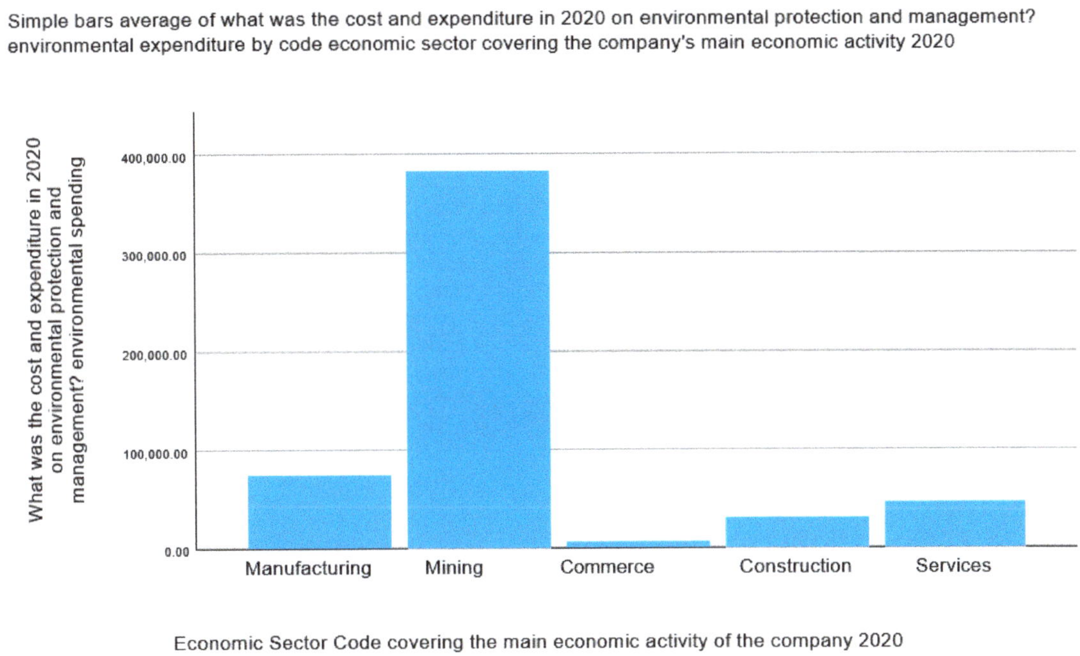

Figure 2. Average environmental expenditure by economic sector.

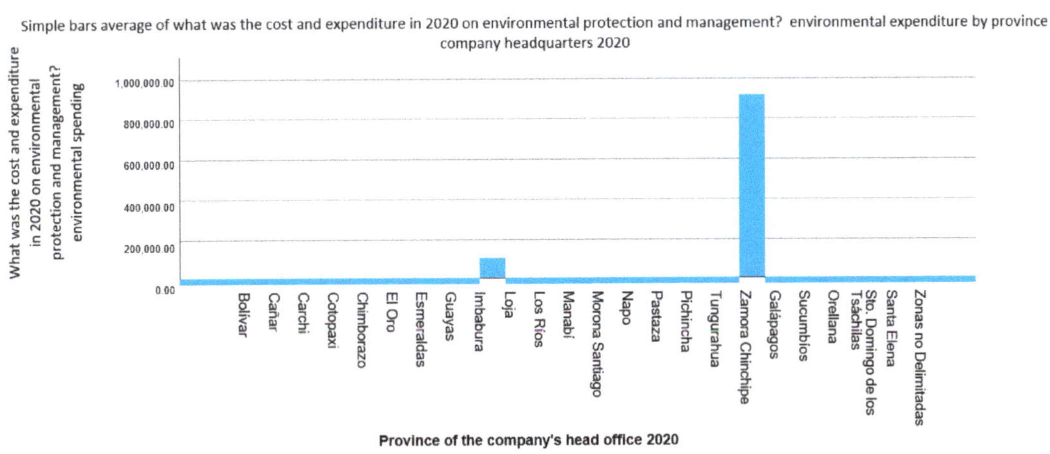

Figure 3. Average environmental expenditure by province.

With respect to the bivariate analysis, to determine associations between the variables, the following hypotheses are established:

H0: The economic sector of the company is not associated with the type of action against climate change (considered as the current expenditure).

H1: The economic sector of the company is associated with the type of action against climate change (considered as the current expenditure).

In this case, the five economic sectors (manufacturing, mining, commerce, construction and services) and seven types of actions against climate change that the companies consider to be the current expenditure are:

1. Reducing emissions or pollutant concentrations in the air (through process modification, gas treatment, measurement, laboratory control, similar activities and others);
2. Preventing surface water pollution by reducing wastewater release (includes wastewater collection and treatment);
3. Preventing the generation of waste and/or residues, the reduction of harmful effects to the environment (includes collection, treatment, recycling and composting, street cleaning and garbage collection);
4. Preventing the infiltration of soil and groundwater pollutants, the cleaning of soils and water bodies and the protection of soil against erosion;
5. Reducing or eliminating the negative consequences of radiation emitted from any source (including the handling, transport and treatment of radioactive waste);
6. Minimizing the use of mineral and energy resources by modifying processes, such as the recovery, reuse, recycling and saving of mineral sources;
7. Minimizing the extraction of water resources through the modification of processes, such as the reuse, recycling, saving and use of fresh water.

Table 3 shows the crossing of the variables analyzed and Cramer's V coefficient.

Table 3. Economic sector/reduction of emissions or air pollutant concentrations.

Reduce Emissions or Pollutant Concentrations in the Air (Through Process Modification, Gas Treatment, Measurement, Control, Laboratory, etc.) Current Costs: YES/NO				**Total**
		Yes	**No**	
Economic Sector Code covering the main economic activity of the company 2020	Manufacturing	237	466	703
	Mining	42	62	104
	Commerce	32	1630	1662
	Construction	20	161	181
	Services	81	1046	1127
Total		412	3365	3777
Symmetrical Measurements				
			Value	Approximate significance
Nominal by Nominal		Phi	0.406	0.000
		Cramer's V	0.406	0.000
N of valid cases			3777	

The highest proportion of companies that reduce emissions into the air through process modification, gas treatment, measurement, laboratory control, similar activities and others, which include this as a current expense, are those in manufacturing and mining sectors with 34% and 40%, respectively.

When examining Cramer's V index, which is 0.406, it can be pointed out that there is a moderate association between the variables. Therefore, the null hypothesis is rejected, meaning that there is an association between the economic sector of the company and the reduction of emissions into the air considered as a current expense, which is observed in the proportion of manufacturing and mining companies; these are the two sectors that are most associated with this action.

The majority of companies that prevent surface water pollution by reducing wastewater release, which include this as a current expense, are manufacturing and mining sectors with 72% and 74%, respectively (in Table 4).

Table 4. Economic sector / Prevention of surface water contamination.

Prevent Surface Water Pollution by Reducing Wastewater Release (Includes Wastewater Collection and Treatment) Current Costs: YES/NO				**Total**
		Yes	**No**	
Code of the Economic Sector that covers the main economic activity of the company 2020	Manufacturing	373	330	703
	Mining	60	44	104
	Commerce	95	1567	1662
	Construction	17	164	181
	Services	133	994	1127
Total		678	3099	3777
Symmetrical Measurements				
			Value	Approximate significance
Nominal by Nominal		Phi	0.490	<0.001
		Cramer's V	0.490	<0.001
N of valid cases			3777	

When examining Cramer's V index, which is 0.490, it can be noted that there is a moderate association between the variables. Therefore, the null hypothesis is rejected, meaning that there is an association between the economic sector of the company and the prevention of surface water pollution by reducing the release of wastewater, included as a current expense, which is visualized in the proportion of manufacturing and mining companies; these are the two sectors that are most associated with this action.

The highest volume of companies that prevent the generation of waste and/or residues, the reduction of harmful effects to the environment, which include this as a current expense, are those in manufacturing and mining sectors with 53% and 58%, respectively (Table 5).

Table 5. Economic sector/prevention of waste generation and/or wastes.

Preventing the Generation of Waste and/or Residues, Reducing the Harmful Effects on the Environment (Includes Collection, Treatment, Recycling and Composting, Street Cleaning and Garbage Collection) Current Costs: YES/NO				**Total**
		Yes	**No**	
Economic Sector Code covering the main economic activity of the company 2020	Manufacturing	503	200	703
	Mining	77	27	104
	Commerce	387	1275	1662
	Construction	76	105	181
	Services	375	752	1127
Total		1418	2359	3777
Symmetrical Measurements				
			Value	Approximate significance
Nominal by Nominal		Phi	0.385	0.000
		Cramer's V	0.385	0.000
N of valid cases			3777	

When examining Cramer's V index, which is 0.385, it can be pointed out that there is a moderate association between the variables. Therefore, the null hypothesis is rejected, which means that there is an association between the economic sector of the company and the prevention of waste generation, considered as a current expenditure, which is seen in the proportion of manufacturing and mining companies; these are the two sectors that are most associated with this action.

The largest number of companies that prevent the infiltration of soil and groundwater pollutants, cleaning of soils and water bodies and soil protection against erosion, which include this as a current expense, are in the mining sector with 28% (Table 6).

Table 6. Economic sector/prevention of infiltration of soil and groundwater contaminants.

Preventing Infiltration of Soil and Groundwater Pollutants, Cleaning of Soils and Water Bodies and Protection of Soil against Erosion Current Costs: YES/NO				**Total**
		Yes	**No**	
Code of the Economic Sector that covers the main economic activity of the company 2020	Manufacturing	54	649	703
	Mining	29	75	104
	Commerce	18	1644	1662
	Construction	12	169	181
	Services	23	1104	1127
Total		136	3641	3777
Symmetrical Measurements				
Value				Approximate significance
Nominal by Nominal		Phi	0.259	0.000
		Cramer's V	0.259	0.000
N of valid cases			3777	

When examining the Cramer's V index, which is 0.259, it can be noted that there is a moderate association between the variables. Therefore, the null hypothesis is rejected, which means that there is an association between the economic sector of the company and the prevention of infiltration of soil and groundwater contaminants, the cleaning of soils and bodies of water and the protection of soil against erosion, included as a current expense, which is observed in the proportion of mining companies; this is the sector that is most associated with this action.

The largest fraction of companies that reduce or eliminate the negative consequences of radiation emitted from any source, which include this as a current expense, are from the mining sector with 1.9% (Table 7).

When examining the Cramer's V index, which is 0.053, it can be noted that there is no association between the variables. Therefore, the null hypothesis is accepted, meaning that there is no association between the economic sector of the company and the reduction or elimination of the negative consequences of radiation, considered as a current expense; this can be seen in the proportion of companies in all sectors, since in general, they oscillate in very even percentages, which do not exceed 2 percent.

The largest portion of companies that minimize the use of mineral and energy resources by modifying processes, such as the recovery, reuse, recycling and saving of mineral sources, which include this as a current expense, come from the manufacturing sector with 2.6% (Table 8).

Table 7. Economic sector/reduction or elimination of the adverse effects of radiation.

To Reduce or Eliminate the Negative Consequences of Radiation Emitted from Any Source (Including the Handling, Transportation and Treatment of Radioactive Waste) Current Costs: YES/NO				**Total**
		Yes	**No**	
Code of the Economic Sector that covers the main economic activity of the company 2020	Manufacturing	4	699	703
	Mining	2	102	104
	Commerce	2	1660	1662
	Construction	1	180	181
	Services	7	1120	1127
Total		16	3761	3777
Symmetrical Measurements				
Value				Approximate significance
Nominal by Nominal		Phi	0.053	0.031
		Cramer's V	0.053	0.031
N of valid cases			3777	

Table 8. Economic sector and minimization of the use of mineral and energy resources.

Minimize the Use of Mineral and Energy Resources by Modifying Processes, such as the Recovery, Reuse, Recycling and Saving of Mineral Sources Current Costs: YES/NO				**Total**
		Yes	**No**	
Code of the Economic Sector that covers the main economic activity of the company 2020	Manufacturing	18	685	703
	Mining	1	103	104
	Commerce	4	1685	1662
	Construction	0	181	181
	Services	12	1115	1127
Total		35	3742	3733
Symmetrical Measurements				
Value				Approximate significance
Nominal by Nominal		Phi	0.090	0.000
		Cramer's V	0.090	0.000
N of valid cases			3777	

When examining the Cramer's V index, which is 0.090, it can be pointed out that there is no association between the variables. Therefore, the null hypothesis is accepted, meaning that there is no association between the economic sector of the company and the minimization of the use of mineral and energy resources, included as a current expense. This can be seen in the proportion of companies in all sectors, since in general, they oscillate in very similar percentages that do not exceed 3 percent.

The largest number of companies that minimize the extraction of water resources by modifying processes, such as the reuse, recycling, saving and use of fresh water, which include this as a current expense, are from the mining sector with 9.6% (Table 9).

Table 9. Economic sector/minimization of water resource extraction.

Minimize the Extraction of Water Resources by Modifying Processes, such as the Reuse, Recycling, Saving and Use of Fresh Water. Current Costs: YES/NO				**Total**
		Yes	**No**	
Code of the Economic Sector that covers the main economic activity of the company 2020	Manufacturing	16	687	703
	Mining	10	94	104
	Commerce	1	1661	1662
	Construction	0	181	181
	Services	7	1120	1127
Total		34	3747	3777
Symmetrical Measurements				
Value				Approximate significance
Nominal by Nominal		Phi	0.178	0.001
		Cramer's V	0.178	0.001
N of valid cases			3777	

When examining the Cramer's V index, which is 0.178, it can be noted that there is no association between the variables. Therefore, the null hypothesis is accepted, meaning that there is no association between the economic sector of the company and the minimization of water resource extraction, considered as a current expense, which is seen in the proportion of companies in all sectors; since in general, they oscillate in very equivalent percentages that do not exceed 10 percent.

General Consideration

It can be summarized that there is a moderate association between the economic sector and the following types of action against climate change, namely the reduction of emissions or pollutant concentrations in the air, prevention of surface water pollution, prevention of the generation of waste and/or residues, and the prevention of infiltration of soil and groundwater pollutants, which implies that knowing the economic sector helps to predict whether a company considers these actions in its current expenditure.

On the other hand, there is no association between the economic sector and the following types of action against climate change, namely the reduction or elimination of the negative consequences of radiation emitted from any source, the minimization of the use of mineral or energy resources, and the minimization of the extraction of water resources, which means that knowing the economic sector does not help predict whether a company considers these actions in its current expenditure.

4. Discussion

The results of the EIAEE reveal that in terms of ISO 14001:2015 certifications, the number of companies that have this certification is very small; of the companies surveyed, only 199 out of 3777 had this environmental management certification, which only represents a little more than 5%. Although, worldwide it is noted that "emissions are increasing, but fortunately, so are the number of certifications globally not only in ISO 14001:2015 but also in other standards such as ISO 9001, ISO 45001 or ISO 27001" [18].

The fact that companies are ISO 14001 certified is relevant because they are standardized norms, which allow them to apply the same principles and practices to improve their environmental performance. It also helps them to identify, control and continuously improve their environmental impact, prevent costly environmental incidents, regulatory fines and damage to the company's reputation, as well as promoting operational efficiency.

On average, companies make environmental expenditure of a little more than USD 40.00, and it is observed that three quarters of these companies spend below USD 15,000. Hence, it is importance, not only that companies increase this expenditure, but also that more companies are incorporated into caring for the environment, which involves making the necessary disbursements. According to the United Nations, to mitigate the climate crisis and other environmental damage, it is necessary to triple the resources currently devoted to it by 2030 and quadruple them by 2050. A new report has been revealed, "which calls on the 20 largest economies in the world to promote private capital investment in this area" [19].

It is also very important that environmental spending is considered as a current expense, since it represents regular and recurring expenditure to maintain a company's operational activities.

Companies' environmental spending has a significant impact on the fight against climate change, which is also reflected in corporate sustainability, since investment in technologies and practices that reduce greenhouse gas emissions, in clean energy and technologies, in the implementation of water management systems and the diversification of supply chains will enable the mitigation and adaptation to climate risks.

The Organic Environmental Code (COA) establishes environmental incentives for conservation, sustainable use and management, and ecosystem restoration, in order to promote the sustainable use of biological resources, promote a culture of pollution prevention and reduction, and compliance with environmental regulations.

"The breakdown of current investment marks 86% public funds and 14% private investment (...) Private financing, (...) is basically allocated to biodiversity offsets, sustainable supply chains and impact investments".

In other words, obtaining measurable biodiversity conservation results that compensate for what has been impacted, obtaining a zero net loss of biodiversity, involves a chain that includes an established strategy and process to identify negative environmental impacts, and a plan to diminish these impacts, and investments that are made with the specific intention of providing a positive environmental impact that is measurable, in addition to a financial return.

The results of the research by [9], who points out that expenditure on environmental management represents approximately 0.34% of the average gross domestic product (GDP) during the study period in Ecuador, show that the country's environmental performance is in an increasing phase of the Kuznets curve, in which environmental degradation tends to increase, but there comes a turning point where economic growth begins to lead to environmental improvement.

An important point from the point of view of the sustainability of companies is the importance that this environmental expenditure not only contributes to the care of the environment, but also to strengthen their permanence in the market, since "The study of environmental costs is a factor that becomes relevant since there are tax incentives for companies that invest in a more ecological and environmentally friendly production" [20].

The study analyzes the relationship between the economic sector and environmental spending. The results obtained through Cramer's V test reveal a moderate association between the economic sector and environmental spending on air care, soil, wastewater and waste treatment. This association suggests that certain economic sectors are contributing significantly to environmental management spending in these specific areas. In this case, when observing the figures, it can be seen that the manufacturing and mining sectors represent the highest proportion in terms of their response, as they do consider these four actions as a current expenditure.

On the other hand, the lack of association between the economic sector and the treatment of radiation, the use of mineral or energy resources and water resources indicates that, at least according to the data analyzed in this study, none of these sectors excels in regard to their contribution to these challenges, as they do not consider these three actions as a current expenditure.

In any case, in the seven questions examined, the highest proportion of companies answered NO to many of them, this proportion far exceeded the number of companies that answered YES, which implies that, despite the fact that the business sector contributes to Ecuador's economic growth, it is also failing to include environmental spending as a current expense, causing further deterioration to the environment. This finding underscores the need to implement sustainable development policies and strategies that balance economic growth with environmental protection, so that companies implement strategies that not only favor economic and social practices, but also environmental practices, as pointed out by [11].

As for the seven actions aligned to the fight against climate change considered in this study, companies assume them by considering them as a current expense, which implies that it does so on an ongoing basis. In theory, the strategies proposed to do so include: increasing the generation of renewable energy through reducing the current dependence on fossil fuels; improving energy efficiency through the use of energy-efficient electrical equipment (Type A), heat and energy recovery, and recycling; And Monitoring, reporting and verifying emissions through the constant monitoring of emissions and the implementation of corrective measures, when necessary [21].

5. Conclusions

The global pact encourages organizations to voluntarily join together to implement strategies, among which is care for the environment, including climate change. In Ecuador, the Climate Ambition Accelerator program has involved the participation of several large companies and training within the framework of Ecuador Verde.

From the EIAA survey conducted by the INEC in Ecuador, the results show that only a little more than 5% of companies have the ISO 14001:2015 certification. This provides room to adopt strategies for more companies to achieve certification, since implementing standardized standards helps to identify and evaluate the environmental impact and reducing costs, while promoting the operational efficiency and the reputation of the company.

The results show that only a little more than 5% of companies have the ISO 14001:2015 certification, the province with the highest environmental expenditure is Zamora Chinchipe, and the mining sector, with the average of USD 40,000, although there is a very large gap between 75% of the companies that spend the least and 25% of companies that spend the most, with an increase of almost a thousand times, ranging from companies that do have zero expenditure to those that spend more than USD 16 million per year. At the world level, the United Nations establishes that private capital for environmental care should be tripled by 2030 and quadrupled by 2050.

Concerning the seven actions against climate change considered in this study, the reduction of emissions or pollutant concentrations in the air, the prevention of surface water pollution, the prevention of the generation of waste and/or residues, the prevention of infiltration of soil and groundwater pollutants, the reduction or elimination of the negative consequences of radiation, the minimization of the use of mineral and energy resources and the minimization of the extraction of water resources, there is a predominance of companies that do NOT consider them as current expenses. Only in the treatment of wastewater and waste was there a majority of companies in the manufacturing and mining sectors that indicated that these areas are considered as current expenses.

As for the association by means of the Cramer's V test, it was determined that there is a moderate association between the economic sector and environmental spending on air, soil, wastewater and waste treatment, showing that the manufacturing and mining sectors have a significant contribution in relation to the other economic sectors in regard to these four actions and that there is no association between the economic sector and the treatment of radiation, the use of mineral or energy resources and water resources, which implies that none of the economic sectors has a significant contribution in relation to another economic sector in regard to these three actions.

Author Contributions: M.A., M.R., M.C.C.M. and R.M.T. have contributed equally to all steps in the development of this article. All authors have read and agreed to the published version of the manuscript.

Funding: This research received no external funding.

Institutional Review Board Statement: Not applicable.

Informed Consent Statement: Not applicable.

Data Availability Statement: The data were taken from the Environmental Economic Information Module in Companies 2020, published by the National Institute of Statistics and Census of Ecuador as data available for public use.

Conflicts of Interest: The authors declare that they have no conflicts of interest.

References

1. Xie, X.M.; Zang, Z.P.; Qi, G.Y. Assessing the environmental management efficiency of manufacturing sectors: Evidence from emerging economies. *J. Clean. Prod.* **2016**, *112*, 1422–1431. [CrossRef]
2. Rasche, A.; Kell, G. *The United Nations Global Compact: Achievements, Trends and Challenges*; Cambridge University Press: Cambridge, UK, 2010.
3. Sethi, S.P.; Schepers, D.H. United Nations Global Compact: The Promise-Performance Gap. *J. Bus. Ethics* **2014**, *122*, 193–208. [CrossRef]
4. Diez Principios de la ONU | ¿Qué Puedes Hacer Tú?·Pacto Mundial. Available online: https://www.pactomundial.org/que-puedes-hacer-tu/diez-principios/ (accessed on 27 November 2023).
5. Da Fonseca, L.M.C.M. ISO 14001:2015: An improved tool for sustainability. *J. Ind. Eng. Manag.* **2015**, *8*, 37–50. [CrossRef]
6. Andrade, J.E.O.; Crespo, J.D.O.; Valarezo, F.R.; Vázquez, J.O.Q. Responsabilidad social empresarial en el Ecuador: Abordaje desde la Agenda 2030. *Rev. Cienc. Soc.* **2020**, *26*, 175–193.
7. Pacto Global Red Ecuador, Informe de Gestión. 2022. Available online: https://pactoglobal-ecuador.org/wp-content/uploads/2022/12/INFORME-PACTO-GLOBAL-ECUADOR-2022.pdf (accessed on 10 January 2024).
8. Nations, U. Energías Renovables: Energías para un Futuro más Seguro | Naciones Unidas. Available online: https://www.un.org/es/climatechange/raising-ambition/renewable-energy (accessed on 4 February 2024).
9. Ochoa Bósquez, J.I.; Senmache Chillagano, T.S. Incidencia de la actividad económica en la gestión ambiental de las empresas del sector privado en Ecuador durante el periódo 2010–2015. *Espirales Rev. Multidiscip. Investig.* **2018**, *2*, 86–104.
10. Cueva, F.D. Efectos del cambio climático en la economía, el comercio internacional y la estrategia empresarial. *Contab. Neg.* **2014**, *9*, 75–98.
11. España-Merchán, A.Y. Responsabilidad Social Empresarial hacia la implementación de prácticas ambientales en Ecuador. *Rev. Amaz. Cienc. Econ.* **2023**, *2*, e475. [CrossRef]
12. Ecuador—Módulo de Información Ambiental Económica en Empresas 2019—Información General. Available online: https://anda.inec.gob.ec/anda/index.php/catalog/881 (accessed on 4 January 2024).
13. Bui, B.; de Villiers, C. Business strategies and management accounting in response to climate change risk exposure and regulatory uncertainty. *Br. Account. Rev.* **2017**, *49*, 4–24. [CrossRef]
14. Wu, B.; Zhang, L.; Zhao, Y. Feature selection via Cramer's V-test discretization for remote-sensing image classification. *IEEE Trans. Geosci. Remote Sens.* **2014**, *52*, 2593–2606. [CrossRef]
15. Sadriddinovich, J.T. Capabilities of SPSS Software in High Volume Data Processing Testing. *Am. J. Public Dipl. Int. Stud.* **2023**, *1*, 82–86.
16. Häberer, H. *Guía de Manejo Ambiental para Minería no Metálica*; Gobierno del Peru: Lima, Peru, 2014.
17. Banco Central del Ecuador. *Resultados al Tercer Trimestre de 2020 Enero 2021*; Subgerencia de Programación y Regulación: Quito, Ecuador, 2021.
18. Gomez, A.; Rodriguez, M.A. The effect of ISO 14001 certification on toxic emissions: An analysis of industrial facilities in the north of Spain. *J. Clean. Prod.* **2011**, *19*, 1091–1095. [CrossRef]
19. Sörgel, B.; Kriegler, E.; Weindl, I.; Rauner, S.; Dirnaichner, A.; Ruhe, C.; Hofmann, M.; Bauer, N.; Bertram, C.; Bodirsky, B.L.; et al. A sustainable development pathway for climate action within the UN 2030 Agenda. *Nat. Clim. Chang.* **2021**, *11*, 656–664. [CrossRef]
20. Ekins, P.; Zenghelis, D. The costs and benefits of environmental sustainability. *Sustain. Sci.* **2021**, *16*, 949–965. [CrossRef] [PubMed]
21. Shahbaz, M.; Li, J.; Dong, X.; Dong, K. How financial inclusion affects the collaborative reduction of pollutant and carbon emissions: The case of China. *Energy Econ.* **2022**, *107*, 105847. [CrossRef]

Disclaimer/Publisher's Note: The statements, opinions and data contained in all publications are solely those of the individual author(s) and contributor(s) and not of MDPI and/or the editor(s). MDPI and/or the editor(s) disclaim responsibility for any injury to people or property resulting from any ideas, methods, instructions or products referred to in the content.

climate

Systematic Review

A Systematic Review of Agroecology Strategies for Adapting to Climate Change Impacts on Smallholder Crop Farmers' Livelihoods in South Africa

Mashford Zenda * and Michael Rudolph

Centre for Ecological Intelligence, Faculty of Engineering and the Build Environment (FEBE), University of Johannesburg, Electrical and Electronic Engineering Science, Auckland Park Campus, Auckland Park, P.O. Box 524, Johannesburg 2006, South Africa; michaelr@uj.ac.za
* Correspondence: mzenda@uj.ac.za; Tel.: +27-738-193-999

Abstract: This systematic review identified the prevalence, effectiveness, and potential benefits of agroecology strategies in promoting sustainable agriculture practices implemented by smallholder crop farmers in South Africa. The review carried out a comprehensive literature search across various academic databases, including PubMed, Scopus, and Web of science. The relevant studies were screened and selected based on predetermined inclusion criteria where a total of 262 articles were extracted and reduced to 30 articles for this systematic review. Data were extracted and synthesised to classify patterns and trends in the adoption of agroecology elements. The results obtained from the review of this study highlights the identification of specific strategies such as indigenous crop varieties, conservation agriculture, intercropping, agroforestry, drought-tolerant crop varieties, and water management strategies. These outcomes demonstrated insights into the prevalence of different strategies applied by smallholder crop farmers in South Africa. Furthermore, the review determined the reported benefits, such as increased crop resilience, improved soil fertility, and enhanced water use efficiency. These benefits were assessed on the available evidence from the selected studies. This review contributes to a better understanding of agroecology practices in South African. The results can inform policymakers, researchers, and farmers in developing appropriate strategies to enhance sustainable agricultural practices.

Keywords: agroecological strategies; sustainable agriculture; smallholder crop farmers; South Africa

1. Introduction

Agriculture plays a vital role in global food security and the livelihoods of millions of people. However, the agricultural sector faces numerous challenges, including climate change, diminishing natural resources, and the need to feed a growing population (Dlamini et al., 2023) [1]. Additionally, climate change is increasing the occurrence of floods, drought and other climate extremes (Gemeda et al., 2023) [2]. These pressures may compromise the agricultural sector's ability to meet the needs of a growing population and ensure food security. Without rapid and effective adaptation, climate change will continue to have devastating consequences (Vögt et al., 2023) [3]. In this context, the concept of agroecology has gained significant attention as a promising approach to address these challenges and promote sustainable and resilient agricultural systems.

Climate change is a pressing global issue that presents significant threats to various sectors, including agriculture (IPCC, 2014; Gomez-Zavaglia et al., 2020; Dlamini et al., 2023; Maluleke et al., 2024 [1,4–6]. The African continent has a higher proportion of people who are vulnerable to the effects of climate change than any other region in the world (Weber et al., 2018; Gobir et al., 2021) [7,8]. Africa's continent is vulnerable to the effects of climate change due to its reliance on natural resources. The continent is heavily dependent on rain-fed agriculture, which is sensitive to changes in rainfall patterns. In addition, the continent's vulnerability to climate is further exacerbated by its low adaptive capacity (Weber et al., 2018; Mthembu and Hlophe, 2021; [7,9].

In South Africa, smallholder crop farmers are vulnerable to the adverse impacts of climate change on their livelihoods (Nkambule et al., 2019; Thinda et al., 2020; Okolie et al., 2023; Serote et al., 2023; Ubisi et al., 2023) [10–14]. The negative effects of climate change affect smallholder farmers mostly in developing countries, who depend on natural resources for their livelihoods (Ndlovu et al., 2021; Serote et al., 2023; Okolle et al., 2023) [12,13,15]. Changes in rainfall patterns, increased temperatures, and extreme weather events can disrupt crop production, reduce yields, and jeopardise food security for these farmers (Nhemachena and Hassan, 2007; Thinda et al., 2020) [11,16]. In response, the adoption of agroecology strategies has gained attention as a promising approach to enhance the resilience and adaptive capacity of smallholder farmers in the face of climate change (Altieri, 2018; Odusola, 2021; Kerr et al., 2023) [17–19].

Agroecology represents a holistic and interdisciplinary approach that integrates ecological principles and practices into agricultural systems. It underscores the importance of biodiversity, ecological processes, and local knowledge in enhancing productivity, resilience, and sustainability. Agroecology is the application of science of ecology to agricultural systems (Zenda and Heyns, 2017; Rudolph, Muchesa and Kroll, 2020; Kesselman, Ngcoya and Casale, 2021) [20–22]. Agroecology promotes the integration of ecological principles into agricultural systems (FAO, 2018) [23]. It accentuates sustainable farming practices that enhance biodiversity, soil health, and ecosystem services while minimising external inputs (Gliessman, 2014) [24]. Agroecology strategies include the use of indigenous and drought-tolerant crop varieties, conservation agriculture practices, intercropping, agroforestry, and water management techniques (Tittonell, 2014 [25]). These strategies aim to improve crop resilience, soil fertility, water use efficiency, and overall farm productivity, thereby enhancing the livelihoods of smallholder crop farmers (Akanmu et al., 2023) [26]. In addition, the adoption of agroecological techniques by smallholder farmers is crucial to ensuring food security (Akanmu et al., 2023) [26].

While agroecology has shown promise in the context of climate change adaptation, there is a need to consolidate, analyse and interrogate the results reported in the existing literature. Numerous studies have investigated the effectiveness, challenges, and outcomes of agro-ecology strategies for smallholder farmers in South Africa (Mugwanya, 2019; Mutenje et al., 2020) [27,28]. However, the findings from these studies are scattered across various sources, making it difficult to obtain a comprehensive understanding of the overall picture.

The effectiveness of agroecology strategies in addressing climate change is highly dependent on the local agroclimatic conditions. Different regions in South Africa experience different agroclimatic conditions and these conditions are changing over time due to climate change. This variability creates challenges for implementing effective agroecology strategies due to ongoing climate change (Trnka et al., 2021) [29]. Studies have shown that climate change is accelerating significantly, impacting local agroclimatic conditions and subsequently influencing the effectiveness of agroecology strategies (Calleja-Cabrera et al., 2020) [30]. Furthermore, the link between climate variability and food insecurity emphasises the need to consider local agroclimatic conditions when implementing agroecology strategies to address climate change (Thornton et al., 2014) [31]. Farmers' perceptions of climate change also vary across different agroecologies, highlighting the localised nature of agroclimatic conditions and their influence on the adoption and effectiveness of agroecology strategies (Esayas et al., 2019) [32]. Additionally, the prevalence of diseases such as bovine trypanosomosis and the density of its vectors are influenced by agroclimatic conditions, further emphasising the need to tailor agroecology strategies to local contexts (Kahsay et al., 2019) [33].

Moreover, the genetic diversity of crops within specific agroclimatic conditions indicates adaptation to local climate, suggesting that agroecology strategies should be context-specific to effectively mitigate climate change impacts (Amante and Tesgera, 2020) [34]. Local genetic resources and traditional practices gain importance in sustainable agricultural production, emphasising the significance of considering agroclimatic conditions in agroecology strategies (Izzatullayeva et al., 2014) [35]. The observed poor adaptation of some cultivars to local agroclimatic conditions further underscores the necessity of aligning agroecology strategies with specific local conditions to enhance their effectiveness in addressing climate change (Jaćimović et al., 2020) [36].

A systematic review offers a rigorous and comprehensive approach to synthesise the available evidence. By systematically searching, selecting, and analysing relevant studies, such a systematic review can provide a consolidated and unbiased summary of the results, identify key patterns and trends, and assess the overall effectiveness of agroecology strategies in South Africa (Haddaway et al., 2018) [37]. These results can contribute to evidence-based decision-making, inform policy development.

This review aims to address the research gap regarding the utilisation of agroecology strategies for climate change adaptation among smallholder crop farmers in South Africa and will provide a comprehensive and current understanding of the effectiveness, challenges, and outcomes of agroecology practices in enhancing the livelihoods and resilience of smallholder farmers. The results of this review will offer important insights for policymakers, researchers, and practitioners seeking to promote sustainable agricultural approaches and support the adaptation of smallholder farmers to the impacts of climate change guide future research efforts, facilitate the adoption and scaling-up of successful agroecological practices.

The FAO's 10 Elements of Agroecology framework is a holistic tool that helps countries achieve transformative change towards sustainable agriculture and food systems (Barrios et al., 2020) [38]. It describes how different elements interact and how they can create desirable outcomes. The elements can be used by practitioners and decision-makers to plan, implement, manage, and evaluate agroecological transitions. This review paper is based on climate change resilience; an element of agroecological framework. It refers to the capacity of agroecological systems to adapt to and cope with the effects of climate change. This includes the ability to maintain food production, protect natural resources and ensure livelihoods in the face of climate change. It also includes the ability to bounce back and adapt to changing climate conditions.

2. Materials and Methods

A protocol was developed to outline the objectives, inclusion and exclusion criteria, search strategy, data extraction methods, and quality assessment procedures (Figure 1). The inclusion criteria specified that the studies must be peer-reviewed articles, dissertations, or reports focused on agroecology strategies utilised by smallholder crop farmers in South Africa to adapt to climate change impacts on their livelihoods. The publication date range for the search was set between January 2010 and December 2023.

A systematic search was conducted in electronic databases, including Scopus, Web of Science, and PubMed. Relevant keywords such as "agroecology", "climate change", "smallholder farmers", and "South Africa" were used to identify potentially relevant studies. The search strategy included combining these keywords using appropriate Boolean operators (e.g., AND, OR).

The retrieved studies were screened based on their titles and abstracts to identify potentially relevant studies. The full texts of the selected studies were obtained for further evaluation. The inclusion criteria were applied to determine the final set of studies to be included in the systematic review.

A standardised data extraction form was developed to collect relevant information from the included studies. Key data elements were extracted, including study characteristics (e.g., study design, sample size), agroecology strategies employed, outcomes measured, and findings reported. Multiple reviewers were involved in the data extraction process to ensure consistency and accuracy.

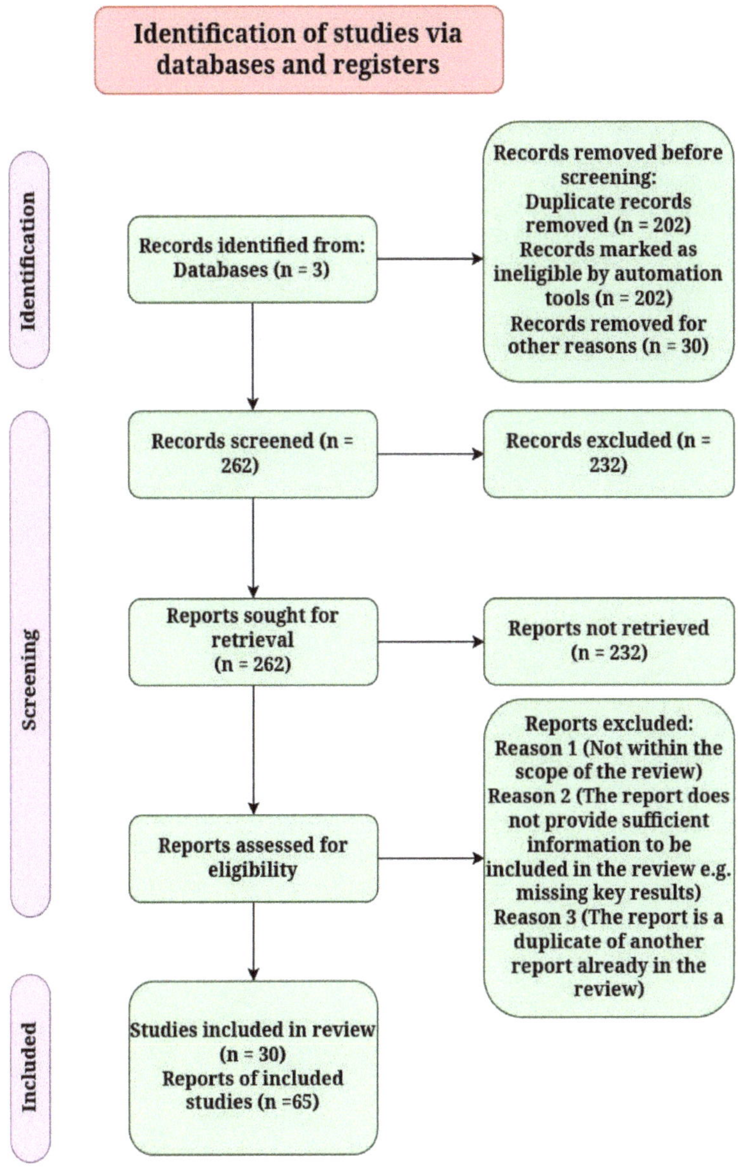

Figure 1. Prisma flow diagram for the current study.

The included studies were critically appraised to assess their quality and risk of bias. Appropriate tools such as critical appraisal skills programme (CASP) Tools or checklists were used to evaluate the methodological rigour, internal validity, and external validity of the studies.

The extracted data were analysed and synthesised to identify common themes, patterns, and trends. The findings of the included studies were summarised in a clear and structured manner, highlighting the main outcomes, challenges, and benefits of agroecology strategies.

3. Results

The initial search yielded a total of 262 articles. After screening and applying inclusion criteria, 30 studies were selected for detailed analysis. From this review, the findings regarding the use of agroecological strategies by smallholder farmers in South Africa were categorised into different themes such as conservation agriculture, drought tolerant crop varieties, water management, indigenous crop varieties, crop management and agroforestry. The findings indicate that many smallholder crop farmers in South Africa have utilised various agroecology strategies to adapt to climate change impacts. These strategies include the use of indigenous and drought-tolerant crop varieties, conservation agriculture practices, intercropping, agroforestry, crop management and water management techniques. The studies show that these practices can have many benefits, including increasing crop resilience, improving soil fertility and using water more efficiently. This could lead to improved livelihoods for smallholder farmers (Figure 2).

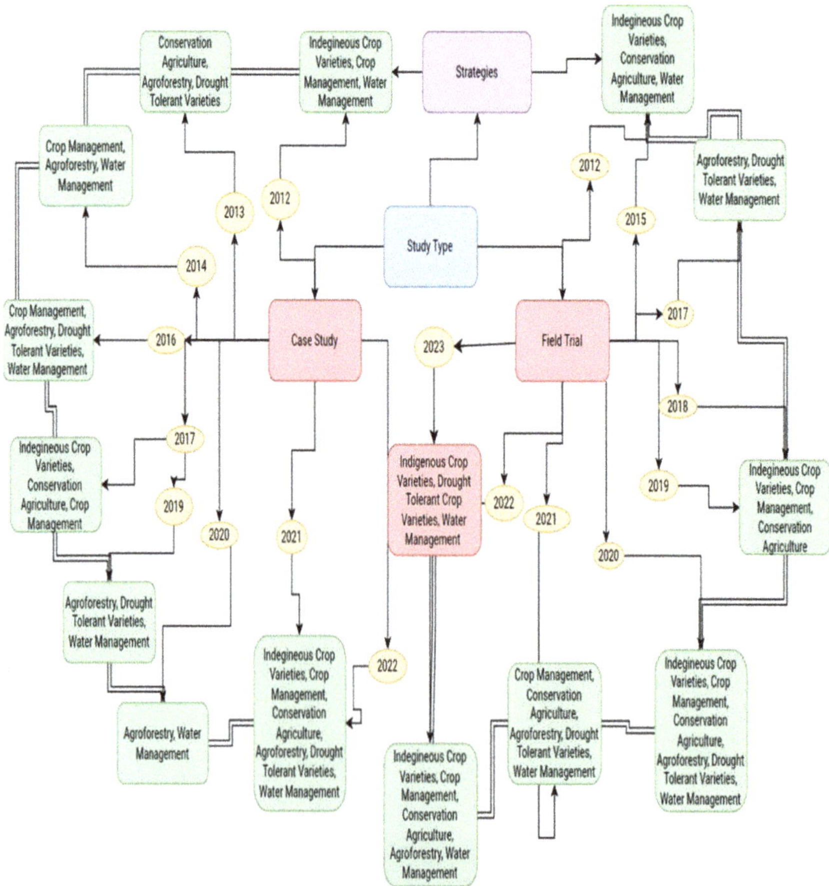

Figure 2. Agroecology strategies utilised by smallholder crop farmers in South Africa.

Agroecology is a holistic approach that combines ecological and social perspectives to create sustainable and equitable food systems. It aims to optimise the interactions between plants, animals, humans, and the environment, while addressing the social dimensions that are essential for a fair and sustainable food system.

It is a bottom-up approach that emphasises local, place-based solutions to address specific problems. It draws on the knowledge and experience of farmers and communities, combining this practical and traditional knowledge with scientific research. This co-creation of knowledge empowers farmers and communities to be key agents of change, enhancing their autonomy and ability to adapt to changing conditions. In this way, agroecology fosters resilience, sustainability, and social and economic justice.

While conventional agricultural practices often focus on fixing specific problems within an unsustainable system, agroecology aims to transform the system as a whole. It takes a holistic approach, addressing root causes and providing long-term solutions that address the social, economic, and environmental dimensions of food systems. This includes recognising and promoting the rights of marginalised groups, such as women, youth, and indigenous peoples. Ultimately, agroecology seeks to create more just sustainable, and resilient food systems.

Table 1 provides a comprehensive overview of the agroecology strategies employed by smallholder crop farmers in South Africa, based on the selected studies. It presents a breakdown of the strategies studied in each research article, allowing us to observe the prevalence and combination of different approaches.

Table 1. Agroecology strategies utilized by smallholder crop farmers in South Africa.

Study	
Study 1	Indigenous crop varieties, intercropping, water management
Study 2	Conservation agriculture, agroforestry, drought-tolerant crop varieties
Study 3	Intercropping, water management, agroforestry
Study 4	Conservation agriculture, indigenous crop varieties, water management
Study 5	Drought-tolerant crop varieties, agroforestry, water management
Study 6	Indigenous crop varieties, conservation agriculture, intercropping
Study 7	Agroforestry, water management, drought-tolerant crop varieties
Study 8	Conservation agriculture, intercropping, indigenous crop varieties
Study 9	Water management, agroforestry, drought-tolerant crop varieties
Study 10	Intercropping, indigenous crop varieties, conservation agriculture
Study 11	Drought-tolerant crop varieties, water management, agroforestry
Study 12	Conservation agriculture, intercropping, drought-tolerant crop varieties
Study 13	Indigenous crop varieties, water management, agroforestry
Study 14	Intercropping, agroforestry, conservation agriculture
Study 15	Drought-tolerant crop varieties, water management, indigenous crop varieties
Study 16	Planting of improved seed, application of chemical fertilizer, changing planting date, application of pesticide/herbicides, mixed cropping, nulching, rainwater harvesting, irrigation, planting of trees
Study 17	Alterations to soil and crop management, crop rotations, and water conservation techniques.
Study 18	Drought-tolerant seeds, shorter cycle crops, diversification of crops, changing planting dates, small-scale irrigation
Study 19	Planting early-maturing plants and drought-tolerant crops, altering planting dates, crop diversification, and irrigating
Study 20	Early maturing variety, soil and water conservation, tree planting, migration, varying planting dates, mulching, improved variety
Study 21	Water harvesting, change of crop variety, soil conservation, improved soil fertility, crop diversification and changing of crop type
Study 22	Improved crop varieties, varying planting dates, soil and water conservation, tree planting
Study 23	Soil and water management
Study 24	Changing crop types, planting schedule and crop rotation, water management and zero tillage
Study 25	Changing planting dates, planted drought resistant or short season crops, soil conservation, diversification, planting trees and irrigation
Study 26	Drought tolerant varieties, shifting planting dates, irrigation, conservation agriculture
Study 27	Drought resistant crops, changing planting dates, planting crop varieties, mixed cropping, conservation agriculture, adjusting fertilise inputs, irrigation systems, rain harvesting and drilling boreholes
Study 28	Planting drought tolerant variety and changing of planting time
Study 29	Shifting planting dates, reduced cultivated area, drought-resistant varieties, crop diversification, and intercropping
Study 30	Changing fertilizer, water harvesting and planting drought resistant varieties

Table 2 shows the presence or absence of specific agroecology strategies in each study. It reveals that the selected studies encompass a range of strategies, including the use of indigenous crop varieties, conservation agriculture practices, intercropping, agroforestry, drought-tolerant crop varieties, and water management techniques. This indicates that smallholder farmers in South Africa are employing diverse approaches to adapt to the impacts of climate change on their crop production.

Table 2. The presence or absence of specific agroecology strategies in each study.

Study	Indigenous Crop Varieties	Conservation Agriculture	Intercropping	Agroforestry	Drought-Tolerant Crop Varieties	Water Management
1	Yes	No	Yes	No	No	Yes
2	No	Yes	No	Yes	Yes	No
3	No	No	Yes	Yes	No	Yes
4	Yes	Yes	No	No	No	Yes
5	No	No	No	Yes	Yes	Yes
6	Yes	Yes	Yes	No	No	No
7	No	No	No	Yes	Yes	Yes
8	Yes	Yes	Yes	No	No	No
9	No	No	No	Yes	Yes	Yes
10	Yes	Yes	Yes	No	No	No
11	No	No	No	Yes	Yes	Yes
12	No	Yes	Yes	No	Yes	No
13	Yes	No	No	Yes	No	Yes
14	No	Yes	Yes	Yes	No	No
15	Yes	No	No	No	Yes	Yes
16	No	No	Yes	Yes	No	Yes
17	No	Yes	Yes	No	No	Yes
18	Yes	Yes	No	No	Yes	Yes
19	Yes	No	Yes	No	Yes	Yes
20	No	Yes	No	Yes	No	Yes
21	No	Yes	Yes	No	No	Yes
22	No	Yes	Yes	Yes	No	Yes
23	Yes	Yes	No	No	No	Yes
24	Yes	Yes	Yes	No	Yes	Yes
25	No	Yes	Yes	Yes	Yes	Yes
26	No	Yes	Yes	No	Yes	Yes
27	No	Yes	Yes	No	Yes	Yes
28	No	No	Yes	No	Yes	No
29	No	Yes	Yes	No	No	No
30	No	No	Yes	No	Yes	Yes

Table 3 highlights the reported benefits of these agroecology strategies. The findings indicate that many of these strategies contribute positively to agricultural outcomes. Increased crop resilience, improved soil fertility, and enhanced water use efficiency are among the benefits observed in various studies. These findings suggest that the adoption of agroecology practices can potentially improve the livelihoods of smallholder farmers by increasing their agricultural productivity and sustainability.

Figure 3 illustrates the benefits of agroecology strategies for smallholder farmers in South Africa. A review of the literature found that water use efficiency was the most frequently reported benefit of agroecology strategies, followed by soil fertility benefits and increased crop resilience. This suggests that agroecology practices may have the greatest impact on water use efficiency in the short term, while other benefits such as increased crop resilience may take longer to be realised.

Table 3. Benefits of Agroecology strategies utilised by smallholder crop farmers in South Africa.

Study	Increased Crop Resilience	Improved Soil Fertility	Enhance Water Use Efficiency
1	Yes	Yes	Yes
2	No	Yes	Yes
3	Yes	No	Yes
4	Yes	Yes	Yes
5	No	No	Yes
6	Yes	Yes	No
7	No	Yes	Yes
8	Yes	Yes	No
9	No	Yes	Yes
10	Yes	No	Yes
11	No	No	Yes
12	Yes	Yes	No
13	Yes	No	Yes
14	No	Yes	No
15	No	Yes	Yes
16	Yes	Yes	Yes
17	Yes	Yes	No
18	Yes	Yes	Yes
19	No	No	Yes
20	No	Yes	Yes
21	Yes	Yes	Yes
22	Yes	No	Yes
23	No	Yes	Yes
24	Yes	No	Yes
25	Yes	No	No
26	No	No	Yes
27	Yes	No	Yes
28	No	Yes	No
29	Yes	Yes	No
30	No	Yes	Yes

Table 4 provides insights into the characteristics of the selected studies. It includes information such as the year of the study, study design, and sample size. This information helps to contextualise the findings and provides a basis for assessing the reliability and generalisability of the results. The studies span 10 years different years, indicating that research on agroecology strategies in South Africa. The inclusion of diverse study designs, such as field trials and case studies, indicates a comprehensive approach to understanding the effectiveness of these strategies. Diverse study designs are critical to ensuring that a research project is robust and comprehensive. A single study design may not be able to fully capture the nuances of a complex issue. By including a range of study designs, the researchers can obtain a more holistic view of the issue and understand how different strategies work in different settings. The variation in sample sizes reflects the different scales at which the studies were conducted, ranging from small-scale investigations to larger-scale assessments.

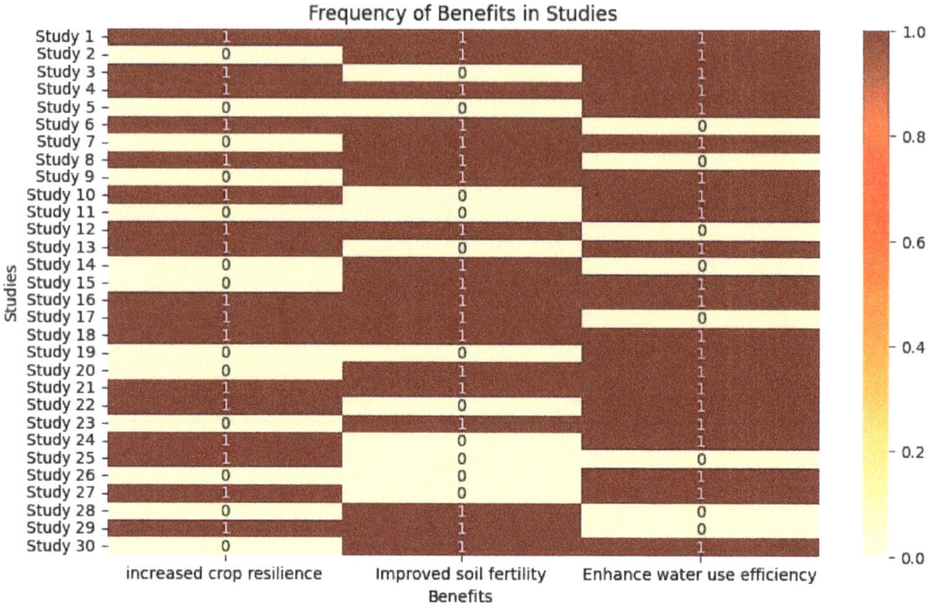

Figure 3. Heat map of benefits of agroecology strategies utilisation by smallholder crop farmers in South Africa.

Table 4. Study characteristics of selected studies.

Study	Year	Study Design	Sample Size
1	2012	Case study	20
2	2013	Case study	30
3	2014	Case study	25
4	2015	Field trial	100
5	2016	Case study	35
6	2017	Case study	22
7	2017	Field trial	50
8	2018	Field trial	60
9	2019	Case study	28
10	2019	Field trial	80
11	2020	Field trial	70
12	2021	Field trial	90
13	2022	Field Trial	45
14	2022	Case study	22
15	2023	Field trial	90
16	2021	Field trial	16

Table 4. *Cont.*

Study	Year	Study Design	Sample Size
17	2021	Field Trial	120
18	2020	Field trial	224
19	2022	Case study	200
20	2020	Case study	183
21	2020	Field trial	391
22	2022	Case study	183
23	2012	Field	300
24	2021	Case study	90
25	2021	Case study	328
26	2021	Field	160
27	2022	Case study	18
28	2016	Case study	70
29	2021	Field	200
30	2021	Case study	235

4. Discussion

The review findings suggest that agroecology strategies have the potential to enhance the adaptive capacity of smallholder crop farmers in South Africa. However, the adoption and scaling-up of these practices face several challenges, including economic challenges faced by smallholder farmers, social and cultural factors, and institutional barriers.

One of the key reasons for the limited adoption of agroecology strategies is the economic challenges faced by smallholder farmers. Agroecology often requires initial investments in sustainable farming practices, such as organic inputs, crop diversification, and agroforestry. These initial investments may pose financial barriers for small-scale farmers who may lack access to credit or financial resources. Additionally, the transition to agroecology may involve a period of adjustment and lower yields during the initial phases, which can be financially challenging for farmers who rely on their agricultural production for their livelihoods.

Social and cultural factors also play a significant role in the adoption of agroecology strategies. Traditional farming practices and knowledge systems may be deeply ingrained in farming communities, making it challenging for farmers to transition to new and unfamiliar agroecological methods. Moreover, social norms and perceptions within farming communities may influence the willingness of farmers to adopt new practices, particularly if there is a lack of awareness or understanding of the potential benefits of agroecology.

Institutional barriers, such as limited access to extension services, technical support, and market opportunities for agroecological products, also contribute to the low adoption of agroecology strategies among farmers in South Africa. The lack of supportive policies and incentives for agroecological practices may further hinder farmers from transitioning to sustainable farming methods.

Additionally, the socio-economic context and local agroecological conditions influence the effectiveness of these strategies. The socio-economic context and local agroecological conditions can have a significant impact on the effectiveness of adaptation strategies. For example, in some communities, traditional agricultural practices, such as subsistence farming, may not be compatible with the types of adaptation strategies that are being promoted by outside organisations. One major challenge related to the socio-economic context is the high level of poverty among smallholder farmers in South Africa. Many farmers lack the financial resources to invest in agroecological practices, such as organic fertilisers or drought-tolerant crops. In addition, farmers may not have access to the information or training needed to successfully adopt agroecology strategies. Furthermore, farmers may be reluctant to adopt new practices if they believe they will be less productive or profitable than their current practices.

It is important to note that these challenges are not static, and that they can change over time. For example, climate change is expected to have a significant impact on the agroecological conditions in South Africa, with hotter and drier conditions predicted in the future. This could lead to additional challenges for farmers, such as increased water stress and the need for new crop varieties that are adapted to changing conditions. As such, it is important to regularly assess the local agroecological conditions and the broader context in which farmers are operating when developing adaptation strategies to climate change. The review emphasises the need for tailored approaches, capacity building, supportive policies, and multi-stakeholder engagement to promote the widespread adoption of agroecology practices among smallholder farmers.

There are various agroecology strategies that are being utilised by smallholder crop farmers in South Africa such as the use of indigenous crop varieties, conservation agriculture, intercropping, agroforestry, planting drought-tolerant crop varieties and water management strategies. These strategies will be explained in the following paragraphs.

Indigenous crop varieties are often well-adapted to the local climate and environmental conditions, making them a valuable resource for building resilience to climate change. For example, many indigenous crops are drought-resistant, can withstand high temperatures, and are able to tolerate poor soils. They may also have other beneficial traits, such as disease resistance, high nutrient content, or resistance to pests. By using indigenous crop varieties, farmers can maintain yields even in the face of climate change.

Conservation agriculture is a farming system that relies on practices such as minimal soil disturbance, crop rotation, and the use of cover crops. It is a particularly effective adaptation strategy for climate change since it helps to improve soil health and fertility, enhance water retention, and reduce the risk of soil erosion. By implementing these practices, farmers can increase the resilience of their crops and maintain productivity even in the face of extreme weather events such as droughts or floods. It also has the added benefits of improving water quality, biodiversity, and carbon sequestration.

Intercropping is another agroecological practice that can help farmers adapt to climate change. It involves planting two or more crops together in the same area, rather than growing them separately. This can help to improve soil health, increase yields, and diversify farm income. It also has the benefit of reducing the impact of pests and diseases, since different crops have different susceptibilities to pests and diseases. Additionally, intercropping can help to reduce the need for inputs such as fertilizer and water.

Agroforestry is a practice that combines agriculture and forestry to create productive, diverse, and sustainable land-use systems. It is a particularly effective adaptation strategy for climate change since it can help to improve soil health, increase biodiversity, and provide shade and wind protection for crops. It can also reduce soil erosion and improve water quality. Additionally, agroforestry can help to mitigate climate change by sequestering carbon.

Planting drought-tolerant crop varieties is another effective adaptation strategy for climate change. Drought-tolerant crops are able to withstand periods of low water availability, which is becoming more common due to climate change. They can maintain their productivity even when there is less rainfall, making them a valuable tool for farmers. Some examples of drought-tolerant crops include sorghum, millet, and cowpeas.

Water management strategies are crucial for adapting to climate change, especially in regions where water is already scarce. These strategies can include rainwater harvesting, the use of efficient irrigation techniques, and the use of mulch to reduce water evaporation. Additionally, it is important to conserve water resources by reducing water loss through leakage and evaporation and promoting water-saving behaviors among farmers.

5. Conclusions

This systematic review highlights the importance of agroecology strategies in enabling smallholder crop farmers in South Africa to adapt to the impact of climate change on their livelihoods. The findings underscore the potential benefits of agroecology practices in enhancing resilience, promoting sustainable agriculture, and improving smallholder farmers' well-being. However, addressing the challenges associated with adoption and scaling-up is crucial. Policymakers, researchers, and practitioners need to collaborate to develop context-specific strategies, provide necessary resources, and create an enabling environment for the widespread adoption of agroecology practices among smallholder crop farmers in South Africa. Future studies should conduct a meta-analysis of existing studies on the topic, synthesizing the results of multiple studies to identify the most effective agroecological strategies for adapting to climate change.

Author Contributions: M.Z.—idea conceptualization, methodology, software, writing—original draft; writing—reviewing and editing. M.R.—Editing and proof reading—reviewing and editing original draft. All authors have read and agreed to the published version of the manuscript.

Funding: This research received no external funding.

Data Availability Statement: Anonymised data are available by contacting the authors.

Conflicts of Interest: The authors declare no known competing financial interests or personal relationships that could have appeared to influence the work reported in this paper.

References

1. Dlamini, L.; Crespo, O.; Van Dam, J.; Kooistra, L. A global systematic review of improving crop model estimations by assimilations remote sensing data: Implications for small-scale agricultural systems. *Remote Sens.* **2023**, *15*, 4066. [CrossRef]
2. Gemeda, D.O.; Korecha, D.; Garedew, W. Climate change perception and vulnerability assessment of the farming communities in the Southwest Parts of Ethiopia. *Climate* **2023**, *11*, 183. [CrossRef]
3. Vögt, V.; Harrs, J.A.; Reinhart, V.; Hollenbach, P.; Bühler, M.M.; Tewes, T. Implementing agile data workflows to unlock climate-resilient urban planning. *Climate* **2023**, *11*, 174. [CrossRef]
4. Intergovernmental Panel on Climate Change (IPCC). *Climate Change 2014: Impacts, Adaptation, and Vulnerability. Part A: Global and Sectoral Aspects. Contribution of Working Group II to the Fifth Assessment Report of the Intergovernmental Panel on Climate Change*; Cambridge University Press: Cambridge, UK, 2014.
5. Gomez-Zavaglia, A.; Mejuto, J.C.; Simal-Gandara, J. Mitigation of emerging implications of climate change on food production systems. *Food Res. Int.* **2020**, *134*, 109256. [CrossRef]
6. Maluleke, P.; Moeletsi, M.E.; Tsubo, M. Analysis of climate variability and its implications on rangelands in the Limpopo Province. *Climate* **2024**, *12*, 2. [CrossRef]
7. Weber, T.; Haensler, A.; Rechid, D.; Pfeifer, S.; Eggert, B.; Jacob, D. Analyzing regional climate change in africa in a 1.5, 2, and 3 °C global warming world. *Earth's Future* **2018**, *6*, 643–655. [CrossRef]
8. Gobir, A.A.; Aliyu, A.; Abubakar, A.; Esekhaigbe, C.; Joshua, I.; Adagba, K.; Nmadu, A. Climate change awareness and related tree planting practices in a rural community in north-western nigeria. *J. Community Med. Prim. Health Care* **2021**, *33*, 41–49. [CrossRef]
9. Mthembu, A.; Hlophe, S. Building resilience to climate change in vulnerable communities: A case study of umkhanyakude district municipality. *Town Reg. Plan.* **2021**, *77*, 42–56.
10. Nkambule, N.P. Climate change adaptation in smallholder agriculture: Evidence from two Southern African countries. *Clim. Dev.* **2019**, *11*, 144–158.
11. Thinda, K.T.; Ogundeji, A.A.; Belle, J.A.; Ojo, T. Understanding the adoption of climate change adaptation strategies among smallholder farmers: Evidence from land reform beneficiaries in South Africa. *Land Use Policy* **2020**, *99*, 104858. [CrossRef]
12. Okolie, C.C.; Danso-Abbeam, G.; Ogundeji, A. Livelihood vulnerability to the changing climate: The experiences of smallholder farming households in the Free State Province, South Africa. *Clim. Serv.* **2023**, *30*, 100371. [CrossRef]
13. Serote, B.; Mokgehle, S.; Senyolo, G.; du Plooy, C.; Hlophe-Ginindza, S.; Mpandeli, S.; Nhamo, L.; Araya, H. Exploring the barriers to the adoption of climate-smart irrigation technologies for sustainable crop productivity by smallholder farmers: Evidence from South Africa. *Agriculture* **2023**, *13*, 246. [CrossRef]
14. Ubisi, N.R.; Kolanisi, U.; Jiri, O. The Role of Indigenous Knowledge Systems in Rural Smallholder Farmers' Response to Climate Change: Case Study of Nkomazi Local Municipality, Mpumalanga, South Africa. *J. Asian Afr. Stud.* **2023**, *55*, 273–284. [CrossRef]
15. Ndlovu, M.; Clulow, A.D.; Savage, M.J.; Nhamo, L.; Magidi, J.; Mabhaudhi, T. An assessment of the impacts of climate variability and change in KwaZulu-Natal Province, South Africa. *Atmosphere* **2021**, *2*, 427. [CrossRef]

16. Nhemachena, C.; Hassan, R. *Micro-Level Analysis of Farmers' Adoption of Conservation Farming Practices in Africa: A Review of Constraints and Effective Interventions*; Environment for Development Discussion Paper Series; International Food Policy Research Institute: Washington, DC, USA, 2007; pp. 1–29.
17. Altieri, M.A. *Agroecology: The Science of Sustainable Agriculture*, 4th ed.; CRC Press: Boca Raton, FL, USA, 2018.
18. Odusola, A. Agriculture as the fulcrum of inclusive development in Africa. In *Africa's Agricultural Renaissance*; Palgrave Macmillan: Cham, Switzerland, 2021.
19. Kerr, R.B.; Postigo, J.C.; Smith, P.; Cowie, A.; Singh, P.K.; Rivera-Ferre, M.; Tirado-von der Pahlen, M.C.; Campbell, D.; Neufeldt, H. Agroecology as a transformative approach to tackle climatic, food, and ecosystemic crises. *Environ. Sustain.* **2023**, *62*, 101275.
20. Zenda, M.; Heyns, S. *Advancing Agroecology through Soil Fertility*; Soil booklet; Surplus Peoples Project: Cape Town, South Africa, 2017.
21. Rudolph, M.; Muchesa, E. A Review of the Agroecological Farming System as a Viable Alternative Food Production Approach in South Africa. *S. Afr. J. Agric. Ext.* **2023**, *51*, 43–67.
22. Kesselman, B.; Ngcoya, M.; Casale, D. The challenge posed by urban dietary norms to the practice of urban agroecology. *Agroecol. Sustain. Food Syst.* **2021**, *45*, 480–498. [CrossRef]
23. Food and Agriculture Organisation of the United Nations (FAO). Agroecology Knowledge Hub. Food and Agriculture Organization of the United Nations. 2018. Available online: http://www.fao.org/agroecology/overview/en (accessed on 23 October 2023).
24. Gliessman, S.R. *Agroecology: The Ecology of Sustainable Food Systems*, 3rd ed.; CRC Press: Boca Raton, FL, USA, 2014.
25. Tittonell, P. Ecological intensification of agriculture—Sustainable by nature. *Curr. Opin. Environ. Sustain.* **2014**, *8*, 53–61. [CrossRef]
26. Akanmu, A.O.; Akol, A.M.; Ndolo, D.O.; Kutu, F.R.; Babalola, O.O. Agroecological techniques: Adoption of safe and sustainable agricultural practices among the smallholder farmers in Africa. Front. *Sustain. Food Syst.* **2023**, *7*, 1143061. [CrossRef]
27. Mugwanya, N. Why agroecology is a dead end for Africa. *Outlook Agric.* **2019**, *48*, 113–116. [CrossRef]
28. Mutenje, M.J. Effects of conservation agriculture on crop yield, crop income, and household food security in southern Africa: A systematic review. *Agronomy* **2020**, *10*, 1177.
29. Trnka, M.; Brázdil, R.; Balek, J.; Dubrovský, M.; Eitzinger, J.; Hlavinka, P.; Žalud, Z. Observed changes in the agroclimatic zones in the czech republic between 1961 and 2019. *Plant Soil Environ.* **2021**, *67*, 154–163. [CrossRef]
30. Calleja-Cabrera, J.; Botër, M.; Oñate-Sánchez, L.; Pernas, M. Root growth adaptation to climate change in crops. *Front. Plant Sci.* **2020**, *11*, 544. [CrossRef]
31. Thornton, P.; Ericksen, P.; Herrero, M.; Challinor, A. Climate variability and vulnerability to climate change: A review. *Glob. Change Biol.* **2014**, *20*, 3313–3328. [CrossRef] [PubMed]
32. Esayas, B.; Simane, B.; Teferi, E.; Ongoma, V.; Tefera, N. Climate variability and farmers' perception in southern ethiopia. *Adv. Meteorol.* **2019**, *2019*, 7341465. [CrossRef]
33. Kahsay, H.; Guta, D.; Simane, B.; Gidey, T. Farmers' perceptions of climate change trends and adaptation strategies in semiarid highlands of eastern tigray, northern ethiopia. *Adv. Meteorol.* **2019**, *2019*, 3849210. [CrossRef]
34. Amante, M.; Tesgera, H. Prevalence of cattle trypanosomosis and apparent density of its fly vectors in bambasi district of benishangul-gumuz regional state, western ethiopia. *Vet. Med. Int.* **2020**, *2020*, 8894188. [CrossRef]
35. Izzatullayeva, V.; Akparov, Z.; Babayeva, S.; Ojaghi, J.; Abbasov, M. Efficiency of using rapd and issr markers in evaluation of genetic diversity in sugar beet. *Turk. J. Biol.* **2014**, *38*, 429–438. [CrossRef]
36. Jaćimović, V.; Božović, Đ.; Ercisli, S.; Bosančić, B.; Nečas, T. Sustainable cornelian cherry production in montenegro: Importance of local genetic resources. *Sustainability* **2020**, *12*, 8651. [CrossRef]
37. Haddaway, N.R. The role of systematic reviews in evidence-informed decision-making in environmental management. *Environ. Evid.* **2018**, *7*, 22.
38. Barrios, E.; Gemmill-Herren, B.; Bicksler, A.; Siliprandi, E.; Brathwaite, R.; Moller, S.; Batello, C.; Tittonell, P. The 10 Elements of Agroecology: Enabling transitions towards sustainable agriculture and food systems through visual narratives. *Ecosyst. People* **2020**, *16*, 230–247. [CrossRef]

Disclaimer/Publisher's Note: The statements, opinions and data contained in all publications are solely those of the individual author(s) and contributor(s) and not of MDPI and/or the editor(s). MDPI and/or the editor(s) disclaim responsibility for any injury to people or property resulting from any ideas, methods, instructions or products referred to in the content.

sustainability

MDPI

Article

Understanding Constraints and Enablers of Climate Risk Management Strategies: Evidence from Smallholder Dairy Farmers in Regional South India

Anupama Shantharaju [1,*], Md Aminul Islam [2,3], Jarrod M. Kath [1,4], Shahbaz Mushtaq [1], Arun Muniyappa [5] and Lila Singh-Peterson [1,4,6]

1 Centre for Applied Climate Sciences, Institute for Life Sciences and the Environment, University of Southern Queensland, Toowoomba, QLD 4350, Australia; jarrod.kath@unisq.edu.au (J.M.K.)
2 Southern Queensland and Northern New South Wales Drought Resilience Adoption and Innovation Hub, Future Drought Fund (Research), Institute for Life Sciences and the Environment, Institute for Resilient Regions, University of Southern Queensland, Toowoomba, QLD 4350, Australia; amin.islam@unisq.edu.au
3 School of Environmental Science, University of New England, Armidale, NSW 2351, Australia
4 School of Agriculture and Environmental Science, Faculty of Health, Engineering and Sciences, University of Southern Queensland, Toowoomba, QLD 4350, Australia
5 College of Horticulture, UHS-Bagalkot, Bengaluru 560065, India; arun.muniyappa@gmail.com
6 Centre for Heritage and Culture, University of Southern Queensland, Toowoomba, QLD 4350, Australia
* Correspondence: anu.shantharaju@unisq.edu.au

Abstract: The adoption of effective coping strategies is crucial for successful adaptation to the impacts of climate change in the dairy sector. However, little attention has been paid to understanding the perceived constraints and motivations toward such strategies. A survey was conducted among 104 dairy farmers from three semi-arid regions of South India. The aim of the survey was to explore the dairy farmers' perception of climate risk, how it impacts their dairy farming system, the coping strategies they employ, and the barriers they face when implementing these strategies. The survey also investigated the factors that facilitate the adoption of adaptation measures. The results indicate dairy farmers in the region perceive drought, pests and diseases, and high temperatures as the major risks associated with climate change, which has resulted in decreased dairy income, animal health problems, reduced fertility, and food intake problems for their cattle. In response to climate variability, dairy farmers have adopted various coping strategies. The most important strategies include buying livestock insurance, keeping low debt obligations, and growing drought-tolerant grass varieties. However, most farmers face significant constraints in adopting these and other strategies including a lack of climate forecast data, the high cost of adaptation activities, and weak institutional support. On the other hand, the key enabling factors that support the adoption of these strategies include milk production security, suitable feed growing conditions, and family interest. Most importantly, the study found that certain factors such as age, education, number of earning family members, annual milk production, monthly cattle expenses, and landholdings significantly influenced dairy farmers' strategies for adapting to climate change. The study recommends that providing timely climate forecasts, implementing improved policies such as vaccination and cattle health services, and establishing strong institutional support systems can help dairy farmers become more resilient to climate change and protect their livelihoods.

Keywords: risk perceptions; adaptation constraints; adaptation facilitation; institutional support; access climate forecast

1. Introduction

India has been the largest milk-producing country in the world for nearly two-and-a-half decades. India contributes 23% to global milk production [1]. At present, global milk production is growing at a rate of two percent per annum, whereas in India, its growth rate is over six percent per annum [1]. The per capita availability of milk in India is also much higher than the world average. In the last three decades, the daily milk consumption in the country rose from 107 g per person in 1970 to 427 g per person in 2020–2021, compared to a world average of 322 g per day during 2021 [1].

The dairy sector in India plays a crucial role in supporting livelihood opportunities for 81 million dairy farmers [2], who are predominantly smallholder dairy farmers (i.e., 2–5 cattle per farm family) and the landless, particularly in the rural areas of the country. These smallholder dairy farmers contribute 62% of milk production in the country [3]. In India, dairy farming systems are most often benevolent due to the complementary nature of crop and livestock production. Much of the feed to animals is derived from agricultural remains and by-products. Although dairy farming has been reported as an age-old and complementary agricultural practice, particularly for the marginal and landless farmers in India [4], it has gained much popularity and attracted attention of the large farming households in recent decades [5–7].

However, this vital dairy industry in India is facing tremendous pressure from climate change impacts, which ultimately threatens sustainable livelihoods in the rural areas of most of the regions of the country [8]. Dairy farmers are similarly exposed to uncertainty regarding changing climatic conditions that produce abrupt changes in environmental factors. For example, the southern semi-arid regions of the country have been hit hard by changes in rainfall patterns, and widespread drought for many years, which has greatly affected dairy farming operations in the last few decades [9].

Climate-change-induced abrupt precipitation patterns and rising temperatures have heightened risks to the health and welfare of livestock [8,10]. For instance, foot-and-mouth disease (FMD), which is a dominant livestock disease in India, has been reported to be more prevalent under extreme and abrupt changes in rainfall and temperature [11–13]. Moreover, health risks of livestock animals are predicted to be accelerated due to a likely increase in average temperature by 2 °C by 2050 in the coming decades [14], as heat stress can inhibit the immune system of livestock leading to an increase in the potential outbreak of infectious diseases [12,15–17].

The quality and quantity of milk, wool, and meat associated with climate variability and extreme weather have also led to a drop in the market prices of final products [8,10,18], resulting in variable production supply that threatens food security [19–23]. However, managing the multi-faceted risks associated with climate change impacts (e.g., health risks, feed availability, milk prices) in the dairy farming industry is becoming more challenging due to a lack of relevant climate information [24–27]. Thus, there is a need for a thoughtful understanding of the prospects of decision making that are site-specific to help smallholder dairy farmers battling the wave of climate risks [28–30].

Numerous studies have been conducted worldwide on farmers' perceptions of climate change and their adaptation strategies in the cropping sector. These studies highlight the significance of understanding farmers' perceptions of climate change and their coping mechanisms. However, most existing studies have focused on the impacts of climate change on the livelihoods of farming communities in the cropping sector and their locally preferred adaptation strategies [31–34]. Limited research has been conducted on farmers' perceptions of climate change and their adaptation strategies in the livestock sector.

For example, a study by Montcho et al. [35] found that farmers adopted local measures to adapt to perceived climate change, which had negative impacts on livestock productivity, such as reduced herd size, milk production, and fodder availability. However, existing studies have paid little attention to the autonomous adaptation choices of smallholder dairy farmers and the factors that motivate them to adopt these strategies [36,37].

A comprehensive investigation of farmers' autonomous adaptation choices is crucial for developing climate-responsive adaptation planning on a large scale. It is also important to document the susceptibility of dairy farmers and animal production to climate risks in the Indian context. Furthermore, limited research has been conducted to assess the knowledge and perceptions of dairy farmers in southern India, highlighting the need for additional research [28,30,38]. Therefore, it is imperative to understand smallholder dairy farmers' perceptions of climate change and their coping strategies while considering the factors that affect their adaptation choices.

This study aims to investigate how smallholder dairy farmers perceive climate risks and how these risks affect their dairy activities. The study will also examine the current risk management strategies used by farmers, as well as the obstacles and opportunities for adaptation. Furthermore, the study will identify the factors that influence dairy farmers' choice of coping strategies for climate-change-related risks in their dairy farming system.

2. Material and Methods

2.1. Selection and Description of the Study Area

Karnataka is the second most vulnerable state in India to be impacted by climate variability and change as it comprises the aridest and driest regions [39]. Karnataka's agriculture and socio-economic settings act as an appropriate representative unit for South India's semi-arid regions. Dairy farming is one of the major livelihood activities besides agriculture in most Indian states. The state has variable rainfall, diverse soil types, and cropping patterns, and the state is divided into 10 agro-climatic zones [40]. The climate-change-induced risks and impacted zones can cause distress to cattle health and other dairying activities. Hence, it is important to study the dairy farmers' perception of various climate risks and impacts, adaptation, and mitigation strategies in the semi-arid pastoral regions in southern India (Bengaluru Rural/Urban and Chikkaballapur).

The reason for choosing these regions is because of their vulnerability to drought, livestock feed constraints, and water availability issues [39]. Bengaluru and Kolar-Chikkaballapur milk unions are the top two in the state, contributing around 25% to total dairy cooperatives and milk producers in Karnataka while adding about 34% to average milk procurement [41]. The study area receives erratic rainfall averaging 661–1086 mm per annum and high temperatures of over 29–39 °C in summer. Figures S1 and S2 show average rainfall and temperature trends from 2000 to 2022. From the Karnataka state, the study regions (e.g., districts) were selected based on the secondary sources where climate change impacts on the dairy sectors were reported in the literature. Then, from the districts, eight villages were selected to administer the survey. The selected villages were Adde Vishwanathapura, Byatha, Chalahalli, Kadathanamalli, Marelenahalli, Neralaghatta, Tharabanahalli, and Bachuvarihalli (Figure 1).

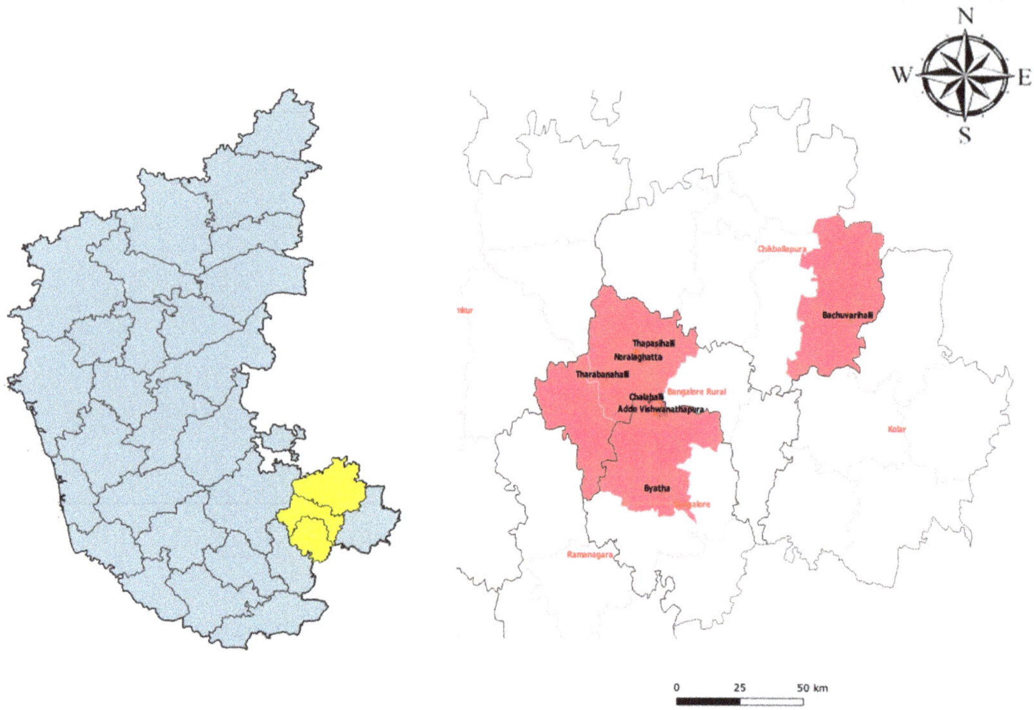

Figure 1. The location of the study areas (red polygons—**right** insert) on the regional map was selected in the state of Karnataka, covering the districts of Bengaluru (rural and urban) and Chikkaballapur (yellow polygons—**left** insert).

2.2. Survey and Data Collection

Primary data were collected through a survey using a pre-tested and structured multiple-choice questionnaire where respondents were able to select different answer options for each question. The initial questionnaire of the study was prepared following Phellas et al. [42]. Having preliminary contact with some of the key informants in the study area, the lead author organized several phone calls with several smallholder dairy farmers to pre-test the questionnaire. By addressing the comments returned during the pre-testing, the initial questionnaire was then restructured and finalized, and employed for data collection. A similar procedure of pre-testing a structured questionnaire was also used in several studies exploring farmers' perceptions of environmental changes [43,44].

The questionnaire was divided into sections to collect information on dairy farmers' socio-economic characteristics; cattle information and consumption patterns; climate risks and impacts (peril) that farmers experienced in the last decade in the region; the climate risks impact on dairy activities that farmers consider (relevant/nonrelevant); a list of barriers and enablers, that farmers need toward their adaption and mitigation measures; and the climate risk management strategies that are (important/not important) to dairy farmers.

Before administering the survey in the studied area, human resources ethics clearance was sought, and approval was gained from the University of Southern Queensland (H22REA088). Following ethics approval, a field trip was organized to administer the survey. The lead author carried out the surveys in a face-to-face interview with each of the 104 respondents.

To operationalize the survey, a stratified sampling procedure was followed to select the respondents, where the individual dairy farming household was considered a primary sampling unit. Firstly, the study regions (e.g., districts) were selected based on a literature review of reported climate change impacts on the dairy sectors. Then, eight villages were selected from the districts to administer the survey. The selected villages were Adde Vishwanathapura, Byatha, Chalahalli, Kadathanamalli, Marelenahalli, Neralaghatta, Tharabanahalli, and Bachuvarihalli. From each village, the smallholder dairy farmers (i.e., respondents) were selected based on recommendations from dairy heads in each village and the availability of participants, to obtain a representative (in terms of gender, cattle ownership) sample of farmers. Thus, a total of 104 diverse socio-economic-centric rural dairy and agricultural farmers were selected from the study region. A similar stratified sampling procedure was also reported in a study that explored farmers' perceptions of climate change and adaptation strategies [45]. In addition to primary data collected from dairy farmers, this study also utilized secondary data (documents and reports) from public institutions such as the Department of Animal Husbandry and Dairying to support its findings.

Farmers' perceptions regarding the climate risks, impacts, coping strategies, barriers of adaptation, and enablers of adaptation were gathered on the predefined (structured) statements focusing on whether they agree with the statements (i.e., yes/no, important/not important). The details of various dimensions relevant to climate risk perception, adaptation, etc., elicited through farmer's survey are depicted in the flow chart (Figure 2) for a clear understanding.

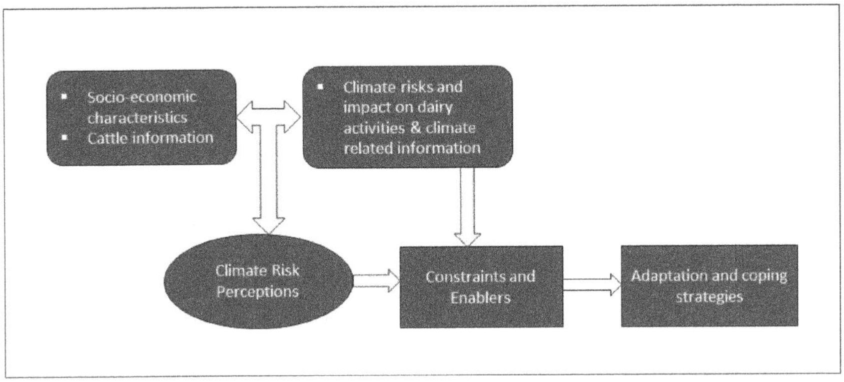

Figure 2. Conceptual framework of data collection process on various factors by dairy farmers.

2.3. Statistical Analysis

Farmers' perceived responses to different issues of climate risks affecting dairy farming, and adaptation choices were presented in terms of "yes" and "important" to demonstrate that the farmers agreed with the statement asked through a structured questionnaire. Similar analysis and results presentation were also reported in studies of farmers' perceptions of climate change in the agricultural sector [35,46].

Following gathering responses on the different variables, these "yes/important" were then coded with a number, and analyzed using SPSS statistical software (version 28.0.1.0 (142)) [47]. Non-parametric chi-square tests were conducted to show the relationship between some of the independent variables and farmers' knowledge of climate change. Furthermore, a chi-square test was carried out to demonstrate the differences among the farmers' perceptions in different dimensions of climate change and adaptation strategies, the constraints of adopting these strategies, as well as facilitating factors in their adoption of these coping strategies, and expressed in percentages.

Additionally, a logistic regression test was conducted to explore the factors affecting farmers' choices of coping strategies using whether each of the selected adaptation strategies (e.g., cash in hand, using drought tolerant grass varieties, buying livestock insurance) was important or not important using the dependent variable with the socio-economic characteristics of the respondents (e.g., gender, age, education, family member, land holding) as independent variables. This analysis identified whether socio-economic, sociopsychological, and related variables contributed to the adaptation strategies chosen, leading to meaningful inferences from the study.

3. Results

3.1. Socio-Economic Characteristics of Respondents

The socio-economic characteristics of the smallholder dairy farmers surveyed are summarized in Table 1. The mean age of the respondents was 50.55 years, having 8.32 years of schooling. The average landholding capacity of the respondents was 3.26 acres across the eight surveyed villages in Karnataka. The average annual income was 108,928 INR and the annual milk production per cattle was 3499 L.

Table 1. Socio-economic characteristics of the respondents surveyed in this study. Number of surveyed farmers = 104.

Socio-Economic Characteristics (Unit of Measurement)	Mean	Std. Deviation	S.E
Age (Number)	50.55	13.55	1.32
Education (Schooling years)	8.32	4.19	0.41
Landholding capacity (Acres)	3.26	2.89	0.28
Annual Income (INR)	108,928	60,751	5957
Herd size (Number of animals)	2.21	1.19	0.11
Annual milk production per cattle (Liter)	3499	1319.07	129.34
Annual expenses per cattle (INR)	45,450	15,625.24	1532.18

3.2. Relationship of Socio-Economic Features of the Smallholder Dairy Farmers and Their Knowledge of Climate Change

The majority of the socio-economic characteristics of the respondents had a significant relationship with the dairy farmer's knowledge of climate change, except for the gender and age of the respondents. The results of these relationships among the socio-economic characteristics and knowledge of climate change in terms of "yes" responses are presented in Table 2. Among the respondents, 56% of males and 44% of females reported knowledge of climate change. In terms of age of the respondents, the vast majority (51.5%) of respondents who held knowledge of climate change were between "31 and 50" years of age, followed by the 51–70 years age group (35.4%). In connection to the educational qualifications, the vast majority (53.5%) of the respondents who mentioned "yes" have a high school level of education. Results also revealed that the highest "yes" responses (96%) indicating having knowledge of climate change were recorded from the respondents whose primary occupation was agriculture. Furthermore, in terms of landholding capacity, the highest "yes" responses (41.4%), indicating that they have knowledge of climate change, were reported from the marginal landholding group of farmers followed by small landholding farmers (34.3%), as presented in Table 2.

Table 2. Relationship among the socio-economic features of the respondents and their knowledge of climate change.

Socio-Economic Features		% of Respondents Mentioned "Yes"	Chi-Square
Gender	Male	55.6	2.42 ^{ns}
	Female	44.4	
Age	<30 years	5.1	4.13 ^{ns}
	31–50 years	51.5	
	51–70 years	35.4	
	>70 Years	8.1	
Education	Illiterate	11.1	8.75 **
	Primary	14.1	
	Middle school	11.1	
	High School	53.5	
	University	10.1	
Primary occupation	Agriculture (Dairy and crops)	96.0	14.23 **
	Other business	1.0	
	Dairy only	3.0	
Landholding	Marginal (<2.47 acre)	41.4	3.14 *
	Small (2.47–4.94 acre)	34.3	
	Medium (4.94–9.88 acre)	20.2	
	Large (>9.88 acre)	4.0	

^{ns} = not significant (i.e., $p > 0.05$); * $p < 0.10$ ** $p < 0.05$.

3.3. Farmers' Perception of the Risks of Climate Change on the Dairy Farming Systems

Smallholder dairy farmers' perceptions of climate change risks on their dairy farming system are presented in Figure 3. Risks related to drought and to pests and disease were the two most highly ranked risks. When asked about drought, 91% of the respondents mentioned "yes" as drought was a severe climate change risk. Similarly, 89% of the respondents reported an increase in pests and diseases as a result of climate change, and 81% of the farmers identified heat stress as a climate change risk. Regarding erratic rainfall, hailstorms, and floods, 78%, 68%, and 64% of the farmers recognized these as climate change risks, respectively.

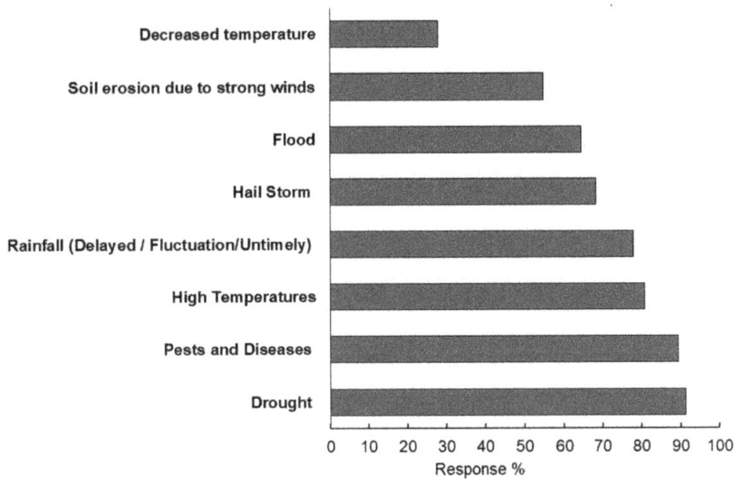

Figure 3. Main climate risks identified during the survey and corresponding farmers' responses (in percentage). Number of farmers surveyed = 104.

3.4. *Farmers' Perceptions of the Impacts of Climate Risks on the Dairy Farming System*

Smallholder dairy farmers were asked whether they agreed (yes) or not (no) with the impacts of each of the climate change risks on their dairy farming system, and their responses indicating "yes" they agreed are presented in Figure 4. Most farmers agreed that climate change has a negative impact on their dairy farming activities, especially in terms of labor and income. They attributed this to a decrease in milk production. Additionally, 99% of the farmers noted that animal health issues were a direct result of high temperatures, particularly during the summer months (March to May). The farmers also reported experiencing severe drought in the region, which has resulted in a decrease in food intake (98%) and feed issues (96%) due to the high temperatures (see Figure 4).

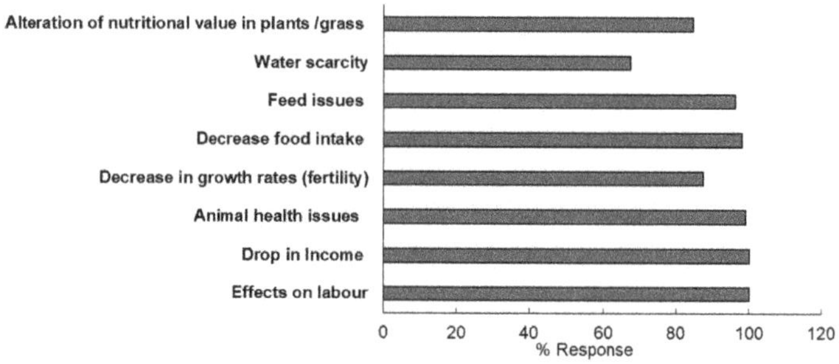

Figure 4. Farmers' perceptions of the impacts of climate risks on the dairy farming system. Number of farmers surveyed = 104.

3.5. *Dairy Farmers Adaptation Strategies*

Farmers were asked to share their thoughts on the impact of climate risks on dairy activities and which adaptation strategies they deemed important to mitigate these effects. These responses were expressed in percentages of farmers who mentioned that the specific adaptation strategy was important. The study revealed a range of adaptation strategies that are commonly used in the dairy industry to combat climate-change-induced risks. Results indicated that in terms of buying livestock insurance and low debt obligations, 100% of the respondents mentioned these coping strategies are important, followed by using drought-tolerant variety grass (99%), value addition of dairy products, and keeping cash in hand (98%), off-farm employment (88%), and selling livestock (80%) as the most important adaptation strategies to climate change impacts, However, regarding income diversification, crop insurance, and increasing herd size, farmers' responses were divided by both important and not important, which indicated by 61%, 60%, and 58% toward "yes" the adaptation strategies are important respectively (Figure 5).

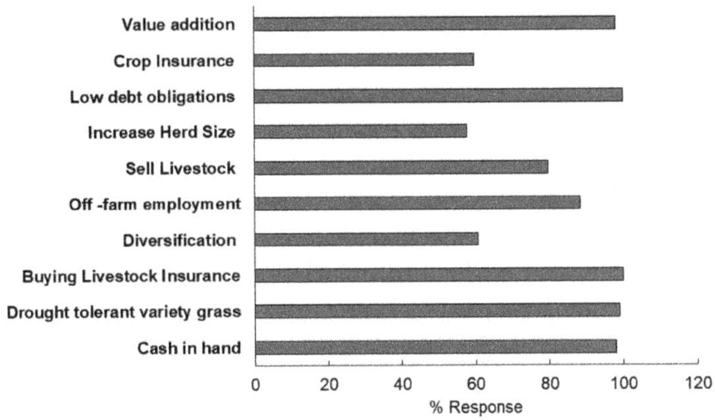

Figure 5. Coping/adaptation strategies adopted/practiced by the smallholder dairy farmers. Number of farmers surveyed = 104.

Following an understanding of the perceived coping strategies for dealing with climate change impacts, this study then investigated how these perceptions varied according to the salient socio-economic features of the respondents (Table 3). The study found that farmers were divided on the importance of adaptation strategies based on the gender of the respondents. However, in terms of the primary occupation of the respondents, more than 95% of those farmers whose occupation was agriculture (livestock and dairy) mentioned all the perceived adaptation strategies were important to them, while only 5% of them perceived coping strategies were not important. Similar trends were also observed in the case of the educational qualifications of the respondents. The findings demonstrated that many of the respondents who had high school-level educational qualifications (~50%) mentioned that adaptation strategies were important to them (Table 3).

Table 3. Farmers' perceived coping strategies for climate change impacts as influenced by different socio-economic features.

Adaptation Strategies	% of the Respondents Mentioned the Adaptation Strategies as Important										Chi-Square
	Gender		Occupation			Educational Qualification					
	Male	Female	Agriculture	Business	Dairy	Illiterate	Primary	Middle School	High School	University	
Value addition	53.8	46.2	95.7	1.1	3.2	14.0	16.1	9.7	50.5	9.7	64.65 **
Crop Insurance	54.8	45.2	93.5	1.6	4.8	14.5	11.3	9.7	58.1	6.5	3.84 *
Low debt obligation	53.8	46.2	94.5	1.1	4.4	11.0	16.5	13.2	49.5	9.9	58.50 **
Increase herd size	53.3	46.7	93.3	1.7	5.0	11.7	16.7	11.7	51.7	8.3	2.46 ns
Sell livestock	51.8	48.2	94.0	1.2	4.8	12.0	15.7	10.8	49.4	12.0	36.96 **
Off-farm employment	54.3	45.7	95.7	1.1	3.3	12.0	15.2	10.9	51.1	10.9	61.53 **
Diversification	46.0	54.0	93.7	1.6	4.8	15.9	12.7	9.5	52.4	9.5	4.6 **
Buying livestock insurance	54.0	46.0	94.0	1.0	5.0	13.0	16.0	12.0	50.0	9.0	88.61 **
Cash in hand	52.9	47.1	94.1	1.0	4.9	12.7	15.7	11.8	51.0	8.8	95.15 **
Drought-tolerant grass cultivation	47.1	52.9	94.1	1.0	4.9	12.7	14.7	11.8	51.0	9.8	96.15 **
Alternative income sources	55.1	44.9	94.9	0.0	5.1	12.2	16.3	12.2	50.0	9.2	81.38 **

ns = not significant (i.e., $p > 0.05$); * $p < 0.10$, ** $p < 0.05$.

3.5.1. Barriers to Adaptation

Apart from perceived important adaptation strategies and factors influencing adaptation in this study, we further assessed the barriers to their adaptation by dairy farmers. Findings indicated that most of the farmers agreed with the perceived statements of the adaptation barriers, such as "lack of climate forecast", "adaptation strategies were expensive", and "weak institutional support". The "yes" responses toward these statements were 100%, 99%, and 84%, respectively, while the case of "policies of dairy co-operatives" yielded a similar proposition of the responses toward "yes" it is important, and "no" it is not as important (Figure 6).

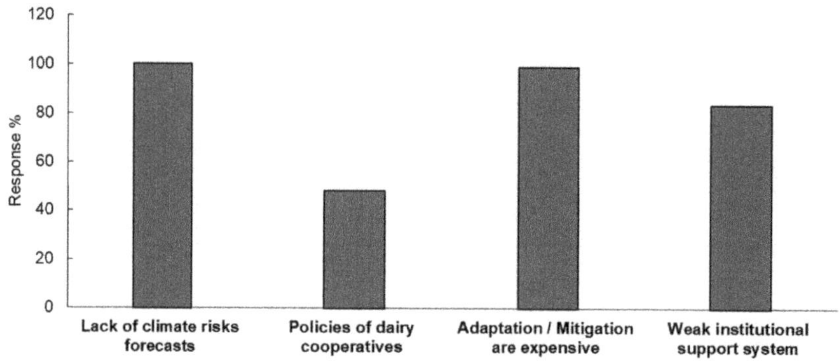

Figure 6. Barriers to adaptation reported by dairy farmers. Number of farmers surveyed = 104.

3.5.2. Motivating Factors for Adopting the Climate Impact Adaptation Strategies

Despite the barriers to adaptation, some factors drove the local farmers to act upon and adopt appropriate and timely measures. The findings of this study revealed that in cases of "family interests", "milk production security", and "economic interests", and "suitable growing feed condition" were reported as the most important motivating factors, as almost 99% of the respondents mentioned "yes" toward these statements. In comparison, 90% of the respondents mentioned "no" toward the statement of a motivating factor as "taking collective action for climate risks" (Figure 7).

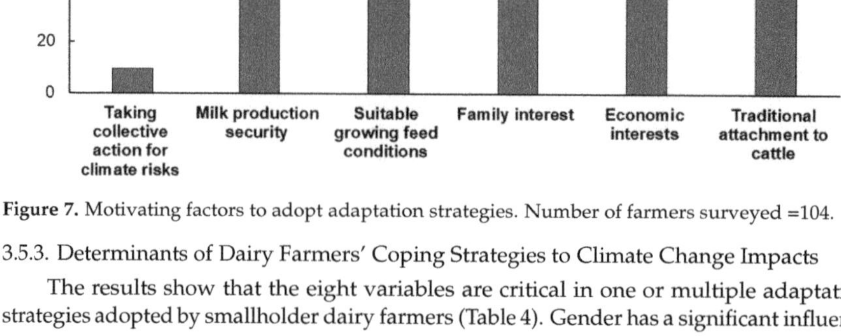

Figure 7. Motivating factors to adopt adaptation strategies. Number of farmers surveyed =104.

3.5.3. Determinants of Dairy Farmers' Coping Strategies to Climate Change Impacts

The results show that the eight variables are critical in one or multiple adaptation strategies adopted by smallholder dairy farmers (Table 4). Gender has a significant influence

($p < 0.05$) when considering diversification in dairy farming, and female farmers preferred this over male dairy farmers. Age affected the adoption of cash in hand, the decision to grow drought-resistant grass, increasing the herd size, and low debt obligation. Farmers with more experience in the dairy industry tend to prefer conservative adaptation strategies such as diversification and using drought-tolerant grass varieties as livestock feed. Education also had a positive effect on similar adaptation decisions. The more years spent in education, the more farmers chose diverse adaptation measures to counter the risks of climate change. Higher education positively affected the consideration of crop insurance as an adaptation strategy. More earning family members opted for off-farm employment and this positively affected crop insurance adaptation by 6%, which shows the additional income makes the farmer capable of buying risk-sharing strategies like insurance. Annual dairy income influenced the farmers to opt for growing drought-tolerant varieties of grasses, which makes sense in so far as additional investment is required for such an adaptation measure apart from regular dairy operations. The cattle expenses affected farmers' decisions on the amount to be held in the form of cash for unforeseen circumstances and affected the decisions to adapt diversification and consider off-farm employment significantly ($p < 0.05$). Land holdings of farmers impacted the decision to obtain off-farm employment. It positively influenced the growth of drought-tolerant grass varieties and value addition to dairy farming activities in terms of better hygiene conditions for cattle in place on the relevant farm. In this study context, value addition meant improved conditions for cattle in terms of hygiene and other actions taken for the protection and welfare of animals.

Table 4. Coefficient estimates from a logistic regression on the factors influencing/affecting different adaptation measures (coping strategies) adopted by dairy farmers.

Predictors Variables	Response Variables (i.e., Adaptation Measures)									
	Cash on Hand	Drought-Tolerant Variety Grass	Buying Livestock Insurance	Diversification	Off-farm Employment	Sell Livestock	Increase Herd Size	Low-Debt Obligation	Crop Insurance	Value Addition
Gender (0 for male, 1 for female)	NS	NS	NS	−0.63 **	NS	NS	NS	NS	NS	NS
Age	−0.86 *	0.951 *	NS	NS	NS	NS	0.007 *	0.029 **	NS	NS
Education (y)	NS	NS	NS	NS	NS	NS	0.01 *	0.11 *	0.007 *	NS
Earning family members	NS	NS	NS	NS	1.19 **	NS	NS	NS	0.06 **	NS
Annual income	NS	0.00 **	NS	NS	NS	NS	NS	NS	NS	NS
Annual milk production	−0.018 **	NS	−0.001*	NS	NS	NS	NS	−0.00 **	NS	NS
Expenses per cattle per month	0.016 *	NS	NS	0.00 *	−0.001 **	NS	NS	NS	NS	NS
Landholding	NS	NS	NS	NS	0.021 **	NS	NS	NS	NS	NS

Statistical significance; * indicates $p < 0.1$; ** indicates $p < 0.05$, NS = not significant (and so coefficient not shown). Dependent variable: Not important = 0, Important = 1.

4. Discussions

Exploring the perceived constraints and enabling factors of the coping strategies is imperative to successfully plan for climate change adaptation. However, the existing discourses of perceptions of climate change adaptation studies are overwhelmed by the crop sector studies, which are predominantly focused on risk and adaptation and pay little attention to perceived constraints and facilitating factors. To understand these issues, a structured questionnaire survey was adopted in this study to investigate the smallholder dairy farmers' perceptions of climate change and their coping strategies, as well as barriers and enabling factors to adopt these coping strategies, which could help to plan comprehensive adaptation actions. This section discusses the findings of this study in the following subsections.

4.1. Smallholder Dairy Farmers' Knowledge of Climate Change and Their Perceptions of Climate Risk and Impacts

The socio-economic characteristics of dairy farmers are important because they influence their economic behavior and determine the ability of farmers to adequately adopt climate risk management strategies. The results of this study demonstrated a significant relationship between their knowledge of climate change and most of the socio-economic characteristics of the farmers (Table 2). These results were found to be consistent with previous studies, indicating that dairy farmers understanding and knowledge of climate change are key drivers to their adaptation approaches [46]. Further, socio-economic features play a vital role in farmers' perceptions of climate change issues, which, in this study, was a driving indicator of adopting coping strategies. For example, the existing literature has demonstrated that education levels play a key role in the enactment of climate adaptation measures based on perceptions of farmers and climate-related information [48]. In addition to local indigenous knowledge, climate science education would help farmers acquire further knowledge and skills to read and understand a wide range of climate-related information and determine how it impacts dairy. The level of a farmer's education also influences the quality of decisions made by dairy farmers whenever a climate risk triggers an adaptation response.

Results further indicated significant associations among the respondents in connection to the impacts of educational qualification on their knowledge of climate change. The results showed that the farmers who were more aware of climate change had completed high school education, followed by those who had completed primary education. On the contrary, most respondents who indicated that they did not know about climate change were illiterate or held primary levels of educational qualification. Results also revealed that land holdings of dairy farmers had a significant association with their knowledge of climate change; in particular, those farmers with marginal and small landholdings agreed they had knowledge of climate change, which dominated in this sample study. These findings are consistent with another study, which indicated that farmers' perceptions play a significant role in the adaptation process [49].

The study conducted in the region has shown that farmers consider drought as the most devastating climate risk, followed by increased pests and diseases, erratic rainfall, hailstorms, and floods. Moreover, the farmers' views on climate change have been verified by comparing them with the observed rainfall and temperature data obtained from the meteorological stations in the region between 2000 and 2022, as illustrated in Figure S1. The meteorological records reveal that the region has experienced 23 years of erratic annual average rainfall, with seven years (2011–2014, 2016, 2018, 2019) in the last decade having below-average rainfall, leading to droughts as shown in Figure S2. Additionally, drought periods were characterized by high mean temperatures ranging from 34 and 39 degrees, which were reported frequently. Therefore, the perceived climate risks by farmers in the region are consistent with the reported data in the existing literature.

In addition, farmers reported heat stress and increased pest and disease incidence in cattle could be a result of drought and high mean temperatures. Rainfall data show a fluctuation trend notably in the years (2020–2022) with heavy rainfall posing different challenges to dairy activities, particularly the landless and small land-holding farmers who depend on pasture grazing around water bodies. Thus, the findings of this study regarding farmers' perceptions of climate change issues were consistent with the meteorological records, signifying that farmers were closely connected to weather and climate cycles and were able to describe environmental factors affecting their dairy farming activities. As most farmers were involved in agriculture production, they were mainly concerned about drought and above-average rainfall years (2020–2022) in the region since they caused a series of crop failures. This experience created further worry for future cultivation and cattle feed security. The foot and mouth disease (FMD) outbreak in the years 2013 and 2018 was confirmed [11] as an extreme challenge by dairy farmers in the study. Unfortunately, stress on cattle is just one way that extreme heat is challenging dairy farmers with small herd sizes.

The reported list of climate change impacts on dairy activities in this study was found to be consistent with previous studies that also indicated similar implications on dairy farming [46,50–52]. The findings of this study are consistent with [35,53–55], who reported that climate change impacted risks, particularly, heat stress affects milk production in cattle. Persistent dry spells and severe rain were also found to affect the availability of fodder [56] and, in turn, the cost of feed. Thus, the findings of this study were consistent with previous research on the pressures of soaring cattle feed costs and water scarcity issues that led to severe distress from season to season in terms of climate-associated risks [39,57,58], compelling farmers to review their decisions. Previous research has mainly concentrated on the environmental effects that affect livestock productivity, such as the availability of food, shelter, and milk production [59,60]. However, this study has taken into consideration animal welfare concerns, particularly, nutrition, which were previously overlooked.

4.2. Smallholder Dairy Farmers' Coping Strategies, Constraints, and Facilitating Factors in Adopting These Strategies

The findings of this study have shown that a vast majority of the smallholder dairy farmers opined that buying livestock insurance is the most important adaptation strategy to cope with the climate change impacts, followed by growing drought-tolerant variety grass and off-farm employment (Figure 4). Farmers' primary occupations and educational qualifications were found to be the major factors influencing their coping strategies for climate change impacts. In fact, a range of socio-economic, political, institutional, and region-specific environmental factors have a bearing on the adaptive capability of the livestock sector toward climate change. The findings also demonstrated that those farmers with marginal land holdings subject to droughts and unpredictable rainfall are often required to seek off-farm employment, which can involve migration to nearby towns to cope with the financial distress caused by climate change.

The study found that farmers agreed to keep cash in hand as an important adaptation measure to climate risk. Saving money has been a long-standing practice among farmers to tackle tough times by making lifestyle changes when required. Keeping low debt obligations and not depending on money lenders was preferred due to the exorbitant interest rates they charged, which has led fellow dairy farmers to fall into a debt trap. In small-scale mixed crop–livestock systems, farmers have limited economic opportunities compared to cropping systems [61]. Previous studies have shown that keeping cash on hand is the most important risk management strategy for dairy farmers [62,63].

Previous studies also indicated that frequent and persistent droughts lead to the drying of natural pastures, which causes feeding problems for animals and affects the availability of water and the quality of fodder resources in the region [35,64]. Thus, those farmers facing fodder deficiencies made sure to stock some seasonal-based crop residue, purchasing from nearby localities. In addition to diversified feedstock (e.g., drought-tolerant grass, drumstick tree leaves), farmers also introduced diversified livestock to keep their livelihood sustainable through alternative income generation activities and to enhance their livelihood resilience by spreading risk across a number of income channels. For example, farmers reared goats/sheep to gain extra income, particularly to support the period when milk production is usually reduced due to the gestation period. Some women farmers have started chicken farming and selling eggs to earn extra income when their land is not used for agriculture.

Many farmers chose to take advantage of cattle insurance offered by dairy co-operatives, who subsidized the insurance premium by 50%. Although insurance was purchased regularly over the years, some farmers reported losing their animals if they missed a year. A few well-educated farmers were aware that the insurance policy was put in place to deliver protection mechanisms to the cattle rearers against any eventual loss of their animals due to death and to validate the benefit of cattle insurance. They also knew the scheme was designed to achieve improvements in the health of cattle and their products. However, the illiterate farmers were not aware of this scheme in place to support their business. According to [65,66], cattle holding size positively affected the adaptation decisions of dairy farmers. Those farmers in this study who had experienced a series of losses in agriculture operations preferred to increase their herd size to sustain their living standards.

Farmers may face crop failures, loss of income, and additional expenses during climate change. To offset these challenges, they may be forced to sell some of their livestock [49,64]. Additionally, farmers concerned about the potential impact of pest and disease outbreaks on their animals or reduced milk production due to diseases may prefer to sell some of their livestock. This situation is further exacerbated by the threat of increased livestock mortality due to extreme temperature events in the region.

Another strategy adopted by some farmers was to plant additional shade trees in the study regions, particularly for those farmers who lacked an animal shelter. This was considered an excellent adaptation strategy to prevent animals from the impacts of excessive heat stress. Planting trees also served as supplementary fodder sources. An example of a tree that is suitable for the region is *Moringa olifera,* also known as the 'drumstick tree' [67]. It is drought-resistant, can withstand varying temperature ranges, and grows rapidly and vigorously. Leaves of the tree have high protein content with other essential amino acids [67]. Thus, this kind of tree provides shade to the smallholder dairy farmers and provides nutritious feed to the livestock. This tells us how farmers keep indigenous practices in place and how these traditionally practiced mechanisms hold their importance even after multiple generations. Their perception of its effectiveness was based on the belief that it can lower the negative effects of rising temperatures on cattle health and quality milk production.

Moreover, various, green-dried fodder mixes offered to cattle are an experimental strategy to fight climate change, particularly during hot and rainy months, as reported by farmers in the study region. Besides averting the adverse impact of climate change on agricultural and cattle milk production in the future, planting fodder trees for cattle and ruminants is emerging in the region as a practical solution. This approach not only helps with sustainability but also generates income through growth and harvesting. While crop insurance was another potential strategy, some farmers deemed other strategies as important in the study region to their farming approach.

Findings of this study have shown that dairy farmers in the studied region perceived "lack of climate forecast", "high cost of adaptation strategies", and "weak institutional support" as the most important constraints of adopting the coping strategies. In addition to the lack of farmers' knowledge, easy accessibility to available climate forecasting is also a barrier to dairy operations mentioned by the respondents of this study. This was particularly relevant for the landless farmers who were entirely dependent on natural pastures for animal feed and were worried that along with cattle's health, the operator's health was at stake in extreme weather conditions. Therefore, it is of the utmost importance to establish institutions that will help coordinate and implement farmer-friendly measures and provide educational support to access and interpret climate-related information.

In addition, farmers who had experienced crop failures in agriculture operations were solely dependent on income from milk sales and the sale of manure from the cattle owned. This being the case with most farmers in the study region, adaptation activities became expensive, requiring farmers to make tough choices daily considering the best interest of their animals. Farmers in this study reported that feed costs had jumped 30-40%, while the milk price had not changed to keep pace with additional costs in dairy operations.

Farmers appreciated vaccination programs, but those who lost cattle did not receive support, except for insurance claims. Dairy-cooperative-associated doctors were unable to provide timely information. The required help did not reach farmers on time, resulting in conflicts with policies. Some suggested that more money was needed to buy healthy cattle and that a lack of communication between policy changes and farmers resulted in a lack of support. Lack of support and relevant information affected farmers, particularly when rainfall fell short of expectations. The study found that fundamentally improving access to essential services can effectively protect farmers' livelihoods and enhance their resilience to climate-change-induced risks, but there were some challenges around communication and access to critical services remaining that needed to be addressed.

Despite these barriers to adaptation discussed above, there were some factors that motivated the local farmers to act upon and adopt timely coping strategies. This study found that in cases of "family interests", "milk production security", "economic interests", and "suitable feed growing condition" are reported as the most important motivating factors that could facilitate farmers' adoption of climate risk management. Since climate-change-induced risks gravely affect dairy farmers' operations, they adopt various measures to minimize the impact when encountered. Furthermore, this study found that many farmers had recently built cattle housing on their farms, which was influenced by their strong bond with the animals they raised. The idea of taking collective action at the village level did not seem to drive them to adaptation as only 10% of farmers agreed it may be workable because of income differences, but other aspects took precedence over this. Although a labor-intensive activity, many women preferred to utilize their time in other ways than to become involved in low-productive activities, especially for those who were solely managing every dairy activity. These findings provide an understanding of dairy farmers' requirements and priorities, which can help guide researchers and policymakers in their effort to develop and align comprehensive strategies to tackle the climate change impacting dairy in Karnataka, India, and other countries with a similar set of dairy systems.

4.3. Determinants of Dairy Farmers' Coping Strategies for Climate Change Adaptation

In this study, the factors influencing the adaptation strategies were examined by logistic regression analysis. Results revealed a significant positive relationship between most of the socio-economic variables and climate-induced adaptation measures of dairy farmers in the studied region (Table 4). The study results were consistent with previous findings indicating dairy farmers' socio-economic variables and other factors associated with dairy related to the location, which affected adaptation decisions and coping mechanisms. For instance, Abbas et al. [49] reported that farmers' education, farming experience, herd size, and access to extension services influenced adaptation strategies such as selling weak/deceased animals, migration, and off-farm income activities in their study of dairy farmers in Punjab, Pakistan.

Furthermore, the age of the farmer, tropical livestock unit, type of animal breed, perceived benefits of the technology, access to extension, and farmer group membership influenced the adoption of climate-smart Brachiaria grass among dairy farmers in Eastern and Western regions of Kenya [68]. Farming experience, cattle herd size, non-agricultural income, membership in an organization, number of farm assets, level of education, and climate zone were the major variables affecting farmers' adaptation strategies among cattle farmers in Benin and Turkey [34,69]. Thus, to make the policies effective aiming at climate change adaptation, it is, therefore, necessary to take into account the local farmers' understanding of how dairy farmers perceive climate change, the differences in perception, and what factors influence them to adopt various strategies and decision making in a regional setting.

5. Conclusions

India's smallholder dairy farmers are facing significant challenges from climate change, which not only threaten their dairy activities but also the well-being of their livestock, which are valuable possessions. Understanding smallholder farmers' perceptions of climate change and coping strategies could help better plan and ultimately protect their livelihood options. This study in regional South India investigated dairy farmers' perceptions of climate change risk impact on dairy, as well as adaptation strategies. First, dairy farmers perceived significant changes in the local climate, and impacts such as drought, lengthy periods of high temperatures, pests, and diseases in cattle, and changing rainfall patterns affected dairy production in the study area. Second, farmers indicated how this has affected animal health, with exceptionally high temperatures followed by heavy rainfall for those who are dependent on natural pastures grazing their livestock, highlighted as being particularly important.

Third, in response to the changing local climate, dairy farming households are adopting multiple response measures to avoid hazards in milk production and cattle health. Among diverse adaptation strategies, buying livestock insurance, keeping low debt obligations, and growing drought-tolerant varieties of grass appeared to be the most adopted adaptation measures, while crop insurance, diversification, and increasing herd size were reported as moderately adopted strategies. Financial constraints and lack of information were regarded as the greatest barriers to dairy farm-level adaptation. Regression analysis further revealed that dairy farmers' age, education, land holdings, annual milk production, and cattle expenses influenced the types of adaptation strategies undertaken. The findings of this study also demonstrated that dairy farmers' adaptation to climate risks is largely associated with their access to essential institution-led services such as climate forecasts and extension services.

The use of a survey based on the recall method of eliciting information from farmers itself was a limitation. In addition, the study results may not generalize to other areas well as the data were confined to only two districts of southern India. Nonetheless, based on our study findings, we suggest that relevant institutions, policymakers, and stakeholders should improve farmers' access to essential services that could enhance their ability to prepare for and adapt to climate hazards in order to increase their climate resilience. Finally, the study findings highlight where future studies should focus to better understand dairy farmer perspectives, priorities, and needs, and importantly how to incorporate these into climate change adaptation.

Supplementary Materials: The following supporting information can be downloaded at: https://www.mdpi.com/article/10.3390/su16052018/s1, Figure S1. Mean monthly rainfall and temperature in the study area (data presented is the average of 2000–2022). Figure S2. Annual total rainfall in the study area.

Author Contributions: A.S. conducted survey, literature review, analyzed the data, and drafted the manuscript. M.A.I. helped improve data analysis and made essential edits on multiple versions. J.M.K. advised on conceptualization and made essential edits. S.M. advised on conceptualization and made comprehensive edits and critical revisions to the manuscript. A.M. helped in conceptualization and essential edits. L.S.-P. advised on conceptualization and essential edits with grammar in the manuscript. All authors have read and agreed to the published version of the manuscript.

Funding: This research was funded by the University of Southern Queensland's International Fees Research Scholarship and the Australian Government Department of Industry, Science and Resources through the Australia-India Strategic Research Fund (AISRF) Round 12 grant number DISER-AIRXII000075.

Institutional Review Board Statement: The study was conducted in accordance with the Declaration of Helsinki, and approved by the Ethics Committee of the University of Southern Queensland (H22REA088; 17/06/2022).

Informed Consent Statement: Informed consent was obtained from all subjects involved in the study.

Data Availability Statement: Data will be made available upon request to the corresponding author.

Acknowledgments: We thank all the dairy farmers who participated in the interviews and kindly provided information about their farming systems. We gratefully acknowledge the funding received by the first author from the University of Southern Queensland's International Fees Research Scholarship. We gratefully acknowledge the funding received from the Australian Government Department of Industry, Science and Resources through the Australia–India Strategic Research Fund (AISRF)

Round 12 project. The lead author also thanks Suresh Babu for his support in the preparation of the study map.

Conflicts of Interest: The authors declare no conflicts of interest.

References

1. DAHD. Basic Animal Husbandry Statistics, Department of Animal Husbandry and Dairying, Government of India. Available online: https://dahd.nic.in/schemes/programmes/animal-husbandry-statistics (accessed on 7 February 2024).
2. DAHD. Annual Report 2019–2020, Departmenrt of Animal Husbandry, Ministry of Fisheries (DAHD), Government of India. Available online: https://dahd.nic.in/sites/default/filess/Annual2019-20.pdf (accessed on 17 June 2021).
3. Jatwani, M.; Swain, S. Is small scale dairy farming dying out? An In-depth study. *Indian J. Community Med. Off. Publ. Indian Assoc. Prev. Soc. Med.* **2020**, *45*, S47.
4. Aubron, C.; Bainville, S.; Vignes, M.; Philippon, O.; Dervillé, M.; Daviron, B.; Dorin, B. Milk and dairy in India's development path: Lessons from a multi-location research. *VKCoE Dig.* **2020**, *1*, hal-03658334.
5. Kishore, A.; Birthal, P.S.; Joshi, P.K.; Shah, T.; Saini, A. Patterns and drivers of dairy development in India: Insights from analysis of household and district-level data. *Agric. Econ. Res. Rev.* **2016**, *29*, 1–14. [CrossRef]
6. Das, A.K.; Saha, C.K.; Alam, M.M. Greenhouse gas emissions from dairy farming in Bangladesh. *World* **2017**, *1*, 92–101.
7. Tiwari, S. Contribution of Dairy Farming on Livelihood: A Case Study of Madhyabindu Municipality, Nawalpur. Ph.D. Thesis, Department of Rural Development, Nagaland, India, 2023.
8. Pawar, S.S.; Sajjanar, B.; Lonkar, V.; Kurade, N.; Kadam, A.; Nirmal, A.; Brahmane, M.; Bal, S. Assessing and mitigating the impact of heat stress in poultry. *Adv. Anim. Vet. Sci* **2016**, *4*, 332–341. [CrossRef]
9. Muralikrishnan, L.; Padaria, R.N.; Choudhary, A.K.; Dass, A.; Shokralla, S.; El-Abedin, T.K.Z.; Abdelmohsen, S.A.M.; Mahmoud, E.A.; Elansary, H.O. Climate change-induced drought impacts, adaptation and mitigation measures in semi-arid pastoral and agricultural watersheds. *Sustainability* **2022**, *14*, 6. [CrossRef]
10. Ahmed, S.; Iqbal, Z.; Antahal, P.C. Impact of climate change on nomadic herders' livelihoods: Evidence from Bakarwal community in the Western Himalayas in India. *Environ. Dev.* **2023**, *48*, 100930. [CrossRef]
11. Subramaniam, S.; Mohapatra, J.K.; Sahoo, N.R.; Sahoo, A.P.; Dahiya, S.S.; Rout, M.; Biswal, J.K.; Ashok, K.S.; Mallick, S.; Ranjan, R. Foot-and-mouth disease status in India during the second decade of the twenty-first century (2011–2020). *Vet. Res. Commun.* **2022**, *46*, 1011–1022. [CrossRef]
12. Eskdale, A.; Tholth, M.; Paul, J.D.; Desphande, J.; Cole, J. Climate stress impacts on livestock health: Implications for farming livelihoods and animal disease in Karnataka, India. *CABI One Health* **2022**, *2022*. [CrossRef]
13. Nejash, A.; Kula, J. Impact of climate change on livestock health: A review. *Glob. Vet.* **2016**, *16*, 419–424.
14. World Bank Group. *Mean Projections—CMIP6, IPCC Sixth Assessment Report*; World Bank Group: Washington, DC, USA, 2022.
15. Lara, L.J.; Rostagno, M.H. Impact of heat stress on poultry production. *Animals* **2013**, *3*, 356–369. [CrossRef]
16. Cole, J.; Desphande, J. Poultry farming, climate change, and drivers of antimicrobial resistance in India. *Lancet Planet. Health* **2019**, *3*, e494–e495. [CrossRef]
17. Vinke, K.; Gabrysch, S.; Paoletti, E.; Rockström, J.; Schellnhuber, H.J. Corona and the climate: A comparison of two emergencies. *Glob. Sustain.* **2020**, *3*, e25. [CrossRef]
18. Cramer, W.; Guiot, J.; Fader, M.; Garrabou, J.; Gattuso, J.-P.; Iglesias, A.; Lange, M.A.; Lionello, P.; Llasat, M.C.; Paz, S. Climate change and interconnected risks to sustainable development in the Mediterranean. *Nat. Clim. Chang.* **2018**, *8*, 972–980. [CrossRef]
19. Barnes, A.P.; Toma, L. A typology of dairy farmer perceptions towards climate change. *Clim. Chang.* **2012**, *112*, 507–522. [CrossRef]
20. Thornton, P.; Nelson, G.; Mayberry, D.; Herrero, M. Impacts of heat stress on global cattle production during the 21st century: A modelling study. *Lancet Planet. Health* **2022**, *6*, e192–e201. [CrossRef] [PubMed]
21. Birthal, P. Climate Change and Risk Management in Indian Agriculture. *Environ. Chall.* **2022**, *8*, 100543.
22. Moran, J. *Business Management for Tropical Dairy Farmers*; CSIRO Publishing: Clayton, Australia; Landlinks Press: Collingwood, Australia, 2009.
23. Douphrate, D.I.; Hagevoort, G.R.; Nonnenmann, M.W.; Lunner Kolstrup, C.; Reynolds, S.J.; Jakob, M.; Kinsel, M. The dairy industry: A brief description of production practices, trends, and farm characteristics around the world. *J. Agromedicine* **2013**, *18*, 187–197. [CrossRef] [PubMed]
24. Patel, S.; Patel, M.; Patel, J.; Patel, A.; Gelani, R. Role of women gender in livestock sector: A review. *J. Livest. Sci.* **2016**, *7*, 92–96.
25. Birthal, P.; Taneja, V.; Thorpe, W. Smallholder livestock production in India-Opportunities and challenges. In Proceedings of the ICAR-ILRI International Workshop, New Delhi, India, 31 January–1 February 2006.
26. Sarkar, A.; Dutta, A. Challenges and opportunities of dairy sector in India vis-à-vis world: A critical review. *Explor. Anim. Med. Res.* **2020**, *10*, 9–17.
27. Rohith, G. Performance of Cattle Insurance in Select Districts of Karnataka. *J. Anim. Res.* **2019**, *9*, 431–438. [CrossRef]
28. Ahmed, M.; Suphachalasai, S. *Assessing the Costs of Climate Change and Adaptation in South Asia*; Asian Development Bank: Mandaluyong, Philippines, 2014.
29. Byrareddy, V.; Kouadio, L.; Mushtaq, S.; Kath, J.; Stone, R. Coping with drought: Lessons learned from robusta coffee growers in Vietnam. *Clim. Serv.* **2021**, *22*, 100229. [CrossRef]

30. Singh, S.; Kagweza, O.O. *The Impacts of Climate Change in Lwengo, Uganda*; Intechopen: London, UK, 2021.
31. Islam, M.A.; Warwick, N.; Koech, R.; Amin, M.N.; de Bruyn, L.L. The importance of farmers' perceptions of salinity and adaptation strategies for ensuring food security: Evidence from the coastal rice growing areas of Bangladesh. *Sci. Total Environ.* **2020**, *727*, 138674. [CrossRef]
32. Alam, G.M.; Alam, K.; Mushtaq, S. Climate change perceptions and local adaptation strategies of hazard-prone rural households in Bangladesh. *Clim. Risk Manag.* **2017**, *17*, 52–63. [CrossRef]
33. Below, T.B.; Schmid, J.C.; Sieber, S. Farmers' knowledge and perception of climatic risks and options for climate change adaptation: A case study from two Tanzanian villages. *Reg. Environ. Chang.* **2015**, *15*, 1169–1180. [CrossRef]
34. Koç, G.; Uzmay, A. Determinants of dairy farmers' likelihood of climate change adaptation in the Thrace Region of Turkey. *Environ. Dev. Sustain.* **2022**, *24*, 9907–9928. [CrossRef]
35. Montcho, M.; Padonou, E.A.; Montcho, M.; Mutua, M.N.; Sinsin, B. Perception and adaptation strategies of dairy farmers towards climate variability and change in West Africa. *Clim. Chang.* **2022**, *170*, 38. [CrossRef]
36. Sejian, V.; Bhatta, R.; Malik, K.; Madiajagan, B.; Ali, Y.; Al-Hosni, Y.; Sullivan, M.; Gaughan, J. Livestock as Sources of Greenhouse Gases and Its Significance to Climate Change. In *Greenhouse Gases*; Intechopen: London, UK, 2016; pp. 243–260.
37. Singhal, K.K.; Mohini, M.; Jha, A.K.; Gupta, P.K. Methane emission estimates from enteric fermentation in Indian livestock: Dry matter intake approach. *Curr. Sci.* **2005**, *88*, 119–127.
38. Lane, D.; Murdock, E.; Genskow, K.; Rumery Betz, C.; Chatrchyan, A. Climate change and dairy in New York and Wisconsin: Risk perceptions, vulnerability, and adaptation among farmers and advisors. *Sustainability* **2019**, *11*, 3599. [CrossRef]
39. DDMA. Chikkaballapur District Disaster Management Plan, District Disaster Management Authority (DDMA), Chikkaballapur. Available online: https://chikkaballapur.nic.in/en/departments/ddma_cbp_en/ (accessed on 29 March 2021).
40. Garg, K.K.; Anantha, K.; Nune, R.; Akuraju, V.R.; Singh, P.; Gumma, M.K.; Dixit, S.; Ragab, R. Impact of land use changes and management practices on groundwater resources in Kolar district, Southern India. *J. Hydrol. Reg. Stud.* **2020**, *31*, 100732. [CrossRef]
41. Geetha, R.S.; Murthy, P.S.S. Evolution of Dairy Cooperatives in Karnataka: A Special Focus on Women's Dairy Cooperatives. *Asian J. Agric. Ext. Econ. Sociol.* **2021**, *39*, 1–7. [CrossRef]
42. Phellas, C.N.; Bloch, A.; Seale, C. Structured methods: Interviews, questionnaires and observation. *Res. Soc. Cult.* **2011**, *3*, 23–32.
43. Islam, M.A.; de Bruyn, L.L.; Warwick, N.W.; Koech, R. Salinity-affected threshold yield loss: A signal of adaptation tipping points for salinity management of dry season rice cultivation in the coastal areas of Bangladesh. *J. Environ. Manag.* **2021**, *288*, 112413. [CrossRef] [PubMed]
44. Hasan, M.K.; Kumar, L. Perceived farm-level climatic impacts on coastal agricultural productivity in Bangladesh. *Clim. Chang.* **2020**, *161*, 617–636. [CrossRef]
45. Islam, M.A.; Lobry de Bruyn, L.; Koech, R.; Warwick, N. Contextualisation of salinisation and adaptation preferences in the coastal areas of Bangladesh: Bringing together farmers' salinity perspectives into place-based policy initiatives. *Land Degrad. Dev.* **2022**, *33*, 3882–3900. [CrossRef]
46. Wetende, E.; Olago, D.; Ogara, W. Perceptions of climate change variability and adaptation strategies on smallholder dairy farming systems: Insights from Siaya Sub-County of Western Kenya. *Environ. Dev.* **2018**, *27*, 14–25. [CrossRef]
47. Verma, J.P. *Data Analysis in Management with SPSS Software*; Springer Science & Business Media: Berlin, Germany, 2012.
48. Sertse, S.F.; Khan, N.A.; Shah, A.A.; Liu, Y.; Naqvi, S.A.A. Farm households' perceptions and adaptation strategies to climate change risks and their determinants: Evidence from Raya Azebo district, Ethiopia. *Int. J. Disaster Risk Reduct.* **2021**, *60*, 102255. [CrossRef]
49. Abbas, Q.; Han, J.; Adeel, A.; Ullah, R. Dairy Production under Climatic Risks: Perception, Perceived Impacts and Adaptations in Punjab, Pakistan. *Int. J. Environ. Res Public Health* **2019**, *16*, 4036. [CrossRef]
50. Asravor, R. Smallholder farmers' risk perceptions and risk management responses: Evidence from the semi-arid region of Ghana. *Afr. J. Econ. Manag. Stud.* **2018**, *9*, 367–387. [CrossRef]
51. Bishu, K.G.; O'Reilly, S.; Lahiff, E.; Steiner, B. Cattle farmers' perceptions of risk and risk management strategies: Evidence from Northern Ethiopia. *J. Risk Res.* **2018**, *21*, 579–598. [CrossRef]
52. Meraner, M.; Finger, R. Risk perceptions, preferences and management strategies: Evidence from a case study using German livestock farmers. *J. Risk Res.* **2019**, *22*, 110–135. [CrossRef]
53. Bett, B.; Kiunga, P.; Gachohi, J.; Sindato, C.; Mbotha, D.; Robinson, T.; Lindahl, J.; Grace, D. Effects of climate change on the occurrence and distribution of livestock diseases. *Prev. Vet. Med.* **2017**, *137*, 119–129. [CrossRef] [PubMed]
54. Diniso, Y.S.; Zhou, L.; Jaja, I.F. Dairy farmers' knowledge and perception of climate change in the Eastern Cape province, South Africa. *Int. J. Clim. Chang. Strateg. Manag.* **2022**, *14*, 168–179. [CrossRef]
55. Abdelatty, A.M.; Iwaniuk, M.E.; Potts, S.B.; Gad, A. Influence of maternal nutrition and heat stress on bovine oocyte and embryo development. *Int. J. Vet. Sci. Med.* **2018**, *6*, S1–S5. [CrossRef] [PubMed]
56. Radha, M.; Singh, D.S.K.; Beillard, M.J. *Dairy and Products Annual—2022*; USDA: Washington, DC, USA, 2022.
57. KSNDMC. Drought Management Strategies. In Proceedings of the National Conference, Melbourne, Australia, 8–9 March 2019.
58. Rawat, S.; Ganapathy, A.; Agarwal, A. Drought characterization over Indian sub-continent using GRACE-based indices. *Sci. Rep.* **2022**, *12*, 15432. [CrossRef] [PubMed]

59. Herrero, M.; Wirsenius, S.; Henderson, B.; Rigolot, C.; Thornton, P.; Havlík, P.; De Boer, I.; Gerber, P.J. Livestock and the environment: What have we learned in the past decade? *Annu. Rev. Environ. Resour.* **2015**, *40*, 177–202. [CrossRef]
60. Kumar, S.; Raju, B.; Ramarao, C.; Ramilan, T. Sensitivity of livestock production to climatic variability under Indian drylands and future perspective. *Curr. Agric. Res. J.* **2015**, *3*, 142–149. [CrossRef]
61. Rivera-Ferre, M.G.; López-i-Gelats, F.; Howden, M.; Smith, P.; Morton, J.F.; Herrero, M. Re-framing the climate change debate in the livestock sector: Mitigation and adaptation options. *WIREs Clim. Change* **2016**, *7*, 869–892. [CrossRef]
62. Lien, G.D.; Flaten, O.; Ebbesvik, M.; Koesling, M.; Valle, P.S. Risk and risk management in organic and conventional dairy farming: Empirical results from Norway. In Proceedings of the 14th Congress of the International Farm Management Association, Perth, Australia, 10–15 August 2003.
63. Akcaoz, H.; Kizilay, H.; Ozcatalbas, O. Risk management strategies in dairy farming: A case study in Turkey. *Int. J. Anim. Vet. Adv.* **2009**, *8*, 949–958.
64. Godde, C.M.; Mason-D'Croz, D.; Mayberry, D.E.; Thornton, P.K.; Herrero, M. Impacts of climate change on the livestock food supply chain; a review of the evidence. *Glob. Food Secur.* **2021**, *28*, 100488. [CrossRef]
65. Aryal, J.P.; Rahut, D.B.; Maharjan, S.; Erenstein, O. Factors Affecting the Adoption of Multiple Climate-Smart Agricultural Practices in the Indo-Gangetic Plains of India. In *Natural Resources Forum*; Blackwell Publishing Ltd.: Oxford, UK, 2018; pp. 141–158.
66. Maindi, N.; Osuga, I.; Gicheha, M. Advancing climate smart agriculture: Adoption potential of multiple on-farm dairy production strategies among farmers in Murang'a County, Kenya. *Livest. Res. Rural. Dev.* **2020**, *32*, 63.
67. Nouala, F.; Akinbamijo, O.; Adewumi, A.; Hoffman, E.; Muetzel, S.; Becker, K. The influence of Moringa oleifera leaves as substitute to conventional concentrate on the in vitro gas production and digestibility of groundnut hay. *Livest. Res. Rural Dev.* **2006**, *18*, 121.
68. Maina, K.W.; Ritho, C.; Lukuyu, B.A.; Rao, E. Socio-economic determinants and impact of adopting climate-smart Brachiaria grass among dairy farmers in Eastern and Western regions of Kenya. *Heliyon* **2020**, *6*, e04335. [CrossRef] [PubMed]
69. Idrissou, Y.; Assani, A.S.; Baco, M.N.; Yabi, A.J.; Traoré, I.A. Adaptation strategies of cattle farmers in the dry and sub-humid tropical zones of Benin in the context of climate change. *Heliyon* **2020**, *6*, e04373. [CrossRef]

Disclaimer/Publisher's Note: The statements, opinions and data contained in all publications are solely those of the individual author(s) and contributor(s) and not of MDPI and/or the editor(s). MDPI and/or the editor(s) disclaim responsibility for any injury to people or property resulting from any ideas, methods, instructions or products referred to in the content.

sustainability

MDPI

Article

Current Situation of Traditional Architecture Located inside Cultural Mayan Heritage Spaces in Remote Villages of Guatemala: Case of the Black Salt Kitchens

Luis Pablo Yon Secaida [1,*], Suguru Mori [2] and Rie Nomura [2]

[1] Laboratory of Architectural Planning, Division of Architectural and Structural Design, N216, Engineering Faculty, Hokkaido University, Kita 13-Jo, Nishi 8-Chome, Sapporo 060-8628, Japan

[2] Faculty of Engineering, Hokkaido University, Sapporo 060-8628, Japan; suguru-m@eng.hokudai.ac.jp (S.M.); nomurarie@eng.hokudai.ac.jp (R.N.)

* Correspondence: secaidaluispablo.yon.m4@elms.hokudai.ac.jp

Abstract: In the town of Sacapulas, located in the mountainous country of Guatemala, there is a constant risk of natural disasters. Floods and landslides occur frequently, resulting in the loss of human lives and cultural aspects. Important to the region, the creation of the black salt is most affected. This resource has been created since the time of the Mayans on the salt beach surrounding the town. However, from the 1940s onwards, this industry has shrunk. As a result, architectural expressions known as "salt kitchens" have almost disappeared, and there is no information on the subject available. By employing interviews, area survey, and GPS mapping, it was discovered that the location of the salt kitchens is determined by the shape of the beach. However, only one third of the beach area is accessible up to this day. It was discovered that the destruction of the salt kitchens is due to natural elements as well as owners reusing the land for other economically viable functions. To preserve their existence, the first plans of the salt kitchens were created, and will help future researchers if necessary.

Keywords: cultural areas; space use; Mayan architecture

1. Introduction

This research is a continuation of the continuous effort to categorize, analyze, and recover the salt beach area of Sacapulas. Previous research has been conducted regarding the beach's current situation. In this paper, a part of the same original dataset will be used with new information gathered on site.

Culture and its inclusion in sustainable development has been researched lately, with more people being interested in the topic [1–3]. Sustainable development is defined as "development that meets the needs of the present without compromising the ability of future generations to meet their own needs" [4]. Its inclusion is based on three important aspects: social, environmental, and economical [1,2]. Examples can be given, such as the Sendai framework for disaster risk reduction, a joint effort seeking to reduce the impact of disasters over the next 15 years from its creation [5]. It is of the utmost importance to understand how an affected group of people reacts, adapts, and uses its resources to reduce the impact of disasters. Some groups, sadly, do not have the means to properly address such effects and seek external assistance to solve them [6]. This phenomenon has been researched lately, yet the efforts vary by country and region. This is true in European countries where specific care has been given. Countries in development still lack the interest and effort to reduce the loss of cultural heritage, especially the countries with a high degree of natural disasters [7].

A specific aspect of disaster management is related to culture heritage [7,8]. It is understood that climate change can affect culture, both the tangible [9] and intangible [10,11] expressions that a group of people consider culture. A response to counteract this issue is the implementation of new approaches with a diverse set of tools, be it interdisciplinary, multidisciplinary, or transdisciplinary [12]. A four-pillar approach created by the US National Park Service (NPS) is a good example of this. The same could be said of UNESCO's Local and Indigenous Knowledge System (LINKS), which implements Intangible Cultural Heritage (ICH) into disaster management. ICH is knowledge shared by cultures all across the globe, recollected through ages of experience by local people. Another term for this knowledge is "Indigenous Knowledge systems and practices" (IKSP) [9,13]. The implementation of LINKS has been effective during the creation of new science and environmental policies [13].

Even though there has been an international effort towards reducing disasters, this has not been fully developed in the Latin American country of Guatemala. Research conducted in the country shows that there is an erosion of authority figures, resulting in slow decision making [14–16]. While not specifically using the LINKS, Guatemala does have disaster risk governance legislation, which is included in the legislative decree 109-96 called "Law on the National Coordinator for Disaster Reduction" [17]. This law was based on the Yokohama Strategy and Framework for Action, which gives priority attention to developing countries [18]. Currently, due to the lack of enforcement of said law, local disasters are normally addressed with foreign cooperation such as USAID [19]. Yet, the international help is not enough to address less-known communities such as the town of Sacapulas. This town is located in the mountain range of the Chucumatanes, Guatemala. To the north of the village is the Chixoy River, which flows to the south of Mexico. The importance of this town is found in the production of black salt, a traditional product with more than a thousand years of tradition. The creation of black salt is directly linked to three main actors: the people directly involved in the industry, known as salt makers; the architecture created specifically for the industry, known as "salt kitchens"; and the space where they collect the materials for the salt, called "Black Salt Beach", located north of the village, next to the Chixoy River. Historically, the Maya society has always been vulnerable against natural disasters [20–22], but there has been an increase in recent years [23]. Proof of this has been the constant flooding of the Chixoy River in the area. As a consequence, the black salt industry and every aspect linked to it (tangible and intangible) have been affected. Of the countless people that were part of this industry previously, only four people in an advanced stage of life remain [24].

Considering that Guatemala is a country with a high risk of natural disasters, and a high lack of adaptive capabilities [25], swift action must be taken as soon as possible. In fact, research to safeguard the black salt culture has been few and far between, focusing only on the process of black salt creation. Interest in the subject has been diverse, and mainly by foreigners. One of the most important research projects in the area was performed by Anthony P. Andrews in 1983 [26], where he described how the black salt created in Sacapulas has always been of interest both to people outside of the culture (Spanish conquerors and researchers), and people inside the region, such as the nearby kingdom of Quiche, which was interested in the town and its resources. Overall, the town of Sacapulas and the black salt have been an important product of the Guatemalan culture and a point of interest to many people across the years. The previous known research was conducted in 2002 and it focused specifically on the black salt, until 2023, with an effort to understand the current situation of the beach [24,27]. This research focused on understanding how the situation evolved since the 2002 research conducted in the area. It was found that there is a constant lack of information regarding the physical aspects of the space, and also the cultural expressions that were born due to the salt industry. Without a concrete point of comparison, it was decided to instead develop the first spatial information in the form of planimetry, photography of the current state, and activity division of the area based on the accounts of people interviewed. The analysis identified that, compared to the accounts of previous research, the area in fact shrunk and was overrun with more soil than was reported before. This has affected the space and the salt industry in ways that have yet to be researched, such as the architectural elements used for salt cooking. There is also a lack of research focused on the relation between the space, its features, and how the architecture functions. This is because research in the area is usually performed by people not interested in these topics, and also because all of the research does not come from the viewpoint of an architect. This has become an important topic as there is the threat of a resilient culture, and cultural expressions that are valuable due to their historical context, being erased from memory. Understanding how to properly adapt a local LINKS system, it must first identify what is at stake to be lost. At this moment, we know what the current situation is, yet we need to understand how the past impacted and transformed the salt industry and its specific architecture.

1.1. Case Area

Guatemala is a country located in Central America, represented in Figures 1 and 2. It has borders with Mexico to the north, Belize and the Caribbean Sea to the northeast, Honduras to the east, El Salvador to the southeast, and the Pacific Ocean to the south. The country's territory is primarily mountainous due to the "Sierra de Los Cuchumatanes," a mountain range throughout Central America. As a result, topographic elevation varies from 500 m to 3800 m. Rural towns and cities are located all across this terrain variation, resulting in many people living in mountainous, remote areas.

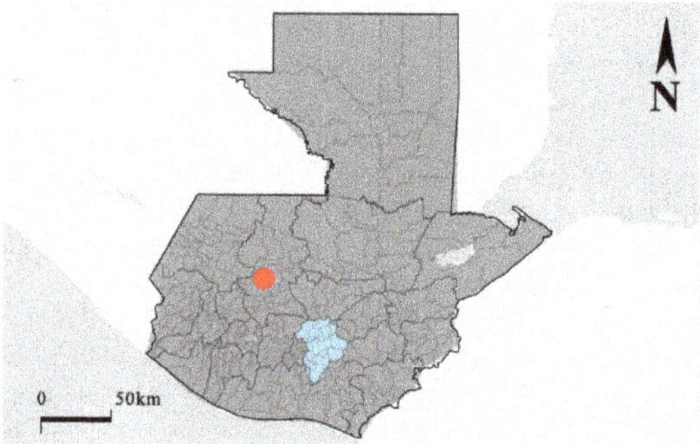

Figure 1. Geographical location of Guatemala. Research area, in red. Capital of Guatemala in blue.

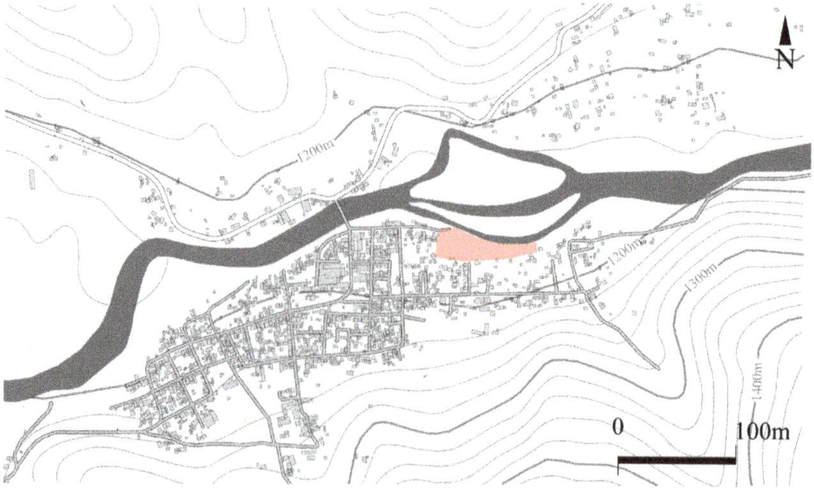

Figure 2. Sacapulas Town in the south of the Chixoy River, with the salt beach in red.

Inside the Cuchumatanes mountains is located the town of Sacapulas. While the region has developed steadily, communication is slow due to its current location, resulting in difficulties sharing the local culture. These difficulties have resulted in almost no current research being conducted in the area. Next to the town is the Chixoy River. This river runs through much of the northwestern territory of Guatemala and flows into southern Mexico. Across the river can be found salt deposits in the form of white spots, while the beach presents a source of thermal water. Both of these resources are used in the process of the black salt, which Figures 3 and 4 depicts. Because of its traditional way of being made, when cooked, it results in a black color with attributed medicinal properties that the local population benefits from [26,27].

Figure 3. Black salt production; the mold is being broken to take the salt out. RAICES NGO (2023).

Figure 4. Black salt product. Author (2022).

Current research in the area has successfully identified the situation of the salt beach. At this moment, the salt beach comprises a surface area of 4873.42 m^2, of which almost 57% is covered by soil. As mentioned before, this soil comes from constant flooding from the Chixoy River. Of this area, only 5% has been excavated and is used for the salt industry.

The rest of the area is used for what seems a small array of activities, such as using the small shore of the beach to bathe due to hot water found in the area; cattle rising occurs all across the covered salt beach; and people use the space to move within the town of Sacapulas.

1.2. Objective

At this moment, research conducted towards the spatial and architectural features of the area has yet to be conducted. The importance of such a point of view resides in the relation between these two subjects and how they adapted due to the natural disasters of recent years. Therefore, this research is aimed specifically towards the black salt kitchens. It seeks to identify their current situation, usage, location, and relation to the black salt beach. Secondarily, it seeks to document the architectural information of the kitchens through architecture planimetry.

1.3. Limits

Considering the amount of information yet to be researched about the subject, several limits were imposed on this research. Firstly, this research will not consider the other cultural expressions that exist in relation to the black salt other than the architectural elements. To locate the kitchens, only people related to the black salt industry are considered, as they present the information firsthand. Finally, the research area is located across the current black salt beach. Previous research suggests that salt kitchens were located at the north of the black salt beach, yet this area is currently flooded and access to it is impossible.

1.4. Current Literature

1.4.1. The Salt in Guatemala

The nature of this research is not about the overall situation and history of the Salinas, but a general understanding of its characteristics will be discussed to understand its context, from an architecture perspective. The traditional salt industry in Guatemala spans the entire Pacific Coast from the Suchiate River on the Mexican border to the border of El Salvador. According to Andrews, several hundred salt sources are located along the shores of the Pacific. According to Andrews, all of the salt produced on the Pacific coast of Guatemala was made by a process known as "cooking salt", which has more than 2000 years of history. Figure 5 depicts the major salt sources according to Andrews, which are located mostly on the west of Guatemala. There have been accounts by Spaniard chroniclers that described in detail the process to cook the salt. It is generally understood that it involves a three-step process: placing the soil impregnated with salt, filtering the soil with estuary water, and then cooking the result inside a clay or ceramic pan (*"olla"* in Spanish). While Andrews mentioned replacement of the pans with a more convenient vessel made from iron called *"perol"*, this eventually transitioned to the usage of ovens. Interestingly enough, Sacapulas uses what is known as a "salt kitchen" as a space where the salt is cooked. Plenty of modern research mentions the salt kitchens as a novelty and an important factor for the fame of the Sacapulas salt. Yet, a specific analysis of this architectural element has been extremely scarce.

Figure 5. Historical salt sources identified by Andrews in his salt research. In red, Sacapulas. Andrews [26].

1.4.2. Salt Industry Formation

Thanks to archeological data found in the region of Ocos and Champerico, it is understood that the industry of salt began around the Formative Mayan period and was underway 2000 [26] years ago. Meanwhile, Spaniard chroniclers mention the existence of the salt industry to a degree of novelty [26,28], being conducted exclusively by natives. While the territories used for the salt industry were eventually divided with Spanish colonizers, the salt industry remained stable until the introduction of new technological innovations in the region. The previously mentioned ceramic pans, iron vessels, and ovens did increase its production, but the introduction of solar technology marked a noticeable shift in the local production. In fact, Andrews mentioned the increased use of solar evaporation as the main reason for the decline of cooked salt in the region. This is true, as, comparatively, solar evaporation yields more product with less effort [26]. The usage of this technology was slow though; it was introduced during the 1920s and it became the dominant method of salt creation until the 1950s.

There are mentions of the complications due to the salt making process, both inside and outside of the salt industry, by Andrews, influencing its disappearance in the future. Firstly, the lack of natural resources such as firewood results in its disappearance. Secondly, the benefit of large agricultural and ranch spaces has led to deforestation. It is important to note that rural areas in Guatemala benefit mostly from these two economical activities, which makes them more attractive to people. In turn, cooked salt is both labor-intense and low yield compared to other salt methods.

1.4.3. Salt Making and Sacapulas Importance

The value of Sacapulas salt goes beyond its traditional method of production, as cooked salt is not specific to the town. Instead, its value comes from the historical background that it represents to Mayan history, the cultural expressions born from said industry, and the attributed value that is given by people that know about this product. Specifically, Sacapulas black salt is considered a famous type of salt due to its attributed health benefits, though no research has been conducted to validate or deny such a claim. Figure 6 depicts the process in which it attains its famous black color: the salt cooking process. Nevertheless, it has been reported by many researchers as an important factor of its notoriety [26,27]. This kind of salt making specifically uses an area known as "Salinas", which comprises a beach located north of the town of Sacapulas, and south of the Chixoy River. Most of this space is currently covered with alluvial deposits due to the constant floods caused by the river, which both Andrews and current research have confirmed as a constant issue. Specifically, Andrews mentions that at the time of his research, the space was covered with "as much as 2 m of alluvial deposits". Historical accounts place Sacapulas in a favorable position due to its location in relation to major trading routes. Its location is along the Chixoy River, which constitutes the east–west salt trade in the region. It is also located in a favorable position towards the Atlantic lowlands due to the Chixoy River being connected to the Usumacinta River. Its location was attractive to the Quiche region, as its location made it an important economical center. This resulted in the kingdom of Quiche eventually incorporating the town into its control until the arrival of Spanish conquerors. Colonial and modern history has shown considerable information on the Sacapulas industry, with recounts such as Titulo de los senores de Sacapulas and Spanish chronicles as important resources to understand the position that the town had for the Spanish crown and local Mayan people. Current research has focused on the actual state of the beach, and understanding how it has evolved due to the decline of salt making in the region. Traces of the importance of Sacapulas salt making still exist. An example of this is the current trading routes in the region. According to Andrews, the trading routes used by Mayas during the precolonial times were adapted by Spaniards and even used by current traders. Such examples exist, like the trading routes towards south and central Quiche, which are a pattern that dated to the XIV century when Sacapulas was integrated into Quiche. It was also mentioned that the salt route to the west was very old. This is due to the fact that salt tributes were found in the western region of Guatemala, tributes paid with Sacapulas salt [26].

Figure 6. Black salt cooking process. RAICES NGO (2023).

Historically, the importance of the black salt industry is found in several mentions by Spanish conquistadors and Mayan mythology related to the region. Specifically, chronicles written after the conquest mention the Spanish conquistadors' interest in salt creation. The earliest mention of salt is a report by three Franciscans in 1579 [28]. This report relates the conquistadors' need for salt and how, for this reason, salt had to be obtained from Sacapulas. The salt makers were referred as "salineros" (Salt workers), and received a part of the salt as payment, which they eventually sold in their communities. The next mention of salt consisted of impressions by the Spanish colonizers. This recollection was written and reported by Martin Alfonso Tovilla, the mayor of the province of Vera Paz, Golfo Dulce, Sacapulas, and Manchen. His report to the Spanish crown was compiled in the book "Historicas Dyscriptivas" (Descriptive stories) in 1629. On this occasion, he mentions the importance of salt in the gastronomy of the locals, as they "do not do or want anything more than corn tortillas and a little chili and salt, which makes them happier than if they were eating turkeys..." [29].

Tovilla also mentions that the area's importance was also perceived as high because it was one of the few regions which produced salt. This importance was critical with the political situation of Sacapulas, as the riches were a point of issue with the Quiche kings and their kingdom located in the ruins of "Gumarkaaj", south of Sacapulas (around one hour of driving). Prisoners of war were commonly captured and forced to work in the salt industry. Tovilla concludes by giving his opinion regarding the low price of black salt compared to the volume of wood necessary for its creation, finding the process too exhaustive for the price it is sold. He said the rate was around fifteen "cakes" (salt molds) for one "*real*" (currency at the time).

Little research has been conducted in recent years, the latest being on the importance of black salt. This is found in the anthropological book "Traditions of Guatemala," which defines black salt as "a resource of great importance" [27]. In addition, there is evidence that various tributes were made in the form of salt during colonial times, proving that salt was considered an essential and valuable product for the Spaniards and the natives [26].

1.4.4. Salt Kitchens—Architecture Specific to the Industry

The area known as the black salt beach has been afflicted by constant flooding. This is due to the location of the Chixoy River in relation to the town of Sacapulas [24]. The result of the constant flooding and loss of space is a reduction in people active in the black salt industry. Consequently, the tools used in salt making have also ceased to be used. In this case, the "traditional salt stove" is one of the essential tools in salt making, known locally as "salt kitchens". Figure 7 is an example of this architecture. These constructions are concentrated above the black salt beach and comprise two essential elements: the main kitchen and the distillation box or dripping area. The kitchen comprises an enclosed space needed for cooking salt. It is a small room excavated underground, surrounded by stone walls [30]. While the construction of the walls is rudimentary (it does not present any internal structure), the gable roof presents a wood structure.

Figure 7. Salt kitchen. Author (2022).

1.4.5. Cajón and Pila de Destilación (Distillation Box and Basin)

In front of the kitchen, on the outside, is the salt preparation area represented by Figure 8. It comprises the distillation box denominated "cajon", an element usually 1.68 m in diameter and 1 meter deep. While the main structure is of wood, the box floor is made of wooden rods or cane and matting, named "petate". The box is placed on top of stone walls, on a mound of earth, at a certain height. Underneath, the distillation area or basin is built. Its function is to distill the salt using the thermal waters found around the beach shore. This step is crucial as it happens before cooking the salt in the kitchen. The basin is made of brick and stone, glued with mortar. The basin has the necessary capacity for eight "tinajas": a traditional plastic jar used in rural areas.

Figure 8. Distillation box and area. Author (2022).

1.4.6. Features of the Salt Kitchen

The architectural language found in the kitchens is of something other than Maya origin, which suggests that they were created after the pre-hispanic period. This fact is alluded to by Tovilla's text, which does not mention the use of specific architecture to create salt. Also, the use of clay tiles in roofs as an architectural element started after the conquest, and did not originate from the region. However, Ruud Van Akkereen mentions in "Titulo de los señores de Sacapulas" that the pre-hispanic name of Sacapulas comes from the name "Tuja" or "Tujalá" [31]. He mentions the possibility that tuja refers to the low houses where people worked the black salt. Thus, there is a possibility that this architecture existed in pre-hispanic times in a different shape. There are discrepancies regarding the height of the salt kitchen. According to Reyna, the total height of a kitchen is 2 to 2.5 m (one meter underground, the rest over ground level), with measurements of 6 × 8 m. Ordoñez mentions between 3 and 4 m in height, with no mention of its measurements. The primary tool found inside the salt kitchen is the "heart of the fire," a slight mound of earth measuring 1.67 × 5.02 × 0.41–0.84 m wide, located in the center. On it, 40 tenamastes are placed: pointed stones taken from the Chixoy River. Over the tenamastes, "cajetes" or pots made from animal feces or clay are placed with the previously distilled water inside. These are usually used only once due to the fragility of the material. Also, the differences in temperature result in many of these pots being broken after they are used (Figure 3). The salt creation process is detailed in Section 1.4.7.

New kitchens are not built due to the small size of the salt industry. For this reason, maintenance must be performed to keep them usable. Maintenance focuses specifically on the roof structure comprising wood. This is replaced every two or three years due to the high intensity of the fire. In contrast, the tiles used on the roof are cleaned, not replaced. Informants estimate that the tiles are over 100 years old [27]. Due to the scarcity of kitchens, Ordoñez reports that salt kitchens can be borrowed or rented. The price is Q10 (USD 1.27) each time they are used or Q100 (USD 12.77) for a season from January to April. Another important fact is that the "heart of the fire" can be shared among salineros, which can help divide the cost of rent between two or more workers. Finally, throughout the off-season, salt makers aim to extract enough soil with salt to cook throughout the year. Thus, the salt kitchen also serves as a warehouse storing potential economical revenue to the salt maker.

1.4.7. Salt Creation Process

Understanding the value of the black salt's architecture and industry requires knowledge of how the product is made. The process initially mentioned by Alfonso Tovilla is the same as the one used in 1978, reported by Reyna and Monaghan, and by Amilcar Ordoñez in his 2001 research [27,30]. It should be noted that the process Tovilla mentions is described in general terms compared to Reyna and Monaghan's work. An element that has not changed has been the tools used in the salt creation process, which implies that the work process has not been optimized with new technologies. The process mentioned by Tovilla, and Reyna and Monaghan consists of two main parts: the "sowing of the salt"; and the cooking and subsequent finalization of the product. The first step of the process ("sowing the salt") is a method of salt germination involving the beach's specific temperature and soil obtained from it. Due to the nature of this research, the process is summarized in the most important steps for its creation. Therefore, this is a compilation of the steps mentioned by both Reyna and Monaghan and Ordoñez in their respective investigations.

As mentioned previously, salt production only occurs during the dry season of the year (November to June). However, the salted soil obtained during these months is stored and produced as salt throughout the year. Usually, the salt kitchens serve as a salt creation area and salt soil storage for future work. The work begins in the morning when the salt maker spreads special, clean soil on his or her part of the beach, no more than one centimeter thick. The documentation does not mention why the soil used is special. The land of the beach is measured by "petaquilla," a specific informal measurement unit that equals 12 square varas (Spanish measure equal to 8.385 sqr m). Then, he fills a tinaja with thermal water obtained from one of the many open wells on the beach. With half of a gourd, he hydrates the soil so that later the sun heats the beach enough to make the salt sprout and impregnate the soil. According to the report, this process is conducted for three consecutive days until the salt maker, from his own experience, decides that the soil has enough salt impregnated. The next step is the mobilization of the soil to the salt kitchens. Figures 9 and 10 depicts this part of the process.

Figure 9. Distillation process. Hot water from the springs is being poured over the soil impregnated with salt. RAICES NGO (2023).

Figure 10. Lower part of the distillation box. RAICES NGO (2023).

The salt is transported to the salt kitchens on foot. Here, the soil will be distilled one petaquilla at a time. The soil is transported to the front of the kitchen, where the distillation box and basin are located. The salt is placed in the box, separated by a few m from the ground. The next day, the hot water from wells on the beach is poured in, a process that takes from 5 a.m. to 3 p.m. (10 h). At the end of the process, the soil is removed, and more is hauled back to start again. The result is distilled water with salt, or "aguasal". With the filtered salt water, the salt maker creates the "heart of fire", a rudimentary oven in which the salt molds are placed on the 40 tenamastes. The cajetes are heated with salted water inside and cooked for 5 to 6 h to evaporate the water. This process is called "burning" the salt. The next day, the pot is broken, and the salt lid is removed, ready to be sold. Figures 11 and 12 depicts this process.

Figure 11. The cooking space known as "The heart of fire". RAICES NGO (2023).

Figure 12. Cooking process. Source: RAICES NGO (2023).

The resulting product of the salt workers is twofold: Atzam and Xuupej. Atzam is small pieces of white salt resulting from the burning of the aguasal three times. However, reintroducing the atzam into the heart of fire results in Xuupej, black salt [27]. At the time of Reyna and Monaghan's work, approximately 250 pounds of salt were produced in two days and sold at high or low prices, depending on the season. This fact does not seem to have changed according to Ordoñez's research, where he mentions that each time salt is made, between 1 and 2 quintals (150 to 300 pounds) are produced. Figures 13 and 14 depict the last hours before the salt is finished.

Figure 13. Final hours until the cooking is finished. RAICES NGO (2023).

Figure 14. Finished black salt. RAICES NGO (2023).

1.4.8. Number of Salt Kitchens

The total number of kitchens is a mystery; in 1981, Reyna and Monaghan mentioned the existence of 11 salt kitchens. Meanwhile, Andrews mentioned on his 1983 research that there were 14 salt kitchens standing. However, in 2002, according to Ordoñez, only 6 kitchens were still in use during his research. According to Ordoñez's informants, there were around 16 kitchens in operation previously. Ordoñez did not define the specific year the 16 kitchens were in use. It can be hypothesized that the small number is due to two factors: a reduction in the black salt industry results in the reinterpretation of the space of the salt kitchens for another, more economically attractive use. And second, the historical earthquake of February 1976, where most of the buildings suffered damage [32]. According to Ordoñez, many areas of the black salt beach are now used as corrals and pigpens.

2. Materials and Methods

Black salt is a research topic wide enough but yet to be researched to a degree that access to information on it would be considered easy. Therefore, the lack of information has resulted in a complicated starting point for any person interested in the subject. A good way to address this challenge is by implementing not only one research method, but a diverse array of techniques that cover a broader set of research questions. This is known as "triangulation" [33]. The idea of triangulation is obtaining data from difference sources, varying the methodology, researcher, and/or theories. The triangulation approach is not an end in itself, but instead a tool to better understand a complex problem while also being effective in the time employed during the research. It also serves as a way to reduce the amount of personal bias the researcher might have, as different sources of information yield a more complete result. Finally, the relationship between quantitative and qualitative methods benefits from a triangulation approach, as it gives a better reach to the research while confirming or denying information from interviews. In the case of this research, the triangulation approach was used to complete the resulting dataset, with different information groups used for this process. Specifically, using the Denzin approach to triangulation techniques, "Methodological triangulation" was used. Of this group, "between method" triangulation was selected. This triangulation approach focuses on the usage of a variety of methods to collect and interpret data, and is used where two or more distinct methods are employed to research the same phenomenon—in this case, the relation of the salt kitchen and the salt beach. There is evidence of this approach being helpful in understanding disasters and local issues in Guatemala [14]. A summary of the analyses performed is shown in Table 1, while Figure **??** depicts the methodology employed and the workflow conducted.

Table 1. Types of analysis performed in this research.

Method	Data Sources	Time	Space	Researcher	Goal
Interviews	People related to black salt (former workers, traders, and current salt workers).	Interviews were conducted twice; the first was conducted over one week as the pilot survey. The second one was conducted in person for one week, with a visual analysis of the space.	Town of Sacapulas	A principal investigator together with a 2-person working group in the area.	Information about the historical use of the area and current situation.
Salt kitchen measurements	On-site measurements of one salt kitchen and its features	Carried out in the same timeframe as the primary survey	Salt kitchen location	A principal investigator, together with an assistant	General planimetry of the area and analysis
Salt kitchen's location	On-site visit, GPS localization	Carried out in the same timeframe as the primary survey	Beach area surroundings	A principal investigator, together with an assistant	Location of the salt kitchens and spaces where they were built. Documentation of how the architecture was distributed concerning salt beach.

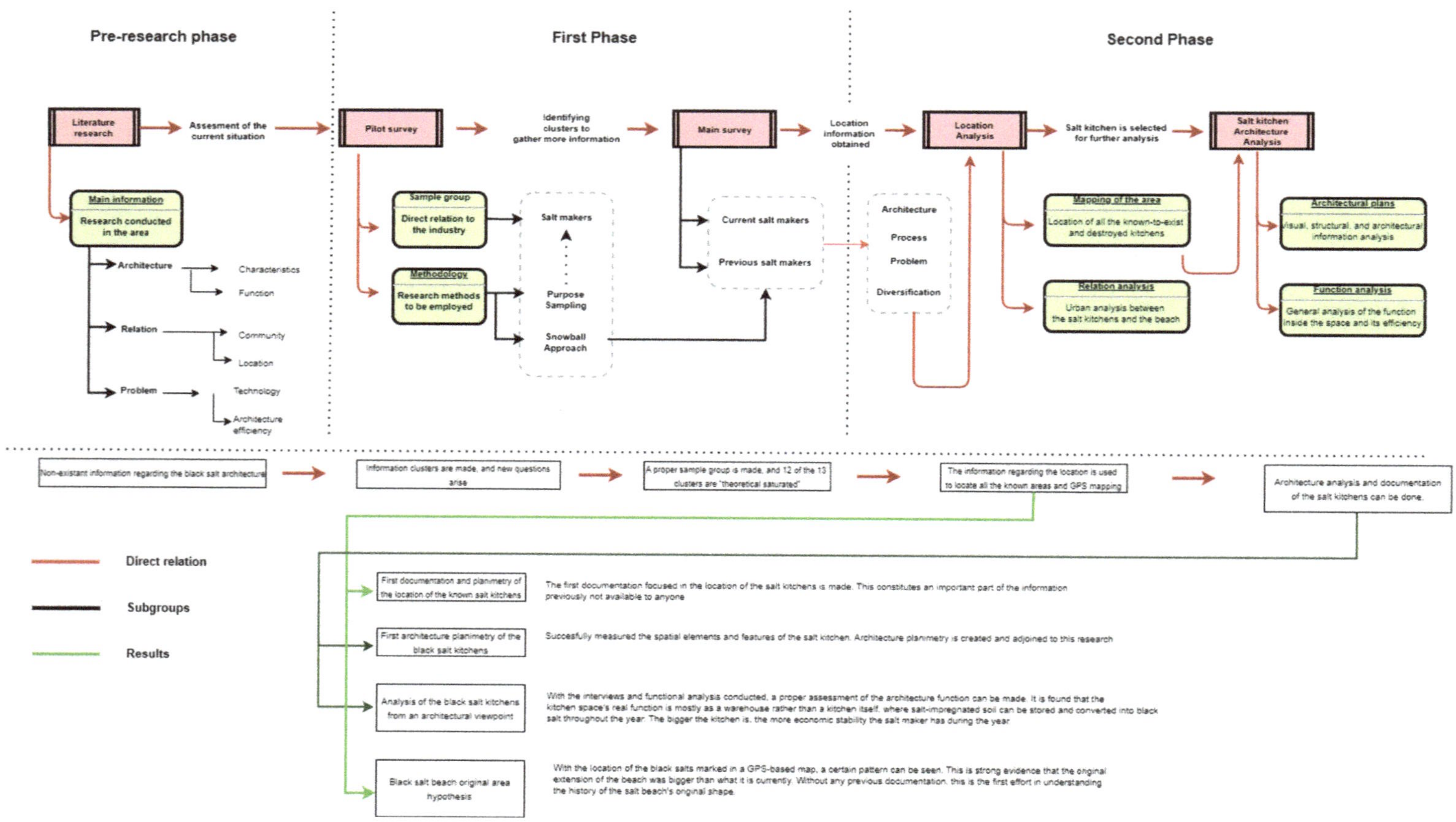

Figure 15. Methodology diagram.

After selecting the triangulation method, we need to define the types of analysis to be conducted in this research. Three analysis types were selected for this research, ranging from qualitive to quantitative. The main analysis and information were obtained by interview. A certain part of the information obtained was used originally in the previous research conducted in the area [24], yet part of the unused information was utilized on this follow up research by implementing new information specifically from the salt kitchen. This analysis was conducted in 2 parts, the first being an approximation of the area conducted in March to April 2022 as a pilot survey, and the second one being conducted in September 2022 as the main survey. The pilot survey was focused on specific interviews of a sample group of people who had a relationship with the black salt, using the "modified ground theory" as the main methodology employed. The "Modified Ground Theory" approach is based on the systematic method developed by Glaser and Strauss called "Ground Theory", in their 1967 book [34]. This qualitative method requires that the researcher start the investigation with pure data and no available theoretical background or knowledge of the phenomena. This is quite a challenge in the case of the black salt due to the complexity of the situation and the lack of information available: it is required to know the value of the industry and its relation to the people of Sacapulas. Therefore, a method which implements rough definitions and theories previous to the start of the research is needed while using the general "ground theory" methodology. This is called the "Modified Ground Theory", and it facilitates the researcher by taking into account the context of the black salt beach and its kitchens [35]. The result is a main dataset comprising qualitative information in the form of interviews, which was further completed in the second phase. The second phase was completed in September 2022 and focused on deepening the information obtained in the pilot survey by conducting a main survey. A further description of this method is found in Section 2.2.1.

The next analyses are considered secondary due to their being dependent on the information retrieved by the interview phase. These are the "location analysis of the salt kitchens" and "Salt Kitchen measurements". Location analysis is based on validating the information given by the interviewees regarding their knowledge of the salt kitchens. Due to the inexistence of these data, it required asking and then physically going to the spaces that were mentioned. After this, an inventory of their situation and usage of the terrain was conducted to update the current information of the architectural elements. This was supported by the usage of GPS tools to "map" their locations and eventually compare the data to the physical urban thread and distance between them. This helped not only understand the position of each element, but also to identify if there exists a pattern or a relation with the salt beach.

"Salt Kitchen measurements" refers to the measurement and construction analysis conducted on one of the remaining salt kitchens found in the area. Due to not knowing the specific location of any of the remaining salt kitchens (or their existence), this part of the research also depended on the previous two analyses conducted. This analysis focused on determining the general measurements of all the structural and architectural elements that comprise the salt kitchen. While information regarding these elements roughly exists in previous research [30], it is not concise and does not present the required architectural documents to be used by people interested in the salt kitchens. In contrast, this analysis seeks to provide the measurements and visual information of each element found in the space while also analyzing its function. This is further explained in Sections 2.2.2, 2.2.3 and 3.1.

After defining the analysis to be conducted, the methodology was developed as a step-by-step process, dividing the overall research into two steps: the first analysis phase and the second analysis phase. The first analysis phase consists of the literature research and the interview stage. The literature research comprises the research of all available data on the subject on the basis of previous research in the area. All this has been discussed in Sections 1 and 2. Meanwhile, the interview analysis comprises a pilot interview and a subsequent main interview. The resulting information was catalogued and analyzed, which helped move to the second phase. The secondary phase comprises the location analysis and salt kitchen analysis. This is further explained in Sections 2.1.2, 2.2.2 and 2.2.3.

2.1. Materials Descriptions

2.1.1. Interview

As previously mentioned, the same sampling group that was part of the previous research will be used in this one. Yet, a different dataset extracted from the result will be used in this part of the research with additional data obtained from the secondary analyses conducted in the space. The original sampling conducted in this study is based on 17 people with a relationship with the black salt industry. They are divided into those who have maintained or had a direct relationship. This group is divided by gender, resulting in seven men and ten women. Due to the initial lack of information on the subject, it was decided to use the snowball approach, which helped increase the group of correspondents in the main interview. The snowball approach comprises asking a starting sample group for more people with knowledge of the subject of the research. This "snowballs" into more and more people while retaining the same level of relationship to the research question [36]. This results in two surveys (one conducted in the first phase, or the "pilot survey", and a second one in the second phase, or the "main survey"), with a noticeable difference in size. The pilot survey resulted in four correspondents. Meanwhile, the main survey correspondents increased in number up to 11, shown in Table 2. Figure 16 shows the total division of gender between both surveys conducted. The name for the subjects is the "Main survey subject" (MSS).

Table 2. Main survey sample subjects.

No.	Ref. Name	Age	Relationship with Salt	Current Job
1	MSS-A	75 years	It is carried out	Currently in the industry, diversifies his income by farming in the wet season.
2	MSS-B	72–75 years	It is carried out	Diversifies the income with local candy named "alpinique"
3	MSS-C	75 years	It is carried out	-
4	MSS-D	46 years old	Father performed	Bakes cakes and bread thanks to the oven his father built.
5	MSS-E	49 years old	Father performed	Makes ice cream and works with her husband in metal plating and jewelry.
6	MSS-F	71 years old	Mother made	No information
7	MSS-G	74 years old	Worked on it	Lives with her husband. She did not share the information
8	MSS-H	48 years old	Worked on it	No information
9	MSS-I	50–55 years	Worked on it	Works in a taco shop near the central plaza.
10	MSS-J	57 years old	Worked on it	Along with other products, they are small merchants/resellers.
11	MSS-K	84 years old	It is marketed	Along with other products, they are small traders/retailers.

Table 2. *Cont.*

No.	Ref. Name	Age	Relationship with Salt	Current Job
12	MSS-L	78 years old	The marketer/family realized	Along with other products, they are small traders/retailers.
13	MSS-M	85 years old	It was marketed	Switched to sugar candy because it is easier to sell.
14	MSS-N	71 years old	I grew up with her	Not shared

Figure 16. (**a**) Pie diagram of the total percentage of respondents on the second survey. (**b**) Pie diagram of the relationship of the respondents with the black salt industry on the second survey.

2.1.2. Salt Kitchen Measurements and Area Location

Analyzing all the data obtained from past research, it was concluded that information related to salt kitchens and their architecture has been extremely scarce. The only data found by Monaghan and Reina are relatively general and function as a description of the area rather than an architectural analysis. It was decided to obtain, catalog, and digitize the architectural information of the salt kitchens, inside and outside of them. The process involved the researcher, his assistant, and the owner of the salt kitchen. Obtaining the information comprised mostly a site visit to the MSS-A salt kitchen, where the measurements of the external and internal elements were identified. It was initially expected to analyze the three kitchens that are still being used. Yet, due to the privacy of rural Mayan communities and the researcher being an outsider, only one of the salt makers (MSS-A) agreed to this. Though according to MSS-A, the construction and space elements are the same in every one of the kitchens. This analysis was conducted with a research assistant, and took place during the wet season; thus, there was no inconvenience in disturbing the daily salt cooking of MSS-A.

After understanding the general aspects of the architecture, the research was expanded to the surrounding areas to find their whereabouts. Based on the information previously obtained in the interviews, the information collected in both the interviews was used as a basis to find the ruins of the salt kitchens. A subsequent field survey of the area near the salt beach was conducted. The objective was to understand the terrain and find if there was any relation between their distribution on the salt beach, between each salt kitchen, and the current usage of the spaces where salt kitchens were built.

2.2. Methodology

2.2.1. Interview

The first pilot survey was conducted over one month due to the location and availability of the people involved. The goal of this survey was to understand the actual situation of the beach, how many people are still working in the black salt industry, and what kind of information can be obtained from the interviewees. As mentioned before, a snowball approach was initially used to help achieve these goals, which helped further deepen the understanding of the black salt beach in the main survey. In total, it resulted in 4 h of conversation. There were several challenges in gathering information due to the area's location and respondents' age. To overcome this issue, questions were reinterpreted in the second survey and a temporary stay was conducted in the town. The questionnaire used is presented in Appendix B.

During the pilot survey, the information obtained was analyzed line by line using microanalysis and separated using the methodology employed by Kambaru [37]. For this purpose, all the answers obtained in the interviews were transcribed verbatim. Then, each of the responses was analyzed to find similar patterns. This resulted in "information groups" that present the same meanings that relate to similar concepts which are depicted in Table 3. These groups will be referred as "information clusters" and the aim of this method is to reach "theoretical saturation". This means that once enough information is obtained from any of these clusters, the concept is considered to have reached "theoretical saturation" and no more information is needed to be gathered. The "Theoretical saturation" is a limit that has to be decided by the researcher, though. In the case of the pilot survey, there is no prior knowledge of which information clusters will be made. The only information gathered was from the literature research available at the moment, and even then, it has been mentioned that the amount of information on black salt is still scarce on topics outside the salt industry. Therefore, saturation was considered when the information gathered was similar to the previous literature research. As such, only the "Process" information cluster was too well researched that new information was null. In total, 10 information clusters were identified, ranging from the relation of the black salt industry with the interviewee, to the influence of black salt beyond the town of Sacapulas (such as clusters "Religion" and "Trade"). An example of the microanalysis tool can be found in Appendix C.

Table 3. Pre-survey information clusters.

No.	Concept	Definition	No. of Iterations Pre-Survey	Theoretical Saturation?
1	Relation	What is the relationship between salt and the people of Sacapulas?	11	NO
2	Product	Any product related to black salt (such as black salt, salt water, and hot springs.)	5	NO
3	Future	What is the future you envision for black salt?	4	NO
4	Usage	The use of black salt as a product	7	YES
5	Religion	Religion-related aspects performed	8	NO
6	Architecture	Everything related to architecture related to production, or people working in black salt.	1	NO
7	Terrain	Related information on the spatial area used for the creation of the black salt	3	NO
8	Authorities	Information about the authorities' perspective	4	NO
9	Process	Information related to the process of creation of black salt	Theoretical saturation	YES
10	Trade	Price information and sales locations	1	NO

During the main survey, the results were analyzed in the same way as in the pilot survey. In total, the main dataset comprised 171 new entries divided into 13 information clusters, and 3 of the new information clusters were identified by recurring mentions in the interviews conducted. Their importance was also noted to the relevancy of the investigation effort conducted in the area. Thanks to the pilot survey, it was understood what information was necessary to deepen and obtain about the different clusters previously identified. For this, the main survey consisted of pinpoint questions that sought to deepen the understanding and successfully reach a logical conclusion or perspective. This is the "theoretical saturation" that was considered. After the microanalysis conducted on the information, 12 of the 13 reached "theoretical saturation". The only one that could not reach this point was "authorities", which is a cluster focused on the intervention of local authorities to manage the salt beach. At the point of developing this research, communication with the authorities has been unsuccessful. The summary of the new information is found in Table 4. Due to the nature and objectives of this research, only the relevant information clusters will be used in the analysis and discussion. As mentioned before, the nature of this research paper focuses on a specific topic matter; thus, only a part of the dataset is relevant and will be employed. The first research paper made about the subject used "Terrain", "Usage", "Religion", and "Ownership" clusters in the research process. This leaves 9 information clusters that could be triangulated with new information about different aspects of the subject matter. In this paper, it was decided to use the information recapitulated in the "Architecture," "Process", "Problem," and "Diversification" clusters.

Table 4. Main survey information clusters.

No.	Concept	Definition	No. of Iterations Pre-Survey	No. of Iterations Primary Survey	Theoretical Saturation?
1	Relation	What is the relationship between salt and the people of Sacapulas?	11	31	YES
2	Product	Any product related to the black salt process (such as white salt, black salt, salt water, and thermal water).	6	12	YES, extended with triangulation
3	Future	What is the future you envision for black salt?	4	6	YES, interviewees mentioned the same desire
4	Usage	The use of black salt as a product	Theoretical saturation	-	YES
5	Religion	Religion-related aspects performed	8	20	YES
6	Architecture	Everything related to architecture related to production, or people working in black salt.	1	12	YES, extended with triangulation
7	Terrain	Related information on the spatial area used for the creation of black salt	3	32	YES, extended with triangulation
8	Authorities	Information about the authorities' perspective	4	1	Authorities could not be contacted
9	Process	Information related to the process of creation of black salt	Theoretical saturation	-	YES
10	Trade	Price information and sales locations	1	17	YES
11	Problem	Any problems encountered when making black salt	0	15	YES
12	Ownership	Current land ownership by salt producers or outsiders	0	20	YES, extended with triangulation
13	Diversification	Changes in economic income due to salt	0	5	YES

The objective of this research paper is to fully understand the situation of the salt kitchens and its relation to the salt beach. Therefore, the cluster "Architecture" was selected due to it containing all the relevant information about the salt kitchens not previously mentioned by other research. At this point, the information gathered on this cluster mainly comprises the known location of the active salt beaches and the known location of previous salt kitchens (currently destroyed salt kitchens). "Process" was also selected due to the analysis of the overall function of the kitchen. This helped understand how the architecture works towards the end goal of salt cooking, and a further analysis was made with the secondary analysis conducted. "Problem" and "Diversification" are clusters that present both the issues that the salt industry is facing, and what the response by salt maker families to this phenomenon is. Specially, how it affected the cultural expressions that this research is interested on. All of this information was eventually triangulated with the information from the secondary analysis performed in the area, giving a bigger and more complete outlook and analysis.

2.2.2. Architecture Measurements and Planimetry

In this research, we chose to analyze in detail the measurement of the black salt kitchens. This type of work has yet to be conducted in detail before, the mention of it by Reyna and Monaghan being the only available data on the subject. This has resulted in a lack of critical information of certain aspects about the process: What is the main tool used for salt cooking, what are its dimensions, and what is its overall function? Specifically, the information obtained in this analysis comes from the viewpoint of an architect. Therefore, the information obtained corresponds to the following: construction elements used, architectural elements, and spatial measurements inside and outside the architectural elements. After that, the data were entered into the Revit 2023 architectural modeling program, where the architecture's views, plans, sections, and general details were created. The usage of this program was due to its convenience and easiness to complete the task at hand, and because it presents the results not only in 2D (planimetry), but also in 3D (visualization of the overall building). The results can be found in Appendix A with a detailed analysis of its features and spatial use in Sections 3.1 and 3.1.1.

2.2.3. On-Site Architecture Localization

Analyzing the location of each salt kitchen, two essential things could be determined. First, there is no record of the locations of any salt kitchens; only the knowledge of the former salt makers provides a general idea of their location. Second, a record of the salt kitchens currently in use is needed. Since this information has not yet been developed, it is not easy to understand the current state of the black salt industry. By obtaining the location of salt kitchens and the spaces where they existed, it is possible to understand their relationship to the salt beach, which would help define the shape of the beach before the floods buried the majority of it. Two sets of information were used at this stage:

- The collective memory of the people related to black salt. This has been obtained in the interviews and analyzed previously. The results were several areas in which salt kitchens had been built but are now destroyed.
- Field analysis of the spaces to determine if there are residues of previous constructions. The excavation of a rectangular space determines the construction of a salt kitchen. This excavation is not backfilled and can be easily identified unless the space is reused for growing crops. An example of such a space is found in Figure 17.

The result of the field visit was the location of fifteen areas where salt kitchens were built and three existing ones in use. Due to privacy concerns, some of the space owners did not share information other than the location of the salt kitchen. Meanwhile, other locations were found run over with vegetation. A summary of these data is presented in Table 5. By knowing the location, it was then pinned in a GPS and mapped in Google Maps, resulting in a semblance of an orderly distribution across the area. With these results, it can be much better understood if there is a relationship between the space, the kitchens, and the salt beach. The locations of all the elements found is represented in Figure 18 while a discussion is conducted in Section 3.2.

Figure 17. Area where a salt kitchen was built, currently destroyed. Author (2022).

Table 5. Summary of salt kitchens' situation and distance between each of them.

No.	Situation	Current Situation	Distance between the Closest Salt Kitchen
1	Destroyed	Idle	Point 0
2	Destroyed	Idle	58.72 m (1–2)
3	Destroyed	Planting	20.03 m (2–3)
4	Destroyed	Idle	11.98 m (3–4)
5	Destroyed	Planting	48.44 m (4–5)
6	Existing	-	14.54 m (5–6)
7	Destroyed	Construction	26.40 m (6–7)
8	Destroyed	Construction	26.75 m (7–8)
9	Existing	-	30.25 m (8–9)
10	Destroyed	Idle	10.03 m (9–10)
11	Destroyed	Idle	9.98 m (10–11)
12	Existing	-	10.10 mts (11–12)
13	Destroyed	Idle, private land	22.14 mts (12–13)
14	Destroyed	Idle, private land	13.75 m (13–14)
15	Destroyed	Cattle breeding (?)	49.75 m (14–15)
16	Destroyed	Idle	41.86 m (15–16)
17	Destroyed	Idle	17.27 mts (16–17)
18	Destroyed	Idle	8.43 m (17–18)
19	Destroyed	Idle	8.72 m (18–19)

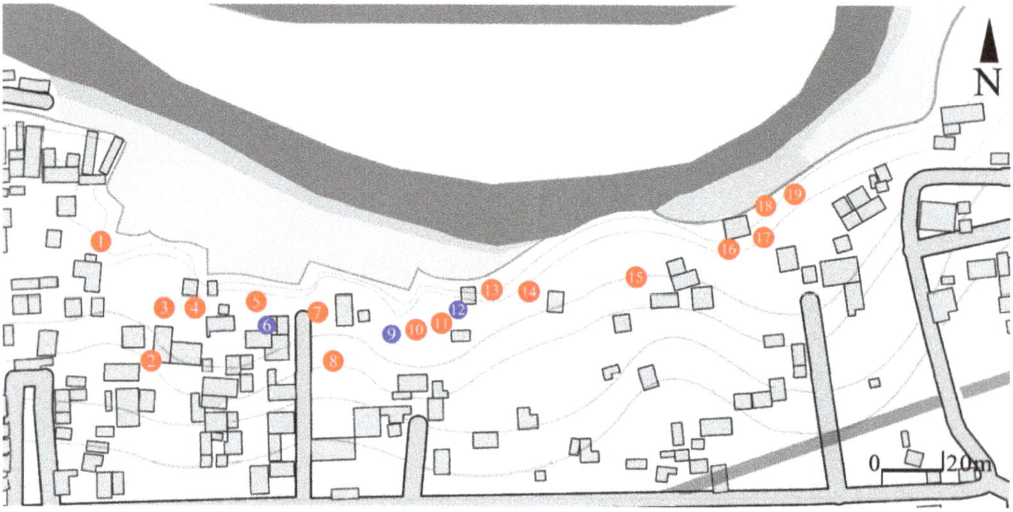

Figure 18. Locations of the salt kitchens. In blue, the existing ones, in red, the empty spaces of previous kitchens. The numbers depict the salt kitchens mentioned in Table 5.

There is some difference in the number of salt kitchens found compared to the information obtained in the existing literature and the interviews conducted with the people involved. According to MSS-A, the number of salt kitchens in 1970 was around 44. However, it was mentioned that, due to the earthquake in 1976, most of the salt kitchens were destroyed due to their structural composition. Of the mentioned salt kitchens, at this moment it is hard to know their location due to the loss of the collective memory. However, the 19 found present a particular distribution and relationship with the beach space, which clarifies how they interacted with the beach to some extent. Figures 19–21 depict the last salt kitchens standing.

Figure 19. Salt kitchen No. 6. Author (2022).

Figure 20. Salt kitchen No. 9. Author (2022).

Figure 21. Salt kitchen No. 12. Author (2022).

3. Results

3.1. Architecture Features of the Salt Kitchen

Spanish chroniclers mentioned the architecture of the salt kitchen. The information coming from *"Titulo de los señores de Sacapulas"* (Title of the Sacapulas Lords) determined that the pre-hispanic name of Sacapulas comes from the name "**Tuja**" or "**Tujalá**." As mentioned before, Van Akkereen mentions the possibility that tuja refers to the low houses where people worked the black salt. This supposes that this architectural element has existed since before colonial times. However, the kitchen analysis presents both colonial elements and construction systems. This is defined as masonry walls: stone and mortar of semi-regular shape with a gable roof of wood structure. On the roof, a Spanish-type clay tile is used. The selection of these materials assumes the original function of the kitchen was maintained. However, its construction has been modernized at least once. This is because both the tile roof and the walls show no similarities with the pre-colonial Maya construction types; the most typical system of Mayan construction of houses was based on load-bearing walls and vegetation cover. However, this type of Mayan construction is the hardest to obtain information on, according to research conducted by Gaspar Muñoz Cosme and Cristina Vidal Lorenzo [38].

Upon analyzing the disposition of each element, it is evident that the arrangement of the materials used on each element is irregular. The walls of the kitchen, for example, are made of stone and mortar. However, the stones are selected for size rather than shape. This results in a certain degree of irregularity, reducing the structure's sturdiness, which translates to a high probability of collapse. MSS-A mentions that several of the buildings were affected due to the 1976 earthquake, demonstrating the construction's fragility. However, there is some sort of arrangement of the stones on the front wall, specifically near the entrance door. Meanwhile, in the exterior, stones delimit the area of the outside patio. These stones do not present regularity and are only selected for their size. Within this perimeter, the space serves as a storage area for the equipment used in the salt cooking. It is also the space where the distillation basin is found. The arrangement of the basin does not follow a rule among the three kitchens, and it is assumed that the location is decided simply by the proximity to the salt kitchen. All the kitchens analyzed have in their outside patio the *"tinajas"* used to carry the thermal waters to the basin, and the *"ocote"* (pine) firewood used in the salt burning. Around the kitchen, there are also Spanish clay tiles used to exchange used tiles deteriorated from the heat inside the kitchen as well as the climate of the region. Figures 22 and 23 depicts the clay tiles location.

Figure 22. MSS-2 salt kitchen; the wall made from stucco and stones can be seen. It shows a rudimentary construction process without any internal structure. Author (2022).

Figure 23. The Spanish roof tiles replacements can be seen next to the kitchen due to space constraints. Author (2022).

The salt kitchen analyzed (kitchen No. 6) has a semi-regular roof structure. This means that although it presents a typical wooden ceiling, its structure does not present regular spacing between elements. Specifically, the roof is composed of three beams and one ridge beam. Two of the three beams are circular sections, while the remaining one is a square section. The reason for this variation has yet to be discovered, but it can be assumed that it is due to the use of whatever material was available to the owner. The ridge beam rests on a large rock with a wedge in the upper area to fit the structure. All beams extend about 0.55 m outward on the edges. Because wood tends to rot due to external factors, the extended areas are covered by Spanish tiles. Following the beams are the rafters and battens. The rafters' layout is regular, with a spacing of 0.50 m between each end. The battens do not have a regular distribution, varying their spacing distance between rafters. However, it was possible to define between 5 and 6 battens arranged between the rafter spacing. The last layer of the roof is the Spanish tile. The tiles are 0.46 m by 0.28 m, and have no unique characteristics. When MSS-A was asked about the number of tiles found around the kitchen, he mentioned that due to the constant deterioration, a large sum was needed at all times. According to him, it is common to replace them every two or three years. Figure 24 depicts the wood frame, while Figure 25 depicts the front of the kitchen and location of the "*tinajas*".

Figure 24. Salt kitchen extended roof structure, covered by Spanish tiles. The wear can be seen due to the black spots from both moisture and constant cooking inside the kitchen. Author (2022).

Figure 25. Front of the kitchen where the "tinajas" can be found, on the left side.

3.1.1. Interior of the Space

According to the interviews conducted and the literature research, the salt kitchen serves primarily as an encapsulation of space for cooking salt. This still holds true up to this time, as the process of salt making has not changed at all. The salt kitchen interior has no finishes on the walls or floor, which is made of earth. In fact, the interior of the space is relatively simple and has no significant qualities, being an enclosed space during the salt cooking process due to the temperature difference between inside and outside, which can "crack" the mold used to make the black salt. The result is an area that can maintain the high temperatures of the flame constantly. There are openings between the walls and the wooden beams, previously mentioned as a storage area for various materials used in the process, which can further be seen in Appendix A. This is the only area where fresh air can enter or leave the space indirectly. As mentioned in Section 1.4.7, the heart of the fire is inside the space, which must be kept in flames the entire time the black salt is made. There is no information about the temperature needed to cook the salt, which assumes this knowledge is empirical. However, the interviewees mentioned that wood from "*ocote*" (pine) is needed because it "burns better," strengthening the hypothesis that this knowledge is not scientific.

Inside, the "heart of the fire" is created: a rudimentary oven of about 2.5 m in diameter, according to the research conducted. It comprises oval stones called "tenamastes" that raise the salt molds. Previously, the use of 40 of them was mentioned. However, the result of the research was that nowadays, 15–18 tenamastes are used due to the reduction in salt production. This results in a much smaller "heart of fire" than reported by Reyna and Monaghan during their research. The size of the heart seems to be governed by its functionality and convenience to the salt maker. During the observed process, it was identified that eight salt molds are made per burn. The function of the salt maker within the space is to maintain this flame at a specific temperature, introducing "*cargas*" (loads) of firewood into the fire during a process that lasts about 6 to 8 h.

There are some discrepancies between what was mentioned in the interview and the overall function of the salt kitchen. Specifically, it was determined by the interviews that the space is used mostly a warehouse saving soil impregnated with salt. According to MSS-A, this soil will be used throughout the whole year, even the wet season when it is impossible to impregnate the soil with salt. This in turn determines how economically flexible one salt maker would be during the whole year, as more soil equals more potential economic benefit. In fact, the salt cooking process presents a rudimentary way of cooking, effectively transforming the space into a stove. This process is something that both salt makers and ex-salt makers mentioned as a cause of people's disinterest in continuing with this tradition. As it turns out, the subject cannot open the space until the end of the cooking, a process in which several noxious gases are generated.

3.2. *Relationship between Architectural Elements and Space*

Beyond the number of kitchens, essential characteristics of how the architecture interacts with the space were found. During the field visit, it was possible to define three significant areas represented in Figure 19: A (current salt beach), B (intermediate plain), and C (land near the "pit"). General pictures of the space are found in Section 2 and Figures 16–18. Area A is currently used, Area B is a plain used for cattle, and Area C is abandoned. Two areas (A and C) are connected by main roads (pink). The three areas are also connected by a makeshift road created by the people of Sacapulas (yellow). This road also connects all the spaces of the salt kitchens. The salt beach is connected by two roads, which are unsafe due to neglect. However, they are the only means of access. Finally, areas A and C have roads that do not connect to any place (green color).

At first glance, it could be hypothesized that the distribution of the salt kitchens is a function of the proximity to the salt beach, as it was impossible to determine specific factors of their construction other than the proximity to the area. Two interviewees mentioned that several kitchens existed close to each other because they belonged to the same family. In this case, kitchens 2–4 and 6 belonged to the same lineage of salt makers. It is unknown if this also happened in other salt kitchens or if there is some relationship resulting in the construction of the architecture. Figure **??** shows the current location of the salt kitchens and roads that connect to the different spaces. Currently, only groups 1–12 are close to the beach. Groups 13–15 are closer to Area B, and groups 16–19 are closer to Area C. Specifically, the roads that connect to Area A (blue color) are relatively close to the existing salt kitchens. Although several salt kitchens were destroyed due to external factors, others were destroyed due to a lack of people interested in salt. While several of the spaces have been reused, many of them are abandoned. Figures 27–29 show the actual situation of the spaces.

As resumed in Table 5 of all the salt kitchen areas, only No. 3, 5, and 15 were reused for planting (16%). No. 7 and 8 were reused for new construction (10.5%). No. 7 was used for constructing a private house, while No. 8 was reused for constructing a public, synthetic grass soccer field. This new use is probably due to its location in front of a main street compared to the other kitchen spaces.

If it is assumed that the salt beach has always been Area A and its surroundings, questions arise about its accessibility. Considering kitchens 13–19 had access to the beach, it would be challenging for a salt maker to travel from A to the salt kitchens as it demands too much energy. Therefore, the location of these kitchens is not logical if it relates to Area A. In addition, Area B is a plain, and has a direct connection to the Chixoy River; this is important because it does not present fences like other terrains next to the area: this space must not have a current owner. Its topology supposes it was an access road to the area in front of the Chixoy River. This is supported by the informal roads found in both A and C. Area C especially presents a connection to the town of Sacapulas, utilizing a main road and two informal roads to the north and the west. The northern road connects to the outskirts of the town of Sacapulas. Meanwhile, the road to the west only connects to somewhere currently buried by the river. A previous connection to some existing area in front of B must have existed. However, this area is currently inaccessible due to the Chixoy River.

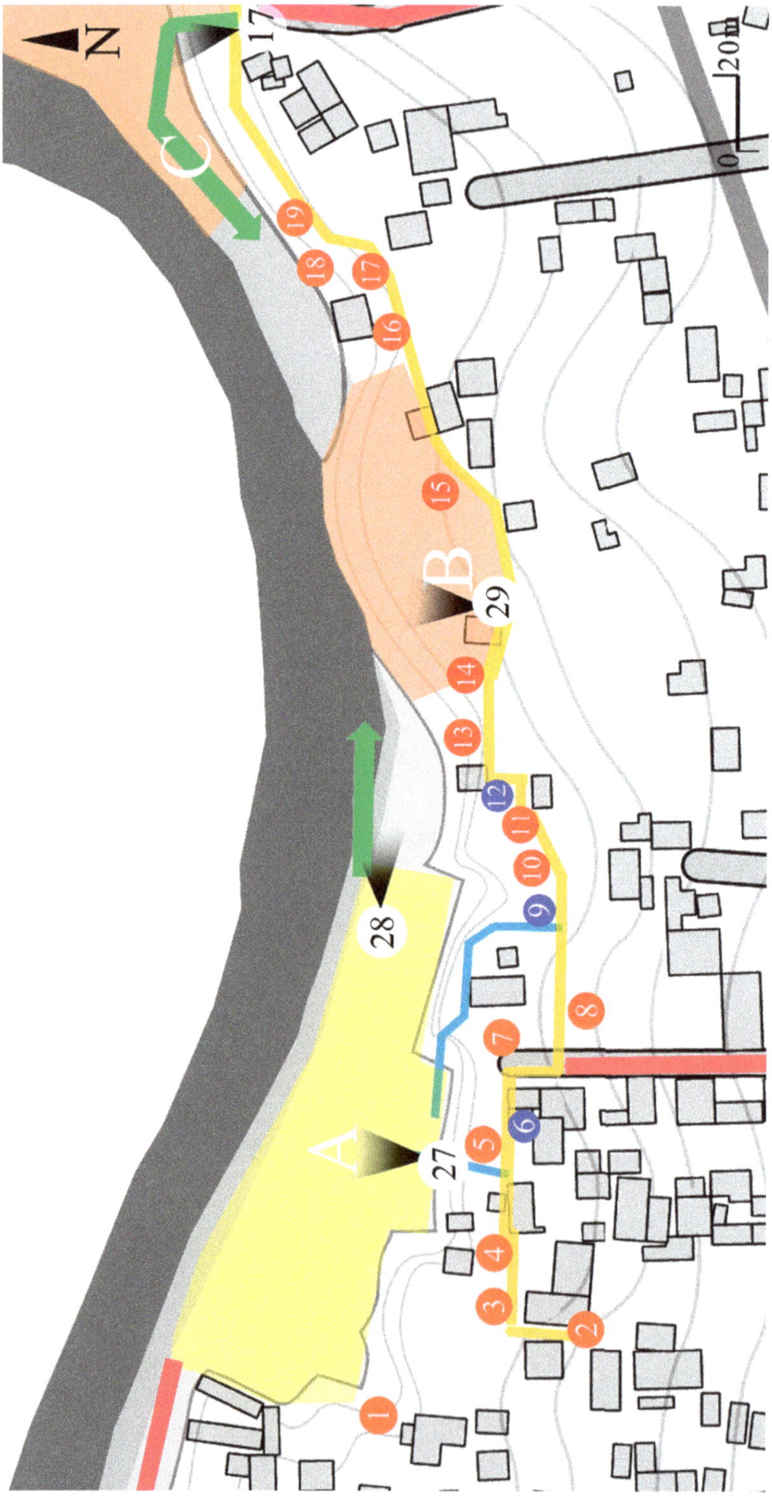

Figure 26. Area analysis. In red, destroyed kitchens. In blue, existing kitchens. In white, location of the pictures taken (Figures 27 and 28).

Figure 27. Pictures from Area A and its current state. This is the actual "black salt" beach, as it is known. Author (2022).

(**a**) (**b**)

Figure 28. Picture (**a**) represents the road going across Area B. Picture (**b**) presents the current state of Area B. Author (2022).

Figure 29. Area C. Author (2022).

4. Discussion and Conclusions

4.1. Salt Kitchens, Their Function, and Current Issue

Continuous natural disasters and lack of interest have impacted the traditional Mayan knowledge of black salt. This research successfully created the first planimetry of the traditional salt kitchen, safekeeping its knowledge for further research and understanding. It also analyzed its relationship towards the space and its current usage. Salt kitchens are a rudimentary way of creating salt. This was confirmed by two crucial factors: the research conducted by Andrews [26], and how the salt kitchen functions. In general terms, the kitchen shows evidence of being fit for local people rather than outsiders. The salt kitchen presents measurements that only seem logical if we consider the height of a person is around 1.60 m. Considering that the height of people in Guatemala is around 1.60 m [39], it does not follow an international standard, rather a local one. This suggests a much greater importance beyond the black salt industry, and being connected to the people and culture of the town.

During the analysis, it was found that the function of the kitchen is mostly as a space for storing earth impregnated with salt used during the off season, representing an annual economic income. Without presenting a specific aesthetic, the space is a highly functional area, as the bigger the internal space, the more money salt makers could get during the year. Considering the price of maintaining the structure and salt cooking materials, salt farmers are encouraged to resort to better incomes. Evidence of this is that many ex-salt makers preferred to demolish the kitchen and turn it into a crop-growing space. By destroying the salt kitchen, the possibility of creating salt is also destroyed. It also breaks the generational tradition that this specific family has been performing for years. Although the loss of interest in black salt has been a progressive event due to both human and natural factors, it is the 1976 earthquake that was the determining factor in the loss of the industry. According to MSS-A, most of the salt kitchens were destroyed in this event due to the construction system used in them. Due to the lack of internal structure, any natural catastrophe could destroy the kitchen, earthquakes being cited as a common occurrence. Considering the regularity of these natural events and their magnitude, it is self-evident that the kitchen slowly deteriorates to a point where structural fractures develop.

The reason why the structural system was not modernized up to this point was unknown to the interviewees. However, it can be hypothesized in terms of two reasons:

- Age of the constructions: When interviewing the salt makers, it was evident that the exact date of construction of the salt kitchens is unknown, as they are inherited among family members. Taking into account this, together with the construction system used (adobe masonry construction), and the lack of mention by Spaniards in their chronicles, it is hypothesized that the date of the construction was at least during the colonial era (1524 onwards).
- No new kitchens built: Due to the loss of interest in the industry and personal economy of the salt makers, there are no data on new salt kitchens built. The industry depends on the three existing kitchens to subsist, resulting in an extremely delicate sustainability and resilience towards any natural or man-made catastrophe.

Creating black salt also harms the salt maker, and is a process yet to be modernized. Instead, the same tradition has been maintained since before the conquest. The salt makers understand that the process is highly body-consuming, but there has not been any action or interest in improving it. When asked for their opinion, all of the salt makers mentioned that it is the "*costumbre*" (social custom) that dictates how salt is made, regardless of its negative impact. It can be hypothesized that the lack of development resulted in an industry that is not very competitive in the market and a cultural aspect at risk of complete loss.

4.2. Location and Relationship with the Black Salt Beach

Considering that there are no records of the salt kitchens and planimetry of the space, it can be hypothesized that the original shape of the salt beach was different from the current situation [30], to such an extent that the beach area was much more prominent to the north than it is today. The location of kitchens 13–19 is more logical if one assumes an area in front of B, connecting Area A to C. The chronicler Martin Alfonso Tovilla makes a mention regarding the size of the black salt beach in his recount [29]. Hypothesizing that he meant that the total area was around 0.687 kilom, by measuring the land from Area A to the east, this results in an extension slightly in front of Area C, towards the mouth of the Chixoy River. If we take a hypothesis of how the original river was, an affluence on the river flow would mean a progressive crooking of the river towards the black salt beach. This is due to the area being at an extremely low height and being primed for modifications by the river. Due to the shape of the current meander located west of Area A, the increase in the river current could result in slowly burying Area B. The hypothetical original shape of the beach is presented by Figure 18. There are still new questions regarding this hypothesis:

- What was the size of the area in 1979, when Ruben E. Reina and John Monaghan conducted their research in the area? Although they took photos, only one presents the black salt beach, and the shape of it is barely visible. In their research, there is mention that the floods in the 1940s and early 1950s buried most of the beaches, while only one part was unburied. It also mentions that the salt makers informed them that only one-fifth of the original space was being used. However, they do not mention where the other four parts of the space were located.
- What was the size of the area in 2002, when Ordoñez Chicano conducted his research in the area? Although he took a photo, his research never delved into the exact measurements of the salt beach. There is no mention of the size during that year or before the floods. Regarding floods, he mentions one in 1949 that buried half of the present salt flats. Like Reina and Monaghan, there are no documents that can determine the total size of the area.

In order to obtain information regarding the buried beach, it is necessary to analyze the composition of the currently buried soil. However, this is beyond the scope of the research carried out.

The results show a close relationship between the architecture distribution and the beach space. If we take the original area of the hypothesis, the relationship between the urban fabric (which does not present urban planning), the shape of the beach, and the black salt industry is evident. Proof of this is how the salt kitchens were distributed in space, with the urban fabric arranged with the need for access to the kitchens. However, all of the roads connecting the salt kitchens are informal, both in shape and materials used (earthen, defined specifically by function rather than aesthetic or safeness). This implies lesser interest from the Sacapulas government, resulting in little investment towards the access roads. The only part of the road that uses concrete is the steps that connect the southern area where the salt kitchens are located, with the existing space of the black salt beach [24]. However, evidence shows that the measurements of these elements are slightly irregular and are not in an optimal state. Considering that the people dedicated to black salt are at an advanced age (between 50 and older), this poses an extremely high risk to their safety. Moreover, transporting the black salt to the kitchen proves a difficult endeavor for someone of this age.

The location of the kitchens has helped them be sheltered from any flood or natural disaster occurred during the years, as they are located in elevated terrain. However, the loss of salt harvesting areas is the main reason the salt industry shrank considerably, affecting their usage. In addition, the difficulties of preparing black salt have alienated new generations. The interviews showed that many people prefer to repurpose the kitchen spaces into crop-growing areas, as it is much more profitable and demonstrates an alienation from the Mayan customs rooted in the black salt. Today, many kitchen spaces are abandoned or within large plots of land belonging to the children of the original salt maker families. However, the urban thread and salt kitchen locations present a history where this was different which is depicted in Figure 30. An era where the people, the salt workers, and the architecture were important aspects of their identity.

Figure 30. The hypothesis of the original black salt beach. Blue: Chixoy River, Green: salt beaches hypothesis. Red defines the destroyed salt kitchens, in blue, the existing ones.

5. Conclusions

By taking into account the research conducted since Andrews and the latest researchers, it is understood that the local salt industry is effectively dying. As discussed in Section 1, Andrews mentioned certain causes that will eventually lead to the loss of the black salt. This and the previous research conducted on the subject have confirmed it to be true, and also identified new reasons that have accelerated this loss. As a consequence of the industry dying, cultural expressions such as the architecture have also been affected. Black salt is a product with close ties to the people of Sacapulas. This has been evidenced by how the black salt beach has influenced the architecture, town, and urban fabric. In addition, Maya texts and chronicles present an era when the native people of Sacapulas celebrated their roots and black salt work in religious demonstrations [31], evidence of the importance of the space. Today, this is different; the celebrations are no longer held and a part of the culture has effectively died. However, a specific group of people still practice this profession. By developing this research, a documented account of the architecture has been successfully completed. Now, several new steps must be taken to properly understand to what extent the salt beach space and industry can be recovered. How we can adapt such traditional methods to make them economically feasible, or at least interesting enough for the people, is an important point to discuss. Plenty of research has yet to be conducted in the area, which could yield a better understanding of the current intangible value it presents to the people of Sacapulas. By developing the current research, it was found that people outside of the salt industry still use the beach, even knowing that the space presents an actual danger. A first step should be to understand what the usage and frequency of the salt beach are, and what activities are currently conducted in the space. This research will yield the necessary information to decide what the best direction to recover the space is, and the feasibility of such action. Information about the usage outside of the salt industry does not exist yet, and no research conducted up to this point has ever been interested in activities other than the salt creation. There is an important need due to such historical lack of information, and prompt action is required.

After understanding its current usage, future research may focus on two aspects: the black salt product, and the recovering of the salt beach. As discussed previously, one of the most attractive reasons for the black salt is the medicinal properties attributed to it, which have yet to be proved by research. This goes beyond the expertise of this researcher, yet the subject might prove attractive to researchers interested in Mayan culture who desire its safeguarding. Meanwhile, this research has hypothesized the original shape of the black salt beach due to the relation between known architecture elements, urban thread, and the current black salt beach. This should be further researched as research and locals suggest that more salt beaches existed beyond the ones found. It is also noted that soil analysis should be conducted around the hypothesized area to understand if, in effect, the flooded area presents similar properties as the current salt beach. Beyond this, the most important decision is the call for action. The benefits of such work go beyond saving a culturally important product; it impacts the livelihood of people living in constant danger. All the research conducted in the area has mentioned this, and yet there has been a holistic approach towards danger reduction. By managing, planning, and executing, the region of Sacapulas can capitalize on the natural resources that the salt beach has given to them for a long time. And by saving part of the black salt industry, an opportunity for cultural celebration will be possible for the Guatemalan people. A proper plan to safeguard in any way the culture of the area while also helping the community can be conducted in many ways, yet the economic feasibility of said action will shape decision making. Implementing KIPS for an auto-sustainable approach, recovering the "costumbre", and cataloguing and creating urban and architectural planimetry all help raise awareness and will eventually revitalize the black salt beach.

Regardless of the final outcome, decisions are being made right now. And hopefully in the future, both the town of Sacapulas and the black salt can become an attractive and important part of Guatemala, as they were long ago.

Author Contributions: Conceptualization, S.M. and L.P.Y.S.; methodology, L.P.Y.S.; software, L.P.Y.S.; validation, S.M. and R.N.; formal analysis, L.P.Y.S.; investigation, L.P.Y.S.; resources, S.M., R.N. and L.P.Y.S.; data curation, L.P.Y.S.; writing—original draft preparation, L.P.Y.S.; writing—review and editing, S.M. and R.N.; visualization, L.P.Y.S.; supervision, S.M. and R.N.; project administration, S.M. and R.N.; funding acquisition, S.M. All authors have read and agreed to the published version of the manuscript.

Funding: This work was supported by the Japan Society for the Promotion of Science Grant-in-Aid for Scientific Research (A) Grant Number 24H00341.

Institutional Review Board Statement: Ethical review and approval were waived for this study since no data and information related to the ethical guidelines were at the discretion of the committee at Hokkaido University.

Informed Consent Statement: Informed consent was obtained from all subjects involved in this study.

Data Availability Statement: Data is contained within the article.

Acknowledgments: A big acknowledgement to the non-profit organization RAICES for their help during the development of this research. A big thanks to the people of Sacapulas who helped to conduct this research.

Conflicts of Interest: The authors declare no conflicts of interest.

Appendix A. Planimetry from the Salt Kitchens

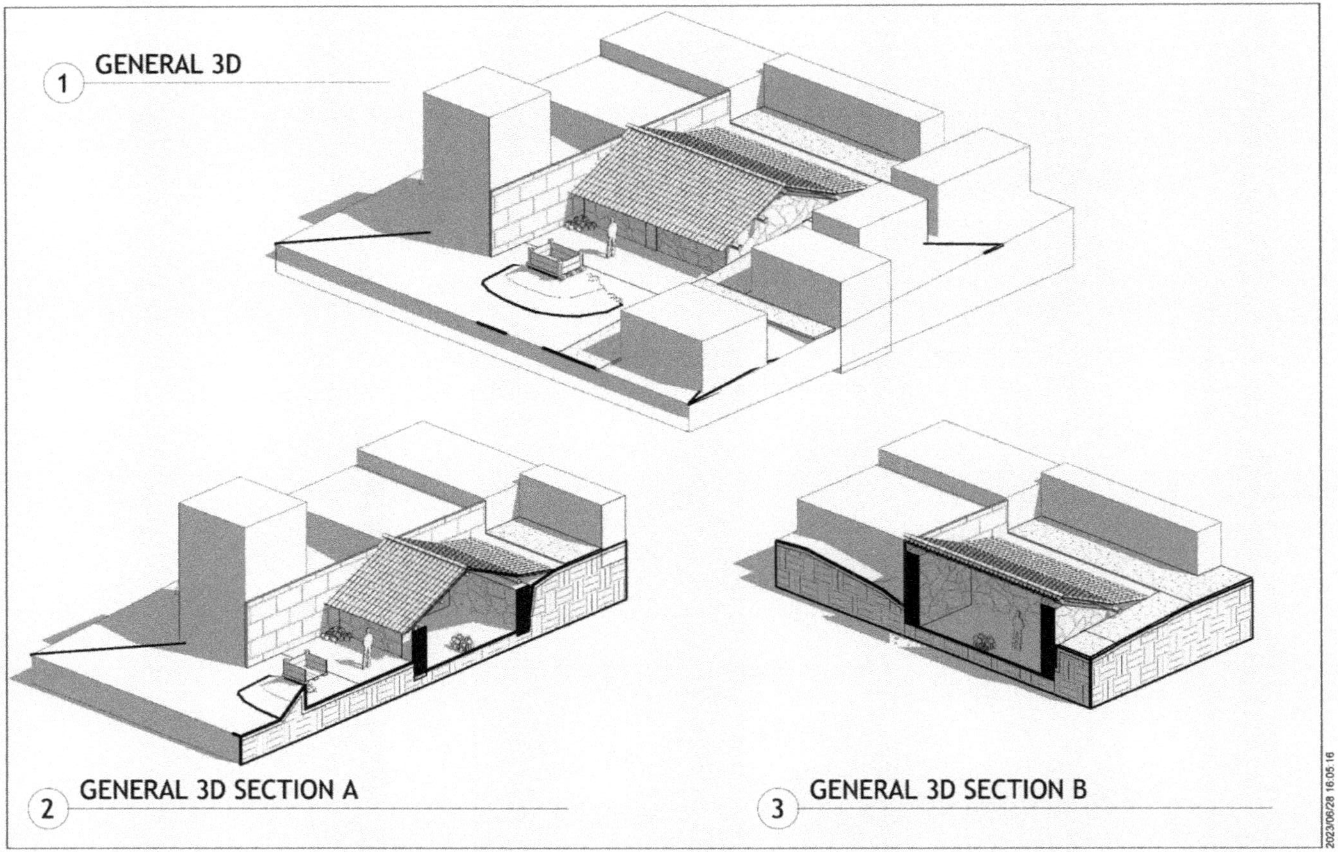

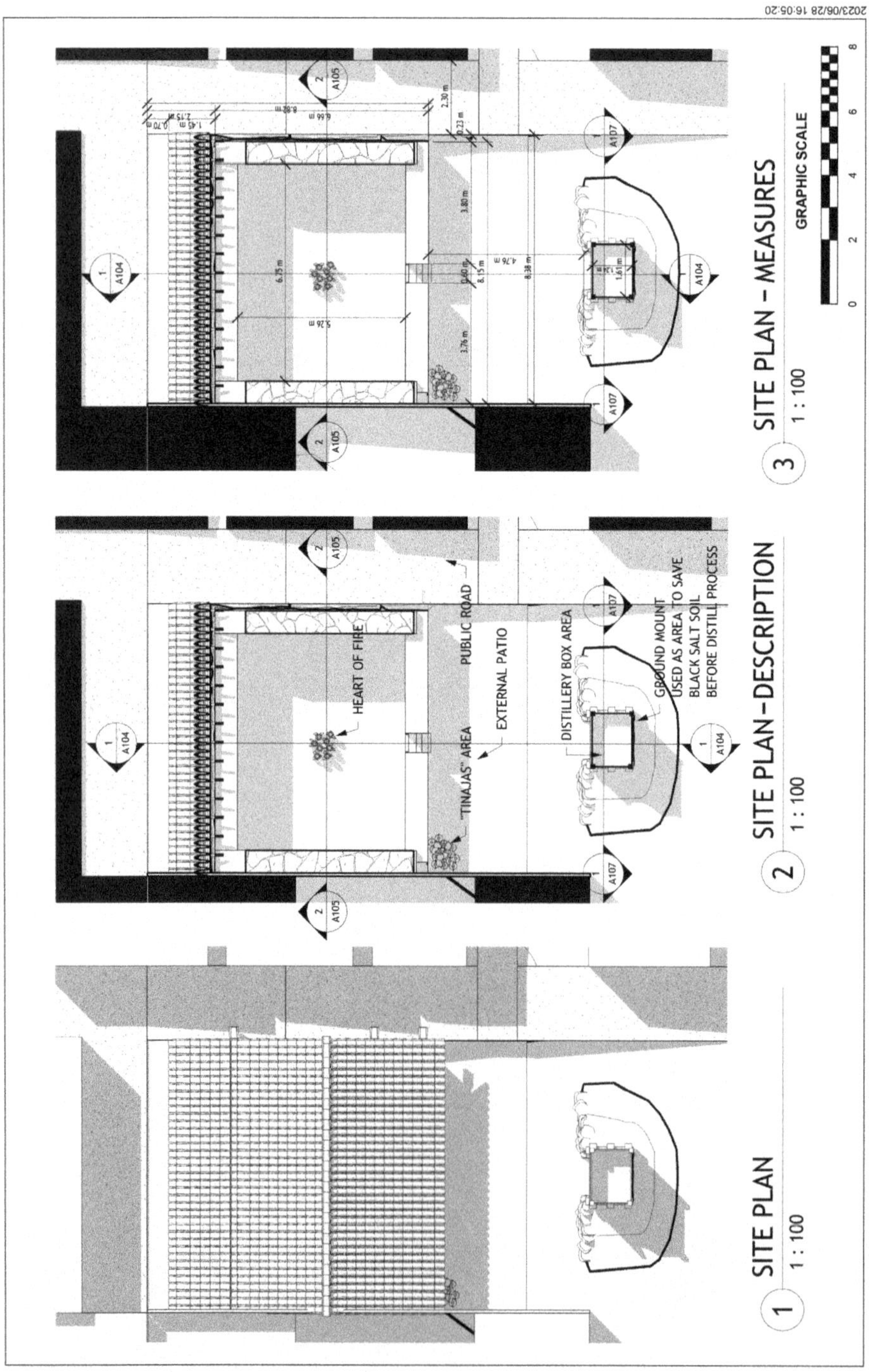
SITE PLAN
1 : 100
SITE PLAN-DESCRIPTION
1 : 100
HEART OF FIRE
"TINAJAS" AREA
EXTERNAL PATIO
PUBLIC ROAD
DISTILLERY BOX AREA
GROUND MOUNT USED AS AREA TO SAVE BLACK SALT SOIL BEFORE DISTILL PROCESS
SITE PLAN – MEASURES
1 : 100
GRAPHIC SCALE
0
2
4
6
8

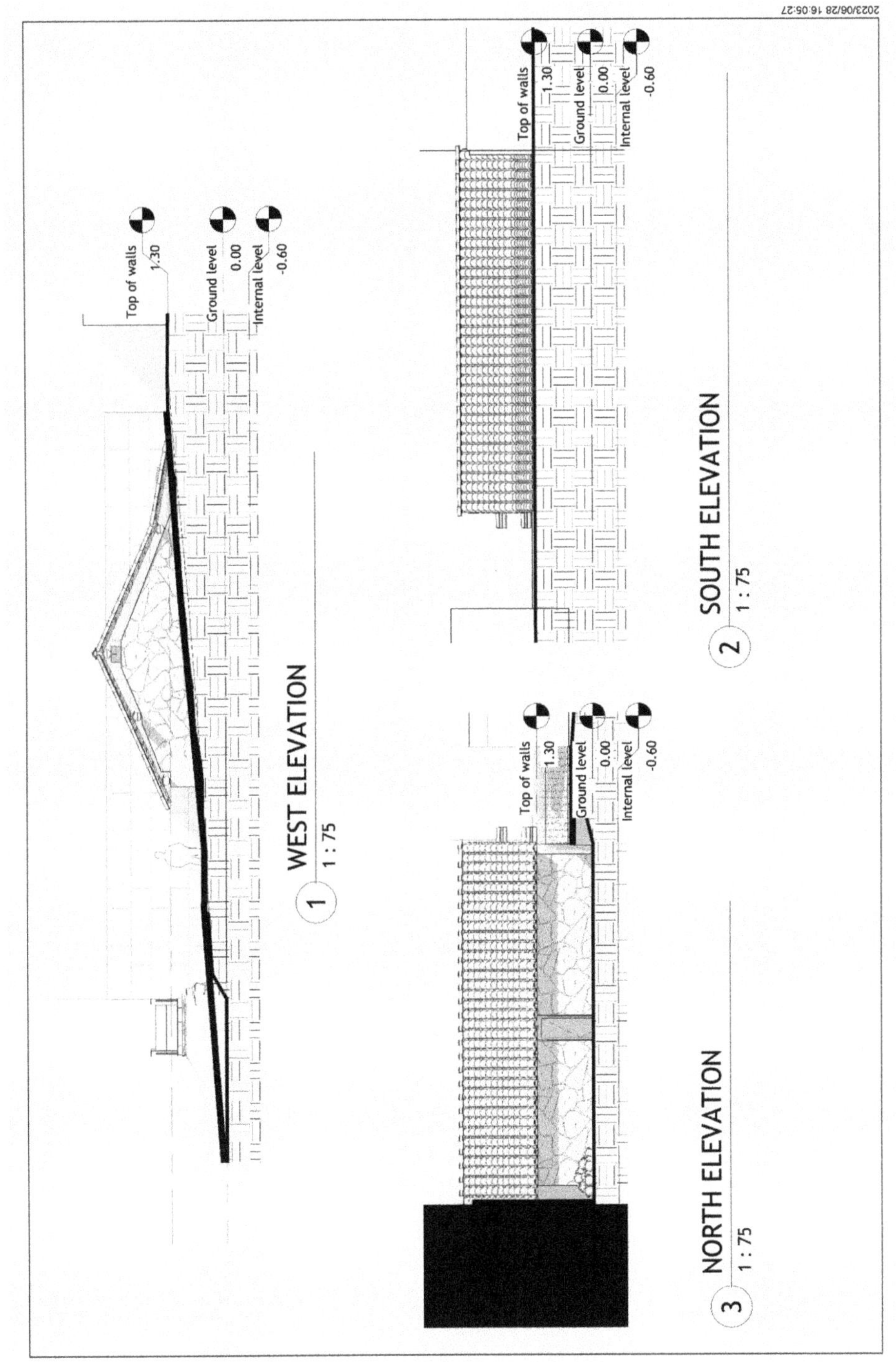
2023/06/28 16:05:27
Top of walls
1.30
Ground level
0.00
Internal level
-0.60
1
WEST ELEVATION
1 : 75
2
SOUTH ELEVATION
1 : 75
3
NORTH ELEVATION
1 : 75

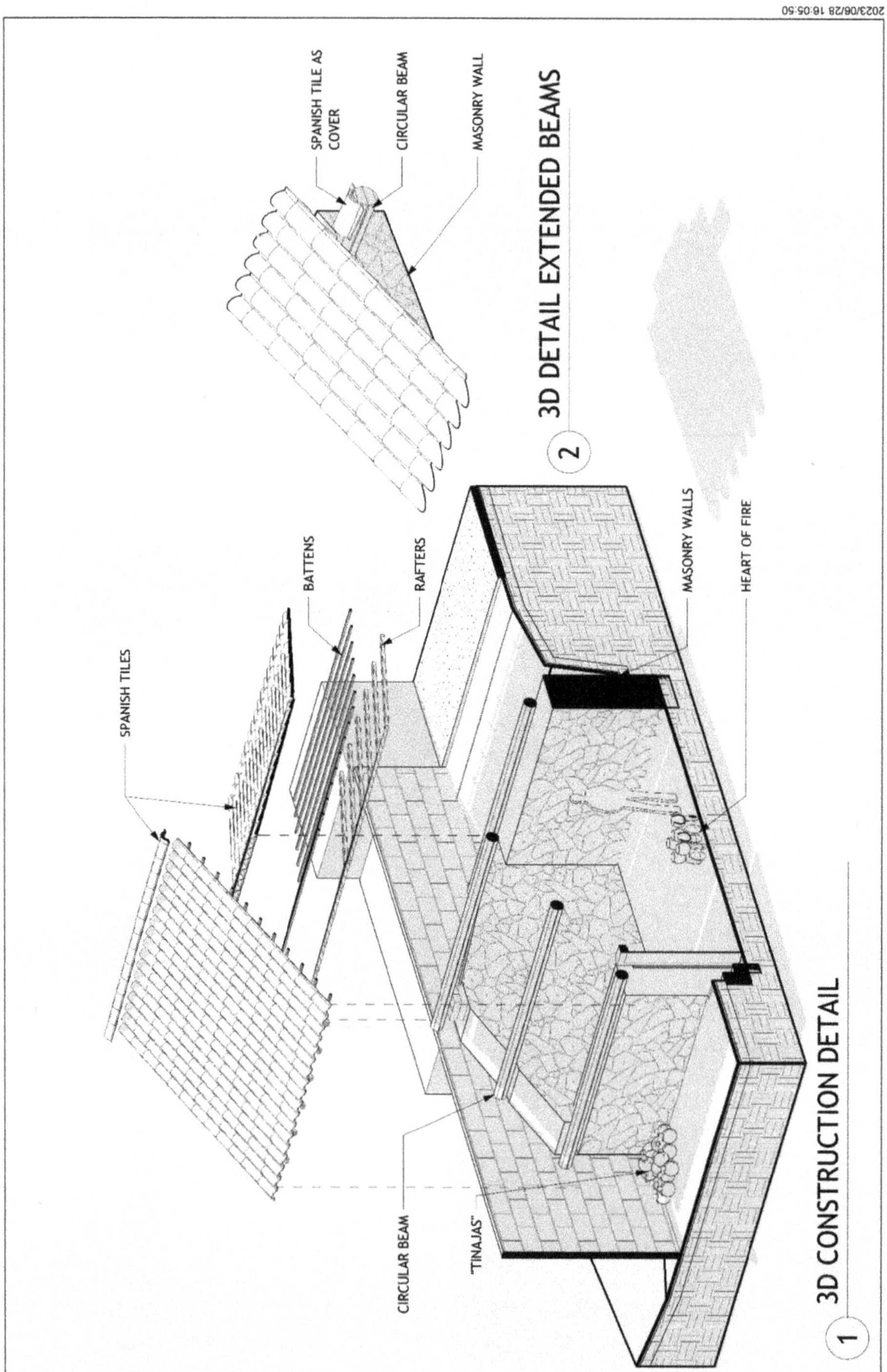
SPANISH TILES
BATTENS
RAFTERS
CIRCULAR BEAM
"TINAJAS"
MASONRY WALLS
HEART OF FIRE
1
3D CONSTRUCTION DETAIL
SPANISH TILE AS COVER
CIRCULAR BEAM
MASONRY WALL
2
3D DETAIL EXTENDED BEAMS

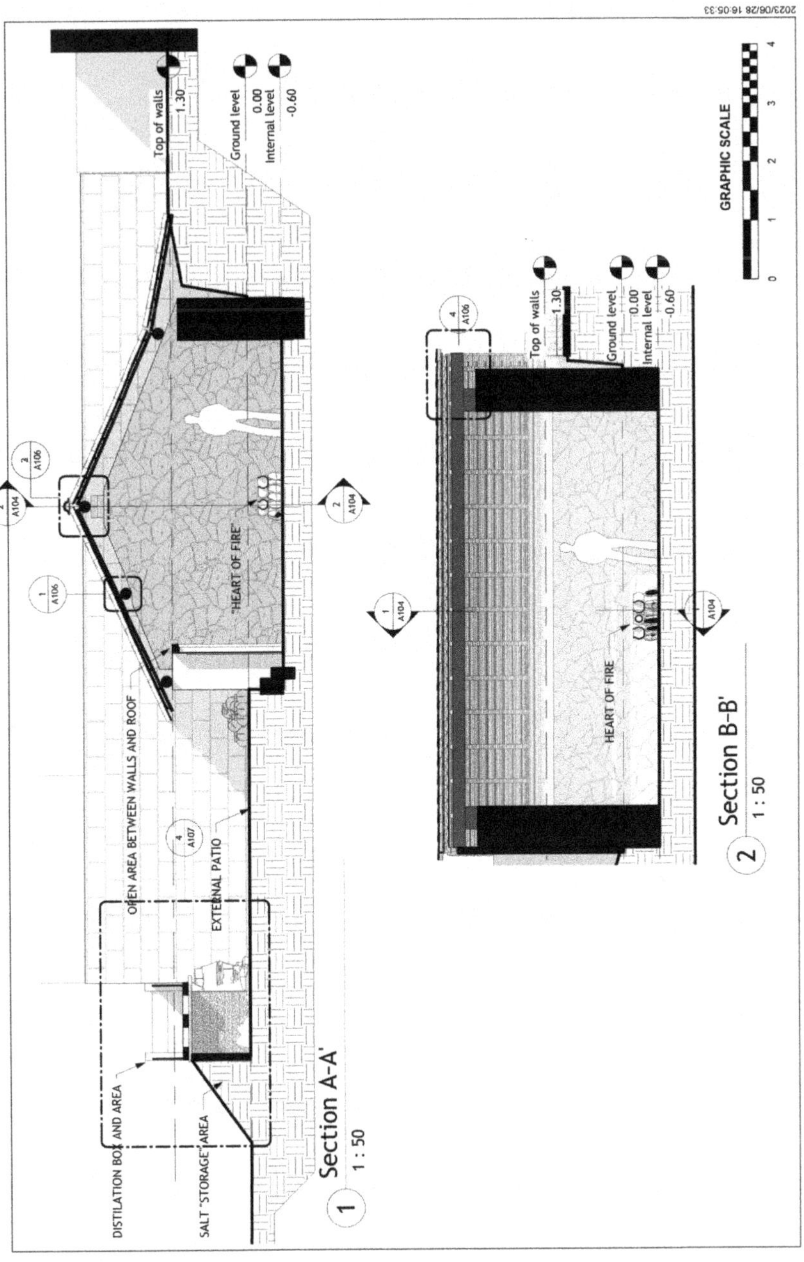

2023/06/28 16:05:33
Top of walls
1.30
Ground level
0.00
Internal level
-0.60
GRAPHIC SCALE
0
1
2
3
4
"HEART OF FIRE"
OPEN AREA BETWEEN WALLS AND ROOF
EXTERNAL PATIO
DISTILATION BOX AND AREA
SALT "STORAGE" AREA
Section A-A'
1 : 50
Section B-B'
1 : 50
HEART OF FIRE

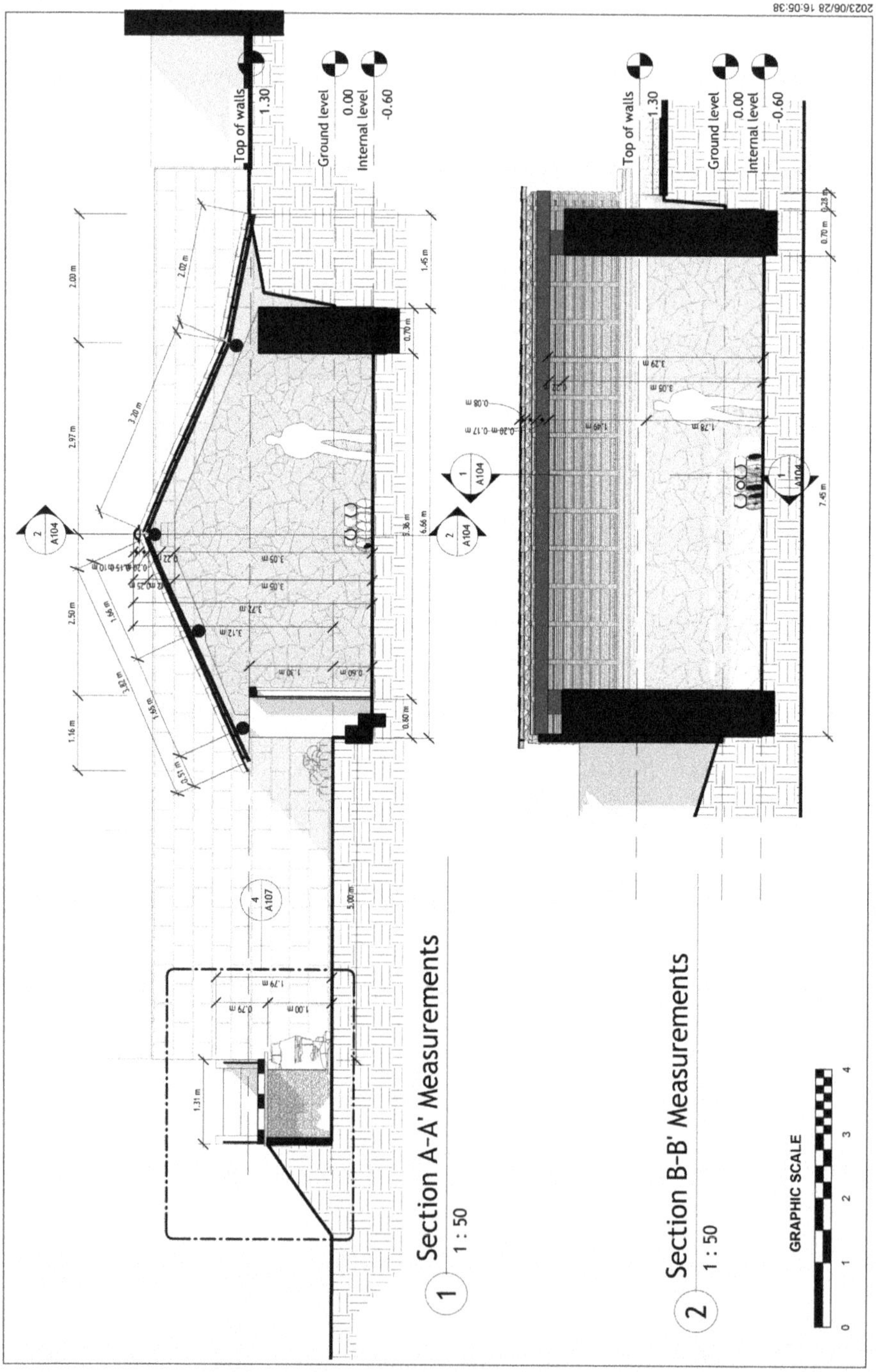
Section A-A' Measurements
1 : 50
Section B-B' Measurements
1 : 50
Top of walls
1.30
Ground level
0.00
Internal level
-0.60
GRAPHIC SCALE
2023/06/28 16:05:38

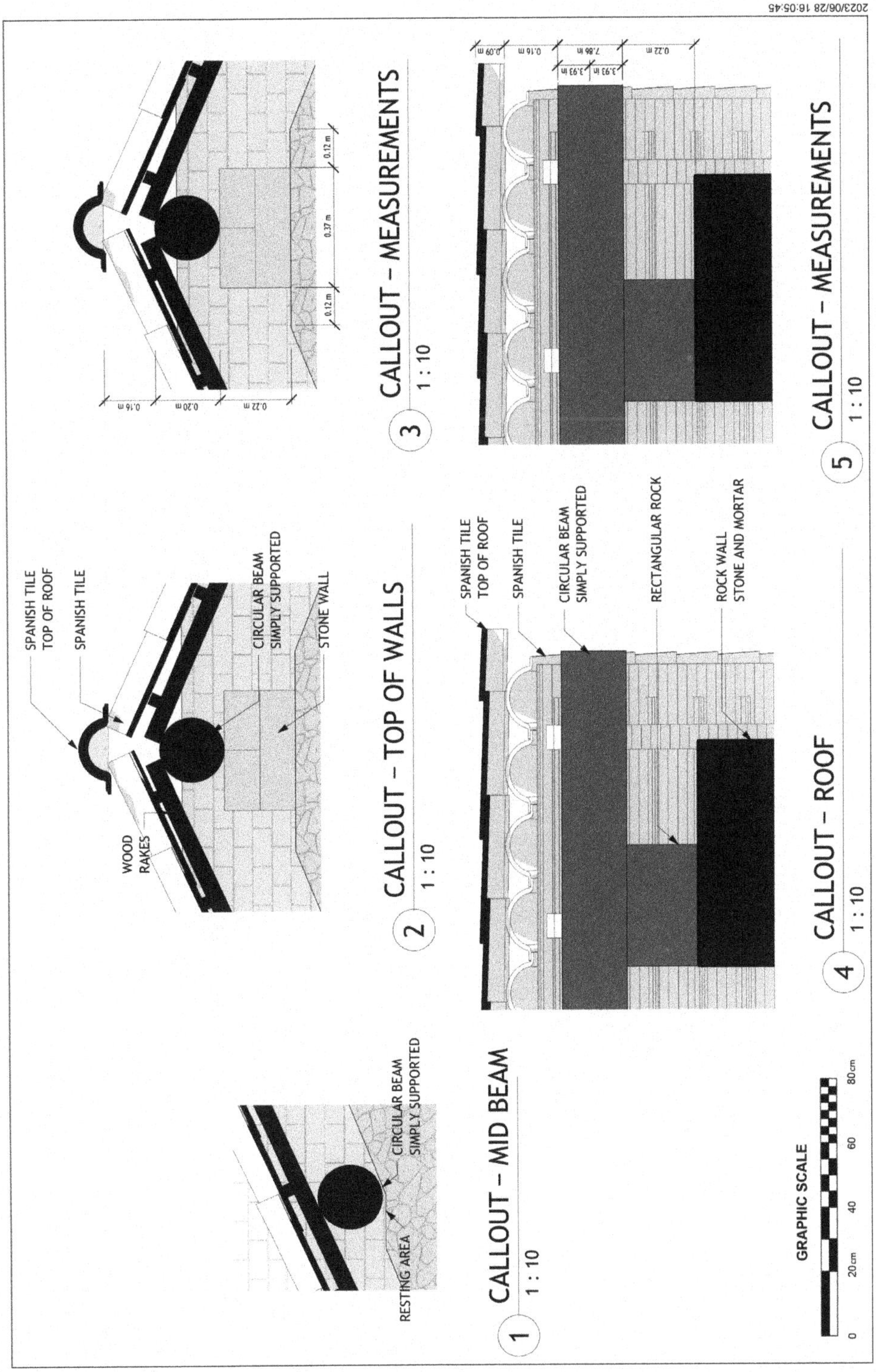
2023/06/28 16:05:45
RESTING AREA
CIRCULAR BEAM SIMPLY SUPPORTED
1 CALLOUT – MID BEAM 1 : 10
GRAPHIC SCALE
0 20cm 40 60 80cm
WOOD RAKES
SPANISH TILE TOP OF ROOF
SPANISH TILE
CIRCULAR BEAM SIMPLY SUPPORTED
STONE WALL
2 CALLOUT – TOP OF WALLS 1 : 10
SPANISH TILE TOP OF ROOF
SPANISH TILE
CIRCULAR BEAM SIMPLY SUPPORTED
RECTANGULAR ROCK
ROCK WALL STONE AND MORTAR
4 CALLOUT – ROOF 1 : 10
0.16 m
0.20 m
0.22 m
0.12 m
0.37 m
0.12 m
3 CALLOUT – MEASUREMENTS 1 : 10
0.09 m
0.16 m
7.86 in
3.93 in
3.93 in
0.22 m
5 CALLOUT – MEASUREMENTS 1 : 10

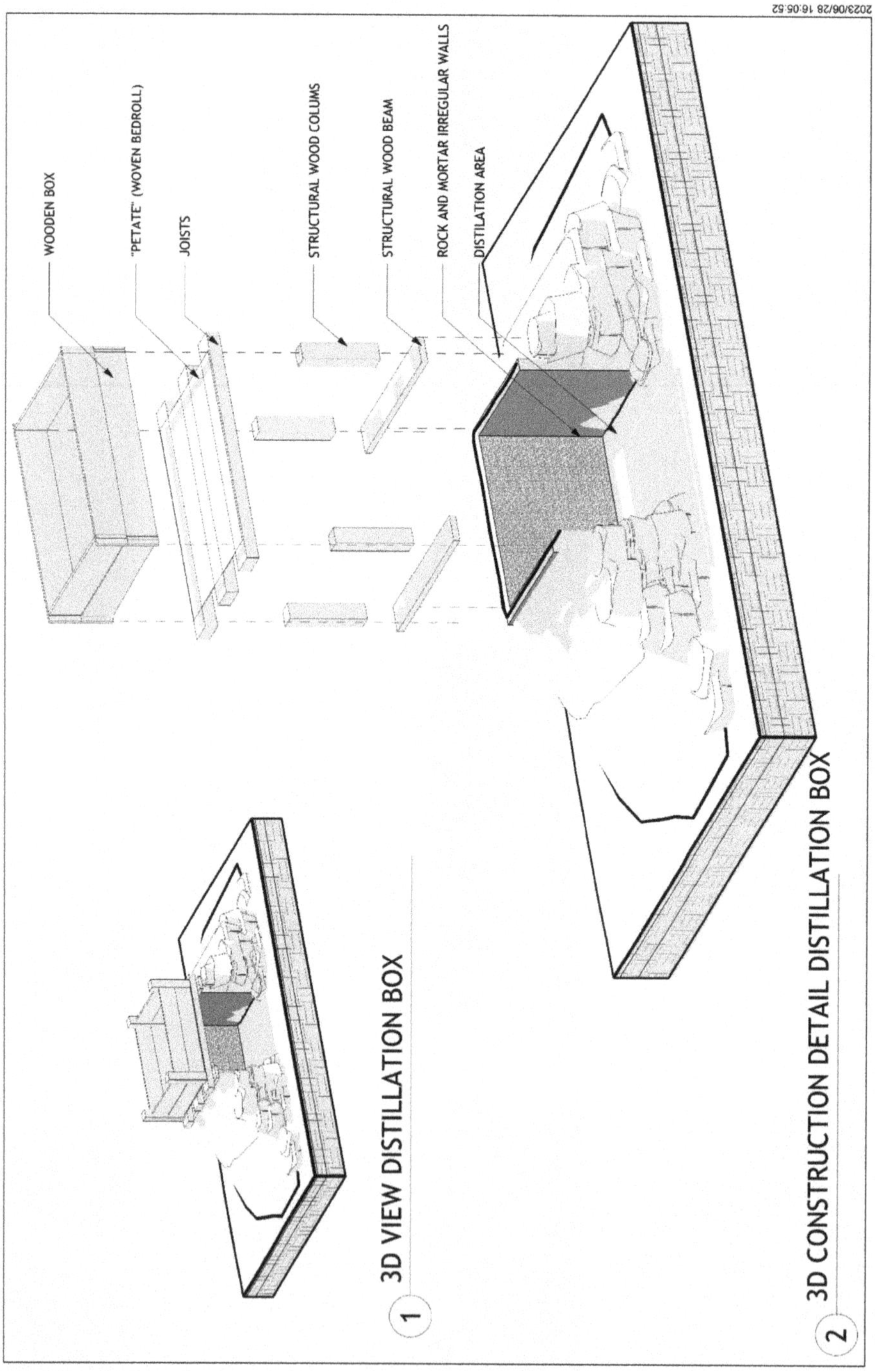
2023/06/28 16:05:52
WOODEN BOX
"PETATE" (WOVEN BEDROLL)
JOISTS
STRUCTURAL WOOD COLUMS
STRUCTURAL WOOD BEAM
ROCK AND MORTAR IRREGULAR WALLS
DISTILATION AREA
1
3D VIEW DISTILLATION BOX
2
3D CONSTRUCTION DETAIL DISTILLATION BOX

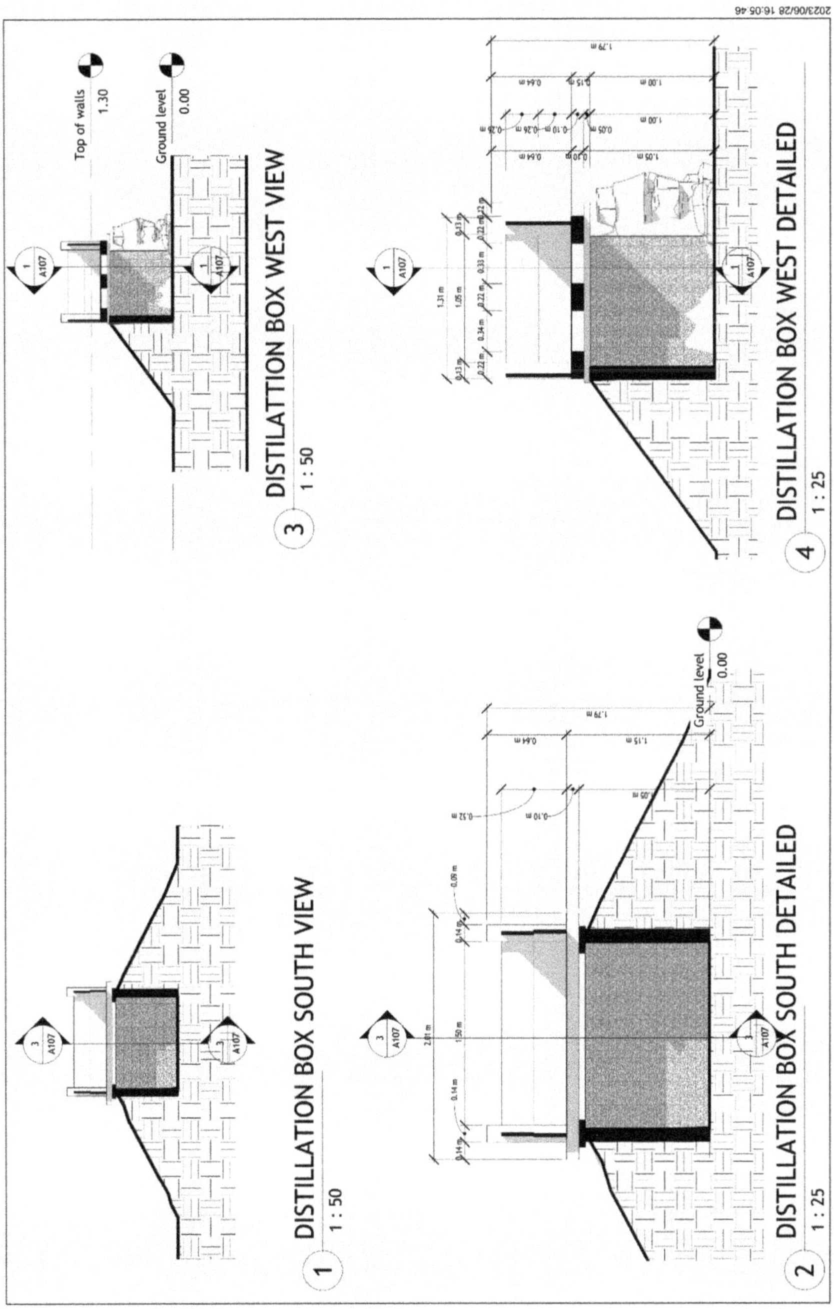
DISTILLATION BOX SOUTH VIEW
1 : 50
DISTILLATION BOX SOUTH DETAILED
1 : 25
DISTILATTION BOX WEST VIEW
1 : 50
DISTILLATION BOX WEST DETAILED
1 : 25
Top of walls
1.30
Ground level
0.00
A107
2023/06/28 16:05:46

Appendix B. Questionnaire Employed in the On-Site Survey

Black Salt in Guatemala

The objective of the interview: to understand the current situation of black salt, the use of the space around the salt beaches, people's perception of this space, and the future of this industry.

Therefore, the following questions are given, which are necessary to obtain the necessary data from the pre-survey. They don't necessarily have to be the specific question, but related enough to get the desired answer.

For the interviewer, read the following statement to the interviewee:

Dear Mr./Mrs. (enter name of interviewee) Nice to greet you, my name is *(interviewer name, present identification document)*, and I come on behalf of Mr. Luis Pablo Yo Secaida. I understand that according to what has been discussed with Mr. Yo Secaida, I would be carrying out the agreed interview with him. **I will proceed to read you the following clarification:** The following interview is conducted in order to obtain information for strictly academic purposes. This information will be processed in order to obtain quantitative and qualitative statistical data, and will serve as the basis for other interviews and surveys that will delve into the influence and impact of the black salt industry in the Department of Quiché. In addition, you would be requested to record this interview, if possible with video and audio, in order to transfer it to Mr. Luis Pablo Yo Secaida, so that he can carry out the corresponding analysis. If you agree with this, I will turn on the recorder right now. *(Interviewer starts recording)* . *(Continuous reading, which must be recorded on video and audio / audio)* By agreeing to answer this interview, you authorize Mr. Luis Pablo Yon Secaida to use the answers and information for the preparation of his postgraduate thesis, as well as inputs for the formulation of a methodological guide for other interviews and surveys. At no time during the other interviews and surveys will your personal information or responses be revealed. However, your name may be mentioned in the thesis to be carried out. At the end of the thesis, a copy of it will be sent to you. Would you agree?

interview Ballot

Section 1	**Personal data and housing**

To the interviewer:

In this part of the interview, the main objective is to know a little about his life and the space in which he grew up. We want to understand how it changed, and if this influenced their curreent life. It is extremely necessary that the subject draws the spaces where he lived, as well as the space outside and around. Could it be that the activity of the black salt influenced his day-to-day development? If possible, it would be good if he drew a ***sketch****, or, description of the entire space. Each of the houses in which he has lived must be described, along with a sketch if possible*

1 **Where exactly were you born?**
(Department, municipality, name of the community)

2 **How many people reside in your household?**
(Enter amount)

3 **Who resides in your household?**
(Write down the roles, understand marital status, children, grandparents, in-laws, grandchil

4 **What do you do?**
(Ask for each member mentioned above)

5 **Have you lived in this house all your life?**
(If you answer yes, skip to question 7)

6 **Can you describe the homes where you have previously lived?**
(Write down the amount of dwelling and its characteristics: number of rooms and what they were (example 1 kitchen, 1 bedroom, a bathroom outside the dwelling, a patio).

7 **What are the things you remember from that stage of your life?**
(This applies to events that marked your childhood WITHIN the territory of your home)

8 **Why did you decide to move out of the house?**
(Ask for each of the dwellings described in questions 5 and 6)

9 **Describe your current dwelling?**
(Here you can make a drawing (sketch) of the house. You must describe, or draw a sketch of each house that you have lived in your life)

Section 2A	Black Salt Industry
To the interviewer: This section only applies to current black salt workers. *In this part of the interview, the main objective is to learn as much as possible about black salt. Information related not so much about its realization (creation process), but about the space where it is created, the people involved, how the people perceive this product, etc... It is necessary to know exactly where the salt beaches are, and where they were (it will be given a general map so that the subject can mark the locations of the beaches)*	

10 How was black salt introduced?

11 What role did your family have in teaching you the process of creating black salt?

12 How many people previously made black salt?

13 Was it possible to live only on black salt? Because otherwise?

14 How has its price/value changed?

15 Previously, how was the interest of people towards black salt?

16 In your opinion, what had the biggest impact on the black salt industry?
(Important question, we want to know what influenced the loss of culture)

17 This graphic shows several events that were documented about black salt, is it true? Could you expand it? Is information missing?
(The graph is attached to this document)

18 Are there other influences NOT investigated, that influenced the loss of culture?

19 Has this space used (for the creation of black salt) changed over time?

20 Are you the owner of said space?
(Mark your exact location on a map)

21 How was it before, how is it now?

22 Is there a risk of losing the beach where you make the black salt?
(If you answer no, skip to question 25).

23 Is this risk constant?
(If you answer no, skip to question 25).

24 If it were lost, what would be the consequences?

25 How do you see the future of black salt?

Section 2B	Black Salt Industry
*For the interviewer: This section only applies to **former workers** of black salt.* *In this part of the interview, the main objective is to learn as much as possible about black salt. Information not so much about its realization (creation process), but about the space where it is created, the people involved, how the people perceive this product, etc... It is necessary to know exactly where the salt beaches are, and where they were (it will be given a general map so that the subject can mark the locations of the beaches)*	

10 How was black salt introduced?

11 What role did your family have in teaching you the process of creating black salt?

12 How many people previously made black salt?

13 Was it possible to live only on black salt? Because otherwise?

14 How has its price/value changed?

15 Previously, how was the interest of people towards black salt?

16 In your opinion, what had the biggest impact on the black salt industry?
(Important question, we want to know what influenced the loss of culture)

17 This graphic shows several events that were documented about black salt, is it true? Could you expand it? Is information missing?
(The graph is attached to this document)

18 Are there other influences NOT investigated, that influenced the loss of culture?

19 Are you or were you the owner of a salt beach?

(Mark your exact location on a map)

20 How did your life change when you left the black salt industry?

21 Was it financially beneficial to leave the black salt industry?

22 Have your daily activities changed in any way?

23 Do you miss working in the black salt?

24 How do you see the future of black salt?

Section 3	People related to the industry
In this part of the investigation, the objective is to be able to find more information about people who ***currentlyis making*** *the black salt. Documentation found online mentions only Don Max, however photos and small comments describe other people involved. It is also necessary to know the point of view of the people who were involved in black salt in the past. What is the reality of these people? How did your life change due to the change of profession?*	

25 Are you aware of other people who make black salt?
(Inquire if it is possible to find these people and thus carry out this survey)

26 Are you aware of other people who MADE black salt?
(Inquire if it is possible to find these people and thus carry out this survey)

27 Are these people still in Sacapulas or did they move out of town?

28 Are your salt beaches still usable, or is it too late?

29 Can we contact these people in any way?

Section 4	Black Salt Beaches
In this part of the interview, we want to know everything that exists ***around*** *of the salt beaches. Due to the loss of these spaces for various reasons (questions from the previous section), we want to know what the new use of these spaces is, or if they are just idle. It is important to catalog the type of activity around these spaces. Are they used by children as a play area? Are they used for x,y,z activity?*	

30 Can you remember the black salt beaches that were used in the past?
(Mark the location of the salt beaches on the map)

31 Of the salt beaches that you point out, which ones are no longer used?
(Mark the location of the salt beaches on the map)

32 How are these spaces used?
(Ask specifically if the town has reclaimed these spaces in the form of informal use, or if they are simply idle spaces)

33 Around the black salt beaches, do the owners live?

34 The spaces around of the black salt beaches were used for its production?
(Ask if the industry or other activity was developed in the surroundings of the beaches)

35 The spaces around of the salt beaches, are they used today?

Section 5	Religion and the Rio Negro
In this part of the interview, what we are looking for specifically is the ***Relationship between Mayan religion, ancient product and current reality*** *. Due to the introduction of the Christian religion, the Mayan religion had to be adapted in one way or another. Today, documentation presents information about a certain religious value to the river. However, this is scarce. If possible, this aspect should be investigated deeply.*	

36 Does the Rio Negro and the elaboration of salt have any religious significance?
(Ask about a specifically Mayan religious value, syncretism.)

37 Documents (report from the 70s) mention 2 nahuales that protect the river. Are you aware of
(Find your exact description of the Mayan mythology around the lake. Extend if necessary)

38 Are there annual celebrations in relation to the river?
(If YES, go to question 39. If NO, go to question 45.)

39 Please describe in as much detail as possible the celebration in question.

40 •Has this celebration changed in size and in people who celebrate it?

41 What religion professes this celebration?

42 Do people have any attachment to these religious festivals?

43 Is the whole town part of these festivities?

44 What do you think is the opinion of the people towards this celebration?

45 Is there any other type of celebration related to black salt?

Appendix C. An Example of the Microanalysis Tool Used in the Interview, Where the Time, Interviewee and Concept Are Organized

Concept	< Terrain >	
Definition	**Information**	
Iterations	PSS-A-1,2	00:28 I: At what point was all that space filled? A-1: When they came, they left the area...; the bosses didn't come anymore... to plant or something like that. And others left the place.
		24:18 I: Do you remember more or less where the salt was before? A-1: Yes, there on the beach, where that hole was (area C). There were salt pans belonging to an owner that was there. And that's where it started, where there was a mangal.
	PSS-D	00:07 D: The earth stayed and go, who is going to activate That's why we needed to study more, we needed to work more.
Theoretical notes		Those involved do not have exact knowledge of where each person's land is located. No specific dates were mentioned, so there is a gap at this time.

References

1. Sabatini, F. Culture as fourth pillar of sustainable development: Perspectives for integration, Paradigms of Action. *Eur. J. Sustain. Dev.* **2019**, *8*, 31. [CrossRef]
2. Durrant, L.J.; Vadher, A.N.; Teller, J. Disaster Risk Management and cultural heritage: The perceptions of european world heritage site managers on disaster risk management. *Int. J. Disaster Risk Reduct.* **2023**, *89*, 103625. [CrossRef]
3. Durrant, L.J.; Vadher, A.N.; Sarač, M.; Başoğlu, D.; Teller, J. Using organigraphs to map disaster risk management governance in the field of Cultural Heritage. *Sustainability* **2022**, *14*, 1002. [CrossRef]
4. Brown, M.L. Sustainable development. In *Oxford Research Encyclopedia of International Studies*; Oxford University Press: Oxford, UK, 2017. [CrossRef]
5. United Nations Office for Disaster Risk Reduction. Sendai Framework for Disaster Risk Reduction 2015–2030—PreventionWeb. 18 March 2015. Available online: https://www.preventionweb.net/files/43291_sendaiframeworkfordrren.pdf (accessed on 12 June 2023).
6. United Nations. Risks and Disasters. Available online: https://www.un-spider.org/risks-and-disasters (accessed on 12 June 2023).
7. Fatorić, S.; Seekamp, E. Are cultural heritage and resources threatened by climate change? A systematic literature review. *Clim. Chang.* **2017**, *142*, 227–254. [CrossRef]
8. Daly, P.; Ninglekhu, S.; Hollenbach, P.; McCaughey, J.W.; Lallemant, D.; Horton, B.P. Rebuilding Historic Urban Neighborhoods After disasters: Balancing disaster risk reduction and heritage conservation after the 2015 earthquakes in Nepal. *Int. J. Disaster Risk Reduct.* **2023**, *86*, 103564. [CrossRef]
9. Lawangen, A.; Roberts, J.K. Interactions between disaster risk reduction and intangible culture among indigenous communities in Benguet, Philippines. *Int. J. Disaster Risk Reduct.* **2023**, *94*, 103801. [CrossRef]
10. Vhurumuku, E.; Mokeleche, M. The nature of science and indigenous knowledge systems in South Africa, 2000–2007: A critical review of the research in Science Education. *Afr. J. Res. Math. Sci. Technol. Educ.* **2009**, *13*, 96–114. [CrossRef]
11. Rosa, A.; Santangelo, A.; Tondelli, S. Investigating the integration of cultural heritage disaster risk management into urban planning tools. The Ravenna Case Study. *Sustainability* **2021**, *13*, 872. [CrossRef]
12. Hambrecht, G.; Rockman, M. International approaches to climate change and cultural heritage. *Am. Antiq.* **2017**, *82*, 627–641. [CrossRef]
13. UNESCO. Local and Indigenous Knowledge Systems (Links). UNESCO.org. 1 January 1970. Available online: https://www.unesco.org/en/links (accessed on 12 June 2023).
14. Fernandez, M. Risk perceptions and management strategies in a post-disaster landscape of Guatemala: Social conflict as an opportunity to understand disaster. *Int. J. Disaster Risk Reduct.* **2021**, *57*, 102153. [CrossRef]
15. Lemus, V.M.G. Disaster risk governance in the midst of the COVID-19 pandemic in Central America: The case of Guatemala. Elsevier: Amsterdam, The Netherlands, 2022; Chapter 5; pp. 51–60. [CrossRef]
16. Castillo, J.G.; Hernandez, M.A. The unintended consequences of confinement: Evidence from the rural area in Guatemala. *J. Econ. Psychol.* **2023**, *95*, 102587. [CrossRef]

17. Congreso de Guatemala. Decreto Legislativo ley 109-96. Decreto Legislativo 109-96. 7 November 1996. Available online: https://www2.congreso.gob.pe/sicr/cendocbib/con2_uibd.nsf/52B034338E20D4570525770C0060034B/$FILE/Decreto_109.pdf (accessed on 12 June 2023).
18. United Nations Office for Disaster Risk. Yokohama Strategy and Plan of Action for a Safer World: Guidelines for Natural Disaster Prevention, Preparedness and Mitigation. PreventionWeb. 4 February 2009. Available online: https://www.preventionweb.net/publication/yokohama-strategy-and-plan-action-safer-world-guidelines-natural-disaster-prevention# (accessed on 12 June 2023).
19. U.S. Agency for International Development. Guatemala: Humanitarian Assistance. 31 May 2023. Available online: https://www.usaid.gov/humanitarian-assistance/guatemala (accessed on 12 June 2023).
20. Me-Bar, Y.; Valdez, F. On the vulnerability of the ancient maya society to natural threats. *J. Archaeol. Sci.* **2005**, *32*, 813–825. [CrossRef]
21. Sioui, M.P. Drought in the Yucatan: Maya perspectives on tradition, change, and adaptation. *Drought Chall.* **2019**, *2*, 67–75. [CrossRef]
22. Turner, B.L.; Sabloff, J.A. Classic period collapse of the Central Maya Lowlands: Insights about human–environment relation-ships for Sustainability. *Proc. Natl. Acad. Sci. USA* **2012**, *109*, 13908–13914. [CrossRef] [PubMed]
23. SGCCC. Primer Reporte de Evaluacion del Conocimiento Sobre CAMBIO CLIMATICO en Guatemala. 1er Reporte de Cambio Climatico en Guatemala. 2019. Available online: https://www.sgccc.org.gt/wp-content/uploads/2019/07/1erRepCCGua_Introduccion.pdf (accessed on 12 June 2023).
24. Yon Secaida, L.P.; Mori, S.; Nomura, R. Assessment of Natural Disasters Impact on Cultural Mayan Heritage Spaces in Remotes Villages of Guatemala: Case of Black Salt. *Sustainability* **2023**, *15*, 12591. [CrossRef]
25. World Risk Report 2022. Bündnis Entwicklung Hilft, Ruhr University Bochum—Institute for International Law of Peace and Conflict 2022. Available online: https://weltrisikobericht.de/weltrisikobericht-2022-e (accessed on 12 June 2023).
26. Andrews, A. *Maya Salt Production and Trade*; University of Arizona Press: Tucson, AZ, USA, 1983.
27. Ordóñez Chocano, A. (2005). Estudio Histórico Antropológico de la sal Negra en Sacapulas, Departamento del Quiché, Guatemala [Online]. In *Tradiciones de Guatemala* (Vol. 59, pp. 159–192). CEFOL, Universidad San Carlos de Guatemala. Available online: https://ls3.usac.edu.gt/revindex/articulos/editor5-r86_pi159_pfi193_ra42612.pdf (accessed on 10 August 2023).
28. De Fuentes y Guzmán, F.A. *Recordación Florida*; Dirección General del Diario de Centro América y Tipografía Nacional: Ciudad de Guatemala, Guatemala, 2012.
29. Tovilla, M.A.; Scholes, F.V.; Adams, E.B. *Relaciones Histórico-Descriptivas de la Verapaz, el Manché y Lacandón en Guatemala*; Editorial Universitaria: Santiago, Chile, 1960.
30. Reina, R.; Monaghan, J. *The Ways of the Maya Salt Production in Sacapulas, Guatemala Expedition*; University of Pennsylvania: Philadelphia, PA, USA, 1981; Volume 23, pp. 13–33.
31. Akkeren, R. Título de Los Señores de Sacapulas. In *Crónicas Mesoamericanas*; Cabezas Carcache, H., Akkeren, R., Eds.; Universidad Mesoamericana: Guatemala City, Guatemala, 2008; pp. 59–91.
32. Orlando, O.; Moreno, R.; Ibarra, F. *The Guatemala Earthquake Disaster of 1976: A Review of Its Effects and of the Contribution of the United Nations Family*; UNDP: Guatemala City, Guatemala, 1977.
33. Webb, E.J.; Campbell, D.T.; Schwartz, R.D.; Sechrest, L. *Unobtrusive Measures: Nonreactive Research in the Social Sciences*; Rand Mcnally: Chicago, IL, USA, 1966.
34. Glaser, B.; Strauss, A. *Discovery of Grounded Theory: Strategies for Qualitative Research*, 1st ed.; Routledge: Milton Park, UK, 1999. [CrossRef]
35. Alnsour, M.A. Using modified grounded theory for conducting systematic research study on sustainable project management field. *MethodsX* **2022**, *9*, 101897. [CrossRef] [PubMed]
36. Arksey, H.; Knight, P.T. *Interviewing for Social Scientists*; SAGE Publications, Ltd.: Thousand Oaks, CA, USA, 1999. [CrossRef]
37. Kambaru, A. Qualitative Research and a Modified Grounded Theory Approach. *Tsuru Univ. Rev.* **2018**, *88*, 47–58. [CrossRef]
38. Gaspar, M.C.; Lorenzo, C.V. Análisis Comparativo de Los Diferentes Sistemas Constructivos en el Área Maya. In *XVII Simposio de Investigaciones Arqueológicas en Guatemala, 2003*; Laporte, J.P., Arroyo, B., Escobedo, H., Mejía, H., Eds.; Museo Nacional de Arqueología y Etnología: Guatemala City, Guatemala, 2004; pp. 736–748.
39. Instituto Nacional de Estadistica. *Informe Final, V Encuesta Nacional de Salud Materno Infantil, 2008–2009*; National Institute of Statistics: Guatemala City, Guatemala, 2011; pp. 16–358.

Disclaimer/Publisher's Note: The statements, opinions and data contained in all publications are solely those of the individual author(s) and contributor(s) and not of MDPI and/or the editor(s). MDPI and/or the editor(s) disclaim responsibility for any injury to people or property resulting from any ideas, methods, instructions or products referred to in the content.

climate MDPI

Article

The Effectiveness of Climate Adaptation Finance and Readiness on Vulnerability in African Economies

Purity Maina [1,2,*] and Anett Parádi-Dolgos [3]

1 Doctoral School of Economic and Regional Sciences, The Hungarian University of Agriculture and Life Sciences, Páter Károly u. 1, 2100 Gödöllő, Hungary
2 Business School, Strathmore University, Nairobi 0200, Kenya
3 Department of Investment, Finance, and Accounting, Institute of Rural Development and Sustainable Economy, Kaposvár Campus, The Hungarian University of Agriculture and Life Sciences, 7400 Kaposvár, Hungary; paradi-dolgos.anett.katalin@uni-mate.hu
* Correspondence: maina.purity.watetu@phd.uni-mate.hu

Abstract: Addressing climate vulnerability remains a priority for economies globally. This study used the panel-corrected standard error (PCSE) methodology to investigate the impact of adaptation financing on climate vulnerability. This analysis examined 52 African countries from 2012 to 2021 while considering their climate adaptation readiness. The impact was also assessed based on the Human Development Index (HDI) categories to reflect different levels of development. The findings showed that adaptation finance considerably influenced climate vulnerability reduction in Africa, particularly in nations with a moderate HDI. However, most countries still need higher levels of adaptation financing, resulting in a small impact on vulnerability reduction. Furthermore, the impact of readiness measures differed by HDI category. Economic and social climate readiness strongly impacted climate vulnerability in high-HDI nations, but governance preparedness was more critical in low-HDI countries. Based on the empirical facts, two policy proposals emerge. First, it is critical to reconsider the distribution of adaptation financing to reduce disparities and effectively alleviate climate vulnerability. Moreover, African economies should consider implementing innovative localized financing mechanisms to mobilize extra adaptation finance. Second, African governments should customize climate readiness interventions based on their HDI levels to improve the achievement of a positive impact on climate vulnerability.

Keywords: climate adaptation finance; climate readiness; climate vulnerability; panel-corrected standard error analysis; Africa

1. Introduction

In light of the pressing issue of climate change, it is imperative to swiftly and consistently execute measures that address climate change adaptation on a continental scale [1]. The implemented adaptation measures' long-term effectiveness partly depends on the climate vulnerability of the respective regions [2]. Climate vulnerability, referred to as the susceptibility to adverse impacts caused by climate-driven hazards [3], remains a subject of considerable interest as policymakers work toward enhancing climate risk management [4]. The far-reaching consequences of climate change risks, which threaten economic and social advancement [5,6], have garnered significant attention across the globe, including in Africa.

The African continent, categorized as the most vulnerable to climate change [7], continues to experience disruptions, particularly in sectors susceptible to the adverse effects of climate change [8–10]. The need to address climate vulnerability remains an area of focus, given the crucial role these sectors play in the countries' socioeconomic landscapes. The agriculture sector, for example, significantly influences the macro- and micro-economies [11]. A considerable proportion of Africa's mixed economies notably rely on the rainfed agricultural sector for employment and livelihood, directly or indirectly.

In 2021, the agriculture sector employed 52% of the workforce, according to World Bank employment statistics [12].

Moreover, the sector's significant contribution to the gross domestic production, averaging nearly 20% across African countries, cannot be undermined [13]. The services sector, which is the highest contributor to the economic output of most African countries, is also sensitive to extreme climate-related events like drought and floods [14–16]. The impact of climate change across the key sectors has stimulated more discussion regarding proactive adaptation, notwithstanding the obstacles faced, particularly on confined financial, infrastructural, and technological resources [17].

Improving adaptation readiness toward tackling the complexities of climate change and strengthening resilience is crucial for Africa. Lagging in implementing proper coping measures will undeniably increase the number of Africans living in poverty by 2030 [18]. The Notre Dame Global Adaptation Index (ND-GAIN), an open-source index [19] adopted by scholars [20–22], defines readiness as the capacity to attract adaptation investment from various avenues. Its measurement is based on proxies that could encourage or discourage public and private sector investment.

A country's readiness is evaluated based on economic, governance, and social metrics. Economic readiness is assessed through the business environment since this is expected to attract or discourage investment in adaptation, particularly from the private sector. Governance readiness focuses more on the political risks and their possible impact on foreign capital flows. Social readiness proxies revolve around social attributes, including inequalities, education, access to knowledge, and innovation that could derail the community's ability to identify adaptation opportunities, invest, and yield returns. Adaptation readiness is pivotal in effectively habituating climate change's impacts [17,23]

In addition to readiness, actual progress in adaptation requires massive finance disbursement for related investments in Africa. The developed economies' commitment toward Africa's climate finance basket has seen more distributions toward mitigation than the adaptation course. This fact has been attributed to mitigation being considered a public good that directly benefits all nations, whereas adaptation indirectly benefits the fund contributors [24]. Disbursements on adaptation finance are also lagging compared to the commitments made [25]. The estimated deficit in financing based on the annual average is above 70% for 2020–2030 [26]. This sluggish adaptation financing trend attracts interest, given that climate adaptation financing initiatives have been laid out globally to support climate adaptation projects since the 1990s. The initiatives include the establishment of the Global Environment Facility in 1991, the Kyoto Protocol adaptation fund in 2001, and the Green Climate Fund in 2010. Generally, the formation of adaptation finance initiatives has mainly focused on vulnerable developing countries since adaptation finance has been acknowledged as having the potential to address climate vulnerability in these regions [27,28].

The influence of human development on climate adaptation and mitigation measures in different countries is significant [29]. Analyzing and understanding the disparities in countries' levels of development is crucial for effective policy design [30]. The Human Development Index (HDI) is a comprehensive metric systematically evaluating a country's development [31]. African countries exhibit a diverse range of HDI rankings and trends. Several African countries have achieved notable advancements in human development and exhibit relatively high HDI scores.

Nevertheless, a significant proportion of the HDI values in this region are relatively low compared to other areas. Specifically, more than 50% of the nations in this region have an average HDI below 0.55 (Figure 1), which falls within the low category according to the United Nations Development Programme (UNDP) classification ([32], p. 1). This results from countries with poverty, inequality, and limited access to quality education and healthcare. Countries with higher levels of human development are anticipated to possess the resources and capability to allocate and execute more ambitious adaptation readiness policies in contrast to countries with lower levels, owing to divergent priorities [33].

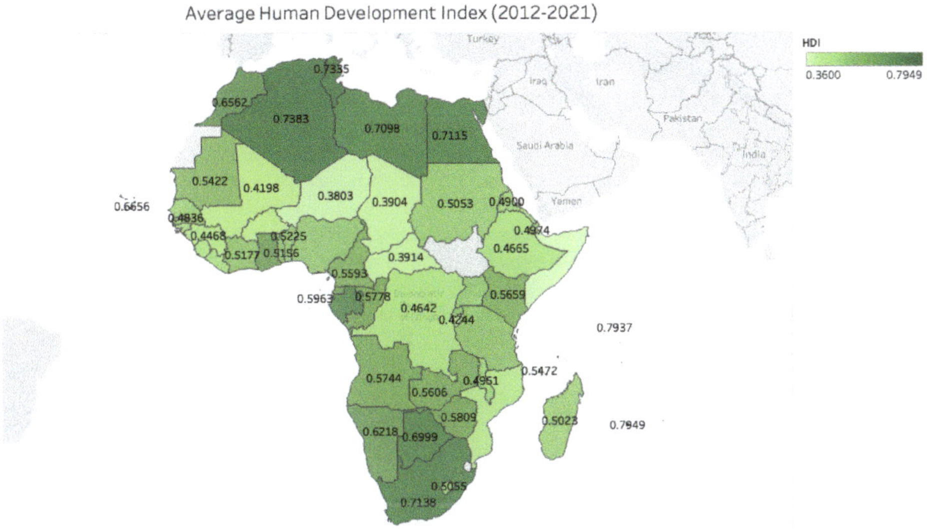

Figure 1. Average Human Development Index (HDI) of African countries for the period of 2012–2021. The Human Development Index is a United Nations country-ranking composite index comprising social and economic development indicators.

Given the interaction between adaptation financing, readiness, and vulnerability, we aimed to establish if the adaptation finance to Africa by bilateral, multilateral, and private philanthropic organizations has achieved the intended objective of reducing climate vulnerability while considering the various countries' readiness and HDI categories. The novelty of our research model is derived from the fact that we (i) evaluate the impact of traceable adaptation finance on Africa's climate vulnerability while taking into account the individual countries' readiness levels, (ii) employ the Prais–Winsten regression and panel-corrected standard error analysis (PCSE) to predict this impact, and (iii) analyze the impact across three African countries categories based on the level of development as measured by the Human Development Index (HDI). The rest of this paper is organized as follows: Section 2 discusses the relevant literature, Section 3 presents materials and methods, Section 4 presents the results and discussion of our findings, and we conclude the paper in Section 5.

2. Climate Vulnerability, Adaptation Readiness, and Adaptation Finance Literature

The climate vulnerability aspect has been examined through empirical research, resulting in interconnected themes defining the policy and scholarly investigations. The three main subjects of interest related to the topic are vulnerability assessment methodologies, factors influencing vulnerability, and their association with adaptation.

Climate vulnerability assessment entails modeling based on specific factors and risks to evaluate the levels at which climate impacts affect societies. The vulnerability assessment process of communities comprises the measurement of four components: climate change exposure, sensitivity to exposure, implications on points of sensitivity, and adaptive capabilities based on steps taken to address the impacts [34]. Moreover, it is imperative to acknowledge the importance of complex adaptive systems theory to assess vulnerability effectively. Considerations of community modulation, feedback mechanisms, redundancy, and susceptibility to change, extending from the theory, are essential incorporations in vulnerability models [35].

The approach adopted to assess vulnerability differs depending on the scale of analysis, emphasizing single risks impacting a specific industry or society or those affecting multiple sectors [36,37]. Integrating more indicators can strengthen the vulnerability indexes and provide unique insights to guide the strategies adopted to reduce the vulnerability [38]. The efficacy of climate vulnerability assessment is crucial in informing and implementing suitable local strategies ([39], p. 12). Our study builds upon vulnerability measurement papers by going beyond assessment and determining the role of adaptability on the vulnerability trend using a preexisting ND-GAIN vulnerability measure.

The second broad category of researchers explores the factors influencing climate vulnerability, including socioeconomic and governance factors, considered components of adaptation readiness in ND-GAIN metrics [19,40]. A notable interaction between these factors and vulnerability is evident, though with mixed findings, as reported by [22,41–44]. One specific focus is the association between socioeconomic factors, including wealth and education levels, and climate vulnerability. Examining the correlation between wealth and vulnerability has established a clear negative relationship between low-income populations and those facing greater vulnerability risk. While assessing the dynamics of vulnerability to climate-related hazards, [45] reported a decreasing trend in global socioeconomic vulnerability, suggesting that societies worldwide have become less vulnerable to climate-related hazards over time. Moreover, the study revealed a strong negative relationship between population income levels and climate vulnerability.

Relatedly, wealth increases the sense of control, leading individuals to be climate change conscious and take climate action [46]. From the perspective of educational levels, there are two possible scenarios, especially for future generations [47]. Educational factors could improve the Human Development Index (HDI), a primary indicator of climate resilience, thus reducing vulnerability. However, population education attainments could increase energy consumption resulting from labor productivity and slow progress in reducing environmental pressures. Empirical evidence provided for the association between a country's economic level and climate vulnerability has indicated mixed findings. The contributors to a country's economic growth determine whether climate vulnerability reduces or accelerates [48]. Furthermore, there is a correlation between economic development and climate vulnerability [49]. The findings above underscore the significance of giving due consideration to socioeconomic factors in a bid to mitigate the impacts of climate change effectively.

The discussions on governance factors encompass political factors and institutional and societal frameworks that determine climate finance resource allocation and regulation. The efficacy of national and local governments, non-government institutions, and relevant international institutions in climate governance has been assessed [6,50–53]. While examining adaptation readiness from a climate governance perspective, [23] examined the effectiveness of governance across various countries and found that it impacts climate vulnerability. Participation, collaboration, and knowledge sharing among the various regulating bodies are essential to enhancing climate governance [54,55]. Regarding the influence of political factors, [56] acknowledges the possible self-interest of powerful political–economic institutions, politicians, and other interested actors in the vulnerability measurement and policymaking role. This creates challenges, including conflicting ideologies on the definition and indicators of climate vulnerability and implementing transformative and practical approaches that benefit the most affected.

Empirical studies on the adaptation readiness index comprising a combination of economic, social, and governance factors have indicated that countries with a high adaptation readiness, majorly the developed economies, are less vulnerable compared to their counterparts [17,57]. In essence, most studies on all the above factors focus on regions considered most vulnerable. This stems from the generally accepted ideology that climate vulnerability distribution is unequal [58,59], with developed economies being less vulnerable than developing economies [60]. The importance of reviewing the climate vulnerability influencing factors for our paper was the appreciation that they could contribute to aggravating or reducing climate vulnerability in addition to the distribution of adaptation finance [61,62]. This was a critical aspect to consider in our discussion of findings.

The climate adaptation topic has also been emphasized, with empirical evaluations focusing mainly on countries' adaptive capacities, adaptive interventions, and adaptation finance distributions. While developed countries have greater climate adaptation capacities due to their adaptation readiness, developing countries struggle to counter climate-related risks due to adaptation constraints [23,63]. The adaptive capacities of various communities have been assessed using different frameworks, with the strength varying across sectors and contexts [64–66]. For example, from the perspective of households, evidence suggests disparities in adaptive capacity scores across different age groups, genders, education levels, and locations. The technological, financial, social, infrastructural, governance, and human capital capabilities further determine the adaptive capacities [67–70].

Adaptation interventions to facilitate effective responses to climate change effects have been implemented from different levels, including national and local governments, international communities, and community-based initiatives. The intervention cruciality is highest in the most vulnerable regions [71,72]. Deep engagement with the communities in vulnerable regions to understand the different local contexts could lead to more adaptation success [71]. However, a lack of proper adaptation interventions may reinforce, redistribute, or create new vulnerabilities, thus hindering making substantial progress. From a related angle, [73] note that interventions may create losers and winners, just like the climate-related crisis. The main losers could be vulnerable communities if they are misrepresented, stereotyped, or excluded from adaptation discussions.

One essential adaptation intervention effort is the distribution of adaptation financing to fund essential adaptation projects. International funds are crucial in supporting adaptation for low-income states, given that adaptation finance is a significant constraint [74,75]. So far, examinations of adaptation finance have focused on the allocation levels across various sectors and countries, as well as the associated opportunities and challenges. The discrepancy between countries that need adaptation funding and those that are allocated exists [76,77]. The extent to which adaptation finance has reached vulnerable countries through the Green Climate Fund is low, particularly for the least developed countries that are stricken by conflict and have sluggish government institutions [78,79]. To this end, climate justice in adaptation financing has been explored, with inequities attributed to donors' use of aid to foster their political and economic interest in recipient countries [6,80].

Although scholars have explored the link between vulnerability and adaptation finance, focusing on whether vulnerability determines the adaptation finance received, [21, 81] the findings are still controversial. Moreover, while it is recognized that investing in adaptation measures is crucial for reducing vulnerability and building resilience to the adverse effects of climate change, scholarly emphasis on trends in achieving this goal is minimal. Accordingly, we expect climate vulnerability to decrease in countries with higher adaptation finance over time and a higher adaptation readiness index. We therefore formulate the following hypothesis:

Alternative hypotheses:

H1: *In the case of African economies, climate adaptation finance positively impacts the reduction of climate vulnerability when readiness factors are considered.*

Further, given the different levels of development across African countries, as measured by the HDI, we expect the impact of adaptation finance on vulnerability to be different across the low, moderate, and high HDI classifications. We hence formulate the second alternative hypothesis:

H2: *There is a significant difference in the impact of adaptation finance and climate readiness on climate vulnerability across Africa based on the Human Development Index (HDI) categories.*

In essence, testing the hypothesis and answering the questions could contribute to a deeper understanding of climate adaptation finance, readiness, and vulnerability dynamics and enhance evidence-based decision-making in policy formulation. This can, in turn, promote the development of more effective policies and empower policymakers to fine-tune policy interventions relating to climate adaptation.

3. Materials and Methods

3.1. Scope

The first hypothesis, regarding whether adaptation finance reduces climate vulnerability in Africa, was evaluated using panel data analysis. We used panel data since it has several advantages, including controlling for cross-section heterogeneity and providing a greater capacity to find and quantify effects that are not discernible just through cross-sectional or time-series data [82]. We analyzed balanced panel data comprising 52 of 54 African countries over ten years (2012–2021) using a combination of R version 4.3.2, Stata/SE 17, and Microsoft Excel version 16.78.3. South Sudan and Eswatini were excluded from the analysis due to missing data cases.

To test the second hypothesis, we stratified African economies into three classifications based on their average HDI as reported by the UNDP data repository over ten years. According to the data, no African country has an HDI level that is significantly higher than 0.8. Further, less than five countries have an average high HDI above 0.7. Since most countries fall within the medium and low UNDP HDI classifications, we categorized the countries into three classifications [83]. The first category included countries with an average HDI of 0.6 or higher, indicating a relatively high level of human development. The second category encompassed countries with an average HDI ranging between 0.5 and 0.6, representing moderate human development. Lastly, the third category consisted of countries with an average HDI below 0.5, suggesting a lower level of human development.

The stratification allowed for a comprehensive analysis of the impact across varying levels of countries' development. We selected the ten years following the establishment of the Green Climate Fund, including 2015, the year that saw the ratification of the Paris Agreement. Focusing on this period, we aimed to capture the recent developments and advancements in climate finance adaptation and vulnerability.

3.2. Data

The data on climate adaptation finance originated from the Organization for Economic Cooperation and Development (OECD). OECD monitors climate finance using OECD DAC Rio Markers for Climate. This encompasses bilateral public climate finance from institutions in developed countries, multilateral public climate finance, climate-related officially supported export credits, and bilateral and multilateral private finance attributed to developed countries [84].

The climate vulnerability and adaptation readiness data were sourced from the University of Notre Dame Global Adaptation Index (ND-GAIN). This index serves as a comprehensive measure of countries' readiness and vulnerability to the impacts of climate change. It assesses various factors such as exposure to climate hazards, sensitivity to climate change, and adaptive capacity. The ND-GAIN index is widely recognized and used by researchers, policymakers, and organizations worldwide to understand climate vulnerabilities better and support effective adaptation strategies. Moreover, the HDI and greenhouse gas emissions (GHG) data, included as a control variable, were obtained from the World Bank Data Bank and UNDP databases [85]

Data Limitation

The challenges associated with establishing adaptation financing components and acquiring accurate data to differentiate commitments from disbursements hindered our ability to estimate adaptation finance requirements precisely. Nevertheless, the OECD has actively monitored climate development finance since 2000 and offers comprehensive data explorer tools and methodological documents. The distinction between adaptation financing and mitigation financing was of particular interest since our study focuses specifically on climate adaptation.

4. Findings and Interpretation

This section presents the empirical findings derived from the study's data analysis. The results are then systematically discussed, providing insightful interpretations and implications.

4.1. Descriptive Statistic Analysis

The descriptive statistics below provide a comprehensive overview of the relevant data. The statistics play a crucial role in understanding the current state and trends in vulnerability to climate-related risks, adaptation finance, the adaptation readiness of different countries, and the classification based on levels of development as measured by HDI in Africa.

4.1.1. Climate Change Vulnerability

Climate vulnerability, as assessed by ND-GAIN, monitors a country's exposure, sensitivity, and mechanisms to adapt to the negative impact of climate change. In the visualization presented in Figure 2, Somalia, Chad, Guinea Bissau, and Niger stand out as the most climate-vulnerable countries. At the same time, Algeria, Tunisia, Morocco, and South Africa are identified as among the least climate vulnerable. On average, the vulnerability across the entire continent is 0.513. Further analysis revealed that highly developed countries based on HDI exhibit an average vulnerability of 0.427, moderately developed nations average 0.524, and the least developed countries demonstrate an average of 0.574. This indicated that vulnerability averages are not significantly different across the development classifications, especially for the moderate and less developed nations.

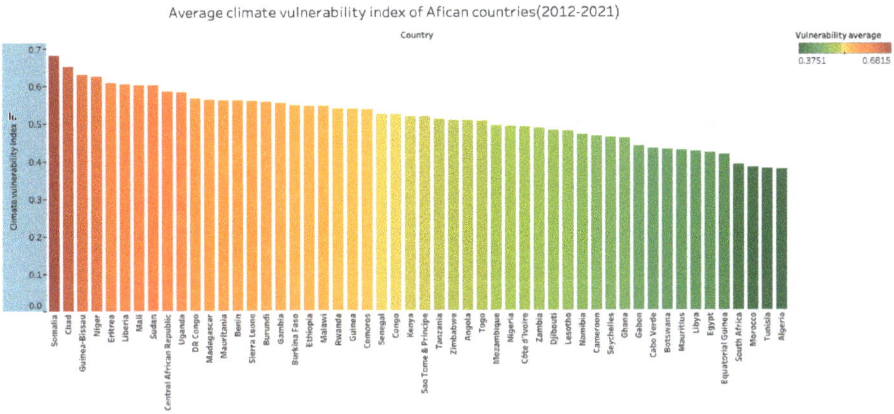

Figure 2. African countries' climate vulnerability. The climate vulnerability averages for the 2012–2021 period are sourced from the University of Notre Dame Global Adaptation Index (ND-GAIN).

4.1.2. Climate Adaptation Development Finance

Climate Adaptation Development Finance reported by OECD comprises bilateral public climate funding from institutions in developed nations, multilateral public climate finance, climate-related publicly supported export credits, and bilateral and multilateral private financing attributed to developed countries.

To understand the level of climate adaptation finance across African countries, the study presents a map in Figure 3. The data are based on a scale of 28,018 to 10,035,018 USD thousand, with the blue color intensifying for countries with higher amounts of adaptation finance. Kenya emerged as the top African country in terms of total commitments in adaptation-related development financing recorded by the OECD within the period. Additionally, the Eastern Africa region has received more commitments than other regions, with Ethiopia, Somalia, Uganda, and Tanzania among the top ten countries in this category.

The West African countries have also received significant climate adaptation commitments. A comparison of the top ten most vulnerable countries and those that have received the highest commitments indicates that only two countries, Somalia and Niger, fall into both categories. This suggests there may be a disconnect between the level of vulnerability and the amount of adaptation support provided, highlighting a potential area for further analysis and targeted intervention.

Figure 3. Africa Climate Adaptation Development Finance. The data are the totals in USD thousands from the period of 2012–2021 and are based on the OECD Climate-related Development Finance datasets.

On the other hand, the countries that have received minimum commitments include Libya, Seychelles, Equatorial Guinea, Sao Tome and Principe, and Guinea-Bissau. Of this, Guinea-Bissau is among the top ten most vulnerable countries. The other four countries are among the top 30% of countries in terms of HDI-measured development. There is a significant discrepancy in the distribution of commitment allocation, as the bottom 52% of countries' allocation is only 12% of the total share.

The global adaptation finance gap is increasing despite the increase in total climate finance [86,87]. Globally, the regions that are considered the most climate-vulnerable and thus might compete for climate adaptation finance include Africa, South Asia, the Pacific region, and Latin America [88,89]. A comparison of the adaptation finance across the regions during the period of 2021–2022 indicates that Asia and the Pacific region received 45% of global adaptation financing as compared to 20% in Africa and 10% in Latin America [90]. This means that there is an opportunity for enhancement to optimize the allocation of limited adaptation finance across vulnerable regions.

4.1.3. Climate Adaptation Overall Readiness

ND-GAIN measure of the overall climate adaptation readiness focuses on a nation's capacity to leverage investments and convert them to adaptation actions. The overall readiness comprises economic, governance, and social readiness.

The readiness of Mauritius for climate adaptation surpasses that of other African economies, as illustrated in Figure 4. The islands of Seychelles and Cabo Verde are also highly climate adaptation-ready from an overall perspective, along with Tunisia, Botswana, Morocco, and Rwanda. The least adaptation-ready economies include the Central African Republic, Chad, DR Congo, Zimbabwe, and Eritrea. Central African Republic and Chad's economic readiness was very low compared to the continent's average, leading to a low overall readiness. Zimbabwe and Eritrea lagged on social readiness, contributing to their low overall readiness. This indicated a low community's ability to tap into climate adaptation opportunities.

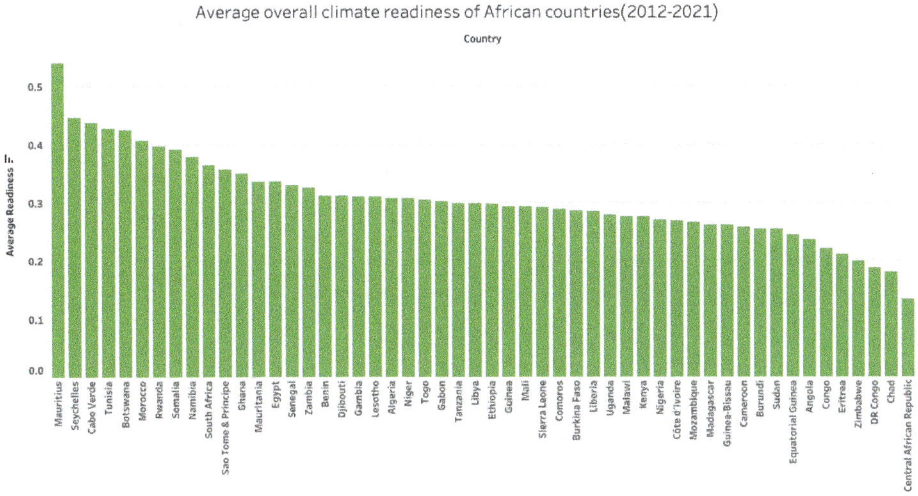

Figure 4. Africa's overall climate adaptation readiness by country. The overall climate readiness index is an average of the period of 2012–2022 and is a factor of economic, social, and governance readiness as reported by ND-GAIN.

4.1.4. Time Trend of Variables Analysis

The visuals presented in this section illustrate African countries' average governance, social, and economic readiness from 2012 to 2021. The ND-GAIN governance readiness, as illustrated in Figure 5, focuses on the institutional capabilities that enhance adaptation investment applications.

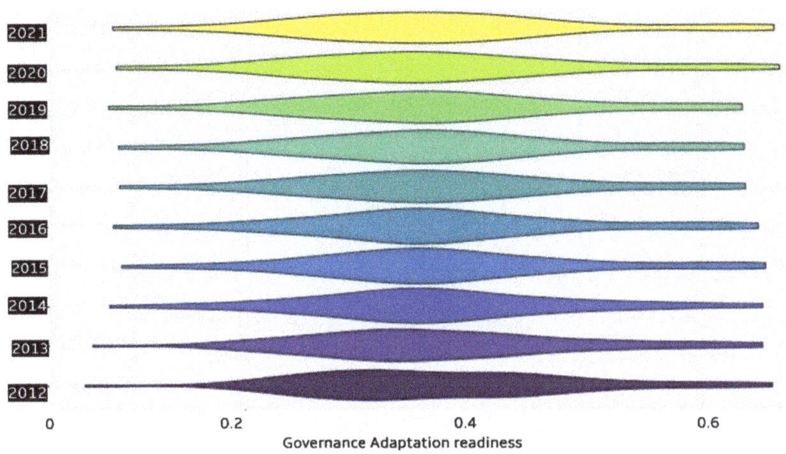

Figure 5. Average governance adaptation readiness of African countries for the period of 2012–2021.

The governance climate adaptation readiness metric has exhibited relative consistency over time. Throughout the years, certain countries constantly demonstrated high governance readiness, with index values exceeding 0.6, while others remained below 0.1, leading to an average index of 0.37

The social readiness index featured in Figure 6 incorporates indicators of the social and technological capabilities of the society that influence adaptation investment mobility and actions.

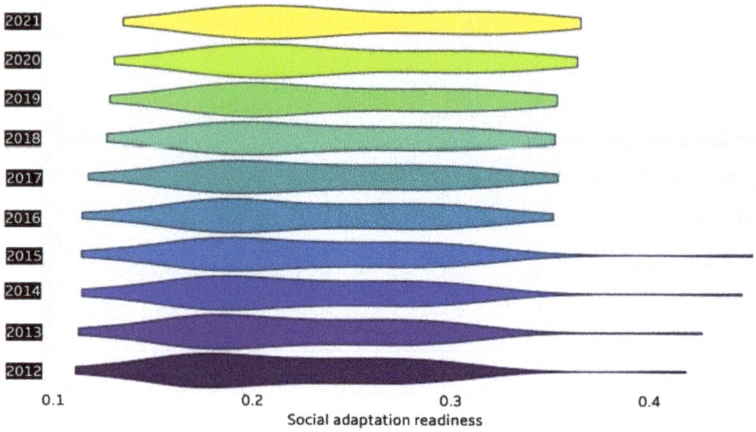

Figure 6. Average social adaptation readiness of African countries for the period of 2012–2021.

The social climate adaptation readiness, presents a skewed distribution, including notable positive outliers, particularly in 2012–2015. However, during the period from 2016 to 2021, the majority of indices are concentrated between 0.1 and 0.35 across countries. In the same period, some countries previously considered more socially prepared experienced a decline in readiness, resulting in a decreased overall mean index of 0.237, lower than the governance and economic readiness. Moreover, the countries with a lower index below 0.2 are still notable, but the trend has reduced over the years.

The economic readiness index presented in Figure 7 assesses a country's business environment that could influence its adaptation investments. The economic readiness trend shows some similarity with the social readiness in the period from 2012 to 2015. However, in 2016–2021, the countries' indices are more centered around the mean of 0.313, with some countries experiencing a decline from the previous years.

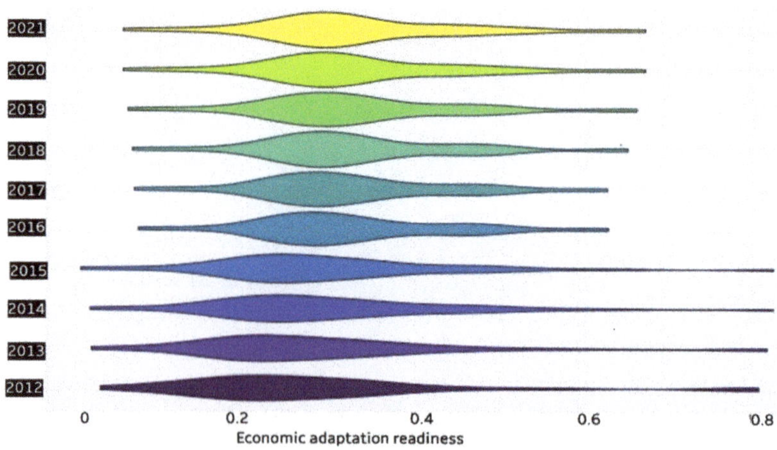

Figure 7. Average economic adaptation readiness of African countries for the period of 2012–2021.

The climate adaptation finance for most countries, depicted in Figure 8, has been very low averaging less than 500,000 USD thousand every year. However, the weight on the lower side has reduced over the years, with more countries receiving more adaptation funding recently.

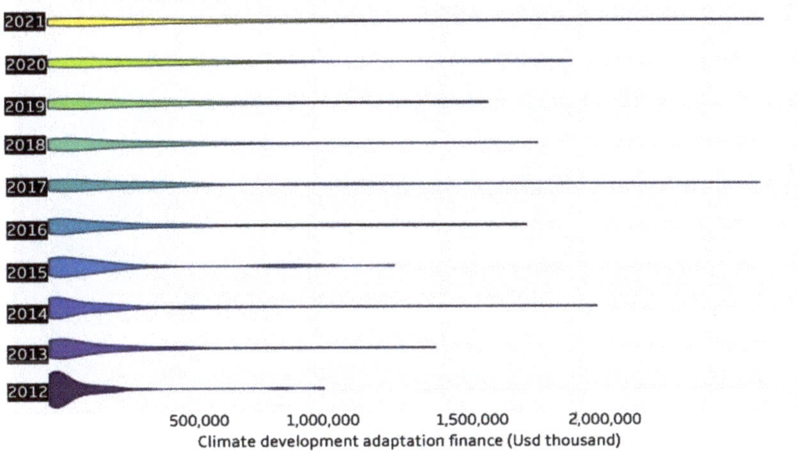

Figure 8. Climate adaptation finance of African countries for the period of 2012–2021.

Finally, Figure 9 shows the trend of overall climate readiness and vulnerability. The vulnerability has reduced as readiness has improved over the years. The increasing trend in adaptation readiness is notable as compared to the reduction in climate vulnerability.

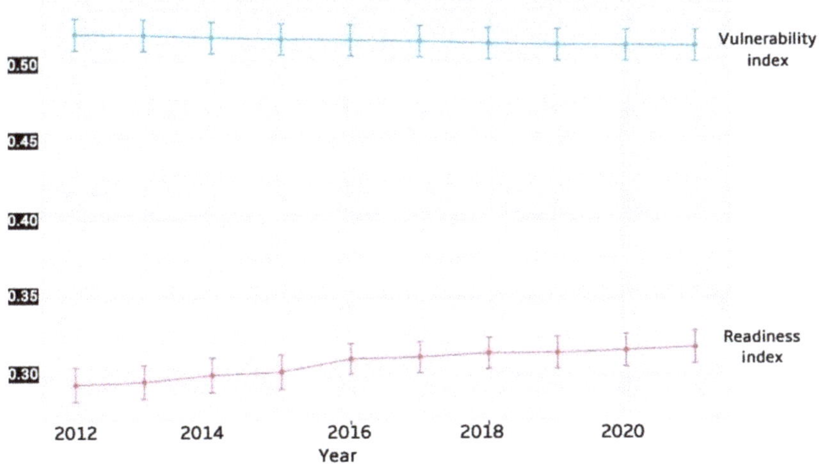

Figure 9. Overall African adaptation readiness vs. climate vulnerability trend.

4.2. Diagnostic Test

In this section, we discuss the results of five diagnostic tests. These tests played a crucial role in detecting potential issues in our panel data and, hence, selecting a robust analysis model in line with the problems identified.

4.2.1. Normality Test

Normal distribution of residuals is required for the test to be reliable while running a regression analysis [91]. The normality test on residuals was conducted at a 5% significance level. We utilized the sktest skewness and kurtosis normality test to test the null hypothesis that variable residuals were normally distributed against the alternative hypothesis that variable residuals were not normally distributed.

Table 1 indicates that the probability of skewness of the residuals is less than 0.05, meaning that the residuals are not normally distributed. Nonetheless, the probability of kurtosis implies that the residuals approach an asymptotic normal distribution with a 95% confidence level. Additionally, the joint test probability of Chi-square is less than 0.05, which provides sufficient evidence to conclude that the residuals are not normally distributed within a 5% level of significance. To address the non-normality issue, we use natural logs of the variables climate adaptation finance and GHG emissions that exhibited non-normality.

Table 1. Normality test output.

Skewness and Kurtosis Test for Normality Joint Test					
Variable	0 bs	Pr (Skewness)	Pr (Kurtosis)	Adj chi2 (2)	Prob > chi2
resid	520	0.0000	0.6644	25.42	0.0000

4.2.2. Pesaran CD Cross-Sectional Dependence Test

Cross-sectional dependence of error terms could lead to biased test results and is problematic. Ignoring sufficient cross-sectional dependency in panel analysis could greatly diminish estimate efficiency because the panel estimator may not differ significantly from the single equation ordinary least squares estimator [92].

Using the Pesaran CD test, we ran the pcdtest R command to examine the possibility of cross-sectional dependence in the residuals across countries. This test was appropriate for our data set, given that the number of cross-sectional units was sizeable compared to the small number of periods. The CD test is advantageous when many cross-sectional units are recorded over periods, unlike the conventional Breusch–Pagan LM test of independence [93,94]. The null hypothesis assumes no correlation between cross-section residuals at a 5% significance level, while the alternative hypothesis assumes a correlation between cross-section residuals at a 5% significance level.

As shown in Table 2, Pesaran's test probability of 0.0161 is less than 0.05; hence, we fail to accept the null hypothesis and conclude that the cross-section residuals exhibit contemporaneous correlation.

Table 2. Cross-sectional dependence test output.

Z Statistic	*p*-Value
2.4075	0.0161

4.2.3. Heteroscedasticity Test

Heteroscedasticity occurs when the errors do not exhibit a constant variance, meaning that var(ut) = σ2 < ∞ [91]. Testing and tackling the heteroscedasticity of the residuals is vital since its presence may result in inadequate least squares estimates and inconsistent covariance matrix estimates. We used the Aod R package module by [95] to compute the Wald statistic for groupwise heteroskedasticity in the residuals of a fixed effect regression model. The test null hypothesis (H0) was that the residuals are derived from homoscedastic distributions, against an alternative hypothesis that the residuals are not derived from homoscedastic distributions, meaning they are expected to have unequal variance.

The results in Table 3 indicate the chi2 *p*-value of the test was 0.000, less than $\alpha = 0.05$; hence, we rejected the null hypothesis and concluded that heteroscedasticity was present in the panel regression model residuals.

Table 3. Heteroscedasticity test output.

Chi2	DF	*p*-Value
68.6	5	0.0000

4.2.4. Serial Correlation

Serial correlation is the correlation between matched data sets that are not obtained at a single point in time but rather observed progressively through time. Autocorrelation is a type of serial correlation where the second data set matches the first set, albeit with a temporal lag [96]. Testing and solving for serial correlation, also referred to as autocorrelation in panel data, are essential to avoid a false high R squared and erroneous more minor standard errors of the coefficients [97–99]. We utilized the pbgtest command of the Lm R package to run the Breusch–Godfrey–Wooldridge test on the null hypothesis of no serial correlation in idiosyncratic errors as proposed by [100]. The test is appropriate since it permits heteroscedasticity, which was present in our data.

The significant *p*-value of 0.000 in Table 4 below indicated the presence of serial correlation in the panels. Moreover, Figure 10 below further demonstrates a correlation between the temporal mean of the dependent variable, climate vulnerability.

Table 4. Serial correlation output.

Chi2	DF	*p*-Value
118.93	1	0.0000

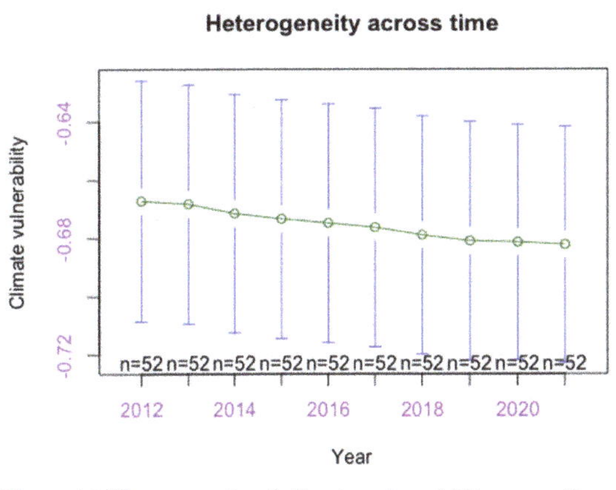

Figure 10. Heterogeneity of climate vulnerability across time.

4.2.5. Unit Root and Cointegration Tests

The unit root test is conducted to determine whether the panels are stationary. Stationarity is crucial to ensure no spurious relationships from the analysis [101,102]. We employed the Phillips–Peron test of the Tseries R package to examine the following hypothesis:

Null hypothesis (H0): *All panels contain a unit root.*

Alternative hypothesis (HA): *None of the panels have a unit root.*

The test yields more reliable results when dealing with disturbance processes that may exhibit heteroscedasticity and autocorrelation in our dataset [103,104]. Most panel unit root tests, including the Breitung and Das test and Levin–Lin–Chu unit root test, assume homoskedasticity [105] and hence were not applicable.

Table 5 shows that the *p*-values are not significant at 5%. We, therefore, fail to reject the null hypothesis and conclude that the panels contain a unit root, implying they are not stationary.

Table 5. Phillips–Perron unit root test output per variable.

Variable	Dickey–Fuller	*p*-Value
Climate vulnerability	−6.016	0.01
Climate Adaptation Finance	−7.2769	0.01
Economic readiness	−6.3659	0.01
Governance readiness	−5.4711	0.01
Social readiness	−5.3766	0.01
GHG emissions	−5.8296	0.01

Consequently, we conducted the Kao cointegration test to ascertain the existence of a long-run association between the non-stationary variables. Cointegration is crucial for avoiding spurious outcomes in variables that lack stationarity [106]. The test results, presented in Table 6, support the alternative hypothesis that panels are cointegrated. Therefore, we conclude that there is a long-term relationship between the variables and hence proceeded with regression analysis.

Table 6. Kao-cointegration test results.

	Static	*p*-Value
Modified Dickey–Fuller t	−5.0399	0.0000
Dickey–Fuller t	−11.9640	0.0000
Augmented Dickey–Fuller t	−6.2431	0.0000
Unadjusted modified Dickey–Fuller t	−11.8750	0.0000
Unadjusted Dickey–Fuller t	−14.7455	0.0000

From the diagnostic tests conducted, the standard error estimates exhibit the problems of non-normal distribution, cross-sectional dependence, heteroscedasticity, and autocorrelation. To solve the non-normality issue, we transformed the data into logs. Additionally, we ran and interpreted the panel-corrected standard error (PCSE) estimation models. This error-corrected model introduced by [107] solves for standard error estimates that are not robust due to the disturbances above in panel models. The advantages of using the model include its accountability for serial correlation, heteroskedasticity, and contemporaneous correlation. The model was also appropriate given that the data comprised many cross-sections (n) compared to the periods (t), a pre-requisite of using the model. The model employment has limitations in cases where the cross-sections are fewer than the periods [107,108].

4.3. PCSE Empirical Outputs

This study estimated four panel-corrected standard error (PCSE) econometric models to answer the research hypothesis. The results of Model 1 relating to hypothesis 1 are presented in Section 4.3.1 below, while those of Model 2–4 testing hypothesis 2 are presented in Section 4.3.2.

4.3.1. Hypothesis 1 Estimation Results

We applied the first PCSE regression model (Equation (1) below) to test the null hypothesis that climate adaptation finance does not positively impact the reduction of climate vulnerability in Africa when readiness factors are taken into account. We examined the effect of adaptation finance; economic, governance, and social readiness; GHG emissions; and vulnerability, the value of which was lagging in terms of climate vulnerability. The output is presented in Table 7 below.

Table 7. Overall Prais–Winsten regression output with panel-corrected standard error (PCSE): Dependent variable:climate vulnerability.

Variable	Coefficient	Std. Err.	Z	$p > Z$	[95% Conf. Interval]	
CAFlog	−0.0207	0.0004	−2.07	0.039	−0.0014	0.0003
Economic readiness	−0.1282	0.0170	−7.52	0.000	−0.1616	−0.0948
Governance readiness	−0.1878	0.0285	−6.59	0.000	−0.2437	−0.1319
Social readiness	0.1198	0.0181	6.62	0.000	0.0843	0.1552
GHGlog	0.0044	0.0008	5.25	0.000	0.0028	0.0060
LagV	0.3883	0.0697	4.86	0.000	0.2022	0.4755
Constant	0.3560	0.0363	9.82	0.000	0.2849	0.4271
rhos=	1	−0.2923	0.4447	−0.3651	1	−0.0269
Wald chi2(6)=	2810.32					
Prob > chi2=	0.0000					

CAF log is the natural log of the climate adaptation finance. GHG log is the natural log of greenhouse gas emissions. Lag V is the lagged value of the vulnerability index.

Equation (1): PCSE estimation output

$$\begin{aligned} Xtpcse\ Climate\ vulnerability_{it} &= Climate\ adaptation\ finance_{it} + Economic\ readiness_{it} + Governance\ readiness_{it} \\ &+ social\ readiness_{it} + GHG_{it} + Lag\ vulnerability_{it} + \varepsilon_{it} \end{aligned} \quad (1)$$

At a 95% level of confidence, all the variables were statistically significant. Climate adaptation finance, economic readiness, and governance readiness had a statistically negative relationship with the vulnerability index, while social readiness, GHG emissions, and the lagged value of the vulnerability index had a positive relationship. These results imply that climate adaptation finance has a positive impact on reducing vulnerability, given the negative relationship. However, based on the beta coefficient, the decrease is minimal since a 1% increase in climate adaptation finance led to an average reduction of 0.02% in climate vulnerability. The slight decline may be attributed to most nations' insufficient aggregate adaptation funds. According to the data depicted in the adaptation finance map, 52% of the nations have received total commitments of less than 1,000,000 USD thousands. The disparity in financing between the highest- and lowest-ranking countries is also substantial.

Concerning adaptation readiness, countries' economic and governance readiness reduces climate vulnerability, with governance readiness exhibiting a higher impact. This inferred that a good adaptation investment environment that defines the economic readiness and governance factors contributing to political stability and stability is important in minimizing climate vulnerability. Unexpectedly, contrary to prior research [17], the social readiness variable exhibited a positive relationship with vulnerability, implying that as the overall African social structures of communities become more solid, the climate vulnerability is likely to increase. The Notre Dame social readiness indicators include social inequality, innovation, information communication technology (ICT), and education.

Prior research suggests that the impact of ICT on the environment could be two-way, depending on the ICT quality and interventions [109,110]. This could suggest a possible explanation for this relationship, considering that African nations have had a higher rate of ICT technology adoption over the past decade compared to previous years. Furthermore, as argued by [111] innovation can potentially increase climate challenges, depending on the nature of the innovation, its accessibility, and its environmental implications. A significant impact of green technological innovations on carbon productivity in developing economies is difficult to find [112]. Moreover, education enhancement is expected to result in improved HDI but could also increase emissions resulting from economic growth that negatively impact the environment in the short run [47]. The results further imply that prioritizing the reduction of GHG gas emissions and addressing climate vulnerability from previous years are essential steps toward achieving overall climate vulnerability reduction.

In summary, we reject the null hypothesis 1 and conclude that at a 95% confidence level, climate adaptation finance positively influences the reduction of climate vulnerability in Africa. Nevertheless, it is important to take note of the minimal nature of the decrease and the unique characteristics of individual countries in the African continent.

4.3.2. Hypothesis 2 Empirical Output

Given that African countries vary in development, we tested the null hypothesis 2 that there is no significant difference in the impact of adaptation finance and climate readiness on climate vulnerability across African economies based on HDI 3 classifications. The results of the three empirical models are presented in Tables 8–10 below based on low HDI, average HDI, and high HDI ranking.

Table 8. Low-HDI countries' PCSE empirical output. Dependent variable: climate vulnerability.

Variable	Coefficient	Std. Err.	Z	$p > Z$	[95% Conf.	Interval]
CAFlog	−0.0106	0.0005	−1.24	0.214	−0.0015	0.0003
Economic readiness	−0.0735	0.0125	−5.87	0.000	−0.0980	−0.0489
Governance readiness	−0.2219	0.0161	−13.74	0.000	−0.2536	−0.1903
Social readiness	0.0561	0.0435	1.29	0.198	−0.0293	0.1415
GHGlog	0.0174	0.0016	11.23	0.000	0.01438	0.0205
Lag V	0.3033	0.0545	5.57	0.000	0.1966	0.4101
cons	0.1805	0.0153	11.82	0.000	0.1506	0.2105
rhos=	1	1	1	−0.3687	1	−1
Wald chi2(6)=	9692.34					
Prob > chi2=	0.0000					

Table 9. Average-HDI countries' PCSE empirical output. Dependent variable: climate vulnerability.

V	Coefficient	Std. Err.	Z	$p > Z$	[95% Conf.	Interval]
CAFlog	−0.0322	0.0005	−4.48	0.000	−0.0031	−0.0012
Economic readiness	0.0328	0.0119	2.75	0.006	0.0094	0.0561
Governance readiness	−0.0856	0.0222	−3.86	0.000	−0.1291	−0.0421
Social readiness	0.0544	0.0297	1.83	0.067	−0.0039	0.1127
GHGlog	0.0038	0.0007	−5.57	0.000	−0.0052	−0.0025
lagV	0.1556	0.0570	2.73	0.006	0.0438	0.2674
cons	0.5394	0.0371	14.53	0.000	0.4667	0.6122
rhos=	0.4035	0.0303	1	0.0454	0.5301	1
Wald chi2(6)=	73.82					
Prob > chi2=	0.0000					

Table 10. High-HDI countries PCSE empirical output. Dependent variable: climate vulnerability.

Variable	Coefficient	Std. Err.	Z	$p > Z$	[95% Conf.	Interval]
CAFlog	−0.0107	0.0003	−2.43	0.015	−0.0013	−0.0001
Economic readiness	−0.0960	0.0188	−5.09	0.000	−0.1329	−0.0590
Governance readiness	−0.0704	0.01732	4.07	0.000	0.0365	0.1044
Social readiness	−0.0180	0.01769	−1.02	0.039	−0.0527	0.0167
GHGlog	0.0044	0.0020	2.18	0.029	0.0004	0.0084
lagV	0.4693	0.0800	5.87	0.000	0.3126	0.6261
cons	0.1678	0.0494	3.40	0.001	0.0711	0.2646
rhos	0.4749	1	1	0.3313	0.7167	0.6119
Wald chi2(6)=	705.5					
Prob > chi2=	0.0000					

Results in Table 8 were based on African countries whose average HDI index was below 0.5, which signifies a lower level of human development. These countries include

Somalia, Niger, Chad, Central African Republic, and Mali. All the variables except social readiness were statistically significant at a 5% level of significance. A 1% increase in climate adaptation finance is expected to reduce vulnerability by 0.01% in low-HDI African countries. From the readiness perspective, a unit increase in economic preparedness might lower climate vulnerability by 0.07, whereas a unit increase in governance readiness could reduce climate vulnerability by 0.22. Similar to the overall continent results, the findings also suggest that it is crucial to tackle climate vulnerability from past years and prioritize the decrease of greenhouse gas emissions in low-HDI economies to effectively reduce overall climate vulnerability.

Table 9 presents the output of category 2, comprising countries with an average HDI between 0.5 and 0.6. Some countries in this category included Senegal, Uganda, Nigeria, Rwanda, Zambia, Kenya, and Angola. Similar to the low HDI category, all the variables except social readiness were statistically significant at a 95% level of confidence. Climate adaptation finance was negatively related to climate vulnerability, with a 1% increase in the finance expected to lower the vulnerability by 0.03%. The reason for the greater influence, in comparison to other categories, may be attributed to the fact that countries with an average HDI, including Kenya and Nigeria, ranked among the top nations in terms of climate adaptation finance.

In contrast to the lower HDI group, the economic readiness variable exhibited a positive relationship with climate vulnerability. This finding revealed that as the ease of doing business index improves in the middle-developed countries, the climate vulnerability is likely to increase. The ease of doing business in middle-HDI countries, the economic readiness indicator, is associated with the prospect of infrastructure development, industrial growth, and increased energy intensity [113,114] as the countries expand. These activities may heighten climate vulnerability if not well planned out. Vulnerability was still positively related to the GHG and lagging vulnerability variables. However, the beta coefficients indicated a lesser potential impact on vulnerability.

The third classification was comprised of countries categorized as high-HDI countries with a value higher than 0.6. These countries included Mauritius, Seychelles, South Africa, Egypt, Namibia, and Ghana. The findings, as depicted in Table 10 above, indicated that all variables were statistically significant at 5%, similar to the overall continent's output presented in Table 7. The relationship of climate adaptation finance, economic readiness, and governance readiness with climate vulnerability was negative, signifying a positive impact similar to that of the low HDI category. The social readiness variable exhibited a statistically significant negative relationship in this category, indicating that social readiness is essential in reducing the climate vulnerability of highly human-developed African countries. The GHG emissions and lagged value of vulnerability maintained the same results as previous models.

In summary, the impact of variables across the various HDI categories is different. The relationship between the readiness variables and climate vulnerability and the magnitude of change in the climate vulnerability index caused by change in the independent variables varied across the classifications, indicating that the influence of the variables varies with HDI levels. According to the findings, the impact of adaptation finance is higher among average HDI countries than the lower and higher HDI countries. Economic and social readiness has more influence in high-HDI countries, while governance readiness is more important in low-HDI countries.

As a result, we rejected the null hypothesis 2. We concluded that the impact of adaptation funding and climate preparation on climate vulnerability varies significantly among African economic classifications based on the Human Development Index.

5. Conclusions and Policy Recommendations

The cruciality of promptly and consistently implementing actions to minimize climate vulnerability cannot be undermined in aiding the prevention and alleviation of the detrimental impacts of climate change [115]. African countries' socioeconomic, political, geographical, and environmental processes contribute to their heightened vulnerability to climate impacts [116]. In recent years, the continent has experienced a growing number of climate-related calamities, prompting increased discourse on the significance of adaptation.

Using Prais–Winsten regression and panel-corrected standard error analysis, this study empirically tested the effect of adaptation finance on climate change vulnerability while factoring in adaptation readiness in 52 African countries over the 2012 to 2021 period. Further, the paper explicitly assessed the effect across three different HDI categories, given the various levels of development across African countries. The empirical findings from all models validate the contribution of adaptation finance in reducing climate vulnerability to climate change in Africa. The influence is more significant in countries with average HDI than those with lower and higher HDI. However, the levels of adaptation finance are still low, with significant disparities across African countries leading to a small vulnerability reduction impact between 0.01% and 0.03% across the various HDI categories. On the other hand, the effect of adaptation readiness indicators varies across the categories. Economic and social preparedness significantly impacts vulnerability in countries with high HDI, while governance preparedness plays a more critical role in countries with low HDI.

Utilizing the empirical data, we make the following two policy recommendations. First, adaptation funding needs to be reassessed. While recognizing the critical role of adaptation finance in reducing climate vulnerability, it is essential to tackle the notable disparities in the allocation across African countries. Given that just 12% of the overall adaptation funding share to Africa is allocated to the 52% of bottom-HDI-ranked nations, it is crucial to prioritize improving the equitable distribution of funds to maximize its effectiveness in reducing vulnerability. The effective allocation process requires collaboration from stakeholders to conduct a comprehensive adaptation investment needs assessment of all countries while factoring in their risk susceptibility. Subsequently, the allocation can be carried out equitably while factoring in the assessment findings and the available resources.

Bridging the climate finance gap requires African countries' innovation of localized adaptation financing mechanisms, continuous international cooperation, and related partnerships. The establishment of adaptation project financing mechanisms at both the national and local government levels could ensure a comprehensive reach. To foster a sense of inclusivity and community engagement, these mechanisms should aim to provide favorable terms to attract contributory financing, even from local residents and corporations in vulnerable areas. By structuring financing mechanisms in this way, a wider range of adaptation projects can be supported, contributing to greater resilience across diverse geographic and demographic contexts.

To enhance international cooperation and partnerships, especially with private investors, African countries should focus on adaptation projects that have clearly defined and measurable outcomes. Further, maintaining consistent and high-level transparency and accountability in the usage of the adaptation finances is crucial. The assurance will not only reinforce existing partners' trust but could also encourage new international partnerships, attracting a wider pool of adaptation finance.

Second, African countries' governments should continue enhancing their adaptation readiness. Given the varying impact of readiness indicators on vulnerability across different HDI categories, adaptation readiness interventions should be tailored to the specific needs of countries based on their HDI levels. For countries with high HDI, the focus should be on strengthening economic and social preparedness, which has been shown to have a more significant impact. This entails enhancing the business environment to attract additional investment in adaptation and leveraging societal preparedness capabilities, such as education and technology adoption. For countries with low HDI, improving governance preparedness first is crucial. This may involve capacity building, institutional reforms, and policy frameworks to enhance governance readiness, specifically in these countries.

We acknowledge that obstacles to implementing policies may arise as a result of factors such as political will, institutional capacity, and resource constraints, which may impact the effectiveness of proposed interventions. Political will plays a pivotal role in facilitating coordination among key players involved in the implementation process and in establishing important legal frameworks. Conversely, the absence of the will could pose implementation challenges. Moreover, strong and effective institutions are critical in implementation coordination and management. The absence of such institutions in African countries could challenge the implementation. Additionally, the policy recommendations also necessitate resources, particularly financial resources, and therefore, limited resources could derail the process.

This study's conclusions are limited to a certain extent. First, this study focused on the effect of adaptation financing, readiness, and the country's classification of HDI on climate vulnerability. Other variables outside this scope were not included in this study, which determined the overall explanatory level of the model. Other researchers could further investigate other factors to enhance knowledge of their impact on the climate vulnerability aspect, especially in regions considered to be the most vulnerable.

Second, this study excluded South Sudan and Eswatini due to the lack of sufficient data, which impacts the generalizability of our findings to these countries. The exclusion, however, was not very significant since it represented 3.85% of the sample. The limitation highlights the need for improved data collection and reporting standards in Africa to ensure more inclusive and accurate research outcomes.

Third, the research was based on Africa, a continent mostly composed of developing and low-income nations with unique socioeconomic and developmental features. The findings may, therefore, only apply to other regions with relatable similar characteristics. It is hence important to exercise caution when extrapolating the results to developed nations or regions with different socioeconomic systems and development levels.

The assumptions underlying the analysis and which might have influenced the results included a presumption that data reported for the various variables from the sources was accurate. Given that biases in data collection could influence the robustness of results, validation of data sources and collection methods is important. Further, this study assumed a causal relationship where the independent variables impact climate vulnerability. The possible reverse causation of climate vulnerability influencing the independent variables was not tested and can be considered in future research.

Author Contributions: Conceptualization, P.M.; methodology, P.M.; software, P.M.; validation, A.P.-D.; formal analysis, P.M.; investigation, P.M.; resources, P.M. and A.P.-D.; data curation, P.M.; writing—original draft preparation, P.M.; writing—review and editing, P.M. and A.P.-D.; visualization, P.M.; supervision, A.P.-D. All authors have read and agreed to the published version of the manuscript.

Funding: This research received no external funding.

Data Availability Statement: The data presented in this study are available upon request from the corresponding author.

Acknowledgments: We acknowledge the support offered by the Hungarian University of Agriculture and Life Sciences (MATE) and the Stipendium Hungaricum Scholarship.

Conflicts of Interest: The authors declare no conflicts of interest.

References

1. Tilleard, S.; Ford, J. Adaptation readiness and adaptive capacity of transboundary river basins. *Clim. Chang.* **2016**, *137*, 575–591. [CrossRef]
2. Dhar, T.; Moncada, S.; Briguglio, L.; Bambrick, H.; Kelman, I.; Iorns, C.; Nurse, L. Identifying Climate Change Vulnerability and Adaptation Challenges in the Caribbean SIDS: An Urban Morphological Approach. In *Small Island Developing States: Vulnerability and Resilience under Climate Change*; The World of Small States; Springer International Publishing: Cham, Switzerland, 2021; pp. 329–350, ISBN 978-3-030-82774-8.
3. UNFCCC. Glossary of Key Terms. Available online: https://www4.unfccc.int/sites/NAPC/Pages/glossary.aspx (accessed on 14 September 2023).
4. Klein, R.; Mukhala, E.; Stephen, L.; Winograd, M.; Ziervogel, G. *Assessing Vulnerability for Climate Adaptation*; Technical Paper 3; UNFCCC: New York, NY, USA, 2005.

5. Thangjam, U.; Thong, P.; Sahoo, U.K. Climate change threat on socio-economic condition of agroforestry managers: A vulnerability study in eastern Himalayan state of Mizoram, Northeast India. *Clim. Risk Manag.* **2023**, *40*, 100519. [CrossRef]
6. Weiler, F.; Klöck, C.; Dornan, M. Vulnerability, good governance, or donor interests? The allocation of aid for climate change adaptation. *World Dev.* **2018**, *104*, 65–77. [CrossRef]
7. UNFCCC. *Report of the Conference of the Parties on Its Thirteenth Session, Held in Bali from 3 to 15 December 2007*; United Nations Framework Convention on Climate Change: New York, NY, USA, 2008.
8. African Development Bank Climate Change in Africa. Available online: https://www.afdb.org/en/cop25/climate-change-africa (accessed on 31 July 2021).
9. Adhikari, U.; Nejadhashemi, A.P.; Woznicki, S.A. Climate change and eastern Africa: A review of impact on major crops. *Food Energy Secur.* **2015**, *4*, 110–132. [CrossRef]
10. Nhemachena, C.; Nhamo, L.; Matchaya, G.; Nhemachena, C.R.; Muchara, B.; Karuaihe, S.T.; Mpandeli, S. Climate Change Impacts on Water and Agriculture Sectors in Southern Africa: Threats and Opportunities for Sustainable Development. *Water* **2020**, *12*, 2673. [CrossRef]
11. Eugenio, D.-B. *Macroeconomics, Agriculture, and Food Security: A Guide to Policy Analysis in Developing Countries*; International Food Policy Research Institute: Washington, DC, USA, 2015; ISBN 978-0-89629-859-0.
12. World Bank Open Data. Available online: https://data.worldbank.org (accessed on 13 September 2023).
13. Statista Africa: Agriculture as GDP Share by Country. 2021. Available online: https://www.statista.com/statistics/1265139/agriculture-as-a-share-of-gdp-in-africa-by-country/ (accessed on 18 September 2023).
14. Attoh, E.M.N.A.N.; de Bruin, K.; Goosen, H.; van Veldhoven, F.; Ludwig, F. Making physical climate risk assessments relevant to the financial sector—Lessons learned from real estate cases in the Netherlands. *Clim. Risk Manag.* **2022**, *37*, 100447. [CrossRef]
15. Burton, A.J.; Bambrick, H.J.; Friel, S. Is enough attention given to climate change in health service planning? An Australian perspective. *Glob. Health Action* **2014**, *7*, 23903. [CrossRef] [PubMed]
16. Dlugolecki, A.; Loster, T. Climate Change and the Financial Services Sector: An Appreciation of the UNEPFI Study. *Geneva Pap. Risk Insur. Issues Pract.* **2003**, *28*, 382–393. [CrossRef]
17. Amegavi, G.B.; Langnel, Z.; Ofori, J.J.Y.; Ofori, D.R. The impact of adaptation on climate vulnerability: Is readiness relevant? *Sustain. Cities Soc.* **2021**, *75*, 103325. [CrossRef]
18. Jafino, B.; Walsh, B.; Rozenberg, J.; Hallegatte, S. *Revised Estimates of the Impact of Climate Change on Extreme Poverty by 2030*; World Bank Group: Washington, DC, USA, 2020.
19. Chen, C.; Noble, I.; Hellmann, J.; Coffee, J.; Murillo, M.; Chawla, N. *The Notre Dame-Global Adaptation Index (ND-GAIN) Country Index*; ND-GAIN: South Bend, IN, USA, 2015.
20. Nagy, G.J.; Gutiérrez, O.; Brugnoli, E.; Verocai, J.E.; Gómez-Erache, M.; Villamizar, A.; Olivares, I.; Azeiteiro, U.M.; Leal Filho, W.; Amaro, N. Climate vulnerability, impacts and adaptation in Central and South America coastal areas. *Reg. Stud. Mar. Sci.* **2019**, *29*, 100683. [CrossRef]
21. Chen, C.; Hellmann, J.; Berrang-Ford, L.; Noble, I.; Regan, P. A global assessment of adaptation investment from the perspectives of equity and efficiency. *Mitig. Adapt. Strateg. Glob. Chang.* **2018**, *23*, 101–122. [CrossRef]
22. Halkos, G.; Skouloudis, A.; Malesios, C.; Jones, N. A Hierarchical Multilevel Approach in Assessing Factors Explaining Country-Level Climate Change Vulnerability. *Sustainability* **2020**, *12*, 4438. [CrossRef]
23. Sarkodie, S.A.; Strezov, V. Economic, social and governance adaptation readiness for mitigation of climate change vulnerability: Evidence from 192 countries. *Sci. Total Environ.* **2019**, *656*, 150–164. [CrossRef] [PubMed]
24. Skovgaard, J. *The Economisation of Climate Change: How the G20, the OECD and the IMF Address Fossil Fuel Subsidies and Climate Finance*, 1st ed.; Cambridge University Press: Cambridge, UK, 2021; ISBN 978-1-108-68804-8.
25. Savvidou, G.; Atteridge, A.; Omari-Motsumi, K.; Trisos, C.H. Quantifying international public finance for climate change adaptation in Africa. *Clim. Policy* **2021**, *21*, 1020–1036. [CrossRef]
26. Global Center on Adaptation. *Adaptation Finance Flows in Africa*; Global Center on Adaptation: Rotterdam, The Netherlands, 2022.
27. Scandurra, G.; Thomas, A.; Passaro, R.; Bencini, J.; Carfora, A. Does climate finance reduce vulnerability in Small Island Developing States? An empirical investigation. *J. Clean. Prod.* **2020**, *256*, 120330. [CrossRef]
28. Mungai, E.M.; Ndiritu, S.W.; da Silva, I. Unlocking Climate Finance Potential for Climate Adaptation: Case of Climate Smart Agricultural Financing in Sub Saharan Africa. In *African Handbook of Climate Change Adaptation*; Oguge, N., Ayal, D., Adeleke, L., da Silva, I., Eds.; Springer International Publishing: Cham, Switzerland, 2021; pp. 2063–2083. ISBN 978-3-030-45106-6.
29. Lombardi, D. Climate crisis mitigation and adaptation: Educational and developmental psychology's responsibility in helping face this threat. *Educ. Dev. Psychol.* **2022**, *39*, 1–4. [CrossRef]
30. Ahmad, Z.; Saleem, A. Predicting Level of Development for Different Countries. *J. Sustain. Dev.* **2012**, *5*, 15. [CrossRef]
31. Morse, S. Quality of Life, Well-Being and the Human Development Index: A Media Narrative for the Developed World? *Soc. Indic. Res.* **2023**, *170*, 1035–1058. [CrossRef]
32. United Nations Development Programme. *Technical Note 1. Human Development Index*; UNDP: New York, NY, USA, 2018; pp. 1–6.
33. Osakede, U.A.; Aramide, V.O.; Adesipo, A.E.; Akunna, L.C. Correlates of human development in Africa: Evidence across gender and income group. *Res. Glob.* **2023**, *6*, 100135. [CrossRef]

34. Boswell, M.R.; Greve, A.I.; Seale, T.L. Climate Change Vulnerability Assessment. In *Climate Action Planning: A Guide to Creating Low-Carbon, Resilient Communities*; Boswell, M.R., Greve, A.I., Seale, T.L., Eds.; Island Press/Center for Resource Economics: Washington, DC, USA, 2019; pp. 172–191. ISBN 978-1-61091-964-7.
35. Naylor, A.; Ford, J.; Pearce, T.; Van Alstine, J. Conceptualizing Climate Vulnerability in Complex Adaptive Systems. *One Earth* **2020**, *2*, 444–454. [CrossRef]
36. Michalak, J.L.; Withey, J.C.; Lawler, J.J.; Case, M.J. Future climate vulnerability—Evaluating multiple lines of evidence. *Front. Ecol. Environ.* **2017**, *15*, 367–376. [CrossRef]
37. Warren, R.F.; Wilby, R.L.; Brown, K.; Watkiss, P.; Betts, R.A.; Murphy, J.M.; Lowe, J.A. Advancing national climate change risk assessment to deliver national adaptation plans. *Philos. Trans. R. Soc. Math. Phys. Eng. Sci.* **2018**, *376*, 20170295. [CrossRef]
38. Papathoma-Köhle, M.; Cristofari, G.; Wenk, M.; Fuchs, S. The importance of indicator weights for vulnerability indices and implications for decision making in disaster management. *Int. J. Disaster Risk Reduct.* **2019**, *36*, 101103. [CrossRef]
39. Fakhruddin, B.; Boylan, K.; Wild, A.; Robertson, R. Chapter 12—Assessing vulnerability and risk of climate change. In *Climate Extremes and Their Implications for Impact and Risk Assessment*; Sillmann, J., Sippel, S., Russo, S., Eds.; Elsevier: Amsterdam, The Netherlands, 2020; pp. 217–241. ISBN 978-0-12-814895-2.
40. Wang, P.; Qiao, W.; Wang, Y.; Cao, S.; Zhang, Y. Urban drought vulnerability assessment—A framework to integrate socio-economic, physical, and policy index in a vulnerability contribution analysis. *Sustain. Cities Soc.* **2020**, *54*, 102004. [CrossRef]
41. Mboya, A. Vulnerability and the Climate Change Regime. *UCLA J. Environ. Law Policy* **2018**, *36*, 79–103. [CrossRef]
42. Santos-Lacueva, R.; Ariza, E.; Romagosa, F.; Saladié, Ò. The vulnerability of destinations to climate change: A comparative analysis of contextual socio-political factors. *J. Sustain. Tour.* **2019**, *27*, 1217–1238. [CrossRef]
43. Sarun, S.; Ghermandi, A.; Sheela, A.M.; Justus, J.; Vineetha, P. Climate change vulnerability in a tropical region based on environmental and socio-economic factors. *Environ. Monit. Assess.* **2018**, *190*, 727. [CrossRef]
44. Surjan, A.; Kudo, S.; Uitto, J.I. Risk and Vulnerability. In *Sustainable Development and Disaster Risk Reduction*; Uitto, J.I., Shaw, R., Eds.; Disaster Risk Reduction; Springer: Tokyo, Japan, 2016; pp. 37–55. ISBN 978-4-431-55078-5.
45. Formetta, G.; Feyen, L. Empirical evidence of declining global vulnerability to climate-related hazards. *Glob. Environ. Chang.* **2019**, *57*, 101920. [CrossRef]
46. Nauges, C.; Wheeler, S.A.; Fielding, K.S. The relationship between country and individual household wealth and climate change concern: The mediating role of control. *Environ. Dev. Sustain.* **2021**, *23*, 16481–16503. [CrossRef]
47. O'Neill, B.C.; Jiang, L.; Kc, S.; Fuchs, R.; Pachauri, S.; Laidlaw, E.K.; Zhang, T.; Zhou, W.; Ren, X. The effect of education on determinants of climate change risks. *Nat. Sustain.* **2020**, *3*, 520–528. [CrossRef]
48. Bowen, A.; Cochrane, S.; Fankhauser, S. Climate change, adaptation and economic growth. *Clim. Chang.* **2012**, *113*, 95–106. [CrossRef]
49. Wu, J.; Han, G.; Zhou, H.; Li, N. Economic development and declining vulnerability to climate-related disasters in China. *Environ. Res. Lett.* **2018**, *13*, 034013. [CrossRef]
50. Kim, H.; Marcouiller, D.; Woosnam, K. Multilevel Climate Governance, Anticipatory Adaptation, and the Vulnerability-Readiness Nexus. *Rev. Policy Res.* **2021**, *38*, 222–242. [CrossRef]
51. Jubeh, G.; Mimi, Z. Governance and Climate Vulnerability Index. *Water Resour. Manag.* **2012**, *26*, 4147–4162. [CrossRef]
52. Thomas, K.; Hardy, R.D.; Lazrus, H.; Mendez, M.; Orlove, B.; Rivera-Collazo, I.; Roberts, J.T.; Rockman, M.; Warner, B.P.; Winthrop, R. Explaining differential vulnerability to climate change: A social science review. *WIREs Clim. Chang.* **2019**, *10*, e565. [CrossRef]
53. Muiderman, K.; Gupta, A.; Vervoort, J.; Biermann, F. Four approaches to anticipatory climate governance: Different conceptions of the future and implications for the present. *WIREs Clim. Chang.* **2020**, *11*, e673. [CrossRef]
54. Hamilton, M.; Lubell, M. Collaborative Governance of Climate Change Adaptation Across Spatial and Institutional Scales. *Policy Stud. J.* **2018**, *46*, 222–247. [CrossRef]
55. Trell, E.-M.; Van Geet, M.T. The Governance of Local Urban Climate Adaptation: Towards Participation, Collaboration and Shared Responsibilities. *Plan. Theory Pract.* **2019**, *20*, 376–394. [CrossRef]
56. Barnett, J. Global environmental change II: Political economies of vulnerability to climate change. *Prog. Hum. Geogr.* **2020**, *44*, 1172–1184. [CrossRef]
57. Salamanca, A.; Nguyen, H. *Climate Change Adaptation Readiness in the ASEAN Countries*; Stockholm Environment Institute: Stockholm, Sweden, 2016.
58. Cappelli, F. Investigating the Origins of Differentiated Vulnerabilities to Climate Change and their Effects on Wellbeing 2020. *Econ. Politica* **2023**, *40*, 1051–1074. [CrossRef]
59. Turner, M.D. Climate vulnerability as a relational concept. *Geoforum* **2016**, *68*, 29–38. [CrossRef]
60. Scott, D.; Hall, C.M.; Gössling, S. Global tourism vulnerability to climate change. *Ann. Tour. Res.* **2019**, *77*, 49–61. [CrossRef]
61. Doshi, D.; Garschagen, M. Understanding Adaptation Finance Allocation: Which Factors Enable or Constrain Vulnerable Countries to Access Funding? *Sustainability* **2020**, *12*, 4308. [CrossRef]
62. Williamson, T.; Hesseln, H.; Johnston, M. Adaptive capacity deficits and adaptive capacity of economic systems in climate change vulnerability assessment. *For. Policy Econ.* **2012**, *15*, 160–166. [CrossRef]
63. Adom, P.K.; Amoani, S. The role of climate adaptation readiness in economic growth and climate change relationship: An analysis of the output/income and productivity/institution channels. *J. Environ. Manag.* **2021**, *293*, 112923. [CrossRef]

64. Banerjee, S.; Hussain, A.; Tuladhar, S.; Mishra, A. Building capacities of women for climate change adaptation: Insights from migrant-sending households in Nepal. *Clim. Chang.* **2019**, *157*, 587–609. [CrossRef]
65. Glaas, E.; Hjerpe, M.; Wihlborg, E.; Storbjörk, S. Disentangling municipal capacities for citizen participation in transformative climate adaptation. *Environ. Policy Gov.* **2022**, *32*, 179–191. [CrossRef]
66. Mesfin, D.; Simane, B.; Belay, A.; Recha, J.W.; Schmiedel, U. Assessing the Adaptive Capacity of Households to Climate Change in the Central Rift Valley of Ethiopia. *Climate* **2020**, *8*, 106. [CrossRef]
67. Abdul-Razak, M.; Kruse, S. The adaptive capacity of smallholder farmers to climate change in the Northern Region of Ghana. *Clim. Risk Manag.* **2017**, *17*, 104–122. [CrossRef]
68. Chepkoech, W.; Mungai, N.W.; Stöber, S.; Lotze-Campen, H. Understanding adaptive capacity of smallholder African indigenous vegetable farmers to climate change in Kenya. *Clim. Risk Manag.* **2020**, *27*, 100204. [CrossRef]
69. Nguyen, Q.A.; Miller, F.; Bowen, K.; Tan Sinh, B. Evaluating capacity for climate change adaptation in the health and water sectors in Vietnam: Constraints and opportunities. *Clim. Dev.* **2017**, *9*, 258–273. [CrossRef]
70. Ranasinghe, R.D.A.K.; Korale-Gedara, P.M.; Weerasooriya, S.A. Climate change adaptation and adaptive capacities of dairy farmers: Evidence from village tank cascade systems in Sri Lanka. *Agric. Syst.* **2023**, *206*, 103609. [CrossRef]
71. Eriksen, S.; Schipper, E.L.F.; Scoville-Simonds, M.; Vincent, K.; Adam, H.N.; Brooks, N.; Harding, B.; Khatri, D.; Lenaerts, L.; Liverman, D.; et al. Adaptation interventions and their effect on vulnerability in developing countries: Help, hindrance or irrelevance? *World Dev.* **2021**, *141*, 105383. [CrossRef]
72. Leal Filho, W.; Azeiteiro, U.M.; Balogun, A.-L.; Setti, A.F.F.; Mucova, S.A.R.; Ayal, D.; Totin, E.; Lydia, A.M.; Kalaba, F.K.; Oguge, N.O. The influence of ecosystems services depletion to climate change adaptation efforts in Africa. *Sci. Total Environ.* **2021**, *779*, 146414. [CrossRef] [PubMed]
73. Marino, E.; Ribot, J. Special Issue Introduction: Adding insult to injury: Climate change and the inequities of climate intervention. *Glob. Environ. Chang.* **2012**, *22*, 323–328. [CrossRef]
74. Conevska, A.; Ford, J.; Lesnikowski, A.; Harper, S. Adaptation financing for projects focused on food systems through the UNFCCC. *Clim. Policy* **2019**, *19*, 43–58. [CrossRef]
75. Theokritoff, E.; Thomas, A.; Lissner, T.; Schleussner, C.-F. Interacting adaptation constraints in the Caribbean highlight the importance of sustained adaptation finance. *Clim. Risk Manag.* **2023**, *39*, 100483. [CrossRef]
76. Rahman, S.M.; Ahmad, M.M. Climate finance towards vulnerable countries: Evidence from the Global Environment Facility (GEF). *Int. J. Green Econ.* **2015**, *9*, 199. [CrossRef]
77. Das, M. Allocation of Climate Funds and the Adaptation Infrastructure Gap. *Sustain. Dev. Res.* **2022**, *4*, p42. [CrossRef]
78. Omukuti, J.; Barrett, S.; White, P.C.L.; Marchant, R.; Averchenkova, A. The green climate fund and its shortcomings in local delivery of adaptation finance. *Clim. Policy* **2022**, *22*, 1225–1240. [CrossRef]
79. Saeed, S.; Makhdum, M.S.A.; Anwar, S.; Yaseen, M.R. Climate Change Vulnerability, Adaptation, and Feedback Hypothesis: A Comparison of Lower-Middle, Upper-Middle, and High-Income Countries. *Sustainability* **2023**, *15*, 4145. [CrossRef]
80. Barrett, S. Subnational Adaptation Finance Allocation: Comparing Decentralized and Devolved Political Institutions in Kenya. *Glob. Environ. Polit.* **2015**, *15*, 118–139. [CrossRef]
81. Basty, N.; Azouz Ghachem, D. A Sectoral Approach of Adaptation Finance in Developing Countries: Does Climate Justice Apply? *Sustainability* **2022**, *14*, 10835. [CrossRef]
82. Baltagi, B.H. Introduction. In *Econometric Analysis of Panel Data*; Baltagi, B.H., Ed.; Springer Texts in Business and Economics; Springer International Publishing: Cham, Switzerland, 2021; pp. 1–14. ISBN 978-3-030-53953-5.
83. Mabaso, M.; Zama, T.; Mlangeni, L.; Mbiza, S.; Mkhize-Kwitshana, Z. Association between the Human Development Index and Millennium Development Goals 6 Indicators in Sub-Saharan Africa from 2000 to 2014: Implications for the New Sustainable Development Goals. *J. Epidemiol. Glob. Health* **2018**, *8*, 77–81. [CrossRef]
84. Climate Finance and the USD 100 Billion Goal—OECD. Available online: https://www.oecd.org/climate-change/finance-usd-100-billion-goal/ (accessed on 8 November 2023).
85. World Development Indicators | DataBank. Available online: https://databank.worldbank.org/source/world-development-indicators/Series/EN.ATM.GHGT.KT.CE (accessed on 14 November 2023).
86. Malik, I.H.; Ford, J.D. Addressing the Climate Change Adaptation Gap: Key Themes and Future Directions. *Climate* **2024**, *12*, 24. [CrossRef]
87. United Nations Environment Programme. *Adaptation Gap Report 2022: Too Little, Too Slow—Climate Adaptation Failure Puts World at Risk*; United Nations Environment Programme: Nairobi, Kenya, 2022.
88. Asian Development Bank. *Climate Finance Landscape of Asia and the Pacific*; Asian Development Bank: Manila, Philippines, 2023.
89. UNFCCC. *Climate Change: Impacts, Vulnerabilities and Adaptation in Developing Countries*; UNFCCC: New York, NY, USA, 2007.
90. Global Center on Adaptation. *State and Trends in Climate Adaptation Finance 2023*; Global Center on Adaptation: Rotterdam, The Netherlands, 2023.
91. Phillips, P.C.B.; Sul, D. Dynamic panel estimation and homogeneity testing under cross section dependence. *Econom. J.* **2003**, *6*, 217–259. [CrossRef]
92. Breusch, T.S.; Pagan, A.R. The Lagrange Multiplier Test and its Applications to Model Specification in Econometrics. *Rev. Econ. Stud.* **1980**, *47*, 239–253. [CrossRef]

93. Baltagi, B.H.; Feng, Q.; Kao, C. A Lagrange Multiplier test for cross-sectional dependence in a fixed effects panel data model. *J. Econom.* **2012**, *170*, 164–177. [CrossRef]
94. Brooks, C. *Introductory Econometrics for Finance*; Cambridge University Press: Cambridge, UK, 2019; ISBN 978-1-108-42253-6.
95. Lancelot, M.L. Raod: Analysis of Overdispersed Data. R Package Version 0.4-1.2 2022. Available online: https://cran.r-project.org/web/packages/aod/aod.pdf (accessed on 10 April 2024).
96. Riffenburgh, R.H. Chapter 25—Epidemiology. In *Statistics in Medicine*, 3rd ed.; Riffenburgh, R.H., Ed.; Academic Press: San Diego, CA, USA, 2012; pp. 535–549. ISBN 978-0-12-384864-2.
97. Born, B.; Breitung, J. Testing for Serial Correlation in Fixed-Effects Panel Data Models. *Econom. Rev.* **2016**, *35*, 1290–1316. [CrossRef]
98. Drukker, D.M. Testing for Serial Correlation in Linear Panel-data Models. *Stata J.* **2003**, *3*, 168–177. [CrossRef]
99. Bai, J.; Choi, S.H.; Liao, Y. Feasible generalized least squares for panel data with cross-sectional and serial correlations. *Empir. Econ.* **2021**, *60*, 309–326. [CrossRef]
100. Hothorn, T.; Zeileis, A.; Farebrother, R.W.; Cummins, C.; Millo, G.; Mitchell, D. lmtest: Testing Linear Regression Models. R Package Version 0.9-40 2022. Available online: https://cran.r-project.org/web/packages/lmtest/lmtest.pdf (accessed on 10 April 2024).
101. Shrestha, M.B.; Bhatta, G.R. Selecting appropriate methodological framework for time series data analysis. *J. Finance Data Sci.* **2018**, *4*, 71–89. [CrossRef]
102. Lin, Z.; Brannigan, A. Advances in the Analysis of Non-stationary Time Series: An Illustration of Cointegration and Error Correction Methods in Research on Crime and Immigration. *Qual. Quant.* **2003**, *37*, 151–168. [CrossRef]
103. Jalil, A.; Rao, N.H. Chapter 8—Time Series Analysis (Stationarity, Cointegration, and Causality). In *Environmental Kuznets Curve (EKC)*; Özcan, B., Öztürk, I., Eds.; Academic Press: Cambridge, MA, USA, 2019; pp. 85–99. ISBN 978-0-12-816797-7.
104. Phillips, P.; Peron, R. Testing for a Unit Root in Time Series Regression. *Oxf. Univ. Press Biom. Trust.* **1988**, *75*, 335–346. Available online: https://www.jstor.org/stable/2336182 (accessed on 10 April 2024). [CrossRef]
105. Herwartz, H.; Maxand, S.; Raters, F.H.C.; Walle, Y.M. Panel Unit-root Tests for Heteroskedastic Panels. *Stata J.* **2018**, *18*, 184–196. [CrossRef]
106. Banerjee, A. Panel Data Unit Roots and Cointegration: An Overview. *Oxf. Bull. Econ. Stat.* **1999**, *61*, 607–629. [CrossRef]
107. Beck, N.; Katz, J.N. What to do (and not to do) with Time-Series Cross-Section Data. *Am. Polit. Sci. Rev.* **1995**, *89*, 634–647. [CrossRef]
108. Maureen Tobechukwu, N.; Maxwell Azubuike, I. Panel Data Estimators in the Presence of Serial and Spatial Correlation with Panel Heteroscedasticity: A Simulation Study. *Q. J. Econom. Res.* **2020**, *6*, 1–11. [CrossRef]
109. Appiah-Otoo, I.; Acheampong, A.O.; Song, N.; Chen, X. The impact of information and communication technology (ICT) on carbon dioxide emissions: Evidence from heterogeneous ICT countries. *Energy Environ.* **2023**, *34*, 3080–3102. [CrossRef]
110. Freitag, C.; Berners-Lee, M.; Widdicks, K.; Knowles, B.; Blair, G.S.; Friday, A. The real climate and transformative impact of ICT: A critique of estimates, trends, and regulations. *Patterns* **2021**, *2*, 100340. [CrossRef]
111. Dauda, L.; Long, X.; Mensah, C.N.; Salman, M. The effects of economic growth and innovation on CO_2 emissions in different regions. *Environ. Sci. Pollut. Res.* **2019**, *26*, 15028–15038. [CrossRef] [PubMed]
112. Du, K.; Li, J. Towards a green world: How do green technology innovations affect total-factor carbon productivity. *Energy Policy* **2019**, *131*, 240–250. [CrossRef]
113. Bétila, R.R. The impact of Ease of Doing Business on economic growth: A dynamic panel analysis for African countries. *SN Bus. Econ.* **2021**, *1*, 144. [CrossRef]
114. Alaeddine, H. The Mediating Effect of Ease of Doing Business on the Relationship Between Public Debt and infrastructure In Arab Countries. *BAU J. Creat. Sustain. Dev.* **2023**, *4*, 10. [CrossRef]
115. Fernández Lopera, C.C.; Mendes, J.M.; Barata, E.J. The Differential Risk Transfer: A new approach for reducing vulnerability to climate-related hazards. *Disaster Prev. Manag. Int. J.* **2022**, *31*, 550–564. [CrossRef]
116. Ayanlade, A.; Smucker, T.A.; Nyasimi, M.; Sterly, H.; Weldemariam, L.F.; Simpson, N.P. Complex climate change risk and emerging directions for vulnerability research in Africa. *Clim. Risk Manag.* **2023**, *40*, 100497. [CrossRef]

Disclaimer/Publisher's Note: The statements, opinions and data contained in all publications are solely those of the individual author(s) and contributor(s) and not of MDPI and/or the editor(s). MDPI and/or the editor(s) disclaim responsibility for any injury to people or property resulting from any ideas, methods, instructions or products referred to in the content.

forests

MDPI

Article

Traces of Local Adaptive Acclimatization Response in the Tracheid Anatomical Traits between Dry and Wet Mesic Norway Spruce (*Picea abies*) Forests in Moravia, Czech Republic?

Dimitrios Tsalagkas [1,*], Tomáš Novák [1], Marek Fajstavr [1,2], Hanuš Vavrčík [1], Vladimír Gryc [1], Petr Horáček [1,2] and Kyriaki Giagli [1]

[1] Department of Wood Science, Faculty of Forestry and Wood Technology, Mendel University in Brno, Zemědělská 3, 613 00 Brno, Czech Republic; tom.novak.olsi@seznam.cz (T.N.); fajstavr.m@czechglobe.cz (M.F.); hanus.vavrcik@mendelu.cz (H.V.); vladimir.gryc@mendelu.cz (V.G.); horacek.p@czechglobe.cz (P.H.); kyriaki.giagli@mendelu.cz (K.G.)

[2] Department of Xylogenesis and Biomass Allocation, Czechglobe—Global Change Research Institute, The Czech Academy of Sciences, Bělidla 4a, 603 00 Brno, Czech Republic

* Correspondence: dimitrios.tsalagkas@mendelu.cz

Abstract: Norway spruce (*Picea abies*) forests in temperate zones are already reacting to short-term extreme summer heatwaves, threatening the vitality of trees and forest productivity, and can even lead to local and regional dieback events. Examining quantitative wood anatomy can provide helpful information in terms of understanding the physiology mechanisms and related responses of conifer trees to local environmental interactions in relation to tracheid adaptive capacity. This study analysed the tracheid functional anatomical traits (FATs) plasticity of six young Norway spruce trees growing in two mesic research plots with high annual precipitation (~43%) and air temperature differences during 2010–2017. The research plots are located in the sub-mountainous (Rájec Němčice) and mountainous (Bílý Kříž) belts of the Moravia region, Czech Republic. Vapour pressure deficit and cell wall reinforcement index (CWRI) were shown to be the most representative environmental parameters as proxies of dry conditions. Tracheid FATs indicated latewood phenological plasticity sensitivity, with more pronounced variability in the warmer and drier plots. Latewood tracheids of Norway spruce trees grown in the RAJ formed significantly thicker cell walls than BK during the studied period. The observed differences between the two research plots indicate additional support for tracheid cells' hydraulic safety against cavitation and potential traces of adaptive acclimatization response.

Keywords: quantitative wood anatomy; climatic response; cavitation resistance; vapour pressure deficit; cell wall reinforcement index

1. Introduction

Trees' growth and forest ecosystems are closely interlinked with climate forcing and variability at short- and long-term scales [1]. Changes in any environmental factor trigger a cascade of modifications at the physiological and morphoanatomical levels of plants, affecting plant growth (meristem activity and morphological development) and productivity [2]. As the most fundamental attributes of woody plants over the life of a tree are the result of short- to long-term physiological responses, xylem tissues are under intensive selective pressure to respond to abiotic and biotic challenges [3,4] adaptively.

The variability of xylem cell anatomy within a population can be very high, reflecting a genetic richness that leads to the selection of new genotypes of the same species (genetic change adaptation fitting into the new environment) and/or phenotypic abilities in terms of acclimatization (potentially reversible physiological responses to environmental variables), which is fundamental for adaptive processes [5]. Phenotypic plasticity is genetically controlled and allows plastic reactions of the functional anatomical traits (FATs), which permit tree species to withstand a broad range of environmental conditions in which trees live [6]. Phenotypic plasticity is essential in trees' hydraulic strategies concerning water availability, suggesting that tracheid FATs may have allowed conifers to radiate into different habitats [7].

Quantitative wood anatomy, i.e., intra-annual resolved xylem cell anatomical traits along tree ring series, is a methodological approach that operates at the cellular level [8]. This method provides new possibilities in terms of providing better insights into the mechanisms related to year-to-year xylem FATs and the related consequences of environmental interactions on conifers' tree physiology and growth at highly resolved temporal and spatial scales [9,10]. This is achieved by quantitatively investigating the variability of the tracheid FATs within specific tree-ring zones and can provide useful information concerning tracheid cell anatomy and environment [11,12]. Several studies have emphasized a close interconnection between conifer tracheid FATs and seasonal environmental and weather fluctuations, reflecting changes in regional conditions related to these short, sub-annual intervals [8,13–19]. Climate change is expected to intensify regional-scale droughts; therefore, quantifying the extent of plasticity driven by the FATs of the xylem is critical for predicting the tolerance ranges of different species and their resilience to dry conditions [20].

Norway spruce (*Picea abies* (L.) Karst) a late-successional species with slow, early growth and late mean annual increment culmination; is considered to grow better in mesic sites and intermediate-to-high-fertility soils [21]. In addition, Norway spruce, as an intermediately shade-tolerant species, is sensitive to weather conditions, for instance, summer water deficits and soil water recharge in the preceding autumn and early spring [22]. It is also predicted to suffer from warmer and wetter winters [23]. Since 2000, parts of Central Europe have experienced persistent hot and dry periods in summer and non-summer periods [24], leading to a greater risk of precipitation deficits combined with increased temperatures and evapotranspiration [25]. Precipitation deficits can lead to a tight hydrological balance and, potentially, plant water stress [26]; thus, concerns have been raised concerning the ecological stability and adaptation of drought-sensitive Norway spruce trees [22,27,28]. Boden et al. [29] indicated a loss of growth resilience and stress response to decreasing water availability and a relatively limited short-term adaptive capacity to changing climate conditions in tree-ring width (TRW) series of Norway spruce trees located in southwestern Germany in sub-mountainous and mountainous belts.

However, information relating to tree species' adaptive capacity and regional growth resilience to climate change in European forests is complex. Moreover, individual site characteristics may buffer or boost the impacts of heat, drought, and storm events [30]. Focused research is required because trees adapt to local water availability [25] at various temporal and spatial scales [31]. Considering the current environmental changes and their impact on tracheid cells, a better understanding of the environmental effects on dry- and wet-mesic Norway spruce forests with significant precipitation differences and how the anatomy of Norway spruce tracheid cells cope with the harsher climate conditions is decisive.

Cavitation resistance is associated with the 'thickness to span ratio', i.e., the ratio between the double cell wall thickness (CWT) and the conduit lumen radial diameter width (LRD) of tracheid cells. Higher resistance to cavitation requires stronger tracheids with a higher 'thickness to span ratio' to withstand mechanical stress and lower negative pressures [7,32–34]. Therefore, we hypothesized that the Norway spruce trees growing in warmer and drier environments would display greater hydraulic safety, i.e., thicker cell walls and narrower conduits, than in colder and wetter environments due to the increased unfavourable conditions in response to water stress and drought intensity. We also hypothesized that the FATs of the Norway spruce trees in the selected research plots were different concerning local environmental conditions and precipitation regimes, suggesting an adaptive acclimatization response [35–37]. The objectives of this study were (i) to assess the phenotypic plasticity adjustments of tracheid FATs on young Norway spruce trees to environmental variation and (ii) to analyse the correlations between the examined environmental parameters (EPs) and tracheid FATs to assess which parameters determine the observed plasticity of tracheid cellular anatomy.

2. Materials and Methods

2.1. Study Sites and Characteristics

Two Norway spruce research plots, located on two experimental ecosystem stations (Figure A1) in the Czech Republic (CZ), part of the Czech long-term ecosystem research network (https://lter.cz/en/homepage, accessed on 2 April 2024), the Czech Carbon Observation System (CzeCOS) network (http://www.czecos.cz/en.html, accessed on 2 April 2024) and the Integrated Carbon Observation System (ICOS) International Network (https://www.icos-cp.eu, accessed on 2 April 2024) [38,39] were studied to account for local variability. Rájec Němčice (RAJ) is situated at a lower altitude (600–660 m a.s.l.) on a hilly slope with NEE exposure in the South Moravian region, while Bílý Kříž (BK) is situated at a higher altitude (800–900 m a.s.l.) on an SSW-oriented planar slope with a mountain ridge terrain [40]. The ecosystem type can be classified as coniferous evergreen forests, and both forest research plots are dominated by a monoculture of Norway spruce trees [41]. Cambisol (RAJ) and podzol (BK) soils crosslinked with several bedrock types and altitudinal zones represent the two major forest soil and associated soil groups, covering over 80% of the CZ territory [42].

From a climatic point of view (Table 1), the RAJ sub-mountainous region is relatively warm and moderately dry, with frequent high-temperature episodes and dry periods, particularly in the summer months [43]. BK experiences a cold and humid environment with high annual precipitation [39,40] and represents the mountainous region conditions of Central Europe [44]. Therefore, this study will assess RAJ as a dry-mesic Norway spruce forest research plot and BK as a wet-mesic Norway spruce forest research plot.

Table 1. Long-term weather data (annual mean air temperature and sum of precipitation) for the 1990–2022 period, according to the records of two nearby stations belonging to the Czech Hydrometeorological Institute (www.chmi.cz, accessed on 2 April 2024).

Long-Term Weather Data	Rájec Němčice (RAJ)	Bílý Kříž (BK)
Mean annual air temperature, °C	7.3 ± 0.8 [1]	7.0 ± 0.7 [2]
Mean annual sum precipitation, mm	638.5 ± 93.1	945.1 ± 126.5
Mean air temperature Apr–Sept, °C	13.7 ± 0.9	13.1 ± 0.7
Mean sum precipitation Apr–Sept, mm	409.3 ± 86.4	575.1 ± 131.5

[1] Protivanov meteorological station (ID: B1PROT01; ~675 m a.s.l.; 49°29′ N, 16°50′ E); [2] Bílá Konečná meteorological station (ID: O1BILA01; ~720 m a.s.l.; 49°27′ N, 18°31′ E).

2.2. Sample Collection and Preparation

Six healthy, young Norway spruce trees were harvested at the donor research plots during the spring of 2018 before the budburst period. All of the selected trees were uniform, with a conical crown, and without visible damage or abnormality on either the stem or the crown. The trees were comparable in size and age. The mean tree height, breast height diameter, and age of Norway spruce trees harvested in the RAJ were 20 m, 18 cm, and 39 years, respectively. Similarly, the mean tree height, breast height diameter and age of Norway spruce trees in the BK were 18 m, 17 cm, and 33 years, respectively. Wood samples of each tree were collected on-site by cutting stem discs (3–4 cm thickness) at breast height immediately after felling. Subsequently, with a small saw, 1 cm wide strips (with a direction from the bark to the pith) of wood were cut from the stem discs. Transversion sections of wood strips were prepared by using a WSL core-microtome [45] and coated with cornstarch-water-glycerol solution (10 g of cornstarch, 8 mL of distilled water, and 7 g of glycerol) to stabilize and prevent the cells from being distorted or broken, as suggested by Schneider and Gärtner [46]. The prepared microsections were stained using Safranin

(Waldeck GmbH & Co., Münster, Germany) and permanently mounted on the microscope slides with Euparal (Waldeck GmbH & Co., Münster, Germany).

2.3. Measurements and Statistical Analysis

2.3.1. Tree-Ring Width and Tracheid Anatomical Measurements

The TRW and FATs measurements were conducted cell-by-cell along three radial files within each ring, per year and tree, in each research plot, avoiding areas with visible defects (e.g., reaction wood, traumatic areas). The TRW tracheid FATs used in the study were direct measurements derived from the image analysis program WinCELL Pro 2010 (Regent Instruments, Canada). The analysis was restricted to the tree rings formed in 2010–2017 at the stems' northern alignment to avoid juvenile wood in the young Norway spruce trees [47]. For analysing the xylem structure at the sub-annual time scale, each tree ring was divided into earlywood (EW) and latewood (LW) tree-ring zones, according to Mork's index, which is the ratio between the double CWT and the LRD [48]. The tracheid measurements were assembled into three functional groups related to tree-ring growth, cell growth, and hydraulic safety (Table 2). Within the tree ring, TRW, earlywood width (EWW), latewood width (LWW), and latewood width percentage (LWW%) were calculated. Within each cell, we measured the following FATs: (1) CWT, (2) LRD, (3) the tracheid radial diameter width (TRD), (4) the cell lumen area (CLA), and (5) the cell wall reinforcement index (CWRI). The CWRI is defined as the 'thickness to span' ratio ($[t/b]^2$) [15,33], where t is the double CWT and b is the LRD as the side of a hypothetical square conduit of CLA [49]. During xylem formation and plant secondary growth, inter- and intra-annual changes in the anatomical structure of cells (such as the size of tracheid lumens and walls) in response to EPs also represent the trees' hydraulic architecture, which balances the optimisation of water transport (hydraulic efficiency) against the risk of hydraulic failure (or hydraulic safety) [37]. Xylem hydraulic safety estimated by the CWRI is considered a good proxy for xylem resistance to embolism and cell wall implosion since this ratio is believed to be a surrogate of the tracheid FATs linked to the mechanical and hydraulic properties of trees [19,37,50].

Table 2. A descriptive summary of Norway spruce (*Picea abies*) tree variables used in this study, divided by the functional groups.

Functional Groups	Parameters	Acronym	Unit
Ring growth	Tree-ring width	TRW	mm
	Earlywood width	EWW	mm
	Latewood width	LWW	mm
	Percentage of latewood width	LWW%	%
Cell growth	Cell wall thickness	CWT	µm
	Lumen radial diameter	LRD	µm
	Tracheid radial diameter	TRD	µm
	Cell lumen area	CLA	μm^2
Hydraulic safety	Conduit cell wall reinforcement index	CWRI	-

To calculate the *TRD* based on the relative position (*RP*) of each tracheid cell within a tree ring, we used the following equations [51]:

$$TRD = \frac{1}{2}(2CWT) + LRD + \frac{1}{2}(2CWT) \quad (1)$$

and

$$RP = \frac{X_n}{N} \quad (2)$$

where $2CWT$ is the thickness of a double cell wall, X_n is the rank of a cell in the tree ring increment, and N is the total number of cells. The $LWW\%$ was calculated as the ratio of LWW to TRW [52]:

$$LWW\% = \frac{LWW}{TRW} \times 100 \tag{3}$$

Since the number of tracheids varied among radial files among trees and within tree rings of each tree ring, tracheid FATs were standardized and then averaged based on the relative position of each tracheid within the ring [53].

2.3.2. Meteorological Parameters and Vapour Pressure Deficit

A multi-year dataset of half–hourly eddy covariance measurements made by the CzeCOS network instruments per research plot, recorded above the canopy, was used for this study. The air temperature (Tair) and relative humidity (RH) profile records were taken with EMS33 temperature and humidity sensors (EMS Brno, CZ). In contrast, the amount of precipitation (PRCP) was recorded using 386C Met One precipitation gauge instruments (Met One Instruments, USA). The daily-resolved values for mean Tair, RH and the sum of PRCP for the studied period of 2010–2017 were calculated. The mean annual Tair at the RAJ during 2010–2017 was 7.8 ± 0.7 °C, i.e., 1.2 °C greater than the BK (mean annual Tair was 6.6 °C). Accordingly, the mean annual sum PRPC at the RAJ was 660 ± 94 mm, i.e., 42.81% less than the mean annual sum PRPC at the BK (1154 ± 178 mm) (Figure 1).

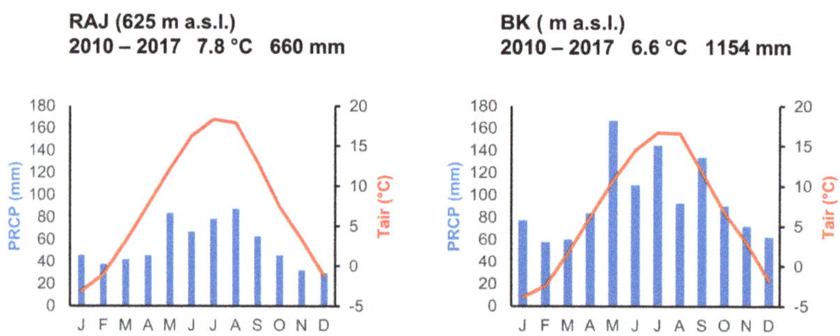

Figure 1. Mean monthly weather data of the studied research plots during 2010–2017 (Tair, air temperature; PRCP, sum of precipitation).

The vapour pressure deficit (VPD) reflects the effect of Tair and PRCP on the RH and transpiration demand. It is often monitored as a proxy for plant water stress, stimulating stomatal closure and photosynthetic carbon fixation [54]. Above-ground atmospheric demand for moisture was represented by the VPD (kPa), according to the equation:

$$\boldsymbol{VPD} = \left(\mathbf{1} - \frac{\boldsymbol{RH}}{\mathbf{100}}\right) \times \boldsymbol{SVP} \tag{4}$$

where SVP is the saturated vapour pressure for a given temperature. SVP was estimated according to Tetens equation:

$$\boldsymbol{SVP} = \mathbf{0.61078exp}\left(\frac{\mathbf{17.27}\boldsymbol{T_{air}}}{\boldsymbol{T_{air}} + \mathbf{237.3}}\right) \tag{5}$$

where *Tair* is in °C and SVP in kPa [55]. A VPD value of 1.5 kPa (VPD1.5) was chosen as an 'extreme' threshold, above which stomata close, representing the extremely dry air [56].

2.3.3. Standardized Precipitation Evapotranspiration Index (SPEI)

Variations in drought severity, duration, and timing complicate the situation, and it is challenging to quantify the tree species' physiological and anatomical adjustments to cope with drought [22]. Frequency, duration, and severity are the most critical characteristics of drought, and they depend on the established time scales as a function of the period. As time scales become larger, each new month has less impact, and drought becomes less frequent and of longer duration [57]. A one-month SPEI (SPEI–1) resolution was chosen to cover short drought episodes since these affect drought-sensitive tree species on shallow soils, as Fonti and Babushkina [17] suggested. Short-term interactions are preferable in assessing meteorological drought to soil moisture drought because they capture drought onsets and short-term severe bouts of drying that are potentially blunt in more prolonged interactions [58].

The impact of meteorological droughts is a complex, multisectoral, spatially variant and context-dependent phenomenon with different responses as a function of hydrology subsystems, vegetation, water resources management and seasonal time scales, making it difficult to quantify its characteristics and may lead to over- or underestimating reliable thresholds [59,60]. Thus, it is more appropriate to use an objective, location-specific method when defining drought thresholds to properly quantify drought seasonality, severity, and duration [59]. We adopted the drought definition of [57] as proposed by Parente et al. [61] and Spinoni et al. [62] to identify the drought events (DEs), which started once the analyzed SPEI–1 indicator drops below the corresponding value to a given negative standard deviation (SPEI–1 ≤ -1.0) for at least two consecutive months and finished when this indicator rises above zero (SPEI–1 ≥ 0). If, during the DE, at least two successive months show extreme drought conditions, the event would be considered extreme. Drought frequency displays the number of events, while drought duration demonstrates the number of months that a DE lasts. The assessment of meteorological DE intensity was classified into four classes, according to the increasing negative SPEI–1 values of less or equal to -1.0 [61]: (i) mild ($-1.0 <$ SPEI–1 ≤ 0), (ii) moderate ($-1.5 <$ SPEI–1 ≤ -1.0), (iii) severe ($-2.0 <$ SPEI–1 ≤ -1.5), and (iv) extreme (SPEI–1 ≤ -2.0).

2.3.4. Statistical Analysis

Based on the monthly Tair and PRCP data, we quantified the meteorological drought according to the SPEI climatic water balance drought index using the Thornthwaite R v. 4.1.1. package SPEI [63]. Daily records of mean Tair, RH, PRCP, and VPD values for the period 2010–2017 obtained from the meteorological stations located at the research plots were used to test the relationships between the FATs (i.e., CWT, TRD, CLA, and CWRI) and the EPs. The tested EPs (Tair, RH, PRCP, and VPD) were arranged in fixed intervals of 15-day time windows from the beginning of April till the end of July in the year of tree-ring formation and from the beginning of July till the end of October for the EW and LW tree-ring zones, respectively. The relationships between the EPs between the research plots and medial values of Norway spruce tracheid FATs in each tree-ring zone were computed using the non-parametric Kendall's tau rank correlation coefficient, which accounts for a smaller sample size [64]. Short time intervals of 15-day long time scales were chosen to capture better the dependency of the tracheid of FATs on short-time responses [9]. Correlations between the EPs and FATs were considered significant if the p-value was <0.05.

Finally, we adopted the non-parametric Mann–Whitney test to compare EPs between the two research plots, which is more robust to possible outliers and distribution skewness [65]. Similarly, to compare dry-mesic (RAJ) vs. wet-mesic (BK) FATs on the tracheids of Norway spruce trees without considering the time dimension, we used the non-parametric Mann–Whitney test, a method used to contrast two groups without considering any distribution information on the population [54]. All non-parametric tests were performed using the Statistica 13.4.0.14 analysis package (TIBCO Software Inc., Palo Alto, USA).

3. Results

3.1. Environmental Parameters

The trend agreement among the examined EPs between the two research plots (Figure 2) showed that the variations in time series resolutions of 15-day intervals follow a similar pattern across the investigated period of 2010–2017. Apart from the existing PRCP differences, we can also detect the low RH and high VPD values that occurred at both research plots during the summer months (July–August) of 2015 and, to a lesser extent, in 2013.

The interannual variability of the meteorological parameters (Tair and PRCP) during 2010–2017 portrayed a warmer and drier research plot in RAJ compared to the wetter and slightly colder conditions in the BK (Table A1). The number of VPD1.5 days, i.e., the days in which the extreme threshold value of VPD1.5 occurred, as was expected mainly occurred during the summer months of July and August. Interestingly, the years (2012 and 2014 during the spring period and 2013 and 2015 during the summer period) and the number of days with the increased VPD1.5 were almost identical for both research plots. However, during the most severe year of 2015, the number of VDP1.5 days in the driest RAJ was double when compared to the wettest BK, i.e.,18, and 9 days, respectively.

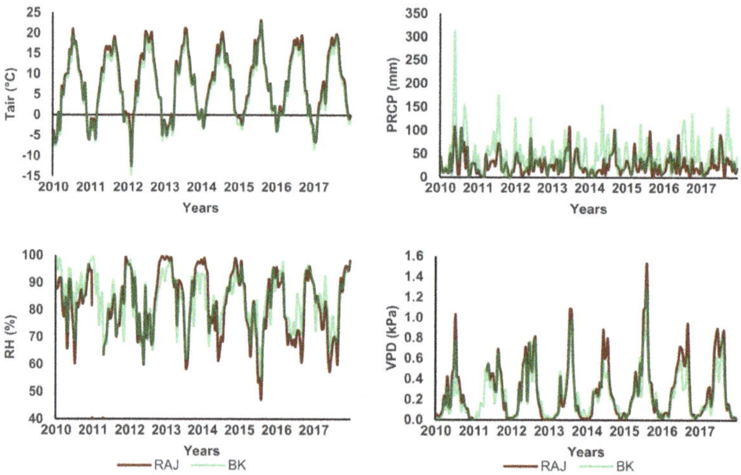

Figure 2. Time series of the mean values of the environmental parameters (EPs) for the 2010–2017 period for the Rájec Němčice (RAJ) and the Bílý Kříž (BK) plot (Tair, air temperature; PRCP, precipitation; RH, relative humidity; VPD, vapour pressure deficit).

Seasonal Mann–Whitney U test analyses during the growing season (April–September), as well as between the spring (April–June) and summer (July–September) sub-annual intervals, were performed to assess statistical differences in the EPs between the two studied research plots (Table 3, Figure A2). Tair, PRCP, and VPD were determined to be statistically significantly different during the growing season. Accordingly, only Tair and PRCP of the sub-annual periods were statistically significant, while RH and VPD were not. Yet, we should not overlook the higher calculated mean VPD value in the RAJ (0.56 $\pm$ 0.15 kPa) when compared to BK (0.45 $\pm$ 0.13 kPa). Since the variability of EPs during the 2010–2017 period was indicated between the two research plots, we expect that any differences in the phenotypic plasticity of the tracheid FATs of Norway spruce trees in the research plots would reflect the adjustment of the anatomy of tracheids' to environmental variations.

3.2. Drought Assessment

We compared drought frequency, duration, and intensity during the 2010–2017 reference period to investigate the spatial and temporal distributions between the RAJ and BK plots. During this period, we counted two DEs for RAJ and one DE for the BK plot, respectively (Figure 3). RAJ's first shorter duration DE occurred from January to February 2011 (two months), while a prolonged DE of seven months was detected from December 2013 to June 2014. Similarly, a quite dry period of three months, but not a DE in response to the definition, was also identified for the BK plot from January to March 2014. The only

DE that occurred in BK was during the spring of 2012 for two months (April to May 2012). No extreme conditions were detected, and approximately 3% and 15% of the total months were detected as severe and moderate drought months (Table 4).

Table 3. Seasonal environmental parameters (EPs) for the studied research plots during the 2010–2017 period. Mean air temperature, relative humidity, and vapour pressure deficit ($Tair_{2010–2017}$, $RH_{2010–2017}$, $VPD_{2010–2017}$), and mean sum precipitation ($PRCP_{2010–2017}$) for each duration of the growing and sub-annual period (April–September, April–June, and July–September) are presented. The ± symbol depicts the standard deviation values. Asterisks denote significant differences between the two research plots (ns—not significant, * $p < 0.05$, ** $p < 0.01$).

	Annual		April–September			April–June			July–September		
	RAJ	BK	RAJ	BK	*p*-Value	RAJ	BK	*p*-Value	RAJ	BK	*p*-Value
$Tair_{2010–2017}$ (°C)	7.8 (±0.7)	6.6 (±0.6)	14.2 (±0.6)	12.7 (±0.4)	**	12.0 (±0.8)	10.5 (±0.6)	**	16.4 (±0.7)	15.0 (±0.7)	**
$RH_{2010–2017}$ (%)	82.3 (±3.2)	83.7 (±3.0)	74.4 (±4.5)	78.1 (±4.1)	ns	75.0 (±5.8)	78.2 (±4.5)	ns	73.8 (±6.0)	78.1 (±5.4)	ns
$PRCP_{2010–2017}$ (mm)	660.3 (±94)	1153.7 (±178.6)	425.8 (±104.5)	732.1 (±239.2)	**	196.9 (±67.7)	360.4 (±132.5)	*	228.8 (±71.5)	371.7 (±113.6)	*
$VPD_{2010–2017}$ (kPa)	0.29 (±0.05)	0.25 (±0.05)	0.49 (±0.09)	0.39 (±0.08)	*	0.41 (±0.10)	0.34 (±0.07)	ns	0.56 (±0.15)	0.45 (±0.13)	ns

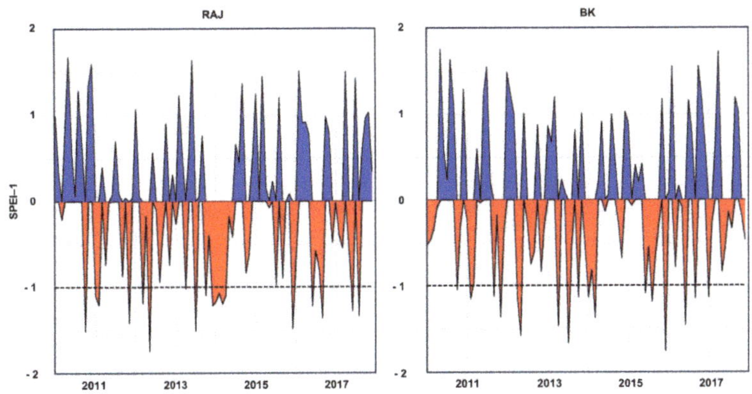

Figure 3. Running 1-month time scale Standardized Precipitation Evapotranspiration Index (SPEI–1) values from January 2010 to December 2017 in the RAJ and BK research plots. Positive values (blue) indicate cooler and wetter conditions, and negative values (red) indicate warmer and drier conditions. SPEI–1 was calculated with a Thornthwaite-type water balance. Dashed lines indicate the −1.0 threshold as the benchmark of drought stress conditions (moderate, severe, and extreme).

Table 4. General drought conditions (GDCs) were assessed with SPEI–1 and drought classes (mild, moderate, severe, and extreme) for the reference period of 2010–2017. Relative values are equal to the ratio between the number of drought months and the total number of months.

Plots	GDCs	Drought Class			
		Mild	Moderate	Severe	Extreme
RAJ	46.8%	27 (28%)	16 (17%)	2 (2%)	0 (0%)
BK	53.1%	34 (35%)	14 (15%)	3 (3%)	0 (0%)

3.3. Tree-Ring Growth Variability

Norway spruce trees grown in the dry-mesic (RAJ) and wet-mesic (BK) plots did not differ in diameter, height and age. In general, drier conditions of the RAJ plot during the 2010–2017 growing seasons promoted wider LWW, apart from the 2015 summer drought of July–August (based on VPD values and the number of VPD1.5 days), in which LWW sharply dropped. Specifically, the mean TRW and LWW% values in the RAJ were 1.9 ± 0.6 mm and 30.0 ± 7.5%, while in the BK 2.13 ± 0.4 mm and 19.5 ± 4.2%. Mann–Whitney *t*-tests demonstrated statistically significant differences in the EWW ($p < 0.05$) and LWW% ($p < 0.01$). At the same time, TRW and LWW were found not to be significantly different (Table 5).

Table 5. Intra-annual mean values tree-ring width (TRW), earlywood width (EWW), latewood width (LWW), and latewood width percentage (LWW%).

	TRW (mm)		EWW (mm)		LWW (mm)		LWW%	
Years	**RAJ**	**BK**	**RAJ**	**BK**	**RAJ**	**BK**	**RAJ**	**BK**
2010	2.27	2.00	1.47	1.64	0.80	0.36	35.2%	18.0%
2011	2.94	2.66	1.96	2.36	0.98	0.3	33.3%	11.2%
2012	2.17	2.47	1.58	1.86	0.59	0.61	27.2%	24.7%
2013	2.02	1.92	1.52	1.55	0.50	0.37	24.7%	19.3%
2014	1.62	2.65	1.10	2.05	0.52	0.60	32.7%	22.6%
2015	1.04	1.60	0.88	1.25	0.16	0.35	15.4%	21.9%
2016	1.88	2.17	1.30	1.80	0.58	0.37	31.4%	17.0%
2017	1.22	1.53	0.72	1.53	0.50	0.40	40.1%	21.1%
2010–2017 (st. dev.)	1.9 (±0.6)	2.13 (±0.4)	1.32 (a) (±0.4)	1.76 (b) (±0.3)	0.58 (±0.2)	0.42 (±0.1)	30.0% (a) (±7.5%)	19.5% (b) (±4.2%)

Letters denote statistically significant differences.

Although detecting long-term trends on such a relatively short time series of an 8-year analysis is challenging, most of the phenotypic plasticity variability was primarily related to changes occurring at the LW tree-ring zone. Based on the standardized tracheidograms and Mork's index, we observed a more considerable year-to-year variability in the LW tree-ring zone, which was more prominent in the drier RAJ. We also estimated that LW formation occurred earlier in the RAJ within the summer months, as illustrated in Figure 4.

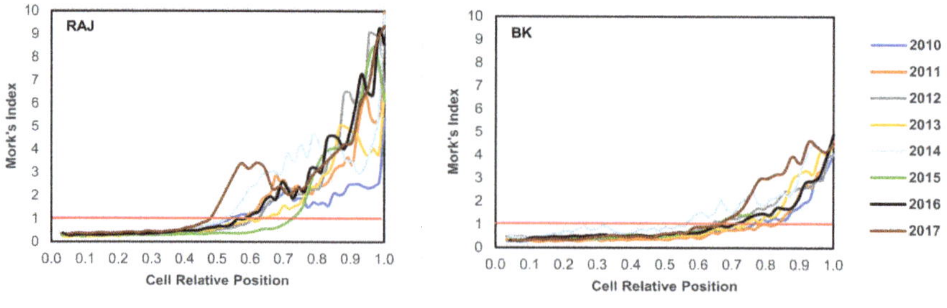

Figure 4. Interannual variability (2010–2017) of earlywood (EW) and latewood (LW) of the standardized tracheid cells within the Norway spruce (*Picea abies*) tree ring. The red line shows the boundary between the EW and LW tracheids as determined by Mork's index and the tracheid cells' relative position.

3.4. Sub-Annual Tracheid Anatomical Variability

The results indicate substantial changes in the sub-annual distributions and median values of tracheid FATs within the tree-ring zones between the two research plots (Table 6). Tracheid size (TRD and CLA) was found to be statistically significantly different in the EW tree-ring zone and CWT only in the LW tree-ring zone. Median CWRI, despite not being statistically significant (presumably due to the substantial variable values), was more than double in the RAJ (2.44 ± 1.9) compared to the CWRI (1.01 ± 0.5) in the BK. The results of this study suggest a downscaling of cell size (i.e., TRD and CLA) and the upscaling of CWT in the tracheid cells relating to Norway spruce tree ring growth in both research plots.

Table 6. Statistics describing the median values of measured and derived functional anatomical traits (FATs) of the Norway spruce trees in the dry-mesic RAJ (Rájec Němčice) and wet-mesic BK (Bílý Kříž) research plots.

FATs	Earlywood (EW)		Latewood (LW)	
	RAJ	BK	RAJ	BK
CWT (μm)	3.3 ± 0.4	3.3 ± 0.3	6.9 ± 0.7 (a)	5.3 ± 0.3 (b)
TRD (μm)	41.1 ± 3.1 (a)	35.9 ± 1.7 (b)	25.2 ± 3.9	22.5 ± 2.0
CLA (μm^2)	920.9 ± 131.8 (a)	746.7 ± 94.2 (b)	150.1 ± 57.0	200.9 ± 54.9
CWRI	0.04 ± 0.01	0.05 ± 0.02	2.44 ± 1.9	1.01 ± 0.5

Values are presented separately for EW and LW. Data are medians ± SD and correspond to the 2010–2017 reference period. Different letters indicate significant ($p < 0.05$) differences based on Mann–Whitney tests. Abbreviations of variables are CWT, cell wall thickness; TRD, Tracheid radial diameter; CLA, cell lumen area; CWRI, cell wall reinforcement index.

Figure 5 depicts the interannual variation in the FATs of the Norway spruce trees grown in the relatively warm and moderately dry RAJ area that has more frequent summer heatwaves compared to BK, with its cooler and higher annual precipitation. We observed that LW CWT and/or LW LRD in the Norway spruce trees located at the RAJ was reduced when unfavourable environmental conditions occurred, such as during short-term summer heatwaves with several days of high evapotranspiration demand (the year 2015, 18 days VPD1.5 threshold value) or after a prolonged DE during the 2014 winter-spring season.

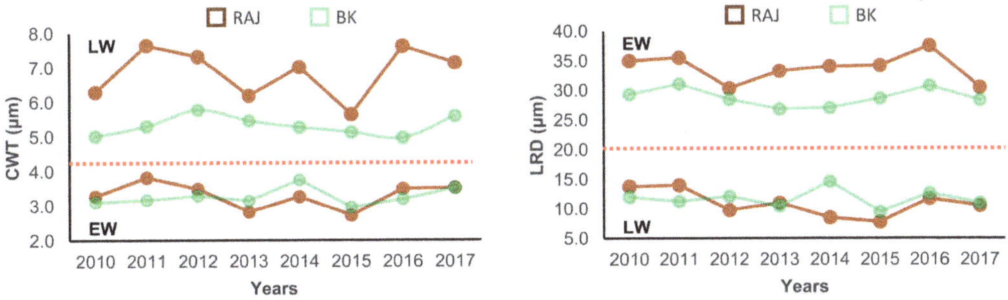

Figure 5. Interannual changes in median CWT and LRD were grouped per tree-ring zone (EW, earlywood; LW, latewood) and research plot (RAJ, Rájec Němčice; BK, Bílý Kříž) from 2010–2017. Horizontal red dotted lines indicate the boundary within each tree-ring zone.

3.5. Environment Parameters–Tracheid Functional Anatomical Traits Relationships

Within this study, we considered the April–June period more relevant to the EW tree-ring zone and the July–September period for the LW tree-ring zone, respectively. Furthermore, based on the lag in the EW–LW transition of the Norway spruce tracheids between the two plots, as can be observed in Figure 4 (according to Mork's index and cells' relative position), we considered July as the EW–LW transition month, yet also representing partly the LW tree-ring zone in the RAJ plot and the EW tree-ring zone in the BK plot. We also considered October since it is usually a month in which LW tracheids of Norway spruce trees are still forming.

The FATs in response to the EPs in the Norway spruce trees growing in the two plots showed seasonal and sub-annual variations, with VPD and PRCP being the most important EPs during tracheid formation. The contrasting positive or negative correlations illustrate the impact of EPs in each tree-ring zone by enlarging or narrowing CWT and TRD size, especially during the hottest summer months of July and August. Mainly, EPs were more influential in the warmer and drier RAJ rather than the colder and wetter BK (Figures 6 and 7).

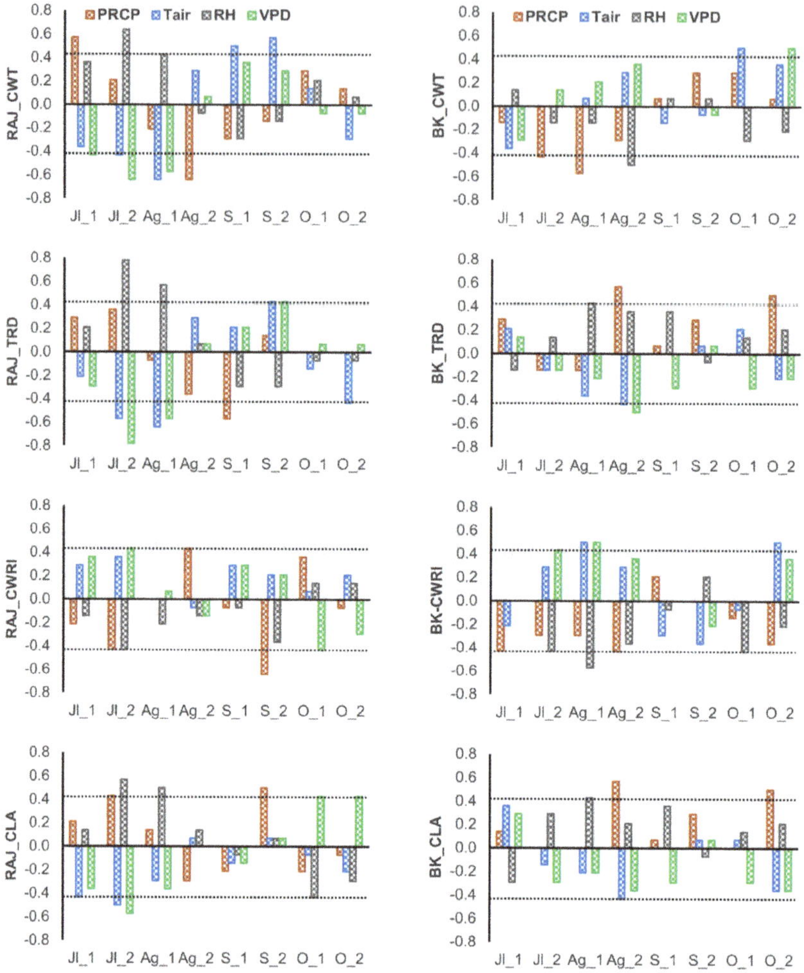

Figure 6. Kendall tau correlation coefficients between the studied EPs (PRCP, Tair, RH, and VPD) and the median values of the tracheid FATs (CWT, TRD, CLA, and CWRI) in the EW tree-ring zone of the Norway spruce trees in the RAJ and BK plots for the period 2010–2017. Months are abbreviated by letters (Ap, April; M, May; Jn, June; Jl, July). Numbers 1 and 2 denote the first half (1–15 days) and second half (16–30 days) time intervals of each month, respectively. Horizontal dotted lines indicate a threshold for significant correlation at $p < 0.05$.

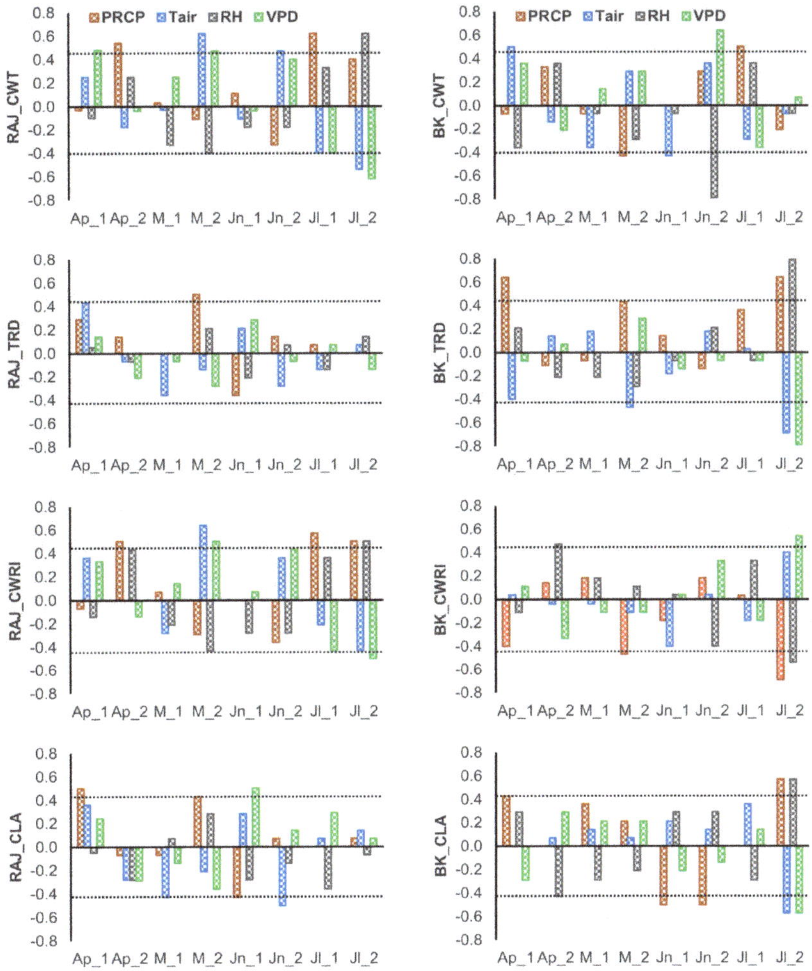

Figure 7. Kendall tau correlation coefficients between the studied EPs (PRCP, Tair, RH, and VPD) and the median values of the tracheid FATs (CWT, TRD, CLA, and CWRI) in the LW tree-ring zone of the Norway spruce trees in the RAJ and BK plots for the period 2010–2017. Months are abbreviated by letters (Ap, April; M, May; Jn, June; Jl, July). Numbers 1 and 2 denote the first half (1–15 days) and second half (16–30 days) time intervals of each month, respectively. Horizontal dotted lines indicate a threshold for significant correlation at $p < 0.05$.

4. Discussion

4.1. Tree-Ring Growth Traits

The amount of EWW and LWW has been demonstrated to play a prominent role in regulating xylem hydraulic safety and efficiency, wood density, and the mechanical functioning of cells. The climate also strongly affects LWW% [19]. In our study, LWW% in the drier RAJ was approximately 42% higher than in the wetter BK and was statistically significant. In addition, the EWW of RAJ was significantly shorter than the EWW measured in BK. Luostarinen et al. [66] report that a higher growth rate is believed to increase the EWW, while the amount of LWW remains relatively constant in Norway spruce. Further, a wider EWW indicates more effective tree water use [67].

4.2. Tracheid Anatomy Traits and Latewood Plasticity Sensitivity

Short-term heatwaves during the summer months of July–August or long-term drought events had a perceptible effect on LW plasticity and sensitivity in the warmer and drier RAJ. The diverse and statistically significant differences in the tracheid FATs of the Norway spruce trees between the two research plots of RAJ and BK demonstrate the phenotypic plasticity and adaptability to the local environmental conditions. EW tracheids of the Norway spruce trees in the dry-mesic RAJ formed larger CLA and almost identical CWT compared to the Norway spruce trees in the wet-mesic BK, while LW tracheids presented narrower sizes and thicker cell walls. Our findings on Norway spruce trees seem to support the conclusions reported by Björklund et al. [14], Piermattei et al. [19], and Rathgeber [68] that conifers may enhance the following: (i) conduction efficiency during the spring period by influencing EW TRD, (ii) water conduction security during the summer period by influencing LW CWT, and (iii) LW is probably the tree-ring zone involved in adaptive responses to climate and environmental changes.

Despite presenting higher year-to-year variability, the CWT of the LW tracheid cells in the dry RAJ was always thicker than that of the LW tracheid cells in the wet BK during the 2010–2017 period. An experimental drought treatment in Norway spruce trees demonstrated changes in xylem cell anatomy traits, with drought-treated trees presenting new cells with a high proportion of thick cell walls and narrower lumen diameters after the experiment's onset. On the contrary, irrigated trees had generally thinner cell walls and wider lumen diameters [34]. Thicker cell walls in conifer species, including Norway spruce grown in drier environments, have also been suggested in other studies as being a reinforcement strategy relating to cavitation resistance to water stress [33,49,67]. Thick cell walls reduce the water flow due to a reduced CLD and increase the water flow path through pits, consequently increasing cavitation resistance [7].

Higher cavitation resistance is associated with lower negative sap pressure, requiring tracheids with a higher CWRI to resist mechanical stress [4]. Pittermann et al. [69] concluded that variation in the CWRI was determined by narrowing lumen diameter rather than thickening cell walls. In contrast, Bouche et al. [32] and Rosner et al. [70] demonstrated that CWT—which is genetically largely determined—and not the LRD, is the critical factor in the collapse of tracheid walls. This study further confirms the usefulness of this index as an indicator of the dry conditions that occurred on the research plots. We believe that the LW differences observed at CWRI values between the two research plots, as well as the interannual observations during the summer months of 2015 or after a prolonged dry winter-spring period is a combined result of the interaction performed in both these anatomical traits (i.e., CWT and LRD). Reductions in both tracheid size and CWT and increased hydraulic safety were explained as being plastic responses to the drought experienced by Norway spruce trees in a mountainous, dry environment in the alpine Swiss region [15].

Although a reduction in the tracheid cell lumina of LW tracheids is beneficial to increasing hydraulic safety, signs of reduced LWW% (~16%) and thin-walled tracheids (5.65 μm) of the Norway spruce trees at RAJ during 2015 compared to the other years, according to Liang et al. [71], can be a mark of water stress in conifer trees. The FATs results of our study also confirmed this conclusion. The summer drought of 2015 was documented in CZ during a record year with high annual fluctuations of anomalous warm days [65] and seasonal dry atmospheric circulation during the April–September vegetation period [72], which was also the case in the Central Europe region [73,74].

4.3. Tracheid Anatomy Traits and Environmental Parameters

Trnka et al. [75] identified that the trends in actual evapotranspiration from July to September are highly dependent on PRCP but are also sensitive to reference evapotranspiration driven by the evaporative demand of the atmosphere. In our study, these EPs were statistically significant during the spring (April–June) and summer (July–September) sub-annual periods between the RAJ and BK plots. In addition, Mensah et al. [76] observed reduced forest productivity at both research plots and the sensitivity of both spruce forest plots to the VPD and Tair, especially at the drier RAJ. These outcomes agree with our findings since (i) VPD was found to be highly correlated with the FATs of Norway spruce tracheids, and (ii) mean VPD was higher in the dry-mesic RAJ compared to the wet-mesic BK; however, there were no statistically significant differences between each sub-annual period.

Despite the fact that, in the context of climate change, meteorological factors such as Tair and PRCP significantly influence evapotranspiration, they are not the only driving factors [77,78]. The effect of evapotranspiration is more complex, as seen in the dry RAJ, since the number of VPD1.5 days during the summer period of 2012 and 2016, two years with low PRCP and high Tair, were 2 and 0 days, respectively. The adequacy of water and the extent of vegetation factors also directly affect the actual evapotranspiration besides the meteorological factors [78]. Moreover, it is generally accepted that evapotranspiration is limited by available soil moisture when the soil moisture content is low, and it responds more to variability in meteorological and atmospheric conditions when there is sufficient soil moisture [77].

Along with its direct impact on plant physiology, high VPD increases water loss rates from moist soils. A drier atmosphere naturally demands more water from the soil surface [2]. Low PRCP is associated with an increased probability of hot days and higher Tair, which can further deplete soil moisture [58]. In this sense, we could consider that summer VPD also partly depicts PRCP and soil water deficits when soil moisture data are unavailable. In this study, we did not consider soil type characteristics. Soil characteristics also significantly affect moisture availability and soil water deficits [22], which could not be captured through the investigated EPs. The relatively high proportion of Norway spruce trees grown in cambisol-type soils (RAJ) instead of podzol-type soils (BK) can be partly associated with their higher clay content and more frequent hydromorphic properties [79].

5. Conclusions

Intra-annual analysis and high phenotypic plasticity on the xylem traits of Norway spruce in the studied research plots potentially point to adaptive acclimatization response mechanisms and a more safety-oriented, water-conducting system influenced by the local environmental conditions. Specifically, the young Norway spruce trees growing in the warmer and drier RAJ presented the following xylem structure differences: (i) on the size of the CWRI by adjusting CWT and LRD dimensions and (ii) by forming tree rings with increased LWW% and higher number of LW tracheids, relevant to the BK. Moreover, the presence of thicker CWT on LW cells to withstand higher rates of negative xylem pressure and, thus, increased cavitation resistance against the harsh summer conditions potentially provides an additional indication of adaptation to the local environmental conditions. Additionally, the interannual analysis revealed that in 2015 and 2013 years, which were characterized by disturbance regimes of heatwaves during the summer months (July–August) and days with high evapotranspiration demand or after prolonged drought events, such as the winter-spring in 2014, were formed LW tracheids with thinner cell walls in the warmer and drier RAJ.

Norway spruce is Europe's most widespread conifer species, with a high economic and ecological value. The practical outcome that the seasonal RH and VPD values were not statistically significant between the two research plots might indicate that the long-term resilience and function of Norway spruce trees growing in wet-mesic forest ecosystems might also be affected if days with high evapotranspiration demand and water deficit will continue to rise. However, understanding species' adaptive capacity and response adaptation to warmer and drier environmental conditions is multidisciplinary and still challenging.

Author Contributions: Conceptualization, H.V., V.G. and P.H.; Methodology, H.V., V.G. and P.H.; Validation, D.T., P.H. and K.G.; Investigation, T.N. and D.T.; Formal analysis, P.H., K.G. and D.T.; Resources, M.F., H.V., V.G. and P.H.; Data curation, D.T.; Writing—Original draft preparation, D.T.; Writing—Review and editing, D.T., V.G., H.V., M.F., K.G. and P.H.; Visualization, D.T. and H.V.; Supervision, V.G., P.H. and K.G.; Project administration D.T. and K.G.; Funding acquisition, K.G. All the authors critically reviewed and approved the final version of the manuscript submission. All authors have read and agreed to the published version of the manuscript.

Funding: This study was supported by funding from the Internal Grant Agency (IGA) of the Faculty of Forestry and Wood Technology at Mendel University in Brno under Grant Agreement No. LDF_VP_2021027. This study is also part of the ASFORCLIC project that has received funding from the European Union's Horizon 2020 research and innovation programme under Grant Agreement No. 952314.

Data Availability Statement: The datasets used or analyzed during the current study are available from the corresponding author upon reasonable request.

Conflicts of Interest: The authors declare that they have no known competing financial interests or financial relationships that could have appeared to influence the work reported in this study.

Abbreviation

FATs	Functional anatomical traits
TRW	Tree-ring width
CWT	Cell wall thickness
LRD	Lumen radial diameter width
EPs	Environmental parameters
CZ	Czech Republic
RAJ	Rájec Němčice research plot
BK	Bílý Kříž research plot
EW	Earlywood
LW	Latewood
EWW	Earlywood width
LWW	Latewood width
LWW%	Latewood width percentage
TRD	Tracheid radial diameter width
CLA	Cell lumen area
CWRI	Cell wall reinforcement index
Tair	Air temperature
RH	Relative humidity
PRCP	Precipitation
VPD	Vapour pressure deficit
VPD1.5	threshold value of VPD above 1.5 kPa
SPEI	Standardized Precipitation-Evapotranspiration Index
SPEI–1	one month Standardized Precipitation-Evapotranspiration Index
DE	Drought Event

Appendix A

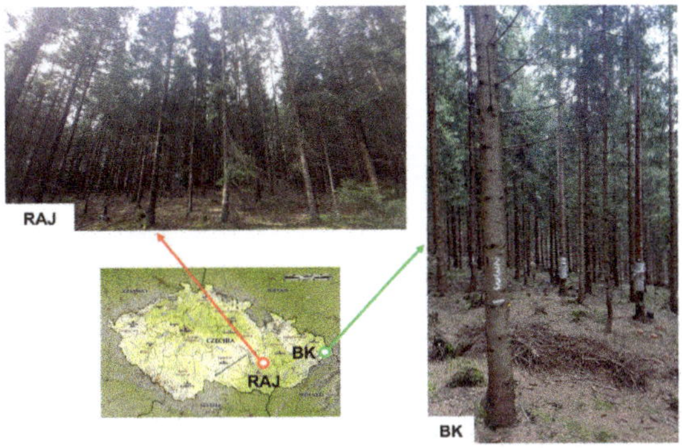

Site characteristics	Rájec-Němčice (RAJ)	Bílý Kříž (BK)
Location	Drahanská Highlands	Moravian-Silesian Beskydy Mountains
Latitude, N	49°26′N	49°30′N
Longtitude, E	16°41′E	18°32′E
Elevation, m a.s.l.	~ 625	~ 875
Soil type	loam-clay Modal Cambisol	Sandy-loam Haplic and Entic Podzol
Soil depth (m)	0.40-0.70	0.60-0.80

Figure A1. Location of the studied research plots in the territory of the Czech Republic (map source: www.bluegreenatlas.com, license: https://creativecommons.org/licenses/by/4.0/), including the main site characteristics of the studied research plots [42,44]. View into the inner part of the studied Norway spruce stands. RAJ, Rájec-Němčice; BK, Bílý Kříž.

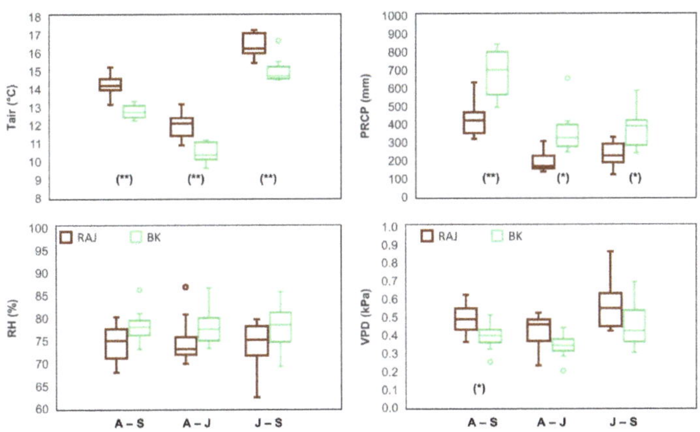

Figure A2. Box plots of the EPs: mean Tair, PRCP, RH, VPD for the growing season (April–September, A–S), and the sub-annual periods of April–June (A–J) and July–September (J–S) in the RAJ and BK plots during the 2010–2017 period. The thick horizontal line in each box plot represents the median value. Each box plot also indicates the upper and lower quartiles, with the vertical lines representing the minimum and maximum values and the outlier points with open circles. Asterisks denote significant differences between the two research plots (* $p < 0.05$, ** $p < 0.01$).

Table A1. Interannual meteorological variability along the Rájec Němčice (RAJ) and Bílý Kříž (BK) plots during the study period 2010–2017. For each time duration (annual, April–September, April–June, and July–September), we calculated the mean air temperature (Tair, °C) and the sum of precipitation (PRCP, mm). We also report the number of days with the extreme threshold value (VPD1.5), with a vapour pressure deficit above 1.5 kPa during the sub-annual periods of April–June and July–September.

RAJ	Annual		April–September		April–June			July–September		
Year	Tair	PRCP	Tair	PRCP	Tair	PRCP	VPD1.5	Tair	PRCP	VPD1.5
2010	6.4	847.1	13.1	628.8	10.9	302.8	0	15.4	326.0	1
2011	8.3	552.7	14.6	399.4	13.0	162.8	0	16.3	236.6	0
2012	8.0	622.1	15.2	316.3	13.2	155.7	3	17.2	160.6	2
2013	7.3	740.9	13.7	497.3	11.3	303.2	0	16.2	194.1	7
2014	8.5	597.1	14.0	435.8	12.2	135.9	3	15.9	299.9	0
2015	8.3	636.4	14.3	359.3	11.4	149.6	0	17.2	209.8	18
2016	7.8	603.1	14.6	315.6	12.1	196.1	0	17.0	119.6	0
2017	7.8	682.6	14.0	453.5	12.1	169.3	2	16.0	284.2	1
BK	**Annual**		**April–September**		**April–June**			**July–September**		
Year	**Tair**	**PRCP**	**Tair**	**PRCP**	**Tair**	**PRCP**	**VPD1.5**	**Tair**	**PRCP**	**VPD1.5**
2010	5.6	1518.9	12.2	1235.9	9.9	652.3	0	14.4	583.6	0
2011	6.7	1011.4	13.0	783.6	11.2	387.0	0	14.8	396.6	0
2012	6.4	1064.0	13.3	491.2	11.1	243.4	2	15.5	247.8	3
2013	6.2	1043.2	12.4	570.6	10.3	279.6	0	14.5	291.0	6
2014	7.6	1199.8	12.4	841.2	10.4	415.2	3	14.5	426.0	0
2015	7.4	968.6	13.1	536.4	9.6	299.2	0	16.6	237.2	9
2016	6.6	1268.0	13.1	638.0	11.0	261.2	0	15.1	376.8	0
2017	6.5	1155.4	12.4	759.6	10.1	345.2	0	14.6	414.4	1

References

1. Carrer, M.; Motta, R.; Nola, P. Significant Mean and Extreme Climate Sensitivity of Norway Spruce and Silver Fir at Mid-Elevation Mesic Sites in the Alps. *PLoS ONE* **2012**, *7*, e50755. [CrossRef]
2. Amitrano, C.; Arena, C.; Rouphael, Y.; De Pascale, S.; De Micco, V. Vapour Pressure Deficit: The Hidden Driver Behind Plant Morphofunctional Traits in Controlled Environments. *Ann. Appl. Biol.* **2019**, *175*, 313–325. [CrossRef]
3. Fonti, P.; von Arx, G.; García-González, I.; Eilmann, B.; Sass-Klaassen, U.; Gärtner, H.; Eckstein, D. Studying Global Change Through Investigation of the Plastic Responses of Xylem Anatomy in Tree Rings. *New Phytol.* **2010**, *185*, 42–53. [CrossRef]
4. Lachenbruch, B.; McCulloh, K.A. Traits, Properties and Performance: How Woody Plants Combine Hydraulic and Mechanical Functions in a Cell, Tissue or Whole Plant. *New Phytol.* **2014**, *204*, 747–764. [CrossRef]
5. Bussotti, F.; Pollastrini, M.; Holland, V.; Brüggemann, W. Functional Traits and Adaptive Capacity of European Forests to Climate Change. *Environ. Exp. Bot.* **2015**, *111*, 91–113. [CrossRef]
6. Rathgeber, C.B.K.; Fonti, P.; Shishov, V.V.; Rozenberg, P. Wood Formation and Tree Adaptation to Climate. *Ann. For. Sci.* **2019**, *76*, 109. [CrossRef]
7. Song, Y.; Poorter, L.; Horsting, A.; Delzon, S.; Sterck, F. Pit and Tracheid Anatomy Explain Hydraulic Safety but not Hydraulic Efficiency of 28 Conifer Species. *J. Exp. Bot.* **2022**, *73*, 1033–1048. [CrossRef]
8. Matisons, R.; Krišāns, O.; Kārkliņa, A.; Adamovičs, A.; Jansons, Ā.; Gärtner, H. Plasticity and Climatic Sensitivity of Wood Anatomy Contribute to Performance of Eastern Baltic Provenances of Scots Pine. *For. Ecol. Manag.* **2019**, *452*, 117568. [CrossRef]
9. Carrer, M.; Castagneri, D.; Prendin, A.L.; Petit, G.; von Arx, G. Retrospective Analysis of Wood Anatomical Traits Reveals a Recent Extension in Tree Cambial Activity in Two High-Elevation Conifers. *Front. Plant Sci.* **2017**, *8*, 737. [CrossRef]
10. Ziaco, E.; Biondi, F.; Rossi, S.; Deslauriers, A. Intra-Annual Wood Anatomical Features of High-Elevation Conifers in the Great Basin, USA. *Dendrochronologia* **2014**, *32*, 303–312. [CrossRef]
11. Gebregeorgis, E.G.; Boniecka, J.; Piątkowski, M.; Robertson, I.; Rathgeber, C.B.K. SabaTracheid 1.0: A Novel Program for Quantitative Analysis of Conifer Wood Anatomy—A Demonstration on African Juniper from the Blue Nile Basin. *Front. Plant Sci.* **2021**, *12*, 595258. [CrossRef]
12. von Arx, G.; Crivellaro, A.; Prendin, A.L.; Čufar, K.; Carrer, M. Quantitative Wood Anatomy—Practical Guidelines. *Front. Plant Sci.* **2016**, *1*, 781. [CrossRef]
13. Belokopytova, L.V.; Babushkina, E.A.; Zhirnova, D.F.; Panyushkina, I.P.; Vaganov, E.A. Pine and Larch Tracheids Capture Seasonal Variations of Climatic Signal at Moisture-Limited Sites. *Trees* **2019**, *33*, 227–242. [CrossRef]

14. Björklund, J.; Seftigen, K.; Schweingruber, F.; Fonti, P.; von Arx, G.; Bryukhanova, M.V.; Cuny, H.E.; Carrer, M.; Castagneri, D.; Frank, D.C. Cell Size and Wall Dimensions Drive Distinct Variability of Earlywood and Latewood Density in Northern Hemisphere Conifers. *New Phytol.* **2017**, *216*, 728–740. [CrossRef]
15. Bryukhanova, M.; Fonti, P. Xylem Plasticity Allows Rapid Hydraulic Adjustment to Annual Climatic Variability. *Trees* **2013**, *27*, 485–496. [CrossRef]
16. Castagneri, D.; Fonti, P.; von Arx, G.; Carrer, M. How Does Climate Influence Xylem Morphogenesis Over the Growing Season? Insights from Long-Term Intra-Ring Anatomy in *Picea abies*. *Ann. Bot.* **2017**, *119*, 1011–1020.
17. Fonti, P.; Babushkina, E.A. Tracheid Anatomical Responses to Climate in a Forest-Steppe in Southern Siberia. *Dendrochronologia* **2016**, *39*, 32–41. [CrossRef]
18. Olano, J.M.; Eugenio, M.; García-Cervigón, A.I.; Folch, M.; Rozas, V. Quantitative Tracheid Anatomy Reveals a Complex Environmental Control of Wood Structure in Continental Mediterranean Climate. *Int. J. Plant Sci.* **2012**, *173*, 137–149. [CrossRef]
19. Piermattei, A.; von Arx, G.; Avanzi, C.; Fonti, P.; Gärtner, H.; Piotti, A.; Urbinati, C.; Vendramin, G.G.; Büntgen, U.; Crivellaro, A. Functional Relationships of Wood Anatomical Traits in Norway Spruce. *Front. Plant Sci.* **2020**, *11*, 683. [CrossRef]
20. Choat, B.; Brodribb, T.J.; Brodersen, C.R.; Duursma, R.A.; López, R.; Medlyn, B.E. Triggers of Tree Mortality under Drought. *Nature* **2018**, *558*, 531–539. [CrossRef]
21. Nilsson, O.; Hjelm, K.; Nilsson, U. Early Growth of Planted Norway Spruce and Scots Pine After Site Preparation in Sweden. *Scand. J. For. Res.* **2019**, *34*, 678–688. [CrossRef]
22. Lévesque, M.; Saurer, M.; Siegwolf, R.; Eilmann, B.; Brang, P.; Bugmann, H.; Rigling, A. Drought Response of Five Conifer Species Under Contrasting Water Availability Suggests High Vulnerability of Norway Spruce and European Larch. *Glob. Chang. Biol.* **2013**, *19*, 3184–3199. [CrossRef] [PubMed]
23. Maringer, J.; Stelzer, A.-S.; Paul, C.; Albrecht, A.T. Ninety-Five Years of Observed Disturbance-Based Tree Mortality Modeled with Climate-Sensitive Accelerated Failure Time Models. *Eur. J. For. Res.* **2021**, *140*, 255–272. [CrossRef]
24. Hanel, M.; Rakovec, O.; Markonis, Y.; Máca, P.; Samaniego, L.; Kyselý, J.; Kumar, R. Revisiting the Recent European Droughts from a Long-Term Perspective. *Sci. Rep.* **2018**, *8*, 9499. [CrossRef] [PubMed]
25. Lindner, M.; Maroschek, M.; Netherer, S.; Kremer, A.; Barbati, A.; Garcia-Gonzalo, J.; Seidl, R.; Delzon, S.; Corona, P.; Kolström, M.; et al. Climate Change Impacts, Adaptive Capacity and Vulnerability of European Forest Ecosystems. *For. Ecol. Manag.* **2010**, *259*, 698–709. [CrossRef]
26. Kuželková, M.; Jačka, L.; Kovář, M.; Hradilek, V.; Máca, P. Tree Trait-Mediated Differences in Soil Moisture Regimes: A Comparative Study of Beech, Spruce and Larch in a Drought-Prone Area. *Eur. J. For. Res.* **2024**, *143*, 319–332. [CrossRef]
27. Jevšenak, J.; Tychkov, I.; Gričar, J.; Levanič, T.; Tumajer, J.; Prislan, P.; Arnič, D.; Popkova, M.; Shishov, V.V. Growth-Limiting Factors and Climate Response Variability in Norway Spruce (*Picea abies* L.) Along an Elevation and Precipitation Gradients in Slovenia. *Int. J. Biometeorol.* **2021**, *65*, 311–324. [CrossRef]
28. Rukh, S.; Poschenrieder, W.; Heym, M.; Pretzsch, H. Drought Resistance of Norway Spruce (*Picea abies* [L.] Karst) and European Beech (*Fagus sylvatica* [L.]) in Mixed vs. Monospecific Stands and on Dry vs. Wet Sites. From Evidence at the Tree Level to Relevance at the Stand Level. *Forests* **2020**, *11*, 639. [CrossRef]
29. Boden, S.; Kahle, H.-P.; von Wilpert, K.; Spiecker, H. Resilience of Norway Spruce (*Picea abies* (L.) Karst) Growth to Changing Climatic Conditions in Southwest Germany. *For. Ecol. Manag.* **2014**, *315*, 12–21. [CrossRef]
30. Spathelf, P.; van der Maaten, E.; van der Maaten-Theunissen, M.; Campioli, M.; Dobrowolska, D. Climate Change Impacts in European Forests: The Expert Views of Local Observers. *Ann. For. Sci.* **2014**, *71*, 131–137. [CrossRef]
31. Tumajer, J.; Altman, J.; Štěpánek, P.; Treml, V.; Doležal, J.; Cienciala, E. Increasing Moisture Limitation of Norway Spruce in Central Europe Revealed by Forward Modelling of Tree Growth in Tree-Ring Network. *Agric. For. Meteorol.* **2017**, *247*, 56–64. [CrossRef]
32. Bouche, P.S.; Larter, M.; Domec, J.-C.; Burlett, R.; Gasson, P.; Jansen, S.; Delzon, S. A Broad Survey of Hydraulic and Mechanical Safety in the Xylem of Conifers. *J. Exp. Bot.* **2014**, *65*, 4419–4431. [CrossRef]
33. Hacke, U.G.; Sperry, J.S. Functional and Ecological Xylem Anatomy. *Perspect. Plant Ecol. Evol. Syst.* **2001**, *4*, 97–115. [CrossRef]
34. Montwé, D.; Spiecker, H.; Hamann, A. An Experimentally Controlled Extreme Drought in a Norway Spruce Forest Reveals Fast Hydraulic Response and Subsequent Recovery of Growth Rates. *Trees* **2014**, *28*, 891–900. [CrossRef]
35. Arnič, D.; Gričar, J.; Jevšenak, J.; Božič, G.; von Arx, G.; Prislan, P. Different Wood Anatomical and Growth Responses in European Beech (*Fagus sylvatica* L.) at Three Forest Sites in Slovenia. *Front. Plant Sci.* **2021**, *12*, 669229. [CrossRef]
36. Lange, J.; Carrer, M.; Pisaric, M.F.J.; Porter, T.J.; Seo, J.-W.; Trouillier, M.; Wilmking, M. Moisture-Driven Shift in the Climate Sensitivity of White Spruce Xylem Anatomical Traits is Coupled to Large-Scale Oscillation Patterns Across Northern Treeline in Northwest North America. *Clob. Chang. Biol.* **2020**, *26*, 1842–1856. [CrossRef] [PubMed]
37. Ziaco, E.; Liu, X.; Biondi, F. Dendroanatomy of Xylem Hydraulics in Two Pine Species: Efficiency Prevails on Safety for Basal Area Growth in Drought-Prone Conditions. *Dendrochronologia* **2023**, *81*, 126116. [CrossRef]
38. Chawla, S.; Nguyen, V.X.; Torres, C.P.G.; Pavelka, M.; Trusina, J.; Marek, M.V. Wind Characteristics Recorded at the Czech Carbon Observation System (CZECOS) Site Rajec. *Eur. J. Environ. Sci.* **2018**, *8*, 117–123. [CrossRef]
39. Urban, O.; Klem, K.; Ač, A.; Havránková, K.; Holišová, P.; Navrátil, M.; Zitová, M.; Kozlová, K.; Pokorný, R.; Šprtová, M.; et al. Impact of Clear and Cloudy Sky Conditions on the Vertical Distribution of Photosynthetic CO_2 Uptake Within a Spruce Canopy. *Funct. Ecol.* **2012**, *26*, 46–55. [CrossRef]

40. McGloin, R.; Šigut, L.; Fischer, M.; Foltýnová, L.; Chawla, S.; Trnka, M.; Pavelka, M.; Marek, M.V. Available Energy Partitioning During Drought at two Norway Spruce Forests and a European Beech Forest in Central Europe. *J. Geophys. Res. Atmos.* **2019**, *124*, 3726–3742. [CrossRef]
41. Sedlák, P.; Aubinet, M.; Heinesch, B.; Janouš, D.; Pavelka, M.; Potužníková, K.; Yernaux, M. Night-Time Airflow in a Forest Canopy near a Mountain Crest. *Agric. For. Meteorol.* **2010**, *150*, 736–744. [CrossRef]
42. Samec, P.; Voženílek, V.; Vondráková, A.; Macků, J. Diversity of Forest Soils and Bedrock in Soil Regions of the Central-European Highlands (Czech Republic). *Catena* **2018**, *160*, 95–102. [CrossRef]
43. Kolář, T.; Giagli, K.; Trnka, M.; Bednářová, E.; Vavrčík, H.; Rybníček, M. Response of the Leaf Phenology and Tree-Ring Width of European Beech to Climate Variability. *Silva Fenn.* **2016**, *50*, 1520. [CrossRef]
44. Zapletal, M.; Cudlín, P.; Chroust, P.; Urban, O.; Pokorný, R.; Edwards-Jonášová, M.; Czerný, R.; Janouš, D.; Taufarová, K.; Večeřa, Z.; et al. Ozone Flux over a Norway Spruce Forest and Correlation with Net Ecosystem Production. *Environ. Pollut.* **2011**, *159*, 1024–1034. [CrossRef] [PubMed]
45. Gärtner, H.; Nievergelt, D. The Core-Microtome: A new Tool for Surface Preparation on Cores and Time Series Analysis of Varying Cell Parameters. *Dendrochronologia* **2010**, *28*, 85–92. [CrossRef]
46. Schneider, L.; Gärtner, H. The Advantage of Using a Starch Based Non-Newtonian Fluid to Prepare Micro Sections. *Dendrochronologia* **2013**, *31*, 175–178. [CrossRef]
47. Pritzkow, C.; Heinrich, I.; Grudd, H.; Helle, G. Relationship Between Wood Anatomy, Tree-Ring Widths and Wood Density of *Pinus sylvestris* L. and Climate at High Latitudes in Northern Sweden. *Dendrochronologia* **2014**, *32*, 295–302. [CrossRef]
48. Denne, M.P. Definition of Latewood According to Mork (1928). *IAWA J.* **1989**, *10*, 59–62. [CrossRef]
49. Martin-Benito, D.; Anchukaitis, K.J.; Evans, M.N.; del Río, M.; Beeckman, H.; Cañellas, I. Effects of Drought on Xylem Anatomy and Water-Use Efficiency of two Co-Occuring Pine Species. *Forests* **2017**, *8*, 332. [CrossRef]
50. Pellizzari, E.; Camarero, J.J.; Gazol, A.; Sangüesa-Barreda, G.; Carrer, M. Wood Anatomy and Carbon-Isotope Discrimination Support Long-Term Hydraulic Deterioration as a Major Cause of Drought-Induced Dieback. *Glob. Chang. Biol.* **2016**, *22*, 2125–2137. [CrossRef]
51. Fajstavr, M.; Giagli, K.; Vavrčík, H.; Gryc, V.; Horáček, P.; Urban, J. The Cambial Response of Scots Pine Trees to Girdling and Water Stress. *IAWA J.* **2020**, *41*, 159–185. [CrossRef]
52. Filion, L.; Cournoyer, L. Variation in Wood Structure of Eastern Larch Defoliated by the Larch Sawfly in Subarctic Quebec, Canada. *Can. J. For. Res.* **1995**, *25*, 1263–1268. [CrossRef]
53. Campelo, F.; Nabais, C.; Carvalho, A.; Vieira, J. TracheideR- An R Package to Standardize Tracheidograms. *Dendrochronologia* **2016**, *37*, 64–68. [CrossRef]
54. Broz, A.; Retallack, G.J.; Maxwell, T.M.; Silva, L.C.R. A Record of Vapour Pressure Deficit Preserved in Wood and Soil Across Biomes. *Sci. Rep.* **2021**, *11*, 662. [CrossRef] [PubMed]
55. Hoylman, Z.H.; Jencso, K.G.l; Hu, J.; Holden, Z.A.; Martin, J.T.; Gardner, W.P. The Climatic Water Balance and Topography Control Spatial Patterns of Atmospheric Demand, Soil Moisture and Shallow Subsurface Flow. *Water Resour. Res.* **2019**, *55*, 2370–2389. [CrossRef]
56. de Cárcer, P.S.; Vitasse, Y.; Peñuelas, J.; Jassey, V.E.J.; Buttler, A.; Signarbieux, C. Vapor-Pressure Deficit and Extreme Climatic Variables Limit Tree Growth. *Glob. Chang. Biol.* **2018**, *24*, 1108–1122. [CrossRef] [PubMed]
57. McKee, T.B.; Doesken, N.J.; Kleist, J. The Relationship of Drought Frequency and Duration to Time Scales. In Proceedings of the 8th Conference on Applied Climatology, Anaheim, CA, USA, 17–22 January 1993; pp. 179–183.
58. Manning, C.; Widmann, M.; Bevacqua, E.; Van Loon, A.F.; Maraun, D.; Vrac, M. Soil Moisture Drought in Europe: A Compound Event of Precipitation and Potential Evapotranspiration on Multiple Time Scales. *J. Hydrometerol.* **2018**, *19*, 1255–1271. [CrossRef]
59. Quiring, S.M. Developing Objective Operational Definitions for Monitoring Drought. *J. Appl. Meteorol. Climatol.* **2009**, *48*, 1217–1229. [CrossRef]
60. Vicente-Serrano, S.M.; Peña-Angulo, D.; Murphy, C.; López-Moreno, J.I.; Tomas-Burguera, M.; Domínguez-Castro, F.; Tian, F.; Eklundh, L.; Cai, Z.; Alvarez-Farizo, B.; et al. The Complex Multisectoral Impacts of Drought: Evidence from a Mountainous Basin in the Central Spanish Pyrenees. *Sci. Total. Environ.* **2021**, *769*, 144702. [CrossRef]
61. Parente, J.; Amraoui, M.; Menezes, I.; Pereira, M.G. Drought in Portugal: Current Regime, Comparison of Indices and Impacts on Extreme Wildfires. *Sci. Total Environ.* **2019**, *685*, 150–173. [CrossRef]
62. Spinoni, J.; Naumann, G.; Vogt, J.; Barbosa, P. European Drought Climatologies and Trends Based on a Multi-Indicator Approach. *Glob. Planet. Chang.* **2015**, *127*, 50–57. [CrossRef]
63. Vicente-Serrano, S.M.; Beguería, S.; López-Moreno, J.I. A Multiscalar Drought Index Sensitive to Global Warming: The Standardized Precipitation Evapotranspiration Index. *J. Clim.* **2010**, *23*, 1696–1718. [CrossRef]
64. Newson, R. Parameters Behind 'Non-Parametric' Statistics: Kendall's tau, Somers' D and median differences. *Stata J.* **2002**, *2*, 45–64. [CrossRef]
65. Zahradníček, P.; Brázdil, R.; Řehoř, J.; Lhotka, O.; Dobrovolný, P.; Štěpánek, P.; Trnka, M. Temperature Extremes and Circulation Types in the Czech Republic, 1961–2020. *Int. J. Climatol.* **2022**, *42*, 4808–4829. [CrossRef]
66. Luostarinen, K.; Pikkarainen, L.; Ikonen, V.-P.; Gerendiain, A.Z.; Pulkkinen, P.; Peltola, H. Relationships of Wood Anatomy with Growth and Wood Density in Three Norway Spruce Clones of Finnish Origin. *Can. J. For. Res.* **2017**, *47*, 1184–1192. [CrossRef]

67. Gričar, J.; Prislan, P.; de Luis, M.; Gryc, V.; Hacurová, J.; Vavrčík, H.; Čufar, K. Plasticity in Variation of Xylem and Phloem Cell Characteristics of Norway Spruce Under Different Local Conditions. *Front. Plant Sci.* **2015**, *6*, 730. [CrossRef] [PubMed]
68. Rathgeber, C.B.K. Conifer Tree-Ring Density Inter-Annual Variability–Anatomical, Physiological and Environmental Determinants. *New Phytol.* **2017**, *216*, 621–625. [CrossRef] [PubMed]
69. Pittermann, J.; Sperry, J.S.; Wheeler, J.K.; Hacke, U.G.; Sikkema, E.H. Mechanical Reinforcement of Tracheids Compromises the Hydraulic Efficiency of Conifer Xylem. *Plant Cell Environ.* **2006**, *29*, 1618–1628. [CrossRef] [PubMed]
70. Rosner, S.; Gierlinger, N.; Klepsch, M.; Karlsson, B.; Evans, R.; Lundqvist, S.-O.; Světlík, J.; Børja, I.; Dalsgaard, L.; Andreassen, K.; et al. Hydraulic and Mechanical Dysfunction of Norway Spruce Sapwood due to Extreme Summer Drought in Scandinavia. *For. Ecol. Manag.* **2018**, *409*, 527–540. [CrossRef]
71. Liang, C.; Filion, L.; Cournoyer, L. Wood Structure of Biotically and Climatically Induced Light Rings in Eastern Larch (*Larix laricina*). *Can. J. For. Res.* **1997**, *27*, 1538–1547. [CrossRef]
72. Lhotka, O.; Trnka, M.; Kyselý, J.; Markonis, Y.; Balek, J.; Možný, M. Atmospheric Circulation as a Factor Contributing to Increasing Drought Severity in Central Europe. *J. Geophys. Res. Atmos.* **2020**, *125*, e2019JD032269. [CrossRef]
73. Buras, A.; Schunk, C.; Zeiträg, C.; Herrmann, C.; Kaiser, L.; Lemme, H.; Straub, C.; Taeger, S.; Gößwein, S.; Klemmt, H.-J.; et al. Are Scots Pine Forest Edges Particularly Prone to Drought-Induced Mortality? *Environ. Res. Lett.* **2018**, *13*, 025001. [CrossRef]
74. Burri, S.; Haeler, E.; Eugster, W.; Haeni, M.; Etzold, S.; Walthert, L.; Braun, S.; Zweifel, R. How did Swiss Forest Trees Respond to the Hot Summer 2015? *Die Erde* **2019**, *150*, 214–229.
75. Trnka, M.; Brázdil, R.; Balek, J.; Semerádová, D.; Hlavinka, P.; Možný, M.; Štěpánek, P.; Dobrovolný, P.; Zahradníček, P.; Dubrovský, M.; et al. Drivers of Soil Drying in the Czech Republic Between 1961 and 2012. *Int. J. Climatol.* **2015**, *35*, 2664–2675. [CrossRef]
76. Mensah, C.; Šigut, L.; Fischer, M.; Foltýnová, L.; Jocher, G.; Urban, O.; Wemegah, C.S.; Nyantakyi, E.K.; Chawla, S.; Pavelka, M.; et al. Environmental Effects on Normalized Gross Primary Productivity in Beech and Norway Spruce Forests. *Atmosphere* **2021**, *12*, 1128. [CrossRef]
77. Cui, Z.; Wang, Y.; Zhang, G.J.; Yang, M.; Liu, J.; Wei, L. Effects of Improved Simulation of Precipitation on Evapotranspiration and Its Partitioning Over Land. *Geophys. Res. Lett.* **2022**, *49*, e2021GL097353. [CrossRef]
78. Liu, Y.; Jiang, Q.; Wang, Q.; Jin, Y.; Yue, Q.; Yu, J.; Zheng, Y.; Jiang, W.; Yao, X. The Divergence Between Potential and Actual Evapotranspiration: An Insight from Climate, Water and Vegetation Change. *Sci. Total Environ.* **2022**, *807*, 150648. [CrossRef]
79. Daněk, P.; Šamonil, P.; Vrška, T. Four Decades of the Coexistence of Beech and Spruce in a Central European Old-Growth Forest. Which Succeeds on what Soils and Why? *Plant Soil* **2019**, *437*, 257–272. [CrossRef]

Disclaimer/Publisher's Note: The statements, opinions and data contained in all publications are solely those of the individual author(s) and contributor(s) and not of MDPI and/or the editor(s). MDPI and/or the editor(s) disclaim responsibility for any injury to people or property resulting from any ideas, methods, instructions or products referred to in the content.

forests

MDPI

Article

The Influence of Spatial Heterogeneity of Urban Green Space on Surface Temperature

Mengru Zhang [1], Jianguo Wang [1] and Fei Zhang [2,*]

[1] College of Geography and Remote Sensing Sciences, Xinjiang University, Urumqi 830017, China; zhangmr981114@163.com (M.Z.); jianguowang@ksu.edu.cn (J.W.)
[2] College of Geography and Environmental Sciences, Zhejiang Normal University, Jinhua 321004, China
* Correspondence: zhangfei3s@163.com or zhangfei3s@zjnu.edu.cn

Abstract: Urban green space (UGS) has been recognized as a key factor in enhancing the urban ecosystem balance, particularly in arid areas. It is often considered an effective means to mitigate the urban heat island (UHI) effect. In this study, the reference comparison method was utilized to optimize the process of nighttime lighting data; the random forest classification method was employed to extract UGS data; and the radiative transfer method was applied in land surface temperature (LST) inversion. Additionally, moving window analysis was conducted to assess the robustness of the results. The objective of this research was to analyze the spatial distribution characteristics of UGS and LST and to explore their bivariate local spatial autocorrelations by calculating four landscape metrics, including the aggregation index (AI), edge density (ED), patch density (PD), and area-weighted mean shape index (Shape_am). It was found that the distribution of UGS in the study area was uneven, with higher temperatures in the eastern and western regions and lower temperatures in the central and southern regions. The results also revealed that ED, PD, and Shape_am were negatively correlated with LST, with correlation coefficients being −0.469, −0.388, and −0.411, respectively, indicating that UGS in these regions were more effective in terms of cooling effect. Conversely, AI was found to be positively correlated with LST (Moran' I index of 0.449), indicating that surface temperatures were relatively higher in regions of high aggregation. In essence, the fragmented, complex, and evenly distributed green patches in the study area provided a better cooling effect. These findings should persuade decision makers and municipal planners to allocate more UGS in cities for UHI alleviation to improve quality of life and enhance recreational opportunities.

Keywords: urban green space (UGS); land surface temperature (LST); Gaofen-2; thermal comfort; bivariate local spatial autocorrelation

1. Introduction

The United Nations 2030 Agenda for Sustainable Development has proposed 17 Sustainable Development Goals (SDGs), of which SDG 11 is "to build inclusive, safe, resilient and sustainable cities and human settlements" [1]. This goal underscores the global and national focus on sustainable urban development, which was intensified by rising personal incomes and public demand for higher-quality living environments. Consequently, the Chinese government escalated the requirements for living environments of residents in its pursuit of ecological civilization [2]. Urban green space (UGS) is a vital component of the urban environment, playing significant roles in the ecological and social realms [3]. Therefore, rationalizing the allocation of UGS to enhance thermal comfort in urban areas has become a pressing issue.

Since the 20th century, the rapid development of cities has brought many ecological and environmental problems, among which the urban heat environment is particularly prominent. Large amounts of construction, industrial emissions, and other human activities lead to a continuous increase in surface temperature, causing cities to be hotter than surrounding zones. This phenomenon is known as the urban heat island (UHI) effect [4]. In general, two types of UHI are recognized: the air temperature UHI, which affects atmospheric conditions, and the surface temperature UHI, sometimes referred to as the surface urban heat island (SUHI), which impacts the temperatures of urban surfaces [5]. Air temperature UHI can be extracted from infrared remote sensing images with land surface temperature (LST) and has been extensively utilized to study the UHI phenomenon [6]. LST threatens the urban ecological environment, and there are natural disasters such as heat waves and floods, and it is increasingly used to study UHI. A number of researchers have indicated that UGSs perform a very valuable function in improving urban thermal environment problems [7–9] and has important ecological landscape and health value for cities and residents, not only improving air quality and reducing noise pollution and extreme weather impacts on people's lives, but also reducing stress on humans by controlling the development of UHI. It can improve air quality, reduce the impact of noise pollution and extreme weather on people's lives, as well as promote psychological health by reducing stress, thus making residents more comfortable. With the growing urbanization and the increasing expectation to have a higher quality of life in urban environments, the impact of UGS on human settlements has proven to be increasingly evident and has become a key enabling aspect for all governments and societies in urban planning. However, different green space patterns have different relationships with urban temperature, and exploring their patterns and distributions is important for planners to optimize urban patterns in the future.

Recent studies have consistently highlighted the crucial role of UGS in mitigating UHI effects [10–12]. These studies illustrate various mechanisms, through which UGSs lower LST, such as shading and evapotranspiration [13,14]. Despite the limited availability of land for greening in urban planning, more attention has been given to the effect of the spatial configuration of green space on LST, aiming to maximize cooling benefits by modifying its spatial distribution. It has been demonstrated that the spatial configuration of green space significantly influences LST [11,15], with the area, shape, and type of UGS being crucial factors. While some studies indicate a strong cooling effect related to the proportion of area covered by UGS [8,16], there is no consensus on the impact of UGS shape and type on temperature, potentially due to variations in study areas, data spatial resolution, etc. [6,17,18]. Furthermore, comparative studies across different urban settings have shown that the effectiveness of UGS in reducing LST varies significantly and depends on the local climatic conditions, urban morphology, and vegetation types [19,20]. However, while there is a growing understanding of these processes, significant gaps remain in specific areas such as semi-arid regions, where unique climatic challenges exacerbate the UHI effect [21]. Moreover, the impact of UGS configuration—such as the size, shape, and connectivity of green spaces—on urban thermal comfort is still not fully understood, with studies reporting mixed outcomes based on the spatial resolution of the data used and the methodological approaches employed [22,23].

This study delineates the influence of various configurations of UGS on LST, providing crucial insights for strategic UGS planning and optimization within urban settings. Notably, variations in the geographic scope of study areas, analytical scales, and the resolution of the data employed can significantly affect temperature outcomes. A review of the literature highlights a significant lack of empirical investigations into arid regions [24], which face unique challenges such as acute water shortages that could be exacerbated by global warming [25]. These gaps underscore the need for a comprehensive exploration of the direct interactions between UGS and LST. Additionally, variations in findings regarding the associations between landscape metrics and LST across different studies call for comparative analyses with other regional case studies.

This investigation aims to provide a robust scientific basis for UGS optimization by addressing pivotal questions including the following: (1) What are the spatial distributions of UGS landscape patterns in different regions? (2) How does LST vary spatially across these regions? (3) What are the differences in the correlation between the spatial distribution of

UGS and LST in different regions? To our knowledge, this is the inaugural study in Urumqi employing GIS technology and spatial metrics to analyze the correlation between UGS and LST. The outcomes of this analysis are anticipated to furnish the governmental policy makers and urban planners with enhanced understanding of the UGS spatial patterns and facilitate more profound investigations into the correlation between UGS and LST across the entire urban community.

2. Materials and Methods

2.1. Study Area

Urumqi, the capital of Xinjiang, is situated in the northwest border of China, encompassed by coordinates ranging from 86°37′33″ to 88°58′24″ E and 42°45′32″ to 45°00′00″ N. It is enclosed by mountain ranges on three sides, and the city boasts an average altitude of 800 m (Figure 1). Characterized by a mid-temperate semi-arid continental climate, Urumqi is distinguished as the city farthest from any ocean globally, receiving an average annual precipitation of only 236 mm. According to the Köppen-Geiger climate classification system, this climatic condition categorizes Urumqi in the Cold Semi-Arid (BSk) zone. Summers exhibit high temperatures, with peak heat occurring in July and August, boasting an average temperature of 25.7 °C. Winters, on the other hand, are characterized by mild conditions, with an average temperature of −15.2 °C. This moderation is attributed to the barrier effect of the Tien Shan Mountains, which effectively entraps cold air within the basin.

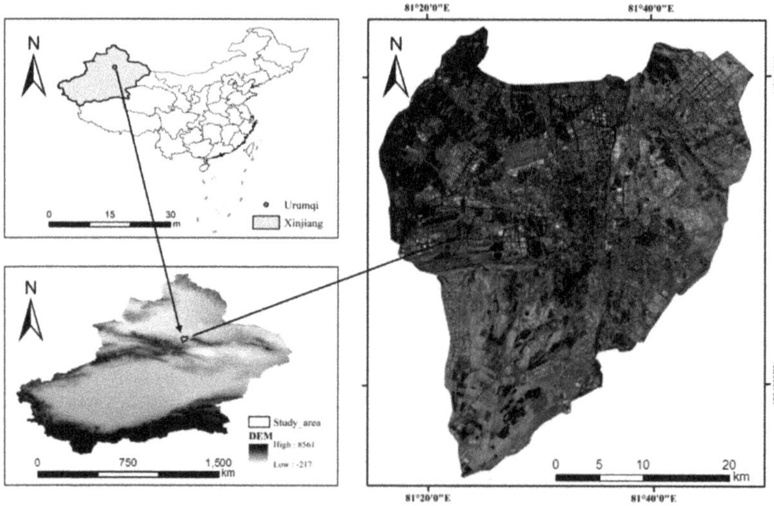

Figure 1. Location map of the study area.

As reported in the 2021 National Economic and Social Development Statistical Bulletin of Urumqi (http://www.urumqi.gov.cn/fjbm/tjj/tjgb/509211.htm, accessed on 14 September 2022), the city covers an area of approximately 1.38 × 10^4 km^2, with the built-up area having expanded by 14.60 km^2 from the previous year to 536.20 km^2. With a resident population of 4.05 million, Urumqi stands as Xinjiang's major urban center and the most populous and economically dynamic city in Central Asia. The UGS in Urumqi faces significant challenges due to the harsh dry climate, climate change, and rapid urban expansion, which emphasize the necessity for strategic environmental planning and management. Thus, Urumqi, the capital of the Xinjiang Uyghur Autonomous Region, is selected as the focal study area to investigate the impact of green space composition and configuration on LST.

2.2. Research Data

UGSs, distributed within urban areas, are the focus of this paper. For the precise definition of urban built-up areas, nighttime light data from the Lujia-1 (LJ1-01) satellite [26,27], developed by Wuhan University, were employed. As the world's first satellite dedicated to luminescence remote sensing, LJ1-01 has been widely validated to monitor and extract economic activities in urban built-up areas. Data acquired on 30 October 2018, with a spatial resolution of 130 m, were utilized to define urban built-up areas accurately.

The Gaofen-2 (GF-2) satellite, launched by China, represents a significant advancement in civil optical remote sensing technology. With a spatial resolution exceeding 1 m, it is the highest resolution civil remote sensing satellite developed by China to date. Its extensive observation area and long operational lifespan [28] make GF-2 an exemplary data source for urban studies [29]. In this research, GF-2 PMS images captured on 21 July 2021 and 1 August 2021 were used to comprehensively cover the designated study area. The image processing was meticulously executed using ENVI 5.3 software, which involved several critical stages. Radiometric calibration was performed with settings adjusted to "Radiance", converting raw data to accurate radiance values, and atmospheric correction was applied using the FLAASH module, configured for an urban aerosol model and mid-latitude summer atmospheric conditions. Orthogonal correction was performed using a cubic convolutional resampling method to correct geometric distortions. Image fusion was conducted using NNDiffuse Pan Sharpening to improve spatial resolution. Finally, the images were mosaicked and cropped to accurately define the study area, ensuring high fidelity and consistency of the processed images. These steps are essential to maintain the integrity and usability of the data for detailed urban structure analysis.

Landsat 8, the eighth installment in the U.S. Landsat program, is equipped with the Operational Land Imager (OLI) and the Thermal Infrared Sensor (TIRS), which provide spatial resolutions of 30 m and 100 m, respectively. This study utilizes Landsat 8 OLI_TIRS data collected on 2 August 2021, notable for its minimal cloud coverage of 0.01%, to derive surface temperature estimates. The accuracy of these derived temperatures was assessed using the Landsat Collection 2 Level-2 product data. This product, available from the USGS (https://www.usgs.gov/landsat-missions/landsat-collection-2-surface-temperature, accessed on 12 September 2022), includes multispectral band surface reflectance and thermal infrared band surface temperatures, which have been processed using the enhanced Land Surface Reflectance Code (LaSRC) with an improved atmospheric correction algorithm [30]. The use of Collection 2 Level-2 data is critical in validating the accuracy of inverted surface temperatures due to its proven efficacy in surface temperature analysis [31]. This validation process ensures that the surface temperature data derived from the Landsat 8 images are robust and reliable for subsequent analyses.

2.3. Methodology

2.3.1. Reference Comparison Method

In this study, the delineation of built-up areas in Urumqi was performed using the reference comparison method, which is a widely recognized technique for processing nighttime lighting data [32]. This method involves setting various lighting thresholds and comparing the extracted built-up area results against official built-up area statistics released by the government. The optimal threshold, identified as the one that minimizes discrepancies between these datasets, was selected to ensure the robustness of the findings. Supported by quantitative analysis, this method allows the accurate delineation of built-up areas and ensures high precision and consistency in the spatial data utilized [33].

2.3.2. Random Forest Classification Method

For the extraction of UGS, hyperspectral data were analyzed using a supervised classification technique under ENVI 5.3 software, employing the Random Forest (RF) algorithm. This algorithm was chosen due to its proven efficacy in handling similar spectral categories [34,35], demonstrating superior performance over other classification methods. The RF algorithm, configured to default settings with a number of trees set to 100 and the number of features determined by the square root of the total number of features, was utilized. Classification samples were meticulously selected based on visual interpretations of Google Earth and GF-2 true color imagery, ensuring the relevance and accuracy of the training data. The effectiveness of this method was quantitatively assessed, achieving an overall classification accuracy of 92.6% with a kappa coefficient of 0.82, which affirms the robustness of the classification process and validates the suitability of the data for subsequent analytical applications.

2.3.3. Radiative Transfer (Atmospheric Correction) Method

Currently, there are three main surface temperature inversion algorithms, with the radiative transfer method (also known as the atmospheric correction method) having an advantage over a single-channel and split-window algorithms for inversion of surface temperature [36]. The thermal infrared radiation brightness value L_λ received by the satellite sensor in the radiative transfer method consists of three components: the atmospheric upward radiation brightness $L\uparrow$; the energy of the real radiation brightness of the ground reaching the satellite sensor after passing through the atmosphere; and the energy $L\downarrow$ reflected by the atmospheric downward radiation reaching the ground. The expression of the thermal infrared radiation brightness value L_λ received by the satellite sensor is shown as follows (radiation transmission equation):

$$L_\lambda = [\varepsilon B(T_S) + (1-\varepsilon)L\downarrow]\tau + L\uparrow \tag{1}$$

$$T_s = \frac{K_2}{\ln\left(\left(\frac{K_1}{B(T_S)}\right) + 1\right)} \tag{2}$$

where B is the Planck function; ε is the surface specific emissivity; τ is the atmospheric transmittance in the thermal infrared band; T_S is the true surface temperature (K); K_1 and K_2 are the radiation constants, for TIRS Band10, K_1 = 774.89(W·m^{-2}·sr^{-1}·μm^{-1}), K_2 = 1321.08 (K); TIRS Band11, K_1 = 480.89(W·m^{-2}·sr^{-1}·μm^{-1}), K_2 = 1201.14 (K).

2.3.4. Statistical Analysis of the Sampling

In this paper, we focus on the spatial distribution of UGS and LST. So, it is necessary to determine the scale of the analysis. The sample method is an important method to study the inner city [37,38]. Therefore, 527 sample grids were obtained after a 1 km × 1 km grid sampling of the study area. Then, the UGS landscape metrics and mean surface temperature of each grid were extracted.

2.3.5. Moving Window Analysis

The moving window analysis employed FRAGSTATS 4.2 (http://www.umass.edu/landeco/research/fragstats/fragstats.html, accessed on 26 September 2022) software landscape metrics are sensitive to changes in scale. If the window is too small, the local characteristics of the landscape will cover the overall characteristics, and the generated image will appear to have no continuity. If the window is too large, details will be lost, and the resulting image will be blurry [39]. To avoid the increase in the spatial heterogeneity of the landscape pattern due to the unreasonable setting of the moving window size, the variation maps of landscape indices (AI, ED, PD and Shape_am) under the corresponding windows are generated by setting different moving window edge lengths. Most of the current studies based on semi-variance function and block–base ratio in determining the analysis scale of landscape patterns assume that the variation degree of the landscape index is the same in all directions [40,41], that is, isotropic. This is followed by determining the analysis magnitude of the study area with the help of omnidirectional semi-variance function to quantitatively describe the spatial heterogeneity of the landscape pattern. A total of 735 random points were randomly generated under ArcGIS 10.4.1, and the landscape index values of the spatially randomly distributed points were extracted. GS+ 10.0 software was used to simulate the landscape index semi-variance function under different moving window radii. This was followed by calculating the block–base ratio to analyze and determine the appropriate moving window size for the study area, and the results are shown in Figure 2.

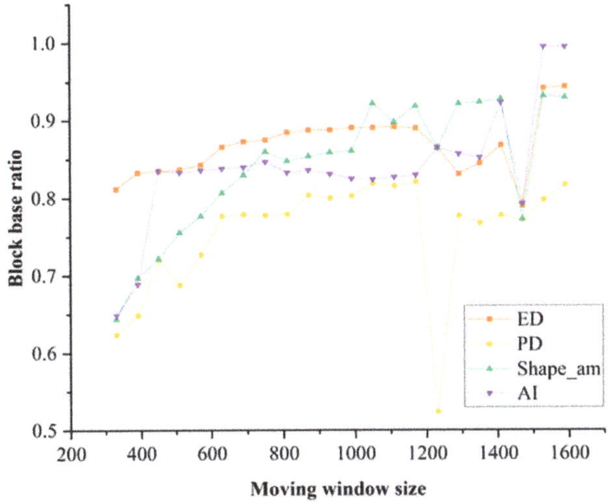

Figure 2. Spatial variation characteristics of landscape index.

In a moving window, the window is always rounded to the nearest odd cell so that the focal cell can be located in the center of the window. In this study, an odd multiple of 30 m was used as the test moving window size, which is a multiple of the resolution of Landsat 8 data. A total of 22 moving window sizes were set in intervals from 300 to 1600 m. From Figure 2, the fluctuation amplitude starts to decrease when the moving window edge length is in the interval from 810 to 990 m. Finally, 870 m was determined as the window edge length for the moving window analysis, and the shape was chosen as square, and the moving window was moved from the top left of the study area to analyze the whole study area.

2.3.6. Bivariate Local Indicator of Spatial Association (LISA)

The analysis in this study incorporates the bivariate local spatial autocorrelation (Bivariate Moran's I), a methodological advancement over the traditional Moran's I index for spatial autocorrelation, facilitated by GeoDa 1.18 software (http://geodacenter.github.io/, accessed on 28 March 2022). It extends the utility of Moran's I by enabling the examination of the correlation between a specific attribute of a spatial unit and a corresponding attribute of its neighboring units. Developed to provide deeper insights into spatial interdependencies, the Bivariate Moran's I allows a more nuanced exploration of spatial patterns by quantifying the degree of correlation between two distinct but spatially contiguous attributes [42,43], with the following equation:

$$I_{ab}^{i} = \frac{X_a^i - \overline{X_a}}{\sigma_a} \times \sum_{j=1}^{n} W_{ij} \frac{X_b^j - \overline{X_b}}{\sigma_b} \quad (3)$$

where x_a^i is the value of attribute a of cell i; x_b^i is the value of attribute b of cell j; σ_a and σ_b are the variances of attributes a and b, respectively.

The analysis of the spatial distribution data from regions i and j revealed distinct patterns of correlation between the independent and dependent variables, which can be categorized into four primary types of clustering including High–High (H-H), Low–Low (L-L), High–Low (H-L), and Low–High (L-H). The H-H and L-L clusters signify a positive correlation, indicating that high values of the independent variable in region i are associated with high values of the dependent variable in region j and similarly for low–low correlations. Conversely, H-L and L-H clusters represent a negative correlation, where high values of the independent variable in one region correspond to low values of the dependent variable in another, and vice versa.

Figure 3 shows the conceptual framework of this study divided into four stages: (1) collection of data and pre-processing; (2) extraction of built-up area, UGS and inversion of urban LST; (3) spatial distribution of LST and UGS landscape indices; and (4) analyzing the relationship between the spatial distribution of LST and UGS landscape indices.

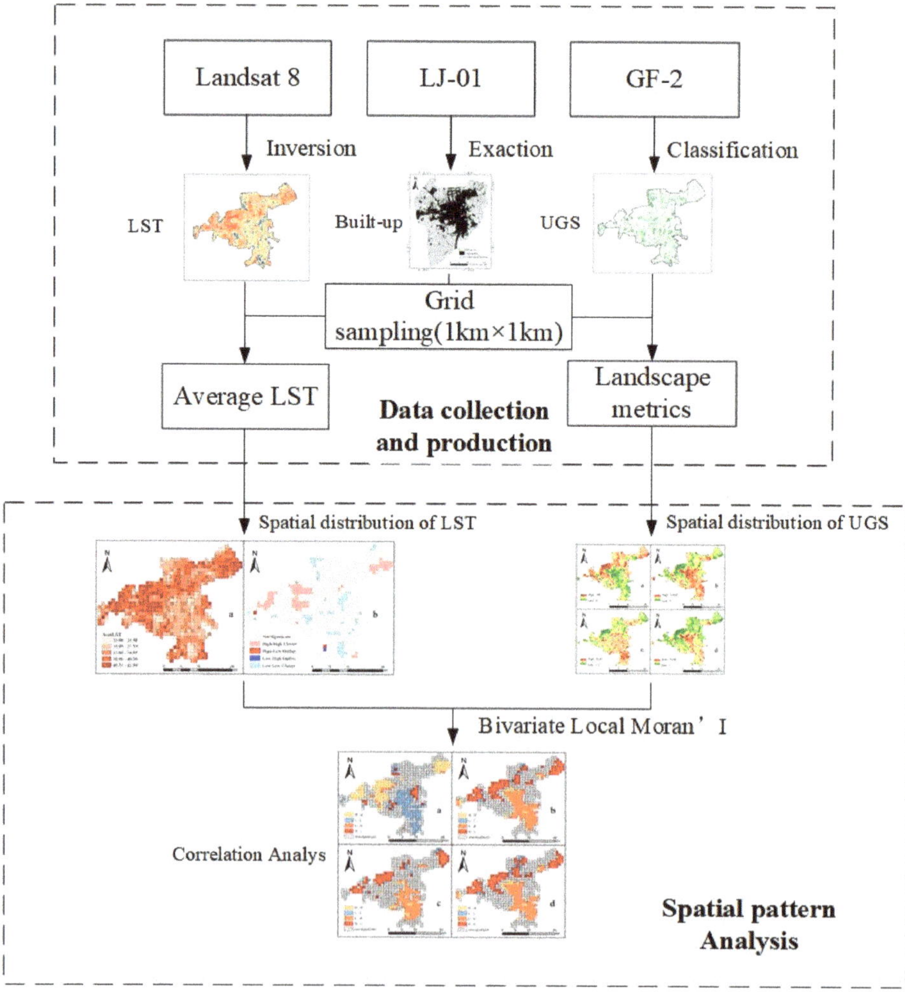

Figure 3. A flow chart showing the study's conceptual framework, divided into four principal stages to be implemented in sequence.

3. Results

3.1. Extracting Built-Up Area and UGS as Well as Inversing Land Surface Temperature

Built-up areas of Urumqi were delineated using the reference comparison method with LJ-01 satellite nighttime lighting data. This approach involved layering nighttime lighting data from high to low intensity to align with officially announced built-up area statistics as of 15 June 2022, accessible via the Urumqi municipal website (http://www.urumqi.gov.cn/fjbm/tjj/tjgb/509211.htm, accessed on 14 September 2022). This method ensured accurate and current representations of urban extents. The results, depicted in Figure 4, indicate that the built-up area encompasses approximately 536.20 km^2, primarily distributed across six districts including Tianshan, Saybag, Shuimogou, Xinshi, Toutunhe, and Midong.

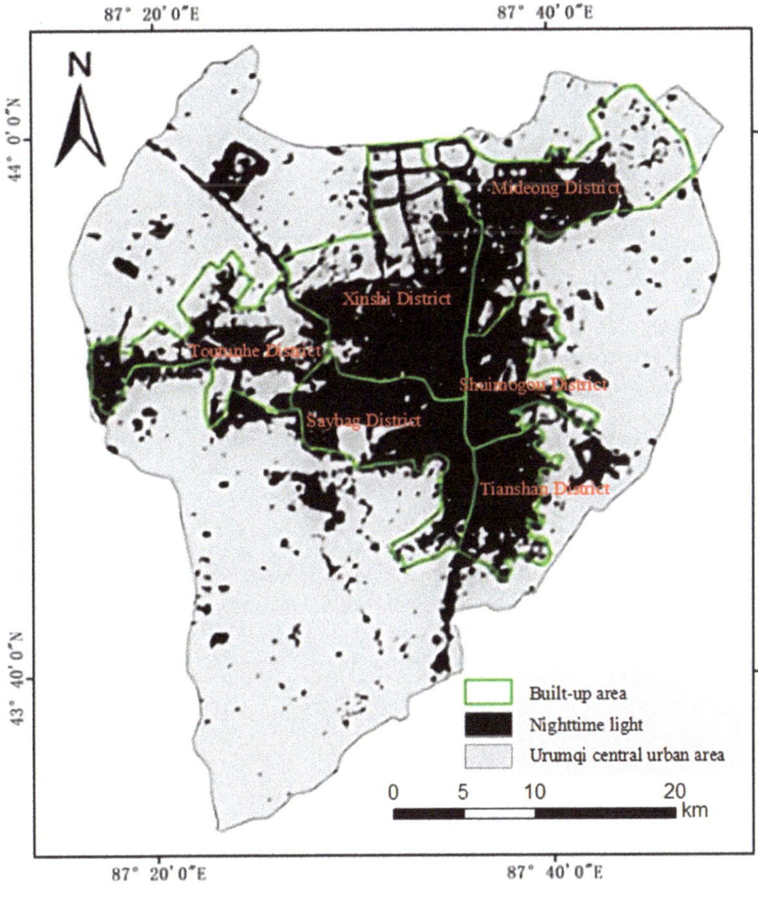

Figure 4. Schematic diagram of the built-up area.

Secondly, the UGS within the study area were delineated using high-resolution HMS-2 satellite imagery, as depicted in Figure 5. The map illustrates a varied distribution of UGS, with sparse coverage in the northern regions and more extensive coverage in central and southern sectors of the study area. Due to the 1 m spatial resolution of the HMS-2 imagery, which demands substantial memory for processing, it was necessary to resample these data. Therefore, Landsat imagery with a 30 m resolution was employed for resampling purposes. This resolution is acknowledged as optimal for examining the relationship between LST and landscape patterns [44]. The resampled UGS map at 30 m resolution facilitates more efficient subsequent analytical processes, ensuring the study maintains computational feasibility without compromising the spatial integrity of the environmental data.

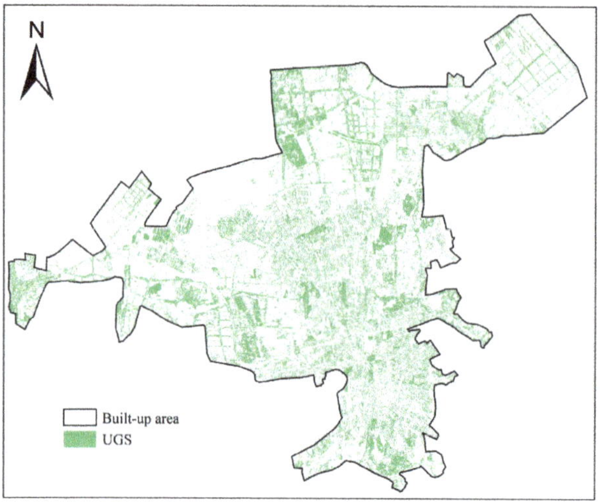

Figure 5. Extraction of urban green spaces (UGSs) for the study area.

Surface temperature was derived using a radiative transfer method applied to Landsat 8 imagery. The inversion results, displayed in Figure 6, reveal a spatial temperature distribution with higher temperatures predominantly located in the eastern and western parts of the study area, whereas the central and southern parts exhibit lower temperatures. To ensure the reliability of these results, a rigorous validation process was conducted using Landsat Collection 2 Level-2 data. Specifically, 500 random points were selected for a comparative analysis between the surface temperatures obtained from the Landsat 8 inversion and the corresponding data from the Collection 2 Level-2 data. The analysis, visualized in Figure 7, indicated a strong correlation and alignment between the two datasets, substantiating the accuracy of the inversion process. These validated results provide a robust basis for supporting subsequent studies and analyses.

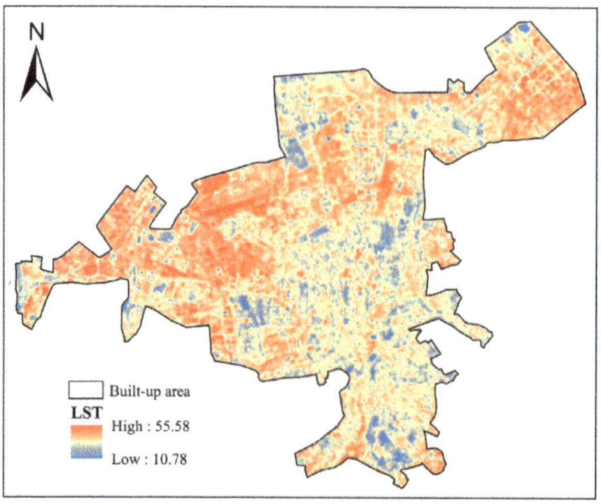

Figure 6. Surface temperature inversion diagram.

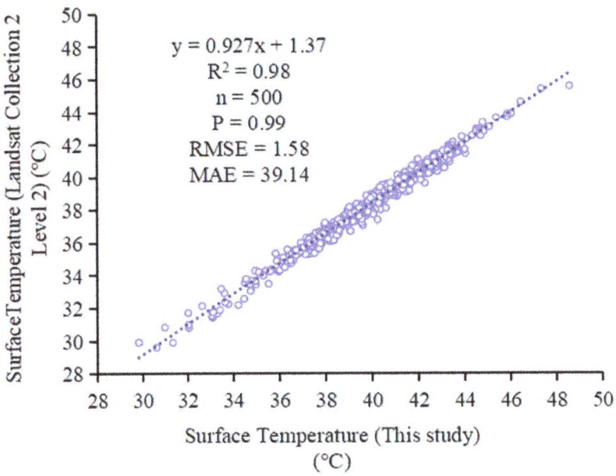

Figure 7. LST inversion accuracy verification.

3.2. Spatial Distribution of UGS Landscape Metrics

Figure 8 visualizes the distribution of four key landscape indices: AI, ED, PD, and Shape_am, which are instrumental in analyzing UGS configuration for urban thermal environment studies.

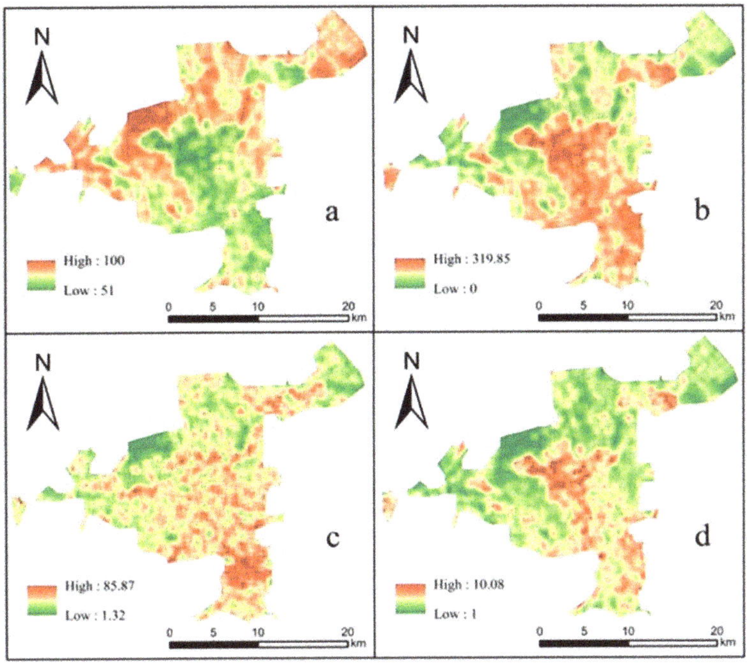

Figure 8. Spatial distribution of urban green space (UGS) landscape indies: (**a**) aggregation index (AI); (**b**) edge density (ED); (**c**) patch density (PD); and (**d**) area-weighted mean shape index (Shape_am).

High aggregation values, ranging from 0.51 to 1.00, were observed in Midong, Toutunhe, and northern Xinshi, suggesting that there is a great deal of connectivity between green space, and that large, contiguous green spaces are critical for ecological corridors and effective urban heat mitigation (Figure 8a). Lower edge densities in these areas correspond to the reduced fragmentation, supporting cohesive green infrastructure, which is advantageous for maintaining microclimatic stability and reducing edge effects that negatively impact flora and fauna by exposing them to urban environmental stresses (Figure 8b). Elevated patch densities in southern Xinshi, Saybag, Shuimogou, and Tianshan demonstrate a fragmented arrangement of green patches (Figure 8c). More irregularly shaped patches in southern Xinshi, with Shape_am values above 1, demonstrate a complex green space structure (Figure 8d), which is in contrast to more regular and simpler green space shapes that dominate in Toutunhe and Midong. These indices collectively provide a quantitative foundation for evaluating the structural and functional characteristics of UGSs within Urumqi, supporting targeted urban planning initiatives aimed at optimizing the ecological and thermal benefits of urban green spaces.

3.3. Spatial Distribution of LST

Empirical studies have established a significant correlation between LST and key urban surface features, notably for impervious surfaces and vegetation cover. High-temperature areas are primarily concentrated in Midong District, Toutunhe District, and the western part of the new urban area, while low-temperature areas cluster in the southern part of the new urban area, Saybag District, Shuimogou District, and Tianshan District (Figure 9a).

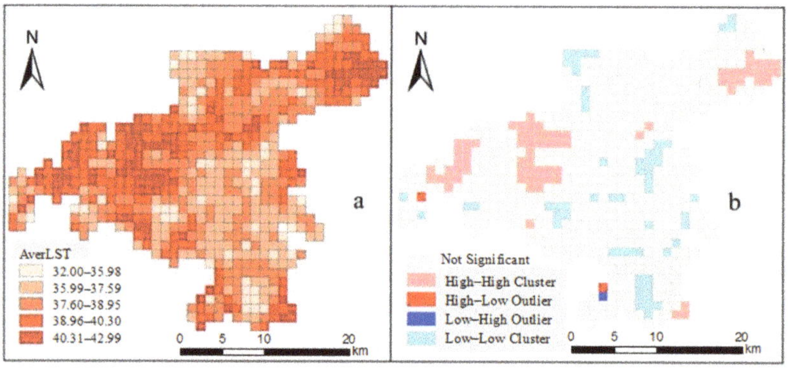

Figure 9. Spatial distribution of surface temperature: (**a**) AverLST; (**b**) cluster analysis.

Figure 9b shows that the high values are mainly concentrated in the western part of Xinshi District, the central part of Toutunhe District, and the eastern part of Midong District, indicating that not only their own temperature is high, but also the temperature of the adjacent units is equally high. The low values are mainly concentrated in Tianshan District, Saybag District, and Shuimogou District, with less distribution in the north and south of Xinshi District, the north of Midong District, and the west and south of Toutunhe District. The two cells with high values surrounded by low values are located in Toutunhe District and Tianshan District, indicating that these two cells are in high temperatures in spatial distribution and the cells next to them are in low-temperature conditions. The only cell with low values surrounded by high values is located in Tianshan District, indicating that this cell is at a low temperature in spatial distribution, but the cells next to it are at a high temperature. The conclusion drawn in Figure 9b coincides with Figure 9a.

3.4. Spatial Distribution Relationship between UGS and LST

It has been consistently indicated by previous research that an increase in UGS leads to a reduction in LST. Building on this foundation, the current study employs bivariate local spatial correlation analysis to investigate the influence of urban landscape structure indicators on regional surface temperatures, which is depicted in Figure 10.

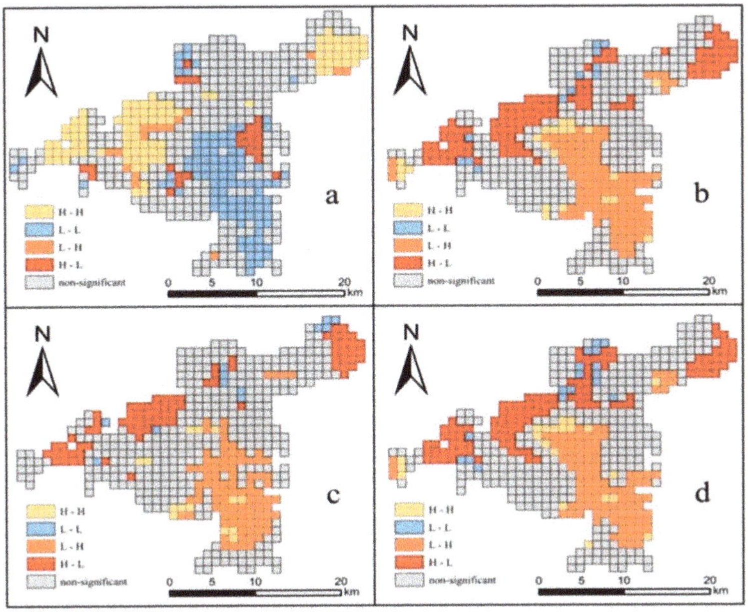

Figure 10. Bivariate LISA clustering diagram of UGS and LST: (**a**) aggregation index (AI); (**b**) edge density (ED); (**c**) patch density (PD); and (**d**) area-weighted mean shape index (Shape_am).

The landscape indexes were all significantly correlated with LST to different degrees, and AI, ED, PD, and Shape_am were all significantly correlated with LST at the 0.001 level. The results indicated that the degree of aggregation, edge density, patch density, and shape complexity index all played a role in LST. Among them, ED, PD and Shape_am were all negatively correlated with LST, and the Moran' I index was −0.469, −0.388, and −0.411, respectively, indicating that plaque fragmentation, UGS with a complex shape and better cooling effect, and ED had more significant effect on LST than PD. AI was positively correlated with LST, and the Moran' I index was 0.449, indicating that the higher the aggregation, the higher the LST.

Based on these results, planners can logically assume that a uniform distribution of street trees within the city may be an effective way to reduce the urban heat island effect. This suggests that when planning for cities in the future with cooling benefits from UGS as an option, decision makers should focus more on the pattern of UGS rather than their number to bring maximum benefits with the same number of UGS. This is especially true for cities with severe spatial constraints due to the competing land use demands, where adding more UGS remains a major challenge.

4. Discussion

4.1. Policy Influence on Spatial Distribution of LST

The spatial distribution of LST in Urumqi has been notably influenced by the national planning policies, especially during the 14th Five-Year Plan period. Development was primarily focused in the Toutunhe district, potentially overlooking environmental considerations, thereby impacting the ecological balance. In contrast, traditional districts like Tianshan, Saybag, and Shuimogou have achieved an equilibrium between development and environmental benefits (thanks to mature urban planning). Notably, areas such as the Botanical Garden and Liyu Mountain Park, located in the south of the new urban area, exhibit lower average surface temperatures, underlining the cooling effects of well-planned green spaces. Conversely, the Midong District, further from the city center, exhibits higher temperatures due to its lower population density, minimal green coverage, and proximity to industrial zones.

4.2. Influence of Spatial Resolution and Landscape Metrics on LST

The relationship between the landscape metrics and LST appears to vary significantly depending on the spatial resolution of the remote sensing imagery used. For instance, ED was found to be negatively correlated with LST in studies utilizing Landsat TM imagery with a resolution of 30 m [15,45]. However, this relationship was reversed in higher resolution imagery from SPOT (10 m) and QuickBird (2.44 m), where a positive correlation between ED and LST was observed, and it intensifies with increased resolution [46]. PD has predominantly shown a negative correlation with LST, which is aligning with our findings [47]. An analysis using Sentinel 2A data to assess cooling effects in green spaces indicated a significant positive correlation between LST and PD [16]. Similarly, using GF-6 imagery and a spatial analysis grid of 1 km × 1 km, researchers found a positive correlation between PD and LST, and a negative correlation between ED, AI, and LST, which diverges from the results reported by Bao et al. [48] where AI was positively correlated with LST. A comprehensive study by Xiang et al. [49] explored the impact of various landscape metrics on LST using a 9000 m × 9000 m analysis grid. The findings indicated that AI was negatively correlated with LST, whereas PD, ED, and Shape_am were positively correlated, utilizing land cover data at a 30 m resolution and LST data from MODIS MYDLT1D. These findings underscore the complexity of the interactions between landscape structure and LST, suggesting that spatial resolution and methodological nuances significantly influence these relationships.

4.3. Methodological Considerations and Future Directions

Discrepancies observed in the correlation between landscape metrics and LST across various studies may largely stem from the differential spatial scales, at which these metrics were calculated, coupled with the inherent sensitivity of the metrics to scale changes. In our study, UGS model metrics were computed using a 1 km × 1 km grid, primarily focusing on smaller urban patches. Conversely, studies by Li et al. [46] and Xiang et al. [49] concentrating on larger UGS such as parks, often overlook the smaller patches within urban settings. This variation in scale fundamentally affects the cooling distances attributed to different UGS, thereby leading to the conflicting results noted across studies. Additionally, the significant variation in overall urban form and existing UGS patterns across different cities elucidates the observed contradictions [50]. For instance, a comprehensive assessment of urban morphology's impact on UGS models in 262 cities across China revealed that factors such as the UGS perimeter-area ratio, road density, and a composite topographic complexity index exert significant influences on landscape metrics [51]. These findings highlight the critical role that both the scale of landscape metrics and urban structural complexities play a role in influencing LST outcomes.

Recognizing several inherent limitations could impact the generalizability and precision of our findings. Firstly, the spatial resolution of the Landsat 8 data used to derive LST is relatively low (30 m for OLI and 100 m for TIRS), which may affect the granularity of our UGS and LST analysis. While this resolution is adequate for large-scale urban analysis, it might not capture finer landscape details crucial for micro-scale urban studies. To overcome this, it is recommended that future studies utilize higher-resolution satellite data, such as from the Landsat 9 mission or commercial high-resolution satellites, which can provide more detailed information and thus enhance the accuracy and reliability of the analysis. Moreover, our study's focus on a single city, Urumqi, limits the ability to generalize our findings across different urban forms and climatic regions. Urumqi's unique geographic and climatic conditions may not represent other urban environments where different UGS configurations and broader climatic factors might influence the urban heat island effect differently. To address this limitation, future research should aim to include multiple cities that vary significantly in size, climatic conditions, and urban morphology. Conducting comparative studies across these diverse settings will help identify robust UGS evaluation indicators that are applicable under varying urban conditions. Expanding the research scope to include a broader array of urban settings will also allow for the development of generalized urban green space strategies tailored to specific climatic zones or urban configurations. This approach is crucial for devising effective urban planning and management strategies that can mitigate the urban heat island effect and contribute to climate mitigation and urban sustainability more broadly. Additionally, integrating advanced spatial analysis techniques and exploring the potential of new analytical methods, such as machine learning models that can handle complex variable interactions in large datasets, might provide deeper insights into the dynamics of UGS and LST. In conclusion, while the current study provides valuable insights into the correlation between UGS and LST in Urumqi, acknowledging and addressing these limitations through methodological enhancements and broader geographic studies will greatly enhance the reliability and applicability of future urban ecological research.

5. Conclusions

The degradation of urban environmental quality, evidenced by deteriorating air quality and increasing UHI effects, is one of the adverse consequences of uncontrolled urbanization. Nonetheless, these negative effects can be mitigated. Urban vegetation plays a pivotal role in preserving the ecological equilibrium within cities, with the ecological services offered by UGS exhibiting notable diversity across varying landscape configurations. Consequently, this study analyzed the spatial distribution characteristics of UGS and LST in Urumqi, situated in an arid zone, and explored the bivariate local spatial autocorrelation between UGS distribution and LST as a reference for urban planners to design UGS scientifically. The following conclusions are drawn from the results of this study:

(1) Patches with a high aggregation index are predominantly located in the study area of Midong District, Toutunhe District, and the northern part of Xinshi District, whereas the distribution of edge density contrasts with that of the aggregation index. The southern part of Xinshi District, Saybag District, Shuimogou District, and Tianshan District exhibit higher densities of patches, with Xinshi District encompassing the largest distribution area of UGS, albeit non-uniformly distributed.

(2) The high-temperature zones are primarily concentrated in Midong District, Toutunhe District, and the western part of Xinshi District, while the low-temperature zones predominantly cluster in the southern part of Xinshi District, Saybag District, Shuimogou District, and Tianshan District.

(3) AI, ED, PD, and Shape_am were all found to be significantly correlated with LST at the 0.001 level. Conversely, ED, PD, and Shape_am exhibited negative correlations with LST, while AI was positively correlated with LST.

This study provides a more scientific reference for urban planners to evaluate UGS by studying the influence of different landscape metrics of UGS on LST. However, further studies and in-depth exploration of different cities are needed to reveal the impact of the UGS model.

Author Contributions: Conceptualization, M.Z.; methodology, M.Z. and J.W.; formal analysis, M.Z.; resources, F.Z.; data curation, M.Z.; writing—original draft, M.Z.; visualization, M.Z.; supervision, F.Z.; funding acquisition, F.Z. All authors have read and agreed to the published version of the manuscript.

Funding: This research was supported by the Strategic Priority Program of the Chinese Academy of Sciences, Pan-Third Pole Environment Study for a Green Silk Road (XDA20040400), and the Tianshan Talent Project (Phase III) of the Xinjiang Uygur Autonomous region, China.

Data Availability Statement: The datasets used and analyzed during the current study are available from the corresponding author on reasonable request.

Acknowledgments: The authors would like to thank Cheng Chen of Zhejiang Normal University, Daosheng Chen of Xinjiang University, Ngai Weng Chan and Mou Leong Tan of Universiti Sains Malaysia, Jinchao Shi of the University of Memphis, and Cheng Li of Yangzhou University for reviewed and improved our paper.

Conflicts of Interest: The authors declare no competing interests.

References

1. Vaidya, H.; Chatterji, T. SDG 11 sustainable cities and communities: SDG 11 and the new urban agenda: Global sustainability frameworks for local action. In *Actioning the Global Goals for Local Impact: Towards Sustainability Science, Policy, Education and Practice*; Springer: Singapore, 2020; pp. 173–185.
2. Xue, B.; Han, B.; Li, H.; Gou, X.; Yang, H.; Thomas, H.; Stückrad, S. Understanding ecological civilization in China: From political context to science. *Ambio* **2023**, *52*, 1895–1909. [CrossRef] [PubMed]
3. Ngom, R.; Gosselin, P.; Blais, C. Reduction of disparities in access to green spaces: Their geographic insertion and recreational functions matter. *Appl. Geogr.* **2016**, *66*, 35–51. [CrossRef]
4. Corburn, J. Cities, climate change and urban heat island mitigation: Localising global environmental science. *Urban Stud.* **2009**, *46*, 413–427. [CrossRef]
5. Arnfield, A. Two decades of urban climate research: A review of turbulence, exchanges of energy and water, and the urban heat island. *Int. J. Climatol. J. R. Meteorol. Soc.* **2003**, *23*, 1–26. [CrossRef]
6. Li, X.; Zhou, W.; Ouyang, Z.; Xu, W.; Zheng, H. Spatial pattern of greenspace affects land surface temperature: Evidence from the heavily urbanized Beijing metropolitan area, China. *Landsc. Ecol.* **2012**, *27*, 887–898. [CrossRef]
7. Ke, X.; Men, H.; Zhou, T.; Li, Z.; Zhu, F. Variance of the impact of urban green space on the urban heat island effect among different urban functional zones: A case study in Wuhan. *Urban For. Urban Green* **2021**, *62*, 127159. [CrossRef]
8. Wang, X.; Meng, Q.; Zhang, L.; Hu, D. Evaluation of urban green space in terms of thermal environmental benefits using geographical detector analysis. *Int. J. Appl. Earth Obs. Geoinf.* **2021**, *105*, 102610. [CrossRef]
9. Chan, N.; Tan, M.; Ghani, A.; Zakaria, N. Sustainable urban drainage as a viable measure of coping with heat and floods due to climate change. *IOP Conf. Ser. Earth Environ. Sci.* **2019**, *257*, 012013. [CrossRef]
10. Aram, F.; García, E.H.; Solgi, E.; Mansournia, S. Urban green space cooling effect in cities. *Heliyon* **2019**, *5*, e01339. [CrossRef]
11. Masoudi, M.; Tan, P.; Fadaei, M. The effects of land use on spatial pattern of urban green spaces and their cooling ability. *Urban Clim.* **2021**, *35*, 100743. [CrossRef]
12. Shah, A.; Garg, A.; Mishra, V. Quantifying the local cooling effects of urban green spaces: Evidence from Bengaluru, India. *Landsc. Urban Plan.* **2021**, *209*, 104043. [CrossRef]
13. Zhou, W.; Cao, W.; Wu, T.; Zhang, T. The win-win interaction between integrated blue and green space on urban cooling. *Sci. Total Environ.* **2023**, *863*, 160712. [CrossRef] [PubMed]
14. Proutsos, N.; Tigkas, D.; Tsevreni, I.; Alexandris, S.; Solomou, A.; Bourletsikas, A.; Stefanidis, S.; Nwokolo, S. A thorough evaluation of 127 potential evapotranspiration models in two mediterranean urban green sites. *Remote Sens.* **2023**, *15*, 3680. [CrossRef]
15. Maimaitiyiming, M.; Ghulam, A.; Tiyip, T.; Pla, F.; Latorre-Carmona, P.; Halik, Ü.; Sawut, M.; Caetano, M. Effects of green space spatial pattern on land surface temperature: Implications for sustainable urban planning and climate change adaptation. *ISPRS J. Photogramm. Remote Sens.* **2014**, *89*, 59–66. [CrossRef]
16. Pramanik, S.; Punia, M. Environment. Assessment of green space cooling effects in dense urban landscape: A case study of Delhi, India. *Model. Earth Syst. Environ.* **2019**, *5*, 867–884. [CrossRef]
17. Guo, L.; Liu, R.; Men, C.; Wang, Q.; Miao, Y.; Zhang, Y. Quantifying and simulating landscape composition and pattern impacts on land surface temperature: A decadal study of the rapidly urbanizing city of Beijing, China. *Sci. Total Environ.* **2019**, *654*, 430–440. [CrossRef] [PubMed]
18. Rakoto, P.; Deilami, K.; Hurley, J.; Amati, M.; Sun, Q.; Greening, U. Revisiting the cooling effects of urban greening: Planning implications of vegetation types and spatial configuration. *Urban For. Urban Green.* **2021**, *64*, 127266. [CrossRef]
19. Yuan, B.; Zhou, L.; Hu, F.; Wei, C. Effects of 2D/3D urban morphology on land surface temperature: Contribution, response, and interaction. *Urban Clim.* **2024**, *53*, 101791. [CrossRef]
20. Huang, S.; Xiao, X.; Tian, T.; Che, Y. Seasonal influences on preferences for urban blue-green spaces: Integrating land surface temperature into the assessment of cultural ecosystem service value. *Sustain. Cities Soc.* **2024**, *102*, 105237. [CrossRef]
21. Kumar, B.; Anusha, B.; Babu, k.; Sree, P. Identification of climate change impact and thermal comfort zones in semi-arid regions of AP, India using LST and NDBI techniques. *J. Clean. Prod.* **2023**, *407*, 137175. [CrossRef]
22. Cheng, Y.; Wang, W.; Ren, Z.; Zhao, Y.; Liao, Y.; Ge, Y.; Wang, J.; He, J.; Gu, Y.; Wang, Y.; et al. Multi-scale Feature Fusion and Transformer Network for urban green space segmentation from high-resolution remote sensing images. *Int. J. Appl. Earth Obs. Geoinf.* **2023**, *124*, 103514. [CrossRef]
23. Guan, J.; Wang, R.; Berkel, D.; Liang, Z. How spatial patterns affect urban green space equity at different equity levels: A bayesian quantile regression approach. *Landsc. Urban Plan.* **2023**, *233*, 104709. [CrossRef]

24. Connors, J.; Galletti, C.; Chow, W. Landscape configuration and urban heat island effects: Assessing the relationship between landscape characteristics and land surface temperature in Phoenix, Arizona. *Landsc. Ecol.* **2013**, *28*, 271–283. [CrossRef]
25. El-Beltagy, A.; Madkour, M. Impact of climate change on arid lands agriculture. *Agric. Food Secur.* **2012**, *1*, 3. [CrossRef]
26. Chen, D.; Zhang, F.; Zhang, M.; Meng, Q.; Jim, C.Y.; Shi, J.; Tan, M.L.; Ma, X. Landscape and vegetation traits of urban green space can predict local surface temperature. *Sci. Total Environ.* **2022**, *825*, 154006. [CrossRef] [PubMed]
27. Li, X.; Zhao, L.; Li, D.; Xu, H. Mapping urban extent using Luojia 1-01 nighttime light imagery. *Sensors* **2018**, *18*, 3665. [CrossRef] [PubMed]
28. Yun, C. GF-2 satellite. *Satell. Appl.* **2014**, *9*, 65. (In Chinese)
29. Wang, X.; Meng, Q.; Zhao, S.; Li, J.; Zhang, L.; Chen, X. Urban green space classification and landscape pattern measurement based on GF-2 image. *J. Geoinf. Sci* **2020**, *22*, 1971–1982.
30. Duan, S.; Ru, C.; Li, Z.; Wang, M.; Xu, H.; Li, H.; Wu, P.; Zhan, W.; Zhou, J.; Zhao, W. Reviews of methods for land surface temperature retrieval from Landsat thermal infrared data. *Natl. Remote Sens. Bull.* **2021**, *25*, 1591–1617. [CrossRef]
31. Onačillová, K.; Gallay, M.; Paluba, D.; Péliová, A.; Tokarčík, O.; Laubertová, D. Combining landsat 8 and sentinel-2 data in google earth engine to derive higher resolution land surface temperature maps in urban environment. *Remote Sens.* **2022**, *14*, 4076. [CrossRef]
32. Zhang, M.; Zhang, F.; Chen, D.; Tan, M.L.; Chan, N. Urban local surface temperature prediction using the urban gray-green space landscape and vegetation indices. *Build. Environ.* **2022**, *226*, 109723. [CrossRef]
33. Li, G.; Fan, J.; Zhou, Y.; Zhang, Y. Development characteristics estimation of Shandong peninsula urban agglomeration using VIIRS night light data. *Remote Sens. Technol. Appl.* **2020**, *35*, 1348–1359.
34. Naidoo, L.; Cho, M.A.; Mathieu, R.; Asner, G. Classification of savanna tree species, in the Greater Kruger National Park region, by integrating hyperspectral and LiDAR data in a Random Forest data mining environment. *ISPRS J. Photogramm. Remote Sens.* **2012**, *69*, 167–179. [CrossRef]
35. Pu, R.; Landry, S.; Yu, Q. Assessing the potential of multi-seasonal high resolution Pléiades satellite imagery for mapping urban tree species. *Int. J. Appl. Earth Obs. Geoinf.* **2018**, *71*, 144–158.
36. Liu, H.; Weng, Q. Scaling effect on the relationship between landscape pattern and land surface temperature. *Photogramm. Eng. Remote Sens.* **2009**, *75*, 291–304. [CrossRef]
37. Yu, X.; Guo, X.; Wu, Z. Land surface temperature retrieval from Landsat 8 TIRS—Comparison between radiative transfer equation-based method, split window algorithm and single channel method. *Remote Sens.* **2014**, *6*, 9829–9852. [CrossRef]
38. Mutani, G.; Todeschi, V. Roof-integrated green technologies, energy saving and outdoor thermal comfort: Insights from a case study in urban environment. *Int. J. Sustain. Dev. Plan* **2021**, *16*, 13–23. [CrossRef]
39. Vega, K.A.; Küffer, C. Promoting wildflower biodiversity in dense and green cities: The important role of small vegetation patches. *Urban For. Urban Green.* **2021**, *62*, 127165. [CrossRef]
40. McGarigal, K. *FRAGSTATS Help*; University of Massachusetts: Amherst, MA, USA, 2015; 182p.
41. Mou, Y.; Chen, Q.; Li, L.; Tu, C.; Lu, X. Spatial differentiation and influencing factors of surface soil selenium in Huaxi District, Guiyang. *Acta Sci. Circumstantiae* **2022**, *42*, 415–424. (In Chinese)
42. Song, S.; Wang, Y.; Shi, M.; Hu, S.; Xu, D. Effects of landscape pattern type on soil erosion. *Soil Water Conserv. Res.* **2022**, *29*, 85–92. (In Chinese)
43. Gutiérrez, J.; García-Palomares, J.C.; Romanillos, G.; Salas-Olmedo, M. The eruption of Airbnb in tourist cities: Comparing spatial patterns of hotels and peer-to-peer accommodation in Barcelona. *Tour. Manag.* **2017**, *62*, 278–291. [CrossRef]
44. Yao, X.; Zeng, J.; Li, W. Spatial correlation characteristics of urbanization and land ecosystem service value in Wuhan Urban Agglomeration. *Trans. Chin. Soc. Agric. Eng.* **2015**, *31*, 249–256.
45. Li, B.; Shi, X.; Wang, H.; Qin, M. Analysis of the relationship between urban landscape patterns and thermal environment: A case study of Zhengzhou City, China. *Environ. Monit. Assess.* **2020**, *192*, 540. [CrossRef] [PubMed]
46. Li, X.; Zhou, W.; Ouyang, Z. Relationship between land surface temperature and spatial pattern of greenspace: What are the effects of spatial resolution? *Landsc. Urban Plan.* **2013**, *114*, 1–8. [CrossRef]
47. Li, J.; Song, C.; Cao, L.; Zhu, F.; Meng, X.; Wu, J. Impacts of landscape structure on surface urban heat islands: A case study of Shanghai, China. *Remote Sens. Environ.* **2011**, *115*, 3249–3263. [CrossRef]
48. Bao, T.; Li, X.; Zhang, J.; Zhang, Y.; Tian, S. Assessing the distribution of urban green spaces and its anisotropic cooling distance on urban heat island pattern in Baotou, China. *ISPRS Int. J. Geo-Inf.* **2016**, *5*, 12. [CrossRef]
49. Xiang, Y.; Ye, Y.; Peng, C.; Teng, M.; Zhou, Z. Seasonal variations for combined effects of landscape metrics on land surface temperature (LST) and aerosol optical depth (AOD). *Ecol. Indic.* **2022**, *138*, 108810. [CrossRef]
50. Huang, J.; Lu, X.; Sellers, J. A global comparative analysis of urban form: Applying spatial metrics and remote sensing. *Landsc. Urban Plan.* **2007**, *82*, 184–197. [CrossRef]
51. Huang, C.; Yang, J.; Jiang, P. Assessing impacts of urban form on landscape structure of urban green spaces in China using Landsat images based on Google Earth Engine. *Remote Sens.* **2018**, *10*, 1569. [CrossRef]

Disclaimer/Publisher's Note: The statements, opinions and data contained in all publications are solely those of the individual author(s) and contributor(s) and not of MDPI and/or the editor(s). MDPI and/or the editor(s) disclaim responsibility for any injury to people or property resulting from any ideas, methods, instructions or products referred to in the content.

climate MDPI

Article

Assessment of the Vulnerability of Households Led by Men and Women to the Impacts of Climate-Related Natural Disasters in the Coastal Areas of Myanmar and Vietnam

Aung Tun Oo [1,*], Ame Cho [2] and Dao Duy Minh [3]

[1] Department of Economics, Society and Politics, University of Urbino Carlo Bo, 61029 Urbino, Italy
[2] Gran Sasso Science Institute, 67100 L'Aquila, Italy; ame.cho@gssi.it
[3] Faculty of Economics and Development Studies, University of Economics-Hue University, Thua Thien Hue 530000, Vietnam; daoduyminh@hueuni.edu.vn
* Correspondence: aung.oo@uniurb.it or aunghtunoo717.yau@gmail.com

Abstract: Farm households along the coastlines of Myanmar and Vietnam are becoming increasingly vulnerable to flooding, saltwater intrusion, and rising sea levels. There is little information available on the relative vulnerability of men- and women-headed households, and the governments of Myanmar and Vietnam have not identified or implemented any adaptive measures aimed specifically at vulnerable peoples. This study aims to fill these gaps and assess the relative climate change vulnerability of men- and women-headed farm households. This study considers 599 farm households from two regions of Myanmar and 300 households from Thua Thien Hue province of Vietnam for the period 2021–2022. We offer a livelihood vulnerability index (LVI) analysis of men- and women-headed farm households using 46 indicators arranged into seven major components. The aggregate LVI scores indicate that farm households in Myanmar are more vulnerable (scores of 0.459 for men and 0.476 for women) to climate-related natural disasters than farm households in Vietnam (scores of 0.288 for men and 0.292 for women), regardless of the gender of the head of household. Total vulnerability indexing scores indicate that women-headed households are more vulnerable than men-headed households in both countries. Poor adaptive capacity and highly sensitive LVI dimensional scores explain the greater vulnerability of women-headed farm households. The findings also highlight the importance of the adaptive capacity components reflected in the LVI analysis in reducing farm households' vulnerability.

Keywords: coastal vulnerability; gender-treatment effect; adaptive capacity; livelihood vulnerability index (LVI)

1. Introduction

Climate change has exacerbated the rise of sea levels and poses an existential threat to some low-lying coastal areas, affecting the socioeconomic vulnerability of communities in coastline areas [1–3]. Sea-level rise, floods, and saltwater intrusion are becoming more common in the Mekong region and wreaking havoc on farming communities, livelihoods, and food and nutrition security. Nevertheless, little is known about the vulnerability to climate change in farming households in the coastal Mekong region, where there are scarce resources for creating and implementing adaptation strategies [4,5]. National adaptation strategies have focused on hard strategies such as infrastructure rebuilding; few have focused on identifying the vulnerability of various groups and communities and how climate adaptation funds might be better allocated to meet their needs. An assessment of the vulnerability of the communities and regions in the Mekong region has been particularly encouraged by development agencies and public and private organisations [4,6,7].

The countries in the Mekong region are extremely exposed to the effects of extreme climate events such as rising temperatures, erratic rainfall, flooding, and sea-level rise [4–7].

The Global Climate Risk Index report ranks Myanmar as the second most climate-hazardous country in the world [2]. The coastline areas of Myanmar lie just 3 m above sea level and are densely populated. Therefore, these areas are vulnerable to flooding, saltwater intrusion, and other adverse effects of climate change [6,8]. Since the main source of income in the delta areas is agriculture, the adverse impacts of climate change and the rise of sea level will put an additional burden on the people in these areas. Recent data show that rice farming in coastline areas is frequently impacted by saltwater intrusion and coastal flooding along with rising sea levels [9–11]. Vietnam is similarly vulnerable; it is currently sixth among countries that have suffered the most from climate change over the last two decades [2]. It is one of nine countries in which at least 50 million people are at risk of exposure to, among other things, increasing sea levels, saltwater intrusion, and powerful storms [12,13]. With over 3200 km long of coastline, shrimp aquaculture, rice-based cropping, and fisheries are three distinct systems in Vietnam, and these sectors are particularly vulnerable to adverse effects of climate change [14,15]. We have thus selected Vietnam and Myanmar as countries in the Mekong region with a high climate risk to assess the varying effects of farm households' vulnerability to flooding and saltwater intrusion.

The effects of climate change are not evenly distributed, and marginalised groups, including women, are disproportionately affected. Female farmers frequently face gender-specific barriers to resources and decision-making power, exacerbating their vulnerability to climate risks [16,17]. In many regions, women play an important role in food production, processing, and marketing, and their vulnerability has consequences for agricultural productivity and food security. Understanding the gendered differences in the impacts of climate change is therefore crucial for developing effective and gender-sensitive adaptation strategies that address the specific needs and challenges of women farmers [16–18]. Women constitute the majority of the world's poor and rely on natural resources that are most vulnerable to climate change [19]. There are nevertheless few research assessments of the relative level of climate change vulnerability experienced by women- and men-headed households where the causes of women's and men's vulnerability to climate change are many and varied [19–21].

As the United Nations Framework Convention on Climate Change (UNFCCC) [19] points out, simply adding women to existing enquiry frameworks and approaches is insufficient. Rather, the root causes of vulnerability need to be addressed by using a more nuanced approach to capture multiple and intersecting social pressures related to gender. Therefore, we conceptualise the root causes of vulnerability in men- and women-headed households by identifying multiple and intersecting indicators, social pressures, and climate-related natural hazards. The remainder of the study is structured as follows. In Section 2, we present theoretical conceptualisation, followed by our research methodology in Section 3, research findings and discussion in Section 4, and our conclusions and recommendations are presented in Section 5.

2. Theoretical Conceptualisation

The literature highlights that although climate change is a global process, there is a need to test climate change vulnerability at the community level [6,22,23]. Therefore, many scholars advise conducting localised assessments of farm households' vulnerability and evaluating the vulnerability of communities and regions [10,23,24]. There are many methods available to assess the vulnerability and impacts of climate change on specific communities or societies; the most common are econometrics and indicator-based approaches [6,23]. Indicator methods are easy to use, and indications can be easily compiled into an index to measure various vulnerabilities [10,23,24].

Climate change impacts male and female farmers differently, depending on their social, economic, and cultural contexts, and it is important to identify and address the differential vulnerabilities of men and women farmers [18,19,25,26]. Aryal et al. [16] find that even when exposed to the same climate risks, households in Bangladesh's coastal region face different levels of livelihood impact depending on whether they are led by men or women; these disparate livelihood impacts are largely explained by differences in access to assets and the availability of adaptation options. Thus, this study considers indicators reported across the scientific literature to assess the dimensions and contexts of climate-related vulnerability of women- and men-headed households in the coastline areas of Myanmar and Vietnam.

Since the combined impacts of climate change and water insecurity may be severe for communities in remote locations with limited infrastructure, we consider access to market and infrastructure accessibility to estimate the adverse effects of climate change on communities with or without limited infrastructure [27]. This may capture the effect of public investment in infrastructure as climate resilience strategies in Myanmar and Vietnam. Time spent travelling to source water is a well-known determinant of quality of life and determines the overall burden of fetching water [6,10,28]. Women often play a greater role than men in managing water supplies as they are the primary users of water in the household and are more likely to detect water insecurities. Women are also more burdened than men by household caregiving, including having larger roles in household food preparation [6,29,30]. This study explicitly considers some norms and cultural indicators related to fetching water, time taken for such collection, and caregiving of family members to better estimate the differences in climate change vulnerability of farm households in the coastline areas of Myanmar and Vietnam.

Some studies divide livelihood vulnerability to climate change into five major components based on five livelihood capitals identified in sustainable livelihood framework analysis. However, this would not as accurately reflect the intricacies of household vulnerability as would the seven major components of livelihood analysis that we apply here [6,10,31,32]. The seven major components of livelihood analysis are the following: socio-demographics, social networks, livelihood strategy, food, health, water, and natural hazards. According to the Intergovernmental Panel on Climate Change (IPCC)-defined vulnerability standard, the first three components can be constituted in the adaptive capacity dimension, while health, food, and water components can be constituted in the sensitivity dimension. The exposure dimension is used to characterise the natural hazard component [1,3].

This study introduces new indicators based on the outcomes of focus group discussions. These indicators reflect support from private, community-based organisations and public organisations during the COVID-19 pandemic. This study also considers new indicators related to the conflict between water users and the effect of this conflict on local communities in water resource utilisation and management at the farm level in Myanmar and Vietnam. The study assesses the difference in vulnerability of women and men households by indexing livelihood.

We adopt the indexing method applied in earlier studies [6,33–35]. However, our innovative contribution is that we employ indicators reported across various studies as well as local-specific and community-validated indicators. Our study offers concrete findings and has implications for policymakers and development planners in Myanmar and Vietnam. Second, using the 46 indicators, we apply the livelihood vulnerability index (LVI) and a gender-treatment effect to examine the climate change vulnerability of men- and women-led households in two countries.

3. Research Methodology

3.1. Study Areas and Data

This research is conducted in Vietnam and Myanmar. Due to their vast geographies and their populations' limited knowledge and understanding regarding the influence of climate change, this study is focused on two regions of Myanmar and one region of Vietnam. A randomised sampling method is considered because of the complications and impacts of the COVID-19 pandemic in Myanmar and Vietnam; samples are mostly from areas in which travel was less restricted. For instance, Myanmar had fewer travel restrictions within communities at the time of the survey, but there were strict travel restrictions throughout

Vietnam due to the COVID-19 pandemic. Initially, villages are selected based on particular characteristics, such as villages located near coastline areas and in which both agriculture and aquaculture farmers are located. Data are collected in four distinct steps: (i) field observation, (ii) validation of selected indicators with key stakeholders through focus group discussions, (iii) final questionnaire preparation, and (iv) farm household surveys or interviews conducted from September to October 2021 in Myanmar and from February to March 2022 in Vietnam (see Figure 1).

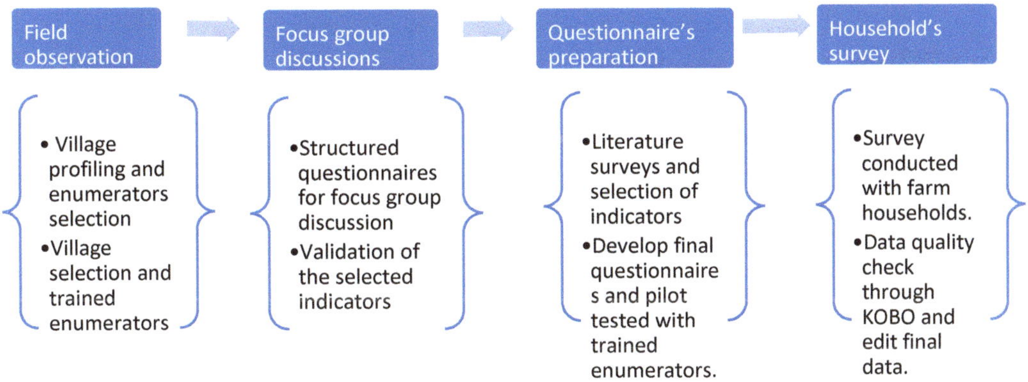

Figure 1. Four distinct steps of data collection.

Taking into account the factors noted above, we select 12 sampling communities in Myanmar and six sampling communities in Vietnam along the coastline and delta areas. We select a sample of 300 individuals in each region for interviews. In Myanmar, our final sample, from which some observations were missing, comprises 599 respondents, and that for Vietnam is 300. Despite inconsistencies in the available dataset, the LVI method that we apply in this study uses a balanced weighted average approach to minimise the variance. To ensure the indicators selected are appropriate, these are tested and validated with the key stakeholders during the focus group discussions. Nineteen indicators are selected to represent the components of socio-demographic factors, livelihood strategy, and social network. Twenty indicators are considered for inclusion in the sensitivity dimensions (i.e., health, food, and water components), and seven indicators are selected for inclusion in the exposure dimension. The LVI analysis includes a final total of 46 indicators divided into seven major components.

In Myanmar, we selected six communities in Dedaye and Kyaiklatt Township in the Ayeyarwaddy region (Ka Lar Su, Kyon Thin, Hmyaw Sin, Kunpa Laing, Kyee Chaung, and Hpoe Shan Gyi villages), and six communities in Kyauktan Township in the Yangon region (Kamarmat, Oke Pho, Thet Kel Kone, Yae Kyaw, Shwe Pyi Thit, and Zwe Bar villages). The three townships that were chosen are all low-lying areas with elevations under 5 m; see Figure 2.

In Vietnam, we select six communities in three districts of the Tam Giang–Cai Hai lagoon: Quang Dien (Quang Loi and Quang Thai), Phu Vang (Vinh Xuan and Vinh Phu), and Phu Loc (Vinh Hung and Vinh Giang); see Figure 3. For the village profiles in the sampling areas, we developed gender- and ethnicity-related questions and worked to ensure the participation of women and minority groups in the research assessment.

Enumerators from the selected villages are chosen and trained through virtual training and pilot surveys; this approach is taken because of the impacts of travel restrictions during the COVID-19 pandemic and the political situation in Myanmar. The data collection tool, KOBO, is used to collect data and check the quality of the dataset. In Vietnam, the enumerator training is conducted at Hue University, and field-based data collection is carried out. All potential interviewees were asked to sign an ethical consent form to participate in the survey; interviews are only conducted with those willing to participate. The research team paid attention to the norms, values, and ethics of community participants and considered their ethical principles, such as following COVID-19 coping measures, keeping participant data anonymous, and practising secure data management during the research assessment, despite the development of COVID-19 pandemic in the region.

Although our study is in line with the proposed research objectives, it has certain limitations. The enumerators were trained and tested in advance of the survey using a set of questionnaires. Their level of understanding of the concepts of climate change and the types of questionnaires may not be perfect. This study may also be associated with the specificities of each of two different contexts or countries in one single questionnaire. However, we confirm that our data are of high quality through a daily data quality check. Moreover, this study focuses on the impacts of climate change on farm households and does not take into account the impacts on the ecosystem, biodiversity, livestock, and animals, and so on.

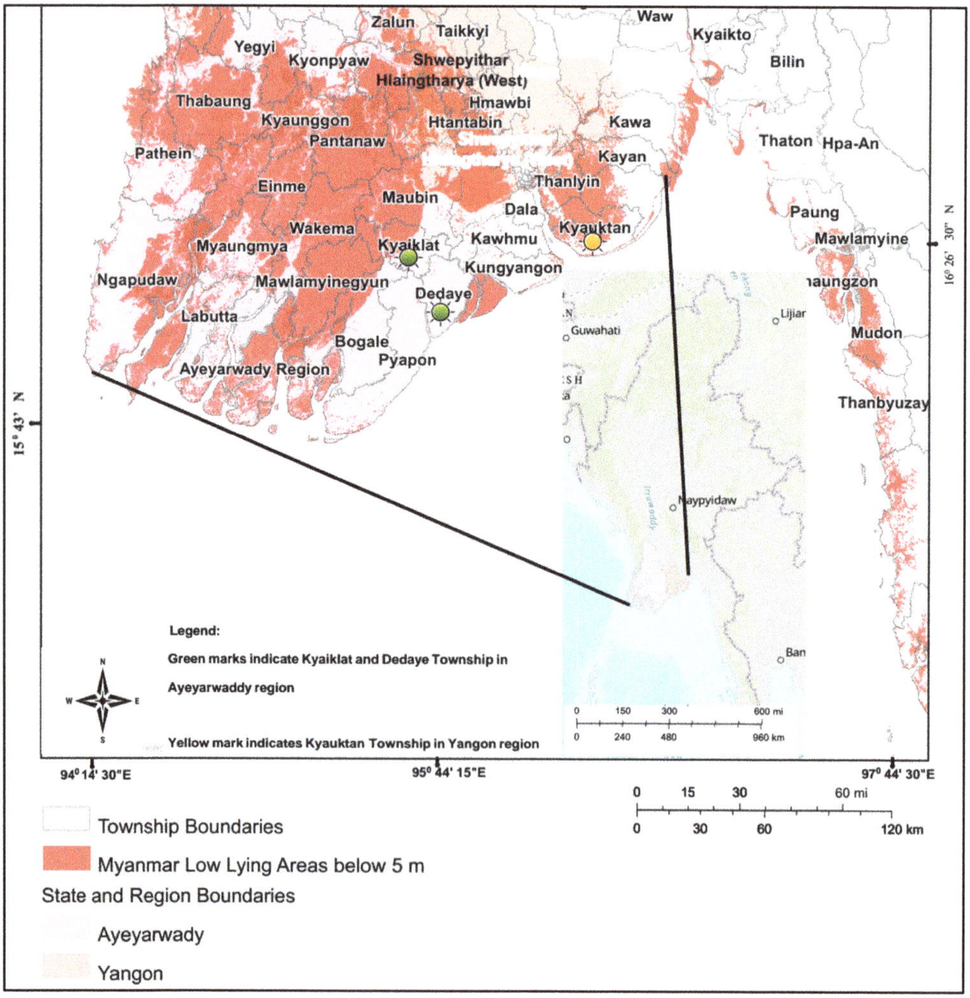

Figure 2. The study area's location map, Myanmar. (Source: Author's work based on MIMU map maker).

3.2. Development of Livelihood Vulnerability Index

Although climate change is happening globally, vulnerability is time- and location-specific. Deressa [23] and Tun Oo [6] recommend comparing and testing climate change vulnerability at the community level. However, others argue that the concepts used to understand vulnerability remain vague and are inconsistently defined [36,37], leading to different indexes. Others state that, in principle, vulnerability cannot be measured [38]. Hinkel [39] nevertheless presents four arguments for developing vulnerability indicators. The indicator approach has weaknesses and strengths; however, due to the simplicity of constructing an index, it is applied in many types of research related to climate change vulnerability assessment [39].

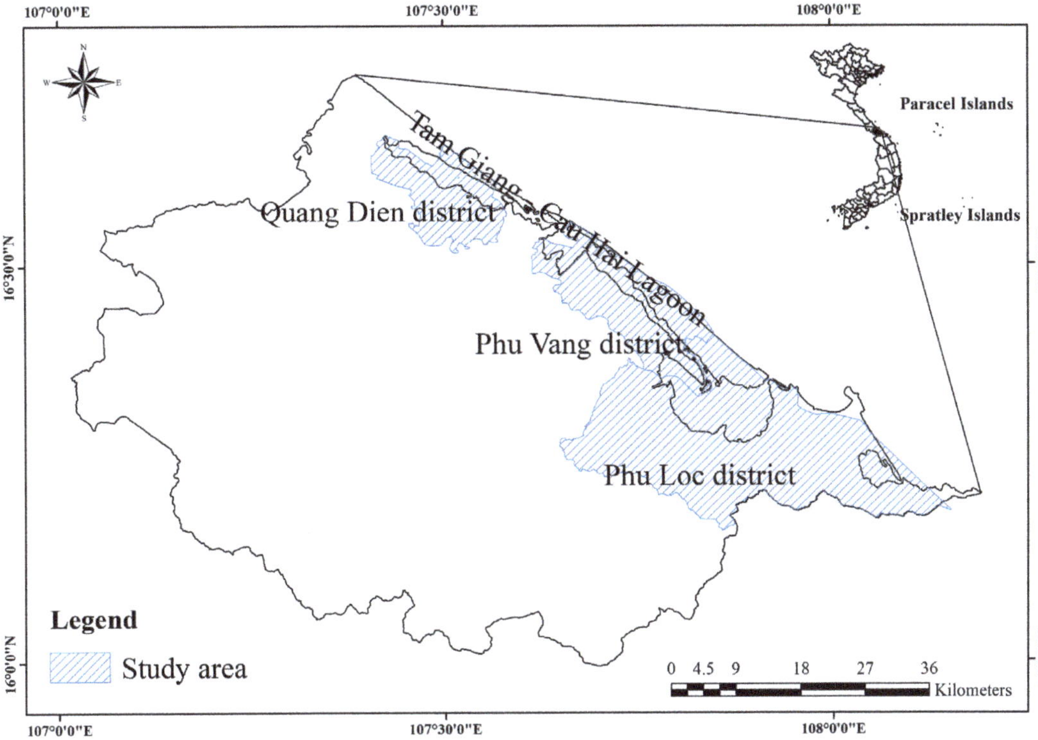

Figure 3. The study area's location map, Vietnam. (Source: author's work).

In this study, we selected and validated indicators to convey farm household vulnerability to flooding and saltwater intrusion in both countries (see detail in Appendix A). Each component comprises several indicators, and these are standardised using a balanced average approach [10,22]. The major component index of the men- and women-headed households can be calculated after each index has been standardised following Equation (1):

$$M_h = \frac{\sum_{i=1}^{n} index\ S_h\ i}{n} \quad (1)$$

where M_h denotes one of the seven major components of men- or women-headed households 'h'; $\boldsymbol{indexS_h}$ represents the subcomponent index or indicator denoted by 'i' of households 'h' and 'n' the number of indicators applied in each major component. After the major component of men- and women-headed households 'h' for each of the seven major components is deduced, the LVI of men- and women-headed households is calculated as follows in Equation (2):

$$LVI_h = \frac{\sum_{i=1}^{n} M_h}{n}, \quad (2)$$

where LVI_h is the livelihood vulnerability indexing scores for the man- or woman-led household 'h'; 'M_h' represents one of the seven major components by household type 'h', 'i' is the index of farm households; and 'n' is the number of major components. As the balanced average approach was applied, the LVI was aggregated with the index derived from the average scale of each indicator. The contribution factor such as exposure, sensitivity, and adaptive capacity scores can be calculated as the following Equation (3):

$$CF_h = \frac{\sum_{i=1}^{n} M_{ih}}{n} \quad (3)$$

where CF_h is an IPCC-defined contribution factor for households 'h', M_{ih} are the major components for households 'h', and 'n' is the number of major components in each contributing factor. Once the exposure, sensitivity, and adaptive capacity dimensions of vulnerability are calculated, the IPCC-defined livelihood vulnerability index (IPCC_LVI) can be calculated as the following Equation (4):

$$\text{IPCC\_LVI} = (e_h - a_h) \times s_h \quad (4)$$

where e_h denotes the exposure dimension of the farm households 'h', a_h is the adaptive capacity of farm household 'h', and s_h is the sensitivity dimension of the farm households 'h'. According to the IPCC [1], adaptive capacity is the ability of a system to adjust or respond to climate change impacts and cope with consequences. Climate change exposure refers to the extent to which a system is subjected to climate change impacts such as floods, drought, heat waves, and sea-level rise. Climate change sensitivity refers to the degree to which a system is affected by climate-related factors [6,10,40].

4. Results and Discussion

In this study, out of 899 respondents, 603 men-headed households and 296 women-headed households participated in the survey during the survey periods in 2021 and 2022. Women-headed households are defined as farm households currently under the supervision of women in the absence of (or as a result of the death of) a male household head and with farm activities mostly undertaken by the women in the household. While we intended to include an equal number of men- and women-headed households in the survey, only around one-third of women-headed farmers could be included. As a result, only 222 (37.1%) women farmers participated (out of 599) in Myanmar, and 74 (24.67%) women farmers participated out of 300 respondents in Vietnam. In terms of ethnicity, 509 respondents (85%) are of the Bamar ethnic group, and the remaining respondents are of other ethnicities, including 41 Hindu (6.8%) and 43 Kayin (7.2%) farmers in Myanmar.

4.1. Component-Based Vulnerability: LVI

The aggregate LVI score is the distributed output of the seven major components for a total of 46 indicators. Table 1 shows the balanced average scores from the indicators making up the LVI, spread across the seven components. In the following section, we consider the climate change vulnerability of men- and women-headed households for each component, with a brief discussion of each.

Table 1. Calculated major components and subcomponent index scores of LVI for men- and women-headed households in Myanmar and Vietnam.

Major Component	Indicators	Unit	Men-VN (226)	Women-VN (74)	Men-MM (377)	Women-MM (222)
Socio-demographic	Average Age (years)	Year	0.548	0.411	0.445	0.432
	A household without electricity (0: yes; 1: no)	%	0.013	0.092	0.491	0.586
	Households without secondary education (0: yes; 1: no)	%	0.237	0.303	0.485	0.631
	Without work for family men (0: yes; 1: no)	%	0.158	0.25	0.013	0.135
	Without work for family women (0: yes; 1: no)	%	0.171	0.276	0.223	0.135
	HH with family members less than 15 years (dependent children=number)	%	0.263	0.237	0.585	0.61
	Balanced Average Score (St.dev)		**0.232 (0.03)**	**0.262 (0.017)**	**0.374 (0.036)**	**0.422 (0.039)**
Livelihood strategies	Household agriculture as the main income source (1: yes; 0: no)	%	0.408	0.513	0.393	0.356
	Household aquaculture as the main income source (1: yes; 0: no)	%	0.355	0.224	0.255	0.217
	Households without having income jobs at the moment (0: yes; 1: no)	%	0.526	0.474	0.549	0.559
	Households without income from non-farm sources (0: yes; 1: no)	%	0.526	0.421	0.533	0.541
	Households without incomes from remittance (0: yes; 1: no)	%	0.763	0.303	0.822	0.788
	Households without income from off-farm sources (0: yes; 1: no)	%	0.5	0.724	0.714	0.73
	Households difficult to repay loans (1: yes; 0: no)	%	0.355	0.421	0.676	0.73
	Household currently without access to any credit (0: yes; 1: no)	%	0.316	0.263	0.581	0.59
	Balanced Average Score (St.dev)		**0.469 (0.018)**	**0.418 (0.02)**	**0.565 (0.023)**	**0.564 (0.025)**
Social network	Average distance to nearest market (mile=number)	miles	0.408	0.41	0.445	0.150
	Households without/not receiving social/gov help (0: yes; 1: no)	%	0.039	0.211	0.905	0.896
	Households without/not receiving community help (0: yes; 1: no)	%	0.145	0.224	0.785	0.77
	Without gov support during COVID-19 (0: yes; 1: no)	%	0.132	0.066	0.864	0.91
	Without private support during COVID-19 (0: Yes; 1: no)	%	0.618	0.25	0.612	0.68
	Balanced Average Score (St.dev)		**0.268 (0.048)**	**0.232 (0.025)**	**0.722 (0.038)**	**0.681 (0.062)**
Health	Average distance to health facilities (miles=number)	miles	0.459	0.52	0.355	0.346
	Average time to health facilities (minutes=number)	Minute	0.355	0.388	0.247	0.243
	Households without sanitary latrine/toilet (0: yes; 1: no)	%	0.026	0.105	0.085	0.185
	Household with missed work or school due to illness (1: yes; 0: no)	%	0.053	0.276	0.560	0.572
	Households with members having chronic illness (1: yes; 0: no)	%	0.053	0.118	0.303	0.365
	Households not receiving good health facilities (0: yes; 1: no)	%	0.184	0.197	0.279	0.212
	Balanced Average Score (St.dev)		**0.188 (0.03)**	**0.267 (0.027)**	**0.305 (0.026)**	**0.321 (0.024)**

Table 1. *Cont.*

Major Component	Indicators	Unit	Men-VN (226)	Women-VN (74)	Men-MM (377)	Women-MM (222)
Food	Household food strategy—take loans (1: yes; 0: no)	%	0.118	0.237	0.828	0.867
	Household food strategy—sell property (1: yes; 0: no)	%	0.066	0.211	0.535	0.467
	Households consume fewer foods (1: yes; 0: no)	%	0.184	0.263	0.263	0.244
	Households did not save food (0: yes; 1: no)	%	0.158	0.224	0.694	0.775
	Households did not save seed (0: yes; 1: no)	%	0.289	0.145	0.586	0.653
	Households consume non-cash food items (1: yes; 0: no)	%	0.316	0.211	0.578	0.523
	Average food insecure months (months=number)	Number	0.026	0.211	0.848	1.254
	Balanced Average Score (St.dev)		**0.165 (0.015)**	**0.215 (0.005)**	**0.619 (0.034)**	**0.683 (0.034)**
Water	Households report water conflict problems (1: yes; 0: no)	%	0.474	0.316	0.096	0.059
	Slept very few hours due to water collection duty (1: yes; 0: no)	%	0.237	0.289	0.117	0.153
	Reduced time for daily work/income generative activities due to water collection (1: yes; 0: no)	%	0.263	0.355	0.147	0.222
	Reduced time for studies/missed school due to water collection (1:yes; 0:no)	%	0.237	0.303	0.061	0.086
	Collected water from an undesirable/dirty source (1: yes; 0: no)	%	0.224	0.184	0.152	0.162
	Households that do not have private water facilities (0: yes; 1: no)	%	0.25	0.25	0.56	0.626
	Climate and hydrological changes adversely affecting households and their economy (1: yes; 0: no)	%	0.724	0.579	0.979	0.973
	Balanced Average Score (St.dev)		**0.344 (0.027)**	**0.325 (0.018)**	**0.302 (0.049)**	**0.326 (0.049)**
Natural Hazards	Household report no early warning (0: yes; 1: no)	%	0.066	0.342	0.316	0.36
	Loss of family members due to disasters and/or incurring injury (1: yes, 0:no)	%	0.013	0.066	0.037	0.063
	Household loss of housing (asset) as a result of floods/disasters (1: yes; 0: no)	%	0.289	0.184	0.324	0.371
	Households with loss of farming products as a result of flooding/disaster events (1: yes; 0: no)	%	0.316	0.263	0.629	0.724
	Past 10 years. Households with water insecurity problems (1: yes; 0: no)	%	0.803	0.763	0.654	0.716
	Within 1 year. Households with water insecurity problems (1: yes; 0: no)	%	0.355	0.276	0.263	0.284
	Flood destroys farmlands with debris (1: yes; 0: no)	%	0.592	0.395	0.739	0.822
	Balanced Average Score (St.dev)		**0.348 (0.04)**	**0.327 (0.031)**	**0.423 (0.036)**	**0.477 (0.04)**

4.1.1. Socio-Demographic Component

As shown in Table 1, women-headed households achieve higher scores on the socio-demographic components than men-headed households, with an average score of 0.262 (±0.017) and 0.232 (±0.03) in Myanmar and an average score of 0.422 (±0.03) and 0.374 (±0.017) in Vietnam. The contributing indicator scores—households without secondary education and family members having income-earning jobs—are higher in women-headed than in men-headed households in both countries. This suggests that during the survey period, women-headed households were less likely to obtain a good education and have income from their male family members. In Vietnam, access to basic education has been more difficult to access for women farmers than for male farmers in the past several decades. The study also demonstrates that women-headed farm households have a larger percentage of children than men-headed farm households. This is consistent with the findings that children are almost exclusively reared by women, who shoulder the large burden of providing unpaid care for sick or injured children [41,42].

Additionally, women-headed households had less access to electricity compared to men-headed households. The lack of reliable and affordable electricity beyond key urban centres has meant that many communities rely on biomass for cooking and other needs [26]. The scores for this component show that vulnerable women-headed households should be prioritised to enhance socio-demographic development. Further, addressing the norms and cultures that place additional responsibilities on women-headed households is crucial to mitigating the negative impact of climate change in the Mekong region.

4.1.2. Livelihood Component

Based on the average vulnerability score for the livelihood component, in Myanmar, there is little difference between households led by men, with a score of 0.565 (±0.025), and those led by women, with a score of 0.564 (±0.023). However, in Vietnam, men-headed households with a livelihood component score of 0.469 (±0.018) are more likely to be vulnerable than women-headed households with a score of 0.418 (±0.02). Men-headed households also report that they are currently less likely to access credit (0.316) as compared to women-headed farm households (0.263). However, in Vietnam, women-headed households receive less income from off-farm jobs (0.724) than men-headed households (0.5). Unlike the livelihood score for farm households in Vietnam, in Myanmar, men-headed households receive less income from remittance transfers from their migrant family members, while women-headed households receive less non-farm and off-farm income.

Compared to male-headed households, in both countries, women-headed farm households report that they are facing increasing difficulty repaying their loans. Access to financial services is crucial for men- and women-headed farm households to invest in farming. However, in both countries, there is a need to consider possible ways for farm households to escape the debt trap and extend loan repayments with lower interest rates. This is similar to the finding of Aryal et al. [18]: Women farmers are more vulnerable to climate risks due to their subsistence-based livelihoods and access to credit. Therefore, the livelihood component indexing scores suggest the need for access to credit, the creation of non-farm and off-farm employment opportunities, and livelihood activities in Myanmar and Vietnam.

4.1.3. Social-Network Component

In terms of social network component scores, men-headed farm households had higher indexing scores than women-headed households in both countries. The COVID-19 pandemic has had differential impacts on men and women, and this indicator is included to examine the strengths and weaknesses of the social network component. In Vietnam, men-headed households report receiving less support from private institutions (0.618) and the government (0.132) during the COVID-19 pandemic than women-headed households, with a score of (0.25) and (0.066), respectively. In the case of Myanmar, we observe similar scores. These findings show the importance of an equal and effective COVID-19 response and the need for better reporting of data by sex and social status [43]. Access to markets for selling farm products is also essential for social well-being and community income generation. In the case of Myanmar, men-headed households report an average distance to the nearest market that is slightly higher than for women-headed households. Better access to market information can be ensured by enhanced social capital and information-sharing platforms [10,43]. For instance, women with less access to resources, such as information, credit, and market services, are less likely to adopt climate change adaptation strategies, increasing vulnerability to climate risks [17,18]. Therefore, the scores for this component highlight the need for more concerted action and emergency COVID-19 responses from public and private organisations to meet the needs of vulnerable farm households in the coastal areas of Myanmar and Vietnam.

4.1.4. Health Component

In terms of the health component, women-headed households had higher scores than men-headed households in both countries. In Myanmar, the balanced average health indicator scores indicate that women-headed households, with a score of 0.321 (±0.024), are more vulnerable to climate-related natural disasters than those headed by men, with a score of 0.305 (±0.026). The same finding is observed in Vietnam, with a score of 0.267 (0.027) for women-headed households and a score of 0.188 (0.03) for men-headed households. This study reveals that family members in women-headed households suffer from more chronic health conditions (with a score of 0.118 in Vietnam and 0.365 in Myanmar) than those in men-headed households (with a score of 0.053 in Vietnam and 0.303 in Myanmar). This is because men-headed households take more initiative and are more active in utilising the health care services provided than women-headed households [44]. Women-headed households in Vietnam are also less likely to access good health facilities than men-headed households and are thus more likely to suffer from water-related diseases in both countries. WHO [43] and World Bank Group [45] highlighted that women are especially vulnerable regarding access to hygienic water and basic health facilities. This study highlights the importance of providing efficient and effective healthcare services and facilities to all people, especially children and the elderly, in the coastal areas of Myanmar and Vietnam.

4.1.5. Food Component

The vulnerability of farm households is also assessed using a seven-indicator food component. The balanced average indicator scores show that in Myanmar and Vietnam, women-headed households are more vulnerable than men-headed households. For instance, with respect to food, women-headed farm households, with a score of 0.215 (±0.005), are more vulnerable than men-headed farm households, with a score of 0.165 (±0.015). Women-headed households in both countries also report that they can no longer retain seeds for the next growing seasons and food for family members in the off-season, instead having to sell all their farm products after harvesting. This study thus shows that women-headed farm households are vulnerable to the adverse effects of climate change due to a lack of food for their families and seeds for the next growing seasons. This is in line with existing findings that farm households who do not keep food or save seeds are the most vulnerable to climate shocks in the next growing seasons [6,10].

The average food-insecure months is higher for women-headed households, with a score of 0.266 in Myanmar and a score of 0.211 in Vietnam. Men-headed households reported food insecurity conditions with a score of 0.154 in Myanmar and 0.026 in Vietnam, respectively. In food insecure situations, farm households reported that they had to rely on non-cash food items such as fishing, eel collecting, foods from forests, etc. In Vietnam, the women-headed households report that their food strategies are mostly selling their household's properties, such as draught cattle and motorcycles, with an average score of 0.211 in women-headed households, compared to men-headed households, for which the average score is 0.066. Also, this study observes that the family members consumed fewer foods in women-headed households (0.263) compared to men-headed farm households (0.184). The food security strategies of men-headed households in Myanmar differ from those in Vietnam. Men-headed households are less likely to use loans as a means of securing food (0.828) than women-headed households (0.867). On the other hand, women-headed households exercise other options, such as selling gold and draught cattle. Therefore, the indexing scores reveal the need to address food insecurity issues and provide an affordable supply of inputs such as rice seeds, particularly to women-headed households in the region.

4.1.6. Water Component

Men-headed households reported more water conflict issues on the farm (0.474) than women-headed households (0.316) in Vietnam. The finding is similar for Myanmar, with an indexing score of 0.096 for men-headed households and 0.059 for women-headed households. It seems likely that men-headed households report more water conflict as aquaculture operations are mostly carried out by men, and it requires a higher level of technical expertise, knowledge, and capacity. Furthermore, aquaculture and agriculture livelihood activities are the major sources of water conflict issues among farm households in both countries. Aquaculture is impacted by water contamination that is caused by chemical residues and runoff produced during agricultural activities. In addition, aquaculture households sometimes have conflicts in the management of wastewater discharge within the communities, and these conflicts are frequently observed in men-headed households. Women-headed households in Myanmar and Vietnam also report sleeping less than men-headed households to allow time for water collection. We show that, compared to households headed by men, women-headed households have reduced time for daily work or income-generation activities due to fetching water for household consumption. Furthermore, climate and hydrological changes have adversely affected the income generation and social activities of the farm households surveyed. Men-headed households report that their livelihood activities are being impacted by hydrological change and climate change (0.724 in Vietnam and 0.979 in Myanmar) at a higher rate than women-headed households (0.579 in Vietnam and 0.973 in Myanmar). Therefore, the indicator scores in this water component suggest that additional action is required to address water insecurity issues and ensure affordable, clean, and hygienic water for all farm households in both countries.

4.1.7. Natural Hazard Component

The natural hazard component is considered an exposure dimension of the LVI. This component's average scores indicate that women-headed households in Myanmar are more vulnerable than men-headed households. The average score on this dimension for men-headed farm households is 0.423 ($\pm$0.036) and 0.477 ($\pm$0.04) for women-headed households in Myanmar. However, men-headed households in Vietnam are more vulnerable, with an average score of 0.348 ($\pm$0.04), than women-headed households, with a score of 0.327 ($\pm$0.031). Factors such as a loss of assets and farm equipment during climate-related disasters are major factors contributing to the vulnerability of men-headed households in Vietnam.

Conversely, in Myanmar, all indicators contributing to these subcomponent scores indicate the higher vulnerability of women-headed than men-headed farm households. This study reveals that women-headed farm households in Myanmar are increasingly vulnerable to climate-related natural disasters as they are generally less likely to adapt or cope with the adverse impacts of climate change and natural hazards. Moreover, women-headed households also report limited (or no) access to early information and warning systems to cope with climate-related natural disasters. Aryal et al. [18] note that due to their limited access to resources and climate information, women are disproportionately impacted by climate change. Most men- and women-headed households in both countries report experiencing water insecurity over the previous ten years. Also, men-headed households report an increasing occurrence of flooding and saltwater intrusion in the region. Due to flood inundation, both farm households report that their farmlands have been destroyed by debris from floods, and they lack the capacity to cope with flooding due to their lack of access to early warning information. Therefore, there is a need for projects or programmes to improve access to early warning information and boost awareness among farm households. In addition, tackling the problems of water insecurity is also a critical strategy for dealing with the adverse effects of climate-related natural disasters in Myanmar and Vietnam.

4.2. IPCC-Defined Livelihood Vulnerability Index Assessment

The scores for the major components are presented in spider diagrams, and the contributing factor or dimensional scores are presented in a triangular diagram (see Figure 4 and also Figure ??). In Vietnam, women-headed households are more vulnerable in three out of the seven major components: health, food, and socio-demographic factors. Therefore, women-headed households are especially vulnerable in terms of their sensitivity and adaptive capacity. Conversely, men-headed households in Vietnam are more vulnerable than women-headed households in respect of the livelihood, social, water, and natural hazard components. The total calculated vulnerability score indicates that women-headed farm households are slightly more vulnerable (0.292) than men-headed farm households (0.288) in Vietnam. In Myanmar, women-headed households are more vulnerable for five out of seven components, contributing to a higher index score for vulnerability (0.476) than men-headed households scores (0.459). In Myanmar, the IPCC-defined dimensional indexing score is also higher for women-headed than for men-headed households in the exposure dimension, possibly due to their high sensitivity to the impacts of climate-related natural disasters. Therefore, this study shows that women-headed farm households are more vulnerable according to the vulnerability indexing scores than men-headed farm households in Myanmar and Vietnam.

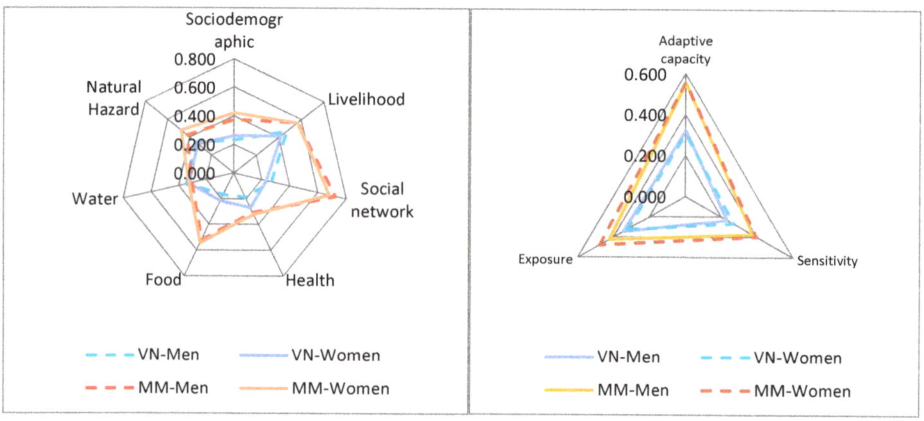

Figure 4. Major component LVI scores for farm households in Myanmar and Vietnam.

The IPCC-defined livelihood vulnerability indexing scores of farm households in Myanmar and Vietnam are also presented in Figure 5. Overall, IPCC-defined LVI scores indicate that women-headed households (−0.472) are more impacted by climate-related natural disasters than men-headed households (−0.433) in Vietnam. Our findings are similar to those for Myanmar, where the scores for women-headed farm households (−1.414) suggest these are more affected by climate-related natural disasters than men-headed farm households (−1.394). Therefore, we suggest that it is imperative to strengthen the adaptive capacity of farm households and address the adverse effects of climate change on the sensitivity dimension of farm households, particularly regarding health, food, and water. More concerted efforts are required in both countries to improve the adaptive capacity of farm households through market innovation, investment in soft and hard infrastructures, sharing of early warning information, and affordable supply of clean water.

Table 2 explains the degree of vulnerability of men- and women-led households to climate-related natural disasters based on the IPCC-defined LVI scores. Compared to those in Myanmar, farm households in Vietnam are moderately vulnerable to climate-related natural disasters. This could be due to the mid-level scores that farm households in Vietnam achieved for the exposure and sensitivity dimensions; adaptive capacity also achieves a moderate score.

By contrast, farm households in Myanmar are highly vulnerable to climate-related disasters like flooding and saltwater intrusion. This is because livelihood activities such as fish farming and rice farming are also highly sensitive to climate change. Farm households in Myanmar have a very high adaptive capacity, but their overall LVI score indicates that they are vulnerable to the impacts of climate-related natural disasters due to the high scores on the exposure dimension. This exposure results in the sensitivity of the farming sector in both countries and the continued vulnerability of farm households. Therefore, these findings are consistent with the previous finding that farm households are adjusting their farming systems and improving their capacities if they face many external shocks or natural disasters [6].

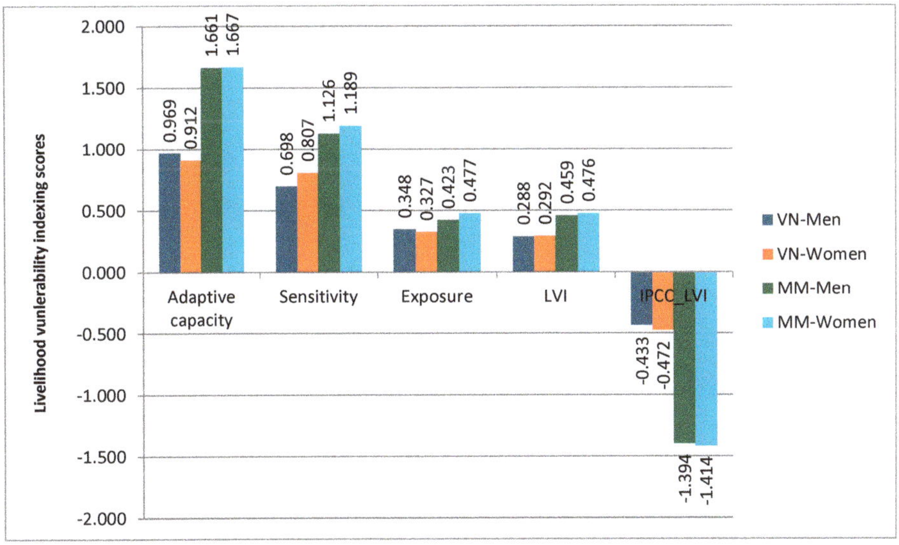

Figure 5. Overall LVI component scores for farm households in Myanmar and Vietnam.

Table 2. Comparison of the degree of LVI of men- and women-headed households in Myanmar and Vietnam.

LVI Components	Exposure [1]	Adaptive Capacity [2]	Sensitivity [3]	Over LVI [4]
VN-Men (226)	Moderate	Moderate	Moderate	Moderate
VN-Women (74)	Moderate	Moderate	Moderate	Moderate
MM-Men (377)	High	Very High	High	High
MM-Women (222)	High	Very High	High	High

[1,2] Low (<0.3), moderate (0.3–0.4), high (0.4–0.55), and very high (>0.55); [3,4] low (<0.2), moderate (0.2–0.35), high (0.35–0.5), and very jigh (>0.5).

In Myanmar, greater exposure to natural hazards leads to greater sensitivity of the farm household, even though there is little difference in the adaptive capacity dimension of households headed by men and women. Therefore, to reduce or limit the impacts of climate-related natural hazards, it is critical to pay attention to the indicators used in the exposure dimension of the study, such as the early provision of weather and climate change information, and to address the water insecurity challenges faced by farm households in the coastal areas of Myanmar and Vietnam.

5. Conclusions

In this study, we have examined the difference in climate change vulnerability of men-headed and women-headed households in Myanmar and Vietnam by using the LVI approach. This study assesses the climate change vulnerability of farm households in selected case-study areas in Myanmar and Vietnam using an LVI comprising 46 selected indicators during the period 2021–2022. Although the individual indicators for the LVI's major components show different vulnerability trends for men- and women-headed farm households, the overall balanced indexing scores indicate that in Myanmar and Vietnam, women-headed farm households are more vulnerable to climate-related natural disasters than men-headed farm households. The low adaptive capacity of women-headed households explains their high vulnerability, even though those of exposure dimension are weak. This will lead to increased sensitivity of farm households to climate-related natural disasters in the study areas. Following this assessment, we conclude that the lower adaptive capacity of farm households leads to greater sensitivity to climate change when external shocks like flooding and saltwater intrusion affect agriculture and aquaculture farming households in the coastline areas of Myanmar and Vietnam.

As vulnerability is time- and location-specific, the vulnerability of farm households in the study areas can fluctuate over time or within a different period, and thus, additional assessments can be conducted over a different time and or longer period. This study provides a snapshot of the context and location-specific information for policymakers and development planners in these areas. We, therefore, recommend additional assessments in different regions of Myanmar and Vietnam using additional indicators such as impacts on the ecosystem, biodiversity, etc., as well as advanced vulnerability assessment methods to gain a deeper understanding of the relative climate change vulnerability of men- and women-headed households in Myanmar and Vietnam. In addition, the adaptive capacity of farm households in both countries, particularly women-headed households, should be strengthened through the creation of livelihood income opportunities, provision of formal and informal training programmes, and early warning information, as revealed in this study. Reducing the sensitivity of farm households requires additional investment in both soft and hard infrastructures, market innovation and information sharing, equitable supply of farm inputs and foods, as well as improved management of water for clean and affordable supply in the coastline areas of Myanmar and Vietnam. Moreover, there is a need for private and public organisations to make a more concerted effort to improve the adaptive capacity of farm households.

Author Contributions: Conceptualisation, A.T.O. and D.D.M.; methodology, A.C. and D.D.M.; validation, A.T.O., D.D.M., and A.C.; formal analysis, A.T.O. and D.D.M.; investigation, A.C.; data curation, A.C. and D.D.M.; writing—original draft preparation, A.T.O. and D.D.M.; writing—review and editing, A.T.O., A.C., and D.D.M.; visualisation, A.T.O. and D.D.M.; supervision, A.T.O.; funding acquisition, A.T.O. All authors have read and agreed to the published version of the manuscript.

Funding: This research was funded by Stiftelsen the Stockholm Environmental Institute—Sustainable Mekong Research Network for all (SUMERNET 4 All) collaborative research grant SEI Project Np. 100099/Work Order No. 100099205.

Data Availability Statement: The data used are confidential.

Acknowledgments: Wholehearted thanks to the farmers who participated in the study. The authors would like to express their sincere gratitude to the field-based enumerators, key farmers, and community-based organisations for their support of the field survey.

Conflicts of Interest: The authors declare no conflicts of interest. The funders had no role in the design of the study; in the collection, analyses, or interpretation of data; in the writing of the manuscript; or in the decision to publish the results.

Appendix A

Table A1. Major component, subcomponent indexes of LVI for men- and women-headed households in Vietnam and Myanmar.

Major Components	Indicators	Unit	Men-VN (226)	Women-VN (74)	Men-MM (377)	Women-MM (222)
Socio-demographic (6)	Average Age (years)	Year	53.0	56.37	44.5	43.2
	A household without electricity (0: yes; 1: no)	%	1.3	9.21	49.1	58.6
	Households without secondary education (0: yes; 1: no)	%	23.7	30.26	48.5	63.1
	Without work for family men (0: yes; 1: no)	%	15.8	25.00	1.3	13.5
	Without work for family-women (0: yes; 1: no)	%	17.1	27.63	22.3	13.5
	HH with family members less than 15 years (dependent children = number)	%	26.3	23.68	58.5	61
Livelihood strategies (8)	Household agriculture as the main income source (1: yes; 0: no)	%	40.8	51.32	39.3	35.6
	Household aquaculture as the main income source (1: yes; 0: no)	%	35.5	22.37	25.5	21.7
	Households without having income jobs at the moment (0: yes; 1: no)	%	52.6	47.37	54.9	55.9
	Households without income from non-farm sources (0: yes; 1: no)	%	52.6	42.11	53.3	54.1
	Households without incomes from remittance (0: yes; 1: no)	%	76.3	30.26	82.2	78.8
	Households without income from off-farm sources (0: yes; 1: no)	%	50.0	72.37	71.4	73
	Households difficult to repay loans (1: yes; 0: no)	%	35.5	42.11	67.6	73
	Household currently without access to any credit (0: yes; 1: no)	%	31.6	26.32	58.1	59
Social network (5)	Average Distance to nearest market (mile = number)	miles	3.2	3.23	44.5	15
	Households without/not receiving social/gov help (0: yes; 1: no)	%	3.9	21.05	90.5	89.6
	Households without/not receiving community help (0: yes; 1: no)	%	14.5	22.37	78.5	77
	Without Gov support during COVID-19 (0: yes; 1: no)	%	13.2	6.58	86.4	91
	Without Private support during COVID-19 (0: Yes; 1: no)	%	61.8	25.00	61.2	68
Health (6)	Average Distance to health facilities (miles = number)	miles	2.8	3.08	35.5	34.6
	Average Time to health facilities (minutes = number)	Minute	8.1	8.33	24.7	24.3
	Households without sanitary latrine/toilet (0: yes; 1: no)	%	2.6	10.53	8.5	18.5
	Household with missed work or school due to illness (1: yes; 0: no)	%	5.3	27.63	56	57.2
	Households with members having chronic illness (1: yes; 0: no)	%	5.3	11.84	30.3	36.5
	Households not receiving good health facilities (0: yes; 1: no)	%	18.4	19.74	27.9	21.2

Table A1. *Cont.*

Major Components	Indicators	Unit	Men-VN (226)	Women-VN (74)	Men-MM (377)	Women-MM (222)
Food (7)	Household food strategy take loans (1: yes; 0: no)	%	11.8	23.68	82.8	86.7
	Household food strategy sell property (1: yes; 0: no)	%	6.6	21.05	53.5	46.7
	Households consume fewer foods (1: yes; 0: no)	%	18.4	26.32	26.3	24.4
	Households did not save food (0: yes; 1: no)	%	15.8	22.37	69.4	77.5
	Households did not save seed (0: yes; 1: no)	%	28.9	14.47	58.6	65.3
	Households consume non-cash food items (1: yes; 0: no)	%	31.6	21.05	57.8	52.3
	Average Food insecure months (months = number)	Number	2.6	1.05	15.4	26.6
Water (7)	Households report water conflict problems (1: yes; 0: no)	%	47.4	31.58	9.6	5.9
	Slept very few hours due to water collection duty (1: yes; 0: no)	%	23.7	28.95	11.7	15.3
	Reduced time for daily work/income generative activities due to water collection (1: yes; 0: no)	%	26.3	35.53	14.7	22.2
	Reduced time for studies/missed school due to water collection (1:yes; 0:no)	%	23.7	30.26	6.1	8.6
	Collected water from an undesirable/dirty source (1: yes; 0: no)	%	22.4	18.42	15.2	16.2
	Households that do not have private water facilities (0: yes; 1: no)	%	25.0	25.00	56	62.6
	Climate and hydrological changes adversely affecting households and their economy (1: yes; 0: no)	%	72.4	57.89	97.9	97.3
Natural Hazards (7)	Household report no early warning (0: yes; 1: no)	%	6.6	34.21	31.6	36
	Loss of family members due to disasters and/or incurring injury (1: yes, 0:no)	%	1.3	6.58	3.7	6.3
	Household loss of housing (asset) as a result of floods/disasters (1: yes; 0: no)	%	28.9	18.42	32.4	37.1
	Households with loss of farming products as a result of flooding/disaster events (1: yes; 0: no)	%	31.6	26.32	62.9	72.4
	Past 10 years. Households with water insecurity problems (1: yes; 0: no)	%	80.3	76.32	65.4	71.6
	Within 1 year. Households with water insecurity problems (1: yes; 0: no)	%	35.5	27.63	26.3	28.4
	Flood destroys farmlands with debris (1: yes; 0: no)	%	59.2	39.47	73.9	82.2

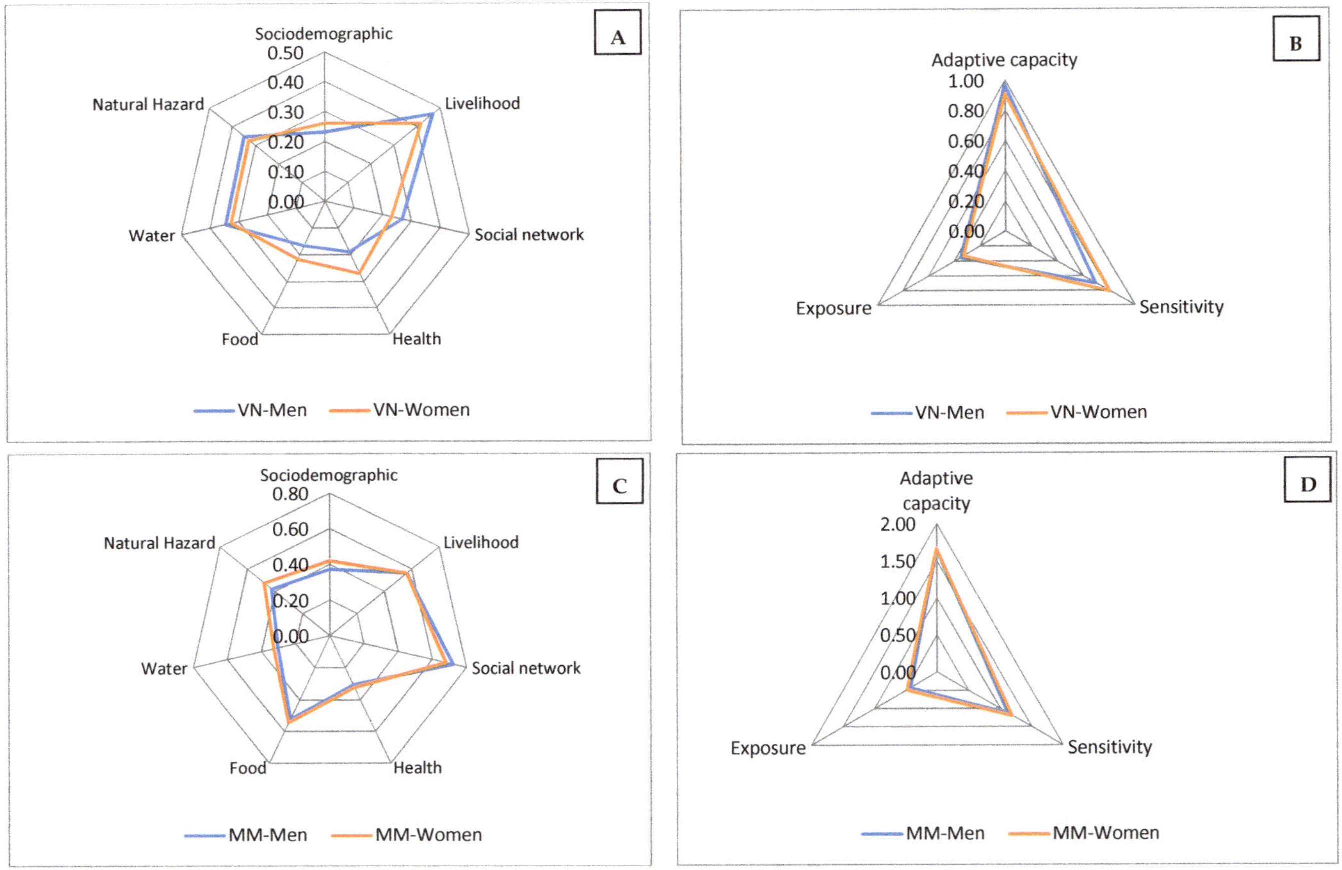

Figure A1. Major components (spider diagram) and contributing factors (triangular diagram) of LVI for Vietnam (**A**,**B**) and Myanmar (**C**,**D**).

References

1. IPCC. *Climate Change 2014: Impacts, Adaptation, and Vulnerability*; Contribution of Working Group II to the Fifth Assessment Report of the Intergovernmental Panel on Climate Change; Field, C.B., Barros, V.R., Dokken, D.J., Mach, K.J., Mastrandrea, M.D., Bilir, T.E., Chatterjee, M., Ebi, K.L., Estrada, Y.O., Genova, R.C., et al., Eds.; Cambridge University Press: Cambridge, UK; New York, NY, USA, 2014; 1132p.
2. Eckstein, D.; Künzel, V.; Schäfer, L. *Global Climate Risk Index 2021: Who Suffers Most from Extreme Weather Events? Weather-Related Loss Events in 2019 and 2000 to 2019*; Germanwatch: Berlin, Germany, 2021. Available online: https://www.germanwatch.org/en/19777 (accessed on 25 January 2021).
3. IPCC. *Climate Change 2022: Impacts, Adaptation and Vulnerability*; Contribution of Working Group II to the Sixth Assessment Report of the Intergovernmental Panel on Climate Change; Po, H.-O., Roberts, D.C., Tignor, M., Poloczanska, E.S., Mintenbeck, K., Alegri, A., Craig, M., Langsdorf, S., Lo, S., Mo, V., Eds.; Cambridge University Press: Cambridge, UK; New York, NY, USA, 2022; 3056p.
4. Krittasudthacheewa, C.; Navy, H.; Tinh, B.D.; Voladet, S. *Development and Climate Change in the Mekong Region—Case Studies*; Strategic Infromation and Research Development Centre (SIRD): Petaling Jaya, Malaysia, 2019. Available online: https://www.sei.org/wp-content/uploads/2019/08/climate-change-mekong-sumernet-2019-smallsize.pdf (accessed on 30 November 2019).
5. Tran, D.D.; Bich Thuc, P.T.; Park, E.; Hang, P.T.T.; Duong, B.M.; Wang, J. Extent of saltwater intrusion and freshwater exploitability in the coastal Vietnamese Mekong Delta assessed by gauging records and numerical simulations. *J. Hydrol.* **2024**, *630*, 130655. [CrossRef]
6. Tun Oo, A. Characterizing Farm Households' Vulnerability and Adaptation to Climate Change in Myanmar. Ph.D. Thesis, Ghent University, Ghent, Belgium, 2018. Available online: http://hdl.handle.net/1854/LU-8581606 (accessed on 26 November 2018).
7. Tran, D.D.; Park, E.; Ngoc Tuoi, H.T.; Thien, N.D.; Hong Tu, V.; Anh Ngoc, P.T.; Van, C.T.; Long, P.K.; Ho, H.L.; Chau Quang, N.X. Climate change impacts on rice-based livelihood vulnerability in the lower Vietnamese Mekong Delta: Empirical evidence from Can Tho City and Tra Vinh Province. *Environ. Technol. Innov.* **2022**, *28*, 102834. [CrossRef]
8. Horton, R.; De Mel, M.; Peters, D.; Lesk, C.; Bartlett, R.; Helsingen, H.; Bader, D.; Capizzi, P.; Martin, S.; Rosenzweig, C. *Assessing Climate Risk in Myanmar: Summary for Policymakers and Planners*; Center for Climate Systems Research at Columbia University, WWF-US and WWF-Myanmar, UNHabitat Myanmar: New York, NY, USA, 2017.
9. Driel, W.F.V.; Nauta, T.A. *Vulnerability and Resilience Assessment of the Ayeyarwaddy Delta in Myanmar, Scoping Phase*; Bay of Bengal Large Marine Ecosystem (BOBLME) Project, Global Water Partnership (GWP) and Delta Alliance: Delft-Wageningen, The Netherlands, 2013; pp. 1–148. Available online: https://themimu.info/sites/themimu.info/files/documents/Report_Vulnerability_Resilience_Assessment_Ayeyarwady_Delta_Feb2015.pdf (accessed on 20 February 2015).
10. Tun Oo, A.; Van Huylenbroeck, G.; Speelman, S. Assessment of climate change vulnerability of farm households in Pyapon District, a delta region in Myanmar. *Int. J. Disaster Risk Reduct.* **2018**, *28*, 10–21.
11. Ministry of Natural Resources and Environmental Conservation (MONREC). *Myanmar Climate Change Strategy (2018–2030)*; MONREC: Yangon, Myanmar, 2019. Available online: https://myanmar.un.org/sites/default/files/2019-11/MyanmarClimateChangeStrategy_2019.pdf (accessed on 12 January 2019).
12. Ministry of Natural Resource and Environment (MONRE). Climate Change and Sea Level Rise Scenarios for Vietnam. Available online: https://www.preventionweb.net/files/11348_ClimateChangeSeaLevelScenariosforVi.pdf (accessed on 30 June 2009).
13. Gustafson, S.; Joehl Cadena, A.; Ngo, C.C.; Kawash, A.; Saenghkaew, I.; Hartman, P. Merging science into community adaptation planning processes: A cross-site comparison of four distinct areas of the Lower Mekong Basin. *Clim. Change* **2018**, *149*, 91–106. [CrossRef]
14. Gowing, J.; Tuong, T.; Hoanh, C.T. *Land and Water Management in Coastal Zones: Dealing with Agriculture, Aquaculture, Fishery Conflicts*; CAB International: Wallingford, UK, 2006.
15. Nguyen, T.T.N.; Tran, H.C.; HO, T.M.H.; Burny, P.; Lebailly, P. Dynamics of Farming Systems under the Context of Coastal Zone Development: The Case of Xuan Thuy National Park, Vietnam. *Agriculture* **2019**, *9*, 138. [CrossRef]
16. Aryal, J.P.; Sapkota, T.B.; Rahut, D.B.; Krupnik, T.J.; Shahrin, S.; Jat, M.L.; Stirling, C.M. Major Climate Risks and Adaptation Strategies of Smallholder Farmers in Coastal Bangladesh. *Environ. Manag.* **2020**, *66*, 105–120. [CrossRef] [PubMed]
17. Ayanlade, A.; Oluwatimilehin, I.A.; Ayanlade, O.S.; Adeyeye, O.; Abatemi-Usman, S.A. Gendered vulnerabilities to climate change and farmers' adaptation responses in Kwara and Nassarawa States, Nigeria. *Humanit. Soc. Sci. Commun.* **2023**, *10*, 911. [CrossRef]
18. Aryal, J.P.; Sapkota, T.B.; Rahut, D.B.; Gartaula, H.N.; Stirling, C. Gender and climate change adaptation: A case of Ethiopian farmers. *Nat. Resour. Forum* **2022**, *46*, 263–288. [CrossRef]
19. UNFCCC. Differentiated Impacts of Climate Change on Women and Men; the Integration of Gender Considerations in Climate Policies, Plans, and Actions; and Progress in Enhancing Gender Balance in National Climate Delegations. In Proceedings of the Synthesis Report by the Secretariat, Subsidiary Body for Implementation, Fiftieth Session, Bonn, Germany, 17–27 June 2019. Available online: https://coe.insuresilience.org/knowledge/differentiated-impacts-of-climate-change-on-women-and-men-the-integration-of-gender-considerations-in-climate-policies-plans-and-actions-and-progress-in-enhancing-gender-balance-in-national-climate/ (accessed on 12 June 2019).

20. Nellemann, C.; Verma, R.; Hislop, L. (Eds.) *Women at the Frontline of Climate Change: Gender Risks and Hopes*; A Rapid Response Assessment. United Nations Environment Programme, GRID-Arendal; 2011. Available online: https://genderandenvironment.org/women-at-the-frontline-of-climate-change-gender-risks-and-hopes-a-rapid-response-assessment/ (accessed on 12 December 2011).
21. Mujere, N. Gendered vulnerability and adaptation to climate change. In *Natural Resources Management: Concepts, Methodologies, Tools, and Applications*; Information Resources Management Association, IGI Global: Hershey, PA, USA, 2017; 14p, pp. 1133–1146.
22. Hahn, M.B.; Riederer, A.M.; Foster, S.O. The Livelihood Vulnerability Index: A pragmatic approach to assessing risks from climate variability and change-A case study in Mozambique. *Glob. Environ. Change* **2009**, *19*, 74–88. [CrossRef]
23. Deressa, T.T. Assessment of the Vulnerability of Ethiopian Agriculture to Climate Change and Farmers' Adaptation Strategies. Ph.D. Thesis, University of Pretoria, Pretoria, South Africa, 2010. Available online: http://hdl.handle.net/2263/28969 (accessed on 24 October 2010).
24. Ahsan, M.N.; Warner, J. The socioeconomic vulnerability index: A pragmatic approach for assessing climate change led risks—A case study in the south-western coastal Bangladesh. *Intern. J. Disaster Risk Reduc.* **2014**, *8*, 32–49. [CrossRef]
25. World Bank Group; FAO; IFAD. Gender in Climate-Smart Agriculture. *Module 18 for Gender in Agriculture Sourcebook.* 2013. Available online: https://www.cac.int/sites/default/files/BM_FAO_IFAD._2015._Gender_in_climate_smart_agriculture.pdf (accessed on 25 August 2015).
26. Simpson, A.; South, A. *Evolving Climate Change Governance in Myanmar: Limitations and Opportunities in a Political Crisis*; Chapter 7 in Governing Climate Change in Southeast Asia: Critical Perspective; Routledge: Abingdon-on-Thames, UK, 2022.
27. Eichelberger, L.; Dev, S.; Howe, T.; Barnes, D.L.; Bortz, E.; Briggs, B.R.; Cochran, P.; Dotson, A.D.; Drown, D.M.; Hahn, M.B.; et al. Implications of inadequate water and sanitation infrastructure for community spread of COVID-19 in remote Alaskan communities. *Sci. Total Environ.* **2021**, *776*, 145842. [CrossRef]
28. Zuin, V.; Ortolano, L.; Alvarinho, M.; Russel, K.; Thebo, A.; Muximpua, O.; Davis, J. Water supply services for Africa's urban poor: The role of resale. *J. Water Health* **2011**, *9*, 773–784. [CrossRef] [PubMed]
29. Adams, E.A.; Adams, Y.J.; Koki, C. Water, sanitation, and hygiene (WASH) insecurity will exacerbate the toll of COVID-19 on women and girls in low-income countries sustainability: Science. *Pract. Policy* **2021**, *17*, 86–90. [CrossRef]
30. Lebel, L.; Navy, H.; Siharath, P.; Long, C.T.M.; Aung, N.; Lebel, P.; Hoanh, C.T.; Lebel, B. COVID-19 and household water insecurities in vulnerable communities in the Mekong Region. *Environ. Dev. Sustain.* **2022**, *25*, 3503–3522. [CrossRef] [PubMed]
31. Duy Can, N.; Hong Tu, V.; Thai Hoanh, C. Application of livelihood vulnerability index to assess risks from flood vulnerability and climate variability case study in the Mekong Delta of Vietnam. *J. Environ. Sci. Eng.* **2013**, *2*, 476–486.
32. Adu, D.T.; Kuwornu, J.K.M.; Anim-Somuah, H.; Sasaki, N. Application of livelihood vulnerability index in assessing smallholder maize farming households' vulnerability to climate change in Brong-Ahafo region of Ghana. *Kasetsart J. Soc. Sci.* **2018**, *39*, 22–32. [CrossRef]
33. Paavola, J. Livelihoods, vulnerability and adaptation to climate change in Morogoro, Tanzania. *Environ. Sci. Policy* **2008**, *11*, 642–654. [CrossRef]
34. Shah, K.U.; Dulal, H.B.; Johnson, C.; Baptiste, A. Understanding livelihood vulnerability to climate change: Applying the livelihood vulnerability index in Trinidad and Tobago. *Geoforum* **2013**, *47*, 125–137. [CrossRef]
35. Zhang, Q.; Zhao, X.; Tang, H. Vulnerability of communities to climate change: Application of the livelihood vulnerability index to an environmentally sensitive region of China. *Clim. Dev.* **2019**, *11*, 525–542. [CrossRef]
36. Hinkel, J. Trans-Disciplinary Knowledge Integration. Cases from Integrated Assessment and Vulnerability Assessment. Ph.D. Thesis, Wageningen University, Wageningen, The Netherlands, 2008. Available online: https://ciret-transdisciplinarity.org/biblio/biblio_pdf/Hinkel.pdf (accessed on 7 March 2008).
37. Wolf, S.; Hinkel, J.; Hallier, M.; Bisaro, A.; Lincke, D.; Ionescu, C.; Klein, R.J.T. Clarifying vulnerability definitions and assessments using formalization. *Int. J. Clim. Change Strateg. Manag.* **2013**, *5*, 54–70. [CrossRef]
38. Patt, A.; Schroeter, D.; de la Vega-leinert, A.C.; Klein, R.J.T. Vulnerability research and assessment to support adaptation and mitigation: Common themes from the diversity of approaches. In *Assessing Vulnerability to Global Environmental Change*; Patt, A., Schroeter, D., Klein, R.J.T., de la Vega-Leinert, A.C., Eds.; Earthscan: Oxford, UK, 2009; pp. 1–26.
39. Hinkel, J. Indicators of vulnerability and adaptive capacity, towards a clarification of the science-policy interface. *Glob. Environ. Change* **2011**, *21*, 198–208. [CrossRef]
40. Deressa, T.T.; Hassan, R.M.; Ringler, C.; Alemu, T.; Yesuf, M. Determinants of farmers' choice of adaptation methods to climate change in the Nile Basin of Ethiopia. *Glob. Environ. Change* **2009**, *19*, 248–255. [CrossRef]
41. UN Women; World Bank; UNEP; UNDP. *The Cost of the Gender Gap in Agricultural Productivity in Malawi, Tanzania, and Uganda*; UN Women, UNDP, UNEP, and the World Bank Group Publication: Nairobi, Kenya, 2015; p. 32. Available online: https://africa.unwomen.org/en/digital-library/publications/2015/10/the-cost-of-the-gender-gap-in-agricultural-productivity-in-malawi-tanzania-and-uganda (accessed on 22 October 2015).
42. UN Women; UNDP. *Factors Driving the Gender Gap in Agricultural Productivity: Malawi*; UN Women, UNDP and UN Environment: Nairobi, Kenya, 2018; pp. 1–30. Available online: https://africa.unwomen.org/sites/default/files/Field%20Office%20Africa/Attachments/Publications/2019/the20gender20gapcompressed.pdf (accessed on 21 March 2018).
43. World Health Organization (WHO). Gender and COVID-19. Advocacy Brief. 14 May 2020. Available online: https://www.who.int/publications/i/item/WHO-2019-nCoV-Advocacy_brief-Gender-2020.1 (accessed on 14 May 2020).

44. Tessema, Y.A.; Aweke, C.S.; Endris, G.S. Understanding the process of adaptation to climate change by smallholder farmers: The case of east Hararghe Zone, Ethiopia. *J. Agric. Food Econ.* **2013**, *1*, 13. [CrossRef]
45. World Bank Group. *Gender Dimensions of the COVID-19 Pandemic*; Policy Note by World Bank Group, April 26; World Bank Group: Washington, DC, USA, 2020. Available online: https://documents1.worldbank.org/curated/en/618731587147227244/pdf/Gender-Dimensions-of-the-COVID-19-Pandemic.pdf (accessed on 16 April 2020).

Disclaimer/Publisher's Note: The statements, opinions and data contained in all publications are solely those of the individual author(s) and contributor(s) and not of MDPI and/or the editor(s). MDPI and/or the editor(s) disclaim responsibility for any injury to people or property resulting from any ideas, methods, instructions or products referred to in the content.

sustainability

MDPI

Article

Urban Vulnerability under the Extreme High Temperatures in the Chengdu-Chongqing Area, Western China

Zhaoqi Yin [1], Weipeng Li [1,2,*], Zhongsheng Chen [1,2], Li Zhu [3] and Panheng Shui [1]

1 School of Geographical Sciences, China West Normal University, Nanchong 637002, China; yinzhaoqicwnu@163.com (Z.Y.)
2 Sichuan Provincial Engineering Laboratory of Monitoring and Control for Soil Erosion on Dry Valleys, China West Normal University, Nanchong 637002, China
3 Severe Weather in Northeast Sichuan Key Laboratory of Nanchong City, Nanchong 637000, China
* Correspondence: wpengli@cwnu.edu.cn

Abstract: The frequent occurrence of extreme high-temperature events in the summer under global climate change poses a serious threat to Chinese society. An urban vulnerability evaluation system for counties in the Chengdu-Chongqing Area was constructed to calculate the urban vulnerability and distribution characteristics of each district. In this study, a vulnerability-contribution model was used to analyze the types of urban vulnerability in the Chengdu-Chongqing Area. Additionally, combined with the optimal parameter geographic detector (OPGD) model, the main influencing factors and interactions of urban vulnerability were explored. The results show that: ① The urban vulnerability of the Chengdu-Chongqing Area is high in the east and low in the west, with vulnerability degree mostly below the medium degree. ② Exposure contributes more than 50% to severe and general urban vulnerability in the region, while adaptability contributes the highest proportion to mild urban vulnerability, reaching 47.53%. ③ From the factor perspective, the impact ratio of high-temperature days on urban vulnerability is 39.1%, and the interaction between various meteorological factors and social factors produces an enhancement effect, with the highest interaction q-value reaching 0.7863.

Keywords: urban vulnerability; contribution rate; optimal parameter geographic detector; cities in western China

1. Introduction

The sixth assessment report of the Intergovernmental Panel on Climate Change (IPCC) noted that in the past 70 years, the frequency of heat waves and droughts has increased, and the future probability of potential hazards of heat waves, droughts, and their combined disasters has also increased [1,2]. High-temperature and heat wave-related disasters threaten the sustainable development of economy, health, and natural ecology [3–5]. It is pointed out that taking urgent action to address climate change and the impacts is one of the goals of sustainable development in 2015 [6]. In recent years, the scope of the impact has expanded, with more areas being exposed to high-temperature and hot environments [7,8]. After 2007, the increase in the number of related official documents in China, such as the "Notice on Further Strengthening Summer Heat Prevention and Cooling Work in the Workplace", indirectly confirms increasingly severe high-temperature events.

The number of studies on group health vulnerability under extreme high-temperature conditions has increased in recent years [9–12]. It has been shown that excessive environmental temperature increases the risk of death [13]. Racial genetic differences may have different adaptive capacities to high temperatures, while linguistic barriers may make it more difficult to access assistance during a disaster [14]. The research on the vulnerability of social operational capacity and sustainable development management under heat waves is also improving constantly. Urbanization and increased population density have led to an increase in the vulnerability [15], and there are significant differences in heat waves between urban and rural areas [16], so the population exposure under the high temperatures should be taken seriously [17]. The studies on the vulnerability of high-temperature heat waves in China include different scales such as the whole country [18], the Yangtze River Economic Belt [19], and specific cities [20]. With the normalization of extreme heat disasters, engineering technology, organizational systems, and politics should be combined to improve capabilities of pre-disaster warning, disaster rescue, and post-disaster resettlement [21]. A high-precision and dynamic high-temperature health stress warning system needs to be continuously improved, and the goals of cooling and health should be included in the national spatial planning goals. International, domestic, and community cooperation should be strengthened to alleviate the threat of high temperatures [22]. Overall, scholars have made rich research achievements on the vulnerability of regional extreme high temperatures, which can provide prevention suggestions for large-scale heat wave disasters.

The Chengdu-Chongqing region is economically developed in western China with a complex terrain and unique climate [23]. It is similar to the Yangtze River Delta urban agglomeration [24] and some urban belts in Western countries [25] in population, economy, and scale. The Chengdu-Chongqing region is close to the southern edge of the North-South Transitional Zone, bordering Gansu Province, Shaanxi Province, Hubei Province, and Guizhou Province. The latitude of the region is around 30° N, which is similar to the important hubs such as the Wuhan urban agglomeration and the Tokyo–Osaka transportation line in climate and latitude. Meanwhile, it is an important stronghold of the ancient Three Kingdoms culture and a hub area of the South and North Silk Road, with an important cultural background. Studies have shown that the heat island area in the downtown area of Chengdu increases during the daytime, and the variation rate of air temperature may be higher than the global average while the intensity of the heat island effect is positively correlated with population [26,27]. The index related to minimum temperature is more sensitive to urbanization [28]. Similarly, the central and southwestern parts of Chongqing were the areas with the most severe heat wave disasters in the past 60 years [29]. Luo et al. [30] pointed out that the impact of three-dimensional variables (altitude) on surface temperature in mountainous cities is 2.3% higher than that in plain cities. Yang et al. [31] and Yin et al. [32] explored the risk and response measures of high-temperature disasters in the Chengdu-Chongqing Economic Zone and Nanchong, which have considered socio-economic indicators, social environmental indicators, and climate characteristics when conducting risk assessments, which can reflect social vulnerability.

With the development of society, the amount of refrigeration equipment such as air conditioning is increasing, and the aging population is also intensifying. However, at the same time, the global average annual temperature is also constantly rising, and most cities around the world have been hit by heat waves. In the dynamic pattern of intensified climate change and socio-economic development in the new era, the potential risk of heat wave disasters may differ. Therefore, it is necessary to determine new indicators based on these changes to analyze the high-temperature response capacity of key cities and regions. The Chengdu-Chongqing region holds an important strategic position in the development and economic construction of western China [33], with high representativeness in geographical location and economic scale. In summary, this study considered the potential driving effect of the Chengdu-Chongqing Economic Circle on the surrounding areas. The Chengdu-Chongqing Economic Circle was expanded to cover some surrounding areas (Chengdu-Chongqing Area) and has been used as the research object.

2. Research Areas and Methods

2.1. Overview of the Research Area and Data Sources

The Chengdu-Chongqing Area (27°43′ N–32°54′ N, 101°48′ E–110°6′ E) defined in the study includes Sichuan Province and Chongqing. The total area is nearly 2.4×10^5 km^2. The area is mainly composed of basins, surrounded by mountains, with an average elevation below 1500 m. Some areas in the west are plateau mountains, while the east is parallel valleys. Additionally, it has a subtropical monsoon climate, with a vegetation type of subtropical evergreen broad-leaved forest, but extreme high temperatures have occurred frequently in recent years [34].

The population of Chengdu Chongqing region is about 100 million, and the regional GDP in 2021 has exceeded 7 trillion yuan and it is showing an upward trend, which plays an important role in strategic economic and ecological status (shown in Figure 1).

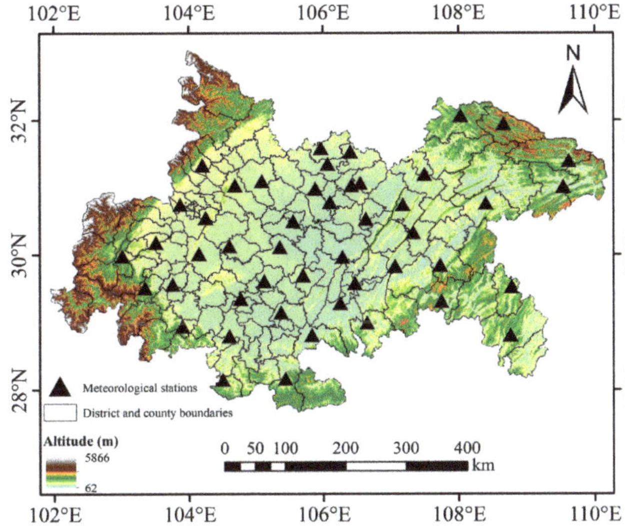

Figure 1. Overview of the study area (GS (2022) No. 1873). The research area spans Sichuan and Chongqing, and the black triangle in the picture shows the location of the meteorological stations.

Meteorological data were obtained from the National Meteorological Science Data Center (http://data.cma.cn/, accessed on 20 September 2023). The daily maximum temperature data from 48 meteorological stations in the Chengdu-Chongqing Area (including 33 in Sichuan and 15 in Chongqing) from 1961 to 2020 were selected for this research. The outliers were eliminated by using the 3σ method and the original data quality control code, and the missing values were supplemented by linear interpolation. Finally, the daily maximum temperature data for 60 years were obtained. The distribution of the meteorological stations and terrain overview are shown in Figure 1. The socioeconomic data are taken from the statistical yearbook (2020) and social questionnaire survey of each area.

2.2. Research Methods for Urban Vulnerability

2.2.1. Identification of High-Temperature Heat Waves and High-Temperature Days

High-temperature heat waves refer to weather processes where the temperature remains abnormally high during a certain period. A general definition of a high-temperature day (a day with a maximum temperature higher than 35 °C is considered a high-temperature day) [1] is adopted because of the small spatial span and non-significant elevation differences. During heat waves (a duration of 3 or more days with high temperature), the severity and degree of harm caused by high-temperature disasters are often expressed in terms of high temperature intensity. It can be expressed as the cumulative number of temperatures in the part where the daily highest temperature is higher than the absolute threshold (35 °C) during the heat wave process [35] (in degrees Celsius). The formula for calculating high temperature intensity is as follows:

$$Q = \frac{\sum_{i=1}^{m}\left[\left(\overline{T_i} - 35\right) \times (d_i)\right]}{m} \tag{1}$$

where Q is the intensity of high-temperature heat waves in a certain year (°C); $\overline{T_i}$ is the daily average of the highest temperature during the i-th heat wave of a certain year; m is the total number of heat wave frequencies that occur in the region in a certain year; and d_i is the duration of the i-th heat wave.

In addition, Nearest Neighbor is used for interpolation of temperature-related data in various districts and counties. Previous studies have shown that this method has high accuracy in interpolation of meteorological elements.

2.2.2. Construction of the Indicator System and Data Processing

The definition of vulnerability in catastrophology refers to the (potential) loss caused by natural disasters to the disaster-bearing entity. Usually, the loss is the result of the joint action of various factors such as nature, society, and economy [36]. This study focuses more on the impact of high temperature heat waves on social operations, and the framework proposed by IPCC [37] provides guidance for the construction of the indicator system in this article. Xie et al. [38] noted that indicators should include both small-scale residents and large-scale regions as much as possible. Therefore, the 13 indicators (shown in Table 1) were comprehensively considered for vulnerability measurement. The weights of each factor in Table 1 are determined by the entropy method to avoid subjectivity caused by artificial settings.

Table 1. Urban vulnerability evaluation index system in the Chengdu-Chongqing Area.

Target	Dimensional Layers	Weight	Index (Driving Factors)
urban vulnerability	Exposure	0.388	Number of hot days (X_1)
		0.343	Intensity of heat wave (°C) (X_2)
		0.032	Green space for each inhabitant (X_3)
		0.237	Frequency of heat waves (X_4)
	Susceptibility	0.083	Proportion of rural population (X_5)
		0.744	Population density (X_6)
		0.054	Proportion of the elderly population (X_7)
		0.119	Gender ratio (X_8)
	Adaptability	0.099	Number of air conditioners at home (X_9)
		0.410	Quantity of regular heatstroke medication/food (X_{10})
		0.189	Proportion of medical staff (X_{11})
		0.226	Per capita number of medical treatments (X_{12})
		0.076	Per Capita GDP (X_{13})

Due to dimensional differences in various statistical indicators, range standardization is used here to perform dimensionless processing on the data of each indicator. The forward formula is as follows:

$$b_{\text{ij}} = \frac{a_{ij} - a_{jmin}}{a_{jmax} - a_{jmin}} \tag{2}$$

The calculation formula for negative indicators is:

$$b_{\text{ij}} = \frac{a_{jmax} - a_{ij}}{a_{jmax} - a_{jmin}} \tag{3}$$

Above, b_{ij}, a_{ij}, a_{jmax}, a_{jmin} are the standardized value, original value, maximum value, and minimum value of the j-th indicator of the i-th research object, respectively.

The multiplication and division method can better reflect the combined changes in various dimensional layers [38], and the urban vulnerability index (*UVI*) is calculated using the multiplication and division method to characterize urban vulnerability. The formula is as follows:

$$UVI = \frac{EI \times SI}{AI} \tag{4}$$

$$EI = \sum\nolimits_{i,j=1}^{n} W_{ij} I_{ij} \quad SI = \sum\nolimits_{i,j=1}^{n} W_{ij} I_{ij} \quad AI = \sum\nolimits_{i,j=1}^{n} W_{ij} I_{ij} \tag{5}$$

The *UVI* represents the urban vulnerability index, while the *EI*, *SI*, and *AI* represent the exposure index, sensitivity index, and adaptability index, respectively. W_{ij} represents the weight of the *j*-th indicator in the *i*-th region, and I_{ij} represents the standardized value of the *j*-th indicator in the *i*-th region (each index is categorized into 5 degrees by the natural breakpoint method).

2.2.3. The Division of Vulnerability Levels

The sample is smaller when the districts were selected as the subject of the study. So, the severe, general, and mild urban vulnerability districts and counties were defined here (severe, general, and mild vulnerability levels are different from the above vulnerability degree). The specific division is shown in Table 2.

Table 2. Interval division of vulnerability levels in the Chengdu-Chongqing Area.

Vulnerability Level	Division Basis
serious	$1.9001 < UVI$
general	$1.0465 < UVI \leq 1.9001$
mild	$0 \leq UVI \leq 1.0465$

2.3. *Research Methods for Determining the Influencing Factors of Urban Vulnerability*

2.3.1. Contribution Analysis Method

This article aims to reveal the impact of exposure, sensitivity, and adaptability on urban vulnerability, clarify the vulnerability factors of each district and county, and further quantify the contribution of the three dimensions to urban vulnerability based on vulnerability assessment, referring to the methods of Zhang et al. [20]. The calculation formula is as follows:

$$C_{mn} = \frac{W_n \times I_{mn}}{\sum_{n=1}^{3} W_n \times I_{mn}} \times 100\% \quad (6)$$

In the formula, C_{mn} represents the contribution of n dimensions in the *m*-th region, and W_n represents the weight of the *n*-th dimension. I_{mn} is the standardized value of the *n* indicator in layer *m*. The weight of the dimensions is determined by using the entropy method, with the weights of the exposure, sensitivity, and adaptability dimensions being 0.1697, 0.2834, and 0.5469, respectively.

2.3.2. Optimal Parameter-Based Geographical Detector

To further explore the impact of various factors and their combinations on vulnerability in different dimensions, the optimal parameter geographic detector (OPGD) [39] was used to eliminate subjectivity in partitioning continuous variables and detecting spatial heterogeneity, revealing the driving factors behind vulnerability [40]. First, the *q*-value ($q \in [0, 1]$) of each continuous factor is calculated under different grading methods and different numbers of discontinuities. The larger the value is, the stronger the spatial differentiation of regional vulnerability, and the stronger the explanatory power of the influencing factor. Then, equal breaks, natural breaks, quantile breaks, geometric breaks, and standard deviation breaks are used to classify and set the classification level as 3 to 9. The parameter combination with the largest *q* value was selected as the discretization standard to detect and identify the influence of a single factor on urban vulnerability.

$$q = 1 - \frac{\sum_{h=1}^{L} N_h \sigma_h^2}{N\sigma^2} = 1 - \frac{SSW}{SST} \quad (7)$$

In the equation, *q* represents the explanatory power of the factor, *h* represents the stratification of the explanatory variable or the dependent variable, N_h and N represent the number of units in layer *h* and the entire region, respectively, and σ_h^2 and σ^2 represent the variance of layer *h* and the whole region value, respectively.

The joint effects of two factors on urban vulnerability are analyzed through interaction detection. The main types include nonlinear attenuation $q(X_1 \cap X_2) < min(q(X_1), q(X_2))$, single-factor nonlinear attenuation $min(q(X_1), q(X_2)) < q(X_1 \cap X_2) < max(q(X_1), q(X_2))$, double-factor enhancement $q(X_1 \cap X_2) > max(q(X_1), q(X_2))$, independent $q(X_1 \cap X_2) = q(X_1) + q(X_2)$, and nonlinearity enhancement $q(X_1 \cap X_2) > q(X_1) + q(X_2)$ [41]. The specific discrimination method can be found in reference [27].

3. The Results

3.1. Spatial Pattern of Regional Vulnerability

The exposure degree, sensitivity degree, and adaptability degree of each district were quantified in the Chengdu-Chongqing Area through the *EI*, *SI*, and *AI*. The calculation of each indicator is shown in Formula (5), and the final calculation results are brought into Formula (4) to obtain the urban vulnerability index of each district and county. This article uses the natural breakpoint method to divide the calculation results of each dimension into five degrees: low, medium-low, medium, medium-high, and high (Figure 2).

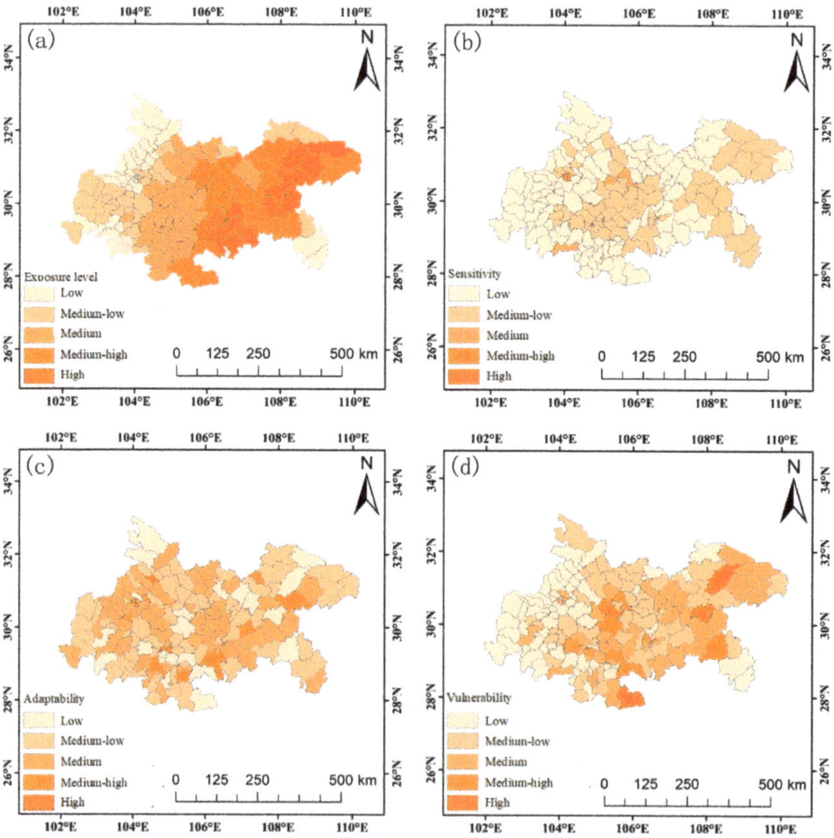

Figure 2. Distribution of high-temperature heat wave exposure, sensitivity, adaptability, and vulnerability in the Chengdu-Chongqing Area. (**a**) shows the spatial distribution of high temperature exposure levels, (**b**) shows the spatial distribution of regional sensitivity, (**c**) shows the spatial distribution of regional adaptability, and (**d**) shows the spatial distribution of overall vulnerability.

Overall, the percentages of districts with exposure indices ranging from high to low were 20, 32, 42, 35, and 27, accounting for 12.82%, 20.51%, 26.92%, 22.44%, and 17.31%, of the total number of districts, respectively. From the spatial distribution of the exposure index, the areas with high exposure indices gradually increase from northwest to southeast in a ladder shape, mainly concentrated in Chongqing, showing a trend of relative aggregation, in which the southwest of Chongqing (such as Yuzhong District, Yubei District, Jiulongpo District, Banan District, etc.) is the most concentrated, and the northeast of Chongqing also has high exposure index districts (such as Wanzhou District, Wuxi County, etc.). This finding is highly consistent with the conclusions of Mao and Wang [42]. There is no area with a high exposure index in Sichuan, and the area with a high exposure index is mainly located in the border area between Sichuan and Chongqing (such as southern Nanchong, Dazhou, southern Guang'an, and some districts in Luzhou). Areas and counties with medium and low exposure are mainly located in Sichuan, and there are only three areas and counties with low or lower exposure indices in Chongqing (Qianjiang District, Youyang Tujia and Miao Autonomous County, and Xiushan Tujia and Miao Autonomous County). The previous methods were uesd to measure the exposure under high-temperature conditions [17,18] and the exposure in green space areas [21]. It has been noted that topography, atmospheric circulation, and vegetation are important factors affecting high-temperature exposure [42,43]. Compared with Sichuan, Chongqing is closer to the east and is more easily controlled by the Western Pacific Horse latitudes in summer. It is sunny, hot, and less rainy [44]. In addition, the parallel mountains and valleys in eastern Sichuan in the north prevent some cold air in the north from flowing southward [42], resulting in strong high-temperature sustainability.

From the perspective of the sensitivity exponential distribution, the Chengdu-Chongqing Area has a low sensitivity overall. The areas with high sensitivity indices are Wuhou District and Yuzhong District, which are highly consistent with the locations of Chengdu and the main urban areas of Chongqing. The areas with higher sensitivity indices were distributed only in the main urban area of Chengdu (mainly the Qingyang District, Jinniu District, Jinjiang District, and Chenghua District) and not in other locations in the Chengdu-Chongqing Area. The districts with medium or low vulnerability indices account for a significant proportion of the total area. From a quantitative perspective, the number of districts with vulnerability indices ranging from high to low is 2, 4, 3, 57, and 90, accounting for 1.28%, 2.56%, 1.92%, 36.54%, and 57.69%, respectively, of the total number of districts.

From the exponential distribution of adaptability, there is no obvious trend in the horizontal distribution of the overall adaptability, showing a broken patch distribution. Among them, the number of districts with below-medium adaptability indices is the largest in the Chengdu-Chongqing Area. The districts with medium adaptability indices are mostly around the city center, while the districts with medium-high and high adaptability indices are mainly the main urban areas of cities at various levels (such as Luojiang District and Da'an District). Notably, there are no districts with high adaptability indices in Chengdu and Chongqing, which are all districts with high adaptability indices. In terms of quantity, the numbers of districts with high to low adaptability indices are 2, 13, 59, 61, and 21, accounting for 1.28%, 8.33%, 37.82%, 39.10%, and 13.46% of the total, respectively.

In terms of the spatial characteristics of the vulnerability index, the overall distribution is high in the east and low in the west. The vulnerability indices are arranged from high to low, with proportions of 2.56%, 6.41%, 25.00%, 31.41%, and 34.62%, respectively. There were only 33.97% of the districts with moderate or greater vulnerability indices, of which there were only four districts with high vulnerability indices, which were distributed in both Sichuan and Chongqing, including Anju District of Suining city, Gulin County of Luzhou city, Kaizhou District of Chongqing, and Zhongxian County of Chongqing city. There are 10 districts with high vulnerability indices, 7 of which are located in Sichuan (such as Daying County, Pengxi County, and Anyue County) and 3 of which are located in Chongqing (Shapingba District, Yuzhong District, Pengshui Miao, and Tujia Autonomous County). The areas and counties with medium vulnerability image indices are mostly located in the eastern part of the Chengdu Plain, while the areas and counties with medium-low and low vulnerability degrees are mainly located on the western edge of the Chengdu Plain.

3.2. *Main Causes of Regional Vulnerability*

3.2.1. Statistics on Different Causes of Vulnerability

The three dimensions of exposure, sensitivity, and adaptability comprehensively determine the urban vulnerability of the Chengdu-Chongqing Area. Here, the causes of vulnerability in each district and county are classified into three categories from a dimensional perspective, namely, high-frequency vulnerability caused by high exposure, vulnerability caused by high social sensitivity, and vulnerability caused by insufficient adaptability. The specific discrimination method calculates the contribution degree of different dimensions of each district and county according to Formula (6). If the contribution degree of the corresponding dimension accounts for the largest proportion, it is the corresponding cause of vulnerability.

Figure 3 shows that the high vulnerability in the central and eastern parts of the Chengdu-Chongqing Area is almost entirely caused by high-temperature exposure. The changes in the western region, especially the Chengdu Plain, Ya'an, and its surrounding areas, are mainly due to insufficient adaptability. However, there are very few districts with high sensitivity that lead to vulnerability, which are mainly scattered in the main urban areas of Chengdu and Chongqing. In terms of quantity, there are a total of 96 vulnerable districts caused by high exposure, accounting for 61.54%; 6 vulnerable districts caused by high sensitivity, accounting for 3.85%; and 54 vulnerable districts due to insufficient adaptability, accounting for 34.61%.

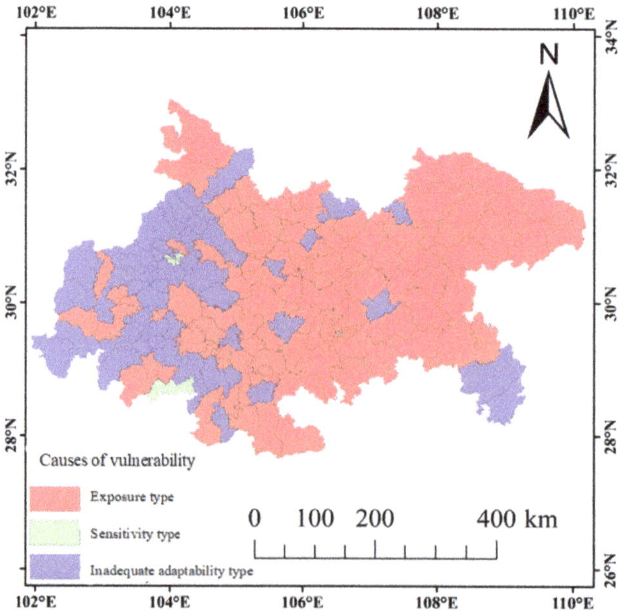

Figure 3. Main causes of vulnerability in various districts in the Chengdu-Chongqing Area.

3.2.2. Regional Distribution of Different Causes of Vulnerability

From the perspective of districts with different levels of high-temperature urban vulnerability, out of the 14 severely vulnerable districts, 13 have high-temperature exposure, and only 1 has high social sensitivity. Out of 39 general urban vulnerability districts, a total of 36 have high-temperature exposure, and 2 have high social sensitivity. Out of 103 mild urban vulnerability districts, 47 have high-temperature exposure, 2 have high social sensitivity, and 54 have insufficient adaptability. According to the average contribution of each district and county type (Figure 4), the average contribution of the exposure index in both the severe and general urban vulnerability districts exceeded 50%, reaching 62.02% and 54.67%, respectively. Among the mild urban vulnerability districts, the average contribution degree of adaptability was the highest, accounting for 47.53%.

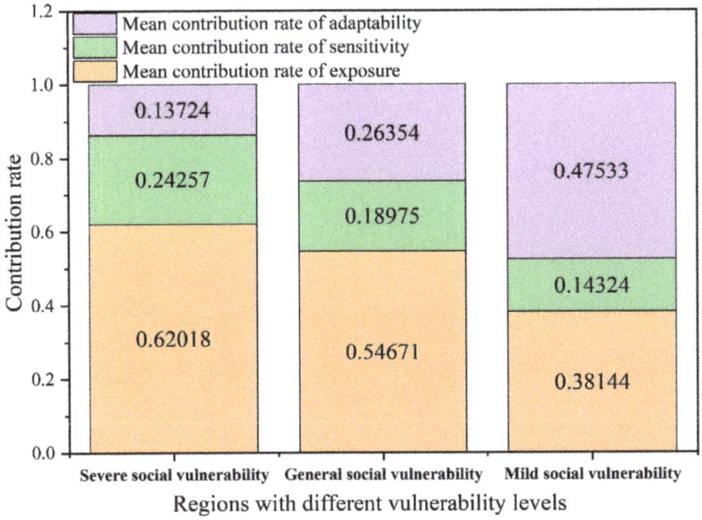

Figure 4. Mean contributions of the three indices to the different levels of urban vulnerability in the districts.

High and medium-high vulnerability (severe urban vulnerability level) districts and counties, as well as medium vulnerability (general urban vulnerability level) districts, are mostly affected by high temperature exposure, while the main cause of low and medium-low vulnerability (mild urban vulnerability level) districts is insufficient adaptability. The number of vulnerable areas in the Chengdu-Chongqing Area due to high sensitivity is relatively low.

3.3. Exploration of the Driving Factors of Urban Vulnerability

3.3.1. Discretization of the Continuity Factor

The academic community has conducted many studies using geographic detectors (GDs) in areas such as ecological vulnerability, urban waterlogging risk, and human habitat vulnerability, proving the feasibility of using evaluation factors as independent variables to detect the aforementioned objects [45–50]. The Chengdu-Chongqing Area covers mountains, plains, and other major geomorphic areas, has a subtropical monsoon climate, has stable socioeconomic growth, and is representative of both natural and social conditions. Therefore, the selection of independent variable factors can be found in the relevant research on other urban agglomerations in China. The *UVI* is taken as the dependent variable, and 13 evaluation indicators under each district and county are selected as the independent variables. The specific processing results are shown as follows (Figure 5).

As an example, for X_1, the geometric interval method has a significantly higher q-value when classified as 5 compared to the other methods. Therefore, the optimal parameter selection for high-temperature days (X_1) in geographical detectors should be based on the geometric interval divided into five categories. Different driving factors have different spatial discretization methods to maximize their q-values, and this parameter combination that maximizes q-values is the optimal parameter for geographic detectors. Given the impact of space, this article provides parameter selection for the top eight impact factors with the highest q-values. The specific parameter division and setting results are shown in Table 3.

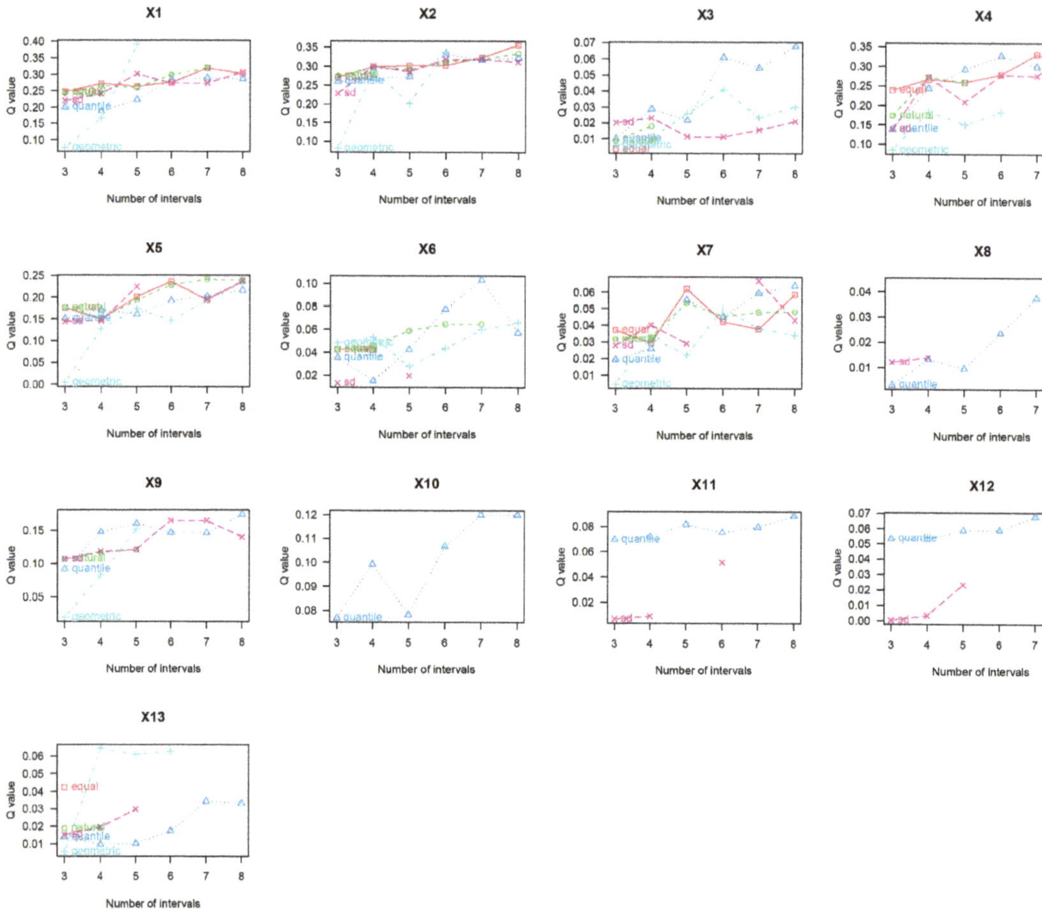

Figure 5. Parameter matching occurs when the q-values of each factor are at their maximum (where the abscissa indicate different ways of categorization. The vertical coordinates show the factor interpretations of the different categorizations under the different classifications. Different colored lines represent different classification methods. The purple line segment is sd, the deep blue line segment is quantile, the light blue line segment is geometric, the green line segment is natural, and the red line segment is equal).

Table 3. Classification method and number of stages for impact factors.

Factors	Methods	Intervals	Factors	Methods	Intervals
X_1	geometric	5	X_8	quantile	8
X_2	equal	8	X_9	natural	8
X_3	quantile	8	X_{10}	quantile	9
X_4	sd	8	X_{11}	quantile	9
X_5	quantile	9	X_{12}	natural	8
X_6	quantile	9	X_{13}	natural	9
X_7	sd	9	-	-	-

3.3.2. Analysis of a Single Driving Factor

By using OPGD to conduct single factor-driven detection of urban vulnerability in the Chengdu-Chongqing Area to identify its explanatory power, we find that the significance of each factor corresponds to the q-value as follows (Table 4). After removing the factors with insufficient significance, the explanatory power of the remaining factors is as follows (Figure 6):

Table 4. The correspondence between q values and the significance of each factor.

Variable	q-Value	Sig.	Variable	q-Value	Sig.
X_1	0.390771	0.000	X_8	0.043767	0.493
X_2	0.356134	0.000	X_9	0.173176	0.001
X_3	0.067794	0.194	X_{10}	0.123695	0.030
X_4	0.348757	0.000	X_{11}	0.102308	0.067
X_5	0.247505	0.000	X_{12}	0.076069	0.162
X_6	0.119297	0.031	X_{13}	0.070002	0.539
X_7	0.072545	0.300	-	-	-

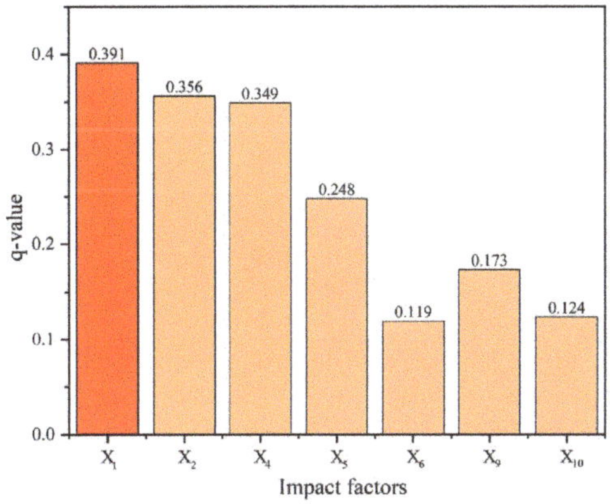

Figure 6. Single factors with significant explanatory power for urban vulnerability (the vertical coordinate is the explanatory power of each single factors for urban vulnerability, the bar chart filled in red represents the factor with the highest q-value).

Overall, the q-value of X_1 is 0.391, indicating that the explanatory power of the high-temperature days factor for overall urban vulnerability reaches 39.1%, which is the highest among all factors. The q-values of X_2 and X_4 are 0.356 and 0.349, respectively, both reaching a high level of explanatory power. At the socioeconomic level, the explanatory power of each factor in descending order is the proportion of the township population (X_5), the number of air conditioners at home (X_9), the number of regular heatstroke medications/food (X_{10}), and the regional population density (X_6), with corresponding q values of 0.248, 0.173, 0.124, and 0.119, respectively. Single-factor detection revealed that natural factors such as high-temperature days, intensity, and frequency of heat waves play a dominant role in the level of urban vulnerability in the Chengdu-Chongqing Area, especially high-temperature days, which have the strongest explanatory power for regional urban vulnerability, while socioeconomic factors such as population density have relatively weak explanatory power for regional urban vulnerability. This conclusion also maintains high consistency with the results of the contribution in the previous text.

3.3.3. Analysis of Interaction Driving Factors

To further explore the interactions between various factors, the OPGD interaction detector was used for detection. The results are as follows (Figure 7). Most factors show a nonlinear enhancement trend (interaction types are identified with "2" in Figure 7a). The factor interactions with explanatory powers exceeding 70% are $X_4 \cap X_6$, $X_2 \cap X_6$, $X_4 \cap X_9$, and $X_2 \cap X_9$. Among them, the interaction between average heat wave frequency and regional population density has the strongest explanatory power, and the trend shows nonlinear enhancement, with a q-value of 0.7863, indicating that the synergy of the two factors can explain 78.63% of urban vulnerability. Therefore, it is necessary to focus on adjusting population density and arranging heat relief facilities in this area to reduce urban vulnerability.

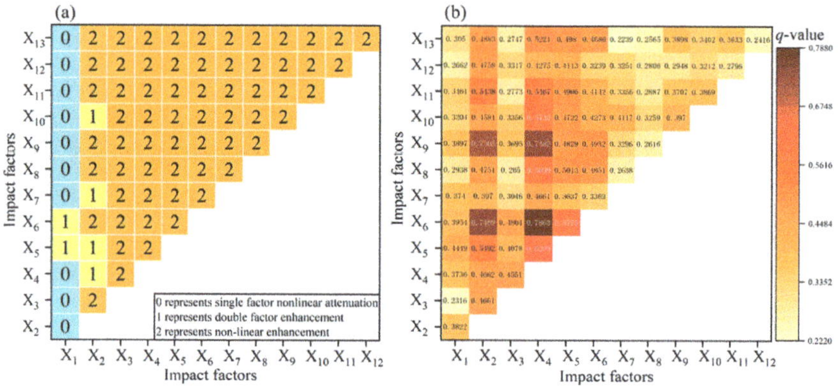

Figure 7. Double factor interaction detection results ((**a**) shows the form of factor interaction and (**b**) shows the q-value after factor interaction; the horizontal and vertical coordinates represent the corresponding relationships between different interaction factors).

In addition, after the interaction between various socio-economic factors (X_5~X_{13}) with the intensity and frequency of heat waves (X_2, X_4), the q-value of the interaction ranged from 0.397 to 0.7863 (higher than the q-value under a single socio-economic factor), the interaction between socio-economic factors (X_5~X_{13}) with X_2 and X_4 showed an enhanced signal. In other words, the differences in social factors alone have less impact on the divergence of urban vulnerability in the context of high temperatures, but the divergence of urban vulnerability is more pronounced with the addition of heat wave intensity (X_2) and heat wave frequency (X_4).

It is worth noting that the interaction between high-temperature days (X_1) and the most factors show single-factor nonlinear weakening, showing only a double-factor enhancement with X_5 and X_6. Except for X_5 and X_6, the q-values of the interactions between X_1 and the other factors ranged from 0.2316 to 0.3897, which was lower than its explanatory power. The reason may be that the number of high-temperature days in a region increases beyond a certain threshold, and the impact of high-temperature days on urban vulnerability decreases due to the increased operational intensity of cooling equipment and medical systems in the region. In addition, it is also possible that the increase in the number of high-temperature days may cause urban vulnerability to reach a peak in the region, resulting in paralysis and no longer increasing.

4. Discussion

The evaluation indicators are improved in this study to make the indicator system more in line with the current development status of the area. In the exploration of influencing factors, a more objective OPGD model was used to determine the explanatory power of each influencing factor on urban vulnerability, and their interaction modes were accurately identified. Within the study area, there are differences in exposure levels, sensitivity, and adaptability among different counties. Overall, the vulnerability is higher in the east and lower in the west. In addition, extreme high temperatures make a greater contribution to severe social vulnerability, while mild social vulnerability is often associated with insufficient adaptability. Therefore, for the eastern part of the research area, reducing direct solar thermal radiation is very important. This requires the government (meteorological department) to carry out artificial rainfall when necessary and take measures for high temperature warnings. From the results of factor detection, in addition to natural factors, the proportion of rural population and the equipment of heat dissipation facilities will affect extreme high temperatures in urban systems. Therefore, it is necessary to continue to promote the rural revitalization plan, strengthen rural infrastructure construction, and increase the deployment scope of heat dissipation facilities. In fact, the threshold for cooling demand varies among different regions, which leads to differences in energy consumption (especially electricity) between regions under extreme high temperatures [51,52]. So, the energy reserves will also affect the adaptability indicators to a certain extent (areas with fewer backup energy sources are more sensitive to high-temperature attacks, such as power outages, resulting in an increase in *UVI*). In addition, the number of days with high temperatures has the strongest explanatory power for the spatial distribution of vulnerability. However, some factors (e.g., X_4, X_5) weakened their explanatory power after the interaction with X_1, which may be related to the increase in high-temperature days leading to the peak of urban vulnerability in the region.

The main influencing factors of urban vulnerability under high temperature include natural factors such as high-temperature exposure and sunshine duration, as well as socioeconomic factors such as population density, the urban green space ratio, and heatstroke prevention capacity [53]. The scope of extreme climate events caused by ocean surface temperature anomalies [53,54] is also highly likely to increase, so the differentiation of urban vulnerability under the combined effects of socioeconomic and natural factors may be more significant. There is research indicating that Sichuan Province lacks adaptability in the face of flood disasters [55]. This is similar to the conclusion of this article, which is that the number of counties with increased urban vulnerability due to insufficient adaptability exceeds one-third. In terms of spatial distribution of vulnerability, the study points out that highly vulnerable areas are concentrated in the center areas of Chengdu and Chongqing, as well as the rurales in the northwest of Chongqing [31]. The results are similar to the above conclusions. However, this study found that the distribution area of counties with exposure-induced vulnerability is larger. On the one hand, this difference may be due to the longer meteorological data in this article (frequent occurrence of severe high temperature weather in recent years). On the other hand, differences in evaluation indicators and methods may also be one of the reasons for the differences. So, if the method needs to be applied to other regions, it is necessary to consider the special geographical differences of other regions (including natural and social factors).

At present, there are still limitations in the indicators of exposure, sensitivity, and adaptability in the study of urban heat wave vulnerability [56]. The evaluation system here may not always emphasize the mitigating effect and the concern for outdoor workers [57,58]. However, these indicators can meet the research needs at the county level. To make research more refined, human activities (e.g., increased CO_2 emissions due to human activities) as well as land use types could be considered. In addition, the smallest unit of this study is the county. The data accuracy is slightly lower than that of the kilometer grid dataset (1 km × 1 km), and the temperature data can only be obtained from some national stations and the site density and coverage area are limited, which may make it difficult to match the mountainous terrain. Therefore, the impact of mountainous terrain on temperature and heat has not been fully considered, which leads to a reduction in the impact of ventilation [29] and altitude [30] on temperature. However, considering that the population concentration areas in most counties are at low altitudes and close to plain areas, the impact of errors on the results is negligible on the county scale.

In subsequent research, optimization can be made from aspects such as evaluation systems and data to compensate for the above shortcomings. The studies on the risk of high-temperature heat wave disasters still mainly focused on the number of high-temperature days and the intensity of heat waves [59–61]. However, the indicator of humidity may be an important influencing factor of high-temperature heat wave disasters in some areas. Moreover, land use types also continue to be altered [62]. The ability to absorb heat and long-wave radiation is different between built-up land (e.g., hard concrete surfaces) and agricultural land (e.g., paddy fields). Therefore, different land use types have different impacts on regional warming. In the future, improving the selection of influencing factors can comprehensively evaluate the threat of high-temperature heat waves in the context of global climate change [63]. In addition, against the backdrop of economic recovery and the gradual control of the COVID-19 pandemic, the increase in outdoor population mobility has the potential to expand the population base suffering from sensitive diseases such as hypertension. Therefore, groups with high exposure and sensitivity, such as the tourism industry [56], high-temperature sensitive patients [64], and outdoor workers [65], should also be considered at the sensitivity level. Finally, with the establishment of more automatic weather stations and the development of real-time weather information sharing systems, the problem of insufficient station density and the lack of consideration for the impact of terrain on temperature in future research can gradually be solved.

5. Conclusions

With the vulnerability evaluation index system, contribution model, and OPGD, the high-temperature disasters and the impacts in the Chengdu-Chongqing Area were analyzed. The results are as follows:

① The exposure index increases from northwest to southeast. The sensitivity index is low in most areas, but the sensitivity indices of central Chengdu and central Chongqing are approximately 0.79. The adaptability index exhibited significant regional differences and low concentrations. The vulnerability index is high in the east and low in the west, with significant differences among districts and counties.

② The vulnerable districts caused by high exposure have the greatest proportion and the largest area, with a total of 96, followed by the vulnerable counties with insufficient adaptability, accounting for 34.61%, and the vulnerable counties with insufficient sensitivity have a relatively small number, accounting for only 3.85%. The average contribution of exposure levels to general and severe urban vulnerability areas and counties exceeds half, with the highest proportion being the average contribution of adaptability in mild urban vulnerability areas and counties.

③ The explanatory power of climate factors on the spatial differentiation of urban vulnerability is stronger than that of socioeconomic factors, with the highest explanatory power being the number of high-temperature days, reaching 39.1%. The combined effect of climate factors and regional socioeconomic factors had a greater impact on the urban vulnerability of the Chengdu-Chongqing Area.

Author Contributions: Conceptualization, W.L.; Methodology, W.L. and L.Z.; Software, Z.C. and P.S.; Validation, Z.C. and P.S.; Resources, Z.Y.; Data curation, Z.C. and P.S.; Writing—original draft, Z.Y.; Writing—review & editing, Z.Y. and W.L.; Visualization, Z.Y.; Supervision, W.L., Z.C. and L.Z. All authors have read and agreed to the published version of the manuscript.

Funding: This research was funded by the Regional Cultural Research Center of the Sichuan Provincial Social Science Key Research Base Annual Project, grant number CQYYJC2101; the National Innovation and Entrepreneurship Training Program for College Students, grant number 201910638025, S202010638090, 202110638013; the Educational Reform Project of China West Normal University, grant number 403995; the Scientific Research Fund of Sichuan Provincial Education Department, grant number 18ZA0476; the Meritocracy Research Funds of China West Normal University, grant number 17YC112; the Fundamental Research Funds of China West Normal University, grant number 16C003.

Institutional Review Board Statement: Not applicable.

Informed Consent Statement: Not applicable.

Data Availability Statement: Due to privacy and related policy factors, we will not provide relevant data now.

Conflicts of Interest: The authors declare no conflicts of interest.

References

1. IPCC. *Climate Change 2021: The Physical Science Basis;* Contribution of Working Group I to the Sixth Assessment Report of the Intergovernmental Panel on Climate Change; Masson-Delmotte, V., Zhai, P., Pirani, A., Connors, S.L., Clotild, P., Chen, Y., Goldfarb, L., Gomis, M.I., Matthews, J.B.R., Berger, S., et al., Eds.; Cambridge University Press: Cambridge, UK; New York, NY, USA, 2021.
2. Mei, M.; Gao, G.; Li, Y.; Wang, G.; Dai, T.; Chen, Y. Change characteristics in compound high temperature and drought extreme events over Yangtze River Basin from 1961 to 2022. *Yangtze River* **2023**, *54*, 12–20. [CrossRef]
3. Wang, F.; Zhang, J.; Ge, Q.; Hao, Z. Projected changes in risk of heat waves throughout Belt and Road Region in the 21st century. *Chin. Sci. Bull.* **2021**, *66*, 3045–3058. (In Chinese) [CrossRef]
4. Zhang, J.; Qian, C. Linear Trends in Occurrence of High Temperature and Heat Waves in China for the 1960–2018 Period: Method and Analysis Results. *Clim. Environ. Res.* **2020**, *25*, 225–239.
5. Sherwood, S.C.; Huber, M. An adaptability limit to climate change due to heat stress. *Proc. Natl. Acad. Sci. USA* **2010**, *107*, 9552–9555. [CrossRef] [PubMed]
6. United Nations. *Transforming Our World: The 2030 Agenda for Sustainable Development;* UN: New York, NY, USA, 2015.
7. Shen, H.; You, Q.; Wang, P.; Kong, L. Analysis on heat waves variation features in China during 1961–2014. *J. Meteorol. Sci.* **2018**, *38*, 28–36. (In Chinese) [CrossRef]
8. Renée, M.M.; Diana, B.; Alessandro, O.; Michelle, R.L.; Mark, G.T.; David, S.E. Extreme heat increases stomatal conductance and drought-induced mortality risk in vulnerable plant species. *Glob. Change Biol.* **2021**, *28*, 1133–1146. [CrossRef]
9. Shu, Z.; Li, W.; Zhang, J.; Jin, J.; Xue, Q.; Wang, Y.; Wang, G. Historical Changes and Future Trends of Extreme Precipitation and High Temperature in China. *Strateg. Study CAE* **2022**, *24*, 116–125. (In Chinese) [CrossRef]
10. Jia, J.; Hu, Z. Spatial and Temporal Features and Trend of Different Level Heat Waves Over China. *Adv. Earth Sci.* **2017**, *32*, 546–559. (In Chinese) [CrossRef]
11. Reid, C.E.; O'Neill, M.S.; Gronlund, C.J.; Brines, S.J.; Brown, D.G.; Diez-Roux, A.V.; Schwartz, J. Mapping Community Determinants of Heat Vulnerability. *Environ. Health Perspect.* **2009**, *117*, 1730–1736. [PubMed]
12. Xue, Q.; Xie, M.; Guo, Q.; Wang, Y.; Wu, R.; Liu, Q. Research progress on urban heat wave vulnerability assessments geographical perspective. *Prog. Geogr.* **2020**, *39*, 685–694. (In Chinese)
13. Benmarhnia, T.; Grenier, P.; Brand, A.; Fournier, M.; Deguen, S.; Smargiassi, A. Quantifying Vulnerability to Extreme Heat in Time Series Analyses: A Novel Approach Applied to Neighborhood Social Disparities under Climate Change. *Int. J. Environ. Res. Public Health* **2015**, *12*, 11869–11879. [CrossRef]
14. Cutter, S.L.; Barnes, L.; Berry, M.; Burton, C.; Evans, E.; Tate, E.; Webb, J. A Place-Based Model for Understanding Community Resilience to Natural Disasters. *Glob. Environ. Change* **2008**, *18*, 598–606. [CrossRef]
15. Setiawati, M.D.; Jarzebski, M.P.; Fukushi, K. *Heat Vulnerability Index Development and Application in Medan City, Indonesia;* Integrated Research on Disaster Risks; Springer: Cham, Switzerland, 2021; pp. 87–110. [CrossRef]
16. Li, D.; Wang, L.; Liao, W.; Sun, T.; Katul, G.; Bou-Zeid, E.; Maronga, B. Persistent urban heat. *Sci. Adv.* **2024**, *10*, eadj7398. [CrossRef]
17. Lemonsu, A.; Viguié, V.; Daniel, M.; Masson, V. Vulnerability to heat waves: Impact of urban expansion scenarios on urban heat island and heat stress in Paris. *Urban Clim.* **2015**, *14*, 586–605. (In France) [CrossRef]
18. Huang, X.; Wang, B.; Liu, M.; Guo, Y.; Li, Y. Characteristics of urban extreme heat and assessment of urban vulnerability in China. *Geogr. Res.* **2020**, *39*, 1534–1547. (In Chinese)
19. Zhang, Y.; Huang, X.; Zheng, D. Temporal and Spatial Characteristics of Heat Waves and Assessment of Vulnerability in the Yangtze River Economic Belt. *Resour. Environ. Yangtze Basin* **2023**, *32*, 440–450.
20. Huang, X.; Qi, M.; Zhao, K.; Zheng, D.; Liu, M. Assessment of population vulnerability to heat stress and spatial differentiation in Xi'an. *Geogr. Res.* **2021**, *40*, 1684–1700. (In Chinese)
21. Peng, B.; Zhang, H. Research Progress and Future Topics on Integrated Governance for Urban Extreme Heat Disasters: A Review of Multi-disciplinary Perspectives. *J. Public Adm.* **2024**, *17*, 4–25+196. (In Chinese)
22. Xiao, H.; Guo, Y.; Wang, Y.; Xu, Y.; Shi, J. Community-Scale Mitigation and Adaptation Pathways for Coping with High-Temperature Health Stressors—Experiences and Insights from Cool Communities Program in New York. *Planners* **2022**, *38*, 151–158. (In Chinese)
23. Sun, X.; Ge, F.; Luo, H.; Feng, C. Projected changes of temperature extremes in Chengdu-Chongqing Economic Zone and its surrounding areas under 1.5 °C and 2.0 °C global warming. *Trans. Atmos. Sei* **2021**, *44*, 875–887. (In Chinese)
24. Zhang, X.; Wang, X. Equilibrium and Accessibility: Equalization of Basic Public Services in Urban Agglomeration. *Price Theory Pract.* **2024**, 1–6. (In Chinese)
25. Qin, P.; Liu, H. The theoretical logic and path exploration of the coordinated development of Chengdu-Chongqing Economic Circle: Based on the perspective of the functional series theory. *J. Chongqing Univ. (Soc. Sci. Ed.)* **2021**, *27*, 44–54. (In Chinese) [CrossRef]
26. Cheng, Z.; Yang, X.; Sun, C.; Xu, Y. The Trend of Summer Urban Heat Island Effect and Its Relationship with Urban Development in Chengdu. *Clim. Change Res.* **2016**, *12*, 322–331. (In Chinese) [CrossRef]
27. Yuan, J.; Lin, Q.; Chen, S.; Zhao, H.; Xie, X.; Cai, Z.; Zhang, J.; Cheng, T.; Hua, M.; Zhang, R. Influence of global warming and urbanization on regional climate of Megacity: A case study of Chengdu, China. *Urban Clim.* **2022**, *44*, 101227. [CrossRef]

28. Zhang, X.; Liu, M.; Qing, Q.; Deng, Z.; Hua, M.; Zhang, Y. Extreme Temperature Change and Urbanization Contribution in Chengdu from 1960 to 2018. *Plateau Mt. Meteorol. Res.* **2022**, *42*, 10–17. (In Chinese)
29. Huang, H.; Jie, P.; Yang, Y.; Mi, S. Spatial and Temporal Characteristics of High-Temperature Heat Wave Disasters in Chongqing. *Atmosphere* **2023**, *13*, 1396. [CrossRef]
30. Luo, P.; Yu, B.; Li, P.; Liang, P.; Zhang, Q.; Yang, L. Understanding the relationship between 2D/3D variables and land surface temperature in plain and mountainous cities: Relative importance and interaction effects. *Build. Environ.* **2023**, *245*, 110959. [CrossRef]
31. Yang, M.; Liu, R.; Gou, X.; Li, G.; Xu, Y. Spatial-temporal analysis and risk assessment of heat waves in the Chengdu-Chongqing economic circle. *Comput. Tech. Geophys. Geochem. Explor.* **2022**, *44*, 392–402. (In Chinese)
32. Yin, Z.; Li, W.; Chen, Z.; Shui, P.; Li, X.; Qin, C. The Spatiotemporal Characteristics of Extreme High Temperatures and Urban Vulnerability in Nanchong, China. *Atmosphere* **2023**, *14*, 1318. [CrossRef]
33. Li, B. Strategic Thinking on Coordinated Development of Chengyu Economic Zone. *Macroecon. Manag.* **2012**, *7*, 54–55. (In Chinese)
34. Qi, D.; Li, Y.; Wang, Y.; Deng, M.; Ren, Q. Temporal—Spatial Abnormity Characteristics of Drought in Sichuan Province Based on Z Index. *J. Arid Meteorol.* **2017**, *35*, 734–744.
35. Deng, T.; Geng, G.; Yang, R.; Zhang, B. Temporal and spatial variation characteristics of high temperature and heat wave in the Weihe River Basin from 1980 to 2020. *Arid Land Geogr.* **2023**, *46*, 211–221. (In Chinese)
36. Shang, Y. Review on Concept Model of Disaster Vulnerability. *J. Catastrophology* **2013**, *28*, 112–116. (In Chinese)
37. IPCC. *Climate Change 2014*; Working Group II Contribution to the Fifth Assessment Report of the Intergovernmental Panel on Climate Change; Cambridge University Press: Cambridge, UK; New York, NY, USA, 2014.
38. Xie, P.; Wang, Y.; Peng, J.; Liu, Y. Health related urban heat wave vulnerability assessment research progress and framework. *Prog. Geogr.* **2015**, *34*, 165–174. (In Chinese) [CrossRef]
39. Song, Y.; Wang, J.; Ge, Y.; Xu, C. An optimal parameters-based geographical detector model enhances geographic characteristics of explanatory variables for spatial heterogeneity analysis: Cases with different types of spatial data. *GIScience Remote Sens.* **2020**, *57*, 593–610. [CrossRef]
40. Wang, J.; Xu, C. Geodetector: Principle and prospective. *Acta Geogr. Sin.* **2017**, *72*, 116–134. (In Chinese) [CrossRef]
41. Shi, Z.; Xie, H.; Wang, Z.; Hu, X.; Wang, X.; Xie, X.; Lin, H.; Liu, X. Analysis of spatiotemporal heterogeneity of habitat quality and their driving factors based on optimal parameters-based geographic detector for Fuzhou City, China. *J. Environ. Eng. Technol.* **2023**, *13*, 1921–1930. [CrossRef]
42. Mao, Y.; Wang, Y. Characteristics and Trend Analysis of Extreme High Temperature in Chongqing in Recent 30 Years. *Bull. Sci. Technol.* **2019**, *35*, 60–68. (In Chinese)
43. Zhou, C.; Li, Y.; Pu, Q.; Peng, J. Features of Drought-Flood Coexistence in West and East of Sichuan-Chongqing Basin in Midsummer and Its Relating Background of Atmospheric Circulation. *Plateau Meteorol.* **2011**, *30*, 620–627. (In Chinese)
44. Chen, H. *Primary Character of Climate Change and the Research on Extreme Weather-Climate Events in Chongqing*; Lanzhou University: Lanzhou, China, 2011.
45. Niyazihainimu, X.; Ma, Q.; Bao, A.; Wusiman, S.; Jiang, Z. Ecological Vulnerability Evaluation and Driving Force Analysis of Ebinur Lake Based on Geodetector. *J. Hydroecology* **2022**, 1–15. (In Chinese) [CrossRef]
46. Hu, W.; Zhang, Z.; Zhao, X.; Qi, X.; Wang, Y. VSD model and geodetector-based assessment on urban waterlogging vulnerability in Beijing and analysis of its driving mechanism. *Water Resour. Hydropower Eng.* **2022**, *53*, 86–100. (In Chinese) [CrossRef]
47. Zhao, B.; Zhu, L.; Wang, L.; Sun, Z.; Zhang, Z.; Dong, Q.; Zhu, W. Coupling Mechanisms of Eco-environmental Quality and Human Activities in China and Their Influencing Factors. *Environ. Sci.* **2023**, *13*, 808–816. (In Chinese) [CrossRef]
48. Feng, Y.W.; Zhen, J.H.; Tian, Y.T.; Cao, Y.; Zhang, L.; Zhang, S.; Zhang, Y.F. Change in Vulnerability of Human Settlement Environment and Its Mechanism in Hohhot. *Econ. Geogr.* **2020**, *40*, 91–99. (In Chinese) [CrossRef]
49. Xu, C.; Lu, C.; Huang, S. Study on ecological vulnerability and its influencing factors in Zhangjiakou area. *J. Nat. Resour.* **2020**, *35*, 1288–1300. (In Chinese) [CrossRef]
50. Wang, Y.; Hu, B. Spatial and Temporal Differentiation of Ecological Vulnerability of Xijiang River in Guangxi and Its Driving Mechanism. *J. Geo-Inf. Sci.* **2018**, *20*, 947–956. (In Chinese) [CrossRef]
51. Li, W.; Wen, X.; Zhan, L.; Wang, J.; Chen, Z.; Li, W.H.; You, Y. The Variation of Cooling and Heating Degree Days and Its Response to Climate Warming in Nanchong. *J. China West Norm. Univ. (Nat. Sci.)* **2017**, *38*, 334–339. (In Chinese) [CrossRef]
52. You, Y.; Shui, P.; Li, W. The Variation of Apparent Temperature and Its Influence on Building Energy Consumption for Temperature Regulation in Nanchong City. *J. China West Norm. Univ. (Nat. Sci.)* **2018**, *39*, 195–201. (In Chinese) [CrossRef]
53. Jin, F.; Cheng, Z.; Jin, L.; Zhang, B. Study on the Relationship between Thermal Environment Effect and Vegetation Coverage in Chengyu Urban Agglomeration. *J. Chengdu Univ. Inf. Technol.* **2019**, *34*, 308–315. (In Chinese) [CrossRef]
54. Sun, B.; Wang, H.; Huang, Y.; Yin, Z.; Zhou, B.; Duan, M. Characteristics and causes of the hot-dry climate anomalies in China during summer of 2022. *Trans. Atmos. Sci.* **2023**, *46*, 1–8. (In Chinese) [CrossRef]
55. He, L.; Chen, J.; Cheng, C.; Yuan, J. Assessment and spatiotemporal analysis of flood vulnerability in the Yangtze River Economic Belt. *Water Resour. Hydropower Eng.* **2024**, *55*, 24–39. (In Chinese) [CrossRef]
56. Wu, Z.; Shui, W.; Huang, Z.; Wang, C.; Qiao, L.; Wu, Y. Construction and Application of Urban Heat Vulnerability Early Warning System. *Trop. Geogr.* **2024**, *44*, 121–142. (In Chinese) [CrossRef]

57. Yao, H.; Li, Q.; Zhao, L.; Wu, X.; Shen, X.; Duan, C.; Li, C. Evolution characteristics of compound drought and heat events during the warm season in the Huaihe River Basin and their relationship with climate and vegetation. *Acta Ecol. Sin.* **2024**, *13*, 1–13. (In Chinese) [CrossRef]
58. Gao, J.; Liu, W. Suggestions for high-quality development of express delivery industry in the era of "billion level" in China. *China Econ. Trade Her.* **2022**, *39*, 47–49.
59. Zheng, D.; Huang, X. Population Exposure Risk of Heat wave and Its Influencing Factors at County Level in China. *Areal Res. Dev.* **2022**, *41*, 143–149. (In Chinese)
60. Chen, J.; Tao, H.; Zhai, J.; Su, B.; Jiang, T. Risk assessment of extreme high temperature events over the China-Pakistan Economic Corridor. *J. Nat. Disasters* **2022**, *31*, 65–74. (In Chinese) [CrossRef]
61. Kamal, N.I.A.; Ash'Aari, Z.H.; Abdullah, A.M.; Kusin, F.M.; Yusuff, F.M.; Sharaai, A.H.; Muharam, F.M.; Ariffin, N.A.M. Extreme heat vulnerability assessment in tropical region: A case study in Malaysia. *Clim. Dev.* **2022**, *14*, 472–486. [CrossRef]
62. Xu, H.; Zhang, F.; Li, W.; Shi, J.; Johnson, B.A.; Tan, M.L. Spatial-temporal pattern of change in production-living-ecological space of Nanchong City from 2000 to 2020 and underlying factors. *Environ. Monit. Assess* **2024**, *196*, 94. [CrossRef]
63. Sylla, M.B.; Faye, A.; Giorgi, F.; Diedhiou, A.; Kunstmann, H. Projected Heat Stress Under 1.5 °C and 2 °C Global Warming Scenarios Creates Unprecedented Discomfort for Humans in West Africa. *Earth's Future* **2018**, *6*, 1029–1044. [CrossRef]
64. Yin, P.; Chen, R.; Wang, L.; Liu, C.; Niu, Y.; Wang, W.; Jiang, Y.; Liu, Y.; Liu, J.; Qi, J.; et al. The Added Effects of Heatwaves on Cause-Specific Mortality: A Nationwide Analysis in 272 Chinese Cities. *Environ. Int.* **2018**, *121*, 898–905. [CrossRef]
65. Fu, W.; Guo, D.; Wang, Y.; Yi, G. Epidemiological Characteristics and Meteorological Effects of Heatstroke in Wuhan City from 2009 to 2017. *Occup. Health* **2020**, *36*, 1087–1090. (In Chinese)

Disclaimer/Publisher's Note: The statements, opinions and data contained in all publications are solely those of the individual author(s) and contributor(s) and not of MDPI and/or the editor(s). MDPI and/or the editor(s) disclaim responsibility for any injury to people or property resulting from any ideas, methods, instructions or products referred to in the content.

sustainability

MDPI

Review

Meta-Analysis of Life Cycle Assessment Studies for Polyethylene Terephthalate Water Bottle System

Yoo-Jin Go [1,2,†], Dong-Ho Kang [2,†], Hyun-Jin Park [1], Jun-Hyuk Lee [2] and Jin-Kie Shim [2,*]

1 Department of Biotechnology, College of Life Sciences and Biotechnology, Korea University, 145 Anam-ro, Seongbuk-gu, Seoul 02841, Republic of Korea; yoojinp0210@korea.ac.kr (Y.-J.G.); hjpark@korea.ac.kr (H.-J.P.)

2 Korea Packaging Center, Korea Institute of Industrial Technology, Bucheon 14449, Republic of Korea; kangppp@kitech.re.kr (D.-H.K.); junhyuklee@kitech.re.kr (J.-H.L.)

* Correspondence: jkshim@kitech.re.kr; Tel.: +82-32-624-477

† These authors contributed equally to this work.

Abstract: The life cycle assessment (LCA) serves as a crucial tool for assessing the environmental impact of products, with recent emphasis on polyethylene terephthalate (PET) bottles. Our meta-analytical review of 14 LCA research papers (2010–2022) on PET bottles, aligned with PRISMA guidelines, spans six phases: raw material production (MP), bottle production (BP), distribution and transportation (DT), collection and transport (CT), waste management (WM), and environmental benefits (EB). Utilizing the global warming potential (GWP) as the indicator, our study harmonized data into a consistent functional unit, revealing an average emission of 5.1 kg CO_2 equivalent per 1 kg of PET bottles. Major contributors to global warming were identified across the MP, BP, and DT phases. While the MP and BP phases exhibited low variability due to uniform processes, the CT, WM, and EB phases displayed higher variability due to scenario considerations. A comparison with Korean environmental product declaration data affirmed the methodology's practical utility. Our approach offers potential applicability in diverse product category assessments, emphasizing its relevance for informed decision-making in sustainable product development.

Keywords: life cycle assessment; global warming potential; meta-analysis; PET bottle; packaging

1. Introduction

The expansion of e-commerce channels, rising purchasing power, and population growth have increased packaging waste-associated environmental burdens [1]. Consequently, there is an increase in demand for eco-friendly packaging solutions. In this context, the life cycle assessment (LCA) methodology has gained prominence in evaluating the environmental performance of packaging solutions [2]. LCA is a systematic process to assess the environmental factors associated with products and services throughout their life cycles [3]. It is commonly referred to as the "cradle-to-grave" approach, encompassing the entire life cycle of a product, from the extraction of raw materials to its disposal. Standards such as ISO 14040:2006 [4] and 14044:2006 [5] provide a uniform framework for conducting the LCA analysis while keeping it a flexible tool for research [6].

The LCA research field has grown exponentially, leading to various attempts to integrate and summarize the studies [1]. Previous LCA reviews have been conducted in a narrative style, focusing on specific topics or product categories. Furberg et al. (2021) [7] conducted a narrative review of reusable bottle LCAs and explored the impact of the packaging size, composition, transportation distance, and end-of-life scenarios on the reuse system. Gomes et al. (2019) [8] presented an overview of the LCA studies of polyethylene terephthalate (PET) packaging, comparing different end-of-life scenarios and evaluating alternative materials, aiming to identify the trends and challenges within the two life phases. Bishop et al. (2021) [9] conducted a review of LCA studies that compared the environmental efficiency of bioplastics with conventional petrochemical plastics to explore the strengths and limitations of the environmental performance of bioplastics. Despite the existence of numerous review papers on LCA studies pertaining to PET bottles, there has been no direct comparison utilizing quantitative values.

In the LCA procedure, Ortiz et al. (2009) [10] analyzed the significance of methodological choices, such as impact category selection, inventory completeness, boundary definition, representation of biogenic carbon, and choice of end-of-life scenarios. Therefore, there is an issue of incomparability in the direct comparison of LCA results across different studies due to the methodological choices and assumptions made by LCA practitioners, as well as the differences in datasets [11]. Notably, when different authors evaluate the same product, significant variability in the results can arise owing to the use of different functional units, assessment methods, and scenarios [11,12]. Guidelines to minimize such variability have been created, including the Product Category Rule for Environmental Product Declarations (EPDs) and the Product Environmental Footprint by the European Commission. These initiatives aim to standardize the evaluation methodologies and datasets for specific product categories [13].

Meta-analysis is a review method used to obtain impartial and quantitative results on a specific topic [14]. It is a statistical technique that combines the results of multiple independent studies on the same research question or topic [15]. By synthesizing data from multiple studies, meta-analysis provides a comprehensive understanding of the research question, allowing for the identification of patterns, trends, and discrepancies in the findings [16]. In the field of medicine, meta-analysis is an essential tool to keep the research knowledge constantly updated and to quantitatively assess the heterogeneity of the results [17]. For instance, multiple clinical trials have been conducted to evaluate the efficacy and safety of inhibitors, and using meta-analysis of the randomized clinical trial data increases the precision of estimates [18].

Thus, to enhance the reliability of environmental impact assessments given the variability in LCA study results, it is essential to develop a methodology that can derive comprehensive conclusions through meta-analysis [19,20]. To provide a detailed explanation of the methodology, the proposed systematic review approach was applied to a comparative LCA of PET-bottled water. Among the quantifiable impact categories expressing environmental impact, we seek to establish an environmental impact assessment procedure for PET water bottles, with a focus on indicators such as the global warming potential (GWP) and greenhouse gas (GHG), which are representative of carbon footprints [21].

PET water bottles were chosen for the case study because of their role in the beverage industry, recording the fastest growth rate in the global plastic market [22]. The increasing consumption of bottled water raises concerns over its environmental footprint owing to the increased amount of PET bottle waste. This issue has been addressed through the use of recycled materials, source reduction, and bio-PET [23]. Consequently, plastic-bottled water has become a subject of many LCA studies, making it a suitable subject for meta-analysis [24].

In this study, we aimed to conduct a meta-analysis of the LCA studies to statistically compare the environmental impacts over its full life cycle. A systematic assessment of multiple studies can offer valuable insights into its environmental impacts. This, in turn, can contribute to the development of a methodology for LCA studies and enable evidence-based decision-making and policy evaluation.

By adhering to the ISO framework, the LCA ensures that all studies clearly define the FU and scope while following a standardized four-step procedure [25]. This standardized assessment tool facilitates a seamless harmonization of diverse research findings through meta-analysis [26]. Meta-analysis facilitates the calculation of the mean estimates and identification of GWP variation factors across the LCA results of different PET bottles. Moreover, through a comparative analysis of the average GWP value derived from meta-analysis and the actual GWP value of commercially available products, one can evaluate the extent to which it faithfully represents typical carbon footprint results. The findings of this study provide detailed guidelines for the systematic and practical evaluation of the environmental impacts of the PET bottle life cycle through LCA.

2. Materials and Methods

This study encompasses a sequential five-step methodology, delineated in Figure 1, which guides the systematic execution of the research process to derive an integrated conclusion. The methodology can be adapted to both products and services by employing an iterative approach that allows for continuous data updates or modifications during the study. The prescribed iterative procedure ensures a systematic and refined evaluation, thereby enhancing the accuracy and comprehensiveness of the assessment.

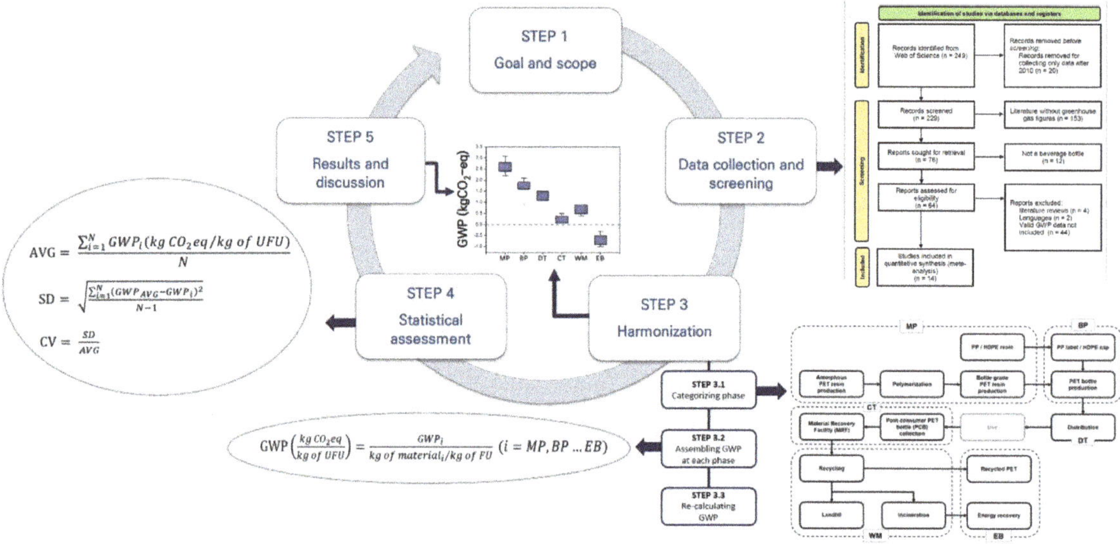

Figure 1. Five-step overview of the research method.

2.1. Goal and Scope

To systematically integrate the results of LCA studies, it is essential to establish a target for meta-analysis. This begins with a clear definition of the study's goal and scope. In this study, the identified goal is to analyze the environmental performance across the entire life cycle of the PET-bottled water system. The defined scope, in alignment with this objective, adopts a cradle-to-cradle perspective, covering from raw material collection to the post-disposal process. The focus of the analysis parameter is specifically on evaluating the GWP.

2.2. Data Collection and Screening Process

LCA studies were compiled from electronic bibliographic databases of scientific literature and web search engines. The references included papers and project reports published in international journals and EPDs. The search method utilizes a specific query to effectively target the subject matter. For this review, LCA research literature was collected from the Web of Science using the following query:

Queries: LCA OR life cycle assessment (All Fields) and PET OR polyethylene terephthalate (All Fields) and bottle OR bever* OR pack* OR water system (All Fields) and GWP OR global warming OR GHG OR CO2 (All Fields) 2010–2023.

To perform the meta-analysis, the selected studies were screened using the PRISMA guidelines. PRISMA is a scientifically grounded essential item for reporting systematic reviews and meta-analyses. It was developed in an attempt to increase the clarity, transparency, and quality of reports [27]. The screening process involved searching for documents in the Web of Science database that specifically addressed LCA and PET bottles. Studies published before 2010 were excluded to ensure that the data is up-to-date. Additionally, only studies focusing on carbon emissions, including terms such as GWP, GHG, and CO_2, were considered for the analysis, whereas studies unrelated to PET bottles or packaging were omitted.

After the initial screening, 64 LCA research articles were selected for further review. The examination involved identifying documents that lack life-cycle phase-specific GWP data, as well as those not adhering to the desired system boundaries or those that primarily consisted of literature reviews that were unsuitable for quantitative analysis. Figure 2 shows a visual representation of the screening process. The final set of 14 selected studies served as the subjects for this meta-analysis. A comprehensive summary of the 14 studies is provided in Table S1 within the Supplementary Materials.

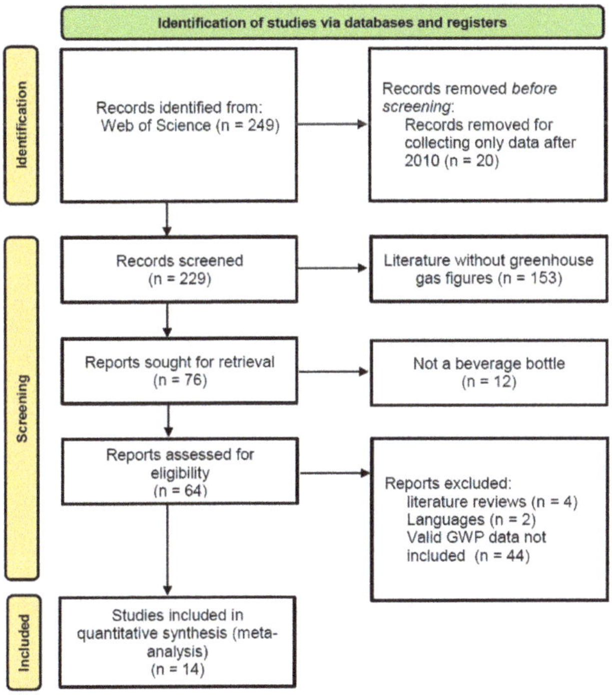

Figure 2. A flow chart of the PRISMA-based screening process for study purposes.

By adhering to the PRISMA guidelines, a systematic and rigorous screening process was ensured, laying the groundwork for a robust and reliable meta-analysis.

Following the literature selection for the review, the differentiation of geographical boundaries, system boundaries, data sources, and FUs for each document is essential for identifying significant parameters. Chapter 3 presents the results of this differentiation process for each element.

2.3. Harmonization of Data

2.3.1. Categorizing Phases for Research Result Collection

The process of synthesizing the literature results is essential to facilitate a quantitative literature review via meta-analysis. To consistently harmonize the GWP results from each study within the determined system boundaries, the life cycle of the PET bottle was subdivided into six phases: Material production (MP), bottle production (BP), distribution transportation (DT), post-consumer PET bottle collection transportation (CT), waste management (WM), and environmental benefits (EB). A flowchart is shown in Figure 3.

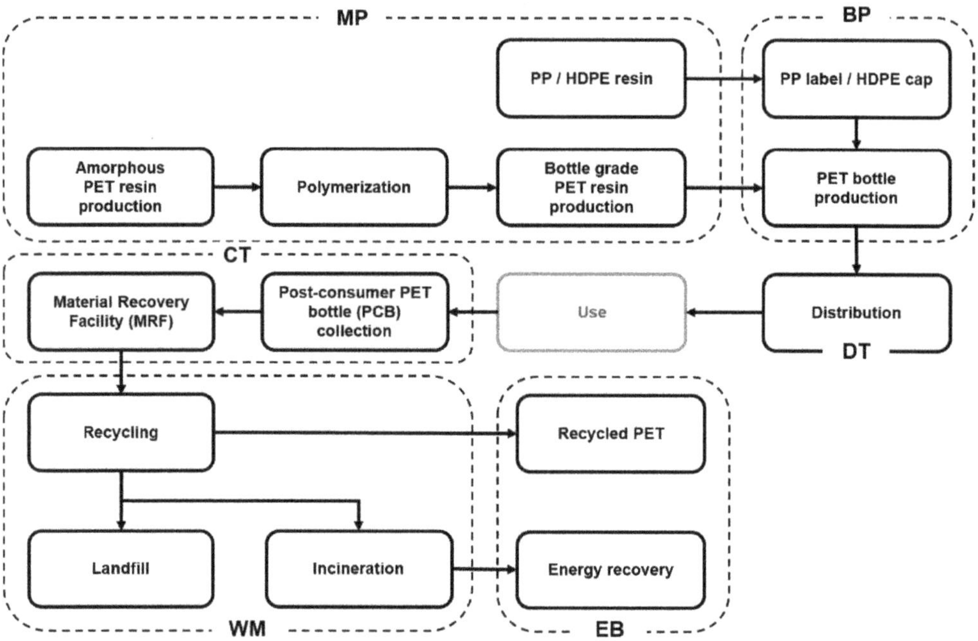

Figure 3. The life phases of a PET bottle within the study system boundaries were used for the harmonization process.

The MP phase involves the production of bottle-grade PET resin from the extraction of crude oil and natural gas through polymerization. It also includes the production of polypropylene (PP), which is used as a label material, and high-density polyethylene (HDPE), which is used as a cap material. The BP phase involves the creation of preforms by injecting PET resin, forming PET bottles through stretch-blow molding, producing labels from the extruded PP film, and making caps through the injection molding of HDPE resin. DT refers to consumer distribution, whereas CT encompasses the transportation process that collects post-consumer PET bottles and moves them to a post-collection treatment facility, such as a material recovery facility. WM consists of measures taken to minimize the environmental impacts through the recycling, incineration, and landfill of collected PET bottles. Recycling involves mechanical and chemical processes. Finally, EB refers to the benefits achieved by utilizing recycled raw materials obtained from recycling during the WM phase, leading to a reduction in the production of new plastic raw materials. Additionally, it includes the heat recovery resulting from incineration. Recycled raw materials can be categorized into open loops, which are recycled into fibers or strings, and closed loops, which are recycled back into PET bottles.

2.3.2. Assembling GWP at Each Phase

At this stage, the 'global warming potential per functional unit (GWP/FU)' values for each phase were extracted from the selected literature. The functional unit of LCA defines the specific quantity and function of a product being analyzed, serving as the reference point for assessing environmental impacts. These values were categorized into six predefined phases. It is important to note that not all studies provided a comprehensive coverage of each phase. Some studies focused solely on specific phases, such as waste disposal, or excluded certain phases, such as transport from system boundaries. In instances where some phases were missing from the literature, only the results from the existing phases were utilized, and the GWP values for the omitted phases were not considered. This approach allows the inclusion of partial results to extend the parameters used in the analysis.

The presentation of the GWP results varied across studies. Some studies had explicitly stated the GWP values for each phase based on a well-defined classification. Conversely, other studies presented the GWP values of the entire product system and displayed the contribution of each phase as a percentage or in a graphical format. When the contribution of each phase was not explicitly provided, the GWP value of a specific phase was derived by multiplying the total GWP value by the contribution of the corresponding phase. Additionally, when data were illustrated as a stacked bar graph, software programs, such as Illustrator, were used to calculate the length of each bar, ensuring the proportional representation of each phase's contribution.

2.3.3. Re-Calculating GWP per Unified Functional Unit

The analyzed literature shared the same product, yet directly comparing the GWP values proved challenging because of the use of different FUs. To address this issue, each functional unit in the references was labeled as "FU", and a unified functional unit that was used in the meta-analysis was established as "UFU".

The literature on LCA defines FUs based on their weight or volume. For instance, with PET beverage bottles, even for the same 0.5 L product, the actual amount of PET raw material can vary. Therefore, the UFU was defined as "1 kg of PET bottles".

To standardize the GWP values, an equation was determined for each document, enabling the conversion from "GWP per FU" to "GWP per UFU" (i.e., UFU = 1 kg of PET bottles). Using this Equation (1), the GWP per UFU value was calculated. The resulting values represented the GWP per UFU and served as quantitative parameters for the meta-analysis.

$$\mathrm{GWP}\left(\frac{\mathrm{kg\ CO_2eq}}{\mathrm{kg\ of\ UFU}}\right) = \frac{\mathrm{GWP_i}}{\mathrm{kg\ of\ material_i / kg\ of\ FU}}\ (i = \mathrm{MP, BP \ldots EB}) \quad (1)$$

2.4. *Statistical Assessment*

The harmonized parameters were subjected to statistical analyses. First, the average (AVG), standard deviation (SD), and coefficient variation (CV) of the GWP were analyzed for each phase. AVG, SD, and CV adhere to Equations (2)–(4). By comparing the average GWP estimates for each phase, the hotspot phase with the highest environmental impact could be identified. Additionally, the comparison and analysis of CV revealed the factors contributing to the variation in GWP values and briefly described the resulting tendencies across the environmental impacts. Building upon the insights gained from the factors influencing environmental impact, we propose technologies and strategies designed to effectively reduce the carbon footprint in each phase.

$$\mathrm{AVG} = \frac{\sum_{i=1}^{N} \mathrm{GWP_i(kg\ CO_2eq/\ kg\ of\ UFU)}}{\mathrm{N}}\ (\mathrm{N = The\ number\ of\ data\ for\ each\ phase}) \quad (2)$$

$$\mathrm{SD} = \sqrt{\frac{\sum_{i=1}^{N} (\mathrm{GWP_{AVG}} - \mathrm{GWP_i})^2}{\mathrm{N} - 1}}\ (\mathrm{N = The\ number\ of\ data\ for\ each\ phase}) \quad (3)$$

$$\mathrm{CV} = \frac{\mathrm{SD}}{\mathrm{AVG}} \quad (4)$$

Second, to validate whether the GWP values extracted from the literature represented integrated and representative results, they were compared with the GWP values of commercially available products. The EPD system quantifies and discloses the environmental impact of products throughout their life cycle, including raw material collection, production, transportation, use, and disposal. Products with EPD certification in Korea disclose their evaluation scopes, methodologies, and results. By comparing the GWP of EPD-certified products, the validity and usefulness of the GWP estimation methodology for the PET bottle's life cycle were verified through a meta-analysis. In addition, this comparison provides an opportunity to evaluate the quality and authenticity of Korea's environmental evaluation methodology.

In order to compare the statistical difference of product volumes (0.33 L, 0.5 L, 1 L, 2 L), the Games–Howell post hoc test was conducted with a confidence level ($\alpha = 0.05$) using the IBM Statistical Package for Social Sciences (SPSS) version 29. This test is often used when the assumption of homogeneity of variances of a group is violated [28].

2.5. Assumptions and Limitations

(1) In statistical analysis, a higher number of parameters allows for more accurate conclusions. To increase the number of parameters, only partial values from the PET bottle's life cycle were included in this study, as the entire life cycle was not considered.
(2) To increase the number of parameters, the literature review also included studies on subjects other than bottled water, such as PET flakes. However, cases related to carbonated beverages that exhibited significant differences in PET raw material consumption were excluded from data collection.
(3) While some studies explicitly specified disposal rates, there were cases in which information about how disposal scenarios were set was not adequately provided. In such instances, the environmental impact of recycling, incineration, and landfill processes was collectively termed "waste management" and calculated by aggregating the effects of these processes.

3. Results and Discussion

3.1. Analysis of References

In this study, a total of 14 LCA studies were meticulously selected through a PRISMA screening process. These studies were scrutinized for various parameters, including the study year, geographic location, system boundary, FU, and content type, to comprehensively analyze their implications for PET bottle systems. It is important to note that the year of the study is significant, as the environmental impact results can vary depending on the technology prevalent at the time the study was conducted. Similarly, the geographic location exerts a distinct influence on the environmental load owing to disparities in technological advancements across different countries. Factors such as production technology, energy efficiency, transportation distance, and disposal methods exhibit substantial variations based on geographical considerations.

In terms of system boundaries, our approach aims to consider the entire life cycle comprehensively, encompassing a cradle-to-grave perspective, to collect data to the fullest extent possible. Furthermore, we sought to compile a diverse database that included studies that conducted detailed assessments of specific life-cycle phases. It is worth noting that within the domain of LCA, the definition of the FU not only significantly influences the quantity of raw materials used but also impacts the results of the assessment, which underscores their importance as key parameters to be identified. These parameters are summarized in Table 1 to facilitate meaningful cross-comparisons among the literature sources.

Data were distributed evenly across the years from 2011 to 2022, with a notable concentration of literature in 2021, comprising five articles. Geographically, Europe emerged as the predominant source, contributing six articles. In most cases, the system boundaries were delineated as cradle-to-grave, reflecting a comprehensive life cycle perspective. The definition of the FU is primarily based on volume or weight metrics, often centered on the capacity and quantity of PET beverage bottles. In addition, the literature presents the weights of the PET bottles. Certain instances involved specifying the weight of both the cap and label of a beverage bottle, whereas others concentrated solely on quantifying the weight of the bottle's body.

Table 1. The LCA research literature was selected for the systematic review.

No.	Reference	Year	Geographical Boundary	System Boundaries	Functional Unit (FU)	PET Bottle Weight
1	Gileno and Turci [29]	2021	Brazil	cradle-to-gate	1 t of post-consumer recycled PET resin or 1 t of rPolyester fiber	NA
2	Bataineh [30]	2020	Jordan	cradle-to-grave	1 t of PET flake	NA
3	Kouloumpis, et al. [31]	2020	England	cradle-to-grave	2468 t of PET bottles	NA
4	Tamburini, et al. [32]	2021	Italy	cradle-to-grave	1 year of use: 1095 bottles of PET	19.1 g
5	Grisales, et al. [33]	2022	Italy	cradle-to-grave	The distribution of 100 L mineral water	15.22 g
6	Olatayo, et al. [34]	2021	South Africa	cradle-to-grave	10 single-use 0.5 L PET bottles	20.3 g
7	Lonca, et al. [35]	2020	USA	cradle-to-grave	2.78 t PET bottles (in a year in the US)	NA
8	Gursel, et al. [36]	2021	Europe	cradle-to-grave	Packaging water in one hundred 0.5 L bottles (1 kg PET bottles)	10 g
9	Benavides, et al. [37]	2018	USA	cradle-to-grave	One 26 g 500 mL PET bottle	26 g
10	Martin, et al. [38]	2021	Brazil	cradle-to-grave	1 t of PET waste	NA
11	Papong, et al. [39]	2014	Thailand	cradle-to-gate	1000 units of 250 mL drinking water bottles	NA
12	Kuczenski and Geyer [40]	2013	California	cradle-to-gate	1 L beverages in single-use PET bottles consumed by Californians during the years 2007–2009	40.8 g
13	Gironi and Piemonte [41]	2011	Italy	cradle-to-grave	1000 units of 500 mL PET bottles to be used for drinking water	12.2 g
14	Kim, et al. [42]	2022	Korea	cradle-to-grave	A PET water bottle, 500 mL	17 g

3.2. Harmonization

3.2.1. Categorizing and Extracting GWP

The GWP values were systematically obtained for each of the six phases of the study. Given the diverse unit expressions employed for the GWP values, encompassing both kilograms of CO_2 equivalent (kg CO_2eq) and grams of CO_2 equivalent (g CO_2eq), meticulous efforts were undertaken to standardize all data by converting them into a unified unit of kilograms of CO_2 equivalent (kg CO_2eq).

This comprehensive data collection effort yielded 48 GWP extraction datasets distributed across various phases, including 10 in MP, 10 in BP, 2 in DT, 7 in CT, 12 in WM, and 7 in EB. These datasets were consistently organized and presented as "kg CO_2eq per FU" and are meticulously summarized in Table 2.

Among the distinct phases, the WM phase emerged as the most extensively covered, boasting a higher volume of data, whereas the distribution and transportation phase exhibited a comparatively limited dataset. This can be attributed to the predominant focus of LCA studies on scenarios associated with PET bottle disposal. Furthermore, some studies have incorporated the EB stemming from the waste treatment process, thereby complicating the independent selection of GWP values for the EB phase and potentially yielding lower GWP values for the WM phase compared to other studies.

Table 2. Global warming potential (GWP) values of the chosen studies.

No.	MP	BP	DT	CT	WM	EB
1				134	856	
2	2746			155.560	660.610	−1800
3		8.920×10^6		1.160×10^5	2.300×10^6	-3.500×10^5
4	98.018	25.766			1.675	−7.300
5	6.809	3.450	3.950		1.155	−1.127
6					0.012	−0.011
7	7822	6532		912	1960	
8	2.237	0.920			0.047	
9	0.073	0.046				−0.026
10				26.633	73.445	−56.715
11	48.240	19.640				
12	98.300	80.500		15.700	32.100	
13	33.306	4.636		0.122	0.910	
14	3.910×10^{-2}	1.860×10^{-2}	3.029×10^{-3}		1.860×10^{-2}	

3.2.2. Normalized GWP by Unified FU

To enable a meaningful comparison of the GWP values extracted with various FUs, a harmonization process was implemented. This involved aligning them with the UFU, representing 1 kg of PET bottles. The resultant harmonized GWP values, expressed in kg CO_2eq/1 kg of PET bottles, are presented in Table 3.

Table 3. Data items for GWP (kg CO_2eq) unification by phase in UFU (1 kg of PET bottle).

No.	MP	BP	DT	CT	WM	EB
1				0.134	0.856	
2	2.746			0.156	0.661	−1.800
3		3.614		0.047	0.932	−0.142
4	4.687	1.232			0.080	−0.349
5	2.237	1.133	1.298		0.379	−0.370
6					0.059	−0.056
7	2.813	2.349		0.328	0.705	
8	2.237	0.920			0.047	
9	2.805	1.772				−0.991
10				0.027	0.073	−0.057
11	2.950	1.201				
12	2.409	1.973		0.079	0.787	
13	2.730	0.380		0.010	0.075	
14	2.300	1.094	1.094		0.178	

Previously, the diversity of the GWP values was hindered by variations in the FUs. However, through the systematic organization of these values based on UFU, quantitative comparisons are now attainable. The ensuing step of our analysis involves a comprehensive examination and statistical assessment of the harmonized GWP results.

3.3. *Statistical Assessment*

3.3.1. Assessment of Harmonized GWP

In the domain of statistical analysis, box plots serve as valuable tools for visually comparing multiple datasets and categories. They offer a succinct overview of the distribution of the environmental impacts among the assessed groups. The plot boxes indicate the interquartile range, with the median represented by the inner line. The lower and upper whiskers extend to values within a predefined range, excluding the outliers. Interpreting the box plot facilitated the identification of central tendencies and the spread of the GWP (kg CO_2eq/1 kg of PET bottle) values for each phase, providing insights into comparative environmental performance. Figure 4 presents a boxplot that excludes outliers to enable a comprehensive analysis.

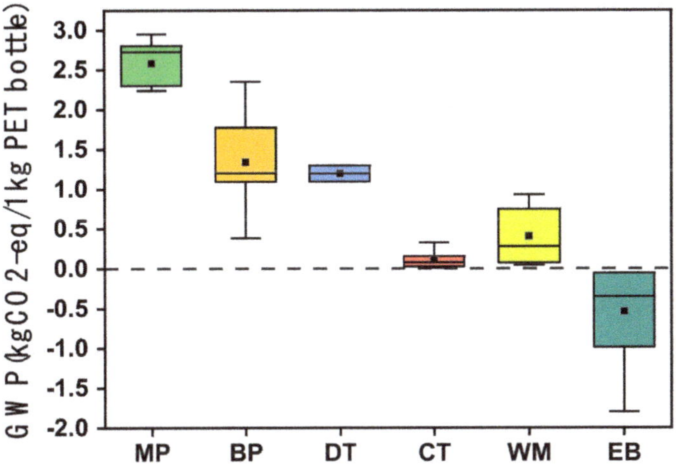

Figure 4. Box plot analysis of GWP values for six phases of the PET bottle.

Two outliers were identified: one in the MP and one in BP. The outliers in the MP phase, as demonstrated by Tamburini et al. (2021) [32], result from meticulous calculations performed by individual companies. These calculations encompass raw material production yields along with the electricity and heat consumption for each process, starting from the synthesis of monoethylene glycol and purified terephthalic acid. This level of scrutiny reveals the distinct regional characteristics that increase the likelihood of energy overlap. Therefore, it is reasonable to designate this data point as an outlier.

The outlier of the BP phase is attributed to the inherent challenge of isolating the PET resin production from the bottle stretch-blow molding process, as detailed in the study by Kouloumpis et al. (2020) [31]. Due to the inseparability of these two processes, they were combined into a single value, resulting in the removal of outliers before subsequent analyses. Additionally, the statistical analysis encompassed key metrics, including the AVG, SD, coefficient of variation (CV), and number of parameters (N) for all life phases of a PET bottle. The analytical findings are summarized in Table 4 for a comprehensive evaluation.

Table 4. The summary of the statistical analyses of the GWP values of each phase.

Phase	MP	BP	DT	CT	WM	EB
AVG	2.581	1.340	1.196	0.111	0.403	−0.538
SD	0.282	0.596	0.144	0.110	0.358	0.643
CV	0.109	0.445	0.120	0.983	0.888	1.196
N	9	9	2	7	12	7

The aggregation of the average values from each phase serves as an indicator of the total greenhouse gas emissions across the life cycle of the PET bottle. The calculated life-cycle carbon emissions, based on data derived from LCA research literature, amounted to 5.093 kg CO_2eq. It is evident that the MP contributes most significantly to the overall global warming impact, followed by the BP, DT, WM, and CT phases.

The CV emerged as a valuable relative dispersion metric that facilitates comparisons among phases with different AVG, SD, and measurement units. This aided in assessing the concentration of data around the mean, with a smaller CV indicating a higher degree of concentration. Notably, the CT, WM, and EB phases exhibit higher CV values. These phases involve a multitude of scenarios within the LCA, signifying a higher degree of variability in the data compared to the MP and BP phases, which are characterized by more consistent processes.

3.3.2. Comprehensive Evaluation of GWP in Each Phase

The environmental impact contributions and factors contributing to the variation were analyzed for each phase. The significant contribution of MP to the overall GWP can be attributed to the energy-intensive nature of plastic raw material production. This process involves high-temperature procedures such as melt polymerization, solid-state polymerization, extrusion, and refining. PET production consists of the polymerization of ethylene glycol and terephthalic acid monomers. Post-polymerization, PET pellets are obtained in a dissolved form, necessitating a solid-state polymerization process to yield crystalline pellets known as bottle-grade granules [23]. The production of PET resin involves multiple high-temperature and high-energy steps, resulting in a substantial environmental burden.

The MP phase can be improved by utilizing recycled PET and bio-PET materials. This choice not only mitigates the environmental impact of raw material production by reducing the need for virgin PET—thereby curbing carbon emissions—but also offers cost-saving opportunities. Moreover, a recent study by Benavides et al. (2018) [37] has indicated that landfilling bio-PET and recycled PET bottles can substantially reduce greenhouse gas emissions from a cradle-to-grave perspective. These reductions can range from 12% to 82% in comparison with fossil-fuel-based virgin PET bottles. This outcome was attributed to the carbon-absorbing capacity of the plants used for the bio-PET raw material supply, creating a carbon dioxide cycle structure.

The BP phase encompasses several distinct processes, including the production of PET bottles, HDPE caps, and PP films. Initially, the PET resin underwent injection molding to create preforms, which were subsequently transformed into PET bottles through stretch-blow molding. Simultaneously, the HDPE resin was subjected to injection molding to craft caps, and the PP film was extruded. Notably, these processes involve the use of high-temperature equipment operating at approximately 280 °C. This results in increased energy consumption, primarily electricity, which contributes to an elevated GWP. In the context of bottled water production, there are instances in which bottles are manufactured within cleanroom facilities. This further amplifies energy consumption and subsequently leads to a heightened GWP [43].

Minimizing energy consumption using lightweight PET bottles contributes to a decrease in carbon emissions. An example involves reducing the bottle weight by 20%

by incorporating approximately 100 ppm of inorganic substances into the PET preform. Consequently, the power consumption during stretch-blow molding was decreased by 18%, resulting in a 21% reduction in carbon emissions. Moreover, enhancing process efficiency, implementing steam recycling, and transitioning from thermal power generation to renewable energy sources can further mitigate the environmental impacts.

Despite having a limited database, the DT phase ranks as the third largest contributor to the GWP among the six phases considered. The environmental impact of transportation is measured in kgkm, allowing for variations in geography, transport scenarios, and types.

To reduce the environmental impact of transportation, optimizing resource logistics can reduce exhaust emissions by enhancing the energy efficiency of transportation modes, such as transitioning from truck transport with high exhaust emissions to mass transport options such as railroads and shipping. Second, it is vital to actively promote logistics pooling to improve the utilization and loading capacity, necessitating the selection of appropriate vehicle types based on the logistics volume and optimization of transportation distances. Furthermore, reducing the bottle weight contributes fundamentally to a reduction in the greenhouse gas emissions associated with transportation.

The CT phase exhibited the lowest average GWP among the considered phases. Out of the seven research studies providing GWP data for the CT phase, six specified the collection and transportation distances. Table 5 presents a comprehensive summary, including the GWP (kg CO_2eq per 1 kg of PET bottles), distance (km), and GWP/km (kg CO_2eq per 1 kgkm) for the CT phase. Notably, Kuczenski and Geyer (2013) [40] conducted a comprehensive study covering the entire collection system, encompassing curbside and drop-off methods, in the eastern US, where collection distances were notably extensive. Similarly, Gileno and Turci (2021) [29] modeled the collection conditions in Brazil as a case study. The remaining literature primarily reported simplified distances from curbside collection to recycling and incineration facilities, with the majority falling within 300 km.

Table 5. Collection route distance (km), GWP (kg CO_2eq/1 kg of PET bottles), and GWP/km (kg CO_2eq/1 kgkm) in the CT phase, per reference.

No.	Reference	Distance (km)	GWP (kgCO_2eq)	GWP/km
(1)	Bataineh (2020) [30]	60	0.016	0.000259
(2)	Gironi and Piemonte (2011) [41]	100	0.010	0.000100
(3)	Martin et al. (2021) [38]	160	0.027	0.000169
(4)	Kouloumpis et al. (2020) [31]	286	0.047	0.000164
(5)	Kuczenski and Geyer (2013) [40]	570	0.079	0.000139
(6)	Gileno and Turci (2021) [29]	750	0.134	0.000179

A trend emerged where the GWP increased with longer transportation distances, as shown in Figure 5. The calculation of R-squared, a robust measure of the relationship between independent and dependent variables, yielded a value of 0.964, signifying the strong explanatory power of distance (the independent variable) in relation to the GWP (the dependent variable). LCA employs the unit of kgkm to evaluate the environmental impact during transport, and when normalized to 1 kg, it is directly proportional to kilometers.

The GWP/km, which represents the comparison of GWP for 1 km, is shown in Figure 6 as a scatter plot. Among the six datasets, cases (1) and (2) were determined to be outside of the 95% confidence interval. For instance, the GWP value in Bataineh (2020) [30], which was set at 60 km, was comparatively high. However, this discrepancy arises because of various factors with regard to transportation, such as fuel type, vehicle category, and distance. Similarly, Gironi and Piemonte (2011) [41] found that GWP values can be influenced by factors such as the EURO emission standards and truck types in the transport datasets of the DB. For instance, EURO1's GWP value is 68% that of EURO6, leading to potential variations in the results within the transportation sector [38]. Consequently, the significant CV observed in the CT phase can be attributed to disparities resulting from factors like EURO standards, truck types, truck sizes, country-specific road conditions, and the diversity of waste recovery systems.

In the CT phase, it is imperative to establish an efficient collection system that minimizes the number of collection routes. Ultimately, consumers must proficiently separate and dispose of their products. This approach facilitates the collection of substantial quantities of PET waste, leading to an increased recycling rate.

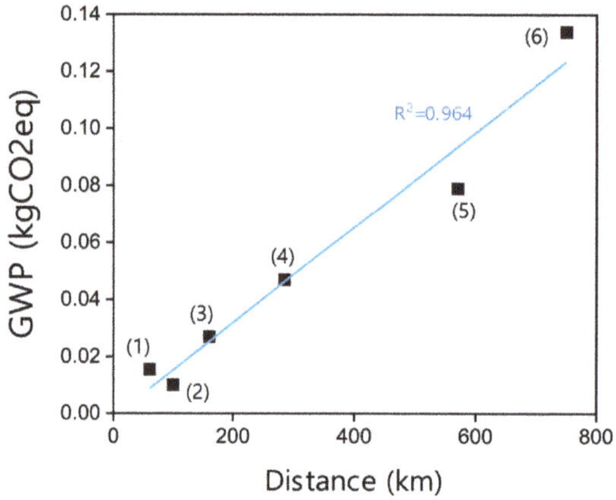

Figure 5. Scatter plot and R-squared values for CT phase GWP converted to "kg CO_2eq per 1 kg of PET bottle". The numbers in parentheses within the figure correspond to the numbers listed in Table 5.

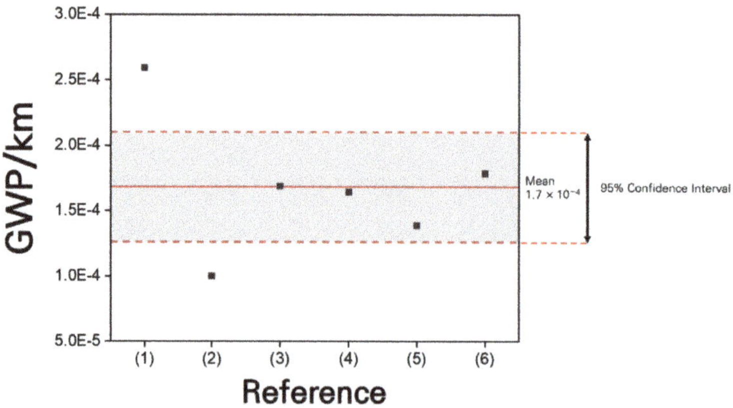

Figure 6. Scatter plot of CT phase GWP values normalized to "kg CO_2eq per 1 kgkm". The numbers in parentheses within the figure correspond to the numbers listed in Table 5.

The WM phase ranked fourth in terms of the GWP and exhibited a high CV. These variations in GWP values were mainly attributed to disparities in the study years and geographical regions. This phase is intricately linked to the evolving landscape of recycling and incineration technologies, wherein improvements in energy efficiency and increased yields in recycled raw material production through recycling play pivotal roles. Moreover, regional differences in GWP values within the WM phase can be attributed to the waste collection systems, disposal technologies, and environmental policies adopted by each country.

Major waste disposal methods include recycling, incineration, and landfill. Recycling is further categorized into physical, chemical, and open-loop or closed-loop processes. Due to the variability in the proportions of recycling, incineration, and landfill in different studies, these factors were treated as variables when calculating the GWP of the WM phase. Table 6 provides a summary of the specific disposal scenarios utilized in the cited references, with instances of unspecified scenarios marked as "N/A".

Table 6. Rate and type of recycling, incineration, and landfill for disposal scenarios by reference in the WM phase.

No.	Reference	Recycling	Recycling Type	Incineration	Landfill
1	Gileno and Turci (2021) [29]	100%	Closed-loop	N/A	N/A
2	Bataineh (2020) [30]	100%	Open-loop	N/A	N/A
3	Kouloumpis et al. (2020) [31]	52.78%	N/A	47.17%	N/A
4	Tamburini et al. (2021) [32]	100%	Closed-loop	N/A	100%
5	Grisales et al. (2022) [33]	55%	N/A	45%	N/A
6	Olatayo et al. (2021) [34]	46.3%	N/A	N/A	53.7%
7	Lonca et al. (2020) [35]	4.6%	Closed-loop	92%	
8	Gursel et al. (2021) [36]	60%	Open-loop	20%	20%
9	Benavides et al. (2018) [37]	35%	Closed-loop	N/A	65%
10	Martin et al. (2021) [38]	3.6%	N/A	N/A	96.4%
12	Kuczenski and Geyer (2013) [40]	N/A	N/A	N/A	N/A
13	Gironi and Piemonte (2011) [41]	N/A	Closed-loop	N/A	N/A
14	Kim et al. (2022) [42]	36.60%	N/A	48.50%	14.90%

To enable a quantitative comparison of the environmental impacts across various disposal scenarios, we arranged the GWP of the WM phases in ascending order and visually presented them as a heat map, as shown in Table 7. Three discernible trends emerge from the analysis. First, scenarios characterized by elevated landfill rates tend to exhibit lower carbon emissions. When petrochemical-based plastics are disposed of in landfills, they remain non-decomposable for at least 100 years, resulting in minimal greenhouse gas emissions. The fact that the default time frame for calculating the GWP is based on 100 years is the reason why the GWP is calculated low due to the landfill.

Table 7. Ratios of recycling, incineration, and landfill for each reference disposal scenario in the WM phase, organized in ascending order of resulting GWP (kg CO_2eq per 1 kg of PET bottles).

No.	Reference	Recycling	Incineration	Landfill	WM GWP (kg CO_2eq)
8	Gursel et al. (2021) [36]	60%	20%	20%	0.047
6	Olatayo et al. (2021) [34]	46%		54%	0.059
10	Martin et al. (2021) [38]	4%		96%	0.073
13	Gironi and Piemonte (2011) [41]	N/A	N/A	N/A	0.075
4	Tamburini et al. (2021) [32]			100%	0.080
14	Kim et al. (2022) [42]	37%	49%	15%	0.178
5	Grisales et al. (2022) [33]	55%	45%		0.397
2	Bataineh (2020) [30]	100%			0.661
7	Lonca et al. (2020) [35]	4.60%	92%		0.705
12	Kuczenski and Geyer (2013) [40]	N/A	N/A	N/A	0.787
1	Gileno and Turci (2021) [29]	100%			0.856
3	Kouloumpis et al. (2020) [31]	53%	47%		0.932

Second, when comparing the recycling strategies in the WM phase, closed-loop recycling, which involves the conversion of PET to bottle-grade PET, exhibited a higher GWP value than open-loop recycling, which resulted in the production of lower-quality products. Consequently, from the perspective of the WM phase, closed-loop recycling appears to have a more substantial environmental impact.

Finally, scenarios characterized by high incineration rates tend to yield larger GWP values owing to the significant release of carbon dioxide during the combustion of plastic materials. Within the WM-phase GWP dataset, the highest-impact scenario was observed in a study conducted by Kouloumpis et al. (2020) [31]. This scenario combines recycling and incineration, resulting in increased energy consumption and a heightened environmental impact contribution compared with other studies.

Some studies have combined the WM and EB phases to draw conclusions. Consequently, it is crucial to recognize that when solely considering the WM phase, the GWP values may appear relatively low compared to studies that separate these phases.

Increasing the recycling rate of waste plastics is highly effective for reducing greenhouse gas emissions during disposal. The primary goal is to promote mechanical recycling and concurrently enhance recycling technologies to complement chemical recycling methods. Colorless PET is a clean source that is recognized as suitable for contact with food; hence, it must be meticulously segregated during the recovery and sorting processes. PET obtained through mechanical recycling can then be employed in products intended for food contact.

Chemical recycling is helpful in areas where clean separation and screening are challenging. The focus should be on boosting the recycling rate rather than debating the environmental merits of mechanical versus chemical recycling. Developing technologies for application in both regional and environmental policies is a more pragmatic approach [44].

The EB phase encompasses recycled raw materials and heat derived from recycling and incineration during the WM process. Notably, it exhibited the largest CV among the six phases, signifying a broader range of data compared to other categories. This variance likely arises from the diverse disposal scenarios considered in each study, as all LCA investigations modeled hypothetical scenarios when assessing the EB. The GWP values and associated disposal scenarios for the EB phase are listed in Table 8, have been systematically arranged in ascending order and visually represented as a heat map for comparison.

Table 8. Ratios of recycling, incineration, and landfill for each reference disposal scenario in the EB phase, organized in ascending order of resulting avoided GWP (kg CO_2eq per 1 kg of PET bottles).

No.	Reference	Recycling	Incineration	Landfill	EB GWP (kg CO_2eq)
2	Bataineh (2020) [30]	100%			−1.800
9	Benavides et al. (2018) [37]	35%		65%	−0.991
5	Grisales et al. (2022) [33]	55%	45%		−0.370
4	Tamburini et al. (2021) [32]	100%			−0.349
3	Kouloumpis et al. (2020) [31]	53%	47%		−0.142
10	Martin et al. (2021) [38]	4%		96%	−0.057
6	Olatayo et al. (2021) [34]	46%		54%	−0.056

A limitation of this study pertains to the level of detail provided in the collected references regarding the disposal scenarios. Specifically, some studies did not differentiate between the open-loop or closed-loop methods and mechanical or chemical recycling methods, hindering a precise quantitative comparison between these disposal classifications. Nevertheless, when assessing the EBs of various disposal methods and their combinations, it became evident that scenarios with higher recycling and incineration rates outperformed landfills in terms of the EB.

Closed-loop recycling is expected to yield the most significant EB, primarily because it enables the acquisition of high-quality bottle-grade PET raw materials via chemical recycling. This approach mitigates the environmental burden associated with the production of virgin PET materials. This approach has the potential to mitigate substantial environmental loads compared to the use of low-quality PET products.

However, in the case of Tamburini et al. (2021) [32], despite adopting a 100% closed-loop recycling scenario, it demonstrated fewer benefits than the open-loop 100% and closed-loop 35% disposal scenarios. This discrepancy arises because the study relies on data derived from the external reports available in the literature. It is important to note that this information represents a simplified dataset rather than values derived from comprehensive LCA procedures. Consequently, the introduction of such data, which are not based on research modeling data, introduces uncertainty into the environmental assessment, as it may not fully capture intricate environmental nuances.

Bataineh (2020) [30] reported that the EBs were the greatest despite open-loop recycling through mechanical recycling. This study was conducted by comparing the GWP value calculated by entering the input and output of the PET recycling process in detail with the GWP of PET raw material production established in the database. As a result, it was concluded that PET flake recycling provides a net benefit for greenhouse gas emissions of 1.8 kg of CO_2 equivalent per 1 kg of recycled PET flakes.

Chemical recycling is associated with greenhouse gas emissions and is on par with the production of virgin PET, resulting in a GWP that is four times higher than that of mechanical recycling [45]. However, it offers the advantages of reducing virgin PET usage and solid waste generation by 56% and 64%, respectively. Therefore, when assessing the EBs of closed-loop recycling via chemical recycling alone in terms of the GWP, it may be less efficient than open-loop recycling through mechanical means. It is essential to consider the broader context of circular economies in such evaluations.

In a study conducted by Olatayo et al. (2021) [34], which demonstrated the lowest EB, an LCA was carried out in South Africa, where approximately 90% of plastic bottle waste enters landfills. This study aims to quantify the potential environmental advantages achievable if the waste slated for landfill disposal is redirected for recycling, assuming a recycling rate of 46.3%. The conclusion of the study suggests that even with a 36.3% increase in recycling rates, the resulting reduction in environmental impact remains relatively modest. However, the need to increase waste collection and recycling rates has been emphasized to reduce plastic generation.

While various methodologies are available for enhancing the environmental benefits, the industry for products reliant on fossil fuels faces challenges in implementing strategies for carbon emission reduction [46]. Hence, there is a growing demand for carbon dioxide reduction through carbon capture, utilization, and storage [47]. Captured carbon dioxide can be applied to the production of bio-based plastic raw materials or mineralized for manufacturing composite materials.

3.3.3. Validation and Utility Assessment: Comparing Meta-Analysis with Korea's EPD

This study presents the total GWP value derived from a comprehensive meta-analysis of LCA studies concerning the life cycle of PET bottles in accordance with the ISO 14040 [4] and 14044 [5] standards. This step aims to demonstrate the utility of a methodology that integrates data from various sources encompassing different years, countries, and scenarios to arrive at quantitative conclusions. Korea's EPD system serves as an eco-labeling initiative that delivers standardized and verified environmental impact information for products across their entire life cycle, adhering to the ISO 14025 specifications [48]. To assign labels, the EPD conducts a thorough LCA, encompassing data related to greenhouse gas emissions, energy consumption during production, resource utilization, and water consumption. Accredited by a professional LCA organization to meet international standards, the EPD system assigns labels that contribute to sustainability goals and mitigate a company's environmental footprint [49].

In the second quarter of 2023, an environmental impact assessment encompassing 32 products within Korea's bottled water bottle classification was conducted, leading to the public disclosure of environmental performance results, including the CO_2 equivalent values, as part of the Korean EPD. These data served as the basis for calculating the average kg CO_2eq per 1 kg of PET bottles in the Korean market. Considering the diverse capacities of bottled water products, ranging from 0.33 L, 0.5 L, 1 L, to 2 L, the "kg CO_2eq per 1 kg of PET bottle" metric was computed using the average weight of the PET bottles corresponding to each capacity. Figure 7 illustrates the resulting scatter and average GWP values for each product capacity. The GWP for the entire life cycle of the Korean PET bottle was 5.424 kg CO_2eq. These values were approximately 6.5% higher than the average GWP estimate of 5.093 kg CO_2eq, which was calculated after removing the outliers through a meta-analysis of the literature. These findings indicate a notable alignment of the average estimates between this study and existing research.

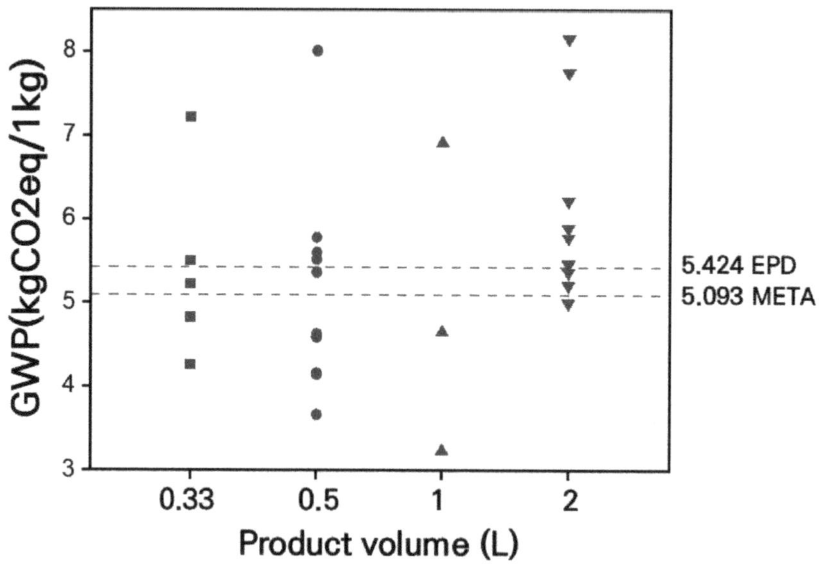

	0.33L	0.5L	1L	2L
AVG	5.34[a]	5.06[a]	4.94[a]	5.98[a]
SD	0.94	1.21	1.85	1.04
CV	0.18	0.24	0.38	0.17

Figure 7. Scatter plot of EPD product values grouped by volume (0.33 L, 0.5 L, 1 L, and 2 L) with GWP normalized to 1 kg PET bottle. Below the scatter plot, the following statistics are provided for each group: average (AVG), standard deviation (SD), and coefficient of variation (CV). Significant differences between groups, determined through statistical techniques, are indicated by superscripts alongside the AVG values. All product volume groups are marked with the superscript "a" indicating the absence of significant differences among them. This is further supported by the post-test results presented in Table S2 within the Supplementary Materials.

Using post hoc analysis, we converted the GWP for each product-volume group to a basis per 1 kg of PET, followed by an assessment of intergroup differences. Underneath Figure 7, the intergroup relationships are depicted, which are indicated by the AVG number superscripts. The post-test results are presented in Table S2 within the Supplementary Materials. There were no significant differences among the four groups, suggesting that when standardized to 1 kg, they could all be considered the same group.

In the subsequent analysis, the GWP for various volumes of EPD products was converted to a standardized measure of 100 mL, and the findings are presented in Figure 8. The average of the EPD products' GWP was calculated at 17.3 g $CO_2eq/100$ mL, with all 2 L products exhibiting GWP values below the average, whereas the 0.33 L variant was observed to exceed the average emissions load. A cursory examination of Figure 8 reveals a consistent trend in which an increase in product volume leads to a consistent reduction in greenhouse gas emissions per 100 mL. This phenomenon can be attributed to the reduction in the amount of raw plastic materials required to carry water, resulting in a more efficient environmental impact.

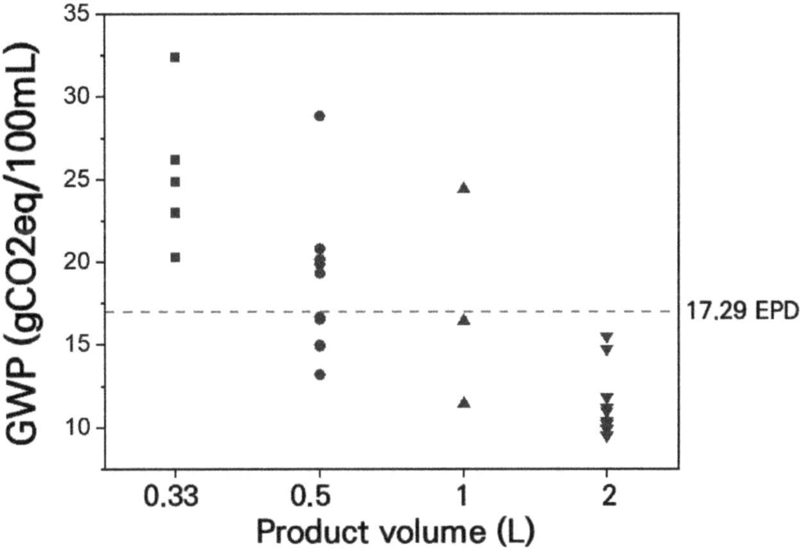

	0.33L	0.5L	1L	2L
AVG	25.11[a]	18.20[b]	17.42[a,b,c]	11.36[c]
SD	3.82	4.36	6.54	1.97
CV	0.15	0.24	0.38	0.17

Figure 8. Scatter plot of EPD product values grouped by volume (0.33 L, 0.5 L, 1 L, and 2 L) with GWP normalized to 100 mL PET bottle. Below the scatter plot, the following statistics are provided for each group: average (AVG), standard deviation (SD), and coefficient of variation (CV). Significant differences between product volume groups are denoted by superscripts "a", "b" and "c" corresponding to AVG values, as revealed by Games–Howell post hoc tests.

The Games–Howell post hoc tests showed that there was a significant difference between the 0.33, 0.5, and 2 L product groups. However, no statistically significant difference was observed between the 1 L product group and the other groups. Following the post hoc test, significant relationships among the groups were revealed by examining their respective average values, as depicted in Figure 8. The post hoc results are presented in Table S3 within the Supplementary Materials.

The scatter plots for the 0.33 L, 0.5 L, and 1 L groups revealed that the highest values were consistently associated with products from a specific company. This suggests that these products contain substantial amounts of PET raw materials, leading to increased input and output within the system and resulting in a higher GWP compared with products from other companies. This company exhibited a similar GWP pattern in the scatter plots when considering GWP per 1 kg. Furthermore, companies manufacturing top-tier 2 L products emit substantial quantities of greenhouse gases during the production of plastic raw materials. This implies that the company's 2 L product may have a greater weight than the other 2 L products.

Although the classification method for the product life cycle and GWP units differed between the EPD and meta-analysis, both approaches share the commonality of quantitatively assessing the environmental impacts from raw material collection to disposal using the same LCA technique. Consequently, by standardizing and comparing the meta-analysis and EPD result data on a "kg CO_2eq per 1 kg of PET bottle" basis, similar GWP averages were obtained. This underscores the efficacy of the method, which employs meta-analysis to harmonize the data and derive comprehensive insights into the environmental impacts of the PET bottle's life cycle. While this study serves as a case study for PET bottles, the methodology is applicable to other products, particularly various packaging materials, enabling a holistic evaluation of the environmental impact throughout their life cycles.

4. Conclusions

We systematically collected a broad array of LCA research literature to comprehensively evaluate the environmental impacts of PET bottle life cycles. Subsequently, we performed a meta-analysis to standardize the GWP data by categorizing the PET bottle life into six distinct phases. This systematic review included 14 research studies conducted in eight countries across various study years from 2010 to 2022.

The research conducted involved unifying the units of GWP data extracted from 14 studies, facilitating their comparability. Through comprehensive statistical analysis, it was established that within the six phases of the PET bottle system, the MP phase emerged as the hot spot for the greenhouse effect. Subsequently, the BP, DT, WM, and CT phases exhibited significant environmental impacts. Moreover, the phases of EB, CT, and WM displayed noteworthy coefficients of variation, signifying substantial fluctuations in GWP data. This variance suggests a significant disparity between environmentally superior and inferior data, implying a high potential for environmental improvement in these phases.

The MP phase yields the highest GWP value owing to the presence of energy-intensive processes, such as solid-state polymerization, which are essential for the production of fossil fuel-based plastic resins. In the BP phase, numerous processes involve the use of high-temperature equipment, particularly during injection and stretch-blow molding, which significantly contribute to global warming. In the distribution and collection transportation phases, the GWP was found to be high and dependent on factors such as the weight (kg) and travel distance (km) of the PET bottles.

Waste management is an active area of research, with various countries incorporating disposal scenarios or assumptions. Generally, a high incineration ratio is associated with higher GWP values, whereas a higher recycling and landfill ratio leads to lower GWP values. The EB phase is characterized by the GWP value determined from renewable raw materials and the heat gain from incineration. This phase has a more pronounced impact on disposal scenarios in which the incineration and recycling ratio exceeds that of landfills.

To assess the similarity between the GWPs derived from meta-analysis and the carbon footprint results of commercially available products, a comparative analysis was conducted using Korea's EPD system, the comparison of AVG estimates, presented in kg CO_2eq per 1 kg of PET bottles, revealed a close match, differing by only approximately 6.5%. This demonstrates that meta-analysis serves as an efficient method for inferring the average carbon footprint of commercial products.

This methodology can be considered a suitable technique for deriving comprehensive conclusions, as it facilitates quantitative comparisons independent of the research year, region, or functional unit. This approach can be extended to various product groups, establishing itself as a valuable tool for identifying and assessing greenhouse effects. Furthermore, the application of this methodology supports evidence-based decision-making in the development of sustainable solutions for product systems.

Supplementary Materials: The following supporting information can be downloaded at: https://www.mdpi.com/article/10.3390/su16020535/s1, Figure S1: Illustration of how to extract the GWP from stacked bar chart; Figure S2: Box plot analysis that encompasses all outliers of GWP values for six phases of the PET bottle system; Table S1: The literature screening forms with the references that passed the final screening criteria. The screening forms contain the type of reference with titles, author, year, and information of references; Table S2: Organization of capacities and GWP values for Korean EPD products; Table S3: Results of the Games–Howell post hoc test for GWP values of EPD products per 1 kg; Table S4: Results of the Games–Howell post hoc test for GWP values of EPD products per 100 mL.

Author Contributions: Conceptualization, Y.-J.G. and D.-H.K.; methodology, Y.-J.G. and D.-H.K.; validation, Y.-J.G. and D.-H.K.; investigation, Y.-J.G. and D.-H.K.; formal analysis, Y.-J.G. and D.-H.K.; data curation, Y.-J.G. and D.-H.K.; writing—original draft, Y.-J.G.; visualization, Y.-J.G.; review and editing, D.-H.K. and J.-H.L.; supervision, H.-J.P. and J.-K.S. All authors have read and agreed to the published version of the manuscript.

Funding: This research was funded by the Ministry of Agriculture, Food and Rural Affairs: 321045-3.

Institutional Review Board Statement: Not applicable.

Data Availability Statement: Data are contained within the article and Supplementary Materials.

Acknowledgments: This work was supported by the Korea Institute of Planning and Evaluation for Technology in Food, Agriculture and Forestry (IPET) through the 'High Value-added Food Technology Development Program'.

Conflicts of Interest: The authors declare no conflicts of interest.

References

1. Kang, D. *Environmental Evaluation of Non-Alcoholic Single-Serve PET Beverage Bottles in the State of California Using Life Cycle Assessment and System Dynamics*; Michigan State University: East Lansing, MI, USA, 2015.
2. Burkhardt III, J.J.; Heath, G.; Cohen, E. Life cycle greenhouse gas emissions of trough and tower concentrating solar power electricity generation: Systematic review and harmonization. *J. Ind. Ecol.* **2012**, *16*, S93–S109. [CrossRef]
3. Arvanitoyannis, I.S. ISO 14040: Life cycle assessment (LCA)–principles and guidelines. In *Waste Management for the Food Industries*; Elsevier: Amsterdam, The Netherlands, 2008; pp. 97–132.
4. *ISO 14040:2006*; Environmental Management–Life Cycle Assessment–Principles and Framework. ISO: Geneva, Switzerland, 2006.
5. *ISO 14044:2006*; International Organization of Standardization. Environmental Management—Life Cycle Assessment—Requirements and Guidelines. ISO: Geneva, Switzerland, 2006.
6. Pesonen, H.-L.; Ekvall, T.; Fleischer, G.; Huppes, G.; Jahn, C.; Klos, Z.S.; Rebitzer, G.; Sonnemann, G.W.; Tintinelli, A.; Weidema, B.P. Framework for scenario development in LCA. *Int. J. Life Cycle Assess.* **2000**, *5*, 21–30. [CrossRef]
7. Furberg, A.; Moum, A.; Nørsterud, S.; Raadal, H.L. *Review of Life Cycle Assessments of Reuse Systems for Bottles*; NORSUS: Kråkerøy, Norway, 2021.
8. Gomes, T.S.; Visconte, L.L.; Pacheco, E.B. Life cycle assessment of polyethylene terephthalate packaging: An overview. *J. Polym. Environ.* **2019**, *27*, 533–548. [CrossRef]
9. Bishop, G.; Styles, D.; Lens, P.N. Environmental performance comparison of bioplastics and petrochemical plastics: A review of life cycle assessment (LCA) methodological decisions. *Resour. Conserv. Recycl.* **2021**, *168*, 105451. [CrossRef]
10. Ortiz, O.; Castells, F.; Sonnemann, G. Sustainability in the construction industry: A review of recent developments based on LCA. *Constr. Build. Mater.* **2009**, *23*, 28–39. [CrossRef]
11. Fantin, V.; Scalbi, S.; Ottaviano, G.; Masoni, P. A method for improving reliability and relevance of LCA reviews: The case of life-cycle greenhouse gas emissions of tap and bottled water. *Sci. Total Environ.* **2014**, *476*, 228–241. [CrossRef] [PubMed]
12. Weidema, B.; Wenzel, H.; Petersen, C.; Hansen, K. The product, functional unit and reference flows in LCA. *Environ. News* **2004**, *70*, 1–46.
13. Hunsager, E.A.; Bach, M.; Breuer, L. An institutional analysis of EPD programs and a global PCR registry. *Int. J. Life Cycle Assess.* **2014**, *19*, 786–795. [CrossRef]
14. Lee, Y.H. An overview of meta-analysis for clinicians. *Korean J. Intern. Med.* **2018**, *33*, 277. [CrossRef]
15. Rosenthal, R. Meta-analysis: A review. *Psychosom. Med.* **1991**, *53*, 247–271. [CrossRef]
16. Aromataris, E.; Pearson, A. The systematic review: An overview. *AJN Am. J. Nurs.* **2014**, *114*, 53–58. [CrossRef]

17. Ownby, R.L.; Crocco, E.; Acevedo, A.; John, V.; Loewenstein, D. Depression and risk for Alzheimer disease: Systematic review, meta-analysis, and metaregression analysis. *Arch. Gen. Psychiatry* **2006**, *63*, 530–538. [CrossRef] [PubMed]
18. Song, G.G.; Bae, S.-C.; Lee, Y.H. Efficacy and safety of tofacitinib for active rheumatoid arthritis with an inadequate response to methotrexate or disease-modifying antirheumatic drugs: A meta-analysis of randomized controlled trials. *Korean J. Intern. Med.* **2014**, *29*, 656. [CrossRef] [PubMed]
19. Asdrubali, F.; Baldinelli, G.; D'Alessandro, F.; Scrucca, F. Life cycle assessment of electricity production from renewable energies: Review and results harmonization. *Renew. Sustain. Energy Rev.* **2015**, *42*, 1113–1122. [CrossRef]
20. Razman, K.K.; Hanafiah, M.M.; Mohammad, A.W. An overview of LCA applied to various membrane technologies: Progress, challenges, and harmonization. *Environ. Technol. Innov.* **2022**, *27*, 102803. [CrossRef]
21. Clune, S.; Crossin, E.; Verghese, K. Systematic review of greenhouse gas emissions for different fresh food categories. *J. Clean. Prod.* **2017**, *140*, 766–783. [CrossRef]
22. El Essawy, N.A.; Konsowa, A.H.; Elnouby, M.; Farag, H.A. A novel one-step synthesis for carbon-based nanomaterials from polyethylene terephthalate (PET) bottles waste. *J. Air Waste Manag. Assoc.* **2017**, *67*, 358–370. [CrossRef] [PubMed]
23. Welle, F. Twenty years of PET bottle to bottle recycling—An overview. *Resour. Conserv. Recycl.* **2011**, *55*, 865–875. [CrossRef]
24. Zhang, R.; Ma, X.; Shen, X.; Zhai, Y.; Zhang, T.; Ji, C.; Hong, J. PET bottles recycling in China: An LCA coupled with LCC case study of blanket production made of waste PET bottles. *J. Environ. Manag.* **2020**, *260*, 110062. [CrossRef]
25. Pieragostini, C.; Mussati, M.C.; Aguirre, P. On process optimization considering LCA methodology. *J. Environ. Manag.* **2012**, *96*, 43–54. [CrossRef]
26. Menten, F.; Chèze, B.; Patouillard, L.; Bouvart, F. A review of LCA greenhouse gas emissions results for advanced biofuels: The use of meta-regression analysis. *Renew. Sustain. Energy Rev.* **2013**, *26*, 108–134. [CrossRef]
27. Sohrabi, C.; Franchi, T.; Mathew, G.; Kerwan, A.; Nicola, M.; Griffin, M.; Agha, M.; Agha, R. *PRISMA 2020 Statement: What's New and the Importance of Reporting Guidelines*; Elsevier: Amsterdam, The Netherlands, 2021; Volume 88, p. 105918.
28. Ruxton, G.D.; Beauchamp, G. Time for some a priori thinking about post hoc testing. *Behav. Ecol.* **2008**, *19*, 690–693. [CrossRef]
29. Gileno, L.A.; Turci, L.F.R. Life cycle assessment for PET-bottle recycling in Brazil: B2B and B2F routes. *Clean. Environ. Syst.* **2021**, *3*, 100057. [CrossRef]
30. Bataineh, K.M. Life-cycle assessment of recycling postconsumer high-density polyethylene and polyethylene terephthalate. *Adv. Civ. Eng.* **2020**, *2020*, 8905431. [CrossRef]
31. Kouloumpis, V.; Pell, R.; Correa-Cano, M.; Yan, X. Potential trade-offs between eliminating plastics and mitigating climate change: An LCA perspective on Polyethylene Terephthalate (PET) bottles in Cornwall. *Sci. Total Environ.* **2020**, *727*, 138681. [CrossRef]
32. Tamburini, E.; Costa, S.; Summa, D.; Battistella, L.; Fano, E.A.; Castaldelli, G. Plastic (PET) vs bioplastic (PLA) or refillable aluminium bottles–What is the most sustainable choice for drinking water? A life-cycle (LCA) analysis. *Environ. Res.* **2021**, *196*, 110974. [CrossRef]
33. Grisales, V.; Tua, C.; Rigamonti, L. Life cycle assessment of bottled mineral water for the hospitality industry in Northern Italy. *Packag. Technol. Sci.* **2022**, *35*, 301–314. [CrossRef]
34. Olatayo, K.I.; Mativenga, P.T.; Marnewick, A.L. Life cycle assessment of single-use and reusable plastic bottles in the city of Johannesburg. *S. Afr. J. Sci.* **2021**, *117*, 1–10. [CrossRef]
35. Lonca, G.; Lesage, P.; Majeau-Bettez, G.; Bernard, S.; Margni, M. Assessing scaling effects of circular economy strategies: A case study on plastic bottle closed-loop recycling in the USA PET market. *Resour. Conserv. Recycl.* **2020**, *162*, 105013. [CrossRef]
36. Gursel, I.V.; Moretti, C.; Hamelin, L.; Jakobsen, L.G.; Steingrimsdottir, M.M.; Junginger, M.; Høibye, L.; Shen, L. Comparative cradle-to-grave life cycle assessment of bio-based and petrochemical PET bottles. *Sci. Total Environ.* **2021**, *793*, 148642. [CrossRef]
37. Benavides, P.T.; Dunn, J.B.; Han, J.; Biddy, M.; Markham, J. Exploring comparative energy and environmental benefits of virgin, recycled, and bio-derived PET bottles. *ACS Sustain. Chem. Eng.* **2018**, *6*, 9725–9733. [CrossRef]
38. Martin, E.J.; Oliveira, D.S.; Oliveira, L.S.; Bezerra, B.S. Life cycle comparative assessment of pet bottle waste management options: A case study for the city of Bauru, Brazil. *Waste Manag.* **2021**, *119*, 226–234. [CrossRef]
39. Papong, S.; Malakul, P.; Trungkavashirakun, R.; Wenunun, P.; Chom-in, T.; Nithitanakul, M.; Sarobol, E. Comparative assessment of the environmental profile of PLA and PET drinking water bottles from a life cycle perspective. *J. Clean. Prod.* **2014**, *65*, 539–550. [CrossRef]
40. Kuczenski, B.; Geyer, R. PET bottle reverse logistics—Environmental performance of California's CRV program. *Int. J. Life Cycle Assess.* **2013**, *18*, 456–471. [CrossRef]
41. Gironi, F.; Piemonte, V. Life cycle assessment of polylactic acid and polyethylene terephthalate bottles for drinking water. *Environ. Prog. Sustain. Energy* **2011**, *30*, 459–468. [CrossRef]
42. Kim, J.; Lee, C.; Lim, N.-H.; Han, Y.; Han, I. A Study on the Environmental Assessment of Bottled Water using Life Cycle Assessment Methodology. *KSEE J. Korean Soc. Environ. Eng.* **2022**, *44*, 345–353. [CrossRef]
43. Marathe, K.; Chavan, K.R.; Nakhate, P. Life cycle assessment (LCA) of PET bottles. In *Recycling of Polyethylene Terephthalate Bottles*; Elsevier: Amsterdam, The Netherlands, 2019; pp. 149–168.
44. Jeswani, H.; Krüger, C.; Russ, M.; Horlacher, M.; Antony, F.; Hann, S.; Azapagic, A. Life cycle environmental impacts of chemical recycling via pyrolysis of mixed plastic waste in comparison with mechanical recycling and energy recovery. *Sci. Total Environ.* **2021**, *769*, 144483. [CrossRef]

45. Gracida-Alvarez, U.R.; Xu, H.; Benavides, P.T.; Wang, M.; Hawkins, T.R. Circular Economy Sustainability Analysis Framework for Plastics: Application for Poly (ethylene Terephthalate)(PET). *ACS Sustain. Chem. Eng.* **2023**, *11*, 514–524. [CrossRef]
46. von der Assen, N.; Jung, J.; Bardow, A. Life-cycle assessment of carbon dioxide capture and utilization: Avoiding the pitfalls. *Energy Environ. Sci.* **2013**, *6*, 2721–2734. [CrossRef]
47. Von Der Assen, N.; Voll, P.; Peters, M.; Bardow, A. Life cycle assessment of CO_2 capture and utilization: A tutorial review. *Chem. Soc. Rev.* **2014**, *43*, 7982–7994. [CrossRef]
48. Durão, V.; Silvestre, J.D.; Mateus, R.; de Brito, J. Assessment and communication of the environmental performance of construction products in Europe: Comparison between PEF and EN 15804 compliant EPD schemes. *Resour. Conserv. Recycl.* **2020**, *156*, 104703. [CrossRef]
49. Fet, A.M.; Skaar, C. Eco-labeling, product category rules and certification procedures based on ISO 14025 requirements (6 pp). *Int. J. Life Cycle Assess.* **2006**, *11*, 49–54. [CrossRef]

Disclaimer/Publisher's Note: The statements, opinions and data contained in all publications are solely those of the individual author(s) and contributor(s) and not of MDPI and/or the editor(s). MDPI and/or the editor(s) disclaim responsibility for any injury to people or property resulting from any ideas, methods, instructions or products referred to in the content.

remote sensing

MDPI

Article

Enhancing Maize Yield Simulations in Regional China Using Machine Learning and Multi-Data Resources

Yangfeng Zou [1], Giri Raj Kattel [1,2] and Lijuan Miao [1,*]

1 School of Geographical Sciences, Nanjing University of Information Science & Technology, Nanjing 210044, China; 15308093142@163.com (Y.Z.); gkattel@unimelb.edu.au (G.R.K.)
2 Department of Infrastructure Engineering, The University of Melbourne, Parkville, VIC 3052, Australia
* Correspondence: miaolijuan1111@gmail.com

Abstract: Improved agricultural production systems, together with increased grain yield, are essential to feed the growing global population in the 21st century. Global gridded crop models (GGCMs) have been extensively used to assess crop production and yield simulation on a large geographical scale. However, GGCMs are less effective when they are used on a finer scale, significantly limiting the precision in capturing the yearly maize yield. To address this issue, we propose a relatively more advanced approach that downsizes GGCMs by combining machine learning and crop modeling to enhance the accuracy of maize yield simulations on a regional scale. In this study, we combined the random forest algorithm with multiple data sources, trained the algorithm on low-resolution maize yield simulations from GGCMs, and applied it to a finer spatial resolution on a regional scale in China. We evaluated the performance of the eight GGCMs by utilizing a total of 1046 county-level maize yield data available over a 30-year period (1980–2010). Our findings reveal that the downscaled models created for maize yield simulations exhibited a remarkable level of accuracy ($R^2 \geq 0.9$, MAE < 0.5 t/ha, RMSE < 0.75 t/ha). The original GGCMs performed poorly in simulating county-level maize yields in China, and the improved GGCMs in our study captured an additional 17% variability in the county-level maize yields in China. Additionally, by optimizing nitrogen management strategies, we identified an average maize yield gap at the county level in China ranging from 0.47 to 1.82 t/ha, with the south maize region exhibiting the highest yield gap. Our study demonstrates the high effectiveness of machine learning methods for the spatial downscaling of crop models, significantly improving GGCMs' performance in county-level maize yield simulations.

Keywords: maize yield; global gridded crop models; random forest; multiple data sources; county-level

1. Introduction

By the year 2050, the global population is projected to exceed 9 billion, which is an increase of 2 billion people compared to the baseline year of 2010 [1]. As the population continues to expand, global food demand is also rapidly growing. The 25% increase in the global food demand by 2010 is projected to increase to 70% by 2050 [2,3] under ongoing climate warming and population growth, presenting serious food security challenges worldwide [4,5]. The Intergovernmental Panel on Climate Change (IPCC) Special Report on Climate Change and Land (SRCCL) highlights the fact that climate warming will trigger a series of cascading effects that will adversely affect food security [6]. A reduction in food production and the quality of grain will, in turn, pose threats to the development of agriculture-related industries and may exacerbate food shortages worldwide [7]. Given these concerns, achieving the "Zero Hunger" goal of the United Nations Sustainable Development Goals has become a top priority of national governments and other international communities [8]. In such a context, increasing crop production and accurately estimating crop yields by advancing appropriate models and technology are becoming increasingly crucial for global and regional food security [9,10].

Global gridded crop models (GGCMs) are some of the most significant tools being used extensively for estimating crop yields under various environmental and management conditions worldwide [11,12]. GGCMs have the ability to assess the impact of climate change on crop yields more effectively [13–15]. Researchers have parameterized crop growth processes using GGCMs by, for example, combining information such as weather, soil, and management parameters, followed by simulating the dynamics of crop growth and yield on regional and global scales [16,17]. Through analysis of the crop yield simulations of GGCMs, it has been revealed that climate change has been found to be the major driver causing a negative impact on global agricultural production [18]. A recent study by Yin and Leng [19] suggests that GGCMs effectively captured the adverse effects of climate change on maize production when using the 30-year (1980–2010) time series data on global climate change and maize production. Furthermore, GGCMs have been found to show varying performance in simulating major grain crop production in China [20]. GGCMs commonly use a 50 km $\times$ 50 km spatial resolution, which is suitable for robust crop yield assessments on a national scale [13]. However, there has been a rising concern that GGCM estimates at lower spatial resolutions may ignore real geographical factors, such as soil and topography [21]. One of the major concerns raised by researchers is that missing regional-level heterogeneity may lead to discrepancies between sub-grid level information and data and farm land and agricultural practices, consequently biasing the simulation model [22]. Conducting county-level maize yield studies poses challenges due to the lower resolution of GGCMs, making it challenging to identify spatial variations in maize yield [23]. Additionally, the utilization of global-scale datasets in the model results in considerable errors when simulating county-level maize yield, thereby hindering the effectiveness of local government efforts in risk prevention and decision making for food security [24]. Improved simulation accuracy of GGCMs at the county level holds the potential to significantly enhance research on crop yield loss risks spanning from local to regional scales; studies on potential crop yields will benefit significantly from such improvements [25,26]. On the other hand, applying gridded crop models at very high resolutions significantly increases computational demands and is often constrained by data availability [21]. GGCM crop simulations are primarily determined by climate data, but the global-scale higher-resolution climatic data needed have only been available at coarser spatial resolutions until recently [27].

An approach that can help address these issues efficiently and flexibly is downscaling the GGCM results, which has become significant in addressing unresolved issues of regional and global crop yield simulations and allows for obtaining high-resolution simulations without setting up high-resolution crop model infrastructures and comprehensive datasets [21]. Studies suggest that a machine learning-based downscaling approach could utilize correlations between dependent and independent predictor variables to refine low-resolution crop yield data into finer spatial-resolution data [28]. For instance, combining the GPM model and random forest increased the prediction potential of land productivity by accurately showing the reduction in the spatial-scale crop yield data from the United States [29]. Recently, machine learning and remote sensing variables have been employed to downscale global gridded crop yield data in southern Asia, which produced highly accurate results in crop yield at a 1 km resolution [30]. This suggests that when constrained by technical limitations and data accessibility, using machine learning to downscale low-resolution data into high-resolution data has become a viable approach for predicting crop production at both the regional and global scales [21]. However, currently, within agricultural sciences, this application has primarily been limited to the processing and analysis of remote sensing data [31,32]. Limited research on downscaling crop simulation yields from GGCMs has hindered generating unbiased results on global and regional crop yield simulations [21,33].

China is the world's second-largest maize producer, contributing to 23% of global production in recent years [34]. Conducting research with a focus on ensuring maize yield in China is significant for global food security. We utilized the random forest algorithm and integrated climate data, soil data, and topographical data to downscale low-resolution data into high-resolution data to achieve the best possible model outcome. While downscaling the data, we used eight different GGCMs (e.g., CLM-CROP, GEPIC, EPIC-BOKU, EPIC-TAMU, EPIC-IIASA, PDSSAT, PAPSIM, PEGASUS) and calculated the ensemble mean of multiple GGCMs (ENSEMBLE). This enabled us to compare the model outputs, which assist in making better decisions. Taking China's maize plantation regions as the research area, the primary objectives of this research were: (1) to develop precise downscaling models for simulating maize yields in regional China; (2) to evaluate the effectiveness of GGCMs in simulating maize yield, both before and after downscaling; (3) to compare and estimate maize yield disparities at the county level in China.

2. Methods and Datasets

2.1. Study Area

China's maize plantation regions are classified into five distinct zones based on their cultivation characteristics, management practices, and geographical environments [35]. As shown in Figure 1, these zones contain the north spring maize region (Zone I), the Huang-Huai-Hai summer maize region (Zone II), the southwest maize region (Zone III), the northwest maize region (Zone IV), and the south maize region (Zone V). The mean annual temperature across these regions falls within approximately 9–25 °C, which is within the optimal range for maize growth, spanning from 9 to 29 °C [36]. Maize cultivation requires adequate water supply throughout its lifecycle [37]. Therefore, Zone IV accounts for only 3% of China's annual maize cultivation area due to the low precipitation levels, while the remaining regions, with relatively higher precipitations, together account for 93% of the nation's maize cultivation area.

2.2. Data Sources

2.2.1. Climate Data

In the Global Gridded Crop Model Intercomparison Project (GGCMI), all participating GGCMs adhered to a standardized protocol and were constrained by the same climate dataset [16,38]. We especially focused on one of the enforced simulations, namely the AgMERRA dataset, and thus, the gridded climate data were obtained from the publicly available AgMERRA climate dataset [27]. It features a spatial resolution of 25 km × 25 km and spans the period from 1980 to 2010. AgMERRA incorporates a comprehensive retrospective analysis of modern research and practical applications [39]. This dataset has been bias-corrected using station data and remote sensing data for agricultural land use. The selected key variables include the maximum temperature (TMAX), minimum temperature (TMIN), average temperature (TAVG), total precipitation (PRATE), potential evapotranspiration (PET), downward surface shortwave radiation (SRAD), wind speed (WS), and vapor pressure deficit (VPD) (Table 1). Several statistics were calculated for each climate variable, resulting in a total of 76 features.

2.2.2. Soil Data

Soil information was extracted from the Harmonized World Soil Database v1.2, featuring a spatial resolution of 1 km [40]. The dominant soil type was selected for each grid cell at a spatial resolution of 50 km from the largest soil mapping unit. We chose a total of 11 variables from the soil dataset, including the soil reference depth (DEPTH), bulk density in topsoil (BD), carbonate content in topsoil (CARB), cation exchange capacity in topsoil (CEC), clay content in topsoil (CLAY), electrical conductivity in topsoil (EC), organic carbon content in topsoil (OC), pH in topsoil (PH), coarse fragment (rock) content in topsoil (ROK), sum of bases in topsoil (SB), and soil sand content (SAND) (Table 1).

Figure 1. The five maize planting regions in China. Notes: Zone I: the north maize region; Zone II: the Huang-Huai-Hai maize region; Zone III: the southwest maize region; Zone IV: the northwest maize region; and Zone V: the south maize region.

Table 1. Features and target variables for machine learning.

Variables	Variable Descriptions
	Climate variables (VARs)
TMAX	Maximum temperature (°C)
TMIN	Minimum temperature (°C)
TAVG	Average temperature (°C)
PRATE	Total precipitation (mm)
SRAD	Solar radiation (MJ/m^2)
PET	Potential evapotranspiration (mm)
WS	Wind speed (m/s)
VPD	Vapor pressure deficit (h PA)

Table 1. *Cont.*

Variables	Variable Descriptions
Temporal aggregates of climate variables	
VAR_X	Monthly value for month X in calendar year (e.g., "TMAX_1")
VARsumGS	Sum of climate variables in growing season (e.g., "TMAXsumGS")
VARavgGS	Average of climate variables in growing season (e.g., "TMAXavgGS")
Soil and topography variables	
DEPTH	Total soil depth (m)
OC	Organic carbon content in topsoil (%)
SAND	Sand content in topsoil (%)
SB	Sum of bases in topsoil (cmol/kg)
ROK	Coarse fragment (rock) content in topsoil (%)
CLAY	Clay content in topsoil (%)
EC	Electric conductivity in topsoil (mmho/cm)
BD	Bulk density in topsoil (g/cm^3)
CEC	Cation exchange capacity in topsoil (cmol/kg)
PH	pH in topsoil
CARB	Carbonate content in topsoil (%)
DEM	Digital elevation model
Target variables	
YIELD	Simulated maize yield (t/ha)

2.2.3. Topography Data

Topographical information was provided by digital elevation models (DEMs). DEMs are digital mapping datasets that consist of three-dimensional coordinates, typically derived from contour lines or photogrammetric methods [41]. We utilized a 1 km-resolution DEM sourced from the Space Shuttle Radar Topography Mission (SRTM) data. This dataset was derived from the latest SRTM V4.1 data (http://www.resdc.cn, accessed on 15 August 2023).

2.2.4. Irrigation Mask Data

To accurately evaluate the performance of each GGCM in simulating maize yields under both rain-fed conditions and irrigated conditions, we employed the mask data (irrigated and rain-fed crop harvested areas), to process the maize yield simulation results. The 1 km-resolution mask data were obtained from MIRCA2000 (https://www.uni-frankfurt.de/45218031/data_download, accessed on 15 August 2023), providing detailed information on the crop-specific irrigated and rain-fed harvested areas for each grid cell [42].

2.2.5. Maize Yield Data

To assess the performance of GGCMs, we used China's county-level maize yield data. They were sourced from the Agricultural Statistical Yearbook compiled by the Ministry of Agriculture of China (http://www.stats.gov.cn, accessed on 15 August 2023), and its unit is kilograms per hectare (kg/ha). These data represent the average crop yield for each county and include both the rain-fed and irrigated maize yields.

2.2.6. Date Processing

Since maize yield data was available at the county-level, we aggregated the simulation results to the county scale using an area-weighted average, as described below:

$$Y_{county} = \frac{\sum_{i=1}^{n} A_{irr,i} \times Y_{irr,i} + \sum_{i=1}^{n} A_{noirr,i} \times Y_{noirr,i}}{\sum_{i=1}^{n} (A_{irr,i} + A_{noirr,i})} \tag{1}$$

where n is the number of grid cells within each county, i is the index of grid cell. $A_{irr,i}$ represents the irrigated maize harvested area (ha) in grid cell i, and $A_{noirr,i}$ is the rain-fed maize harvested area (ha) in grid cell i. $Y_{irr,i}$ represents the simulated yield (t·ha^{-1}) of irrigated maize in grid cell i, and $Y_{noirr,i}$ represents the simulated yield (t·ha^{-1}) of rain-fed maize in grid cell i.

Since variations in maize yields are mainly driven by management factors [43,44], and due to the limitations of accessible data, these influences should be removed from the maize yield data for an accurate evaluation of the GGCMs' simulations. We employed a moving window approach spanning 5 years to eliminate trends in both the observed and simulated maize yields. This involved detrending the annual maize yield data by subtracting the yield mean in a 5-year window. Compared to other methods, this detrending approach has demonstrated greater effectiveness in mitigating the impact of trend effects [45].

2.3. Experiments Design

The primary objective of this research was to assess the feasibility of achieving high-resolution predictions using low-resolution simulated maize yield data. We employed a downscaling model constructed by the random forest algorithm. All the data were resampled to a uniform 50 km resolution, consistent with the spatial resolution of the maize yields simulated by GGCMs. Climate data, soil data, and topographical data were used as feature variables for model development, while the simulated maize yields served as the target variable. The data from 1980 to 2009 were randomly split into training and test sets, containing 75% and 25% of the samples, respectively. Data for the year 2010 were kept separate for data validation. Random forest exhibits greater flexibility in fitting training data, but it is also more prone to overfitting. To avoid overfitting, we employed various techniques, including monitoring out-of-bag error, performing n-fold cross-validation, and applying regularization to make the training procedure more conservative. We evaluated the high-resolution GGCM maize yield simulations after downscaling using observed county-level maize yield data from 1980–2010. As an example, we estimated the maize yield disparities for the year 2010.

2.3.1. Description for the Gridded Crop Model

The GGCMI (Global Gridded Crop Model Intercomparison Project) Phase 1 dataset comprises outputs from fourteen modeling groups, spanning various time periods and four major staple crops: soybeans, wheat, maize, and rice [16]. The GGCMI defined three distinct model configurations based on different crop management practices [44]. Firstly, each modeling group developed a "default" configuration (default) based on their typical historical period simulation management, technical assumptions, and inputs. Next, they developed a "full harmonization" configuration (Fullharm), which involved harmonized growing seasons (i.e., prescribed grid-cell- and crop-specific sowing and maturity dates) and fertilizer inputs. Lastly, simulations were conducted with the same harmonized growing seasons but with unlimited nutrient supply, termed the "Harmnon" configuration (Harmnon). All simulations were conducted under both rain-fed (noirr) and fully irrigated (irr) conditions to facilitate a closer approximation of crop yields to actual production in subsequent processing [46,47]. Due to the unavailability of simulation results for all model configurations from some groups, we utilized the data from only eight models (e.g., CLM-CROP, EPIC-BOKU, EPIC-IIASA, EPIC-TAMU, GEPIC, PAPSIM, PDSSAT, and PEGASUS). These GGCMs were driven by historical weather datasets, such as WFDEI, AgMERRA, WATCH (WFD), GRASP, AgCFSR, and Princeton GF (Table S1). Given that these weather datasets are all based on station data or reanalysis data, we assumed that the choice of different weather datasets had minimal impact on the simulation results.

2.3.2. Random Forest

The random forest (RF) algorithm stands as a well-established ensemble learning algorithm that performs regression or classification tasks by combining a large number of decision trees [48]. RF is particularly suitable for capturing both linear and nonlinear relationships between crop yields and various environmental factors [49,50]. RF offers a valuable feature by providing a metric for assessing the relative importance of different predictor variables. This capability aids researchers in better understanding how climate and soil variables influence maize yield, contributing to a more comprehensive understanding of the underlying factors driving crop production [51]. Numerous studies have demonstrated the good performance of RF in predicting county-level maize yields [52,53]. RF exhibits robustness to variations in data distribution and demonstrates less sensitivity to hyperparameter tuning [54]. Its ensemble learning approach helps mitigate overfitting concerns [54]. Additionally, RF's default settings commonly yield satisfactory results, making it a practical choice. The model's effectiveness may also stem from its ability to capture specific patterns or relationships in maize yield data and from its generally strong performance in diverse applications [55,56]. Since RF is less prone to overfitting, we struck a balance between its computational performance and efficiency by adjusting the global parameters, as its computational demands increase linearly with the number and depth of trees. The key parameters that required tuning for RF were as follows: the number of trees (n_estimators), the tree depth (max_depth), and the minimum number of samples required to split a node (min_samples_split). To optimize the downscaling model, we selected the parameters that minimized the RMSE through ten-fold cross-validation.

2.3.3. Feature Engineering and Feature Importance

Our features were derived from the model's input data, as described in Section 2.2; they were climate, soil, and topographical data. Depending on the variable type, the daily climate data were averaged to the monthly average climate data, with the number following the variable name representing the month. Since maize in China is typically planted in April and harvested in October, we defined April to September as the growing season for the simulations [57,58]. The daily climate data were aggregated into seasonal averages and seasonal totals based on the maize growing season timeline. Soil data mainly consist of surface soil variables (0–30 cm, divided according to the HWSD), as surface soil mainly affects maize growth [21]. Additionally, relevant features affecting maize yield include DEM data, which serve as site-specific characteristics. This process is referred to as feature engineering, which involves normalizing raw data based on domain knowledge. RF can internally determine feature importance by initially splitting on a fixed feature, calculating the total reduction in the sum of squared residuals achieved by each tree as a result of this split, and averaging it over all trees, where larger averages indicate greater variable importance [51]. Here, we present the relative importance of each feature as a percentage and rank all features accordingly.

2.3.4. Calculation of Regional Maize Yield Gap

In the GGCMI, the input data for GGCMs in the Fullharm scenario were the same as that in the Harmnon scenario, except for the nitrogen fertilizer input data. In the Harmnon scenario, characterized by an unlimited nutrient supply, the simulated maize yield is referred to as the potential yield. This potential yield serves as the baseline for defining the yield gap and related analyses [16]. To quantify the difference in the simulated maize yields between the Harmnon and the Fullharm scenarios (t/ha), we calculated the maize yield gap by computing the multi-model ensemble mean using Equation (2):

$$Y_{g,e} = \frac{1}{n}\sum_{i=1}^{n} Y_{h,i} - Y_{f,i} \tag{2}$$

where $Y_{g,e}$ is the relative maize yield gap as the multi-model ensemble mean, n is the number of GGCMs, i is the member of GGCMs, $Y_{h,i}$ is the simulated maize yield in the Harmnon scenario, and $Y_{f,i}$ is the simulated maize yield in the Fullharm scenario.

2.4. Metrics for Model Performance Evaluations

The coefficient of determination (R^2), the root-mean-square error (RMSE), and the mean absolute error (MAE) were applied to assess the effectiveness of the downscaling model developed in this study. Furthermore, we utilized the correlation coefficient (R), RMSE, and MAE to evaluate the performance of GGCMs in simulating the spatiotemporal variability of maize yield both before and after downscaling. The calculations for these metrics are provided in Equations (3)–(6):

$$R^2 = 1 - \frac{\sum_{i=1}^{n}\left(Y_{s,i} - \overline{Y}_p\right)^2}{\sum_{i=1}^{n}\left(Y_{s,i} - \overline{Y}_s\right)^2} \tag{3}$$

$$RMSE = \sqrt{\frac{1}{n}\sum_{i=1}^{n}\left(Y_{s,i} - Y_{p,i}\right)^2} \tag{4}$$

$$MAE = \frac{1}{n}\sum_{i=1}^{n}\left|Y_{s,i} - Y_{p,i}\right| \tag{5}$$

$$R = \frac{\sum_{i=1}^{n}\left(Y_{s,i} - \overline{Y}_s\right)\left(Y_{r,i} - \overline{Y}_r\right)}{\sqrt{\sum_{i=1}^{n}\left(Y_{s,i} - \overline{Y}_s\right)^2\sum_{i=1}^{n}\left(Y_{r,i} - \overline{Y}_r\right)^2}} \tag{6}$$

where n (i = 1, 2, …, n) is the number of samples, $Y_{s,i}$ is the simulated maize yield, $Y_{p,i}$ is the predicted maize yield, $Y_{r,i}$ is the observed maize yield, and $\overline{Y}$ is the corresponding mean value.

3. Results

3.1. Performance of Simulated Maize Yield Downscaling Model

3.1.1. General Performance of Downscaling Model and Spatial Patterns of Maize Yield

All GGCMs except PAPSIM showed exceptionally high goodness-of-fit and demonstrated low errors under both the irrigated and rain-fed conditions ($R^2 \geq 0.9$, RMSE < 0.75 t/ha, MAE < 0.5 t/ha) (Figures S1–S7). Notably, EPIC-BOKU exhibited the highest goodness-of-fit, with R^2 = 0.93, RMSE = 0.65 t/ha, and MAE = 0.42 t/ha under irrigated conditions, and R^2 = 0.94, RMSE = 0.66 t/ha, and MAE = 0.42 t/ha under rain-fed conditions (Figure 2a,b). Predominantly, maize yield overestimations occurred when the simulated yields were low, while underestimations were observed when the simulated yields were high (Figure 2c,d). Despite some deviations, the overall density distribution of the predicted yields closely resembles that of the simulated yields (Figure 2c,d). These deviations vary within different model-specific maize yield intervals. For all models, the range of simulated maize yields corresponding to the peak density in irrigated maize consistently surpassed that of the simulated rain-fed maize (Figures S1–S7). This indicates a significant improvement in maize yield following adequate irrigation. Due to the significantly low accuracy exhibited by PAPSIM's downscaling model, it has been omitted from this study for further consideration.

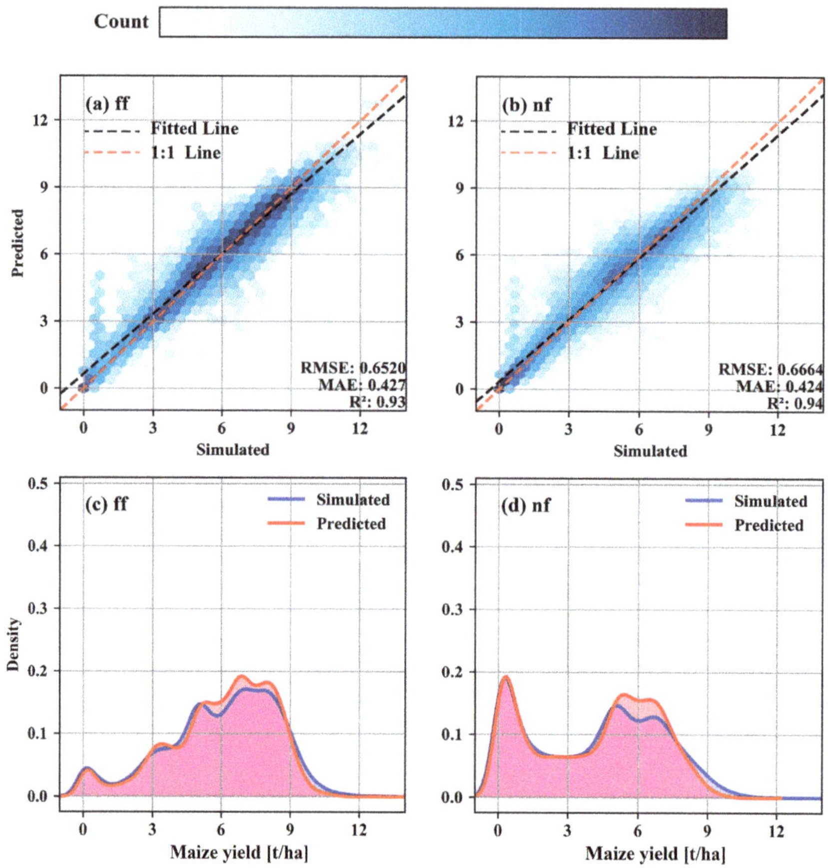

Figure 2. Scatter and density plots for GGCM simulations and RF-predicted maize yields in the validation dataset. Taking EPIC-BOKU as an example, the other models are detailed in the Supplementary Materials (Figures S1–S7). Notes: "ff" represents the irrigated condition, while "nf" represents the rain-fed condition. (**a**) Maize yield under irrigated conditions; (**b**) maize yield under rain-fed conditions; (**c**) density distribution of maize yield under irrigated conditions; (**d**) density distribution of maize yield under rain-fed conditions.

We compared the low-resolution simulations with the high-resolution machine learning predictions for maize yield in China in 2010 (Figures S8–S13). And here, taking EPIC-BOKU as an example (Figure 3), it was evident that for both the rain-fed and irrigated conditions, there was a consistent spatial distribution pattern observed in both the low-resolution simulations and the high-resolution machine learning predictions. The maize yields exhibited significant spatial heterogeneity under both the irrigated and rain-fed conditions, with the maize yield under the irrigated conditions being significantly higher than that under the rain-fed conditions (Figures 3 and S8–S13). However, the high-resolution machine learning predictions outperformed the low-resolution simulations in accurately capturing the heterogeneity of the maize yield across China (Figures 3 and S8–S13). This suggests that high-resolution machine learning predictions are more suitable for county-level maize yield studies in China compared to low-resolution simulations.

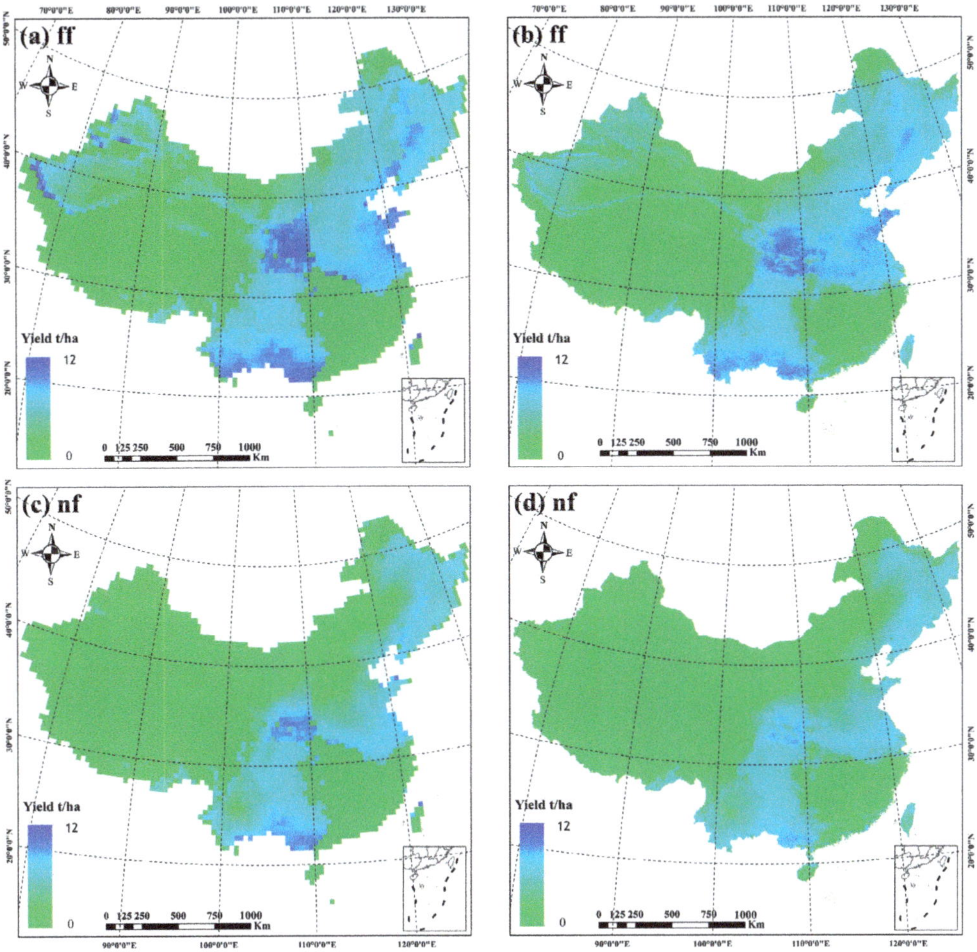

Figure 3. Spatial distributions of maize yield in China before and after downscaling. Taking EPIC-BOKU as an example, the other models are detailed in the Supplementary Materials (Figures S8–S13). (**a**) Irrigated maize yield at a spatial resolution of 50 km; (**b**) irrigated maize yield at a spatial resolution of 1 km; (**c**) rain-fed maize yield at a spatial resolution of 50 km; and (**d**) rain-fed maize yield at a spatial resolution of 1 km.

3.1.2. Feature Importance from Maize Yield Simulations

We calculated the feature importance for all models (Figures S14–S19), taking EPIC-BOKU as an example (Figure 4). In the rain-fed conditions, the most critical indicators for predicting maize yield consistently revolved around precipitation-related features. Following closely were features related to PET, temperature, and soil–terrain characteristics (Figures 4b and S14b–S19b). Under adequate irrigation conditions, temperature-related features and VPD were the predominant indicators for maize yield, whereas the significance of precipitation-related features diminished (Figures 4a and S14a–S19a). Whether the maize was rain-fed or irrigated, the climate characteristics during the maize growing season (e.g., PRATEsumGS, PRATEavgGS, TMINsumGS, TMAXsumGS, TMINavgGS, TMAXavgGS, TAVGsumGS, and TAVGavgGS) consistently ranked as pivotal indicators. This emphasizes that drought or excessive moisture during the growing season significantly impacts maize yield. Features related to soil and terrain characteristics played a relatively lesser role in maize yield predictions. DEM was the most relevant feature among the soil and terrain characteristics. Furthermore, features related to wind speed rarely ranked among the top-ranked features in terms of feature importance when predicting maize yield.

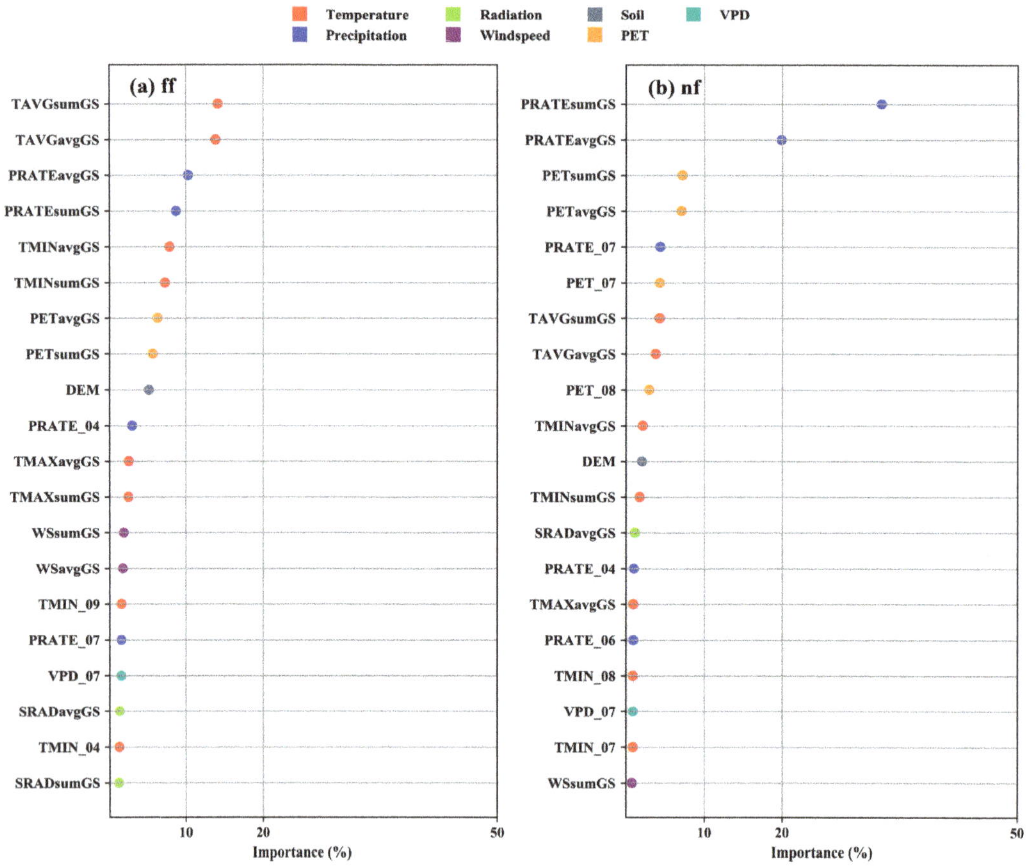

Figure 4. Ranking of feature importance. Taking EPIC-BOKU as an example, the other models are detailed in the Supplementary Materials (Figures S14–S19). (**a**) Feature importance under irrigated conditions; (**b**) feature importance under rain-fed conditions.

3.2. Evaluation of County-Level Maize Yields in China

The downscaling models were applied to spatially downscale the county-level maize yield grids simulated by seven GGCMs (EPIC-BOKU, EPIC-TAMU, EPIC-IIASA, GEPIC, PDSSAT, PEGASUS, CLM-CROP) from a 50 km resolution to 1 km for the period of 1980–2010. The ensemble model was created by averaging the outputs of the seven global crop models, and all models were masked using the maize planting area grid provided by MIRCA2000. The data were then aggregated at the county level. Subsequently, we computed the correlation coefficients (R) between the simulated maize yields and the observed maize yields before and after the downscaling process. The density curves in Figure 5 demonstrate a significant improvement in the R values for all crop models after downscaling. The original 50 km-resolution maize yield simulations exhibited R values primarily within the range of 0 to 0.4, signifying poor performance at the county level. Conversely, following downscaling, the R values for the 1 km-resolution maize yield simulations were concentrated within the range of 0.2 to 0.6. This suggests that downscaling effectively enhanced the county-level maize yield simulations for all the GGCMs.

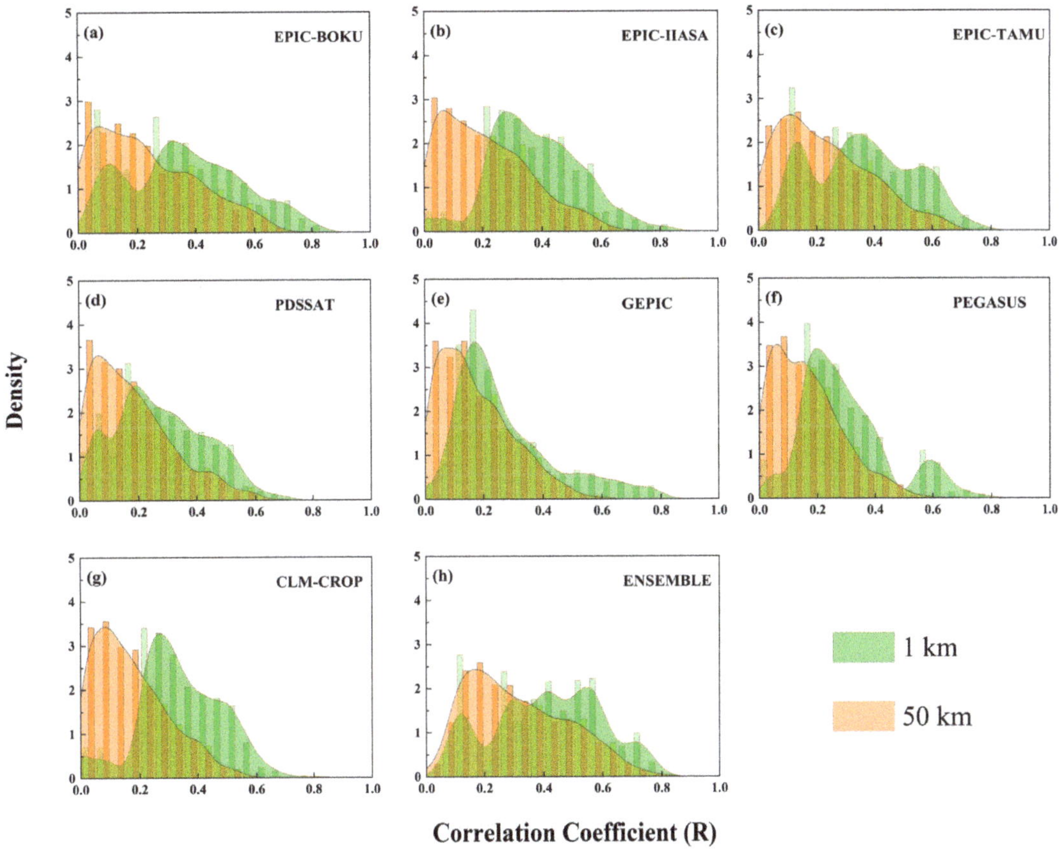

Figure 5. Density distribution of the correlation coefficients between 1 km-resolution simulated maize yields and 50 km-resolution simulated maize yields, in comparison to maize yield observations in China from 1980 to 2010. (**a**) Density distribution of EPIC-BOKU; (**b**) density distribution of EPIC-IIASA; (**c**) density distribution of EPIC-TAMU; (**d**) density distribution of PDSSAT; (**e**) density distribution of GEPIC; (**f**) density distribution of PEGASUS; (**g**) density distribution of CLM-CROP; (**h**) density distribution of ENSEMBLE.

To determine the best-performing model in simulating county-level maize yields in China, we assessed the performance of seven GGCMs before and after downscaling, as well as the ensemble mean of the multi-model, using the indices of R, RMSE, and MAE. The GGCMs demonstrated an improved capacity to elucidate maize yield variability after downscaling (Figure 6). The pre-improvement crop models were only able to explain 16 to 31% of the maize yield variability across a minimum of 1046 counties. However, following improvement, they could explain 30 to 48% of the maize yield variability (average R value across 1046 counties). Simultaneously, both the RMSE and MAE exhibited substantial reductions, with the RMSE decreasing by 0.4 to 0.7 t/ha, and the MAE decreasing by 0.3 to 0.5 t/ha. Among the individual models, EPIC-BOKU performed the best in simulating the county-level maize yield variability (R = 0.38, RMSE = 2.2 t/ha, MAE = 1.7 t/ha), while PDSSAT performed the worst (R = 0.29, RMSE = 2.8 t/ha, MAE = 2.3 t/ha) among the seven models. The multi-model ensemble approach further enhanced the model performance, with ENSEMBLE demonstrating the highest explanatory power for maize yield variability and the lowest errors among all models (R = 0.48, RMSE = 1.8 t/ha, MAE = 1.3 t/ha).

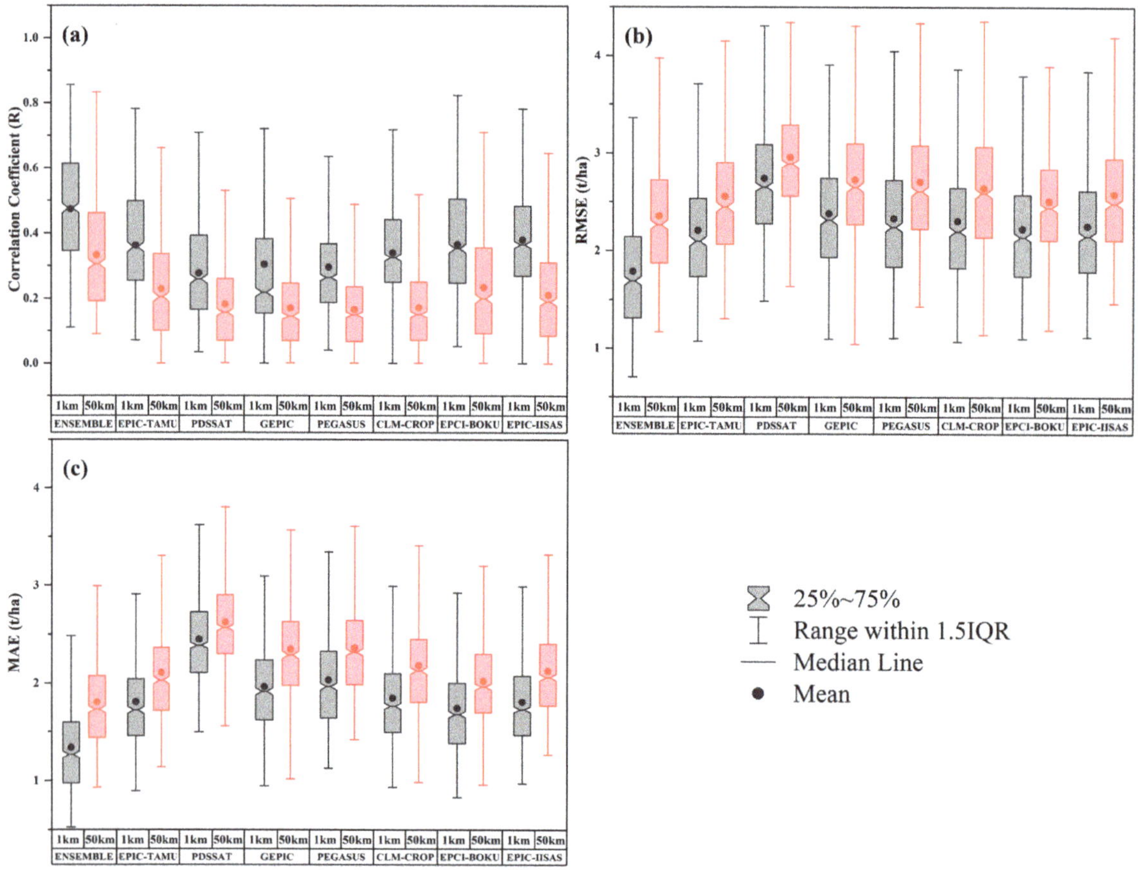

Figure 6. The performance of GGCMs' simulated maize yields before and after downscaling. (**a**) The correlation coefficient (R) between simulated maize yields and observed county-level maize yields before and after downscaling; (**b**) the root mean square error (RMSE) between simulated maize yields and observed county-level maize yields before and after downscaling; (**c**) the mean absolute error (MAE) between simulated maize yields and observed county-level maize yields before and after downscaling. Notes: Red represents GGCMs before downscaling (50 km); brown represents GGCMs after downscaling (1 km).

3.3. Evaluation of County-Level Maize Yield Gap in China

We quantified the maize yield gap in China's major maize-producing regions (1046 counties). There was still substantial room for improving the maize yield in most counties (Figure 7a). Only 7.1% of all the counties had maize yields that exceeded their potential maize yields, while the remaining 92.9% of counties exhibited room for enhancement in their maize yield (Figure 7b). Counties with maize yield gaps ranging from 0 to 1 t/ha constituted the largest share at 38.4%; counties with maize yield gaps ranging from 1 to 2 t/ha made up 32% of the total; those with maize yield gaps ranging from 2 to 3 t/ha accounted for 15.3% of the counties; and counties with maize yield gaps exceeding 3 t/ha represented a proportion of 7.2%. The spatial distribution pattern indicates that counties with maize yield gaps exceeding 2 t/ha are primarily located in the south and southwest maize regions (Figure 7a). Figure 7c highlights the substantial disparities in the maize yield gaps among different regions, with a substantial average maize yield gap of 1.82 t/ha in the south maize region. The northwest maize region, the Huang-Huai-Hai maize region, and the southwest maize region had maize yield gaps of 0.85 t/ha, 0.65 t/ha, and 1.34 t/ha, respectively. In contrast, the north maize region featured the smallest maize yield gap, at only 0.47 t/ha. These findings emphasize the considerable potential for enhancing maize yields in most counties across China, particularly in the south maize region.

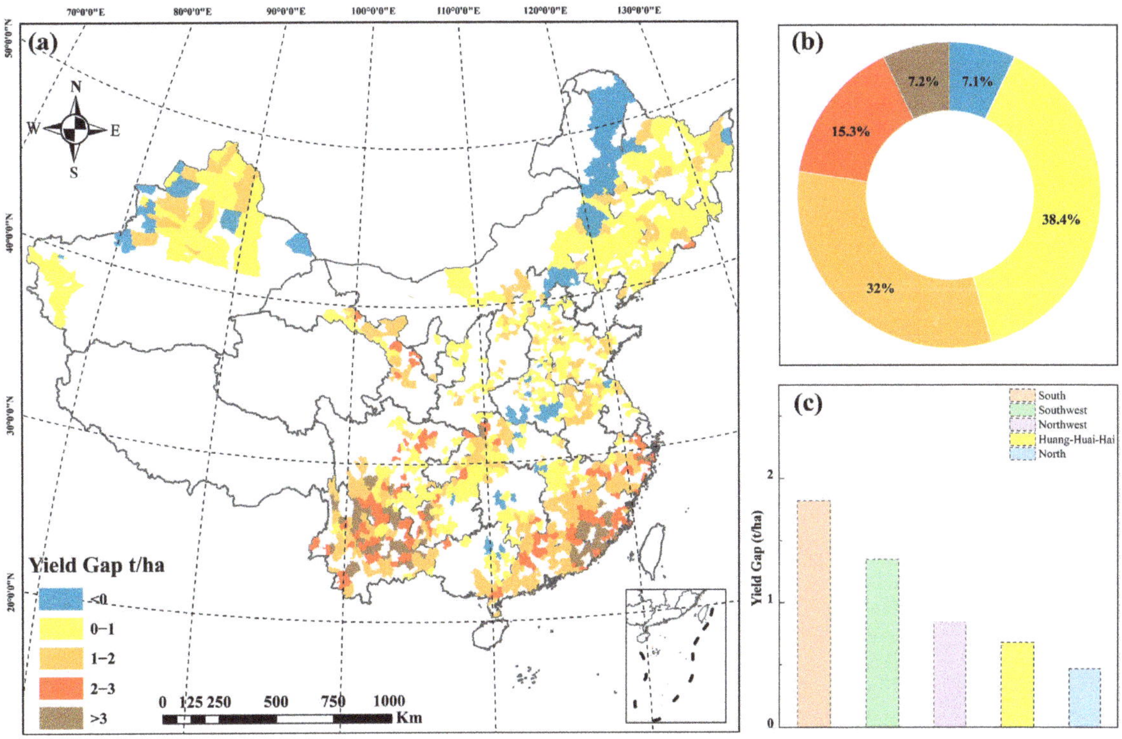

Figure 7. Maize yield disparity at the county level in China. (**a**) Maize yield gaps in various counties; (**b**) percentage of counties falling into each interval (<0 t/ha, 0–1 t/ha, 1–2 t/ha, 2–3 t/ha, >3 t/ha); (**c**) average maize yield discrepancy across five maize plantation regions in China (the south maize region, the southwest maize region, the northeast maize region, the Huang-Huai-Hai maize region, and the northwest maize region).

4. Discussion

Among the eight GGCMs models that provided complete configurations for maize yield simulation scenarios, the combination of these models with random forest and multiple data sources yielded outstanding results. Specifically, seven of these models (EPIC-TAMU, EPIC-IIASA, EPIC-BOKU, GEPIC, PDSSAT, PEGASUS, and CLM-CROP) have demonstrated exceptional performance in downscaling simulated maize yields ($R^2 \geq 0.9$, RMSE < 0.75 t/ha, MAE < 0.5 t/ha). However, due to the use of high-yielding hybrid maize varieties and different maize cultivation practices within PAPSIM, the accuracy of the downscaling models for PAPSIM faced significant challenges [59,60]. Previous studies have shown that GGCMs encounter difficulties in accurately simulating maize yield variability at the county level in China [19]. However, their performance in simulating maize yield variability at the county level in China significantly improved following downscaling, for example, yield variability increased by 14 to 17%, the RMSE decreased by 0.4 to 0.7 t/ha, and the MAE decreased by 0.3 to 0.5 t/ha (Figure 6). The performance of integrated models using the multiple GGCMs (ENSEMBLE, R = 0.48) in our study was found to be very close to that of using the GGCMs for crop yield simulation studies at larger regional scales elsewhere (at the global scale, R = 0.65; at the national scale, R = 0.60; at the provincial scale, R = 0.50) [19,20,61], indicating the increased strength of the model when ensembled. This not only underscores the effectiveness of the downscaling approach but also highlights the potential for more accurate and detailed assessments of crop yield at the county level in China. In comparison to continuously evolving grid-based model simulators, our method significantly improves the computational efficiency and reduces the data processing requirements without sacrificing the simulation performance [62,63]. This allows for a comprehensive assessment of future variations in the yield variability and the associated distribution of extreme yield levels.

When the water supply was insufficient, precipitation and other water-related factors consistently ranked as the most important factors affecting maize yield in China (Figures 4b and S14b–S19b). Temperature and solar radiation were considered the primary factors influencing maize growth only when the maize received adequate irrigation (Figures 4a and S14a–S19a). This reaffirms the predominant influence of precipitation over temperature in determining maize yield variability in most regions [64]. The cumulative or mean climate variables throughout the growing season consistently outweigh the importance of the individual monthly climate variables (Figures 4 and S14–S19). This indicates that the overall levels of high/low temperature, drought, and excessive soil moisture regimes during the maize growing season can significantly impact the maize yield [35,65]. When considering only monthly climate variables, July and August, for example, were found to exert a substantial influence on maize yield in China (Figures 4 and S14–S19), aligning with the notion that climatic conditions during the peak growing period would have been more critical for maize growth [66–68]. Interestingly, the importance of soil and topographic features remained consistently low for both the rain-fed and irrigated conditions, possibly due to the inherent insensitivity of GGCMs to soil data [69]. However, despite the low importance of soil and topography, the spatial patterns of maize yield, as shown in Figure 3, indicate that the downscaling model effectively captured information from the soil and topographic data.

Our study has uncovered the untapped potential for maize yields by revealing the maize yield gap in China. In 2010, only 7.1% of the counties achieved their potential maize yield, while the remaining 92.9% still had significant room for increased yields (Figure 7b). Our study also ranked the maize yield gaps in regional China (Figure 7c); the south maize region ranked the highest, and the north maize region ranked the lowest. These rankings of maize yield gaps in regional China can profoundly impact food distribution within the country. Rankings of the maize yield gaps in regional China could also provide a focus in management practices in maize production across different regions. For instance, despite having lower average maize yields, the south maize region exhibited substantial potential for improving their maize yield in the future. One of the contributing factors to the substantially higher maize yield gap in this region is excessive moisture, which can lead to a rapid rate of nitrogen depletion in the soil [17,70,71]. Similarly, the southwest maize region was also found to be affected by excessive moisture availability, experiencing the second-largest yield gaps following the south maize region (Figure 7c) [35,72]. The south maize region and the southwest maize region experienced excessive annual precipitation, leading to soil oversaturation and intense rainfall that could physically damage maize growth [73]. Excessive moisture reduced maize resistance to lodging and nitrogen utilization, impacting regional irrigation, fertilization practices, and soil characteristics [74,75]. Through the optimization of nitrogen management levels, simulated maize yields significantly increased in the south maize region and the southwest maize region (Figure 7c). This indicates that improving nitrogen utilization in these regions can extend the maize planting potential. Our approach used in this study successfully identified maize yield gaps in regional China, and this has become increasingly useful for providing policy strategies to enhance maize yields and reduce yield gaps through better crop management practices, including adequate irrigation, appropriate plantation, and optimized crop fertilization [76,77]. Our findings suggest that China may need to focus on addressing the impact of climate change and improving nitrogen utilization to increase maize yields in the future [72], particularly in the south maize region and the southwest maize region.

The method introduced in this paper demonstrates the tremendous potential of machine learning for building readily applicable downscaling models for GGCMs in spatiotemporal applications. However, certain limitations exist. For instance, the feature variables do not incorporate regional crop management conditions or crop varieties. Crop management conditions often vary based on the local circumstances, and crop varieties need to adapt to regional conditions, such as temperature and precipitation [78]. Therefore, manually combining and systematically training regional crop varieties and management conditions would enhance crop yield simulations at a finer resolution. Additionally, research has shown that remote sensing data can provide supplementary information describing crop growth to further improve crop yield simulations [79,80]. Hence, this could serve as a blueprint for expanding the range and dimensions of feature variables, making the model applicable to a broader range of applications.

5. Conclusions

Our study addresses the important food security issue that China has been facing under rapid climate warming in the 21st century. The advancement of numerical techniques could potentially enhance food security in regional China. In this study, we have demonstrated the effectiveness of integrated machine learning techniques and the downscaling of GGCMs to improve model performance when simulating the county-level maize yield in China. By employing the random forest algorithm with seven GGCMs (EPIC-TAMU, EPIC-IIASA, EPIC-BOKU, GEPIC, PDSSAT, PEGASUS, and CLM-CROP), we have successfully established high-performance downscaled models that can inform China's crop management practices. We have clearly indicated that the original GGCMs face challenges in accurately simulating county-level maize yields, mainly due to their low spatial resolution, which could only capture approximately 16 to 31% of the yield variability across 1046 counties in China from 1980 to 2010. However, the improved GGCMs significantly enhanced the model performance, explaining 30 to 48% of the maize yield variability at the county level. The model ensemble we made through integration of GGCMs further enhanced the performance and overall accuracy of our results. The maize yield gap quantification we made at the county level in China has successfully revealed that the south maize plantation region has the highest gap, requiring major management attention. We argue that the machine learning approach offers greater spatial advantages for simulations by opening new avenues for the application of GGCMs in regional China and providing valuable insights into improving crop yield simulations.

Supplementary Materials: The following supporting information can be downloaded at: https://www.mdpi.com/article/10.3390/rs16040701/s1, Table S1: Basic introduction to the selected GGCMs. Figure S1: Scatter and density plots for EPIC-IIASA simulations and RF-predicted maize yields in the validation dataset. Notes: "ff" represents the irrigated condition, while "nf" represents the rain-fed condition. (a) Maize yield under irrigated conditions; (b) maize yield under rain-fed conditions; (c) density distribution of maize yield under irrigated conditions; (d) density distribution of maize yield under rain-fed conditions. Figure S2: Scatter and density plots for EPIC-TAMU simulations and RF-predicted maize yields in the validation dataset. Figure S3: Scatter and density plots for PDSSAT simulations and RF-predicted maize yields in the validation dataset. Figure S4: Scatter and density plots for GEPIC simulations and RF-predicted maize yields in the validation dataset. Figure S5: Scatter and density plots for PEGASUS simulations and RF-predicted maize yields in the validation dataset. Figure S6: Scatter and density plots for CLM-CROP simulations and RF-predicted maize yields in the validation dataset. Figure S7: Scatter and density plots for PAPSIM simulations and RF-predicted maize yields in the validation dataset. Figure S8: Spatial distributions of maize yield in China before and after downscaling for EPIC-IIASA. (a) Irrigated maize yield at a spatial resolution of 50 km; (b) irrigated maize yield at a spatial resolution of 1 km; (c) rain-fed maize yield at a spatial resolution of 50 km; and (d) rain-fed maize yield at a spatial resolution of 1 km. Figure S9: Spatial distributions of maize yield in China before and after downscaling for EPIC-TAMU. Figure S10: Spatial distributions of maize yield in China before and after downscaling for PDSSAT. Figure S11: Spatial distributions of maize yield in China before and after downscaling for GEPIC. Figure S12: Spatial distributions of maize yield in China before and after downscaling for PEGASUS. Figure S13: Spatial distributions of maize yield in China before and after downscaling for CLM-CROP. Figure S14: Ranking of feature importance for EPIC-IIASA. (a) Feature importance under irrigated conditions; (b) feature importance under rain-fed conditions. Figure S15: Ranking of feature importance for EPIC-TAMU. Figure S16: Ranking of feature importance for PDSSAT. Figure S17: Ranking of feature importance for GEPIC. Figure S18: Ranking of feature importance for PEGASUS. Figure S19: Ranking of feature importance for CLM-CROP.

Author Contributions: Conceptualization, Y.Z. and L.M.; data curation, Y.Z.; formal analysis, Y.Z.; investigation, Y.Z. and L.M.; methodology, Y.Z.; software, Y.Z.; validation, L.M.; resources, Y.Z.; writing—original draft preparation, Y.Z.; writing—review and editing, G.R.K. and L.M.; visualization, Y.Z.; supervision, L.M.; project administration, Y.Z. and L.M.; funding acquisition, Y.Z. and L.M. All authors have read and agreed to the published version of the manuscript.

Funding: This research was supported by the Postgraduate Research and Practice Innovation Program of Jiangsu Province (KYCX23_1294), the National Natural Science Foundation of China (4210010673), and the Natural Science Foundation of Jiangsu Province (SBK2020040616).

Data Availability Statement: All data resources are provided in the manuscript with links.

Acknowledgments: Yangfeng Zou, Giri Raj Kattel, and Lijuan Miao would like to acknowledge the School of Geographical Sciences, Nanjing University of Information Science and Technology.

Conflicts of Interest: The authors declare no conflicts of interest.

References

1. Hunter, M.C.; Smith, R.G.; Schipanski, M.E.; Atwood, L.W.; Mortensen, D.A. Agriculture in 2050: Recalibrating Targets for Sustainable Intensification. *BioScience* **2017**, *67*, 386–391. [CrossRef]
2. Keating, B.A.; Herrero, M.; Carberry, P.S.; Gardner, J.; Cole, M.B. Food wedges: Framing the global food demand and supply challenge towards 2050. *Glob. Food Secur.* **2014**, *3*, 125–132. [CrossRef]
3. Van Dijk, M.; Morley, T.; Rau, M.L.; Saghai, Y. A meta-analysis of projected global food demand and population at risk of hunger for the period 2010–2050. *Nat. Food* **2021**, *2*, 494–501. [CrossRef]
4. Tilman, D.; Balzer, C.; Hill, J.; Befort, B.L. Global food demand and the sustainable intensification of agriculture. *Proc. Natl. Acad. Sci. USA* **2011**, *108*, 20260–20264. [CrossRef]
5. Trnka, M.; Rötter, R.P.; Ruiz-Ramos, M.; Kersebaum, K.C.; Olesen, J.E.; Žalud, Z.; Semenov, M.A. Adverse weather conditions for European wheat production will become more frequent with climate change. *Nat. Clim. Chang.* **2014**, *4*, 637–643. [CrossRef]
6. IPCC. Climate Change and Land: An IPCC Special Report on Climate Change, Desertification, Land Degradation, Sustainable Land Management, Food Security, and Greenhouse Gas Fluxes in Terrestrial Ecosystems. Available online: https://www.ipcc.ch/srccl-report-download-page/ (accessed on 23 August 2019).
7. Gomez-Zavaglia, A.; Mejuto, J.C.; Simal-Gandara, J. Mitigation of emerging implications of climate change on food production systems. *Food Res. Int.* **2020**, *134*, 109256. [CrossRef] [PubMed]
8. Godfray, H.C.J.; Beddington, J.R.; Crute, I.R.; Haddad, L.; Lawrence, D.; Muir, J.F.; Pretty, J.; Robinson, S.; Thomas, S.M.; Toulmin, C. Food Security: The Challenge of Feeding 9 Billion People. *Science* **2010**, *327*, 812–818. [CrossRef] [PubMed]
9. Lobell, D.B.; Burke, M.B. On the use of statistical models to predict crop yield responses to climate change. *Agric. For. Meteorol.* **2010**, *150*, 1443–1452. [CrossRef]
10. Zhao, C.; Liu, B.; Piao, S.; Wang, X.; Lobell, D.B.; Huang, Y.; Huang, M.; Yao, Y.; Bassu, S.; Ciais, P.; et al. Temperature increase reduces global yields of major crops in four independent estimates. *Proc. Natl. Acad. Sci. USA* **2017**, *114*, 9326–9331. [CrossRef] [PubMed]
11. Heinicke, S.; Frieler, K.; Jägermeyr, J.; Mengel, M. Global gridded crop models underestimate yield responses to droughts and heatwaves. *Environ. Res. Lett.* **2022**, *17*, 044026. [CrossRef]
12. Mistry, M.N.; Sue Wing, I.; De Cian, E. Simulated vs. empirical weather responsiveness of crop yields: US evidence and implications for the agricultural impacts of climate change. *Environ. Res. Lett.* **2017**, *12*, 075007. [CrossRef]
13. Rosenzweig, C.; Elliott, J.; Deryng, D.; Ruane, A.C.; Müller, C.; Arneth, A.; Boote, K.J.; Folberth, C.; Glotter, M.; Khabarov, N.; et al. Assessing agricultural risks of climate change in the 21st century in a global gridded crop model intercomparison. *Proc. Natl. Acad. Sci. USA* **2014**, *111*, 3268–3273. [CrossRef] [PubMed]

14. Rosenzweig, C.; Jones, J.W.; Hatfield, J.L.; Ruane, A.C.; Boote, K.J.; Thorburn, P.; Antle, J.M.; Nelson, G.C.; Porter, C.; Janssen, S.; et al. The Agricultural Model Intercomparison and Improvement Project (AgMIP): Protocols and pilot studies. *Agric. For. Meteorol.* **2013**, *170*, 166–182. [CrossRef]
15. Warszawski, L.; Frieler, K.; Huber, V.; Piontek, F.; Serdeczny, O.; Schewe, J. The Inter-Sectoral Impact Model Intercomparison Project (ISI–MIP): Project framework. *Proc. Natl. Acad. Sci. USA* **2014**, *111*, 3228–3232. [CrossRef] [PubMed]
16. Elliott, J.; Müller, C.; Deryng, D.; Chryssanthacopoulos, J.; Boote, K.J.; Büchner, M.; Foster, I.; Glotter, M.; Heinke, J.; Iizumi, T.; et al. The Global Gridded Crop Model Intercomparison: Data and modeling protocols for Phase 1 (v1.0). *Geosci. Model Dev.* **2015**, *8*, 261–277. [CrossRef]
17. Folberth, C.; Elliott, J.; Müller, C.; Balkovic, J.; Chryssanthacopoulos, J.; Izaurralde, R.C.; Jones, C.D.; Khabarov, N.; Liu, W.; Reddy, A.; et al. Uncertainties in global crop model frameworks: Effects of cultivar distribution, crop management and soil handling on crop yield estimates. *Biogeosciences Discuss.* **2016**, *2016*, 1–30. [CrossRef]
18. Van Meijl, H.; Havlik, P.; Lotze-Campen, H.; Stehfest, E.; Witzke, P.; Domínguez, I.P.; Bodirsky, B.L.; van Dijk, M.; Doelman, J.; Fellmann, T.; et al. Comparing impacts of climate change and mitigation on global agriculture by 2050. *Environ. Res. Lett.* **2018**, *13*, 064021. [CrossRef]
19. Yin, X.; Leng, G. Modelling global impacts of climate variability and trend on maize yield during 1980–2010. *Int. J. Climatol.* **2021**, *41*, E1583–E1596. [CrossRef]
20. Li, Z.; Zhan, C.; Hu, S.; Ning, L.; Wu, L.; Guo, H. Evaluation of global gridded crop models (GGCMs) for the simulation of major grain crop yields in China. *Hydrol. Res.* **2022**, *53*, 353–369. [CrossRef]
21. Folberth, C.; Baklanov, A.; Balkovič, J.; Skalský, R.; Khabarov, N.; Obersteiner, M. Spatio-temporal downscaling of gridded crop model yield estimates based on machine learning. *Agric. For. Meteorol.* **2019**, *264*, 1–15. [CrossRef]
22. Reidsma, P.; Ewert, F.; Boogaard, H.; Diepen, K.v. Regional crop modelling in Europe: The impact of climatic conditions and farm characteristics on maize yields. *Agric. Syst.* **2009**, *100*, 51–60. [CrossRef]
23. Ewert, F.; van Ittersum, M.K.; Heckelei, T.; Therond, O.; Bezlepkina, I.; Andersen, E. Scale changes and model linking methods for integrated assessment of agri-environmental systems. *Agric. Ecosyst. Environ.* **2011**, *142*, 6–17. [CrossRef]
24. Rosenzweig, C.; Ruane, A.C.; Antle, J.; Elliott, J.; Ashfaq, M.; Chatta, A.A.; Ewert, F.; Folberth, C.; Hathie, I.; Havlik, P.; et al. Coordinating AgMIP data and models across global and regional scales for 1.5 °C and 2.0 °C assessments. *Philos. Trans. R. Soc. A Math. Phys. Eng. Sci.* **2018**, *376*, 20160455. [CrossRef]
25. Prodhan, F.A.; Zhang, J.; Pangali Sharma, T.P.; Nanzad, L.; Zhang, D.; Seka, A.M.; Ahmed, N.; Hasan, S.S.; Hoque, M.Z.; Mohana, H.P. Projection of future drought and its impact on simulated crop yield over South Asia using ensemble machine learning approach. *Sci. Total Environ.* **2022**, *807*, 151029. [CrossRef] [PubMed]
26. Li, X.; Fang, S.; Wu, D.; Zhu, Y.; Wu, Y. Risk analysis of maize yield losses in mainland China at the county level. *Sci. Rep.* **2020**, *10*, 10684. [CrossRef]
27. Ruane, A.C.; Goldberg, R.; Chryssanthacopoulos, J. Climate forcing datasets for agricultural modeling: Merged products for gap-filling and historical climate series estimation. *Agric. For. Meteorol.* **2015**, *200*, 233–248. [CrossRef]
28. Witten, I.H.; Frank, E. Data mining: Practical machine learning tools and techniques with Java implementations. *ACM SIGMOD Rec.* **2002**, *31*, 76–77. [CrossRef]
29. Yang, P.; Zhao, Q.; Cai, X. Machine learning based estimation of land productivity in the contiguous US using biophysical predictors. *Environ. Res. Lett.* **2020**, *15*, 074013. [CrossRef]
30. Mohanasundaram, S.; Kasiviswanathan, K.S.; Purnanjali, C.; Santikayasa, I.P.; Singh, S. Downscaling Global Gridded Crop Yield Data Products and Crop Water Productivity Mapping Using Remote Sensing Derived Variables in the South Asia. *Int. J. Plant Prod.* **2023**, *17*, 1–16. [CrossRef]
31. Duro, D.C.; Franklin, S.E.; Dubé, M.G. A comparison of pixel-based and object-based image analysis with selected machine learning algorithms for the classification of agricultural landscapes using SPOT-5 HRG imagery. *Remote Sens. Environ.* **2012**, *118*, 259–272. [CrossRef]
32. Shin, J.-Y.; Kim, K.R.; Ha, J.-C. Seasonal forecasting of daily mean air temperatures using a coupled global climate model and machine learning algorithm for field-scale agricultural management. *Agric. For. Meteorol.* **2020**, *281*, 107858. [CrossRef]
33. Shahhosseini, M.; Hu, G.; Huber, I.; Archontoulis, S.V. Coupling machine learning and crop modeling improves crop yield prediction in the US Corn Belt. *Sci. Rep.* **2021**, *11*, 1606. [CrossRef]
34. Yang, Y.; Xu, W.; Hou, P.; Liu, G.; Liu, W.; Wang, Y.; Zhao, R.; Ming, B.; Xie, R.; Wang, K.; et al. Improving maize grain yield by matching maize growth and solar radiation. *Sci. Rep.* **2019**, *9*, 3635. [CrossRef]
35. Liu, W.; Li, Z.; Li, Y.; Ye, T.; Chen, S.; Liu, Y. Heterogeneous impacts of excessive wetness on maize yields in China: Evidence from statistical yields and process-based crop models. *Agric. For. Meteorol.* **2022**, *327*, 109205. [CrossRef]
36. Butler, E.E.; Huybers, P. Adaptation of US maize to temperature variations. *Nat. Clim. Chang.* **2013**, *3*, 68–72. [CrossRef]
37. Liu, H.; Zhang, X.; Wang, Y.; Guo, Y.; Shen, Y. Spatio-temporal characteristics of the hydrothermal conditions in the growth period and various gro wth stages of maize in China from 1960 to 2018. *Chin. J. Eco-Agric.* **2021**, *29*, 1417–1429. [CrossRef]
38. Müller, C.; Elliott, J.; Chryssanthacopoulos, J.; Deryng, D.; Folberth, C.; Pugh, T.A.M.; Schmid, E. Implications of climate mitigation for future agricultural production. *Environ. Res. Lett.* **2015**, *10*, 125004. [CrossRef]

39. Gelaro, R.; McCarty, W.; Suárez, M.J.; Todling, R.; Molod, A.; Takacs, L.; Randles, C.A.; Darmenov, A.; Bosilovich, M.G.; Reichle, R.; et al. The Modern-Era Retrospective Analysis for Research and Applications, Version 2 (MERRA-2). *J. Clim.* **2017**, *30*, 5419–5454. [CrossRef] [PubMed]
40. Wieder, W. *Regridded Harmonized World Soil Database v1.2*; Oak Ridge National Laboratory: Roane County, TN, USA, 2014. [CrossRef]
41. Gandhi, S.M.; Sarkar, B.C. Chapter 3—Reconnaissance and Prospecting. In *Essentials of Mineral Exploration and Evaluation*; Gandhi, S.M., Sarkar, B.C., Eds.; Elsevier: Amsterdam, The Netherlands, 2016; pp. 53–79.
42. Portmann, F.T.; Siebert, S.; Döll, P. MIRCA2000—Global monthly irrigated and rainfed crop areas around the year 2000: A new high-resolution data set for agricultural and hydrological modeling. *Glob. Biogeochem. Cycles.* **2010**, *24*, 24. [CrossRef]
43. Ray, D.K.; Ramankutty, N.; Mueller, N.D.; West, P.C.; Foley, J.A. Recent patterns of crop yield growth and stagnation. *Nat. Commun.* **2012**, *3*, 1293. [CrossRef]
44. Müller, C.; Elliott, J.; Chryssanthacopoulos, J.; Arneth, A.; Balkovic, J.; Ciais, P.; Deryng, D.; Folberth, C.; Glotter, M.; Hoek, S.; et al. Global gridded crop model evaluation: Benchmarking, skills, deficiencies and implications. *Geosci. Model Dev.* **2017**, *10*, 1403–1422. [CrossRef]
45. Lobell, D.B.; Field, C.B. Global scale climate–crop yield relationships and the impacts of recent warming. *Environ. Res. Lett.* **2007**, *2*, 014002. [CrossRef]
46. Villoria, N.B.; Elliott, J.; Müller, C.; Shin, J.; Zhao, L.; Song, C. Rapid aggregation of global gridded crop model outputs to facilitate cross-disciplinary analysis of climate change impacts in agriculture. *Environ. Model. Softw.* **2016**, *75*, 193–201. [CrossRef]
47. Porwollik, V.; Müller, C.; Elliott, J.; Chryssanthacopoulos, J.; Iizumi, T.; Ray, D.K.; Ruane, A.C.; Arneth, A.; Balkovič, J.; Ciais, P.; et al. Spatial and temporal uncertainty of crop yield aggregations. *Eur. J. Agron.* **2017**, *88*, 10–21. [CrossRef]
48. Breiman, L. Random Forests. *Mach. Learn.* **2001**, *45*, 5–32. [CrossRef]
49. Elavarasan, D.; Vincent, D.R.; Sharma, V.; Zomaya, A.Y.; Srinivasan, K. Forecasting yield by integrating agrarian factors and machine learning models: A survey. *Comput. Electron. Agric.* **2018**, *155*, 257–282. [CrossRef]
50. Cai, Y.; Guan, K.; Peng, J.; Wang, S.; Seifert, C.; Wardlow, B.; Li, Z. A high-performance and in-season classification system of field-level crop types using time-series Landsat data and a machine learning approach. *Remote Sens. Environ.* **2018**, *210*, 35–47. [CrossRef]
51. Strobl, C.; Boulesteix, A.-L.; Zeileis, A.; Hothorn, T. Bias in random forest variable importance measures: Illustrations, sources and a solution. *BMC Bioinform.* **2007**, *8*, 25. [CrossRef] [PubMed]
52. Zhang, L.; Zhang, Z.; Luo, Y.; Cao, J.; Tao, F. Combining Optical, Fluorescence, Thermal Satellite, and Environmental Data to Predict County-Level Maize Yield in China Using Machine Learning Approaches. *Remote Sens.* **2020**, *12*, 21. [CrossRef]
53. Li, M.; Zhao, J.; Yang, X. Building a new machine learning-based model to estimate county-level climatic yield variation for maize in Northeast China. *Comput. Electron. Agric.* **2021**, *191*, 106557. [CrossRef]
54. Roy, M.-H.; Larocque, D. Robustness of random forests for regression. *J. Nonparametric Stat.* **2012**, *24*, 993–1006. [CrossRef]
55. Li, Z.; Zhang, Z.; Zhang, L. Improving regional wheat drought risk assessment for insurance application by integrating scenario-driven crop model, machine learning, and satellite data. *Agric. Syst.* **2021**, *191*, 103141. [CrossRef]
56. Li, L.; Zhang, Y.; Wang, B.; Feng, P.; He, Q.; Shi, Y.; Liu, K.; Harrison, M.T.; Liu, D.L.; Yao, N.; et al. Integrating machine learning and environmental variables to constrain uncertainty in crop yield change projections under climate change. *Eur. J. Agron.* **2023**, *149*, 126917. [CrossRef]
57. Luo, Y.; Zhang, Z.; Chen, Y.; Li, Z.; Tao, F. ChinaCropPhen1km: A high-resolution crop phenological dataset for three staple crops in China during 2000–2015 based on leaf area index (LAI) products. *Earth Syst. Sci. Data* **2020**, *12*, 197–214. [CrossRef]
58. Liu, Y.; Qin, Y.; Ge, Q. Spatiotemporal differentiation of changes in maize phenology in China from 1981 to 2010. *J. Geogr. Sci.* **2019**, *29*, 351–362. [CrossRef]
59. Keating, B.A.; Carberry, P.S.; Hammer, G.L.; Probert, M.E.; Robertson, M.J.; Holzworth, D.; Huth, N.I.; Hargreaves, J.N.G.; Meinke, H.; Hochman, Z.; et al. An overview of APSIM, a model designed for farming systems simulation. *Eur. J. Agron.* **2003**, *18*, 267–288. [CrossRef]
60. Müller, C.; Elliott, J.; Kelly, D.; Arneth, A.; Balkovic, J.; Ciais, P.; Deryng, D.; Folberth, C.; Hoek, S.; Izaurralde, R.C.; et al. The Global Gridded Crop Model Intercomparison phase 1 simulation dataset. *Sci. Data* **2019**, *6*, 50. [CrossRef] [PubMed]
61. Ringeval, B.; Müller, C.; Pugh, T.A.M.; Mueller, N.D.; Ciais, P.; Folberth, C.; Liu, W.; Debaeke, P.; Pellerin, S. Potential yield simulated by global gridded crop models: Using a process-based emulator to explain their differences. *Geosci. Model Dev.* **2021**, *14*, 1639–1656. [CrossRef]
62. Franke, J.A.; Müller, C.; Elliott, J.; Ruane, A.C.; Jägermeyr, J.; Snyder, A.; Dury, M.; Falloon, P.D.; Folberth, C.; François, L.; et al. The GGCMI Phase 2 emulators: Global gridded crop model responses to changes in CO_2, temperature, water, and nitrogen (version 1.0). *Geosci. Model Dev.* **2020**, *13*, 3995–4018. [CrossRef]
63. Blanc, É. Statistical emulators of maize, rice, soybean and wheat yields from global gridded crop models. *Agric. For. Meteorol.* **2017**, *236*, 145–161. [CrossRef]
64. Frieler, K.; Schauberger, B.; Arneth, A.; Balkovič, J.; Chryssanthacopoulos, J.; Deryng, D.; Elliott, J.; Folberth, C.; Khabarov, N.; Müller, C.; et al. Understanding the weather signal in national crop-yield variability. *Earth's Future* **2017**, *5*, 605–616. [CrossRef]
65. Zhu, X.; Xu, K.; Liu, Y.; Guo, R.; Chen, L. Assessing the vulnerability and risk of maize to drought in China based on the AquaCrop model. *Agric. Syst.* **2021**, *189*, 103040. [CrossRef]

66. Benincasa, P.; Reale, L.; Tedeschini, E.; Ferri, V.; Cerri, M.; Ghitarrini, S.; Falcinelli, B.; Frenguelli, G.; Ferranti, F.; Ayano, B.E.; et al. The relationship between grain and ovary size in wheat: An analysis of contrasting grain weight cultivars under different growing conditions. *Field Crops Res.* **2017**, *210*, 175–182. [CrossRef]
67. Kim, N.; Lee, Y.-W. Machine Learning Approaches to Corn Yield Estimation Using Satellite Images and Climate Data: A Case of Iowa State. *J. Korean Soc. Surv. Geodesy Photogramm. Cartogr.* **2016**, *34*, 383–390. [CrossRef]
68. Zhao, Y.; Lobell, D.B. Assessing the heterogeneity and persistence of farmers' maize yield performance across the North China Plain. *Field Crops Res.* **2017**, *205*, 55–66. [CrossRef]
69. Folberth, C.; Skalský, R.; Moltchanova, E.; Balkovič, J.; Azevedo, L.B.; Obersteiner, M.; van der Velde, M. Uncertainty in soil data can outweigh climate impact signals in global crop yield simulations. *Nat. Commun.* **2016**, *7*, 11872. [CrossRef] [PubMed]
70. Lobell, D.B.; Deines, J.M.; Tommaso, S.D. Changes in the drought sensitivity of US maize yields. *Nat. Food* **2020**, *1*, 729–735. [CrossRef] [PubMed]
71. Lesk, C.; Coffel, E.; Horton, R. Net benefits to US soy and maize yields from intensifying hourly rainfall. *Nat. Clim. Chang.* **2020**, *10*, 819–822. [CrossRef]
72. Yu, Y.; Jiang, Z.; Wang, G.; Kattel, G.R.; Chuai, X.; Shang, Y.; Zou, Y.; Miao, L. Disintegrating the impact of climate change on maize yield from human management practices in China. *Agric. For. Meteorol.* **2022**, *327*, 109235. [CrossRef]
73. Li, Y.; Guan, K.; Schnitkey, G.D.; DeLucia, E.; Peng, B. Excessive rainfall leads to maize yield loss of a comparable magnitude to extreme drought in the United States. *Glob. Chang. Biol.* **2019**, *25*, 2325–2337. [CrossRef] [PubMed]
74. Haqiqi, I.; Grogan, D.S.; Hertel, T.W.; Schlenker, W. Quantifying the impacts of compound extremes on agriculture. *Hydrol. Earth Syst. Sci.* **2021**, *25*, 551–564. [CrossRef]
75. Lu, Y.; Wang, E.; Zhao, Z.; Liu, X.; Tian, A.; Zhang, X. Optimizing irrigation to reduce N leaching and maintain high crop productivity through the manipulation of soil water storage under summer monsoon climate. *Field Crops Res.* **2021**, *265*, 108110. [CrossRef]
76. Cui, Z.; Chen, X.; Miao, Y.; Zhang, F.; Sun, Q.; Schroder, J.; Zhang, H.; Li, J.; Shi, L.; Xu, J.; et al. On-Farm Evaluation of the Improved Soil Nmin–based Nitrogen Management for Summer Maize in North China Plain. *Agron. J.* **2008**, *100*, 517–525. [CrossRef]
77. Chlingaryan, A.; Sukkarieh, S.; Whelan, B. Machine learning approaches for crop yield prediction and nitrogen status estimation in precision agriculture: A review. *Comput. Electron. Agric.* **2018**, *151*, 61–69. [CrossRef]
78. Jin, Z.; Azzari, G.; Burke, M.; Aston, S.; Lobell, D.B. Mapping Smallholder Yield Heterogeneity at Multiple Scales in Eastern Africa. *Remote Sens.* **2017**, *9*, 931. [CrossRef]
79. Zhou, W.; Liu, Y.; Ata-Ul-Karim, S.T.; Ge, Q.; Li, X.; Xiao, J. Integrating climate and satellite remote sensing data for predicting county-level wheat yield in China using machine learning methods. *Int. J. Appl. Earth Obs. Geoinf.* **2022**, *111*, 102861. [CrossRef]
80. Cai, Y.; Guan, K.; Lobell, D.; Potgieter, A.B.; Wang, S.; Peng, J.; Xu, T.; Asseng, S.; Zhang, Y.; You, L.; et al. Integrating satellite and climate data to predict wheat yield in Australia using machine learning approaches. *Agric. For. Meteorol.* **2019**, *274*, 144–159. [CrossRef]

Disclaimer/Publisher's Note: The statements, opinions and data contained in all publications are solely those of the individual author(s) and contributor(s) and not of MDPI and/or the editor(s). MDPI and/or the editor(s) disclaim responsibility for any injury to people or property resulting from any ideas, methods, instructions or products referred to in the content.

sustainability

MDPI

Article

Developing a New ANN Model to Estimate Daily Actual Evapotranspiration Using Limited Climatic Data and Remote Sensing Techniques for Sustainable Water Management

Halil Karahan [1,*], Mahmut Cetin [2], Muge Erkan Can [2] and Omar Alsenjar [2]

1 Department of Civil Engineering, Pamukkale University, Denizli 20160, Turkey
2 Department of Agricultural Structures and Irrigation, Cukurova University, Adana 01250, Turkey; mcet64@cu.edu.tr (M.C.); merkan@cu.edu.tr (M.E.C.); omarsenjar@yahoo.com (O.A.)
* Correspondence: hkarahan@pau.edu.tr

Abstract: Accurate estimations of actual evapotranspiration (ETa) are essential to various environmental issues. Artificial intelligence-based models are a promising alternative to the most common direct ETa estimation techniques and indirect methods by remote sensing (RS)-based surface energy balance models. Artificial Neural Networks (ANNs) are proven to be suitable for predicting reference evapotranspiration (ETo) and ETa based on RS data. This study aims to develop a methodology based on ANNs for estimating daily ETa values using NDVI and land surface temperature, coupled with limited site-specific climatic variables in a large irrigation catchment. The ANN model has been applied to the two different scenarios. Data from only the 38 days of satellite overpass dates was selected in Scenario I, while in Scenario II all datasets, i.e., the 769-day data were used. An irrigation scheme, located in the Mediterranean region of Turkiye, was selected, and a total of 38 Landsat images and local climatic data collected in 2021 and 2022 were used in the ANN model. The ETa results by the ANN model for Scenarios I and II showed that the R^2 values for training (0.79 and 0.86), testing (0.75 and 0.81), and the entire dataset (0.76 and 0.84) were all remarkably high. Moreover, the results of the new ANN model in two scenarios showed an acceptable agreement with ETa-METRIC values. The proposed ANN model demonstrated the potential for obtaining daily ETa using limited climatic data and RS imagery. As a result, the suggested ANN model for daily ETa computation offers a trustworthy way to determine crop water usage in real time for sustainable water management in agriculture. It may also be used to assess how crop evapotranspiration in drought-prone areas will be affected by climate change in the 21st century.

Keywords: evapotranspiration; artificial neural networks (ANNs); remote sensing (RS); METRIC model; climate change; sustainability

1. Introduction

Scientists, as well as practitioners, have been seeking appropriate, integrated, and sustainable approaches to determine reference and actual evapotranspiration amounts in both irrigation and rainfed catchments. Actual evapotranspiration (ETa) measurements in the field or estimation by any respective conventional method [1] are needed to ensure the sustainability of irrigation water management and irrigation scheduling at large-scale irrigation schemes, as well as in farmyards, water-balance works in hydrologic studies, accurate designs of hydraulic structures related to irrigation schemes, etc. By introducing a two-step procedure [2], ETa, in other words, crop evapotranspiration ETc, is estimated under standard conditions [2], i.e., well-watered and optimal agronomic conditions. On the other hand, ETa is an essential component of the water budget, hydrological modeling, and irrigation water management in arid and semi-arid regions of the world. Nevertheless, it is not an easy task to acquire a representative ETa value for it can be determined directly using either a lysimeter or water-balance work/approach which are rather labor-intensive, time-consuming, and very expensive, thereby threatening the sustainability of freshwater resources. As stated clearly by Allen et al. [2], Rawat et al. [3], Gharbia et al. [4], and Alsenjar et al. [5], among others, ETa describes the physical processes of the amount of water that can occur either through evaporation or transpiration according to climatic conditions, crop types, and soil status. Since direct methods of ETa calculation focus only on one point or parcel, they have limitations in showing the variation in ETa spatially and temporally at a large scale [6], particularly in areas with large irrigation schemes of >100,000 ha as in Turkiye. Therefore, accurately estimating ETa is of the utmost importance, as well as a critical issue in water-balance methods and agricultural water management at the irrigation scheme level. In this regard, the remote sensing (RS)-based surface energy balance models are one of the indirect methods to estimate spatially ETa over large-scale irrigation catchments with a high spatial and temporal resolution [7].

Several researchers have estimated ETa based on the RS-based surface energy balance models, such as the "Surface Energy Balance Algorithm for Land" (SEBAL) [8,9], "Mapping EvapoTranspiration at High Resolution and with Internalized Calibration", i.e., METRIC [10,11], and "Surface Energy Balance System" (SEBS) [12]. These models normally use Landsat satellite imagery or Moderate Resolution Imaging Spectroradiometer (MODIS) which include more required RS data alongside weather parameters [7,10], among others. However, as pointed out by Bachour et al. [13], among others, RS technology has some limitations. Therefore, some of these data cannot be provided due to cloud cover and the unavailability of all relevant climatic data. For example, the implemented METRIC model to estimate ETa requires more of the input parameters of Landsat satellite data (net radiation flux (Rn), soil heat flux (G), sensible heat flux (H), latent heat flux (LE), leaf area index (LAI), land surface temperature (LST), surface albedo (α), and normalized difference vegetation index (NDVI), to name but a few).

In addition, H calculation in the METRIC model is based on the anchoring pixels, i.e., hot and cold, so there are complexities involved in sensitivity selecting anchor pixels for improved estimation of H and LE fluxes [14]. Moreover, Landsat satellite data with a 30 m by 30 m spatial resolution are normally available every 16 days for satellite scenes [7]. As such, the methodology obtained by Allen et al. [10] has been used to estimate ETa based on the METRIC model for each pixel during the satellite image overpass date. In this regard, to determine ETa values based on the METRIC model for a predetermined period, i.e., month or season, Allen et al. [10] applied the procedure by making the interpolation of daily reference evapotranspiration fraction (ETrF) or crop coefficient (Kc) values between two satellite images and multiplying by reference evapotranspiration, i.e., ETo, for each day and then integrated for a given or specific period. However, the accepted method by Allen et al. [10] has some gaps or limitations in showing the effect of precipitation or irrigation practices precisely on the daily growth stages and NDVI of crops between two satellite image overpasses. As known, both ETa and ETo have a nonlinear character in nature [15], and in turn, is a complex phenomenon. Therefore, more nonlinearity exists in the evapotranspiration process due to its stochastic behavior [16]. Thus, over the past few decades, artificial neural networks (ANNs) have been successfully utilized in modeling reference evapotranspiration, i.e., ETo [1,15,16]. As such, the idea of artificial neural networks usage in engineering applications goes back to the 1940s [15]. Due to these reasons, it has been widely used in hydrological practices, particularly in the estimations of reference evapotranspiration, since the early nineties. Contrary to this, the literature review revealed that there have been hardly any studies on estimating METRIC-based actual evapotranspiration values by ANN models.

The most significant merit of the ANN models is to solve complex problems using fewer inputs by adjusting the weights to be able to predict the correct output of the input parameters. Recently, the ANN algorithm has been applied to estimate reference evapotranspiration (ETo) and crop evapotranspiration (ETc) for wheat, maize, and potato in different regions of the world [17–23]. As for ANN-based ET estimation, studies rarely used land surface parameters calculated from RS data, including but not limited to land surface temperature, vegetation indices, etc. and limited meteorological parameters as inputs in the ANN model [24].

ETa estimation by the ANN model can be beneficial and powerful for using it as an input factor in water-balance calculations at large-scale irrigation catchments since, as claimed by Kumar et al. [1], theoretically, ANN is expected to produce better results than a regression model for the same data length. The novelty of this study is to establish a new methodology for generating daily actual evapotranspiration (ETa) series based on the ANN model using some of the parameters of MODIS data coupled with availably daily moderate spatial and less input of weather variables as compared to the existing methods of ETa-based surface energy balance models estimation. Therefore, this study aims to use an artificial neural network (ANN) approach to estimate daily ETa values in large-scale irrigation catchments using two parameters of MODIS data, i.e., NDVI and LST, coupled with limited site-specific climatic variables at a large-scale irrigation scheme located in the Lower Seyhan Plain irrigation project area with >210,000 ha of land [7]. Furthermore, this paper is the first attempt to generate a new model using the ANN algorithm as an alternative to the existing methods of actual evapotranspiration estimation in a large-scale irrigation district in the Eastern Mediterranean Region of Turkiye. Moreover, this methodology can be generalized for estimating daily ETa using the ANN model to different climate regions and zones of the world.

2. Materials and Methods

2.1. Study Area and Its Characteristics

The research area that has gained popularity under the name Akarsu Irrigation District (hereafter, AID, A = 9495 ha ≈ 95 km^2, Figure 1) in the studies carried out so far is located in the Lower Seyhan Plain (LSP) in the southeastern part of the Mediterranean region of Turkiye. The LSP shows typical characteristics of a deltaic plain with a rather flat topography (a slope of 1% or less) and a large-scale irrigation and drainage network [5,7]. The Mediterranean climate, characterized by warm and rainy in the winter season, whereas dry and hot in the summer season, prevails utterly in the LSP and, in turn, in the study area. The average annual precipitation of the basin is approximately 650 mm [5]. In the LSP of Turkiye, there is a very remarkable difference in temperature and evaporation values in the irrigation season, i.e., in July and August in particular, compared to those in the winter season, i.e., in December, January, and February [5,7]. As reported by Cetin et al. [7], due to meteorological and geographical factors, the definition of the water year, i.e., hydrological year, varies from region to region. It has been defined as the period, with a length of 365 days, between 1 October of one year and 30 September of the next, as late September to early October is the time for many drainage areas or catchments in Turkiye to have the lowest stream flows and consistent groundwater levels.

2.2. Remote Sensing Data Used

2.2.1. Landsat Satellite Imagery

To run the METRIC model and for the actual evapotranspiration (ETa) estimations, a total of 38 clear-sky Landsat satellite images were downloaded from the USGS website (http://earthexplorer.usgs.gov, accessed on 16 February 2023) (path 175, row 34) and used in this research (Table 1). These images are Landsat 7, Landsat 8, and Landsat 9 with 30 m by 30 m spatial resolution. General characteristics of the Landsat satellite images are given in Table 1. The "Environment for Visualizing Images (ENVIs 5.3)" software program was applied to perform a cloud mask for one satellite image on 1 May 2022, i.e., DOY 121, of grayscale fill, as shown in Table 1. Scan Lines Error for Landsat 7 ETM$^+$ was corrected and/or the gaps were filled before using this data through the implementation of Quantum Geographic Information System (QGIS 3.24.3) and Landsat toolbox by ArcGIS 10.4.1.

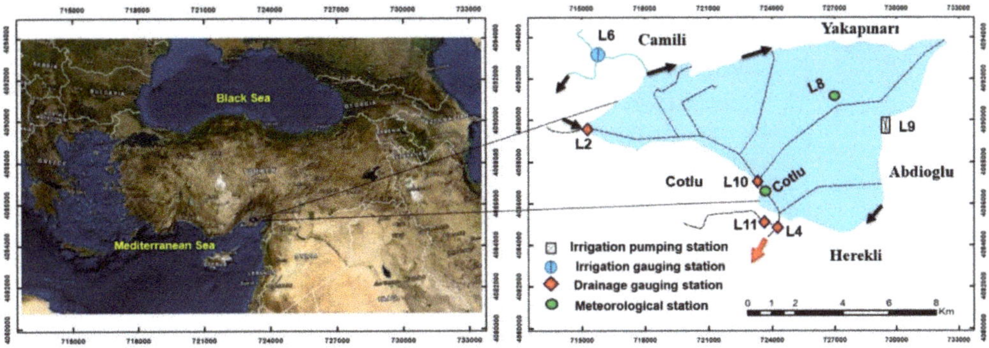

Figure 1. The study area is located in the southeastern Mediterranean region of Turkiye. Dotted lines, red and black arrows represent the drainage canals, and water flow directions, respectively. Meteorological stations are located at L8 and Cotlu. Irrigation water is diverted from L6 and L9 locations into the AID; L2 and L11 stand for drainage water inputs and L4 is the drainage outlet of the catchment.

Table 1. Availability of Landsat 7, Landsat 8, and Landsat 9 scene information in the 2021 and 2022 water years: names of scenes, acquisition dates, and overpass time.

Image	Day of the Year (DOY)	Landsat Scene-ID	Satellite Type	Cloud Cover (%)	Acquisition Dates	Overpass Local Time (AM)
1	260	LC81750342020260LGN00	Landsat 8	1	16 September 2020	11:15:56.5028510
2	300	LE71750342020300SG100	Landsat 7	8	26 October 2020	10:38:56.1154274
3	316	LE71750342020316NPA00	Landsat 7	3	11 November 2020	10:37:51.0228172
4	364	LE71750342020364NPA00	Landsat 7	1	29 December 2020	10:34:21.8153233
5	22	LC81750342021022LGN00	Landsat 8	9	22 January 2021	11:15:49.9861710
6	54	LC81750342021054LGN00	Landsat 8	7	23 February 2021	11:15:43.2139690
7	79	LE71750342021078SG100	Landsat 7	5	19 March 2020	10:28:24.8443048
8	118	LC81750342021118LGN00	Landsat 8	8	28 April 2021	11:15:15.8809360
9	134	LC81750342021134LGN00	Landsat 8	1	14 May 2021	11:15:15.9098560
10	158	LE71750342021158SG100	Landsat 7	1	7 June 2021	10:21:43.6865663
11	182	LC81750342021182LGN00	Landsat 8	4	1 July 2021	11:15:35.1871370
12	190	LE71750342021190SG100	Landsat 7	3	9 July 2021	10:19:04.5579664
13	198	LC81750342021198LGN00	Landsat 8	5	17 July 2021	11:15:36.9021800
14	214	LC81750342021214LGN00	Landsat 8	0	2 August 2021	11:15:45.3409259
15	230	LC81750342021230LGN00	Landsat 8	2	18 August 2021	11:15:51.0643500
16	262	LC81750342021262LGN00	Landsat 8	9	19 September 2021	11:15:59.0216650
17	278	LC81750342021278LGN00	Landsat 8	1	5 October 2021	11:16:04.4435930
18	294	LC81750342021294LGN00	Landsat8	0	21 October 2021	11:16:07.4309270
19	326	LC81750342021326LGN00	Landsat 8	6	22 November 2021	11:16:02.0432969
20	358	LC81750342021358LGN00	Landsat 8	4	24 December 2021	11:15:59.4307040
21	001	LE71750342022001NPA00	Landsat 7	13	1 January 2022	10:03:01.7753300
22	017	LE71750342022017NPA00	Landsat 7	3	17 January 2022	10:01:30.1854341
23	049	LE71750342022049NPA00	Landsat 7	6	18 February 2022	09:58:18.1083401
24	081	LE71750342022081NPA00	Landsat 7	19	22 March 2022	09:55:10.7155745
25	089	LC81750342022089LGN00	Landsat 8	7	30 March 2022	11:15:25.6965150
26	113	LC91750342022113LGN00	Landsat 9	2	23 April 2022	11:15:26.7327310
27	121	LC81750342022121LGN00	Landsat 8	60	1 May 2022	11:15:28.7371250

Table 1. *Cont.*

Image	Day of the Year (DOY)	Landsat Scene-ID	Satellite Type	Cloud Cover (%)	Acquisition Dates	Overpass Local Time (AM)
28	140	LE71750342022140SG100	Landsat 7	12	20 May 2022	09:50:56.6332014
29	157	LE71750342022157SG100	Landsat 7	23	6 June 2022	09:50:06.2292853
30	169	LC81750342022169LGN00	Landsat 8	1	18 June 2022	11:15:53.3072380
31	186	LE71750342022186SG100	Landsat 7	0	5 July 2022	09:42:29.8002719
32	201	LC81750342022201LGN00	Landsat 8	1	20 July 2022	11:15:58.6410700
33	209	LC91750342022209LGN00	Landsat 9	0	28 July 2022	11:15:42.4933480
34	220	LE71750342022220SG100	Landsat 7	29	8 August 2022	09:39:50.3921524
35	237	LE71750342022237SG100	Landsat 7	9	25 August 2022	09:38:18.9640686
36	249	LC81750342022249LGN00	Landsat 8	8	06 September 2022	11:16:15.2109079
37	271	LE71750342022271SG100	Landsat 7	3	28 September 2022	09:34:51.9536731
38	297	LC81750342022297LGN00	Landsat 8	0	24 October 2022	11:16:18.0041120

2.2.2. Moderate Resolution Imaging Spectroradiometer (MODIS) Products

Two parameters of MODIS data, i.e., normalized difference vegetation index (NDVI) and land surface temperature (LST) were downloaded by the Google Earth engine ('MODIS/MOD09GA_006_NDVI'; 'MODIS/061/MOD11A1_LST') to apply the artificial neural network (ANN) for estimating daily ETa values for the entire study area. Typical characteristics of MODIS data are illustrated in Table 2 along with the spatial and temporal resolutions.

Table 2. MODIS data used in the research area.

MODIS Standard Products	Parameter	Spatial Resolution	Temporal Resolution
MOD09GA-Terra	NDVI	500 m by 500 m	Daily
MOD11A1.061-Terra	LST	1000 m by 1000 m	Daily

Daily NDVI and LST datasets were used as casual variables in the modeling practice since NDVI and LST by Landsat or Sentinel-2 are not available daily [5,7] due to the revisit period of the satellites. However, NDVI and LST by MODIS products are available as daily time series even though they have low spatial resolutions. Furthermore, these indices by MODIS data can be used to monitor daily crop development, surface temperature, and spatial distribution over the study area if compared to other remote sensing data which revisit the time of satellite every 3–5 days for Sentinel-2 and 8–16 days for Landsat. These remote sensing data, however, require an interpolation process to fill the gaps of NDVI and LST between two overpass satellite dates to obtain a daily dataset.

2.3. *In Situ Meteorological Observations*

In this research, hourly and daily climatic variables (minimum and maximum temperatures (Tmin, Tmax), wind speed (U), solar radiation (Rs), minimum and maximum relative humidity values (RHmin and RHmax), and precipitation (P)) acquired from two meteorological stations, i.e., L8 and Cotlu meteorological stations established in the AID (Figure 1), were used in calculations. Before using any climatic data observed in L8 and Cotlu meteorological stations, quality controls, i.e., QC (gaps in the data, outliers, constant values, jumps, etc.) were checked thoroughly. No consistency was found in the meteorological datasets. Climatic datasets cover from 16 September 2020 to 24 October 2022, i.e., 769 daily datasets.

2.4. *Reference Evapotranspiration (ETo) Estimation*

Reference evapotranspiration (ETo in mm day^{-1} unit) is defined and computed using the standard FAO–Penman–Monteith approach given by Allen et al. [2]. Equation (1) was developed for short grass; the albedo was 0.23, whilst the aerodynamic resistance was 70 s m^{-1}.

$$ETo = \frac{0.408\Delta(R_n - G) + \gamma\frac{900}{T+273}u_2(e_s - e_a)}{\Delta + \gamma(1 + 0.34u_2)} \quad (1)$$

where ETo is the reference evapotranspiration (mm day^{-1}), R_n is the net radiation at the crop surface (MJ m^{-2} day^{-1}), G is the soil heat flux density (MJ m^{-2} day^{-1}), T is the mean daily air temperature at 2 m height (°C), u_2 is the wind speed at 2 m height (m s^{-1}), e_s is the saturation vapor pressure (kPa), e_a is the actual vapor pressure (kPa), $e_s - e_a$ is the saturation vapor pressure deficit (kPa), Δ slope (kPa °C^{-1}) is the vapor pressure curve, and γ the psychrometric constant (kPa °C^{-1}).

2.5. METRIC Model

The METRIC model was applied to estimate the following: (a) surface energy balance components (SEBs), i.e., latent heat (LE), net radiation (R_n), sensible heat (H), and soil heat flux (G) in Equation (2), (b) ETa for each pixel, and the whole study area by using Landsat satellite imagery and meteorological stations, i.e., L8 and Cotlu at the time of satellite overpass, primarily based on Allen et al. [10,11] through R-METRIC model using a water package in the *R* program [25] and LandMOD ET mapper-MATLAB [26].

$$LE = R_n - G - H \quad (2)$$

All the energy fluxes are in the unit of watt per meter square (i.e., W m^{-2}). Further information on the METRIC model and its associated equations, i.e., step-by-step ETa calculation, as well as the FAO–Penman–Monteith approach, is given by Allen et al. [10].

2.6. Developing an ANN Model for Actual Evapotranspiration (ETa) Estimation

Artificial Neural Networks (ANNs) are mathematical models that resemble biological neural networks. ANNs can learn from examples and adapt solutions over time by recognizing patterns in data, along with rapidly processing information [27]. In essence, ANNs are tools to mimic the underlying likely relationship between input and output variables in the hand adequately.

Water resources and hydrological processes are often complex, multivariable, and nonlinear. ANNs exhibit a flexible structure to address these complex relationships, making them capable of learning and integrating complex relationships by using various input data. Therefore, in recent times, ANNs have been increasingly utilized in hydrology and water resource management. ANNs might be considered flexible modeling tools and can theoretically model any type of relationship with good accuracy. With ANNs, there is no need to make specific assumptions about the models and the underlying probability distributions or relationships; the underlying relationship is determined solely through data mining procedures.

This data-driven approach is one of the most significant advantages of ANNs in solving various complex real-world prediction problems. ANNs have been used in a whole range of hydrological applications, including reference evapotranspiration estimations and predicting groundwater levels [28,29], flood forecasting simulations [30], rainfall and streamflow modeling [31], and aquifer parameter estimations [32].

Models for calculating ETo and plant water requirements involve a myriad of variables such as meteorological data, soil properties, plant type, and climate conditions. ANNs have a significant advantage in handling these complexities due to their ability to use a large amount of data. In particular, when trained with large datasets, these networks have a better capacity to learn complex relationships and patterns for making logical predictions. Predicting future changes in water resources and plant water requirements due to climate change is becoming increasingly important. In ET modeling, machine learning algorithms are being used more and more as alternatives to traditional methods [33,34]. These algorithms can be used as alternatives to traditional equations for ETc and/or ETo predictions. They also provide insights into how ET behaves over time and space [35,36]. ANNs can be trained and adapted to be used in different geographical areas, allowing for customized predictions based on different plant species and climatic conditions.

Despite their mentioned advantages, ANNs have some important disadvantages. They require a significant amount of data to learn complex relationships and to determine the optimal network architecture [37]. To create the network structure, the number of hidden layers in the model and the optimal number of neurons in each layer need to be determined. In most of the ANN studies in the literature, a trial-and-error procedure is used to determine the network architecture, which is a time-consuming process. According to Maier and Dandy [38], in most water resources problems, using a single hidden layer is sufficient. Therefore, in this study, a single hidden layer is used in ETa estimation.

To determine the optimal number of neurons in the hidden layer, 80% of the data was used for training the network, and 20% was used for testing. This process was repeated for 100 different randomly selected training and testing datasets, and the Mean Squared Error (MSE) value was calculated. This process was repeated in a loop from 1 to the maximum number of neurons, which is 30 in this study, and the number of neurons that yielded the minimum MSE value was selected as the optimal number of neurons, and the analyses were conducted accordingly [37]. Figure 2 illustrates the typical structure of multi-layer ANNs. The input layer consists of 4 nodes, 2 parameters acquired by the two meteorological stations (Figure 1) installed in the study area, and two variables downloaded from MODIS satellite data. In situ climatic observations are solar radiation (Rs), extraterrestrial radiation (Ra), wind speed (u2) at 2 m height, and the reference evapotranspiration values, i.e., ETo, were calculated by the standard FAO–Penman–Monteith approach through following Allen et al. [2]. The connections between the input layers and 30 hidden layers take different weights and are trained depending on the required output of daily ETa.

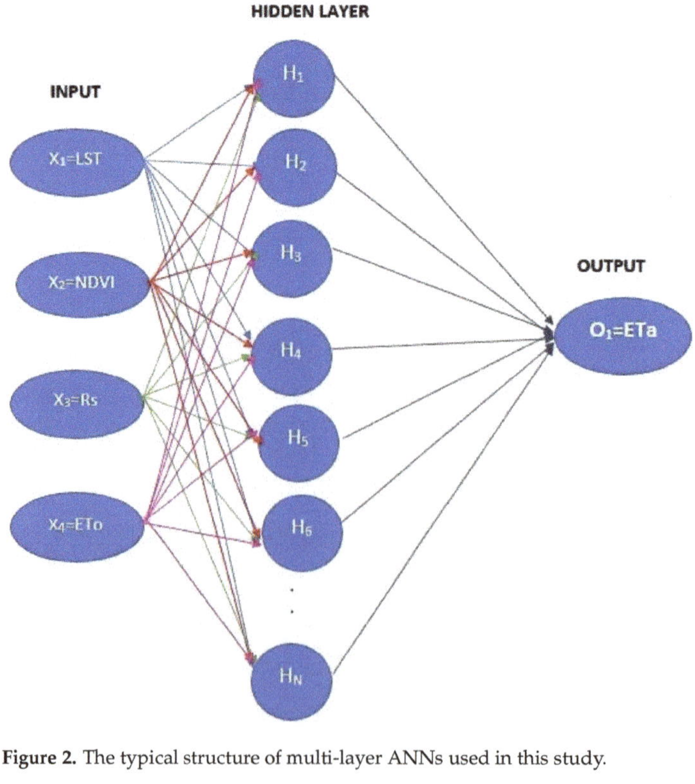

Figure 2. The typical structure of multi-layer ANNs used in this study.

3. Results

3.1. Implementation of the ANN Model

The ANN model has been applied to two different scenarios. In the training of Scenario I, data from only the 38 days of satellite observations listed in Table 1 were used, while the data, which were acquired from the interpolation procedure by Allen et al. [10] from the other days were used as test data. In Scenario II, 80% of the total dataset was used for training, and the remaining 20% was used for testing.

For both scenarios, firstly, the network architecture was created, and the change in the MSE value concerning the number of neurons is presented in Figure 3a for Scenario I and Figure 3b for Scenario II. As seen in Figure 3, the optimum number of neurons obtained for Scenario I and II is 5 and 15, respectively.

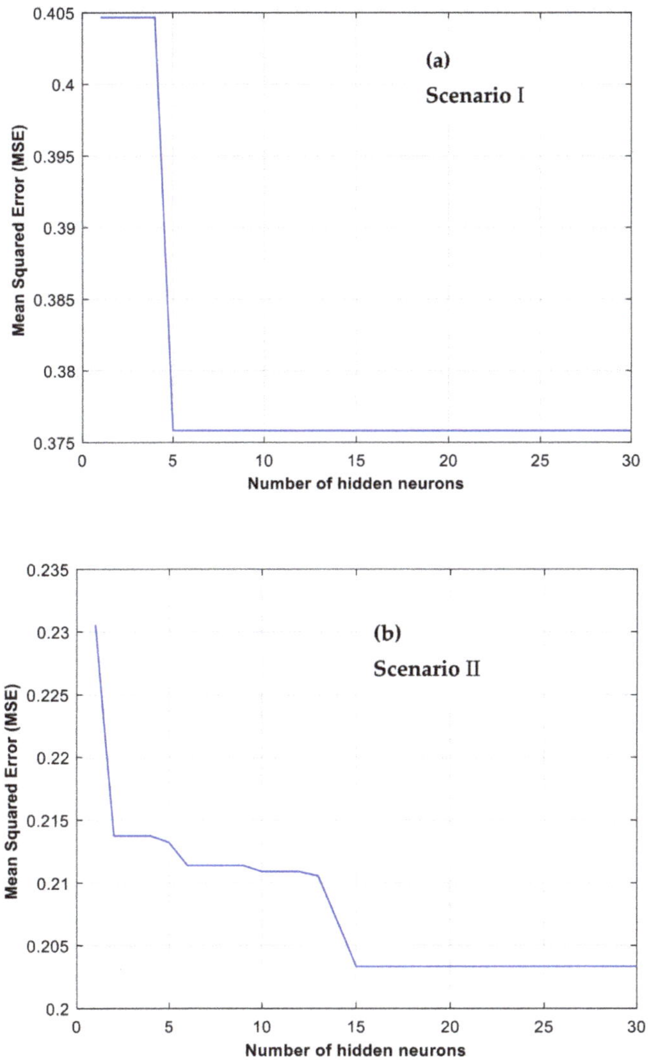

Figure 3. A three-layer feed-forward ANN.

3.1.1. Scenario I

The developed model was applied to Scenario I, and the model results are summarized in Figure 4. As seen in Figure 4, even though the data used in the model training accounts for approximately 5% (N = 38) of the total data, the R-squared value is 0.7547 for the test data (N = 731) and 0.7561 for the total data (N = 769). This indicates that the performance of the developed model is quite good concerning the learning stage and prediction ability; however, it also shows an increased error rate in predicting high ETa values. This result is a natural consequence of the need for a substantial amount of data for ANN models to learn complex relationships in complex problems, as mentioned above. In the following section, this situation will be evaluated in more detail.

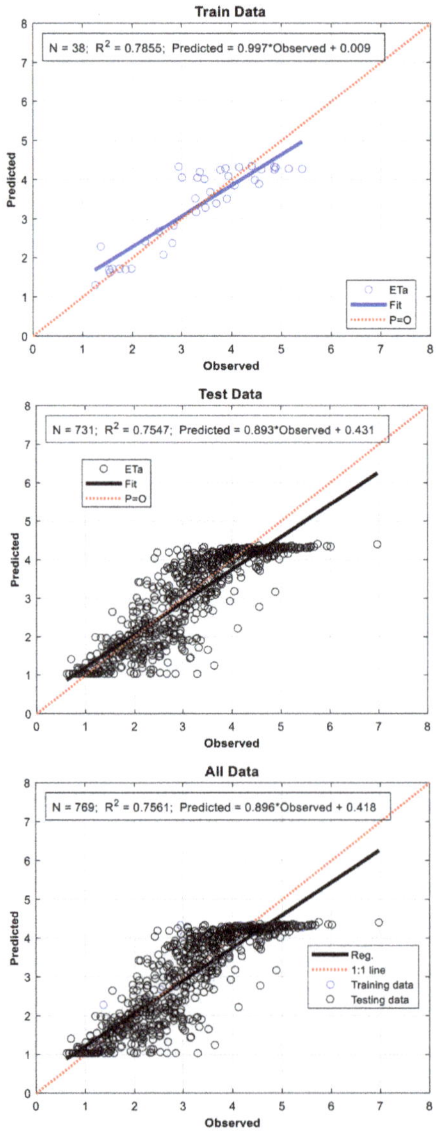

Figure 4. Model results for Scenario I.

3.1.2. Scenario II

Figure 5 shows the model results of the developed model for Scenario II. As seen in Figure 5, the number of data points used in the model training is 615, which constitutes 80% of the total data. During the training phase, the R-squared value is 0.8496, while R-squared values are 0.8055 and 0.8411 for the test data and total data, respectively. These results indicate that the developed model has a very good learning and prediction ability.

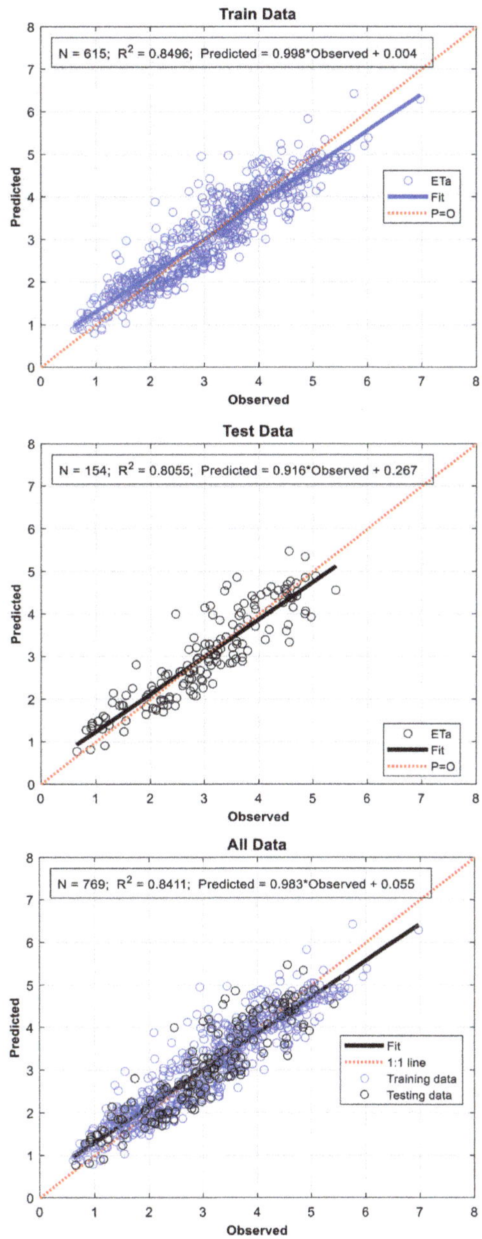

Figure 5. Model results for Scenario II.

4. Discussion

The primary objective of this study was to develop a methodology for estimating daily ETa values for large irrigation areas based on artificial neural networks using some parameters of MODIS data and in situ climatic data with ETo values. It is important to highlight that ETo estimations will be an easy task, provided that in situ climatic data required by the standard FAO–Penman–Monteith approach [2,39–44] are available at the study site. In our study, there are two meteorological stations for collecting data. However, the problem is that Landsat satellite images are available with a 16-day repeat cycle. Therefore, ETa values for the days of no satellite overpass need to be estimated by using an appropriate methodology such as ANNs. To this end, the developed methodology consists of two main parts. In the first stage of the methodology, daily ETo values calculated using the standard FAO–Penman–Monteith approach detailed in Allen et al. [2] are combined with MODIS data to create the input data for the proposed ANN model. These values are based on observations from a meteorological station established in the study area. Additionally, observed ETa values are calculated based on Allen et al. [10,11] using L8 and Cotlu meteorological station data for each pixel at satellite overpass times and other times. This completes the dataset required for the training and testing stages of the developed ANN model.

After completing the dataset, calculations are performed in two different scenarios. In Scenario I, only data acquired for the days with satellite imagery are used for model training, while a randomly selected 80% portion of the total dataset is used in Scenario II. It should be noted that the input and output layers of both scenarios are the same, as shown in Figure 2. As can be seen in Figure 3, the optimal number of neurons in the hidden layer is determined to be 5 for Scenario I and 15 for Scenario II. After determining the appropriate network architecture, detailed analyses were conducted for both scenarios.

In the model training, the goal was to minimize the MSE values as the objective function, and the results obtained for training, testing, and the entire datasets were provided graphically in the previous section. In the relevant graphs, the number of data points used, the R^2 value, and the regression relationships between predicted values and observed values are presented.

As seen in Figure 4, in Scenario I, even though approximately 5% of the total dataset was used for model training, high R^2 values of 0.7855 (N = 38), 0.7547 (N = 731), and 0.7561 (N = 769) were achieved for training, testing, and the entire datasets, respectively. However, it can be observed from the relevant graphs that despite the overall high model performance, the model's prediction ability in Scenario I decreases unexpectedly if the ETa values are approximately 4 mm day^{-1} or greater, indicating conspicuous underestimates of high actual evapotranspiration rates. The underestimation behavior of the model in Scenario I indicates evidently that the model fails to mimic the actual evapotranspiration of 4 mm day^{-1} or greater. This behavior can be seen more clearly in Figure 6, where the model results are presented graphically for both scenarios in comparison with ETa values calculated based on Allen et al. [10,11,42,43]. As clearly visible in the respective Figure, although the general trend is captured in both scenarios, it is noteworthy that the error rate in predicting values, especially those greater than four, increases in Scenario I. It is believed that this is due to the low number of satellite observations (only 38 daily data with very few extreme values) available for training the ANN model. Scatter plots provided in Figure 5 for Scenario II, as well as the temporal ETa values presented in Figure 6, support our postulate. Most studies in the literature also support this argument by allocating 80% of the total dataset for training and the remaining 20% for testing when training ANN models.

However, there are some challenges encountered in both the provision of Landsat satellite data and the implementation of the procedure suggested by Allen et al. [10]. These issues of Landsat satellite data belong to the availability of satellite images every 8–16 days, and cloud coverage, especially for the rainy seasons. Furthermore, using the METRIC model, as suggested by Allen and other scholars, requires well-trained experts with a strong background in handling RS data and choosing the anchor pixels, i.e., hot and cold pixels. Therefore, in order to determine the optimal dataset to be used in training the ANN model, ratios of different training data used to the total data (TRNRatio) were employed, and the change in model performance is presented in Figure 7. As seen clearly from Figure 7, the optimal TRNRatio value was 0.20. Although the model performance, i.e., R-squared values for training, testing, and all datasets, increases, to some extent, up to this value, there is a significant decrease in test performance for larger TRNRatio values, indicating overfitting. Consequently, the model becomes more and more dependent on the data, which is considered a weakness of the ANN models. This situation can be observed from the scatter plots provided in Figure 8. After evaluating the analysis results given in Figures 7 and 8, it can be concluded that a TRNRatio value of 0.10 can yield ETa values that are quite satisfactory from a practical perspective without increasing the number of satellite data.

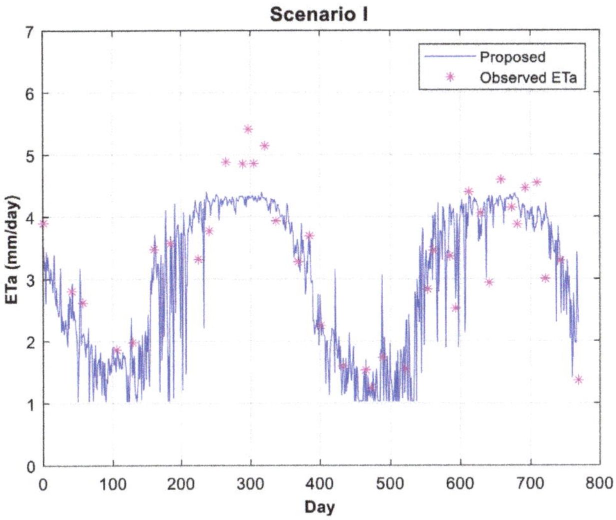

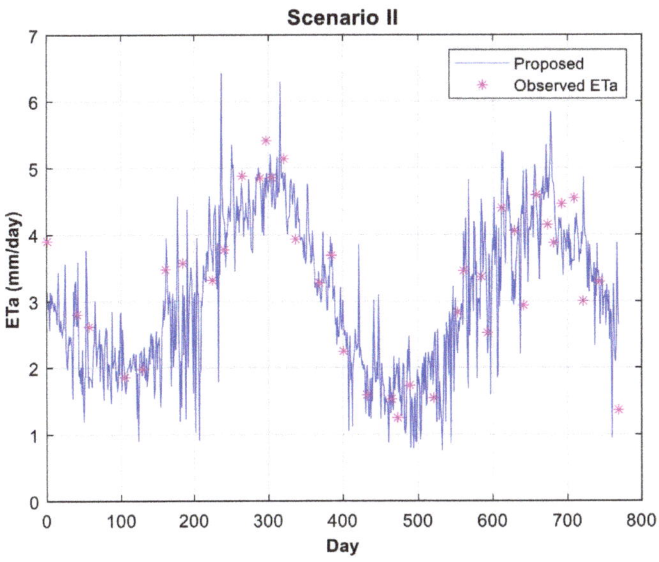

Figure 6. Model results for Scenario I and II.

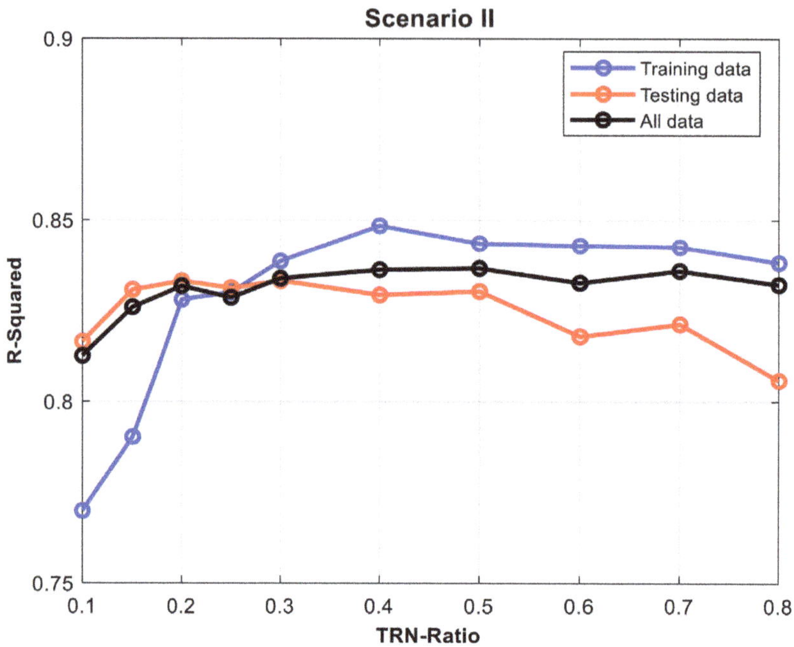

Figure 7. The impact of the data ratio used in the training of the model on the model's performance.

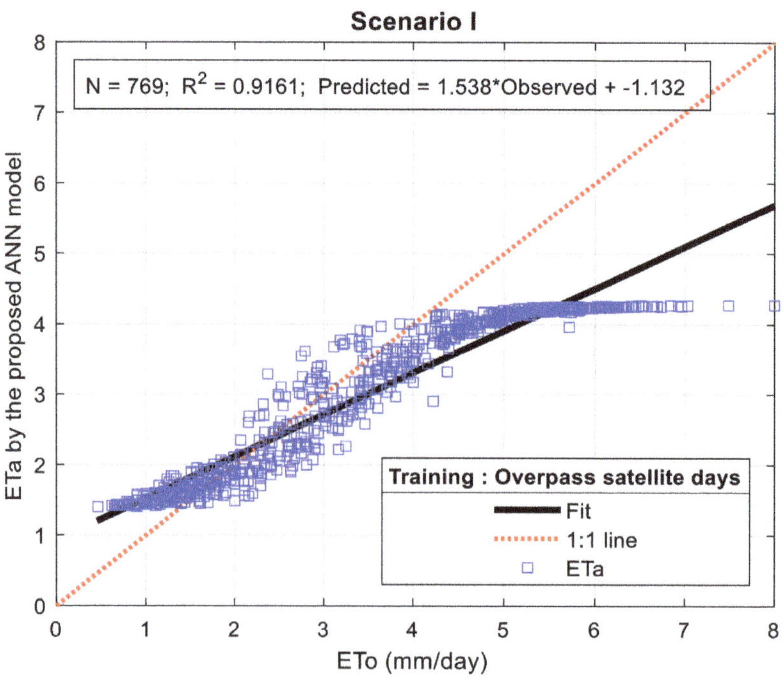

Figure 8. *Cont.*

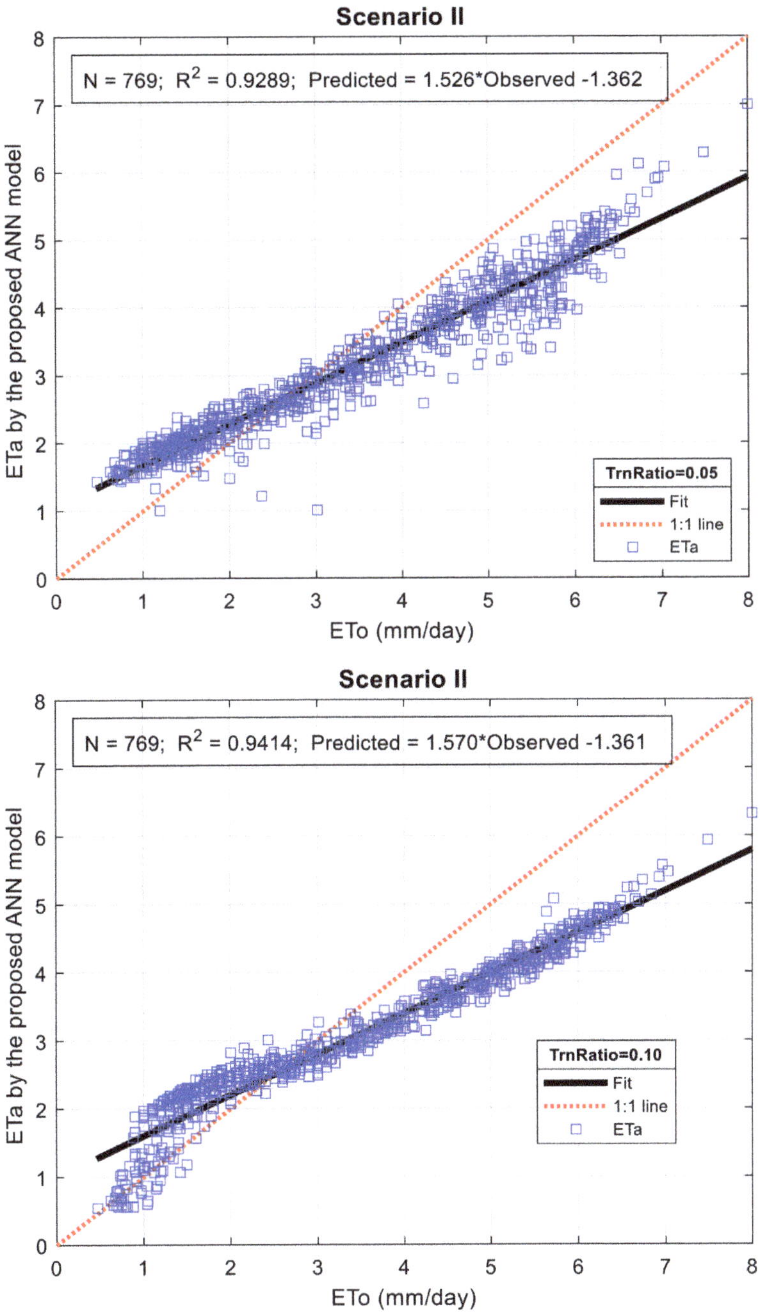

Figure 8. Relationship between ETo by the *FAO–Penman–Monteith* approach and *ETa* by the *ANN model* for Scenario I and II in the Akarsu Irrigation District, i.e., the AID, located in the Eastern Mediterranean Region of Turkiye.

In this study, ETo by FAO–Penman–Monteith method was calculated using the meteorological parameters from two ground-based observations (L8 and Cotlu meteorological stations shown in Figure 1) established in the AID. In addition, the estimated ETa by the ANN model for two scenarios was validated to the reference ET by using a simple linear regression approach (ETa = a + b*ETo) over the AID. As can be seen in Figure 8, the results demonstrate that the R^2 value is 0.9161 for Scenario I if we use the whole dataset (N = 769) for training. On the other hand, in Scenario II with TRNRatios of 0.05 and 0.10, R^2 values reach 0.9289 and 0.9414, respectively. In the Cukurova region, where our study is based, several studies have been conducted to estimate ETa by direct methods [7]. For example, Cetin et al. [7] compared the estimated ETa-METRIC of the second crop of soybean to the ETa-lysimeter. They concluded that the ETa-METRIC values were more closely aligned with the lysimeter observations. Additionally, as indicated by the findings reported by Cetin et al. [7], there is a good agreement between ETc values for the crops of citrus, wheat, potato, lettuce, first-crop corn, and peanut, as well as second-crop soybean, and ETa by the MERIC model during 2021 water year. All these could lead us to conclude that the estimated ETa results by the METRIC model, as well as the proposed ANN model, were consistent with the ETa values by direct and indirect techniques in the same study region in both this research and prior studies.

5. Conclusions

This study presents a novel approach that allows daily ETa estimation using a new ANN model in a large-scale irrigation scheme using limited climatic data and MODIS satellite data. The study demonstrated that the daily ETa predicted values are comparable with the ETa which is estimated by the RS-based surface energy balance models. Furthermore, this methodology is the first attempt to generate a new model using the ANN algorithm as an alternative to the existing methods of ETa estimation in a large-scale irrigation district in the Eastern Mediterranean Region of Turkiye. The daily ETa estimation results of the ANN model in the two implemented scenarios showed a reasonable degree of agreement with ETa values computed by the METRIC model. Thus, this work can be considered a significant contribution to obtaining reliable ETa results without the need for lengthy and labor-intensive processes for complex equations, as in RS-based surface energy balance models. This study revealed that the proposed ANN model is a powerful tool for estimating daily ETa values in large-scale irrigation schemes using limited meteorological observations and some of the parameters of remote sensing in arid and semi-arid regions, as well as in different climate regions and zones of the world. Moreover, the proposed methodology in this research may help water authorities, practitioners, and end users estimate actual evapotranspiration for sustainable water management, especially in Mediterranean countries, where freshwater resources are scarce in an increasingly globalized world.

Author Contributions: Conceptualization, H.K.; Methodology, H.K., M.C. and O.A.; Software, H.K.; Validation, H.K.; Formal analysis, M.E.C. and O.A.; Investigation, H.K. and M.C.; Resources, O.A.; Data curation, M.C., M.E.C. and O.A.; Writing—original draft, H.K., M.C. and O.A.; Writing—review & editing, H.K. and O.A.; Supervision, H.K.; Project administration, H.K. and M.E.C. All authors have read and agreed to the published version of the manuscript.

Funding: The authors thank the Scientific and Technological Research Council of Turkiye [TUBITAK, Project number: 122Y007] and the Scientific Research Projects (BAP) Coordination Unit of Cukurova University [Project Number: FDK-2022-14907] for obtaining financial support for this work.

Institutional Review Board Statement: Not applicable.

Informed Consent Statement: Not applicable.

Data Availability Statement: Data is available on request from the authors.

Conflicts of Interest: The authors declare no conflicts of interest.

References

1. Kumar, M.; Raghuwanshi, N.S.; Singh, R. Artificial Neural Networks Approach in Evapotranspiration Modeling: A Review. *Irrig. Sci.* **2011**, *29*, 11–25. [CrossRef]
2. Allen, R.G.; Pereira, L.S.; Raes, D.; Smith, M. *Crop Evapotranspiration Guidelines for Computing Crop Water Requirements*; FAO Irrigation and Drainage Paper 56; FAO: Rome, Italy, 1998.
3. Rawat, K.S.; Bala, A.; Singh, S.K.; Pal, R.K. Quantification of Wheat Crop Evapotranspiration and Mapping: A Case Study from Bhiwani District of Haryana, India. *Agric. Water Manag.* **2017**, *187*, 200–209. [CrossRef]
4. Gharbia, S.S.; Smullen, T.; Gill, L.; Johnston, P.; Pilla, F. Spatially Distributed Potential Evapotranspiration Modeling and Climate Projections. *Sci. Total Environ.* **2018**, *633*, 571–592. [CrossRef] [PubMed]
5. Alsenjar, O.; Cetin, M.; Aksu, H.; Akgul, M.A.; Golpinar, M.S. Cropping Pattern Classification Using Artificial Neural Networks and Evapotranspiration Estimation in the Eastern Mediterranean Region of Turkey. *J. Agric. Sci.* **2023**, *29*, 677–689. [CrossRef]
6. Zhang, X.C.; Wu, J.W.; Wu, H.Y.; Li, Y. Simplified SEBAL Method for Estimating Vast Areal Evapotranspiration with MODIS Data. *Water Sci. Eng.* **2011**, *4*, 24–35. [CrossRef]
7. Cetin, M.; Alsenjar, O.; Aksu, H.; Golpinar, M.S.; Akgul, M.A. Comparing Actual Evapotranspiration Estimations by METRIC to In-situ Water Balance Measurements Over an Irrigated Field in Turkey. *Hydrol. Sci. J.* **2023**, *68*, 1162–1183. [CrossRef]
8. Bastiaanssen, W.G.M.; Menenti, M.; Feddes, R.A.; Holtslag, A.A.M. A Remote Sensing Surface Energy Balance Algorithm for Land (SEBAL): 1. Formulation. *J. Hydrol.* **1998**, *212–213*, 198–212. [CrossRef]
9. Bastiaanssen, W.G.M.; Pelgrum, H.; Wang, J.; Ma, Y.; Moreno, J.F.; Roerink, G.J.; van der Wal, T. A Surface Energy Balance Algorithm For Land (SEBAL): 2. Validation. *J. Hydrol.* **1998**, *212–213*, 213–229. [CrossRef]
10. Allen, R.; Tasumi, M.; Morse, A.; Trezza, R.; Wright, J.L.; Bastiaanssen, W.; Kramber, W.; Lorite, L.; Robison, C.W. Satellite-Based Energy Balance for Mapping Evapotranspiration with Internalized Calibration, METRIC (applications). *J. Irrig. Drain. Eng.* **2007**, *133*, 395–406. [CrossRef]
11. Allen, R.G.; Tasumi, M.; Trezza, R. Satellite-Based Energy Balance for Mapping Evapotranspiration with Internalized Calibration, METRIC (model). *J. Irrig. Drain. Eng.* **2007**, *133*, 380–394. [CrossRef]
12. Su, Z. The Surface Energy Balance System (SEBS) for Estimation of Turbulent Heat Fluxes. *Hydrol. Earth Syst. Sci.* **2002**, *6*, 85–100. [CrossRef]
13. Bachour, R.; Walker, W.; Ticlavilca, A.; McKee, M.; Maslova, I. Estimation of Spatially Distributed Evapotranspiration Using Remote Sensing and A Relevance Vector Machine. *J. Irrig. Drain. Eng.* **2014**, *140*, 04014029. [CrossRef]
14. Singh, R.K.; Irmak, A. Treatment of Anchor Pixels in the METRIC Model for Improved Estimation of Sensible and Latent Heat Fluxes. *Hydrol. Sci. J.* **2011**, *56*, 895–906. [CrossRef]
15. ASCE Task Committee on Application of Neural Networks in Hydrology. Artificial Neural Network in Hydrology. I: Preliminary Concepts. *J. Hydrol. Eng.* **2000**, *5*, 115–123. [CrossRef]
16. ASCE Task Committee on Application of Neural Networks in Hydrology. Artificial Neural Network in Hydrology. II: Hydrologic Application. *J. Hydrol. Eng.* **2000**, *5*, 124–137. [CrossRef]
17. Bruton, J.M.; McClendon, R.W.; Hoogenboom, G. Estimating Daily Pan Evaporation with Artificial Neural Networks. *Trans.* **2000**, *43*, 491–496. [CrossRef]
18. Odhiambo, L.O.; Yoder, R.E.; Yoder, D.C.; Hines, J.W. Optimization of Fuzzy Evapotranspiration Model Through Neural Training with Input-Output Examples. *Trans. ASAE* **2001**, *44*, 1625–1633. [CrossRef]
19. Kumar, M.; Raghuwanshi, N.; Singh, R.; Wallender, W.W.; Pruitt, W.O. Estimating Evapotranspiration Using Artificial Neural Network. *J. Irrig. Drain. Eng.* **2002**, *128*, 224–233. [CrossRef]
20. Dehbozorgi, F.; Sepaskhah, A.R. Comparison of Artificial Neural Networks and Prediction Models for Reference Evapotranspiration Estimation in a Semi-Arid Region. *Arch. Agron. Soil. Sci.* **2011**, *58*, 477–497. [CrossRef]
21. Khoshhal, J.; Mokarram, M. Model for Prediction of Evapotranspiration Using MLP Neural Network. *Inter. J. Environ. Sci.* **2012**, *3*, 1000–1009.
22. Abrishami, N.; Sepaskhah, A.R.; Shahrokhnia, M.H. Estimating Wheat and Maize Daily Evapotranspiration Using Artificial Neural Network. *Theoret Appl. Climatol.* **2019**, *135*, 945–958. [CrossRef]
23. Yamac, S.S.; Todorovic, M. Estimation of Daily Potato Crop Evapotranspiration Using Three Different Machine Learning Algorithms and Four Scenarios of Available Meteorological Data. *Agric. Water Manag.* **2020**, *228*, 105875. [CrossRef]
24. Virnodkar, S.S.; Pachghare, V.K.; Patil, V.C.; Jha, S.K. Application of machine learning on remote sensing data for sugarcane crop classification: A review. In *ICT Analysis and Applications. Lecture Notes in Networks and Systems*; Fong, S., Dey, N., Joshi, A., Eds.; Springer: Singapore, 2020; Volume 93. [CrossRef]
25. Olmedo, G.F.; Ortega-farias, S.; Fonseca-luengo, D. Tools and Functions to Estimate Actual Evapotranspiration Using Land Water: Toolsand Functions to Estimate Actual Evapotranspiration Using Land Surface Energy Balance Models. *R J.* **2016**, *8*, 352. [CrossRef]
26. Bhattarai, N.; Quackenbush, L.J.; Im, J.; Shaw, S.B. A New Optimized Algorithm for Automating Endmember Pixel Selection in the SEBAL and METRIC Models. *Remote Sens. Environ.* **2017**, *196*, 178–192. [CrossRef]
27. Jain, S.K.; Nayak, P.C.; Sudheer, K.P. Models for Estimating Evapotranspiration Using Artificial Neural Networks, and Their Physical Interpretation. *Hydrol. Process.* **2008**, *22*, 2225–2234. [CrossRef]

28. Coppola, E., Jr.; Szidarovszky, F.; Poulton, M.; Charles, E. Artificial Neural Network Approach for Predicting Transient Water Levels in a Multilayered Groundwater System Under Variable State, Pumping, and Climate Conditions. *J. Hydrol. Eng.* **2003**, *8*, 348–360. [CrossRef]
29. Daliakopoulos, I.N.; Coulibaly, P.; Tsanis, I.K. Groundwater Level Forecasting Using Artificial Neural Networks. *J. Hydrol.* **2005**, *309*, 229–240. [CrossRef]
30. Karahan, H.; Iplikci, S.; Yasar, M.; Gurarslan, G. River Flow Estimation from Upstream Flow Records Using Support Vector Machines. *J. Appl. Math.* **2014**, *2014*, 714213. [CrossRef]
31. Luk, K.C.; Ball, J.E.; Sharma, A. A Study of Optimal Model Lag and Spatial Inputs to Artificial Neural Network for Rainfall Forecasting. *J. Hydrol.* **2000**, *227*, 56–65. [CrossRef]
32. Garcia, L.A.; Shigidi, A. Using Neural Networks for Parameter Estimation in Groundwater. *J. Hydrol.* **2006**, *318*, 215–231. [CrossRef]
33. Granata, F. Evapotranspiration Evaluation Models Based on Machine Learning Algorithms: A Comparative Study. *Agric. Water Manag.* **2019**, *217*, 303–315. [CrossRef]
34. Tikhamarine, Y.; Malik, A.; Kumar, A.; Souag-Gamane, D.; Kisi, O. Estimation of Monthly Reference Evapotranspiration Using Novel Hybrid Machine Learning Approaches. *Hydrol. Sci. J.* **2019**, *64*, 1824–1842. [CrossRef]
35. Madugundu, R.; Al-Gaadi, K.A.; Tola, E.; Hassaballa, A.A.; Patil, V.C. Performance of the METRIC Model in Estimating Evapotranspiration Fluxes Over an Irrigated Field in Saudi Arabia Using Landsat-8 Images. *Hydrol. Earth Syst. Sci.* **2017**, *21*, 6135–6151. [CrossRef]
36. Tang, D.; Feng, Y.; Gong, D.; Hao, W.; Cui, N. Evaluation of Artificial Intelligence Models for Actual Crop Evapotranspiration Modeling in Mulched and Non-mulched Maize Croplands. *Comput. Electron. Agric.* **2018**, *152*, 375–384. [CrossRef]
37. Karahan, H.; Ayvaz, M.T. Simultaneous Parameter Identification of a Heterogeneous Aquifer System Using Artificial Neural Networks. *Hydrogeol. J.* **2008**, *16*, 817–827. [CrossRef]
38. Maier, H.R.; Dandy, G.C. Neural Networks for the Prediction and Forecasting of Water Resources Variables: A Review of Modelling Issues and Applications. *Environ. Model. Softw.* **2000**, *15*, 101–124. [CrossRef]
39. Antonopoulos, V.Z.; Antonopoulos, A.V. Daily Reference Evapotranspiration Estimates by Artificial Neural Networks Technique and Empirical Equations Using Limited Input Climate. *Comput. Electron. Agric.* **2017**, *132*, 86–96. [CrossRef]
40. Ferreira, L.B.; Cunha, F.F. New Approach to Estimate Daily Reference Evapotranspiration Based on Hourly Temperature and Relative Humidity Using Machine Learning and Deep Learning. *Agric. Water Manag.* **2020**, *234*, 106–113. [CrossRef]
41. Mattar, M.A. Using Gene Expression Programming in Monthly Reference Evapotranspiration Modeling: A Case Study in Egypt. *Agric. Water Manag.* **2018**, *198*, 28–38. [CrossRef]
42. Santos, P.A.B.; Schwerz, F.; Carvalho, L.G.; Baptista, V.B.S.; Marin, D.B.; Ferraz, G.A.S.; Rossi, G.; Conti, L.; Bambi, G. Machine Learning and Conventional Methods for Reference Evapotranspiration Estimation Using Limited-Climatic-Data Scenarios. *Agronomy* **2023**, *13*, 2366. [CrossRef]
43. Chen, X.; Yu, S.; Zhang, H.; Li, F.; Liang, C.; Wang, Z. Estimating the Actual Evapotranspiration Using Remote Sensing and SEBAL Model in an Arid Environment of Northwest China. *Water* **2023**, *15*, 1555. [CrossRef]
44. Abdel-Fattah, M.K.; Kotb Abd-Elmabod, S.; Zhang, Z.; Merwad, A.R.M.A. Exploring the Applicability of Regression Models and Artificial Neural Networks for Calculating Reference Evapotranspiration in Arid Regions. *Sustainability* **2023**, *15*, 15494. [CrossRef]

Disclaimer/Publisher's Note: The statements, opinions and data contained in all publications are solely those of the individual author(s) and contributor(s) and not of MDPI and/or the editor(s). MDPI and/or the editor(s) disclaim responsibility for any injury to people or property resulting from any ideas, methods, instructions or products referred to in the content.

sustainability

MDPI

Article

Harmonizing the Development of Local Socioeconomic Scenarios: A Participatory Downscaling Approach Applied in Four European Case Studies

Athanasios Thomas Vafeidis [1], Lena Reimann [1,2,*], Gerald Jan Ellen [3], Gunnel Goransson [4], Gerben Koers [3,5], Lisa Van Well [4], Bente Vollstedt [1], Maureen Tsakiris [1] and Amy Oen [6]

[1] Department of Geography, Kiel University, Ludewig-Meyn-Straße 14, 24118 Kiel, Germany; vafeidis@geographie.uni-kiel.de (A.T.V.); vollstedt@geographie.uni-kiel.de (B.V.); tsakiris@geographie.uni-kiel.de (M.T.)
[2] Institute for Environmental Studies (IVM), De Boelelaan 1111, 1081 HV Amsterdam, The Netherlands
[3] Department of Urban Water and Subsurface, Unit Subsurface and Groundwater Systems, Deltares, Daltonlaan 600, 3584 BK Utrecht, The Netherlands; gerben.koers@deltares.nl (G.K.)
[4] Department of Natural Hazards and Geodata, Swedish Geotechnical Institute, 581 83 Linköping, Sweden; gunnel.goransson@sgi.se (G.G.); lisa.vanwell@sgi.se (L.V.W.)
[5] Department of Geography, Planning and Environment, Institute for Management Research, Radboud University, Heyendaalseweg 141, 6525 AJ Nijmegen, The Netherlands
[6] Department of Risk, Slope Stability and Climate Adaptation, NGI, N-0855 Oslo, Norway; amy.oen@ngi.no
* Correspondence: lena.reimann@vu.nl

Abstract: Scenario analysis is a widely employed method for addressing uncertainties when assessing the physical and socio-economic impacts of climate change. Global scenarios have been extensively used in this context. However, these scenarios are in most cases not suitable for supporting local analyses. On the other hand, locally developed scenarios may lack the global context, thus having limited comparability with or transferability to other locations. The Shared Socioeconomic Pathways (SSP), which have been primarily developed for climate impact research, provide the possibility to extend the existing global narratives and adapt them to local characteristics in order to develop locally relevant scenarios. Here, we propose a methodological framework for producing harmonized scenarios across different case studies. This framework was developed in the EVOKED project and combines elements of top-down and bottom-up approaches to develop local scenarios for four regions in northern Europe. We employ the SSP as boundary conditions and, in cooperation with stakeholders from these four regions, develop local scenarios for a range of SSP. The developed sets of scenarios are consistently informed by global developments and are therefore comparable with other downscaled scenarios developed in different regions. At the same time, they have been based on local participatory processes, thus being locally credible and relevant to the needs of stakeholders. The local scenarios constitute a climate service per se as they can raise stakeholder awareness of the processes that will drive risk, exposure, and adaptive capacity in the future and inform discussions on mitigation strategies and adaptation pathways.

Keywords: Shared Socioeconomic Pathways; local scenarios; extended SSP; co-production

1. Introduction

Scenario analysis is an integral tool in climate change research for exploring uncertainties in the assessment of potential future socioeconomic and physical impacts of climate change-related hazards such as heat stress, sea-level rise, or heavy precipitation [1,2]. Socioeconomic scenarios, in particular, can inform mitigation and adaptation decisions as developments in socioeconomic conditions drive society's adaptive capacity [3,4] as well as potential exposure to climate change hazards [5]. For developing adaptation strategies at the local scale, socioeconomic scenarios need to reflect the local characteristics of the study area [6–8]. Local socioeconomic scenarios that are developed in collaboration with local stakeholders can, for instance, challenge perceptions and attitudes, encourage discussion about current issues, raise awareness, and serve as a tool to bridge knowledge gaps [9–12].

Previous work has proposed a range of approaches for developing locally relevant socioeconomic scenarios, ranging from top-down expert-led approaches to bottom-up stakeholder-led approaches. Top-down approaches are based on the assumption that socioeconomic developments at the local scale are embedded in the developments at regional, national, and global scales [13,14]. Therefore, these approaches use existing global and/or regional socioeconomic scenarios as boundary conditions, which are downscaled based on the characteristics of the study area. These characteristics are established by integrating expert knowledge with a literature review [15–20]. The advantage of top-down approaches is that the downscaled scenarios are consistent with the global scenarios used as boundary conditions, which ensures comparability of studies across different regional contexts. However, scenarios developed exclusively using a top-down approach may not be relevant for local stakeholders, which limits their usability in local decision-making [21].

Bottom-up approaches, on the other hand, do not use existing scenario frameworks to guide the scenario development process, but consult with local stakeholders from the beginning of the process, resulting in an original set of local socioeconomic scenarios [22–27]. Such stakeholder-led approaches ensure that the developed scenarios are tailored to the local characteristics of the study area, thereby creating a sense of ownership among stakeholders. As a result, the scenarios are more likely to be adopted as a decision tool at the local level, for instance as a basis for the development of local adaptation strategies [9,24,27].

This study has been designed as part of the EVOKED (EVOKED: Enhancing the value of climate data—translating risk and uncertainty utilizing a living-labs approach (http://tinyurl.com/3tjwbwbz, accessed on 1 March 2024) project, which has developed climate services in close collaboration with local stakeholders, with the aim to support adaptation planning in four case-study regions across Europe (see Section 2) [28]. The integration of local knowledge is one important precondition for a successful adaptation process [29]. The project has employed a living lab (LL) approach, which puts the end-user (i.e., local stakeholders) at the center of the development process in order to "co-create" and co-develop new knowledge, products or services [30–34]. The development of locally relevant socioeconomic scenarios is an integral component of such climate services as these scenarios provide the local basis for exploring different adaptation strategies. The project case studies are exposed to a variety of climate change-related hazards, which yield specific adaptation needs. Therefore, the socioeconomic scenarios need to reflect the local characteristics of each case study region.

To develop such locally relevant socioeconomic scenarios, we devise and apply a scenario development framework that combines elements of top-down and bottom-up scenario development approaches [35,36]. Applying this framework, we produce local socioeconomic scenarios that integrate local stakeholder knowledge, while at the same time harmonizing the scenario development process across case studies by using the same underlying scenarios as a starting point. Harmonization of the scenario development process ensures that the developed scenarios are internally consistent and coherent with the overarching scenarios used as boundary conditions, i.e., they follow the same scenario logic and the same underlying scenario assumptions, although storylines at the local scale may differ across case studies [21]. In this way, scenarios become locally relevant and can be directly employed in local-level decision-making and planning processes. At the same time, more overarching conclusions can be drawn across case studies and results can be compared to studies in other regions that have used the same scenarios as boundary conditions for the development of local scenarios [37,38].

Our scenario development framework comprises a "multi-scale co-production approach" where the developed local socioeconomic scenarios are coherent across multiple spatial scales (i.e., global—regional—local) [11,39] and are co-produced in collaboration with local stakeholders. We have designed this approach specifically for the development of local socioeconomic scenario narratives, which constitute qualitative descriptions of plausible socioeconomic developments in the form of a story [2,12]. We focus on the co-production of local scenario narratives as narratives ease the integration of local stakeholder knowledge into the scenario development process due to their typically non-scientific language [40]. The local narratives describe long-term developments, until 2100 (following the timeframe of the SSP), and can constitute the basis for producing local-scale projections of key variables that need to be considered in adaptation planning, such as the spatial distribution of population and infrastructure [13].

In the next sections, we provide a brief introduction into each case study region; present our "multi-scale co-production approach" and explain how we use this approach to develop and implement the scenario framework in each case study region by tailoring it to the local stakeholder needs of the respective study area; compare the scenario outcomes across the case studies; and critically reflect upon the scenario development process, its implications for adaptation planning at local level and its potential use in climate impact research.

2. Case Study Regions

To address climate adaptation at local scale and develop climate services, EVOKED focuses on four local- and regional-scale case studies in Norway, Sweden, Germany and the Netherlands. The selected sites (Figure 1) are exposed to a range of natural hazards and local governments seek to initiate, or enhance the value of ongoing, adaptation planning, via climate services.

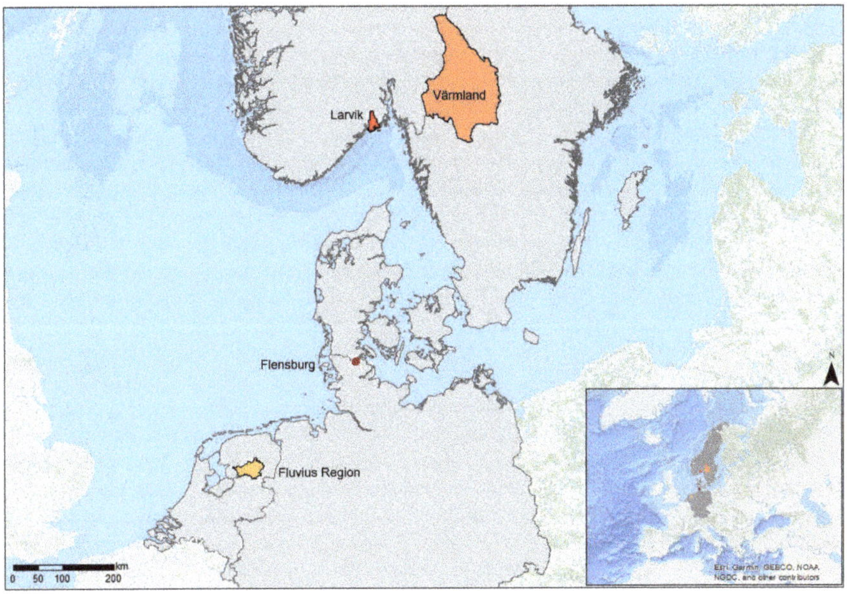

Figure 1. Map of the study region and case study areas.

2.1. City of Larvik, Norway

As a coastal city located in southern Norway, Larvik is exposed to hazards and has historically experienced floods, strong winds, storms and storm surges. These events are expected to become more frequent and more intense with potentially increasing damage costs as a result of climate change [41]. In particular, the expected increase in extreme precipitation could lead to increases in intensity and frequency of urban flooding, erosion, quick clay slides, rockslides, and river flooding; while increases in storm activity in Skagerrak, in combination with a rising sea level, will increase the intensity and frequency of storm surges, coastal flooding, and erosion in Larvik.

According to the Civil Protection Act, all Norwegian municipalities have the obligation to carry out comprehensive risk and vulnerability assessments (RVA) which should be considered in land-use planning and should map potential impacts of climate change [42]. The municipality of Larvik aims to reduce both current and future risks due to climate change, e.g., through actions of its emergency preparedness group, or by incorporating adaptation analysis in all development projects.

As part of the last update of Larvik municipality's RVA, the potential impacts of climate change due to a range of natural hazards including intense rainfall, flooding, erosion, landslides, and storm surges were explored [43]. In this context, exposure maps for Larvik were developed and used for the building development of the Martineåsen area within the municipality, also accounting for a range of suggested local adaptation and mitigation measures; while the knowledge needs and perceptions of stakeholders were explored as input to the development of future socioeconomic scenarios that were relevant not only for the local authorities, but also for the community of Larvik.

2.2. *Värmland, Sweden*

Värmland is an inland county located in Sweden that is exposed to flooding and other water-related hazards from Lake Vänern (the largest lake in the European Union) and river flows. The county capital of Karlstad situated on a delta where the Klarälven river meets Lake Vänern and has experienced several floods and landslides. In Värmlands county there is a need for new housing to deal with urbanization, especially in Karlstad. Smaller municipalities are losing population due to urbanization migration to Karlstad and are using the possibility of living near the water as a way to attract people. This leads to more construction occurring close to the water today than previously, even though recent flooding events have put many areas at risk.

The Värmland County Administrative Board (VCAB) coordinates climate adaptation work in the county. It supports municipalities and other regional actors through awareness raising activities and disseminating knowledge about current and future climate change and climate change impacts. Within Värmland county, VCAB suggests measures which could help increase local and regional resilience to climate change; develops the Regional Climate-Adaption Action Plans [44]; and plays an active part in complying to the European Union Flood Directive. It focuses on adapting society to a changing climate to protect the environment, people, and property in a long-term perspective.

2.3. *Fluvius Region, The Netherlands*

For the Netherlands, climate adaptation is often associated with flood risk (e.g., [45]). However, water scarcity, drought and heat stress are also considered as threats [46], particularly in the rural areas in the sandy east and south of the Netherlands, where drought is expected to become a problem [47,48]; while extreme rainfall and flooding are expected to occur more frequently. These hazards may affect both rural and urban areas [46]. In response, the Deltaplan Spatial Adaptation (Deltaplan Ruimtelijke Adaptatie) was developed to help guide Dutch efforts [46] to such hazards. In this approach, the initial focus is to gain insights in the potential risks of climate change (stress test), followed by the initiation of dialogues between local governmental bodies and regional stakeholders (e.g., agriculture, public healthcare, drinking water, nature) (risk dialogue) leading eventually to co-developing adaptation plans (regional adaptation strategy).

The Fluvius region is located in the North-Eastern part of The Netherlands. The abundant presence of water has driven the social and economic development in this region. The regional and local governments in the region collaborate for establishing a water robust and climate proof region, with the aim to provide for a water safe area in 2100; and have launched the program "Living with Water in the IJssel-Vechtdelta" in 2014. This program aimed, among others, to improve the awareness of the regional communities regarding the effects of climate change. Within this context the region has developed a range of climate services included in Platform Water Fluvius [49] and has completed the process of consultation with regional stakeholders. This process has resulted in a regional adaptation strategy and regional implementation agenda, which acts as a strategic document for informing more localized climate adaptation plans for the Fluvius-partners [49].

2.4. *City of Flensburg, Germany*

The city of Flensburg is located at the German Baltic Sea coast, directly at the national border to Denmark. The city is prone to coastal flooding during storm conditions, when a high-pressure system with winds from the northeast follows a low-pressure system with strong winds from the west, as these wind conditions can potentially result in a seiche over the entire Baltic Sea basin [50]. The intensity and frequency of these flooding events are expected to increase over the next century and beyond due to accelerated sea-level rise [29,51–53]. Currently, no assessment of vulnerability to coastal flooding exists for the city and no adaptation measures are in place apart from measures taken on an individual level [54,55]. These household-level adaptation measures include sandbags and the installation of mobile barriers during flooding events [56,57]. Due to these circumstances, the city of Flensburg has recently initiated the process of developing an adaptation agenda, in cooperation with the local community. Furthermore, local stakeholders do not tend to favor conventional protection strategies in the form of structural solutions, and appear more positive to alternative adaptation options for coping with future flood risk. In this context, the city can benefit from support in exploring potential adaptation options.

3. The Multi-Scale Co-Production Approach for Local Scenario Development

Our method is based on previous work by Rotmans et al., 2000 [12] who developed scenarios for sustainable development in Europe, using a participatory scenario development approach. They proposed five steps to be taken during each stakeholder workshop. We adapted these steps to our approach and introduced an additional one, resulting in the following six steps shown in Figure 2. Beyond the introduction of an additional step, our approach differs from [12] in its implementation. In our study, we employed an existing scenario framework, the SSP, which we downscale to each of the study regions. By defining local scenario elements for each case study, we extended the SSP for the study regions, thus developing new locally relevant scenarios, while at the same time keeping them consistent between all case studies. Our approach is not only workshop-based and requires a longer period for its implementation. This has allowed more interaction with the stakeholders, which is necessary for introducing the SSP framework; and a larger number of iterations before finalizing every step. The six steps that we undertook for the development of local scenarios are described in detail below:

Step 1: Select scenarios as boundary conditions

When compiling socioeconomic scenarios at a local scale, it is important to not only account for local developments and processes, but also to take into consideration that the region is embedded in developments at different spatial scales, ranging from global to national, regional, and local levels [13,14]. Therefore, in order to ensure compatibility between case studies, it is important in the first step to select scenarios for all case study regions that will serve as boundary conditions for the local scenarios. This step needs to be taken jointly (i.e., with all those working on the different case studies) to ensure that the local socioeconomic scenarios are harmonized across case studies.

In EVOKED, we used the global-scale Shared Socioeconomic Pathways (SSP) as a starting point for providing the global context to the local scenario narratives as the SSP are the current state-of-the-art socioeconomic scenarios used in climate change research [58,59]. The SSP describe five broad-scale pathways of plausible socioeconomic development until the end of the 21st century at a global scale, based on societal challenges for mitigation and adaptation [59,60]. SSP1 "Sustainability" describes a sustainable world with low challenges for both mitigation and adaptation; SSP2 "Middle of the Road" is a scenario with moderate challenges; SSP3 "Regional Rivalry" has high challenges; SSP4 "Inequality" is characterized by high challenges for adaptation, but low challenges for mitigation; SSP5 "Fossil-fueled Development" has low challenges for adaptation and high challenges for mitigation [61]. Each SSP consists of a qualitative narrative and projections of several variables that have been produced on a national level, such as population [62], urbanization [63], and GDP [64].

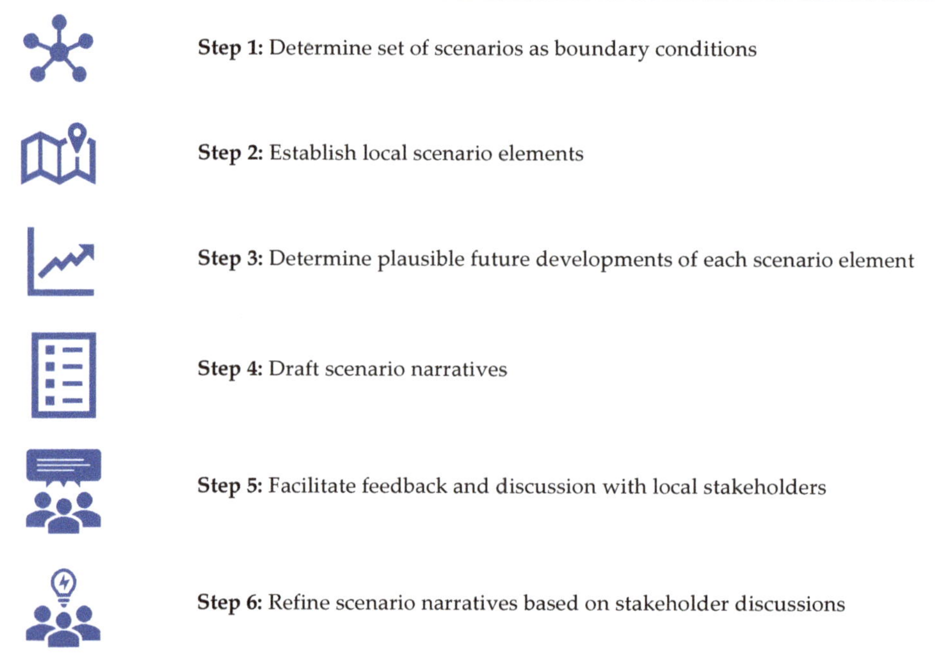

Figure 2. Steps undertaken for the development of the local scenarios.

The SSP have been intentionally designed in a broad manner to allow for the development of so-called "extended SSP" for sectoral and/or regional to local applications [59,65]. More than thirty key elements that have been identified as important drivers of socioeconomic development at the global scale are described in each global SSP narrative [61]. As a result, the global narratives are generic enough to provide the overall socioeconomic context in each SSP while providing elements that are relevant to sub-global analyses as well [61,66]. We use these scenario elements and their characteristics in each SSP as a starting point for the development of extended SSPs for each case study region. We must note that, as the development of our local SSP extensions was completed in 2018, we did not associate the SSP with radiative forcings, as presented in Gidden et al., 2019 [67] and adopted in the IPCC AR6 [68]. Nevertheless, as the scope of our study is the compilation of local narratives for socioeconomic development, relevant future forcings can subsequently be associated to each of the developed local scenarios for impact and vulnerability analysis.

Step 2: Establish local scenario elements

In step two, we extended the global SSP key elements by local scenario elements that are important drivers of societal development in each case study region. To establish local elements, we reviewed the locally relevant policy documents and literature, analyzed data of the local and regional administrations and statistics offices, and/or consulted with local stakeholders of each case study region [69]. In addition, several guiding questions for each case study region have been compiled in order to explore how drivers of socioeconomic development are embedded in the global to regional scale, which is an important aspect of multi-scale scenario development [10,11]:

- What is the demographic structure of the population? What are recent population trends?
- What are the major characteristics of political and socioeconomic importance and/or issues of concern in the case study region?
- How are local politics embedded in and influenced by regional, national, and global politics?
- In which ways is the local economy embedded in global markets? What are the biggest companies in the case study region? Are they national, regional, or global players?

As pointed out by Kok et al., 2007 [10] and Özkaynak and Rodríguez-Labajos 2010 [11], it is important to include local stakeholders at this stage of the process in order to confirm whether the established elements are relevant at the local scale and to establish potentially missing elements.

Once the local elements were established, we further identified the global SSP elements that are most relevant to each case study region and extended them based on the respective local SSP elements. We focused only on central global elements in order to keep the total number of scenario elements manageable, thereby ensuring that the draft narrative is as short as possible to facilitate the communication with stakeholders [11].

Step 3: Determine plausible future developments of each scenario element

Based on current characteristics in socioeconomic development identified through the literature review and analysis of regional socioeconomic data in step two, we proceeded to define for each local scenario the characteristics of the scenario elements, including the newly established local elements as well as the selected global elements. As it is important to concentrate on a small number of scenarios in order to enhance communication and to facilitate discussion with local stakeholders, we propose using a set of three to four scenarios [18,70,71]. This number is sufficient to account for the full range of uncertainty regarding the challenges for mitigation and adaptation, as defined in the global SSP [72].

To maintain consistency with the global SSP, we used the developments of the global SSP as the basis for each respective local scenario; adapted them where necessary to reflect the developments at local scale; and extended them with further socioeconomic context using the information for the local elements. Here, it is particularly important to ensure that the assumptions of the future developments (at local level) in each scenario are plausible, in order to increase the likelihood that local stakeholders will adopt the developed scenario narratives for local decision-making. Plausibility of scenarios is achieved if, based on their current knowledge and understanding of the world, stakeholders agree that the developments described in each scenario "could happen" [73,74]. We would generally propose to assemble the results of step three in a table in order to allow for comparing and contrasting the characteristics of all scenarios in a structured and clear manner.

Step 4: Draft scenario narratives

Using the table outlining the characteristics of the scenario narrative elements, full-text narratives were then drafted for each scenario, qualitatively describing the developments in the form of a story [2,12]. The narratives add further context to the scenario elements and are written in non-scientific language, with the aim to facilitate stakeholders' understanding of each scenario [40]. As pointed out by Kok et al., 2007 [10], local stakeholders will be less overwhelmed by global developments, particularly social and economic developments, if they are not presented as facts but rather as underlying assumptions and changes in world views. In this step, we suggest giving each scenario a new name that reflects the overall socioeconomic developments in each narrative in a concise manner, also referring to the case study context. New scenario names are important for stakeholder identification with the local scenarios. Therefore, they should be developed in collaboration with the local stakeholders, aiming to create a sense of ownership among them [75].

Step 5: Facilitate feedback and discussion with local stakeholders

In step five, the first drafts of the narratives along with the scenario names for every case study were discussed with the respective stakeholders, to ensure plausibility and relevance of the local scenarios. In this step, it is essential to clearly communicate to the stakeholders that scenarios are explorations of plausible futures rather than predictions of what will happen [70,74,76]. Therefore, it is important to note that the developed narratives can also reflect developments that are considered as less likely by the stakeholders [10]. Depending on stakeholders' background knowledge and prior experience with scenarios, sufficient time should be set aside for explaining these basic notions [9,39].

Stakeholder interactions can take place in different communication formats, such as interviews, expert workshops, focus group discussions or stakeholder workshops [10]. Besides the narrative text, other visualization tools can be employed in order to facilitate understanding of each scenario; such tools include, for instance, pictures, collages, comics, graphs, explanatory videos, or a theatrical play [10,77,78]. It is important that the communication formats and visualization tools are selected based on the amount of time available and the previous experience of the stakeholders. Therefore, the project partners select those formats and tools most suitable for their case study region [12,78].

In EVOKED, we aimed at including a broad group of stakeholders in the process as one important component of the living lab approach. An initial list of stakeholders was compiled for each case study, based on web-based search and previous co-operation. Based on suggestions from our key stakeholders, who were already partners in EVOKED, we extended this list using a snowballing method. Each project partner conducted a stakeholder analysis to establish the relevant stakeholders in each case study region and understand their interest and influence in the region, based on Reed et al., 2009 [79]. We aimed to include representatives from business, governments, NGOs, science/experts, and citizens of the case study region, following the recommendations of previous work [10,75,80]. More detailed information on the selection of stakeholders can be found in Reimann et al., 2021 [73].

Step 6: Refine scenario narratives based on stakeholder discussions

Based on the stakeholder feedback in step five, the narrative drafts were revised to include the ideas and local knowledge of the stakeholders. Here, the scenario experts weigh which of the points raised during the discussion to include in the scenario, thereby ensuring coherence across scales (i.e., with the global SSPs) [21] as well as consistency within each scenario narrative [10,12]. Subsequently, the revised narrative drafts were discussed with the local stakeholders again to ensure that the revised scenarios were plausible and relevant at the local level. If necessary, this process should be repeated in several iterations until the scenarios are fully approved by the stakeholders [26,80]. Depending on the time and resources available, this iterative process can take place in workshops or focus group discussions, but also remotely via email or during conference calls.

4. Implementation of the Approach in Each Case Study Region

In this section, we describe how we applied the six-step multi-scale co-production approach in each case study region and present the outcomes of this process. Table 1 provides an overview of how each step of the approach was tailored to the local particularities of each case study.

Table 1. Overview of the implementation of the six-step multi-scale co-production approach in each case study region.

Step #	Norway	Sweden	Netherlands	Germany
	Shared Socioeconomic Pathways (SSPs)—Global narratives			
	-Literature review, emphasis on municipality's planning documents -Discussions with municipality representatives	-Literature review of guiding documents -Discussion with county and Arvika municipal actors	-Literature review -Expert judgement	-Literature review -Focus group discussion
	-Literature review, emphasis on municipality's planning documents -Regional and local statistics (SSB)	-Literature review of guiding documents -Local and regional statistics from web pages and national statistics	-Literature review -Local statistical information -Expert judgement	-Literature review -Regional and local statistics offices
Selected scenarios	SSP 1, 2, 3	VCAB—SSP 1, 2, 3, 5 Arvika—SSP 1, 2, 4, 5	SSP 1 and 5	SSP 1,3,4,5
Methods used for determining current conditions	-Municipality planning documents -Regional and local statistics (SSB)	-Online brainstorming meetings with municipal and regional representatives -Strategy documents	-Literature review	-Literature review -Statistics offices
	An overview text with factual details created into half-page narratives for the illustrator to create drawings	One narrative drafted for the Värmland county (VCAB) but including the Arvika scenarios as descriptive and illustrative examples	-One-page narrative text drafted by Deltares EVOKED project member	-One-page narrative text drafted by scenario experts
	-Workshop with children and youth representatives	Online meetings with stakeholders in Arvika and VCAB	Did not take place. Rather feedback was provided by colleagues at Deltares in the form of expert judgement	-Stakeholder workshop -jigsaw puzzle -group discussions
	-Illustrations revised based on feedback -Presented during open meeting with local politicians	SGI refines text scenarios together with VCAB, Arvika and feedback from CAU. Discussion on potential uses of scenarios.	Focus on SSP 1 and 5 instead of all five scenarios	-replace SSP4 with SSP2 -Email-based feedback -Focus group discussion

4.1. City of Larvik, Norway

The local scenarios for Larvik were based on the five global-scale SSP. First (steps one and two), policy documents, the literature, and data specific to the region of Larvik, were analyzed in order to explore challenges and potential trends, already observed in Larvik today. Three of the global SSP were selected following discussions with the municipality, to ensure that the selected SSP are considered most relevant for Larvik. The selected global scenarios included SSP1, SSP2 and SSP3, and plausible future developments were determined for each scenario element in the third step.

SSP1, "Sustainable Larvik", was primarily based on the municipal plans, such as the land-use plan and energy- and climate plan, among others. This scenario reflects the way that Larvik is currently planning for the future. SSP2, "Business as usual", is based on relevant policy and strategy documents and knowledge bases, depicting the current state of Larvik. SSP3, "Regional rivalry", is based on similar sources, but depicts a scenario that could take place if the existing strategies "failed". Regional cooperation is an important factor for the future of Larvik as all neighboring municipalities in the region are preparing for an anticipated centralization process towards the larger cities, especially in the Oslo region. In a situation where regional municipalities were not able to compete at the same level as the nearest urban metropolitan areas, regional rivalry could erupt, with regional municipalities competing amongst them.

Following a dialogue with the municipality, we decided not to develop dystopic scenarios, since Norway is a welfare state with a fairly robust safety net system for its citizens. The municipality believed that dystopic scenarios would not be considered as plausible by the stakeholders involved in the ongoing urban development plans (followed also by the EVOKED project). Here, legitimacy was an important concern for the municipality and developing scenarios that people could relate to was deemed to be the way forward.

In the fourth step, the scenario table was compiled with partners from the municipality of Larvik; in this process missing information or information that was not relevant was identified. The draft table was then revised and extended into narratives. In the fifth step, the narratives were transformed into illustrations for children and the youth (Figure 3). These illustrations were used during a workshop held at a primary school in Larvik. Feedback from the children and youth participants on these illustrations was then used in step six, for further refining the illustrations. The final illustrations were presented to local politicians as part of a larger discussion on the development of the Martineåsen project. They formed the basis for a further dialog on how to use a blue-green factor tool to ensure a high quality of the blue–green infrastructure and how building developers incorporate stormwater runoff and infiltration in their projects to reduce the impact of urban flooding.

Figure 3. *Cont.*

Figure 3. Illustrations of the three local narratives (illustrations by Bar Bakke), (**a**) SSP1, "Sustainable Larvik"; (**b**) SSP2, Larvik in a "Business as usual" scenario; (**c**) SSP3, "Regional rivalry" for Larvik.

4.2. Värmland, Sweden

In the first step, we compiled a first list of local and regional scenario elements. SSP2 was based on the official statistical data for the years of 2017–2018 of Arvika municipality and Värmland county. SSP1 was based on policy documents for sustainable development: Arvika's Environmental strategy [81], Strategic plan for Arvika 2019–2020 [82], and Värmlandsstrategin 2014–2020 [83]. SSP 3, 4 and 5 were based on assumptions regarding potential developments in case the strategies for sustainable development were only partially (SSP3 and 4) or not at all realized. In step two, the various scenario elements were discussed during an online meeting with representatives from VCAB and Arvika Teknik and Arvika municipality in September 2018 to determine the plausibility of the elements in each scenario and to select the scenarios that were most relevant.

In the third step VCAB and Arvika representatives discussed the possibility of combining SSP3 (regional rivalry) and SPP4 (inequality: A more insular Arvika) into a single scenario and whether two separate scenarios, one for Arvika and one for the entire region (Värmland), were required. SSP3 was relevant in the debate about larger regions in Sweden, but SSP4 was more relevant for the local conditions in Arvika. The decision was taken to develop two separate scenarios, one for Arvika using SSP4 and one for VCAB using SSP3. The other SSP-based scenarios would remain the same, as well as the global and local elements, but the weighting and description of the elements were different. A first draft of the scenario narratives was compiled in step four; this was subsequently discussed and revised by both VCAB and Arvika, who proposed changes and additions to the narratives to enhance their plausibility and relevance. Eventually, the local and county scenarios were merged into one narrative for the entire county of Värmland and use the Arvika scenarios as local examples. In steps five and six additional stakeholders in VCAB and Arvika were brought in during separate meetings to discuss the potential usefulness of the scenarios. For VCAB the scenarios could be used as input in the municipality's work with a regional climate adaptation strategy. For Arvika they could be used in comprehensive plans or risk and vulnerability assessments, particularly for the building and planning section (Samhällsplanering) and less so for the technical sections (Teknik i väst).

4.3. Fluvius-Region, The Netherlands

For the Netherlands, national scenarios with detailed narratives (the "Delta Scenarios") that had been developed in consultation with stakeholders already existed [84]. These scenarios describe potential development paths for the Netherlands along two axes: with regard to climate change ("slow and little amount of climate impact vs. fast and large amount of climate impact") and socio-economic trends ("socio-economic shrinkage vs. socio-economic growth"). As these scenarios match some of the SSP scenarios, and are at the same time relevant for the local studies in the Netherlands, they were also employed for the local scenario development by adjusting the six-step approach presented in Section 3.

In the first step, the five SSP scenarios were translated into potential developmental storylines for the case study areas. Then, we established local scenario elements based on the literature specific to the case study areas, on the "Delta Scenarios" [84] and on (local) statistical data and trends that are available for both regions [85]. We employed the "Delta Scenarios" [84] as a basis for extending the plausible future development of each scenario element and focused on the scenarios "Rest" and "Steam". These correspond with SSP 1 and 5, which we believe best represent the solution space for the future development of the study areas. Based on the local data and the narrative described in the Delta Scenarios, we developed narratives for both selected SSP scenarios. For step five (asking for feedback), we made use of expert judgement from colleagues working at Deltares that are involved in their daily work with scenario development in regard to climate change impacts in the Netherlands. This judgement involved the plausibility of the developments described in the first draft of the local SSP scenarios. Based on the feedback we made final adjustments to the developed scenario drafts for SSP 1 and 5.

4.4. City of Flensburg, Germany

The scenario development process in the city of Flensburg closely followed the six-step approach (see [73]). We established local scenario elements (step two) based on locally relevant literature, such as policy documents, [86–90], as well as quantitative data of the regional and local administrations and statistics offices [91,92]. We discussed the preliminary local scenario elements with Flensburg's city administration in a focus group of four participants. During this discussion, we presented the local elements and asked participants to identify missing elements as well as items that were not relevant for the city of Flensburg. Subsequently, we devised the plausible future developments of each scenario element for four selected SSPs, ensuring coherence of the local-scale scenarios with the global SSPs. We chose SSPs 1, 3, 4 and 5 with the aim to span the full range of uncertainty with respect to challenges for mitigation and adaptation. Step three resulted in a table outlining the future characteristics of each scenario element for each selected SSPs. Based on this table, we drafted a one-page narrative for each scenario (step four). We further devised new scenario names that reflect the local developments described in each narrative.

In step five, we discussed the local scenario narrative drafts in a stakeholder workshop, which we co-organized with Flensburg's city administration. We focused on three scenarios (SSP1, SSP3 and SSP5) due to the number of workshop participants (13 stakeholders) and time constraints (half-day workshop). Following Kok and van Vliet 2011 [78], we used a mix of visualization tools and formats of interaction to address a range of perceptions and tastes. At the beginning of the workshop (phase one), a scenario expert provided general information of the SSP framework to the stakeholders via a flip chart presentation. In order to develop a better understanding for the scenarios, the stakeholders assigned the different scenario elements and their characteristics to the respective SSP in a second phase. Similar to a jigsaw puzzle, the task was to use different pieces of graphical elements of the SSP and organize them on a poster (Figure 4).

Figure 4. Photo of the completed jigsaw puzzle exercise, showing the pieces of the graphical elements and the completed scenario puzzles.

The narratives were then discussed in smaller groups, with each group focusing on one SSP. As our aim was to motivate a lively discussion of the narrative drafts, we decided upon a group size of four to five participants for each scenario. We asked the workshop participants to provide feedback on the individual elements; report on missing ones; and to provide feedback on the narratives' drafts. After the discussion in smaller groups, we reflected upon all comments with all workshop participants. During this phase, the stakeholders identified the need for a scenario reflecting current development trends in Flensburg.

Last, we integrated the stakeholder comments into each narrative. It was not possible to include all suggestions into the local scenarios as some were not compatible with the developments described in the global narrative of the respective SSP. We sent the revised narratives to the stakeholders via email for a second round of comments. We also attached a justification and explanation document in those cases where we could not include specific comments. As we received further minor comments on SSP1 "Sustainable Flensburg", we changed the narrative accordingly. We additionally replaced the SSP4-based narrative, "Flensburg's elites on the rise" (which was not discussed during the workshop), with the newly devised SSP2-based narrative, "Flensburg as we know it", in order not to exceed a total number of four scenarios. We presented this narrative to representatives of Flensburg's city administration for their comments on its plausibility before finalizing it.

5. Discussion and Conclusions

In the context of the EVOKED project, we developed a set of local scenarios, following the common for all four case studies, stepwise approach described in Section 3. By combining a top-down (steps 1–4) with a bottom-up approach (steps 5–6) we have produced groups of scenarios that are locally relevant and accepted by the potential end-users. Further, the developed scenarios are consistent with the global SSP that were selected as boundary conditions and are therefore comparable with, or complementary to, other local and regional scenarios developed in different regions or in different contexts.

Despite the distinct local context of each case study of EVOKED, the scenarios developed for the four case studies are similar in terms of the SSP that were preferred by the stakeholders of every case study. The same trend was also observed in Kok and van Vliet 2011 [78]. In all case studies there was a clear preference for a local SSP1, while three of the case studies developed local extensions for SSP2, SSP3 and SSP5. In particular, the inclusion of SSP2 was, in most cases, included following the demand of the stakeholders. This is possibly due to the nature of the SSP2 which, as a middle of the road scenario, entails less extreme change and its storyline is therefore closer to what stakeholders can envisage and project. The overall choice of scenarios seems to reflect ongoing discussions in climate policy and the general political landscape, with society being torn between sustainable solutions, the use of fossil fuels, and the increasing tendency to protect regional markets. Nevertheless, the participatory approach leads to a shift from "typical" key elements (economic, demographic, technological) to socio-cultural, political and institutional elements [12].

Our approach leads to the development of sets of scenarios that are consistently informed by global developments as well as by local participatory processes, thus being locally credible and relevant to the needs of stakeholders. The methods leave ample room for interpretation in order to develop the scenarios based on the specific context of each case study [93] but are still coherent with the SSPs, thus avoiding a range of very different scenarios that do not allow for cross-case comparisons [13,38]. As such, they are characterized by internal consistency and similarity despite possible differences between them [36]. Such scenarios allow for adaptation options to be explored across plausible futures, trust to be built in the context of participatory processes, local values to be understood by the developers and policies to be more robust under changing climate conditions [69]. The developed scenarios provide the possibility to explore a wide range of societal futures and the resulting climate-change impacts, thus providing a solid basis for informing adaptation planning in each project case study. Importantly, coherence further leads to consistency between spatial scales, which is particularly important in the context of climate change adaptation [78].

The developed narratives also provide the basis for quantifying trends of socioeconomic development and for producing local-scale projections of socioeconomic variables that can directly feed into the analysis of risk and consequently guide the adaptation process. Examples of such variables are spatial population distributions, distribution of income, age groups, and land use changes under each scenario. For example, population projections that can be employed for assessing future exposure and vulnerability of population to potential impacts of climate change were developed for the selected SSP for the case-study of Flensburg, using the respective SSP national population growth projection rates and adjusting them depending on the scenario narrative. Using the same model for all case studies allows for using projection input data from the same baseline scenarios (SSP) [37,94], thus making the quantified projections consistent across the case studies. Further, the narratives constitute a climate service per se as they raise stakeholder awareness of the processes that will drive risk, exposure, and adaptive capacity in the future, and provide context for stakeholder discussions on mitigation and adaptation strategies and pathways [94] by comparing scenarios to determine most desirable pathway and devise policies to achieve the vision [11]. As our work has primarily focused on developing narratives for local socioeconomic development and was completed before the year 2021, we have not associated the local SSP to plausible radiative forcings, as presented in IPCC AR6 [68]. However, we must point out that in some of the case studies, stakeholders expressed interest in pathways that are deemed not plausible [67] according to the new framework. For example, local sustainable development, corresponding to SSP1, was perceived locally plausible in a world with high-end emissions (e.g., RCP5). Such questions, which may be relevant in a local context but seem implausible in a global context, require the use of the scenario matrix architecture [4] and cannot be fully described in the context of the newer global SSP framework, where socioeconomic development is linked to specific radiative forcings.

The process of co-developing local scenarios with stakeholders also has some drawbacks and challenges to overcome. The quality of the stakeholder feedback largely depends on the background knowledge of the stakeholders and their active participation. Regional long-term scenarios did not generally exist in most of our case studies and only a few stakeholders had previous experience in developing scenarios. In this context, exploring plausible long-term future socioeconomic developments (e.g., until 2100) can seem rather abstract to stakeholders, who require time to fully immerse in the storylines discussed. For this reason, thorough preparation of the stakeholder workshop and the use of visualization techniques and material, such as figures, graphs, animations, and posters, is essential for facilitating comprehension. Further, we have observed that exercises in small groups can be particularly useful as these result in more lively discussions that help participants to engage in the topic, especially in those cases where the group of stakeholders is very diverse in terms of roles and background.

Second, achieving coherence between scenarios from different case studies may lead to loss of variety [78] as the use of common scenario elements can lead to the developed scenarios being too rigid or restricted. Kok et al., 2007 [10] propose that formal downscaling should be attempted only if there is particular scientific or policy value; if the aim is primarily to aid decision making constructing the scenarios for each scale using a broad common framework could suffice.

Third, aligning new scenarios with already existing scenarios at a national level that are accepted and used by stakeholders is a complex process. In EVOKED, for the Dutch case studies, we decided to map the already established national scenarios to the SSP. However, this is not a straightforward task and may introduce uncertainties when comparing case studies. Similarly, Hagemann et al. (2020) [37] note that if stakeholders are free to choose scenarios, the outcomes may eventually look very different, reflecting to a greater extent the local logic based on cultural and geomorphological conditions [37].

Further work to tackle some of the above limitations could involve at improved implementation of the participatory part of the process, in line with the elements of a living lab. Moreover, integrating stakeholders right from the start of the scenario development and increasing the iterations with the stakeholders would allow further integration of local knowledge. However, time availability often constitutes a major constraint in this process. Last, we must note that the development of scenarios has only constituted one aspect of the EVOKED project in the process of the production of climate services. Therefore, the proposed method should be considered in this context and adapted in projects or analysis that, for example, focus entirely on scenario development (e.g., MedAction, Visions).

Author Contributions: Conceptualization, A.T.V. and L.R.; Methodology, A.T.V., L.R., B.V. and A.O.; Formal analysis, L.R., G.J.E., G.G., G.K., L.V.W., B.V. and M.T.; Writing—original draft, A.T.V. and L.R.; Writing—review & editing, G.J.E., G.K., L.V.W. and A.O.; Supervision, A.T.V.; Funding acquisition, A.T.V. and A.O. All authors have read and agreed to the published version of the manuscript.

Funding: The project EVOKED is part of ERA4CS, an ERA-NET initiated by JPI Climate, and funded by RCN (NO), FORMAS (SE), NOW (NL), BMBF (DE) with co-funding by the European Union (Grant 690462). Publication of this research is financially supported by DFG within the funding programme Open Access-Publikationskosten.

Institutional Review Board Statement: Not applicable.

Informed Consent Statement: Not applicable.

Data Availability Statement: Data are contained within the article.

Acknowledgments: We acknowledge financial support by DFG within the funding programme Open Access-Publikationskosten.

Conflicts of Interest: The authors declare no conflict of interest.

References

1. Ebi, K.L.; Hallegatte, S.; Kram, T.; Arnell, N.W.; Carter, T.R.; Edmonds, J.; Kriegler, E.; Mathur, R.; O'Neill, B.C.; Riahi, K.; et al. A new scenario framework for climate change research: Background, process, and future directions. *Clim. Chang.* **2014**, *122*, 363–372. [CrossRef]
2. Moss, R.H.; Edmonds, J.A.; Hibbard, K.A.; Manning, M.R.; Rose, S.K.; van Vuuren, D.P.; Carter, T.R.; Emori, S.; Kainuma, M.; Kram, T.; et al. The next generation of scenarios for climate change research and assessment. *Nature* **2010**, *463*, 747–756. [CrossRef]
3. O'Neill, B.C.; Carter, T.R.; Ebi, K.; Harrison, P.A.; Kemp-Benedict, E.; Kok, K.; Kriegler, E.; Preston, B.L.; Riahi, K.; Sillmann, J.; et al. Achievements and needs for the climate change scenario framework. *Nat. Clim. Chang.* **2020**, *10*, 1074–1084. [CrossRef] [PubMed]

4. Riahi, K.; van Vuuren, D.P.; Kriegler, E.; Edmonds, J.; O'Neill, B.C.; Fujimori, S.; Bauer, N.; Calvin, K.; Dellink, R.; Fricko, O.; et al. The Shared Socioeconomic Pathways and their energy, land use, and greenhouse gas emissions implications: An overview. *Glob. Environ. Chang.* **2017**, *42*, 153–168. [CrossRef]
5. Hallegatte, S.; Przyluski, V.; Vogt-Schilb, A. Building world narratives for climate change impact, adaptation and vulnerability analyses. *Nat. Clim. Chang.* **2011**, *1*, 151–155. [CrossRef]
6. Carlsen, H.; Dreborg, K.H.; Wikman-Svahn, P. Tailor-made scenario planning for local adaptation to climate change. *Mitig. Adapt. Strateg. Glob. Chang.* **2013**, *18*, 1239–1255. [CrossRef]
7. Haasnoot, M.; Kwakkel, J.H.; Walker, W.E.; ter Maat, J. Dynamic adaptive policy pathways: A method for crafting robust decisions for a deeply uncertain world. *Glob. Environ. Chang.* **2013**, *23*, 485–498. [CrossRef]
8. Walker, W.; Haasnoot, M.; Kwakkel, J. Adapt or Perish: A Review of Planning Approaches for Adaptation under Deep Uncertainty. *Sustainability* **2013**, *5*, 955–979. [CrossRef]
9. Berkhout, F.; Hertin, J. Foresight Futures Scenarios. *Greener Manag. Int.* **2002**, *37*, 37–52. [CrossRef]
10. Kok, K.; Biggs, R.; Zurek, M. Methods for Developing Multiscale Participatory Scenarios: Insights from Southern Africa and Europe. *Ecol. Soc.* **2007**, *13*, 16p. [CrossRef]
11. Özkaynak, B.; Rodríguez-Labajos, B. Multi-scale interaction in local scenario-building: A methodological framework. *Futures* **2010**, *42*, 995–1006. [CrossRef]
12. Rotmans, J.; van Asselt, M.; Anastasi, C.; Greeuw, S.; Mellors, J.; Peters, S.; Rothman, D.; Rijkens, N. Visions for a sustainable Europe. *Futures* **2000**, *32*, 809–831. [CrossRef]
13. van Ruijven, B.J.; Levy, M.A.; Agrawal, A.; Biermann, F.; Birkmann, J.; Carter, T.R.; Ebi, K.L.; Garschagen, M.; Jones, B.; Jones, R.; et al. Enhancing the relevance of Shared Socioeconomic Pathways for climate change impacts, adaptation and vulnerability research. *Clim. Chang.* **2014**, *122*, 481–494. [CrossRef]
14. van Vuuren, D.P.; Smith, S.J.; Riahi, K. Downscaling socioeconomic and emissions scenarios for global environmental change research: A review. *WIREs Clim Chang.* **2010**, *1*, 393–404. [CrossRef]
15. Absar, S.M.; Preston, B.L. Extending the Shared Socioeconomic Pathways for sub-national impacts, adaptation, and vulnerability studies. *Glob. Environ. Chang.* **2015**, *33*, 83–96. [CrossRef]
16. Kamei, M.; Hanaki, K.; Kurisu, K. Tokyo's long-term socioeconomic pathways: Towards a sustainable future. *Sustain. Cities Soc.* **2016**, *27*, 73–82. [CrossRef]
17. Kok, K.; Pedde, S.; Gramberger, M.; Harrison, P.A.; Holman, I.P. New European socio-economic scenarios for climate change research: Operationalising concepts to extend the shared socio-economic pathways. *Reg. Environ. Chang.* **2019**, *19*, 643–654. [CrossRef]
18. Kok, K.; Rothman, D.S.; Patel, M. Multi-scale narratives from an IA perspective: Part I. European and Mediterranean scenario development. *Futures* **2006**, *38*, 261–284. [CrossRef]
19. Lehtonen, H.S.; Aakkula, J.; Fronzek, S.; Helin, J.; Hildén, M.; Huttunen, S.; Kaljonen, M.; Niemi, J.; Palosuo, T.; Pirttioja, N.; et al. Shared socioeconomic pathways for climate change research in Finland: Co-developing extended SSP narratives for agriculture. *Reg. Environ. Chang.* **2021**, *21*, 7. [CrossRef]
20. Reimann, L.; Merkens, J.-L.; Vafeidis, A.T. Regionalized Shared Socioeconomic Pathways: Narratives and spatial population projections for the Mediterranean coastal zone. *Reg. Environ. Chang.* **2018**, *18*, 235–245. [CrossRef]
21. Zurek, M.B.; Henrichs, T. Linking scenarios across geographical scales in international environmental assessments. *Technol. Forecast. Soc. Chang.* **2007**, *74*, 1282–1295. [CrossRef]
22. Berg, C.; Rogers, S.; Mineau, M. Building scenarios for ecosystem services tools: Developing a methodology for efficient engagement with expert stakeholders. *Futures* **2016**, *81*, 68–80. [CrossRef]
23. Gramberger, M.; Zellmer, K.; Kok, K.; Metzger, M.J. Stakeholder integrated research (STIR): A new approach tested in climate change adaptation research. *Clim. Chang.* **2015**, *128*, 201–214. [CrossRef]
24. McBride, M.F.; Lambert, K.F.; Huff, E.S.; Theoharides, K.A.; Field, P.; Thompson, J.R. Increasing the effectiveness of participatory scenario development through codesign. *Ecol. Soc.* **2017**, *22*. [CrossRef]
25. Palazzo, A.; Vervoort, J.M.; Mason-D'Croz, D.; Rutting, L.; Havlík, P.; Islam, S.; Bayala, J.; Valin, H.; Kadi Kadi, H.A.; Thornton, P.; et al. Linking regional stakeholder scenarios and shared socioeconomic pathways: Quantified West African food and climate futures in a global context. *Glob. Environ. Chang. Hum. Policy Dimens.* **2017**, *45*, 227–242. [CrossRef]
26. Pedde, S.; Kok, K.; Onigkeit, J.; Brown, C.; Holman, I.; Harrison, P.A. Bridging uncertainty concepts across narratives and simulations in environmental scenarios. *Reg. Environ. Chang.* **2018**, *250*, 145. [CrossRef]
27. Reed, M.S.; Kenter, J.; Bonn, A.; Broad, K.; Burt, T.P.; Fazey, I.R.; Fraser, E.D.G.; Hubacek, K.; Nainggolan, D.; Quinn, C.H.; et al. Participatory scenario development for environmental management: A methodological framework illustrated with experience from the UK uplands. *J. Environ. Manag.* **2013**, *128*, 345–362. [CrossRef] [PubMed]
28. Swedish Geotechnical Institute. Living Lab Co-Design Requirements Guiding Paper: EVOKED Deliverable 1.1. 2018.
29. Oppenheimer, M.; Glavovic, B.C.; Hinkel, J.; van de Wal, R.; Magnan, A.K.; Abd-Elgawad, A.; Cai, R.; Cifuentes-Jara, M.; DeConto, R.M.; Ghosh, T.; et al. Sea Level Rise and Implications for Low-Lying Islands, Coasts and Communities. In *IPCC Special Report on the Ocean and Cyrosphere in a Changing Climate*; Pörtner, H.-O., Roberts, D.C., Masson-Delmotte, V., Zhai, P., Tignor, M., Poloczanska, E., Mintenbeck, K., Alegria, A., Nicolai, M., Okem, A., et al., Eds.; Cambridge University Press: Cambridge, UK; New York, NY, USA, 2019; pp. 321–445.

30. Dell'Era, C.; Landoni, P.; Gonzalez, S.J. Investigating the innovation impacts of user-centred and participatory strategies adopted by European Living Labs. *Int. J. Innov. Mgt.* **2019**, *23*, 1950048. [CrossRef]
31. Evans, P.; Schuurmann, D.; Stahlbröst, A.; Vervoort, K. *Living Lab Methodology Handbook*; University of Manchester: Manchester, UK, 2017.
32. Kolstad, E.W.; Sofienlund, O.N.; Kvamsås, H.; Stiller-Reeve, M.A.; Neby, S.; Paasche, Ø.; Pontoppidan, M.; Sobolowski, S.P.; Haarstad, H.; Oseland, S.E.; et al. Trials, Errors, and Improvements in Coproduction of Climate Services. *Bull. Amer. Meteor. Soc.* **2019**, *100*, 1419–1428. [CrossRef]
33. Robles, A.G.; Hirvikoski, T.; Schuurman, D.; Stokes, L. *Introducing ENoLL and Its Living Lab Community*; European Network of Living Labs: Brussels, Belgium, 2015.
34. Tiwari, A.; Rodrigues, L.C.; Lucy, F.E.; Gharbia, S. Building Climate Resilience in Coastal City Living Labs Using Ecosystem-Based Adaptation: A Systematic Review. *Sustainability* **2022**, *14*, 10863. [CrossRef]
35. Alizadeh, M.R.; Adamowski, J.; Inam, A. Integrated assessment of localized SSP–RCP narratives for climate change adaptation in coupled human-water systems. *Sci. Total Environ.* **2022**, *823*, 153660. [CrossRef]
36. Pedde, S.; Harrison, P.A.; Holman, I.P.; Powney, G.D.; Lofts, S.; Schmucki, R.; Gramberger, M.; Bullock, J.M. Enriching the Shared Socioeconomic Pathways to co-create consistent multi-sector scenarios for the UK. *Sci. Total Environ.* **2021**, *756*, 143–172. [CrossRef] [PubMed]
37. Hagemann, N.; van der Zanden, E.H.; Willaarts, B.A.; Holzkämper, A.; Volk, M.; Rutz, C.; Priess, J.A.; Schönhart, M. Bringing the sharing-sparing debate down to the ground—Lessons learnt for participatory scenario development. *Land Use Policy* **2020**, *91*, 104262. [CrossRef]
38. Zandersen, M.; Hyytiäinen, K.; Meier, H.E.M.; Tomczak, M.T.; Bauer, B.; Haapasaari, P.E.; Olesen, J.E.; Gustafsson, B.G.; Refsgaard, J.C.; Fridell, E.; et al. Shared socio-economic pathways extended for the Baltic Sea: Exploring long-term environmental problems. *Reg. Environ. Chang.* **2019**, *19*, 1073–1086. [CrossRef]
39. Biggs, R.; Raudsepp-Haerne, C.; Atkinson-Palombo, C.; Bohensky, E.; Boyd, E.; Cundill, G.; Fox, H.; Ingram, S.; Kok, K.; Spehar, S.; et al. Linking Futures across Scales: A Dialog on Multiscale Scenarios. *Ecol. Soc.* **2007**, *12*, 1–16. [CrossRef]
40. Nilsson, A.E.; Bay-Larsen, I.; Carlsen, H.; van Oort, B.; Bjørkan, M.; Jylhä, K.; Klyuchnikova, E.; Masloboev, V.; van der Watt, L.-M. Towards extended shared socioeconomic pathways: A combined participatory bottom-up and top-down methodology with results from the Barents region. *Glob. Environ. Chang.* **2017**, *45*, 124–132. [CrossRef]
41. KSS Klimaprofil Vestfold—Norsk Klima Service Senter. 2023. Available online: https://klimaservicesenter.no/kss/klimaprofiler/vestfold (accessed on 22 September 2023).
42. Junker, E. Legal requirements for risk and vulnerability assessments in Norwegian land-use planning. *Local Environ.* **2013**, *20*, 474–488. [CrossRef]
43. NGI Lardal og Larvik Kommuner—Tilpasning til Klimaendringer. Vurderinger og Ekstrem Nedbør, Skred, Flom, Stormflo og Havnivåstigning, Report. 2016. Available online: https://www.larvik.kommune.no/media/4850/lardal-og-larvik-kommuner-tilpasning-til-klimaendringer-15042016.pdf (accessed on 18 March 2024).
44. Länsstyrelsen Värmland. Värmland i ett Förändrat Klimat: Plan för vad Länsstyrelsen Värmland Behöver Göra—Handlingsplan och Myndighetsmål 2020–2025. 2023. Available online: https://catalog.lansstyrelsen.se/store/38/resource/4 (accessed on 22 September 2023).
45. Oliver, J.; Qin, X.S.; Madsen, H.; Rautela, P.; Joshi, G.C.; Jorgensen, G. A probabilistic risk modelling chain for analysis of regional flood events. *Stoch. Environ. Res. Risk Assess.* **2019**, *33*, 1057–1074. [CrossRef]
46. Ministerie van Infrastructuur en Waterstaat. NATIONAL DELTA PROGRAMME, Annual Report, The Hague. 2020; 140p.
47. Bartholomeus, R.P.; van der Wiel, K.; van Loon, A.F.; van Huijgevoort, M.H.; van Vliet, M.T.; Mens, M.; Muurling-van Geffen, S.; Wanders, N.; Pot, W. Managing water across the flood–drought spectrum: Experiences from and challenges for the Netherlands. *Camb. Prism. Water* **2023**, *1*, e2. [CrossRef]
48. van den Eertwegh, G.; Bartholomeus, R.; Louw P de Witte, F.; van Dam, J.; van Deijl, D.; Hoefsloot, P.; Clevers, S.; Hendriks, D.; van Huijgevoort, M.; Hunink, J.; et al. *Droogte in Zandgebieden van Zuid-, Midden-en Oost-Nederland: Rapportage Fase 1: Ontwikkeling van Uniforme Werkwijze Voor Analyse van Droogte en Tussentijdse Bevindingen*; KWR: Amsterdam, The Netherlands, 2019.
49. Werkregio Fluvius. Samen Werken aan Klimaatbestendigheid—Regionale Adaptatiestrategie en Uitvoeringsagenda—2022–2027. 2022. Available online: https://simcms.wdodelta.nl/_flysystem/media/ras-en-rua-samenwerkingsregio-fluvius_definitief.pdf (accessed on 10 February 2023).
50. Jensen, J.; Mueller-Navarre, S.H. Storm Surges on the German Coast. *Die Küste* **2008**, *74*, 92–124.
51. Sterr, H. Assessment of Vulnerability and Adaptation to Sea-Level Rise for the Coastal Zone of Germany. *J. Coast. Res.* **2008**, *242*, 380–393. [CrossRef]
52. The BACC II Author Team. *Second Assessment of Climate Change for the Baltic Sea Basin*; Springer International Publishing: Cham, Switzerland, 2015.
53. Weiße, R.; Meinke, I. Meeresspiegelanstieg, Gezeiten, Sturmfluten und Seegang. In *Klimawandel in Deutschland: Entwicklung, Folgen, Risiken und Perspektiven*; Brasseur, G.P., Jacob, D., Schuck-Zöller, S., Eds.; Springer: Berlin/Heidelberg, Germany, 2017; pp. 77–85.
54. Hofstede, J. Coastal Flood Defence and Coastal Protection along the Baltic Sea Coast of Schleswig-Holstein. *Die Küste* **2008**, *74*, 170–178.

55. Landesbetrieb für Küstenschutz, Nationalpark und Meeresschutz Schleswig-Holstein. Fachplan Küstenschutz Ostseeküste, Kiel, Germany. 2015. Available online: https://www.schleswig-holstein.de/DE/fachinhalte/K/kuestenschutz_fachplaene/Dossier/teaser_tabelle_Ostsee.html (accessed on 18 March 2024).
56. NDR Sturmtief "Zeetje" Brachte Hochwasser. 2019. Available online: https://www.ndr.de/nachrichten/schleswig-holstein/Sturmtief-Zeetje-brachte-Hochwasser,sturm2742.html (accessed on 25 October 2019).
57. Ohlsen, H. Keine Deiche, Sperrwerke oder mobile Sicherungseinheiten: "Wir Haben nur Sandsäcke!". Flensburger Tageblatt, Flensburg. 2017. Available online: https://www.shz.de/lokales/flensburg/artikel/keine-deiche-sperrwerke-oder-mobile-sicherungseinheiten-wir-haben-nur-sandsaecke-41614619 (accessed on 18 March 2024).
58. Abram, N.; Gattuso, J.-P.; Prakash, A.; Cheng, L.; Chidichimo, M.P.; Crate, S.; Enomoto, H.; Garschagen, M.; Gruber, N.; Harper, S.; et al. Framing and Context of the Report. In *IPCC Special Report on the Ocean and Cyrosphere in a Changing Climate*; Pörtner, H.-O., Roberts, D.C., Masson-Delmotte, V., Zhai, P., Tignor, M., Poloczanska, E., Mintenbeck, K., Alegria, A., Nicolai, M., Okem, A., et al., Eds.; Cambridge University Press: Cambridge, UK; New York, NY, USA, 2019; pp. 73–129.
59. O'Neill, B.C.; Kriegler, E.; Riahi, K.; Ebi, K.L.; Hallegatte, S.; Carter, T.R.; Mathur, R.; van Vuuren, D.P. A new scenario framework for climate change research: The concept of shared socioeconomic pathways. *Clim. Chang.* **2014**, *122*, 387–400. [CrossRef]
60. Kriegler, E.; O'Neill, B.C.; Hallegatte, S.; Kram, T.; Lempert, R.J.; Moss, R.H.; Wilbanks, T. The need for and use of socio-economic scenarios for climate change analysis: A new approach based on shared socio-economic pathways. *Glob. Environ. Chang.* **2012**, *22*, 807–822. [CrossRef]
61. O'Neill, B.C.; Kriegler, E.; Ebi, K.L.; Kemp-Benedict, E.; Riahi, K.; Rothman, D.S.; van Ruijven, B.J.; van Vuuren, D.P.; Birkmann, J.; Kok, K.; et al. The roads ahead: Narratives for shared socioeconomic pathways describing world futures in the 21st century. *Glob. Environ. Chang.* **2017**, *42*, 169–180. [CrossRef]
62. Kc, S.; Lutz, W. The human core of the shared socioeconomic pathways: Population scenarios by age, sex and level of education for all countries to 2100. *Glob. Environ. Chang. Hum. Policy Dimens.* **2017**, *42*, 181–192. [CrossRef]
63. Jiang, L.; O'Neill, B.C. Global urbanization projections for the Shared Socioeconomic Pathways. *Glob. Environ. Chang.* **2017**, *42*, 193–199. [CrossRef]
64. Dellink, R.; Chateau, J.; Lanzi, E.; Magné, B. Long-term economic growth projections in the Shared Socioeconomic Pathways. *Glob. Environ. Chang.* **2017**, *42*, 200–214. [CrossRef]
65. Hunter, L.M.; O'Neill, B.C. Enhancing engagement between the population, environment, and climate research communities: The shared socio-economic pathway process. *Popul. Environ.* **2014**, *35*, 231–242. [CrossRef]
66. Ebi, K.L.; Kram, T.; van Vuuren, D.P.; O'Neill, B.C.; Kriegler, E. A New Toolkit for Developing Scenarios for Climate Change Research and Policy Analysis. *Environ. Sci. Policy Sustain. Dev.* **2014**, *56*, 6–16. [CrossRef]
67. Gidden, M.J.; Riahi, K.; Smith, S.J.; Fujimori, S.; Luderer, G.; Kriegler, E.; Van Vuuren, D.P.; Van Den Berg, M.; Feng, L.; Klein, D.; et al. Global emissions pathways under different socioeconomic scenarios for use in CMIP6: A dataset of harmonized emissions trajectories through the end of the century. *Geosci. Model Dev.* **2019**, *12*, 1443–1475. [CrossRef]
68. Chen, D.; Rojas, M.; Samset, B.H.; Cobb, K.; Diongue Niang, A.; Edwards, P.; Emori, S.; Faria, S.H.; Hawkins, E.; Hope, P.; et al. Framing, Context, and Methods. In *Climate Change. The Physical Science Basis. Contribution of Working Group I to the Sixth Assessment Report of the Intergovernmental Panel on Climate Change*; Masson-Delmotte, V.P., Zhai, A., Pirani, S.L., Connors, C., Péan, S., Berger, N., Caud, Y., Chen, L., Goldfarb, M.I., Gomis, M., et al., Eds.; Cambridge University Press: Cambridge, UK; New York, NY, USA, 2021; pp. 147–286.
69. Frame, B.; Lawrence, J.; Ausseil, A.-G.; Reisinger, A.; Daigneault, A. Adapting global shared socio-economic pathways for national and local scenarios. *Clim. Risk Manag.* **2018**, *21*, 39–51. [CrossRef]
70. Alcamo, J.; Henrichs, T. Chapter Two towards Guidelines for Environmental Scenario Analysis. In *Environmental Futures: The Practice of Environmental Scenario Analysis, Vol 2. Developments in Integrated Environmental Assessment, v. 2, 1st Edn.*; Alcamo, J., Ed.; Elsevier: Amsterdam, The Netherlands; Boston, MA, USA, 2008; pp. 13–35.
71. van der Heijden, K. Scenarios: The Art of Strategic Conversation, 2nd ed.Wiley: Chichester, UK, 2009.
72. Carlsen, H.; Eriksson, E.A.; Dreborg, K.H.; Johansson, B.; Bodin, Ö. Systematic exploration of scenario spaces. *Foresight* **2016**, *18*, 59–75. [CrossRef]
73. Reimann, L.; Vollstedt, B.; Koerth, J.; Tsakiris, M.; Beer, M.; Vafeidis, A.T. Extending the Shared Socioeconomic Pathways (SSPs) to support local adaptation planning—A climate service for Flensburg, Germany. *Futures* **2021**, *127*, 102691. [CrossRef]
74. Voros, J. A generic foresight process framework. *Foresight* **2003**, *5*, 10–21. [CrossRef]
75. Kok, K.; van Vliet, M.; Bärlund, I.; Dubel, A.; Sendzimir, J. Combining participative backcasting and exploratory scenario development: Experiences from the SCENES project. *Technol. Forecast. Soc. Chang.* **2011**, *78*, 835–851. [CrossRef]
76. Rounsevell, M.D.A.; Metzger, M.J. Developing qualitative scenario storylines for environmental change assessment. *WIREs Clim. Chang.* **2010**, *1*, 606–619. [CrossRef]
77. Ballarini, E.; D'Adamo, R.; Pazienza, G.; Zaggia, L.; Vafeidis, A. Assessing the applicability of a bottom-up or top-down approach for effective management of a coastal lagoon area. *Ocean. Coast. Manag.* **2021**, *200*, 105417. [CrossRef]
78. Kok, K.; van Vliet, M. Using a participatory scenario development toolbox: Added values and impact on quality of scenarios. *J. Water Clim. Chang.* **2011**, *2*, 87–105. [CrossRef]

79. Reed, M.S.; Graves, A.; Dandy, N.; Posthumus, H.; Hubacek, K.; Morris, J.; Prell, C.; Quinn, C.H.; Stringer, L.C. Who's in and why? A typology of stakeholder analysis methods for natural resource management. *J. Environ. Manag.* **2009**, *90*, 1933–1949. [CrossRef]
80. Kok, K.; Bärlund, I.; Flörke, M.; Holman, I.; Gramberger, M.; Sendzimir, J.; Stuch, B.; Zellmer, K. European participatory scenario development: Strengthening the link between stories and models. *Clim. Chang.* **2015**, *128*, 187–200. [CrossRef]
81. Arvika Kommun. Miljöstrategi för Arvika Kommun. 2015. Available online: https://www.arvika.se/kommunochpolitik/planerochstyrandedokument/forfattningssamling/bestammelser/miljostrategi.2478.html (accessed on 22 September 2023).
82. Arvika Kommun. Strategisk Plan 2024–2026—Budget 2024. 2023. Available online: https://www.arvika.se/download/18.39e0dd96188d6b45c534fd/1692092130920/Strategisk%20plan%202024-2026.pdf (accessed on 22 September 2023).
83. Länsstyrelsen Värmland. Värmland i ett Förändrat Klimat—Regional Handlingsplan för Klimatanpassning 2014. 2023.
84. Wolters, H.A.; van den Born, G.J.; Dammers, E.; Reinhard, S. *Deltascenario's Voor de 21e Eeuw: Actualisering 2017*; Hoofdrapport: Utrecht, The Netherlands, 2018.
85. Commission E, Eurostat. *Eurostat Regional Yearbook, 2019 ed.*; Publications Office: Brussels, Belgium, 2019.
86. Beer, M.; Jahn, M.; Kovac, E.; Köster, H.; Laros, S.; Maas, H. Masterplan 100% Klimaschutz Flensburg: CO_2-Neutralität und Halbierung des Energiebedarfs bis zum Jahr 2050, Municipality of Flensburg 2013. Available online: https://www.uni-flensburg.de/fileadmin/content/abteilungen/industrial/dokumente/downloads/veroeffentlichungen/forschungsergebnisse/konzept-masterplan-100-klimaschutz-fuer-flensburg.pdf (accessed on 18 March 2024).
87. City of Flensburg. Stadtportrait. 2017. Available online: https://www.flensburg.de/Leben-Soziales/Stadtportrait (accessed on 18 September 2019).
88. City of Flensburg. Perspektiven für Flensburg: Ein integriertes Stadtentwicklungskonzept. Konzept und Maßnahmen, Municipality of Flensburg. 2018. Available online: https://www.flensburg.de/media/custom/2306_6869_1.PDF?1542120210 (accessed on 18 March 2024).
89. IHR Sanierungsträger. Stadt Flensburg Vorbereitende Untersuchungen Hafen-Ost: Öffentlichkeitsbeteiligung. Ergebnisse des Experten-Workshops am 13. Flensburg, Germany. October 2016. Available online: https://www.ihrsan.de/fileadmin/Content/PDF_und_Images/Hafen-Ost/Hafen-Ost_161013_Experten-WS_Ergebnisdokumentation.pdf (accessed on 18 March 2024).
90. Staudt, M. Serie: Stadtteile in Flensburg. Flensburger Tageblatt. *Flensburg* 2018.
91. City of Flensburg. Statistik. 2019. Available online: https://www.flensburg.de/Politik-Verwaltung/Stadtverwaltung/Statistik (accessed on 18 September 2019).
92. Statistisches Amt für Hamburg und Schleswig-Holstein. Regionaldaten für Flensburg. 2018. Available online: http://region.statistik-nord.de/detail/0010000000000000000/1/0/356/ (accessed on 18 September 2019).
93. Swart, R.J.; Raskin, P.; Robinson, J. The problem of the future: Sustainability science and scenario analysis. *Glob. Environ. Chang.* **2004**, *14*, 137–146. [CrossRef]
94. Kok, K.; Pedde, S. IMPRESSIONS Socio-Economic Scenarios: Deliverable D2.2. 2016. Available online: http://www.impressions-project.eu/getatt.php?filename=D2.2_Socioeconomic_Scenarios_FINAL_14009.pdf (accessed on 18 April 2018).

Disclaimer/Publisher's Note: The statements, opinions and data contained in all publications are solely those of the individual author(s) and contributor(s) and not of MDPI and/or the editor(s). MDPI and/or the editor(s) disclaim responsibility for any injury to people or property resulting from any ideas, methods, instructions or products referred to in the content.

remote sensing

Article

Unveiling Deviations from IPCC Temperature Projections through Bayesian Downscaling and Assessment of CMIP6 General Circulation Models in a Climate-Vulnerable Region

Giovanni-Breogán Ferreiro-Lera [1,*], Ángel Penas [1] and Sara del Río [1,2]

1 Department of Biodiversity and Environmental Management, Faculty of Biological and Environmental Sciences, University of León, Campus de Vegazana s/n, 24071 León, Spain; apenm@unileon.es (Á.P.); sriog@unileon.es (S.d.R.)

2 Mountain Livestock Institute (CSIC-ULE), León-Vega de Infanzones Road (Finca Marzanas-Grulleros), 24346 León, Spain

* Correspondence: gferl@unileon.es; Tel.: +34-987-293-406

Abstract: The European Mediterranean Basin (Euro-Med), a region particularly vulnerable to global warming, notably lacks research aimed at assessing and enhancing the widely used remote climate detection products known as General Circulation Models (GCMs). In this study, the proficiency of GCMs in replicating reanalyzed 1981–2010 temperature data sourced from the ERA5 Land was assessed. Initially, the least data-modifying interpolation method for achieving a resolution match of 0.1° was ascertained. Subsequently, a pixel-by-pixel evaluation was conducted, employing five goodness-of-fit metrics. From these metrics, we compiled a Comprehensive Rating Index (CRI). A Multi-Model Ensemble using Random Forest was constructed and projected across three emission scenarios (SSP1-RCP2.6, SSP2-RCP4.5, and SSP5-RCP8.5) and timeframes (2026–2050, 2051–2075, and 2076–2100). Empirical Bayesian Kriging, selected for its minimal data alteration, supersedes the commonly employed Bilinear Interpolation. The evaluation results underscore MPI-ESM1-2-HR, GFDL-ESM4, CNRM-CM6-1, MRI-ESM2-0, CNRM-ESM2-1, and IPSL-CM6A-LR as top-performing models. Noteworthy geospatial disparities in model performance were observed. The projection outcomes, notably divergent from IPCC forecasts, revealed a warming trend of 1 to over 2 °C less than anticipated for spring and winter over the medium–long term, juxtaposed with heightened warming in mountainous/elevated regions. These findings could substantially refine temperature projections for the Euro-Med, facilitating the implementation of policy strategies to mitigate the effects of global warming in vulnerable regions worldwide.

Keywords: climate change; CMIP6; Empirical Bayesian Kriging; ERA5 Land; Global Circulation Models; global warming; Euro-Med region; Random Forest; temperature

1. Introduction

The Massachusetts Institute of Technology reported in the 1970s a "inadvertent climate disruption" [1]. The analysis of trends in average air temperatures swiftly emerged as the principal evidence of planetary warming [2]. Nearly four decades later, the Intergovernmental Panel on Climate Change (IPCC) deemed it highly probable that over half of the 1951–2010 temperature surge can be ascribed to human activities [3]. This escalation manifested as a 0.36 °C (0.26 to 0.44 °C) anomaly in the 1961–1990 air temperature relative to the reference period of 1850–1900, which escalates to 1.09 °C (0.95 to 1.20 °C) when scrutinizing from 2011 to 2020 [4]. Alongside the transition toward a warmer climate, there has been an intensification in the hydrological cycle; an uptick in the frequency and magnitude of weather extremes, such as droughts in arid regions; and an increase in dry–warm and wet–warm events [5–10]. As a result, primary food production [11], availability of natural resources, energy production [12] and water reservoirs [13], and even human health and behavior [14–16] are poised to be impacted. Hence, global warming is a paramount concern for the scientific community, the media, and the general population [17].

However, it is important to note that warming is not uniform across the Earth's surface, but rather there are significant warming holes [18,19], as well as regions that experience warming well above the global average [20]. The world's Mediterranean climate regions are characterized by strong summer drought conditions [21,22] and fall within these "superwarming" regions [4]. Significant temperature increases have been observed in these regions, along with the prevalence of droughts driven by extreme heat events [4,23,24]. In California, human demand, combined with extremely high temperatures, led to the most significant drought event of the millennium in 2012–2014 [25]. Recent studies warn that, in this state, 1980–2019 air temperatures have increased by about 0.01 °C per year (0.008–0.017 °C) [26]. Similar annual change magnitudes have also been detected in South Africa or Australia [27–29]. However, the region of the globe where the Mediterranean climate reaches its maximum extent is in the European Mediterranean Basin [22]. There, according to the latest report from the Mediterranean Experts on Climate and Environmental Change (MedECC), the temperature increase has been 0.4 °C (0.1–0.5 °C) higher compared to the reference period of 1860–1890 than the global average [30].

In recent decades, General Circulation Models (GCMs), along with satellite products, have become pivotal in simulating and remotely detecting climate trends [19,31–35]. There are numerous reasons that have made GCMs the most widely used tools for working with present and projected climate simulations. Firstly, many regions globally lack readily available in situ climate data that are accessible in open access [36]. Furthermore, past climate models, despite being criticized as alarmist, have proven highly skillful in simulating 1970–2007 global surface warming [37]. In fact, it is this forecasting capability that makes them indispensable in the routine of climate simulation [38]. The latest generation of GCMs, introduced with the release of CMIP6, exhibits a substantial improvement compared to its predecessors [39]. Precision and error rates in simulating climatic types appear to enhance across most model families, particularly those that have increased their spatial resolution compared to those presented in CMIP5 [40]. Similarly, CMIP6 GCMs better capture trends in atmospheric circulation [41,42], resulting in the improved simulation of climatic variables such as temperature or precipitation [43–45]. Specifically, for the Euro-Mediterranean Region, CMIP6 GCMs seem to mitigate the exaggerated warming trends observed in previous generations and show higher dispersion in non-extreme precipitation (percentile 5th–95th) [46]. However, the success of GCMs is not without criticism. Should they be used only for qualitative and probabilistic assessments? Is model bias being considered in the adoption of political measures? Can all variables be predicted with even acceptable accuracy? Cases of contentious variables [47] and regions [48,49] are already known in this regard. Moreover, there is a growing concern in recent years about the discrepancy of models in predicting processes at the mesoscale and microscale, even labeling regional predictions as arbitrary [50,51]. Thus, the evaluation of GCMs at a local scale is absolutely necessary before incorporating them into a Multi-Model Ensemble (MME), aiming to identify possible deviations or abnormal discrepancies in certain GCMs that may undermine the predictive capacity of studies. This locally based evaluation approach, combined with MME, is prevalent in numerous research endeavors [32,35,52–56]. Nevertheless, a notable absence of such a localized assessment for the European Mediterranean region has been identified [57], with few studies conducting a comprehensive comparison of model-performance disparities [58–61]. Furthermore, to the best of our knowledge, the scrutinized works within the Euro-Med lack a geospatial component indicating areas where General Circulation Models (GCMs) achieve optimal performance versus those where they exhibit mediocre simulation outcomes. In the context of the anticipated significant warming within the Euro-Med region, it becomes imperative to possess projections that are both realistic and precise, characterized by high resolution and grounded in an analysis of GCMs renowned for their adept simulation of Mediterranean temperatures. Furthermore, acquiring insights into the spatial domains critical for climate modeling with respect to specific GCMs is of paramount importance.

It was the principal goal of this study to refine current temperature projections in the Euro-Med region through the systematic application of downscaling and a comprehensive assessment of 34 GCMs. Another key aim of our research was to compare the results of the projections obtained using the methodology applied here with the global and regional projections of the IPCC in order to reveal potential differences. IPCC predictions were established using a large MME (>30 GCMs), with no verification of local performance. To this end, both a Simple Averages MME (SAE) and a Random Forest MME (RFE) were constructed, tested, and projected across three distinct future timeframes—short (2026–2050), medium (2051–2075), and long (2076–2100) term—contingent upon three distinct emission scenarios: SSP1-RCP2.6, SSP2-RCP4.5, and SSP5-RCP8.5. Moreover, as a secondary objective, the developed MME was indeed evaluated at a monthly scale to potentially substantiate the hypothesis that the performance of models at longer scales is inherently linked to performance at shorter scales [62].

It is believed that this research brings significant novelties to the field of GCM analysis and evaluation. In recent years, several studies have conducted projections concerning Mediterranean temperature [46,63–67]. As far as we are aware, our study is the first to rely on a pre-established ranking of the optimum downscaled, best-performing GCMs in the Euro-Med. Thus, we endeavor to introduce several advancements to current knowledge in temperature forecasting and GCM assessment, focusing on a climate-vulnerable and poorly studied region. Currently, to our knowledge, no thorough analysis of different interpolation techniques has been carried out alongside a GCM performance evaluation in our research region. Therefore, fine-scale downscaling allows us to retrieve the simulated temperatures of each GCM at the scale of the reference dataset used (ERA5 Land, 0.1°), enabling a high-resolution assessment of the geospatial performance, an approach not previously available in the reviewed scientific literature. In line with this, while ERA5 Land has recently been used in the correction of climate datasets [68], including GCMs [69], this may be the first application as a reference dataset in GCMs assessment. Lastly, an important feature of this research is the wide integration of GCMs (34 total) from top modelling centers, which includes differences in resolution and considers aerosol–oceanic dynamics [70,71]. Although an annual evaluation will be conducted, the authors posit that the presented method could be applied at seasonal and monthly scales, representing a potential avenue for future research. Similarly, the developed method can be applied to other regions worldwide. The decision to focus on the European Mediterranean region was solely due to its vulnerability to climate change and limited research, where temperature variability and differences from previous predictions are expected to be significant. Hence, our regional study can be deemed of global significance. It is expected that, if found, the differences with the IPCC predictions may help inform the development of effective policy measures at regional level in climate adaptation and land management.

2. Materials and Methods

2.1. Study Area

The European Mediterranean Basin (or Euro-Med) is typically defined as the collection of river basins draining into the Mediterranean Sea. However, for the purposes of this research, countries within the region were considered in their entirety to provide comprehensive information in an area particularly vulnerable to climate variability (Figure 1). The selected study area, covering an area of over 3 million km^2, is delimited by the parallels 34°N and 52°N, and by the meridians -10°W and 45°E.

In terms of its topography, the area is notably mountainous due to the significant geological changes during the Alpine orogeny [72]. Prominent mountain systems include the Alps, north of the Italian Peninsula, and the Armenian Highlands, east of Turkey, with peaks exceeding 4800 m. Other relevant mountain ranges include the Sierra Nevada and the Pyrenees in Southeastern and Northwestern Spain, respectively; the Apennines in

Central Italy; the Dinaric Alps along the Adriatic coast; the Carpathians in Romania; and the Balkans in Bulgaria. The major river basins in the region include the Danube, Rhône, and Po, which collect much of the Alpine water. The Ebro in Southeastern Spain and the Kizilirmak in Northern Turkey are also notable for their flow rates.

ANK - Ankara	**BUC** - Bucarest	**MAD** - Madrid	**ROM** - Roma	**TIR** - Tirana
ATH - Athens	**BUD** - Budapest	**NIC** - Nicosia	**SKO** - Skopje	**VIE** - Viena
BEL - Belgrade	**LIS** - Lisbon	**PAR** - Paris	**SOF** - Sofia	**VLT** - Valletta
BRN – Bern	**LJU** – Ljubljana	**POD** - Podgorica	**SRJ** – Sarajevo	**ZGB** - Zagreb

Figure 1. Topographical map of the Euro-Med region, indicating capitals and major rivers.

Regarding its climatology, much of the territory experiences pronounced seasonality in precipitation, resulting in a summer drought lasting for at least two consecutive months. These areas have a subtropical or Mediterranean climate, with average annual temperatures ranging from 18 to 24 °C and summer averages between 28 and 36 °C [21,73]. The northern zones north of the 45° parallel, however, have temperate climates, with a more stable hydrological regime, no summer drought, and milder temperatures throughout the year [21].

2.2. Data Collection

2.2.1. GCMs Temperature Data

Monthly mean temperature data simulated by 34 CMIP6 GCMs were obtained from the Lawrence Livermore National Laboratory node of the World Climate Research Program website (https://esgf-node.llnl.gov/projects/cmip6/, accessed on 16 December 2022). The GCMs selected, along with their supporter institution and their native resolution, are listed in Table 1. Only the first ensemble run (r1i1p1) was considered, shortening the timeframe to 1981–2010, which is the last 30-year period recommended by the World Meteorological Organization (WMO) and included in the CMIP6 historical experiment (1850–2014) [74].

Table 1. The 34 CMIP6 Global Circulation Models (GCMs) selected in this study and their attributes (supporter institution, name, and native resolution).

Supporter Institution	GCM Name	Native Resolution (x, y)
Commonwealth Scientific and Industrial Research Organization (Australia)	ACCESS-CM2 ACCESS-ESM1-5	1.875° × 1.250°
Canadian Centre for Climate Modelling and Analysis (Canada)	CanESM5 CanESM5-CanOE	2.813° × 2.789°

Table 1. *Cont.*

Supporter Institution	GCM Name	Native Resolution (x, y)
Beijing Climate Center (China)	BCC-CSM2-MR BCC-ESM1	1.125° × 1.121° 2.813° × 2.789°
Chinese Academy of Meteorological Sciences (China)	CAMS-CSM1-0	1.125° × 1.121°
Nanjing University of Information Science and Technology (China)	NESM3	1.875° × 1.865°
Institute of Atmospheric Physics Chinese Academy of Sciences (China)	CAS-ESM2-0	1.406° × 1.417°
Institute of Atmospheric Physics, Chinese Academy of Sciences (China)	FGOALS-f3-L	1.250° × 1.000°
The First Institution of Oceanography (China)	FIO-ESM2-0	1.250° × 0.942°
Centre National de Recherches Météorologiques (France)	CNRM-CM6-1 CNRM-ESM2-1	1.406° × 1.400°
Institut Pierre-Simon Laplace/Centre National de Recherche Scientifique (France)	IPSL-CM6A-LR IPSL-CM6A-LR-INCA	2.500° × 1.268°
Max-Planck-Institut fuer Meteorologie, Deutsches Klimarechenzentrum (Germany)	MPI-ESM1-2-HAM MPI-ESM1-2-LR MPI-ESM1-2-HR	1.875° × 1.865° 0.938° × 0.935°
Centro Euro-Mediterraneo sui Cambiamenti Climatici (Italy)	CMCC-ESM2	1.250° × 0.942°
Indian Institute of Tropical Meteorology (India)	IITM-ESM	1.875° × 1.904°
National Institute for Environmental Studies, Japan Agency for Marine–Earth Science and Technology (Japan)	MIROC6 MIROC-ES2L	1.406° × 1.400° 2.813° × 2.789°
Meteorological Research Institute (Japan)	MRI-ESM2-0	1.125° × 1.121°
Bjerknes Centre for Climate Research, Norwegian Meteorological Institute (Norway)	NorESM2-MM	1.250° × 0.942°
Russian Academy of Sciences, Institute of Numerical Mathematics (Russia)	INM-CM5-0	2.000° × 1.500°
National Institute of Meteorological Sciences, Korea Met. Administration (South Korea)	KACE-1-0-G	1.875° × 1.250°
Korea Institute of Ocean Science and Technology (South Korea)	KIOST-ESM	1.875° × 1.895°
Met Office Hadley Centre (United Kingdom)	HadGEM-GC31-LL UKES-M1-1-LL	1.875° × 1.250°
National Center for Atmospheric Research	CESM2-WACCM	1.250° × 0.942°
Geophysical Fluid Dynamics Laboratory/NOAA (United States of America)	GFDL-ESM4	1.250° × 1.000 °
NASA Goddard Institute for Space Studies (United States of America)	GISS-E2-1-G GISS-E2-2-G	2.500° × 2.000°
University of Arizona—Department of Geosciences (United States of America)	MCM-UA-1-0	3.750° × 2.235°

2.2.2. ERA5 Land Temperature Data

ERA5 Land, the latest reanalysis product from the European Centre for Medium-Range Weather Forecasts (ECMWF), has emerged as one of the most widely utilized terrestrial reanalysis products to date [75].

Its remarkable simulation of in situ temperature is positioning it as the forefront product in the field. Several studies have reported a correlation exceeding $r = 0.9$ with observed temperatures, showcasing its exceptional ability to detect thermal fluctuations across various temporal scales (from intraday to intermonth) [76–78]. Moreover, it outperforms its predecessors in spatial resolution ($0.1° \approx 11.1$ km) and simulation capacity [79,80], while also matching or surpassing satellite products. It has been suggested that, on a global scale, ERA5 Land is as proficient as satellite products like AIRS or MODIS in detecting temperature trends [81]. Furthermore, in regions characterized by complex topography, such as the Euro-Mediterranean region, ERA5 Land has been shown to outperform MODIS in trend detection while exhibiting less exaggeration of temperature increases [82]. Analogously, it has demonstrated greater accuracy in capturing both interdecadal and interannual temperature trends compared to NASA POWER, based on MERRA-2, and is more suitable for calculating specific extreme temperature indices [83].

While ERA5 has been used to evaluate the performance of GCMs land surface temperature simulation [84–86], our study would represent, to our understanding, the first use of ERA5 Land with that specific objective in mind. The monthly averaged temperature dataset from ERA5 Land was acquired through the Copernicus Climate Change Service (C3S) website (https://cds.climate.copernicus.eu/, accessed on 16 December 2022). The temporal window was adjusted, following the process conducted in the historical experiment of the GCMs, to the period 1981–2010.

2.3. Cross-Validation of Interpolation Methods

In this paper, GCMs are interpolated to ERA5 Land resolution (0.1°). While this may seem a fine resolution, GCMs have already been interpolated to this or lower scales [56,87,88]. Even downscales of less than one square kilometer have been performed [89]. Bilinear interpolation is undoubtedly the most commonly used method either when matching the spatial resolution of GCMs or when adjusting them to the reference dataset [32,35,52,56,87]. However, this may not be the optimal way to do it, and the data may be undergoing significant transformation. Therefore, it is intended to establish which interpolation method is the least data-modifying in a specific study area.

Six interpolation methods are tested in this paper: Bilinear Interpolation (BI), Inverse Distance Weighting (IDW), Kernel Smoothing (KS) and three Kriging methods: Ordinary Kriging (OK), Cokriging (plus Elevation) (COK+E) and Empirical Bayesian Kriging (EBK). The former two are deterministic methods, whereas Kriging is a geostatistical technique based on the methodical computation of the influence of certain points on other adjacent points [90]. On the other hand, Kernel is a polynomial method based on the obtention of the optimal prediction polynomial for each pixel [91].

The aforementioned interpolation methods were compared using a cross-validation procedure based on the error between raw data and the interpolated output [92,93]. It consists of a leave-one-out cross-validation, which is the default cross-validation procedure provided by the program in the version adopted. From the native resolution of each GCM, which is very divergent from each other, as shown in Table 1, we applied the aforementioned interpolation methods, setting 0.1° as the target (output) resolution. Spatial Analyst and Geostatistical Analyst extensions of ESRI ArcGIS 10.8 were used [94].

2.4. GCMs Performance Assessment and Ranking

An extensive review would be required to explain in detail the profusion of methods available for assessing and ranking General Circulation Models (GCMs): reliability ensemble averaging, relative entropy, symmetric uncertainty, Bayesian approaches, probability density functions, or clustering are merely a sample of this wide diversity [57]. Nevertheless, despite the multitude of existing methods, conventional metrics continue to be used in GCM evaluation, as they offer greater comparability across studies [57]. Furthermore, they vary within known ranges, making them easily interpretable. These conventional metrics, also referred to as goodness-of-fit (GOF) metrics, employ simple calculations to compare the values of GCMs with those of a reference dataset—in this case, ERA5 Land.

Coefficient of Correlation (CC), Normalized Root Mean Squared Error (*NRMSE*), Nash–Sutcliffe Efficiency (*NSE*), Kling–Gupta Efficiency (*KGE*), and modified index of agreement (*md*) were implemented as GOF metrics in this study. All of them were devised to assess various facets of two data distributions to ascertain whether both sets of data could stem from a shared underlying distribution. Indeed, this serves as one of the primary rationales underpinning the utilization of diverse goodness-of-fit metrics. Their mathematical formulas are provided below (Equations (1)–(5)).

$$CC = \frac{\sum_{i=1}^{N}\left(sim_i - \overline{sim}\right)\left(obs_i - \overline{obs}\right)}{\sqrt{\sum_{i=1}^{N}\left(sim_i - \overline{sim}\right)^2}\cdot\sqrt{\sum_{i=1}^{N}\left(sim_i - \overline{sim}\right)^2}} \tag{1}$$

$$NRMSE = \frac{\sqrt{\frac{1}{N}\sum_{i=1}^{N}\left(sim_i - obs_i\right)^2}}{obs_{(max)} - obs_{(min)}} \tag{2}$$

$$NSE = 1 - \frac{\sum_{i=1}^{N}\left(sim_i - obs_i\right)^2}{\sum_{i=1}^{N}\left(obs_i - \overline{obs}\right)^2} \tag{3}$$

$$KGE = 1 - \sqrt{\left(CC-1\right)^2 + \left(\frac{\overline{sim}}{\overline{\overline{obs}}} - 1\right)^2\left(\frac{\sigma_{sim}/\overline{sim}}{\sigma_{obs}/\overline{obs}} - 1\right)^2} \tag{4}$$

$$md = 1 - \frac{\sum_{i=1}^{N}\left|\left(obs_i - sim_i\right)\right|^j}{\sum_{i=1}^{N}\ \left|sim_i - \overline{obs}\right| + \left|\left(obs_i - \overline{obs}\right)\right|^j} \tag{5}$$

where sim_i stands for simulated data, obs_i for observed data, $\overline{sim}$ for mean simulated data, $\overline{obs}$ for mean observed data, σ_{sim} for simulated data variance, σ_{obs} for observed data variance, and N for the number of records.

All metrics presented, except for NRMSE, have a range that reaches its maximum at the value 1, representing the ideal fit between simulation and observation. The only difference is that both *CC* (Equation (1)) and *md* (Equation (5)) have a lower bound of 0, where this value represents total disagreement between simulation and observation, while *NSE* (Equation (3)) and *KGE* (Equation (4)) have no lower limit, so the closer a value is to $-\infty$, the worse the model–reality relationship. On the other hand, *NRMSE* has the opposite interpretation: it varies between 0 and 1, with 0 representing the ideal fit and positive values representing the error between simulation and observation. Examples of the application of the GOF metrics can be found in current research [52,55,60,61,95].

Logically, from each of the GOF metrics, a ranking is obtained that orders the different GCMs from worst to best according to their value for that metric. To obtain an overall ranking, the most common approach is to resort to a method based on a single index, which most studies typically refer to as the Comprehensive Rating Index (*CRI*) [60,96,97]. Its mathematical expression is presented herein (Equation (6)).

$$CRI = 1 - \frac{1}{nm}\sum_{i=1}^{n} Rank_i \tag{6}$$

where n is the number of metrics (5); m is the number of models evaluated (34); and $Rank_i$ represents the ranking position of GCM m based on metric n, ranging from 1 to 34. Hence, the closer the model is to *CRI* = 1, the better it fits ERA5 Land reanalyzed data.

2.5. Multi-Model Ensemble and Future Projections

Multi-Model Ensemble (MME) methods have been classified into unweighted and weighted [98]. Unweighted procedures have been recommended and employed as standard practice for several decades, primarily due to their ease of use [3]. The most widely used among these methods is the Simple Averages Multi-Model Ensemble (SAE). However, it has been found that weighted methods, especially those based on machine learning algorithms, produce better results and reduce the systematic biases inherent in each GCM [99,100].

Within these, the Random Forest Multi-Model Ensemble (RFE) has proven to be one of the most practical, accurate, and popular methods currently [100].

Both SAE and RFE were compared in this research. The SAE was obtained by directly averaging the values of the GCMs intended for ensemble. For a more detailed explanation of how to obtain an RFE, one should refer to the original publication [101]. However, it is worth mentioning that the parameters used were $n_{tree} = 500$, and $m_{try} = 2$, with 80% of the data for calibration, and the reanalyzed temperature was used as the independent variable.

To select the number of models included in the MME, the "best MME" criterion was followed. This involved generating 34 MMEs by adding the 34 GCMs in order of ranking and selecting the MME with the maximum correlation with the reanalyzed temperature. This approach is similar to that used by other authors [61]. However, as an innovation, this study also considers the "explained variance" criterion. This criterion entails determining which of the 34 designed MMEs predominantly explains (has a higher percentage of) the variance present in the ERA5 Land reference database. In other words, for instance, if the variance of ERA5 Land is 1.0, an MME with a variance value of 0.85 would be considered better, initially, than one with a value of 0.80.

Finally, the optimal MME obtained will be used to establish projections for short-term (2026–2050), medium-term (2051–2075), and long-term (2076–2100) periods under different greenhouse gas emission scenarios from CMIP6: optimistic (SSP1-RCP2.6), intermediate (SSP2-RCP4.5), and pessimistic (SSP5-RCP8.5) [102]. Emission scenarios data were also obtained from the Lawrence Livermore National Laboratory node of the World Climate Research Program website (https://esgf-node.llnl.gov/projects/cmip6/, accessed on 16 December 2022)

3. Results

3.1. Cross-Evaluation of Interpolation Methods

Six interpolation methods were applied to match the resolution of the 34 GCMs, with the resolution of the reference dataset being (0.1°): Inverse Distance Weighting (IDW), Bilinear Interpolation (BI), Kernel Smoothing (KS), Ordinary Kriging (OK), Cokriging plus Elevation (COK+E), and Empirical Bayesian Kriging (EBK). Figure 2 shows the Mean Error (ME) and Root Mean Square Error (RMSE) between GCM value in its native resolution and GCM value in its interpolated resolution (0.1°) for the 34 selected GCMs. Concerning the deterministic methods (BI and IDW), subtle discrepancies between them are noted, albeit both exhibiting similar downscaling capabilities, as indicated by their comparable RMSE values (0.9 versus 0.8 as the median value). BI manifests a ME range spanning from +0.08 to −0.013, with certain disregarded outliers reaching extremes of +0.18 and −0.35. Conversely, IDW displays a maximum ME of +0.02 and a minimum of −0.06. In general, IDW demonstrates greater stability as a deterministic method (with lesser intermodel discrepancies) than BI, with its median value in closer proximity to 0, and displaying either a slight or negligible propensity for underestimation. The polynomial KS method mirrors IDW's traits, albeit exhibiting a substantially compressed median value around 0 and a diminished RMSE, with a median hovering around 0.6. Among geostatistical methods, EBK emerges as the exemplar, showcasing the lowest error values, both in terms of ME and RMSE. Its entire ME spectrum spans from +0.005 to −0.004, positioning it closest to 0, while registering scarce absolute under or overestimations. The median value of its RMSE stands at 0.41, unequivocally the most minimal. Conversely, neither KS nor COK+E yield favorable outcomes, yielding even poorer estimates than deterministic counterparts. Particularly, COK+E's ME values exhibit a marked skew towards underestimation (median = −0.05), with standardized error values surpassing 1 for several GCMs, constituting their median. Although OK portrays a more acceptable ME range (−0.06 to +0.06, with outliers disregarded), it yields RMSE values similar to those of COK+E.

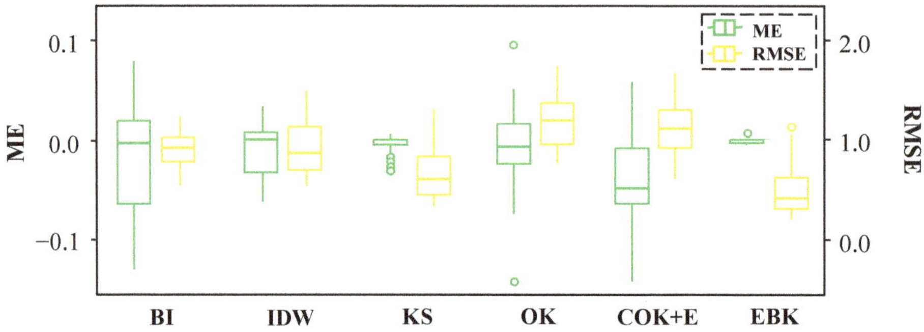

Figure 2. Mean error (ME) and Root Mean Square Error (RMSE) of the applied interpolation methods in downscaling the GCMs temperature data. IDW, Inverse Distance Weighting; BI, Bilinear Interpolation; KS, Kernel Smoothing; OK, Ordinary Kriging; COK+E, Cokriging plus Elevation; EBK, Empirical Bayesian Kriging.

The outcomes regarding individual model deviations are illustrated in Figure 3. It is evident that EBK stands out as the method that minimizes RMSE to the greatest extent, with values equal to or less than 0.5 for the majority of GCMs. Even in models with certain "resistance" to interpolation, such as those from the IPSL and GISS families, along with KIOST, which consistently exhibit errors irrespective of the method employed, the application of EBK appears to mitigate these discrepancies. BI, KS, and to a lesser extent, IDW, also demonstrate considerable consistency. Conversely, both OK and COK+E display substantial errors, with values close to or exceeding 1.5 in RMSE for models from the CNRM, CMCC, and CESM2 families, and significant errors ranging between 1.0 and 1.5 in RMSE for numerous other GCMs (Hadley, MPI, or GFDL families), underscoring once again their inability to yield acceptable downscaling results.

Finally, it is imperative to underscore the geospatial disparities observed in the interpolation of each GCM. To illustrate this, a particular model, the HadGEM-GC31-LL, was chosen as an example, and its spatial distribution of errors in the Euro-Med region is delineated in Figure 4. Notably problematic areas arise across all methods except for EBK. These problematic zones are concentrated in Central and Eastern Turkey, as well as in high-altitude regions such as Northern Spain, Central France, the Alps, the Dinaric Alps, and the Balkans. In these areas, methods such as IDW, OK, and especially COK+E, alter the data by up to +4.4 °C. Additionally, a profound depression is observed in the center of the Czech Republic, where the alteration reaches as low as −4.2 °C. Although BI exhibits more consistency than the previous methods and does not manifest such gross positive errors (maximum of +3.5 °C), it demonstrates negative errors across virtually the entire study area, with areas of significant modification (exceeding −3.8 °C) in Eastern Spain, Southern Turkey, or the Franco-Italian coast.

Regarding EBK, it shows its greatest strength in these findings, as the low error rate (Figure 2) and consistency among models observed previously (Figure 3) translate into remarkable spatial regularity. The entire range of data alteration through interpolation ranges from −1.15 °C (Czech Republic) to +1.41 °C (Eastern Turkey), significantly lower than that of any other method. However, despite specific zones, such as in most of the territory, the range of modification oscillates between +0.19 and +0.35 °C.

The superiority of EBK as the interpolation method with the least data modification was demonstrated on the global error rate, intermodel consistency, and geospatial regularity level. Therefore, this Bayesian approach is implemented to equalize the scale of the 34 selected GCMs to that of the ERA5 Land reference dataset (0.1°).

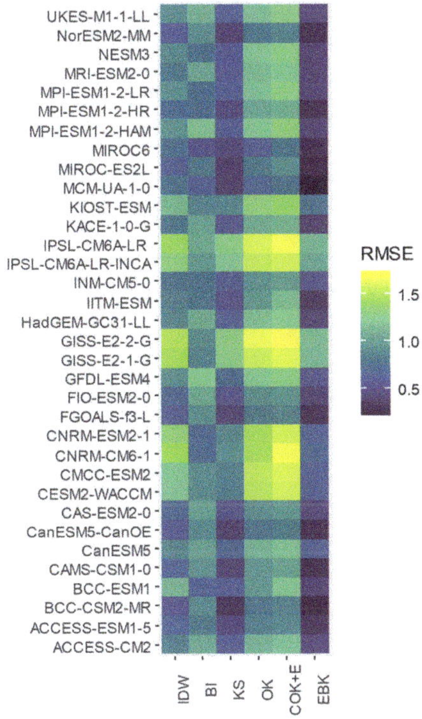

Figure 3. Heatmap of the Root Mean Square Error (RMSE) for each GCM and interpolation method. IDW, Inverse Distance Weighting; BI, Bilinear Interpolation; KS, Kernel Smoothing; OK, Ordinary Kriging; COK+E, Cokriging plus Elevation; EBK, Empirical Bayesian Kriging.

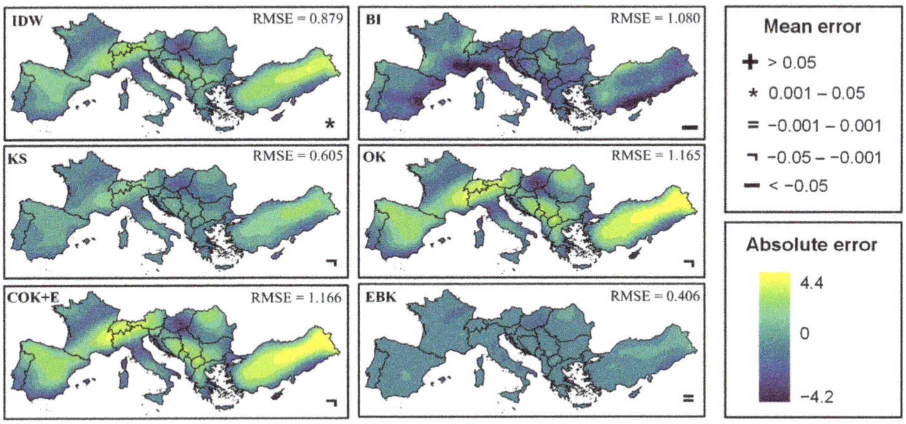

Figure 4. Spatial distribution of the absolute error in temperature (°C) for an example of GCM (HadGEM-GC31-LL). Root Mean Square Error (RMSE) and Mean error (ME) are also shown. IDW, Inverse Distance Weighting; BI, Bilinear Interpolation; KS, Kernel Smoothing; OK, Ordinary Kriging; COK+E, Cokriging plus Elevation; EBK, Empirical Bayesian Kriging.

3.2. Ranking and Assessment of GCMs

Table 2 presents the performance of each of the 34 GCMs for each of the five GOF metrics used. These results stem from a pixel-to-pixel comparison of the annual mean temperature at 0.1° provided by each interpolated GCM versus the observed (reanalyzed) by ERA5 Land. Additionally, the table displays the global ranking based on the Comprehensive Rating Index (CRI).

Table 2. Goodness-of-fit between General Circulation Models (GCMs) and ERA5 Land temperature data. NRMSE, Normalized Root Mean Squared Error; NSE, Nash–Sutcliffe Efficiency; KGE, Kling–Gupta Efficiency; CC, Correlation Coefficient; md, modified index of agreement. Comprehensive rating index (CRI) value and overall rank are also displayed. The rank for each metric is shown between brackets. The top 5 models for each variable are highlighted in bold.

GCM	NRMSE (%)	NSE	KGE	CC	md	CRI
ACCESS-CM2	53.20 (16)	0.72 (16)	**0.85 (5)**	0.87 (20)	0.75 (14)	0.58 (14)
ACCESS-ESM1-5	57.60 (22)	0.67 (22)	0.72 (28)	0.86 (23)	0.72 (20)	0.32 (24)
BCC-CSM2-MR	49.30 (12)	0.76 (12)	0.80 (13)	0.87 (21)	0.77 (11)	0.59 (11)
BCC-ESM1	66.60 (30)	0.56 (30)	0.65 (32)	0.77 (32)	0.67 (30)	0.09 (31)
CAMS-CSM1-0	51.70 (14)	0.73 (14)	0.82 (12)	0.90 (10)	0.73 (19)	0.59 (12)
CanESM5	58.70 (26)	0.66 (26)	0.75 (25)	0.83 (28)	0.72 (23)	0.25 (28)
CanESM5-CanOE	58.20 (24)	0.66 (24)	0.76 (22)	0.83 (29)	0.72 (21)	0.29 (26)
CAS-ESM2-0	57.70 (23)	0.67 (23)	0.77 (20)	0.91 (8)	0.71 (25)	0.42 (20)
CESM2-WACCM	57.40 (21)	0.67 (21)	0.75 (26)	**0.91 (5)**	0.71 (24)	0.43 (19)
CMCC-ESM2	55.60 (19)	0.69 (19)	0.78 (17)	0.89 (13)	0.73 (17)	0.50 (17)
CNRM-CM6-1	**41.70 (3)**	**0.83 (3)**	**0.87 (3)**	**0.91 (4)**	**0.80 (3)**	**0.91 (3)**
CNRM-ESM2-1	44.40 (7)	0.80 (7)	**0.86 (4)**	0.91 (7)	0.78 (7)	**0.81 (5)**
FGOALS-F3-L	47.80 (9)	0.77 (9)	0.74 (27)	0.89 (14)	0.76 (12)	0.58 (15)
FIO-ESM2	59.20 (27)	0.65 (27)	0.77 (21)	0.90 (9)	0.70 (28)	0.34 (23)
GFDL-ESM4	**42.40 (4)**	**0.82 (4)**	**0.89 (1)**	**0.94 (1)**	**0.80 (4)**	**0.92 (2)**
GISS-E2-1-G	46.90 (8)	0.78 (8)	0.78 (15)	0.89 (15)	0.77 (10)	0.67 (10)
GISS-E2-2-G	73.60 (31)	0.46 (31)	0.72 (29)	0.88 (17)	0.61 (32)	0.18 (29)
HadGEM-GC31-LL	48.00 (11)	0.77 (11)	0.84 (6)	0.88 (16)	0.77 (9)	0.69 (8)
IITM-ESM	57.10 (20)	0.67 (20)	0.83 (10)	0.84 (26)	0.73 (18)	0.45 (18)
INM-CM5-0	62.90 (29)	0.60 (29)	0.68 (31)	0.83 (30)	0.70 (26)	0.15 (30)
IPSL-CM6A	44.30 (6)	0.80 (6)	0.84 (7)	0.90 (11)	0.79 (6)	0.79 (6)
IPSL-CM6A-INCA	**44.10 (5)**	**0.81 (5)**	0.83 (9)	0.90 (12)	**0.79 (5)**	0.79 (7)
KACE-1-0-G	51.70 (13)	0.73 (13)	0.78 (18)	0.87 (22)	0.75 (13)	0.54 (16)
KIOST-ESM	60.00 (28)	0.64 (28)	0.78 (16)	0.88 (19)	0.69 (29)	0.29 (27)
MCM-UA-1-0	76.00 (32)	0.42 (32)	0.51 (34)	0.69 (34)	0.59 (33)	0.03 (34)
MIROC6	90.90 (34)	0.17 (34)	0.68 (30)	0.80 (31)	0.58 (34)	0.04 (32)
MIROC-ES2L	78.60 (33)	0.38 (33)	0.61 (33)	0.71 (33)	0.64 (31)	0.04 (33)
MPI-ESM1-2-HAM	58.60 (25)	0.66 (25)	0.79 (14)	0.85 (24)	0.70 (27)	0.32 (25)
MPI-ESM1-2-HR	**41.00 (2)**	**0.83 (2)**	**0.88 (2)**	**0.92 (3)**	**0.82 (1)**	**0.94 (1)**
MPI-ESM1-2-LR	54.20 (17)	0.71 (17)	0.76 (24)	0.85 (25)	0.74 (16)	0.42 (21)
MRI-ESM2-0	**40.20 (1)**	**0.84 (1)**	0.83 (11)	**0.92 (2)**	**0.81 (2)**	**0.90 (4)**
NESM3	55.50 (18)	0.69 (18)	0.76 (23)	0.83 (27)	0.72 (22)	0.36 (22)
NorESM2-MM	52.90 (15)	0.72 (15)	0.77 (19)	0.91 (6)	0.74 (15)	0.59 (13)
UKES-M1-1-LL	47.90 (10)	0.77 (10)	0.83 (8)	0.88 (18)	0.78 (8)	0.68 (9)

Regarding the GOF metrics, both NRMSE and NSE offer identical rankings. They appear to be the ones that show the greatest range of variability among models, as they oscillate, from worst to best GCM, between 90.9 and 40.2 for NRMSE (%) and between 0.17 and 0.84 for NSE. It is noteworthy that NSE does not exhibit negative values, ensuring that no GCM presents an aberrant temperature simulation. Conversely, CC seems to provide the least information, as its interquartile range of values ranges from 0.85 to 0.89, with only three models below 0.8. It could be attributed to the data distribution of the GCMs and to the intrinsic characteristic of the metric. It could be hypothesized that in coarse-resolution models, the inner variance is small and therefore the difference between the simulated value of a pixel and the mean value is minimal. In these particular cases, metrics like CC virtually measure the observed variance of the data distribution, resulting in similar CC values for almost all models. On the other hand, metrics such as NRMSE or NSE compare (subtract) the simulated and observed data distributions, yielding results with a wider range of values. Moreover, models with a high correlation with the reanalyzed temperature can be identified, but their data distribution, according to the other metrics, is significantly distant from the reanalyzed data. For instance, CESM2-WACCM exhibits a correlation with ERA5 Land exceeding 0.9, ranking 5th for this metric, yet occupying a position below 20th for the rest of the metrics. CAS-ESM2-0, FIO-ESM2, or NorESM2-MM present a similar problem.

Concerning the models, none repeat the same ranking position for the five analyzed metrics. Models from the CNRM family and MIROC-ES2L exhibit the most repetitions: CNRM-CM6-1 ranks 3rd on four occasions, CNRM-ESM2-1 occupies the 7th position on four occasions, and MIROC-ES2L is placed 33rd on four occasions. GFDL-ESM4 (4th), MPI-ESM1-2-HR (2nd), MIROC6 (34th), BCC-ESM1 (30th), and IPSL-CM6A (6th) also show consistency in their results and repeat their ranking position at least three times without significant fluctuations or performance losses.

Lastly, regarding the CRI value, the top five models demonstrating overall superior performance, from highest to lowest, are MPI-ESM1-2-HR, GFDL-ESM4, CNRM-CM6-1, MRI-ESM2-0, and CNRM-ESM2-1. Models from the IPSL and Hadley families, along with GISS-E2-1-G, complete the top 10.

3.3. Geospatial Performance Analysis

The aforementioned pixel-by-pixel comparison of model performance, which reveals the geospatial component of this variable, is the main advantage of the developed method. In other words, it allows for the identification of areas where GCM-simulated temperatures are close to those reanalyzed by ERA5 Land, as well as areas where this simulation capacity decreases drastically. To our knowledge, our study is the first to propose a geospatial context for the CRI value, which has previously only been used to rank models based on their overall performance across multiple metrics.

Figure 5 confirms that models do not behave uniformly across the Euro-Med region. The MPI-ESM1-2-HR, which proved to be the most competent, shows areas of decreasing temperature simulation, such as the central regions of Spain and France or the Black Sea coast (Eastern Romania and Bulgaria). Conversely, it is the best at simulating temperatures in high altitude areas, which is remarkable in such a rugged region. Its CRI values are equal to or close to 1 in the Alps, the Dinaric Alps, and the entire Turkish plateau, and very high (0.8–1.0) in the Apennines (Italy) or the Sierra Nevada (Spain). Interestingly, the two models in the CNRM family show similar patterns, although both have deeper dips in performance, particularly the CNRM-ESM2-1, whose CRI values range from 0.15 to 0.40 in Northwest France and the Black Sea Coast. The GFDL-ESM4 and MRI-ESM2-0 models appear to be much more regular among those with competent simulations, as they do not show areas where the simulation drops drastically. The use of GFDL-ESM4 appears to be preferable in Northwest and Southeast Spain and Northern Turkey, while MRI-ESM2-0 appears to simulate exceptionally well along the entire Adriatic and Black Sea coasts, precisely where other better-ranked models fail miserably.

Other models showing poorer overall performance but whose use may be locally justified include: (i) HadGEM-GC31-LL in Northwestern France, (ii) GISS-E2-1-G on the coast of the Black Sea (and in Northwestern France), (iii) UKES-M1-1-LL in the Aegean Sea, and (iv) IPSL models in Central Europe. Regarding the latter, especially in the case of IPSL that does not include the atmospheric chemistry/aerosol microphysics model (INCA), it should be noted that they competently model the Atlantic/temperate zones of the study area but seem to lose capacity in the large Mediterranean/subtropical zones such as the Iberian Peninsula, Italy, Greece, or Eastern Turkey.

Finally, among the models that simulate the temperature in the Euro-Mediterranean region poorly, MCM-UA-1-0, MIROC-ES2L and MIROC6 are clearly the worst performers, with CRI values below 0.25 in almost all points of the region. BCC-ESM1, GISS-E2-2G, INM-CM5-0, NESM3, or the MPI version with the Hamburg Aerosol Model (HAM) also show a globalized low CRI value. However, our analysis corroborates that even these GCMs can have local simulation peaks in certain areas. For instance, BCC-ESM1 correctly reproduces temperatures in Eastern Romania, MPI-ESM1-2-HAM in the Bulgarian Balkans, and INM-CM5 in Hungary and Northern Serbia. Therefore, even GCMs that are considered "disposable" due to their poor overall performance can be used in certain sub-regions as viable alternatives to models with better overall simulation, or as entities to be included in a Multi-Model Ensemble to improve the capture of existing climate variability.

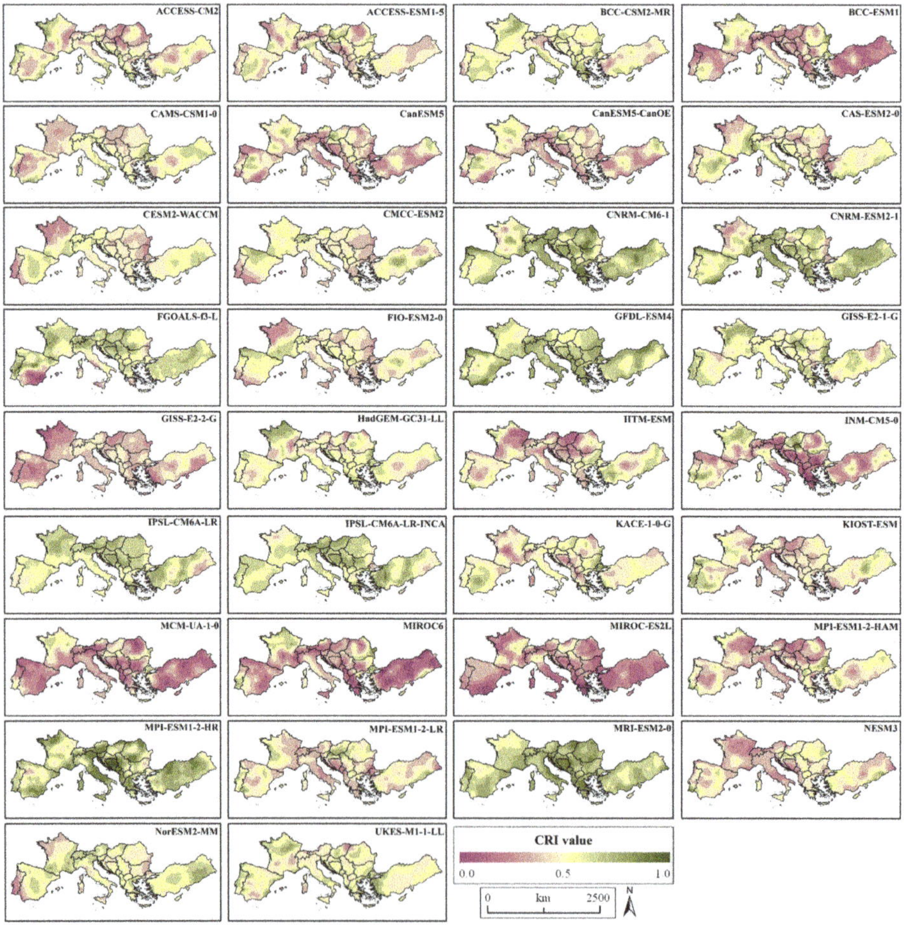

Figure 5. Geospatial analysis of the 34 GCMs in the Euro-Med region through the Comprehensive Rating Index (CRI).

3.4. Multi-Model Ensemble

3.4.1. Determining Ensemble Members

Figure 6 illustrates the correlation and explained variance as a function of the number of additional models added, in strict order of the ranking obtained in Table 2, to the Multi-Model Ensemble (MME). It is evident that with an increasing number of added GCMs, the ensemble value becomes closer to the value of ERA5 Land, resulting in an increase in both correlation and explained variance. With just one GCM (MPI-ESM1-2-HR), the correlation already exceeds 0.9, and the explained variance is greater than 0.8. With three GCMs (MPI-ESM1-2-HR + CNRM-CM6-1 + GFDL-ESM4) the variance explained is greater than 0.9 and the correlation value rises to 0.96. Although at three GCMs the MME has already improved considerably, a slight but constant increase is observed until the value of six GCMs is reached (0.97 in correlation and 0.93 of explained variance). From this value onwards, neither the correlation nor the explained variance improves significantly. Hence, the number of GCMs considered for MME is six GCMs: MPI-ESM1-2-HR, CNRM-CM6-1, GFDL-ESM4, CNRM-ESM2-1, MRI-ESM2-0, and IPSL-CM6A-LR.

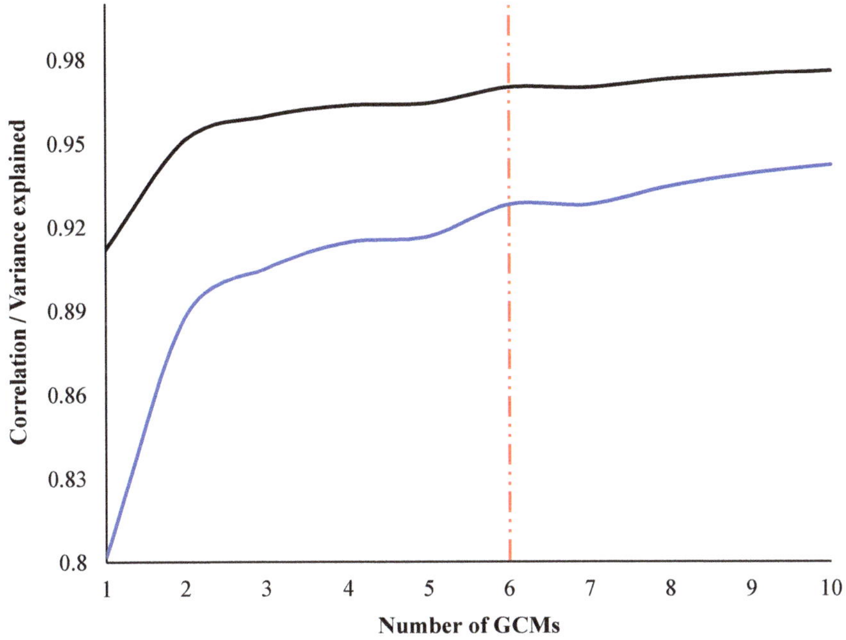

Figure 6. Correlation (black) and variance explained (blue) as a function of the number of models added to Multi-Model Ensemble. The red dotted line indicates the number of GCMs at which both values stabilize.

3.4.2. Simple Averages vs. Random Forest Multi-Model Ensemble

Figure 7 shows how the Random Forest Multi-Model Ensemble (RFE) exhibits significantly improved accuracy in capturing temperature patterns of ERA5 Land on an annual basis, but for a slight overestimation tendency in the southern half of the Iberian Peninsula, the coast of Sardinia, Italy, Southern Turkey, and Cyprus. Having said that, RFE accurately depict the temperatures in the higher foothills of the study area, especially in the Alps, the Carpathians, and the Armenian System. Conversely, Simple Averages Multi-Model Ensemble (SAE), clearly attenuates the contrasts between the coldest and warmest regions, resulting in the loss of extreme values. The RFE superiority in reproducing the observed (reanalyzed) temperature can also be observed at the monthly scale. In this regard, Figure 8 demonstrates that a proper simulation of monthly temperature can be achieved through an annual-scale evaluation of the GCMs that perform best for a specific study area. It can be observed that the RFE not only captures the temperature range of ERA5 Land perfectly for each month but also reproduces an almost identical distribution and provides appropriate detection of extreme values (whiskers of the diagram). In contrast, the SAE tends to overestimate central values and shows a significant inability to capture extreme values. Particularly, it is less effective in detecting the lowest temperatures of each month, as clearly observed during the months of December, April, May, June, September, and October.

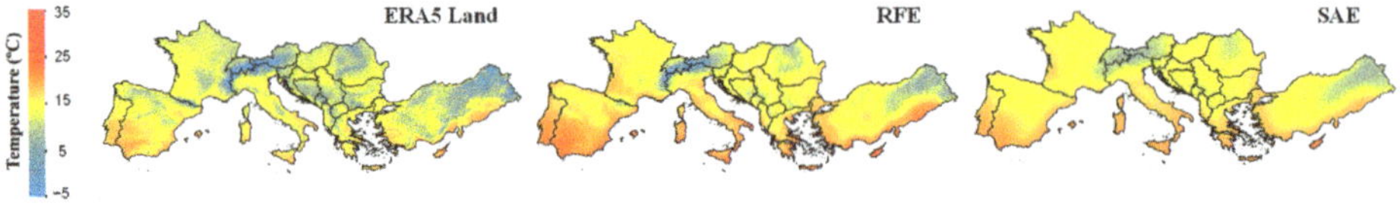

Figure 7. ERA5 Land, Random Forest Multi-Model Ensemble (RFE), and Simple Averages Multi-Model Ensemble (SAE) mean annual temperature in the Euro-Med.

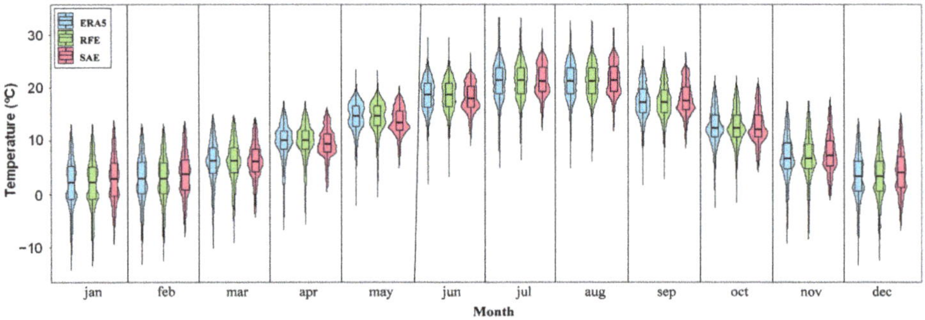

Figure 8. ERA5 Land, Random Forest Multi-Model Ensemble (RFE), and Simple Averages Multi-Model Ensemble (SAE) monthly mean temperature in the Euro-Med.

Finally, Figure 9 illustrates the correlation between the percentage importance assigned by the regression method used in the RFE and the CRI value obtained in the ranking (Table 2). Although a positive correlation detected between these variables (Rho = 0.23), it is not statistically significant ($p > 0.05$). Thus, it is observed that when performing an RFE with the total pool of analyzed GCMs, the regression method automatically discards several of them (GISS-E2-1-G, MPI-ESM1-2-LR, NESM3, etc.) by assigning them 0% importance. However, it also includes, although with little overall importance, models considered as "poor performers" (BCC-ESM1, MIROC6, and MIRPC-ES2L). At the same time, it discards "good performers" (IPSL models, HadGEM-GC31-LL, and UKES-M1-1-LL). In conclusion, we could argue that the majority of the assigned importance is distributed among three of the GCMs that performed best in the ranking: GFDL-ESM4 (14%), MRI-ESM2-0 (6%), and MPI-ESM1-2-HR (3%).

3.5. Future Temperature Projections

In this section, future projections regarding mean seasonal and annual temperature in the Euro-Med region for the short-term (2026–2050), medium-term (2051–2075), and long-term (2076–2100) periods, based on different emission scenarios (SSP1-RCP2.6, SSP2-RCP4.5, and SSP5-RCP8.5), will be established using the previously developed MME. The projections obtained in this study will be contextualized according to the predictions of the IPCC at both global and regional (Euro-Med) scales. The confidence intervals provided by the IPCC were obtained from the Interactive Atlas developed by the Working Group I for AR6 [103], and they are documented in Table A1 of Appendix A. The percentage of territory whose expected warming falls within the IPCC interval for each scenario and time window is presented in Table A2 of Appendix A.

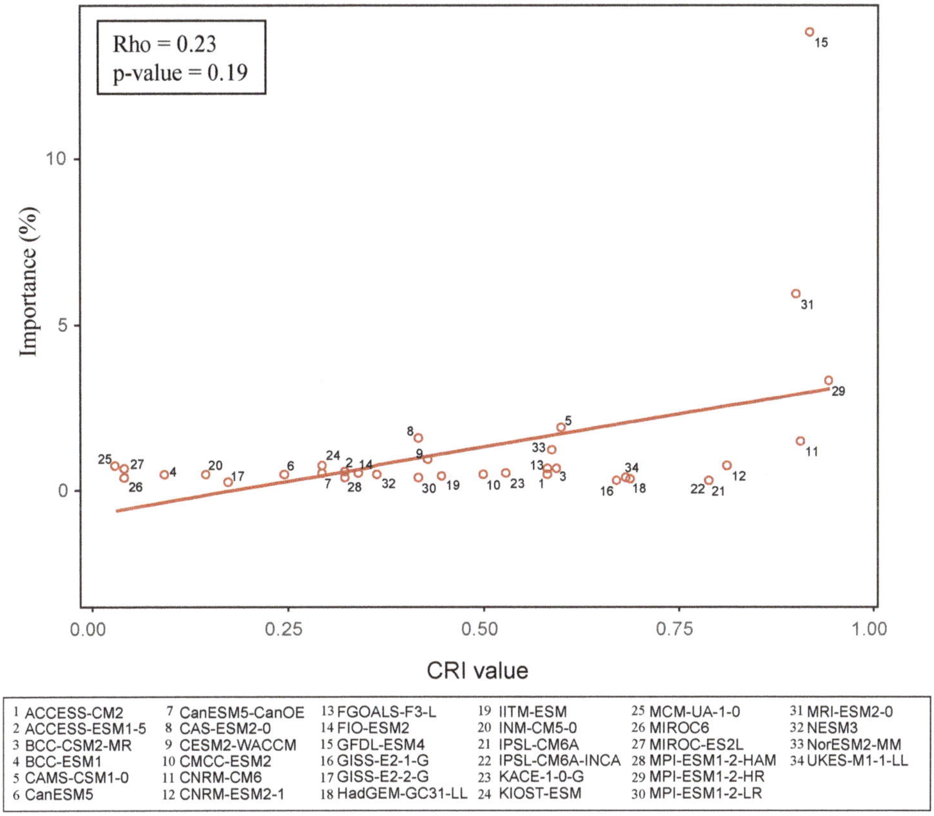

Figure 9. Correlation between CRI value and the percentual importance value assigned by the Random Forest regression method.

Figure 10 shows the seasonal results in the most optimistic scenario: SSP1-RCP2.6 (sustainability or "green road"). As observed, significant differences exist between the outcomes obtained in our study and those predicted by the IPCC at a global or regional scale. The percentage of study area whose increments align with these predictions is lower in winter and spring. In winter, only 20.4% of the study area falls within the interval expected by the IPCC in the short term, a value that expands to 33.4% in the long term. Overall, there is a more significant warming than expected by the IPCC. The areas where this detected warming is most intense (more than 2 °C with respect to the reference interval) correspond to those regions of higher elevation in the Euro-Mediterranean region, as well as the Mediterranean archipelagos. In the Central European plateau, the expected increase is also higher (0–+1 °C) than the global and regional average of the IPCC. In the case of spring, the agreement shown with respect to official predictions is even lower, as it stands at 17.9% in the short term and only increases to 22.8% in the long term. However, in this case, the official projections suggest that the warming will be much greater than that detected in this study, as most of the study area shows an increase between 0 and 2 °C lower than predicted by the IPCC. The Iberian plateau, the entire France, the Po basin (Italy), and large areas of Central Turkey are the main points where this pattern is detected. However, the elevated mountain systems continue to warm more than expected. The wide ranges offered by the IPCC for summer and autumn mean that much of the territory falls within them. In the long term, the agreement is over 51.4% in summer and 39.9% in autumn. In both seasons, except for the aforementioned areas, which are resistant to warming in any season and time period, the pattern indicates a warming greater than expected. Particularly noteworthy, in the case of summer, is that much of the Northwestern French territory expects a warming between 0 and 1 °C higher than expected by the IPCC.

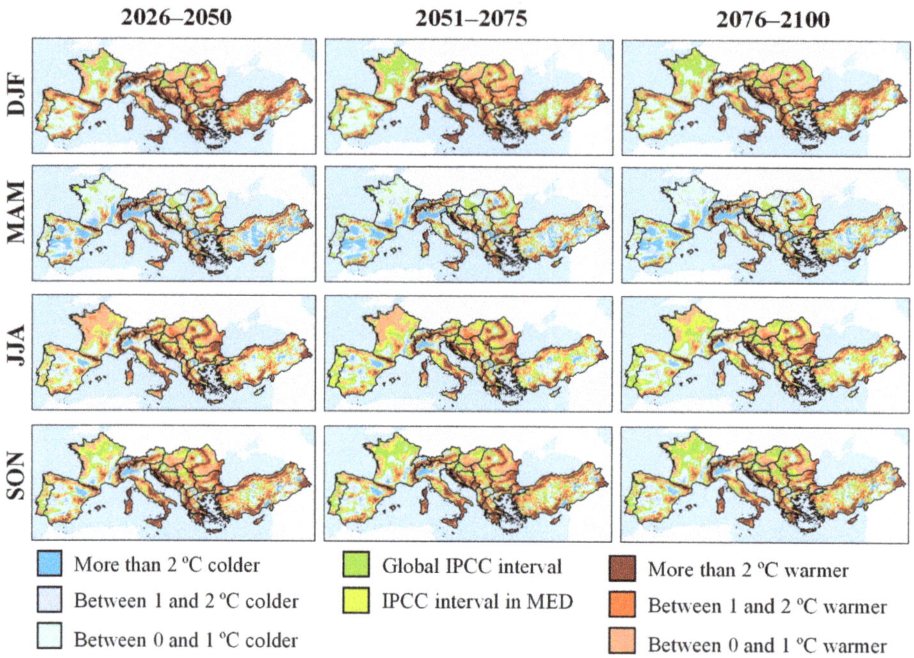

Figure 10. Seasonal temperature anomalies in Mediterranean Europe in the short (2026–2050), medium (2051–2075) and long (2076–2100) term under the SSP1-RCP2.6 emission scenario. Anomalies were established with respect to IPCC predictions globally and for the European Mediterranean region (MED).

Figure 11 depicts the results obtained for the intermediate emission scenario: SSP2-RCP4.5 ("middle of the road"). The patterns and trends are very similar to those discussed in the previous scenario. In winter, however, it is observed that in the short term, most of the region warms more than predicted by the IPCC (concordance of 14.3%). These areas of increase not exceeding 1 °C with respect to the official projections extend to the Portuguese Coast and much of France and Central Europe. However, this pattern seems to reverse over time, as in the long term, the concordance is close to 20%, but due to a reluctance to warm up as much as the IPCC prediction expects (1.8–3.1 °C between the global and regional predictions). However, much of France is close to these values, with an approximate average anomaly of +1.6 °C. In the other seasons, no different patterns are observed compared to those discussed for the pessimistic scenario, although it should be noted that, in the long term, the concordance for summer and autumn is 64.4% and 60.3%, respectively, with respect to the established interval.

Figure 12 presents the results for the pessimistic scenario: SSP5-RCP8.5 (fossil-fuel development or "highway"). Regarding winter, although in the short and medium term the results are similar to those of less pessimistic scenarios, in the long term, it is observed that only a small part of the Mediterranean region will warm as much as expected by the IPCC. The concordance obtained with this interval is 5.2%. Although most of the anomalies detected are positive, they do not reach the +4.1 °C increase, the lower limit of the reference interval. In this regard, we can classify the IPCC projections as more pessimistic than those obtained through our methodical application of a regional assessment with downscaling.

Regarding summer and autumn, there is a noticeable accelerated increase in both cases in the medium term. In the case of summer, the Adriatic coast and much of Eastern Europe will warm by 2 °C above expectations, while other areas of Central Europe or the northwest half of France will warm by over 1 °C compared to the reference. This is particularly concerning because the upper reference limits for summer and autumn are +3.4 °C and over +3.0 °C, respectively, meaning that much of the region could experience increases exceeding +5.0 °C in seasonal average temperature anomalies from the mid-century onwards. In the long term, however, the consequent increase in the upper reference limit to +6.9 °C (summer) and +6.2 °C (autumn) results in a 71.9% concordance for both seasons. However, the Central–Eastern European region continues to experience intense summer overheating.

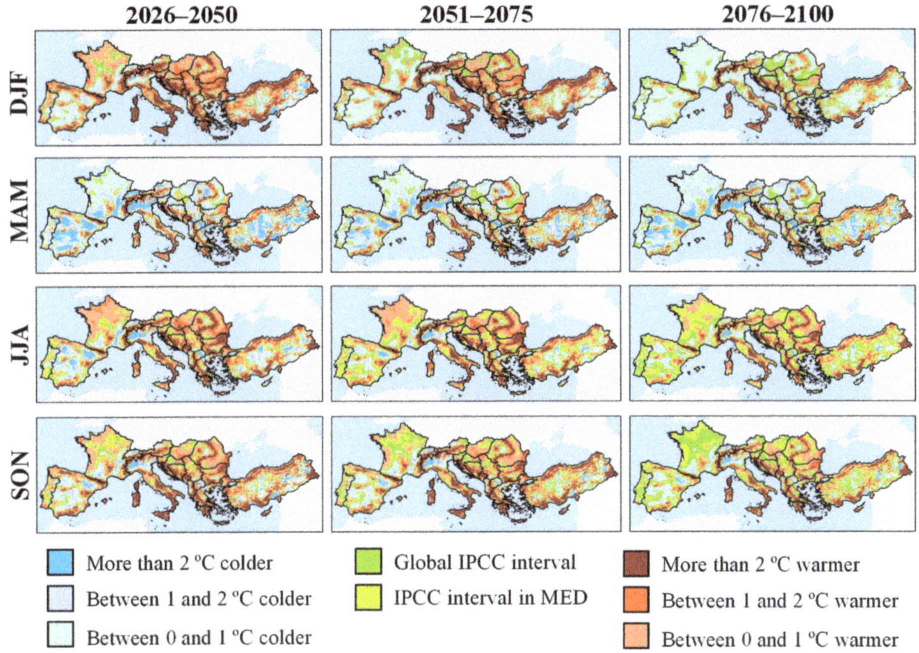

Figure 11. Seasonal temperature anomalies in Mediterranean Europe in the short (2026–2050), medium (2051–2075), and long (2076–2100) term under the SSP2-RCP4.5 emission scenario. Anomalies were established with respect to IPCC predictions globally and for the European Mediterranean region (MED).

Finally, Figure 13 shows the analysis of annual temperature anomalies for the three emission scenarios in the short, medium, and long term. In the long term, the agreement with the IPCC percentages is quite high for all three scenarios: 60.2%, 52.6%, and 66.4% for the optimistic, intermediate, and pessimistic scenarios, respectively. The intermediate scenario SSP2-RCP4.5 shows the greatest discordance with the results obtained in this study. In the medium-to-long term, except for the Iberian Peninsula, Italy, and Central–Eastern Europe, as well as in the higher mountain systems where the detected anomaly is positive, temperature increases are rarely as significant as those predicted by the IPCC. The agreement is also low (32.9%) for the SSP5-RCP8.5 scenario in the medium term, likely influenced by the strong increase in summer mean temperatures detected. In this case, areas with an increase exceeding 2 °C compared to the reference could face a further increase of +4.7 °C in annual mean temperatures from mid-century onwards, which would rise to 4.9–5.8 °C by the end of the century.

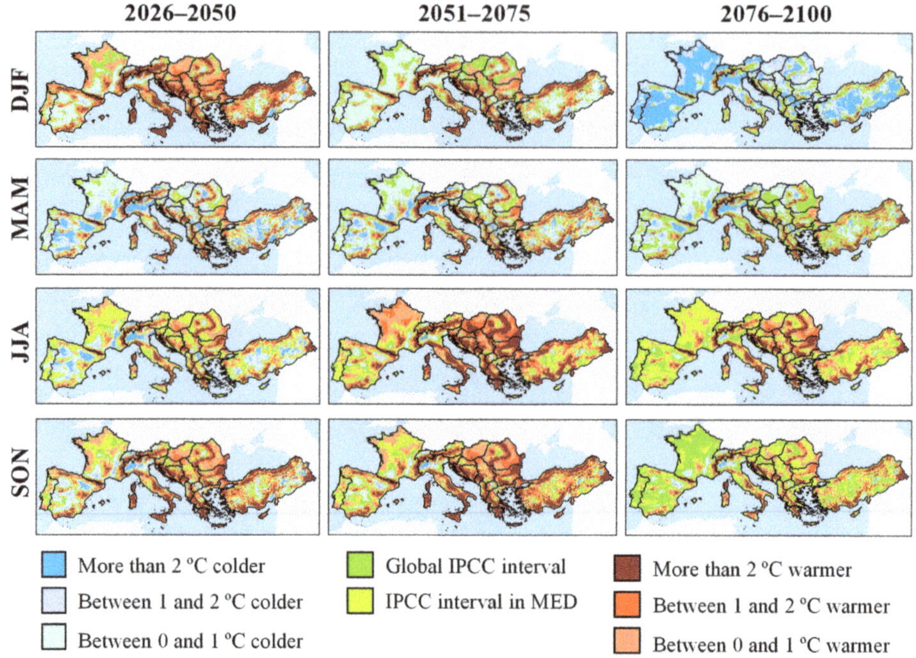

Figure 12. Seasonal temperature anomalies in Mediterranean Europe in the short (2026–2050), medium (2051–2075), and long (2076–2100) term under the SSP5-RCP8.5 emission scenario. Anomalies were established with respect to IPCC predictions globally and for the European Mediterranean region (MED).

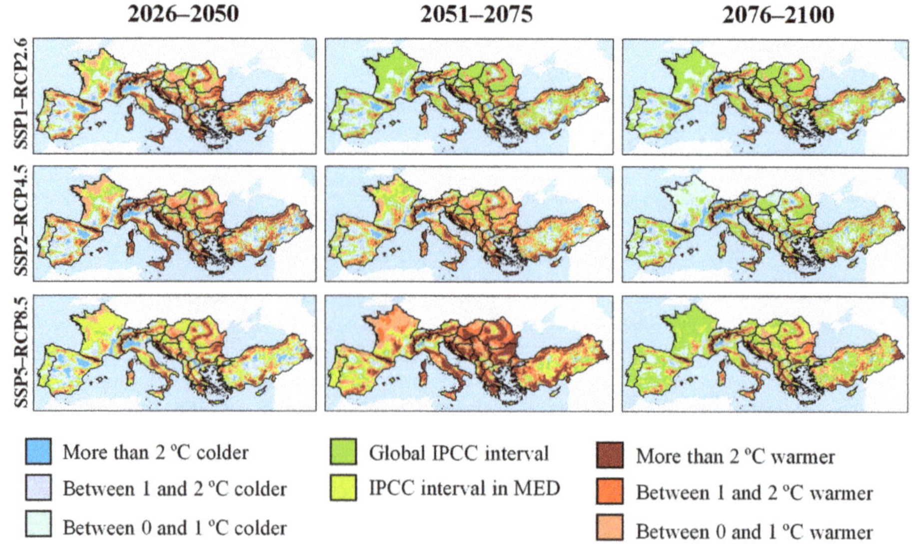

Figure 13. Annual temperature anomalies in the Euro-Med in the short (2026–2050), medium (2051–2075) and long (2076–2100) term under the three emission scenarios (SSP1-RCP2.6, SSP2-RCP4.5, and SSP5-RCP8.5). Anomalies were established with respect to IPCC predictions globally and for the European Mediterranean region (MED).

In conclusion, the results analyzed indicate a clear warming signal in the Euro-Med region. However, the detailed assessment based on downscaling and ranking of models that perform best for the study area reveals certain local differences that diverge, in some cases, greatly from the global and regional predictions established by the IPCC.

4. Discussion

4.1. Downscaling of GCMs Temperature Data

Empirical Bayesian Kriging (EBK) has clearly yielded the best results and introduces the least modifications between raw and interpolated data. The superiority of this method over other interpolation methods has been highlighted in numerous instances in current scientific literature [104–106]. This may be attributed to its less restrictive assumptions regarding data distribution [107]. Bilinear interpolation (BI), although the most commonly used method for downscaling GCMs [32,35,52,56,87], did not yield positive results in our evaluation. This may be because, being a deterministic method, it applies uniformly across all regions of the territory, which may work against it in rugged areas such as the Euro-Med. The use of Inverse Distance Weighting (IDW), another deterministic and simple method, appears preferable, despite overestimation and unfavorable previous results [108,109], at least in the case of temperature in mountainous regions, in the absence of complex geostatistical methods like EBK. Regarding deficiencies in the interpolation of certain GCMs, a clear reason has not been identified. The native resolution of the models could be a possible cause. The internal structure and distribution of the data are more feasible possibilities, as they would explain why models of the same family manifest comparable errors.

Our study underscores the imperative for a thorough and meticulous examination of interpolation techniques that minimize errors when downscaling GCMs. Systematically defaulting to bilinear interpolation may substantially distort the values furnished by GCMs.

4.2. Ranking, Assessment and Ensemble of GCMs

In relation to the goodness-of-fit (GOF) metrics employed, it is noteworthy that both the low variability in the correlation coefficient values and the relative inconsistency shown in the rankings obtained for each measure were detected in several previous studies worldwide [53,61,110,111].

Regarding the GCMs, MPI-ESM1-2-HR (1st) was classified as the best overall performer in our study. Its predecessors (e.g., ECHAM4 and ECHAM5) were highlighted for their superior simulation of climatic variables in Europe, especially along the French coast and the northwest of the Iberian Peninsula [112]. Studies conducted in Southern Italy (Calabria and Sicily) or in Turkey also underscore this accurate temperature modeling [59,60]. Precisely in Southern Italy, the CRI detected in our study exceeds 0.85. However, only the high-resolution (HR) version exhibits such efficacy, as the medium–low-resolution versions (LR and HAM) rank 21st and 25th, respectively. The improvement of the Max Planck family models with increased resolution has been noted in other studies [40,60]. Additionally, studies that do not incorporate this high-resolution version do not highlight MPI or its derivatives as good simulators [58,113].

The MRI-ESM2-0 (fourth) is another model highlighted in our study. However, it is a controversial model since its previous versions were considered incapable of simulating European climate patterns [114]. Recent studies, however, place the CMIP6 version of MRI as one of the best performers in the Mediterranean [60,115]. Regarding other models that have demonstrated great simulation in our analysis, CNRM models (third and fifth) had already shown their superiority over IPSL models (sixth and seventh) ([40]), despite the significant improvement evidenced by the latter compared to their CMIP5 versions [41]. Some studies highlight IPSL models as good predictors of future extreme temperature scenarios in Southern Spain [58] or in simulating temperature in Italy [59]. On the other hand, Regional Circulation Models (RCMs) based on CNRM models had already been highlighted for their good simulation in Mediterranean locations such as the Segura River

Basin (Spain) [113], Turkey [60], and Southern Italy [59]. Specifically, in the latter, the CRI detected by our work for the CNRM-ESM2-1 model exceeds 0.9, with CNRM-CM6-1 showing similar values. Some of these studies also highlight RCMs based on the EC-Earth model, which were not evaluated in our study due to sharing the same climatic component as ERA5 Land [40].

It is noteworthy that, despite the intermediate performance detected in our study, the Hadley Centre models (HadGEM-GC31-LL, eighth; UKES-M1-1-LL, ninth) or their predecessors have been recommended in numerous studies for their excellent simulation of atmospheric circulation and climate types in Europe [40,60,114,116–118]. Some studies even suggest that they outperform ECHAM/MPI models in some territories, such as the Catalan–Provençal coast, the Balkans, or Central Europe [112], a fact not observed in light of our results. However, many of these studies highlight their good reproduction of precipitation, focusing less on the accurate simulation of temperature detected in our research.

Some research has also pointed towards the good temperature simulation of the ACCESS1-0 model in the south of the Iberian Peninsula [58]. In our study, the performance of ACCESS-CM2 (14th) on the Southern and Eastern Spanish Coast is noteworthy (Figure 8), which is consistent with these previous results. The NorESM2-MM (13th) and FGOALS-f3-L (15th) models were not highlighted in our study. However, they show good performance in the south and east of Turkey, a fact highlighted in more local studies [60,61].

Regarding GCMs with poor performance, the MIROC models (32nd and 33rd) are unfortunately the most prominent in this regard. Studies using their previous versions had highlighted them as good predictors in the North Atlantic [114]. However, this study corrected Sea Level Pressure values before evaluating the models, whereas it is now considered that correcting GCM errors should never be performed before evaluation [87,119]. Recent studies implicitly discourage the use of MIROC models in the European region [118]. Their poor performance in the Eastern Mediterranean has also been previously detected [60]. On the other hand, the BCC models, especially the low-resolution version (BCC-ESM1) (31st), are also among the poor performers. Their poor performance had already been detected in other Mediterranean locations [58].

It is also noteworthy that none of the versions of the models that include aerosol sub-models (MPI-ESM1-2-HAM [71]), atmospheric chemistry (IPSL-CM6A-INCA [120]), or ocean circulation (CanESM5-CanOE [70]) perform better than the same versions that do not include these "improvements".

Finally, regarding the Multi-Model Ensemble (MME), it is imperative to underscore the recognized superiority of the Random Forest Multi-Model Ensemble (RFE) over the Simple Averages method (SAE), as well as over Bayesian or machine learning-based approaches, as delineated in prior scholarly investigations ([100,111,121,122]). Furthermore, it merits mention that our study opted for a cohort of six models, a choice aligned with the guidance provided by antecedent research [111]. Nevertheless, the determination of an optimal ensemble size remains a subject of contention within the scientific discourse. While some scholars advocate for the adequacy of a mere three GCMs [123], others contend that a minimum of 25 GCMs is requisite to attain minimal error reduction [98]. It is pertinent to highlight that, in our study, even with two or three GCMs, the correlation and explained variance were deemed satisfactory (>0.9) (Figure 6). The proficient replication of temperature patterns across both annual and monthly timescales by our RFE underscores that exhaustive minimization of error may not be imperative; rather, identifying a suitable local minimum suffices. Additionally, it elucidates the intrinsic linkage between performance at extended temporal scales (annual) and shorter ones (monthly), a conjecture previously posited [62]. Thus, the execution of assessments and ensembles at protracted scales emerges as a pragmatic approach, entailing reduced computational demand and greater expediency under specific circumstances.

4.3. The Sizzling Future of the Euro-Med

Our study highlights certain deviations between the predictions made by the IPCC and those obtained through optimal downscaling, GCM evaluation, and subsequent elimination of inaccurate models (Figures 10–13). Despite the scientific consensus on global warming, criticism and skepticism regarding the models or scenarios employed by the IPCC have continued to grow [124,125]. Some suggest that emission pathways may be exaggerated and divergent from detected realities [126], while other studies indicate that certain temperature-related variables may be underestimated [127,128]. However, studies comparing the CMIP6 scenarios developed for this latest IPCC assessment report (AR6) or for previous reports (AR4, AR5, etc.) conclude that there are no substantial differences between observed and projected or expected data [37,125,129]. In fact, many suggest that, if any, we would be dealing with a "cold bias", meaning that any errors in projections would result in expecting slightly less warming than observed [125,129]. It is worth noting that, in our research, the optimal selection of GCMs implies that, in the long term, for winter and spring, anomalies are much lower than those expected by the IPCC. Recent studies conducted in Europe support our findings in this regard, as they show that, in spring, the season in which we have detected the most disagreement with the IPCC projections, the reanalyzed mean temperature (and mean maximum temperature) would have been much lower than those observed by the models [130]. Therefore, it could be deduced that the IPCC predictions have been conservative, at least until 2020, and scenarios presuming higher levels of greenhouse gas emissions would be more realistic [125]. Our research also suggests that, at least for summer and autumn, the agreement with IPCC predictions is significantly higher in the most pessimistic scenario (SSP5-RCP8.5). Thus, the selection of GCMs that simulate current temperatures at a regional scale with high efficiency (and elimination of less efficient ones) and their subsequent ensemble and projection indicate that the most plausible scenario is the most pessimistic, at least in some specific temporal windows and seasons of the year.

In addition to the "cold bias" referred to in the literature and detected in our research, a significant portion of summer, autumn, and short-to-medium-term scenarios for winter exhibit a "warm bias", meaning a warming greater than expected by the IPCC. Studies that have conducted restricted Multi-Model Ensembles (CMMEs) by eliminating outlier models have concluded that results diverge considerably from expectations using large MME packages, the method followed by the IPCC [131,132]. Some of these studies have detected a slight tendency of MMEs to underestimate (0.2–0.4 °C) temperature projections compared to CMMEs [131]. Furthermore, it has been found that the higher regions of Europe, primarily the Alps, have shown, especially in summer, a reanalyzed temperature much higher (+3 °C) than simulated by the GCMs for the historical period of CMIP6 [130]. It is expected that this "warm bias" will persist in future projections. Therefore, the elimination of less operative GCMs for our study region could reveal a trend towards more significant warming than detected by the IPCC in certain temporal windows, seasons, and areas of the European Mediterranean.

The rates of temperature change detected in the past decades in numerous trend studies carried out in the European Mediterranean Basin are congruent with the expectations of the IPCC and with the anomalies detected in our research [133–140]. The rates of increase per decade are regionally variable: 0.4 °C in Sicily [134], up to 0.5 °C in Spain [136] or Central Europe [137] and even 1 °C in Greece [139] or part of Italy [138]. An increase in mean minimum temperatures of up to +0.04 °C per year across the entire Mediterranean region has also been noticed [140]. This escalation could account for the significant variance observed between our projections and those of the IPCC for the winter season and for elevated regions of the territory, where minimum temperatures tend to be lower. However, several of these studies highlight spring as one of the seasons where a more pronounced temperature change has been observed [133,136,137,139]. Conversely, these studies observe few changes in autumn. Although trend studies contradict the seasonal results obtained in our research, other studies based on future projections seem to ratify them [46,141–144]. Similarly, both the magnitude of the observed changes and the areas suffering the most severe impacts are consistent with previous studies [46,67,145–147]. A full ensemble (34 CMIP6 GCMs) was used to observe, in a pessimistic scenario, end-of-century summertime increases above 8 °C in the interior of the Iberian Peninsula, the Balkans and Central Turkey [46]. With CMIP5 GCMs lower changes for this season (6.5–7 °C) was noticed, although they agree that continental areas will be more affected [67]. A change in winter temperature with an order of magnitude of up to 5 °C in the long term in Eastern Europe and not so in the west is also noted, also with a strong increase in the Alps that is apparent in our study [46,67]. Even summer increases of up to 9 °C at the Serbia–Bulgaria–Romania border were reported [147], somewhat higher than those given in the present study, using the long-term A2 scenario. As could not be otherwise, studies agree that the greater the radiative forcing experienced, the greater the increase in temperature [130].

Last but not least, it is important to dedicate a brief paragraph to the causes (the drivers) of this amplified warming signal in the Euro-Med. Drivers of temperature increase in the Mediterranean Basin are well-studied [148,149]. The majority of the amplification can be attributed to the thermodynamic plus lapse-rate and to the meridional change, as 3D change and high-frequency circulation change have minor effects [147]. Also, large-scale upper tropospheric warming and weakened lapse-rate change has to happen in order to explain the summertime temperature increases [150].

Despite the existence of models or predictions that may not fully align with reality, the adoption of early policy measures to mitigate or adapt to the escalating global warming is of paramount importance [151]. Maintaining our emissions within an optimistic radiative forcing scenario (RCP2.6) can have significant impacts globally, such as controlling polar ice cap melting [152], and regionally, by considerably mitigating heatwaves, which in 2022 resulted in the deaths of over 60,000 European citizens [153].

5. Conclusions

In this study, we conducted a downscaling (to 0.1°) of GCMs' mean annual temperatures over a particularly vulnerable region, the European Mediterranean (Euro-Med), utilizing Empirical Bayesian Kriging (EBK). Subsequently, we evaluated the performance of each General Circulation Model (GCM) by measuring how much its simulated values deviated from the reanalyzed values by ERA5 Land. The top six GCMs, a number maximizing correlation and explained variance, were selected for a Multi-Model Ensemble (MME) using Random Forest (RFE). Lastly, future projections were made for three emission scenarios, namely optimistic (SSP1-RCP2.6), intermediate (SSP2-RCP4.5), and pessimistic (SSP5-RCP8.5), and for three temporal windows: short-term future (2026–2050), medium (2051–2075), and long term (2076–2100). These projections were compared with those predicted by the IPCC at global and regional levels. Through the application of this method, the following conclusions were drawn:

- Empirical Bayesian Kriging (EBK) stands out as the least data-modifying approach for downscaling GCMs and aligning them with ERA5 Land resolution (0.1°). Its Mean Error (ME) is close to 0 °C, while its Root Mean Square Error (RMSE) is less than 0.5 °C. Moreover, its performance is notable across all models and does not exhibit significant errors in specific areas of the study region, even performing well in mountainous regions. Other Kriging methods, particularly Cokriging, yield discouraging results. Bilinear interpolation (BI), the most commonly used GCM downscaling method to date, tends to underestimate temperatures significantly; thus, we do not recommend its use for these purposes.
- MPI-ESM1-2-HR, GFDL-ESM4, CNRM-CM6-1, MRI-ESM2-0, CNRM-ESM2-1, and IPSL-CM6A-LR are the top six GCMs regarding annual temperature simulation in the Euro-Med region across all five metrics used in the assessment. On the other hand, the MCM-UA-1-0 model and the MIROC models demonstrate poor simulation and should not be used in this region of the globe.

- Significant geospatial differences exist in the performance of GCMs across the Euro-Med region. MPI-ESM1-2-HR performs better in elevated regions such as the Alps, Sierra Nevada, and Carpathian Mountains but fails in the Central Iberian and French areas. GFDL-ESM4 and MRI-ESM2-0 demonstrate more consistent performance across the region. The CNRM models excel in simulating the Cantabrian coast and Greece but fall short in the northwestern half of France. Models from the Hadley Center and FGOALS-f3-L perform well in simulating the Northwestern Iberian and French regions. The MIROC, MCM-UA-1-0, GISS-E2-2-G, and BCC-ESM1 models exhibit mean performance values below 0.25 in any area.
- The optimal number of GCMs for ensemble was determined to be six, maximizing the correlation (0.97) and explained variance (0.93). The Random Forest Ensemble (RFE) method effectively replicates the interannual temperature variability. Additionally, it positively correlates with the ranking value assigned to each GCM (rho = 0.23), although it overestimates the relevance of models with high correlation with the reanalyzed temperature, such as the GFDL-ESM4.
- Significant differences arise between the projections made using the method applied in our research and the IPCC projections at both the global and regional scales. Winter temperatures in the medium-to-long term, particularly in spring, warm up between 1 and more than 2 °C less than what the IPCC expects for any emission scenario. On the other hand, winter temperatures in the short term and summer and autumn temperatures in the short-to-medium term warm up between 1 and more than 2 °C more than expected. The warming is especially found in the higher-elevation regions of the study area and in the Mediterranean islands, where temperatures increase by more than 2 °C compared to the official predictions in all scenarios. In the long term, the projections for summer and autumn in our research show broad agreement with the IPCC confidence interval, exceeding 70% in the pessimistic scenario (SSP5-RCP8.5).

Finally, it is important to highlight that reducing the existing uncertainty in temperature projections at a regional scale is crucial for addressing the adaptive challenges posed by climate change [154], especially in regions as climatically vulnerable as the European Mediterranean region. Adopting appropriate adaptive policies to cope with the consequences of climate change requires an analysis that addresses critical aspects in the modeling chain: downscaling, model selection, and selection of future scenarios [151]. We believe that the method and results of our study can support subsequent research, help establish policy measures that address the challenge of global warming from a holistic perspective, and provide information to territorial planners and environmental conservation organizations to continue their efforts to raise awareness about the importance of mitigation and adaptation to Anthropocene climate change.

Author Contributions: Conceptualization, Á.P. and S.d.R.; methodology, G.-B.F.-L.; software, G.-B.F.-L.; validation, Á.P. and S.d.R.; formal analysis, G.-B.F.-L.; investigation, G.-B.F.-L.; resources, Á.P. and S.d.R.; data curation, G.-B.F.-L.; writing—original draft preparation, G.-B.F.-L.; writing—review and editing, G.-B.F.-L., Á.P. and S.d.R.; visualization, G.-B.F.-L., Á.P. and S.d.R.; supervision, Á.P. and S.d.R.; project administration, S.d.R.; funding acquisition, G.-B.F.-L. All authors have read and agreed to the published version of the manuscript.

Funding: This research was funded by the Spanish Ministry of Universities, General Secretary of Universities (Government of Spain), grant number FPU21/03022. The grant was awarded to the first author (G.-B.F.-L.) and included in a Fellowship Scheme for a Doctoral Training Program. The APC was funded by the University of León.

Data Availability Statement: CMIP6 model outputs for hist, ssp126, ssp245, and ssp585 experiments were accessed from the Lawrence Livermore National Laboratory (LLNL) ESGF node (https://esgf-node.llnl.gov/search/cmip6/, accessed on 16 December 2022). ERA5 Land dataset is available from the Copernicus Climate Data Webstore (https://cds.climate.copernicus.eu/, accessed on 16 December 2022). Processed data and R scripts are available upon reasonable request to the correspondence author.

Acknowledgments: Authors would like to thank Barnaby E. Bouchard and Llibertat Cortés for their advice on English terminology.

Conflicts of Interest: The authors declare the following financial interests/personal relationships, which may be considered as potential competing interests: G.-B.F.-L. reports that financial support was provided by the Spanish Ministry of Universities. Funders had no role in the design of the study; in the collection, analyses, or interpretation of data; in the writing of the manuscript; or in the decision to publish the results.

Appendix A

Table A1. Ranges of annual and seasonal temperature increase expected by the IPCC at the global level (GL) and for the Euro-Med (MED) region under different emission scenarios (SSP1-RCP2.5, SSP2-RCP4.5, and SSP5-RCP8.5) in the short (2026–2050), medium (2051–2075), and long (2016–2100) term. Information obtained from the IPCC WGI Interactive Atlas [103].

		SSP1-RCP2.6			SSP2-RCP4.5			SSP5-RCP8.5		
		Short	Medium	Long	Short	Medium	Long	Short	Medium	Long
Annual	GL	0.8–1.2	1.0–2.6	0.9–2.6	0.8–1.2	1.2–1.7	1.8–2.6	0.9–1.2	1.6–2.2	3.3–4.9
	MED	1.0–1.5	1.3–1.9	1.3–2.0	1.3–1.4	1.6–2.1	2.3–3.1	1.4–1.7	2.1–2.7	4.1–5.8
Winter (DJF)	GL	0.8–1.2	1.0–1.6	1.0–1.7	0.8–1.1	1.3–1.8	1.9–2.7	0.9–1.3	1.6–2.3	3.5–5.1
	MED	0.8–1.3	1.0–1.8	1.1–1.8	0.9–1.3	1.3–1.7	1.9–2.9	0.9–1.4	1.6–2.4	3.4–4.9
Spring (MAM)	GL	0.7–1.1	0.9–1.5	0.9–1.6	0.8–1.0	1.1–1.6	1.7–2.5	0.8–1.1	1.5–2.1	3.1–4.7
	MED	0.8–1.3	1.1–1.8	1.2–1.9	1.0–1.4	1.4–1.9	2.0–2.8	0.9–1.6	1.7–2.5	3.7–5.0
Summer (JJA)	GL	0.8–1.1	0.9–1.5	0.9–1.7	0.8–1.1	1.1–1.7	1.7–2.7	0.9–1.2	1.5–2.2	3.2–4.9
	MED	1.3–1.9	1.6–2.4	1.6–2.4	1.4–1.8	1.9–2.7	3.0–3.9	1.5–2.6	2.7–3.4	5.3–6.9
Autumn (SON)	GL	0.8–1.2	1.0–1.6	1.0–1.7	0.8–1.1	1.3–1.8	1.8–2.7	0.9–1.3	1.6–2.3	3.4–5.0
	MED	1.1–1.6	1.3–2.0	1.3–2.0	1.1–1.6	1.6–2.3	2.3–3.5	1.2–1.8	2.2–3.0	4.3–6.2

Table A2. Percentage of agreement between the temperature anomalies identified in this study and those predicted by the IPCC [103] for the three analyzed emission scenarios (SSP1-RCP2.5, SSP2-RCP4.5, and SSP5-RCP8.5) and temporal periods (short, 2026–2050; medium, 2051–2075; and long, 2076–2100) at both annual and seasonal scales.

	SSP1-RCP2.6			SSP2-RCP4.5			SSP5-RCP8.5		
	Short	Medium	Long	Short	Medium	Long	Short	Medium	Long
Annual	35.00	59.55	60.20	29.77	40.57	52.55	39.76	32.92	66.43
Winter (DJF)	20.44	33.22	33.29	14.33	21.40	20.40	20.02	28.11	5.26
Spring (MAM)	17.93	24.70	22.83	18.34	24.77	26.21	25.41	35.53	44.77
Summer (JJA)	30.56	48.19	51.04	31.24	46.74	64.42	57.49	39.36	71.92
Autumn (SON)	31.50	36.10	39.98	25.54	34.53	60.31	29.40	32.69	71.95

References

1. Weart, S. The Discovery of Rapid Climate Change. *Phys. Today* **2003**, *56*, 30–36. [CrossRef]
2. Jones, P.D.; Wigley, T.M.L.; Folland, C.K.; Parker, D.E.; Angell, J.K.; Lebedeff, S.; Hansen, J.E. Evidence for Global Warming in the Past Decade. *Nature* **1988**, *332*, 790. [CrossRef]
3. IPCC. *Climate Change 2014. Impacts, Adaptation, and Vulnerability. Part B: Regional Aspects*; Barros, V.R., Field, C.B., Eds.; Cambridge University Press: New York, NY, USA, 2014.
4. IPCC. *Climate Change 2021: The Physical Science Basis. Contribution of Working Group I to the Sixth Assessment Report of the Intergovernmental Panel on Climate Change*; Masson-Delmotte, V., Zhai, P., Pirani, A., Connors, S.L., Péan, C., Berger, S., Caud, N., Chen, Y., Goldfarb, L., Gomis, M.I., et al., Eds.; Cambridge University Press: New York, NY, USA, 2021.
5. Koutsoyiannis, D. Revisiting the Global Hydrological Cycle: Is It Intensifying? *Hydrol. Earth Syst. Sci.* **2020**, *24*, 3899–3932. [CrossRef]
6. Otto, F.E.L. Attribution of Extreme Events to Climate Change. *Annu. Rev. Environ. Resour.* **2023**, *48*, 813–828. [CrossRef]

7. Cavicchia, L.; von Storch, H.; Gualdi, S. A Long-Term Climatology of Medicanes. *Clim. Dyn.* **2014**, *43*, 1183–1195. [CrossRef]
8. Carvalho, L.M.V. Assessing Precipitation Trends in the Americas with Historical Data: A Review. *Wiley Interdiscip. Rev. Clim. Change* **2020**, *11*, e627. [CrossRef]
9. Meng, Y.; Hao, Z.; Feng, S.; Zhang, X.; Hao, F. Increase in Compound Dry-Warm and Wet-Warm Events under Global Warming in CMIP6 Models. *Glob. Planet. Change* **2022**, *210*, 103773. [CrossRef]
10. Li, M.; Wu, P.; Sexton, D.M.H.; Ma, Z. Potential Shifts in Climate Zones under a Future Global Warming Scenario Using Soil Moisture Classification. *Clim. Dyn.* **2021**, *56*, 2071–2092. [CrossRef]
11. Wakatsuki, H.; Ju, H.; Nelson, G.C.; Farrell, A.D.; Deryng, D.; Meza, F.; Hasegawa, T. Research Trends and Gaps in Climate Change Impacts and Adaptation Potentials in Major Crops. *Curr. Opin. Environ. Sustain.* **2023**, *60*, 101249. [CrossRef]
12. Wang, S.; Zhu, J.; Huang, G.; Baetz, B.; Cheng, G.; Zeng, X.; Wang, X. Assessment of Climate Change Impacts on Energy Capacity Planning in Ontario, Canada Using High-Resolution Regional Climate Model. *J. Clean. Prod.* **2020**, *274*, 123026. [CrossRef]
13. Loucks, D.P. Meeting Climate Change Challenges: Searching for More Adaptive and Innovative Decisions. *Water Resour. Manag.* **2023**, *37*, 2235–2245. [CrossRef]
14. Si, X.; Wang, L.; Mengersen, K.; Hu, W. Epidemiological Features of Seasonal Influenza Transmission among 11 Climate Zones in Chinese Mainland. *Infect. Dis. Poverty* **2024**, *13*, 4. [CrossRef]
15. Cork, S.; Alexandra, C.; Alvarez-Romero, J.G.; Bennett, E.M.; Berbés-Blázquez, M.; Bohensky, E.; Bok, B.; Costanza, R.; Hashimoto, S.; Hill, R.; et al. Exploring Alternative Futures in the Anthropocene. *Annu. Rev. Environ. Resour.* **2023**, *48*, 25–54. [CrossRef]
16. Paravantis, J.; Santamouris, M.; Cartalis, C.; Efthymiou, C.; Kontoulis, N. Mortality Associated with High Ambient Temperatures, Heatwaves, and the Urban Heat Island in Athens, Greece. *Sustainability* **2017**, *9*, 606. [CrossRef]
17. O'Neill, B.C.; Oppenheimer, M.; Warren, R.; Hallegatte, S.; Kopp, R.E.; Pörtner, H.O.; Scholes, R.; Birkmann, J.; Foden, W.; Licker, R.; et al. IPCC Reasons for Concern Regarding Climate Change Risks. *Nat. Clim. Change* **2017**, *7*, 28–37. [CrossRef]
18. Fried, C.; Huybers, P. A Simple Relationship between the Magnitude and Spatial Extent of Global Surface Temperature Anomalies. *Geophys. Res. Lett.* **2024**, *51*, e2023GL106537. [CrossRef]
19. Sun, C.; Zhu, L.; Liu, Y.; Wei, T.; Guo, Z. CMIP6 Model Simulation of Concurrent Continental Warming Holes in Eurasia and North America since 1990 and Their Relation to the Indo-Pacific SST Warming. *Glob. Planet. Change* **2022**, *213*, 103824. [CrossRef]
20. Dai, T.; Dong, W.; Sun, S.; Wang, G. Exploring the Asymmetry and Rate of SAT Warming over the Global Land Area under the 1.5 °C and 2 °C Climate Change Targets. *Meteorol. Atmos. Phys.* **2023**, *135*, 19. [CrossRef]
21. Rivas-Martínez, S.; Rivas-Sáenz, S.; Penas, Á. Worldwide Bioclimatic Classification System. *Glob. Geobot.* **2011**, *1*, 1–634. [CrossRef]
22. Tramblay, Y.; Koutroulis, A.; Samaniego, L.; Vicente-Serrano, S.M.; Volaire, F.; Boone, A.; Le Page, M.; Llasat, M.C.; Albergel, C.; Burak, S.; et al. Challenges for Drought Assessment in the Mediterranean Region under Future Climate Scenarios. *Earth Sci. Rev.* **2020**, *210*, 103348. [CrossRef]
23. Polade, S.D.; Gershunov, A.; Cayan, D.R.; Dettinger, M.D.; Pierce, D.W. Precipitation in a Warming World: Assessing Projected Hydro-Climate Changes in California and Other Mediterranean Climate Regions. *Sci. Rep.* **2017**, *7*, 10783. [CrossRef] [PubMed]
24. Giorgi, F.; Raffaele, F. On the Dependency of GCM-Based Regional Surface Climate Change Projections on Model Biases, Resolution and Climate Sensitivity. *Clim. Dyn.* **2022**, *58*, 2843–2862. [CrossRef]
25. Diffenbaugh, N.S.; Swain, D.L.; Touma, D.; Lubchenco, J. Anthropogenic Warming Has Increased Drought Risk in California. *Proc. Natl. Acad. Sci. USA* **2015**, *112*, 3931–3936. [CrossRef] [PubMed]
26. González-Pérez, A.; Álvarez-Esteban, R.; Penas, Á.; Del Río, S. Analysis of Recent Mean Temperature Trends and Relationships with Teleconnection Patterns in California (U.S.). *Appl. Sci.* **2022**, *12*, 5831. [CrossRef]
27. Adeola, A.M.; Kruger, A.; Makgoale, T.E.; Botai, J.O. Observed Trends and Projections of Temperature and Precipitation in the Olifants River Catchment in South Africa. *PLoS ONE* **2022**, *17*, e0271974. [CrossRef]
28. Ndlovu, M.; Clulow, A.D.; Savage, M.J.; Nhamo, L.; Magidi, J.; Mabhaudhi, T. An Assessment of the Impacts of Climate Variability and Change in Kwazulu-Natal Province, South Africa. *Atmosphere* **2021**, *12*, 427. [CrossRef]
29. Trancoso, R.; Syktus, J.; Toombs, N.; Ahrens, D.; Wong, K.K.H.; Pozza, R.D. Heatwaves Intensification in Australia: A Consistent Trajectory across Past, Present and Future. *Sci. Total Environ.* **2020**, *742*, 140521. [CrossRef] [PubMed]
30. MedECC. *Climate and Environmental Change in the Mediterranean Basin—Current Situation and Risks for the Future. First Mediterranean Assessment Report*; Cramer, W., Guiot, J., Marini, K., Eds.; Mediterranean Experts on Climate and Environmental Change (MedECC): Marseille, France, 2020.
31. Fan, X.; Duan, Q.; Shen, C.; Wu, Y.; Xing, C. Global Surface Air Temperatures in CMIP6: Historical Performance and Future Changes. *Environ. Res. Lett.* **2020**, *15*, 104056. [CrossRef]
32. Almazroui, M.; Ashfaq, M.; Islam, M.N.; Rashid, I.U.; Kamil, S.; Abid, M.A.; O'Brien, E.; Ismail, M.; Reboita, M.S.; Sörensson, A.A.; et al. Assessment of CMIP6 Performance and Projected Temperature and Precipitation Changes Over South America. *ESEV* **2021**, *5*, 155–183. [CrossRef]
33. Zhao, Y.; Qian, C.; Zhang, W.; He, D.; Qi, Y. Extreme Temperature Indices in Eurasia in a CMIP6 Multi-Model Ensemble: Evaluation and Projection. *Int. J. Climatol.* **2021**, *41*, 5368–5385. [CrossRef]
34. Shi, J.; Tian, Z.; Lang, X.; Jiang, D. Projected Changes in the Interannual Variability of Surface Air Temperature Using CMIP6 Simulations. *Clim. Dyn.* **2023**, *62*, 431–446. [CrossRef]
35. Yang, X.; Zhou, B.; Xu, Y.; Han, Z. CMIP6 Evaluation and Projection of Temperature and Precipitation over China. *Adv. Atmos. Sci.* **2021**, *38*, 817–830. [CrossRef]

36. Overpeck, J.T.; Meehl, G.A.; Bony, S.; Easterling, D.R. Climate Data Challenges in the 21st Century. *Science* **2011**, *331*, 700–702. [CrossRef]
37. Hausfather, Z.; Drake, H.F.; Abbott, T.; Schmidt, G.A. Evaluating the Performance of Past Climate Model Projections. *Geophys. Res. Lett.* **2020**, *47*, e2019GL085378. [CrossRef]
38. Balaji, V.; Couvreux, F.; Deshayes, J.; Gautrais, J.; Hourdin, F.; Rio, C. Are General Circulation Models Obsolete? *Proc. Natl. Acad. Sci. USA* **2022**, *119*, e2202075119. [CrossRef] [PubMed]
39. Eyring, V.; Bony, S.; Meehl, G.A.; Senior, C.A.; Stevens, B.; Stouffer, R.J.; Taylor, K.E. Overview of the Coupled Model Intercomparison Project Phase 6 (CMIP6) Experimental Design and Organization. *Geosci. Model. Dev.* **2016**, *9*, 1937–1958. [CrossRef]
40. Brands, S. A Circulation-Based Performance Atlas of the CMIP5 and 6 Models for Regional Climate Studies in the Northern Hemisphere Mid-to-High Latitudes. *Geosci. Model. Dev.* **2022**, *15*, 1375–1411. [CrossRef]
41. Fernandez-Granja, J.A.; Casanueva, A.; Bedia, J.; Fernandez, J. Improved Atmospheric Circulation over Europe by the New Generation of CMIP6 Earth System Models. *Clim. Dyn.* **2021**, *56*, 3527–3540. [CrossRef]
42. Xia, Y.; Hu, Y.; Liu, J. Comparison of Trends in the Hadley Circulation between CMIP6 and CMIP5. *Sci. Bull.* **2020**, *65*, 1667–1674. [CrossRef] [PubMed]
43. Fan, X.; Miao, C.; Duan, Q.; Shen, C.; Wu, Y. The Performance of CMIP6 versus CMIP5 in Simulating Temperature Extremes Over the Global Land Surface. *J. Geophys. Res.-Atmos.* **2020**, *125*, e2020JD033031. [CrossRef]
44. Guo, H.; Zhan, C.; Ning, L.; Li, Z.; Hu, S. Evaluation and Comparison of CMIP6 and CMIP5 Model Performance in Simulating the Runoff. *Theor. Appl. Climatol.* **2022**, *149*, 1451–1470. [CrossRef]
45. Li, J.; Huo, R.; Chen, H.; Zhao, Y.; Zhao, T. Comparative Assessment and Future Prediction Using CMIP6 and CMIP5 for Annual Precipitation and Extreme Precipitation Simulation. *Front. Earth Sci.* **2021**, *9*, 687976. [CrossRef]
46. Cos, J.; Doblas-Reyes, F.; Jury, M.; Marcos, R.; Bretonnière, P.A.; Samsó, M. The Mediterranean Climate Change Hotspot in the CMIP5 and CMIP6 Projections. *Earth Syst. Dyn.* **2022**, *13*, 321–340. [CrossRef]
47. Zhou, C.; Penner, J.E. Why Do General Circulation Models Overestimate the Aerosol Cloud Lifetime Effect? A Case Study Comparing CAM5 and a CRM. *Atmos. Chem. Phys.* **2017**, *17*, 21–29. [CrossRef]
48. Boberg, F.; Christensen, J.H. Overestimation of Mediterranean Summer Temperature Projections Due to Model Deficiencies. *Nat. Clim. Change* **2012**, *2*, 433–436. [CrossRef]
49. Das, R.S.; Rao, S.A.; Pillai, P.A.; Srivastava, A.; Pradhan, M.; Ramu, D.A. Why Coupled General Circulation Models Overestimate the ENSO and Indian Summer Monsoon Rainfall (ISMR) Relationship? *Clim. Dyn.* **2022**, *59*, 2995–3011. [CrossRef]
50. Loehle, C. The Epistemological Status of General Circulation Models. *Clim. Dyn.* **2018**, *50*, 1719–1731. [CrossRef]
51. Maraun, D.; Truhetz, H.; Schaffer, A. Regional Climate Model Biases, Their Dependence on Synoptic Circulation Biases and the Potential for Bias Adjustment: A Process-Oriented Evaluation of the Austrian Regional Climate Projections. *J. Geophys. Res.-Atmos.* **2021**, *126*, e2020JD032824. [CrossRef]
52. Ali, Z.; Hamed, M.M.; Muhammad, M.K.I.; Iqbal, Z.; Shahid, S. Performance Evaluation of CMIP6 GCMs for the Projections of Precipitation Extremes in Pakistan. *Clim. Dyn.* **2023**, *61*, 4717–4732. [CrossRef]
53. Feyissa, T.A.; Demissie, T.A.; Saathoff, F.; Gebissa, A. Evaluation of General Circulation Models CMIP6 Performance and Future Climate Change over the Omo River Basin, Ethiopia. *Sustainability* **2023**, *15*, 6507. [CrossRef]
54. Nguyen-Duy, T.; Ngo-Duc, T.; Desmet, Q. Performance Evaluation and Ranking of CMIP6 Global Climate Models over Vietnam. *J. Water Clim. Change* **2023**, *14*, 1831–1846. [CrossRef]
55. Peng, S.; Wang, C.; Li, Z.; Mihara, K.; Kuramochi, K.; Toma, Y.; Hatano, R. Climate Change Multi-Model Projections in CMIP6 Scenarios in Central Hokkaido, Japan. *Sci. Rep.* **2023**, *13*, 230. [CrossRef] [PubMed]
56. Sa'adi, Z.; Alias, N.E.; Yusop, Z.; Iqbal, Z.; Houmsi, M.R.; Houmsi, L.N.; Ramli, M.W.A.; Muhammad, M.K.I. Application of Relative Importance Metrics for CMIP6 Models Selection in Projecting Basin-Scale Rainfall over Johor River Basin, Malaysia. *Sci. Total Environ.* **2024**, *912*, 169187. [CrossRef] [PubMed]
57. Raju, K.S.; Kumar, D.N. Review of Approaches for Selection and Ensembling of GCMs. *J. Water Clim. Change* **2020**, *11*, 577–599. [CrossRef]
58. Abadie, L.M.; Moral, M.P. A Method for Selecting a Climate Model: An Application for Maximum Daily Temperature in Southern Spain. *Theor. Appl. Climatol.* **2023**, *152*, 773–786. [CrossRef]
59. Peres, D.J.; Senatore, A.; Nanni, P.; Cancelliere, A.; Mendicino, G.; Bonaccorso, B. Evaluation of EURO-CORDEX (Coordinated Regional Climate Downscaling Experiment for the Euro-Mediterranean Area) Historical Simulations by High-Quality Observational Datasets in Southern Italy: Insights on Drought Assessment. *Nat. Hazard Earth Syst.* **2020**, *20*, 3057–3082. [CrossRef]
60. Bağçaci, S.Ç.; Yucel, I.; Duzenli, E.; Yilmaz, M.T. Intercomparison of the Expected Change in the Temperature and the Precipitation Retrieved from CMIP6 and CMIP5 Climate Projections: A Mediterranean Hot Spot Case, Turkey. *Atmos. Res.* **2021**, *256*, 105576. [CrossRef]
61. Seker, M.; Gumus, V. Projection of Temperature and Precipitation in the Mediterranean Region through Multi-Model Ensemble from CMIP6. *Atmos. Res.* **2022**, *280*, 106440. [CrossRef]
62. Jury, M.W.; Prein, A.F.; Truhetz, H.; Gobiet, A. Evaluation of CMIP5 Models in the Context of Dynamical Downscaling over Europe. *J. Clim.* **2015**, *28*, 5575–5582. [CrossRef]
63. Palmer, T.E.; Booth, B.B.B.; McSweeney, C.F. How Does the CMIP6 Ensemble Change the Picture for European Climate Projections? *Environ. Res. Lett.* **2021**, *16*, 094042. [CrossRef]

64. Tuel, A.; Kang, S.; Eltahir, E.A.B. Understanding Climate Change over the Southwestern Mediterranean Using High-Resolution Simulations. *Clim. Dyn.* **2021**, *56*, 985–1001. [CrossRef]
65. Drobinski, P.; Da Silva, N.; Bastin, S.; Mailler, S.; Muller, C.; Ahrens, B.; Christensen, O.B.; Lionello, P. How Warmer and Drier Will the Mediterranean Region Be at the End of the Twenty-First Century? *Reg. Environ. Change* **2020**, *20*, 78. [CrossRef]
66. Todaro, V.; D'Oria, M.; Secci, D.; Zanini, A.; Tanda, M.G. Climate Change over the Mediterranean Region: Local Temperature and Precipitation Variations at Five Pilot Sites. *Water* **2022**, *14*, 2499. [CrossRef]
67. Zittis, G.; Hadjinicolaou, P.; Klangidou, M.; Proestos, Y.; Lelieveld, J.A. Multi-Model, Multi-Scenario, and Multi-Domain Analysis of Regional Climate Projections for the Mediterranean. *Reg. Environ. Change* **2019**, *19*, 2621–2635. [CrossRef]
68. Zhang, L.; Jiapaer, G.; Yu, T.; Umuhoza, J.; Tu, H.; Chen, B.; Liang, H.; Lin, K.; Ju, T.; De Maeyer, P.; et al. Evaluating and Correcting Temperature and Precipitation Grid Products in the Arid Region of Altay, China. *Remote Sens.* **2024**, *16*, 283. [CrossRef]
69. Sun, P.; Zou, Y.; Yao, R.; Ma, Z.; Bian, Y.; Ge, C.; Lv, Y. Compound and Successive Events of Extreme Precipitation and Extreme Runoff under Heatwaves Based on CMIP6 Models. *Sci. Total Environ.* **2023**, *878*, 162980. [CrossRef] [PubMed]
70. Christian, J.R.; Denman, K.L.; Hayashida, H.; Holdsworth, A.M.; Lee, W.G.; Riche, O.G.J.; Shao, A.E.; Steiner, N.; Swart, N.C. Ocean Biogeochemistry in the Canadian Earth System Model Version 5.0.3: CanESM5 and CanESM5-CanOE. *Geosci. Model. Dev.* **2022**, *15*, 4393–4424. [CrossRef]
71. Salzmann, M.; Ferrachat, S.; Tully, C.; Münch, S.; Watson-Parris, D.; Neubauer, D.; Siegenthaler-Le Drian, C.; Rast, S.; Heinold, B.; Crueger, T.; et al. The Global Atmosphere-Aerosol Model ICON-A-HAM2.3–Initial Model Evaluation and Effects of Radiation Balance Tuning on Aerosol Optical Thickness. *J. Adv. Model. Earth Syst.* **2022**, *14*, e2021MS002699. [CrossRef] [PubMed]
72. Schmid, S.M.; Fügenschuh, B.; Kissling, E.; Schuster, R. Tectonic Map and Overall Architecture of the Alpine Orogen. *Eclogae Geol. Helv.* **2004**, *97*, 93–117. [CrossRef]
73. Holdridge, L.R. *Life Zone Ecology*; Tropical Science Center: San Pedro de Montes de Oca, Costa Rica, 1967.
74. WMO. Updated 30-Year Reference Period Reflects Changing Climate. Available online: https://public.wmo.int/en/media/news/updated-30-year-reference-period-reflects-changing-climate (accessed on 5 May 2023).
75. Muñoz-Sabater, J.; Dutra, E.; Agustí-Panareda, A.; Albergel, C.; Arduini, G.; Balsamo, G.; Boussetta, S.; Choulga, M.; Harrigan, S.; Hersbach, H.; et al. ERA5-Land: A State-of-the-Art Global Reanalysis Dataset for Land Applications. *Earth Syst. Sci. Data* **2021**, *13*, 4349–4383. [CrossRef]
76. Zou, J.; Lu, N.; Jiang, H.; Qin, J.; Yao, L.; Xin, Y.; Su, F. Performance of Air Temperature from ERA5-Land Reanalysis in Coastal Urban Agglomeration of Southeast China. *Sci. Total Environ.* **2022**, *828*, 154459. [CrossRef]
77. De Araújo, C.S.P.; Silva, I.A.C.E.; Ippolito, M.; de Almeida, C.D.G.C. Evaluation of Air Temperature Estimated by ERA5-Land Reanalysis Using Surface Data in Pernambuco, Brazil. *Environ. Monit. Assess.* **2022**, *194*, 381. [CrossRef]
78. Zhao, P.; He, Z. A First Evaluation of ERA5-Land Reanalysis Temperature Product over the Chinese Qilian Mountains. *Front. Earth Sci.* **2022**, *10*, 907730. [CrossRef]
79. Barbosa, S.; Scotto, M.G. Extreme Heat Events in the Iberia Peninsula from Extreme Value Mixture Modeling of ERA5-Land Air Temperature. *Weather Clim. Extrem.* **2022**, *36*, 100448. [CrossRef]
80. Pelosi, A.; Terribile, F.; D'Urso, G.; Chirico, G.B. Comparison of ERA5-Land and UERRA MESCAN-SURFEX Reanalysis Data with Spatially Interpolated Weather Observations for the Regional Assessment of Reference Evapotranspiration. *Water* **2020**, *12*, 1669. [CrossRef]
81. Liu, J.; Fiifi, D.; Hagan, T.; Liu, Y.; Liu, J.; Hagan, D.F.T.; Liu, Y. Global Land Surface Temperature Change (2003–2017) and Its Relationship with Climate Drivers: AIRS, MODIS, and ERA5-Land Based Analysis. *Remote Sens.* **2021**, *13*, 44. [CrossRef]
82. Yilmaz, M. Consistency of Spatiotemporal Variability of MODIS and ERA5-Land Surface Warming Trends over Complex Topography. *Environ. Sci. Pollut. R* **2023**, *30*, 94414–94435. [CrossRef]
83. Tan, M.L.; Armanuos, A.M.; Ahmadianfar, I.; Demir, V.; Heddam, S.; Al-Areeq, A.M.; Abba, S.I.; Halder, B.; Cagan Kilinc, H.; Yaseen, Z.M. Evaluation of NASA POWER and ERA5-Land for Estimating Tropical Precipitation and Temperature Extremes. *J. Hydrol.* **2023**, *624*, 129940. [CrossRef]
84. Scafetta, N. Testing the CMIP6 GCM Simulations versus Surface Temperature Records from 1980–1990 to 2011–2021: High ECS Is Not Supported. *Climate* **2021**, *9*, 161. [CrossRef]
85. Collazo, S.; Barrucand, M.; Rusticucci, M. Evaluation of CMIP6 Models in the Representation of Observed Extreme Temperature Indices Trends in South America. *Clim. Change* **2022**, *172*, 21. [CrossRef]
86. Ali, Z.; Hamed, M.M.; Muhammad, M.K.I.; Shahid, S. A Novel Approach for Evaluation of CMIP6 GCMs in Simulating Temperature and Precipitation Extremes of Pakistan. *Int. J. Climatol.* **2024**, *44*, 592–612. [CrossRef]
87. Blanke, J.H.; Lindeskog, M.; Lindström, J.; Lehsten, V. Effect of Climate Data on Simulated Carbon and Nitrogen Balances for Europe. *J. Geophys. Res. Biogeosci.* **2016**, *121*, 1352–1371. [CrossRef]
88. Graux, A.I.; Bellocchi, G.; Lardy, R.; Soussana, J.F. Ensemble Modelling of Climate Change Risks and Opportunities for Managed Grasslands in France. *Agric. For. Meteorol.* **2013**, *170*, 114–131. [CrossRef]
89. Čejka, T.; Trnka, M.; Krusic, P.J.; Stobbe, U.; Oliach, D.; Václavík, T.; Tegel, W.; Büntgen, U. Predicted Climate Change Will Increase the Truffle Cultivation Potential in Central Europe. *Sci. Rep.* **2020**, *10*, 21281. [CrossRef]
90. Chilès, J.-P.; Desassis, N. Fifty Years of Kriging. In *Handbook of Mathematical Geosciences*; Daya Sagar, B.S., Cheng, Q., Agterberg, F., Eds.; Springer: Cham, Switzerland, 2018; pp. 589–612.
91. Mühlenstädt, T.; Kuhnt, S. Kernel Interpolation. *Comput. Stat. Data Anal.* **2011**, *55*, 2962–2974. [CrossRef]

92. Antal, A.; Guerreiro, P.M.P.; Cheval, S. Comparison of Spatial Interpolation Methods for Estimating the Precipitation Distribution in Portugal. *Theor. Appl. Climatol.* **2021**, *145*, 1193–1206. [CrossRef]
93. Katipoğlu, O.M. Spatial Analysis of Seasonal Precipitation Using Various Interpolation Methods in the Euphrates Basin, Turkey. *Acta Geophys.* **2022**, *70*, 859–878. [CrossRef]
94. ESRI ArcGIS 2019. Available online: https://www.esri.com/en-us/arcgis/products/arcgis-desktop (accessed on 28 January 2023).
95. Di Virgilio, G.; Ji, F.; Tam, E.; Nishant, N.; Evans, J.P.; Thomas, C.; Riley, M.L.; Beyer, K.; Grose, M.R.; Narsey, S.; et al. Selecting CMIP6 GCMs for CORDEX Dynamical Downscaling: Model Performance, Independence, and Climate Change Signals. *Earth's Future* **2022**, *10*, e2021EF002625. [CrossRef]
96. Lei, X.; Xu, C.; Liu, F.; Song, L.; Cao, L.; Suo, N. Evaluation of CMIP6 Models and Multi-Model Ensemble for Extreme Precipitation over Arid Central Asia. *Remote Sens.* **2023**, *15*, 2376. [CrossRef]
97. Jiang, Z.; Li, W.; Xu, J.; Li, L. Extreme Precipitation Indices over China in CMIP5 Models. Part I: Model Evaluation. *J. Clim.* **2015**, *28*, 8603–8619. [CrossRef]
98. Wang, B.; Zheng, L.; Liu, D.L.; Ji, F.; Clark, A.; Yu, Q. Using Multi-Model Ensembles of CMIP5 Global Climate Models to Reproduce Observed Monthly Rainfall and Temperature with Machine Learning Methods in Australia. *Int. J. Climatol.* **2018**, *38*, 4891–4902. [CrossRef]
99. Hernanz, A.; García-Valero, J.A.; Domínguez, M.; Rodríguez-Camino, E. Evaluation of Statistical Downscaling Methods for Climate Change Projections over Spain: Present Conditions with Imperfect Predictors (Global Climate Model Experiment). *Int. J. Climatol.* **2022**, *42*, 6793–6806. [CrossRef]
100. Ahmed, K.; Sachindra, D.A.; Shahid, S.; Iqbal, Z.; Nawaz, N.; Khan, N. Multi-Model Ensemble Predictions of Precipitation and Temperature Using Machine Learning Algorithms. *Atmos. Res.* **2020**, *236*, 104806. [CrossRef]
101. Breiman, L. Random Forests. *Mach. Learn.* **2001**, *45*, 5–32. [CrossRef]
102. O'Neill, B.C.; Tebaldi, C.; Van Vuuren, D.P.; Eyring, V.; Friedlingstein, P.; Hurtt, G.; Knutti, R.; Kriegler, E.; Lamarque, J.F.; Lowe, J.; et al. The Scenario Model Intercomparison Project (ScenarioMIP) for CMIP6. *Geosci. Model. Dev.* **2016**, *9*, 3461–3482. [CrossRef]
103. Gutiérrez, J.M.; Jones, R.G.; Narisma, G.T.; Alves, L.M.; Amjad, M.; Gorodetskaya, I.V.; Grose, M.; Klutse, N.A.B.; Krakovska, S.; Li, J.; et al. 2021: Atlas. In *Climate Change 2021: The Physical Science Basis. Contribution of Working Group I to the Sixth Assessment Report of the Intergovernmental Panel on Climate Change*; Masson-Delmotte, V.P., Zhai, A., Pirani, S.L., Connors, C., Péan, S., Berger, N., Caud, Y., Chen, L., Goldfarb, M.I., Gomis, M., et al., Eds.; Cambridge University Press: New York, NY, USA, 2021; pp. 1927–1929.
104. Liu, A.; Qu, C.; Zhang, J.; Sun, W.; Shi, C.; Lima, A.; De Vivo, B.; Huang, H.; Palmisano, M.; Guarino, A.; et al. Screening and Optimization of Interpolation Methods for Mapping Soil-Borne Polychlorinated Biphenyls. *Sci. Total Environ.* **2024**, *913*, 169498. [CrossRef] [PubMed]
105. Krivoruchko, K.; Gribov, A. Distance Metrics for Data Interpolation over Large Areas on Earth's Surface. *Spat. Stat.* **2020**, *35*, 100396. [CrossRef]
106. Biernacik, P.; Kazimierski, W.; Włodarczyk-Sielicka, M. Comparative Analysis of Selected Geostatistical Methods for Bottom Surface Modeling. *Sensors* **2023**, *23*, 3941. [CrossRef] [PubMed]
107. Gribov, A.; Krivoruchko, K. Empirical Bayesian Kriging Implementation and Usage. *Sci. Total Environ.* **2020**, *722*, 137290. [CrossRef]
108. Liu, Z.N.; Yu, X.Y.; Jia, L.F.; Wang, Y.S.; Song, Y.C.; Meng, H.D. The Influence of Distance Weight on the Inverse Distance Weighted Method for Ore-Grade Estimation. *Sci. Rep.* **2021**, *11*, 2689. [CrossRef]
109. Pellicone, G.; Caloiero, T.; Guagliardi, I. The De Martonne Aridity Index in Calabria (Southern Italy). *J. Maps* **2019**, *15*, 788–796. [CrossRef]
110. Iqbal, Z.; Shahid, S.; Ahmed, K.; Ismail, T.; Khan, N.; Virk, Z.T.; Johar, W. Evaluation of Global Climate Models for Precipitation Projection in Sub-Himalaya Region of Pakistan. *Atmos. Res.* **2020**, *245*, 105061. [CrossRef]
111. Ahmed, K.; Sachindra, D.A.; Shahid, S.; Demirel, M.C.; Chung, E.S. Selection of Multi-Model Ensemble of General Circulation Models for the Simulation of Precipitation and Maximum and Minimum Temperature Based on Spatial Assessment Metrics. *Hydrol. Earth Syst. Sci.* **2019**, *23*, 4803–4824. [CrossRef]
112. Dubrovsky, M.; Trnka, M.; Holman, I.P.; Svobodova, E.; Harrison, P.A. Developing a Reduced-Form Ensemble of Climate Change Scenarios for Europe and Its Application to Selected Impact Indicators. *Clim. Change* **2015**, *128*, 169–186. [CrossRef]
113. Senent-Aparicio, J.; Pérez-Sánchez, J.; Carrillo-García, J.; Soto, J.; Abbaspour, K.; Srinivasan, R.; Ashraf Vaghefi, S.; Faramarzi, M.; Chen, L. Using SWAT and Fuzzy TOPSIS to Assess the Impact of Climate Change in the Headwaters of the Segura River Basin (SE Spain). *Water* **2017**, *9*, 149. [CrossRef]
114. Pastor, M.A.; Casado, M.J. Use of Circulation Types Classifications to Evaluate AR4 Climate Models over the Euro-Atlantic Region. *Clim. Dyn.* **2012**, *39*, 2059–2077. [CrossRef]
115. Topál, D.; Hatvani, I.G.; Kern, Z. Refining Projected Multidecadal Hydroclimate Uncertainty in East-Central Europe Using CMIP5 and Single-Model Large Ensemble Simulations. *Theor. Appl. Climatol.* **2020**, *142*, 1147–1167. [CrossRef]
116. Adeyeri, O.E.; Zhou, W.; Wang, X.; Zhang, R.; Laux, P.; Ishola, K.A.; Usman, M. The Trend and Spatial Spread of Multisectoral Climate Extremes in CMIP6 Models. *Sci. Rep.* **2022**, *12*, 21000. [CrossRef] [PubMed]
117. Stryhal, J.; Huth, R. Classifications of Winter Atmospheric Circulation Patterns: Validation of CMIP5 GCMs over Europe and the North Atlantic. *Clim. Dyn.* **2019**, *52*, 3575–3598. [CrossRef]

118. McSweeney, C.F.; Jones, R.G.; Lee, R.W.; Rowell, D.P. Selecting CMIP5 GCMs for Downscaling over Multiple Regions. *Clim. Dyn.* **2015**, *44*, 3237–3260. [CrossRef]
119. McMahon, T.A.; Peel, M.C.; Karoly, D.J. Assessment of Precipitation and Temperature Data from CMIP3 Global Climate Models for Hydrologic Simulation. *Hydrol. Earth Syst. Sci.* **2015**, *19*, 361–377. [CrossRef]
120. Hauglustaine, D.A.; Balkanski, Y.; Schulz, M. A Global Model Simulation of Present and Future Nitrate Aerosols and Their Direct Radiative Forcing of Climate. *Atmos. Chem. Phys.* **2014**, *14*, 11031–11063. [CrossRef]
121. Crawford, J.; Venkataraman, K.; Booth, J. Developing Climate Model Ensembles: A Comparative Case Study. *J. Hydrol.* **2019**, *568*, 160–173. [CrossRef]
122. Dey, A.; Sahoo, D.P.; Kumar, R.; Remesan, R. A Multimodel Ensemble Machine Learning Approach for CMIP6 Climate Model Projections in an Indian River Basin. *Int. J. Climatol.* **2022**, *42*, 9215–9236. [CrossRef]
123. Hussain, M.; Yusof, K.W.; Mustafa, M.R.U.; Mahmood, R.; Jia, S. Evaluation of CMIP5 Models for Projection of Future Precipitation Change in Bornean Tropical Rainforests. *Theor. Appl. Climatol.* **2018**, *134*, 423–440. [CrossRef]
124. Pielke, R., Jr.; Ritchie, J. How Climate Scenarios Lost Touch With Reality. *Issues Sci. Technol.* **2021**, *37*, 74–83.
125. Carvalho, D.; Rafael, S.; Monteiro, A.; Rodrigues, V.; Lopes, M.; Rocha, A. How Well Have CMIP3, CMIP5 and CMIP6 Future Climate Projections Portrayed the Recently Observed Warming. *Sci. Rep.* **2022**, *12*, 11983. [CrossRef] [PubMed]
126. Burgess, M.G.; Ritchie, J.; Shapland, J.; Pielke, R. IPCC Baseline Scenarios Have Over-Projected CO_2 Emissions and Economic Growth. *Environ. Res. Lett.* **2020**, *16*, 014016. [CrossRef]
127. Siegert, M.; Alley, R.B.; Rignot, E.; Englander, J.; Corell, R. Twenty-First Century Sea-Level Rise Could Exceed IPCC Projections for Strong-Warming Futures. *One Earth* **2020**, *3*, 691–703. [CrossRef]
128. Freychet, N.; Hegerl, G.; Mitchell, D.; Collins, M. Future Changes in the Frequency of Temperature Extremes May Be Underestimated in Tropical and Subtropical Regions. *Commun. Earth Environ.* **2021**, *2*, 28. [CrossRef]
129. Hu, D.; Jiang, D.; Tian, Z.; Lang, X. How Skillful Was the Projected Temperature over China during 2002–2018? *Sci. Bull.* **2022**, *67*, 1077–1085. [CrossRef] [PubMed]
130. Carvalho, D.; Cardoso Pereira, S.; Rocha, A. Future Surface Temperatures over Europe According to CMIP6 Climate Projections: An Analysis with Original and Bias-Corrected Data. *Clim. Change* **2021**, *167*, 10. [CrossRef]
131. Ribes, A.; Qasmi, S.; Gillett, N.P. Making Climate Projections Conditional on Historical Observations. *Sci. Adv.* **2021**, *7*, 671–693. [CrossRef] [PubMed]
132. Zhang, J.; Wu, T.; Li, L.; Furtado, K.; Xin, X.; Xie, C.; Zheng, M.; Zhao, H.; Zhou, Y. Constraint on Regional Land Surface Air Temperature Projections in CMIP6 Multi-Model Ensemble. *npj Clim. Atmos. Sci.* **2023**, *6*, 85. [CrossRef]
133. Gönençgil, B.; Acar, Z. Turkey: Clımate Variability, Extreme Temperature, and Precipitation. In *Geographies of Mediterranean Europe*; Lois-González, R.C., Ed.; Springer: Cham, Switzerland, 2021; pp. 167–180.
134. Liuzzo, L.; Bono, E.; Sammartano, V.; Freni, G. Long-Term Temperature Changes in Sicily, Southern Italy. *Atmos. Res.* **2017**, *198*, 44–55. [CrossRef]
135. Ribes, A.; Corre, L.; Gibelin, A.L.; Dubuisson, B. Issues in Estimating Observed Change at the Local Scale—A Case Study: The Recent Warming over France. *Int. J. Climatol.* **2016**, *36*, 3794–3806. [CrossRef]
136. Sandonis, L.; González-Hidalgo, J.C.; Peña-Angulo, D.; Beguería, S. Mean Temperature Evolution on the Spanish Mainland 1916–2015. *Clim. Res.* **2021**, *82*, 177–189. [CrossRef]
137. Spinoni, J.; Szalai, S.; Szentimrey, T.; Lakatos, M.; Bihari, Z.; Nagy, A.; Németh, Á.; Kovács, T.; Mihic, D.; Dacic, M.; et al. Climate of the Carpathian Region in the Period 1961–2010: Climatologies and Trends of 10 Variables. *Int. J. Climatol.* **2015**, *35*, 1322–1341. [CrossRef]
138. Toreti, A.; Desiato, F. Temperature Trend over Italy from 1961 to 2004. *Theor. Appl. Climatol.* **2008**, *91*, 51–58. [CrossRef]
139. Tzanis, C.G.; Koutsogiannis, I.; Philippopoulos, K.; Deligiorgi, D. Recent Climate Trends over Greece. *Atmos. Res.* **2019**, *230*, 104623. [CrossRef]
140. Castellanos, M.; García, M.Á.; Pérez, I.A.; Sánchez, M.L.; Pardo, N.; Fernández-Duque, B. Measuring Temperature Trends in the Mediterranean Basin. *J. Atmos. Sol. Terr. Phys.* **2021**, *222*, 105713. [CrossRef]
141. Fernández, J.; Frías, M.D.; Cabos, W.D.; Cofiño, A.S.; Domínguez, M.; Fita, L.; Gaertner, M.A.; García-Díez, M.; Gutiérrez, J.M.; Jiménez-Guerrero, P.; et al. Consistency of Climate Change Projections from Multiple Global and Regional Model Intercomparison Projects. *Clim. Dyn.* **2019**, *52*, 1139–1156. [CrossRef]
142. Giorgi, F.; Lionello, P. Climate Change Projections for the Mediterranean Region. *Glob. Planet. Change* **2008**, *63*, 90–104. [CrossRef]
143. Lionello, P.; Scarascia, L. The Relation between Climate Change in the Mediterranean Region and Global Warming. *Reg. Environ. Change* **2018**, *18*, 1481–1493. [CrossRef]
144. Ozturk, T.; Ceber, Z.P.; Türkeş, M.; Kurnaz, M.L. Projections of Climate Change in the Mediterranean Basin by Using Downscaled Global Climate Model Outputs. *Int. J. Climatol.* **2015**, *35*, 4276–4292. [CrossRef]
145. Bador, M.; Terray, L.; Boé, J.; Somot, S.; Alias, A.; Gibelin, A.L.; Dubuisson, B. Future Summer Mega-Heatwave and Record-Breaking Temperatures in a Warmer France Climate. *Environ. Res. Lett.* **2017**, *12*, 074025. [CrossRef]
146. Gobiet, A.; Kotlarski, S.; Beniston, M.; Heinrich, G.; Rajczak, J.; Stoffel, M. 21st Century Climate Change in the European Alps—A Review. *Sci. Total Environ.* **2014**, *493*, 1138–1151. [CrossRef] [PubMed]
147. Zittis, G.; Hadjinicolaou, P.; Fnais, M.; Lelieveld, J. Projected Changes in Heat Wave Characteristics in the Eastern Mediterranean and the Middle East. *Reg. Environ. Change* **2016**, *16*, 1863–1876. [CrossRef]

148. Brogli, R.; Kröner, N.; Sørland, S.L.; Lüthi, D.; Schär, C. The Role of Hadley Circulation and Lapse-Rate Changes for the Future European Summer Climate. *J. Clim.* **2019**, *32*, 385–404. [CrossRef]
149. Tuel, A.; Eltahir, E.A.B. Why Is the Mediterranean a Climate Change Hot Spot? *J. Clim.* **2020**, *33*, 5829–5843. [CrossRef]
150. Brogli, R.; Lund Sørland, S.; Kröner, N.; Schär, C. Future Summer Warming Pattern under Climate Change Is Affected by Lapse-Rate Changes. *Weather Clim. Dyn.* **2021**, *2*, 1093–1110. [CrossRef]
151. Noto, L.V.; Cipolla, G.; Pumo, D.; Francipane, A. Climate Change in the Mediterranean Basin (Part II): A Review of Challenges and Uncertainties in Climate Change Modeling and Impact Analyses. *Water Resour. Manag.* **2023**, *37*, 2307–2323. [CrossRef]
152. Wang, X.; Fenech, A.; Farooque, A.A. Possibility of Stabilizing the Greenland Ice Sheet. *Earth's Future* **2021**, *9*, e2021EF002152. [CrossRef]
153. Ballester, J.; Quijal-Zamorano, M.; Méndez Turrubiates, R.F.; Pegenaute, F.; Herrmann, F.R.; Robine, J.M.; Basagaña, X.; Tonne, C.; Antó, J.M.; Achebak, H. Heat-Related Mortality in Europe during the Summer of 2022. *Nat. Med.* **2023**, *29*, 1857–1866. [CrossRef]
154. Qasmi, S.; Ribes, A. Reducing Uncertainty in Local Temperature Projections. *Sci. Adv.* **2022**, *8*, 6872. [CrossRef]

Disclaimer/Publisher's Note: The statements, opinions and data contained in all publications are solely those of the individual author(s) and contributor(s) and not of MDPI and/or the editor(s). MDPI and/or the editor(s) disclaim responsibility for any injury to people or property resulting from any ideas, methods, instructions or products referred to in the content.

remote sensing

MDPI

Technical Note

Modeling with Hysteresis Better Captures Grassland Growth in Asian Drylands

Lijuan Miao [1], Yuyang Zhang [1], Evgenios Agathokleous [2], Gang Bao [3], Ziyu Zhu [4] and Qiang Liu [5,*]

1 School of Geographical Sciences, Nanjing University of Information Science and Technology, Nanjing 210044, China
2 School of Ecology and Applied Meteorology, Nanjing University of Information Science and Technology, Nanjing 210044, China
3 College of Geographical Science, Inner Mongolia Normal University, Hohhot 010028, China
4 Changwang School of Honors, Nanjing University of Information Science and Technology, Nanjing 210044, China
5 School of Remote sensing & Geomatics Engineering, Nanjing University of Information Science and Technology, Nanjing 210044, China
* Correspondence: qiang.liu@nuist.edu.cn

Abstract: Climate warming hampers grassland growth, particularly in dryland regions. To preserve robust grassland growth and ensure the resilience of grassland in these arid areas, a comprehensive understanding of the interactions between vegetation and climate is imperative. However, existing studies often analyze climate–vegetation interactions using concurrent vegetation indices and meteorological data, neglecting time-lagged influences from various determinants. To address this void, we employed the random forest machine learning method to predict the grassland NDVI (Normalized Difference Vegetation Index) in Asian drylands (including five central Asia countries, the Republic of Mongolia, and Parts of China) from 2001 to 2020, incorporating time-lag influences. We evaluated the prediction model's performance using three indexes, namely the coefficient of determination (R^2), root-mean-square error (RMSE), and Mean Absolute Error (MAE). The results underscore the superiority of the model incorporating time-lag influences, demonstrating its enhanced capability to capture the grassland NDVI in Asian drylands ($R^2 \geq 0.915$, RMSE ≤ 0.033, MAE ≤ 0.019). Conversely, the model without time-lag influences exhibited relatively poor performance, notably inferior to the time-lag-inclusive model. The latter result aligns closely with remote sensing observations and more accurately reproduces the spatial distributions of the grassland NDVI in Asian drylands. Over the study period, the grassland NDVI in Asian drylands exhibited a weak decreasing trend, primarily concentrated in the western region. Notably, key factors influencing the grassland NDVI included the average grassland NDVI in the previous month, total precipitation in the current month, and average soil moisture in the previous month. This study not only pioneers a novel approach to predicting grassland growth but also contributes valuable insights for formulating sustainable strategies to preserve the integrity of grassland ecosystems.

Keywords: grassland; NDVI; time lag; random forests; prediction; Asian drylands

1. Introduction

As one of the most widely distributed vegetation types in terrestrial ecosystems [1], grasslands play a vital role in regulating the global carbon cycle and sustaining the structure and function of ecosystems [2,3]. Grassland ecosystems also serve as a key indicator of global climate change, with physiological processes such as photosynthesis and respiration linking natural moisture, atmosphere, and soil [4]. However, the growth of grasslands is strongly influenced by the external environment, especially in dryland areas [5,6]. In the context of climate warming, frequent extreme events profoundly impact grassland growth [7,8]. Recent extensive research on the impacts of climate change on grassland growth indicates that rising temperatures and increased drought could hinder growth or even lead to stagnation [9–12]. Therefore, obtaining reliable and objective information on grassland growth in a timely manner is crucial [13].

To safeguard grassland growth in dryland areas, it is imperative to accurately simulate the vegetation growth process and predict growth conditions. However, predicting vegetation growth using current tools is challenging [14]. Among the available tools, process-based ecosystem models play a vital and beneficial role in predicting vegetation growth [3]. Achieving accurate predictions of vegetation growth requires these models to reproduce multiple processes with greater precision and realism [15,16]. However, process-based ecosystem models face limitations in accurately replicating key ecosystem processes such as photosynthesis and respiration [16,17]. Recognizing these limitations, machine learning methods have emerged as promising alternatives, capable of comprehensively considering influential factors reflecting ecosystem processes. Machine learning methods effectively compensate for the shortcomings of process-based ecosystem models [18]. Techniques such as support vector machines, random forests, and gradient-boosted trees are employed to elucidate relationships between variables and predictor variables [11,18]. However, research on predicting vegetation growth in Asian drylands remains relatively limited [11,19]. Many scholars have investigated the prediction of vegetation NDVI growth using this method. For instance, Li et al. employed the extreme gradient boosting machine learning method to establish a relationship model between vegetation NDVI and meteorological factors in mainland China. They discovered that the machine learning model exhibits strong performance in replicating the spatial and seasonal variations in satellite-derived NDVI throughout China [11]. Peng et al., on the other hand, utilized a novel machine learning method to forecast the NDVI growth of vegetation during the growing season in China under extreme conditions. The findings indicate that the machine learning method demonstrates stability and high accuracy even amidst severe drought conditions [19]. These studies successfully replicated the spatial and temporal distributions of vegetation growth in China, employing machine learning to assess its accuracy in predicting growth under extreme drought conditions. Machine learning was an important tool for predicting vegetation growth, enhancing our overall predictive capabilities. It is conducive to the timely acquisition of reliable, objective, and timely information on vegetation growth.

Asian drylands boast abundant grassland resources, playing a crucial role in the sustainable development of the local livestock economy [20,21]. For centuries, the indigenous communities have relied on the plentiful resources of the grasslands for the growth of animal husbandry, establishing them as a pivotal industry essential for their survival [22]. Consequently, forecasting information regarding Asian dryland grasslands plays a crucial guiding role in shaping local animal husbandry practices. In-depth research has studied the response mechanism of the grassland NDVI and climate change. A discernible time-lag effect was observed between the grassland NDVI and various climate elements, indicating a delayed response [23–26]. In Asian drylands, the grassland NDVI in the current month showed a stronger response to solar radiation and the previous month's soil moisture, without significant lag effects from temperature and total precipitation [25]. This suggests that changes in climatic factors could impact the grassland NDVI within a few months. However, there are few studies that consider the influence of time lag in predicting the grassland growth of Asian drylands, highlighting a notable gap in the current scientific literature. Employing the random forest machine learning method, we developed an NDVI prediction model for Asian dryland grasslands, incorporating both time-lag influences and their absence. Our main research objectives were (1) to evaluate the accuracy of the grassland NDVI predictions from 2001 to 2020, accounting for and excluding time-lag influences; (2) to investigate the spatiotemporal distributions of the NDVI in Asian grasslands from 2001 to 2020 using the two distinct models; and (3) to identify the primary climate factors influencing the NDVI in arid Asian regions based on the outcomes derived from the two models.

2. Materials and Methods

2.1. Study Area

The research area is the Asian drylands, ranging from 46.49° to 126.08°E and 31.54° to 55.43°N, geographically covering six countries: Kazakhstan, Kyrgyzstan, Tajikistan, Turkmenistan, Uzbekistan, and Mongolia, as well as certain regions of China, namely Inner Mongolia, Shaanxi, Ningxia, Gansu, Qinghai, and Xinjiang (refer to Figure 1, adapted from MODIS land cover product, specifically MCD12C1, accessible at https://lpdaac.usgs.gov/pro-ucts/mcd12c1v006/ (accessed on 18 May 2018)). Spanning over 10 million square kilometers, this region features a distinctive continental arid climate characterized by frequent droughts, low precipitation, and escalating temperatures [27,28]. The eastern Asian drylands experience fewer than 200 mm of annual precipitation, featuring a cold and dry winter and high temperatures in summer [29,30]. The western part (the five central Asia countries) holds an annual precipitation of 300–500 mm, and this region witnesses cold winters, hot summers, and substantial day-to-day temperature variations [29,31]. Under arid and semiarid climatic conditions [32], the main land use type is grassland, with sparse vegetation and extensive desert areas in the central region.

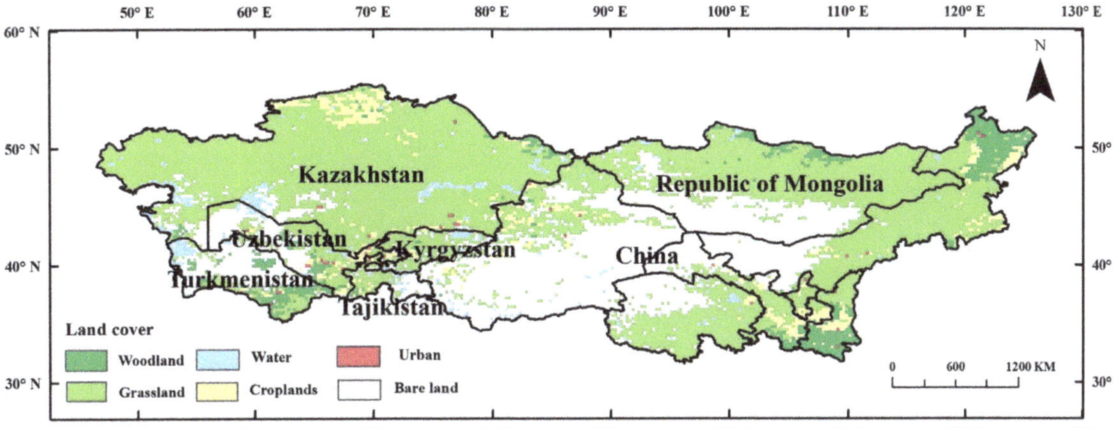

Figure 1. The geographical location of Asian drylands. The land cover information was adapted from MODIS land cover product (i.e., MCD12C1, accessible at https://lpdaac.usgs.gov/pro-ucts/mcd12c1v006/) (accessed on 18 May 2018).

2.2. Data Resources

2.2.1. MODIS NDVI

The MODIS NDVI (Normalized Difference Vegetation Index) data are sourced from the National Aeronautics and Space Administration's (NASA) Moderate Resolution Imaging Spectroradiometer (MODIS) Vegetation Index (VI) product (MOD13C2) (https://ladsweb.modaps.eosdis.nasa.gov) (accessed on 15 April 2022). This dataset spans the period from January 2001 to December 2020, featuring a spatial resolution of 0.05° × 0.05° and a temporal resolution of 1 month. This dataset underwent rigorous preprocessing, including geometric, atmospheric, and radiometric corrections, as well as validation procedures to reduce the impact of atmospheric factors such as cloud shadows and aerosols [33]. The monthly MOD13C2 data are widely recognized for their reliability and are extensively employed for monitoring vegetation conditions at both regional and global scales [34,35].

2.2.2. Climate Dataset

Climate variables include temperature, precipitation, scPDSI (self-calibrating Palmer Drought Severity Index), minimum air temperature, and maximum air temperature. These datasets were obtained from the monthly gridded Climatic Research Unit Time-Series version 4.05 (CRU TS4.05), which is provided by the Climatic Research Unit (CRU) at the University of East Anglia, UK (https://catalogue.ceda.ac.uk/uuid/c26a65020a5e4b80b20018f148556681) (accessed on 21 April 2022) [36]. It features a spatial resolution of 0.5° × 0.5° and a temporal resolution of 1 month, encompassing the period from January 2001 to December 2020. Currently, CRU is one of the most extensively utilized climate datasets. It integrates several well-known existing databases and uses angular distance weighted interpolation to generate global monthly data covering the land surface from 1901 to 2020.

Solar radiation data for model training and testing are sourced from the European Centre for Medium-Range Weather Forecasts (ECMWF) version 5 reanalysis (ERA5) dataset (https://cds.climate.copernicus.eu/cdsapp#!/home) (accessed on 27 April 2022). Representing the latest generation of ECWMF reanalysis data, ERA5 offers improved spatial and temporal resolution, as well as enhanced radiative transfer modeling compared with its predecessor, ERA interim reanalysis [11]. These data can be acquired from 1979 to the present, with a spatial resolution of 0.1° × 0.1° and a temporal resolution of 1 month. For this study, data from January 2001 to December 2020 are utilized.

Vapor Pressure Deficit (VPD) datasets are extracted from the Terra-Climate dataset provided by the University of Idaho (https://climate.northwestknowledge.net/TERRACLIMATE/index_directDownloads.php) (accessed on 5 March 2023). The Terra-Climate dataset provides vpd information at a spatial resolution of 1/24° × 1/24°, with a temporal resolution of 1 month. The data cover the time period from January 2001 to December 2020.

2.2.3. Soil Moisture

The daily soil moisture datasets are obtained from the soil moisture dataset within the European Space Agency (ESA) Climate Change Initiative Program (CCI) (https://www.esa-soilmoisture-cci.org) (accessed on 6 May 2022). It is featured with a spatial resolution of 0.25° × 0.25°, and the data span from January 2001 to December 2020. The soil moisture product was derived by synthesizing an active microwave soil moisture product produced by Bartalis et al. [37] and Wagner et al. [38] and a passive microwave soil moisture product developed by Vrije Universiteit Amsterdam in collaboration with NASA [39,40]. This composite product accurately represents soil moisture levels at a depth ranging from 0 to 10 cm. Due to limitations in satellite sensor coverage, there were significant gaps in the data, resulting in missing pixels. Thus, we removed the missing data points (7%) and computed monthly averages of soil moisture data, ensuring comprehensive coverage of the study area on a month-by-month basis.

2.2.4. Land Use Data

The land cover data were resourced from NASA's MODIS land cover product MCD12C1 (https://lpdaac.usgs.gov/products/mcd12c1v006/) (accessed on 18 May 2018). This dataset features a spatial resolution of 0.05° × 0.05°. This particular research focused on identifying grassland image elements that remained consistent throughout the entire period from 2001 to 2020. This approach was employed to some extent to mitigate the land use impact.

2.3. *Method*

2.3.1. Trend Analyses

We used least squares linear regression to analyze the spatial trends of the grassland NDVI in the dryland areas of Asia from 2001 to 2020. F-test was used to analyze the significance of the trends [41].

$$\boldsymbol{Slope} = \frac{n\sum_{i=1}^{n}(i \times \boldsymbol{NDVI}_i) - \sum_{i=1}^{n} i \times \sum_{i=1}^{n} \boldsymbol{NDVI}_i}{n\sum_{i=1}^{n} i^2 - (\sum_{i=1}^{n} i)^2} \quad (1)$$

where ***Slope*** is the slope of the image element, $\boldsymbol{NDVI}_i$ is the mean value of NDVI in the *i*th year, and n is the length of the study.

Although the observed NDVI indeed provides valuable data on current vegetation conditions over Asia's drylands, there are several compelling reasons for focusing our trend analysis on predicted NDVI. Firstly, the predicted NDVI reveals possible future vegetation patterns under certain environments, allowing model accuracy assessment through comparison with actual observations. Secondly, the reliable NDVI prediction model helps identify links between vegetation growth and the environment, enhancing ecological understanding. Finally, the predicted NDVI supplements observed data, especially in data-scarce or observation-challenging areas, offering a fuller picture of vegetation trends.

2.3.2. Random Forest

The random forest algorithm, proposed and developed by Leo Breiman and Adele Cutler [42], stands out as a powerful tool for addressing both classification and regression problems. Its core concept diverges from the conventional approach of constructing a single, large decision tree using the entire training dataset. Instead, the random forest technique generates multiple decision trees using different subsets of training samples and attributes. These individual decision trees are then combined in a random fashion to create a more robust and effective model, enhancing the overall performance of the random forest model [43,44].

The MODIS NDVI product was used to represent vegetation growth, a choice commonly employed in previous research [4,5]. Previous studies identified a significant time-lag effect on the grassland NDVI in Asian drylands [25]. Because vegetation growth in a current month is influenced by the preceding month, the NDVI from the previous month was also incorporated into the model as an explanatory variable during training. This study utilized MODIS NDVI data from 2001 to 2020 in Asian drylands as the dependent variable, with various other factors serving as independent variables (See Table 1). Each image for each explanatory variable covered an area of 30,528 pixels, for a total of 240 images for a variable. The overall dataset for the random forest encompassed 7,326,720 samples, with 70% allocated for training and 30% for validation.

Table 1. Model details: independent variable, dependent variable, and random forest parameter.

	Independent Variable	Dependent Variable	Temporal Resolution	Spatial Resolution	Time Span	Model Parameter
with time-lag	tmp_0, pre_0, rad_1, sm_1 tmx_0, tmn_0, vpd_0, NDVI_1	NDVI	One month	0.25°	January 2001–December 2020	70% training set, 30% validation set
without time-lag	tmp_0, pre_0, rad_0, sm_0 tmx_0, tmn_0, vpd_0, NDVI_1	NDVI	One month	0.25°	January 2001–December 2020	70% training set, 30% validation set

Notes: NDVI_1 represents NDVI in the previous month; tmp_0 represents temperature in the current month; pre_0 represents precipitation in the current month; sm_0 represents soil moisture in the current month; sm_1 represents soil moisture in the previous month; rad_0 represents solar radiation in the current month; rad_1 represents solar radiation in the previous month; vpd_0 represents vapor pressure deficit in the current month; tmn_0 represents minimum temperature in the current month; tmx_0 represents maximum temperature in the current month.

2.3.3. Statistical Analysis

Two models were developed using the random forest machine learning method to predict the grassland NDVI in Asian drylands. One model excluded time-lag influences, while the other incorporated these influences. The precision differences were then compared between the two models. R^2 was used to measure how effectively each model explained the variation in the data; RMSE served as a metric to measure the standard deviation between the true and predicted values; and MAE was used to measure the size of the gap between the true and predicted values [3,42].

$$R^2 = 1 - \frac{\sum_{i=1}^{n} (y_i - \hat{y}_i)^2}{\sum_{i=1}^{n} (y_i - \overline{y}_i)^2} \quad (2)$$

$$RMSE = \sqrt{\frac{1}{n}\sum_{i=1}^{n}(y_i - \hat{y}_i)^2} \quad (3)$$

$$MAE = \frac{1}{n}\sum_{i=1}^{n}|y_i - \hat{y}_i| \quad (4)$$

3. Results

3.1. Evaluation of Model Performance in Predicting Grassland NDVI in Asian Drylands

The results reveal that the NDVI model incorporating time-lag effects outperforms the model without time-lag across all evaluation indices. The overall coefficient of determination for the model considering time-lag influences was 0.92, whereas the NDVI prediction model without time-lag influences achieved a coefficient of determination of 0.89 (Figure 2a,b). The R^2 values indicated that these models could explain 92% and 89% of the variations in the grassland NDVI, respectively. Further, the RMSE and MAE values of the model considering time-lag influences were 0.033 and 0.019, respectively, while values of the model without time-lag influences were 0.037 and 0.022, respectively (Figure 2a,b). The results from the three evaluation metrics collectively demonstrate that the model considering time-lag influences outperformed the model without time-lag influences in predicting the grassland NDVI in Asian drylands. As depicted in Figure 2, both models fitted well with observations, with the majority of NDVI values clustering between 0 and 0.25. This visual representation affirmed the reliability and accuracy of the developed models in capturing the dynamics of the grassland NDVI in the study region.

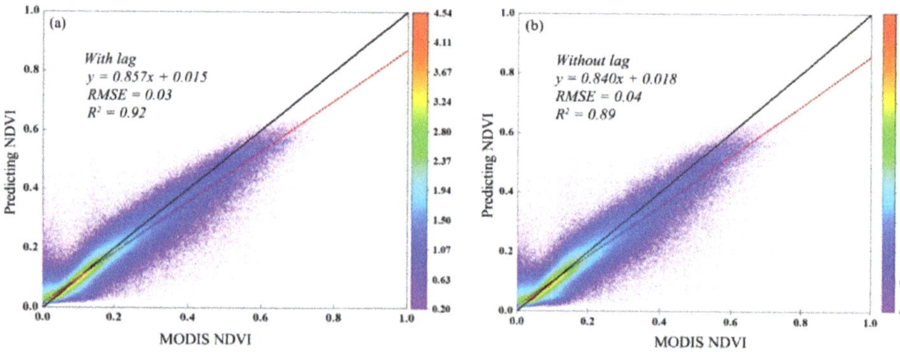

Figure 2. Density scatter plot showing the relationship between the observed MODIS and predicted NDVI values. Notes: (**a**) illustrates MODIS NDVI versus predicted NDVI with time-lag influences considered, while (**b**) depicts MODIS NDVI versus predicted NDVI without time-lag influences. The black line represents the 1:1 correspondence, and the red line indicates the regression line.

According to the spatial distributions of R^2, the developed model for predicting grassland NDVI considering time-lag influences outperformed the model without time-lag influences (Figure 3). Regions exhibiting notably high R^2 values were mainly concentrated in Kazakhstan (Figure 3a,b). Regarding RMSE index, the grassland NDVI model considering time-lag influences demonstrated superior performance compared with the model without time-lag influences, particularly in the northeastern part of the Mongolian Plateau and Kazakhstan (Figure 3c,d). Evaluating the MAE values from the two models, the areas with relatively high MAE of the model considering time-lag influences were mainly distributed in the central part of the Mongolian Plateau and Kazakhstan (Figure 3e,f). The spatial distributions of these three indexes (R^2, RMSE, MAE) collectively indicate that both models were effective in predicting grassland NDVI in Asian drylands. However, the model considering time-lag influences exhibited superior performance, especially in the Kazakhstan region (Figure 3). These findings highlight the significance of considering time-lag influences to enhance the accuracy of predictions, particularly in specific geographical areas.

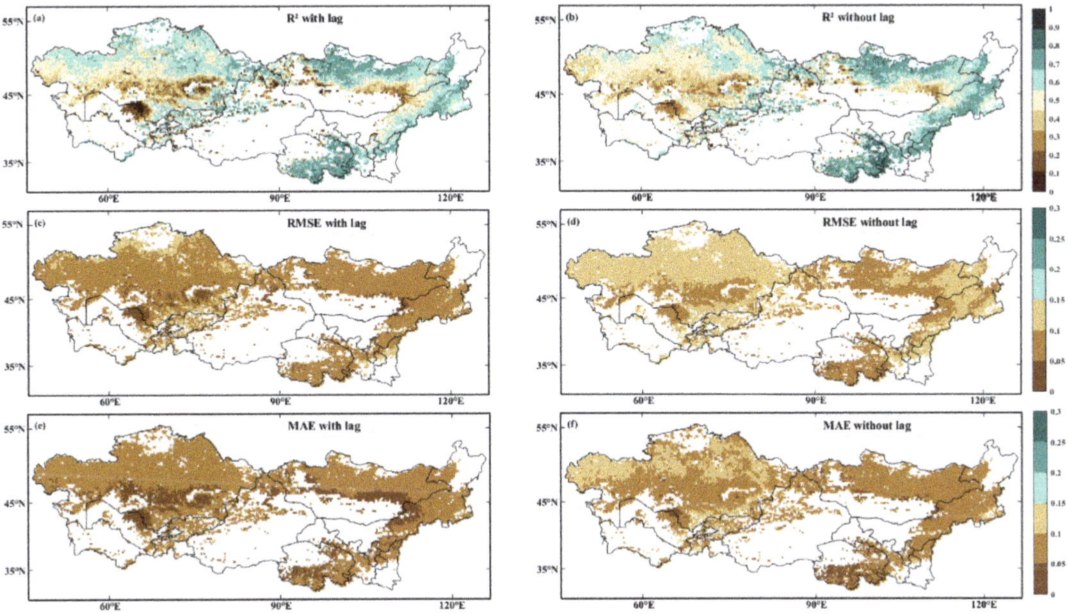

Figure 3. Spatial distributions of three indexes of grassland NDVI without time-lag influences and grassland NDVI with time-lag influences. Notes: (**a**,**b**) coefficient of determination (R^2); (**c**,**d**) root-mean-square error (RMSE); (**e**,**f**) root-mean-square error (MAE).

3.2. Comparing Spatiotemporal Variations between the Two Models

We reproduced the annual variation in the mean grassland NDVI in Asian drylands from 2001 to 2020 (Figure 4). The annual average grassland NDVI with time-lag influences was closer to the observed MODIS NDVI. The average NDVI for the grassland NDVI model, both with a time-lag influence (0.17) and without it (0.21), was higher than the annual mean NDVI derived from MODIS NDVI (0.14). Among the three NDVI values, all had consistent fluctuations in the grassland NDVI change (Figure 4). The time series of MODIS NDVI and NDVI from the two models showed a weak decreasing trend from 2001 to 2020.

We could observe that both models were able to reproduce the spatial distributions of the annual mean grassland NDVI in Asian drylands, and they were consistent with the spatial distributions of the observed MODIS NDVI (Figure 5a–c). The model considering time-lag influences was better than that without time-lag influences, as the former model was similar to the observed annual mean MODIS NDVI (Figure 5b). The annual mean NDVI decreased from the periphery to the center (both for the five central Asia countries and the Mongolia Plateau) (Figure 5a–c). Compared with the annual mean NDVI predicted by the model without time-lag influences, the annual mean NDVI predicted by the model considering time-lag influences was more accurate in the central and southern Kazakhstan and Inner Mongolia regions (Figure 5b,c).

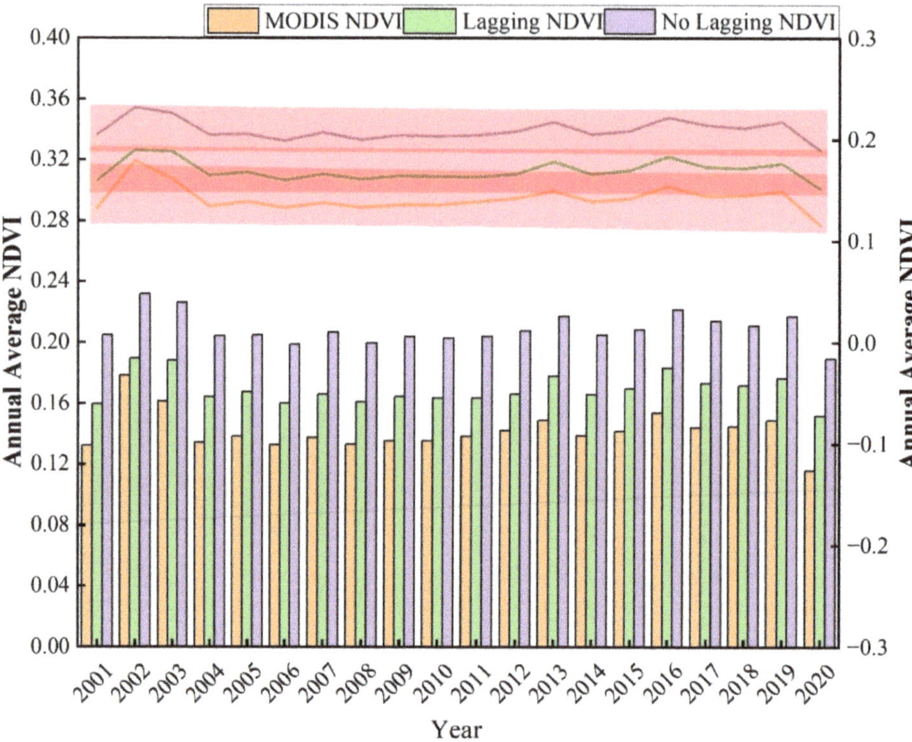

Figure 4. The interannual variability of predicted annual NDVI (with and without lagging effect) compared with MODIS NDVI in Asian grasslands from 2001 to 2020. The left y-axis corresponds to the bar graph, while the right y-axis pertains to the line graph. The pink buffers represent 95% confidence intervals surrounding the annual average NDVI.

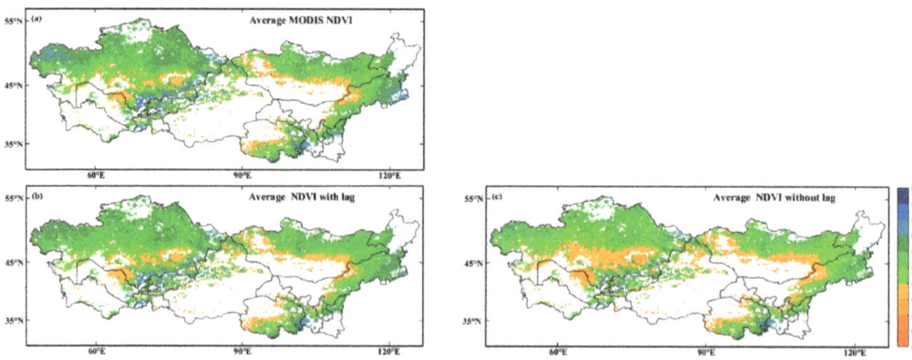

Figure 5. Spatial distributions of the average MODIS NDVI (**a**) and predicted NDVI with (**b**) and without (**c**) time-lag influences.

3.3. Comparing NDVI Trend Spatial Distributions between the Two Models

We analyzed the spatial trends of the grassland NDVI in Asian drylands from 2001 to 2020 using least-squares linear regression. The spatial trends of the predicted grassland NDVI considering time-lag influences were closer to that of the MODIS NDVI (Figure 6a,b) compared with the model simulations without time-lag influences (Figure 6c). From the

MODIS NDVI, 54% of the grassland regions had unchanged NDVI values in Asian drylands (marked in stable), mainly located in Kazakhstan, Inner Mongolia, and Qinghai of China. A fifth (22%) of the grassland areas featured with significantly decreased NDVI values were located in northern Kazakhstan, whereas most of the areas with significant increases were located in the northeastern Republic of Mongolia, Inner Mongolia, and Qinghai (8%) (Figure 6a). The distribution of the five spatial trend types in the model output, taking into account time-lag influences, was superior to that of the model output without considering time-lag influences (Figure 6). In terms of concrete values, the grassland NDVI remained stable in 54%, 48%, and 69% of the grassland areas for (i) MODIS NDVI, (ii) predicted NDVI with time-lag, and (iii) predicted NDVI without time-lag, respectively (Figure 6a). Conversely, the grassland NDVI significantly decreased in 22%, 22%, and 20% (Figure 6b) and significantly increased in 8%, 8%, and 3% (Figure 6c) of the grassland areas in Dryland Asia for (i) MODIS NDVI, (ii) predicted NDVI with time-lag, and (iii) predicted NDVI without time-lag, respectively (Figure 6b).

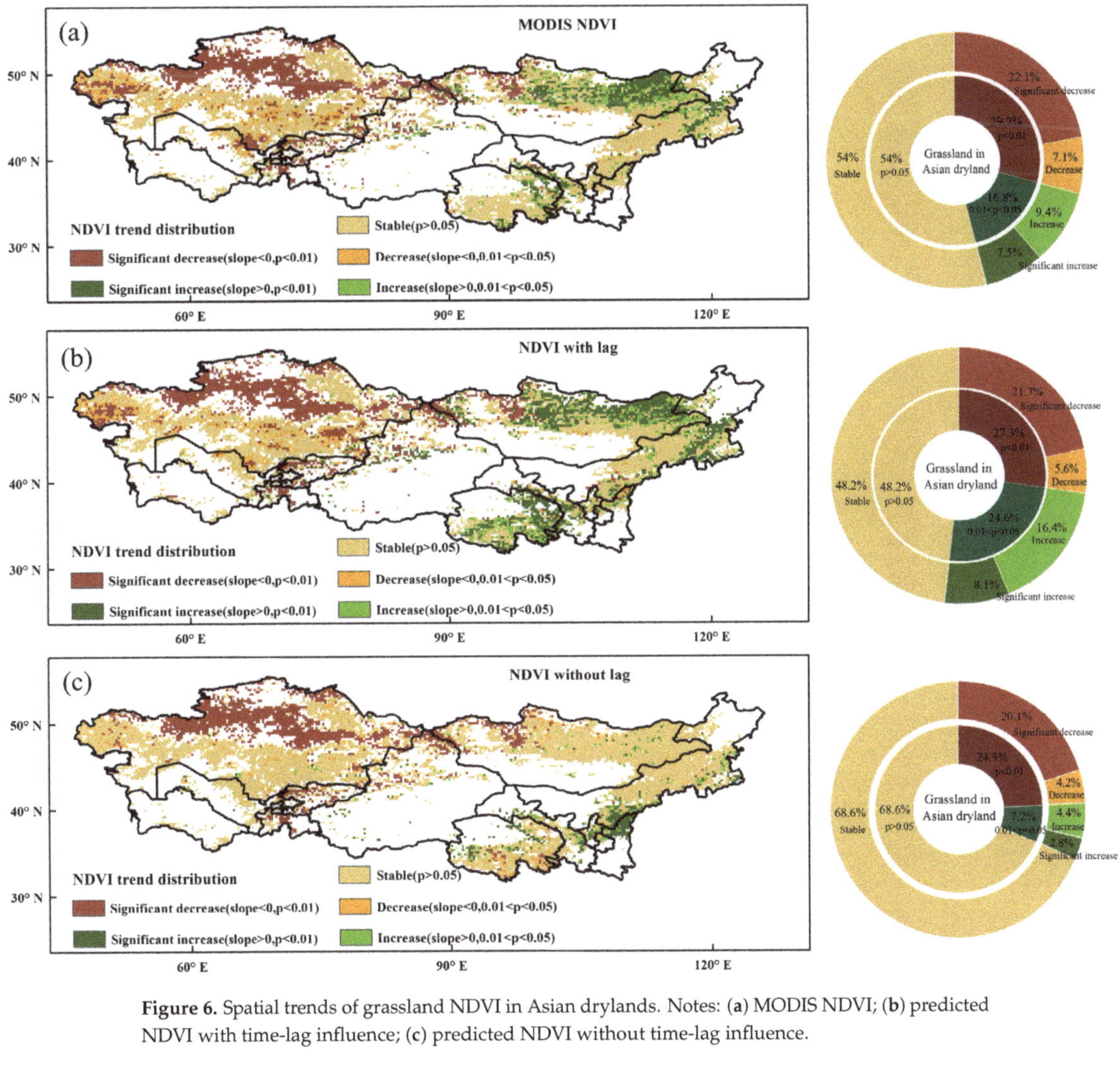

Figure 6. Spatial trends of grassland NDVI in Asian drylands. Notes: (**a**) MODIS NDVI; (**b**) predicted NDVI with time-lag influence; (**c**) predicted NDVI without time-lag influence.

3.4. Importance of the Variables from Random Forest

Figure 7a,b illustrates the relative importance of the two models with and without considering the time-lag influences by model outputs. NDVI_1 was most important for the grassland NDVI in Asian drylands from 2001 to 2020, with importance values of 24.7 and 55.0 for the two models. The second most important factor was pre_0, with importance values of 10.3 and 17.2 for the two models. Except for NDVI_1 and pre_0, sm_1 and vpd_0 were important for the grassland NDVI in Asian drylands. For the model considering time-lag influences, NDVI_1 and moisture variables (pre_0, sm_1, and vpd_0) were the major determinants for the grassland NDVI in Asian drylands, and they accounted for about 86.3% of the overall importance. In comparison, for the model without considering time-lag influences, NDVI_1, pre_0, sm_0, and rad_0 were the major determinants, which accounted for about 87.9% of the overall importance. Hence, soil moisture from the previous month played a crucial role in influencing the current month's grassland NDVI, as depicted in Figure 7a.

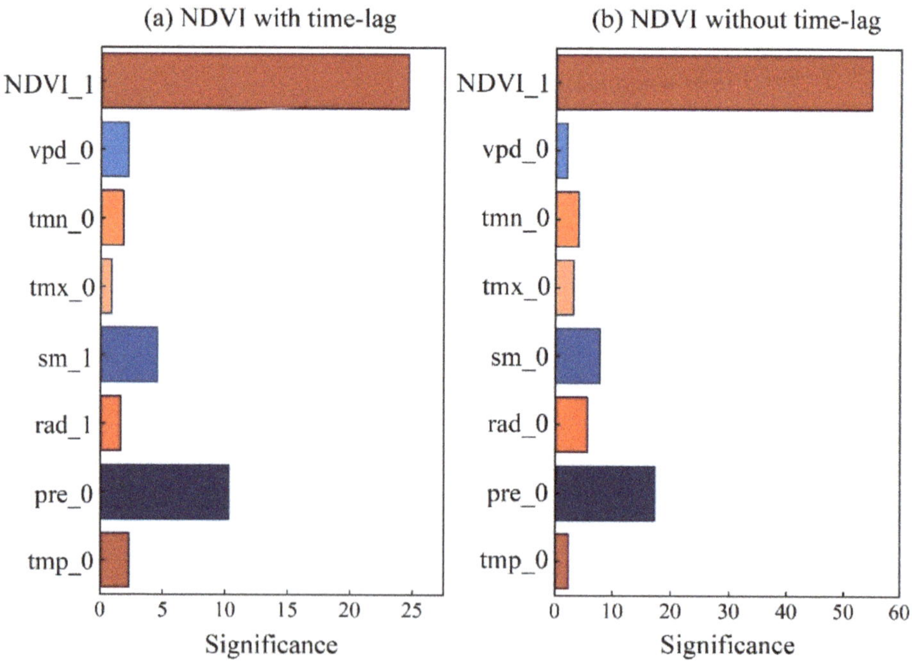

Figure 7. Relative importance of the selected explanatory variables for the predicted grassland NDVI in Asian drylands. Notes: NDVI_1 represents NDVI in the previous month; pre_0 represents precipitation in the current month; sm_0 represents soil moisture in the current month; sm_1 represents soil moisture in the previous month; rad_0 represents solar radiation in the current month; vpd_0 represents vapor pressure deficit in the current month.

4. Discussion

Machine learning methods have been demonstrated in this study to possess remarkable capabilities in predicting the grassland NDVI, consistent with the findings of numerous previous studies [19,45,46]. Additionally, several studies have suggested that lag effects have a substantial impact on vegetation growth. Miao [25] investigated the response of the grassland NDVI to soil moisture and solar radiation with a one-month time lag in Asian drylands, while Chen [47] et al. identified a strong positive relationship between soil moisture and the NDVI, with the NDVI typically lagging behind soil moisture by one month. Two models were developed utilizing the random forest algorithm, with and without considering time-lag influences, for the prediction of the grassland NDVI in Asian drylands. We highlighted the strong performance of both models in capturing the spatial and temporal variations and trends of the grassland NDVI in Asian drylands, while the model considering time-lag influences outperformed the other one. Thus, the integration of time-lag influences and machine learning presents a novel perspective for predicting the grassland NDVI and significantly enhances prediction accuracy.

Whether considering time-lag influence or not, the prediction results for the entire Asian grassland exhibit consistency. Over the period from 2001 to 2020, the grassland NDVI in Asian drylands generally exhibited a weak decreasing trend. Grasslands in eastern Asia generally demonstrate a growth trend in the NDVI, while the NDVI in western Asian dryland grasslands shows a slightly declining trend, with the most pronounced decline observed in northern Kazakhstan. We postulate that the factors driving grassland decline can be attributed to two main factors: climate change and human activities. Research has demonstrated five Central Asia countries witnessing an increase in aridity, marked by the severest drought in the northwestern part of Kazakhstan and a gradual reduction in overall precipitation [4,32,48]. This shift in climate patterns was likely a primary driver behind the declining grassland NDVI in northern Kazakhstan. Meanwhile, the degradation of grassland in northern Kazakhstan could be attributed to human activities, including intensified overgrazing, economic challenges faced by countries after the Soviet Union's dissolution, and suboptimal government policies for grassland preservation [49]. Concerning the regions experiencing grassland NDVI growth, we attribute this phenomenon primarily to climate factors, particularly in northeastern Mongolia, Inner Mongolia, and Qinghai. The results of numerous studies in this region also align with our findings, indicating a consistent increasing trend in the NDVI [50,51]. This was primarily due to precipitation on the Mongolian plateau originating from both the westerly belt and the East Asian monsoon [52]. The northern part of the Mongolian Plateau is influenced by the westerly belt, resulting in higher precipitation in the north and lower levels in the south. The eastern part is influenced by the East Asian monsoon, providing more abundant rainfall, and both rain and warmth coincided, leading to a gradual increase in precipitation from the west to the east [52]. Consequently, the varying trends in the grassland NDVI across different regions of Asian drylands can be attributed to their distinct natural conditions. Therefore, the prediction of grassland growth by machine learning is particularly significant for understanding vegetation physiology and managing human activities. Nonetheless, the prediction of grassland growth in Asian dryland areas was not only related to climatic influences but also to human activities, vegetation physiological processes, and natural hazards. These multifaceted elements might potentially affect the accuracy of grassland growth predictions. Therefore, future endeavors aimed at forecasting grassland growth must comprehensively address the confluence of climatic influences and natural hazards.

The grassland NDVI is typically influenced by various concurrent climatic factors, including temperature and precipitation [53,54]. However, our study reveals a distinctive pattern, as we observed that soil moisture in the preceding month exerts a robust influence on the grassland NDVI, surpassing the impact of temperature on the grassland NDVI in grasslands. Our study identified specific climate variables that played a particularly crucial role in determining the grassland NDVI in the study area, including precipitation in the current month, soil moisture in the previous month, and vapor pressure deficit in the current month. Moisture is often regarded as the primary factor controlling vegetation growth in arid and semiarid areas [55]. Previous studies have commonly used precipitation as a key factor of moisture availability [56,57], but they have often overlooked the underlying mechanisms governing moisture dynamics. In arid and semiarid areas, a significant portion of surface-layer soil moisture originates from precipitation [58], while changes in soil moisture in deeper layers were determined by fluctuations in groundwater depth [59]. When deeper soil layers are well-hydrated, hydraulic redistribution can enhance the efficiency of the root system to absorb and transport water [59], i.e., the conductive movement of water from the wetter part of the soil to the drier part of the soil via the plant's root system [59]. This, in turn, benefits plants by optimizing resource utilization and extending the availability of soil moisture [60]. This mechanism likely explains why soil moisture from the previous month is critical for the grassland NDVI in the current month. Therefore, soil moisture serves as a more accurate indicator of water availability for grasslands in arid regions. In research pertaining to grassland drought, numerous scholars have acknowledged the significance of soil water. They have conducted analyses to examine its impact on vegetation growth, establishing thresholds based on varying soil moisture levels [61,62].

Understanding the interplay between the NDVI and soil properties, climate, and other environmental factors is imperative for predicting grassland growth and productivity. While spectral indexes, notably the Normalized Difference Vegetation Index (NDVI), serve as reliable indicators of vegetation growth and productivity, the augmentation of learning machine models with additional spectral indexes presents an opportunity for a more holistic comprehension of grassland ecosystems. For instance, the incorporation of the Normalized Difference Water Index (NDWI) could elucidate soil moisture content dynamics, whereas the inclusion of the Soil Adjusted Vegetation Index (SAVI) could mitigate soil background effects [59–61]. By incorporating these additional indexes, we can cultivate more realistic models that encompass a broader spectrum of environmental variables. This, in turn, could lead to more accurate predictions of grassland growth and productivity across diverse conditions, such as different temperature scenarios or grazing intensities. Furthermore, the integration of soil properties, climate data, and other environmental factors into these models unveils the potential to reveal previously unrecognized patterns and relationships. For example, factors such as soil nutrients, pH levels, or the presence of specific minerals may exert a substantial influence on grassland growth dynamics. By including these factors in the models, we can attain a deeper understanding of their impacts on vegetation dynamics [5,34].

5. Conclusions

To ensure the resilient growth of grasslands in Asian drylands, understanding the impact of time-lag influences on the grassland NDVI is crucial. Employing a random forest machine learning approach and incorporating seven meteorological elements with MODIS NDVI observations, we predicted the grassland NDVI in Asian drylands, considering or not considering time-lag influences. The grassland NDVI model considering time-lag influences demonstrated relatively higher accuracy, as indicated by R^2, RMSE, and MAE compared with the model without time-lag influences. The annual mean NDVI predicted by the model considering time-lag influences closely aligned with the MODIS NDVI. The annual mean NDVI in Asian drylands demonstrated a weak decreasing trend from 2001 to 2020. In terms of spatial variations, the grassland NDVI remained stable in the majority of areas, significantly decreasing in northern Kazakhstan and significantly increasing in northeastern Republic of Mongolia, Inner Mongolia, and Qinghai. Notably, more areas showed a decrease in the grassland NDVI than those exhibiting an increase. The main factors affecting the grassland NDVI were NDVI in the previous month, the current month's total precipitation, and the previous month's soil moisture. The precise prediction of the grassland NDVI serves as a valuable scientific reference for both grass growth and livestock development in Asian drylands.

Author Contributions: L.M. and Q.L. designed the research; Y.Z. performed the research; L.M., G.B., E.A., Y.Z. and Z.Z. wrote the initial draft, and all authors contributed to the final manuscript. All authors have read and agreed to the published version of the manuscript.

Funding: The research received support from the National Natural Science Foundation of China (42101295), the Natural Science Foundation of Jiangsu Province (BK20210657). Qiang Liu acknowledges additional support from the Jiangsu Distinguished Professor Program of the People's Government of Jiangsu Province, China.

Data Availability Statement: The data that support the findings of this study are available within the manuscript.

Conflicts of Interest: All the authors have no competing interests to declare.

References

1. Chen, C.; Li, G.; Peng, J. Spatio-temporal characteristics of Xinjiang grassland NDVI and its response to climate change from 1981 to 2018. *Acta Ecol. Sin.* **2023**, *43*, 1537–1552.
2. Afuye, G.A.; Kalumba, A.M.; Busayo, E.T.; Orimoloye, I.R. A bibliometric review of vegetation response to climate change. *Environ. Sci. Pollut. Res.* **2021**, 1–13. [CrossRef] [PubMed]
3. Bonan, G. *Climate Change and Terrestrial Ecosystem Modeling*; Cambridge University Press: Cambridge, UK, 2019.
4. Zhang, Q.; Yuan, X.; Chen, X.; LUO, G.; Li, L. Vegetation change and its response to climate change in Central Asia from 1982 to 2012. *Chin. J. Plant Ecol.* **2016**, *40*, 13–23. [CrossRef]
5. Miao, L.; Sun, Z.; Ren, Y.; Schierhorn, F.; Müller, D. Grassland greening on the Mongolian Plateau despite higher grazing intensity. *Land Degrad. Dev.* **2021**, *32*, 792–802. [CrossRef]
6. Geng, Q.; Chen, X.; He, X.; Tian, Z. Vegetation dynamics and its response to climate change and human activities based on different vegetation types in China. *Acta Ecol. Sin.* **2022**, *42*, 3557–3568.
7. Schimel, D.S.; House, J.I.; Hibbard, K.A.; Bousquet, P.; Ciais, P.; Peylin, P.; Braswell, B.H.; Apps, M.J.; Baker, D.; Bondeau, A. Recent patterns and mechanisms of carbon exchange by terrestrial ecosystems. *Nature* **2001**, *414*, 169–172. [CrossRef]
8. Zhang, J.; Yuan, M.; Zhang, J.; HW, L.; Wang, J.; Zhang, X.; Ju, P.; Jiang, H.; Chen, H.; Zhu, Q. Responses of the NDVI of alpine grasslands on the Qinghai-Tibetan Plateau to climate change and human activities over the last 30 years. *Acta Ecol. Sin.* **2020**, *40*, 6269–6281.
9. Liu, Y.Y.; Van Dijk, A.I.; De Jeu, R.A.; Canadell, J.G.; McCabe, M.F.; Evans, J.P.; Wang, G. Recent reversal in loss of global terrestrial biomass. *Nat. Clim. Chang.* **2015**, *5*, 470–474. [CrossRef]
10. Ray, D.; Ramankutty, N.; Mueller, N.; West, P.; Foley, J. Recent patterns of crop yield growth and stagnation. *Nat. Commun.* **2012**, *3*, 1293. [CrossRef] [PubMed]
11. Li, X.; Yuan, W.; Dong, W. A machine learning method for predicting vegetation indices in China. *Remote Sens.* **2021**, *13*, 1147. [CrossRef]
12. Yuan, W.; Zheng, Y.; Piao, S.; Ciais, P.; Lombardozzi, D.; Wang, Y.; Ryu, Y.; Chen, G.; Dong, W.; Hu, Z. Increased atmospheric vapor pressure deficit reduces global vegetation growth. *Sci. Adv.* **2019**, *5*, eaax1396. [CrossRef]
13. Becker-Reshef, I.; Vermote, E.; Lindeman, M.; Justice, C. A generalized regression-based model for forecasting winter wheat yields in Kansas and Ukraine using MODIS data. *Remote Sens. Environ.* **2010**, *114*, 1312–1323. [CrossRef]
14. Friedlingstein, P.; Jones, M.W.; O'sullivan, M.; Andrew, R.M.; Bakker, D.C.; Hauck, J.; Le Quéré, C.; Peters, G.P.; Peters, W.; Pongratz, J. Global carbon budget 2021. *Earth Syst. Sci. Data* **2022**, *14*, 1917–2005. [CrossRef]
15. Li, S.; Yuan, W.; Ciais, P.; Viovy, N.; Ito, A.; Jia, B.; Zhu, D. Benchmark estimates for aboveground litterfall data derived from ecosystem models. *Environ. Res. Lett.* **2019**, *14*, 084020. [CrossRef]
16. Xia, J.; Yuan, W.; Lienert, S.; Joos, F.; Ciais, P.; Viovy, N.; Wang, Y.p.; Wang, X.; Zhang, H.; Chen, Y. Global patterns in net primary production allocation regulated by environmental conditions and forest stand age: A model-data comparison. *J. Geophys. Res. Biogeosci.* **2019**, *124*, 2039–2059. [CrossRef]
17. Jung, N.-C.; Popescu, I.; Kelderman, P.; Solomatine, D.P.; Price, R.K. Application of model trees and other machine learning techniques for algal growth prediction in Yongdam reservoir, Republic of Korea. *J. Hydroinform.* **2010**, *12*, 262–274. [CrossRef]
18. Xia, J.; Ma, M.; Liang, T.; Wu, C.; Yang, Y.; Zhang, L.; Zhang, Y.; Yuan, W. Estimates of grassland biomass and turnover time on the Tibetan Plateau. *Environ. Res. Lett.* **2018**, *13*, 014020. [CrossRef]
19. Peng, Q.; Li, X.; Shen, R.; He, B.; Chen, X.; Peng, Y.; Yuan, W. How well can we predict vegetation growth through the coming growing season? *Sci. Remote Sens.* **2022**, *5*, 100043. [CrossRef]
20. LI, X.; Chen, Y.; Zhang, Y.; Fan, Y.; Zhou, T.; Xie, F. Impact of climate change on desert steppe in northern China. *Adv. Earth Sci.* **2002**, *17*, 254.
21. Yang, S.; Liu, J.; Wang, C.; Zhang, T.; Dong, X.; Liu, Y. Vegetation dynamics influenced by climate change and human activities in the Hanjiang River Basin, central China. *Ecol. Indic.* **2022**, *145*, 109586. [CrossRef]
22. Sneath, D.J.S. State policy and pasture degradation in Inner Asia. *Science* **1998**, *281*, 1147–1148. [CrossRef]
23. Braswell, B.; Schimel, D.S.; Linder, E.; Moore Iii, B. The response of global terrestrial ecosystems to interannual temperature variability. *Science* **1997**, *278*, 870–873. [CrossRef]
24. Wu, D.; Zhao, X.; Liang, S.; Zhou, T.; Huang, K.; Tang, B.; Zhao, W. Time-lag effects of global vegetation responses to climate change. *Glob. Chang. Biol.* **2015**, *21*, 3520–3531. [CrossRef]
25. Miao, L.; Zhang, Y.; Chuai, X.; Bao, G.; He, Y.; Zhu, J. Effects of climatic factors and their time-lag on grassland NDVI in Asian drylands. Chinese Journal of Plant Ecology. *Chin. J. Plant Ecol.* **2022**, *47*, 1375. [CrossRef]
26. Liu, H.; Zhang, A.; Liu, C.; Zhao, Y.; Zhao, A.; Wang, D. Analysis of the time-lag effects of climate factors on grassland productivity in Inner Mongolia. *Glob. Ecol. Conserv.* **2021**, *30*, e01751. [CrossRef]

27. Jin, K.; Wang, F.; Han, J.; Shi, S.; Ding, W. Contribution of climatic change and human activities to vegetation NDVI change over China during 1982–2015. *Acta Geogr. Sin.* **2020**, *75*, 961–974.
28. Hu, R.; Jiang, F.; Wang, Y.; Li, J.; Li, Y.; Abdimijit, A.; Luo, G.; Zhang, J. Arid ecological and geographical conditions in five countries of central Asia. *Arid. Zone Res.* **2014**, *31*, 1–12.
29. Chen, S.; Guo, J.; Guo, Z.; Gao, R.; Shi, Y. Discipline of temporal and spatial variation of annual mean air temperature in arid and semiarid region in northwest China in the recent 46 years. *Arid. Land Geogr.* **2009**, *32*, 364–372.
30. Huang, R.; Xu, L.; Liu, J. Research on spatio-temporal change of temperature in the Northwest Arid Area. *Acta Ecol. Sin.* **2013**, *33*, 4078–4089. [CrossRef]
31. Miao, L.; Jiang, C.; He, B.; Liu, Q.; Zhu, F.; Cui, X. Response of vegetation coverage to climate change in Mongolian Plateau during recent 10 years. *Acta Ecol. Sin.* **2014**, *34*, 1295–1301.
32. Yin, G.; Meng, X.; Wang, H.; Hu, Z.; Sun, Z. Spatial-temporal variation of vegetation and its correlation with climate change in Central Asia during the period of 1982–2012. *Acta Ecol. Sin.* **2017**, *37*, 3149–3163.
33. Pedelty, J.; Devadiga, S.; Masuoka, E.; Brown, M.; Pinzon, J.; Tucker, C.; Vermote, E.; Prince, S.; Nagol, J.; Justice, C. Generating a long-term land data record from the AVHRR and MODIS instruments. In Proceedings of the 2007 IEEE international Geoscience and remote sensing Symposium, Barcelona, Spain, 23–28 July 2007; pp. 1021–1025. [CrossRef]
34. Tian, S.; Van Dijk, A.; Tregoning, P.; Renzullo, L.J. Forecasting dryland vegetation condition months in advance through satellite data assimilation. *Nat. Commun.* **2019**, *10*, 469. [CrossRef]
35. Zhang, H.; Zhang, X.; Shang, Y.; Kattel, G.; Miao, L. Continuously Vegetation Greening over Inner Mongolia for the Past Three Decades. *Remote Sens.* **2021**, *13*, 2446. [CrossRef]
36. Harris, I.; Jones, P.; Osborn, T. *CRU TS4.05: Climatic Research Unit (CRU) Time-Series (TS) Version 4.05 of High-Resolution Gridded Data of Month-by-Month Variation in Climate (Jan. 1901–Dec. 2020)*; NERC EDS Centre for Environmental Data Analysis, University of East Anglia Climatic Research Unit: Norwich, UK, 2021.
37. Bartalis, Z.; Wagner, W.; Naeimi, V.; Hasenauer, S.; Scipal, K.; Bonekamp, H.; Figa, J.; Anderson, C. Initial soil moisture retrievals from the METOP-A Advanced Scatterometer (ASCAT). *Geophys. Res. Lett.* **2007**, *34*. [CrossRef]
38. Wagner, W.; Lemoine, G.; Rott, H. A method for estimating soil moisture from ERS scatterometer and soil data. *Remote Sens. Environ.* **1999**, *70*, 191–207. [CrossRef]
39. De Jeu, R.A.; Wagner, W.; Holmes, T.; Dolman, A.; Van De Giesen, N.; Friesen, J. Global soil moisture patterns observed by space borne microwave radiometers and scatterometers. *Surv. Geophys.* **2008**, *29*, 399–420. [CrossRef]
40. Owe, M.; de Jeu, R.; Holmes, T. Multisensor historical climatology of satellite-derived global land surface moisture. *J. Geophys. Res. Earth Surf.* **2008**, *113*. [CrossRef]
41. Yin, L.; Feng, X.; Fu, B.; Chen, Y.; Wang, X.; Tao, F. Irrigation water consumption of irrigated cropland and its dominant factor in China from 1982 to 2015. *Adv. Water Resour.* **2020**, *143*, 103661. [CrossRef]
42. Leo, B. Random Forests. *Mach. Learn.* **2001**, *45*, 5–32.
43. Amit, Y.; Geman, D. Shape quantization and recognition with randomized trees. *Neural Comput.* **1997**, *9*, 1545–1588. [CrossRef]
44. Breiman, L. Randomizing outputs to increase prediction accuracy. *Mach. Learn.* **2000**, *40*, 229–242. [CrossRef]
45. Dong, J.; Lu, H.; Wang, Y.; Ye, T.; Yuan, W. Estimating winter wheat yield based on a light use efficiency model and wheat variety data. *ISPRS J. Photogramm. Remote Sens.* **2020**, *160*, 18–32. [CrossRef]
46. Johnson, M.D.; Hsieh, W.W.; Cannon, A.J.; Davidson, A.; Bédard, F. Crop yield forecasting on the Canadian Prairies by remotely sensed vegetation indices and machine learning methods. *Agric. For. Meteorol.* **2016**, *218*, 74–84. [CrossRef]
47. Chen, T.; de Jeu, R.; Liu, Y.; van der Werf, G.; Dolman, A. Using satellite based soil moisture to quantify the water driven variability in NDVI: A case study over mainland Australia. *Remote Sens. Environ.* **2014**, *140*, 330–338. [CrossRef]
48. Cheng, X.; Zhou, Z.; Li, W.; Shangguan, D.; Wang, X.; Zhang, X.; Hao, J.; Lin, Q. Monitoring drought situation and analyzing influencing factors in Central Asia using MODIS data. *Trans. Chin. Soc. Agric. Eng.* **2022**, *38*, 128–137.
49. Zou, X.; Song, D.; Chen, S. Ecological carrying capacity of grasslands and animal husbandry sustainability in Central Asia. *J. Resour. Ecol.* **2019**, *10*, 598–604.
50. Du, J.; Bao, G.; Tong, S.; Huang, X.; Wendurina; Meili; Bao, Y. Variations in vegetation cover and its relationship with climate change and human activities in Mongolia during the period 1982–2015. *Acta Pratacult. Sin.* **2021**, *30*, 1–13.
51. Luo, S.; Liu, H.; Gong, H. Nonlinear trends and spatial pattern analysis of vegetation cover change in China from 1982 to 2018. *Acta Ecol. Sin.* **2022**, *42*, 8331–8342.
52. Dong, Y.; Yan, H.; Du, W.; Hu, Y. Spatio-temporal analysis of grassland carrying capacity in Mongolian Plateau based on supply-consumption relationship. *J. Nat. Resour.* **2019**, *34*, 1093–1107. [CrossRef]
53. Pearson, R.G.; Phillips, S.J.; Loranty, M.M.; Beck, P.S.; Damoulas, T.; Knight, S.J.; Goetz, S.J. Shifts in Arctic vegetation and associated feedbacks under climate change. *Nat. Clim. Chang.* **2013**, *3*, 673–677. [CrossRef]
54. Wang, M.; An, Z.; Wang, S. The Time Lag Effect Improves Prediction of the Effects of Climate Change on Vegetation Growth in Southwest China. *Remote Sens.* **2022**, *14*, 5580. [CrossRef]
55. Wang, T.; Wedin, D.A.; Franz, T.E.; Hiller, J. Effect of vegetation on the temporal stability of soil moisture in grass-stabilized semi-arid sand dunes. *J. Hydrol.* **2015**, *521*, 447–459. [CrossRef]
56. Shi, Y.; Jin, N.; Ma, X.; Wu, B.; He, Q.; Yue, C.; Yu, Q. Attribution of climate and human activities to vegetation change in China using machine learning techniques. *Agric. For. Meteorol.* **2020**, *294*, 108246. [CrossRef]

57. Zhang, R.; Zhao, X.; Zuo, X.; Qu, H.; Degen, A.A.; Luo, Y.; Ma, X.; Chen, M.; Liu, L.; Chen, J. Impacts of Precipitation on Ecosystem Carbon Fluxes in Desert-Grasslands in Inner Mongolia, China. *J. Geophys. Res. Atmos.* **2019**, *124*, 1266–1276. [CrossRef]
58. Zou, H.; Gao, G.; FU, B. The relationship between grassland ecosystem and soil water in arid and semi-arid areas: A review. *Acta Ecol. Sin.* **2016**, *36*, 3127–3136.
59. Zhao, S.; Zhao, X.; Li, Y.; Mao, W.; Wang, N.; Chen, Y.; Lu, J.; Chen, X.; Wang, R.; Yang, X. A review on the driving effect of groundwater depth on the evolution of sandy plant soil systems in arid and semi-arid region. *Acta Ecol. Sin.* **2022**, *42*, 9898–9908.
60. Liu, J.; Gao, Q.; Zhu, Y.; Wang, K. Hydraulic Redistribution: Newly Recognized Small Cycle Within the Soil-Plant-Atmosphere Continuum. *Chin. J. Plant Ecol.* **2007**, *31*, 794–803.
61. Li, X.; Piao, S.; Huntingford, C.; Peñuelas, J.; Yang, H.; Xu, H.; Chen, A.; Friedlingstein, P.; Keenan, T.F.; Sitch, S. Global variations in critical drought thresholds that impact vegetation. *Natl. Sci. Rev.* **2023**, *10*, nwad049. [CrossRef]
62. Sun, M.; Li, X.; Xu, H.; Wang, K.; Anniwaer, N.; Hong, S. Drought thresholds that impact vegetation reveal the divergent responses of vegetation growth to drought across China. *Glob. Chang. Biol.* **2024**, *30*, e16998. [CrossRef]

Disclaimer/Publisher's Note: The statements, opinions and data contained in all publications are solely those of the individual author(s) and contributor(s) and not of MDPI and/or the editor(s). MDPI and/or the editor(s) disclaim responsibility for any injury to people or property resulting from any ideas, methods, instructions or products referred to in the content.

MDPI AG
Grosspeteranlage 5
4052 Basel
Switzerland
Tel.: +41 61 683 77 34

MDPI Books Editorial Office
E-mail: books@mdpi.com
www.mdpi.com/books

Disclaimer/Publisher's Note: The title and front matter of this reprint are at the discretion of the Topic Editors. The publisher is not responsible for their content or any associated concerns. The statements, opinions and data contained in all individual articles are solely those of the individual Editors and contributors and not of MDPI. MDPI disclaims responsibility for any injury to people or property resulting from any ideas, methods, instructions or products referred to in the content.

www.ingramcontent.com/pod-product-compliance
Lightning Source LLC
Chambersburg PA
CBHW061446220926
51850CB00004B/29
9783725858194